STUDENT SOLUTIONS M[obscured]
VOLUME ONE, FOR STEWART'S
CALCULUS

THIRD EDITION

JAMES STEWART
McMaster University

DANIEL ANDERSON
University of Iowa

DANIEL DRUCKER
Wayne State University

with the assistance of
Andy Bulman-Fleming

Brooks/Cole Publishing Company

An International Thomson Publishing Company

Pacific Grove • Albany • Bonn • Boston • Cincinnati • Detroit • London • Madrid • Melbourne
Mexico City • New York • Paris • San Francisco • Singapore • Tokyo • Toronto • Washington

Sponsoring Editor: Elizabeth Barelli Rammel
Editorial Associate: Carol Ann Benedict
Production Editor: Tessa A. McGlasson
Cover Design: Katherine Minerva, Vernon T. Boes
Cover Sculpture: Christian Haase
Cover Photo: Ed Young

For more information, contact:

Brooks/Cole Publishing Company
511 Forest Lodge Road
Pacific Grove, CA 93950
USA

International Thomson Editores
Campos Eliseos 385, Piso 7
Col. Polanco
11560 México D. F. México

International Thomson Publishing—Europe
Berkshire House 168-173
High Holborn
London WC1V 7AA
England

International Thomson Publishing GmbH
Königwinterer Strasse 418
53227 Bonn
Germany

Thomas Nelson Australia
102 Dodds Street
South Melbourne, 3205
Victoria, Australia

International Thomson Publishing—Asia
221 Henderson Road #05-10
Henderson Building
Singapore 0315

Nelson Canada
1120 Birchmount Road
Scarborough, Ontario
Canada M1K 5G4

International Thomson Publishing—Japan
Hirakawacho-cho Kyowa Building, 3F
2-2-1 Hirakawacho-cho
Chiyoda-ku, Tokyo 102
Japan

Printed in the United States of America.

5 4 3

ISBN 0-534-21802-4

Preface

I have edited this solutions manual by comparing the solutions provided by Daniel Anderson and Daniel Drucker with my own solutions and those of McGill University students Andy Bulman-Fleming and Alex Taler. Andy also produced this book using EXP Version 3.0 for Windows. I thank him and the staff of TECHarts for producing the diagrams.

JAMES STEWART

CONTENTS

REVIEW AND PREVIEW

EXERCISES 1

1. $f(x) = 2x^2 + 3x - 4$, so $f(0) = 2(0)^2 + 3(0) - 4 = -4$, $f(2) = 2(2)^2 + 3(2) - 4 = 10$,
$f(\sqrt{2}) = 2(\sqrt{2})^2 + 3(\sqrt{2}) - 4 = 3\sqrt{2}$,
$f(1+\sqrt{2}) = 2(1+\sqrt{2})^2 + 3(1+\sqrt{2}) - 4 = 2(1+2+2\sqrt{2}) + 3 + 3\sqrt{2} - 4 = 5 + 7\sqrt{2}$,
$f(-x) = 2(-x)^2 + 3(-x) - 4 = 2x^2 - 3x - 4$,
$f(x+1) = 2(x+1)^2 + 3(x+1) - 4 = 2(x^2 + 2x + 1) + 3x + 3 - 4 = 2x^2 + 7x + 1$,
$2f(x) = 2(2x^2 + 3x - 4) = 4x^2 + 6x - 8$, and
$f(2x) = 2(2x)^2 + 3(2x) - 4 = 2(4x^2) + 6x - 4 = 8x^2 + 6x - 4$.

3. $f(x) = x - x^2$, so $f(2+h) = 2 + h - (2+h)^2 = 2 + h - 4 - 4h - h^2 = -(h^2 + 3h + 2)$,
$f(x+h) = x + h - (x+h)^2 = x + h - x^2 - 2xh - h^2$, and
$$\frac{f(x+h) - f(x)}{h} = \frac{x + h - x^2 - 2xh - h^2 - x + x^2}{h} = \frac{h - 2xh - h^2}{h} = 1 - 2x - h.$$

5. $f(x) = \sqrt{x}$, $0 \le x \le 4$

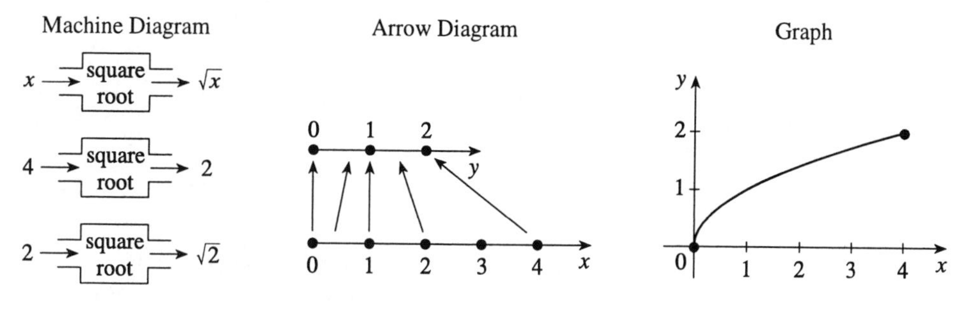

7. The range of f is the set of values of f, $\{0, 1, 2, 4\}$.

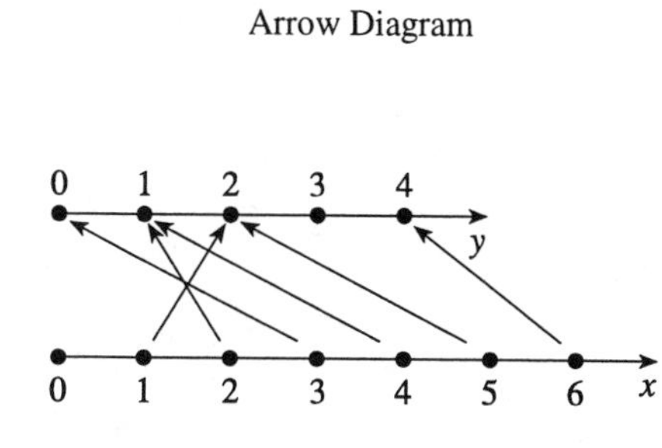

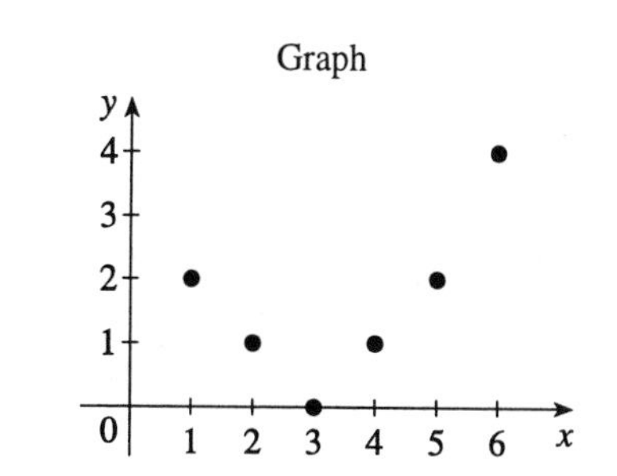

9. $f(x) = 6 - 4x$, $-2 \le x \le 3$. The domain is $[-2, 3]$. If $-2 \le x \le 3$, then
$14 = 6 - 4(-2) \ge 6 - 4x \ge 6 - 4(3) = -6$, so the range is $[-6, 14]$.

11. $h(x) = \sqrt{2x-5}$ is defined when $2x - 5 \geq 0$ or $x \geq \frac{5}{2}$, so the domain is $\left[\frac{5}{2}, \infty\right)$ and the range is $[0, \infty)$.

13. $F(x) = \sqrt{1-x^2}$ is defined when $1 - x^2 \geq 0 \quad \Leftrightarrow \quad x^2 \leq 1 \quad \Leftrightarrow \quad |x| \leq 1 \quad \Leftrightarrow \quad -1 \leq x \leq 1$, so the domain is $[-1, 1]$ and the range is $[0, 1]$.

15. $f(x) = \dfrac{x+2}{x^2-1}$ is defined for all x except when $x^2 - 1 = 0 \quad \Leftrightarrow \quad x = 1$ or $x = -1$, so the domain is $\{x \mid x \neq \pm 1\}$.

17. $g(x) = \sqrt[4]{x^2 - 6x}$ is defined when $0 \leq x^2 - 6x = x(x-6) \quad \Leftrightarrow \quad x \geq 6$ or $x \leq 0$, so the domain is $(-\infty, 0] \cup [6, \infty)$.

19. $\phi(x) = \sqrt{\dfrac{x}{\pi - x}}$ is defined when $\dfrac{x}{\pi - x} \geq 0$. So either $x \leq 0$ and $\pi - x < 0$ ($\Leftrightarrow x > \pi$), which is impossible, or $x \geq 0$ and $\pi - x > 0$ ($\Leftrightarrow x < \pi$), and so the domain is $[0, \pi)$.

21. $f(t) = \sqrt[3]{t-1}$ is defined for every t, since every real number has a cube root. The domain is the set of all real numbers.

23. $f(x) = 3 - 2x$. Domain is $\mathbb{R}$.

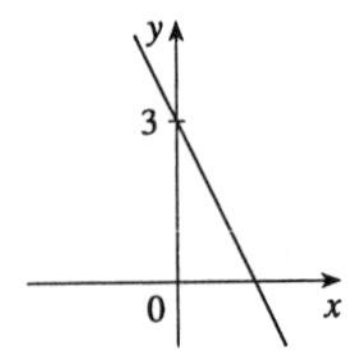

25. $f(x) = x^2 + 2x - 1 = (x^2 + 2x + 1) - 2$
$= (x+1)^2 - 2$, so the graph is a parabola with vertex at $(-1, -2)$. The domain is $\mathbb{R}$.

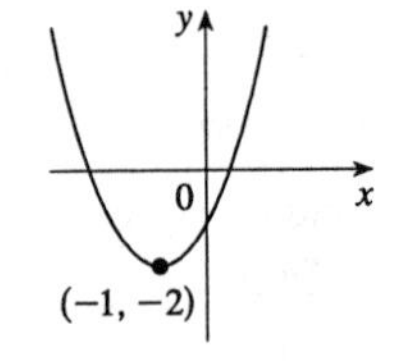

27. $g(x) = \sqrt{-x}$. The domain is $\{x \mid -x \geq 0\} = (-\infty, 0]$.

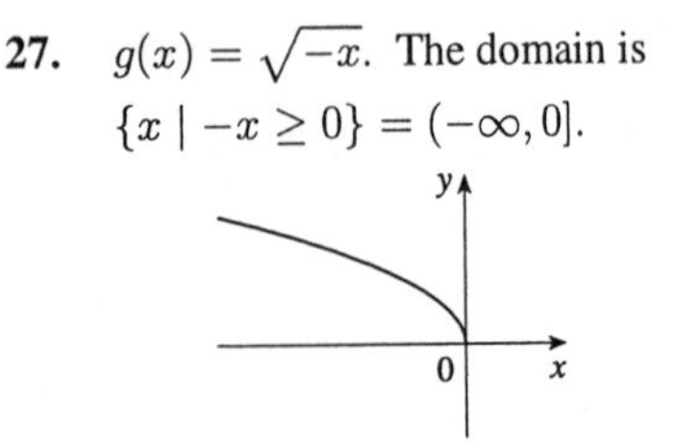

29. $h(x) = \sqrt{4 - x^2}$. Now $y = \sqrt{4 - x^2} \quad \Rightarrow$
$y^2 = 4 - x^2 \quad \Leftrightarrow \quad x^2 + y^2 = 4$, so the graph is the top half of a circle of radius 2.
The domain is $\{x \mid 4 - x^2 \geq 0\} = [-2, 2]$.

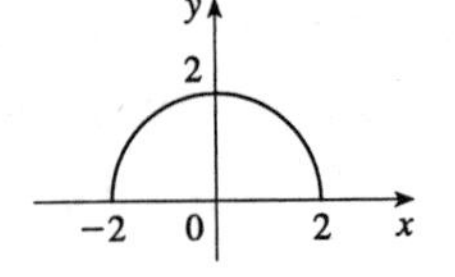

31. $F(x) = \dfrac{1}{x}$. Domain is $\{x \mid x \neq 0\}$.

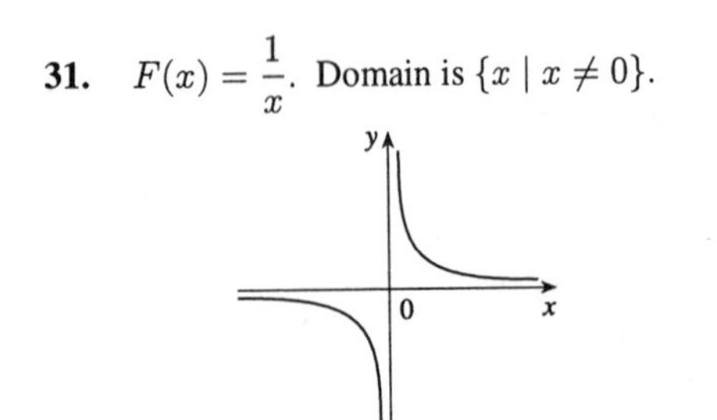

33. $G(x) = |x| + x = \begin{cases} 2x & \text{if } x \geq 0 \\ 0 & \text{if } x < 0 \end{cases}$ Domain is $\mathbb{R}$.

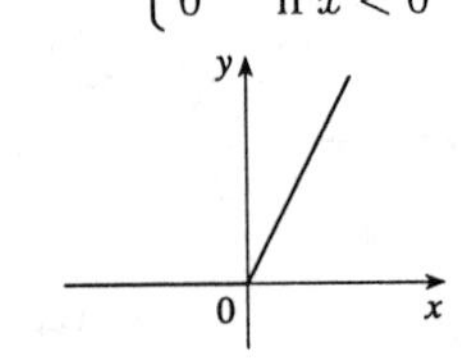

35. $H(x) = |2x| = \begin{cases} 2x & \text{if } x \geq 0 \\ -2x & \text{if } x < 0 \end{cases}$

Domain is $\mathbb{R}$.

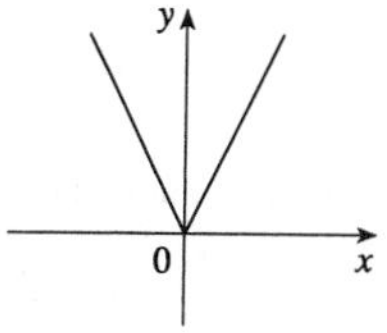

37. $f(x) = \dfrac{x}{|x|} = \begin{cases} 1 & \text{if } x > 0 \\ -1 & \text{if } x < 0 \end{cases}$

Domain is $\{x \mid x \neq 0\}$.

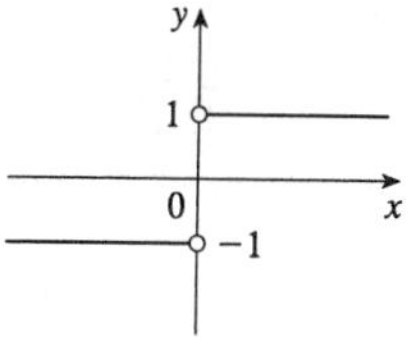

39. $f(x) = \dfrac{x^2 - 1}{x - 1} = \dfrac{(x+1)(x-1)}{x-1}$, so for $x \neq 1$, $f(x) = x + 1$. Domain is $\{x \mid x \neq 1\}$.

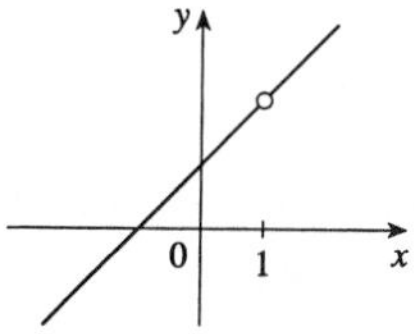

41. $f(x) = \begin{cases} 0 & \text{if } x < 2 \\ 1 & \text{if } x \geq 2 \end{cases}$ Domain is $\mathbb{R}$.

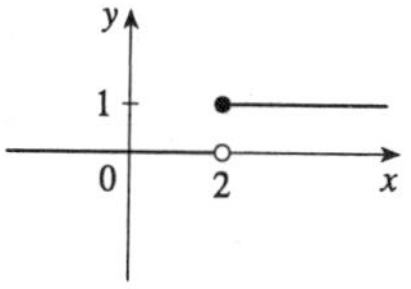

43. $f(x) = \begin{cases} x & \text{if } x \leq 0 \\ x + 1 & \text{if } x > 0 \end{cases}$ Domain is $\mathbb{R}$.

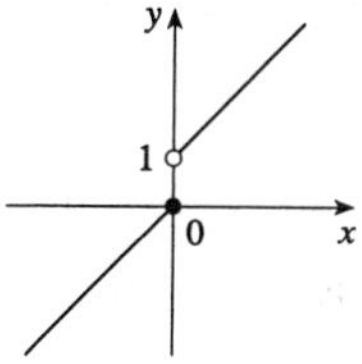

45. $f(x) = \begin{cases} -1 & \text{if } x < -1 \\ x & \text{if } -1 \leq x \leq 1 \\ 1 & \text{if } x > 1 \end{cases}$ Domain is $\mathbb{R}$.

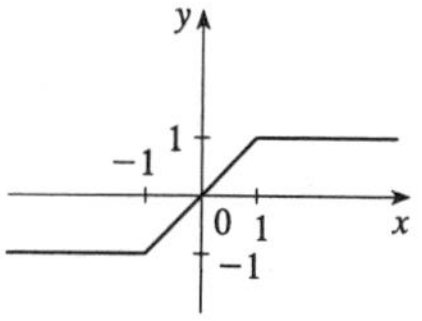

47. $f(x) = \begin{cases} x + 2 & \text{if } x \leq -1 \\ x^2 & \text{if } x > -1 \end{cases}$ Domain is $\mathbb{R}$.

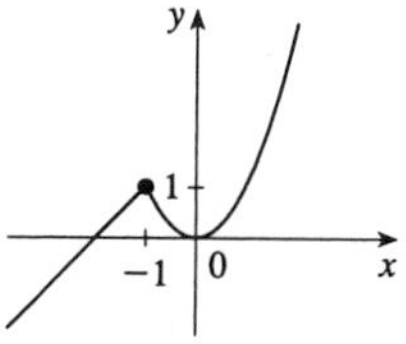

49. $f(x) = \begin{cases} -1 & \text{if } x \leq -1 \\ 3x + 2 & \text{if } -1 < x < 1 \\ 7 - 2x & \text{if } x \geq 1 \end{cases}$ Domain is $\mathbb{R}$.

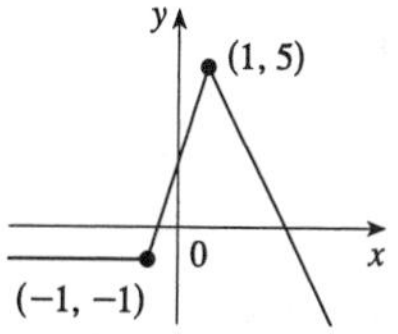

51. Yes, the curve is the graph of a function. The domain is $[-3, 2]$ and the range is $[-2, 2]$.

53. No, this is not the graph of a function since for $x = -1$ there are infinitely many points on the curve.

55. The slope of this line segment is $\dfrac{-6 - 1}{4 - (-2)} = -\dfrac{7}{6}$, so its equation is $y - 1 = -\frac{7}{6}(x + 2)$. The function is $f(x) = -\frac{7}{6}x - \frac{4}{3}$, $-2 \leq x \leq 4$.

57. $x + (y - 1)^2 = 0 \quad \Leftrightarrow \quad y - 1 = \pm\sqrt{-x}$. The bottom half is given by the function $f(x) = 1 - \sqrt{-x}$, $x \leq 0$.

59. For $-1 \le x \le 2$, the graph is the line with slope 1 and y-intercept 1, that is, the line $y = x + 1$. For $2 < x \le 4$, the graph is the line with slope $-\frac{3}{2}$ and x-intercept 4, so $y = -\frac{3}{2}(x - 4) = -\frac{3}{2}x + 6$. So the function is

$$f(x) = \begin{cases} x + 1 & \text{if } -1 \le x \le 2 \\ -\frac{3}{2}x + 6 & \text{if } 2 < x \le 4 \end{cases}$$

61. Let the length and width of the rectangle be L and W respectively. Then the perimeter is $2L + 2W = 20$, and the area is $A = LW$. Solving the first equation for W in terms of L gives $W = \dfrac{20 - 2L}{2} = 10 - L$. Thus $A(L) = L(10 - L) = 10L - L^2$. Since lengths are positive, the domain of A is $0 < L < 10$.

63. Let the length of a side of the equilateral triangle be x. Then by the Pythagorean Theorem, the height y of the triangle satisfies $y^2 + \left(\frac{1}{2}x\right)^2 = x^2$, so that $y = \frac{\sqrt{3}}{2}x$. Thus the area of the triangle is $A = \frac{1}{2}xy$ and so $A(x) = \frac{1}{2}\left(\frac{\sqrt{3}}{2}x\right)x = \frac{\sqrt{3}}{4}x^2$, with domain $x > 0$.

65. Let each side of the base of the box have length x, and let the height of the box be h. Since the volume is 2, we know that $2 = hx^2$, so that $h = 2/x^2$, and the surface area is $S = x^2 + 4xh$. Thus $S(x) = x^2 + 4x(2/x^2) = x^2 + 8/x$, with domain $x > 0$.

67. The height of the box is x and the length and width are $L = 20 - 2x$, $W = 12 - 2x$. Then $V = LWx$ and so $V(x) = (20 - 2x)(12 - 2x)(x) = 4(10 - x)(6 - x)(x) = 4x\left(60 - 16x + x^2\right) = 4x^3 - 64x^2 + 240x$, with domain $0 < x < 6$.

69. **(a)** T is a linear function of h, so $T = mh + b$ with m and b constants. We know two points on the graph of T as a function of h: $(h, T) = (0,20)$ and $(h,T) = (1,10)$. The slope of the line passing through these two points is $(10 - 20)/(1 - 0) = -10$ and the line's T-intercept is 20, so the slope-intercept form of the equation of the line is $T = -10h + 20$.

(b) The slope is $m = -10°\,\text{C/km}$, and it represents the rate of change of temperature with respect to height.

(c)

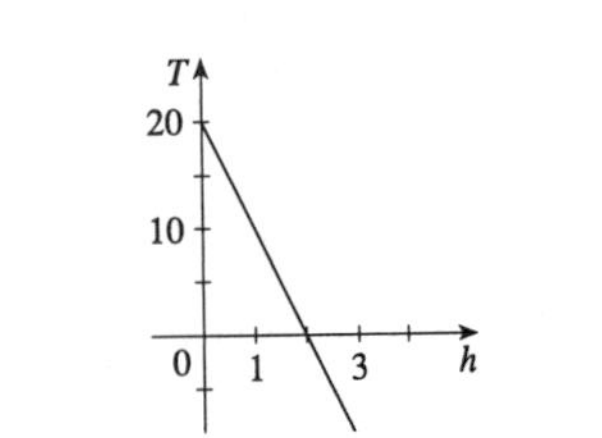

At a height of $h = 2.5$ km, the temperature is

$$\begin{aligned} T &= -10(2.5) + 20 \\ &= -25 + 20 \\ &= -5°\,\text{C}. \end{aligned}$$

71. The water will cool down almost to freezing as the ice melts. Then, when the ice has melted, the water will slowly warm up to room temperature.

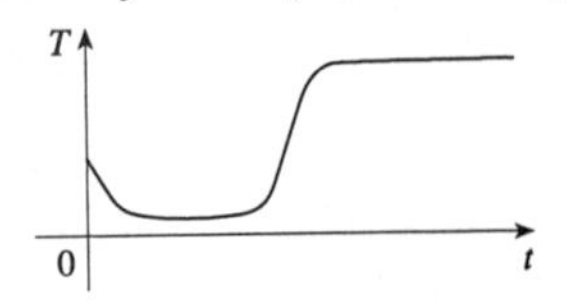

73. Of course, this graph depends strongly on the geographical location!

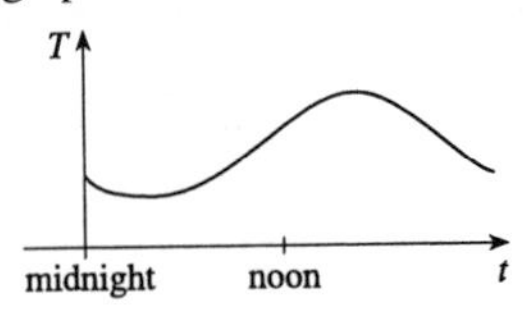

75. **(a)**

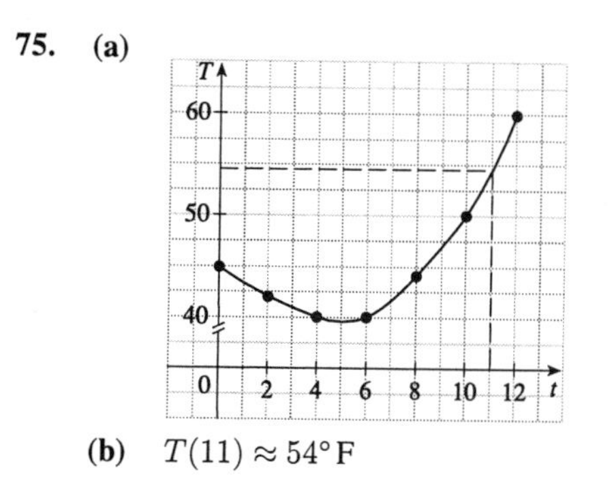

(b) $T(11) \approx 54°\text{F}$

77. $f(-x) = \dfrac{1}{(-x)^2} = \dfrac{1}{x^2} = f(x)$, so

f is an even function.

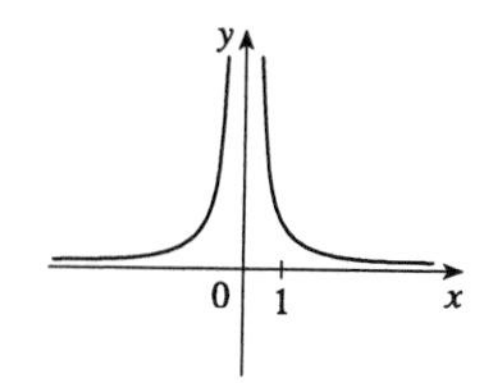

79. $f(-x) = (-x)^2 + (-x) = x^2 - x$.
Since this is neither $f(x)$ nor $-f(x)$, the function f is neither even nor odd.

81. $f(-x) = (-x)^3 - (-x) = -x^3 + x = -(x^3 - x)$
$= -f(x)$, so f is odd.

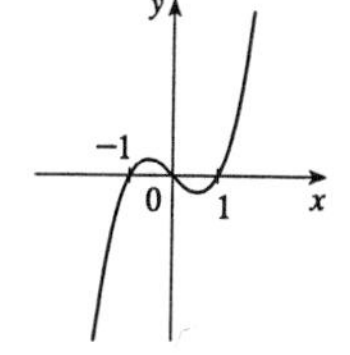

NOTE: For the rest of this section, "$D =$ " stands for "The domain of the function is".

83. $f(x) = x^3 + 2x^2$; $g(x) = 3x^2 - 1$. $D = \mathbb{R}$ for both f and g.
$(f + g)(x) = x^3 + 2x^2 + 3x^2 - 1 = x^3 + 5x^2 - 1$, $D = \mathbb{R}$.
$(f - g)(x) = x^3 + 2x^2 - (3x^2 - 1) = x^3 - x^2 + 1$, $D = \mathbb{R}$.
$(fg)(x) = (x^3 + 2x^2)(3x^2 - 1) = 3x^5 + 6x^4 - x^3 - 2x^2$, $D = \mathbb{R}$.
$(f/g)(x) = (x^3 + 2x^2)/(3x^2 - 1)$, $D = \left\{x \mid x \neq \pm\frac{1}{\sqrt{3}}\right\}$.

85. $f(x) = x$, $g(x) = 1/x$

87. $f(x) = 2x^2 - x$, $g(x) = 3x + 2$. $D = \mathbb{R}$ for both f and g, and hence for their composites.
$(f \circ g)(x) = f(g(x)) = f(3x + 2) = 2(3x + 2)^2 - (3x + 2) = 18x^2 + 21x + 6$.
$(g \circ f)(x) = g(f(x)) = g(2x^2 - x) = 3(2x^2 - x) + 2 = 6x^2 - 3x + 2$.
$(f \circ f)(x) = f(f(x)) = f(2x^2 - x) = 2(2x^2 - x)^2 - (2x^2 - x) = 8x^4 - 8x^3 + x$.
$(g \circ g)(x) = g(g(x)) = g(3x + 2) = 3(3x + 2) + 2 = 9x + 8$.

89. $f(x) = 1/x$, $D = \{x \mid x \neq 0\}$; $g(x) = x^3 + 2x$, $D = \mathbb{R}$.
$(f \circ g)(x) = f(g(x)) = f(x^3 + 2x) = 1/(x^3 + 2x)$, $D = \{x \mid x^3 + 2x \neq 0\} = \{x \mid x \neq 0\}$.
$(g \circ f)(x) = g(f(x)) = g(1/x) = 1/x^3 + 2/x$, $D = \{x \mid x \neq 0\}$.
$(f \circ f)(x) = f(f(x)) = f(1/x) = \dfrac{1}{1/x} = x$, $D = \{x \mid x \neq 0\}$.
$(g \circ g)(x) = g(g(x)) = g(x^3 + 2x) = (x^3 + 2x)^3 + 2(x^3 + 2x) = x^9 + 6x^7 + 12x^5 + 10x^3 + 4x$, $D = \mathbb{R}$.

91. $f(x)=\sqrt[3]{x}$, $D=\mathbb{R}$; $g(x)=1-\sqrt{x}$, $D=[0,\infty)$.

$(f\circ g)(x)=f(g(x))=f(1-\sqrt{x})=\sqrt[3]{1-\sqrt{x}}$, $D=[0,\infty)$.

$(g\circ f)(x)=g(f(x))=g(\sqrt[3]{x})=1-x^{1/6}$, $D=[0,\infty)$.

$(f\circ f)(x)=f(f(x))=f(\sqrt[3]{x})=x^{1/9}$, $D=\mathbb{R}$.

$(g\circ g)(x)=g(g(x))=g(1-\sqrt{x})=1-\sqrt{1-\sqrt{x}}$, $D=\{x\geq 0 \mid 1-\sqrt{x}\geq 0\}=[0,1]$.

93. $f(x)=\dfrac{x+2}{2x+1}$, $D=\{x \mid x\neq -\frac{1}{2}\}$; $g(x)=\dfrac{x}{x-2}$, $D=\{x \mid x\neq 2\}$.

$(f\circ g)(x)=f(g(x))=f\left(\dfrac{x}{x-2}\right)=\dfrac{x/(x-2)+2}{2x/(x-2)+1}=\dfrac{3x-4}{3x-2}$, $D=\{x \mid x\neq 2,\frac{2}{3}\}$.

$(g\circ f)(x)=g(f(x))=g\left(\dfrac{x+2}{2x+1}\right)=\dfrac{(x+2)/(2x+1)}{(x+2)/(2x+1)-2}=\dfrac{-x-2}{3x}$, $D=\{x \mid x\neq 0,-\frac{1}{2}\}$.

$(f\circ f)(x)=f(f(x))=f\left(\dfrac{x+2}{2x+1}\right)=\dfrac{(x+2)/(2x+1)+2}{2(x+2)/(2x+1)+1}=\dfrac{5x+4}{4x+5}$, $D=\{x \mid x\neq -\frac{1}{2},-\frac{5}{4}\}$.

$(g\circ g)(x)=g(g(x))=g\left(\dfrac{x}{x-2}\right)=\dfrac{x/(x-2)}{x/(x-2)-2}=\dfrac{x}{4-x}$, $D=\{x \mid x\neq 2,4\}$.

95. $(f\circ g\circ h)(x)=f(g(h(x)))=f(g(x-1))=f\left(\sqrt{x-1}\right)=\sqrt{x-1}-1$

97. $(f\circ g\circ h)(x)=f(g(h(x)))=f(g(\sqrt{x}))=f(\sqrt{x}-5)=(\sqrt{x}-5)^4+1$

99. Let $g(x)=x-9$ and $f(x)=x^5$. Then $(f\circ g)(x)=(x-9)^5=F(x)$.

101. Let $g(x)=x^2$ and $f(x)=\dfrac{x}{x+4}$. Then $(f\circ g)(x)=\dfrac{x^2}{x^2+4}=G(x)$.

103. Let $h(x)=x^2$, $g(x)=x+1$ and $f(x)=\dfrac{1}{x}$. Then $(f\circ g\circ h)(x)=\dfrac{1}{x^2+1}=H(x)$.

105. Let r be the radius of the ripple in cm. The area of the ripple is $A=\pi r^2$ but, as a function of time, $r=60t$. Thus, $A=\pi(60t)^2=3600\pi t^2$.

107. We need a function g so that $f(g(x))=3(g(x))+5=h(x)=3x^2+3x+2=3(x^2+x)+2$ $=3(x^2+x-1)+5$. So we see that $g(x)=x^2+x-1$.

109. The function $g(x)=x$ has domain $(-\infty,\infty)$. However, the function $f\circ f$, where $f(x)=1/x$, has for its domain $(-\infty,0)\cup(0,\infty)$ even though the rule is the same: $(f\circ f)(x)=f(1/x)=x$.

EXERCISES 2

1. **(a)** $f(x) = \sqrt[5]{x}$ is a root function.

(b) $g(x) = \sqrt{1 - x^2}$ is an algebraic function because it is a root of a polynomial.

(c) $h(x) = x^9 + x^4$ is a polynomial of degree 9.

(d) $r(x) = \dfrac{x^2 + 1}{x^3 + x}$ is a rational function because it is a ratio of polynomials

(e) $s(x) = \tan 2x$ is a trigonometric function.

(f) $t(x) = \log_{10} x$ is a logarithmic function.

3. **(a)** To graph $y = f(2x)$ we compress the graph of f horizontally by a factor of 2.

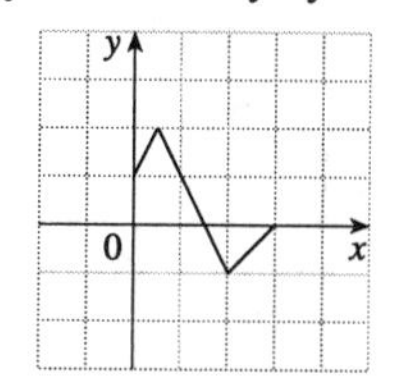

(b) To graph $y = f\left(\frac{1}{2}x\right)$ we expand the graph of f horizontally by a factor of 2.

(c) To graph $y = f(-x)$ we reflect the graph of f about the y-axis.

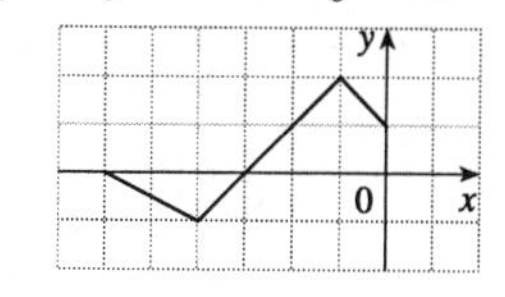

(d) To graph $y = -f(-x)$ we reflect the graph of f about the y-axis, then about the x-axis.

5. $y = -1/x$

7. $y = 2\sin x$

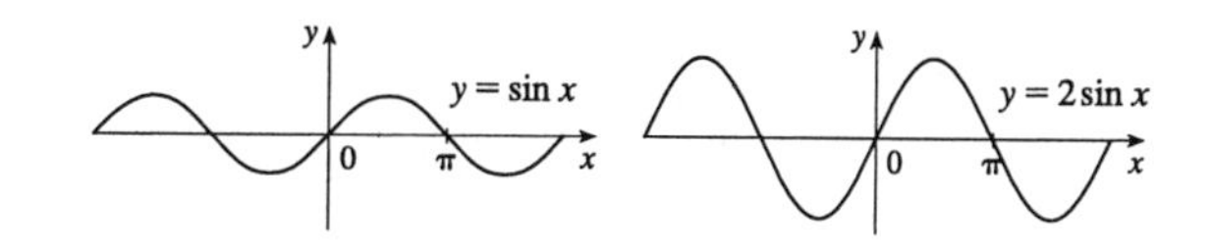

9. 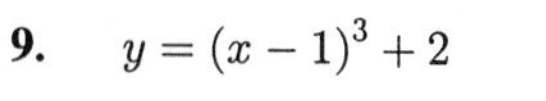
$y = (x - 1)^3 + 2$

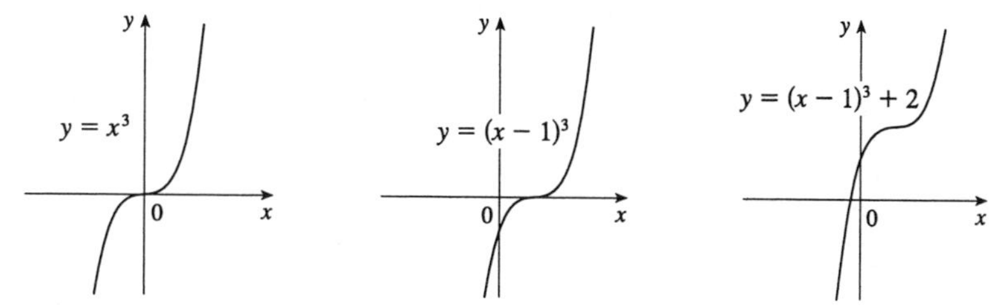

11. $y = \tan 2x$

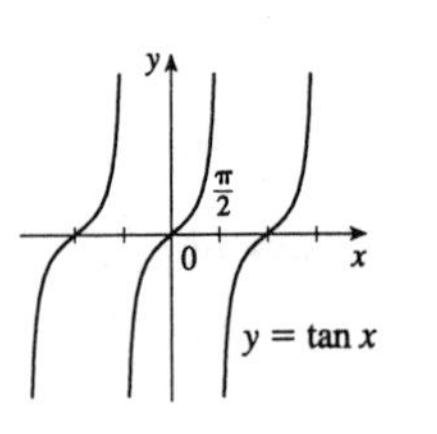

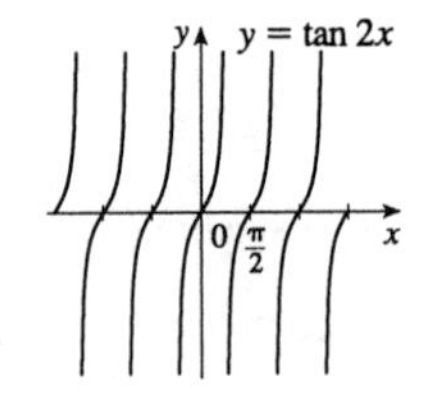

13. $y = \cos(x/2)$

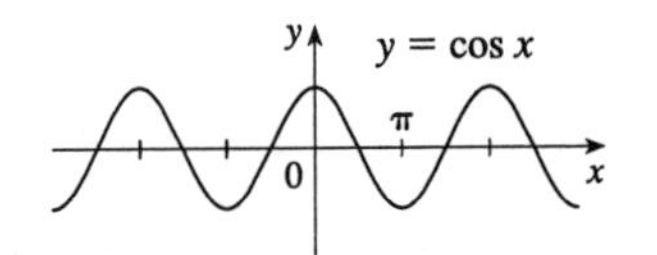

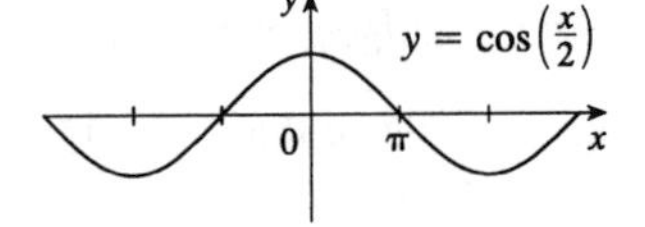

15. $y = \dfrac{1}{x-3}$

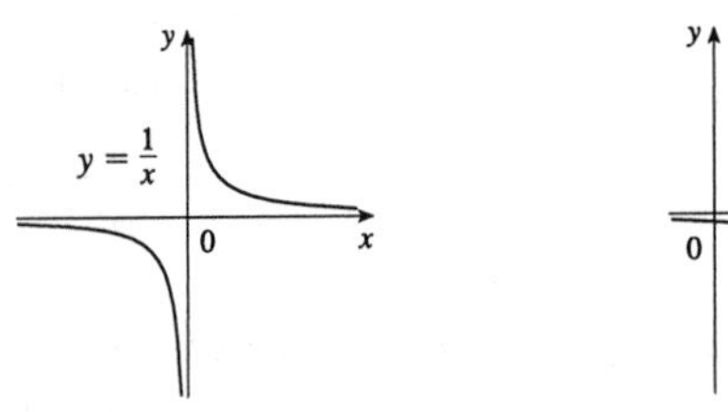

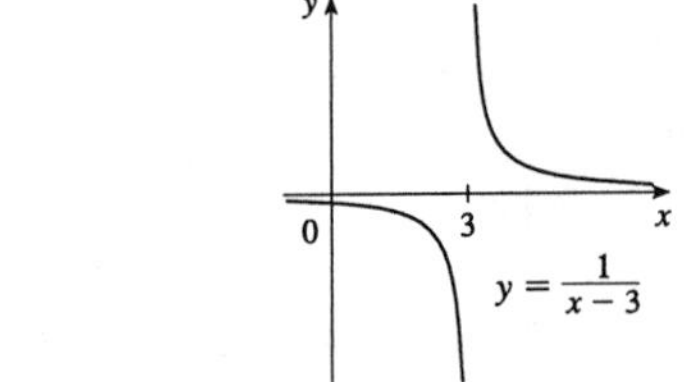

17. $y = \frac{1}{3}\sin\left(x - \frac{\pi}{6}\right)$

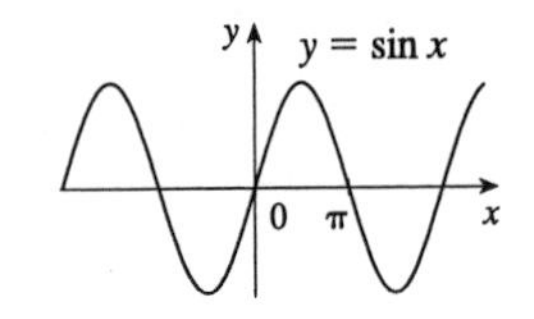

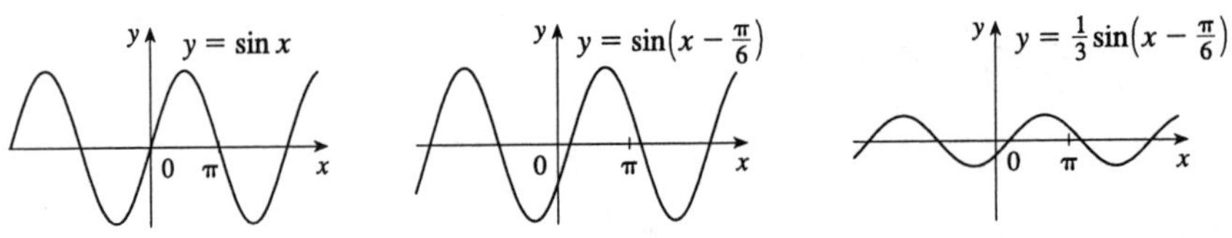

19. $y = 1 + 2x - x^2 = -(x-1)^2 + 2$

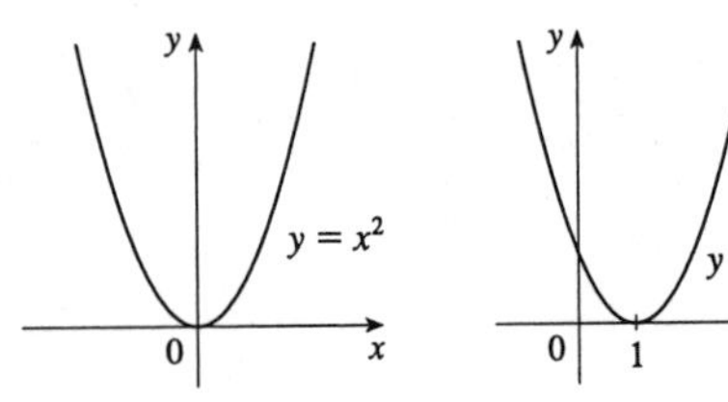

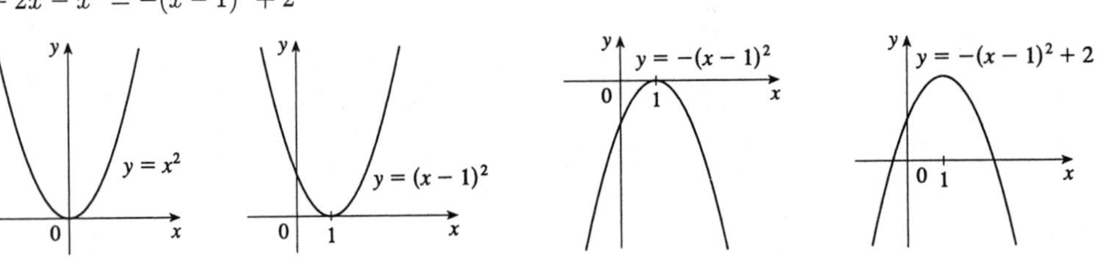

21. $y = 2 - \sqrt{x+1}$

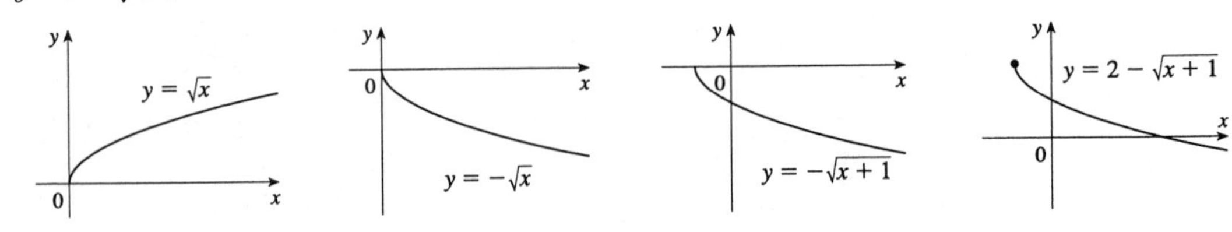

23. $y = x^2 - 2x$
$= (x-1)^2 - 1$

$y = |x^2 - 2x|$

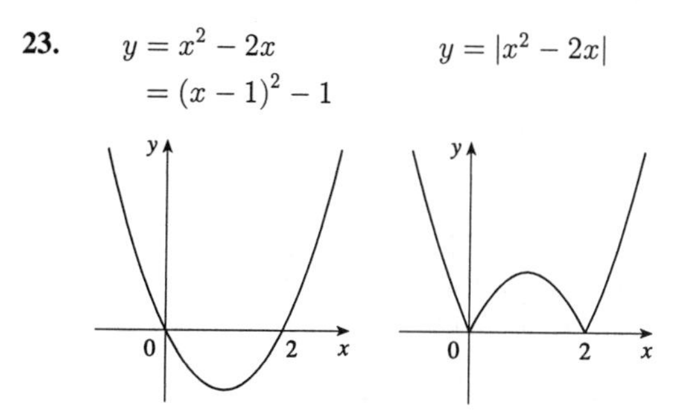

25. $y = ||x| - 1|$

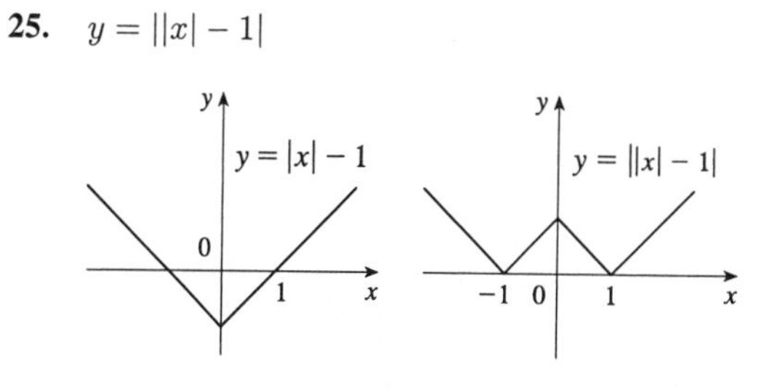

27. **(a)** To obtain $y = f(|x|)$, the portion of $y = f(x)$ right of the y-axis is reflected in the y-axis.

(b) $y = \sin|x|$

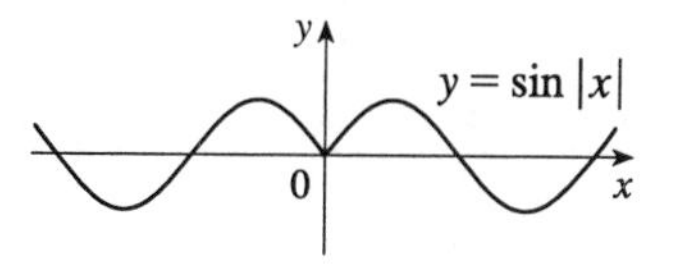

29. Note that there are vertical asymptotes wherever $f(x) = 0$, since division by 0 is impossible.

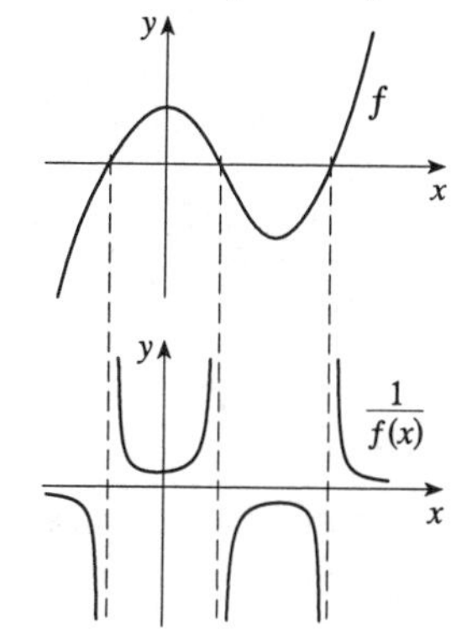

EXERCISES 3

1. $f(x) = x^4 + 2$

(a) $[-2, 2]$ by $[-2, 2]$ **(b)** $[0, 4]$ by $[0, 4]$ **(c)** $[-4, 4]$ by $[-4, 4]$

(d) $[-8, 8]$ by $[-4, 40]$ **(e)** $[-40, 40]$ by $[-80, 800]$

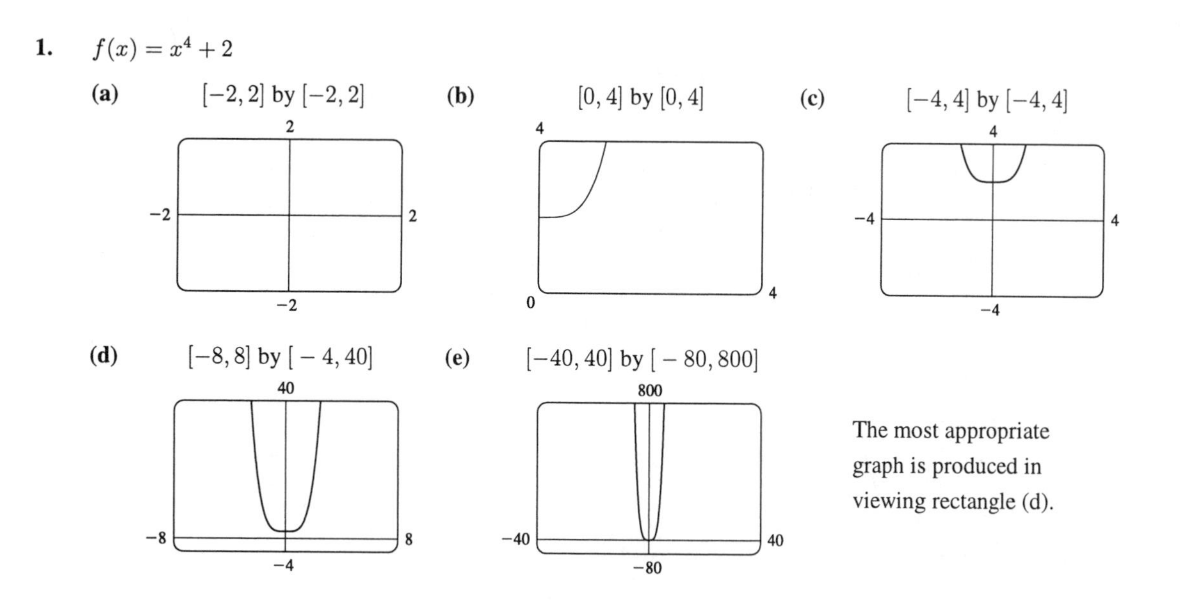

The most appropriate graph is produced in viewing rectangle (d).

3. $f(x) = 10 + 25x - x^3$

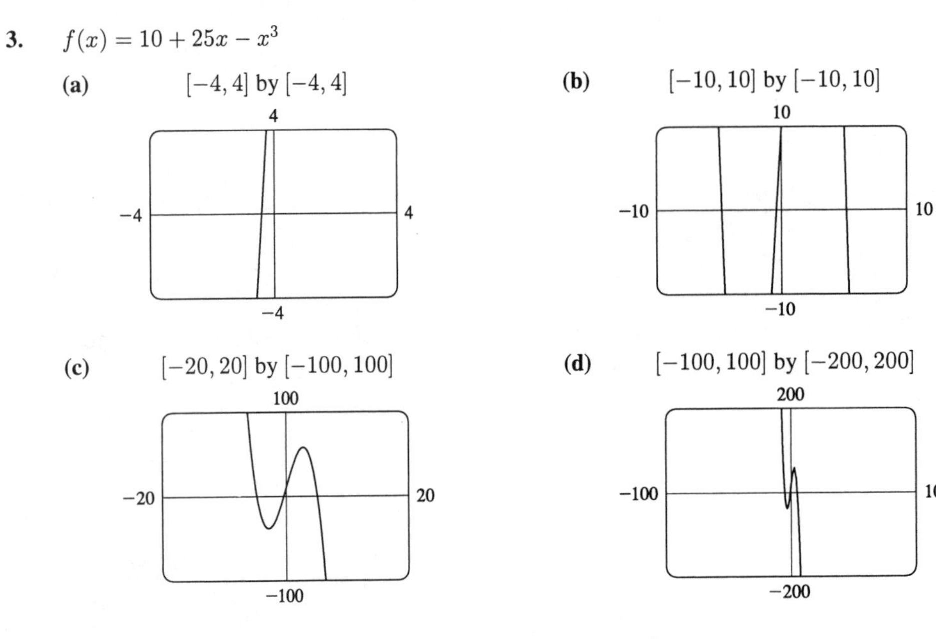

The most appropriate graph is produced in viewing rectangle (c).

5. $f(x) = 4 + 6x - x^2$
Note that many similar rectangles give equally good views of the function.

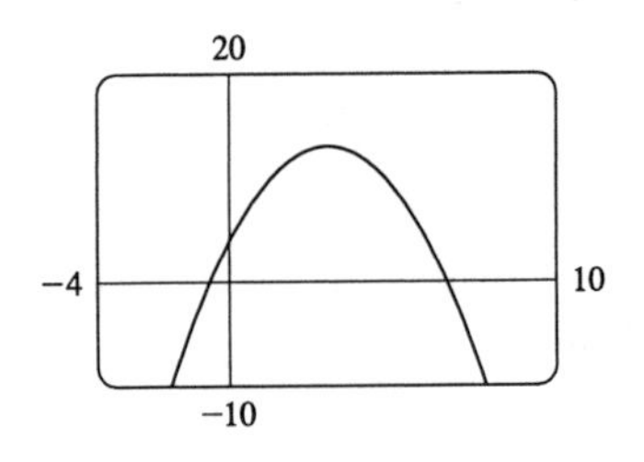

7. $f(x) = \sqrt[4]{256 - x^2}$ To find an appropriate viewing rectangle, we calculate f's domain and range: $256 - x^2 \geq 0 \quad \Leftrightarrow \quad x^2 \leq 256 \quad \Leftrightarrow$ $|x| \leq 16 \quad \Leftrightarrow \quad -16 \leq x \leq 16$, so the domain is $[-16, 16]$. Also, $0 \leq \sqrt[4]{256 - x^2} \leq \sqrt[4]{256} = 4$, so the range is $[0, 4]$. Thus we choose the viewing rectangle to be $[-20, 20]$ by $[-2, 6]$.

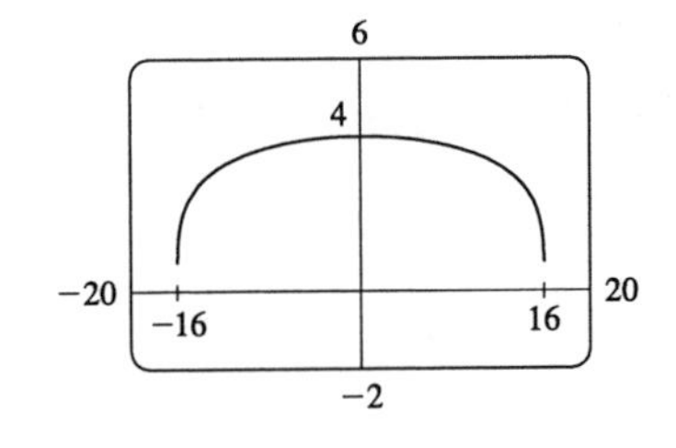

9. $f(x) = 0.01x^3 - x^2 + 5$

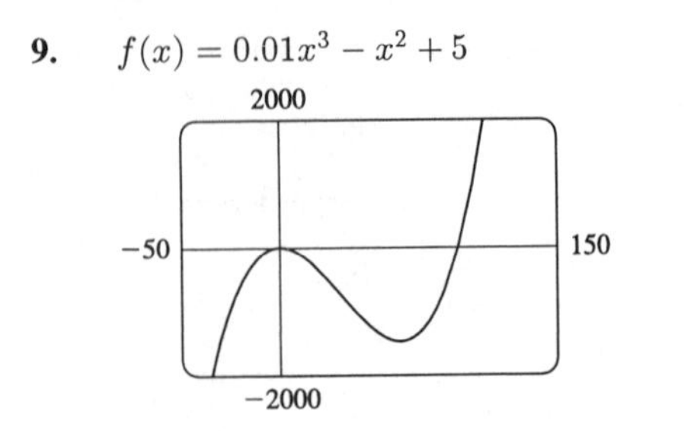

11. $y = \dfrac{1}{x^2 + 25}$

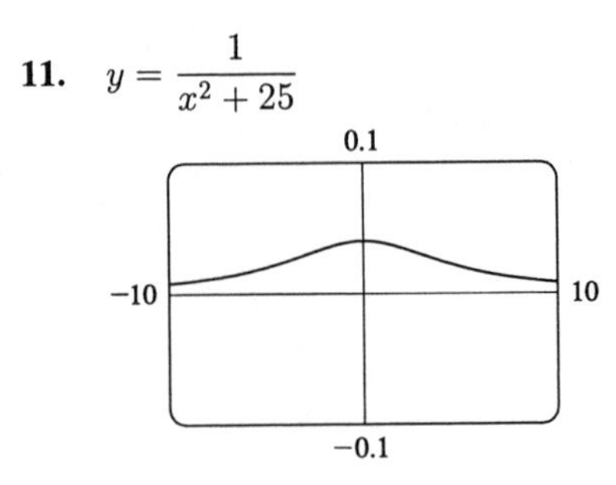

13. $y = x^4 - 4x^3$

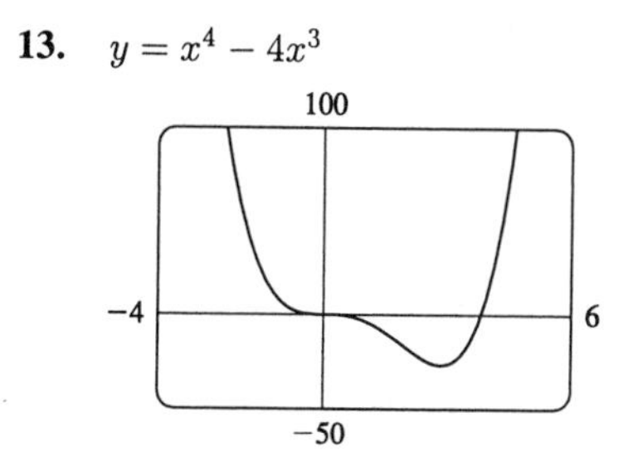

15. $y = \dfrac{2x-1}{x+3}$

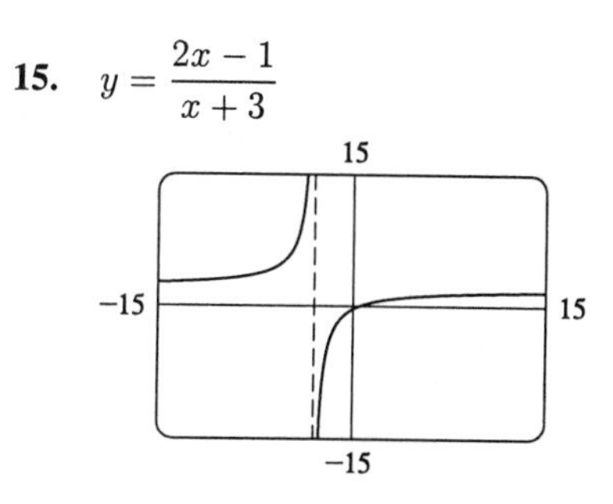

17. $f(x) = \cos(100x)$

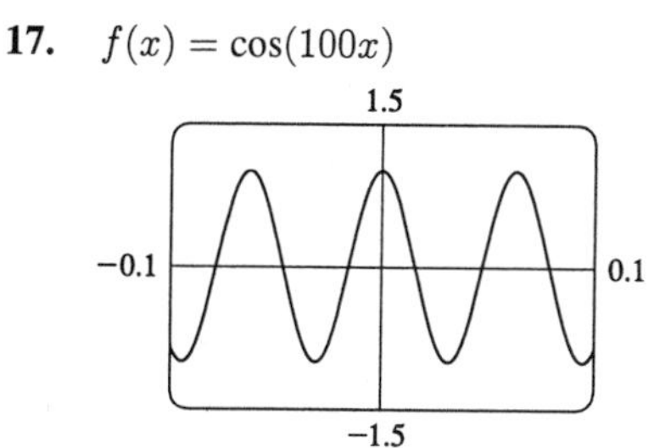

19. $f(x) = \sin(x/40)$

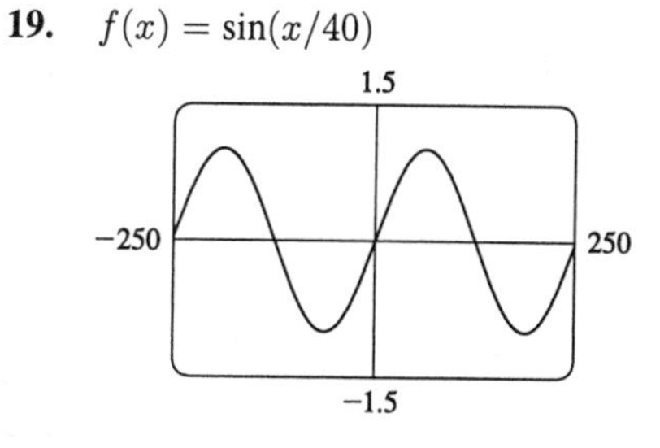

21. $y = 3^{\cos(x^2)}$

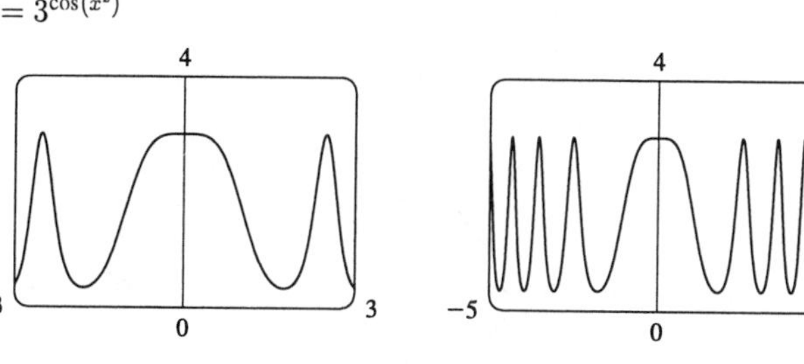

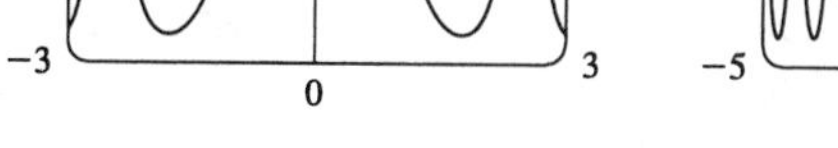

23. $y = \pm\sqrt{\dfrac{1-4x^2}{2}}$

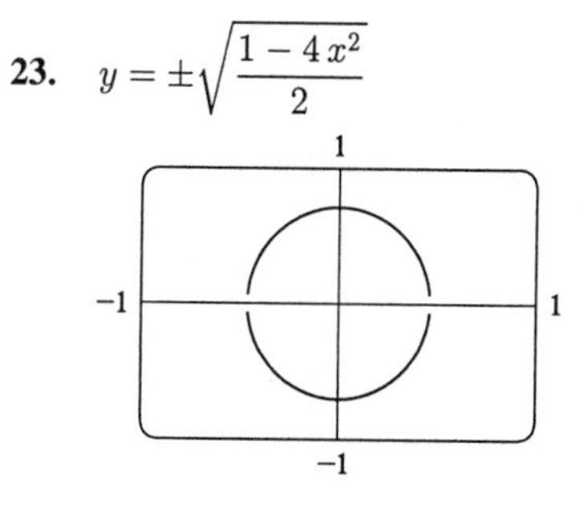

25. In Maple, we can use the procedure

```
f:=proc(x)
    if x<=1 then x^3-2*x+1
    else (x-1)^(1/3) fi
  end;
```

and then `plot(f,-2..3);` to plot the curve.
To define f in Mathematica, we can use
`f[x_]:=If[x<=1,x^3-2*x+1,(x-1)^(1/3)]`.

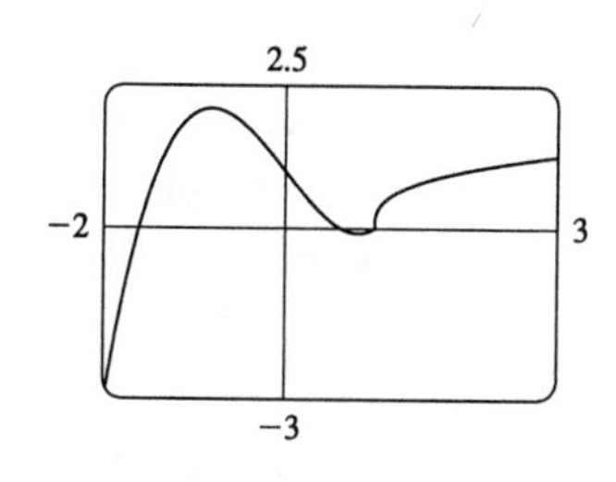

27. We first graph $f(x) = 3x^3 + x^2 + x - 2$ in the viewing rectangle $[-2, 2] \times [-30, 30]$ to find the approximate value of the root. The only root appears to be between $x = 0.5$ and $x = 1$, so we graph f again in the rectangle $[0.5, 1] \times [-1, 1]$ (or use the cursor on a graphing calculator). From the second graph, it appears that the only solution to the equation $f(x) = 0$ is 0.67, to 2 decimal places.

29. From the first graph, it appears that the roots lie near ± 2. We zoom in (or use the cursor) and find that the solutions are -1.90, 0 and 1.90, to 2 decimal places.

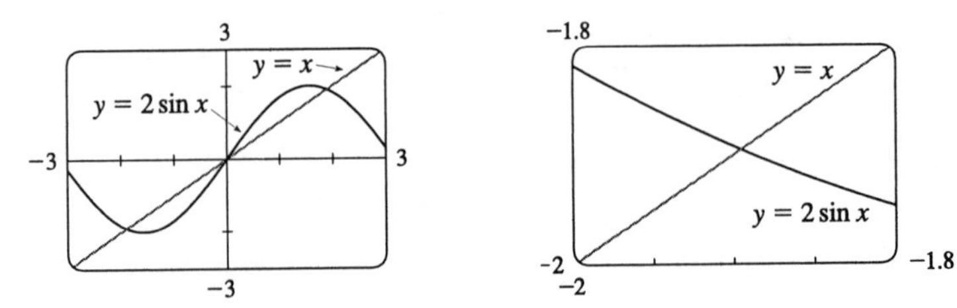

31. $g(x) = x^3/10$ is eventually larger than $f(x) = 10x^2$.

33. **(a)** **(i)** $[0, 5]$ by $[0, 20]$ **(ii)** $[0, 25]$ by $[0, 10^7]$ **(iii)** $[0, 50]$ by $[0, 10^8]$

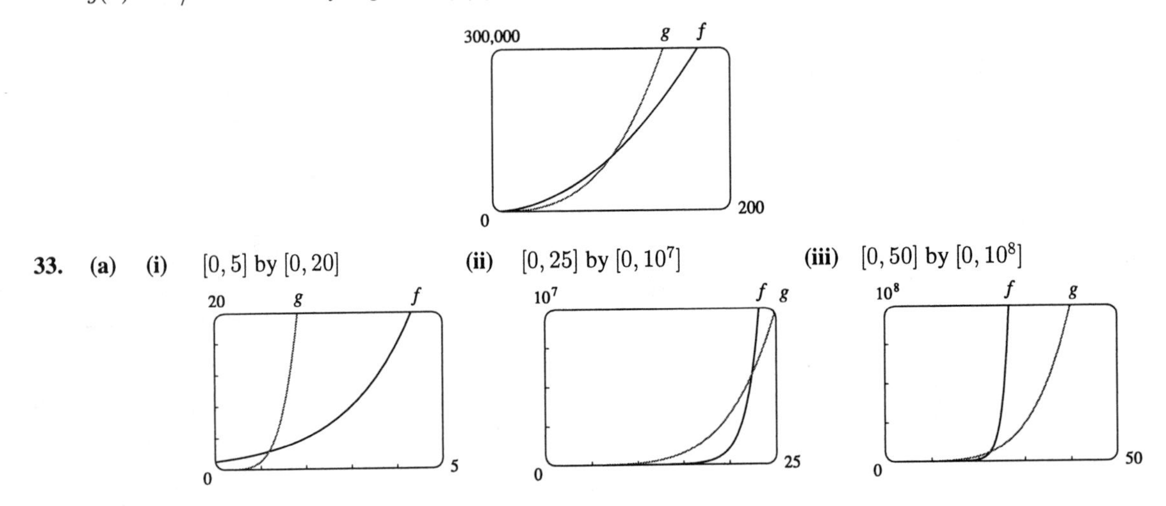

As x gets large, f grows much more quickly than g.

(b) From the graphs in part (a), it appears that the two solutions are $x \approx 1.2$ and 22.4.

35. We see from the graph of $y = |\sin x - x|$ that there are two solutions to the equation $|\sin x - x| = 0.1$: $x \approx -0.85$ and $x \approx 0.85$. The condition $|\sin x - x| < 0.1$ holds for any x lying between these two values.

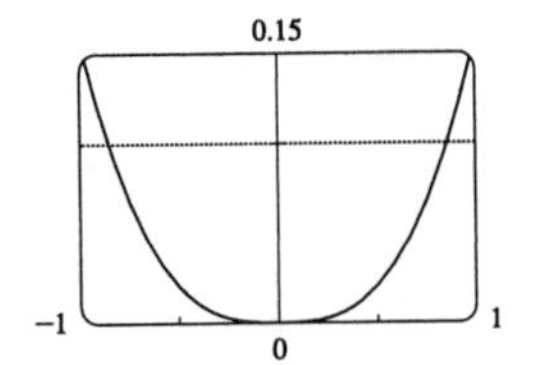

37. **(a)** The root functions $y = \sqrt{x}$, $y = \sqrt[4]{x}$ and $y = \sqrt[6]{x}$

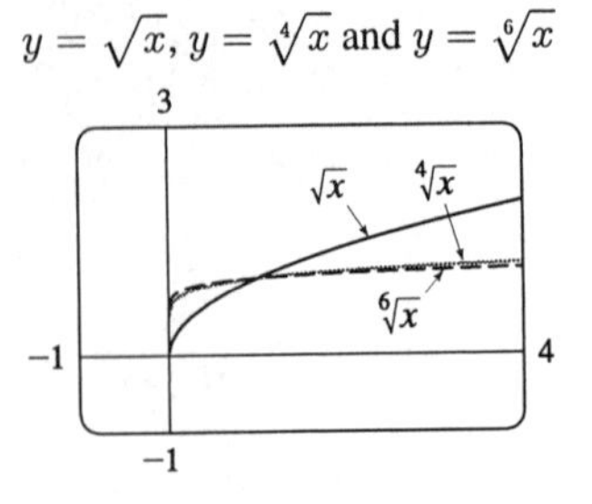

(b) The root functions $y = x$, $y = \sqrt[3]{x}$ and $y = \sqrt[5]{x}$

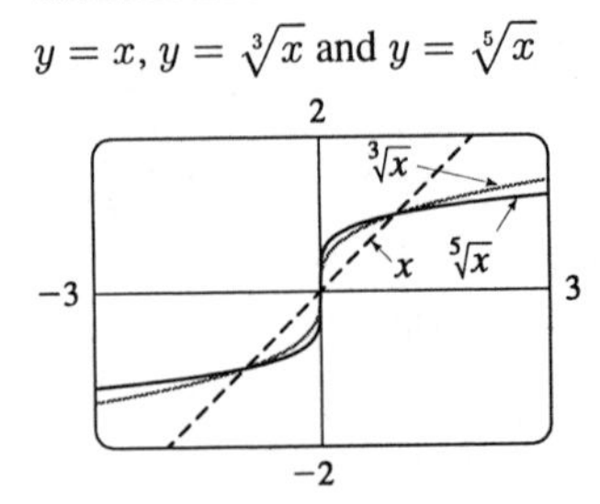

(c) The root functions
$y = \sqrt{x}$, $y = \sqrt[3]{x}$, $y = \sqrt[4]{x}$ and $y = \sqrt[5]{x}$

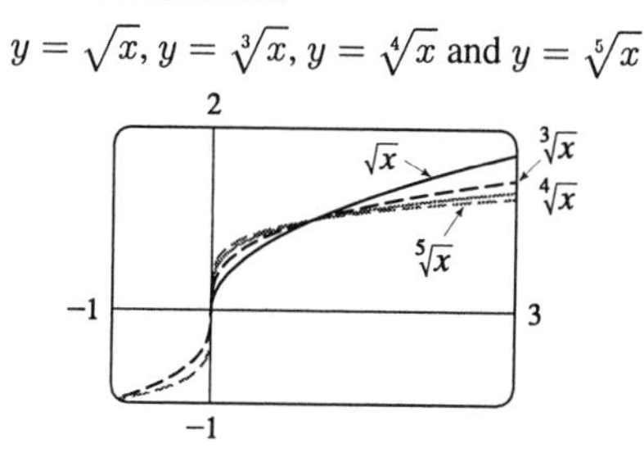

(d)
- For any n, the nth root of 0 is 0 and the nth root of 1 is 1, that is, all nth root functions pass through the points $(0,0)$ and $(1,1)$.
- For odd n, the domain of the nth root function is $\mathbb{R}$, while for even n, it is $\{x \in \mathbb{R} \mid x \geq 0\}$.
- Graphs of even root functions look similar to that of $\sqrt{x}$, while those of odd root functions resemble that of $\sqrt[3]{x}$.
- As n increases, the graph of $\sqrt[n]{x}$ becomes steeper near 0 and flatter for $x > 1$.

39. $f(x) = x^4 + cx^2 + x$
If $c < 0$, there are three humps: two minimum points and a maximum point. These humps get flatter as c increases, until at $c = 0$ two of the humps disappear and there is only one minimum point. This single hump then moves to the right and approaches the origin as c increases.

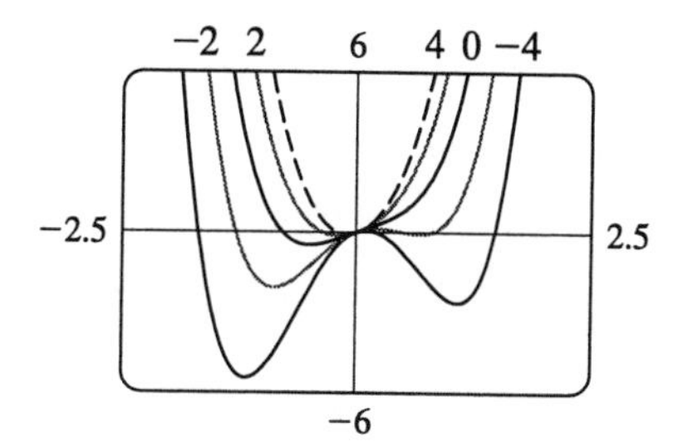

41. $y = x^n 2^{-x}$
As n increases, the maximum of the function moves further from the origin, and gets larger. Note, however, that regardless of n, the function approaches 0 as $x \to \infty$.

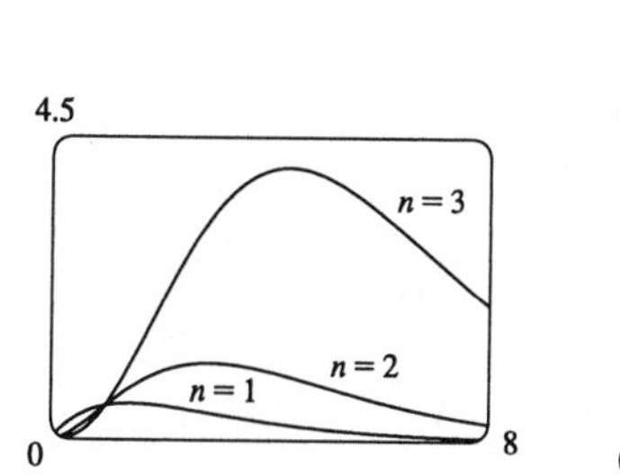

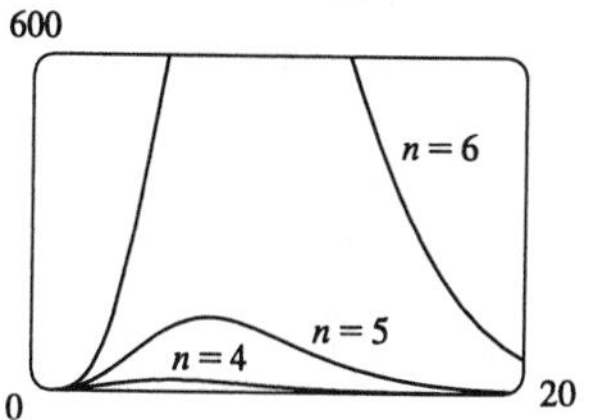

43. $y^2 = cx^3 + x^2$
If $c < 0$, the loop is to the right of the origin, and if c is positive, it is to the left. In both cases, the closer c is to 0, the larger the loop is. (In the limiting case, $c = 0$, the loop is "infinite," that is, it doesn't close.) Also, the larger $|c|$ is, the steeper the slope is on the loopless side of the origin.

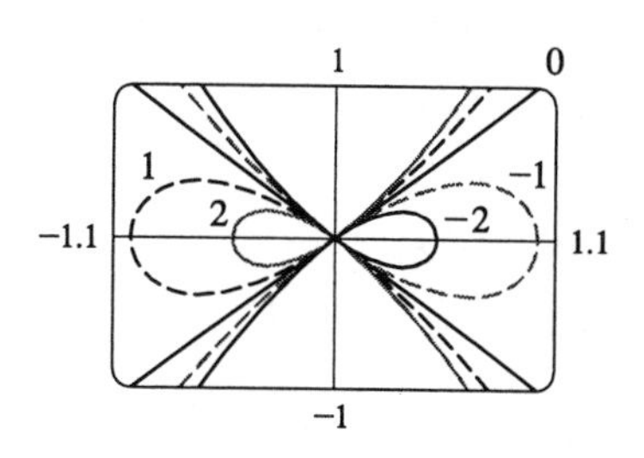

EXERCISES 4

1. As in Example 2, $|x+1| = \begin{cases} -x-1 & \text{if } x < -1 \\ x+1 & \text{if } x \geq 1 \end{cases}$ and $|x+4| = \begin{cases} -x-4 & \text{if } x < -4 \\ x+4 & \text{if } x \geq -4 \end{cases}$

Therefore we consider three cases: $x < -4$, $-4 \leq x < -1$, and $x \geq -1$.

If $x < -4$, we must have $-x-1-x-4 \leq 5 \quad \Leftrightarrow \quad x \geq -5$.

If $-4 \leq x < -1$, we must have $-x-1+x+4 \leq 5 \quad \Leftrightarrow \quad 3 \leq 5$.

If $x \geq -1$, we must have $x+1+x+4 \leq 5 \quad \Leftrightarrow \quad x \leq 0$.

These conditions together imply that $-5 \leq x \leq 0$.

3. $|2x-1| = \begin{cases} 1-2x & \text{if } x < \frac{1}{2} \\ 2x-1 & \text{if } x \geq \frac{1}{2} \end{cases}$ and $|x+5| = \begin{cases} -x-5 & \text{if } x < -5 \\ x+5 & \text{if } x \geq -5 \end{cases}$

Therefore we consider the cases $x < -5$, $-5 \leq x < \frac{1}{2}$, and $x \geq \frac{1}{2}$.

If $x < -5$, we must have $1-2x-(-x-5) = 3 \;\Leftrightarrow\; x = 3$, which is false since we are considering $x < -5$.

If $-5 \leq x < \frac{1}{2}$, we must have $1-2x-(x+5) = 3 \;\Leftrightarrow\; x = -\frac{7}{3}$.

If $x \geq \frac{1}{2}$, we must have $2x-1-(x+5) = 3 \;\Leftrightarrow\; x = 9$.

So the two solutions of the equation are $x = -\frac{7}{3}$ and $x = 9$.

5. The final digit in 947^{362} is determined by 7^{362}, since $947^{362} = (900+40+7)^{362}$, and every term in the expansion of this expression is the product of powers of either 40 or 900 or both, the last digits of which are all zero, except the term 7^{362}. Looking at the first few powers of 7 we see: $7^1 = 7$, $7^2 = 49$, $7^3 = 343$, $7^4 = 2401$, $7^5 = 16{,}807$, $7^6 = 117{,}649$ and it appears that the final digit follows a cyclical pattern, namely $7 \to 9 \to 3 \to 1 \to 7 \to 9$, of length 4. Since $362 \div 4 = 90$ with remainder 2, the final digit is the second in the cycle, that is, 9.

7. $f_0(x) = x^2$ and $f_{n+1}(x) = f_0(f_n(x))$ for $n = 0, 1, 2, \ldots$.
$f_1(x) = f_0(f_0(x)) = f_0(x^2) = (x^2)^2 = x^4$, $\quad f_2(x) = f_0(f_1(x)) = f_0(x^4) = (x^4)^2 = x^8$,
$f_3(x) = f_0(f_2(x)) = f_0(x^8) = (x^8)^2 = x^{16}, \ldots$. Thus a general formula is $f_n(x) = x^{2^{n+1}}$.

9. $f(x) = |x^2 - 4|x| + 3|$. If $x \geq 0$, then $f(x) = |x^2 - 4x + 3| = |(x-1)(x-3)|$.

Case (i): If $0 < x \leq 3$, then $f(x) = x^2 - 4x + 3$.

Case (ii): If $1 < x \leq 3$, then $f(x) = -(x^2 + 4x + 3) = -x^2 - 4x - 3$.

Case (iii): If $x > 3$ then $f(x) = x^2 - 4x + 3$.

This enables us to sketch the graph for $x \geq 0$. Then we use the fact that f is an even function to reflect this part of the graph about the y-axis to obtain the entire graph. Or, we could consider also the cases $x < -3$, $-3 \leq x < -1$, and $-1 \leq x < 0$.

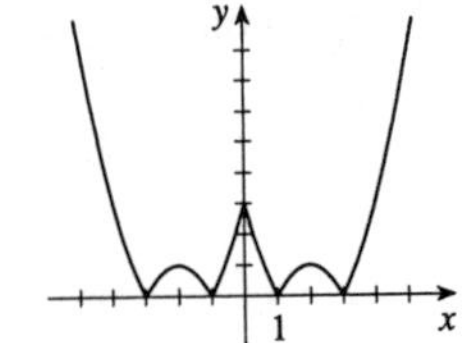

11. $\left[\sqrt{3+2\sqrt{2}}-\sqrt{3-2\sqrt{2}}\right]^2=\left(3+2\sqrt{2}\right)+\left(3-2\sqrt{2}\right)-2\sqrt{\left(3+2\sqrt{2}\right)\left(3-2\sqrt{2}\right)}$

$=6-2\sqrt{9+6\sqrt{2}-6\sqrt{2}-4\sqrt{2}\sqrt{2}}=6-2=4.$ So the given expression is $\sqrt{4}=2$.

13. $|x|+|y|=1+|xy|$

$\Leftrightarrow\quad |xy|-|x|-|y|+1=0$

$\Leftrightarrow\quad (|x|-1)(|y|-1)=0$

$\Leftrightarrow\quad x=\pm1$ or $y=\pm1$.

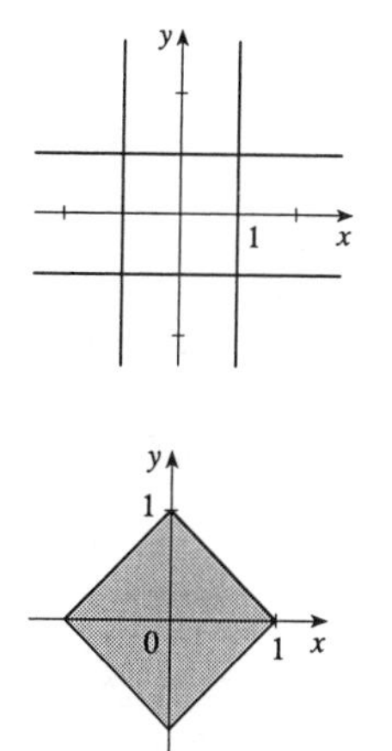

15. $|x|+|y|\le 1$. The boundary of the region has equation $|x|+|y|=1$. In quadrants I, II, III, IV, this becomes the lines $x+y=1$, $-x+y=1$, $-x-y=1$, and $x-y=1$ respectively.

17. **(a)** The amount of ribbon needed is equal to the circumference of the earth, which is about $2\pi r=2\pi(3960)\approx 24{,}880$ mi.

(b) The additional ribbon needed is $2\pi(r+1)-2\pi(r)=2\pi(1)\approx 6.3$ ft. (!)

19. Let x represent the length of the base in cm and h the length of the altitude in cm. By the Pythagorean Theorem, $3^2+x^2=5^2 \quad\Leftrightarrow\quad x^2=16\quad\Leftrightarrow\quad x=4$ and so the area of the triangle is $\frac{1}{2}\cdot 4\cdot 3=6$. But the area of the triangle is also $\frac{1}{2}\cdot 5\cdot h=6\quad\Leftrightarrow\quad h=2.4$ and hence the length of the altitude is 2.4 cm.

21. We use a proof by contradiction. Assume that $\sqrt{3}$ is rational. So $\sqrt{3}=p/q$ for some integers p and q. This fraction is assumed to be in lowest form (that is, p and q have no common factors). Then $3q^2=p^2$. We see that 3 must be one of the prime factors of p^2, and thus of p itself. So $p=3k$ for some integer k. Substituting this into the previous equation, we get $3q^2=(3k)^2=9k^2\quad\Leftrightarrow\quad q^2=3k^2$. So 3 is one of the prime factors of q^2, and thus of q itself. So 3 divides both p and q. But this contradicts our assumption that the fraction p/q was in lowest form So $\sqrt{3}$ is not expressible as a ratio of integers, that is, $\sqrt{3}$ is irrational.

23. The statement is false. Here is one particular counterexample:

	First Half	Second Half	Whole Season
Player A	$1/99$	$1/1$	$2/100=.020$
Player B	$0/1$	$98/99$	$98/100=.980$

25. The odometer reading is proportional to the number of tire revolutions. Let r_1 represent the number of revolutions made by the tire on the 400 mi trip and r_2 represent the number of revolutions made by the tire on the 390 mi trip. Then $r_1=400k$ and $r_2=390k$ for some constant k. Let d be the actual distance traveled, R_1 be the radius of normal tires, and R_2 be the radius of snow tires. Then $d=2\pi R_1 r_1=2\pi R_2 r_2\quad\Leftrightarrow\quad R_1 r_1=R_2 r_2$

$\Leftrightarrow\quad R_2=\dfrac{R_1 r_1}{r_2}=\dfrac{15\cdot 400k}{390k}=\dfrac{15\cdot 40}{39}\approx 15.38$. So the radius of the snow tires is about 15.4 in.

CHAPTER ONE

EXERCISES 1.1

1. **(a)** Slopes of the secant lines:

x	m_{PQ}
0	$\dfrac{2.6-1.3}{0-3}\approx -0.43$
1	$\dfrac{2.0-1.3}{1-3}=-0.35$
2	$\dfrac{1.1-1.3}{2-3}=0.2$
4	$\dfrac{2.1-1.3}{4-3}=0.8$
5	$\dfrac{3.5-1.3}{5-3}=1.1$

(b) The slope of the tangent line at P is about $\dfrac{2.5-0}{5-0.6}\approx 0.57$.

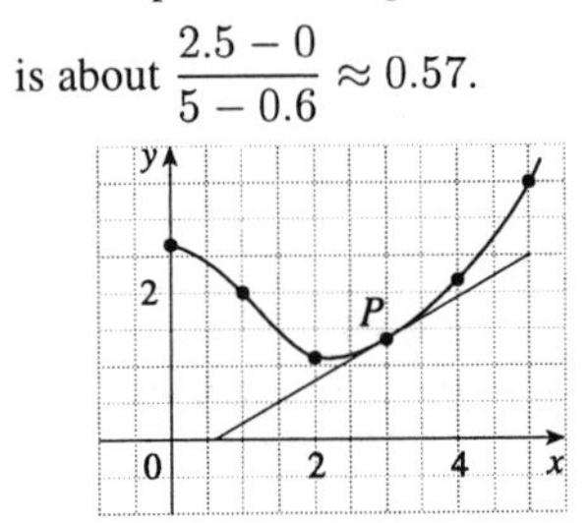

3. For the curve $y=\sqrt{x}$ and the point $P(4,2)$:

(a)

	x	Q	m_{PQ}
(i)	5	$(5, 2.236068)$	0.236068
(ii)	4.5	$(4.5, 2.121320)$	0.242641
(iii)	4.1	$(4.1, 2.024846)$	0.248457
(iv)	4.01	$(4.01, 2.002498)$	0.249844
(v)	4.001	$(4.001, 2.000250)$	0.249984

	x	Q	m_{PQ}
(vi)	3	$(3, 1.732051)$	0.267949
(vii)	3.5	$(3.5, 1.870829)$	0.258343
(viii)	3.9	$(3.9, 1.974842)$	0.251582
(ix)	3.99	$(3.99, 1.997498)$	0.250156
(x)	3.999	$(3.999, 1.999750)$	0.250016

(b) The slope appears to be $\frac{1}{4}$.

(c) $y-2=\frac{1}{4}(x-4)$ or $x-4y+4=0$

5. **(a)** At $t=2$, $y=40(2)-16(2)^2=16$. The average velocity between times 2 and $2+h$ is

$$\frac{40(2+h)-16(2+h)^2-16}{h}=\frac{-24h-16h^2}{h}=-24-16h, \text{ if } h\neq 0.$$

(i) $h=0.5$, -32 ft/s

(ii) $h=0.1$, -25.6 ft/s

(iii) $h=0.05$, -24.8 ft/s

(iv) $h=0.01$, -24.16 ft/s

(b) The instantaneous velocity when $t=2$ is -24 ft/s.

7. Average velocity between times 1 and $1+h$ is

$$\frac{s(1+h)-s(1)}{h}=\frac{(1+h)^3/6-1/6}{h}=\frac{h^3+3h^2+3h}{6h}=\frac{h^2+3h+3}{6} \text{ if } h\neq 0.$$

(a) **(i)** $v_{\text{av}}=\dfrac{2^2+3(2)+3}{6}=\dfrac{13}{6}$ ft/s

(ii) $v_{\text{av}}=\dfrac{1^2+3(1)+3}{6}=\dfrac{7}{6}$ ft/s

(iii) $v_{\text{av}}=\dfrac{(0.5)^2+3(0.5)+3}{6}=\dfrac{19}{24}$ ft/s

(iv) $v_{\text{av}}=\dfrac{(0.1)^2+3(0.1)+3}{6}=\dfrac{331}{600}$ ft/s

(b) As h approaches 0, the velocity approaches $\frac{1}{2}$ ft/s.

(c), (d)

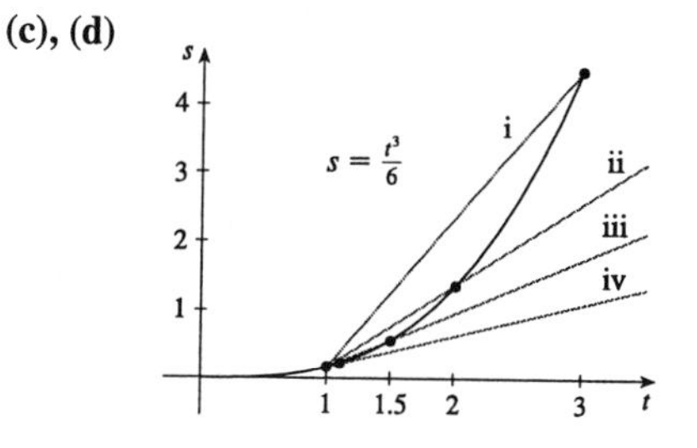

EXERCISES 1.2

1. **(a)** $\lim_{x\to 1} f(x) = 3$ **(b)** $\lim_{x\to 3^-} f(x) = 2$ **(c)** $\lim_{x\to 3^+} f(x) = -2$

(d) $\lim_{x\to 3} f(x)$ doesn't exist **(e)** $f(3) = 1$ **(f)** $\lim_{x\to -2^-} f(x) = -1$

(g) $\lim_{x\to -2^+} f(x) = -1$ **(h)** $\lim_{x\to -2} f(x) = -1$ **(i)** $f(-2) = -3$

3. **(a)** $\lim_{x\to 3} f(x) = 2$ **(b)** $\lim_{x\to 1} f(x) = -1$ **(c)** $\lim_{x\to -3} f(x) = 1$

(d) $\lim_{x\to 2^-} f(x) = 1$ **(e)** $\lim_{x\to 2^+} f(x) = 2$ **(f)** $\lim_{x\to 2} f(x)$ doesn't exist

5. **(a)** $\lim_{x\to 3} f(x) = \infty$ **(b)** $\lim_{x\to 7} f(x) = -\infty$ **(c)** $\lim_{x\to -4} f(x) = -\infty$

(d) $\lim_{x\to -9^-} f(x) = \infty$ **(e)** $\lim_{x\to -9^+} f(x) = -\infty$

(f) The equations of the vertical asymptotes: $x = -9$, $x = -4$, $x = 3$, $x = 7$

7. **(a)**

(b) **(i)** $\lim_{x\to 0^-} f(x) = 1$

(ii) $\lim_{x\to 0^+} f(x) = 1$

(iii) $\lim_{x\to 0} f(x) = 1$

9. For $g(x) = \dfrac{x-1}{x^3-1}$:

x	$g(x)$
0.2	0.806452
0.4	0.641026
0.6	0.510204
0.8	0.409836
0.9	0.369004
0.99	0.336689

x	$g(x)$
1.8	0.165563
1.6	0.193798
1.4	0.229358
1.2	0.274725
1.1	0.302115
1.01	0.330022

It appears that $\lim_{x\to 1} \dfrac{x-1}{x^3-1} = 0.\overline{3} = \dfrac{1}{3}$.

11. For $F(x) = \dfrac{(1/\sqrt{x}) - \frac{1}{5}}{x - 25}$:

x	$F(x)$
26	-0.003884
25.5	-0.003941
25.1	-0.003988
25.05	-0.003994
25.01	-0.003999

x	$F(x)$
24	-0.004124
24.5	-0.004061
24.9	-0.004012
24.95	-0.004006
24.99	-0.004001

It appears that $\lim_{x \to 25} F(x) = -0.004$.

13. For $f(x) = \dfrac{1 - \cos x}{x^2}$:

x	$f(x)$
1	0.459698
0.5	0.489670
0.4	0.493369
0.3	0.496261

x	$f(x)$
0.2	0.498336
0.1	0.499583
0.05	0.499896
0.01	0.499996

It appears that $\lim_{x \to 0} \dfrac{1 - \cos x}{x^2} = 0.5$.

15. $\lim_{x \to 5^+} \dfrac{6}{x - 5} = \infty$ since $(x - 5) \to 0$ as $x \to 5^+$ and $\dfrac{6}{x - 5} > 0$ for $x > 5$.

17. $\lim_{x \to 3} \dfrac{1}{(x - 3)^8} = \infty$ since $(x - 3) \to 0$ as $x \to 3$ and $\dfrac{1}{(x - 3)^8} > 0$.

19. $\lim_{x \to -2^+} \dfrac{x - 1}{x^2(x + 2)} = -\infty$ since $(x + 2) \to 0$ as $x \to 2^+$ and $\dfrac{x - 1}{x^2(x + 2)} < 0$ for $-2 < x < 0$.

21. **(a)**

x	$f(x)$
0.5	-1.14
0.9	-3.69
0.99	-33.7
0.999	-333.7
0.9999	-3333.7
0.99999	$-33,333.7$

x	$f(x)$
1.5	0.42
1.1	3.02
1.01	33.0
1.001	333.0
1.0001	3333.0
1.00001	33,333.3

From these calculations, it seems that $\lim_{x \to 1^-} f(x) = -\infty$ and $\lim_{x \to 1^+} f(x) = \infty$.

(b) If x is slightly smaller than 1, then $x^3 - 1$ will be a negative number close to 0, and the reciprocal of $x^3 - 1$, that is, $f(x)$, will be a negative number with large absolute value. So $\lim_{x \to 1^-} f(x) = -\infty$.

If x is slightly larger than 1, then $x^3 - 1$ will be a small positive number, and its reciprocal, $f(x)$, will be a large positive number. So $\lim_{x \to 1^+} f(x) = \infty$.

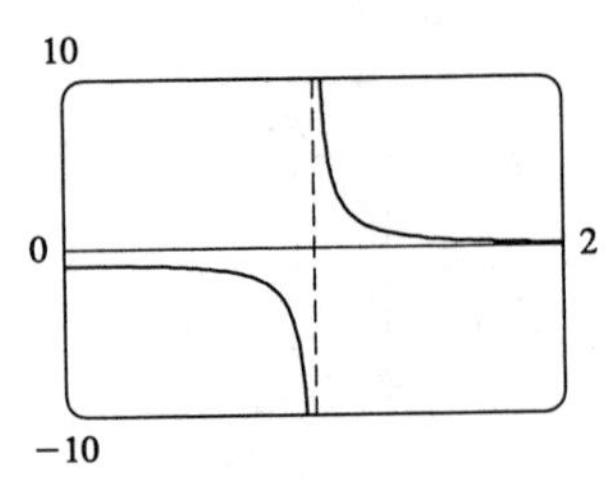

(c) It appears from the graph of f that $\lim_{x \to 1^-} f(x) = -\infty$ and $\lim_{x \to 1^+} f(x) = \infty$.

23. Let $h(x) = (1+x)^{1/x}$.

x	$h(x)$
1.0	2.0
0.1	2.593742
0.01	2.704814
0.001	2.716924
0.0001	2.718146

x	$h(x)$
0.00001	2.718268
0.000001	2.718280
0.0000001	2.718282
0.00000001	2.718282
0.000000001	2.718282

It appears that $\lim_{x\to 0}(1+x)^{1/x} \approx 2.71828$.

25. For $f(x) = x^2 - (2^x/1000)$:

(a)

x	$f(x)$
1	0.998000
0.8	0.638259
0.6	0.358484
0.4	0.158680
0.2	0.038851
0.1	0.008928
0.05	0.001465

It appears that $\lim_{x\to 0} f(x) = 0$.

(b)

x	$f(x)$
0.04	0.000572
0.02	−0.000614
0.01	−0.000907
0.005	−0.000978
0.003	−0.000993
0.001	−0.001000

It appears that $\lim_{x\to 0} f(x) = -0.001$.

27. From the following graphs, it seems that $\lim_{x\to 0} \frac{\tan(4x)}{x} = 4$.

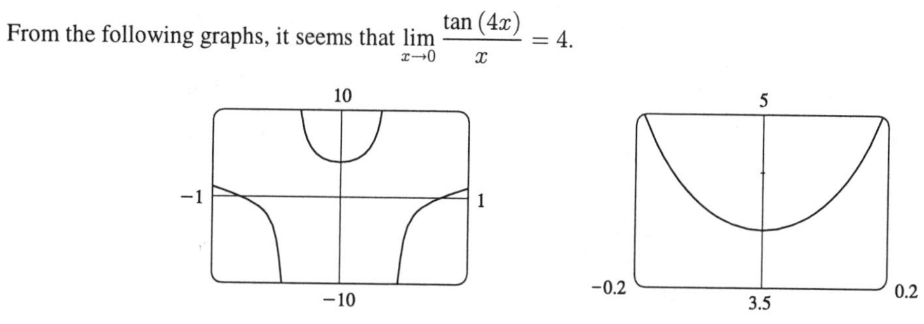

29. No matter how many times we zoom in towards the origin, the graphs appear to consist of almost-vertical lines. This indicates more and more frequent oscillations as $x \to 0$.

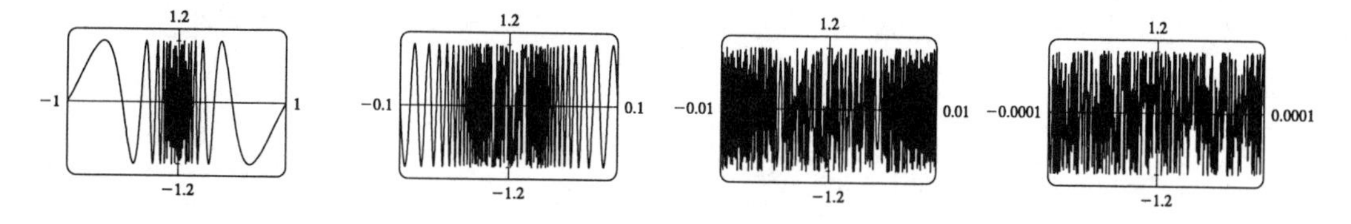

EXERCISES 1.3

1.
$$\begin{aligned}\lim_{x\to4}(5x^2-2x+3) &= \lim_{x\to4}5x^2-\lim_{x\to4}2x+\lim_{x\to4}3 && \text{(Limit Laws 2 \& 1)}\\ &= 5\lim_{x\to4}x^2-2\lim_{x\to4}x+3 && (3\ \&\ 7)\\ &= 5(4)^2-2(4)+3=75 && (9\ \&\ 8)\end{aligned}$$

3.
$$\begin{aligned}\lim_{x\to2}(x^2+1)(x^2+4x) &= \lim_{x\to2}(x^2+1)\lim_{x\to2}(x^2+4x) && (4)\\ &= \left(\lim_{x\to2}x^2+\lim_{x\to2}1\right)\left(\lim_{x\to2}x^2+4\lim_{x\to2}x\right) && (1\ \&\ 3)\\ &= \left[(2)^2+1\right]\left[(2)^2+4(2)\right]=60 && (9,7\ \&\ 8)\end{aligned}$$

5.
$$\begin{aligned}\lim_{x\to-1}\frac{x-2}{x^2+4x-3} &= \frac{\lim_{x\to-1}(x-2)}{\lim_{x\to-1}(x^2+4x-3)} && (5)\\ &= \frac{\lim_{x\to-1}x-\lim_{x\to-1}2}{\lim_{x\to-1}x^2+4\lim_{x\to-1}x-\lim_{x\to-1}3} && (2,1\ \&\ 3)\\ &= \frac{(-1)-2}{(-1)^2+4(-1)-3}=\frac{1}{2} && (8,7\ \&\ 9)\end{aligned}$$

7.
$$\begin{aligned}\lim_{x\to-1}\sqrt{x^3+2x+7} &= \sqrt{\lim_{x\to-1}(x^3+2x+7)} && (11)\\ &= \sqrt{\lim_{x\to-1}x^3+2\lim_{x\to-1}x+\lim_{x\to-1}7} && (1\ \&\ 3)\\ &= \sqrt{(-1)^3+2(-1)+7}=2 && (9,8\ \&\ 6)\end{aligned}$$

9.
$$\begin{aligned}\lim_{t\to-2}(t+1)^9(t^2-1) &= \lim_{t\to-2}(t+1)^9\lim_{t\to-2}(t^2-1) && (4)\\ &= \left[\lim_{t\to-2}(t+1)\right]^9\lim_{t\to-2}(t^2-1) && (6)\\ &= \left[\lim_{t\to-2}t+\lim_{t\to-2}1\right]^9\left[\lim_{t\to-2}t^2-\lim_{t\to-2}1\right] && (1\ \&\ 2)\\ &= [(-2)+1]^9\left[(-2)^2-1\right]=-3 && (8,7\ \&\ 9)\end{aligned}$$

11.
$$\begin{aligned}\lim_{w\to-2}\sqrt[3]{\frac{4w+3w^3}{3w+10}} &= \sqrt[3]{\lim_{w\to-2}\frac{4w+3w^3}{3w+10}} && (11)\\ &= \sqrt[3]{\frac{\lim_{w\to-2}(4w+3w^3)}{\lim_{w\to-2}(3w+10)}} && (5)\\ &= \sqrt[3]{\frac{4\lim_{w\to-2}w+3\lim_{w\to-2}w^3}{3\lim_{w\to-2}w+\lim_{w\to-2}10}} && (1\ \&\ 3)\\ &= \sqrt[3]{\frac{4(-2)+3(-2)^3}{3(-2)+10}}=-2 && (8,9\ \&\ 7)\end{aligned}$$

13. **(a)** $\lim_{x \to a}[f(x) + h(x)] = \lim_{x \to a} f(x) + \lim_{x \to a} h(x) = -3 + 8 = 5$

(b) $\lim_{x \to a}[f(x)]^2 = \left[\lim_{x \to a} f(x)\right]^2 = (-3)^2 = 9$

(c) $\lim_{x \to a} \sqrt[3]{h(x)} = \sqrt[3]{\lim_{x \to a} h(x)} = \sqrt[3]{8} = 2$

(d) $\lim_{x \to a} \dfrac{1}{f(x)} = \dfrac{1}{\lim_{x \to a} f(x)} = \dfrac{1}{-3} = -\dfrac{1}{3}$

(e) $\lim_{x \to a} \dfrac{f(x)}{h(x)} = \dfrac{\lim_{x \to a} f(x)}{\lim_{x \to a} h(x)} = \dfrac{-3}{8} = -\dfrac{3}{8}$

(f) $\lim_{x \to a} \dfrac{g(x)}{f(x)} = \dfrac{\lim_{x \to a} g(x)}{\lim_{x \to a} f(x)} = \dfrac{0}{-3} = 0$

(g) The limit does not exist, since $\lim_{x \to a} g(x) = 0$ but $\lim_{x \to a} f(x) \neq 0$.

(h) $\lim_{x \to a} \dfrac{2f(x)}{h(x) - f(x)} = \dfrac{2 \lim_{x \to a} f(x)}{\lim_{x \to a} h(x) - \lim_{x \to a} f(x)} = \dfrac{2(-3)}{8 - (-3)} = -\dfrac{6}{11}$

15. $\lim_{x \to -3} \dfrac{x^2 - x + 12}{x + 3}$ does not exist since $x + 3 \to 0$ but $x^2 - x + 12 \to 24$ as $x \to -3$.

17. $\lim_{x \to -1} \dfrac{x^2 - x - 2}{x + 1} = \lim_{x \to -1} \dfrac{(x+1)(x-2)}{x+1} = \lim_{x \to -1} (x - 2) = -3$

19. $\lim_{t \to 1} \dfrac{t^3 - t}{t^2 - 1} = \lim_{t \to 1} \dfrac{t(t^2 - 1)}{t^2 - 1} = \lim_{t \to 1} t = 1$

21. $\lim_{h \to 0} \dfrac{(h-5)^2 - 25}{h} = \lim_{h \to 0} \dfrac{(h^2 - 10h + 25) - 25}{h} = \lim_{h \to 0} \dfrac{h^2 - 10h}{h} = \lim_{h \to 0}(h - 10) = -10$

23. $\lim_{h \to 0} \dfrac{(1+h)^4 - 1}{h} = \lim_{h \to 0} \dfrac{(1 + 4h + 6h^2 + 4h^3 + h^4) - 1}{h} = \lim_{h \to 0} \dfrac{4h + 6h^2 + 4h^3 + h^4}{h}$

$= \lim_{h \to 0}(4 + 6h + 4h^2 + h^3) = 4$

25. $\lim_{x \to -2} \dfrac{x + 2}{x^2 - x - 6} = \lim_{x \to -2} \dfrac{x + 2}{(x-3)(x+2)} = \lim_{x \to -2} \dfrac{1}{x - 3} = -\dfrac{1}{5}$

27. $\lim_{t \to 9} \dfrac{9 - t}{3 - \sqrt{t}} = \lim_{t \to 9} \dfrac{(3 + \sqrt{t})(3 - \sqrt{t})}{3 - \sqrt{t}} = \lim_{t \to 9}(3 + \sqrt{t}) = 3 + \sqrt{9} = 6$

29. $\lim_{t \to 0} \dfrac{\sqrt{2-t} - \sqrt{2}}{t} = \lim_{t \to 0} \dfrac{\sqrt{2-t} - \sqrt{2}}{t} \cdot \dfrac{\sqrt{2-t} + \sqrt{2}}{\sqrt{2-t} + \sqrt{2}} = \lim_{t \to 0} \dfrac{-t}{t(\sqrt{2-t} + \sqrt{2})} = \lim_{t \to 0} \dfrac{-1}{\sqrt{2-t} + \sqrt{2}}$

$= -\dfrac{1}{2\sqrt{2}} = -\dfrac{\sqrt{2}}{4}$

31. $\lim_{x \to 9} \dfrac{x^2 - 81}{\sqrt{x} - 3} = \lim_{x \to 9} \dfrac{(x-9)(x+9)}{\sqrt{x} - 3} = \lim_{x \to 9} \dfrac{(\sqrt{x} - 3)(\sqrt{x} + 3)(x + 9)}{\sqrt{x} - 3}$

$= \lim_{x \to 9}(\sqrt{x} + 3)(x + 9) = \lim_{x \to 9}(\sqrt{x} + 3) \lim_{x \to 9}(x + 9) = (\sqrt{9} + 3)(9 + 9) = 108$

33. $\lim_{t \to 0} \left[\dfrac{1}{t\sqrt{1+t}} - \dfrac{1}{t}\right] = \lim_{t \to 0} \dfrac{1 - \sqrt{1+t}}{t\sqrt{1+t}} = \lim_{t \to 0} \dfrac{(1 - \sqrt{1+t})(1 + \sqrt{1+t})}{t\sqrt{t+1}(1 + \sqrt{1+t})} = \lim_{t \to 0} \dfrac{-t}{t\sqrt{1+t}(1 + \sqrt{1+t})}$

$= \lim_{t \to 0} \dfrac{-1}{\sqrt{1+t}(1 + \sqrt{1+t})} = \dfrac{-1}{\sqrt{1+0}(1 + \sqrt{1+0})} = -\dfrac{1}{2}$

35. $\lim\limits_{x\to 0}\dfrac{x}{\sqrt{1+3x}-1}=\lim\limits_{x\to 0}\dfrac{x\left(\sqrt{1+3x}+1\right)}{\left(\sqrt{1+3x}-1\right)\left(\sqrt{1+3x}+1\right)}=\lim\limits_{x\to 0}\dfrac{x\left(\sqrt{1+3x}+1\right)}{3x}$

$$=\lim_{x\to 0}\frac{\sqrt{1+3x}+1}{3}=\frac{\sqrt{1}+1}{3}=\frac{2}{3}$$

37. $\lim\limits_{x\to 2}\dfrac{x-\sqrt{3x-2}}{x^2-4}=\lim\limits_{x\to 2}\dfrac{\left(x-\sqrt{3x-2}\right)\left(x-\sqrt{3x-2}\right)}{\left(x^2-4\right)\left(x-\sqrt{3x-2}\right)}=\lim\limits_{x\to 2}\dfrac{x^2-3x+2}{\left(x^2-4\right)\left(x+\sqrt{3x-2}\right)}$

$$=\lim_{x\to 2}\frac{(x-2)(x-1)}{(x-2)(x+2)\left(x+\sqrt{3x-2}\right)}=\lim_{x\to 2}\frac{(x-1)}{(x+2)\left(x+\sqrt{3x-2}\right)}=\frac{1}{4\left(2+\sqrt{4}\right)}=\frac{1}{16}$$

39. Let $f(x)=-x^2$, $g(x)=x^2\cos 20\pi x$ and $h(x)=x^2$.
Then $-1\le\cos 20\pi x\le 1 \quad\Rightarrow\quad f(x)\le g(x)\le h(x)$.
So since $\lim\limits_{x\to 0}f(x)=\lim\limits_{x\to 0}h(x)=0$, by the Squeeze Theorem we have
$\lim\limits_{x\to 0}\left(x^2\cos 20\pi x\right)=\lim\limits_{x\to 0}g(x)=0$.

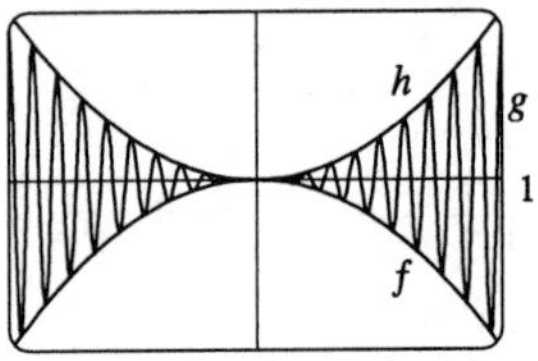

41. $1\le f(x)\le x^2+2x+2$ for all x. But $\lim\limits_{x\to -1}1=1$ and $\lim\limits_{x\to -1}\left(x^2+2x+2\right)=\lim\limits_{x\to -1}x^2+2\lim\limits_{x\to -1}x+\lim\limits_{x\to -1}2$
$=(-1)^2+2(-1)+2=1$. Therefore, by the Squeeze Theorem, $\lim\limits_{x\to -1}f(x)=1$.

43. $-1\le\sin(1/x)\le 1 \quad\Rightarrow\quad -x^2\le x^2\sin(1/x)\le x^2$. Since $\lim\limits_{x\to 0}\left(-x^2\right)=0$ and $\lim\limits_{x\to 0}x^2=0$, we have
$\lim\limits_{x\to 0}x^2\sin(1/x)=0$ by the Squeeze Theorem.

45. $\lim\limits_{x\to 4^-}\sqrt{16-x^2}=\sqrt{\lim\limits_{x\to 4^-}16-\lim\limits_{x\to 4^-}x^2}=\sqrt{16-4^2}=0$

47. If $x>-4$, then $|x+4|=x+4$, so $\lim\limits_{x\to -4^+}|x+4|=\lim\limits_{x\to -4^+}(x+4)=-4+4=0$.
If $x<-4$, then $|x+4|=-(x+4)$, so $\lim\limits_{x\to -4^-}|x+4|=\lim\limits_{x\to -4^-}-(x+4)=4-4=0$.
Therefore $\lim\limits_{x\to -4}|x+4|=0$.

49. If $x>2$, then $|x-2|=x-2$, so $\lim\limits_{x\to 2^+}\dfrac{|x-2|}{x-2}=\lim\limits_{x\to 2^+}\dfrac{x-2}{x-2}=\lim\limits_{x\to 2^+}1=1$.

If $x<2$, then $|x-2|=-(x-2)$ so $\lim\limits_{x\to 2^-}\dfrac{|x-2|}{x-2}=\lim\limits_{x\to 2^-}\dfrac{-(x-2)}{x-2}=\lim\limits_{x\to 2^-}-1=-1$.

The right and left limits are different, so $\lim\limits_{x\to 2}\dfrac{|x-2|}{x-2}$ does not exist.

51. $[\![x]\!]=-2$ for $-2\le x<-1$, so $\lim\limits_{x\to -2^+}[\![x]\!]=\lim\limits_{x\to -2^+}(-2)=-2$

53. $[\![x]\!]=-3$ for $-3\le x<-2$, so $\lim\limits_{x\to -2.4}[\![x]\!]=\lim\limits_{x\to -2.4}(-3)=-3$.

55. $\lim\limits_{x\to 1^+}\sqrt{x^2+x-2}=\sqrt{\lim\limits_{x\to 1^+}x^2+\lim\limits_{x\to 1^+}x-\lim\limits_{x\to 1^+}2}=\sqrt{1^2+1-2}=0$

Notice that the domain of $\sqrt{x^2+x-2}$ is $(-\infty,-2]\cup[1,\infty)$.

57. Since $|x| = -x$ for $x < 0$, we have $\lim_{x\to 0^-}\left(\frac{1}{x} - \frac{1}{|x|}\right) = \lim_{x\to 0^-}\left(\frac{1}{x} - \frac{1}{-x}\right) = \lim_{x\to 0^-}\frac{2}{x}$, which does not exist since the denominator $\to 0$ and the numerator does not.

59. **(a)**

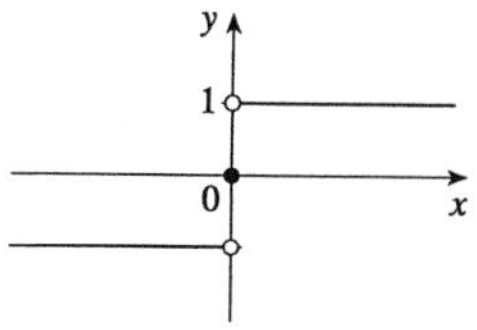

(b) **(i)** Since $\operatorname{sgn} x = 1$ for $x > 0$, $\lim_{x\to 0^+}\operatorname{sgn} x = \lim_{x\to 0^+} 1 = 1$.

(ii) Since $\operatorname{sgn} x = -1$ for $x < 0$, $\lim_{x\to 0^-}\operatorname{sgn} x = \lim_{x\to 0^-} -1 = -1$.

(iii) Since $\lim_{x\to 0^-}\operatorname{sgn} x \neq \lim_{x\to 0^+}\operatorname{sgn} x$, $\lim_{x\to 0}\operatorname{sgn} x$ does not exist.

(iv) Since $|\operatorname{sgn} x| = 1$ for $x \neq 0$, $\lim_{x\to 0}|\operatorname{sgn} x| = \lim_{x\to 0} 1 = 1$.

61. **(a)** $\lim_{x\to -1^-} g(x) = \lim_{x\to -1^-}(-x^3) = -(-1)^3 = 1$,

$\lim_{x\to -1^+} g(x) = \lim_{x\to -1^+}(x+2)^2 = (-1+2)^2 = 1$

(b) By part (a), $\lim_{x\to -1} g(x) = 1$.

(c)

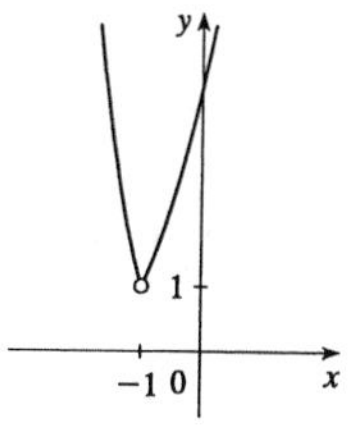

63. **(a)** **(i)** $[\![x]\!] = n - 1$ for $n - 1 \le x < n$, so $\lim_{x\to n^-}[\![x]\!] = \lim_{x\to n^-}(n-1) = n - 1$.

(ii) $[\![x]\!] = n$ for $n \le x < n + 1$, so $\lim_{x\to n^+}[\![x]\!] = \lim_{x\to n^+} n = n$.

(b) $\lim_{x\to a}[\![x]\!]$ exists $\Leftrightarrow$ a is not an integer.

65. **(a)** **(i)** $\lim_{x\to 1^+}\frac{x^2-1}{|x-1|} = \lim_{x\to 1^+}\frac{x^2-1}{x-1} = \lim_{x\to 1^+}(x+1) = 2$

(ii) $\lim_{x\to 1^-}\frac{x^2-1}{|x-1|} = \lim_{x\to 1^-}\frac{x^2-1}{-(x-1)} = \lim_{x\to 1^-} -(x+1) = -2$

(b) No, $\lim_{x\to 1} F(x)$ does not exist since $\lim_{x\to 1^+} F(x) \neq \lim_{x\to 1^-} F(x)$.

(c)

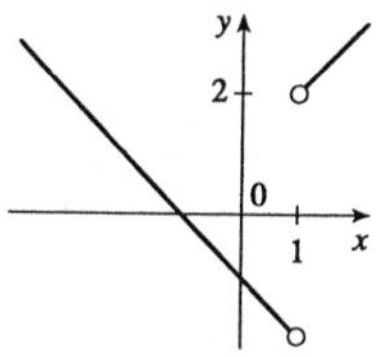

67. Since $p(x)$ is a polynomial, $p(x) = a_0 + a_1 x + a_2 x^2 + \cdots + a_n x^n$. Thus, by the Limit Laws,

$$\lim_{x\to a} p(x) = \lim_{x\to a}\left(a_0 + a_1 x + a_2 x^2 + \cdots + a_n x^n\right) = a_0 + a_1\lim_{x\to a} x + a_2\lim_{x\to a} x^2 + \cdots + a_n\lim_{x\to a} x^n$$

$= a_0 + a_1 a + a_2 a^2 + \cdots + a_n a^n = p(a)$. Thus, for any polynomial p, $\lim_{x\to a} p(x) = p(a)$.

69. Observe that $0 \le f(x) \le x^2$ for all x, and $\lim_{x\to 0} 0 = 0 = \lim_{x\to 0} x^2$. So, by the Squeeze Theorem, $\lim_{x\to 0} f(x) = 0$.

71. Let $f(x) = H(x)$ and $g(x) = 1 - H(x)$, where H is the Heaviside function defined in Example 1.2.6. Then $\lim_{x\to 0} f(x)$ and $\lim_{x\to 0} g(x)$ do not exist but $\lim_{x\to 0}[f(x)g(x)] = \lim_{x\to 0} 0 = 0$.

73. Let $t = \sqrt[3]{1+cx}$. Then $t \to 1$ as $x \to 0$ and $t^3 = 1 + cx$ $\Rightarrow$ $x = (t^3 - 1)/c$. (If $c = 0$, then the limit is obviously 0.) Therefore

$$\lim_{x\to 0}\frac{\sqrt[3]{1+cx}-1}{x} = \lim_{t\to 1}\frac{t-1}{(t^3-1)/c} = \lim_{t\to 1}\frac{c(t-1)}{(t-1)(t^2+t+1)} = \lim_{t\to 1}\frac{c}{t^2+t+1} = \frac{c}{1^2+1+1} = \frac{c}{3}.$$

Another Method: Multiply numerator and denominator by $(1+cx)^{2/3} + (1+cx)^{1/3} + 1$.

75. Since the denominator approaches 0 as $x \to -2$, the limit will exist only if the numerator also approaches 0 as $x \to -2$. In order for this to happen, we need $\lim\limits_{x\to-2}(3x^2+ax+a+3)=0 \quad \Leftrightarrow$

$3(-2)^2+a(-2)+a+3=0 \quad \Leftrightarrow \quad 12-2a+a+3=0 \quad \Leftrightarrow \quad a=15$. With $a=15$, the limit becomes

$\lim\limits_{x\to-2}\dfrac{3x^2+15x+18}{x^2+x-2}=\lim\limits_{x\to-2}\dfrac{3(x+2)(x+3)}{(x-1)(x+2)}=\dfrac{3(-2+3)}{-2-1}=-1.$

77. $y-1<[\![y]\!]\le y$, so $x^2\left(\dfrac{1}{4x^2}-1\right)<x^2\left[\!\!\left[\dfrac{1}{4x^2}\right]\!\!\right]\le x^2\left(\dfrac{1}{4x^2}\right)=\dfrac{1}{4} \quad (x\ne 0)$. But

$\lim\limits_{x\to0}x^2\left(\dfrac{1}{4x^2}-1\right)=\lim\limits_{x\to0}\left(\dfrac{1}{4}-x^2\right)=\dfrac{1}{4}$ and $\lim\limits_{x\to0}\dfrac{1}{4}=\dfrac{1}{4}$. So by the Squeeze Theorem, $\lim\limits_{x\to0}x^2\left[\!\!\left[\dfrac{1}{4x^2}\right]\!\!\right]=\dfrac{1}{4}$.

EXERCISES 1.4

1. **(a)** $|(6x+1)-19|<0.1 \quad \Leftrightarrow \quad |6x-18|<0.1 \quad \Leftrightarrow \quad 6|x-3|<0.1 \quad \Leftrightarrow \quad |x-3|<(0.1)/6=\frac{1}{60}$

(b) $|(6x+1)-19|<0.01 \quad \Leftrightarrow \quad |x-3|<(0.01)/6=\frac{1}{600}$

3. On the left side, we need $|x-2|<\left|\frac{10}{7}-2\right|=\frac{4}{7}$. On the right side, we need $|x-2|<\left|\frac{10}{3}-2\right|=\frac{4}{3}$. For both of these conditions to be satisfied at once, we need the more restrictive of the two to hold, that is, $|x-2|<\frac{4}{7}$. So we can choose $\delta=\frac{4}{7}$, or any smaller positive number.

5. $\left|\sqrt{4x+1}-3\right|<0.5 \quad \Leftrightarrow \quad 2.5<\sqrt{4x+1}<3.5.$

We plot the three parts of this inequality on the same screen and identify the x-coordinates of the points of intersection using the cursor. It appears that the inequality holds for $1.32\le x\le 2.81$. Since $|2-1.32|=0.68$ and $|2-2.81|=0.81$, we choose $0<\delta\le\min\{0.68,0.81\}=0.68$.

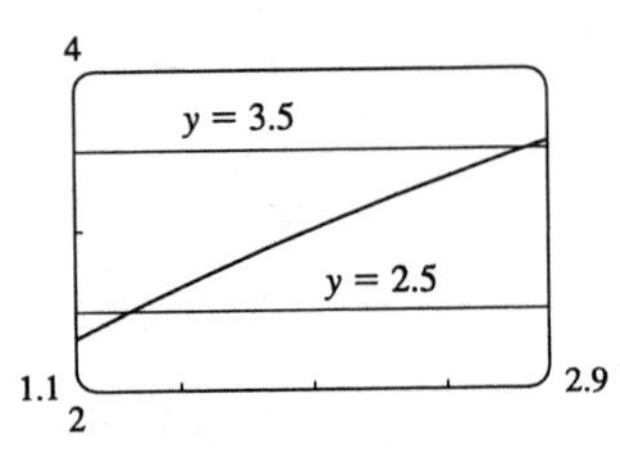

7. For $\epsilon=1$, the definition of a limit requires that we find δ such that $|(4+x-3x^3)-2|<1 \quad \Leftrightarrow$ $1<4+x-3x^3<3$ whenever $|x-1|<\delta$. If we plot the graphs of $y=1$, $y=4+x-3x^3$ and $y=3$ on the same screen, we see that we need $0.86\le x\le 1.11$. So since $|1-0.86|=0.14$ and $|1-1.11|=0.11$, we choose $\delta=0.11$ (or any smaller positive number). For $\epsilon=0.1$, we must find δ such that $|(4+x-3x^3)-2|<0.1 \quad \Leftrightarrow$ $1.9<4+x-3x^3<2.1$ whenever $|x-1|<\delta$. From the graph, we see that we need $0.988\le x\le 1.012$. So since $|1-0.988|=0.012$ and $|1-1.012|=0.012$, we must choose $\delta=0.012$ (or any smaller positive number) for the inequality to hold.

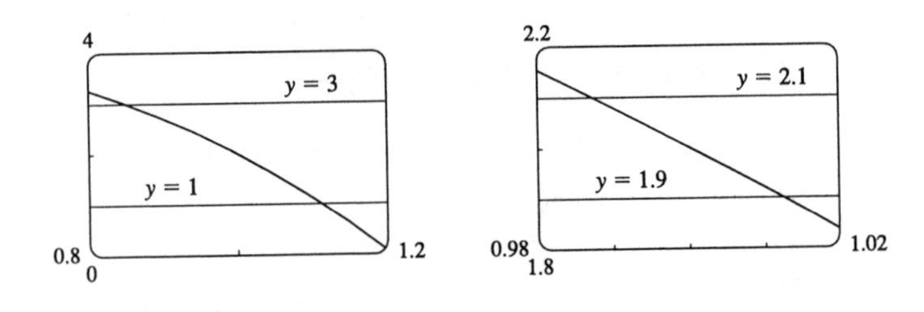

9. From the graph, we see that $\dfrac{x}{(x^2+1)(x-1)^2} > 100$ whenever $0.93 \le x \le 1.07$. So since $|1-0.93| = 0.7$ and $|1-1.07| = 0.7$, we can take $\delta = 0.07$ (or any smaller positive number).

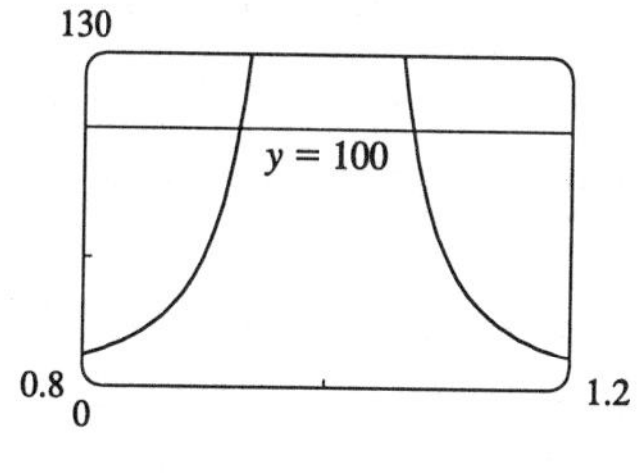

11. Given $\epsilon > 0$, we need $\delta > 0$ such that if $|x-2| < \delta$, then $|(3x-2)-4| < \epsilon \iff |3x-6| < \epsilon \iff 3|x-2| < \epsilon \iff |x-2| < \epsilon/3$. So if we choose $\delta = \epsilon/3$, then $|x-2| < \delta \Rightarrow |(3x-2)-4| < \epsilon$. Thus $\lim_{x\to 2}(3x-2) = 4$ by the definition of a limit.

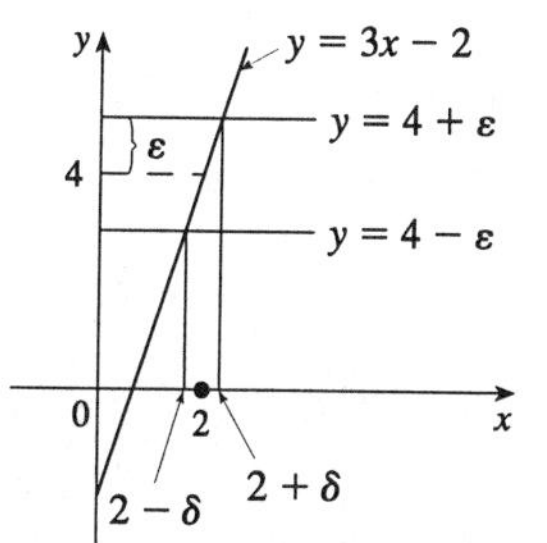

13. Given $\epsilon > 0$, we need $\delta > 0$ such that if $|x-(-1)| < \delta$, then $|(5x+8)-3| < \epsilon \iff |5x+5| < \epsilon \iff 5|x+1| < \epsilon \iff |x-(-1)| < \epsilon/5$. So if we choose $\delta = \epsilon/5$, then $|x-(-1)| < \delta \Rightarrow |(5x+8)-3| < \epsilon$. Thus $\lim_{x\to -1}(5x+8) = 3$ by the definition of a limit.

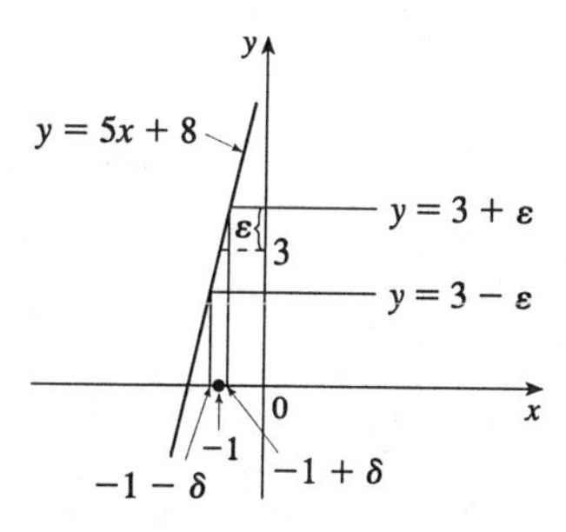

15. Given $\epsilon > 0$, we need $\delta > 0$ such that if $|x-2| < \delta$ then $\left|\dfrac{x}{7} - \dfrac{2}{7}\right| < \epsilon \iff \frac{1}{7}|x-2| < \epsilon \iff |x-2| < 7\epsilon$. So take $\delta = 7\epsilon$. Then $|x-2| < \delta \Rightarrow \left|\dfrac{x}{7} - \dfrac{2}{7}\right| < \epsilon$. Thus $\lim_{x\to 2}\dfrac{x}{7} = \dfrac{2}{7}$ by the definition of a limit.

17. Given $\epsilon > 0$, we need $\delta > 0$ such that if $|x-(-5)| < \delta$ then $\left|\left(4 - \frac{3}{5}x\right) - 7\right| < \epsilon \iff \frac{3}{5}|x+5| < \epsilon \iff |x-(-5)| < \frac{5}{3}\epsilon$. So take $\delta = \frac{5}{3}\epsilon$. Then $|x-(-5)| < \delta \Rightarrow \left|\left(4 - \frac{3}{5}x\right) - 7\right| < \epsilon$. Thus $\lim_{x\to -5}\left(4 - \frac{3}{5}x\right) = 7$ by the definition of a limit.

19. Given $\epsilon > 0$, we need $\delta > 0$ such that if $|x-a| < \delta$ then $|x-a| < \epsilon$. So $\delta = \epsilon$ will work.

21. Given $\epsilon > 0$, we need $\delta > 0$ such that if $|x| < \delta$ then $|x^2 - 0| < \epsilon \iff x^2 < \epsilon \iff |x| < \sqrt{\epsilon}$. Take $\delta = \sqrt{\epsilon}$. Then $|x-0| < \delta \Rightarrow |x^2 - 0| < \epsilon$. Thus $\lim_{x\to 0} x^2 = 0$ by the definition of a limit.

23. Given $\epsilon > 0$, we need $\delta > 0$ such that if $|x-0| < \delta$ then $||x| - 0| < \epsilon$. But $||x|| = |x|$. So this is true if we pick $\delta = \epsilon$.

25. Given $\epsilon > 0$, we need $\delta > 0$ such that if $|x-2| < \delta$, then $|(x^2-4x+5)-1| < \epsilon \quad \Leftrightarrow \quad |x^2-4x+4| < \epsilon$ $\quad \Leftrightarrow \quad |(x-2)^2| < \epsilon$. So take $\delta = \sqrt{\epsilon}$. Then $|x-2| < \delta \quad \Leftrightarrow \quad |x-2| < \sqrt{\epsilon} \quad \Leftrightarrow \quad |(x-2)^2| < \epsilon$. So $\lim\limits_{x\to 2}(x^2-4x+5) = 1$ by the definition of a limit.

27. Given $\epsilon > 0$, we need $\delta > 0$ such that if $|x-(-2)| < \delta$ then $|(x^2-1)-3| < \epsilon$ or upon simplifying we need $|x^2-4| < \epsilon$ whenever $|x+2| < \delta$. Notice that if $|x+2| < 1$, then $-1 < x+2 < 1 \quad \Rightarrow \quad -5 < x-2 < -3$ $\quad \Rightarrow \quad |x-2| < 5$. So take $\delta = \min\{\epsilon/5, 1\}$. Then $|x-2| < 5$ and $|x+2| < \epsilon/5$, so $|(x^2-1)-3| = |(x+2)(x-2)| = |x+2||x-2| < (\epsilon/5)(5) = \epsilon$.
Therefore, by the definition of a limit, $\lim\limits_{x\to -2}(x^2-1) = 3$.

29. Given $\epsilon > 0$, we let $\delta = \min\left\{2, \frac{\epsilon}{8}\right\}$. If $0 < |x-3| < \delta$ then $|x-3| < 2 \quad \Rightarrow \quad 1 < x < 5 \quad \Rightarrow$ $|x+3| < 8$. Also $|x-3| < \frac{\epsilon}{8}$, so $|x^2-9| = |x+3||x-3| < 8 \cdot \frac{\epsilon}{8} = \epsilon$. Thus $\lim\limits_{x\to 3} x^2 = 9$.

31. *1. Guessing a value for* δ Given $\epsilon > 0$, we must find $\delta > 0$ such that $\left|\sqrt{x}-\sqrt{a}\right| < \epsilon$ whenever $0 < |x-a| < \delta$. But $\left|\sqrt{x}-\sqrt{a}\right| = \dfrac{|x-a|}{\sqrt{x}+\sqrt{a}} < \epsilon$ (from the hint). Now if we can find a positive constant C such that $\sqrt{x}+\sqrt{a} > C$ then $\dfrac{|x-a|}{\sqrt{x}+\sqrt{a}} < \dfrac{|x-a|}{C} < \epsilon$, and we take $|x-a| < C\epsilon$. We can find this number by restricting x to lie in some interval centered at a. If $|x-a| < \frac{1}{2}a$, then $\frac{1}{2}a < x < \frac{3}{2}a \quad \Rightarrow$ $\sqrt{x}+\sqrt{a} > \sqrt{\frac{1}{2}a}+\sqrt{a}$, and so $C = \sqrt{\frac{1}{2}a}+\sqrt{a}$ is a suitable choice for the constant. So $|x-a| < \left(\sqrt{\frac{1}{2}a}+\sqrt{a}\right)\epsilon$. This suggests that we let $\delta = \min\left\{\frac{1}{2}a, \left(\sqrt{\frac{1}{2}a}+\sqrt{a}\right)\epsilon\right\}$.

2. Showing that δ *works* Given $\epsilon > 0$, we let $\delta = \min\left\{\frac{1}{2}a, \left(\sqrt{\frac{1}{2}a}+\sqrt{a}\right)\epsilon\right\}$. If $0 < |x-a| < \delta$, then $|x-a| < \frac{1}{2}a \quad \Rightarrow \quad \sqrt{x}+\sqrt{a} > \sqrt{\frac{1}{2}a}+\sqrt{a}$ (as in part 1). Also $|x-a| < \left(\sqrt{\frac{1}{2}a}+\sqrt{a}\right)\epsilon$, so $\left|\sqrt{x}+\sqrt{a}\right| = \dfrac{|x-a|}{\sqrt{x}+\sqrt{a}} < \dfrac{\left(\sqrt{a/2}+\sqrt{a}\right)\epsilon}{\left(\sqrt{a/2}+\sqrt{a}\right)} = \epsilon$. Therefore $\lim\limits_{x\to a}\sqrt{x} = \sqrt{a}$ by the definition of a limit.

33. Suppose that $\lim\limits_{x\to 0} f(x) = L$. Given $\epsilon = \frac{1}{2}$, there exists $\delta > 0$ such that $0 < |x| < \delta \quad \Rightarrow \quad |f(x)-L| < \frac{1}{2}$. Take any rational number r with $0 < |r| < \delta$. Then $f(r) = 0$, so $|0-L| < \frac{1}{2}$, so $L \le |L| < \frac{1}{2}$. Now take any irrational number s with $0 < |s| < \delta$. Then $f(s) = 1$, so $|1-L| < \frac{1}{2}$. Hence $1-L < \frac{1}{2}$, so $L > \frac{1}{2}$. This contradicts $L < \frac{1}{2}$, so $\lim\limits_{x\to 0} f(x)$ does not exist.

35. $\dfrac{1}{(x+3)^4} > 10{,}000 \quad \Leftrightarrow \quad (x+3)^4 < \dfrac{1}{10{,}000} \quad \Leftrightarrow \quad |x-(-3)| = |x+3| < \dfrac{1}{10}$

37. Let $N < 0$ be given. Then, for $x < -1$, we have $\dfrac{5}{(x+1)^3} < N \quad \Leftrightarrow \quad \dfrac{5}{N} < (x+1)^3 \quad \Leftrightarrow$ $\sqrt[3]{\dfrac{5}{N}} < x+1$. Let $\delta = -\sqrt[3]{\dfrac{5}{N}}$. Then $-1-\delta < x < -1 \quad \Rightarrow \quad \sqrt[3]{\dfrac{5}{N}} < x+1 < 0 \quad \Rightarrow \quad \dfrac{5}{(x+1)^3} < N$, so $\lim\limits_{x\to -1^-} \dfrac{5}{(x+1)^3} = -\infty$.

EXERCISES 1.5

1. **(a)** f is discontinuous at -5, -3, -1, 3, 5, 8 and 10.

(b) f is continuous from the left at -5 and -3, and continuous from the right at 8.

It is continuous from neither side at -1, 3, 5, and 10.

3. $\lim\limits_{x\to 3}(x^4 - 5x^3 + 6) = \lim\limits_{x\to 3} x^4 - 5\lim\limits_{x\to 3} x^3 + \lim\limits_{x\to 3} 6 = 3^4 - 5(3^3) + 6 = -48 = f(3)$. Thus f is continuous at 3.

5. $\lim\limits_{x\to 5} f(x) = \lim\limits_{x\to 5}\left(1 + \sqrt{x^2 - 9}\right) = \lim\limits_{x\to 5} 1 + \sqrt{\lim\limits_{x\to 5} x^2 - \lim\limits_{x\to 5} 9} = 1 + \sqrt{5^2 - 9} = 5 = f(5)$. Thus f is continuous at 5.

7. $\lim\limits_{t\to -8} g(t) = \lim\limits_{t\to -8} \dfrac{\sqrt[3]{t}}{(t+1)^4} = \dfrac{\sqrt[3]{\lim\limits_{t\to -8} t}}{\left(\lim\limits_{t\to -8} t + 1\right)^4} = \dfrac{\sqrt[3]{-8}}{(-8+1)^4} = -\dfrac{2}{2401} = g(-8)$. Thus g is continuous at -8.

9. For $-4 < a < 4$ we have $\lim\limits_{x\to a} f(x) = \lim\limits_{x\to a} x\sqrt{16 - x^2} = \lim\limits_{x\to a} x\sqrt{\lim\limits_{x\to a} 16 - \lim\limits_{x\to a} x^2} = a\sqrt{16 - a^2} = f(a)$, so f is continuous on $(-4, 4)$. Similarly, we get $\lim\limits_{x\to 4^-} f(x) = 0 = f(4)$ and $\lim\limits_{x\to -4^+} f(x) = 0 = f(-4)$, so f is continuous from the left at 4 and from the right at -4. Thus f is continuous on $[-4, 4]$.

11. For any $a \in \mathbb{R}$ we have $\lim\limits_{x\to a} f(x) = \lim\limits_{x\to a}(x^2 - 1)^8 = \left(\lim\limits_{x\to a} x^2 - \lim\limits_{x\to a} 1\right)^8 = (a^2 - 1)^8 = f(a)$. Thus f is continuous on $(-\infty, \infty)$.

13. $f(x) = -\dfrac{1}{(x-1)^2}$ is discontinuous at 1 since $f(1)$ is not defined.

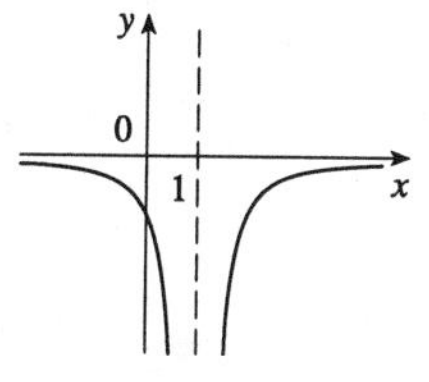

15. $\lim\limits_{x\to 1} f(x) = \lim\limits_{x\to 1}\left[-\dfrac{1}{(x-1)^2}\right]$ does not exist.

Therefore f is discontinuous at 1.

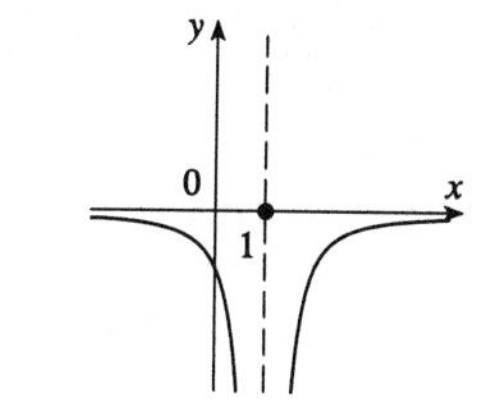

17. Since $f(x) = x^2 - 2$ for $x \neq -3$,

$\lim\limits_{x\to -3} f(x) = \lim\limits_{x\to -3}(x^2 - 2) = (-3)^2 - 2 = 7$.

But $f(-3) = 5$, so $\lim\limits_{x\to -3} f(x) \neq f(-3)$.

Therefore f is discontinuous at -3.

19. $f(x) = (x+1)(x^3 + 8x + 9)$ is a polynomial, so by Theorem 5 it is continuous on $\mathbb{R}$.

21. $g(x) = x + 1$, a polynomial, is continuous (by Theorem 5) and $f(x) = \sqrt{x}$ is continuous on $[0, \infty)$ by Theorem 6, so $f(g(x)) = \sqrt{x+1}$ is continuous on $[-1, \infty)$ by Theorem 8. By Theorem 4 #5, $H(x) = 1/\sqrt{x+1}$ is continuous on $(-1, \infty)$.

23. $g(x) = x - 1$ and $G(x) = x^2 - 2$ are both polynomials, so by Theorem 5 they are continuous. Also $f(x) = \sqrt[5]{x}$ is continuous by Theorem 6, so $f(g(x)) = \sqrt[5]{x-1}$ is continuous on $\mathbb{R}$ by Theorem 8. Thus the product $h(x) = \sqrt[5]{x-1}\,(x^2 - 2)$ is continuous on $\mathbb{R}$ by Theorem 4 #4.

25. Since the discriminant of $t^2 + t + 1$ is negative, $t^2 + t + 1$ is always positive. So the domain of $F(t)$ is $\mathbb{R}$. By Theorem 5 the polynomial $(t^2 + t + 1)^3$ is continuous. By Theorems 6 and 8 the composition $F(t) = \sqrt{(t^2 + t + 1)^3}$ is continuous on $\mathbb{R}$.

27. $g(x) = x^3 - x$ is continuous on $\mathbb{R}$ since it is a polynomial [Theorem 5(a)], and $f(x) = |x|$ is continuous on $\mathbb{R}$ by Example 9(a). So $L(x) = |x^3 - x|$ is continuous on $\mathbb{R}$ by Theorem 8.

29. f is continuous on $(-\infty, 3)$ and $(3, \infty)$ since on each of these intervals it is a polynomial. Also $\lim\limits_{x \to 3^+} f(x) = \lim\limits_{x \to 3^+} (5 - x) = 2$ and $\lim\limits_{x \to 3^-} f(x) = \lim\limits_{x \to 3^-} (x - 1) = 2$, so $\lim\limits_{x \to 3} f(x) = 2$. Since $f(3) = 5 - 3 = 2$, f is also continuous at 3. Thus f is continuous on $(-\infty, \infty)$.

31. f is continuous on $(-\infty, 0)$ and $(0, \infty)$ since on each of these intervals it is a polynomial. Now $\lim\limits_{x \to 0^-} f(x) = \lim\limits_{x \to 0^-} (x - 1)^3 = -1$ and $\lim\limits_{x \to 0^+} f(x) = \lim\limits_{x \to 0^+} (x + 1)^3 = 1$. Thus $\lim\limits_{x \to 0} f(x)$ does not exist, so f is discontinuous at 0. Since $f(0) = 1$, f is continuous from the right at 0.

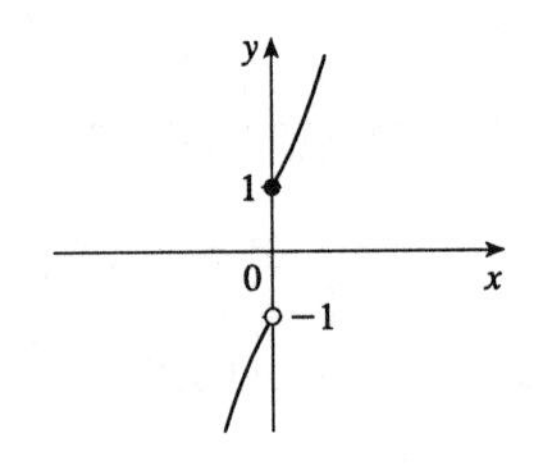

33. f is continuous on $(-\infty, -1)$, $(-1, 1)$ and $(1, \infty)$. Now

$\lim\limits_{x \to -1^-} f(x) = \lim\limits_{x \to -1^-} \dfrac{1}{x} = -1$ and $\lim\limits_{x \to -1^+} f(x) = \lim\limits_{x \to -1^+} x = -1$,

so $\lim\limits_{x \to -1} f(x) = -1 = f(-1)$ and f is continuous at -1.

Also $\lim\limits_{x \to 1^-} f(x) = \lim\limits_{x \to 1^-} x = 1$ and $\lim\limits_{x \to 1^+} f(x) = \lim\limits_{x \to 1^+} \dfrac{1}{x^2} = 1$,

so $\lim\limits_{x \to 1} f(x) = 1 = f(1)$ and f is continuous at 1.

Thus f has no discontinuities.

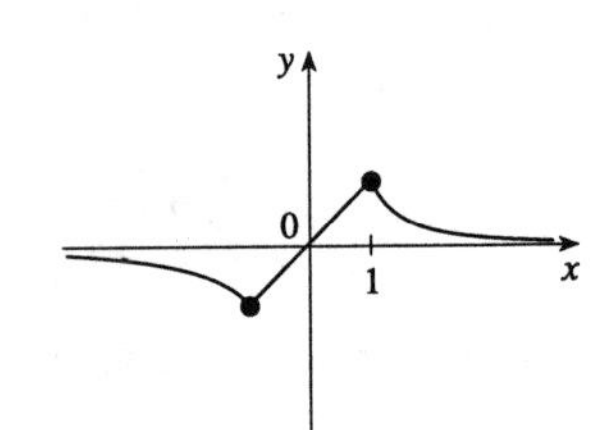

35. $f(x) = [\![2x]\!]$ is continuous except when $2x = n \iff x = n/2$, n an integer. In fact, $\lim\limits_{x \to n/2^-} [\![2x]\!] = n - 1$ and $\lim\limits_{x \to n/2^+} [\![2x]\!] = n = f(n)$, so f is continuous only from the right at $n/2$.

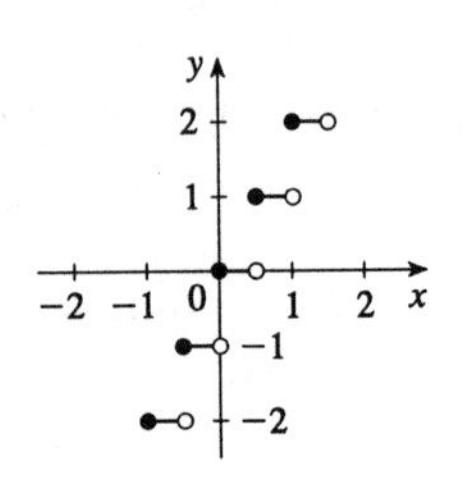

37. f is continuous on $(-\infty, 3)$ and $(3, \infty)$. Now $\lim_{x \to 3^-} f(x) = \lim_{x \to 3^-} (cx + 1) = 3c + 1$ and $\lim_{x \to 3^+} f(x) = \lim_{x \to 3^+} (cx^2 - 1) = 9c - 1$. So f is continuous $\Leftrightarrow$ $3c + 1 = 9c - 1$ $\Leftrightarrow$ $6c = 2$ $\Leftrightarrow$ $c = \frac{1}{3}$. Thus for f to be continuous on $(-\infty, \infty)$, $c = \frac{1}{3}$.

39. The functions $2x$, $cx^2 + d$ and $4x$ are continuous on their own domains, so the only possible problems occur at $x = 1$ and $x = 2$. The left- and right-hand limits at these points must be the same in order for $\lim_{x \to 1} h(x)$ and $\lim_{x \to 2} h(x)$ to exist. So we must have $2 \cdot 1 = c(1)^2 + d$ and $c(2)^2 + d = 4 \cdot 2$. From the first of these equations we get $d = 2 - c$. Substituting this into the second, we get $4c + (2 - c) = 8$ $\Leftrightarrow$ $c = 2$. Back-substituting into the first to get d, we find that $d = 0$.

41. **(a)** $\lim_{x \to 1^-} f(x) = \lim_{x \to 1^-} (1 - x^2) = 0$ and $\lim_{x \to 1^+} f(x) = \lim_{x \to 1^+} (1 + x/2) = \frac{3}{2}$. Thus $\lim_{x \to 1} f(x)$ does not exist, so f is not continuous at 1.

(b) $f(0) = 1$ and $f(2) = 2$. For $0 \le x \le 1$, f takes the values in $[0, 1]$. For $1 < x \le 2$, f takes the values in $(1.5, 2]$. Thus f does not take on the value 1.5 $\big($or any other value in $(1, 1.5]\big)$.

43. $f(x) = x^3 - x^2 + x$ is continuous on $[2, 3]$ and $f(2) = 6$, $f(3) = 21$. Since $6 < 10 < 21$, there is a number c in $(2, 3)$ such that $f(c) = 10$ by the Intermediate Value Theorem.

45. $f(x) = x^3 - 3x + 1$ is continuous on $[0, 1]$ and $f(0) = 1$, $f(1) = -1$. Since $-1 < 0 < 1$, there is a number c in $(0, 1)$ such that $f(c) = 0$ by the Intermediate Value Theorem. Thus there is a root of the equation $x^3 - 3x + 1 = 0$ in the interval $(0, 1)$.

47. $f(x) = x^3 + 2x - (x^2 + 1) = x^3 + 2x - x^2 - 1$ is continuous on $[0, 1]$ and $f(0) = -1$, $f(1) = 1$. Since $-1 < 0 < 1$, there is a number c in $(0, 1)$ such that $f(c) = 0$ by the Intermediate Value Theorem. Thus there is a root of the equation $x^3 + 2x - x^2 - 1 = 0$, or equivalently, $x^3 + 2x = x^2 + 1$, in the interval $(0, 1)$.

49. **(a)** $f(x) = x^3 - x + 1$ is continuous on $[-2, -1]$ and $f(-2) = -5$, $f(-1) = 1$. Since $-5 < 0 < 1$, there is a number c in $(-2, 1)$ such that $f(c) = 0$ by the Intermediate Value Theorem. Thus there is a root of the equation $x^3 - x + 1 = 0$ in the interval $(-2, -1)$.

(b) $f(-1.33) \approx -0.0226$ and $f(-1.32) \approx 0.0200$, so there is a root between -1.33 and -1.32.

51. **(a)** Let $f(x) = x^5 - x^2 - 4$. Then $f(1) = 1^5 - 1^2 - 4 = -4 < 0$ and $f(2) = 2^5 - 2^2 - 4 = 24 > 0$. So by the Intermediate Value Theorem, there is a number c in $(1, 2)$ such that $c^5 - c^2 - 4 = 0$.

(b) We can see from the graphs that, correct to three decimal places, the root is $x \approx 1.434$.

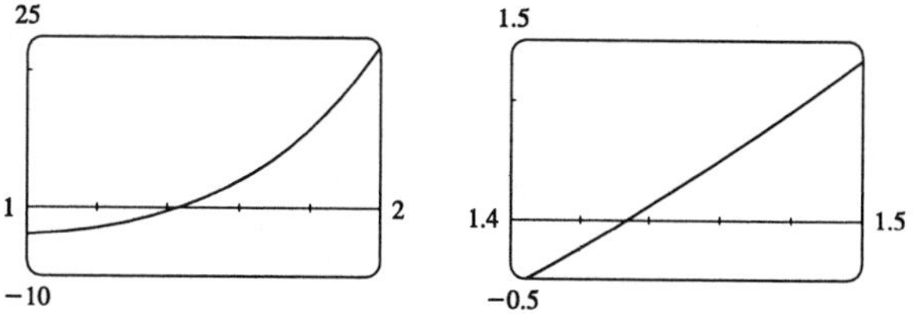

53. ($\Rightarrow$) If f is continuous at a, then by Theorem 7 with $g(h) = a + h$, we have

$$\lim_{h\to 0} f(a+h) = f\left(\lim_{h\to 0}(a+h)\right) = f(a).$$

($\Leftarrow$) Let $\epsilon > 0$. Since $\lim_{h\to 0} f(a+h) = f(a)$, there exists $\delta > 0$ such that $|h| < \delta \quad \Rightarrow$

$|f(a+h) - f(a)| < \epsilon$. So if $|x - a| < \delta$, then $|f(x) - f(a)| = |f(a + (x - a)) - f(a)| < \epsilon$. Thus $\lim_{x\to a} f(x) = f(a)$ and so f is continuous at a.

55. $f(x) = \begin{cases} 0 & \text{if } x \text{ is rational} \\ 1 & \text{if } x \text{ is irrational} \end{cases}$ is continuous nowhere. For, given any number a and any $\delta > 0$, the interval $(a - \delta, a + \delta)$ contains both infinitely many rational and infinitely many irrational numbers. Since $f(a) = 0$ or 1, there are infinitely many numbers x with $|x - a| < \delta$ and $|f(x) - f(a)| = 1$. Thus $\lim_{x\to a} f(x) \neq f(a)$. [In fact, $\lim_{x\to a} f(x)$ does not even exist.]

57. If there is such a number, it satisfies the equation $x^3 + 1 = x \quad \Leftrightarrow \quad x^3 - x + 1 = 0$.

Let the LHS of this equation be called $f(x)$. Now $f(-2) = (-2)^3 - (-2) + 1 = -5 < 0$, and $f(-1) = (-1)^3 - (-1) + 1 = 1 > 0$. Note also that $f(x)$ is a polynomial, and thus continuous. So by the Intermediate Value Theorem, there is a number c between -2 and -1 such that $f(c) = 0$, so that $c = c^3 + 1$.

59. Define $u(t)$ to be the monk's distance from the monastery, as a function of time, on the first day, and define $d(t)$ to be his distance from the monastery, as a function of time, on the second day. Let D be the distance from the monastery to the top of the mountain. From the given information we know that $u(0) = 0$, $d(12) = D$, $u(0) = D$ and $d(12) = 0$. Now consider the function $u - d$, which is clearly continuous (assuming that the monk does not use his mental powers to instantaneously transport himself). We calculate that $(u - d)(0) = -D$ and $(u - d)(12) = D$. So by the Intermediate Value Theorem there must be some time t_0 between 0 and 12 such that $(u - d)(t_0) = 0 \quad \Leftrightarrow \quad u(t_0) = d(t_0)$. So at time t_0 after 7:00 A.M., the monk will be at the same place on both days.

EXERCISES 1.6

1. **(a)** **(i)** $m = \lim_{x \to -3} \dfrac{x^2 + 2x - 3}{x - (-3)} = \lim_{x \to -3} \dfrac{(x+3)(x-1)}{x+3} = \lim_{x \to -3} (x - 1) = -4$

(ii) $m = \lim_{h \to 0} \dfrac{(-3+h)^2 + 2(-3+h) - 3}{h} = \lim_{h \to 0} \dfrac{9 - 6h + h^2 - 6 + 2h - 3}{h} = \lim_{h \to 0} \dfrac{h(h-4)}{h}$
$= \lim_{h \to 0} (h - 4) = -4$

(b) The equation of the tangent line is $y - 3 = -4(x + 3)$ or $y = -4x - 9$.

(c)

3. Using (1), $m = \lim_{x \to -2} \dfrac{1 - 2x - 3x^2 + 7}{x + 2} = \lim_{x \to -2} \dfrac{-3x^2 - 2x + 8}{x + 2} = \lim_{x \to -2} \dfrac{(-3x+4)(x+2)}{x+2}$
$= \lim_{x \to -2} (-3x + 4) = 10$. Thus the equation of the tangent is $y + 7 = 10(x + 2)$ or $y = 10x + 13$.

5. Using (1), $m = \lim_{x \to -2} \dfrac{1/x^2 - \frac{1}{4}}{x + 2} = \lim_{x \to -2} \dfrac{4 - x^2}{4x^2(x+2)} = \lim_{x \to -2} \dfrac{(2-x)(2+x)}{4x^2(x+2)} = \lim_{x \to -2} \dfrac{2 - x}{4x^2} = \dfrac{1}{4}$. Thus the equation of the tangent is $y - \frac{1}{4} = \frac{1}{4}(x + 2)$ or $x - 4y + 3 = 0$.

7. **(a)** $m = \lim_{x \to a} \dfrac{\dfrac{2}{x+3} - \dfrac{2}{a+3}}{x - a} = \lim_{x \to a} \dfrac{2(a - x)}{(x-a)(x+3)(a+3)} = \lim_{x \to a} \dfrac{-2}{(x+3)(a+3)} = \dfrac{-2}{(a+3)^2}$

(b) **(i)** $a = -1 \Rightarrow m = \dfrac{-2}{(-1+3)^2} = -\dfrac{1}{2}$ **(ii)** $a = 0 \Rightarrow m = \dfrac{-2}{(0+3)^2} = -\dfrac{2}{9}$

(iii) $a = 1 \Rightarrow m = \dfrac{-2}{(1+3)^2} = -\dfrac{1}{8}$

9. **(a)** Using (1), $m = \lim_{x \to a} \dfrac{(x^3 - 4x + 1) - (a^3 - 4a + 1)}{x - a} = \lim_{x \to a} \dfrac{(x^3 - a^3) - 4(x - a)}{x - a}$
$= \lim_{x \to a} \dfrac{(x-a)(x^2 + ax + a^2) - 4(x - a)}{x - a} = \lim_{x \to a} (x^2 + ax + a^2 - 4) = 3a^2 - 4$.

(b) At $(1, -2)$: $m = 3(1)^2 - 4 = -1$,
so the equation of the tangent line is
$y - (-2) = -1(x - 1) \quad \Leftrightarrow \quad y = -x - 1$.
At $(2, 1)$: $m = 3(2)^2 - 4 = 8$,
so the equation of the tangent line is
$y - 1 = 8(x - 2) \quad \Leftrightarrow \quad y = 8x - 15$.

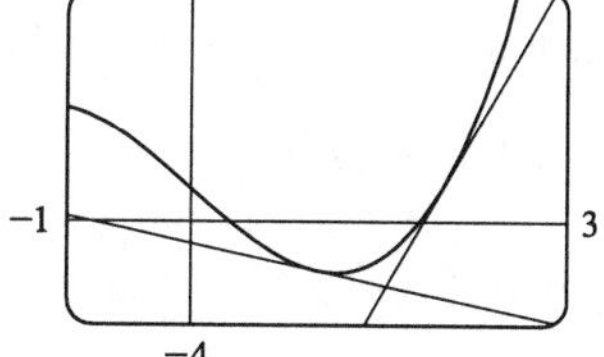

11. Let $s(t) = 40t - 16t^2$. $v(2) = \lim_{t \to 2} \dfrac{s(t) - s(2)}{t - 2} = \lim_{t \to 2} \dfrac{40t - 16t^2 - 16}{t - 2} = \lim_{t \to 2} \dfrac{8(t-2)(-2t+1)}{t - 2}$
$= \lim_{t \to 2} 8(-2t + 1) = -24$. Thus, the instantaneous velocity when $t = 2$ is -24 ft/s.

13. $v(a) = \lim_{h \to 0} \frac{s(a+h) - s(a)}{h} = \lim_{h \to 0} \frac{4(a+h)^3 + 6(a+h) + 2 - (4a^3 + 6a + 2)}{h}$

$= \lim_{h \to 0} \frac{4a^3 + 12a^2h + 12ah^2 + 4h^3 + 6a + 6h + 2 - 4a^3 - 6a - 2}{h}$

$= \lim_{h \to 0} \frac{12a^2h + 12ah^2 + 4h^3 + 6h}{h} = \lim_{h \to 0}\left(12a^2 + 12ah + 4h^2 + 6\right) = 12a^2 + 6$

So $v(1) = 12(1)^2 + 6 = 18\text{ m/s}$, $v(2) = 12(2)^2 + 6 = 54\text{ m/s}$, and $v(3) = 12(3)^2 + 6 = 114\text{ m/s}$.

15. **(a)** Since the slope of the tangent at $s = 0$ is 0, the car's initial velocity was 0.

(b) The slope of the tangent is greater at C than at B, so the car was going faster at C.

(c) Near A, the tangent lines are becoming steeper as x increases, so the velocity was increasing, so the car was speeding up. Near B, the tangent lines are becoming less steep, so the car was slowing down. The steepest tangent near C is the one right at C, so at C the car had just finished speeding up, and was about to start slowing down.

(d) Between D and E, the slope of the tangent is 0, so the car did not move during that time.

17. **(a)** **(i)** $[8, 11]$: $\frac{7.9 - 11.5}{3} = -1.2°/\text{h}$ **(ii)** $[8, 10]$: $\frac{9.0 - 11.5}{2} = -1.25°/\text{h}$

(iii) $[8, 9]$: $\frac{10.2 - 11.5}{1} = -1.3°/\text{h}$

(b) The instantaneous rate of change is approximately $-1.6°/\text{h}$ at 8 P.M.

19. **(a)** **(i)** $\frac{\Delta C}{\Delta x} = \frac{C(105) - C(100)}{5} = \frac{6601.25 - 6500}{5} = \$20.25/\text{unit}.$

(ii) $\frac{\Delta C}{\Delta x} = \frac{C(101) - C(100)}{1} = \frac{6520.05 - 6500}{1} = \$20.05/\text{unit}.$

(b) $\frac{C(100+h) - C(100)}{h} = \frac{5000 + 10(100+h) + 0.05(100+h)^2 - 6500}{h} = 20 + 0.05h,\ h \neq 0.$ So as h approaches 0, the rate of change of C approaches \$20/unit.

REVIEW EXERCISES FOR CHAPTER 1

1. False, since $\lim_{x \to 4} \frac{2x}{x-4}$ and $\lim_{x \to 4} \frac{8}{x-4}$ do not exist.

3. True by Limit Law #5, since $\lim_{x \to 1}(x^2 + 2x - 4) = -1 \neq 0$.

5. False. For example, let $f(x) = \begin{cases} x^2 + 1 & \text{if } x \neq 0 \\ 2 & \text{if } x = 0 \end{cases}$

Then $f(x) > 1$ for all x, but $\lim_{x \to 0} f(x) = \lim_{x \to 0}(x^2 + 1) = 1$.

7. True, by the definition of a limit with $\epsilon = 1$.

9. True. See Exercise 1.3.67.

11. True by Theorem 1.5.7 with $a = 2$, $b = 5$, and $g(x) = 4x^2 - 11$.

13. $\lim\limits_{x\to 4}\sqrt{x+\sqrt{x}} = \sqrt{4+\sqrt{4}} = \sqrt{6}$ since the function is continuous.

15. $\lim\limits_{t\to -1}\dfrac{t+1}{t^3-t} = \lim\limits_{t\to -1}\dfrac{t+1}{t(t+1)(t-1)} = \lim\limits_{t\to -1}\dfrac{1}{t(t-1)} = \dfrac{1}{(-1)(-2)} = \dfrac{1}{2}$

17. $\lim\limits_{h\to 0}\dfrac{(1+h)^2-1}{h} = \lim\limits_{h\to 0}\dfrac{1+2h+h^2-1}{h} = \lim\limits_{h\to 0}\dfrac{2h+h^2}{h} = \lim\limits_{h\to 0}(2+h) = 2$

19. $\lim\limits_{x\to -1}\dfrac{x^2-x-2}{x^2+3x-2} = \dfrac{\lim\limits_{x\to -1}(x^2-x-2)}{\lim\limits_{x\to -1}(x^2+3x-2)} = \dfrac{(-1)^2-(-1)-2}{(-1)^2+3(-1)-2} = \dfrac{0}{-4} = 0$

21. $\lim\limits_{t\to 6}\dfrac{17}{(t-6)^2} = \infty$ since $(t-6)^2 \to 0$ and $\dfrac{17}{(t-6)^2} > 0$.

23. $\lim\limits_{s\to 16}\dfrac{4-\sqrt{s}}{s-16} = \lim\limits_{s\to 16}\dfrac{4-\sqrt{s}}{(\sqrt{s}+4)(\sqrt{s}-4)} = \lim\limits_{s\to 16}\dfrac{-1}{\sqrt{s}+4} = \dfrac{-1}{\sqrt{16}+4} = -\dfrac{1}{8}$

25. $\lim\limits_{x\to 8^-}\dfrac{|x-8|}{x-8} = \lim\limits_{x\to 8^-}\dfrac{-(x-8)}{x-8} = \lim\limits_{x\to 8^-}(-1) = -1$

27. $\lim\limits_{x\to 0}\dfrac{1-\sqrt{1-x^2}}{x}\cdot\dfrac{1+\sqrt{1-x^2}}{1+\sqrt{1-x^2}} = \lim\limits_{x\to 0}\dfrac{1-(1-x^2)}{x\left(1+\sqrt{1-x^2}\right)} = \lim\limits_{x\to 0}\dfrac{x^2}{x\left(1+\sqrt{1-x^2}\right)} = \lim\limits_{x\to 0}\dfrac{x}{1+\sqrt{1-x^2}} = 0$

29. Given $\epsilon > 0$, we need $\delta > 0$ so that if $|x-5| < \delta$ then $|(7x-27)-8| < \epsilon \iff |7x-35| < \epsilon \iff |x-5| < \epsilon/7$. So take $\delta = \epsilon/7$. Then $|x-5| < \delta \Rightarrow |(7x-27)-8| < \epsilon$.

Thus $\lim\limits_{x\to 5}(7x-27) = 8$ by the definition of a limit.

31. Given $\epsilon > 0$, we need $\delta > 0$ so that if $|x-2| < \delta$ then $|x^2-3x-(-2)| < \epsilon$. First, note that if $|x-2| < 1$, then $-1 < x-2 < 1$, so $0 < x-1 < 2 \Rightarrow |x-1| < 2$. Now let $\delta = \min\{\epsilon/2, 1\}$. Then $|x-2| < \delta$ $\Rightarrow |x^2-3x-(-2)| = |(x-2)(x-1)| = |x-2||x-1| < (\epsilon/2)(2) = \epsilon$.

Thus $\lim\limits_{x\to 2}(x^2-3x) = -2$ by the definition of a limit.

33. Since $2x-1 \le f(x) \le x^2$ for $0 < x < 3$ and $\lim\limits_{x\to 1}(2x-1) = 1 = \lim\limits_{x\to 1}x^2$, we have $\lim\limits_{x\to 1}f(x) = 1$ by the Squeeze Theorem.

35. **(a)** $f(x) = \sqrt{-x}$ if $x < 0$, $f(x) = 3-x$ if $0 \le x < 3$, $f(x) = (x-3)^2$ if $x > 3$. So

(i) $\lim\limits_{x\to 0^+}f(x) = \lim\limits_{x\to 0^+}(3-x) = 3$

(ii) $\lim\limits_{x\to 0^-}f(x) = \lim\limits_{x\to 0^-}\sqrt{-x} = 0$

(iii) Because of (i) and (ii), $\lim\limits_{x\to 0}f(x)$ does not exist.

(iv) $\lim\limits_{x\to 3^-}f(x) = \lim\limits_{x\to 3^-}(3-x) = 0$

(v) $\lim\limits_{x\to 3^+}f(x) = \lim\limits_{x\to 3^+}(x-3)^2 = 0$

(vi) Because of (iv) and (v), $\lim\limits_{x\to 3}f(x) = 0$.

(b) f is discontinuous at 0 since $\lim\limits_{x\to 0}f(x)$ does not exist.

f is discontinuous at 3 since $f(3)$ does not exist.

(c)

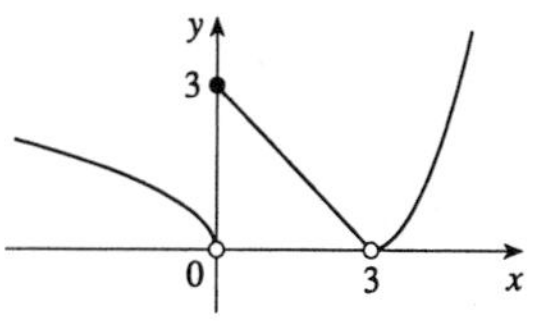

37. $f(x) = \dfrac{x+1}{x^2+x+1}$ is rational so it is continuous on its domain, which is $\mathbb{R}$.

(Note that $x^2 + x + 1 = 0$ has no real roots.)

39. $f(x) = 2x^3 + x^2 + 2$ is a polynomial, so it is continuous on $[-2, -1]$ and $f(-2) = -10 < 0 < 1 = f(-1)$. So by the Intermediate Value Theorem there is a number c in $(-2, -1)$ such that $f(c) = 0$, that is, the equation $2x^3 + x^2 + 2 = 0$ has a root in $(-2, -1)$.

41. **(a)** The slope of the tangent line at $(2, 1)$ is $\displaystyle\lim_{x\to 2} \frac{f(x) - f(2)}{x-2} = \lim_{x\to 2} \frac{9 - 2x^2 - 1}{x-2}$

$$= \lim_{x\to 2} \frac{8 - 2x^2}{x-2} = \lim_{x\to 2} \frac{-2(x^2-4)}{x-2} = \lim_{x\to 2} \frac{-2(x-2)(x+2)}{x-2} = \lim_{x\to 2} -2(x+2) = -8.$$

(b) The equation of this tangent line is $y - 1 = -8(x-2)$ or $8x + y = 17$.

43. **(a)** $s = 1 + 2t + t^2/4$. The average velocity over the time interval $[1, 1+h]$ is

$\dfrac{s(1+h) - s(1)}{h} = \dfrac{1 + 2(1+h) + (1+h)^2/4 - 13/4}{h} = \dfrac{10h + h^2}{4h} = \dfrac{10+h}{4}$. So for the following intervals the average velocities are:

(i) $[1, 3]$: $(10+2)/4 = 3\text{ m/s}$ **(ii)** $[1, 2]$: $(10+1)/4 = 2.75\text{ m/s}$

(iii) $[1, 1.5]$: $(10+0.5)/4 = 2.625\text{ m/s}$ **(iv)** $[1, 1.1]$: $(10+0.1)/4 = 2.525\text{ m/s}$

(b) When $t = 1$ the velocity is $\displaystyle\lim_{h\to 0} \frac{s(1+h) - s(1)}{h} = \lim_{h\to 0} \frac{10+h}{4} = 2.5\text{ m/s}$.

45. The inequality $\left|\dfrac{x+1}{x-1} - 3\right| < 0.2$ is equivalent to the double inequality $2.8 < \dfrac{x+1}{x-1} < 3.2$. Graphing the functions $y = 2.8$, $y = \left|\dfrac{x+1}{x-1}\right|$ and $y = 3.2$ on the interval $[1.9, 2.15]$, we see that the inequality holds whenever $1.91 < x < 2.11$ (approximately). So since $|2 - 1.91| = 0.09$ and $|2 - 2.15| = 0.15$, any positive $\delta \le 0.09$ will do.

47. $|f(x)| \le g(x) \quad \Leftrightarrow \quad -g(x) \le f(x) \le g(x)$ and $\displaystyle\lim_{x\to a} g(x) = 0 = \lim_{x\to a} -g(x)$.

Thus, by the Squeeze Theorem, $\displaystyle\lim_{x\to a} f(x) = 0$.

49. $\displaystyle\lim_{x\to a} f(x) = \lim_{x\to a}\left(\tfrac{1}{2}[f(x)+g(x)] + \tfrac{1}{2}[f(x)-g(x)]\right) = \tfrac{1}{2}\lim_{x\to a}[f(x)+g(x)] + \tfrac{1}{2}\lim_{x\to a}[f(x)-g(x)] = \tfrac{1}{2}\cdot 2 + \tfrac{1}{2}\cdot 1$

$\displaystyle = \tfrac{3}{2}$, and $\displaystyle\lim_{x\to a} g(x) = \lim_{x\to a}([f(x)+g(x)] - f(x)) = \lim_{x\to a}[f(x)+g(x)] - \lim_{x\to a} f(x) = 2 - \tfrac{3}{2} = \tfrac{1}{2}$.

So $\displaystyle\lim_{x\to a}[f(x)g(x)] = \left[\lim_{x\to a} f(x)\right]\left[\lim_{x\to a} g(x)\right] = \tfrac{3}{2}\cdot\tfrac{1}{2} = \tfrac{3}{4}$.

Alternate Solution: Since $\displaystyle\lim_{x\to a}[f(x)+g(x)]$ and $\displaystyle\lim_{x\to a}[f(x)-g(x)]$ exist, we must have

$\displaystyle\lim_{x\to a}[f(x)+g(x)]^2 = \left(\lim_{x\to a}[f(x)+g(x)]\right)^2$ and $\displaystyle\lim_{x\to a}[f(x)-g(x)]^2 = \left(\lim_{x\to a}[f(x)-g(x)]\right)^2$, so

$\displaystyle\lim_{x\to a}[f(x)\,g(x)] = \lim_{x\to a}\tfrac{1}{4}\left([f(x)+g(x)]^2 - [f(x)-g(x)]^2\right)$ (since all of the f^2 and g^2 cancel)

$$= \tfrac{1}{4}\left(\lim_{x\to a}[f(x)+g(x)]^2 - \lim_{x\to a}[f(x)-g(x)]^2\right) = \tfrac{1}{4}(2^2 - 1^2) = \tfrac{3}{4}.$$

CHAPTER TWO

EXERCISES 2.1

1. $f'(2) = \lim_{h\to 0} \dfrac{f(2+h)-f(2)}{h} = \lim_{h\to 0} \dfrac{3(2+h)^2 - 5(2+h) - \left[3(2)^2 - 5(2)\right]}{h}$

$= \lim_{h\to 0} \dfrac{12+12h+3h^2-10-5h-12+10}{h} = \lim_{h\to 0} \dfrac{3h^2+7h}{h} = \lim_{h\to 0}(3h+7) = 7$

So the equation of the tangent line at $(2,2)$ is $y-2=7(x-2)$ or $7x-y=12$.

3. **(a)** $F'(1) = \lim_{x\to 1} \dfrac{F(x)-F(1)}{x-1} = \lim_{x\to 1} \dfrac{x^3-5x+1-(-3)}{x-1} = \lim_{x\to 1} \dfrac{x^3-5x+4}{x-1} = \lim_{x\to 1} \dfrac{(x-1)(x^2+x-4)}{x-1}$

$= \lim_{x\to 1}\left(x^2+x-4\right) = -2.$

So the equation of the tangent line at $(1,-3)$ is $y-(-3) = -2(x-1)$ $\Leftrightarrow$ $y = -2x-1$.

Note: Instead of using Equation 3 to compute $F'(1)$, we could have used Equation 1.

(b)

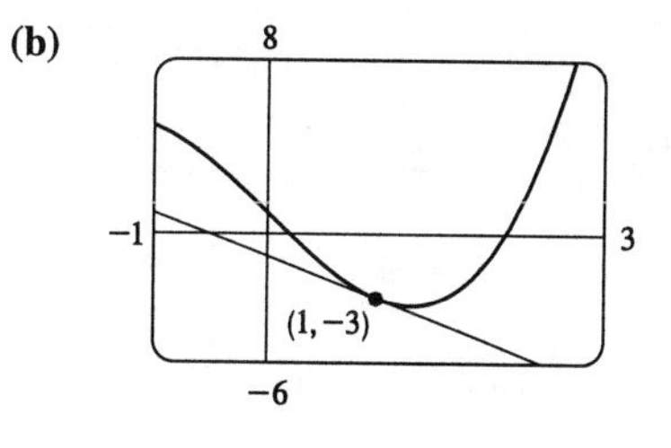

5. $v(2) = f'(2) = \lim_{h\to 0} \dfrac{f(2+h)-f(2)}{h} = \lim_{h\to 0} \dfrac{(2+h)^2 - 6(2+h) - 5 - (2^2 - 6(2) - 5)}{h}$

$= \lim_{h\to 0} \dfrac{4+4h+h^2-12-6h-5-4+12+5}{h} = \lim_{h\to 0} \dfrac{h^2-2h}{h} = \lim_{h\to 0}(h-2) = -2\text{ m/s}$

7. $f'(a) = \lim_{h\to 0} \dfrac{f(a+h)-f(a)}{h} = \lim_{h\to 0} \dfrac{1+(a+h)-2(a+h)^2-(1+a-2a^2)}{h}$

$= \lim_{h\to 0} \dfrac{h-4ah-2h^2}{h} = \lim_{h\to 0}(1-4a-2h) = 1-4a$

9. $f'(a) = \lim_{h\to 0} \dfrac{f(a+h)-f(a)}{h} = \lim_{h\to 0} \dfrac{\dfrac{a+h}{2(a+h)-1} - \dfrac{a}{2a-1}}{h}$

$= \lim_{h\to 0} \dfrac{(a+h)(2a-1) - a(2a+2h-1)}{h(2a+2h-1)(2a-1)} = \lim_{h\to 0} \dfrac{-h}{h(2a+2h-1)(2a-1)}$

$= \lim_{h\to 0} \dfrac{-1}{(2a+2h-1)(2a-1)} = -\dfrac{1}{(2a-1)^2}$

11. $f'(a)=\lim_{h\to 0}\dfrac{f(a+h)-f(a)}{h}=\lim_{h\to 0}\dfrac{\dfrac{2}{\sqrt{3-(a+h)}}-\dfrac{2}{\sqrt{3-a}}}{h}=\lim_{h\to 0}\dfrac{2\left(\sqrt{3-a}-\sqrt{3-a-h}\right)}{h\sqrt{3-a-h}\,\sqrt{3-a}}$

$$=\lim_{h\to 0}\frac{2\left(\sqrt{3-a}-\sqrt{3-a-h}\right)}{h\sqrt{3-a-h}\,\sqrt{3-a}}\cdot\frac{\sqrt{3-a}+\sqrt{3-a-h}}{\sqrt{3-a}+\sqrt{3-a-h}}$$

$$=\lim_{h\to 0}\frac{2[3-a-(3-a-h)]}{h\sqrt{3-a-h}\,\sqrt{3-a}\left(\sqrt{3-a}+\sqrt{3-a-h}\right)}$$

$$=\lim_{h\to 0}\frac{2}{\sqrt{3-a-h}\,\sqrt{3-a}\left(\sqrt{3-a}+\sqrt{3-a-h}\right)}$$

$$=\frac{2}{\sqrt{3-a}\,\sqrt{3-a}\left(2\sqrt{3-a}\right)}=\frac{1}{(3-a)^{3/2}}$$

13. By Equation 1, $\lim_{h\to 0}\dfrac{\sqrt{1+h}-1}{h}=f'(1)$ where $f(x)=\sqrt{x}$. [Or $f'(0)$ where $f(x)=\sqrt{1+x}$; the answers to Exercises 13-18 are not unique.]

15. $\lim_{x\to 1}\dfrac{x^9-1}{x-1}=f'(1)$ where $f(x)=x^9$. (See Equation 3.)

17. $\lim_{t\to 0}\dfrac{\sin\left(\frac{\pi}{2}+t\right)-1}{t}=f'\left(\frac{\pi}{2}\right)$ where $f(x)=\sin x$.

19. $f'(x)=\lim_{h\to 0}\dfrac{f(x+h)-f(x)}{h}=\lim_{h\to 0}\dfrac{5(x+h)+3-(5x+3)}{h}=\lim_{h\to 0}\dfrac{5h}{h}=\lim_{h\to 0}5=5.$

Domain of f = domain of $f'=\mathbb{R}$.

21. $f'(x)=\lim_{h\to 0}\dfrac{f(x+h)-f(x)}{h}=\lim_{h\to 0}\dfrac{(x+h)^3-(x+h)^2+2(x+h)-(x^3-x^2+2x)}{h}$

$$=\lim_{h\to 0}\frac{3x^2h+3xh^2+h^3-2xh-h^2+2h}{h}=\lim_{h\to 0}\left(3x^2+3xh+h^2-2x-h+2\right)=3x^2-2x+2$$

Domain of f = domain of $f'=\mathbb{R}$.

23. $g'(x)=\lim_{h\to 0}\dfrac{g(x+h)-g(x)}{h}=\lim_{h\to 0}\dfrac{\sqrt{1+2(x+h)}-\sqrt{1+2x}}{h}\left[\dfrac{\sqrt{1+2(x+h)}+\sqrt{1+2x}}{\sqrt{1+2(x+h)}+\sqrt{1+2x}}\right]$

$$=\lim_{h\to 0}\frac{1+2x+2h-(1+2x)}{h\left[\sqrt{1+2(x+h)}+\sqrt{1+2x}\right]}=\lim_{h\to 0}\frac{2}{\sqrt{1+2(x+h)}+\sqrt{1+2x}}=\frac{1}{\sqrt{1+2x}}$$

Domain of $g=\left[-\frac{1}{2},\infty\right)$, domain of $g'=\left(-\frac{1}{2},\infty\right)$.

25. $G'(x)=\lim_{h\to 0}\dfrac{G(x+h)-G(x)}{h}=\lim_{h\to 0}\dfrac{\dfrac{4-3(x+h)}{2+(x+h)}-\dfrac{4-3x}{2+x}}{h}$

$$=\lim_{h\to 0}\frac{(4-3x-3h)(2+x)-(4-3x)(2+x+h)}{h(2+x+h)(2+x)}=\lim_{h\to 0}\frac{-10h}{h(2+x+h)(2+x)}$$

$$=\lim_{h\to 0}\frac{-10}{(2+x+h)(2+x)}=\frac{-10}{(2+x)^2}$$

Domain of G = domain of $G'=\{x\mid x\neq -2\}$.

27. $f'(x) = \lim\limits_{h \to 0} \dfrac{f(x+h) - f(x)}{h} = \lim\limits_{h \to 0} \dfrac{(x+h)^4 - x^4}{h} = \lim\limits_{h \to 0} \dfrac{4x^3h + 6x^2h^2 + 4xh^3 + h^4}{h}$

$= \lim\limits_{h \to 0} \left(4x^3 + 6x^2h + 4xh^2 + h^3\right) = 4x^3$

Domain of f = domain of $f' = \mathbb{R}$.

29. $f(x) = x \quad \Rightarrow \quad f'(x) = \lim\limits_{h \to 0} \dfrac{x + h - x}{h} = \lim\limits_{h \to 0} 1 = 1$

$f(x) = x^2 \quad \Rightarrow \quad f'(x) = \lim\limits_{h \to 0} \dfrac{(x+h)^2 - x^2}{h} = \lim\limits_{h \to 0} \dfrac{2xh + h^2}{h} = \lim\limits_{h \to 0} (2x + h) = 2x$

$f(x) = x^3 \quad \Rightarrow \quad f'(x) = \lim\limits_{h \to 0} \dfrac{(x+h)^3 - x^3}{h} = \lim\limits_{h \to 0} \dfrac{3x^2h + 3xh^2 + h^3}{h} = \lim\limits_{h \to 0} \left(3x^2 + 3xh + h^2\right) = 3x^2$

$f(x) = x^4 \quad \Rightarrow \quad f'(x) = 4x^3$ from Exercise 31.

Guess: The derivative of $f(x) = x^n$ is $f'(x) = nx^{n-1}$. Test for $n = 5$: $f(x) = x^5 \quad \Rightarrow$

$f'(x) = \lim\limits_{h \to 0} \dfrac{(x+h)^5 - x^5}{h} = \lim\limits_{h \to 0} \dfrac{5x^4h + 10x^3h^2 + 10x^2h^3 + 5xh^4 + h^5}{h}$

$= \lim\limits_{h \to 0} \left(5x^4 + 10x^3h + 10x^2h^2 + 5xh^3 + h^4\right) = 5x^4$

31. **(a)** $f'(x) = \lim\limits_{h \to 0} \dfrac{f(x+h) - f(x)}{h} = \lim\limits_{h \to 0} \dfrac{\left[x + h - \left(\dfrac{2}{x+h}\right)\right] - \left[x - \left(\dfrac{2}{x}\right)\right]}{h}$

$= \lim\limits_{h \to 0} \left[1 + \dfrac{\dfrac{2}{x} - \dfrac{2}{(x+h)}}{h}\right] = \lim\limits_{h \to 0} \left[1 + \dfrac{2(x+h) - 2x}{h(x)(x+h)}\right]$

$= 1 + 2x^{-2}$

Notice that when f has steep tangent lines, $f'(x)$ is very large. When f is flatter, $f'(x)$ is smaller.

(b)

33. From the graph of f, it appears that

(a) $f'(1) \approx -2$ **(b)** $f'(2) \approx 0.8$

(c) $f'(3) \approx -1$ **(d)** $f'(4) \approx -0.5$

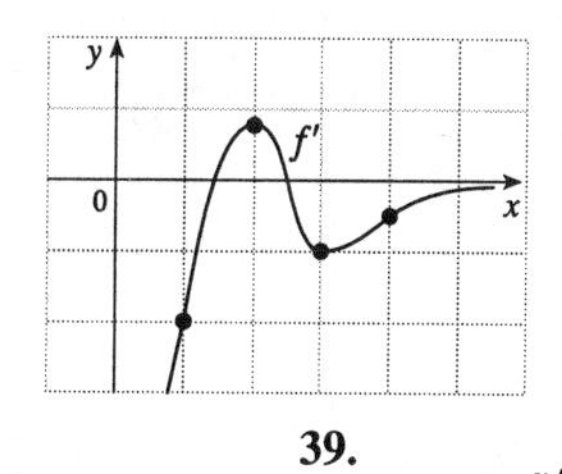

35.

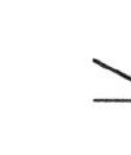

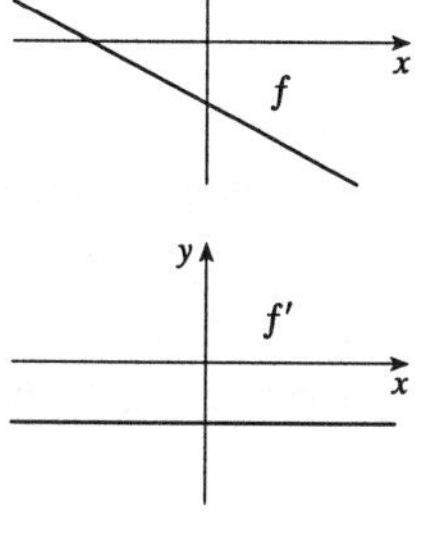

37.

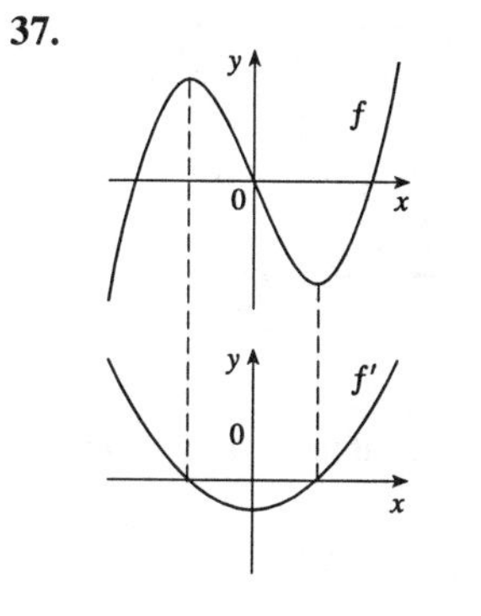

39.

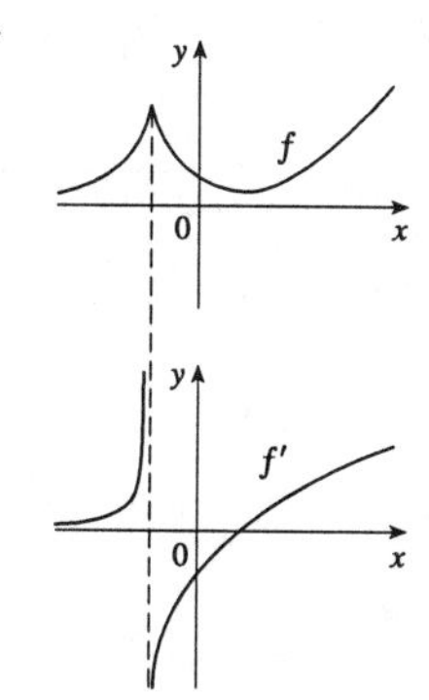

41.

43.

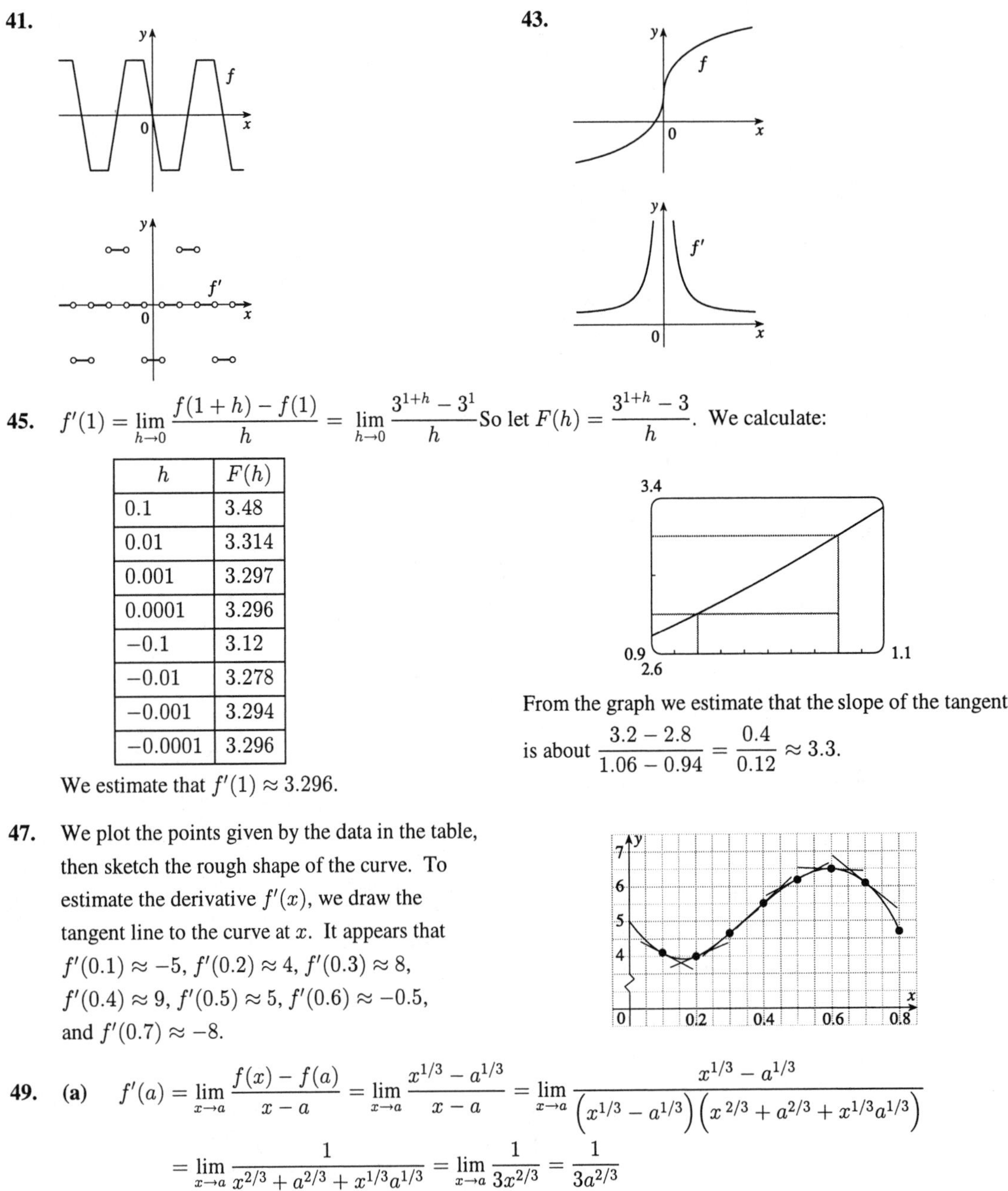

45. $f'(1) = \lim_{h \to 0} \frac{f(1+h) - f(1)}{h} = \lim_{h \to 0} \frac{3^{1+h} - 3^1}{h}$ So let $F(h) = \frac{3^{1+h} - 3}{h}$. We calculate:

h	$F(h)$
0.1	3.48
0.01	3.314
0.001	3.297
0.0001	3.296
−0.1	3.12
−0.01	3.278
−0.001	3.294
−0.0001	3.296

We estimate that $f'(1) \approx 3.296$.

From the graph we estimate that the slope of the tangent is about $\frac{3.2 - 2.8}{1.06 - 0.94} = \frac{0.4}{0.12} \approx 3.3$.

47. We plot the points given by the data in the table, then sketch the rough shape of the curve. To estimate the derivative $f'(x)$, we draw the tangent line to the curve at x. It appears that $f'(0.1) \approx -5$, $f'(0.2) \approx 4$, $f'(0.3) \approx 8$, $f'(0.4) \approx 9$, $f'(0.5) \approx 5$, $f'(0.6) \approx -0.5$, and $f'(0.7) \approx -8$.

49. **(a)** $f'(a) = \lim_{x \to a} \frac{f(x) - f(a)}{x - a} = \lim_{x \to a} \frac{x^{1/3} - a^{1/3}}{x - a} = \lim_{x \to a} \frac{x^{1/3} - a^{1/3}}{\left(x^{1/3} - a^{1/3}\right)\left(x^{2/3} + a^{2/3} + x^{1/3}a^{1/3}\right)}$

$$= \lim_{x \to a} \frac{1}{x^{2/3} + a^{2/3} + x^{1/3}a^{1/3}} = \lim_{x \to a} \frac{1}{3x^{2/3}} = \frac{1}{3a^{2/3}}$$

(b) $f'(0) = \lim_{h \to 0} \frac{f(0+h) - f(0)}{h} = \lim_{h \to 0} \frac{\sqrt[3]{h} - 0}{h} = \lim_{h \to 0} \frac{1}{h^{2/3}}$. This limit does not exist, and therefore $f'(0)$ does not exist.

(c) $\lim_{x \to 0} |f'(x)| = \lim_{x \to 0} \frac{1}{3x^{2/3}} = \infty$. Also f is continuous at $x = 0$ (root function), so f has a vertical tangent at $x = 0$.

51. f is not differentiable at $x = -1$ or at $x = 11$ because the graph has vertical tangents at those points; at $x = 4$, because there is a discontinuity there; and at $x = 8$, because the graph has a corner there.

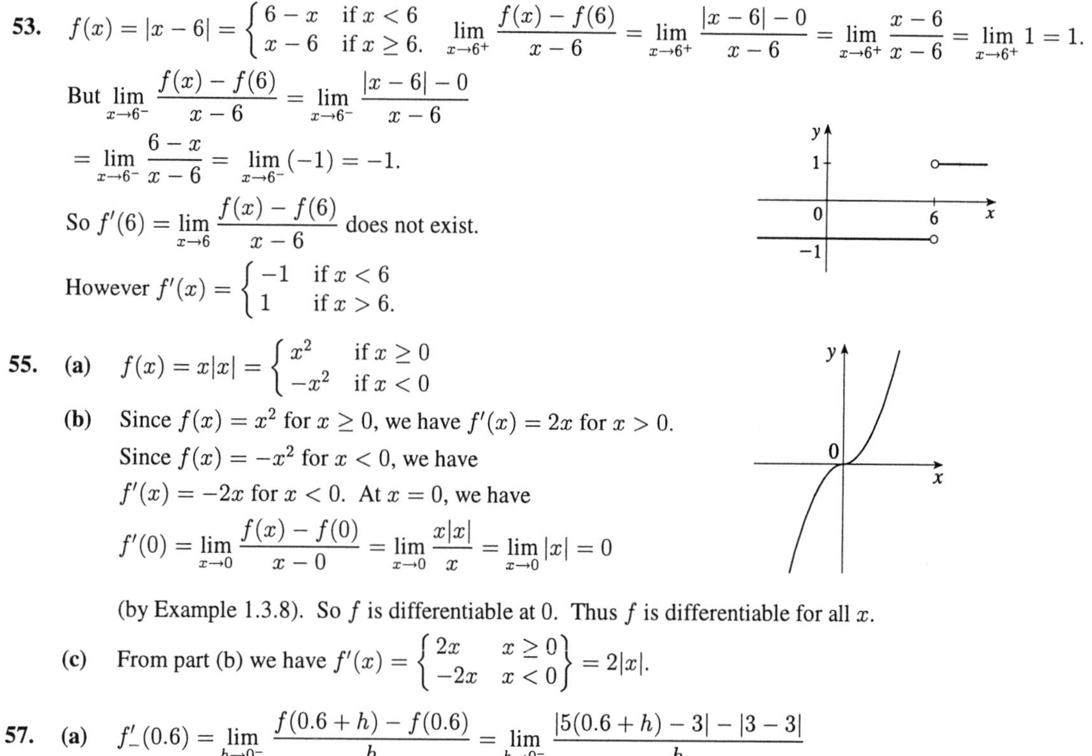

53. $f(x) = |x - 6| = \begin{cases} 6 - x & \text{if } x < 6 \\ x - 6 & \text{if } x \geq 6. \end{cases}$ $\lim\limits_{x \to 6^+} \dfrac{f(x) - f(6)}{x - 6} = \lim\limits_{x \to 6^+} \dfrac{|x - 6| - 0}{x - 6} = \lim\limits_{x \to 6^+} \dfrac{x - 6}{x - 6} = \lim\limits_{x \to 6^+} 1 = 1.$

But $\lim\limits_{x \to 6^-} \dfrac{f(x) - f(6)}{x - 6} = \lim\limits_{x \to 6^-} \dfrac{|x - 6| - 0}{x - 6}$

$= \lim\limits_{x \to 6^-} \dfrac{6 - x}{x - 6} = \lim\limits_{x \to 6^-} (-1) = -1.$

So $f'(6) = \lim\limits_{x \to 6} \dfrac{f(x) - f(6)}{x - 6}$ does not exist.

However $f'(x) = \begin{cases} -1 & \text{if } x < 6 \\ 1 & \text{if } x > 6. \end{cases}$

55. **(a)** $f(x) = x|x| = \begin{cases} x^2 & \text{if } x \geq 0 \\ -x^2 & \text{if } x < 0 \end{cases}$

(b) Since $f(x) = x^2$ for $x \geq 0$, we have $f'(x) = 2x$ for $x > 0$. Since $f(x) = -x^2$ for $x < 0$, we have $f'(x) = -2x$ for $x < 0$. At $x = 0$, we have

$f'(0) = \lim\limits_{x \to 0} \dfrac{f(x) - f(0)}{x - 0} = \lim\limits_{x \to 0} \dfrac{x|x|}{x} = \lim\limits_{x \to 0} |x| = 0$

(by Example 1.3.8). So f is differentiable at 0. Thus f is differentiable for all x.

(c) From part (b) we have $f'(x) = \left\{ \begin{matrix} 2x & x \geq 0 \\ -2x & x < 0 \end{matrix} \right\} = 2|x|.$

57. **(a)** $f'_-(0.6) = \lim\limits_{h \to 0^-} \dfrac{f(0.6 + h) - f(0.6)}{h} = \lim\limits_{h \to 0^-} \dfrac{|5(0.6 + h) - 3| - |3 - 3|}{h}$

$= \lim\limits_{h \to 0^-} \dfrac{|3 + 5h - 3|}{h} = \lim\limits_{h \to 0^-} \dfrac{-5h}{h} = -5$

$f'_+(0.6) = \lim\limits_{h \to 0^+} \dfrac{f(0.6 + h) - f(0.6)}{h} = \lim\limits_{h \to 0^+} \dfrac{|5(0.6 + h) - 3| - |3 - 3|}{h} = \lim\limits_{h \to 0^+} \dfrac{5h}{h} = 5$

(b) Since $f'_-(0.6) \neq f'_+(0.6)$, $f'(0.6)$ does not exist.

59. Since $f(x) = x \sin(1/x)$ when $x \neq 0$ and $f(0) = 0$, we have

$f'(0) = \lim\limits_{h \to 0} \dfrac{f(0 + h) - f(0)}{h} = \lim\limits_{h \to 0} \dfrac{h \sin(1/h) - 0}{h} = \lim\limits_{h \to 0} \sin(1/h)$. This limit does not exist since $\sin(1/h)$ takes the values -1 and 1 on any interval containing 0. (Compare with Example 1.2.4.)

61. **(a)** If f is even, then $f'(-x) = \lim\limits_{h \to 0} \dfrac{f(-x + h) - f(-x)}{h} = \lim\limits_{h \to 0} \dfrac{f(x - h) - f(x)}{h}$

$= -\lim\limits_{h \to 0} \dfrac{f(x - h) - f(x)}{-h}$ [let $\Delta x = -h$] $= -\lim\limits_{\Delta x \to 0} \dfrac{f(x + \Delta x) - f(x)}{\Delta x} = -f'(x)$. Therefore f' is odd.

(b) If f is odd, then $f'(-x) = \lim\limits_{h \to 0} \dfrac{f(-x + h) - f(-x)}{h} = \lim\limits_{h \to 0} \dfrac{-f(x - h) + f(x)}{h}$

$= \lim\limits_{h \to 0} \dfrac{f(x - h) - f(x)}{-h}$ [let $\Delta x = -h$] $= \lim\limits_{\Delta x \to 0} \dfrac{f(x + \Delta x) - f(x)}{\Delta x} = f'(x)$. Therefore f' is even.

63.

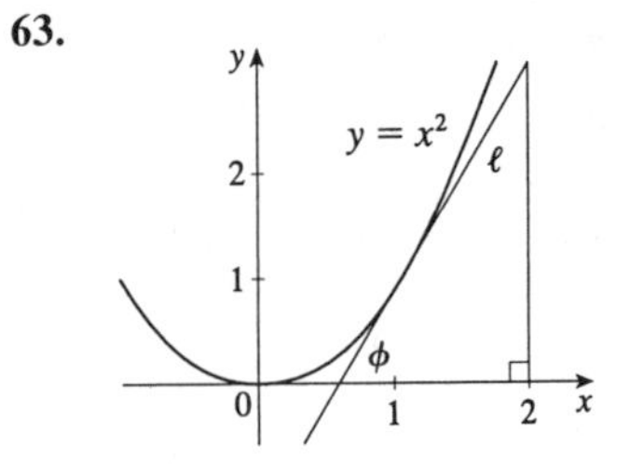

From the diagram, we see that the slope of the tangent is equal to $\tan\phi$, and also that $0 < \phi < \frac{\pi}{2}$. We know (see Exercise 29) that the derivative of $f(x) = x^2$ is $f'(x) = 2x$. So the slope of the tangent to the curve at the point $(1, 1)$ is 2. So ϕ is the angle between 0 and $\frac{\pi}{2}$ whose tangent is 2, that is, $\phi = \tan^{-1} 2 \approx 63°$.

EXERCISES 2.2

1. $f(x) = x^2 - 10x + 100 \quad\Rightarrow\quad f'(x) = 2x - 10$

3. $V(r) = \frac{4}{3}\pi r^3 \quad\Rightarrow\quad V'(r) = \frac{4}{3}\pi(3r^2) = 4\pi r^2$

5. $F(x) = (16x)^3 = 4{,}096x^3 \quad\Rightarrow\quad F'(x) = 4{,}096(3x^2) = 12{,}288x^2$

7. $Y(t) = 6t^{-9} \quad\Rightarrow\quad Y'(t) = 6(-9)t^{-10} = -54t^{-10}$

9. $g(x) = x^2 + \dfrac{1}{x^2} = x^2 + x^{-2} \quad\Rightarrow\quad g'(x) = 2x + (-2)x^{-3} = 2x - \dfrac{2}{x^3}$

11. $h(x) = \dfrac{x+2}{x-1} \quad\Rightarrow\quad h'(x) = \dfrac{(x-1)D(x+2) - (x+2)D(x-1)}{(x-1)^2} = \dfrac{x - 1 - (x+2)}{(x-1)^2} = \dfrac{-3}{(x-1)^2}$

13. $G(s) = (s^2 + s + 1)(s^2 + 2) \quad\Rightarrow\quad G'(s) = (2s+1)(s^2+2) + (s^2+s+1)(2s) = 4s^3 + 3s^2 + 6s + 2$

15. $y = \dfrac{x^2 + 4x + 3}{\sqrt{x}} = x^{3/2} + 4x^{1/2} + 3x^{-1/2} \quad\Rightarrow$

$y' = \frac{3}{2}x^{1/2} + 4\left(\frac{1}{2}\right)x^{-1/2} + 3\left(-\frac{1}{2}\right)x^{-3/2} = \frac{3}{2}\sqrt{x} + \dfrac{2}{\sqrt{x}} - \dfrac{3}{2x\sqrt{x}}$. *Another Method:* Use the Quotient Rule.

17. $y = \sqrt{5x} = \sqrt{5}x^{1/2} \quad\Rightarrow\quad y' = \sqrt{5}\left(\frac{1}{2}\right)x^{-1/2} = \dfrac{\sqrt{5}}{2\sqrt{x}}$

19. $y = \dfrac{1}{x^4 + x^2 + 1} \quad\Rightarrow\quad y' = \dfrac{(x^4 + x^2 + 1)(0) - 1(4x^3 + 2x)}{(x^4 + x^2 + 1)^2} = -\dfrac{4x^3 + 2x}{(x^4 + x^2 + 1)^2}$

21. $y = ax^2 + bx + c \quad\Rightarrow\quad y' = 2ax + b$

23. $y = \dfrac{3t - 7}{t^2 + 5t - 4} \quad\Rightarrow\quad y' = \dfrac{(t^2 + 5t - 4)(3) - (3t - 7)(2t + 5)}{(t^2 + 5t - 4)^2} = \dfrac{-3t^2 + 14t + 23}{(t^2 + 5t - 4)^2}$

25. $y = x + \sqrt[5]{x^2} = x + x^{2/5} \quad\Rightarrow\quad y' = 1 + \frac{2}{5}x^{-3/5} = 1 + \dfrac{2}{5\sqrt[5]{x^3}}$

27. $u = x^{\sqrt{2}} \quad\Rightarrow\quad u' = \sqrt{2}\,x^{\sqrt{2}-1}$

29. $v = x\sqrt{x} + \dfrac{1}{x^2\sqrt{x}} = x^{3/2} + x^{-5/2} \quad\Rightarrow\quad v' = \frac{3}{2}x^{1/2} - \frac{5}{2}x^{-7/2} = \frac{3}{2}\sqrt{x} - \dfrac{5}{2x^3\sqrt{x}}$

31. $f(x) = \dfrac{x}{x + c/x} \quad \Rightarrow \quad f'(x) = \dfrac{(x + c/x)(1) - x(1 - c/x^2)}{(x + c/x)^2} = \dfrac{2cx}{(x^2 + c)^2}$

33. $f(x) = \dfrac{x^5}{x^3 - 2} \quad \Rightarrow \quad f'(x) = \dfrac{(x^3 - 2)(5x^4) - x^5(3x^2)}{(x^3 - 2)^2} = \dfrac{2x^4(x^3 - 5)}{(x^3 - 2)^2}$

35. $P(x) = a_n x^n + a_{n-1}x^{n-1} + \cdots + a_2 x^2 + a_1 x + a_0 \quad \Rightarrow$
$P'(x) = na_n x^{n-1} + (n-1)a_{n-1}x^{n-2} + \cdots + 2a_2 x + a_1$

37. $y = f(x) = x + \dfrac{4}{x} \quad \Rightarrow \quad f'(x) = 1 - \dfrac{4}{x^2}$. So the slope of the tangent line at $(2, 4)$ is $f'(2) = 0$ and its equation is $y - 4 = 0$ or $y = 4$.

39. $y = f(x) = x + \sqrt{x} \quad \Rightarrow \quad f'(x) = 1 + \frac{1}{2}x^{-1/2}$. So the slope of the tangent line at $(1, 2)$ is $f'(1) = 1 + \frac{1}{2}(1) = \frac{3}{2}$ and its equation is $y - 2 = \frac{3}{2}(x - 1)$ or $y = \frac{3}{2}x + \frac{1}{2}$ or $3x - 2y + 1 = 0$.

41. **(a)** $y = f(x) = \dfrac{1}{1 + x^2} \quad \Rightarrow \quad f'(x) = \dfrac{-2x}{(1 + x^2)^2}$.

So the slope of the tangent line at the point $\left(-1, \frac{1}{2}\right)$ is

$f'(-1) = \dfrac{-2(-1)}{\left[1 + (-1)^2\right]^2} = \dfrac{1}{2}$ and its equation is

$y - \frac{1}{2} = \frac{1}{2}(x + 1)$ or $y = \frac{1}{2}x + 1$ or $x - 2y + 2 = 0$.

(b)

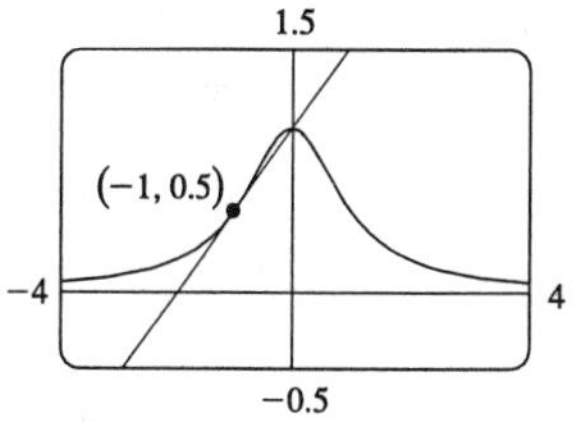

43. **(a)** $f(x) = 3x^{15} - 5x^3 + 3 \quad \Rightarrow$

$f'(x) = 3 \cdot 15x^{14} - 5 \cdot 3x^2$
$= 45x^{14} - 15x^2$

(b)

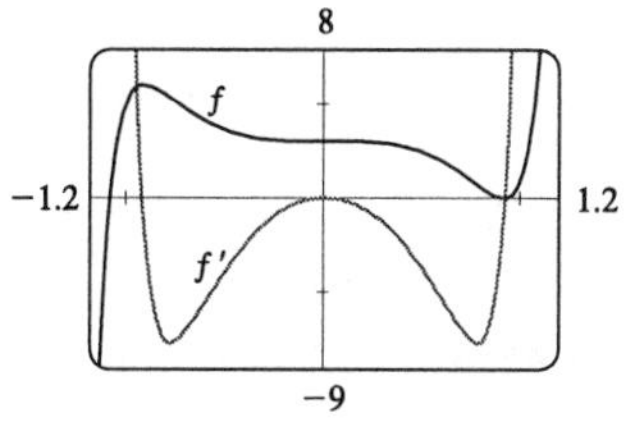

Notice that $f'(x) = 0$ when f has a horizontal tangent.

45. $y = x\sqrt{x} = x^{3/2} \quad \Rightarrow \quad y' = \frac{3}{2}\sqrt{x}$ so the tangent line is parallel to $3x - y + 6 = 0$ when $\frac{3}{2}\sqrt{x} = 3 \quad \Leftrightarrow$ $\sqrt{x} = 2 \quad \Leftrightarrow \quad x = 4$. So the point is $(4, 8)$.

47. $y = x^3 - x^2 - x + 1$ has a horizontal tangent when $y' = 3x^2 - 2x - 1 = 0 \quad (3x + 1)(x - 1) = 0 \Leftrightarrow x = 1$ or $-\frac{1}{3}$. Therefore the points are $(1, 0)$ and $\left(-\frac{1}{3}, \frac{32}{27}\right)$.

49. If $y = f(x) = \dfrac{x}{x + 1}$ then $f'(x) = \dfrac{(x + 1)(1) - x(1)}{(x + 1)^2} = \dfrac{1}{(x + 1)^2}$. When $x = a$, the equation of the tangent line is $y - \dfrac{a}{a + 1} = \dfrac{1}{(a + 1)^2}(x - a)$. This line passes through $(1, 2)$ when $2 - \dfrac{a}{a + 1} = \dfrac{1}{(a + 1)^2}(1 - a) \quad \Leftrightarrow$ $2(a + 1)^2 = a(a + 1) + (1 - a) = a^2 + 1 \quad \Leftrightarrow \quad a^2 + 4a + 1 = 0$. The quadratic formula gives the roots of this equation as $-2 \pm \sqrt{3}$, so there are two such tangent lines, which touch the curve at $\left(-2 + \sqrt{3}, \frac{1 - \sqrt{3}}{2}\right)$ and $\left(-2 - \sqrt{3}, \frac{1 + \sqrt{3}}{2}\right)$.

51. $y = 6x^3 + 5x - 3 \quad \Rightarrow \quad m = y' = 18x^2 + 5$, but $x^2 \geq 0$ for all x so $m \geq 5$ for all x.

53. $y = f(x) = 1 - x^2 \quad \Rightarrow \quad f'(x) = -2x$, so the tangent line at $(2, -3)$ has slope $f'(2) = -4$. The normal line has slope $-1/(-4) = \frac{1}{4}$ and equation $y + 3 = \frac{1}{4}(x - 2)$ or $x - 4y = 14$.

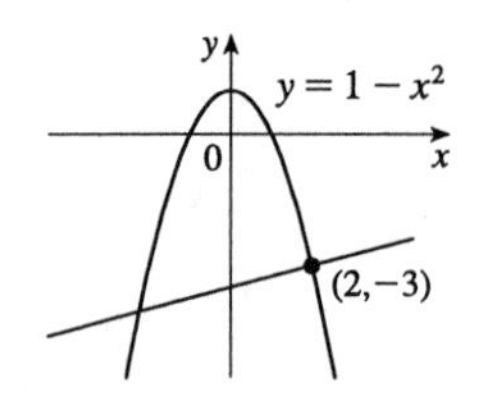

55. $y = f(x) = \sqrt[3]{x} = x^{1/3} \quad \Rightarrow \quad f'(x) = \frac{1}{3}x^{-2/3}$,
so the tangent line at $(-8, -2)$ has slope $f'(-8) = \frac{1}{12}$.
The normal line has slope $-1/\left(\frac{1}{12}\right) = -12$ and equation $y + 2 = -12(x + 8) \quad \Leftrightarrow \quad 12x + y + 98 = 0$.

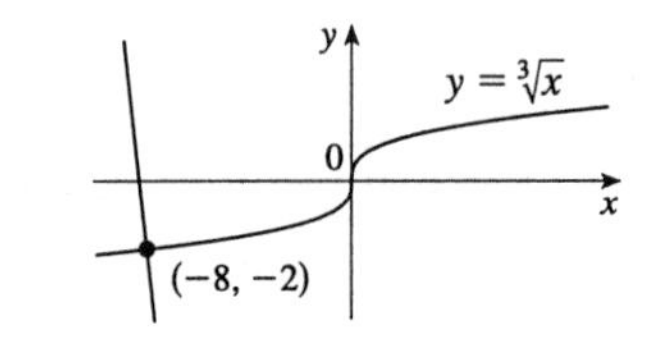

57. If the normal line has slope 16, then the tangent has slope $-\frac{1}{16}$, so $y' = 4x^3 = -\frac{1}{16} \quad \Rightarrow \quad x^3 = -\frac{1}{64} \quad \Rightarrow$
$x = -\frac{1}{4}$. The point is $\left(-\frac{1}{4}, \frac{1}{256}\right)$.

59. **(a)** $(fg)'(5) = f'(5)g(5) + f(5)g'(5) = 6(-3) + 1(2) = -16$

(b) $$\left(\frac{f}{g}\right)'(5) = \frac{f'(5)g(5) - f(5)g'(5)}{(g(5))^2} = \frac{6(-3) - 1(2)}{(-3)^2} = -\frac{20}{9}$$

(c) $$\left(\frac{g}{f}\right)'(5) = \frac{g'(5)f(5) - g(5)f'(5)}{(f(5))^2} = \frac{2(1) - (-3)(6)}{1^2} = 20$$

61. **(a)** $u(x) = f(x)g(x)$, so $u'(1) = f(1)g'(1) + g(1)f'(1) = 2 \cdot (-1) + 1 \cdot 2 = 0$

(b) $v(x) = f(x)/g(x)$, so $v'(5) = \dfrac{g(5)f'(5) - f(5)g'(5)}{[g(5)]^2} = \dfrac{2\left(-\frac{1}{3}\right) - 3 \cdot \frac{2}{3}}{2^2} = -\dfrac{2}{3}$

63. **(a)** $(fgh)' = [(fg)h]' = (fg)'h + (fg)h' = (f'g + fg')h + (fg)h' = f'gh + fg'h + fgh'$

(b) Putting $f = g = h$ in part (a), we have
$$\frac{d}{dx}[f(x)]^3 = (fff)' = f'ff + ff'f + fff' = 3fff' = 3[f(x)]^2 f'(x).$$

65. $y = \sqrt{x}\left(x^4 + x + 1\right)(2x - 3)$. Using Exercise 63(a), we have
$$y' = \frac{1}{2\sqrt{x}}\left(x^4 + x + 1\right)(2x - 3) + \sqrt{x}\left(4x^3 + 1\right)(2x - 3) + \sqrt{x}\left(x^4 + x + 1\right)(2)$$
$$= \left(x^4 + x + 1\right)\frac{2x - 3}{2\sqrt{x}} + \sqrt{x}\left[\left(4x^3 + 1\right)(2x - 3) + 2\left(x^4 + x + 1\right)\right].$$

67. $f(x) = 2 - x$ if $x \leq 1$ and $f(x) = x^2 - 2x + 2$ if $x > 1$. Now we compute the right- and left-hand derivatives defined in Exercise 2.1.57:
$$f'_-(1) = \lim_{h \to 0^-} \frac{f(1 + h) - f(1)}{h} = \lim_{h \to 0^-} \frac{2 - (1 + h) - 1}{h} = \lim_{h \to 0^-} \frac{-h}{h} = \lim_{h \to 0^-} -1 = -1 \text{ and}$$
$$f'_+(1) = \lim_{h \to 0^+} \frac{f(1 + h) - f(1)}{h} = \lim_{h \to 0^+} \frac{(1 + h)^2 - 2(1 + h) + 2 - 1}{h} = \lim_{h \to 0^+} \frac{h^2}{h} = \lim_{h \to 0^+} h = 0.$$

Thus $f'(1)$ does not exist since $f'_-(1) \neq f'_+(1)$, so f is not differentiable at 1. But $f'(x) = -1$ for $x < 1$ and $f'(x) = 2x - 2$ if $x > 1$.

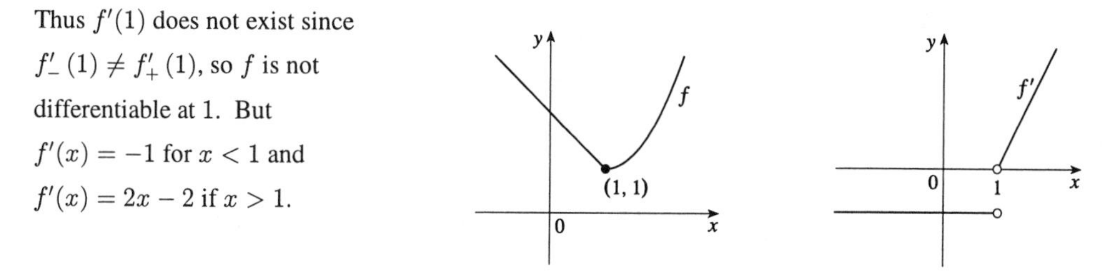

69. **(a)** Note that $x^2 - 9 < 0$ for $x^2 < 9 \quad \Leftrightarrow \quad |x| < 3 \quad \Leftrightarrow \quad -3 < x < 3$. So

$$f(x) = \begin{cases} x^2 - 9 & \text{if } x \leq -3 \\ -x^2 + 9 & \text{if } -3 < x < 3 \\ x^2 - 9 & \text{if } x \geq 3 \end{cases} \quad \Rightarrow \quad f'(x) = \begin{cases} 2x & \text{if } x < -3 \\ -2x & \text{if } -3 < x < 3 \\ 2x & \text{if } x > 3. \end{cases}$$

To show that $f'(3)$ does not exist we investigate $\lim\limits_{h \to 0} \dfrac{f(3+h) - f(3)}{h}$ by computing the left- and right-hand derivatives defined in Exercise 2.1.57.

$$f'_-(3) = \lim_{h \to 0^-} \frac{f(3+h) - f(3)}{h} = \lim_{h \to 0^-} \frac{\left(-(3+h)^2 + 9\right) - 0}{h} = \lim_{h \to 0^-} (-6 + h) = -6 \text{ and}$$

$$f'_+(3) = \lim_{h \to 0^+} \frac{f(3+h) - f(3)}{h} = \lim_{h \to 0^+} \frac{\left[(3+h)^2 + 9\right] - 0}{h} = \lim_{h \to 0^+} \frac{6h + h^2}{h} = \lim_{h \to 0^+} (6 + h) = 6.$$

Since the left and right limits are different, $\lim\limits_{h \to 0} \dfrac{f(3+h) - f(3)}{h}$ does not exist, that is, $f'(3)$ does not exist. Similarly, $f'(-3)$ does not exist. Therefore f is not differentiable at 3 or at -3.

(b)

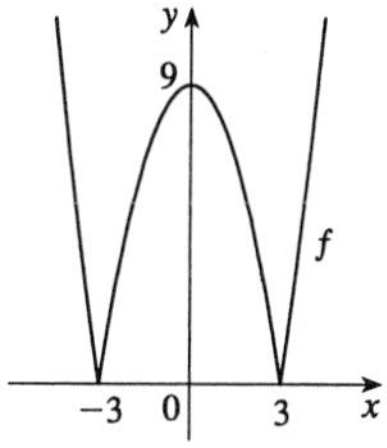

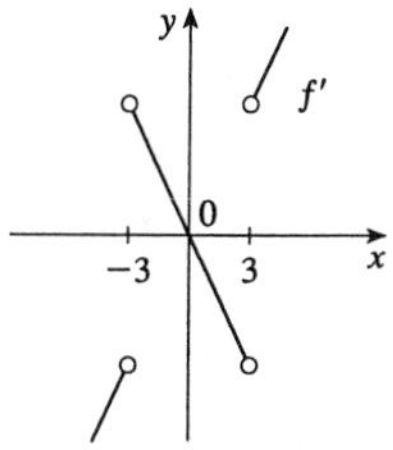

71. $y = f(x) = ax^2 \quad \Rightarrow \quad f'(x) = 2ax$. So the slope of the tangent to the parabola at $x = 2$ is $m = 2a(2) = 4a$. The slope of the given line is seen to be -2, so we must have $4a = -2 \quad \Leftrightarrow \quad a = -\frac{1}{2}$. So the point in question has y-coordinate $-\frac{1}{2} \cdot 2^2 = -2$. Now we simply require that the given line, whose equation is $2x + y = b$, pass through the point $(2, -2)$: $2(2) + (-2) = b \quad \Leftrightarrow \quad b = 2$. So we must have $a = -\frac{1}{2}$ and $b = 2$.

73. $F = \dfrac{f}{g} \quad \Rightarrow \quad f = Fg \quad \Rightarrow \quad f' = F'g + Fg' \quad \Rightarrow \quad F' = \dfrac{f' - Fg'}{g} = \dfrac{f' - (f/g)g'}{g} = \dfrac{f'g - fg'}{g^2}$

75. *Solution 1:* Let $f(x) = x^{1000}$. Then, by the definition of the derivative,

$f'(1) = \lim\limits_{x \to 1} \dfrac{f(x) - f(1)}{x - 1} = \lim\limits_{x \to 1} \dfrac{x^{1000} - 1}{x - 1}$. But this is just the limit we want to find, and we know (from the Power Rule) that $f'(x) = 1000x^{999}$, so $f'(1) = 1000(1)^{999} = 1000$. So $\lim\limits_{x \to 1} \dfrac{x^{1000} - 1}{x - 1} = 1000$.

Solution 2: Note that $(x^{1000} - 1) = (x - 1)\left(x^{999} + x^{998} + x^{997} + \cdots + x^2 + x + 1\right)$. So

$$\begin{aligned} \lim_{x \to 1} \frac{x^{1000} - 1}{x - 1} &= \lim_{x \to 1} \frac{(x-1)\left(x^{999} + x^{998} + x^{997} + \cdots + x^2 + x + 1\right)}{x - 1} \\ &= \lim_{x \to 1}\left(x^{999} + x^{998} + x^{997} + \cdots + x^2 + x + 1\right) = 1 + 1 + 1 + \cdots + 1 + 1 + 1 = 1000 \text{ as above.} \end{aligned}$$

EXERCISES 2.3

1. **(a)** $v(t) = f'(t) = 2t - 6$ **(b)** $v(2) = 2(2) - 6 = -2$ ft/s

(c) It is at rest when $v(t) = 2t - 6 = 0 \quad \Leftrightarrow \quad t = 3$.

(d) It moves in the positive direction when $2t - 6 > 0 \quad \Leftrightarrow \quad t > 3$.

(e) Distance in positive direction $= |f(4) - f(3)| = |1 - 0| = 1$ ft

Distance in negative direction $= |f(3) - f(0)| = |0 - 9| = 9$ ft

Total distance traveled $= 1 + 9 = 10$ ft

(f)

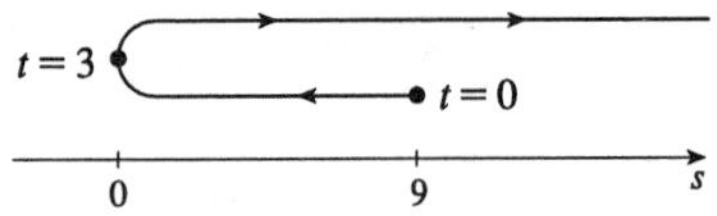

3. **(a)** $v(t) = f'(t) = 6t^2 - 18t + 12$ **(b)** $v(2) = 6(2)^2 - 18(2) + 12 = 0$ ft/s

(c) It is at rest when $v(t) = 6t^2 - 18t + 12 = 6(t-1)(t-2) = 0 \quad \Leftrightarrow \quad t = 1$ or 2.

(d) It moves in the positive direction when $6(t-1)(t-2) > 0 \quad \Leftrightarrow \quad 0 \le t < 1$ or $t > 2$.

(e) Distance in positive direction $= |f(4) - f(2)| + |f(1) - f(0)| = |33 - 5| + |6 - 1| = 33$ ft

Distance in negative direction $= |f(2) - f(1)| = |5 - 6| = 1$ ft

Total distance traveled $= 33 + 1 = 34$ ft

(f)

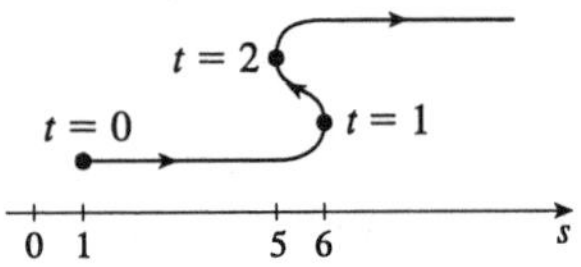

5. **(a)** $v(t) = s'(t) = \dfrac{(t^2+1)(1) - t(2t)}{(t^2+1)^2} = \dfrac{1 - t^2}{(t^2+1)^2}$ **(b)** $v(2) = \dfrac{1 - (2)^2}{(2^2+1)^2} = -\dfrac{3}{25}$ ft/s

(c) It is at rest when $v = 0 \quad \Leftrightarrow \quad 1 - t^2 = 0 \quad \Leftrightarrow \quad t = 1$.

(d) It moves in the positive direction when $v > 0 \quad \Leftrightarrow \quad 1 - t^2 > 0 \quad \Leftrightarrow \quad t^2 < 1 \quad \Leftrightarrow \quad 0 \le t < 1$.

(e) Distance in positive direction $= |s(1) - s(0)| = \left|\frac{1}{2} - 0\right| = \frac{1}{2}$ ft

Distance in negative direction $= |s(4) - s(1)| = \left|\frac{4}{17} - \frac{1}{2}\right| = \frac{9}{34}$ ft

Total distance traveled $= \frac{1}{2} + \frac{9}{34} = \frac{13}{17}$ ft

(f)

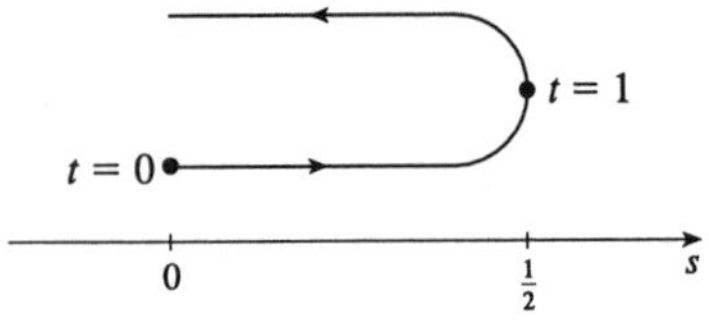

7. $s(t) = t^3 - 4.5t^2 - 7t \quad \Rightarrow \quad v(t) = s'(t) = 3t^2 - 9t - 7 = 5 \quad \Leftrightarrow \quad 3t^2 - 9t - 12 = 0 \quad \Leftrightarrow$
$3(t-4)(t+1) = 0 \quad \Leftrightarrow \quad t = 4$ or -1. Since $t \geq 0$, the particle reaches a velocity of 5 m/s at $t = 4$ s.

9. **(a)** $V(x) = x^3$, so the average rate of change is: **(i)** $\dfrac{V(6) - V(5)}{6-5} = 6^3 - 5^3 = 216 - 125 = 91$

(ii) $\dfrac{V(5.1) - V(5)}{5.1 - 5} = \dfrac{(5.1)^3 - 5^3}{0.1} = 76.51$ **(iii)** $\dfrac{V(5.01) - V(5)}{5.01 - 5} = \dfrac{(5.01)^3 - 5^3}{0.01} = 75.1501$

(b) $V'(x) = 3x^2$, $V'(5) = 75$ **(c)** The surface area is $S(x) = 6x^2$, so $V'(x) = 3x^2 = \frac{1}{2}(6x^2) = \frac{1}{2}S(x)$.

11. After t seconds the radius is $r = 60t$, so the area is $A(t) = \pi(60t)^2 = 3600\pi t^2 \quad \Rightarrow \quad A'(t) = 7200\pi t \quad \Rightarrow$

(a) $A'(1) = 7200\pi$ cm²/s **(b)** $A'(3) = 21{,}600\pi$ cm²/s **(c)** $A'(5) = 36{,}000\pi$ cm²/s

13. $S(r) = 4\pi r^2 \quad \Rightarrow \quad S'(r) = 8\pi r$

(a) $S'(1) = 8\pi$ ft²/ft **(b)** $S'(2) = 16\pi$ ft²/ft **(c)** $S'(3) = 24\pi$ ft²/ft

15. $f(x) = 3x^2$, so the linear density at x is $\rho(x) = f'(x) = 6x$.

(a) $\rho(1) = 6$ kg/m **(b)** $\rho(2) = 12$ kg/m **(c)** $\rho(3) = 18$ kg/m

17. $Q(t) = t^3 - 2t^2 + 6t + 2$, so the current is $Q'(t) = 3t^2 - 4t + 6$.

(a) $Q'(0.5) = 3(0.5)^2 - 4(0.5) + 6 = 4.75$ A **(b)** $Q'(1) = 3(1)^2 - 4(1) + 6 = 5$ A

19. **(a)** $PV = C \quad \Rightarrow \quad V = \dfrac{C}{P} \quad \Rightarrow \quad \dfrac{dV}{dP} = -\dfrac{C}{P^2}$

(b) $\beta = -\dfrac{1}{V}\dfrac{dV}{dP} = -\dfrac{1}{V}\left(-\dfrac{C}{P^2}\right) = \dfrac{C}{(PV)P} = \dfrac{C}{CP} = \dfrac{1}{P}$

21. **(a)** rate of reaction $= \dfrac{d[C]}{dt} = \dfrac{a^2k(akt+1) - (a^2kt)(ak)}{(akt+1)^2} = \dfrac{a^2k(akt+1-akt)}{(akt+1)^2} = \dfrac{a^2k}{(akt+1)^2}$

(b) $a - x = a - \dfrac{a^2kt}{akt+1} = \dfrac{a^2kt + a - a^2kt}{akt+1} = \dfrac{a}{akt+1}$.

So $k(a-x)^2 = k\left(\dfrac{a}{akt+1}\right)^2 = \dfrac{a^2k}{(akt+1)^2} = \dfrac{dx}{dt}$.

23. $m(t) = 5 - 0.02t^2 \quad \Rightarrow \quad m'(t) = -0.04t \quad \Rightarrow \quad m'(1) = -0.04$

25. $v(r) = \dfrac{P}{4\eta\ell}\left(R^2 - r^2\right) \quad \Rightarrow \quad v'(r) = \dfrac{P}{4\eta\ell}(-2r) = -\dfrac{Pr}{2\eta\ell}$.

When $\ell = 3$, $P = 3000$ and $\eta = 0.027$, we have $v'(0.005) = -\dfrac{3000(0.005)}{2(0.027)(3)} \approx -92.6\,\dfrac{\text{cm/s}}{\text{cm}}$.

27. $C(x) = 420 + 1.5x + 0.002x^2 \Rightarrow C'(x) = 1.5 + 0.004x \Rightarrow C'(100) = 1.5 + (0.004)(100) = \$1.90$/item
$C(101) - C(100) = (420 + 151.5 + 20.402) - (420 + 150 + 20) = \$1.902$/item

29. $C(x) = 2000 + 3x + 0.01x^2 + 0.0002x^3 \quad \Rightarrow \quad C'(x) = 3 + 0.02x + 0.0006x^2 \quad \Rightarrow$
$C'(100) = 3 + 0.02(100) + 0.0006(10{,}000) = 3 + 2 + 6 = \$11$/item
$C(101) - C(100) = (2000 + 303 + 102.1 + 206.0602) - (2000 + 300 + 100 + 200)$
$= 11.0702 \approx \$11.07$/item

EXERCISES 2.4

1. $\lim_{x\to 0}(x^2+\cos x) = \lim_{x\to 0} x^2 + \lim_{x\to 0}\cos x = 0^2 + \cos 0 = 0 + 1 = 1$

3. $\lim_{x\to \pi/3}(\sin x - \cos x) = \sin\frac{\pi}{3} - \cos\frac{\pi}{3} = \frac{\sqrt{3}}{2} - \frac{1}{2}$

5. $\lim_{x\to\pi/4}\dfrac{\sin x}{3x} = \dfrac{\sin(\pi/4)}{3\pi/4} = \dfrac{1/\sqrt{2}}{3\pi/4} = \dfrac{2\sqrt{2}}{3\pi}$

7. $\lim_{t\to 0}\dfrac{\sin 5t}{t} = \lim_{t\to 0}\dfrac{5\sin 5t}{5t} = 5\lim_{t\to 0}\dfrac{\sin 5t}{5t} = 5\cdot 1 = 5$

9. $\lim_{\theta\to 0}\dfrac{\sin(\cos\theta)}{\sec\theta} = \dfrac{\sin\left(\lim_{\theta\to 0}\cos\theta\right)}{\lim_{\theta\to 0}\sec\theta} = \dfrac{\sin 1}{1} = \sin 1$

11. $\lim_{x\to\pi/4}\dfrac{\tan x}{4x} = \dfrac{\tan(\pi/4)}{4(\pi/4)} = \dfrac{1}{\pi}$

13. $\lim_{\theta\to 0}\dfrac{\sin^2\theta}{\theta} = \lim_{\theta\to 0}\left(\dfrac{\sin\theta}{\theta}\right)\sin\theta = \lim_{\theta\to 0}\dfrac{\sin\theta}{\theta}\lim_{\theta\to 0}\sin\theta = 1\cdot 0 = 0$

15. $\lim_{x\to 0}\dfrac{\tan 3x}{3\tan 2x} = \lim_{x\to 0}\dfrac{\dfrac{\tan 3x}{3x}}{2\dfrac{\tan 2x}{2x}} = \dfrac{1}{2}\,\dfrac{\lim_{x\to 0}\dfrac{\sin 3x}{3x}\cdot\dfrac{1}{\cos 3x}}{\lim_{x\to 0}\dfrac{\sin 2x}{2x}\cdot\lim_{x\to 0}\dfrac{1}{\cos 2x}} = \dfrac{1}{2}\,\dfrac{1\cdot 1}{1\cdot 1} = \dfrac{1}{2}$

17. $\dfrac{d}{dx}(\csc x) = \dfrac{d}{dx}\left(\dfrac{1}{\sin x}\right) = \dfrac{(\sin x)(0) - 1(\cos x)}{\sin^2 x} = \dfrac{-\cos x}{\sin^2 x} = -\dfrac{1}{\sin x}\cdot\dfrac{\cos x}{\sin x} = -\csc x\cot x$

19. $\dfrac{d}{dx}(\cot x) = \dfrac{d}{dx}\left(\dfrac{\cos x}{\sin x}\right) = \dfrac{(\sin x)(-\sin x) - (\cos x)(\cos x)}{\sin^2 x} = -\dfrac{\sin^2 x + \cos^2 x}{\sin^2 x} = -\dfrac{1}{\sin^2 x} = -\csc^2 x$

21. $y = \sin x + \cos x \;\Rightarrow\; dy/dx = \cos x - \sin x$

23. $y = \csc x\cot x \;\Rightarrow\; dy/dx = (-\csc x\cot x)\cot x + \csc x\,(-\csc^2 x) = -\csc x\,(\cot^2 x + \csc^2 x)$

25. $y = \dfrac{\tan x}{x} \;\Rightarrow\; \dfrac{dy}{dx} = \dfrac{x\sec^2 x - \tan x}{x^2}$

27. $y = \dfrac{x}{\sin x + \cos x} \;\Rightarrow$

$$\frac{dy}{dx} = \frac{(\sin x + \cos x) - x(\cos x - \sin x)}{(\sin x + \cos x)^2} = \frac{(1+x)\sin x + (1-x)\cos x}{\sin^2 x + \cos^2 x + 2\sin x\cos x} = \frac{(1+x)\sin x + (1-x)\cos x}{1 + \sin 2x}$$

29. $y = x^{-3}\sin x\tan x \;\Rightarrow$

$$\frac{dy}{dx} = -3x^{-4}\sin x\tan x + x^{-3}\cos x\tan x + x^{-3}\sin x\sec^2 x = x^{-4}\sin x\left(-3\tan x + x + x\sec^2 x\right)$$

31. $y = \dfrac{x^2\tan x}{\sec x} \;\Rightarrow$

$$\frac{dy}{dx} = \frac{\sec x\,(2x\tan x + x^2\sec^2 x) - x^2\tan x\sec x\tan x}{\sec^2 x} = \frac{2x\tan x + x^2(\sec^2 x - \tan^2 x)}{\sec x} = \frac{2x\tan x + x^2}{\sec x}.$$

Another Method: Write $y = x^2\sin x$. Then $y' = 2x\sin x + x^2\cos x$.

33. $y = \tan x \;\Rightarrow\; y' = \sec^2 x \;\Rightarrow$ The slope of the tangent line at $\left(\frac{\pi}{4}, 1\right)$ is $\sec^2\frac{\pi}{4} = 2$ and the equation is $y - 1 = 2\left(x - \frac{\pi}{4}\right)$ or $4x - 2y = \pi - 2$.

35. **(a)** $y = x\cos x \quad \Rightarrow$
$y' = x(-\sin x) + \cos x\,(1) = \cos x - x\sin x$
So the slope of the tangent at the point $(\pi, -\pi)$ is
$\cos\pi - \pi\sin\pi = -1 - \pi(0) = -1$, and its
equation is $y + \pi = -(x - \pi) \quad \Leftrightarrow \quad y = -x$.

(b)

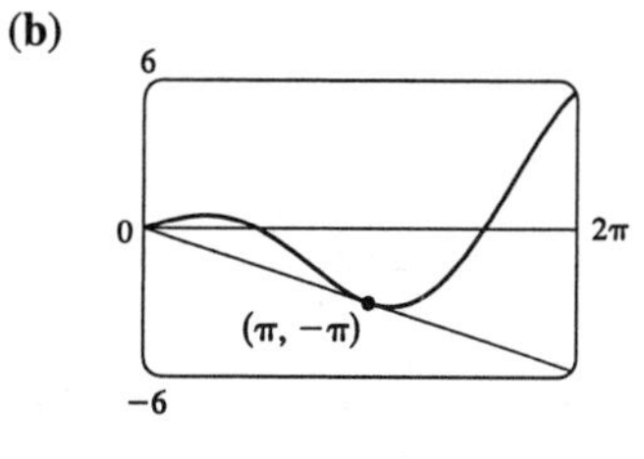

37. $y = x + 2\sin x$ has a horizontal tangent when $y' = 1 + 2\cos x = 0 \quad \Leftrightarrow \quad \cos x = -\frac{1}{2} \quad \Leftrightarrow$
$x = (2n+1)\pi \pm \frac{\pi}{3}$, n an integer.

39.

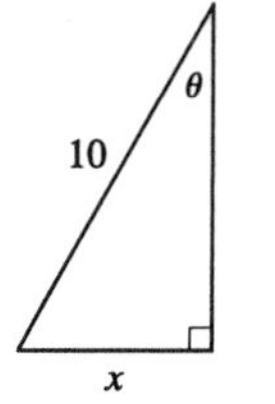

From the diagram we can see that $\sin\theta = 10/x$ $\Leftrightarrow \quad x = 10\sin\theta$. But we want to find the rate of change of x with respect to θ, that is, $dx/d\theta$. Taking the derivative of the above expression, $dx/d\theta = 10(\cos\theta)$. So when $\theta = \frac{\pi}{3}$, $dx/d\theta = 10\cos\frac{\pi}{3} = 10\left(\frac{1}{2}\right) = 5$ ft/rad.

41. $$\lim_{\theta\to 0}\frac{\cos\theta - 1}{\theta} = \lim_{\theta\to 0}\frac{1 - 2\sin^2(\theta/2) - 1}{\theta} = \lim_{\theta\to 0}\frac{-\sin^2(\theta/2)}{\theta/2} = -\lim_{\theta\to 0}\frac{\sin(\theta/2)}{\theta/2}\lim_{\theta\to 0}\sin(\theta/2) = -1\cdot 0 = 0$$

43. $$\lim_{x\to 0}\frac{\cot 2x}{\csc x} = \lim_{x\to 0}\frac{\cos 2x\sin x}{\sin 2x} = \lim_{x\to 0}\cos 2x\left[\frac{(\sin x)/x}{(\sin 2x)/x}\right] = \lim_{x\to 0}\cos 2x\left[\frac{\lim_{x\to 0}[(\sin x)/x]}{2\lim_{x\to 0}[(\sin 2x)/2x]}\right] = 1\cdot\frac{1}{2\cdot 1} = \frac{1}{2}$$

45. $$\lim_{x\to\pi}\frac{\tan x}{\sin 2x} = \lim_{x\to\pi}\frac{\sin x}{\cos x\,(2\sin x\cos x)} = \lim_{x\to\pi}\frac{1}{2\cos^2 x} = \frac{1}{2(-1)^2} = \frac{1}{2}$$

47. Divide numerator and denominator by θ. ($\sin\theta$ also works.)

$$\lim_{\theta\to 0}\frac{\sin\theta}{\theta + \tan\theta} = \lim_{\theta\to 0}\frac{\dfrac{\sin\theta}{\theta}}{1 + \dfrac{\sin\theta}{\theta}\cdot\dfrac{1}{\cos\theta}} = \frac{\lim\limits_{\theta\to 0}\dfrac{\sin\theta}{\theta}}{1 + \lim\limits_{\theta\to 0}\dfrac{\sin\theta}{\theta}\lim\limits_{\theta\to 0}\dfrac{1}{\cos\theta}} = \frac{1}{1 + 1\cdot 1} = \frac{1}{2}$$

49. $$\lim_{x\to 0}\frac{\cos x\sin x - \tan x}{x^2\sin x} = \lim_{x\to 0}\frac{\cos x\sin x - \dfrac{\sin x}{\cos x}}{x^2\sin x} = \lim_{x\to 0}\frac{\cos^2 x\sin x - \sin x}{x^2\sin x\cos x} = \lim_{x\to 0}\frac{\cos^2 x - 1}{x^2\cos x}$$
$$= \lim_{x\to 0}\left(\frac{-\sin^2 x}{x^2}\right)\frac{1}{\cos x} = -\left[\lim_{x\to 0}\frac{\sin x}{x}\right]^2\left[\lim_{x\to 0}\frac{1}{\cos x}\right] = -1$$

51. $\displaystyle\lim_{x\to 0}\frac{\sin(\sin x)}{\sin x} = \lim_{\sin x\to 0}\frac{\sin(\sin x)}{\sin x}$ since as $x\to 0$, $\sin x\to 0$. So we make the substitution $y = \sin x$, and see that $\displaystyle\lim_{x\to 0}\frac{\sin(\sin x)}{\sin x} = \lim_{y\to 0}\frac{\sin y}{y} = 1$.

53. **(a)** $\displaystyle\frac{d}{dx}\tan x = \frac{d}{dx}\frac{\sin x}{\cos x} \quad\Rightarrow\quad \sec^2 x = \frac{\cos x\cos x - \sin x\,(-\sin x)}{\cos^2 x} = \frac{\cos^2 x + \sin^2 x}{\cos^2 x}$. So $\displaystyle\sec^2 x = \frac{1}{\cos^2 x}$.

(b) $\displaystyle\frac{d}{dx}\sec x = \frac{d}{dx}\frac{1}{\cos x} \quad\Rightarrow\quad \sec x\tan x = \frac{(\cos x)(0) - 1(-\sin x)}{\cos^2 x}$. So $\displaystyle\sec x\tan x = \frac{\sin x}{\cos^2 x}$.

(c) $\displaystyle\frac{d}{dx}(\sin x + \cos x) = \frac{d}{dx}\frac{1 + \cot x}{\csc x} \quad\Rightarrow\quad \cos x - \sin x = \frac{\csc x\,(-\csc^2 x) - (1 + \cot x)(-\csc x\cot x)}{\csc^2 x}$

$\displaystyle = \frac{-\csc^2 x + \cot^2 x + \cot x}{\csc x}$. So $\displaystyle\cos x - \sin x = \frac{\cot x - 1}{\csc x}$.

55. By the definition of radian measure, $s = r\theta$, where r is the radius of the circle. By drawing the bisector of the angle θ, we can see that $\sin\frac{\theta}{2} = \frac{d/2}{r} \Rightarrow d = 2r\sin\frac{\theta}{2}$. So
$\lim_{\theta\to 0^+}\frac{s}{d} = \lim_{\theta\to 0^+}\frac{r\theta}{2r\sin(\theta/2)} = \lim_{\theta\to 0^+}\frac{2\cdot(\theta/2)}{2\sin(\theta/2)} = \lim_{\theta\to 0}\frac{\theta/2}{\sin(\theta/2)} = 1.$
[This is just the reciprocal of the limit $\lim_{x\to 0}\frac{\sin x}{x} = 1$ combined with the fact that as $\theta \to 0$, $\frac{\theta}{2} \to 0$ also.]

EXERCISES 2.5

1. $y = u^2$, $u = x^2 + 2x + 3$

(a) $\frac{dy}{dx} = \frac{dy}{du}\frac{du}{dx} = 2u(2x+2) = 4u(x+1)$. When $x = 1$, $u = 1^2 + 2(1) + 3 = 6$, so $dy/dx|_{x=1} = 4(6)(1+1) = 48$.

(b) $y = u^2 = (x^2+2x+3)^2 = x^4 + 4x^2 + 9 + 4x^3 + 6x^2 + 12x = x^4 + 4x^3 + 10x^2 + 12x + 9$, so $dy/dx = 4x^3 + 12x^2 + 20x + 12$ and $dy/dx|_{x=1} = 4(1)^3 + 12(1)^2 + 20(1) + 12 = 48$.

3. $y = u^3$, $u = x + 1/x$

(a) $\frac{dy}{dx} = \frac{dy}{du}\frac{du}{dx} = 3u^2\left(1 - \frac{1}{x^2}\right)$. When $x = 1$, $u = 1 + \frac{1}{1} = 2$, so $\left.\frac{dy}{dx}\right|_{x=1} = 3(2)^2\left(1 - \frac{1}{1^2}\right) = 0$.

(b) $y = u^3 = \left(x + \frac{1}{x}\right)^3 = x^3 + 3x^2\left(\frac{1}{x}\right) + 3x\left(\frac{1}{x}\right)^2 + \left(\frac{1}{x}\right)^3 = x^3 + 3x + 3x^{-1} + x^{-3}$, so $dy/dx = 3x^2 + 3 - 3x^{-2} - 3x^{-4}$ and $dy/dx|_{x=1} = 3(1)^2 + 3 - 3(1)^{-2} - 3(1)^{-4} = 0$.

5. $F(x) = (x^2+4x+6)^5 \Rightarrow$
$F'(x) = 5(x^2+4x+6)^4\frac{d}{dx}(x^2+4x+6) = 5(x^2+4x+6)^4(2x+4) = 10(x^2+4x+6)^4(x+2)$

7. $G(x) = (3x-2)^{10}(5x^2-x+1)^{12} \Rightarrow$
$G'(x) = 10(3x-2)^9(3)(5x^2-x+1)^{12} + (3x-2)^{10}(12)(5x^2-x+1)^{11}(10x-1)$
$= 30(3x-2)^9(5x^2-x+1)^{12} + 12(3x-2)^{10}(5x^2-x+1)^{11}(10x-1)$
[This can be simplified to $6(3x-2)^9(5x^2-x+1)^{11}(85x^2-51x+9)$.]

9. $f(t) = (2t^2-6t+1)^{-8} \Rightarrow f'(t) = -8(2t^2-6t+1)^{-9}(4t-6) = -16(2t^2-6t+1)^{-9}(2t-3)$

11. $g(x) = \sqrt{x^2-7x} = (x^2-7x)^{1/2} \Rightarrow g'(x) = \frac{1}{2}(x^2-7x)^{-1/2}(2x-7) = \frac{2x-7}{2\sqrt{x^2-7x}}$

13. $h(t) = (t-1/t)^{3/2} \Rightarrow h'(t) = \frac{3}{2}(t-1/t)^{1/2}(1+1/t^2)$

15. $F(y) = \left(\frac{y-6}{y+7}\right)^3 \Rightarrow F'(y) = 3\left(\frac{y-6}{y+7}\right)^2\frac{(y+7)(1)-(y-6)(1)}{(y+7)^2} = 3\left(\frac{y-6}{y+7}\right)^2\frac{13}{(y+7)^2} = \frac{39(y-6)^2}{(y+7)^4}$

17. $f(z) = (2z-1)^{-1/5} \Rightarrow f'(z) = -\frac{1}{5}(2z-1)^{-6/5}(2) = -\frac{2}{5}(2z-1)^{-6/5}$

19. $y = (2x-5)^4(8x^2-5)^{-3} \quad \Rightarrow \quad y' = 4(2x-5)^3(2)(8x^2-5)^{-3} + (2x-5)^4(-3)(8x^2-5)^{-4}(16x)$
$$= 8(2x-5)^3(8x^2-5)^{-3} - 48x(2x-5)^4(8x^2-5)^{-4}$$
[This simplifies to $8(2x-5)^3(8x^2-5)^{-4}(-4x^2+30x-5)$.]

21. $y = \tan 3x \quad \Rightarrow \quad y' = \sec^2 3x \, \dfrac{d}{dx}(3x) = 3\sec^2 3x$

23. $y = \cos(x^3) \quad \Rightarrow \quad y' = -\sin(x^3)(3x^2) = -3x^2\sin(x^3)$

25. $y = (1+\cos^2 x)^6 \quad \Rightarrow \quad y' = 6(1+\cos^2 x)^5 2\cos x\,(-\sin x) = -12\cos x \sin x\,(1+\cos^2 x)^5$

27. $y = \cos(\tan x) \quad \Rightarrow \quad y' = -\sin(\tan x)\sec^2 x$

29. $y = \sec^2 2x - \tan^2 2x \quad \Rightarrow \quad y' = 2\sec 2x\,(\sec 2x \tan 2x)(2) - 2\tan 2x \sec^2(2x)(2) = 0$

Easier method: $y = \sec^2 2x - \tan^2 2x = 1 \quad \Rightarrow \quad y' = 0$

31. $y = \csc\dfrac{x}{3} \quad \Rightarrow \quad y' = -\dfrac{1}{3}\csc\dfrac{x}{3}\cot\dfrac{x}{3}$

33. $y = \sin^3 x + \cos^3 x \quad \Rightarrow \quad y' = 3\sin^2 x\cos x + 3\cos^2 x(-\sin x) = 3\sin x\cos x\,(\sin x - \cos x)$

35. $y = \sin\dfrac{1}{x} \quad \Rightarrow \quad y' = \cos\dfrac{1}{x}\left(-\dfrac{1}{x^2}\right) = -\dfrac{1}{x^2}\cos\dfrac{1}{x}$

37. $y = \dfrac{1+\sin 2x}{1-\sin 2x} \quad \Rightarrow \quad y' = \dfrac{(1-\sin 2x)(2\cos 2x) - (1+\sin 2x)(-2\cos 2x)}{(1-\sin 2x)^2} = \dfrac{4\cos 2x}{(1-\sin 2x)^2}$

39. $y = \tan^2(x^3) \quad \Rightarrow \quad y' = 2\tan(x^3)\sec^2(x^3)(3x^2) = 6x^2\tan(x^3)\sec^2(x^3)$

41. $y = \cos^2(\cos x) + \sin^2(\cos x) = 1 \quad \Rightarrow \quad y' = 0.$

43. $y = \sqrt{x+\sqrt{x}} \quad \Rightarrow \quad y' = \frac{1}{2}\left(x+\sqrt{x}\right)^{-1/2}\left(1+\frac{1}{2}x^{-1/2}\right) = \dfrac{1}{2\sqrt{x+\sqrt{x}}}\left(1+\dfrac{1}{2\sqrt{x}}\right)$

45. $f(x) = \left[x^3 + (2x-1)^3\right]^3 \quad \Rightarrow$

$f'(x) = 3\left[x^3+(2x-1)^3\right]^2\left[3x^2+3(2x-1)^2(2)\right] = 9\left[x^3+(2x-1)^3\right]^2\left[9x^2-8x+2\right]$

47. $y = \sin\left(\tan\sqrt{\sin x}\right) \quad \Rightarrow \quad y' = \cos\left(\tan\sqrt{\sin x}\right)\left(\sec^2\sqrt{\sin x}\right)\left(\dfrac{1}{2\sqrt{\sin x}}\right)(\cos x)$

49. $y = f(x) = (x^3-x^2+x-1)^{10} \quad \Rightarrow \quad f'(x) = 10(x^3-x^2+x-1)^9(3x^2-2x+1)$. The slope of the tangent at $(1,0)$ is $f'(1) = 0$ and its equation is $y-0 = 0(x-1)$ or $y = 0$.

51. $y = f(x) = \dfrac{8}{\sqrt{4+3x}} \quad \Rightarrow \quad f'(x) = 8\left(-\frac{1}{2}\right)(4+3x)^{-3/2}(3) = -12(4+3x)^{-3/2}$. The slope of the tangent at $(4,2)$ is $f'(4) = -\frac{3}{16}$ and its equation is $y-2 = -\frac{3}{16}(x-4)$ or $3x+16y = 44$.

53. **(a)** $y = f(x) = \tan\left(\frac{\pi}{4}x^2\right) \quad \Rightarrow$
$f'(x) = \sec^2\left(\frac{\pi}{4}x^2\right)\left(2\cdot\frac{\pi}{4}x\right)$. The slope of the tangent at $(1,1)$ is thus $f'(1) = \sec^2\frac{\pi}{4}\left(\frac{\pi}{2}\right) = 2\cdot\frac{\pi}{2} = \pi$, and its equation is $y-1 = \pi(x-1)$ or $y = \pi x - \pi + 1$.

(b)

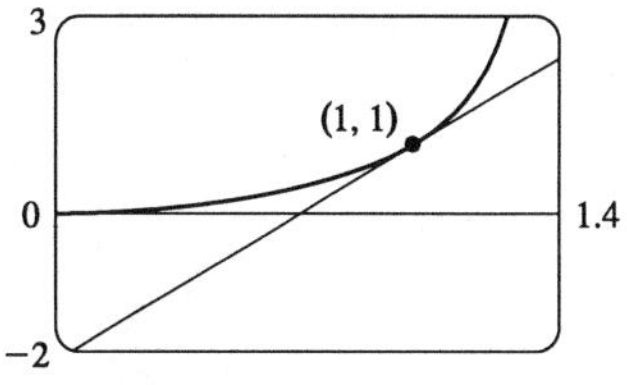

3

(1, 1)

0 1.4

−2

55. **(a)** $f(x) = \dfrac{\sqrt{1-x^2}}{x} \quad \Rightarrow$

$$f'(x) = \frac{x \cdot \frac{1}{2}(1-x^2)^{-1/2}(-2x) - \sqrt{1-x^2}}{x^2}$$
$$= \frac{-1}{\sqrt{1-x^2}} - \frac{\sqrt{1-x^2}}{x^2}$$
$$= \frac{-x^2 - \sqrt{1-x^2}\sqrt{1-x^2}}{x^2\sqrt{1-x^2}} = \frac{-1}{x^2\sqrt{1-x^2}}$$

(b)

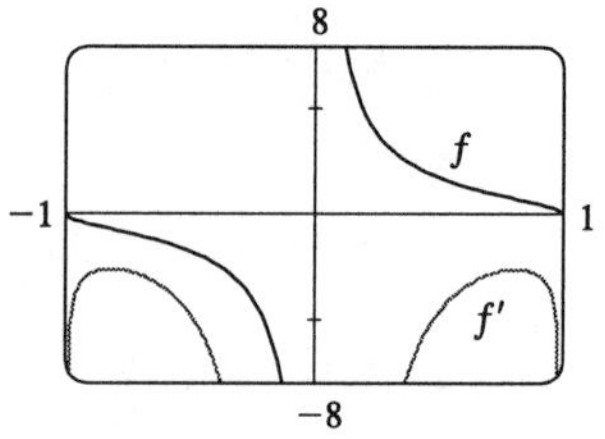

8
−1
1
−8
f
f'

Notice that all tangents to the graph of f have negative slopes and $f'(x) < 0$ always.

57. For the tangent line to be horizontal $f'(x) = 0$. $f(x) = 2\sin x + \sin^2 x \quad \Rightarrow$

$f'(x) = 2\cos x + 2\sin x \cos x = 0 \quad \Leftrightarrow \quad 2\cos x\,(1 + \sin x) = 0 \quad \Leftrightarrow \quad \cos x = 0$

or $\sin x = -1$, so $x = \left(n + \frac{1}{2}\right)\pi$ or $\left(2n + \frac{3}{2}\right)\pi$ where n is any integer. So the points on the curve with a horizontal tangent are $\left(\left(2n + \frac{1}{2}\right)\pi, 3\right)$ and $\left(\left(2n + \frac{3}{2}\right)\pi, -1\right)$ where n is any integer.

59. $F(x) = f(g(x)) \quad \Rightarrow \quad F'(x) = f'(g(x))\,g'(x)$, so $F'(3) = f'(g(3))g'(3) = f'(6)g'(3) = 7 \cdot 4 = 28$.

61. $s(t) = 10 + \frac{1}{4}\sin(10\pi t) \quad \Rightarrow \quad$ the velocity after t seconds is

$v(t) = s'(t) = \frac{1}{4}\cos(10\pi t)(10\pi) = \frac{5\pi}{2}\cos(10\pi t)$ cm/s.

63. **(a)** $\dfrac{dB}{dt} = \left(0.35\cos\dfrac{2\pi t}{5.4}\right)\left(\dfrac{2\pi}{5.4}\right) = \dfrac{7\pi}{54}\cos\dfrac{2\pi t}{5.4}$ **(b)** At $t = 1$, $\dfrac{dB}{dt} = \dfrac{7\pi}{54}\cos\dfrac{2\pi}{5.4} \approx 0.16$.

65. **(a)** Since h is differentiable on $[0, \infty)$ and $\sqrt{x}$ is differentiable on $(0, \infty)$, it follows that $G(x) = h\left(\sqrt{x}\right)$ is differentiable on $(0, \infty)$.

(b) By the Chain Rule, $G'(x) = h'\left(\sqrt{x}\right)\dfrac{d}{dx}\sqrt{x} = \dfrac{h'\left(\sqrt{x}\right)}{2\sqrt{x}}$.

67. **(a)** $F(x) = f(\cos x) \quad \Rightarrow \quad F'(x) = f'(\cos x)\dfrac{d}{dx}(\cos x) = -\sin x\, f'(\cos x)$

(b) $G(x) = \cos(f(x)) \quad \Rightarrow \quad G'(x) = -\sin(f(x))f'(x)$

69. $g(x) = f(b + mx) + f(b - mx) \quad \Rightarrow \quad g'(x) = f'(b + mx)D(b + mx) + f'(b - mx)D(b - mx)$
$= mf'(b + mx) - mf'(b - mx)$. So $g'(0) = mf'(b) - mf'(b) = 0$.

71. **(a)** If f is even, then $f(x) = f(-x)$. Using the Chain Rule to differentiate this equation, we get
$f'(x) = f'(-x)\dfrac{d}{dx}(-x) = -f'(-x)$. Thus $f'(-x) = -f'(x)$, so f' is odd.

(b) If f is odd, then $f(x) = -f(-x)$. Differentiating this equation, we get $f'(x) = -f'(-x)(-1) = f'(-x)$, so f' is even.

73. $\dfrac{d}{dx}(\sin^n x \cos nx) = n\sin^{n-1}x\cos x\cos nx + \sin^n x\,(-n\sin nx)$

$$= n\sin^{n-1}x\,(\cos nx\cos x - \sin nx\sin x) = n\sin^{n-1}x\cos[(n+1)x]$$

75. $f(x) = |x| = \sqrt{x^2} \quad \Rightarrow \quad f'(x) = \frac{1}{2}(x^2)^{-1/2}(2x) = x/\sqrt{x^2} = x/|x|$.

77. Using Exercise 75, we have $h(x) = x|2x-1| \quad \Rightarrow$
$h'(x) = |2x-1| + x\dfrac{2x-1}{|2x-1|}(2) = |2x-1| + \dfrac{2x(2x-1)}{|2x-1|}$.

79. Since $\theta^\circ = \left(\frac{\pi}{180}\right)\theta$ rad, we have $\dfrac{d}{d\theta}(\sin\theta^\circ) = \dfrac{d}{d\theta}\left(\sin\frac{\pi}{180}\theta\right) = \frac{\pi}{180}\cos\frac{\pi}{180}\theta = \frac{\pi}{180}\cos\theta^\circ$.

80. **(a)** $u(x) = f(g(x)) \;\Rightarrow\; u'(x) = f'(g(x))g'(x)$. So $u'(1) = f'(g(1))g'(1) = f'(3)\,g'(1) = \left(-\frac{1}{4}\right)(-3) = \frac{3}{4}$.

(b) $v(x) = g(f(x)) \;\Rightarrow\; v'(x) = g'(f(x))f'(x)$. So $v'(1) = g'(f(1))f'(1) = g'(2)\,f'(1)$, which does not exist because $g'(2)$ does not exist.

(c) $w(x) = g(g(x)) \;\Rightarrow\; w'(x) = g'(g(x))g'(x)$. So $w'(1) = g'(g(1))g'(1) = g'(3)g'(1) = \left(\frac{2}{3}\right)(-3) = -2$.

EXERCISES 2.6

1. **(a)** $x^2 + 3x + xy = 5 \quad\Rightarrow\quad 2x + 3 + y + xy' = 0 \quad\Rightarrow\quad y' = -\dfrac{2x+y+3}{x}$

(b) $x^2 + 3x + xy = 5 \quad\Rightarrow\quad y = \dfrac{5 - x^2 - 3x}{x} = \dfrac{5}{x} - x - 3 \quad\Rightarrow\quad y' = -\dfrac{5}{x^2} - 1$

(c) $y' = -\dfrac{2x+y+3}{x} = \dfrac{-2x - 3 - (-3 - x + 5/x)}{x} = -1 - \dfrac{5}{x^2}$

3. **(a)** $2y^2 + xy = x^2 + 3 \quad\Rightarrow\quad 4yy' + y + xy' = 2x \quad\Rightarrow\quad y' = \dfrac{2x - y}{x + 4y}$

(b) Use the quadratic formula: $2y^2 + xy - (x^2 + 3) = 0 \quad\Rightarrow$

$$y = \frac{-x \pm \sqrt{x^2 + 8(x^2+3)}}{4} = \frac{-x \pm \sqrt{9x^2+24}}{4} \quad\Rightarrow\quad y' = \frac{1}{4}\left(-1 \pm \frac{9x}{\sqrt{9x^2+24}}\right)$$

(c) $$y' = \frac{2x - y}{x + 4y} = \frac{2x - \frac{1}{4}\left(-x \pm \sqrt{9x^2+24}\right)}{x + \left(-x \pm \sqrt{9x^2+24}\right)} = \frac{1}{4}\left(-1 \pm \frac{9x}{\sqrt{9x^2+24}}\right)$$

5. $x^2 - xy + y^3 = 8 \quad\Rightarrow\quad 2x - y - xy' + 3y^2y' = 0 \quad\Rightarrow\quad y' = \dfrac{y - 2x}{3y^2 - x}$

7. $2y^2 + \sqrt[3]{xy} = 3x^2 + 17 \quad\Rightarrow\quad 4yy' + \frac{1}{3}x^{-2/3}y^{1/3} + \frac{1}{3}x^{1/3}y^{-2/3}y' = 6x \quad\Rightarrow$

$$y' = \frac{6x - \frac{1}{3}x^{-2/3}y^{1/3}}{4y + \frac{1}{3}x^{1/3}y^{-2/3}} = \frac{18x - x^{-2/3}y^{1/3}}{12y + x^{1/3}y^{-2/3}}$$

9. $x^4 + y^4 = 16 \quad\Rightarrow\quad 4x^3 + 4y^3y' = 0 \quad\Rightarrow\quad y' = -\dfrac{x^3}{y^3}$

11. $\dfrac{y}{x-y} = x^2 + 1 \quad\Rightarrow\quad 2x = \dfrac{(x-y)y' - y(1-y')}{(x-y)^2} = \dfrac{xy' - y}{(x-y)^2} \quad\Rightarrow\quad y' = \dfrac{y}{x} + 2(x-y)^2$

Another Method: Write the equation as $y = (x-y)(x^2+1) = x^3 + x - yx^2 - y$. Then $y' = \dfrac{3x^2 + 1 - 2xy}{x^2 + 2}$.

13. $\cos(x-y) = y\sin x \quad\Rightarrow\quad -\sin(x-y)(1-y') = y'\sin x + y\cos x \quad\Rightarrow\quad y' = \dfrac{\sin(x-y)+y\cos x}{\sin(x-y)-\sin x}$

15. $xy = \cot(xy) \quad\Rightarrow\quad y + xy' = -\csc^2(xy)(y+xy') \quad\Rightarrow\quad (y+xy')[1+\csc^2(xy)] = 0 \quad\Rightarrow\quad y+xy'=0$
$\Rightarrow\quad y' = -y/x$

17. $y^4 + x^2y^2 + yx^4 = y+1 \;\Rightarrow\; 4y^3 + 2x\,\dfrac{dx}{dy}\,y^2 + 2x^2y + x^4 + 4yx^3\,\dfrac{dx}{dy} = 1 \;\Rightarrow\; \dfrac{dx}{dy} = \dfrac{1-4y^3-2x^2y-x^4}{2xy^2+4yx^3}$

19. $x[f(x)]^3 + xf(x) = 6 \quad\Rightarrow\quad [f(x)]^3 + 3x[f(x)]^2f'(x) + f(x) + xf'(x) = 0 \quad\Rightarrow$
$f'(x) = -\dfrac{[f(x)]^3+f(x)}{3x[f(x)]^2+x} \quad\Rightarrow\quad f'(3) = -\dfrac{(1)^3+1}{3(3)(1)^2+3} = -\dfrac{1}{6}$

21. $\dfrac{x^2}{16} - \dfrac{y^2}{9} = 1 \quad\Rightarrow\quad \dfrac{x}{8} - \dfrac{2yy'}{9} = 0 \quad\Rightarrow\quad y' = \dfrac{9x}{16y}$. When $x=-5$ and $y=\frac{9}{4}$ we have $y' = \dfrac{9(-5)}{16(9/4)} = -\dfrac{5}{4}$
so the equation of the tangent is $y - \frac{9}{4} = -\frac{5}{4}(x+5)$ or $5x+4y+16=0$.

23. $y^2 = x^3(2-x) = 2x^3 - x^4 \quad\Rightarrow\quad 2yy' = 6x^2 - 4x^3 \quad\Rightarrow\quad y' = \dfrac{3x^2-2x^3}{y}$. When $x=y=1$,
$y' = \dfrac{3(1)^2-2(1)^3}{1} = 1$, so the equation of the tangent line is $y-1 = 1(x-1)$ or $y=x$.

25. $2(x^2+y^2)^2 = 25(x^2-y^2) \quad\Rightarrow\quad 4(x^2+y^2)(2x+2yy') = 25(2x-2yy') \quad\Rightarrow\quad y' = \dfrac{25x-4x(x^2+y^2)}{25y+4y(x^2+y^2)}$.
When $x=3$ and $y=1$, $y' = \dfrac{75-120}{25+40} = -\dfrac{9}{13}$ so the equation of the tangent is $y-1 = -\frac{9}{13}(x-3)$ or
$9x+13y=40$.

27. **(a)** $y^2 = 5x^4 - x^2 \quad\Rightarrow\quad 2yy' = 5(4x^3) - 2x \quad\Rightarrow$
$y' = \dfrac{10x^3-x}{y}$. So at the point $(1,2)$ we have
$y' = \dfrac{10(1)^3-1}{2} = \dfrac{9}{2}$, and the equation of
the tangent line is $y-2 = \frac{9}{2}(x-1) \quad\Leftrightarrow\quad y = \frac{9}{2}x - \frac{5}{2}$.

(b)

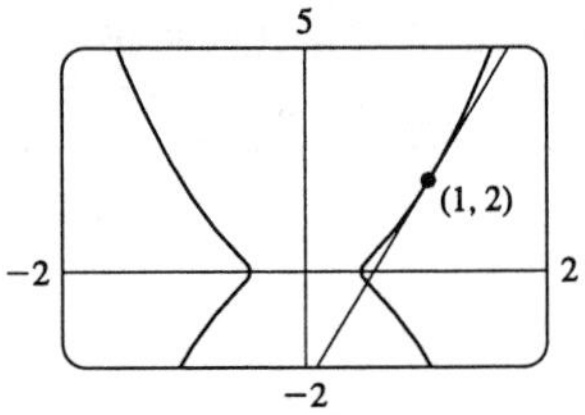

5
(1, 2)
−2
2
−2

29. From Exercise 25, a tangent to the lemniscate will be horizontal $\Rightarrow\; y'=0 \;\Rightarrow\; 25x - 4x(x^2+y^2) = 0$
$\Rightarrow\; x^2+y^2 = \frac{25}{4}$. (Note that $x=0 \;\Rightarrow\; y=0$ and there is no horizontal tangent at the origin.) Putting this in the equation of the lemniscate, we get $x^2 - y^2 = \frac{25}{8}$. Solving these two equations we have $x^2 = \frac{75}{16}$ and $y^2 = \frac{25}{16}$, so the four points are $\left(\pm\frac{5\sqrt{3}}{4}, \pm\frac{5}{4}\right)$.

31. $\dfrac{x^2}{a^2} - \dfrac{y^2}{b^2} = 1 \quad\Rightarrow\quad \dfrac{2x}{a^2} - \dfrac{2yy'}{b^2} = 0 \quad\Rightarrow\quad y' = \dfrac{b^2x}{a^2y} \quad\Rightarrow\quad$ the equation of the tangent at (x_0, y_0) is
$y - y_0 = \dfrac{b^2x_0}{a^2y_0}(x-x_0)$. Multiplying both sides by $\dfrac{y_0}{b^2}$ gives $\dfrac{y_0y}{b^2} - \dfrac{y_0^2}{b^2} = \dfrac{x_0x}{a^2} - \dfrac{x_0^2}{a^2}$. Since (x_0, y_0) lies on the hyperbola, we have $\dfrac{x_0x}{a^2} - \dfrac{y_0y}{b^2} = \dfrac{x_0^2}{a^2} - \dfrac{y_0^2}{b^2} = 1$.

33. If the circle has radius r, its equation is $x^2 + y^2 = r^2 \Rightarrow 2x + 2yy' = 0 \Rightarrow y' = -\dfrac{x}{y}$, so the slope of the tangent line at $P(x_0, y_0)$ is $-\dfrac{x_0}{y_0}$. The slope of OP is $\dfrac{y_0}{x_0} = \dfrac{-1}{-x_0/y_0}$, so the tangent is perpendicular to OP.

35. $2x^2 + y^2 = 3$ and $x = y^2$ intersect when $2x^2 + x - 3 = (2x+3)(x-1) = 0 \Leftrightarrow x = -\frac{3}{2}$ or 1, but $-\frac{3}{2}$ is extraneous. $2x^2 + y^2 = 3 \Rightarrow 4x + 2yy' = 0 \Rightarrow y' = -2x/y$ and $x = y^2 \Rightarrow 1 = 2yy' \Rightarrow$ $y' = 1/(2y)$. At $(1,1)$ the slopes are $m_1 = -2$ and $m_2 = \frac{1}{2}$, so the curves are orthogonal there. By symmetry they are also orthogonal at $(1,-1)$.

37. $x^2 + y^2 = r^2$ is a circle with center O and $ax + by = 0$ is a line through O. By Exercise 35, the curves are orthogonal.

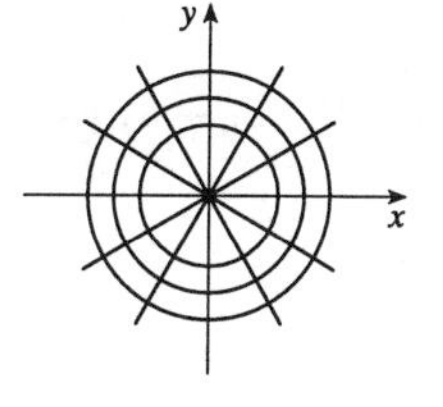

39. $y = cx^2 \Rightarrow y' = 2cx$ and $x^2 + 2y^2 = k$

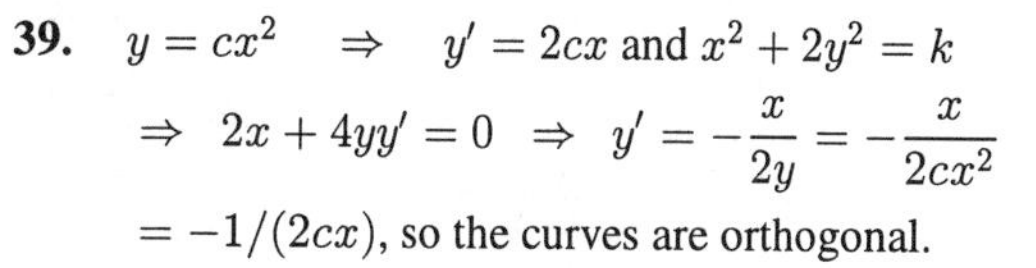

$\Rightarrow 2x + 4yy' = 0 \Rightarrow y' = -\dfrac{x}{2y} = -\dfrac{x}{2cx^2}$ $= -1/(2cx)$, so the curves are orthogonal.

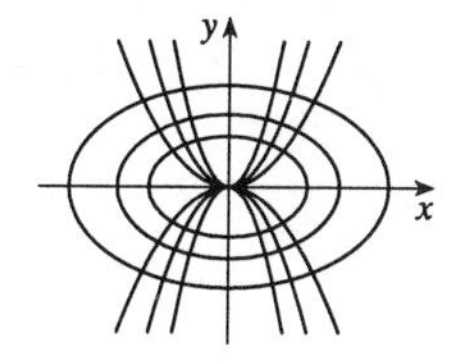

41. $y = 0 \Rightarrow x^2 + x(0) + 0^2 = 3 \Leftrightarrow x = \pm\sqrt{3}$. So the graph of the ellipse crosses the x-axis at the points $(\pm\sqrt{3}, 0)$. Using implicit differentiation to find y', we get $2x - xy' - y + 2yy' = 0 \Rightarrow y'(2y - x) = y - 2x$ $\Rightarrow y' = \dfrac{y - 2x}{2y - x}$. So $y'(\sqrt{3}, 0) = \dfrac{0 - 2\sqrt{3}}{2(0) - \sqrt{3}} = 2$, and $y'(-\sqrt{3}, 0) = \dfrac{0 + 2\sqrt{3}}{2(0) + \sqrt{3}} = 2 = y'(\sqrt{3}, 0)$. So the tangent lines at these points are parallel.

43. $x^2y^2 + xy = 2 \Rightarrow 2xy^2 + 2x^2yy' + y + xy' = 0 \Leftrightarrow y'(2x^2y + x) = -2xy^2 - y \Leftrightarrow$ $y' = -\dfrac{2xy^2 + y}{2x^2y + x}$. So $-\dfrac{2xy^2 + y}{2x^2y + x} = -1 \Leftrightarrow 2xy^2 + y = 2x^2y + x \Leftrightarrow y(2xy+1) = x(2xy+1) \Leftrightarrow$ $(2xy+1)(y-x) = 0 \Leftrightarrow y = x$ or $xy = -\frac{1}{2}$. But $xy = -\frac{1}{2} \Rightarrow x^2y^2 + xy = \frac{1}{4} - \frac{1}{2} \neq 2$ so we must have $x = y$. Then $x^2y^2 + xy = 2 \Rightarrow x^4 + x^2 = 2 \Leftrightarrow x^4 + x^2 - 2 = 0 \Leftrightarrow (x^2+2)(x^2-1) = 0$. So $x^2 = -2$, which is impossible, or $x^2 = 1 \Leftrightarrow x = \pm 1$. So the points on the curve where the tangent line has a slope of -1 are $(-1,-1)$ and $(1,1)$.

45. We use implicit differentiation to find y': $2x + 4(2yy') = 0 \Rightarrow y' = -\dfrac{x}{4y}$. Now let h be the height of the lamp, and let (a,b) be the point of tangency of the line passing through the points $(3,h)$ and $(-5,0)$. This line has slope $(h-0)/[3-(-5)] = \frac{1}{8}h$. But the slope of the tangent line through the point (a,b) can be expressed as $y' = -\dfrac{a}{4b}$, or as $\dfrac{b-0}{a-(-5)} = \dfrac{b}{a+5}$ [since the line passes through $(-5,0)$ and (a,b)], so $-\dfrac{a}{4b} = \dfrac{b}{a+5} \Leftrightarrow$ $4b^2 = -a^2 - 5a \Leftrightarrow a^2 + 4b^2 = -5a$. But $a^2 + 4b^2 = 5$, since (a,b) is on the ellipse, so $5 = -5a \Leftrightarrow$ $a = -1$. Then $4b^2 = -1 - 5(-1) = 4 \Rightarrow b = 1$, since the point is on the top half of the ellipse. So $\dfrac{h}{8} = \dfrac{b}{a+5} = \dfrac{1}{-1+5} = \dfrac{1}{4} \Rightarrow h = 2$. So the lamp is located 2 units above the x-axis.

EXERCISES 2.7

1. $a = f, b = f', c = f''$. We can see this because where a has a horizontal tangent, $b = 0$, and where b has a horizontal tangent, $c = 0$. We can immediately see that c can be neither f nor f', since at the points where c has a horizontal tangent, neither a nor b is equal to 0.

3. $f(x) = x^4 - 3x^3 + 16x \quad \Leftrightarrow \quad f'(x) = 4x^3 - 9x^2 + 16 \quad \Rightarrow \quad f''(x) = 12x^2 - 18x$

5. $h(x) = \sqrt{x^2+1} \quad \Rightarrow \quad h'(x) = \frac{1}{2}\left(x^2+1\right)^{-1/2}(2x) = \dfrac{x}{\sqrt{x^2+1}} \quad \Rightarrow$

$$h''(x) = \frac{\sqrt{x^2+1} - x\left(x/\sqrt{x^2+1}\right)}{x^2+1} = \frac{x^2+1-x^2}{\left(x^2+1\right)^{3/2}} = \frac{1}{\left(x^2+1\right)^{3/2}}$$

7. $F(s) = (3s+5)^8 \;\Rightarrow\; F'(s) = 8(3s+5)^7(3) = 24(3s+5)^7 \;\Rightarrow\; F''(s) = 168(3s+5)^6(3) = 504(3s+5)^6$

9. $y = \dfrac{x}{1-x} \quad \Rightarrow \quad y' = \dfrac{1(1-x) - x(-1)}{(1-x)^2} = \dfrac{1}{(1-x)^2} \quad \Rightarrow \quad y'' = -2(1-x)^{-3}(-1) = \dfrac{2}{(1-x)^3}$

11. $y = (1-x^2)^{3/4} \quad \Rightarrow \quad y' = \frac{3}{4}(1-x^2)^{-1/4}(-2x) = -\frac{3}{2}x(1-x^2)^{-1/4} \quad \Rightarrow$

$y'' = -\frac{3}{2}(1-x^2)^{-1/4} - \frac{3}{2}x\left(-\frac{1}{4}\right)(1-x^2)^{-5/4}(-2x) = -\frac{3}{2}(1-x^2)^{-1/4} - \frac{3}{4}x^2(1-x^2)^{-5/4}$

$= \frac{3}{4}(1-x^2)^{-5/4}(x^2-2)$

13. $H(t) = \tan^3(2t-1) \quad \Rightarrow \quad H'(t) = 3\tan^2(2t-1)\sec^2(2t-1)(2) = 6\tan^2(2t-1)\sec^2(2t-1) \quad \Rightarrow$

$H''(t) = 12\tan(2t-1)\sec^2(2t-1)(2)\sec^2(2t-1) + 6\tan^2(2t-1)2\sec(2t-1)\sec(2t-1)\tan(2t-1)(2)$

$= 24\tan(2t-1)\sec^4(2t-1) + 24\tan^3(2t-1)\sec^2(2t-1)$

15. **(a)** $f(x) = 2\cos x + \sin^2 x \quad \Rightarrow \quad f'(x) = 2(-\sin x) + 2\sin x\,(\cos x) = \sin 2x - 2\sin x \quad \Rightarrow$

$f''(x) = 2\cos 2x - 2\cos x = 2(\cos 2x - \cos x)$

(b)

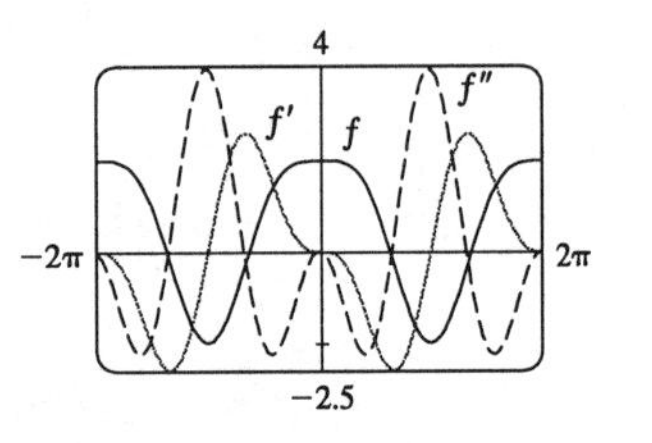

We can see that our answers are plausible, since f has horizontal tangents where $f'(x) = 0$, and f' has horizontal tangents where $f''(x) = 0$.

17. $y = \sqrt{5t-1} \quad \Rightarrow \quad y' = \frac{1}{2}(5t-1)^{-1/2}(5) = \frac{5}{2}(5t-1)^{-1/2} \quad \Rightarrow$

$y'' = -\frac{5}{4}(5t-1)^{-3/2}(5) = -\frac{25}{4}(5t-1)^{-3/2} \quad \Rightarrow \quad y''' = \frac{75}{8}(5t-1)^{-5/2}(5) = \frac{375}{8}(5t-1)^{-5/2}$

19. $f(x) = (2-3x)^{-1/2} \quad \Rightarrow \quad f(0) = 2^{-1/2} = \frac{1}{\sqrt{2}}$

$f'(x) = -\frac{1}{2}(2-3x)^{-3/2}(-3) = \frac{3}{2}(2-3x)^{-3/2} \quad \Rightarrow \quad f'(0) = \frac{3}{2}(2)^{-3/2} = \frac{3}{4\sqrt{2}}$

$f''(x) = -\frac{9}{4}(2-3x)^{-5/2}(-3) = \frac{27}{4}(2-3x)^{-5/2} \quad \Rightarrow \quad f''(0) = \frac{27}{4}(2)^{-5/2} = \frac{27}{16\sqrt{2}}$

$f'''(x) = \frac{405}{8}(2-3x)^{-7/2} \quad \Rightarrow \quad f'''(0) = \frac{405}{8}(2)^{-7/2} = \frac{405}{64\sqrt{2}}$

21. $f(\theta) = \cot\theta \quad \Rightarrow \quad f'(\theta) = -\csc^2\theta \quad \Rightarrow \quad f''(\theta) = -2\csc\theta(-\csc\theta\cot\theta) = 2\csc^2\theta\cot\theta \quad \Rightarrow$
$f'''(\theta) = 2(-2\csc^2\theta\cot\theta)\cot\theta + 2\csc^2\theta(-\csc^2\theta) = -2\csc^2\theta(2\cot^2\theta + \csc^2\theta) \quad \Rightarrow$
$f'''\left(\frac{\pi}{6}\right) = -2(2)^2\left[2\left(\sqrt{3}\right)^2 + (2)^2\right] = -80$

23. $x^3 + y^3 = 1 \quad \Rightarrow \quad 3x^2 + 3y^2y' = 0 \quad \Rightarrow \quad y' = -\dfrac{x^2}{y^2} \quad \Rightarrow$
$y'' = -\dfrac{2xy^2 - 2x^2yy'}{y^4} = -\dfrac{2xy^2 - 2x^2y(-x^2/y^2)}{y^4} = -\dfrac{2xy^3 + 2x^4}{y^5} = -\dfrac{2x(y^3 + x^3)}{y^5} = -\dfrac{2x}{y^5}$, since x and y must satisfy the original equation, $x^3 + y^3 = 1$.

25. $x^2 + 6xy + y^2 = 8 \quad \Rightarrow \quad 2x + 6y + 6xy' + 2yy' = 0 \quad \Rightarrow \quad y' = -\dfrac{x + 3y}{3x + y} \quad \Rightarrow$
$y'' = -\dfrac{(1 + 3y')(3x + y) - (x + 3y)(3 + y')}{(3x + y)^2} = \dfrac{8(y - xy')}{(3x + y)^2} = \dfrac{8\left[y - x(-x - 3y)/(3x + y)\right]}{(3x + y)^2}$
$= \dfrac{8[y(3x + y) + x(x + 3y)]}{(3x + y)^3} = \dfrac{8(x^2 + 6xy + y^2)}{(3x + y)^3} = \dfrac{64}{(3x + y)^3}$, since x and y must satisfy the original equation, $x^2 + 6xy + y^2 = 8$.

27. $f(x) = x - x^2 + x^3 - x^4 + x^5 - x^6 \quad \Rightarrow \quad f'(x) = 1 - 2x + 3x^2 - 4x^3 + 5x^4 - 6x^5 \quad \Rightarrow$
$f''(x) = -2 + 6x - 12x^2 + 20x^3 - 30x^4 \quad \Rightarrow \quad f'''(x) = 6 - 24x + 60x^2 - 120x^3 \quad \Rightarrow$
$f^{(4)}(x) = -24 + 120x - 360x^2 \quad \Rightarrow \quad f^{(5)}(x) = 120 - 720x \quad \Rightarrow \quad f^{(6)}(x) = -720 \quad \Rightarrow$
$f^{(n)}(x) = 0$ for $7 \le n \le 73$.

29. $f(x) = x^n \quad \Rightarrow \quad f'(x) = nx^{n-1} \quad \Rightarrow \quad f''(x) = n(n-1)\,x^{n-2} \quad \Rightarrow \quad \cdots \quad \Rightarrow$
$f^{(n)}(x) = n(n-1)(n-2)\cdots 2 \cdot 1\,x^{n-n} = n!$

31. $f(x) = 1/(3x^3) = \frac{1}{3}x^{-3} \quad \Rightarrow \quad f'(x) = \frac{1}{3}(-3)x^{-4} \quad \Rightarrow \quad f''(x) = \frac{1}{3}(-3)(-4)x^{-5} \quad \Rightarrow$
$f'''(x) = \frac{1}{3}(-3)(-4)(-5)x^{-6} \quad \Rightarrow \quad \cdots \quad \Rightarrow$
$f^{(n)}(x) = \frac{1}{3}(-3)(-4)\cdots[-(n+2)]x^{-(n+3)} = \dfrac{(-1)^n \cdot 3 \cdot 4 \cdot 5 \cdot \cdots \cdot (n+2)}{3x^{n+3}} = \dfrac{(-1)^n (n+2)!}{6x^{n+3}}$

33. In general, $Df(2x) = 2f'(2x)$, $D^2f(2x) = 4f''(2x)$, $\cdots$, $D^nf(2x) = 2^nf^{(n)}(2x)$. Since $f(x) = \cos x$ and $50 = 4(12) + 2$, we have $f^{(50)}(x) = f^{(2)}(x) = -\cos x$, so $D^{50}\cos 2x = -2^{50}\cos 2x$.

35. **(a)** $s = t^3 - 3t \quad \Rightarrow \quad v(t) = s'(t) = 3t^2 - 3 \quad \Rightarrow \quad a(t) = v'(t) = 6t$

(b) $a(1) = 6(1) = 6\text{ m/s}^2$

(c) $v(t) = 3t^2 - 3 = 0$ when $t^2 = 1$, that is, $t = 1$ and $a(1) = 6\text{ m/s}^2$.

37. **(a)** $s = At^2 + Bt + C \quad \Rightarrow \quad v(t) = s'(t) = 2At + B \quad \Rightarrow \quad a(t) = v'(t) = 2A$

(b) $a(1) = 2A\text{ m/s}^2$

(c) The acceleration at these instants is $2A\text{ m/s}^2$, since $a(t)$ is constant.

39. **(a)** $s(t) = t^4 - 4t^3 + 2 \quad \Rightarrow \quad v(t) = s'(t) = 4t^3 - 12t^2 \quad \Rightarrow$

$a(t) = v'(t) = 12t^2 - 24t = 12t(t-2) = 0$ when $t = 0$ or 2.

(b) $s(0) = 2\text{ m}$, $v(0) = 0\text{ m/s}$, $s(2) = -14\text{ m}$, $v(2) = -16\text{ m/s}$

41. **(a)** $y(t) = A\sin\omega t \quad \Rightarrow \quad v(t) = y'(t) = A\omega\cos\omega t \quad \Rightarrow \quad a(t) = v'(t) = -A\omega^2\sin\omega t$

(b) $a(t) = -A\omega^2\sin\omega t = -\omega^2 y(t)$

(c) $|v(t)| = A\omega|\cos\omega t|$ is a maximum when $\cos\omega t = \pm 1 \;\Leftrightarrow\; \sin\omega t = 0 \;\Leftrightarrow\; a(t) = -A\omega^2\sin^2\omega t = 0$.

43. Let $P(x) = ax^2 + bx + c$. Then $P'(x) = 2ax + b$ and $P''(x) = 2a$.

$P''(2) = 2 \quad \Rightarrow \quad 2a = 2 \quad \Rightarrow \quad a = 1.$ $P'(2) = 3 \quad \Rightarrow \quad 4a + b = 4 + b = 3 \quad \Rightarrow \quad b = -1.$

$P(2) = 5 \quad \Rightarrow \quad 2^2 - 2 + c = 5 \quad \Rightarrow \quad c = 3.$ So $P(x) = x^2 - x + 3$.

45. $P(x) = c_n x^n + c_{n-1}x^{n-1} + \cdots + c_1 x + c_0 \quad \Rightarrow \quad P'(x) = nc_n x^{n-1} + (n-1)\,c_{n-1}x^{n-2} + \cdots \quad \Rightarrow$

$P''(x) = n(n-1)c_n x^{n-2} + \cdots \quad \Rightarrow \quad P^{(n)}(x) = n(n-1)(n-2)\cdots(1)\,c_n x^{n-n} = n!\,c_n$ which is a constant.

Therefore $P^{(m)}(x) = 0$ for $m > n$.

47. $f(x) = xg(x^2) \quad \Rightarrow \quad f'(x) = g(x^2) + xg'(x^2)2x = g(x^2) + 2x^2 g'(x^2) \quad \Rightarrow$

$f''(x) = 2xg'(x^2) + 4xg'(x^2) + 4x^3 g''(x^2) = 6xg'(x^2) + 4x^3 g''(x^2)$

49. $f(x) = g(\sqrt{x}) \;\Rightarrow\; f'(x) = \dfrac{g'(\sqrt{x})}{2\sqrt{x}} \;\Rightarrow\; f''(x) = \dfrac{\dfrac{g''(\sqrt{x})}{2\sqrt{x}} \cdot 2\sqrt{x} - \dfrac{g'(\sqrt{x})}{\sqrt{x}}}{4x} = \dfrac{\sqrt{x}\,g''(\sqrt{x}) - g'(\sqrt{x})}{4x\sqrt{x}}$

51. **(a)** $f(x) = \dfrac{1}{x^2 + x} \quad \Rightarrow \quad f'(x) = \dfrac{-(2x+1)}{(x^2+x)^2} \quad \Rightarrow$

$$f''(x) = \frac{(x^2+x)^2(-2) + (2x+1)(2)(x^2+x)(2x+1)}{(x^2+x)^4} = \frac{2(3x^2+3x+1)}{(x^2+x)^3} \quad \Rightarrow$$

$$f'''(x) = \frac{(x^2+x)^3(2)(6x+3) - 2(3x^2+3x+1)(3)(x^2+x)^2(2x+1)}{(x^2+x)^6}$$

$$= \frac{-6(4x^3+6x^2+4x+1)}{(x^2+x)^4} \quad \Rightarrow$$

$$f^{(4)}(x) = \frac{(x^2+x)^4(-6)(12x^2+12x+4) + 6(4x^3+6x^2+4x+1)(4)(x^2+x)^3(2x+1)}{(x^2+x)^8}$$

$$= \frac{24(5x^4+10x^3+10x^2+5x+1)}{(x^2+x)^5}$$

$f^{(5)}(x) = ?$

(b) $f(x) = \dfrac{1}{x(x+1)} = \dfrac{1}{x} - \dfrac{1}{x+1} \quad \Rightarrow \quad f'(x) = -x^{-2} + (x+1)^{-2} \quad \Rightarrow \quad f''(x) = 2x^{-3} - 2(x+1)^{-3}$

$\Rightarrow \; f'''(x) = (-3)(2)x^{-4} + (3)(2)(x+1)^{-4} \;\Rightarrow\; \cdots \;\Rightarrow\; f^{(n)}(x) = (-1)^n n!\left[x^{-(n+1)} - (x+1)^{-(n+1)}\right]$

53. The Chain Rule says that $\dfrac{dy}{dx}=\dfrac{dy}{du}\dfrac{du}{dx}$, so

$$\frac{d^2y}{dx^2}=\frac{d}{dx}\left(\frac{dy}{dx}\right)=\frac{d}{dx}\left(\frac{dy}{du}\frac{du}{dx}\right)=\left[\frac{d}{dx}\left(\frac{dy}{du}\right)\right]\frac{du}{dx}+\frac{dy}{du}\frac{d}{dx}\left(\frac{du}{dx}\right) \quad \text{(Product Rule)}$$

$$=\left[\frac{d}{du}\left(\frac{dy}{du}\right)\frac{du}{dx}\right]\frac{du}{dx}+\frac{dy}{du}\frac{d^2u}{dx^2}=\frac{d^2y}{du^2}\left(\frac{du}{dx}\right)^2+\frac{dy}{du}\frac{d^2u}{dx^2}.$$

EXERCISES 2.8

1. $V=x^3 \quad\Rightarrow\quad \dfrac{dV}{dt}=3x^2\dfrac{dx}{dt}$

3. $xy=1 \quad\Rightarrow\quad x\dfrac{dy}{dt}+y\dfrac{dx}{dt}=0$. If $\dfrac{dx}{dt}=4$ and $x=2$, then $y=\dfrac{1}{2}$, so $\dfrac{dy}{dt}=-\dfrac{y}{x}\dfrac{dx}{dt}=-\dfrac{1/2}{2}(4)=-1$.

5. If the radius is r and the diameter x, then $V=\frac{4}{3}\pi r^3=\frac{\pi}{6}x^3 \quad\Rightarrow\quad -1=\dfrac{dV}{dt}=\dfrac{\pi}{2}x^2\dfrac{dx}{dt} \quad\Rightarrow\quad \dfrac{dx}{dt}=-\dfrac{2}{\pi x^2}$.

When $x=10$, $\dfrac{dx}{dt}=-\dfrac{2}{\pi(100)}=-\dfrac{1}{50\pi}$. So the rate of decrease is $\dfrac{1}{50\pi}\dfrac{\text{cm}}{\text{min}}$.

7. We are given that $dx/dt=5$ ft/s. By similar triangles,

$\dfrac{15}{6}=\dfrac{x+y}{y} \quad\Rightarrow\quad y=\frac{2}{3}x$.

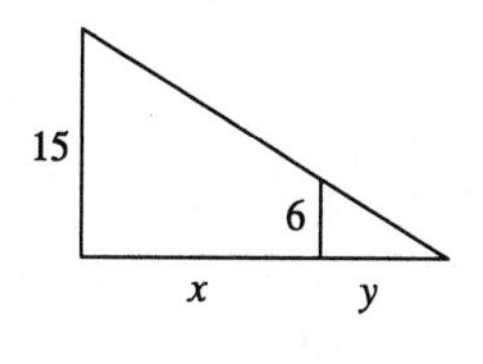

(a) The shadow moves at a rate of

$$\frac{d}{dt}(x+y)=\frac{d}{dt}\left(x+\tfrac{2}{3}x\right)=\frac{5}{3}\frac{dx}{dt}=\tfrac{5}{3}(5)=\tfrac{25}{3}\text{ ft/s}.$$

(b) The shadow lengthens at a rate of $\dfrac{dy}{dt}=\dfrac{d}{dt}\left(\frac{2}{3}x\right)=\dfrac{2}{3}\dfrac{dx}{dt}=\frac{2}{3}(5)=\dfrac{10}{3}$ ft/s.

9. We are given that $dx/dt=500$ mi/h. By the Pythagorean Theorem,

$y^2=x^2+1$, so $2y\dfrac{dy}{dt}=2x\dfrac{dx}{dt} \quad\Rightarrow\quad \dfrac{dy}{dt}=\dfrac{x}{y}\dfrac{dx}{dt}=500\dfrac{x}{y}$.

When $y=2$, $x=\sqrt{3}$, so $\dfrac{dy}{dt}=500\left(\frac{\sqrt{3}}{2}\right)=250\sqrt{3}$ mi/h.

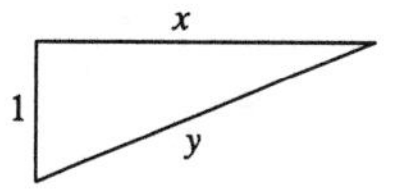

11. We are given that $\dfrac{dx}{dt}=60$ mi/h and $\dfrac{dy}{dt}=25$ mi/h.

$z^2=x^2+y^2 \quad\Rightarrow\quad 2z\dfrac{dz}{dt}=2x\dfrac{dx}{dt}+2y\dfrac{dy}{dt}$. After 2 hours, $x=120$ and $y=50 \quad\Rightarrow\quad z=130$, so

$$\frac{dz}{dt}=\frac{1}{z}\left(x\frac{dx}{dt}+y\frac{dy}{dt}\right)=\frac{120(60)+50(25)}{130}=65\text{ mi/h}.$$

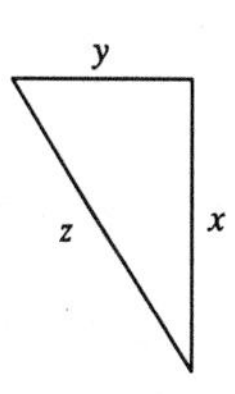

13. We are given that $\frac{dx}{dt} = 35$ km/h and $\frac{dy}{dt} = 25$ km/h.

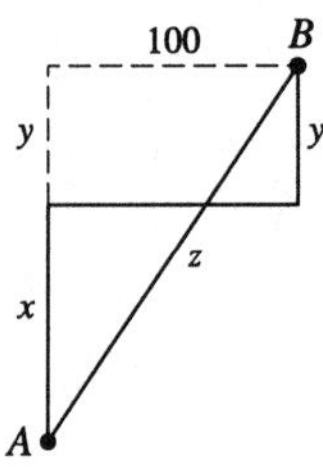

$z^2 = (x+y)^2 + 100^2 \Rightarrow 2z\frac{dz}{dt} = 2(x+y)\left(\frac{dx}{dt} + \frac{dy}{dt}\right)$.

At 4:00 P.M., $x = 140$ and $y = 100 \Rightarrow z = 260$, so

$\frac{dz}{dt} = \frac{x+y}{z}\left(\frac{dx}{dt} + \frac{dy}{dt}\right) = \frac{140+100}{260}(35+25) = \frac{720}{13} \approx 55.4$ km/h.

15. $A = \frac{bh}{2}$, where b is the base and h is the altitude. We are given that $\frac{dh}{dt} = 1$ and $\frac{dA}{dt} = 2$. So

$2 = \frac{dA}{dt} = \frac{b}{2}\frac{dh}{dt} + \frac{h}{2}\frac{db}{dt} = \frac{b}{2} + \frac{h}{2}\frac{db}{dt} \Rightarrow \frac{db}{dt} = \frac{4-b}{h}$. When $h = 10$ and $A = 100$, we have $b = 20$, so

$\frac{db}{dt} = \frac{4-20}{10} = -1.6$ cm/min.

17. If $C =$ the rate at which water is pumped in, then $\frac{dV}{dt} = C - 10{,}000$,

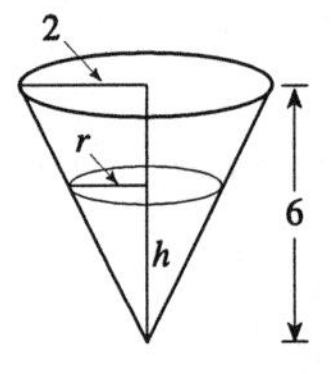

where $V = \frac{1}{3}\pi r^2 h$ is the volume at time t. By similar triangles, $\frac{r}{2} = \frac{h}{6}$

$\Rightarrow r = \frac{1}{3}h \Rightarrow V = \frac{1}{3}\pi\left(\frac{1}{3}h\right)^2 h = \frac{\pi}{27}h^3 \Rightarrow \frac{dV}{dt} = \frac{\pi}{9}h^2\frac{dh}{dt}$.

When $h = 200$, $\frac{dh}{dt} = 20$, so $C - 10{,}000 = \frac{\pi}{9}(200)^2(20) \Rightarrow$

$C = 10{,}000 + \frac{800{,}000}{9}\pi \approx 2.89 \times 10^5$ cm^3/min.

19. $V = \frac{1}{2}[0.3 + (0.3 + 2a)]h(10)$, where $\frac{a}{h} = \frac{0.25}{0.5} = \frac{1}{2}$ so

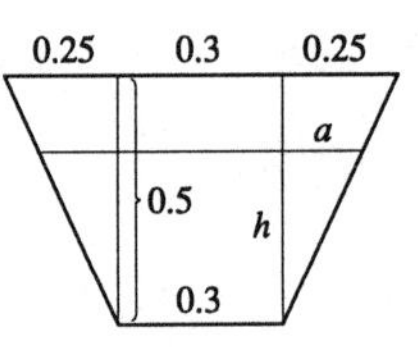

$2a = h \Rightarrow V = 5(0.6 + h)h = 3h + 5h^2 \Rightarrow$

$0.2 = \frac{dV}{dt} = (3 + 10h)\frac{dh}{dt} \Rightarrow \frac{dh}{dt} = \frac{0.2}{3+10h}$. When

$h = 0.3$, $\frac{dh}{dt} = \frac{0.2}{3 + 10(0.3)} = \frac{0.2}{6}$ m/min $= \frac{10}{3}$ cm/min.

21. We are given that $\frac{dV}{dt} = 30$ ft^3/min. $V = \frac{1}{3}\pi\left(\frac{h}{2}\right)^2 h = \frac{h^3\pi}{12}$

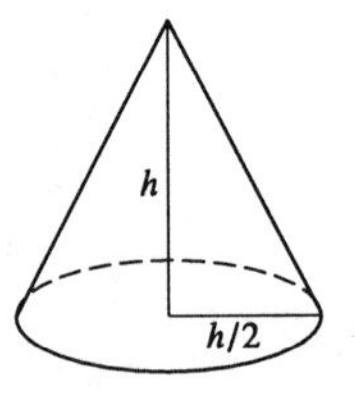

$\Rightarrow 30 = \frac{dV}{dt} = \frac{h^2\pi}{4}\frac{dh}{dt} \Rightarrow \frac{dh}{dt} = \frac{120}{\pi h^2}$.

When $h = 10$ ft, $\frac{dh}{dt} = \frac{120}{10^2\pi} = \frac{6}{5\pi} \approx 0.38$ ft/min.

23. $A = \frac{1}{2}bh$, but $b = 5$ m and $h = 4\sin\theta$ so $A = 10\sin\theta$.

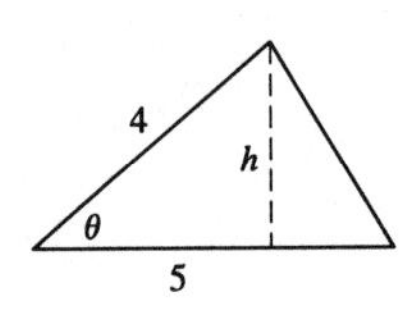

We are given $\frac{d\theta}{dt} = 0.06$ rad/s. $\frac{dA}{dt} = 10\cos\theta\frac{d\theta}{dt} = 0.6\cos\theta$.

When $\theta = \frac{\pi}{3}$, $\frac{dA}{dt} = 10(0.06)\left(\cos\frac{\pi}{3}\right) = (0.6)\left(\frac{1}{2}\right) = 0.3$ m^2/s.

25. $PV = C \quad \Rightarrow \quad P\dfrac{dV}{dt} + V\dfrac{dP}{dt} = 0 \quad \Rightarrow \quad \dfrac{dV}{dt} = -\dfrac{V}{P}\dfrac{dP}{dt}$. When $V = 600$, $P = 150$ and $\dfrac{dP}{dt} = 20$, we have $\dfrac{dV}{dt} = -\dfrac{600}{150}(20) = -80$, so the volume is decreasing at a rate of $80\ \text{cm}^3/\text{min}$.

27. **(a)** By the Pythagorean Theorem, $4000^2 + y^2 = \ell^2$. Differentiating with respect to t, we obtain $2y\dfrac{dy}{dt} = 2\ell\dfrac{d\ell}{dt}$. We know that $\dfrac{dy}{dt} = 600$, so when $y = 3000$ and $\ell = 5000$,

$$\frac{d\ell}{dt} = \frac{y(dy/dt)}{\ell} = \frac{3000(600)}{5000} = \frac{1800}{5} = 360\ \text{ft/s}.$$

(b) Here $\tan\theta = y/4000$, so $\sec^2\theta\,\dfrac{d\theta}{dt} = \dfrac{d}{dt}\tan\theta = \dfrac{1}{4000}\dfrac{dy}{dt} \quad \Rightarrow \quad \dfrac{d\theta}{dt} = \dfrac{\cos^2\theta}{4000}\dfrac{dy}{dt}$. When $y = 3000$, $\dfrac{dy}{dt} = 600$, $z = 5000$ and $\cos\theta = \dfrac{4000}{z} = \dfrac{4000}{5000} = \dfrac{4}{5}$, so $\dfrac{d\theta}{dt} = \dfrac{(4/5)^2}{4000}(600) = 0.096\ \text{rad/s}$.

29. We are given that $\dfrac{dx}{dt} = 2\ \text{ft/s}$. $x = 10\sin\theta \quad \Rightarrow \quad \dfrac{dx}{dt} = 10\cos\theta\,\dfrac{d\theta}{dt}$.

When $\theta = \dfrac{\pi}{4}$, $\dfrac{d\theta}{dt} = \dfrac{2}{10\left(1/\sqrt{2}\right)} = \dfrac{\sqrt{2}}{5}\ \text{rad/s}$.

31. We are given that $\dfrac{dx}{dt} = 30\ \text{km/h}$. By the Law of Cosines,

$y^2 = x^2 + 1 - 2x\cos 120^\circ = x^2 + 1 - 2x\left(-\tfrac{1}{2}\right) = x^2 + x + 1$,

so $2y\dfrac{dy}{dt} = 2x\dfrac{dx}{dt} + \dfrac{dx}{dt} \quad \Rightarrow \quad \dfrac{dy}{dt} = \dfrac{2x+1}{2y}\dfrac{dx}{dt}$. After 1 minute,

$x = \frac{300}{60} = 5 \ \Rightarrow \ y = \sqrt{31} \ \Rightarrow$

$$\frac{dy}{dt} = \frac{2(5)+1}{2\sqrt{31}}(300) = \frac{1650}{\sqrt{31}} \approx 296\ \text{km/h}.$$

33. Let the distance between the runner and the friend be ℓ. Then by the Law of Cosines,

$\ell^2 = 200^2 + 100^2 - 2\cdot 200\cdot 100\cdot\cos\theta = 50{,}000 - 40{,}000\cos\theta \ (\bigstar)$.

Differentiating implicitly with respect to t, we obtain

$2\ell\dfrac{d\ell}{dt} = -40{,}000(-\sin\theta)\dfrac{d\theta}{dt}$. Now if D is the distance run when the angle is θ radians, then $D = 100\theta$, so $\theta = \frac{1}{100}D \quad \Rightarrow$

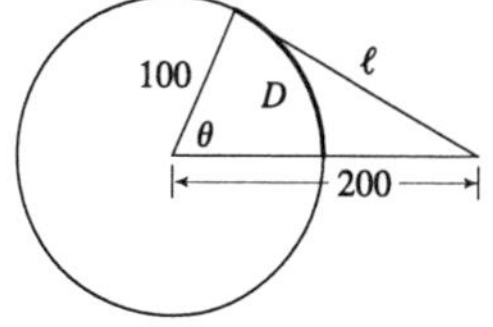

$\dfrac{d\theta}{dt} = \dfrac{1}{100}\dfrac{dD}{dt} = \dfrac{7}{100}$. To substitute into the expression for $\dfrac{d\ell}{dt}$, we must know $\sin\theta$ at the time when $\ell = 200$, which we find from ($\bigstar$): $200^2 = 50{,}000 - 40{,}000\cos\theta \quad \Leftrightarrow \quad \cos\theta = \frac{1}{4} \quad \Rightarrow \quad \sin\theta = \sqrt{1 - \left(\frac{1}{4}\right)^2} = \frac{\sqrt{15}}{4}$.

Substituting, we get $2\ell\dfrac{d\ell}{dt} = 40{,}000\dfrac{\sqrt{15}}{4}\left(\dfrac{7}{100}\right) \quad \Rightarrow \quad d\ell/dt = \dfrac{700\sqrt{15}}{2\cdot 200} = \dfrac{7\sqrt{15}}{4} \approx 6.78\ \text{m/s}$. Whether the distance between them is increasing or decreasing depends on the direction in which the runner is running.

EXERCISES 2.9

1. $y = x^5 \quad \Rightarrow \quad dy = 5x^4 dx$

3. $y = \sqrt{x^4 + x^2 + 1} \quad \Rightarrow \quad dy = \frac{1}{2}\left(x^4 + x^2 + 1\right)^{-1/2}(4x^3 + 2x)dx = \dfrac{2x^3 + x}{\sqrt{x^4 + x^2 + 1}}dx$

5. $y = \sin 2x \quad \Rightarrow \quad dy = 2\cos 2x \, dx$

7. **(a)** $y = 1 - x^2 \quad \Rightarrow \quad dy = -2x\,dx$

(b) When $x = 5$ and $dx = \frac{1}{2}$, $dy = -2(5)\left(\frac{1}{2}\right) = -5$.

9. **(a)** $y = (x^2 + 5)^3 \quad \Rightarrow \quad dy = 3(x^2 + 5)^2 \, 2x \, dx = 6x(x^2 + 5)^2 \, dx$

(b) When $x = 1$ and $dx = 0.05$, $dy = 6(1)(1^2 + 5)^2(0.05) = 10.8$.

11. **(a)** $y = \cos x \quad \Rightarrow \quad dy = -\sin x \, dx$

(b) When $x = \frac{\pi}{6}$ and $dx = 0.05$, $dy = -\frac{1}{2}(0.05) = -0.025$.

13. $y = x^2$, $x = 1$, $\Delta x = 0.5 \quad \Rightarrow$
$\Delta y = (1.5)^2 - 1^2 = 1.25.$
$dy = 2x\,dx = 2(1)(0.5) = 1$

15. $y = 6 - x^2$, $x = -2$, $\Delta x = 0.4 \quad \Rightarrow$
$\Delta y = \left(6 - (-1.6)^2\right) - \left(6 - (-2)^2\right) = 1.44$
$dy = -2x\,dx = -2(-2)(0.4) = 1.6$

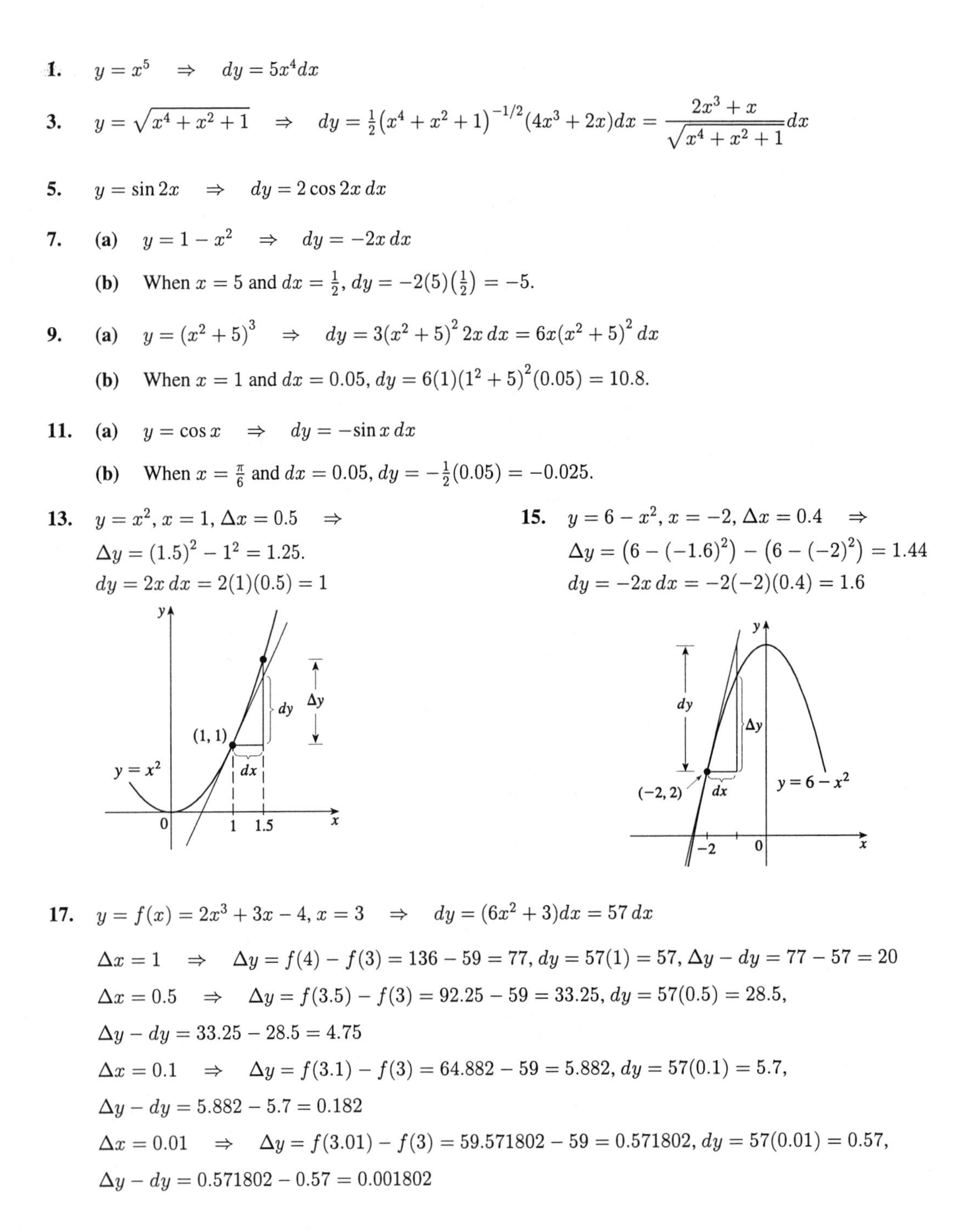

17. $y = f(x) = 2x^3 + 3x - 4$, $x = 3 \quad \Rightarrow \quad dy = (6x^2 + 3)dx = 57\,dx$

$\Delta x = 1 \quad \Rightarrow \quad \Delta y = f(4) - f(3) = 136 - 59 = 77$, $dy = 57(1) = 57$, $\Delta y - dy = 77 - 57 = 20$

$\Delta x = 0.5 \quad \Rightarrow \quad \Delta y = f(3.5) - f(3) = 92.25 - 59 = 33.25$, $dy = 57(0.5) = 28.5$,

$\Delta y - dy = 33.25 - 28.5 = 4.75$

$\Delta x = 0.1 \quad \Rightarrow \quad \Delta y = f(3.1) - f(3) = 64.882 - 59 = 5.882$, $dy = 57(0.1) = 5.7$,

$\Delta y - dy = 5.882 - 5.7 = 0.182$

$\Delta x = 0.01 \quad \Rightarrow \quad \Delta y = f(3.01) - f(3) = 59.571802 - 59 = 0.571802$, $dy = 57(0.01) = 0.57$,

$\Delta y - dy = 0.571802 - 0.57 = 0.001802$

19. $y = f(x) = \sqrt{x} \quad \Rightarrow \quad dy = \dfrac{1}{2\sqrt{x}}\,dx$. When $x = 36$ and $dx = 0.1$, $dy = \frac{1}{2\sqrt{36}}(0.1) = \frac{1}{120}$, so $\sqrt{36.1} = f(36.1) \approx f(36) + dy = \sqrt{36} + \frac{1}{120} \approx 6.0083$.

21. $y = f(x) = 1/x \quad \Rightarrow \quad dy = (-1/x^2)dx$. When $x = 10$ and $dx = 0.1$, $dy = \left(-\frac{1}{100}\right)(0.1) = -0.001$, so $\frac{1}{10.1} = f(10.1) \approx f(10) + dy = 0.1 - 0.001 = 0.099$.

23. $y = f(x) = \sin x \quad \Rightarrow \quad dy = \cos x\,dx$. When $x = \frac{\pi}{3}$ and $dx = -\frac{\pi}{180}$, $dy = \cos\frac{\pi}{3}\left(-\frac{\pi}{180}\right) = -\frac{\pi}{360}$, so $\sin 59^\circ = f\left(\frac{59}{180}\pi\right) \approx f\left(\frac{\pi}{3}\right) + dy = \frac{\sqrt{3}}{2} - \frac{\pi}{360} \approx 0.857$.

25. **(a)** If x is the edge length, then $V = x^3 \quad \Rightarrow \quad dV = 3x^2\,dx$. When $x = 30$ and $dx = 0.1$, $dV = 3(30)^2(0.1) = 270$, so the maximum error is about 270 cm^3.

(b) $S = 6x^2 \quad \Rightarrow \quad dS = 12x\,dx$. When $x = 30$ and $dx = 0.1$, $dS = 12(30)(0.1) = 36$, so the maximum error is about 36 cm^2.

27. **(a)** For a sphere of radius r, the circumference is $C = 2\pi r$ and the surface area is $S = 4\pi r^2$, so $r = C/(2\pi)$ $\Rightarrow \quad S = 4\pi(C/2\pi)^2 = C^2/\pi \quad \Rightarrow \quad dS = \frac{2}{\pi}C\,dC$. When $C = 84$ and $dC = 0.5$, $dS = \frac{2}{\pi}(84)(0.5) = \frac{84}{\pi}$, so the maximum error is about $\frac{84}{\pi} \approx 27\text{ cm}^2$.

(b) Relative error $\approx \dfrac{dS}{S} = \dfrac{84/\pi}{84^2/\pi} = \dfrac{1}{84} \approx 0.012$

29. **(a)** $V = \pi r^2 h \quad \Rightarrow \quad \Delta V \approx dV = 2\pi rh\,dr = 2\pi rh\,\Delta r$

(b) $\Delta V = \pi(r + \Delta r)^2 h - \pi r^2 h$, so the error is $\Delta V - dv = \pi(r + \Delta r)^2 h - \pi r^2 h - 2\pi rh\,\Delta r = \pi(\Delta r)^2 h$

31. $L(x) = f(1) + f'(1)(x - 1)$. $f(x) = x^3 \quad \Rightarrow \quad f'(x) = 3x^2$ so $f(1) = 1$ and $f'(1) = 3$. So $L(x) = 1 + 3(x - 1) = 3x - 2$.

33. $f(x) = 1/x \quad \Rightarrow \quad f'(x) = -1/x^2$. So $f(4) = \frac{1}{4}$ and $f'(4) = -\frac{1}{16}$. So $L(x) = f(4) + f'(4)(x - 4) = \frac{1}{4} + \left(-\frac{1}{16}\right)(x - 4) = \frac{1}{2} - \frac{1}{16}x$.

35. $f(x) = \sqrt{1 + x} \quad \Rightarrow \quad f'(x) = \dfrac{1}{2\sqrt{1 + x}}$ so $f(0) = 1$ and $f'(0) = \frac{1}{2}$.

So $f(x) \approx f(0) + f'(0)(x - 0) = 1 + \frac{1}{2}(x - 0) = 1 + \frac{1}{2}x$.

37. $f(x) = \dfrac{1}{(1 + 2x)^4} \quad \Rightarrow \quad f'(x) = \dfrac{-8}{(1 + 2x)^5}$ so $f(0) = 1$ and $f'(0) = -8$.

So $f(x) \approx f(0) + f'(0)(x - 0) = 1 + (-8)(x - 0) = 1 - 8x$.

39. $f(x) = \sqrt{1 - x} \quad \Rightarrow \quad f'(x) = \dfrac{-1}{2\sqrt{1 - x}}$ so $f(0) = 1$ and $f'(0) = -\frac{1}{2}$.

Therefore $\sqrt{1 - x} = f(x) \approx f(0) + f'(0)(x - 0) = 1 + \left(-\frac{1}{2}\right)(x - 0)$ $= 1 - \frac{1}{2}x$. So $\sqrt{0.9} = \sqrt{1 - 0.1} \approx 1 - \frac{1}{2}(0.1) = 0.95$ and $\sqrt{0.99} = \sqrt{1 - 0.01} \approx 1 - \frac{1}{2}(0.01) = 0.995$.

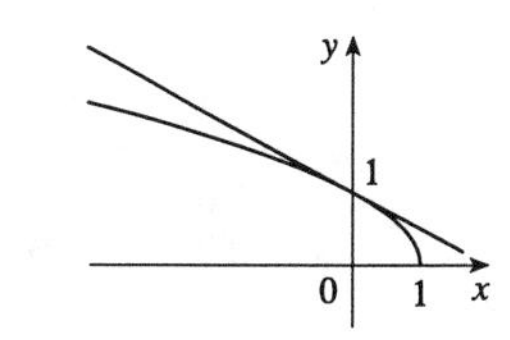

41. We need $\sqrt{1+x} - 0.1 < 1 + \frac{1}{2}x < \sqrt{1+x} + 0.1$. By zooming in or using a cursor, we see that this is true when $-0.69 < x < 1.09$.

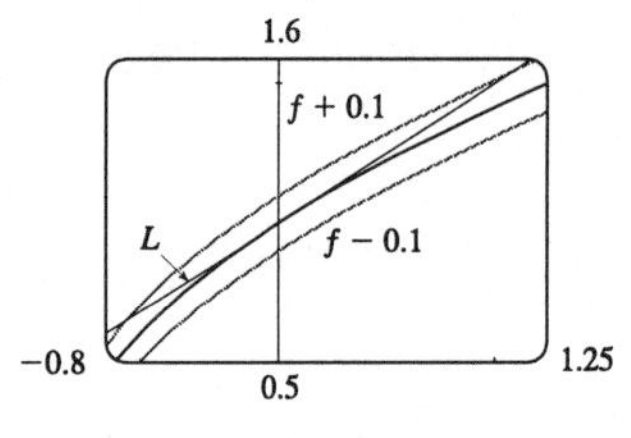

43. We need $1/(1+2x)^4 - 0.1 < 1 - 8x$ and $1 - 8x < 1/(1+2x)^4 + 0.1$, which both hold when $-0.045 < x < 0.055$.

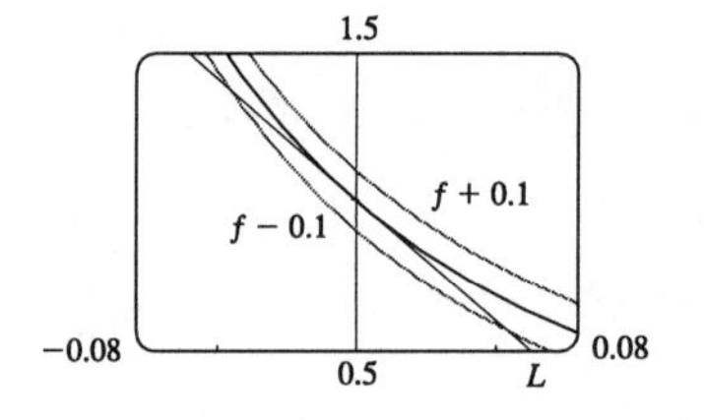

45. Using (10) with $f(x) = 1/x$, $f'(x) = -1/x^2$, and $f''(x) = 2/x^3$,

$$1/x \approx f(4) + f'(4)(x-4) + \tfrac{1}{2} f''(4)(x-4)^2 = \tfrac{1}{4} + (-1)4^{-2}(x-4) + \tfrac{1}{2}(2)4^{-3}(x-4)^2$$
$$= \tfrac{1}{4} - \tfrac{1}{16}(x-4) + \tfrac{1}{64}(x-4)^2$$

47. $f(x) = \sec x$, $f'(x) = \sec x \tan x$, and $f''(x) = \sec x \tan^2 x + \sec^3 x$, so

$$\sec x \approx f(0) + f'(0)(x) + \tfrac{1}{2} f''(x)(x)^2 = \sec 0 + \sec 0 \tan 0(x) + \tfrac{1}{2}[\sec 0(\sec^2 0) + \tan 0(\sec 0 \tan 0)]x^2$$
$$= \tfrac{1}{2}x^2 + 1.$$

49. $f(x) = \sqrt{x}$, $f'(x) = \dfrac{1}{2\sqrt{x}}$, and $f''(x) = \dfrac{1}{-4x\sqrt{x}}$,

so the linear approximation is

$$\sqrt{x} \approx f(1) + f'(1)(x-1) = \sqrt{1} + \tfrac{1}{2\sqrt{1}}(x-1) = 1 + \tfrac{1}{2}(x-1),$$

and the quadratic approximation is

$$\sqrt{x} \approx f(1) + f'(1)(x-1) + \tfrac{1}{2} f''(1)(x-1)^2 = 1 + \tfrac{1}{2}(x-1) - \tfrac{1}{8}(x-1)^2.$$

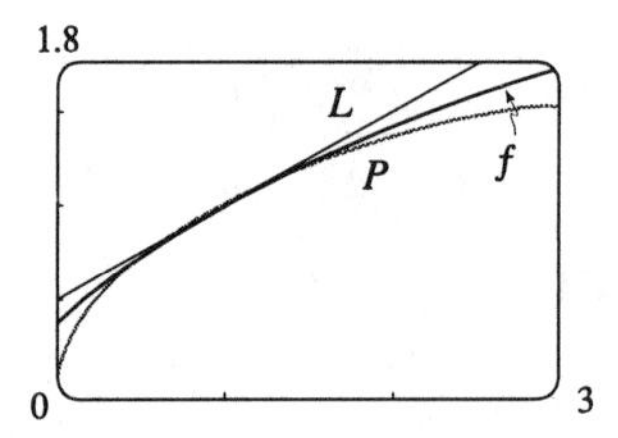

51. **(a)** $f(x) = \cos x \;\Rightarrow\; f'(x) = -\sin x \;\Rightarrow\; f''(x) = -\cos x$.

Thus the linear approximation is

$$\cos x \approx f\left(\tfrac{\pi}{6}\right) + f'\left(\tfrac{\pi}{6}\right)\left(x - \tfrac{\pi}{6}\right) = \cos\tfrac{\pi}{6} - \sin\tfrac{\pi}{6}\left(x - \tfrac{\pi}{6}\right)$$
$$= \tfrac{\sqrt{3}}{2} - \tfrac{1}{2}\left(x - \tfrac{\pi}{6}\right), \text{ and the quadratic approximation is}$$
$$\cos x \approx f\left(\tfrac{\pi}{6}\right) + f'\left(\tfrac{\pi}{6}\right)\left(x - \tfrac{\pi}{6}\right) + \tfrac{1}{2}f''\left(\tfrac{\pi}{6}\right)\left(x - \tfrac{\pi}{6}\right)^2$$
$$= \cos\tfrac{\pi}{6} - \sin\tfrac{\pi}{6}\left(x - \tfrac{\pi}{6}\right) + \tfrac{1}{2}\left(-\cos\tfrac{\pi}{6}\right)\left(x - \tfrac{\pi}{6}\right)^2$$
$$= \tfrac{\sqrt{3}}{2} - \tfrac{1}{2}\left(x - \tfrac{\pi}{6}\right) - \tfrac{\sqrt{3}}{4}\left(x - \tfrac{\pi}{6}\right)^2.$$

(b)

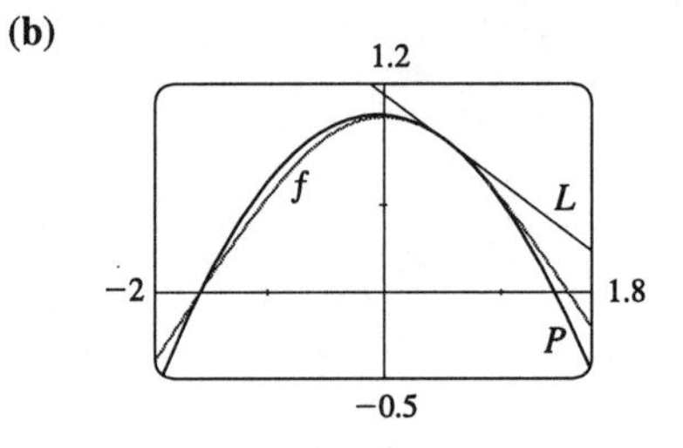

(c) We need $\cos x - 0.1 < \frac{\sqrt{3}}{2} - \frac{1}{2}\left(x - \frac{\pi}{6}\right) < \cos x + 0.1$. From the graph, it appears that the linear approximation has the required accuracy when $0.06 < x < 1.03$.

(d) We need $\cos x - 0.1 < \frac{\sqrt{3}}{2} - \frac{1}{2}\left(x - \frac{\pi}{6}\right) - \frac{\sqrt{3}}{4}\left(x - \frac{\pi}{6}\right)^2 < \cos x + 0.1$. From the graph, it appears that this is true when $-1.82 < x < 1.48$.

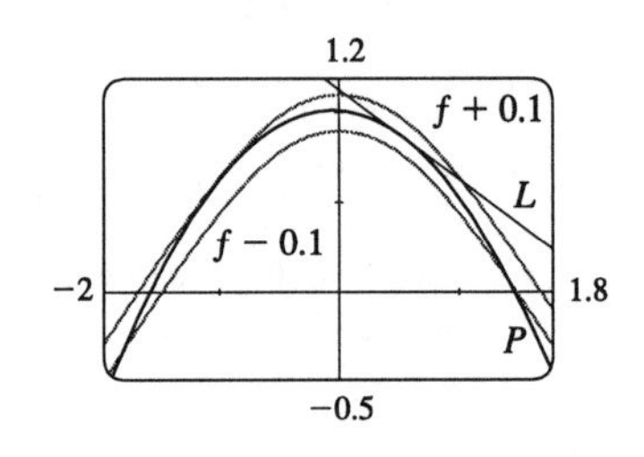

53. **(a)** $dc = \dfrac{dc}{dx}\,dx = 0\,dx = 0$ **(b)** $d(cu) = \dfrac{d}{dx}(cu)dx = c\dfrac{du}{dx}\,dx = c\,du$

(c) $d(u+v) = \dfrac{d}{dx}(u+v)dx = \left(\dfrac{du}{dx} + \dfrac{dv}{dx}\right)dx = \dfrac{du}{dx}\,dx + \dfrac{dv}{dx}\,dx = du + dv$

(d) $d(uv) = \dfrac{d}{dx}(uv)dx = \left(u\dfrac{dv}{dx} + v\dfrac{du}{dx}\right)dx = u\dfrac{dv}{dx}\,dx + v\dfrac{du}{dx}\,dx = u\,dv + v\,du$

(e) $d\left(\dfrac{u}{v}\right) = \dfrac{d}{dx}\left(\dfrac{u}{v}\right)dx = \dfrac{v\dfrac{du}{dx} - u\dfrac{dv}{dx}}{v^2}\,dx = \dfrac{v\dfrac{du}{dx}\,dx - u\dfrac{dv}{dx}\,dx}{v^2} = \dfrac{v\,du - u\,dv}{v^2}$

(f) $d(x^n) = \dfrac{d}{dx}(x^n)dx = nx^{n-1}\,dx$

55. $P(x) = a_0 + a_1x + a_2x^2 + a_3x^3 + \cdots + a_nx^n \quad\Rightarrow$

$P'(x) = a_1 + 2a_2x + 3a_3x^2 + \cdots \quad\Rightarrow$

$P''(x) = 2a_2 + 2\cdot 3a_3x + 3\cdot 4a_4x + \cdots \quad\Rightarrow$

$P'''(x) = 2\cdot 3a_3 + 2\cdot 3\cdot 4a_4x + \cdots \quad\Rightarrow$

$P^{(k)}(x) = 2\cdot 3\cdot 4\cdot\cdots\cdot ka_k + 2\cdot 3\cdot 4\cdot\cdots\cdot k\cdot(k+1)a_{k+1}x + \cdots \quad\Rightarrow$

$P^{(n)}(x) = n!\,a_n$. Therefore $P^{(k)}(0) = f^{(k)}(0) = k!\,a_k$, and so $a_k = \dfrac{f^{(k)}(0)}{k!}$ for $k = 1, 2, \ldots, n$.

Now let $f(x) = \sin x$. Then $f'(x) = \cos x$, $f''(x) = -\sin x$ and $f'''(x) = -\cos x$. So the Taylor polynomial of degree 3 for $\sin x$ is $P(x) = \sin 0 + \dfrac{\cos 0}{1!}x + \dfrac{-\sin 0}{2!}x^2 + \dfrac{-\cos 0}{3!}x^3 = x - \dfrac{x^3}{6}$.

EXERCISES 2.10

1.

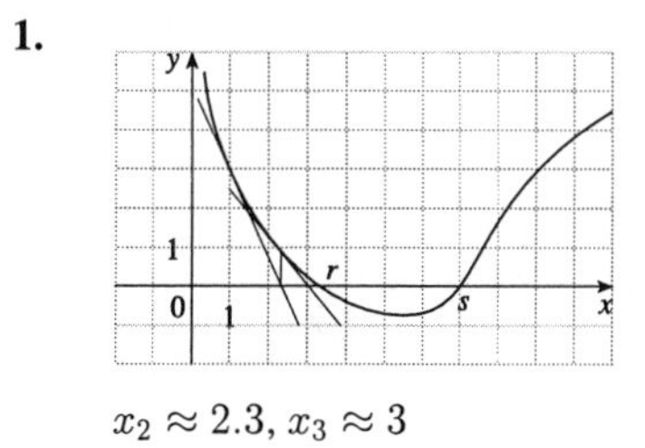

$x_2 \approx 2.3$, $x_3 \approx 3$

3. $f(x) = x^3 + x + 1 \quad\Rightarrow\quad f'(x) = 3x^2 + 1$, so $x_{n+1} = x_n - \dfrac{x_n^3 + x_n + 1}{3x_n^2 + 1}$. $x_1 = -1 \quad\Rightarrow$

$x_2 = -1 - \dfrac{-1 - 1 + 1}{3\cdot 1 + 1} = -0.75 \quad\Rightarrow\quad x_3 = -0.75 - \dfrac{(-0.75)^3 - 0.75 + 1}{3(-0.75)^2 + 1} \approx -0.6860$

53. **(a)** $dc = \dfrac{dc}{dx}\,dx = 0\,dx = 0$ **(b)** $d(cu) = \dfrac{d}{dx}(cu)dx = c\dfrac{du}{dx}\,dx = c\,du$

(c) $d(u+v) = \dfrac{d}{dx}(u+v)dx = \left(\dfrac{du}{dx} + \dfrac{dv}{dx}\right)dx = \dfrac{du}{dx}\,dx + \dfrac{dv}{dx}\,dx = du + dv$

(d) $d(uv) = \dfrac{d}{dx}(uv)dx = \left(u\dfrac{dv}{dx} + v\dfrac{du}{dx}\right)dx = u\dfrac{dv}{dx}\,dx + v\dfrac{du}{dx}\,dx = u\,dv + v\,du$

(e) $d\left(\dfrac{u}{v}\right) = \dfrac{d}{dx}\left(\dfrac{u}{v}\right)dx = \dfrac{v\dfrac{du}{dx} - u\dfrac{dv}{dx}}{v^2}\,dx = \dfrac{v\dfrac{du}{dx}\,dx - u\dfrac{dv}{dx}\,dx}{v^2} = \dfrac{v\,du - u\,dv}{v^2}$

(f) $d(x^n) = \dfrac{d}{dx}(x^n)dx = nx^{n-1}\,dx$

55. $P(x) = a_0 + a_1x + a_2x^2 + a_3x^3 + \cdots + a_nx^n \quad\Rightarrow$

$P'(x) = a_1 + 2a_2x + 3a_3x^2 + \cdots \quad\Rightarrow$

$P''(x) = 2a_2 + 2\cdot 3a_3x + 3\cdot 4a_4x + \cdots \quad\Rightarrow$

$P'''(x) = 2\cdot 3a_3 + 2\cdot 3\cdot 4a_4x + \cdots \quad\Rightarrow$

$P^{(k)}(x) = 2\cdot 3\cdot 4\cdot\cdots\cdot ka_k + 2\cdot 3\cdot 4\cdot\cdots\cdot k\cdot(k+1)a_{k+1}x + \cdots \quad\Rightarrow$

$P^{(n)}(x) = n!\,a_n$. Therefore $P^{(k)}(0) = f^{(k)}(0) = k!\,a_k$, and so $a_k = \dfrac{f^{(k)}(0)}{k!}$ for $k = 1, 2, \ldots, n$.

Now let $f(x) = \sin x$. Then $f'(x) = \cos x$, $f''(x) = -\sin x$ and $f'''(x) = -\cos x$. So the Taylor polynomial of degree 3 for $\sin x$ is $P(x) = \sin 0 + \dfrac{\cos 0}{1!}x + \dfrac{-\sin 0}{2!}x^2 + \dfrac{-\cos 0}{3!}x^3 = x - \dfrac{x^3}{6}$.

EXERCISES 2.10

1.

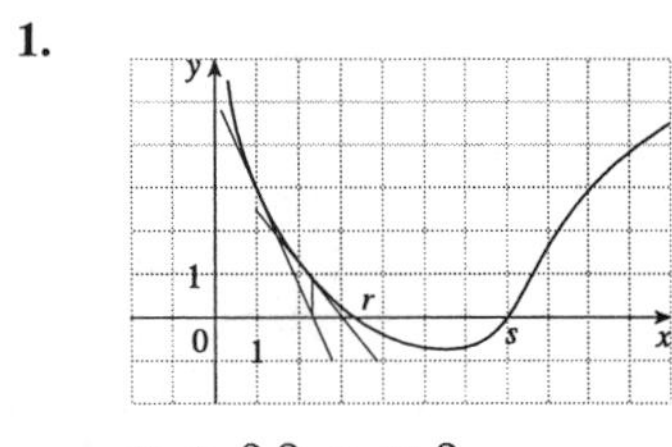

$x_2 \approx 2.3$, $x_3 \approx 3$

3. $f(x) = x^3 + x + 1 \quad\Rightarrow\quad f'(x) = 3x^2 + 1$, so $x_{n+1} = x_n - \dfrac{x_n^3 + x_n + 1}{3x_n^2 + 1}$. $x_1 = -1 \quad\Rightarrow$

$x_2 = -1 - \dfrac{-1 - 1 + 1}{3\cdot 1 + 1} = -0.75 \quad\Rightarrow\quad x_3 = -0.75 - \dfrac{(-0.75)^3 - 0.75 + 1}{3(-0.75)^2 + 1} \approx -0.6860$

17.

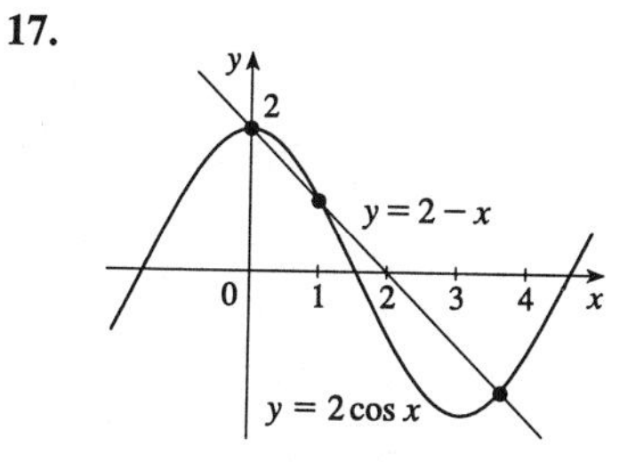

Clearly $x = 0$ is a root. From the sketch, there appear to be roots near 1 and 3.5. Write the equation as $f(x) = 2\cos x + x - 2 = 0$. Then $f'(x) = -2\sin x + 1$, so $x_{n+1} = x_n - \dfrac{2\cos x_n + x_n - 2}{1 - 2\sin x_n}$. Taking $x_1 = 1$, we get $x_2 \approx 1.118026$, $x_3 \approx 1.109188$, $x_4 \approx 1.109144$ and $x_5 \approx 1.109144$. Taking $x_1 = 3.5$, we get $x_2 \approx 3.719159$, $x_3 \approx 3.698331$, $x_4 \approx 3.698154$ and $x_5 \approx 3.698154$. To six decimal places the roots are 0, 1.109144 and 3.698154.

19.

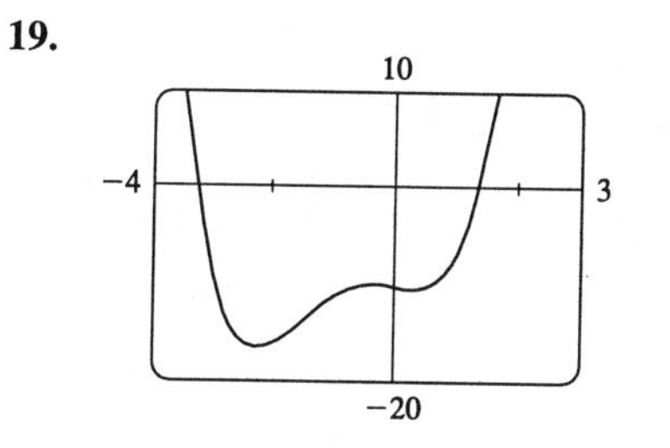

From the graph, there appear to be roots near -3.2 and 1.4. Let $f(x) = x^4 + 3x^3 - x - 10 \quad \Rightarrow \quad f'(x) = 4x^3 + 9x^2 - 1$, so $x_{n+1} = x_n - \dfrac{x_n^4 + 3x_n^3 - x_n - 10}{4x_n^3 + 9x_n^2 - 1}$. Taking $x_1 = -3.2$, we get $x_2 \approx -3.20617358$, $x_3 \approx -3.20614267 \approx x_4$. Taking $x_1 = 1.4$, we get $x_2 \approx 1.37560834$, $x_3 \approx 1.37506496$, $x_4 \approx 1.37506470 \approx x_5$. To eight decimal places, the roots are -3.20614267 and 1.37506470.

21.

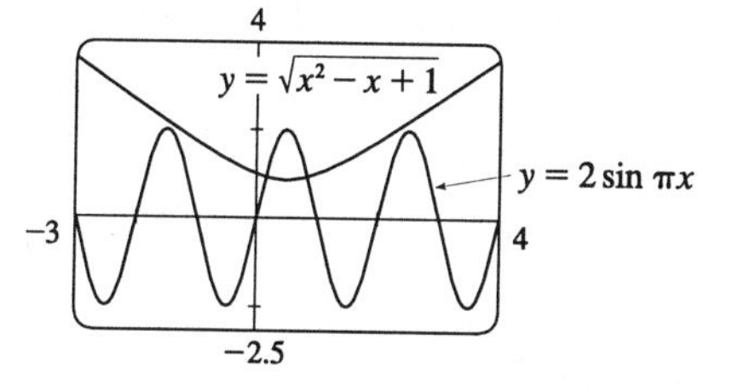

From the graph, we see that there are roots of this equation near 0.2 and 0.8. Let $f(x) = \sqrt{x^2 - x + 1} - 2\sin \pi x \quad \Rightarrow$

$$f'(x) = \frac{2x - 1}{2\sqrt{x^2 - x + 1}} - 2\pi\cos\pi x, \text{ so}$$

$$x_{n+1} = x_n - \frac{\sqrt{x_n^2 - x_n + 1} - 2\sin\pi x_n}{\dfrac{2x_n - 1}{2\sqrt{x_n^2 - x_n + 1}} - 2\pi\cos\pi x_n}$$

Taking $x_1 = 0.2$, we get $x_2 \approx 0.152120155$, $x_3 \approx 0.154380674$, $x_4 \approx 0.154385001 \approx x_5$. Taking $x_1 = 0.8$, we get $x_2 \approx 0.847879845$, $x_3 \approx 0.845619326$, $x_4 \approx 0.845614998 \approx x_5$. So, to eight decimal places, the roots of the equation are 0.15438500 and 0.84561500.

23. **(a)** $f(x) = x^2 - a \quad \Rightarrow \quad f'(x) = 2x$, so Newton's Method gives

$$x_{n+1} = x_n - \frac{x_n^2 - a}{2x_n} = x_n - \tfrac{1}{2}x_n + \frac{a}{2x_n} = \frac{1}{2}\left(x_n + \frac{a}{x_n}\right).$$

(b) Using (a) with $x_1 = 30$, we get $x_2 \approx 31.666667$, $x_3 \approx 31.622807$, $x_4 \approx 31.622777$ and $x_5 \approx 31.622777$. So $\sqrt{1000} \approx 31.622777$.

25. If we attempt to compute x_2 we get $x_2 = x_1 - \dfrac{f(x_1)}{f'(x_1)}$, but $f(x) = x^3 - 3x + 6 \quad \Rightarrow$ $f'(x_1) = 3x_1^2 - 3 = 3(1)^2 - 3 = 0$. For Newton's Method to work $f'(x_n) \neq 0$ (no horizontal tangents).

27. For $f(x) = x^{1/3}$, $f'(x) = \frac{1}{3}x^{-2/3}$ and $x_{n+1} = x_n - \dfrac{f(x_n)}{f'(x_n)} = x_n - \dfrac{x_n^{1/3}}{\frac{1}{3}x_n^{-2/3}} = x_n - 3x_n = -2x_n$. Therefore each successive approximation becomes twice as large as the previous one in absolute value, so the sequence of approximations fails to converge to the root, which is 0.

29. The volume of the silo, in terms of its radius, is
$V(r) = \pi r^2(30) + \frac{1}{2}\left(\frac{4}{3}\pi r^3\right) = 30\pi r^2 + \frac{2}{3}\pi r^3$. From a graph of V, we see that $V(r) = 15{,}000$ at $r \approx 11$ ft. Now we use Newton's Method to solve the equation $V(r) - 15{,}000 = 0$. First we must calculate
$\dfrac{dV}{dr} = 60\pi r + 2\pi r^2$, so $r_{n+1} = r_n - \dfrac{30\pi r_n^2 + \frac{2}{3}\pi r_n^3 - 15{,}000}{60\pi r_n + 2\pi r_n^2}$.
Taking $r_1 = 11$, we get $r_2 = 11.2853$, $r_3 = 11.2807 \approx r_4$. So in order for the silo to hold 15,000 ft^3 of grain, its radius must be about 11.2807 ft.

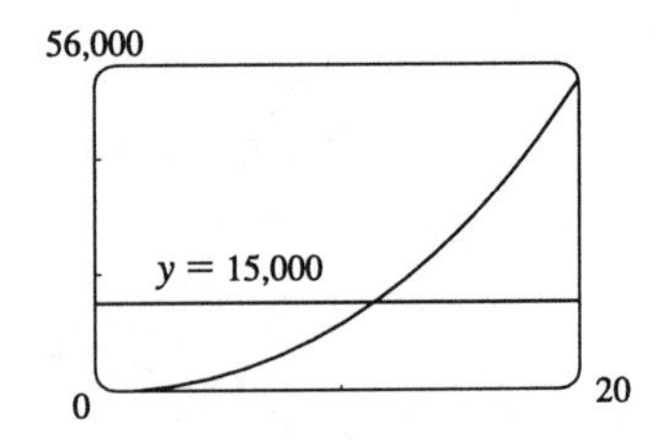

31. In this case, $A = 18{,}000$, $R = 375$, and $n = 60$. So the formula becomes $18{,}000 = \dfrac{375}{x}\left[1 - (1+x)^{-60}\right] \Leftrightarrow$ $48x = 1 - (1+x)^{-60} \Leftrightarrow 48x(1+x)^{60} - (1+x)^{60} + 1 = 0$. Let the LHS be called $f(x)$, so that $f'(x) = 48x(60)(1+x)^{59} + 48(1+x)^{60} - 60(1+x)^{59} = 12(1+x)^{59}(244x - 1)$. So we use Newton's Method with $x_{n+1} = x_n - \dfrac{48x(1+x)^{60} - (1+x)^{60} + 1}{12(1+x)^{59}(244x - 1)}$ and $x_1 = 1\% = 0.01$.

We get $x_2 = 0.0082202$, $x_3 \approx 0.0076802$, $x_4 \approx 0.0076290$, $x_5 \approx 0.0076286 \approx x_6$. So the dealer is charging a monthly interest rate of 0.76286%.

REVIEW EXERCISES FOR CHAPTER 2

1. False; see the warning after Theorem 2.1.8.

3. False. See the discussion before the Product Rule.

5. True, by the Chain Rule.

7. False. $f(x) = |x^2 + x| = x^2 + x$ for $x \geq 0$ or $x \leq -1$ and $|x^2 + x| = -(x^2 + x)$ for $-1 < x < 0$. So $f'(x) = 2x + 1$ for $x > 0$ or $x < -1$ and $f'(x) = -(2x + 1)$ for $-1 < x < 0$. But $|2x + 1| = 2x + 1$ for $x \geq -\frac{1}{2}$ and $|2x + 1| = -2x - 1$ for $x < -\frac{1}{2}$.

9. True. $g(x) = x^5 \Rightarrow g'(x) = 5x^4 \Rightarrow g'(2) = 5(2)^4 = 80$, and by the definition of the derivative, $\lim\limits_{x \to 2} \dfrac{g(x) - g(2)}{x - 2} = g'(2) = 80$.

11. False. A tangent to the parabola has slope $\dfrac{dy}{dx} = 2x$, so at $(-2, 4)$ the slope of the tangent is $2(-2) = -4$ and the equation is $y - 4 = -4(x + 2)$. [The equation $y - 4 = 2x(x + 2)$ is not even linear!]

13. $f(x) = x^3 + 5x + 4 \Rightarrow f'(x) = \lim\limits_{h \to 0} \dfrac{f(x+h) - f(x)}{h} = \lim\limits_{h \to 0} \dfrac{(x+h)^3 + 5(x+h) + 4 - (x^3 + 5x + 4)}{h}$
$= \lim\limits_{h \to 0} \dfrac{3x^2h + 3xh^2 + h^3 + 5h}{h} = \lim\limits_{h \to 0}(3x^2 + 3xh + h^2 + 5) = 3x^2 + 5$

15. $f(x) = \sqrt{3-5x} \quad \Rightarrow$

$$f'(x) = \lim_{h\to 0} \frac{f(x+h) - f(x)}{h} = \lim_{h\to 0} \frac{\sqrt{3-5(x+h)} - \sqrt{3-5x}}{h}$$

$$= \lim_{h\to 0} \frac{\sqrt{3-5x-5h} - \sqrt{3-5x}}{h} \left(\frac{\sqrt{3-5x-5h} + \sqrt{3-5x}}{\sqrt{3-5x-5h} + \sqrt{3-5x}} \right)$$

$$= \lim_{h\to 0} \frac{-5h}{h\left(\sqrt{3-5x-5h} + \sqrt{3-5x}\right)} = \lim_{h\to 0} \frac{-5}{\sqrt{3-5x-5h} + \sqrt{3-5x}} = \frac{-5}{2\sqrt{3-5x}}$$

17. $y = (x+2)^8(x+3)^6 \quad \Rightarrow \quad y' = 6(x+3)^5(x+2)^8 + 8(x+2)^7(x+3)^6 = 2(7x+18)(x+2)^7(x+3)^5$

19. $y = \dfrac{x}{\sqrt{9-4x}} \quad \Rightarrow \quad y' = \dfrac{\sqrt{9-4x} - x\left[-4/\left(2\sqrt{9-4x}\right)\right]}{9-4x} = \dfrac{9-4x+2x}{(9-4x)^{3/2}} = \dfrac{9-2x}{(9-4x)^{3/2}}$

21. $x^2y^3 + 3y^2 = x - 4y \quad \Rightarrow \quad 2xy^3 + 3x^2y^2y' + 6yy' = 1 - 4y' \quad \Rightarrow \quad y' = \dfrac{1-2xy^3}{3x^2y^2 + 6y + 4}$

23. $y = \sqrt{x\sqrt{x\sqrt{x}}} = \left[x\left(x^{3/2}\right)^{1/2}\right]^{1/2} = \left[x\left(x^{3/4}\right)\right]^{1/2} = x^{7/8} \quad \Rightarrow \quad y' = \frac{7}{8}x^{-1/8}$

25. $y = \dfrac{x}{8-3x} \quad \Rightarrow \quad y' = \dfrac{(8-3x) - x(-3)}{(8-3x)^2} = \dfrac{8}{(8-3x)^2}$

27. $y = (x\tan x)^{1/5} \quad \Rightarrow \quad y' = \frac{1}{5}(x\tan x)^{-4/5}(\tan x + x\sec^2 x)$

29. $x^2 = y(y+1) = y^2 + y \quad \Rightarrow \quad 2x = 2yy' + y' \quad \Rightarrow \quad y' = 2x/(2y+1)$

31. $y = \dfrac{(x-1)(x-4)}{(x-2)(x-3)} = \dfrac{x^2-5x+4}{x^2-5x+6} \quad \Rightarrow$

$$y' = \frac{(x^2-5x+6)(2x-5) - (x^2-5x+4)(2x-5)}{(x^2-5x+6)^2} = \frac{2(2x-5)}{(x-2)^2(x-3)^2}$$

33. $y = \tan\sqrt{1-x} \quad \Rightarrow \quad y' = \left(\sec^2\sqrt{1-x}\right)\left(\dfrac{1}{2\sqrt{1-x}}\right)(-1) = -\dfrac{\sec^2\sqrt{1-x}}{2\sqrt{1-x}}$

35. $y = \sin\left(\tan\sqrt{1+x^3}\right) \quad \Rightarrow \quad y' = \cos\left(\tan\sqrt{1+x^3}\right)\left(\sec^2\sqrt{1+x^3}\right)\left[3x^2/\left(2\sqrt{1+x^3}\right)\right]$

37. $y = \cot(3x^2+5) \quad \Rightarrow \quad y' = -\csc^2(3x^2+5)(6x) = -6x\csc^2(3x^2+5)$

39. $y = \cos^2(\tan x) \quad \Rightarrow \quad y' = 2\cos(\tan x)[-\sin(\tan x)]\sec^2 x = -\sin(2\tan x)\sec^2 x$

41. $f(x) = (2x-1)^{-5} \quad \Rightarrow \quad f'(x) = -5(2x-1)^{-6}(2) = -10(2x-1)^{-6} \quad \Rightarrow$

$f''(x) = 60(2x-1)^{-7}(2) = 120(2x-1)^{-7} \quad \Rightarrow \quad f''(0) = 120(-1)^{-7} = -120$

43. $x^6 + y^6 = 1 \quad \Rightarrow \quad 6x^5 + 6y^5y' = 0 \quad \Rightarrow \quad y' = -\dfrac{x^5}{y^5} \quad \Rightarrow$

$$y'' = -\frac{5x^4y^5 - x^5\left(5y^4y'\right)}{y^{10}} = -\frac{5x^4y^5 - 5x^5y^4\left(-x^5/y^5\right)}{y^{10}} = -\frac{5x^4y^6 + 5x^{10}}{y^{11}} = -\frac{5x^4(y^6+x^6)}{y^{11}} = -\frac{5x^4}{y^{11}}$$

45. $\lim\limits_{x\to 0} \dfrac{\sec x}{1-\sin x} = \dfrac{\sec 0}{1-\sin 0} = \dfrac{1}{1-0} = 1$

47. $y = \dfrac{x}{x^2-2} \quad\Rightarrow\quad y' = \dfrac{(x^2-2)-x(2x)}{(x^2-2)^2} = \dfrac{-x^2-2}{(x^2-2)^2}$. When $x=2$, $y' = \dfrac{-2^2-2}{(2^2-2)^2} = -\dfrac{3}{2}$, so the equation of the tangent at $(2,1)$ is $y-1 = -\frac{3}{2}(x-2)$ or $3x+2y-8=0$.

49. $y = \tan x \quad\Rightarrow\quad y' = \sec^2 x$. When $x = \frac{\pi}{3}$, $y' = 2^2 = 4$, so the equation of the tangent line at $\left(\frac{\pi}{3}, \sqrt{3}\right)$ is $y - \sqrt{3} = 4\left(x - \frac{\pi}{3}\right)$ or $y = 4x + \sqrt{3} - \frac{4}{3}\pi$.

51. $y = \sin x + \cos x \quad\Rightarrow\quad y' = \cos x - \sin x = 0 \quad\Leftrightarrow\quad \cos x = \sin x$ and $0 \le x \le 2\pi \quad\Leftrightarrow\quad x = \frac{\pi}{4}$ or $\frac{5\pi}{4}$, so the points are $\left(\frac{\pi}{4}, \sqrt{2}\right)$ and $\left(\frac{5\pi}{4}, -\sqrt{2}\right)$.

53. $f(x) = (x-a)(x-b)(x-c) \quad\Rightarrow\quad f'(x) = (x-b)(x-c)+(x-a)(x-c)+(x-a)(x-b)$. So

$$\frac{f'(x)}{f(x)} = \frac{(x-b)(x-c)+(x-a)(x-c)+(x-a)(x-b)}{(x-a)(x-b)(x-c)} = \frac{1}{x-a}+\frac{1}{x-b}+\frac{1}{x-c}.$$

55. **(a)** $h'(x) = f'(x)g(x) + f(x)g'(x) \quad\Rightarrow\quad h'(2) = f'(2)g(2)+f(2)g'(2) = (-2)(5)+(3)(4) = 2$

(b) $F'(x) = f'(g(x))g'(x) \quad\Rightarrow\quad F'(2) = f'(g(2))g'(2) = f'(5)(4) = 11\cdot 4 = 44$

57. The graph of a has tangent lines with positive slope for $x<0$ and negative slope for $x>0$, and the values of c fit this pattern, so c must be the graph of the derivative of the function for a. The graph of c has horizontal tangent lines to the left and right of the x-axis and b has zeros at these points. Hence b is the graph of the derivative of the function for c. Therefore a is the graph of f, c is the graph of f', and b is the graph of f''.

59. **(a)** $f(x) = x\sqrt{5-x} \quad\Rightarrow\quad f'(x) = \dfrac{-x}{2\sqrt{5-x}} + \sqrt{5-x} = \dfrac{10-3x}{2\sqrt{5-x}}$

(b) At $(1,2)$: $f'(1) = 1\left(\dfrac{-1}{2\sqrt{5-1}}\right) + \sqrt{4} = \dfrac{7}{4}$. So the equation of the tangent is $y-2 = \frac{7}{4}(x-1) \quad\Leftrightarrow$ $y = \frac{7}{4}x + \frac{1}{4}$. At $(4,4)$: $f'(4) = 4\left(\frac{-1}{2\sqrt{1}}\right) + \sqrt{1} = -1$. $y-4 = -(x-4) \quad\Leftrightarrow\quad y = -x+8$.

(c)

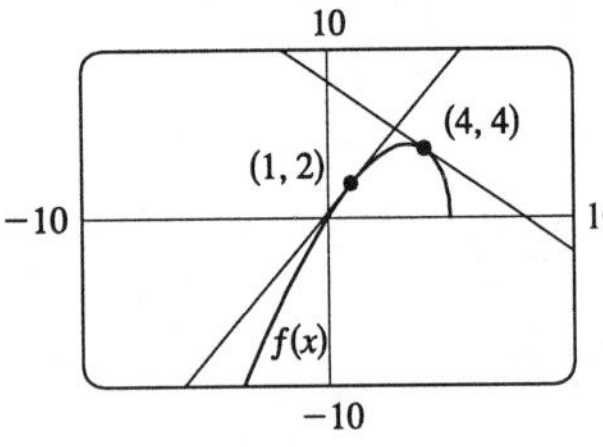

(d)

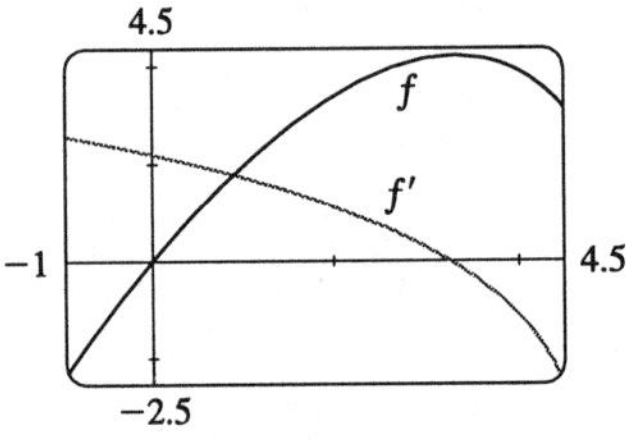

The graphs look reasonable, since f' is positive where f has tangents with positive slope, and f' is negative where f has tangents with negative slope.

61. $f(x) = x^2 g(x) \quad\Rightarrow\quad f'(x) = 2xg(x) + x^2 g'(x)$

63. $f(x) = (g(x))^2 \quad\Rightarrow\quad f'(x) = 2g(x)\,g'(x)$

65. $f(x) = g(g(x)) \quad\Rightarrow\quad f'(x) = g'(g(x))g'(x)$

67. $h(x) = \dfrac{f(x)g(x)}{f(x)+g(x)} \quad \Rightarrow$

$$h'(x) = \frac{[f'(x)g(x) + f(x)g'(x)][f(x)+g(x)] - f(x)g(x)[f'(x)+g'(x)]}{[f(x)+g(x)]^2} = \frac{f'(x)[g(x)]^2 + g'(x)[f(x)]^2}{[f(x)+g(x)]^2}$$

69. Using the Chain Rule repeatedly, $h(x) = f(g(\sin 4x)) \quad \Rightarrow$

$$h'(x) = f'(g(\sin 4x)) \cdot \frac{d}{dx}(g(\sin 4x)) = f'(g(\sin 4x)) \cdot g'(\sin 4x) \cdot \frac{d}{dx}(\sin 4x)$$
$$= f'(g(\sin 4x))g'(\sin 4x)(\cos 4x)(4)$$

71. **(a)** $y = t^3 - 12t + 3 \quad \Rightarrow \quad v(t) = y' = 3t^2 - 12 \quad \Rightarrow \quad a(t) = v'(t) = 6t$

(b) $v(t) = 3(t^2 - 4) > 0$ when $t > 2$, so it moves upward when $t > 2$ and downward when $0 \le t < 2$.

(c) Distance upward $= y(3) - y(2) = -6 - (-13) = 7$.

Distance downward $= y(0) - y(2) = 3 - (-13) = 16$. Total distance $= 7 + 16 = 23$

73. $\rho = x(1+\sqrt{x}) = x + x^{3/2} \quad \Rightarrow \quad d\rho/dx = 1 + \frac{3}{2}\sqrt{x}$, so the density when $x = 4$ is $1 + \frac{3}{2}\sqrt{4} = 4$ kg/m.

75. If $x =$ edge length, then $V = x^3 \quad \Rightarrow \quad dV/dt = 3x^2\,dx/dt = 10 \quad \Rightarrow \quad dx/dt = 10/(3x^2)$ and $S = 6x^2$ $\Rightarrow \quad dS/dt = (12x)dx/dt = 12x\,[10/(3x^2)] = 40/x$. When $x = 30$, $dS/dt = \frac{40}{30} = \frac{4}{3}\text{ cm}^2/\text{min}$.

77. Given $dh/dt = 5$ and $dx/dt = 15$, find dz/dt. $z^2 = x^2 + h^2 \quad \Rightarrow$

$2z\,\dfrac{dz}{dt} = 2x\,\dfrac{dx}{dt} + 2h\,\dfrac{dh}{dt} \quad \Rightarrow \quad \dfrac{dz}{dt} = \dfrac{1}{z}(15x + 5h)$.

When $t = 3$, $h = 45 + 3(5) = 60$ and $x = 15(3) = 45 \quad \Rightarrow$

$z = 75$, so $\dfrac{dz}{dt} = \frac{1}{75}[15(45) + 5(60)] = 13$ ft/s.

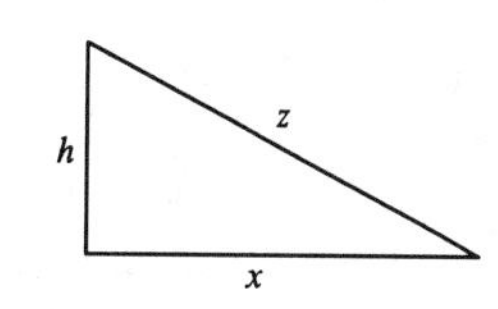

79. We are given $d\theta/dt = -0.25$ rad/h.

$x = 400\cot\theta \quad \Rightarrow \quad \dfrac{dx}{dt} = -400\csc^2\theta\,\dfrac{d\theta}{dt}$.

When $\theta = \dfrac{\pi}{6}$, $\dfrac{dx}{dt} = -400(2)^2(-0.25) = 400$ ft/h.

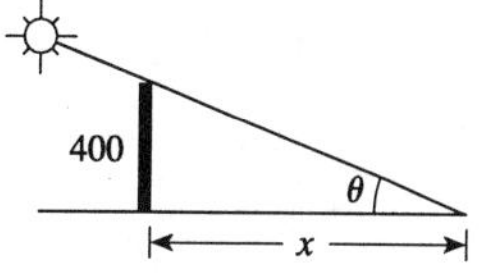

81. $y = x^3 - 2x^2 + 1 \quad \Rightarrow \quad dy = (3x^2 - 4x)dx$. When $x = 2$ and $dx = 0.2$, $dy = \left[3(2)^2 - 4(2)\right](0.2) = 0.8$.

83. $f(x) = \sqrt[3]{1+3x} = (1+3x)^{1/3} \quad \Rightarrow \quad f'(x) = (1+3x)^{-2/3}$ so

$L(x) = f(0) + f'(0)(x - 0) = 1^{1/3} + 1^{-2/3}x = 1 + x$. Thus $\sqrt[3]{1+3x} \approx 1 + x \quad \Rightarrow$

$\sqrt[3]{1.03} = \sqrt[3]{1 + 3(0.01)} \approx 1 + (0.01) = 1.01$.

85. The linear approximation is $\sqrt[3]{1+3x} \approx 1 + x$, so for the required accuracy we want

$\sqrt[3]{1+3x} - 0.1 < 1 + x < \sqrt[3]{1+3x} + 0.1$.

From the graph, it appears that this is true when $-0.23 < x < 0.40$.

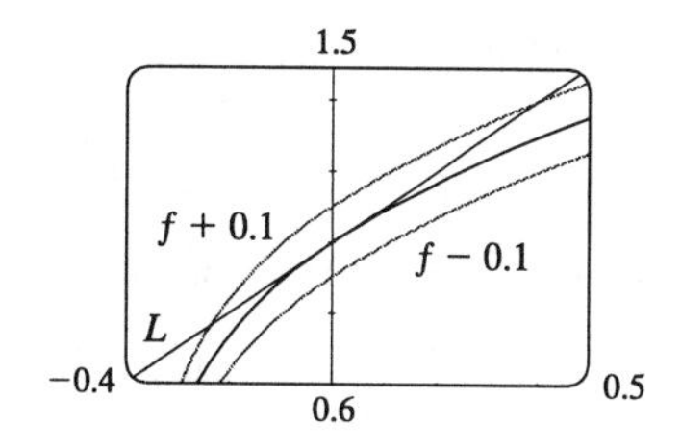

87. $f(x) = x^4 + x - 1 \quad \Rightarrow \quad f'(x) = 4x^3 + 1 \quad \Rightarrow \quad x_{n+1} = x_n - \dfrac{x_n^4 + x_n - 1}{4x_n^3 + 1}$. If $x_1 = 0.5$ then

$x_2 \approx 0.791667$, $x_3 \approx 0.729862$, $x_4 \approx 0.724528$, $x_5 \approx 0.724492$ and $x_6 \approx 0.724492$, so, to six decimal places, the root is 0.724492.

89. $y = x^6 + 2x^2 - 8x + 3$ has a horizontal tangent when $y' = 6x^5 + 4x - 8 = 0$. Let $f(x) = 6x^5 + 4x - 8$. Then $f'(x) = 30x^4 + 4$, so $x_{n+1} = x_n - \dfrac{6x_n^5 + 4x_n - 8}{30\,x_n^4 + 4}$. A sketch shows that the root is near 1, so we take $x_1 = 1$.

Then $x_2 \approx 0.9412$, $x_3 \approx 0.9341$, $x_4 \approx 0.9340$ and $x_5 \approx 0.9340$. Thus, to four decimal places, the point is $(0.9340, -2.0634)$.

91. $\displaystyle\lim_{h\to 0} \frac{(2+h)^6 - 64}{h} = \frac{d}{dx} x^6 \bigg|_{x=2} = 6(2)^5 = 192$

93. Differentiating the expression for $g(x)$ and using the Chain Rule repeatedly, we obtain

$$\begin{aligned}
g(x) &= f(x^3 + f(x^2 + f(x))) \quad \Rightarrow \\
g'(x) &= f'\left(x^3 + f\left(x^2 + f(x)\right)\right)\left(x^3 + f\left(x^2 + f(x)\right)\right)' \\
&= f'\left(x^3 + f\left(x^2 + f(x)\right)\right)\left(3x^2 + f'\left(x^2 + f(x)\right)\left[x^2 + f(x)\right]'\right) \\
&= f'\left(x^3 + f\left(x^2 + f(x)\right)\right)\left(3x^2 + f'\left(x^2 + f(x)\right)\left[2x + f'(x)\right]\right). \text{ So}
\end{aligned}$$

$$\begin{aligned}
g'(1) &= f'\left(1^3 + f\left(1^2 + f(1)\right)\right)\left(3 \cdot 1^2 + f'\left(1^2 + f(1)\right)\left[2 \cdot 1 + f'(1)\right]\right) \\
&= f'(1 + f(1+1))(3 + f'(1+1)[2 + f'(1)]) = f'(1+2)[3 + f'(2)(2+1)] = 3(3 + 2 \cdot 3) = 27.
\end{aligned}$$

95.

$$\begin{aligned}
\lim_{x\to 0} \frac{\sqrt{1+\tan x} - \sqrt{1+\sin x}}{x^3} &= \lim_{x\to 0} \frac{\left(\sqrt{1+\tan x} - \sqrt{1+\sin x}\right)\left(\sqrt{1+\tan x} + \sqrt{1+\sin x}\right)}{x^3\left(\sqrt{1+\tan x} + \sqrt{1+\sin x}\right)} \\
&= \lim_{x\to 0} \frac{(1+\tan x) - (1+\sin x)}{x^3\left(\sqrt{1+\tan x} + \sqrt{1+\sin x}\right)} = \lim_{x\to 0} \frac{\sin x(1/\cos x - 1)\cos x}{x^3\left(\sqrt{1+\tan x} + \sqrt{1+\sin x}\right)\cos x} \\
&= \lim_{x\to 0} \frac{\sin x\,(1 - \cos x)(1 + \cos x)}{x^3\left(\sqrt{1+\tan x} + \sqrt{1+\sin x}\right)\cos x\,(1 + \cos x)} \\
&= \lim_{x\to 0} \frac{\sin x \cdot \sin^2 x}{x^3\left(\sqrt{1+\tan x} + \sqrt{1+\sin x}\right)\cos x\,(1 + \cos x)} \\
&= \left(\lim_{x\to 0} \frac{\sin x}{x}\right)^3 \lim_{x\to 0} \frac{1}{\left(\sqrt{1+\tan x} + \sqrt{1+\sin x}\right)\cos x\,(1 + \cos x)} \\
&= 1^3 \cdot \frac{1}{\left(\sqrt{1} + \sqrt{1}\right) \cdot 1 \cdot (1+1)} = \frac{1}{4}
\end{aligned}$$

97. We are given that $|f(x)| \le x^2$ for all x. In particular, $|f(0)| \le 0$, but $|a| \ge 0$ for all a. The only conclusion is that $f(0) = 0$. Now $\left|\dfrac{f(x) - f(0)}{x - 0}\right| = \left|\dfrac{f(x)}{x}\right| = \dfrac{|f(x)|}{|x|} \le \dfrac{x^2}{|x|} = \dfrac{|x^2|}{|x|} \quad \Rightarrow \quad -|x| \le \dfrac{f(x) - f(0)}{x - 0} \le |x|$. But $\displaystyle\lim_{x\to 0} -|x| = 0 = \lim_{x\to 0} |x|$, so by the Squeeze Theorem, $\displaystyle\lim_{x\to 0} \frac{f(x) - f(0)}{x - 0} = 0$. So by the definition of the derivative, f is differentiable at 0 and, furthermore, $f'(0) = 0$.

PROBLEMS PLUS (page 178)

1. Let a be the x-coordinate of Q. Then $y = 1 - x^2 \quad \Rightarrow \quad$ the slope at $Q = y'(a) = -2a$. But since the triangle is equilateral, $\angle ACB = 60°$, so that the slope at Q is $\tan 120° = -\sqrt{3}$. Therefore we must have that $-2a = -\sqrt{3} \quad \Rightarrow \quad a = \frac{\sqrt{3}}{2}$. Therefore the point Q has coordinates $\left(\frac{\sqrt{3}}{2}, 1 - \left(\frac{\sqrt{3}}{2}\right)^2\right) = \left(\frac{\sqrt{3}}{2}, \frac{1}{4}\right)$ and by symmetry P has coordinates $\left(-\frac{\sqrt{3}}{2}, \frac{1}{4}\right)$.

3. $1 + x + x^2 + \cdots + x^{100} = \dfrac{1 - x^{101}}{1 - x}$ $(x \neq 1)$. If $x = 1$, then the sum is clearly equal to $101 > 0$. If $x \geq 0$, then we have a sum of positive terms which is clearly positive. And if $x < 0$ then $x^{101} < 0 \quad \Rightarrow \quad 1 - x > 0$ and $1 - x^{101} > 0 \quad \Rightarrow \quad \dfrac{1 - x^{101}}{1 - x} > 0$. Therefore $1 + x + x^2 + \cdots + x^{100} = \dfrac{1 - x^{101}}{1 - x} \geq 0$ for all x.

5. For $-\frac{1}{2} < x < \frac{1}{2}$ we have $2x - 1 < 0$, so $|2x - 1| = -(2x - 1)$ and $2x + 1 > 0 \quad \Rightarrow \quad |2x + 1| = 2x + 1$. Therefore, $\displaystyle\lim_{x \to 0} \frac{|2x - 1| - |2x + 1|}{x} = \lim_{x \to 0} \frac{-(2x - 1) - (2x + 1)}{x} = \lim_{x \to 0} \frac{-4x}{x} = \lim_{x \to 0}(-4) = -4$.

7. $V = \frac{4}{3}\pi r^3 \quad \Rightarrow \quad \dfrac{dV}{dt} = 4\pi r^2 \dfrac{dr}{dt}$. But $\dfrac{dV}{dt}$ is proportional to the surface area, so $\dfrac{dV}{dt} = k \cdot 4\pi r^2$ for some constant k. Therefore $4\pi r^2 \dfrac{dr}{dt} = k \cdot 4\pi r^2 \quad \Rightarrow \quad \dfrac{dr}{dt} = k = \text{constant} \quad \Rightarrow \quad r = kt + r_0$. To find k we use the fact that when $t = 3$, $r = 3k + r_0$ and $V = \frac{1}{2}V_0 \quad \Rightarrow \quad \frac{4}{3}\pi(3k + r_0)^3 = \frac{1}{2} \cdot \frac{4}{3}\pi r_0^3 \quad \Rightarrow \quad (3k + r_0)^3 = \frac{1}{2}r_0^3$ $\Rightarrow \quad (3k + r_0) = \frac{1}{\sqrt[3]{2}}r_0 \quad \Rightarrow \quad k = \frac{1}{3}r_0\left(\frac{1}{\sqrt[3]{2}} - 1\right)$. Therefore $r = \frac{1}{3}r_0\left(\frac{1}{\sqrt[3]{2}} - 1\right)t + r_0$. When the snowball has melted completely we have $r = 0 \quad \Rightarrow \quad \frac{1}{3}r_0\left(\frac{1}{\sqrt[3]{2}} - 1\right)t + r_0 = 0$ which gives $t = \dfrac{3\sqrt[3]{2}}{\sqrt[3]{2} - 1}$. Therefore it takes $\dfrac{3\sqrt[3]{2}}{\sqrt[3]{2} - 1} - 3 = \dfrac{3}{\sqrt[3]{2} - 1} \approx 11$ hours and 33 minutes longer.

9. We use mathematical induction. Let S_n be the statement that $\dfrac{d^n}{dx^n}(\sin^4 x + \cos^4 x) = 4^{n-1}\cos(4x + n\pi/2)$.

S_1 is true because
$$\begin{aligned}\frac{d}{dx}(\sin^4 x + \cos^4 x) &= 4\sin^3 x \cos x - 4\cos^3 x \sin x = 4 \sin x \cos x (\sin^2 x - \cos^2 x) \\ &= -4 \sin x \cos x \cos 2x = -2 \sin 2x \cos 2x = -\sin 4x \\ &= \cos\left[\tfrac{\pi}{2} - (-4x)\right] = \cos\left(\tfrac{\pi}{2} + 4x\right) 4^{n-1}\cos\left(4x + n\tfrac{\pi}{2}\right) \text{ when } n = 1.\end{aligned}$$

Now assume S_k is true, that is, $\dfrac{d^k}{dx^k}(\sin^4 x + \cos^4 x) = 4^{k-1}\cos\left(4x + k\frac{\pi}{2}\right)$. Then

$$\begin{aligned}\frac{d^{k+1}}{dx^{k+1}}(\sin^4 x + \cos^4 x) &= \frac{d}{dx}\left[\frac{d^k}{dx^k}(\sin^4 x + \cos^4 x)\right] = \frac{d}{dx}\left[4^{k-1}\cos\left(4x + k\tfrac{\pi}{2}\right)\right] \\ &= -4^{k-1}\sin\left(4x + k\tfrac{\pi}{2}\right) \cdot \frac{d}{dx}\left(4x + k\tfrac{\pi}{2}\right) = -4^k \sin\left(4x + k\tfrac{\pi}{2}\right) \\ &= 4^k \sin\left(-4x - k\tfrac{\pi}{2}\right) = 4^k \cos\left[\tfrac{\pi}{2} - \left(-4x - k\tfrac{\pi}{2}\right)\right] = 4^k \cos\left[4x + (k+1)\tfrac{\pi}{2}\right],\end{aligned}$$

which shows that S_{k+1} is true.

Therefore $\dfrac{d_n}{dx_n}(\sin^4 x + \cos^4 x) = 4^{n-1}\cos\left(4x + n\frac{\pi}{2}\right)$ for every positive integer n by mathematical induction.

Another Proof: First write

$\sin^4 x + \cos^4 x = (\sin^2 x + \cos^2 x)^2 - 2\sin^2 x \cos^2 x = 1 - \frac{1}{2}\sin^2 2x = 1 - \frac{1}{4}(1 - \cos 4x) = \frac{3}{4} + \frac{1}{4}\cos 4x.$

Then we have $\dfrac{d^n}{dx^n}(\sin^4 x + \cos^4 x) = \dfrac{d^n}{dx^n}\left(\frac{3}{4} + \frac{1}{4}\cos 4x\right) = \frac{1}{4} \cdot 4^n \cos\left(4x + n\frac{\pi}{2}\right) = 4^{n-1}\cos\left(4x + n\frac{\pi}{2}\right).$

11. It seems from the figure that as P approaches the point $(0, 2)$ from the right, $x_T \to \infty$ and $y_T \to 2^+$. As P approaches the point $(3, 0)$ from the left, it appears that $x_T \to 3^+$ and $y_T \to \infty$. So we guess that $x_T \in (3, \infty)$ and $y_T \in (2, \infty)$. It is more difficult to estimate the range of values for x_N and y_N. We might perhaps guess that $x_N \in (0, 3)$, and $y_N \in (-\infty, 0)$ or $(-2, 0)$.

In order to actually solve the problem, we implicitly differentiate the equation of the ellipse to find the equation of the tangent line: $\dfrac{x^2}{9} + \dfrac{y^2}{4} = 1 \quad \Rightarrow \quad \dfrac{2x}{9} + \dfrac{2y}{4}y' = 0$, so $y' = -\dfrac{4}{9}\dfrac{x}{y}$. So at the point (x_0, y_0) on the ellipse, the equation of the tangent line is $y - y_0 = -\dfrac{4}{9}\dfrac{x_0}{y_0}(x - x_0)$ or $4x_0 x + 9y_0 y = 4x_0^2 + 9y_0^2$. This can be written as $\dfrac{x_0 x}{9} + \dfrac{y_0 y}{4} = \dfrac{x_0^2}{9} + \dfrac{y_0^2}{4} = 1$, because (x_0, y_0) lies on the ellipse. So an equation of the tangent line is $\dfrac{x_0 x}{9} + \dfrac{y_0 y}{4} = 1$.

Therefore the x-intercept x_T for the tangent line is given by $\dfrac{x_0 x_T}{9} = 1 \quad \Leftrightarrow \quad x_T = \dfrac{9}{x_0}$, and the y-intercept y_T is given by $\dfrac{y_0 y_T}{4} = 1 \quad \Leftrightarrow \quad y_T = \dfrac{4}{y_0}$.

So as x_0 takes on all values in $(0, 3)$, x_T takes on all values in $(3, \infty)$, and as y_0 takes on all values in $(0, 2)$, y_T takes on all values in $(2, \infty)$.

At the point (x_0, y_0) on the ellipse, the slope of the normal line is $-\dfrac{1}{y'(x_0, y_0)} = \dfrac{9}{4}\dfrac{y_0}{x_0}$, and its equation is $y - y_0 = \dfrac{9}{4}\dfrac{y_0}{x_0}(x - x_0)$. So the x-intercept x_N for the normal line is given by $0 - y_0 = \dfrac{9}{4}\dfrac{y_0}{x_0}(x_N - x_0) \quad \Rightarrow$ $x_N = -\dfrac{4x_0}{9} + x_0 = \dfrac{5x_0}{9}$, and the y-intercept y_N is given by $y_N - y_0 = \dfrac{9}{4}\dfrac{y_0}{x_0}(0 - x_0) \quad \Rightarrow$ $y_N = -\dfrac{9y_0}{4} + y_0 = -\dfrac{5y_0}{4}$.

So as x_0 takes on all values in $(0, 3)$, x_N takes on all values in $\left(0, \frac{5}{3}\right)$, and as y_0 takes on all values in $(0, 2)$, y_N takes on all values in $\left(-\frac{5}{2}, 0\right)$.

13. (a)
$$\begin{aligned} D &= \left\{x \mid 3 - x \ge 0, 2 - \sqrt{3 - x} \ge 0, 1 - \sqrt{2 - \sqrt{3 - x}} \ge 0\right\} \\ &= \left\{x \mid 3 \ge x, 2 \ge \sqrt{3 - x}, 1 \ge \sqrt{2 - \sqrt{3 - x}}\right\} = \left\{x \mid 3 \ge x, 4 \ge 3 - x, 1 \ge 2 - \sqrt{3 - x}\right\} \\ &= \left\{x \mid x \le 3, x \ge -1, 1 \le \sqrt{3 - x}\right\} = \{x \mid x \le 3, x \ge -1, 1 \le 3 - x\} \\ &= \{x \mid x \le 3, x \ge -1, x \le 2\} = \{x \mid -1 \le x \le 2\} = [-1, 2] \end{aligned}$$

(b) $f(x) = \sqrt{1 - \sqrt{2 - \sqrt{3 - x}}} \quad \Rightarrow$

$$f'(x) = \frac{1}{2\sqrt{1 - \sqrt{2 - \sqrt{3 - x}}}} \frac{d}{dx}\left(1 - \sqrt{2 - \sqrt{3 - x}}\right)$$

$$= \frac{1}{2\sqrt{1 - \sqrt{2 - \sqrt{3 - x}}}} \cdot \frac{-1}{2\sqrt{2 - \sqrt{3 - x}}} \frac{d}{dx}\left(2 - \sqrt{3 - x}\right)$$

$$= -\frac{1}{8\sqrt{1 - \sqrt{2 - \sqrt{3 - x}}}\sqrt{2 - \sqrt{3 - x}}\sqrt{3 - x}}$$

15. **(a)** For $x \geq 0$,

$$h(x) = |x^2 - 6|x| + 8| = |x^2 - 6x + 8| = |(x-2)(x-4)| = \begin{cases} x^2 - 6x + 8 & \text{if } 0 \leq x \leq 2 \\ -x^2 + 6x - 8 & \text{if } 2 < x < 4 \\ x^2 - 6x + 8 & \text{if } x \geq 4 \end{cases}$$

and for $x < 0$,

$$h(x) = \left|x^2 - 6|x| + 8\right| = |x^2 + 6x + 8|$$

$$= |(x+2)(x+4)| = \begin{cases} x^2 + 6x + 8 & \text{if } x \leq -4 \\ -x^2 - 6x - 8 & \text{if } -4 < x < -2 \\ x^2 + 6x + 8 & \text{if } -2 \leq x < 0 \end{cases}$$

Or: Use the fact that h is an even function and reflect the part of the graph for $x \geq 0$ about the y-axis.

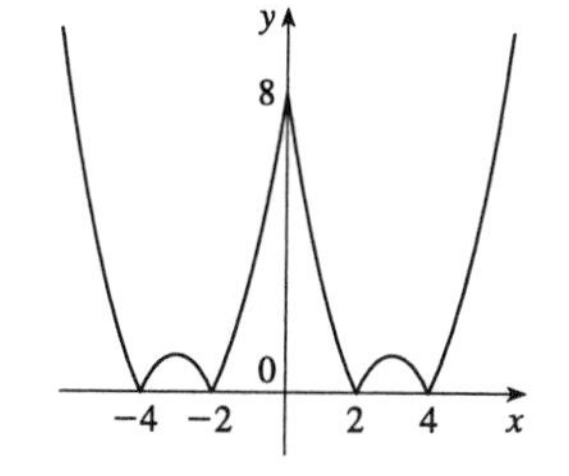

(b) To find where h is differentiable we check the points 0, 2 and 4 by computing the left- and right-hand derivatives: $h'_-(0) = \lim_{t \to 0^-} \frac{h(0+t) - h(0)}{t} = \lim_{t \to 0^-} \frac{(0+t)^2 + 6(0+t) + 8 - 8}{t} = \lim_{t \to 0^-}(t + 6) = 6$

$h'_+(0) = \lim_{t \to 0^+} \frac{h(0+t) - h(0)}{t} = \lim_{t \to 0^+} \frac{(0+t)^2 - 6(0+t) + 8 - 8}{t} = \lim_{t \to 0^+}(t - 6) = -6$
$\neq h'_-(0)$, so h is not differentiable at 0. Similarly h is not differentiable at ± 2 or ± 4. This can also be seen from the graph.

17.

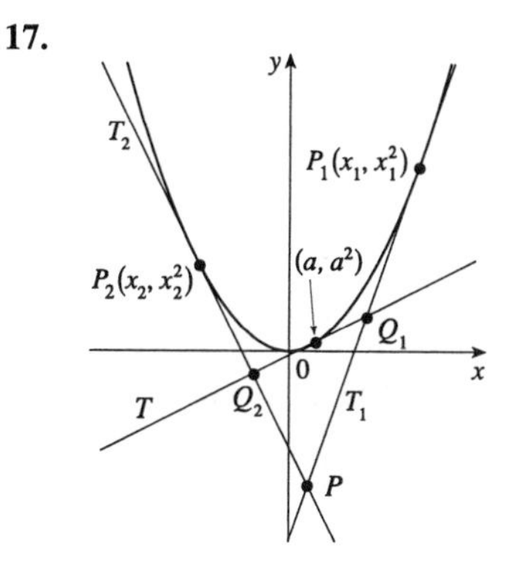

The equation of T_1 is $y - x_1^2 = 2x_1(x - x_1) = 2x_1x - 2x_1^2$ or $y = 2x_1x - x_1^2$. The equation of T_2 is $y = 2x_2x - x_2^2$. Solving for the point of intersection, we get $2x(x_1 - x_2) = x_1^2 - x_2^2$ $\Rightarrow \quad x = \frac{1}{2}(x_1 + x_2)$. Therefore the coordinates of P are $\left(\frac{1}{2}(x_1 + x_2), x_1x_2\right)$. So if the point of contact of T is (a, a^2), then Q_1 is $\left(\frac{1}{2}(a + x_1), ax_1\right)$ and Q_2 is $\left(\frac{1}{2}(a + x_2), ax_2\right)$. Therefore $|PQ_1|^2 = \frac{1}{4}(a - x_2)^2 + x_1^2(a - x_2)^2 = (a - x_2)^2\left(\frac{1}{4} + x_1^2\right)$ and $|PP_1|^2 = \frac{1}{4}(x_1 - x_2^2) + x_1^2(x_1 - x_2)^2 = (x_1 - x_2)^2\left(\frac{1}{4} + x_1^2\right)$.

So $\frac{|PQ_1|^2}{|PP_1|^2} = \frac{(a - x_2)^2}{(x_1 - x_2)^2}$, and similarly $\frac{|PQ_2|^2}{|PP_2|^2} = \frac{(x_1 - a)^2}{(x_1 - x_2)^2}$.

Finally, $\frac{|PQ_1|}{|PP_1|} + \frac{|PQ_2|}{|PP_2|} = \frac{a - x_2}{x_1 - x_2} + \frac{x_1 - a}{x_1 - x_2} = 1$.

19. **(a)** Since f is differentiable at 0, f is continuous at 0 so $f(0) = \lim_{x\to0} f(x) = \lim_{x\to0} \frac{f(x)}{x} \cdot x = \lim_{x\to0} \frac{f(x)}{x} \cdot \lim_{x\to0} x = 4 \cdot 0 = 0$.

(b) $f'(0) = \lim_{x\to0} \frac{f(x) - f(0)}{x - 0} = \lim_{x\to0} \frac{f(x)}{x} = 4$ [since $f(0) = 0$ from (a)]

(c) $\lim_{x\to0} \frac{g(x)}{f(x)} = \lim_{x\to0} \frac{g(x)/x}{f(x)/x} = \frac{\lim_{x\to0}[g(x)/x]}{\lim_{x\to0}[f(x)/x]} = \frac{2}{4} = \frac{1}{2}$

21.

$$\lim_{x\to0} \frac{\sin(a+2x) - 2\sin(a+x) + \sin a}{x^2} = \lim_{x\to0} \frac{\sin a \cos 2x + \cos a \sin 2x - 2\sin a \cos x - 2\cos a \sin x + \sin a}{x^2}$$

$$= \lim_{x\to0} \frac{\sin a(\cos 2x - 2\cos x + 1) + \cos a\,(\sin 2x - 2\sin x)}{x^2}$$

$$= \lim_{x\to0} \frac{\sin a\,(2\cos^2 x - 1 - 2\cos x + 1) + \cos a\,(2\sin x\cos x - 2\sin x)}{x^2}$$

$$= \lim_{x\to0} \frac{\sin a\,(2\cos x)(\cos x - 1) + \cos a\,(2\sin x)(\cos x - 1)}{x^2}$$

$$= \lim_{x\to0} \frac{2(\cos x - 1)[\sin a\cos x + \cos a\sin x](\cos x + 1)}{x^2(\cos x + 1)} = \lim_{x\to0} \frac{-2\sin^2 x\,[\sin(a+x)]}{x^2(\cos x + 1)}$$

$$= -2\lim_{x\to0}\left(\frac{\sin x}{x}\right)^2 \cdot \frac{\sin(a+x)}{\cos x + 1} = -2(1)^2 \frac{\sin(a+0)}{\cos 0 + 1} = -\sin a$$

23. **(a)** If the two lines L_1 and L_2 have slopes m_1 and m_2 and angles of inclination ϕ_1 and ϕ_2, then $m_1 = \tan\phi_1$ and $m_2 = \tan\phi_2$. The figure shows that $\phi_2 = \phi_1 + \alpha$ and so $\alpha = \phi_2 - \phi_1$. Therefore, using the identity for $\tan(x - y)$, we have

$\tan\alpha = \tan(\phi_2 - \phi_1) = \frac{\tan\phi_2 - \tan\phi_1}{1 + \tan\phi_2\tan\phi_1}$ and so $\tan\alpha = \frac{m_2 - m_1}{1 + m_1 m_2}$.

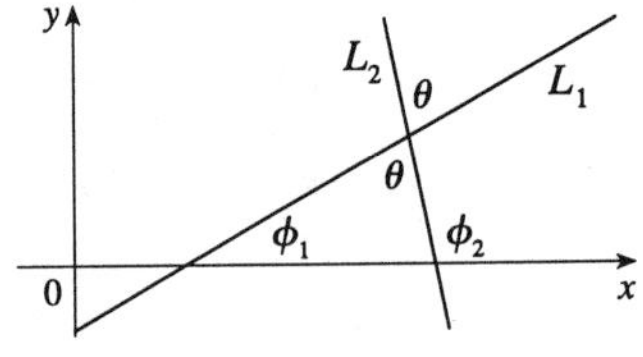

(b) **(i)** The parabolas intersect when $x^2 = (x-2)^2 \Rightarrow x = 1$. If $y = x^2$, then $y' = 2x$, so the slope of the tangent to $y = x^2$ at $(1,1)$ is $m_1 = 2(1) = 2$. If $y = (x-2)^2$, then $y' = 2(x-2)$, so the slope of the tangent to $y = (x-2)^2$ at $(1,1)$ is $m_2 = 2(1-2) = -2$. Therefore

$\tan\alpha = \frac{m_2 - m_1}{1 + m_1 m_2} = \frac{-2-2}{1+2(-2)} = \frac{4}{3}$ and so $\alpha = \tan^{-1}\frac{4}{3} \approx 53°$.

(ii) $x^2 - y^2 = 3$ and $x^2 - 4x + y^2 + 3 = 0$ intersect when $x^2 - 4x + x^2 = 0 \Leftrightarrow 2x(x-2) = 0 \Rightarrow x = 0$ or 2, but 0 is extraneous. If $x^2 - y^2 = 3$ then $2x - 2yy' = 0 \Rightarrow y' = x/y$ and $x - 4x + y^2 + 3 = 0 \Rightarrow 2x - 4 + 2yy' = 0 \Rightarrow y' = \frac{2-x}{y}$. At $(2,1)$ the slopes are $m_1 = 2$ and $m_2 = 0$, so $\tan\alpha = \frac{0-2}{1+2\cdot0} = -2 \Rightarrow \alpha \approx 117°$. At $(2,-1)$ the slopes are $m_1 = -2$ and $m_2 = 0$, so $\tan\alpha = \frac{0-(-2)}{1+(-2)(0)} = 2 \Rightarrow \alpha \approx 63°$.

25. Since $\angle ROQ = \angle OQP = \theta$, the triangle QOR is isosceles, so $|QR| = |RO| = x$. By the Law of Cosines, $x^2 = x^2 + r^2 - 2rx\cos\theta$. Hence $2rx\cos\theta = r^2$, so $x = \dfrac{r^2}{2r\cos\theta} = \dfrac{r}{2\cos\theta}$. Note that as $y \to 0^+$, $\theta \to 0^+$ (since $\sin\theta = y/r$), and hence $x \to \dfrac{r}{2\cos 0} = \dfrac{r}{2}$. Thus as P is taken closer and closer to the x-axis, the point R approaches the midpoint of the radius AO.

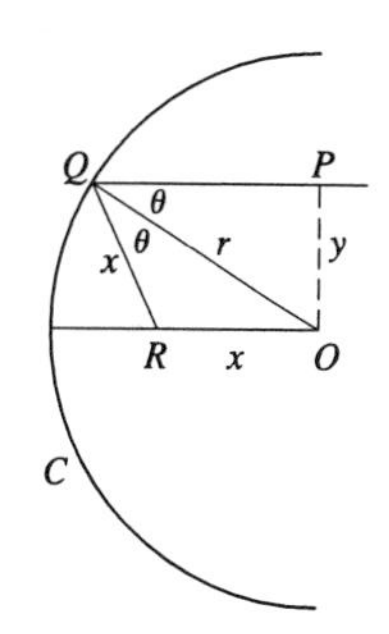

27. $y = x^4 - 2x^2 - x \quad\Rightarrow\quad y' = 4x^3 - 4x - 1$. The equation of the tangent line at $x = a$ is $y - (a^4 - 2a^2 - a) = (4a^3 - 4a - 1)(x - a)$ or $y = (4a^3 - 4a - 1)x + (-3a^4 + 2a^2)$ and similarly for $x = b$. So if at $x = a$ and $x = b$ we have the same tangent line, then $4a^3 - 4a - 1 = 4b^3 - 4b - 1$ and $-3a^4 + 2a^2 = -3b^4 + 2b^2$. The first equation gives $a^3 - b^3 = a - b \quad\Rightarrow\quad (a-b)(a^2 + ab + b^2) = (a - b)$. Assuming $a \neq b$, we have $1 = a^2 + ab + b^2$. The second equation gives $3(a^4 - b^4) = 2(a^2 - b^2) \quad\Rightarrow$ $3(a^2 - b^2)(a^2 + b^2) = 2(a^2 - b^2)$ which is true if $a = -b$. Substituting into $1 = a^2 + ab + b^2$ gives $1 = a^2 - a^2 + a^2 \quad\Rightarrow\quad a = \pm 1$ so that $a = 1$ and $b = -1$ or vice versa. It is easily verified that the points $(1, -2)$ and $(-1, 0)$ have a common tangent line.

As long as there are only two such points, we are done. So we show that these are in fact the only two such points. Suppose that $a^2 - b^2 \neq 0$. Then $3(a^2 - b^2)(a^2 + b^2) = 2(a^2 - b^2)$ gives $3(a^2 + b^2) = 2$ or $a^2 + b^2 = \frac{2}{3}$. Thus $ab = (a^2 + ab + b^2) - (a^2 + b^2) = 1 - \frac{2}{3} = \frac{1}{3}$, so $b = \dfrac{1}{3a}$. Hence $a^2 + \dfrac{1}{9a^2} = \dfrac{2}{3}$, so $9a^4 + 1 = 6a^2 \quad\Rightarrow\quad 0 = 9a^4 - 6a^2 + 1 = (3a^2 - 1)^2$. So $3a^2 - 1 = 0$, so $a^2 = \dfrac{1}{3} \;\Rightarrow\; b^2 = \dfrac{1}{9a^2} = \dfrac{1}{3} = a^2$, contradicting our assumption that $a^2 \neq b^2$.

29. Because of the periodic nature of the lattice points, it suffices to consider the points in the 5×2 grid shown. We can see that the minimum value of r occurs when there is a line with slope $\frac{2}{5}$ which touches the circle centered at $(3, 1)$ and the circles centered at $(0, 0)$ and $(5, 2)$. To find P, the point at which the line is tangent to the circle at $(0, 0)$, we simultaneously solve $x^2 + y^2 = r^2$ and $y = -\frac{5}{2}x \quad\Rightarrow\quad x^2 + \frac{25}{4}x^2 = r^2 \quad\Rightarrow$ $x^2 = \frac{4}{29}r^2 \quad\Rightarrow\quad x = \frac{2}{\sqrt{29}}r$, $y = -\frac{5}{\sqrt{29}}r$. To find Q, we either use symmetry or solve $(x - 3)^2 + (y - 1)^2 = r^2$ and $y - 1 = -\frac{5}{2}(x - 3)$. As above, we get $x = 3 - \frac{2}{\sqrt{29}}r$, $y = 1 + \frac{5}{\sqrt{29}}r$. Now the slope of the line PQ is $\frac{2}{5}$, so

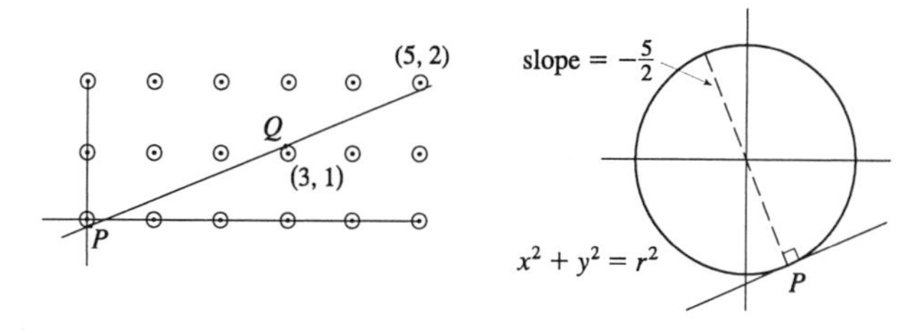

$$m_{PQ} = \frac{1 + \frac{5}{\sqrt{29}}r - \left(-\frac{5}{\sqrt{29}}r\right)}{3 - \frac{2}{\sqrt{29}}r - \frac{2}{\sqrt{29}}r} = \frac{1 + \frac{10}{\sqrt{29}}r}{3 - \frac{4}{\sqrt{29}}r} = \frac{\sqrt{29} + 10r}{3\sqrt{29} - 4r} = \frac{2}{5} \quad\Rightarrow\quad 5\sqrt{29} + 50r = 6\sqrt{29} - 8r \quad\Leftrightarrow$$

$58r = \sqrt{29} \quad\Leftrightarrow\quad r = \frac{\sqrt{29}}{58}$. So the minimum value of r for which any line with slope $\frac{2}{5}$ intersects circles with radius r centered at the lattice points on the plane is $r = \frac{\sqrt{29}}{58}$.

CHAPTER THREE

EXERCISES 3.1

1. Absolute maximum at b; absolute minimum at d; local maxima at b, e; local minima at d, s.

3. Absolute maximum value is $f(4) = 4$; absolute minimum value is $f(7) = 0$; local maximum values are $f(4) = 4$ and $f(6) = 3$; local minimum values are $f(2) = 1$ and $f(5) = 2$.

5. $f(x) = 1 + 2x$, $x \geq -1$.
Absolute minimum $f(-1) = -1$; no local minimum. No local or absolute maximum.

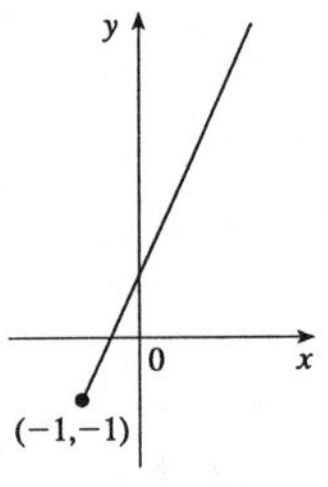

7. $f(x) = 1 - x^2$, $0 < x < 1$. No extremum.

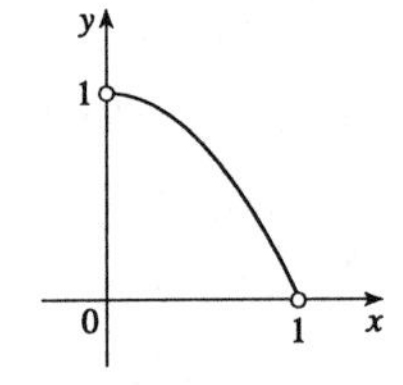

9. $f(x) = 1 - x^2$, $0 \leq x < 1$.
Absolute maximum $f(0) = 1$; no local maximum. No absolute or local minimum.

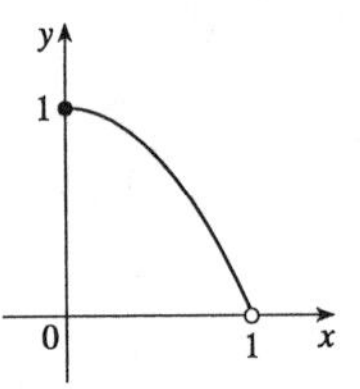

11. $f(x) = 1 - x^2$, $-2 \leq x \leq 1$.
Absolute and local maximum $f(0) = 1$.
Absolute minimum $f(-2) = -3$;
no local minimum.

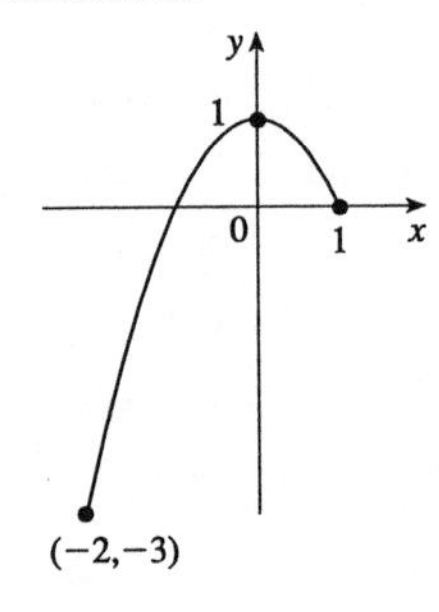

13. $f(t) = 1/t$, $0 < t < 1$. No extremum.

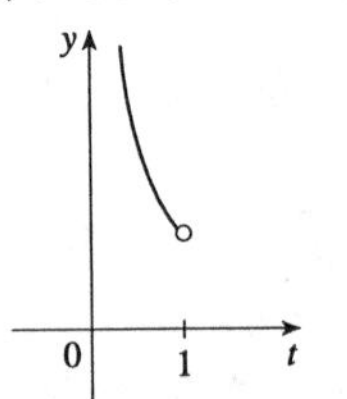

15. $f(\theta) = \sin\theta$, $-2\pi \leq \theta \leq 2\pi$. Absolute and local maxima $f\left(-\frac{3\pi}{2}\right) = f\left(\frac{\pi}{2}\right) = 1$. Absolute and local minima $f\left(-\frac{\pi}{2}\right) = f\left(\frac{3\pi}{2}\right) = -1$.

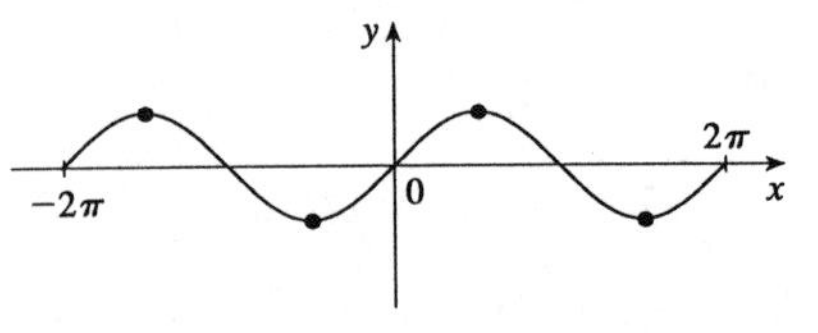

17. $f(x) = x^5$. No extremum.

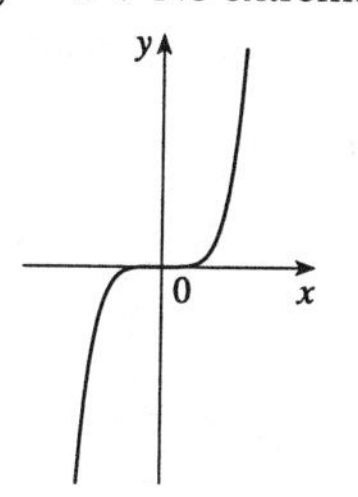

19. $f(x) = \begin{cases} 2x & \text{if } 0 \le x < 1 \\ 2 - x & \text{if } 1 \le x \le 2. \end{cases}$

Absolute minima $f(0) = f(2) = 0$; no local minimum. No absolute or local maximum.

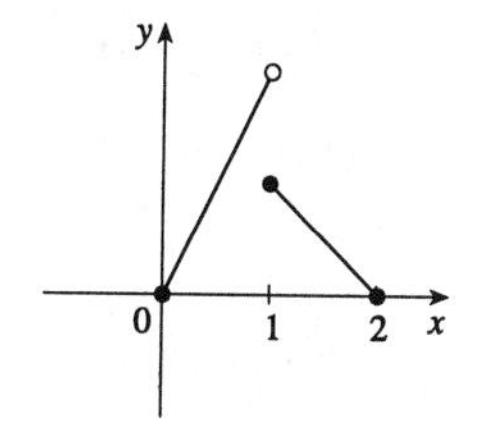

21. $f(x) = 2x - 3x^2 \quad \Rightarrow \quad f'(x) = 2 - 6x = 0 \quad \Leftrightarrow \quad x = \frac{1}{3}$. So the critical number is $\frac{1}{3}$.

23. $f(x) = x^3 - 3x + 1 \quad \Rightarrow \quad f'(x) = 3x^2 - 3 = 3(x^2 - 1) = 3(x + 1)(x - 1)$. So the critical numbers are ± 1.

25. $f(t) = 2t^3 + 3t^2 + 6t + 4 \quad \Rightarrow \quad f'(t) = 6t^2 + 6t + 6$. But $t^2 + t + 1 = 0$ has no real solutions since $b^2 - 4ac = 1 - 4(1)(1) = -3 < 0$. No critical number.

27. $s(t) = 2t^3 + 3t^2 - 6t + 4 \quad \Rightarrow \quad s'(t) = 6t^2 + 6t - 6 = 6(t^2 + t - 1)$. By the quadratic formula, the critical numbers are $t = \left(-1 \pm \sqrt{5}\right)/2$.

29. $g(x) = \sqrt[9]{x} = x^{1/9} \quad \Rightarrow \quad g'(x) = \frac{1}{9}x^{-8/9} = 1\Big/\left(9\sqrt[9]{x^8}\right) \neq 0$, but $g'(0)$ does not exist, so $x = 0$ is a critical number.

31. $g(t) = 5t^{2/3} + t^{5/3} \quad \Rightarrow \quad g'(t) = \frac{10}{3}t^{-1/3} + \frac{5}{3}t^{2/3}$. $g'(0)$ does not exist, so $t = 0$ is a critical number. $g'(t) = \frac{5}{3}t^{-1/3}(2 + t) = 0 \quad \Leftrightarrow \quad t = -2$, so $t = -2$ is also a critical number.

33. $f(r) = \dfrac{r}{r^2 + 1} \quad \Rightarrow \quad f'(r) = \dfrac{1(r^2 + 1) - r(2r)}{(r^2 + 1)^2} = \dfrac{-r^2 + 1}{(r^2 + 1)^2} = 0 \quad \Leftrightarrow \quad r^2 = 1 \quad \Leftrightarrow \quad r = \pm 1$, so these are the critical numbers.

35. $F(x) = x^{4/5}(x - 4)^2 \quad \Rightarrow \quad F'(x) = \frac{4}{5}x^{-1/5}(x - 4)^2 + 2x^{4/5}(x - 4) = \dfrac{(x - 4)(7x - 8)}{5x^{1/5}} = 0$ when $x = 4, \frac{8}{7}$ and $F'(0)$ does not exist. Critical numbers are $0, \frac{8}{7}, 4$.

37. $f(\theta) = \sin^2(2\theta) \quad \Rightarrow \quad f'(\theta) = 2\sin(2\theta)\cos(2\theta)(2) = 2\sin 4\theta = 0 \quad \Leftrightarrow \quad \sin 4\theta = 0 \quad \Leftrightarrow \quad 4\theta = n\pi$, n an integer. So $\theta = n\pi/4$ are the critical numbers.

39. $f(x) = x^2 - 2x + 2$, $[0, 3]$. $f'(x) = 2x - 2 = 0 \quad \Leftrightarrow \quad x = 1$. $f(0) = 2$, $f(1) = 1$, $f(3) = 5$. So $f(3) = 5$ is the absolute maximum and $f(1) = 1$ is the absolute minimum.

41. $f(x) = x^3 - 12x + 1$, $[-3, 5]$. $f'(x) = 3x^2 - 12 = 3(x^2 - 4) = 3(x + 2)(x - 2) = 0 \quad \Leftrightarrow \quad x = \pm 2$. $f(-3) = 10$, $f(-2) = 17$, $f(2) = -15$, $f(5) = 66$. So $f(2) = -15$ is the absolute minimum and $f(5) = 66$ is the absolute maximum.

43. $f(x) = 2x^3 + 3x^2 + 4$, $[-2, 1]$. $f'(x) = 6x^2 + 6x = 6x(x+1) = 0 \quad \Leftrightarrow \quad x = -1, 0$. $f(-2) = 0$, $f(-1) = 5$, $f(0) = 4$, $f(1) = 9$. So $f(1) = 9$ is the absolute maximum and $f(-2) = 0$ is the absolute minimum.

45. $f(x) = x^4 - 4x^2 + 2$, $[-3, 2]$. $f'(x) = 4x^3 - 8x = 4x(x^2 - 2) = 0 \quad \Leftrightarrow \quad x = 0, \pm\sqrt{2}$. $f(-3) = 47$, $f\left(-\sqrt{2}\right) = -2$, $f(0) = 2$, $f\left(\sqrt{2}\right) = -2$, $f(2) = 2$, so $f\left(\pm\sqrt{2}\right) = -2$ is the absolute minimum and $f(-3) = 47$ is the absolute maximum.

47. $f(x) = x^2 + \dfrac{2}{x}$, $\left[\frac{1}{2}, 2\right]$. $f'(x) = 2x - \dfrac{2}{x^2} = 2\dfrac{x^3 - 1}{x^2} = 0 \quad \Leftrightarrow \quad x = 1$. $f\left(\frac{1}{2}\right) = \frac{17}{4}$, $f(1) = 3$, $f(2) = 5$. So $f(1) = 3$ is the absolute minimum and $f(2) = 5$ is the absolute maximum.

49. $f(x) = x^{4/5}$, $[-32, 1]$. $f'(x) = \frac{4}{5}x^{-1/5} \quad \Rightarrow \quad f'(x) \neq 0$ but $f'(0)$ does not exist, so 0 is the only critical number. $f(-32) = 16$, $f(0) = 0$, $f(1) = 1$. So $f(0) = 0$ is the absolute minimum and $f(-32) = 16$ is the absolute maximum.

51. $f(x) = \sin x + \cos x$, $\left[0, \frac{\pi}{3}\right]$. $f'(x) = \cos x - \sin x = 0 \quad \Leftrightarrow \quad x = \frac{\pi}{4}$. $f(0) = 1$, $f\left(\frac{\pi}{4}\right) = \sqrt{2}$, $f\left(\frac{\pi}{3}\right) = \frac{\sqrt{3}+1}{2}$. So $f(0) = 1$ is the absolute minimum and $f\left(\frac{\pi}{4}\right) = \sqrt{2}$ the absolute maximum.

53.

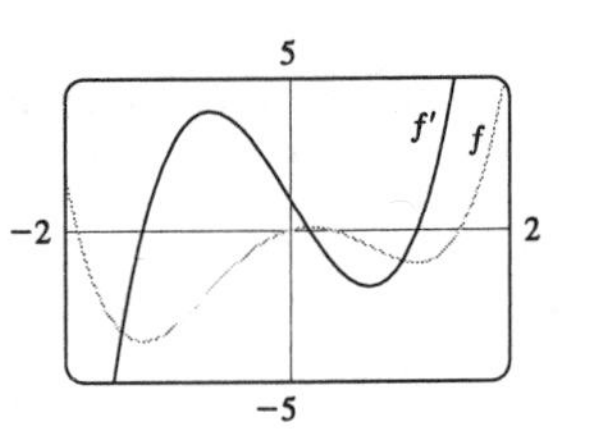

We see that $f'(x) = 0$ at about $x = -1.3, 0.2$, and 1.1. Since f' exists everywhere, these are the only critical numbers.

55. **(a)**

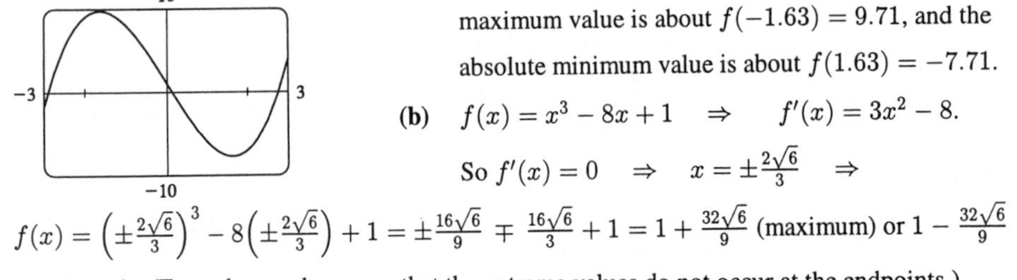

From the graph, it appears that the absolute maximum value is about $f(-1.63) = 9.71$, and the absolute minimum value is about $f(1.63) = -7.71$.

(b) $f(x) = x^3 - 8x + 1 \quad \Rightarrow \quad f'(x) = 3x^2 - 8$. So $f'(x) = 0 \quad \Rightarrow \quad x = \pm\frac{2\sqrt{6}}{3} \quad \Rightarrow$

$f(x) = \left(\pm\frac{2\sqrt{6}}{3}\right)^3 - 8\left(\pm\frac{2\sqrt{6}}{3}\right) + 1 = \pm\frac{16\sqrt{6}}{9} \mp \frac{16\sqrt{6}}{3} + 1 = 1 + \frac{32\sqrt{6}}{9}$ (maximum) or $1 - \frac{32\sqrt{6}}{9}$ (minimum). (From the graph, we see that the extreme values do not occur at the endpoints.)

57. **(a)**

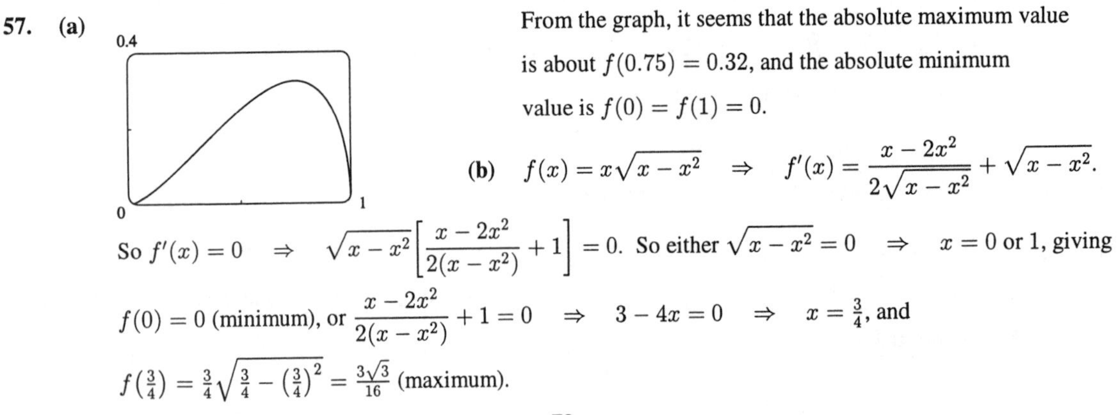

From the graph, it seems that the absolute maximum value is about $f(0.75) = 0.32$, and the absolute minimum value is $f(0) = f(1) = 0$.

(b) $f(x) = x\sqrt{x - x^2} \quad \Rightarrow \quad f'(x) = \dfrac{x - 2x^2}{2\sqrt{x - x^2}} + \sqrt{x - x^2}$.

So $f'(x) = 0 \quad \Rightarrow \quad \sqrt{x - x^2}\left[\dfrac{x - 2x^2}{2(x - x^2)} + 1\right] = 0$. So either $\sqrt{x - x^2} = 0 \quad \Rightarrow \quad x = 0$ or 1, giving $f(0) = 0$ (minimum), or $\dfrac{x - 2x^2}{2(x - x^2)} + 1 = 0 \quad \Rightarrow \quad 3 - 4x = 0 \quad \Rightarrow \quad x = \frac{3}{4}$, and

$f\left(\frac{3}{4}\right) = \frac{3}{4}\sqrt{\frac{3}{4} - \left(\frac{3}{4}\right)^2} = \frac{3\sqrt{3}}{16}$ (maximum).

59. $f(x) = [\![x]\!]$ is discontinuous at every integer n (See Example 11 and Exercise 63 in Section 1.3), so that $f'(n)$ does not exist. For all other real numbers a, $[\![x]\!]$ is constant on an open interval containing a, and so $f'(a) = 0$. Therefore every real number is a critical number of $f(x) = [\![x]\!]$.

61. The density is defined as $\rho = \dfrac{\text{mass}}{\text{volume}} = \dfrac{1000}{V(T)}$ (in g/cm^3). But a critical point of ρ will also be a critical point of V $\left[\text{since } \dfrac{d\rho}{dT} = -1000V^{-2}\dfrac{dV}{dT} \text{ and } V \text{ is never } 0\right]$, and V is easier to differentiate than ρ.

$V(T) = 999.87 - 0.06426T + 0.0085043T^2 - 0.0000679T^3 \quad \Rightarrow$

$V'(T) = -0.06426 + 0.0170086T - 0.0002037T^2$. Setting this equal to 0 and using the quadratic formula to find T, we get $T = \dfrac{-0.0170086 \pm \sqrt{0.0170086^2 - 4 \cdot 0.0003037 \cdot 0.06426}}{2(-0.0002037)} \approx 3.9665°$ or $79.5318°$. Since we are only interested in the region $0° \leq T \leq 30°$, we check the density ρ at the endpoints and at $3.9665°$:

$\rho(0) \approx \dfrac{1000}{999.87} \approx 1.00013$; $\rho(30) \approx \dfrac{1000}{1003.7641} \approx 0.99625$; $\rho(3.9665) \approx \dfrac{1000}{999.7447} \approx 1.000255$. So water has its maximum density at about $3.9665°$ C.

63. $f(x) = x^5$. $f'(x) = 5x^4 \quad \Rightarrow \quad f'(0) = 0$ so 0 is a critical number. But $f(0) = 0$ and f takes both positive and negative values in any open interval containing 0, so f does not have a local extremum at 0.

65. $f(x) = x^{101} + x^{51} + x + 1 \quad \Rightarrow \quad f'(x) = 101x^{100} + 51x^{50} + 1 \geq 1$ for all x, so $f'(x) = 0$ has no solution. Thus $f(x)$ has no critical number, so $f(x)$ can have no local extremum.

67. If f has a local minimum at c, then $g(x) = -f(x)$ has a local maximum at c, so $g'(c) = 0$ by the case of Fermat's Theorem proved in the text. Thus $f'(c) = -g'(c) = 0$.

69.

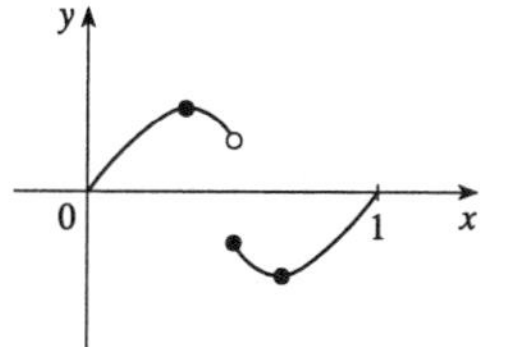

EXERCISES 3.2

1. $f(x) = x^3 - x$, $[-1, 1]$. f, being a polynomial, is continuous on $[-1, 1]$ and differentiable on $(-1, 1)$. Also $f(-1) = 0 = f(1)$. $f'(c) = 3c^2 - 1 = 0 \quad \Rightarrow \quad c = \pm\frac{1}{\sqrt{3}}$.

3. $f(x) = \cos 2x$, $[0, \pi]$. f is continuous on $[0, \pi]$ and differentiable on $(0, \pi)$. Also $f(0) = 1 = f(\pi)$. $f'(c) = -2\sin 2c = 0 \quad \Rightarrow \quad \sin 2c = 0 \quad \Rightarrow \quad 2c = \pi \quad \Rightarrow \quad c = \frac{\pi}{2}$ [since $c \in (0, \pi)$].

5. $f(x)=1-x^{2/3}$. $f(-1)=1-(-1)^{2/3}=1-1=0=f(1)$. $f'(x)=-\frac{2}{3}x^{-1/3}$, so $f'(c)=0$ has no solutions. This does not contradict Rolle's Theorem, since $f'(0)$ does not exist, and so f is not differentiable on $[-1,1]$.

7. $\dfrac{f(8)-f(0)}{8-0}=\dfrac{6-4}{8}=\dfrac{1}{4}$. The values of c which satisfy $f'(c)=\frac{1}{4}$ seem to be about $c=0.8, 3.2, 4.4$, and 6.1.

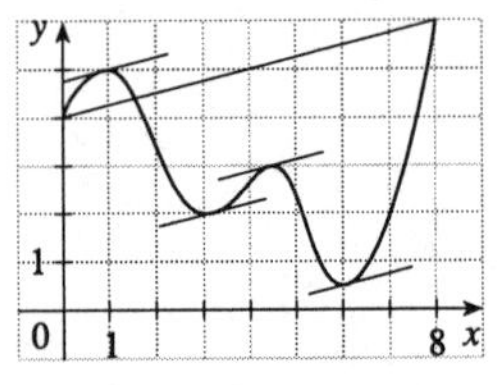

9. (a), (b) The equation of the secant line is

$$y-5=\frac{8.5-5}{8-1}(x-1)\iff y=\tfrac{1}{2}x+\tfrac{9}{2}.$$

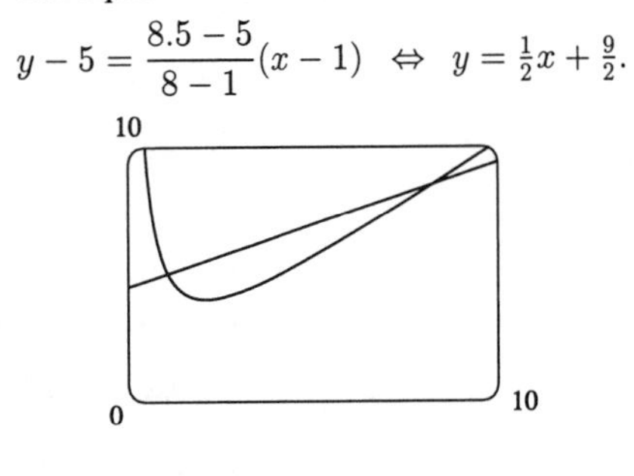

(c) $f(x)=x+4/x \Rightarrow f'(x)=1-4/x^2$. So $f'(c)=\frac{1}{2} \Rightarrow c^2=8\Rightarrow c=2\sqrt{2}$, and $f(c)=2\sqrt{2}+\frac{4}{2\sqrt{2}}=3\sqrt{2}$. Thus the equation of the tangent is $y-3\sqrt{2}=\frac{1}{2}\left(x-2\sqrt{2}\right)$ $\iff y=\frac{1}{2}x+2\sqrt{2}$.

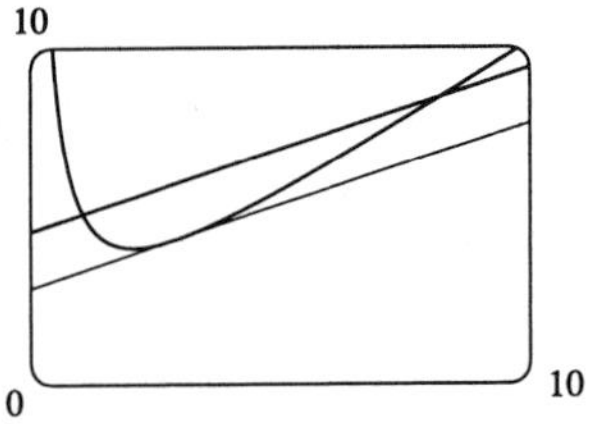

10

0 10

11. $f(x)=1-x^2$, $[0,3]$. f, being a polynomial, is continuous on $[0,3]$ and differentiable on $(0,3)$. $\dfrac{f(3)-f(0)}{3-0}=\dfrac{-8-1}{3}=-3$ and $-3=f'(c)=-2c \Rightarrow c=\frac{3}{2}$.

13. $f(x)=1/x$, $[1,2]$. f, being a rational function, is continuous on $[1,2]$ and differentiable on $(1,2)$. $\dfrac{f(2)-f(1)}{2-1}=\dfrac{\frac{1}{2}-1}{1}=-\dfrac{1}{2}$ and $-\dfrac{1}{2}=f'(c)=-\dfrac{1}{c^2} \Rightarrow c^2=2 \Rightarrow c=\sqrt{2}$ (since c must lie in $[1,2]$).

15. 1 and $x-1$ are continuous on $\mathbb{R}$ by Theorem 1.5.5, $\sqrt[3]{x}$ is continuous on $\mathbb{R}$ by Theorem 1.5.6; therefore $f(x)=1+\sqrt[3]{x-1}$ is continuous on $\mathbb{R}$ by Theorems 1.5.8 and 1.5.4(1), and hence continuous on $[2,9]$. $f'(x)=\frac{1}{3}(x-1)^{-2/3}$, so that f is differentiable for all $x\neq 1$ and so f is differentiable on $(2,9)$. By the Mean Value Theorem, there exists a number c such that $f'(c)=\frac{1}{3}(c-1)^{-2/3}=\dfrac{f(9)-f(2)}{9-2}=\dfrac{3-2}{7}=\dfrac{1}{7} \Rightarrow$ $\frac{1}{3}(c-1)^{-2/3}=\frac{1}{7} \Rightarrow (c-1)^2=\left(\frac{7}{3}\right)^3 \Rightarrow c=\pm\left(\frac{7}{3}\right)^{3/2}+1 \Rightarrow c=\left(\frac{7}{3}\right)^{3/2}+1\approx 4.564$ since $c\in[2,9]$.

17. $f(x)=|x-1|$. $f(3)-f(0)=|3-1|-|0-1|=1$. Since $f'(c)=-1$ if $c<1$ and $f'(c)=1$ if $c>1$, $f'(c)(3-0)=\pm 3$ and so is never equal to 1. This does not contradict the Mean Value Theorem since $f'(1)$ does not exist.

19. $f(x) = x^5 + 10x + 3 = 0$. Since f is continuous and $f(-1) = -8$ and $f(0) = 3$, the equation has at least one root in $(-1, 0)$ by the Intermediate Value Theorem. Suppose that the equation has more than one root; say a and b are both roots with $a < b$. Then $f(a) = 0 = f(b)$ so by Rolle's Theorem $f'(x) = 5x^4 + 10 = 0$ has a root in (a, b). But this is impossible since clearly $f'(x) \geq 10 > 0$ for all real x.

21. $f(x) = x^5 - 6x + c = 0$. Suppose that $f(x)$ has two roots a and b with $-1 \leq a < b \leq 1$. Then $f(a) = 0 = f(b)$, so by Rolle's Theorem there is a number d in (a, b) with $f'(d) = 0$. Now $0 = f'(d) = 5d^4 - 6 \quad \Rightarrow \quad d = \pm\sqrt[4]{\frac{6}{5}}$, which are both outside $[-1, 1]$ and hence outside (a, b). Thus $f(x)$ can have at most one root in $[-1, 1]$.

23. **(a)** Suppose that a cubic polynomial $P(x)$ has roots $a_1 < a_2 < a_3 < a_4$, so $P(a_1) = P(a_2) = P(a_3) = P(a_4)$. By Rolle's Theorem there are numbers c_1, c_2, c_3 with $a_1 < c_1 < a_2$, $a_2 < c_2 < a_3$ and $a_3 < c_3 < a_4$ and $P'(c_1) = P'(c_2) = P'(c_3) = 0$. Thus the second-degree polynomial $P'(x)$ has 3 distinct real roots, which is impossible.

(b) We prove by induction that a polynomial of degree n has at most n real roots. This is certainly true for $n = 1$. Suppose that the result is true for all polynomials of degree n and let $P(x)$ be a polynomial of degree $n + 1$. Suppose that $P(x)$ has more than $n + 1$ real roots, say $a_1 < a_2 < a_3 < \cdots < a_{n+1} < a_{n+2}$. Then $P(a_1) = P(a_2) = \cdots = P(a_{n+2}) = 0$. By Rolle's Theorem there are real numbers $c_1, \ldots, c_{n+1}$ with $a_1 < c_1 < a_2, \ldots, \; a_{n+1} < c_{n+1} < a_{n+2}$ and $P'(c_1) = \cdots = P'(c_{n+1}) = 0$. Thus the nth degree polynomial $P'(x)$ has at least $n + 1$ roots. This contradiction shows that $P(x)$ has at most $n + 1$ real roots.

25. By the Mean Value Theorem, $f(4) - f(1) = f'(c)(4 - 1)$ for some $c \in (1, 4)$. But for every $c \in (1, 4)$ we have $f'(c) \geq 2$. Putting $f'(c) \geq 2$ into the above equation and substituting $f(1) = 10$, we get $f(4) = f(1) + f'(c)(4 - 1) = 10 + 3f'(c) \geq 10 + 3 \cdot 2 = 16$. So the smallest possible value of $f(4)$ is 16.

27. Suppose that such a function f exists. By the Mean Value Theorem there is a number $0 < c < 2$ with $f'(c) = \dfrac{f(2) - f(0)}{2 - 0} = \dfrac{5}{2}$. But this is impossible since $f'(x) \leq 2 < \frac{5}{2}$ for all x, so no such function can exist.

29. We use Exercise 28 with $f(x) = \sqrt{1 + x}$, $g(x) = 1 + \frac{1}{2}x$, and $a = 0$. Notice that $f(0) = 1 = g(0)$ and $f'(x) = \dfrac{1}{2\sqrt{1 + x}} < \dfrac{1}{2} = g'(x)$ for $x > 0$. So by Exercise 28, $f(b) < g(b) \;\Rightarrow\; \sqrt{1 + b} < 1 + \frac{1}{2}b$ for $b > 0$.

Another Method: Apply the Mean Value Theorem directly to either $f(x) = 1 + \frac{1}{2}x - \sqrt{1 + x}$ or $g(x) = \sqrt{1 + x}$ on $[0, b]$.

31. Let $f(x) = \sin x$ and let $b < a$. Then $f(x)$ is continuous on $[b, a]$ and differentiable on (b, a). By the Mean Value Theorem, there is a number $c \in (b, a)$ with $\sin a - \sin b = f(a) - f(b) = f'(c)(a - b) = (\cos c)(a - b)$. Thus $|\sin a - \sin b| \leq |\cos c||b - a| \leq |a - b|$. If $a < b$, then $|\sin a - \sin b| = |\sin b - \sin a| \leq |b - a| = |a - b|$. If $a = b$, both sides of the inequality are 0.

33. For $x > 0$, $f(x) = g(x)$, so $f'(x) = g'(x)$. For $x < 0$, $f'(x) = (1/x)' = -1/x^2$ and $g'(x) = (1 + 1/x)' = -1/x^2$, so again $f'(x) = g'(x)$. However, the domain of $g(x)$ is not an interval $\left[\text{it is } (-\infty, 0) \cup (0, \infty)\right]$ so we cannot conclude that $f - g$ is constant (in fact it is not).

35. Let $g(t)$ and $h(t)$ be the position functions of the two runners and let $f(t) = g(t) - h(t)$. By hypothesis $f(0) = g(0) - h(0) = 0$ and $f(b) = g(b) - h(b) = 0$ where b is the finishing time. Then by Rolle's Theorem, there is a time $0 < c < b$ with $0 = f'(c) = g'(c) - h'(c)$. Hence $g'(c) = h'(c)$, so at time c, both runners have the same velocity $g'(c) = h'(c)$.

EXERCISES 3.3

1. **(a)** $f'(x) > 0$ for $x < 0$ and $x > 3$, so f is increasing on $(-\infty, 0]$ and $[3, \infty)$. $f'(x) < 0$ for $0 < x < 3$, so f is decreasing on $[0, 3]$.

(b) f has a local maximum where f' changes from positive to negative, at $x = 0$, and a local minimum where f' changes from negative to positive, at $x = 3$.

3. $f(x) = 20 - x - x^2$, $f'(x) = -1 - 2x = 0 \quad \Rightarrow \quad x = -\frac{1}{2}$ (the only critical number)

(a) $f'(x) > 0 \quad \Leftrightarrow \quad -1 - 2x > 0 \quad \Leftrightarrow \quad x < -\frac{1}{2}$, $f'(x) < 0 \quad \Leftrightarrow \quad x > -\frac{1}{2}$, so f is increasing on $\left(-\infty, -\frac{1}{2}\right]$ and decreasing on $\left[-\frac{1}{2}, \infty\right)$.

(b) By the First Derivative Test, $f\left(-\frac{1}{2}\right) = 20.25$ is a local maximum.

(c)

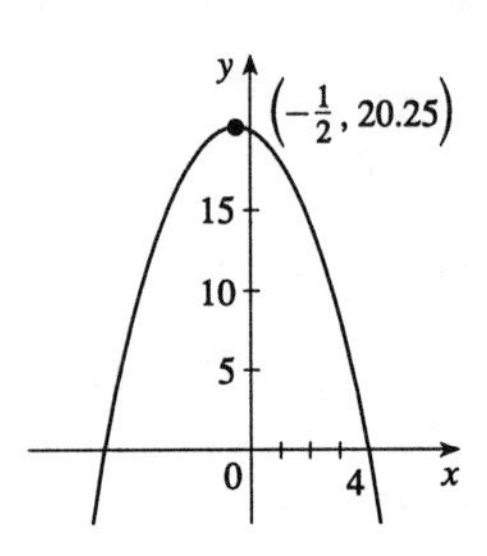

5. $f(x) = x^3 + x + 1 \quad \Rightarrow \quad f'(x) = 3x^2 + 1 > 0$ for all $x \in \mathbb{R}$.

(a) f is increasing on $\mathbb{R}$.

(b) f has no local extremum.

(c)

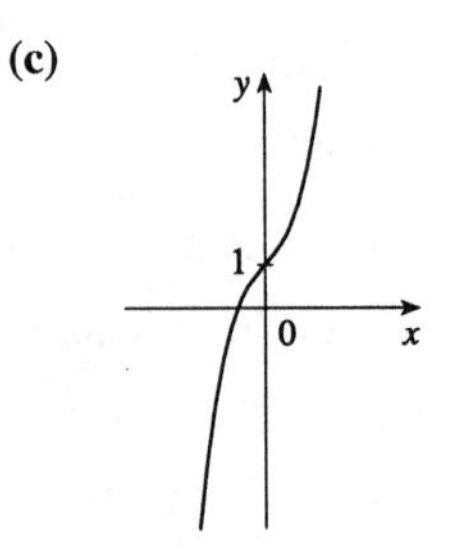

7. $f(x) = 2x^2 - x^4$. $f'(x) = 4x - 4x^3 = 4x(1 - x^2) = 4x(1 + x)(1 - x)$, so the critical numbers are $x = 0, \pm 1$.

(a)

Interval	$4x$	$1+x$	$1-x$	$f'(x)$	f
$x < -1$	$-$	$-$	$+$	$+$	increasing on $(-\infty, -1]$
$-1 < x < 0$	$-$	$+$	$+$	$-$	decreasing on $[-1, 0]$
$0 < x < 1$	$+$	$+$	$+$	$+$	increasing on $[0, 1]$
$x > 1$	$+$	$+$	$-$	$-$	decreasing on $[1, \infty)$

(b) Local maximum $f(-1) = 1$, local minimum $f(0) = 0$, local maximum $f(1) = 1$.

(c)

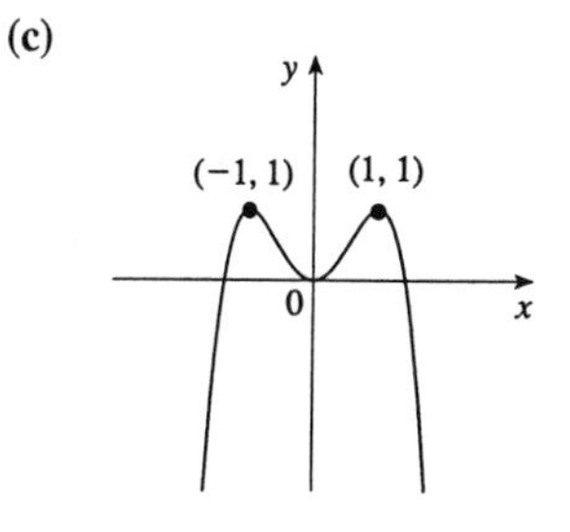

9. $f(x) = x^3(x-4)^4$. $f'(x) = 3x^2(x-4)^4 + x^3\left[4(x-4)^3\right] = x^2(x-4)^3(7x - 12)$. The critical numbers are $x = 0, 4, \frac{12}{7}$.

(a) $x^2(x-4)^2 \geq 0$ so $f'(x) \geq 0 \quad \Leftrightarrow \quad (x-4)(7x-12) \geq 0$ $\Leftrightarrow \quad x \leq \frac{12}{7}$ or $x \geq 4$. $f'(x) \leq 0 \quad \Leftrightarrow \quad \frac{12}{7} \leq x \leq 4$. So f is increasing on $\left(-\infty, \frac{12}{7}\right]$ and $[4, \infty)$ and decreasing on $\left[\frac{12}{7}, 4\right]$.

(b) Local maximum $f\left(\frac{12}{7}\right) = 12^3 \cdot \dfrac{16^4}{7^7} \approx 137.5$, local minimum $f(4) = 0$.

(c)

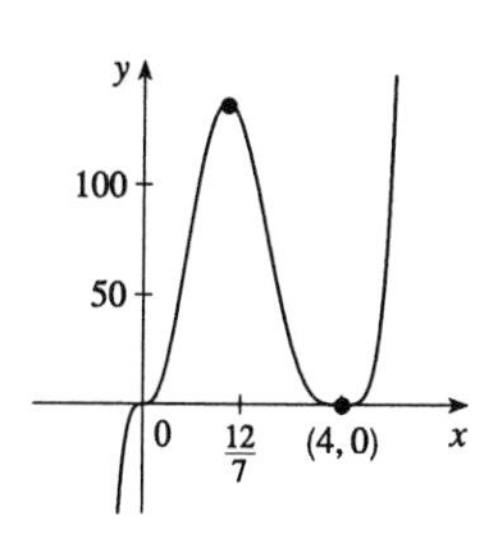

11. $f(x) = x\sqrt{6-x}$. $f'(x) = \sqrt{6-x} + x\left(-\dfrac{1}{2\sqrt{6-x}}\right) = \dfrac{3(4-x)}{2\sqrt{6-x}}$. Critical numbers are $x = 4, 6$.

(a) $f'(x) > 0 \quad \Leftrightarrow \quad 4 - x > 0$ (and $x < 6$) $\quad \Leftrightarrow \quad x < 4$ and $f'(x) < 0 \quad \Leftrightarrow \quad 4 - x < 0$ (and $x < 6$) $\quad \Leftrightarrow \quad 4 < x < 6$. So f is increasing on $(-\infty, 4]$ and decreasing on $[4, 6]$.

(b) Local maximum $f(4) = 4\sqrt{2}$

(c)

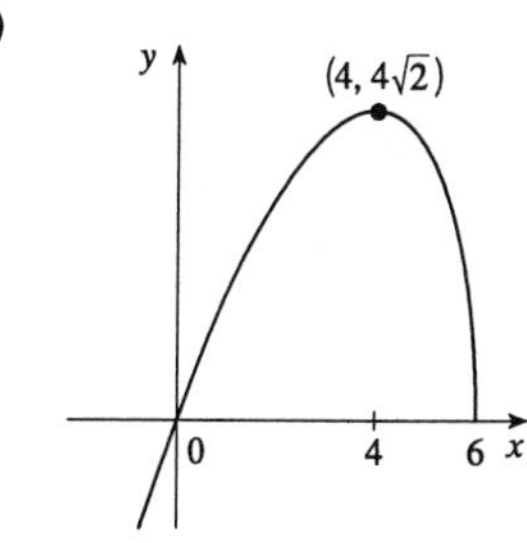

13. $f(x) = x^{1/5}(x+1)$. $f'(x) = \frac{1}{5}x^{-4/5}(x+1) + x^{1/5} = \frac{1}{5}x^{-4/5}(6x+1)$. The critical numbers are $x = 0, -\frac{1}{6}$.

(a) $f'(x) > 0 \;\Leftrightarrow\; 6x + 1 > 0\ (x \neq 0) \;\Leftrightarrow\; x > -\frac{1}{6}\ (x \neq 0)$ and $f'(x) < 0 \quad \Leftrightarrow \quad x < -\frac{1}{6}$. So f is increasing on $\left[-\frac{1}{6}, \infty\right)$ and decreasing on $\left(-\infty, -\frac{1}{6}\right]$.

(b) Local minimum $f\left(-\frac{1}{6}\right) = -\dfrac{5}{6^{6/5}} \approx -0.58$

(c)

15. $f(x) = x\sqrt{x - x^2}$. The domain of f is $\{x \mid x(1-x) \geq 0\} = [0, 1]$.

$f'(x) = \sqrt{x - x^2} + x\dfrac{1 - 2x}{2\sqrt{x - x^2}} = \dfrac{x(3 - 4x)}{2\sqrt{x - x^2}}$. So the critical numbers are $x = 0, \frac{3}{4}, 1$.

(a) $f'(x) > 0 \quad \Leftrightarrow \quad 3 - 4x > 0 \quad \Leftrightarrow$ $0 < x < \frac{3}{4}$. $f'(x) < 0 \quad \Leftrightarrow \quad \frac{3}{4} < x < 1$. So f is increasing on $\left[0, \frac{3}{4}\right]$ and decreasing on $\left[\frac{3}{4}, 1\right]$.

(b) Local maximum $f\left(\frac{3}{4}\right) = \frac{3\sqrt{3}}{16}$

(c)

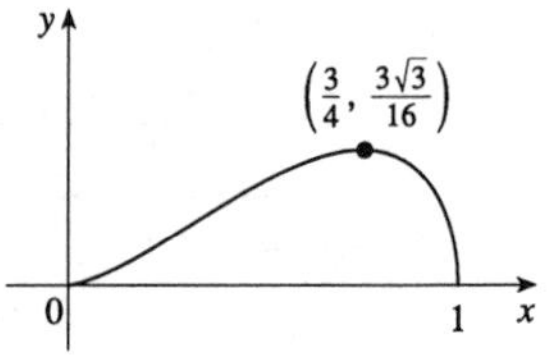

17. $f(x) = x - 2\sin x, 0 \leq x \leq 2\pi$. $f'(x) = 1 - 2\cos x$. So $f'(x) = 0 \quad \Leftrightarrow \quad \cos x = \frac{1}{2} \quad \Leftrightarrow \quad x = \frac{\pi}{3}$ or $\frac{5\pi}{3}$.

(a) $f'(x) > 0 \quad \Leftrightarrow \quad 1 - 2\cos x > 0 \quad \Leftrightarrow \quad \frac{1}{2} > \cos x \quad \Leftrightarrow$ $\frac{\pi}{3} < x < \frac{5\pi}{3}$. $f'(x) < 0 \quad \Leftrightarrow \quad 0 \leq x < \frac{\pi}{3}$ or $\frac{5\pi}{3} < x \leq 2\pi$. So f is increasing on $\left[\frac{\pi}{3}, \frac{5\pi}{3}\right]$ and decreasing on $\left[0, \frac{\pi}{3}\right]$ and $\left[\frac{5\pi}{3}, 2\pi\right]$.

(b) Local minimum $f\left(\frac{\pi}{3}\right) = \frac{\pi}{3} - \sqrt{3} \approx -0.68$, local maximum $f\left(\frac{5\pi}{3}\right) = \sqrt{3} + \frac{5\pi}{3} \approx 6.97$

(c)

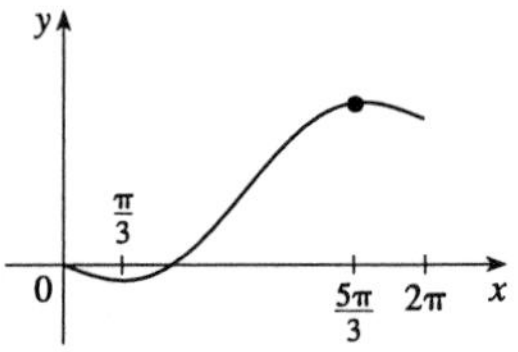

19. $f(x) = \sin^4 x + \cos^4 x, 0 \leq x \leq 2\pi$.

$f'(x) = 4\sin^3 x\cos x - 4\cos^3 x\sin x = -4\sin x\cos x(\cos^2 x - \sin^2 x) = -2\sin 2x\cos 2x = -\sin 4x$.

$f'(x) = 0 \Leftrightarrow \sin 4x = 0 \Leftrightarrow 4x = n\pi \Leftrightarrow x = n\frac{\pi}{4}$. So the critical numbers are $0, \frac{\pi}{4}, \frac{\pi}{2}, \frac{3\pi}{4}, \pi, \frac{5\pi}{4}, \frac{3\pi}{2}, \frac{7\pi}{4}, 2\pi$.

(a) $f'(x) > 0 \quad \Leftrightarrow \quad \sin 4x < 0 \quad \Leftrightarrow \quad \frac{\pi}{4} < x < \frac{\pi}{2}$ or $\frac{3\pi}{4} < x < \pi$ or $\frac{5\pi}{4} < x < \frac{3\pi}{2}$ or $\frac{7\pi}{4} < x < 2\pi$. f is increasing on these intervals. f is decreasing on $\left[0, \frac{\pi}{4}\right], \left[\frac{\pi}{2}, \frac{3\pi}{4}\right], \left[\pi, \frac{5\pi}{4}\right], \left[\frac{3\pi}{2}, \frac{7\pi}{4}\right]$.

(b) Local maxima $f\left(\frac{\pi}{2}\right) = f(\pi) = f\left(\frac{3\pi}{2}\right) = 1$, local minima $f\left(\frac{\pi}{4}\right) = f\left(\frac{3\pi}{4}\right) = f\left(\frac{5\pi}{4}\right) = f\left(\frac{7\pi}{4}\right) = \frac{1}{2}$.

(c)

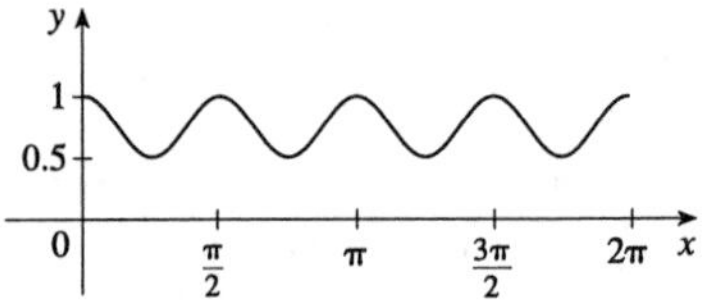

21. $f(x) = x^3 + 2x^2 - x + 1$. $f'(x) = 3x^2 + 4x - 1 = 0 \quad \Rightarrow \quad x = \frac{-4 \pm \sqrt{28}}{6} = \frac{-2 \pm \sqrt{7}}{3}$. Now $f'(x) > 0$ for $x < \frac{-2 - \sqrt{7}}{3}$ or $x > \frac{-2 + \sqrt{7}}{3}$ and $f'(x) < 0$ for $\frac{-2 - \sqrt{7}}{3} < x < \frac{-2 + \sqrt{7}}{3}$. f is increasing on $\left(-\infty, \frac{-2 - \sqrt{7}}{3}\right]$ and $\left[\frac{-2 + \sqrt{7}}{3}, \infty\right)$ and decreasing on $\left[\frac{-2 - \sqrt{7}}{3}, \frac{-2 + \sqrt{7}}{3}\right]$.

23. $f(x) = x^6 + 192x + 17$. $f'(x) = 6x^5 + 192 = 6(x^5 + 32)$. So $f'(x) > 0 \quad \Leftrightarrow \quad x^5 > -32 \quad \Leftrightarrow \quad x > -2$ and $f'(x) < 0 \quad \Leftrightarrow \quad x < -2$. So f is increasing on $[-2, \infty)$ and decreasing on $(-\infty, -2]$.

25. $f(x) = x + \sqrt{1 - x}, 0 \leq x \leq 1$. $f'(x) = 1 - \dfrac{1}{2\sqrt{1 - x}} = \dfrac{2\sqrt{1 - x} - 1}{2\sqrt{1 - x}}$, so $f'(x) = 0$ when $2\sqrt{1 - x} - 1 = 0 \Rightarrow \sqrt{1 - x} = \frac{1}{2} \Rightarrow 1 - x = \frac{1}{4}$ $\Rightarrow \quad x = \frac{3}{4}$. For $0 < x < \frac{3}{4}$, $f'(x) > 0$ and for $\frac{3}{4} < x < 1$, $f'(x) < 0$. So the local maximum is $f\left(\frac{3}{4}\right) = \frac{5}{4}$. Also $f(0) = 1$ and $f(1) = 1$ are the absolute minima and $f\left(\frac{3}{4}\right) = \frac{5}{4}$ is the absolute maximum.

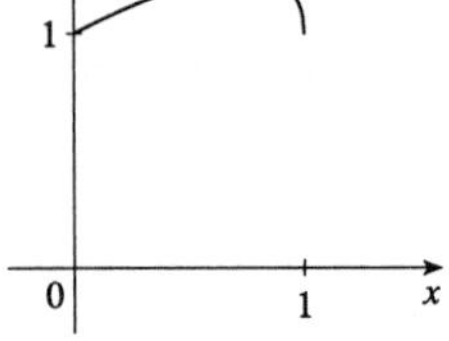

27. $g(x) = \dfrac{x}{x^2+1}$, $-5 \le x \le 5$. $g'(x) = \dfrac{(x^2+1) - x(2x)}{(x^2+1)^2} = \dfrac{1-x^2}{(x^2+1)^2}$.

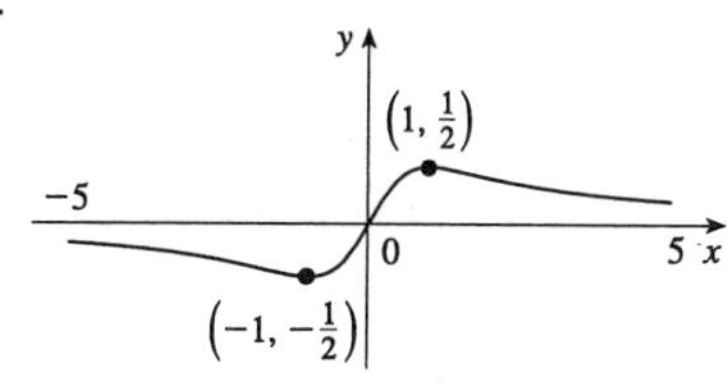

The critical numbers are $x = \pm 1$. $g'(x) > 0 \quad \Leftrightarrow \quad x^2 < 1 \quad \Leftrightarrow$ $-1 < x < 1$ and $g'(x) < 0 \quad \Leftrightarrow \quad x < -1$ or $x > 1$. So $g(-1) = -\frac{1}{2}$ is a local minimum and $g(1) = \frac{1}{2}$ is a local maximum. Also $g(-5) = -\frac{5}{26}$ and $g(5) = \frac{5}{26}$. So $g(-1) = -\frac{1}{2}$ is the absolute minimum and $g(1) = \frac{1}{2}$ is the absolute maximum.

29. **(a)** It seems from the graph that f is increasing on $(-\infty, -0.67]$ and $[0.67, \infty)$ and decreasing on $[-0.67, 0.67]$, that the local maximum at $x \approx -0.67$ is $f(-0.67) \approx 2.53$, and that the local minimum at $x \approx 0.67$ is $f(0.67) \approx 1.47$.

(b)

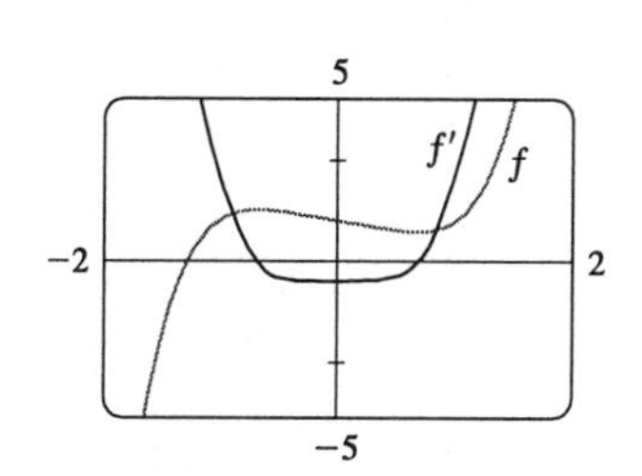

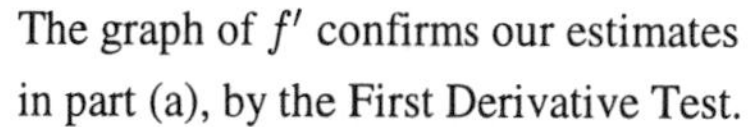
The graph of f' confirms our estimates in part (a), by the First Derivative Test.

(c) $f(x) = x^5 - x + 2 \quad \Rightarrow \quad f'(x) = 5x^4 - 1$. So there are critical points where $f'(x) = 0 \quad \Rightarrow$ $5x^4 = 1 \quad \Rightarrow \quad x = \pm 5^{-1/4}$. Now f' changes from positive to negative at $x = -5^{-1/4}$, and then back to positive at $x = 5^{-1/4}$, so f is increasing on $\left(-\infty, -5^{-1/4}\right]$ and $\left[5^{-1/4}, \infty\right)$ and decreasing on $\left[-5^{-1/4}, 5^{-1/4}\right]$. The local maximum is $f\left(-5^{-1/4}\right) = 4 \cdot 5^{-5/4} + 2 \approx 2.535$, and the local minimum is $f\left(5^{-1/4}\right) = -4 \cdot 5^{-5/4} + 2 \approx 1.465$.

31. Let $f(x) = x + \dfrac{1}{x}$, so $f'(x) = 1 - \dfrac{1}{x^2} = \dfrac{x^2-1}{x^2}$. Thus $f'(x) > 0$ for $x > 1 \quad \Rightarrow \quad f$ is increasing on $[1, \infty)$. Hence for $1 < a < b$, $a + \dfrac{1}{a} = f(a) < f(b) = b + \dfrac{1}{b}$.

33. Let $f(x) = 2\sqrt{x} - 3 + \dfrac{1}{x}$. Then $f'(x) = \dfrac{1}{\sqrt{x}} - \dfrac{1}{x^2} > 0$ for $x > 1$ since for $x > 1$, $x^2 > x > \sqrt{x}$. Hence f is increasing, so for $x > 1$, $f(x) > f(1) = 0$ or $2\sqrt{x} - 3 + \dfrac{1}{x} > 0$ for $x > 1$. Hence $2\sqrt{x} > 3 - \dfrac{1}{x}$ for $x > 1$.

35. Let $f(x) = \sin x - x + \frac{1}{6}x^3$. Then $f'(x) = \cos x - 1 + \frac{1}{2}x^2$. By Exercise 34, $f'(x) > 0$ for $x > 0$, so f is increasing for $x > 0$. Thus $f(x) > f(0) = 0$ for $x > 0$. Therefore $\sin x - x + \frac{1}{6}x^3 > 0$ or $\sin x > x - \frac{1}{6}x^3$ for $x > 0$.

37. **(a)** Let $f(x) = x + \dfrac{1}{x}$, so $f'(x) = 1 - \dfrac{1}{x^2} > 0 \quad \Leftrightarrow \quad x^2 < 1 \quad \Leftrightarrow \quad 0 < x < 1$ (since $x > 0$), and $f'(x) > 0$ for $x > 1$. By the First Derivative Test, there is an absolute minimum for $f(x)$ on $(0, \infty)$ where $x = 1$. Thus $f(x) = x + 1/x \ge f(1) = 2$ for $x > 0$.

(b) Let $y = \dfrac{1}{x}$. Then $\left(\sqrt{x} - \sqrt{y}\right)^2 \ge 0 \Rightarrow \left(\sqrt{x} - \dfrac{1}{\sqrt{x}}\right)^2 \ge 0 \Rightarrow x - 2 + \dfrac{1}{x} \ge 0 \Rightarrow x + \dfrac{1}{x} \ge 2$.

39. $f(x) = ax^3 + bx^2 + cx + d \quad \Rightarrow \quad f(1) = a + b + c + d = 0$ and $f(-2) = -8a + 4b - 2c + d = 3$. Also $f'(1) = 3a + 2b + c = 0$ and $f'(-2) = 12a - 4b + c = 0$ by Fermat's Theorem. Solving these four equations, we get $a = \frac{2}{9}$, $b = \frac{1}{3}$, $c = -\frac{4}{3}$, $d = \frac{7}{9}$, so the function is $f(x) = \frac{1}{9}(2x^3 + 3x^2 - 12x + 7)$.

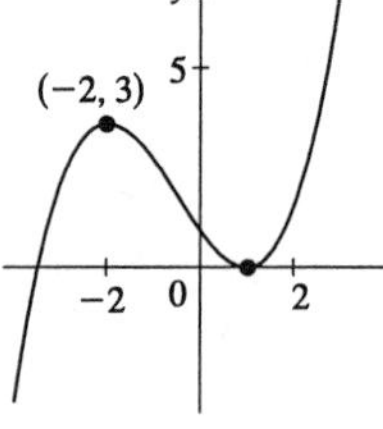

41. There are many possible functions which satisfy all of the conditions.

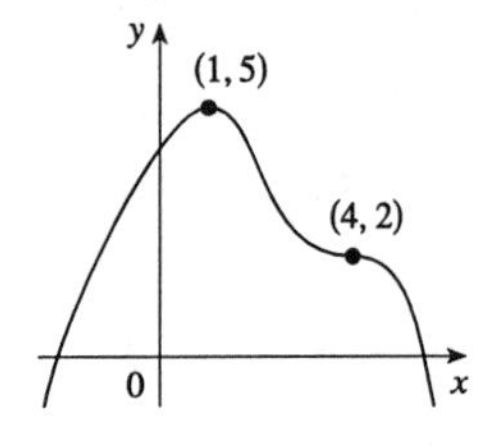

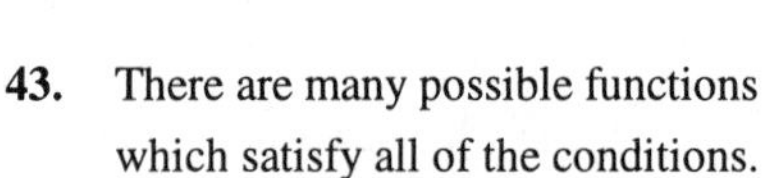

43. There are many possible functions which satisfy all of the conditions.

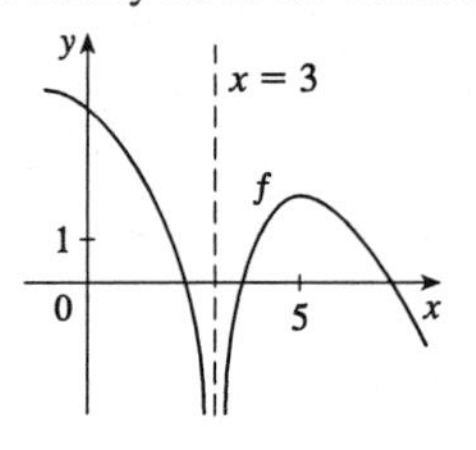

45. Let $x_1, x_2 \in I$ with $x_1 < x_2$. Then $f(x_1) < f(x_2)$ and $g(x_1) < g(x_2)$ (since f and g are increasing on I), so $(f + g)(x_1) = f(x_1) + g(x_1) < f(x_2) + g(x_2) = (f + g)(x_2)$. Therefore $f + g$ is increasing on I.

47. **(a)** Let x_1 and $x_2 \in \mathbb{R}$ with $x_1 < x_2$. Then $g(x_1) < g(x_2)$ since g is increasing on I. So $(f \circ g)(x_1) = f(g(x_1)) < f(g(x_2)) = f \circ g(x_2)$ since f is increasing on $\mathbb{R}$. So for any $x_1 < x_2$ we have $(f \circ g)(x_1) < (f \circ g)(x_2)$, which shows that h is increasing on $\mathbb{R}$.

(b) Let x_1 and $x_2 \in \mathbb{R}$ with $x_1 < x_2$. Then $g(x_1) > g(x_2)$ since g is decreasing on I. So $(f \circ g)(x_1) = f(g(x_1)) < f(g(x_2)) = (f \circ g)(x_2)$ since f is decreasing on $\mathbb{R}$. So for any $x_1 < x_2$ we have $h(x_1) = (f \circ g)(x_1) < (f \circ g)(x_2) = h(x_2)$, which shows that h is increasing on $\mathbb{R}$.

(c) Let x_1 and $x_2 \in \mathbb{R}$ with $x_1 < x_2$. Then $g(x_1) > g(x_2)$ since g is decreasing on $\mathbb{R}$, which implies that $f(g(x_1)) = (f \circ g)(x_1) > (f \circ g)(x_2) = f(g(x_2))$, since f is increasing. This shows that h is decreasing on $\mathbb{R}$.

49. Let x_1 and x_2 be any two numbers in $[a, b]$ with $x_1 < x_2$. Then f is continuous on $[x_1, x_2]$ and differentiable on (x_1, x_2), so by the Mean Value Theorem there is a number c between x_1 and x_2 such that $f(x_2) - f(x_1) = f'(c)(x_2 - x_1)$. Now $f'(c) < 0$ by assumption and $x_2 - x_1 > 0$ because $x_1 < x_2$. Thus $f(x_2) - f(x_1) = f'(c)(x_2 - x_1)$ is negative, so $f(x_2) - f(x_1) < 0$ or $f(x_2) < f(x_1)$. This shows that f is decreasing on $[a, b]$.

EXERCISES 3.4

1. The derivative f' is increasing when the slopes of the tangent lines are becoming larger as x increases. This seems to be the case on the interval $[2, 5]$. The derivative is decreasing when the slopes of the tangent lines are becoming smaller as x increases, and this seems to be the case on $(-\infty, 2]$ and $[5, \infty)$. So f' is increasing on $[2, 5]$ and decreasing on $(-\infty, 2]$ and $[5, \infty)$.

3. **(a)** $f(x) = x^3 - x \;\Rightarrow\; f'(x) = 3x^2 - 1 = 0 \;\Leftrightarrow\; x^2 = \frac{1}{3} \;\Leftrightarrow\; x = \pm\frac{1}{\sqrt{3}}$. $f'(x) > 0 \;\Leftrightarrow\; x^2 > \frac{1}{3}$ $\;\Leftrightarrow\; |x| > \frac{1}{\sqrt{3}} \;\Leftrightarrow\; x > \frac{1}{\sqrt{3}}$ or $x < -\frac{1}{\sqrt{3}}$. $f'(x) < 0 \;\Leftrightarrow\; |x| < \frac{1}{\sqrt{3}} \;\Leftrightarrow\; -\frac{1}{\sqrt{3}} < x < \frac{1}{\sqrt{3}}$. So f is increasing on $\left(-\infty, -\frac{1}{\sqrt{3}}\right]$ and $\left[\frac{1}{\sqrt{3}}, \infty\right)$, and decreasing on $\left[-\frac{1}{\sqrt{3}}, \frac{1}{\sqrt{3}}\right]$.

(b) Local maximum $f\left(-\frac{1}{\sqrt{3}}\right) = \frac{2}{3\sqrt{3}} \approx 0.38$, local minimum $f\left(\frac{1}{\sqrt{3}}\right) = -\frac{2}{3\sqrt{3}} \approx -0.38$.

(c) $f''(x) = 6x \;\Rightarrow\; f''(x) > 0 \;\Leftrightarrow\; x > 0$, so f is CU on $(0, \infty)$ and CD on $(-\infty, 0)$.

(d) Point of inflection at $x = 0$

5. **(a)** $f(x) = x^4 - 6x^2 \;\Rightarrow\; f'(x) = 4x^3 - 12x = 4x(x^2 - 3) = 0$ when $x = 0, \pm\sqrt{3}$.

Interval	$4x$	$x^2 - 3$	$f'(x)$	f
$x < -\sqrt{3}$	$-$	$+$	$-$	decreasing on $\left(-\infty, -\sqrt{3}\right]$
$-\sqrt{3} < x < 0$	$-$	$-$	$+$	increasing on $\left[-\sqrt{3}, 0\right]$
$0 < x < \sqrt{3}$	$+$	$-$	$-$	decreasing on $\left[0, \sqrt{3}\right]$
$x > \sqrt{3}$	$+$	$+$	$+$	increasing on $\left[\sqrt{3}, \infty\right)$

(b) Local minima $f\left(\pm\sqrt{3}\right) = -9$, local maximum $f(0) = 0$

(c) $f''(x) = 12x^2 - 12 = 12(x^2 - 1) > 0 \;\Leftrightarrow\; x^2 > 1$ $\;\Leftrightarrow\; |x| > 1 \;\Leftrightarrow\; x > 1$ or $x < -1$, so f is CU on $(-\infty, -1)$, $(1, \infty)$ and CD on $(-1, 1)$.

(d) Inflection points when $x = \pm 1$

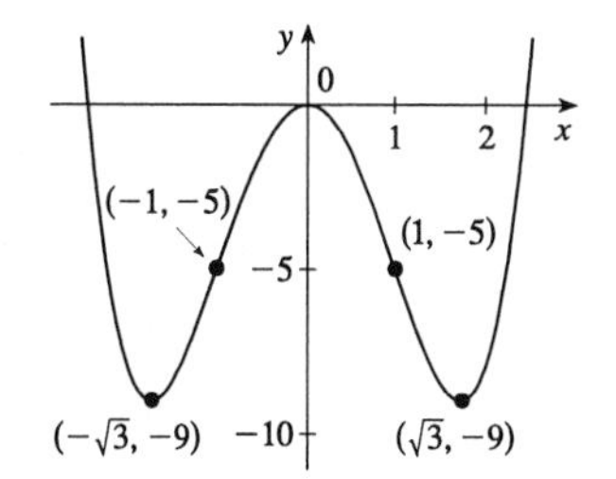

7. **(a)** $h(x) = 3x^5 - 5x^3 + 3 \Rightarrow h'(x) = 15x^4 - 15x^2 = 15x^2(x^2 - 1) = 0$ when $x = 0, \pm 1$. $h'(x) > 0$ $\Leftrightarrow x^2 > 1 \Leftrightarrow x > 1$ or $x < -1$, so h is increasing on $(-\infty, -1]$ and $[1, \infty)$ and decreasing on $[-1, 1]$.

(b) Local maximum $h(-1) = 5$, local minimum $h(1) = 1$

(c) $h''(x) = 60x^3 - 30x = 30x(2x^2 - 1) = 60x\left(x + \frac{1}{\sqrt{2}}\right)\left(x - \frac{1}{\sqrt{2}}\right)$ $\Rightarrow h''(x) > 0$ when $x > \frac{1}{\sqrt{2}}$ or $-\frac{1}{\sqrt{2}} < x < 0$, so h is CU on $\left(-\frac{1}{\sqrt{2}}, 0\right)$ and $\left(\frac{1}{\sqrt{2}}, \infty\right)$ and CD on $\left(-\infty, -\frac{1}{\sqrt{2}}\right)$ and $\left(0, \frac{1}{\sqrt{2}}\right)$.

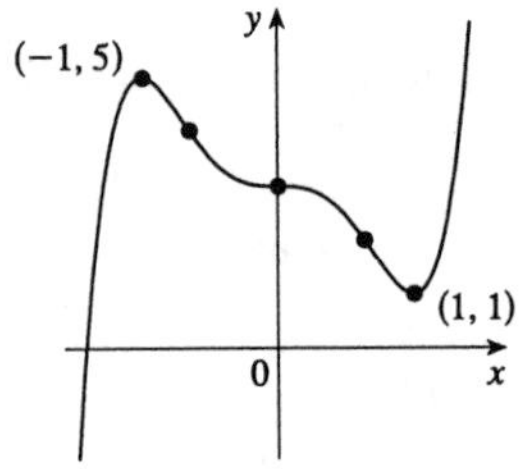

(d) Inflection points at $x = \pm\frac{1}{\sqrt{2}}$ and 0

9. **(a)** $P(x) = x\sqrt{x^2 + 1} \Rightarrow P'(x) = \sqrt{x^2 + 1} + \dfrac{x^2}{\sqrt{x^2 + 1}} = \dfrac{2x^2 + 1}{\sqrt{x^2 + 1}} > 0$, so P is increasing on $\mathbb{R}$.

(b) No extremum

(c) $$P''(x) = \frac{4x\sqrt{x^2 + 1} - (2x^2 + 1)\dfrac{x}{\sqrt{x^2 + 1}}}{x^2 + 1} = \frac{x(2x^2 + 3)}{(x^2 + 1)^{3/2}} > 0$$ $\Leftrightarrow x > 0$ so P is CU on $(0, \infty)$ and CD on $(-\infty, 0)$.

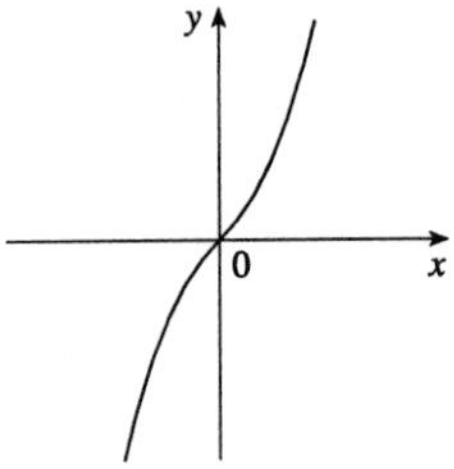

(d) IP at $x = 0$

11. **(a)** $Q(x) = x^{1/3}(x + 3)^{2/3} \Rightarrow Q'(x) = \frac{1}{3}x^{-2/3}(x + 3)^{2/3} + x^{1/3}\left(\frac{2}{3}\right)(x + 3)^{-1/3} = \dfrac{x + 1}{x^{2/3}(x + 3)^{1/3}}$. The critical numbers are -3, -1, and 0. Note that $x^{2/3} \geq 0$ for all x. So $Q'(x) > 0$ when $x < -3$ or $x > -1$ and $Q'(x) < 0$ when $-3 < x < -1 \Rightarrow Q$ is increasing on $(-\infty, -3]$ and $[-1, \infty)$ and decreasing on $[-3, -1]$.

(b) $Q(-3) = 0$ is a local maximum and $Q(-1) = -4^{1/3} \approx -1.6$ is a local minimum.

(c) $Q''(x) = -\dfrac{2}{x^{5/3}(x + 3)^{4/3}} \Rightarrow Q''(x) > 0$ when $x < 0$, so Q is CU on $(-\infty, -3)$ and $(-3, 0)$ and CD on $(0, \infty)$.

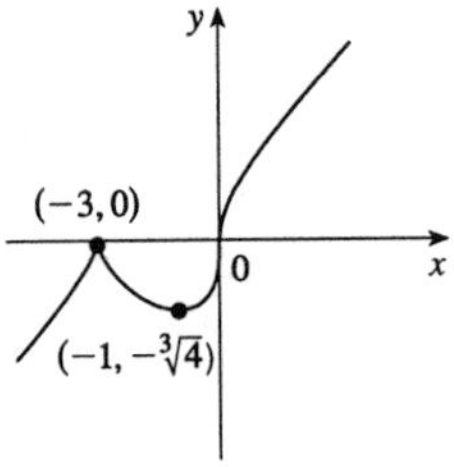

(d) IP at $x = 0$

13. **(a)** $f(\theta) = \sin^2\theta \Rightarrow f'(\theta) = 2\sin\theta\cos\theta = \sin 2\theta > 0 \Leftrightarrow 2\theta \in (2n\pi, (2n + 1)\pi) \Leftrightarrow$ $\theta \in \left(n\pi, n\pi + \frac{\pi}{2}\right)$, n an integer. So f is increasing on $\left[n\pi, n\pi + \frac{\pi}{2}\right]$ and decreasing on $\left[n\pi + \frac{\pi}{2}, (n + 1)\pi\right]$.

(b) Local minima $f(n\pi) = 0$, local maxima $f\left(n\pi + \frac{\pi}{2}\right) = 1$

(c) $f''(\theta) = 2\cos 2\theta > 0 \Leftrightarrow 2\theta \in \left(2n\pi - \frac{\pi}{2}, 2n\pi + \frac{\pi}{2}\right)$ $\Leftrightarrow \theta \in \left(n\pi - \frac{\pi}{4}, n\pi + \frac{\pi}{4}\right)$, so f is CU on these intervals and CD on $\left(n\pi + \frac{\pi}{4}, n\pi + \frac{3\pi}{4}\right)$.

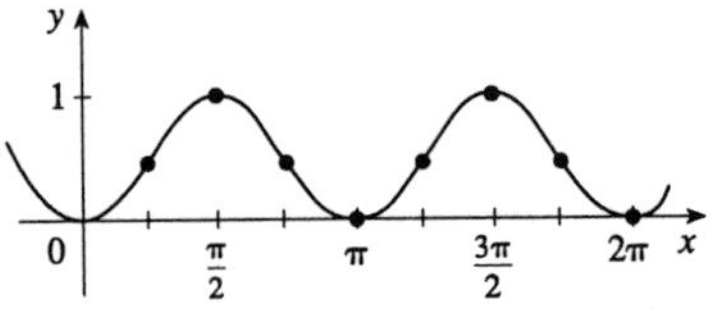

(d) IP at $\theta = n\pi \pm \frac{\pi}{4}$, n an integer

15. $f(x) = 6x^2 - 2x^3 - x^4 \quad \Rightarrow \quad f'(x) = 12x - 6x^2 - 4x^3 \quad \Rightarrow \quad f''(x) = 12 - 12x - 12x^2 = 0 \quad \Leftrightarrow$ $x^2 + x - 1 = 0 \quad \Rightarrow \quad x = \frac{-1 \pm \sqrt{5}}{2}$. For $x < \frac{-1-\sqrt{5}}{2}$, $f''(x) < 0$. For $\frac{-1-\sqrt{5}}{2} < x < \frac{-1+\sqrt{5}}{2}$, $f''(x) > 0$, and if $x > \frac{-1+\sqrt{5}}{2}$ then $f''(x) < 0$. Therefore f is CU on $\left(\frac{-1-\sqrt{5}}{2}, \frac{-1+\sqrt{5}}{2}\right)$.

17. $f(x) = x(1+x)^{-2} \quad \Rightarrow \quad f'(x) = (1+x)^{-2} - 2x(1+x)^{-3} = (1+x)^{-3}(1-x) \quad \Rightarrow$ $f''(x) = -3(1+x)^{-4}(1-x) - (1+x)^{-3} = (1+x)^{-4}(2x-4) > 0 \quad \Leftrightarrow \quad (2x-4) > 0 \quad \Leftrightarrow \quad x > 2$. Therefore f is CU on $(2, \infty)$.

19. (a)

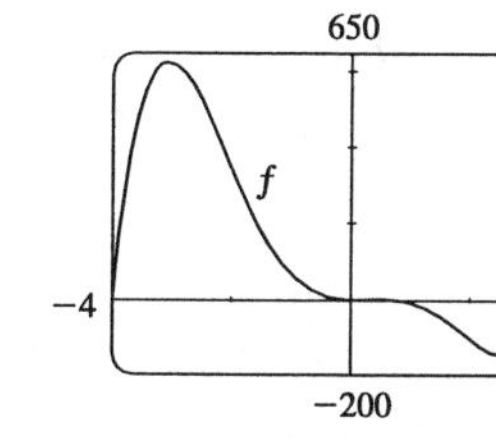

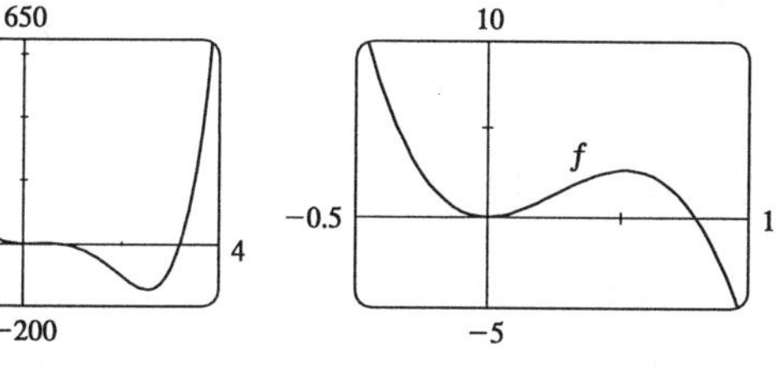

From the graphs of f it seems that f is concave upward on $(-2, 0.25)$ and $(2, \infty)$, and concave downward on $(-\infty, -2)$ and $(0.25, 2)$, with inflection points at about $(-2, 350)$, $(0.25, 1)$, and $(2, -100)$.

(b)

From the graph of f'' it seems that f is CU on $(-2.1, 0.25)$ and $(1.9, \infty)$, and CD on $(-\infty, -2.1)$ and $(0.25, 2)$, with inflection points at about $(-2.1, 386)$, $(0.25, 1.3)$ and $(1.9, -87)$. (We have to check back on the graph of f to find the y-coordinates of the inflection points.)

21. There are many functions which satisfy the given conditions.

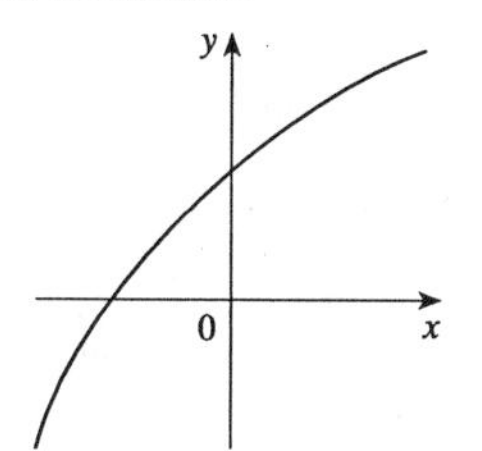

23.

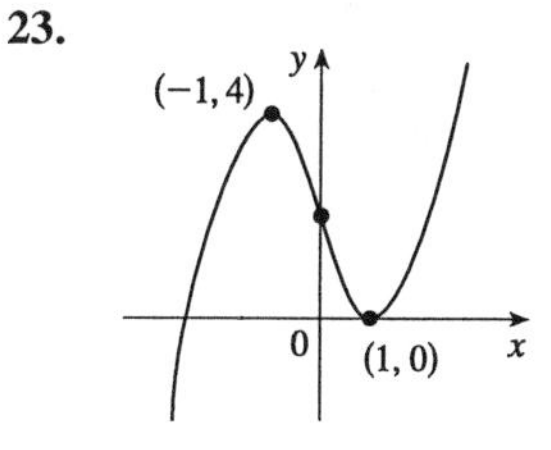

25.

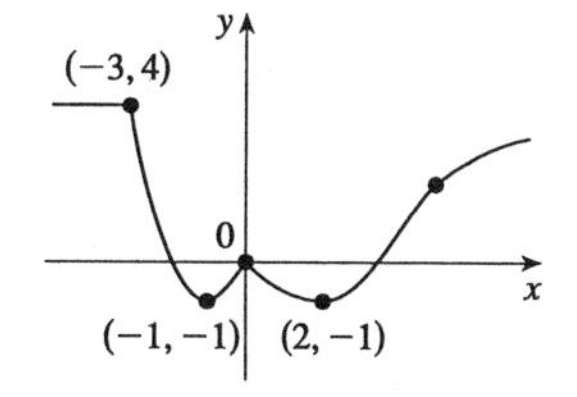

27. (a) f is increasing where f' is positive, that is, on $[0, 2]$, $[4, 6]$, and $[8, \infty)$; and decreasing where f' is negative, that is, on $[2, 4]$ and $[6, 8]$.

(b) f has local maxima where f' changes from positive to negative, at $x = 2$ and at $x = 6$, and local minima where f' changes from negative to positive, at $x = 4$ and at $x = 8$.

(c) f is concave upward where f' is increasing, that is, on $(3, 6)$ and $(6, \infty)$, and concave downward where f' is decreasing, that is, on $(0, 3)$.

(d) There is a point of inflection where f changes from being CD to being CU, that is, at $x = 3$.

(e)

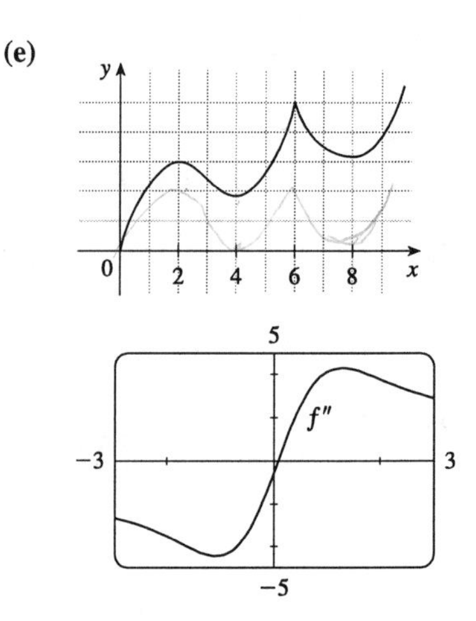

29. In Maple, we define f and then use the command

```
plot(diff(diff(f,x),x),x=-3..3);.
```

In Mathematica, we define f and then use

```
Plot[Dt[Dt[f,x],x],{x,-3,3}].
```

We see that $f'' > 0$ for $x > 0.1$ and $f'' < 0$ for $x < 0.1$. So f is concave up on $(0.1, \infty)$ and concave down on $(-\infty, 0.1)$.

31. By hypothesis $g = f'$ is differentiable on an open interval containing c. Since $(c, f(c))$ is a point of inflection, the concavity changes at $x = c$, so $f'(x)$ changes signs at $x = c$. Hence, by the First Derivative Test, f' has a local extremum at $x = c$. Thus by Fermat's Theorem $f''(c) = 0$.

33. Using the fact that $|x| = \sqrt{x^2}$ (see Exercises 2.5.79-81), we have that $g(x) = x\sqrt{x^2} \quad\Rightarrow$ $g'(x) = \sqrt{x^2} + \sqrt{x^2} = 2\sqrt{x^2} = 2|x| \quad\Rightarrow\quad g''(x) = 2x(x^2)^{-1/2} = \dfrac{2x}{|x|} < 0$ for $x < 0$ and $g''(x) > 0$ for $x > 0$, so $(0, 0)$ is an inflection point. But $g''(0)$ does not exist.

35. If f and g are CU on I, then $f'' > 0$ and $g'' > 0$ on I, so $(f + g)'' = f'' + g'' > 0$ on I $\quad\Rightarrow\quad f + g$ is CU on I.

37. Since f and g are positive, increasing, and CU on I, we have $f > 0$, $f' > 0$, $f'' > 0$, $g > 0$, $g' > 0$, $g'' > 0$ on I. Then $(fg)' = f'g + fg' \quad\Rightarrow\quad (fg)'' = f''g + 2f'g' + fg'' > 0 \quad\Rightarrow\quad fg$ is CU on I.

39. Let the cubic function be $f(x) = ax^3 + bx^2 + cx + d \quad\Rightarrow\quad f'(x) = 3ax^2 + 2bx + c \quad\Rightarrow$ $f''(x) = 6ax + 2b$. So f is CU when $6ax + 2b > 0 \quad\Leftrightarrow\quad x > -\dfrac{b}{3a}$, and CD when $x < -\dfrac{b}{3a}$, and so the only point of inflection occurs when $x = -\dfrac{b}{3a}$. If the graph has three x-intercepts x_1, x_2 and x_3, then the equation of $f(x)$ must factor as

$f(x) = a(x - x_1)(x - x_2)(x - x_3) = a[x^3 - (x_1 + x_2 + x_3)x^2 + (x_1x_2 + x_1x_3 + x_2x_3)x - x_1x_2x_3]$.

So $b = -a(x_1 + x_2 + x_3)$. Hence the x-coordinate of the point of inflection is

$$-\frac{b}{3a} = -\frac{-a(x_1 + x_2 + x_3)}{3a} = \frac{x_1 + x_2 + x_3}{3}.$$

41. There must exist some interval containing c on which f''' is positive, since $f'''(c)$ is positive and f''' is continuous. On this interval, f'' is increasing (since f''' is positive), so $f'' = (f')'$ changes from negative to positive at c. So by the First Derivative Test, f' has a local minimum at $x = c$ and thus cannot change sign there, so f has no extremum at c. But since f'' changes from negative to positive at c, f has a point of inflection at c (it changes from concave down to concave up).

EXERCISES 3.5

1. $\lim\limits_{x\to\infty} \dfrac{1}{x\sqrt{x}} = \lim\limits_{x\to\infty} \dfrac{1}{x^{3/2}} = 0$ by Theorem 4.

3. $$\lim_{x\to\infty} \frac{x+4}{x^2-2x+5} = \lim_{x\to\infty} \frac{\frac{1}{x}+\frac{4}{x^2}}{1-\frac{2}{x}+\frac{5}{x^2}} \overset{(5)}{=} \frac{\lim\limits_{x\to\infty}\left(\frac{1}{x}+\frac{4}{x^2}\right)}{\lim\limits_{x\to\infty}\left(1-\frac{2}{x}+\frac{5}{x^2}\right)} \overset{(1,2,3)}{=} \frac{\lim\limits_{x\to\infty}\frac{1}{x}+4\lim\limits_{x\to\infty}\frac{1}{x^2}}{\lim\limits_{x\to\infty}1-2\lim\limits_{x\to\infty}\frac{1}{x}+5\lim\limits_{x\to\infty}\frac{1}{x^2}}$$
$$= \frac{0+4(0)}{1-2(0)+5(0)} = 0 \text{ by (7) and Theorem 4.}$$

5. $$\lim_{x\to-\infty} \frac{(1-x)(2+x)}{(1+2x)(2-3x)} = \lim_{x\to-\infty} \frac{\left[\frac{1}{x}-1\right]\left[\frac{2}{x}+1\right]}{\left[\frac{1}{x}+2\right]\left[\frac{2}{x}-3\right]} = \frac{\left[\lim\limits_{x\to-\infty}\frac{1}{x}-1\right]\left[\lim\limits_{x\to-\infty}\frac{2}{x}+1\right]}{\left[\lim\limits_{x\to-\infty}\frac{1}{x}+2\right]\left[\lim\limits_{x\to-\infty}\frac{2}{x}-3\right]}$$
$$\overset{(5,4,1,2,7)}{=} \frac{(0-1)(0+1)}{(0+2)(0-3)} = \frac{1}{6}$$

7. $$\lim_{x\to\infty} \frac{1}{3+\sqrt{x}} = \lim_{x\to\infty} \frac{1/\sqrt{x}}{(3/\sqrt{x})+1} \overset{(5,1,3)}{=} \frac{\lim\limits_{x\to\infty}(1/\sqrt{x})}{3\lim\limits_{x\to\infty}(1/\sqrt{x})+\lim\limits_{x\to\infty}1} = \frac{0}{3(0)+1} = 0 \text{ (by Theorem 4 with } r=\tfrac{1}{2}.)$$

Or: Note that $0 < \dfrac{1}{3+\sqrt{x}} < \dfrac{1}{\sqrt{x}}$ and use the Squeeze Theorem.

9. $$\lim_{r\to\infty} \frac{r^4-r^2+1}{r^5+r^3-r} = \lim_{r\to\infty} \frac{\frac{1}{r}-\frac{1}{r^3}+\frac{1}{r^5}}{1+\frac{1}{r^2}-\frac{1}{r^4}} = \frac{\lim\limits_{r\to\infty}\frac{1}{r}-\lim\limits_{r\to\infty}\frac{1}{r^3}+\lim\limits_{r\to\infty}\frac{1}{r^5}}{\lim\limits_{r\to\infty}1+\lim\limits_{r\to\infty}\frac{1}{r^2}-\lim\limits_{r\to\infty}\frac{1}{r^4}} = \frac{0-0+0}{1+0-0} = 0$$

11. $$\lim_{x\to\infty} \frac{\sqrt{1+4x^2}}{4+x} = \lim_{x\to\infty} \frac{\sqrt{(1/x^2)+4}}{(4/x)+1} = \frac{\sqrt{0+4}}{0+1} = 2$$

13. $$\lim_{x\to\infty} \frac{1-\sqrt{x}}{1+\sqrt{x}} = \lim_{x\to\infty} \frac{(1/\sqrt{x})-1}{(1/\sqrt{x})+1} = \frac{0-1}{0+1} = -1$$

15. $$\lim_{x\to\infty}\left(\sqrt{x^2+1}-\sqrt{x^2-1}\right) = \lim_{x\to\infty}\left(\sqrt{x^2+1}-\sqrt{x^2-1}\right)\frac{\sqrt{x^2+1}+\sqrt{x^2-1}}{\sqrt{x^2+1}+\sqrt{x^2-1}}$$
$$= \lim_{x\to\infty} \frac{(x^2+1)-(x^2-1)}{\sqrt{x^2+1}+\sqrt{x^2-1}} = \lim_{x\to\infty} \frac{2}{\sqrt{x^2+1}+\sqrt{x^2-1}}$$
$$= \lim_{x\to\infty} \frac{2/x}{\sqrt{1+(1/x^2)}+\sqrt{1-(1/x^2)}} = \frac{0}{\sqrt{1+0}+\sqrt{1-0}} = 0$$

17. $$\lim_{x\to\infty}\left(\sqrt{1+x}-\sqrt{x}\right) = \lim_{x\to\infty}\left(\sqrt{1+x}-\sqrt{x}\right)\left(\frac{\sqrt{1+x}+\sqrt{x}}{\sqrt{1+x}+\sqrt{x}}\right) = \lim_{x\to\infty} \frac{(1+x)-x}{\sqrt{1+x}+\sqrt{x}}$$
$$= \lim_{x\to\infty} \frac{1}{\sqrt{1+x}+\sqrt{x}} = \lim_{x\to\infty} \frac{1/\sqrt{x}}{\sqrt{(1/x)+1}+1} = \frac{0}{\sqrt{0+1}+1} = 0$$

19. $\lim_{x\to-\infty}\left(\sqrt{x^2+x+1}+x\right)=\lim_{x\to-\infty}\left(\sqrt{x^2+x+1}+x\right)\left[\frac{\sqrt{x^2+x+1}-x}{\sqrt{x^2+x+1}-x}\right]$

$=\lim_{x\to-\infty}\frac{x+1}{\left(\sqrt{x^2+x+1}-x\right)}=\lim_{x\to-\infty}\frac{1+(1/x)}{-\sqrt{1+(1/x)+(1/x^2)}-1}=\frac{1+0}{-\sqrt{1+0+0}-1}=-\frac{1}{2}$

21. $\sqrt{x}$ is large when x is large, so $\lim_{x\to\infty}\sqrt{x}=\infty$.

23. $\lim_{x\to\infty}(x-\sqrt{x})=\lim_{x\to\infty}\sqrt{x}(\sqrt{x}-1)=\infty$ since $\sqrt{x}\to\infty$ and $\sqrt{x}-1\to\infty$ as $x\to\infty$.

25. $\lim_{x\to-\infty}(x^3-5x^2)=-\infty$ since $x^3\to-\infty$ and $-5x^2\to-\infty$ as $x\to-\infty$.

Or: $\lim_{x\to-\infty}(x^3-5x^2)=\lim_{x\to-\infty}x^2(x-5)=-\infty$ since $x^2\to\infty$ and $x-5\to-\infty$.

27. $\lim_{x\to\infty}\frac{x^7-1}{x^6-1}=\lim_{x\to\infty}\frac{1-1/x^7}{(1/x)-(1/x^7)}=\infty$ since $1-\frac{1}{x^7}\to1$ while $\frac{1}{x}-\frac{1}{x^7}\to0^+$ as $x\to\infty$.

Or: Divide numerator and denominator by x^6 instead of x^7.

29. $\lim_{x\to\infty}\frac{\sqrt{x}+3}{x+3}=\lim_{x\to\infty}\frac{(1/\sqrt{x})+(3/x)}{1+3/x}=\frac{0+0}{1+0}=0$

31. If $t=1/x$ then $\lim_{x\to\infty}\cos(1/x)=\lim_{t\to0^+}\cos t=\cos 0=1$.

33. If $f(x)=x^2/2^x$, then a calculator gives $f(0)=0$, $f(1)=0.5$, $f(2)=1$, $f(3)=1.125$, $f(4)=1$, $f(5)=0.78125$, $f(6)=0.5625$, $f(7)=0.3828125$, $f(8)=0.25$, $f(9)=0.158203125$, $f(10)=0.09765625$, $f(20)\approx0.00038147$, $f(50)\approx2.2204\times10^{-12}$, $f(100)\approx7.8886\times10^{-27}$. It appears that $\lim_{x\to\infty}(x^2/2^x)=0$.

35. $\lim_{x\to\pm\infty}\frac{x}{x+4}=\lim_{x\to\pm\infty}\frac{1}{1+4/x}=\frac{1}{1+0}=1$, so $y=1$ is a horizontal asymptote.

$\lim_{x\to-4^-}\frac{x}{x+4}=\infty$ and $\lim_{x\to-4^+}\frac{x}{x+4}=-\infty$, so $x=-4$ is a vertical asymptote. The graph confirms these calculations.

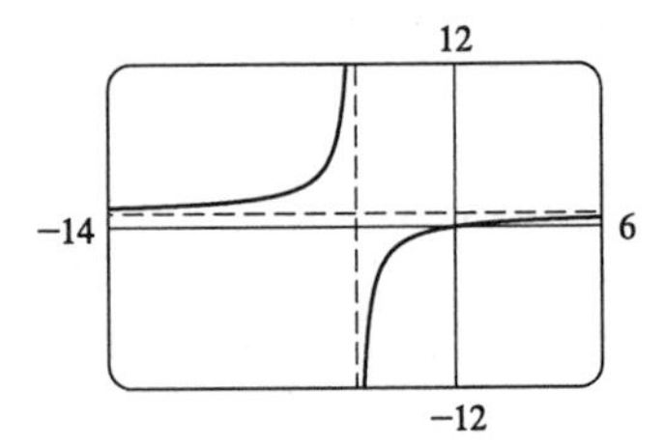

37. $\lim_{x\to\pm\infty}\frac{x^3}{x^2+3x-10}=\lim_{x\to\pm\infty}\frac{x}{1+(3/x)-(10/x^2)}=\pm\infty$, so there is no horizontal asymptote. $\lim_{x\to2^+}\frac{x^3}{x^2+3x-10}=\lim_{x\to2^+}\frac{x^3}{(x+5)(x-2)}=\infty$, since $\frac{x^3}{(x+5)(x-2)}>0$ for $x>2$. Similarly, $\lim_{x\to2^-}\frac{x^3}{x^2+3x-10}=-\infty$ and $\lim_{x\to-5^-}\frac{x^3}{x^2+3x-10}=-\infty$, $\lim_{x\to-5^+}\frac{x^3}{x^2+3x-10}=\infty$, so $x=2$ and $x=-5$ are vertical asymptotes. The graph confirms these calculations.

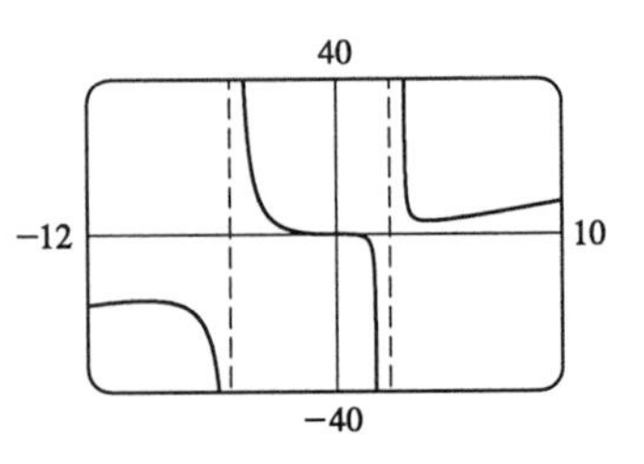

39. $\lim_{x\to\infty} \frac{x}{\sqrt[4]{x^4+1}} = \lim_{x\to\infty} \frac{1}{\sqrt[4]{1+(1/x^4)}} = \frac{1}{\sqrt[4]{1+0}} = 1$ and

$\lim_{x\to-\infty} \frac{x}{\sqrt[4]{x^4+1}} = \lim_{x\to-\infty} \frac{1}{-\sqrt[4]{1+(1/x^4)}} = \frac{1}{-\sqrt[4]{1+0}} = -1,$

so $y = \pm 1$ are horizontal asymptotes. There are no vertical asymptotes.

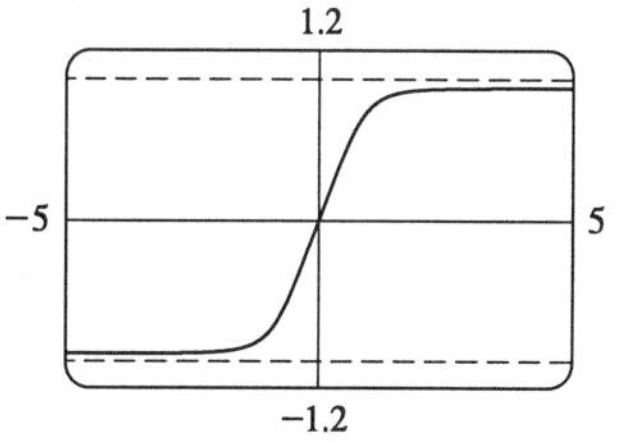

41. $\lim_{x\to 4^+} \frac{4}{x-4} = \infty$ and $\lim_{x\to 4^-} \frac{4}{x-4} = -\infty$, so $x = 4$ is a vertical asymptote. $\lim_{x\to\pm\infty} \frac{4}{x-4} = \lim_{x\to\pm\infty} \frac{4/x}{1-4/x} = 0$, so $y = 0$ is a horizontal asymptote. $y' = -\frac{4}{(x-4)^2} < 0$ $(x \neq 4)$ so y is decreasing on $(-\infty, 4)$ and $(4, \infty)$. $y'' = \frac{8}{(x-4)^3} > 0$ for $x > 4$, so y is CU on $(4, \infty)$ and CD on $(-\infty, 4)$.

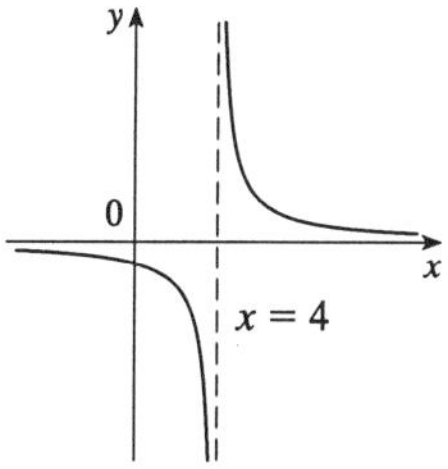

43. $\lim_{x\to\pm\infty} \frac{x}{x^2+1} = \lim_{x\to\pm\infty} \frac{1/x}{1+1/x^2} = \frac{0}{1+0} = 0$, so $y = 0$ is a horizontal asymptote. $y' = \frac{x^2+1-x(2x)}{(x^2+1)^2} = \frac{1-x^2}{(x^2+1)^2} = 0$ when $x = \pm 1$ and $y' > 0 \Leftrightarrow x^2 < 1 \Leftrightarrow -1 < x < 1$, so y is increasing on $[-1, 1]$ and decreasing on $(-\infty, -1]$ and $[1, \infty)$.

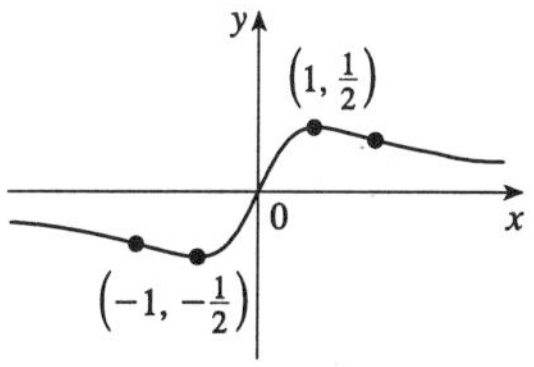

$y'' = \frac{(1+x^2)^2(-2x) - (1-x^2)\,2\,(x^2+1)\,2x}{(1+x^2)^4} = \frac{2x\,(x^2-3)}{(1+x^2)^3} > 0 \Leftrightarrow x > \sqrt{3}$ or $-\sqrt{3} < x < 0$, so y is CU on $(\sqrt{3}, \infty)$ and $(-\sqrt{3}, 0)$ and CD on $(-\infty, -\sqrt{3})$ and $(0, \sqrt{3})$.

45. $\lim_{x\to\pm\infty} \left(1 - 1/\sqrt{x^2+1}\right) = 1 - 0 = 1$, so $y = 1$ is a horizontal asymptote.

$y' = -\left(-\tfrac{1}{2}\right)(x^2+1)^{-3/2}(2x) = \frac{x}{(x^2+1)^{3/2}} > 0 \Leftrightarrow$

$x > 0$, so y is increasing on $[0, \infty)$ and decreasing on $(-\infty, 0]$.

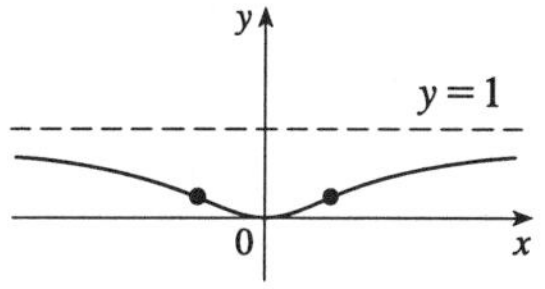

$y'' = \frac{(x^2+1)^{3/2} - x\left(\frac{3}{2}\right)(x^2+1)^{1/2}(2x)}{(x^2+1)^3} = \frac{1-2x^2}{(x^2+1)^{5/2}} > 0 \Leftrightarrow x^2 < \frac{1}{2} \Leftrightarrow |x| < \frac{1}{\sqrt{2}}$, so y is CU on $\left(-\frac{1}{\sqrt{2}}, \frac{1}{\sqrt{2}}\right)$, and CD on $\left(-\infty, -\frac{1}{\sqrt{2}}\right)$ and $\left(\frac{1}{\sqrt{2}}, \infty\right)$.

47. **(a)** If $t = \frac{1}{x}$ then

$$\lim_{x\to\infty} x \sin\frac{1}{x} = \lim_{t\to 0^+} \frac{1}{t}\sin t = \lim_{t\to 0^+} \frac{\sin t}{t} = 1$$

(b)

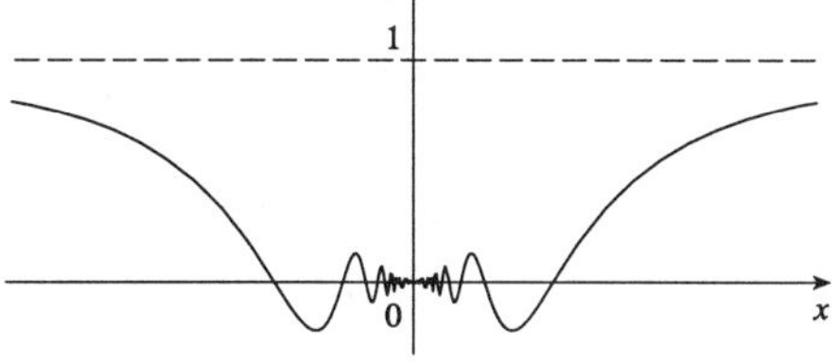

49. $y = f(x) = x^2(x-2)(1-x)$. The y-intercept is $f(0) = 0$, and the x-intercepts occur when $y = 0 \Rightarrow x = 0, 1, 2$. Notice (as in Example 9) that, since x^2 is always positive, the graph does not cross the x-axis at 0, but does cross the x-axis at 1 and 2.
$\lim_{x\to\infty} x^2(x-2)(1-x) = -\infty$, since the first two factors are large positive and the third large negative when x is large positive.
$\lim_{x\to-\infty} x^2(x-2)(1-x) = -\infty$ because the first and third factors are large positive and the second large negative as $x \to -\infty$.

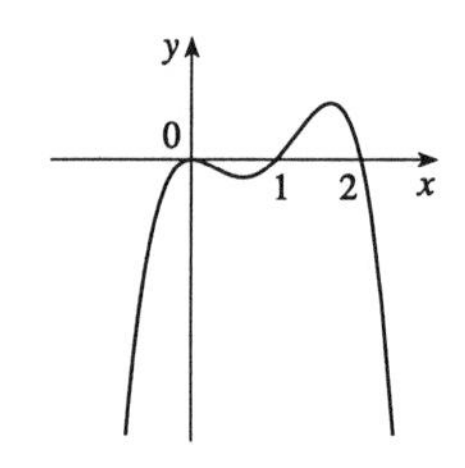

51. $y = f(x) = (x+4)^5(x-3)^4$. The y-intercept is $f(0) = 4^5(-3)^4 = 82{,}944$. The x-intercepts occur when $y = 0 \Rightarrow x = -4, 3$. Notice (as in Example 9) that the graph does not cross the x-axis at 3 because $(x-3)^4$ is always positive, but does cross the x-axis at -4. $\lim_{x\to\infty}(x+4)^5(x-3)^4 = \infty$ since both factors are large positive when x is large positive.
$\lim_{x\to-\infty}(x+4)^5(x-3)^4 = -\infty$ since the first factor is large negative and the second factor is large positive when x is large negative.

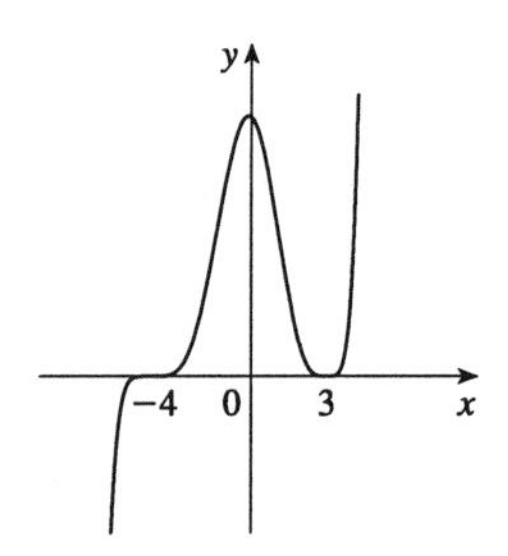

53. First we plot the points which are known to be on the graph: $(2, -1)$ and $(0, 0)$. We can also draw a short line segment of slope 0 at $x = 2$, since we are given that $f'(2) = 0$. Now we know that $f'(x) < 0$ (that is, the function is decreasing) on $(0, 2)$, and that $f''(x) < 0$ on $(0, 1)$ and $f''(x) > 0$ on $(1, 2)$. So we must join the points $(0, 0)$ and $(2, -1)$ in such a way that the curve is concave down on $(0, 1)$ and concave up on $(1, 2)$. The curve must be concave up and increasing on $(2, 4)$, and concave down and increasing on $(4, \infty)$. Now we just need to reflect the curve in the y-axis, since we are given that f is an even function. The diagram shows one possible function satisfying all of the given conditions.

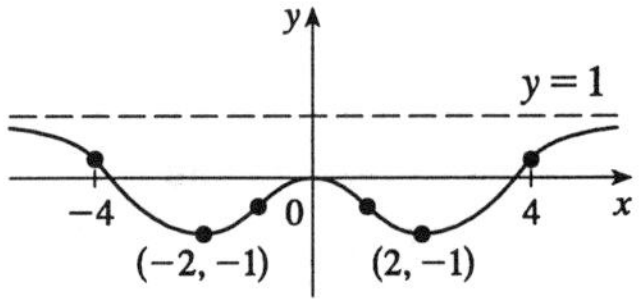

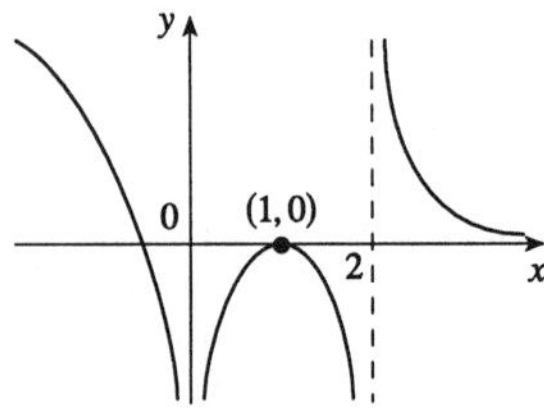

55. We are given that $f(1) = f'(1) = 0$. So we can draw a short horizontal line at the point $(1, 0)$ to represent this situation. We are given that $x = 0$ and $x = 2$ are vertical asymptotes, with $\lim_{x\to 0} f(x) = -\infty$, $\lim_{x\to 2^+} f(x) = \infty$ and $\lim_{x\to 2^-} f(x) = -\infty$, so we can draw the parts of the curve which approach these asymptotes.
On the interval $(-\infty, 0)$, the graph is concave down, and $f(x) \to \infty$ as $x \to -\infty$. Between the asymptotes the graph is concave down. On the interval $(2, \infty)$ the graph is concave up, and $f(x) \to 0$ as $x \to \infty$, so $y = 0$ is a horizontal asymptote.
The diagram shows one possible function satisfying all of the given conditions.

57. Divide numerator and denominator by the highest power of x in $Q(x)$.

(a) If $\deg(P) < \deg(Q)$, then numerator $\to 0$ but denominator doesn't. So $\lim\limits_{x\to\infty}[P(x)/Q(x)] = 0$.

(b) If $\deg(P) > \deg(Q)$, then numerator $\to \pm\infty$ but denominator doesn't, so $\lim\limits_{x\to\infty}[P(x)/Q(x)] = \pm\infty$ (depending on the ratio of the leading coefficients of P and Q.)

59. $\lim\limits_{x\to\infty}\dfrac{4x-1}{x} = \lim\limits_{x\to\infty}\left(4-\dfrac{1}{x}\right) = 4$, and $\lim\limits_{x\to\infty}\dfrac{4x^2+3x}{x^2} = \lim\limits_{x\to\infty}\left(4+\dfrac{3}{x}\right) = 4$. Therefore by the Squeeze Theorem, $\lim\limits_{x\to\infty} f(x) = 4$.

61. $\left|\dfrac{6x^2+5x-3}{2x^2-1} - 3\right| < 0.2 \quad\Leftrightarrow\quad 2.8 < \dfrac{6x^2+5x-3}{2x^2-1} < 3.2$. So we graph the three parts of this inequality on the same screen, and find that the curve $y = \dfrac{6x^2+5x-3}{2x^2-1}$ seems to lie between the lines $y = 2.8$ and $y = 3.2$ whenever $x > 12.8$. So we can choose $N = 13$ (or any larger number), so that the inequality holds whenever $x \geq N$.

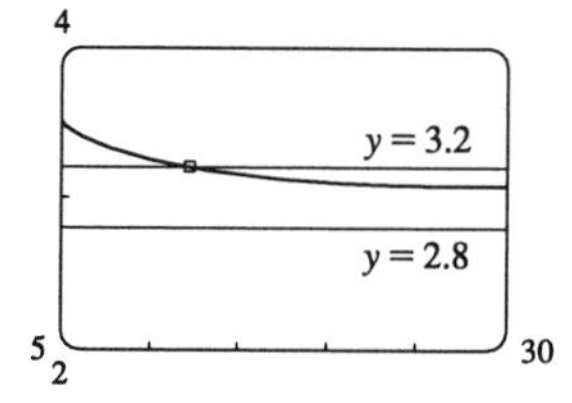

63. For $\epsilon = 0.5$, we need to find N such that

$$\left|\frac{\sqrt{4x^2+1}}{x+1} - (-2)\right| < 0.5 \quad\Leftrightarrow\quad -2.5 < \frac{\sqrt{4x^2+1}}{x+1} < -1.5$$

whenever $x \leq N$. We graph the three parts of this inequality on the same screen, and see that the inequality holds for $x \leq -6$. So we choose $N = -6$ (or any smaller number).

For $\epsilon = 0.1$, we need $-2.1 < \dfrac{\sqrt{4x^2+1}}{x+1} < -1.9$ whenever $x \leq N$. From the graph, it seems that this inequality holds for $x \leq -22$. So we choose any $N = -22$ (or any smaller number).

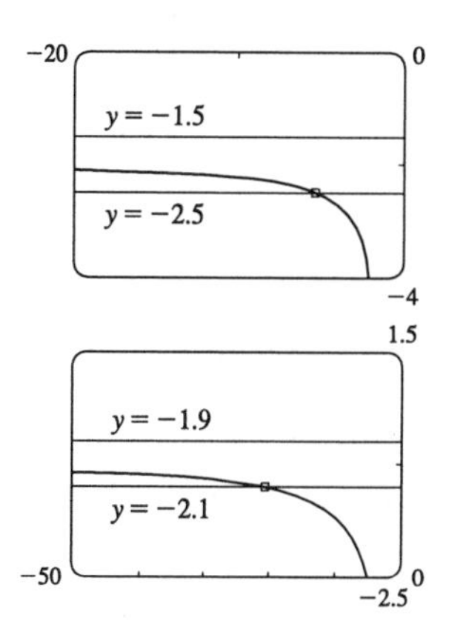

65. **(a)** $1/x^2 < 0.0001 \quad\Leftrightarrow\quad x^2 > 1/0.0001 = 10{,}000 \quad\Leftrightarrow\quad x > 100 \quad (x > 0)$

(b) If $\epsilon > 0$ is given, then $1/x^2 < \epsilon \quad\Leftrightarrow\quad x^2 > 1/\epsilon \quad\Leftrightarrow\quad x > 1/\sqrt{\epsilon}$. Let $N = 1/\sqrt{\epsilon}$. Then $x > N \quad\Rightarrow\quad x > 1/\sqrt{\epsilon} \quad\Rightarrow\quad |1/x^2 - 0| = 1/x^2 < \epsilon$, so $\lim\limits_{x\to\infty}(1/x^2) = 0$.

67. For $x < 0$, $|1/x - 0| = -1/x$. If $\epsilon > 0$ is given, then $-1/x < \epsilon \quad\Leftrightarrow\quad x < -1/\epsilon$. Take $N = -\dfrac{1}{\epsilon}$. Then $x < N \quad\Rightarrow\quad x < -\dfrac{1}{\epsilon} \quad\Rightarrow\quad \left|\dfrac{1}{x} - 0\right| = -\dfrac{1}{x} < \epsilon$, so $\lim\limits_{x\to-\infty}\dfrac{1}{x} = 0$.

69. Suppose that $\lim\limits_{x\to\infty} f(x) = L$ and let $\epsilon > 0$ be given. Then there exists $N > 0$ such that $x > N \quad\Rightarrow\quad |f(x) - L| < \epsilon$. Let $\delta = 1/N$. Then $0 < t < \delta \quad\Rightarrow\quad t < 1/N \quad\Rightarrow\quad 1/t > N \quad\Rightarrow\quad |f(1/t) - L| < \epsilon$. So $\lim\limits_{t\to0^+} f(1/t) = L = \lim\limits_{x\to\infty} f(x)$. Now suppose that $\lim\limits_{x\to-\infty} f(x) = L$ and let $\epsilon > 0$ be given. Then there exists $N < 0$ such that $x < N \quad\Rightarrow\quad |f(x) - L| < \epsilon$. Let $\delta = -1/N$. Then $-\delta < t < 0 \quad\Rightarrow\quad t > \dfrac{1}{N} \quad\Rightarrow\quad \dfrac{1}{t} < N \quad\Rightarrow\quad \left|f\left(\dfrac{1}{t}\right) - L\right| < \epsilon$. So $\lim\limits_{x\to0^-} f\left(\dfrac{1}{t}\right) = L = \lim\limits_{x\to-\infty} f(x)$.

EXERCISES 3.6

Abbreviations:	**D**	**the domain of f**	**VA**	**vertical asymptote(s)**
	HA	**horizontal asymptote**	**IP**	**inflection point(s)**
	CU	**concave up**	**CD**	**concave down**

1. $y = f(x) = 1 - 3x + 5x^2 - x^3$ **A.** $D = \mathbb{R}$ **B.** y-intercept $= f(0) = 1$ **C.** No symmetry **D.** No asymptotes **E.** $f'(x) = -3 + 10x - 3x^2 = -(3x-1)(x-3) > 0 \Leftrightarrow (3x-1)(x-3) < 0 \Leftrightarrow \frac{1}{3} < x < 3$. $f'(x) < 0 \Leftrightarrow x < \frac{1}{3}$ or $x > 3$. So f is increasing on $\left[\frac{1}{3}, 3\right]$ and decreasing on $\left(-\infty, \frac{1}{3}\right]$ and $[3, \infty)$. **F.** The critical numbers occur when $f'(x) = -(3x-1)(x-3) = 0 \Leftrightarrow x = \frac{1}{3}, 3$. The local minimum is $f\left(\frac{1}{3}\right) = \frac{14}{27}$ and the local maximum is $f(3) = 10$. **G.** $f''(x) = 10 - 6x > 0 \Leftrightarrow x < \frac{5}{3}$, so f is CU on $\left(-\infty, \frac{5}{3}\right)$ and CD on $\left(\frac{5}{3}, \infty\right)$. IP $\left(\frac{5}{3}, \frac{142}{27}\right)$ **H.**

3. $y = f(x) = x^4 - 6x^2$ **A.** $D = \mathbb{R}$ **B.** y-intercept $= f(0) = 0$, x-intercepts occur when $f(x) = 0 \Rightarrow x^4 - 6x^2 = 0 \Leftrightarrow x^2(x^2 - 6) = 0 \Leftrightarrow x = 0, \pm\sqrt{6}$. **C.** Since $f(-x) = (-x)^4 - 6(-x^2) = x^4 - 6x^2 = f(x)$, f is an even function and its graph is symmetric about the y-axis. **D.** No asymptotes. **E.** $f'(x) = 4x^3 - 12x = 4x(x^2 - 3) = 0$ when $x = 0, \pm\sqrt{3}$. $f'(x) < 0$ for $x < -\sqrt{3}$ and $0 < x < \sqrt{3}$. $f'(x) > 0$ for $-\sqrt{3} < x < 0$ and $x > \sqrt{3}$, so that f is increasing on $\left[-\sqrt{3}, 0\right]$ and $\left[\sqrt{3}, \infty\right)$ and decreasing on $\left(-\infty, -\sqrt{3}\right]$ and $\left[0, \sqrt{3}\right]$. **F.** Local minima $f\left(\pm\sqrt{3}\right) = -9$, local maximum $f(0) = 0$. **G.** $f''(x) = 12x^2 - 12 = 12(x^2 - 1) > 0 \Leftrightarrow x^2 > 1 \Leftrightarrow |x| > 1 \Leftrightarrow x > 1$ or $x < -1$, so f is CU on $(-\infty, -1)$, $(1, \infty)$ and CD on $(-1, 1)$. IP $(1, -5)$ and $(-1, -5)$. **H.**

5. $y = f(x) = x/(x-1)$ **A.** $D = \{x \mid x \neq 1\} = (-\infty, 1) \cup (1, \infty)$ **B.** x-intercept $= 0$, y-intercept $= f(0) = 0$ **C.** No symmetry **D.** $\displaystyle\lim_{x \to \pm\infty} \frac{x}{x-1} = 1$, so $y = 1$ is a HA. $\displaystyle\lim_{x \to 1^-} \frac{x}{x-1} = -\infty$, $\displaystyle\lim_{x \to 1^+} \frac{x}{x-1} = \infty$, so $x = 1$ is a VA. **E.** $f'(x) = \dfrac{(x-1) - x}{(x-1)^2} = \dfrac{-1}{(x-1)^2} < 0$ for $x \neq 1$, so f is decreasing on $(-\infty, 1)$ and $(1, \infty)$. **F.** No extremum **G.** $f''(x) = \dfrac{2}{(x-1)^3} > 0 \Leftrightarrow x > 1$, so f is CU on $(1, \infty)$ and CD on $(-\infty, 1)$. No IP **H.**

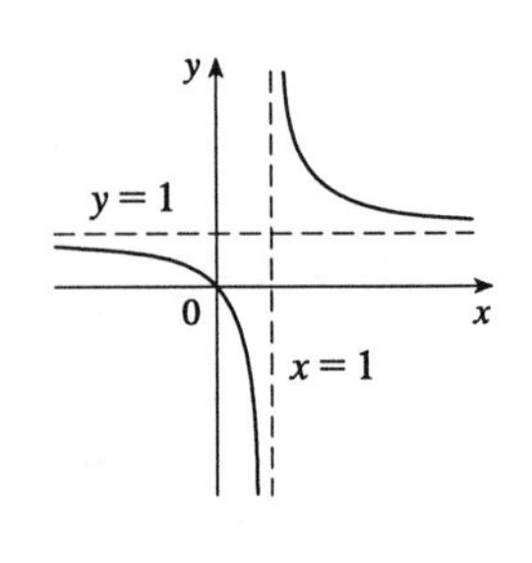

7. $y = f(x) = 1/(x^2 - 9)$ **A.** $D = \{x \mid x \neq \pm 3\} = (-\infty, -3) \cup (-3, 3) \cup (3, \infty)$ **B.** y-intercept $= f(0) = -\frac{1}{9}$, no x-intercept **C.** $f(-x) = f(x) \Rightarrow f$ is even; the curve is symmetric about the y-axis. **D.** $\lim_{x \to \pm\infty} \frac{1}{x^2 - 9} = 0$, so $y = 0$ is a HA. $\lim_{x \to 3^-} \frac{1}{x^2 - 9} = -\infty$, $\lim_{x \to 3^+} \frac{1}{x^2 - 9} = \infty$, $\lim_{x \to -3^-} \frac{1}{x^2 - 9} = \infty$, $\lim_{x \to -3^+} \frac{1}{x^2 - 9} = -\infty$, so $x = 3$ and $x = -3$ are VA. **E.** $f'(x) = -\frac{2x}{(x^2 - 9)^2} > 0 \Leftrightarrow x < 0$ $(x \neq -3)$ so f is increasing on $(-\infty, -3)$ and $(-3, 0]$ and decreasing on $[0, 3)$ and $(3, \infty)$.

F. Local maximum $f(0) = -\frac{1}{9}$.

G. $y'' = \frac{-2(x^2 - 9)^2 + (2x)\,2\,(x^2 - 9)\,(2x)}{(x^2 - 9)^4}$

$= \frac{6(x^2 + 3)}{(x^2 - 9)^3} > 0 \Leftrightarrow x^2 > 9 \Leftrightarrow$

$x > 3$ or $x < -3$, so f is CU on $(-\infty, -3)$ and $(3, \infty)$ and CD on $(-3, 3)$. No IP

H.

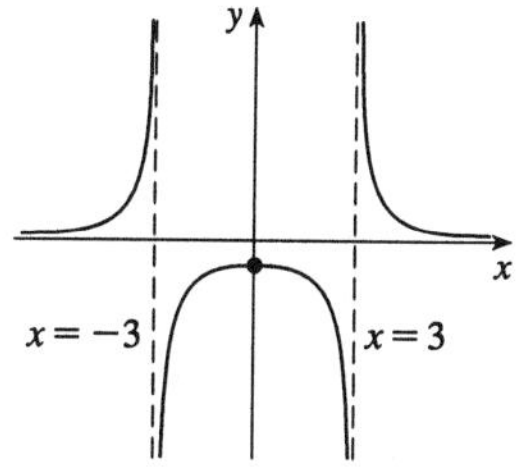

9. $y = f(x) = \frac{1}{(x - 1)(x + 2)}$ **A.** $D = \{x \mid x \neq 1, -2\} = (-\infty, -2) \cup (-2, 1) \cup (1, \infty)$ **B.** No x-intercept, y-intercept $= f(0) = -\frac{1}{2}$ **C.** No symmetry **D.** $\lim_{x \to \pm\infty} \frac{1}{(x - 1)(x + 2)} = 0$, so $y = 0$ is a HA.

$\lim_{x \to 1^-} \frac{1}{(x - 1)(x + 2)} = -\infty$, $\lim_{x \to 1^+} \frac{1}{(x - 1)(x + 2)} = \infty$, $\lim_{x \to -2^-} \frac{1}{(x - 1)(x + 2)} = \infty$,

$\lim_{x \to -2^+} \frac{1}{(x - 1)(x + 2)} = -\infty$. So $x = 1$ and $x = -2$ are VA.

E. $f'(x) = -\frac{2x + 1}{[(x - 1)(x + 2)]^2} \Rightarrow f'(x) > 0 \Leftrightarrow$ $x < -\frac{1}{2}$ $(x \neq -2)$, so f is increasing on $(-\infty, -2)$ and $\left(-2, -\frac{1}{2}\right]$ and decreasing on $\left[-\frac{1}{2}, 1\right)$ and $(1, \infty)$. **F.** $f\left(-\frac{1}{2}\right) = -\frac{4}{9}$ is a local maximum. **G.** $f''(x) = \frac{6(x^2 + x + 1)}{[(x - 1)(x + 2)]^3}$. Now $x^2 + x + 1 > 0$ for all x, so $f''(x) > 0 \Leftrightarrow (x - 1)(x + 2) > 0 \Leftrightarrow x < -2$ or $x > 1$. Thus f is CU on $(-\infty, -2)$ and $(1, \infty)$ and CD on $(-2, 1)$. No IP

H.

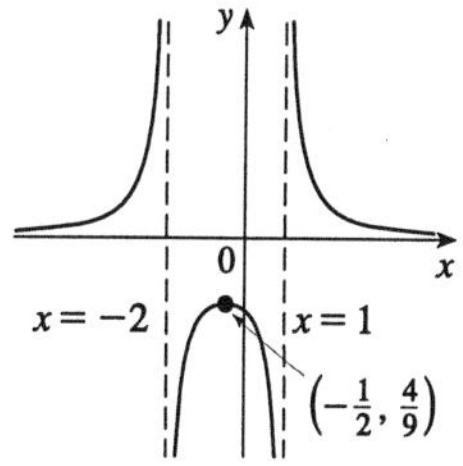

11. $y = f(x) = \frac{1 + x^2}{1 - x^2} = -1 + \frac{2}{1 - x^2}$ **A.** $D = \{x \mid x \neq \pm 1\}$ **B.** No x-intercept, y-intercept $= f(0) = 1$

C. $f(-x) = f(x)$, so f is even and the curve is symmetric about the y-axis.

D. $\lim_{x \to \pm\infty} \frac{1 + x^2}{1 - x^2} = \lim_{x \to \pm\infty} \frac{(1/x^2) + 1}{(1/x^2) - 1} = -1$, so $y = -1$ is a HA. $\lim_{x \to 1^-} \frac{1 + x^2}{1 - x^2} = \infty$, $\lim_{x \to 1^+} \frac{1 + x^2}{1 - x^2} = -\infty$,

$\lim_{x \to -1^-} \frac{1 + x^2}{1 - x^2} = -\infty$, $\lim_{x \to -1^+} \frac{1 + x^2}{1 - x^2} = \infty$. So $x = 1$ and $x = -1$ are VA.

E. $f'(x) = \dfrac{4x}{(1-x^2)^2} > 0 \quad \Leftrightarrow \quad x > 0 \ (x \neq 1)$, so f increases on $[0,1)$, $(1,\infty)$ and decreases on $(-\infty,-1)$, $(-1,0]$.

F. $f(0) = 1$ is a local minimum.

G. $y'' = \dfrac{4(1-x^2)^2 - 4x \cdot 2(1-x^2)(-2x)}{(1-x^2)^4} = \dfrac{4(1+3x^2)}{(1-x^2)^3} > 0 \quad \Leftrightarrow$

$x^2 < 1 \quad \Leftrightarrow \quad -1 < x < 1$, so f is CU on $(-1,1)$ and CD on $(-\infty,-1)$ and $(1,\infty)$. No IP

H.

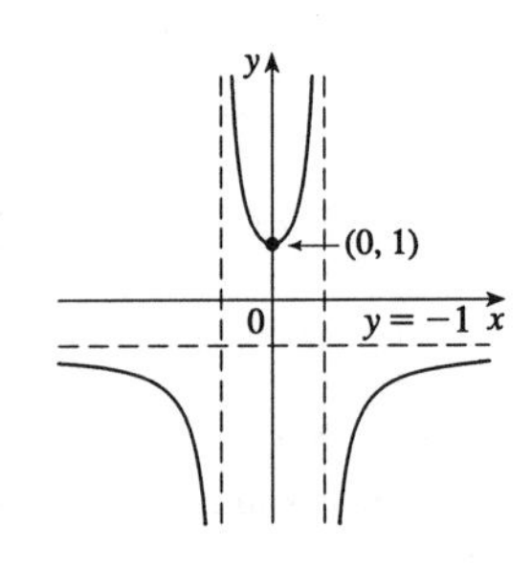

13. $y = f(x) = \dfrac{1}{x^3 - x} = \dfrac{1}{x(x-1)(x+1)}$ **A.** $D = \{x \mid x \neq 0, \pm 1\}$ **B.** No intercepts **C.** $f(-x) = -f(x)$, symmetric about $(0,0)$ **D.** $\lim\limits_{x\to\pm\infty} \dfrac{1}{x^3 - x} = 0$, so $y = 0$ is a HA. $\lim\limits_{x\to 0^-} \dfrac{1}{x^3-x} = \infty$, $\lim\limits_{x\to 0^+} \dfrac{1}{x^3-x} = -\infty$, $\lim\limits_{x\to 1^-} \dfrac{1}{x^3-x} = -\infty$, $\lim\limits_{x\to 1^+} \dfrac{1}{x^3-x} = \infty$, $\lim\limits_{x\to -1^-} \dfrac{1}{x^3-x} = -\infty$, $\lim\limits_{x\to -1^+} \dfrac{1}{x^3-x} = \infty$. So $x = 0$, $x = 1$, and $x = -1$ are VA. **E.** $f'(x) = \dfrac{1 - 3x^2}{(x^3-x)^2} \quad \Rightarrow \quad f'(x) > 0 \quad \Leftrightarrow \quad x^2 < \frac{1}{3} \quad \Leftrightarrow \quad -\frac{1}{\sqrt{3}} < x < \frac{1}{\sqrt{3}} \ (x \neq 0)$, so f is increasing on $\left[-\frac{1}{\sqrt{3}}, 0\right)$, $\left(0, \frac{1}{\sqrt{3}}\right]$ and decreasing on $(-\infty,-1)$, $\left(-1, -\frac{1}{\sqrt{3}}\right]$, $\left[\frac{1}{\sqrt{3}}, 1\right)$, and $(1,\infty)$.

F. Local minimum $f\left(-\frac{1}{\sqrt{3}}\right) = \frac{3\sqrt{3}}{2}$, local maximum $f\left(\frac{1}{\sqrt{3}}\right) = -\frac{3\sqrt{3}}{2}$ **G.** $f''(x) = \dfrac{2(6x^4 - 3x^2 + 1)}{(x^3-x)^3}$.

Since $6x^4 - 3x^2 + 1$ has negative discriminant as a quadratic in x^2, it is positive, so $f''(x) > 0 \quad \Leftrightarrow \quad x^3 - x > 0 \quad \Leftrightarrow \quad x > 1$ or $-1 < x < 0$. f is CU on $(-1,0)$ and $(1,\infty)$, and CD on $(-\infty,-1)$ and $(0,1)$. No IP

H.

15. $y = f(x) = x\sqrt{x+3}$ **A.** $D = \{x \mid x \geq -3\} = [-3,\infty)$ **B.** x-intercepts $0, -3$, y-intercept $= f(0) = 0$ **C.** No symmetry **D.** $\lim\limits_{x\to\infty} \sqrt{x+3} = \infty$, no asymptotes **E.** $f'(x) = \sqrt{x+3} + \dfrac{x}{2\sqrt{x+3}} = \dfrac{3(x+2)}{2\sqrt{x+3}} > 0$ $\Leftrightarrow \quad x > -2$ and $f'(x) < 0 \quad \Leftrightarrow \quad -3 < x < -2$.

So f is increasing on $[-2,\infty)$, decreasing on $[-3,-2]$.

F. $f(-2) = -2$ is a local minimum.

G. $f''(x) = \dfrac{6\sqrt{x+3} - 3(x+2)\left(1/\sqrt{x+3}\right)}{4(x+3)} = \dfrac{3(x+4)}{4(x+3)^{3/2}} > 0$

for all $x > -3$, so f is CU on $(-3,\infty)$.

H.

17. $y = f(x) = \sqrt{x^2+1} - x$ **A.** $D = \mathbb{R}$ **B.** No x-intercept, y-intercept $= 1$ **C.** No symmetry

D. $\lim\limits_{x\to-\infty}\left(\sqrt{x^2+1}-x\right) = \infty$ and $\lim\limits_{x\to\infty}\left(\sqrt{x^2+1}-x\right) = \lim\limits_{x\to\infty}\left(\sqrt{x^2+1}-x\right)\dfrac{\sqrt{x^2+1}+x}{\sqrt{x^2+1}+x}$

$= \lim\limits_{x\to\infty}\dfrac{1}{\sqrt{x^2+1}+x} = 0$, so $y = 0$ is a HA.

E. $f'(x) = \dfrac{x}{\sqrt{x^2+1}} - 1 = \dfrac{x-\sqrt{x^2+1}}{\sqrt{x^2+1}} \quad\Rightarrow\quad f'(x) < 0$, so f is decreasing on $\mathbb{R}$. **F.** No extremum

G. $f''(x) = \dfrac{1}{(x^2+1)^{3/2}} > 0$, so f is CU on $\mathbb{R}$. No IP

H.

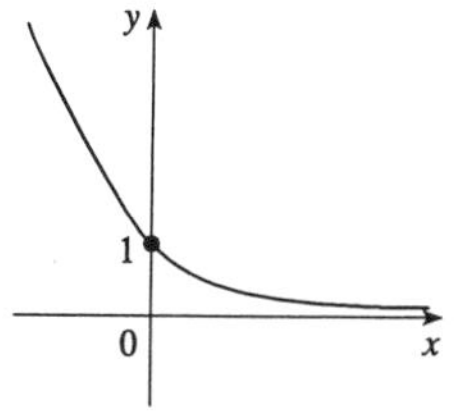

19. $y = f(x) = \sqrt[4]{x^2-25}$ **A.** $D = \{x \mid x^2 \geq 25\} = (-\infty, -5] \cup [5, \infty)$ **B.** x-intercepts are ± 5, no y-intercept **C.** $f(-x) = f(x)$, so the curve is symmetric about the y-axis. **D.** $\lim\limits_{x\to\pm\infty}\sqrt[4]{x^2-25} = \infty$, no asymptotes

E. $f'(x) = \frac{1}{4}(x^2-25)^{-3/4}(2x) = \dfrac{x}{2(x^2-25)^{3/4}} > 0$ if $x > 5$, so f is increasing on $[5, \infty)$ and decreasing on $(-\infty, -5]$. **F.** No local extremum

G. $y'' = \dfrac{2(x^2-25)^{3/4} - 3x^2(x^2-25)^{-1/4}}{4\,(x^2-25)^{3/2}} = -\dfrac{x^2+50}{4(x^2-25)^{7/4}} < 0$,

so f is CD on $(-\infty, -5)$ and $(5, \infty)$. No IP

H.

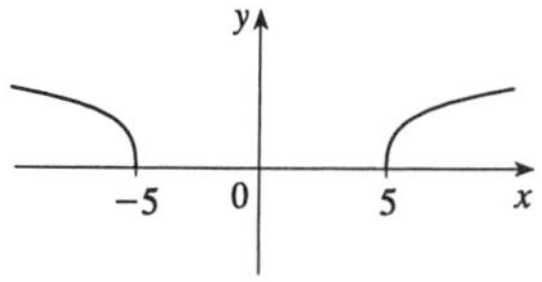

21. $y = f(x) = \dfrac{\sqrt{1-x^2}}{x}$ **A.** $D = \{x \mid |x| \leq 1, x \neq 0\} = [-1, 0) \cup (0, 1]$ **B.** x-intercepts ± 1, no y-intercept

C. $f(-x) = -f(x)$, so the curve is symmetric about $(0, 0)$. **D.** $\lim\limits_{x\to 0^+}\dfrac{\sqrt{1-x^2}}{x} = \infty$, $\lim\limits_{x\to 0^-}\dfrac{\sqrt{1-x^2}}{x} = -\infty$,

so $x = 0$ is a VA. **E.** $f'(x) = \dfrac{\left(-x^2/\sqrt{1-x^2}\right) - \sqrt{1-x^2}}{x^2} = -\dfrac{1}{x^2\sqrt{1-x^2}} < 0$,

so f is decreasing on $[-1, 0)$ and $(0, 1]$. F. No extremum

G. $f''(x) = \dfrac{2-3x^2}{x^3(1-x^2)^{3/2}} > 0 \quad\Leftrightarrow\quad -1 < x < -\sqrt{\frac{2}{3}}$ or

$0 < x < \sqrt{\frac{2}{3}}$, so f is CU on $\left(-1, -\sqrt{\frac{2}{3}}\right)$ and $\left(0, \sqrt{\frac{2}{3}}\right)$

and CD on $\left(-\sqrt{\frac{2}{3}}, 0\right)$ and $\left(\sqrt{\frac{2}{3}}, 1\right)$. IP $\left(\pm\sqrt{\frac{2}{3}}, \pm\frac{1}{\sqrt{2}}\right)$.

H.

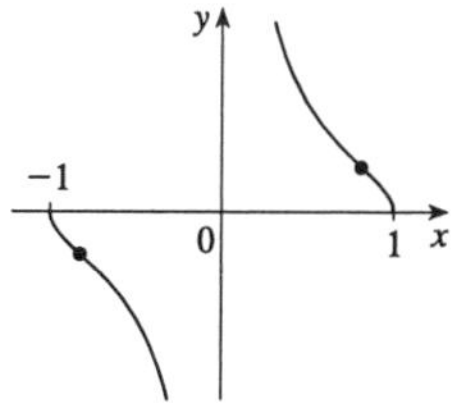

23. $y = f(x) = x + 3x^{2/3}$ **A.** $D = \mathbb{R}$ **B.** $y = x + 3x^{2/3} = x^{2/3}\left(x^{1/3}+3\right) = 0$ if $x = 0$ or -27 (x-intercepts), y-intercept $= f(0) = 0$ **C.** No symmetry **D.** $\lim\limits_{x\to\infty}\left(x + 3x^{2/3}\right) = \infty$,

$\lim\limits_{x\to-\infty}\left(x + 3x^{2/3}\right) = \lim\limits_{x\to-\infty} x^{2/3}\left(x^{1/3}+3\right) = -\infty$, no asymptotes

E. $f'(x) = 1 + 2x^{-1/3} = \left(x^{1/3}+2\right)/x^{1/3} > 0 \quad\Leftrightarrow$ $x > 0$ or $x < -8$, so f increases on $(-\infty, -8]$, $[0, \infty)$ and decreases on $[-8, 0]$. **F.** Local maximum $f(-8) = 4$, local minimum $f(0) = 0$ **G.** $f''(x) = -\frac{2}{9}x^{-4/3} < 0$ $(x \neq 0)$, so f is CD on $(-\infty, 0)$ and $(0, \infty)$. No IP

H.

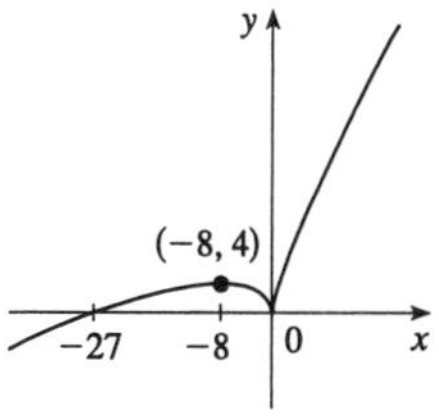

25. $y = f(x) = x + \sqrt{|x|}$ **A.** $D = \mathbb{R}$ **B.** x-intercepts $= 0, -1$, y-intercept 0 **C.** No symmetry **D.** $\lim_{x\to\infty}\left(x + \sqrt{|x|}\right) = \infty$, $\lim_{x\to-\infty}\left(x + \sqrt{|x|}\right) = -\infty$. No asymptotes **E.** For $x > 0$, $f(x) = x + \sqrt{x}$ $\Rightarrow$ $f'(x) = 1 + \dfrac{1}{2\sqrt{x}} > 0$, so f increases on $[0, \infty)$.

For $x < 0$, $f(x) = x + \sqrt{-x}$ $\Rightarrow$ $f'(x) = 1 - \dfrac{1}{2\sqrt{-x}} > 0$ $\Leftrightarrow$ $2\sqrt{-x} > 1$ $\Leftrightarrow$ $-x > \frac{1}{4}$ $\Leftrightarrow$ $x < -\frac{1}{4}$, so f increases on $\left(-\infty, -\frac{1}{4}\right]$ and decreases on $\left[-\frac{1}{4}, 0\right]$. **F.** $f\left(-\frac{1}{4}\right) = \frac{1}{4}$ is a local maximum, $f(0) = 0$ is a local minimum. **G.** For $x > 0$, $f''(x) = -\frac{1}{4}x^{-3/2}$ $\Rightarrow$ $f''(x) < 0$, so f is CD on $(0, \infty)$. For $x < 0$, $f''(x) = -\frac{1}{4}(-x)^{-3/2}$ $\Rightarrow$ $f''(x) < 0$, so f is CD on $(-\infty, 0)$. No IP

H.

27. $y = f(x) = \cos x - \sin x$ **A.** $D = \mathbb{R}$ **B.** $y = 0$ $\Leftrightarrow$ $\cos x = \sin x$ $\Leftrightarrow$ $x = n\pi + \frac{\pi}{4}$, n an integer (x-intercepts), y-intercept $= f(0) = 1$. **C.** Periodic with period 2π **D.** No asymptotes **E.** $f'(x) = -\sin x - \cos x = 0$ $\Leftrightarrow$ $\cos x = -\sin x$ $\Leftrightarrow$ $x = 2n\pi + \frac{3\pi}{4}$ or $2n\pi + \frac{7\pi}{4}$. $f'(x) > 0$ $\Leftrightarrow$ $\cos x < -\sin x$ $\Leftrightarrow$ $2n\pi + \frac{3\pi}{4} < x < 2n\pi + \frac{7\pi}{4}$, so f is increasing on $\left[2n\pi + \frac{3\pi}{4}, 2n\pi + \frac{7\pi}{4}\right]$ and decreasing on $\left[2n\pi - \frac{\pi}{4}, 2n\pi + \frac{3\pi}{4}\right]$. **F.** Local maximum $f\left(2n\pi - \frac{\pi}{4}\right) = \sqrt{2}$, local minimum $f\left(2n\pi + \frac{3\pi}{4}\right) = -\sqrt{2}$. **G.** $f''(x) = -\cos x + \sin x > 0$ $\Leftrightarrow$ $\sin x > \cos x$ $\Leftrightarrow$ $x \in \left(2n\pi + \frac{\pi}{4}, 2n\pi + \frac{5\pi}{4}\right)$, so f is CU on these intervals and CD on $\left(2n\pi - \frac{3\pi}{4}, 2n\pi + \frac{\pi}{4}\right)$. IP $\left(n\pi + \frac{\pi}{4}, 0\right)$

H.

29. $y = f(x) = x\tan x$, $-\frac{\pi}{2} < x < \frac{\pi}{2}$ **A.** $D = \left(-\frac{\pi}{2}, \frac{\pi}{2}\right)$ **B.** Intercepts are 0 **C.** $f(-x) = f(x)$, so the curve is symmetric about the y-axis. **D.** $\lim_{x\to\pi/2^-} x\tan x = \infty$ and $\lim_{x\to-\pi/2^+} x\tan x = \infty$, so $x = \frac{\pi}{2}$ and $x = -\frac{\pi}{2}$ are VA. **E.** $f'(x) = \tan x + x\sec^2 x > 0$ $\Leftrightarrow$ $0 < x < \frac{\pi}{2}$, so f increases on $\left[0, \frac{\pi}{2}\right)$ and decreases on $\left(-\frac{\pi}{2}, 0\right]$. **F.** Absolute minimum $f(0) = 0$. **G.** $y'' = 2\sec^2 x + 2x\tan x\sec^2 x > 0$ for $-\frac{\pi}{2} < x < \frac{\pi}{2}$, so f is CU on $\left(-\frac{\pi}{2}, \frac{\pi}{2}\right)$. No IP

H.

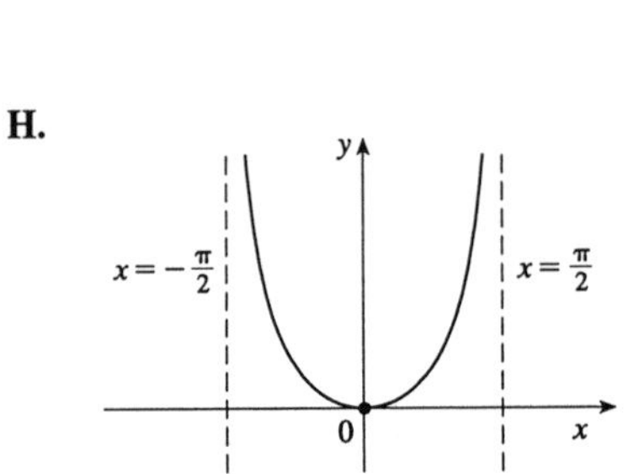

31. $y = f(x) = x/2 - \sin x,\ 0 < x < 3\pi$ **A.** $D = (0, 3\pi)$ **B.** No y-intercept. The x-intercept can be found approximately by Newton's Method (see Exercise 2.10.11). **C.** No symmetry **D.** No asymptotes.
E. $f'(x) = \frac{1}{2} - \cos x > 0 \iff \cos x < \frac{1}{2} \iff \frac{\pi}{3} < x < \frac{5\pi}{3}$ or $\frac{7\pi}{3} < x < 3\pi$, so f is increasing on $\left[\frac{\pi}{3}, \frac{5\pi}{3}\right]$ and $\left[\frac{7\pi}{3}, 3\pi\right)$ and decreasing on $\left(0, \frac{\pi}{3}\right]$ and $\left[\frac{5\pi}{3}, \frac{7\pi}{3}\right]$.
F. $f\left(\frac{\pi}{3}\right) = \frac{\pi}{6} - \frac{\sqrt{3}}{2}$ is a local minimum,
$f\left(\frac{5\pi}{3}\right) = \frac{5\pi}{6} + \frac{\sqrt{3}}{2}$ is a local maximum,
$f\left(\frac{7\pi}{3}\right) = \frac{7\pi}{6} - \frac{\sqrt{3}}{2}$ is a local minimum.
G. $f''(x) = \sin x > 0 \iff 0 < x < \pi$ or $2\pi < x < 3\pi$, so f is CU on $(0, \pi)$ and $(2\pi, 3\pi)$ and CD on $(\pi, 2\pi)$.
IP $\left(\pi, \frac{\pi}{2}\right)$ and $(2\pi, \pi)$.

H.

33. $y = f(x) = 2\cos x + \sin^2 x$ **A.** $D = \mathbb{R}$ **B.** y-intercept $= f(0) = 2$ **C.** $f(-x) = f(x)$, so the curve is symmetric about the y-axis. Periodic with period 2π **D.** No asymptotes
E. $f'(x) = -2\sin x + 2\sin x \cos x = 2\sin x\,(\cos x - 1) > 0 \iff \sin x < 0 \iff$ $(2n-1)\pi < x < 2n\pi$, so f is increasing on $[(2n-1)\pi, 2n\pi]$ and decreasing on $[2n\pi, (2n+1)\pi]$.
F. $f(2n\pi) = 2$ is a local maximum.
$f((2n+1)\pi) = -2$ is a local minimum.
G. $f''(x) = -2\cos x + 2\cos 2x = 2(2\cos^2 x - \cos x - 1)$
$= 2(2\cos x + 1)(\cos x - 1) > 0$
$\iff \cos x < -\frac{1}{2} \iff x \in \left(2n\pi + \frac{2\pi}{3}, 2n\pi + \frac{4\pi}{3}\right)$, so f is CU on these intervals and CD on $\left(2n\pi - \frac{2\pi}{3}, 2n\pi + \frac{2\pi}{3}\right)$.
IP when $x = 2n\pi \pm \frac{2\pi}{3}$

H.

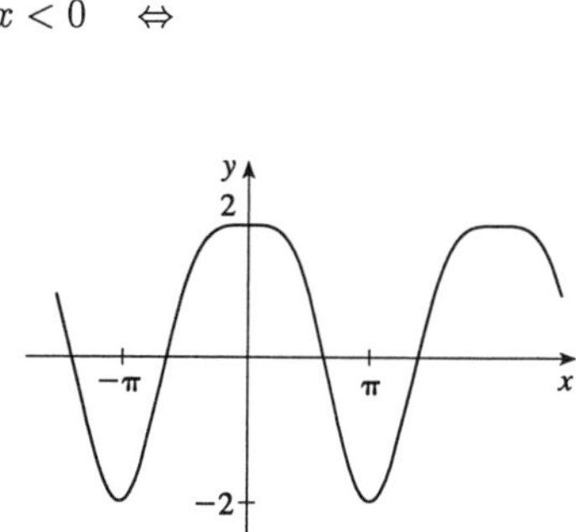

35. $y = f(x) = \sin 2x - 2\sin x$ **A.** $D = \mathbb{R}$ **B.** y-intercept $= f(0) = 0$. $y = 0 \iff$ $2\sin x = \sin 2x = 2\sin x \cos x \iff \sin x = 0$ or $\cos x = 1 \iff x = n\pi$ (x-intercepts)
C. $f(-x) = -f(x)$, so the curve is symmetric about $(0,0)$. *Note:* f is periodic with period 2π, so we determine D-G for $-\pi \le x \le \pi$. **D.** No asymptotes **E.** $f'(x) = 2\cos 2x - 2\cos x$. As in Exercise 33G, we see that $f'(x) > 0 \iff -\pi < x < -\frac{2\pi}{3}$ or $\frac{2\pi}{3} < x < \pi$, so f is increasing on $\left[-\pi, -\frac{2\pi}{3}\right]$ and $\left[\frac{2\pi}{3}, \pi\right]$ and decreasing on $\left[-\frac{2\pi}{3}, \frac{2\pi}{3}\right]$. **F.** $f\left(-\frac{2\pi}{3}\right) = \frac{3\sqrt{3}}{2}$ is a local maximum, $f\left(\frac{2\pi}{3}\right) = -\frac{3\sqrt{3}}{2}$ is a local minimum.
G. $f''(x) = -4\sin 2x + 2\sin x$
$= 2\sin x\,(1 - 4\cos x) = 0$ when $x = 0, \pm\pi$ or $\cos x = \frac{1}{4}$. If $\alpha = \cos^{-1}\frac{1}{4}$, then f is CU on $(-\alpha, 0)$ and (α, π) and CD on $(-\pi, -\alpha)$ and $(0, \alpha)$. IP $(0,0)$, $(\pi, 0)$, $\left(\alpha, -\frac{3\sqrt{15}}{8}\right)$, $\left(-\alpha, \frac{3\sqrt{15}}{8}\right)$.

H.

37. $y = f(x) = x^3/(x^2-1)$ **A.** $D = \{x \mid x \neq \pm 1\} = (-\infty,-1) \cup (-1,1) \cup (1,\infty)$ **B.** x-intercept $= 0$, y-intercept $= 0$ **C.** $f(-x) = -f(x) \Rightarrow$ f is odd, so the curve is symmetric about the origin.

D. $\lim\limits_{x\to\infty} \dfrac{x^3}{x^2-1} = \infty$ but long division gives $\dfrac{x^3}{x^2-1} = x + \dfrac{x}{x^2-1}$ so $f(x) - x = \dfrac{x}{x^2-1} \to 0$ as $x \to \pm\infty$ $\Rightarrow$ $y = x$ is a slant asymptote. $\lim\limits_{x\to 1^-} \dfrac{x^3}{x^2-1} = -\infty$, $\lim\limits_{x\to 1^+} \dfrac{x^3}{x^2-1} = \infty$, $\lim\limits_{x\to -1^-} \dfrac{x^3}{x^2-1} = -\infty$, $\lim\limits_{x\to -1^+} \dfrac{x^3}{x^2-1} = \infty$, so $x = 1$ and $x = -1$ are VA. **E.** $f'(x) = \dfrac{3x^2(x^2-1) - x^3(2x)}{(x^2-1)^2} = \dfrac{x^2(x^2-3)}{(x^2-1)^2}$ $\Rightarrow$ $f'(x) > 0 \Leftrightarrow x^2 > 3 \Leftrightarrow x > \sqrt{3}$ or $x < -\sqrt{3}$, so f is increasing on $\left(-\infty, -\sqrt{3}\right]$ and $\left[\sqrt{3}, \infty\right)$ and decreasing on $\left[-\sqrt{3}, -1\right)$, $(-1,1)$, and $\left(1, \sqrt{3}\right]$.

F. $f\left(-\sqrt{3}\right) = -\frac{3\sqrt{3}}{2}$ is a local maximum and $f\left(\sqrt{3}\right) = \frac{3\sqrt{3}}{2}$ is a local minimum.

G. $y'' = \dfrac{2x\left(x^2+3\right)}{\left(x^2-1\right)^3} > 0 \Leftrightarrow x > 1$ or $-1 < x < 0$, so f is CU on $(-1,0)$ and $(1,\infty)$ and CD on $(-\infty,-1)$ and $(0,1)$. IP $(0,0)$

H.

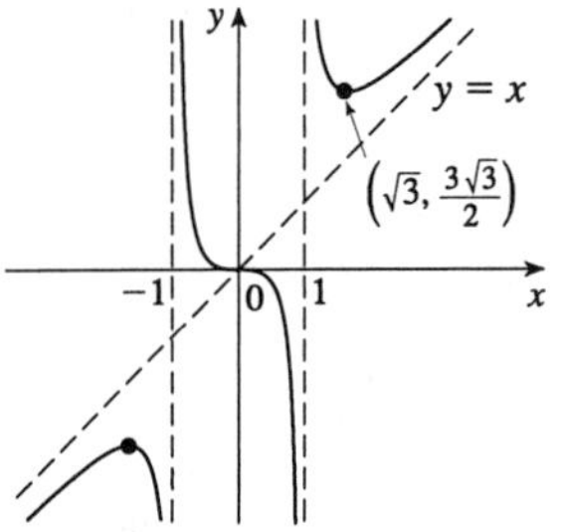

39. $y = f(x) = (x^2+4)/x = x + 4/x$ **A.** $D = \{x \mid x \neq 0\} = (-\infty,0) \cup (0,\infty)$ **B.** No intercept **C.** $f(-x) = -f(x) \Rightarrow$ symmetry about the origin **D.** $\lim\limits_{x\to\infty} (x + 4/x) = \infty$ but $f(x) - x = 4/x \to 0$ as $x \to \pm\infty$, so $y = x$ is a slant asymptote. $\lim\limits_{x\to 0^+} (x + 4/x) = \infty$ and $\lim\limits_{x\to 0^-} (x + 4/x) = -\infty$, so $x = 0$ is a VA.

E. $f'(x) = 1 - 4/x^2 > 0 \Leftrightarrow x^2 > 4 \Leftrightarrow x > 2$ or $x < -2$, so f is increasing on $(-\infty,-2]$ and $[2,\infty)$ and decreasing on $[-2,0)$ and $(0,2]$.

F. $f(-2) = -4$ is a local maximum and $f(2) = 4$ is a local minimum. **G.** $f''(x) = 8/x^3 > 0 \Leftrightarrow x > 0$ so f is CU on $(0,\infty)$ and CD on $(-\infty,0)$. No IP

H.

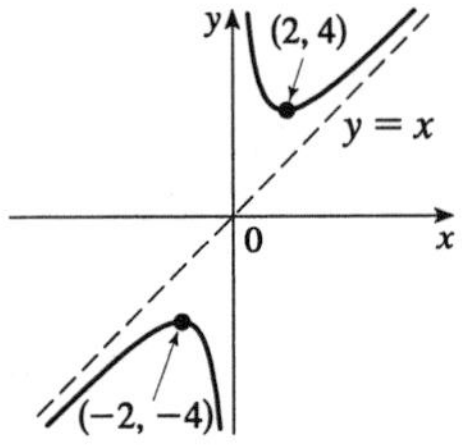

41. $y = \dfrac{1}{x-1} - x$ **A.** $D = \{x \mid x \neq 1\}$ **B.** $y = 0 \Leftrightarrow x = \dfrac{1}{x-1} \Leftrightarrow x^2 - x - 1 = 0 \Rightarrow x = \frac{1 \pm \sqrt{5}}{2}$ (x-intercepts), y-intercept $= f(0) = -1$ **C.** No symmetry **D.** $y - (-x) = \dfrac{1}{x-1} \to 0$ as $x \to \pm\infty$, so $y = -x$ is a slant asymptote.

$\lim\limits_{x\to 1^+} \left(\dfrac{1}{x-1} - x\right) = \infty$ and $\lim\limits_{x\to 1^-} \left(\dfrac{1}{x-1} - x\right) = -\infty$, so $x = 1$ is a VA. **E.** $f'(x) = -1 - 1/(x-1)^2 < 0$ for all $x \neq 1$, so f is decreasing on $(-\infty,1)$ and $(1,\infty)$.

F. No local extremum **G.** $f''(x) = \dfrac{2}{(x-1)^3} > 0 \Leftrightarrow x > 1$, so f is CU on $(1,\infty)$ and CD on $(-\infty,1)$. No IP

H.

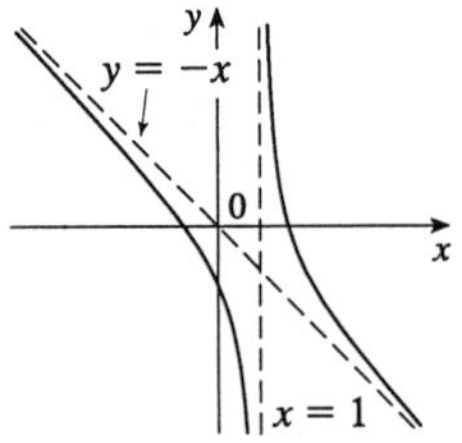

43. $\dfrac{x^2}{a^2} - \dfrac{y^2}{b^2} = 1 \;\Rightarrow\; y = \pm\dfrac{b}{a}\sqrt{x^2 - a^2}$. Now

$$\lim_{x\to\infty}\left[\frac{b}{a}\sqrt{x^2-a^2} - \frac{b}{a}x\right] = \frac{b}{a}\cdot\lim_{x\to\infty}\left(\sqrt{x^2-a^2} - x\right)\frac{\sqrt{x^2-a^2}+x}{\sqrt{x^2-a^2}+x}$$

$$= \frac{b}{a}\cdot\lim_{x\to\infty}\frac{-a^2}{\sqrt{x^2-a^2}+x} = 0,$$

which shows that $y = \dfrac{b}{a}x$ is a slant asymptote. Similarly,

$\displaystyle\lim_{x\to\infty}\left[-\frac{b}{a}\sqrt{x^2-a^2} - \left(-\frac{b}{a}x\right)\right] = -\frac{b}{a}\cdot\lim_{x\to\infty}\frac{-a^2}{\sqrt{x^2-a^2}+x} = 0$, showing that $y = -\dfrac{b}{a}x$ is a slant asymptote.

45. $\displaystyle\lim_{x\to\pm\infty}\left[f(x) - x^3\right] = \lim_{x\to\pm\infty}\frac{x^4+1}{x} - \frac{x^4}{x} = \lim_{x\to\pm\infty}\frac{1}{x} = 0$, so the graph of f is asymptotic to that of $y = x^3$.

A. $D = \{x \mid x \neq 0\}$ **B.** No intercepts **C.** f is symmetric about the origin.

D. $\displaystyle\lim_{x\to 0^-}\left(x^3 + \frac{1}{x}\right) = -\infty$ and $\displaystyle\lim_{x\to 0^+}\left(x^3 + \frac{1}{x}\right) = \infty$, so $x = 0$ is a vertical asymptote, and as shown above, the graph of f is asymptotic to that of $y = x^3$.

E. $f'(x) = 3x^2 - \dfrac{1}{x^2} > 0 \;\Leftrightarrow\; x^4 > \frac{1}{3} \;\Leftrightarrow |x| > \frac{1}{\sqrt[4]{3}}$, so f is increasing on $\left(-\infty, -\frac{1}{\sqrt[4]{3}}\right)$ and $\left(\frac{1}{\sqrt[4]{3}}, \infty\right)$ and decreasing on $\left(-\frac{1}{\sqrt[4]{3}}, 0\right)$ and $\left(0, \frac{1}{\sqrt[4]{3}}\right)$.

F. Local maximum $f\left(-\frac{1}{\sqrt[4]{3}}\right) = -4\cdot 3^{-5/4}$, local minimum $f\left(\frac{1}{\sqrt[4]{3}}\right) = 4\cdot 3^{-5/4}$

G. $f''(x) = 6x + \dfrac{2}{x^3} > 0 \;\Leftrightarrow x > 0$, so f is CU on $(0,\infty)$ and CD on $(-\infty, 0)$.

H.

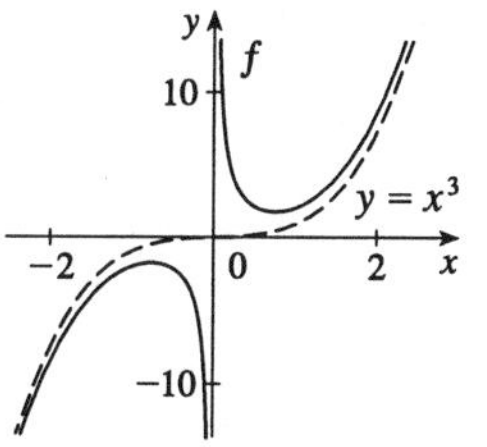

EXERCISES 3.7

Abbreviations:	HA	horizontal asymptote(s)	VA	vertical asymptote(s)
	CU	concave up	CD	concave down
	IP	inflection point(s)	FDT	First Derivative Test

1. $f(x) = 4x^4 - 7x^2 + 4x + 6 \quad \Rightarrow \quad f'(x) = 16x^3 - 14x + 4 \quad \Rightarrow \quad f''(x) = 48x^2 - 14$

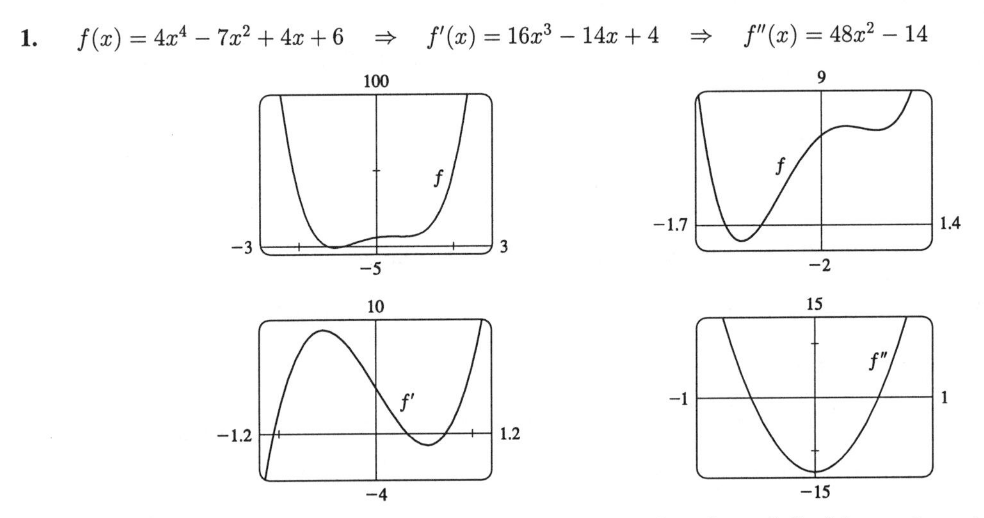

After finding suitable viewing rectangles (by ensuring that we have located all of the x-values where either $f' = 0$ or $f'' = 0$) we estimate from the graph of f' that f is increasing on $[-1.1, 0.3]$ and $[0.7, \infty)$ and decreasing on $(-\infty, -1.1]$ and $[0.3, 0.7]$, with a local maximum of $f(0.3) \approx 6.6$ and minima of $f(-1.1) \approx -1.0$ and $f(0.7) \approx 6.3$. We estimate from the graph of f'' that f is CU on $(-\infty, -0.5)$ and $(0.5, \infty)$ and CD on $(-0.5, 0.5)$, and that f has inflection points at about $(-0.5, 2.0)$ and $(0.5, 6.5)$.

3. $f(x) = \sqrt[3]{x^2 - 3x - 5} \;\Rightarrow\; f'(x) = \dfrac{1}{3}\dfrac{2x - 3}{(x^2 - 3x - 5)^{2/3}} \quad \Rightarrow \quad f''(x) = -\dfrac{2}{9}\dfrac{x^2 - 3x + 24}{(x^2 - 3x - 5)^{5/3}}$

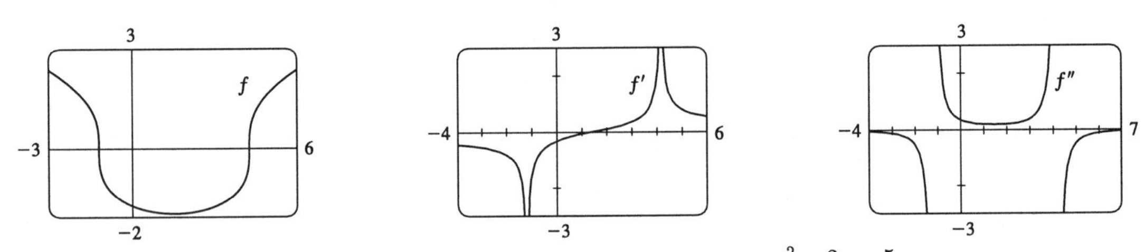

Note: With some CAS's, including Maple, it is necessary to define $f(x) = \dfrac{x^2 - 3x - 5}{|x^2 - 3x - 5|}\left|x^2 - 3x - 5\right|^{1/3}$, since the CAS does not compute real cube roots of negative numbers. (See Example 7 in Section 3 of Review and Preview.) We estimate from the graph of f' that f is increasing on $[1.5, 4.2)$ and $(4.2, \infty)$, and decreasing on $(-\infty, -1.2)$ and $(-1.2, 1.5]$. f has no maximum. Minimum: $f(1.5) \approx -1.9$. From the graph of f'', we estimate that f is CU on $(-1.2, 4.2)$ and CD on $(-\infty, -1.2)$ and $(4.2, \infty)$. IP $(-1.2, 0)$ and $(4.2, 0)$.

5. 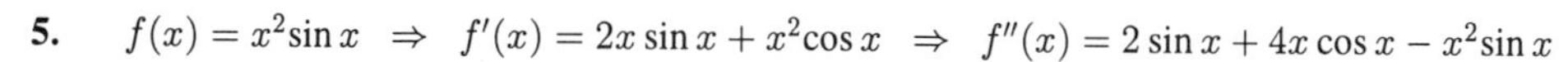
$f(x) = x^2 \sin x \;\Rightarrow\; f'(x) = 2x\sin x + x^2\cos x \;\Rightarrow\; f''(x) = 2\sin x + 4x\cos x - x^2\sin x$

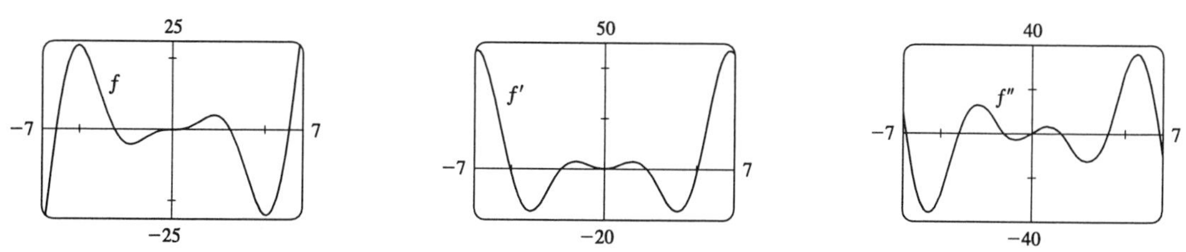

We estimate from the graph of f' that f is increasing on $[-7, -5.1]$, $[-2.3, 2.3]$, and $[5.1, 7]$ and decreasing on $[-5.1, -2.3]$, and $[2.3, 5.1]$. Local maxima: $f(-5.1) \approx 24.1$, $f(2.3) \approx 3.9$. Local minima: $f(-2.3) \approx -3.9$, $f(5.1) \approx -24.1$. From the graph of f'', we estimate that f is CU on $(-7, -6.8)$, $(-4.0, -1.5)$, $(0, 1.5)$, and $(4.0, 6.8)$, and CD on $(-6.8, -4.0)$, $(-1.5, 0)$, $(1.5, 4.0)$, and $(6.8, 7)$. f has IP at $(-6.8, -24.4)$, $(-4, 12.0)$, $(-1.5, -2.3)$, $(0, 0)$, $(1.5, 2.3)$, $(4.0, -12.0)$ and $(6.8, 24.4)$.

7.

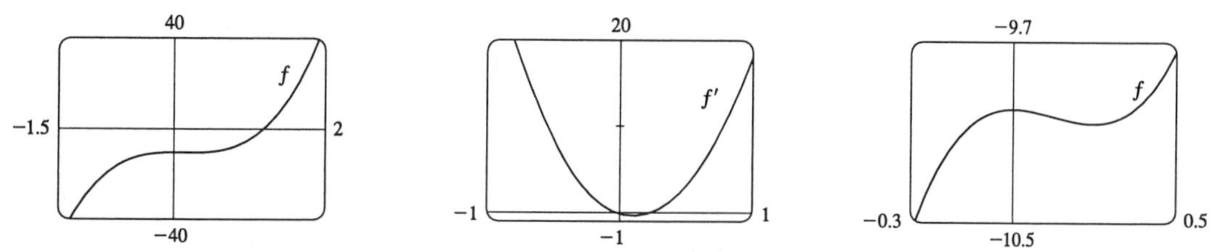

From the graphs, it appears that $f(x) = 8x^3 - 3x^2 - 10$ increases on $(-\infty, 0]$ and $[0.25, \infty)$ and decreases on $[0, 0.25]$; that f has a local maximum of $f(0) = -10.0$ and a local minimum of $f(0.25) \approx -10.1$; that f is CU on $(0.1, \infty)$ and CD on $(-\infty, 0.1)$; and that f has an IP at $(0.1, -10)$. $f(x) = 8x^3 - 3x^2 - 10 \;\Rightarrow$ $f'(x) = 24x^2 - 6x = 6x(4x - 1)$, which is positive ($f$ is increasing) for $(-\infty, 0]$ and $\left[\frac{1}{4}, \infty\right)$, and negative ($f$ is decreasing) on $\left[0, \frac{1}{4}\right]$. By the FDT, f has a local maximum at 0: $f(0) = 8(0)^3 - 3(0)^2 - 10 = -10$; and f has a local minimum at $\frac{1}{4}$: $f\left(\frac{1}{4}\right) = \frac{1}{8} - \frac{3}{16} - 10 = -\frac{161}{16}$. $f'(x) = 24x^2 - 6x \;\Rightarrow\; f''(x) = 48x - 6 = 6(8x - 1)$, which is positive ($f$ is CU) on $\left(\frac{1}{8}, \infty\right)$, and negative ($f$ is CD) on $\left(-\infty, \frac{1}{8}\right)$. f has an IP at $\left(\frac{1}{8}, -\frac{321}{32}\right)$.

9.

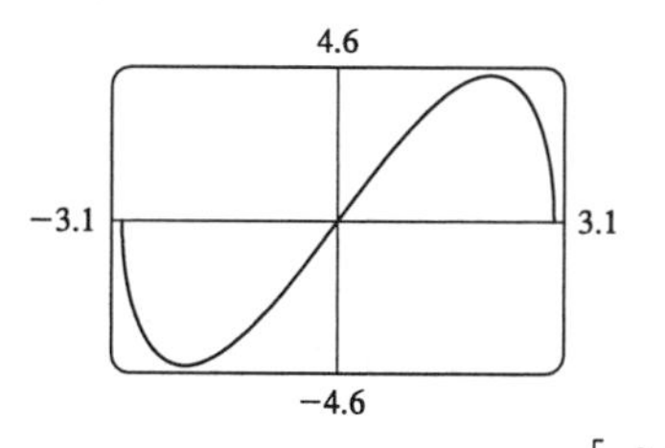

From the graph, it appears that f increases on $[-2.1, 2.1]$ and decreases on $[-3, -2.1]$ and $[2.1, 3]$; that f has a local maximum of $f(2.1) \approx 4.5$ and a local minimum of $f(-2.1) \approx -4.5$; that f is CU on $(-3.0, 0)$ and CD on $(0, 3.0)$, and that f has an IP at $(0, 0)$. $f(x) = x\sqrt{9 - x^2} \;\Rightarrow$ $f'(x) = -x^2/\sqrt{9 - x^2} + \sqrt{9 - x^2} = \left(9 - 2x^2\right)/\sqrt{9 - x^2}$, which is positive ($f$ is increasing) on $\left[\frac{-3\sqrt{2}}{2}, \frac{3\sqrt{2}}{2}\right]$ and negative (f is decreasing) on $\left[-3, \frac{-3\sqrt{2}}{2}\right]$ and $\left[\frac{3\sqrt{2}}{2}, 3\right]$. By the FDT, f has a local maximum of $f\left(\frac{3\sqrt{2}}{2}\right) = \frac{3\sqrt{2}}{2}\sqrt{9 - \left(\frac{3\sqrt{2}}{2}\right)^2} = \frac{9}{2}$; and f has a local minimum of $f\left(\frac{-3\sqrt{2}}{2}\right) = -\frac{9}{2}$ (since f is an odd function.) $f'(x) = -x^2/\sqrt{9 - x^2} + \sqrt{9 - x^2} \;\Rightarrow$

$$f''(x) = \frac{\sqrt{9 - x^2}(-2x) + x^2\left(\frac{1}{2}\right)\left(9 - x^2\right)^{-1/2}(-2x)}{9 - x^2} - x\left(9 - x^2\right)^{-1/2} = \frac{-2x - x^3\left(9 - x^2\right)^{-1} - x}{\sqrt{9 - x^2}}$$

$$= \frac{-3x}{\sqrt{9 - x^2}} - \frac{x^3}{\left(9 - x^2\right)^{3/2}} = \frac{x\left(2x^2 - 27\right)}{\left(9 - x^2\right)^{3/2}},$$

which is positive (f is CU) on $(-3, 0)$, and negative (f is CD) on $(0, 3)$. f has an IP at $(0, 0)$.

11.

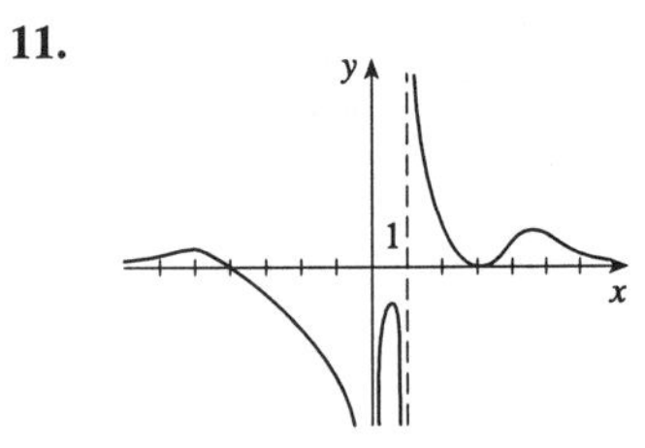

$f(x) = \dfrac{(x+4)(x-3)^2}{x^4(x-1)}$ has VA at $x = 0$ and at $x = 1$ since $\lim\limits_{x \to 0} f(x) = -\infty$, $\lim\limits_{x \to 1^-} f(x) = -\infty$ and $\lim\limits_{x \to 1^+} f(x) = \infty$.

$f(x) = \dfrac{(1+4/x)(1-3/x)^2}{x(x-1)} \to 0^+$ as $x \to \pm\infty$, so f is asymptotic to the x-axis.

Since f is undefined at $x = 0$, it has no y-intercept. $f(x) = 0 \quad \Rightarrow \quad (x+4)(x-3)^2 = 0 \quad \Rightarrow \quad x = -4$ or $x = 3$, so f has x-intercepts -4 and 3. Note, however, that the graph of f is only tangent to the x-axis and does not cross it at $x = 3$, since f is positive as $x \to 3^-$ and as $x \to 3^+$.

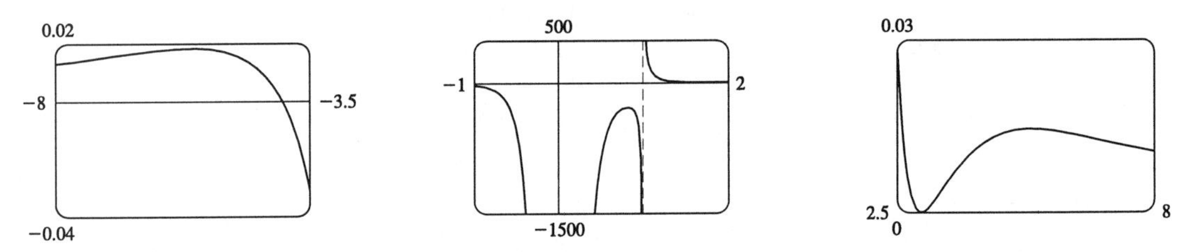

From these graphs, it appears that f has three maxima and one minimum. The maxima are approximately $f(-5.6) = 0.0182$, $f(0.82) = -281.5$ and $f(5.2) = 0.0145$ and we know (since the graph is tangent to the x-axis at $x = 3$) that the minimum is $f(3) = 0$.

13. We use `diff(f,x);` (in Maple) or `Dt[f,x]` (in Mathematica) on the function $f(x) = \dfrac{x^2(x+1)^3}{(x-2)^2(x-4)^4}$, and get $f'(x) = 2\dfrac{x(x+1)^3}{(x-2)^2(x-4)^4} + 3\dfrac{x^2(x+1)^2}{(x-2)^2(x-4)^4} - 2\dfrac{x^2(x+1)^3}{(x-2)^3(x-4)^4} - 4\dfrac{x^2(x+1)^3}{(x-2)^2(x-4)^5}$.

If we then use a CAS to simplify this expression, we get $f'(x) = -\dfrac{x(x+1)^2(x^3+18x^2-44x-16)}{(x-2)^3(x-4)^5}$.

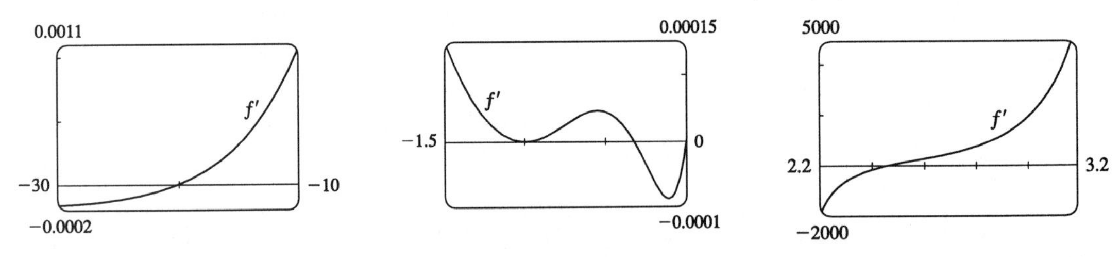

From the graphs of f', it seems that the critical points which indicate extrema occur at $x \approx -20$, -0.3, and 2.5, as estimated in Example 3. (There is another critical point at $x = -1$, but the sign of f' does not change there.) We differentiate again, and after simplifying, we find that

$$f''(x) = 2\frac{(x+1)(x^6+36x^5+6x^4-628x^3+684x^2+672x+64)}{(x-2)^4(x-4)^6}.$$

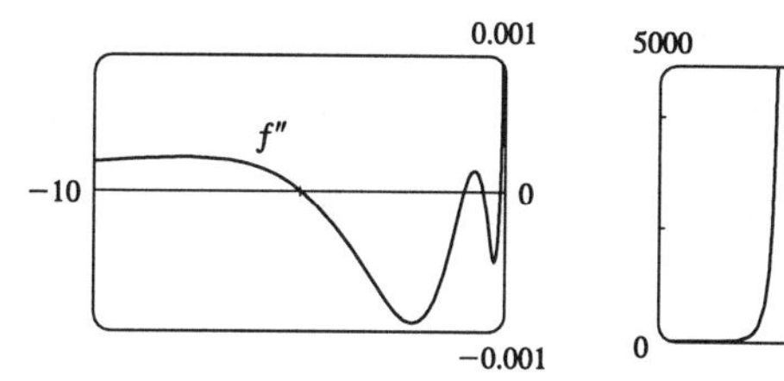

From the graphs of f'', it appears that f is CU on $(-\infty, -5.0)$, $(-1.0, -0.5)$, $(-0.1, 2.0)$, $(2.0, 4.0)$ and $(4.0, \infty)$ and CD on $(-5.0, -1.0)$ and $(-0.5, -0.1)$. We check back on the graphs of f to find the y-coordinates of the inflection points, and find that these are approximately $(-5, -0.005)$, $(-1, 0)$, $(-0.5, 0.00001)$, and $(-0.1, 0.0000066)$.

15.

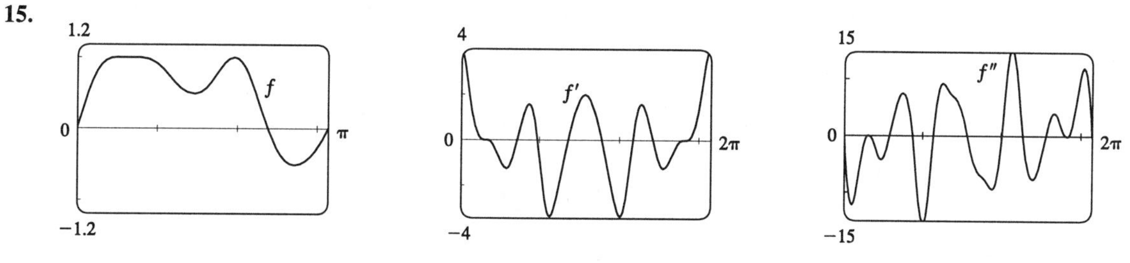

We consider the function only on the interval $[0, \pi]$ and use symmetry to extend. From the graph of f in the viewing rectangle $[0, \pi]$ by $[-1.2, 1.2]$, it looks like f has two maxima and two minima. If we calculate and graph $f'(x) = [\cos(x + \sin 3x)](1 + 3\cos 3x)$ on the same x-interval, we see that the graph of f' appears to be almost tangent to the x-axis at about $x = 0.7$. The graph of $f'' = -[\sin(x + \sin 3x)](1 + 3\cos 3x)^2 + \cos(x + \sin 3x)(-9\sin 3x)$ is even more interesting near this x-value: it seems to just touch the x-axis.

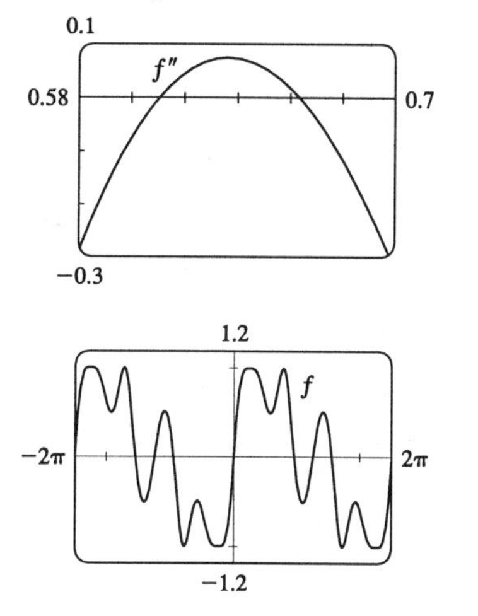

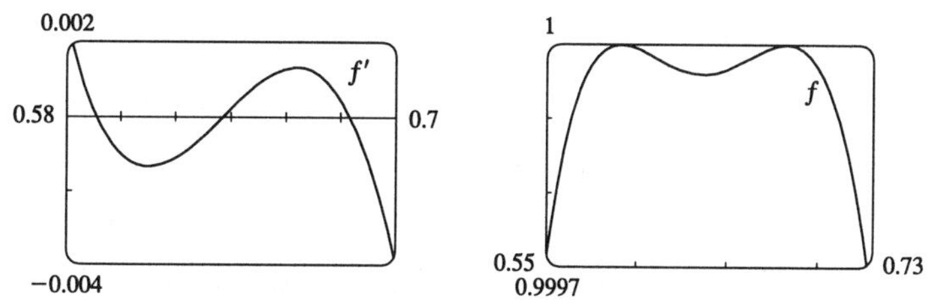

If we zoom in on this place on the graph of f'', we see that f'' actually does cross the axis twice near $x = 0.65$, indicating a change in concavity for a very short interval.

If we look at the graph of f' on the same interval, we see that it changes sign three times near $x = 0.65$, indicating that what we had thought was a broad extremum at about $x = 0.7$ actually consists of three extrema (two maxima and a minimum). These maxima are roughly $f(0.59) = 1$ and $f(0.68) = 1$, and the minimum is roughly $f(0.64) = 0.99996$. There are also a maximum of about $f(1.96) = 1$ and minima of about $f(1.46) = 0.49$ and $f(2.73) = -0.51$. The points of inflection are roughly $(0.61, 0.99998)$, $(0.66, 0.99998)$, $(1.17, 0.72)$, $(1.75, 0.77)$, and $(2.28, 0.34)$. Note that the function is odd and periodic with period 2π, and it is also rotationally symmetric about all points of the form $((2n + 1)\pi, 0)$, n an integer.

17. Note that $c = 0$ is a transitional value at which the graph consists of the x-axis. Also, we can see that if we substitute $-c$ for c, the function $f(x) = \dfrac{cx}{1 + c^2x^2}$ will be reflected in the x-axis, so we investigate only positive values of c (except $c = -1$, as a demonstration of this reflective property). Also, f is an odd function. $\lim\limits_{x\to\pm\infty} f(x) = 0$, so $y = 0$ is a horizontal asymptote for all c. We calculate

$f'(x) = \dfrac{c(1 + c^2x^2) - cx(2c^2x)}{(1 + c^2x^2)^2} = -\dfrac{c(c^2x^2 - 1)}{(1 + c^2x^2)^2}$. So there is an absolute maximum of $f(1/c) = \dfrac{1}{2}$ and an absolute minimum of $f(-1/c) = -\frac{1}{2}$. These extrema have the same value regardless of c, but the maximum points move closer to the y-axis as c increases.

$$f''(x) = \frac{(-2c^3x)(1 + c^2x^2)^2 - (-c^3x^2 + c)[2(1 + c^2x^2)(2c^2x)]}{(1 + c^2x^2)^4} = \frac{(-2cx)(1 + c^2x^2) + (c^3x^2 - c)(4c^2x)}{(1 + c^2x^2)^3}$$

$= \dfrac{2c^3x(c^2x^2 - 3)}{(1 + c^2x^2)^3}$, so there are inflection points at $(0, 0)$ and at $\left(\pm\dfrac{\sqrt{3}}{c}, \pm\dfrac{\sqrt{3}}{4}\right)$.

Again, the y-coordinate of the inflection points does not depend on c, but as c increases, both inflection points approach the y-axis.

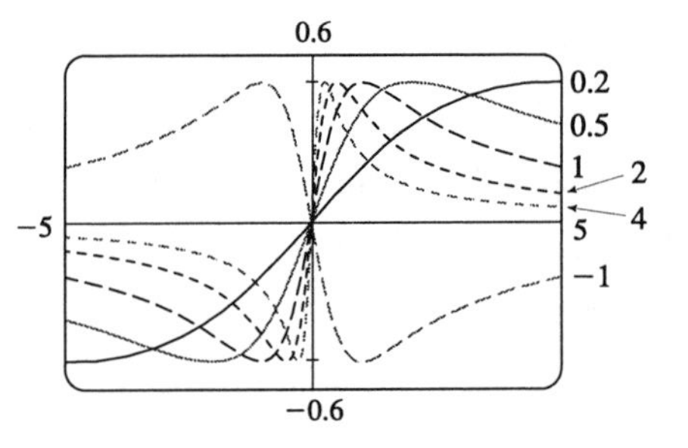

19. $f(x) = x^4 + cx^2 = x^2(x^2 + c)$. Note that f is an even function.

For $c \geq 0$, the only x-intercept is the point $(0, 0)$. We calculate $f'(x) = 4x^3 + 2cx = 4x\left(x^2 + \frac{1}{2}c\right) \Rightarrow$ $f''(x) = 12x^2 + 2c$. If $c \geq 0$, $x = 0$ is the only critical point and there are no inflection points. As we can see from the examples, there is no change in the basic shape of the graph for $c \geq 0$; it merely becomes steeper as c increases. For $c = 0$, the graph is the simple curve $y = x^4$.

For $c < 0$, there are x-intercepts at 0 and at $\pm\sqrt{-c}$. Also, there is a maximum at $(0, 0)$, and there are minima at $\left(\pm\sqrt{-\frac{1}{2}c}, -\frac{1}{4}c^2\right)$. As $c \to -\infty$, the x-coordinates of these minima get larger in absolute value, and the minimum points move downward.

There are inflection points at $\left(\pm\sqrt{-\frac{1}{6}c}, -\frac{5}{36}c^2\right)$, which also move away from the origin as $c \to -\infty$.

EXERCISES 3.8

1. If x is one number, the other is $100-x$. Maximize $f(x)=x(100-x)=100x-x^2$. $f'(x)=100-2x=0$ $\Rightarrow$ $x=50$. Now $f''(x)=-2<0$, so there is an absolute maximum at $x=50$. The numbers are 50 and 50.

3. The two numbers are x and $\dfrac{100}{x}$ where $x>0$. Minimize $f(x)=x+\dfrac{100}{x}$. $f'(x)=1-\dfrac{100}{x^2}=\dfrac{x^2-100}{x^2}$. The critical number is $x=10$. Since $f'(x)<0$ for $0<x<10$ and $f'(x)>0$ for $x>10$, there is an absolute minimum at $x=10$. The numbers are 10 and 10.

5. Let p be the perimeter and x and y the lengths of the sides, so $p=2x+2y$ $\Rightarrow$ $y=\frac{1}{2}p-x$. The area is $A(x)=x\left(\frac{1}{2}p-x\right)=\frac{1}{2}px-x^2$. Now $0=A'(x)=\frac{1}{2}p-2x$ $\Rightarrow$ $x=\frac{1}{4}p$. Since $A''(x)=-2<0$, there is an absolute maximum where $x=\frac{1}{4}p$. The sides of the rectangle are $\frac{1}{4}p$ and $\frac{1}{2}p-\frac{1}{4}p=\frac{1}{4}p$, so the rectangle is a square.

7.

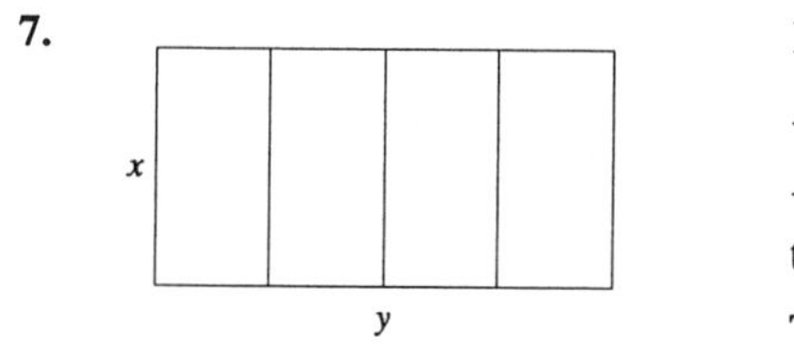

Here $5x+2y=750$ so $y=(750-5x)/2$. Maximize $A=xy=x(750-5x)/2=375x-\frac{5}{2}x^2$. Now $A'(x)=375-5x=0$ $\Rightarrow$ $x=75$. Since $A''(x)=-5<0$ there is an absolute maximum when $x=75$. Then $y=\frac{375}{2}$. The largest area is $75\left(\frac{375}{2}\right)=14{,}062.5\text{ ft}^2$.

9. Let b be the base of the box and h the height. The surface area is $1200=b^2+4hb$ $\Rightarrow$ $h=(1200-b^2)/(4b)$. The volume is $V=b^2h=b^2(1200-b^2)/4b=300b-b^3/4$ $\Rightarrow$ $V'(b)=300-\frac{3}{4}b^2$. $V'(b)=0$ $\Rightarrow$ $b=\sqrt{400}=20$. Since $V'(b)>0$ for $0<b<20$ and $V'(b)<0$ for $b>20$, there is an absolute maximum when $b=20$. Then $h=10$, so the largest possible volume is $(20)^2(10)=4000\text{ cm}^3$.

11.

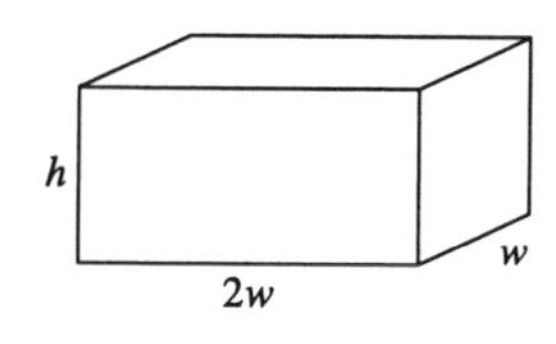

$10=(2w)(w)\,h=2w^2h$, so $h=5/w^2$. The cost is $C(w)=10(2w^2)+6[2(2wh)+2hw]+6(2w^2)$ $=32w^2+36wh=32w^2+180/w$.

$C'(w)=64w-180/w^2=4(16w^3-45)/w^2$ $\Rightarrow$ $w=\sqrt[3]{\frac{45}{16}}$ is the critical number. $C'(w)<0$ for $0<w<\sqrt[3]{\frac{45}{16}}$ and $C'(w)>0$ for $w>\sqrt[3]{\frac{45}{16}}$. The minimum cost is $C\left(\sqrt[3]{\frac{45}{16}}\right)=32(2.8125)^{2/3}+180/\sqrt[3]{2.8125}\approx\$191.28$.

13. For (x,y) on the line $y=2x-3$, the distance to the origin is $\sqrt{(x-0)^2+(2x-3)^2}$. We minimize the square of the distance, that is, $x^2+(2x-3)^2=5x^2-12x+9=D(x)$. $D'(x)=10x-12=0$ $\Rightarrow$ $x=\frac{6}{5}$. Since there is a point closest to the origin, $x=\frac{6}{5}$ and hence $y=-\frac{3}{5}$. So the point is $\left(\frac{6}{5},-\frac{3}{5}\right)$.

15. By symmetry, the points are (x,y) and $(x,-y)$, where $y>0$. The square of the distance is $D(x)=(x-2)^2+y^2=(x-2)^2+4+x^2=2x^2-4x+8$. So $D'(x)=4x-4=0$ $\Rightarrow$ $x=1$ and $y=\pm\sqrt{4+1}=\pm\sqrt{5}$. The points are $\left(1,\pm\sqrt{5}\right)$.

17.

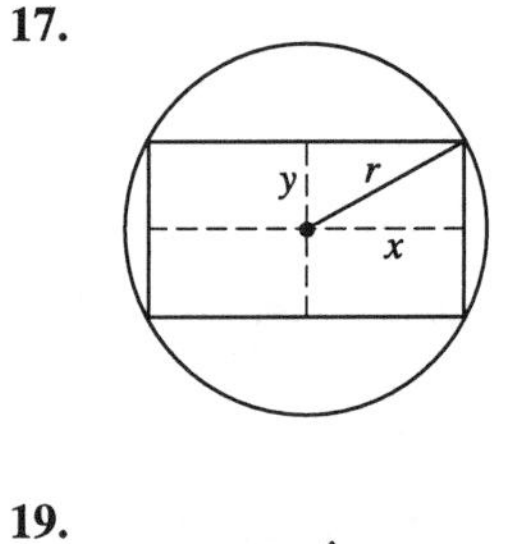

Area of rectangle is $4xy$. Also $r^2 = x^2 + y^2$ so $y = \sqrt{r^2 - x^2}$, so the area is $A(x) = 4x\sqrt{r^2 - x^2}$. Now

$$A'(x) = 4\left(\sqrt{r^2 - x^2} - x^2/\sqrt{r^2 - x^2}\right) = 4(r^2 - 2x^2)/\sqrt{r^2 - x^2}.$$

The critical number is $x = r/\sqrt{2}$. Clearly this gives a maximum. The dimensions are $2x = \sqrt{2}r$ and $2y = \sqrt{2}r$.

19.

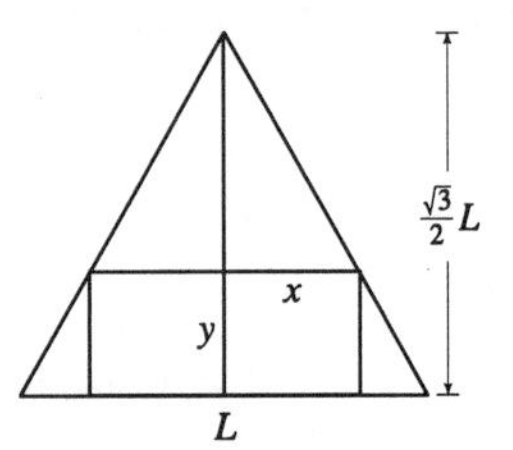

$$\frac{\frac{\sqrt{3}}{2}L - y}{x} = \frac{\frac{\sqrt{3}}{2}L}{L/2} = \sqrt{3} \text{ (similar triangles)} \Rightarrow$$

$\sqrt{3}x = \frac{\sqrt{3}}{2}L - y \quad \Rightarrow \quad y = \frac{\sqrt{3}}{2}(L - 2x)$. The area of the inscribed rectangle is $A(x) = (2x)y = \sqrt{3}x(L - 2x)$ where $0 \le x \le L/2$. Now $0 = A'(x) = \sqrt{3}L - 4\sqrt{3}x \quad \Rightarrow$ $x = \sqrt{3}L/(4\sqrt{3}) = L/4$. Since $A(0) = A(L/2) = 0$, the maximum occurs when $x = L/4$, and $y = \frac{\sqrt{3}}{2}L - \frac{\sqrt{3}}{4}L = \frac{\sqrt{3}}{4}L$, so the dimensions are $L/2$ and $\frac{\sqrt{3}}{4}L$.

21.

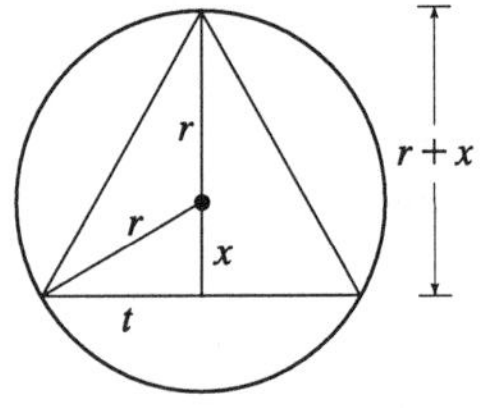

The area of the triangle is

$A(x) = \frac{1}{2}(2t)(r + x) = t(r + x) = \sqrt{r^2 - x^2}(r + x)$. Then

$$0 = A'(x) = r\frac{-2x}{2\sqrt{r^2 - x^2}} + \sqrt{r^2 - x^2} + x\frac{-2x}{2\sqrt{r^2 - x^2}}$$

$$= -\frac{x^2 + rx}{\sqrt{r^2 - x^2}} + \sqrt{r^2 - x^2} \quad \Rightarrow \quad \frac{x^2 + rx}{\sqrt{r^2 - x^2}} = \sqrt{r^2 - x^2}$$

$$\Rightarrow \quad x^2 + rx = r^2 - x^2 \quad \Rightarrow \quad 0 = 2x^2 + rx - r^2 = (2x - r)(x + r)$$

$\Rightarrow \quad x = \frac{1}{2}r$ or $x = -r$. Now $A(r) = 0 = A(-r) \quad \Rightarrow \quad$ the maximum occurs where $x = \frac{1}{2}r$, so the triangle has height $r + \frac{1}{2}r = \frac{3}{2}r$ and base $2\sqrt{r^2 - \left(\frac{1}{2}r\right)^2} = 2\sqrt{\frac{3}{4}r^2} = \sqrt{3}r$.

23.

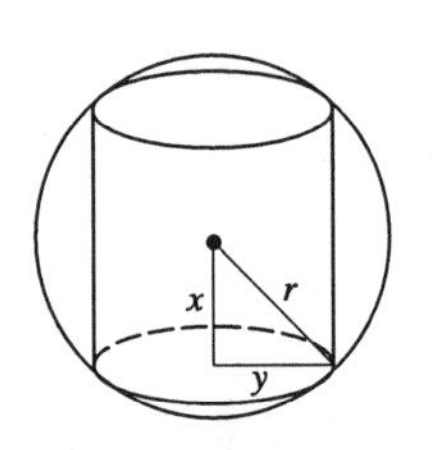

The cylinder has volume $V = \pi y^2(2x)$. Also $x^2 + y^2 = r^2 \quad \Rightarrow$ $y^2 = r^2 - x^2$, so $V(x) = \pi(r^2 - x^2)(2x) = 2\pi\left(r^2x - x^3\right)$, where $0 \le x \le r$. $V'(x) = 2\pi(r^2 - 3x^2) = 0 \quad \Rightarrow \quad x = r/\sqrt{3}$. Now $V(0) = V(r) = 0$, so there is a maximum when $x = r/\sqrt{3}$ and $V\left(r/\sqrt{3}\right) = 4\pi r^3/\left(3\sqrt{3}\right)$.

25.

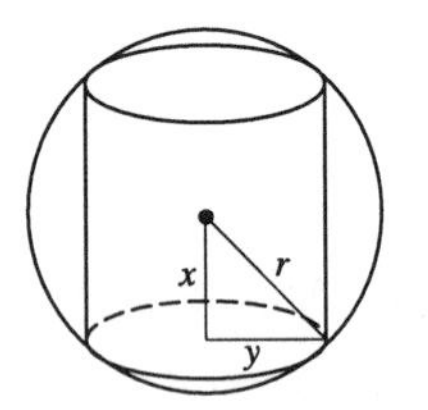

The cylinder has surface area $2\pi y^2 + 2\pi y(2x)$. Now $x^2 + y^2 = r^2 \quad \Rightarrow$ $y = \sqrt{r^2 - x^2}$, so the surface area is $S(x) = 2\pi(r^2 - x^2) + 4\pi x\sqrt{r^2 - x^2}$, $0 \le x \le r$. $S'(x) = -4\pi x + 4\pi\sqrt{r^2 - x^2} - 4\pi x^2/\sqrt{r^2 - x^2}$

$$= \frac{4\pi\left(r^2 - 2x^2 - x\sqrt{r^2 - x^2}\right)}{\sqrt{r^2 - x^2}} = 0 \quad \Rightarrow \quad x\sqrt{r^2 - x^2} = r^2 - 2x^2 \quad (\bigstar)$$

$\Rightarrow \quad x^2(r^2 - x^2) = r^4 - 4r^2x^2 + 4x^4 \quad \Rightarrow \quad 5x^4 - 5r^2x^2 + r^4 = 0$. By the quadratic formula, $x^2 = \frac{5 \pm \sqrt{5}}{10}r^2$, but we reject the root with the $+$ sign since it doesn't satisfy ($\bigstar$). So $x = \sqrt{\frac{5 - \sqrt{5}}{10}}r$. Since $S(0) = S(r) = 0$, the maximum occurs at the critical number and $x^2 = \frac{5 - \sqrt{5}}{10}r^2 \quad \Rightarrow \quad y^2 = \frac{5 + \sqrt{5}}{10}r^2 \quad \Rightarrow \quad$ the surface area is

$$2\pi\left(\frac{5 + \sqrt{5}}{10}\right)r^2 + 4\pi\sqrt{\frac{5 - \sqrt{5}}{10}}\sqrt{\frac{5 + \sqrt{5}}{10}}r^2 = \pi r^2\left(1 + \sqrt{5}\right).$$

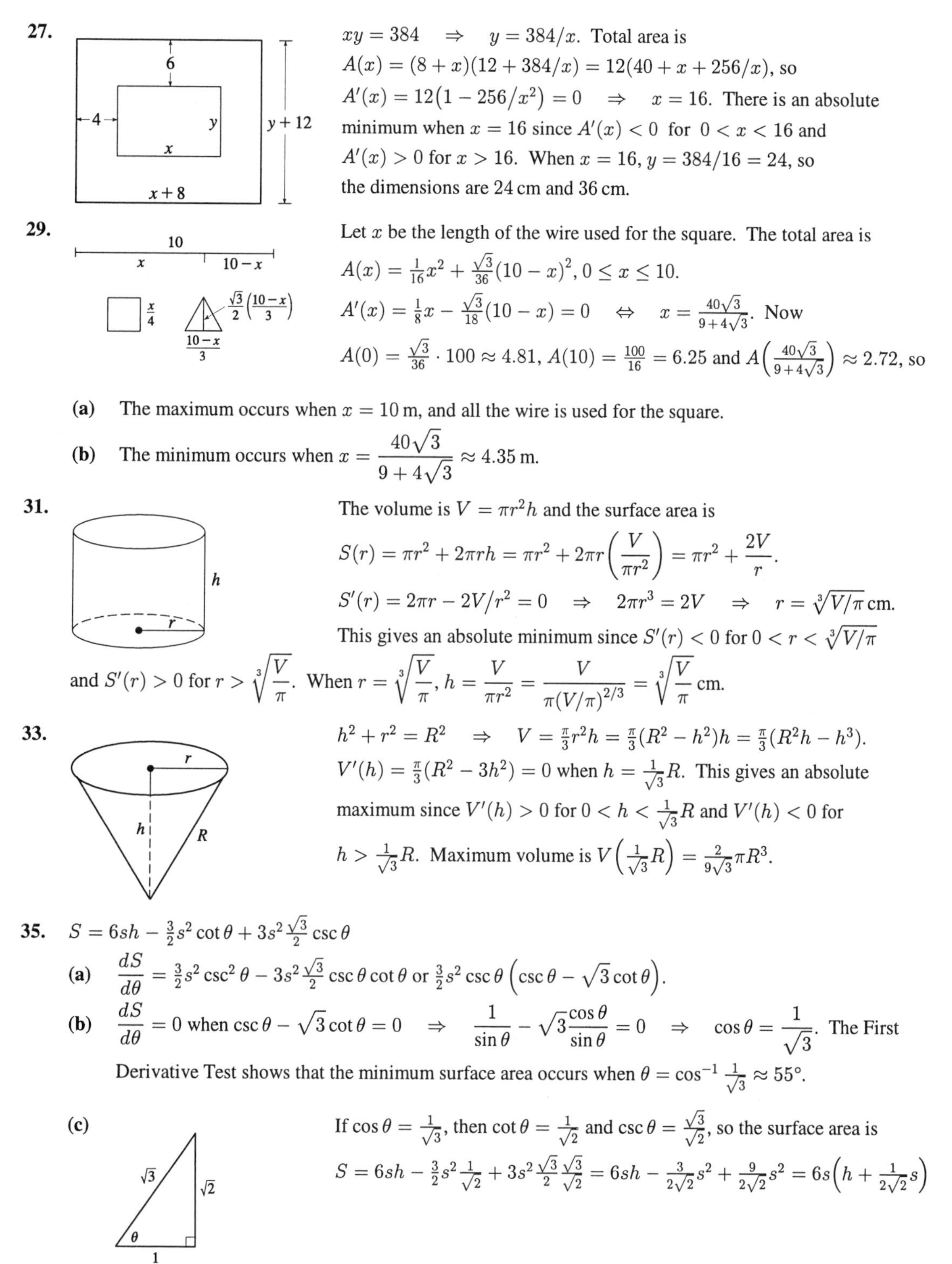

27. $xy = 384 \Rightarrow y = 384/x$. Total area is
$A(x) = (8 + x)(12 + 384/x) = 12(40 + x + 256/x)$, so
$A'(x) = 12(1 - 256/x^2) = 0 \Rightarrow x = 16$. There is an absolute minimum when $x = 16$ since $A'(x) < 0$ for $0 < x < 16$ and $A'(x) > 0$ for $x > 16$. When $x = 16$, $y = 384/16 = 24$, so the dimensions are 24 cm and 36 cm.

29. Let x be the length of the wire used for the square. The total area is
$A(x) = \frac{1}{16}x^2 + \frac{\sqrt{3}}{36}(10 - x)^2, 0 \le x \le 10$.
$A'(x) = \frac{1}{8}x - \frac{\sqrt{3}}{18}(10 - x) = 0 \Leftrightarrow x = \frac{40\sqrt{3}}{9+4\sqrt{3}}$. Now
$A(0) = \frac{\sqrt{3}}{36} \cdot 100 \approx 4.81$, $A(10) = \frac{100}{16} = 6.25$ and $A\left(\frac{40\sqrt{3}}{9+4\sqrt{3}}\right) \approx 2.72$, so

(a) The maximum occurs when $x = 10$ m, and all the wire is used for the square.

(b) The minimum occurs when $x = \dfrac{40\sqrt{3}}{9 + 4\sqrt{3}} \approx 4.35$ m.

31. The volume is $V = \pi r^2 h$ and the surface area is
$$S(r) = \pi r^2 + 2\pi rh = \pi r^2 + 2\pi r\left(\frac{V}{\pi r^2}\right) = \pi r^2 + \frac{2V}{r}.$$
$S'(r) = 2\pi r - 2V/r^2 = 0 \Rightarrow 2\pi r^3 = 2V \Rightarrow r = \sqrt[3]{V/\pi}$ cm.
This gives an absolute minimum since $S'(r) < 0$ for $0 < r < \sqrt[3]{V/\pi}$ and $S'(r) > 0$ for $r > \sqrt[3]{\dfrac{V}{\pi}}$. When $r = \sqrt[3]{\dfrac{V}{\pi}}$, $h = \dfrac{V}{\pi r^2} = \dfrac{V}{\pi(V/\pi)^{2/3}} = \sqrt[3]{\dfrac{V}{\pi}}$ cm.

33. $h^2 + r^2 = R^2 \Rightarrow V = \frac{\pi}{3}r^2 h = \frac{\pi}{3}(R^2 - h^2)h = \frac{\pi}{3}(R^2 h - h^3)$.
$V'(h) = \frac{\pi}{3}(R^2 - 3h^2) = 0$ when $h = \frac{1}{\sqrt{3}}R$. This gives an absolute maximum since $V'(h) > 0$ for $0 < h < \frac{1}{\sqrt{3}}R$ and $V'(h) < 0$ for $h > \frac{1}{\sqrt{3}}R$. Maximum volume is $V\left(\frac{1}{\sqrt{3}}R\right) = \frac{2}{9\sqrt{3}}\pi R^3$.

35. $S = 6sh - \frac{3}{2}s^2 \cot\theta + 3s^2 \frac{\sqrt{3}}{2} \csc\theta$

(a) $\dfrac{dS}{d\theta} = \frac{3}{2}s^2 \csc^2\theta - 3s^2\frac{\sqrt{3}}{2}\csc\theta\cot\theta$ or $\frac{3}{2}s^2\csc\theta\left(\csc\theta - \sqrt{3}\cot\theta\right)$.

(b) $\dfrac{dS}{d\theta} = 0$ when $\csc\theta - \sqrt{3}\cot\theta = 0 \Rightarrow \dfrac{1}{\sin\theta} - \sqrt{3}\dfrac{\cos\theta}{\sin\theta} = 0 \Rightarrow \cos\theta = \dfrac{1}{\sqrt{3}}$. The First Derivative Test shows that the minimum surface area occurs when $\theta = \cos^{-1}\frac{1}{\sqrt{3}} \approx 55°$.

(c) If $\cos\theta = \frac{1}{\sqrt{3}}$, then $\cot\theta = \frac{1}{\sqrt{2}}$ and $\csc\theta = \frac{\sqrt{3}}{\sqrt{2}}$, so the surface area is
$$S = 6sh - \tfrac{3}{2}s^2\tfrac{1}{\sqrt{2}} + 3s^2\tfrac{\sqrt{3}}{2}\tfrac{\sqrt{3}}{\sqrt{2}} = 6sh - \tfrac{3}{2\sqrt{2}}s^2 + \tfrac{9}{2\sqrt{2}}s^2 = 6s\left(h + \tfrac{1}{2\sqrt{2}}s\right)$$

37. Here $T(x) = \dfrac{\sqrt{x^2+25}}{6} + \dfrac{5-x}{8}$, $0 \le x \le 5$, $\Rightarrow$ $T'(x) = \dfrac{x}{6\sqrt{x^2+25}} - \dfrac{1}{8} = 0$ $\Leftrightarrow$ $8x = 6\sqrt{x^2+25}$ $\Leftrightarrow$ $16x^2 = 9(x^2+25)$ $\Leftrightarrow$ $x = \frac{15}{\sqrt{7}}$. But $\frac{15}{\sqrt{7}} > 5$, so T has no critical number. Since $T(0) \approx 1.46$ and $T(5) \approx 1.18$, he should row directly to B.

39.

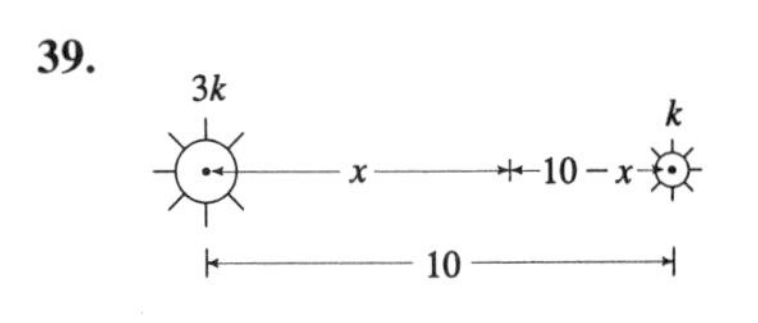

The total illumination is $I(x) = \dfrac{3k}{x^2} + \dfrac{k}{(10-x)^2}$, $0 < x < 10$.

Then $I'(x) = \dfrac{-6k}{x^3} + \dfrac{2k}{(10-x)^3} = 0$ $\Rightarrow$ $6k(10-x)^3 = 2kx^3$

$\Rightarrow$ $\sqrt[3]{3}(10-x) = x$ $\Rightarrow$ $x = 10\sqrt[3]{3}/(1+\sqrt[3]{3}) \approx 5.9$ ft.

This gives a minimum since there is clearly no maximum.

41.

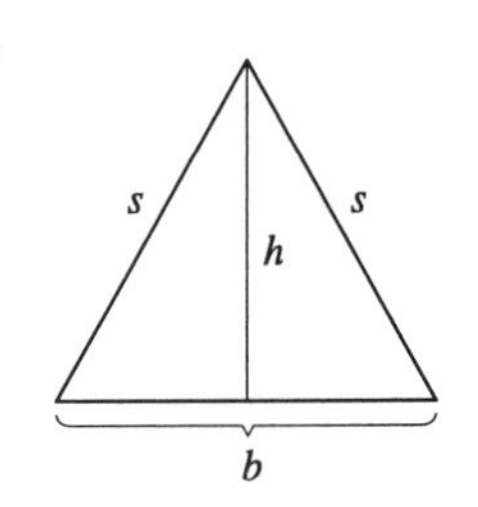

Here $s^2 = h^2 + b^2/4$ so $h^2 = s^2 - b^2/4$. The area is $A = \frac{1}{2}b\sqrt{s^2 - b^2/4}$.

Let the perimeter be p, so $2s + b = p$ or $s = (p-b)/2$ $\Rightarrow$

$A(b) = \frac{1}{2}b\sqrt{(p-b)^2/4 - b^2/4} = b\sqrt{p^2 - 2pb}/4$. Now

$A'(b) = \dfrac{\sqrt{p^2-2pb}}{4} - \dfrac{bp/4}{\sqrt{p^2-2pb}} = \dfrac{-3pb+p^2}{4\sqrt{p^2-2pb}}$. Therefore $A'(b) = 0$

$\Rightarrow$ $-3pb + p^2 = 0$ $\Rightarrow$ $b = p/3$. Since $A'(b) > 0$ for $b < p/3$ and $A'(b) < 0$ for $b > p/3$, there is an absolute maximum when $b = p/3$. But then $2s + p/3 = p$ so $s = p/3$ $\Rightarrow$ $s = b$ $\Rightarrow$ the triangle is equilateral.

43.

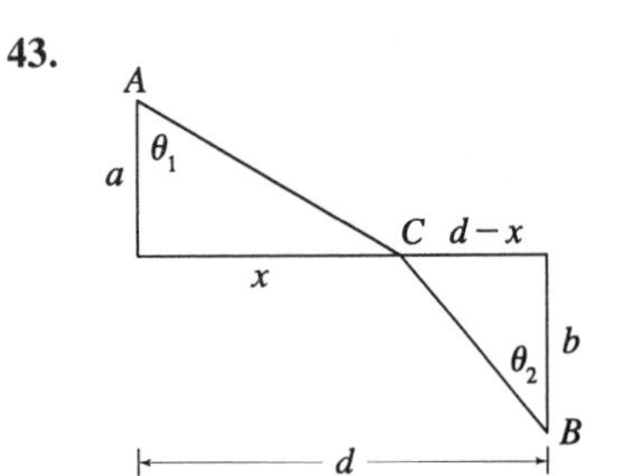

The total time is $T(x) =$ (time from A to C) $+$ (time from C to B)

$= \sqrt{a^2+x^2}/v_1 + \sqrt{b^2+(d-x)^2}/v_2$, $0 < x < d$.

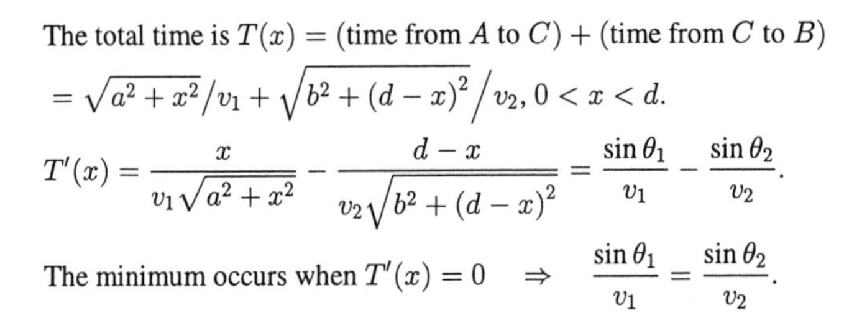

$T'(x) = \dfrac{x}{v_1\sqrt{a^2+x^2}} - \dfrac{d-x}{v_2\sqrt{b^2+(d-x)^2}} = \dfrac{\sin\theta_1}{v_1} - \dfrac{\sin\theta_2}{v_2}$.

The minimum occurs when $T'(x) = 0$ $\Rightarrow$ $\dfrac{\sin\theta_1}{v_1} = \dfrac{\sin\theta_2}{v_2}$.

45.

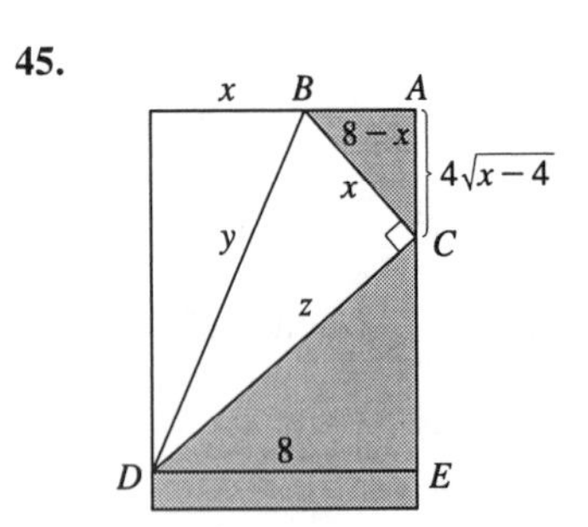

$y^2 = x^2 + z^2$, but triangles CDE and BCA are similar, so $z/8 = x/(4\sqrt{x-4})$. Thus we minimize

$f(x) = y^2 = x^2 + 4x^2/(x-4) = x^3/(x-4)$, $4 < x \le 8$.

$f'(x) = \dfrac{3x^2(x-4) - x^3}{(x-4)^2} = \dfrac{2x^2(x-6)}{(x-4)^2} = 0$ when $x = 6$.

$f'(x) < 0$ when $x < 6$, $f'(x) > 0$ when $x > 6$, so the minimum occurs when $x = 6$ in.

47.

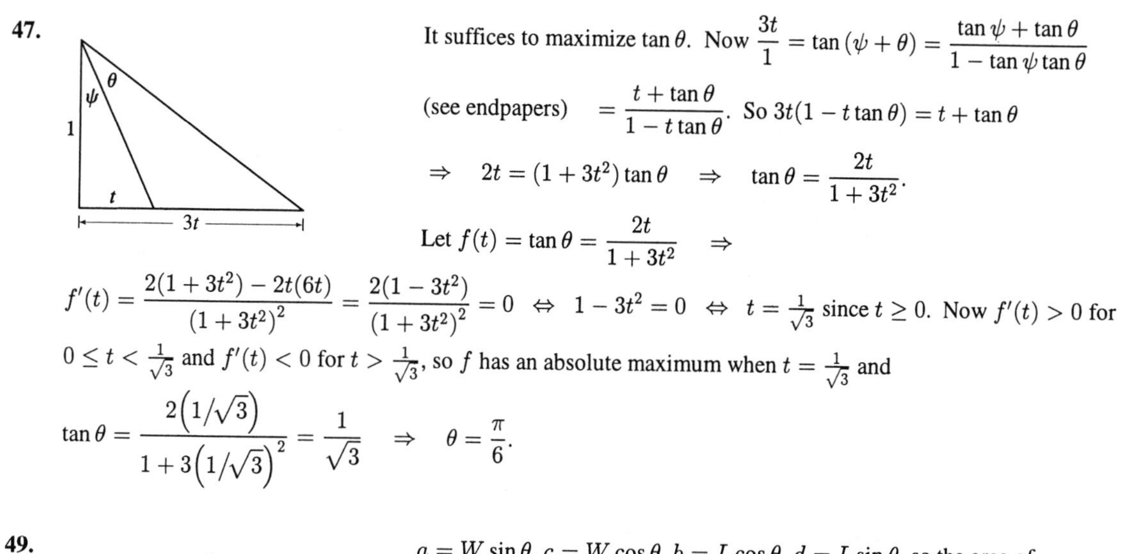

It suffices to maximize $\tan\theta$. Now $\dfrac{3t}{1} = \tan(\psi+\theta) = \dfrac{\tan\psi + \tan\theta}{1 - \tan\psi\tan\theta}$ (see endpapers) $= \dfrac{t + \tan\theta}{1 - t\tan\theta}$. So $3t(1 - t\tan\theta) = t + \tan\theta$

$\Rightarrow \quad 2t = (1 + 3t^2)\tan\theta \quad \Rightarrow \quad \tan\theta = \dfrac{2t}{1 + 3t^2}$.

Let $f(t) = \tan\theta = \dfrac{2t}{1+3t^2} \quad \Rightarrow$

$f'(t) = \dfrac{2(1+3t^2) - 2t(6t)}{(1+3t^2)^2} = \dfrac{2(1-3t^2)}{(1+3t^2)^2} = 0 \;\Leftrightarrow\; 1 - 3t^2 = 0 \;\Leftrightarrow\; t = \frac{1}{\sqrt{3}}$ since $t \geq 0$. Now $f'(t) > 0$ for $0 \leq t < \frac{1}{\sqrt{3}}$ and $f'(t) < 0$ for $t > \frac{1}{\sqrt{3}}$, so f has an absolute maximum when $t = \frac{1}{\sqrt{3}}$ and

$$\tan\theta = \frac{2\left(1/\sqrt{3}\right)}{1 + 3\left(1/\sqrt{3}\right)^2} = \frac{1}{\sqrt{3}} \quad \Rightarrow \quad \theta = \frac{\pi}{6}.$$

49.

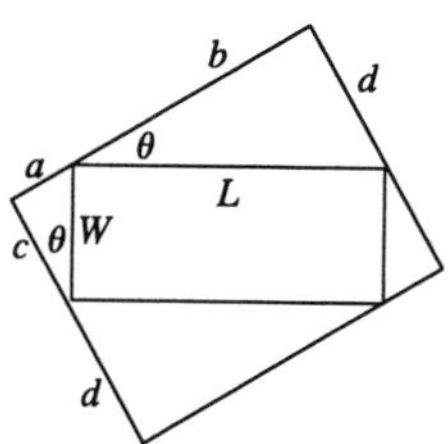

$a = W\sin\theta$, $c = W\cos\theta$, $b = L\cos\theta$, $d = L\sin\theta$, so the area of the circumscribed rectangle is

$$\begin{aligned} A(\theta) &= (a+b)(c+d) = (W\sin\theta + L\cos\theta)(W\cos\theta + L\sin\theta) \\ &= LW\sin^2\theta + LW\cos^2\theta + \left(L^2 + W^2\right)\sin\theta\cos\theta \\ &= LW + \tfrac{1}{2}(L^2+W^2)\sin 2\theta,\; 0 \leq \theta \leq \tfrac{\pi}{2}. \end{aligned}$$

This expression shows, without calculus, that the maximum value of $A(\theta)$ occurs when $\sin 2\theta = 1 \;\Leftrightarrow\; 2\theta = \frac{\pi}{2} \;\Leftrightarrow\; x = \frac{\pi}{4}$. So the maximum area is $A\left(\frac{\pi}{4}\right) = LW + \frac{1}{2}(L^2 + W^2) = \frac{1}{2}(L+W)^2$.

51. $L(x) = |AP| + |BP| + |CP| = x + \sqrt{(5-x)^2 + 2^2} + \sqrt{(5-x)^2 + 3^2}$

$= x + \sqrt{x^2 - 10x + 29} + \sqrt{x^2 - 10x + 34} \quad \Rightarrow$

$$L'(x) = 1 + \frac{x-5}{\sqrt{x^2 - 10x + 29}} + \frac{x-5}{\sqrt{x^2 - 10x + 34}}$$

From the graphs of L and L', it seems that the minimum value of L is about $L(3.59) = 9.35$ m.

EXERCISES 3.9

1. **(a)** $C(0)$ represents the fixed costs of production, such as rent, utilities, machinery etc., which are incurred even when nothing is produced.

(b) The inflection point is the point at which $C''(x)$ changes from negative to positive, that is, the marginal cost $C'(x)$ changes from decreasing to increasing. So the marginal cost is minimized.

(c) The marginal cost function is $C'(x)$.

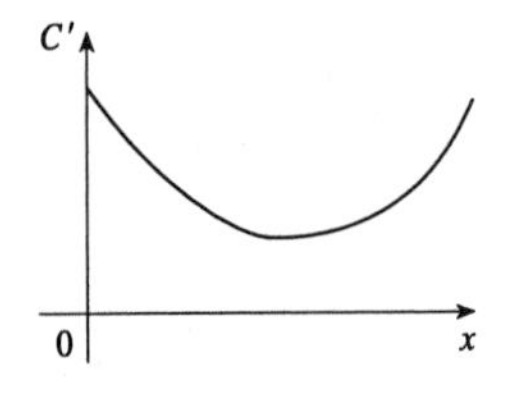

3. **(a)** $C(x) = 10{,}000 + 25x + x^2$, $C(1000) = \$1{,}035{,}000$, $c(x) = \dfrac{C(x)}{x} = \dfrac{10{,}000}{x} + 25 + x$,

$c(1000) = \$1035$. $C'(x) = 25 + 2x$, $C'(1000) = \$2025/\text{unit}$.

(b) We must have $c(x) = C'(x) \Rightarrow 10{,}000/x + 25 + x = 25 + 2x \Rightarrow 10{,}000/x = x \Rightarrow$ $x^2 = 10{,}000 \Rightarrow x = 100$. This is a minimum since $c''(x) = 20{,}000/x^3 > 0$.

(c) The minimum average cost is $c(100) = \$225$.

5. **(a)** $C(x) = 45 + \dfrac{x}{2} + \dfrac{x^2}{560}$, $C(1000) = \$2330.71$. $c(x) = \dfrac{45}{x} + \dfrac{1}{2} + \dfrac{x}{560}$, $c(1000) = \$2.33$.

$C'(x) = \dfrac{1}{2} + \dfrac{x}{280}$, $C'(1000) = \$4.07/\text{unit}$

(b) We must have $C'(x) = c(x) \Rightarrow \dfrac{1}{2} + \dfrac{x}{280} = \dfrac{45}{x} + \dfrac{1}{2} + \dfrac{x}{560} \Rightarrow \dfrac{45}{x} = \dfrac{x}{560} \Rightarrow$ $x^2 = (45)(560) \Rightarrow x = \sqrt{25{,}200} \approx 159$. This is a minimum since $c''(x) = 90/x^2 > 0$.

(c) The minimum average cost is $c(159) = \$1.07$.

7. **(a)** $C(x) = 2\sqrt{x} + \dfrac{x^2}{8000}$, $C(1000) = \$188.25$. $c(x) = \dfrac{2}{\sqrt{x}} + \dfrac{x}{8000}$, $c(1000) = \$0.19$.

$C'(x) = \dfrac{1}{\sqrt{x}} + \dfrac{x}{4000}$, $C'(1000) = \$0.28/\text{unit}$.

(b) We must have $C'(x) = c(x) \Rightarrow \dfrac{1}{\sqrt{x}} + \dfrac{x}{4000} = \dfrac{2}{\sqrt{x}} + \dfrac{x}{8000} \Rightarrow \dfrac{x}{8000} = \dfrac{1}{\sqrt{x}} \Rightarrow$ $x^{3/2} = 8000 \Rightarrow x = (8000)^{2/3} = 400$. This is a minimum since $c''(x) = \frac{3}{2}x^{-5/2} > 0$.

(c) The minimum average cost is $c(400) = \$0.15$.

9. $C(x) = 680 + 4x + 0.01x^2$, $p(x) = 12 \Rightarrow R(x) = xp(x) = 12x$. If the profit is maximum, then $R'(x) = C'(x) \Rightarrow 12 = 4 + 0.02x \Rightarrow 0.02x = 8 \Rightarrow x = 400$. Now $R''(x) = 0 < 0.02 = C''(x)$, so $x = 400$ gives a maximum.

11. $C(x) = 1200 + 25x - 0.0001x^2$, $p(x) = 55 - x/1000$. Then $R(x) = xp(x) = 55x - x^2/1000$. If the profit is maximum, then $R'(x) = C'(x) \quad \Leftrightarrow \quad 55 - x/500 = 25 - 0.0002x \quad \Rightarrow \quad 30 = 0.0018x \quad \Rightarrow$ $x = 30/0.0018 \approx 16{,}667$. Now $R''(x) = -\frac{1}{500} < -0.0002 = C''(x)$, so $x = 16{,}667$ gives a maximum.

13. $C(x) = 1450 + 36x - x^2 + 0.001x^3$, $p(x) = 60 - 0.01x$. Then $R(x) = xp(x) = 60x - 0.01x^2$. If the profit is maximum, then $R'(x) = C'(x) \quad \Leftrightarrow \quad 60 - 0.02x = 36 - 2x + 0.003x^2 \quad \Rightarrow \quad 0.003x^2 - 1.98x - 24 = 0$.

By the quadratic formula, $x = \dfrac{1.98 \pm \sqrt{(-1.98)^2 + 4(0.003)(24)}}{2(0.003)} = \dfrac{1.98 \pm \sqrt{4.2084}}{0.006}$. Since $x > 0$, $x \approx (1.98 + 2.05)/0.006 \approx 672$. Now $R''(x) = -0.02$ and $C''(x) = -2 + 0.006x \quad \Rightarrow \quad C''(672) = 2.032$ $\Rightarrow \quad R''(672) < C''(672) \quad \Rightarrow \quad$ there is a maximum at $x = 672$.

15. $C(x) = 0.001x^3 - 0.3x^2 + 6x + 900$. The marginal cost is $C'(x) = 0.003x^2 - 0.6x + 6$. $C'(x)$ is increasing when $C''(x) > 0 \quad \Leftrightarrow \quad 0.006x - 0.6 > 0 \quad \Leftrightarrow \quad x > 0.6/0.006 = 100$. So $C'(x)$ starts to increase when $x = 100$.

17. **(a)** We are given that the demand function p is linear and $p(27{,}000) = 10$, $p(33{,}000) = 8$, so the slope is $\dfrac{10 - 8}{27{,}000 - 33{,}000} = -\dfrac{1}{3000}$ and the equation of the graph is $y - 10 = \left(-\frac{1}{3000}\right)(x - 27{,}000) \quad \Rightarrow$ $p(x) = 19 - x/3000$.

(b) The revenue is $R(x) = xp(x) = 19x - x^2/3000 \quad \Rightarrow \quad R'(x) = 19 - x/1500 = 0$ when $x = 28{,}500$. Since $R''(x) = -1/1500 < 0$, the maximum revenue occurs when $x = 28{,}500 \quad \Rightarrow \quad$ the price is $p(28{,}500) = \$9.50$.

19. **(a)** $p(x) = 450 - \frac{1}{10}(x - 1000) = 550 - x/10$.

(b) $R(x) = xp(x) = 500x - x^2/10$. $R'(x) = 550 - x/5 = 0$ when $x = 5(550) = 2750$. $p(2750) = 275$, so the rebate should be $450 - 275 = \$175$.

(c) $P(x) = R(x) - C(x) = 550x - x^2/10 - 6800 - 150x = 400x - x^2/10 - 6800$, $P'(x) = 400 - x/5 = 0$ when $x = 2000$. $p(2000) = 550 - 200 = 350$. Therefore the rebate to maximize profits should be $450 - 350 = \$100$.

EXERCISES 3.10

1. $f(x) = 12x^2 + 6x - 5 \quad \Rightarrow \quad F(x) = 12\left(\frac{1}{3}x^3\right) + 6\left(\frac{1}{2}x^2\right) - 5x + C = 4x^3 + 3x^2 - 5x + C$

3. $f(x) = 6x^9 - 4x^7 + 3x^2 + 1 \quad \Rightarrow$
$F(x) = 6\left(\frac{1}{10}x^{10}\right) - 4\left(\frac{1}{8}x^8\right) + 3\left(\frac{1}{3}x^3\right) + x + C = \frac{3}{5}x^{10} - \frac{1}{2}x^8 + x^3 + x + C$

5. $f(x) = \sqrt{x} + \sqrt[3]{x} = x^{1/2} + x^{1/3} \quad \Rightarrow \quad F(x) = \frac{1}{3/2}x^{3/2} + \frac{1}{4/3}x^{4/3} + C = \frac{2}{3}x^{3/2} + \frac{3}{4}x^{4/3} + C$

7. $f(x) = 6/x^5 = 6x^{-5} \quad \Rightarrow \quad F(x) = \begin{cases} 6x^{-4}/(-4) + C_1 = -3/(2x^4) + C_1 & \text{if } x < 0 \\ -3/(2x^4) + C_2 & \text{if } x > 0 \end{cases}$

9. $g(t) = \dfrac{t^3 + 2t^2}{\sqrt{t}} = t^{5/2} + 2t^{3/2} \quad \Rightarrow \quad G(t) = \dfrac{t^{7/2}}{7/2} + \dfrac{2t^{5/2}}{5/2} + C = \frac{2}{7}t^{7/2} + \frac{4}{5}t^{5/2} + C$

11. $h(x) = \sin x - 2\cos x \quad \Rightarrow \quad H(x) = -\cos x - 2\sin x + C$

13. $f(t) = \sec^2 t + t^2 \quad \Rightarrow \quad F(t) = \tan t + \frac{1}{3}t^3 + C_n$ on the interval $\left(n\pi - \frac{\pi}{2}, n\pi + \frac{\pi}{2}\right)$.

15. $f''(x) = x^2 + x^3 \quad \Rightarrow \quad f'(x) = \frac{1}{3}x^3 + \frac{1}{4}x^4 + C \quad \Rightarrow \quad f(x) = \frac{1}{12}x^4 + \frac{1}{20}x^5 + Cx + D$

17. $f''(x) = 1 \quad \Rightarrow \quad f'(x) = x + C \quad \Rightarrow \quad f(x) = \frac{1}{2}x^2 + Cx + D$

19. $f'''(x) = 24x \quad \Rightarrow \quad f''(x) = 12x^2 + C \quad \Rightarrow \quad f'(x) = 4x^3 + Cx + D \quad \Rightarrow$
$f(x) = x^4 + \frac{1}{2}Cx^2 + Dx + E$

21. $f'(x) = 4x + 3 \quad \Rightarrow \quad f(x) = 2x^2 + 3x + C \quad \Rightarrow \quad -9 = f(0) = C \quad \Rightarrow \quad f(x) = 2x^2 + 3x - 9$

23. $f'(x) = 3\sqrt{x} - 1/\sqrt{x} = 3x^{1/2} - x^{-1/2} \quad \Rightarrow \quad f(x) = 3\left(\frac{1}{3/2}\right)x^{3/2} - \frac{1}{1/2}x^{1/2} + C \quad \Rightarrow$
$2 = f(1) = 2 - 2 + C = C \quad \Rightarrow \quad f(x) = 2x^{3/2} - 2x^{1/2} + 2$

25. $f'(x) = 3\cos x + 5\sin x \quad \Rightarrow \quad f(x) = 3\sin x - 5\cos x + C \quad \Rightarrow \quad 4 = f(0) = -5 + C \quad \Rightarrow \quad C = 9$
$\Rightarrow \quad f(x) = 3\sin x - 5\cos x + 9.$

27. $f''(x) = x \quad \Rightarrow \quad f'(x) = \frac{1}{2}x^2 + C \quad \Rightarrow \quad 2 = f'(0) = C \quad \Rightarrow \quad f'(x) = \frac{1}{2}x^2 + 2 \quad \Rightarrow$
$f(x) = \frac{1}{6}x^3 + 2x + D \quad \Rightarrow \quad -3 = f(0) = D \quad \Rightarrow \quad f(x) = \frac{1}{6}x^3 + 2x - 3$

29. $f''(x) = x^2 + 3\cos x \quad \Rightarrow \quad f'(x) = \frac{1}{3}x^3 + 3\sin x + C \quad \Rightarrow \quad 3 = f'(0) = C \quad \Rightarrow$
$f'(x) = \frac{1}{3}x^3 + 3\sin x + 3 \quad \Rightarrow \quad f(x) = \frac{1}{12}x^4 - 3\cos x + 3x + D \quad \Rightarrow \quad 2 = f(0) = -3 + D \quad \Rightarrow$
$D = 5 \quad \Rightarrow \quad f(x) = \frac{1}{12}x^4 - 3\cos x + 3x + 5$

31. $f''(x) = 6x + 6 \quad \Rightarrow \quad f'(x) = 3x^2 + 6x + C \quad \Rightarrow \quad f(x) = x^3 + 3x^2 + Cx + D \quad \Rightarrow \quad 4 = f(0) = D$
and $3 = f(1) = 1 + 3 + C + D = 4 + C + 4 \quad \Rightarrow \quad C = -5 \quad \Rightarrow \quad f(x) = x^3 + 3x^2 - 5x + 4$

33. $f''(x) = x^{-3} \quad \Rightarrow \quad f'(x) = -\frac{1}{2}x^{-2} + C \quad \Rightarrow \quad f(x) = \frac{1}{2}x^{-1} + Cx + D \quad \Rightarrow \quad 0 = f(1) = \frac{1}{2} + C + D$
and $0 = f(2) = \frac{1}{4} + 2C + D$. Solving these equations, we get $C = \frac{1}{4}$, $D = -\frac{3}{4}$, so $f(x) = 1/(2x) + \frac{1}{4}x - \frac{3}{4}$.

35. We have that $f'(x) = 2x + 1 \quad \Rightarrow \quad f(x) = x^2 + x + C$. But f passes through $(1, 6)$ so that $6 = f(1) = 1^2 + 1 + C \quad \Rightarrow \quad C = 4$. Therefore $f(x) = x^2 + x + 4 \quad \Rightarrow \quad f(2) = 2^2 + 2 + 4 = 10$.

37. b is the antiderivative of f. For small x, f is negative, so the graph of its antiderivative must be decreasing. But both a and c are increasing for small x, so only b can be f's antiderivative. Also, f is positive where b is increasing, which supports our conclusion.

39. The graph of F will have a minimum at 0 and a maximum at 2, since $f = F'$ goes from negative to positive at $x = 0$, and from positive to negative at $x = 2$.

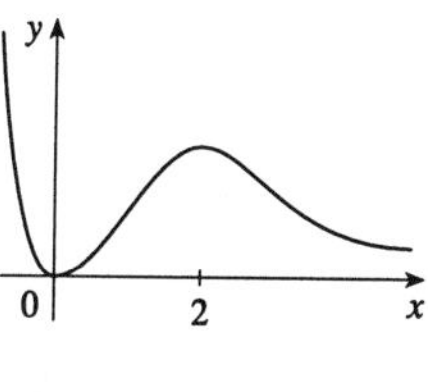

41.

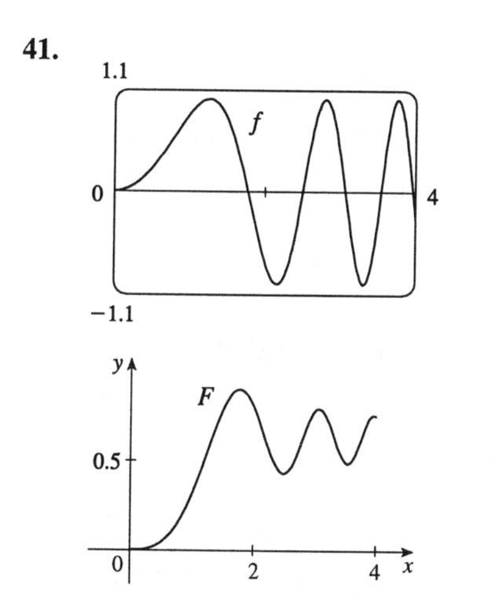

43.

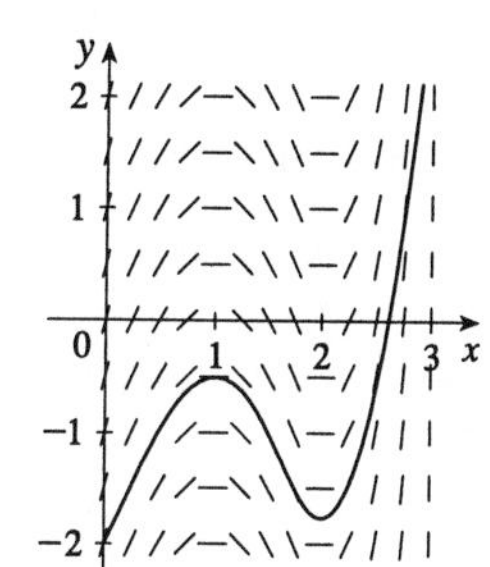

45.

x	$f(x)$
0	1
0.5	0.959
1.0	0.841
1.5	0.665
2.0	0.455
2.5	0.239
3.0	0.470

x	$f(x)$
3.5	−0.100
4.0	−0.189
4.5	−0.217
5.0	−0.192
5.5	−0.128
6.0	−0.047

We compute slopes as in the table and draw a direction field as in Example 6. Then we use the direction field to graph F starting at $(0, 0)$.

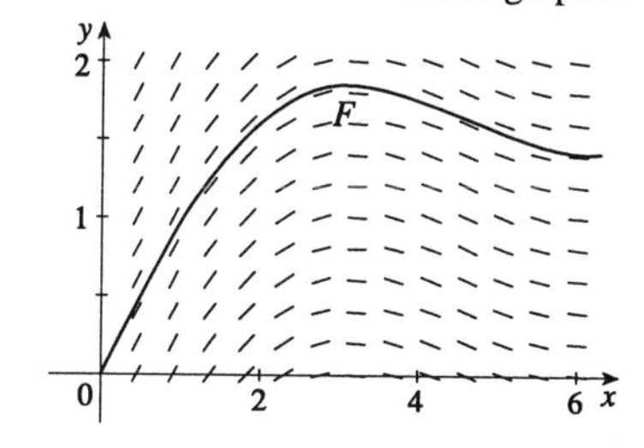

47.

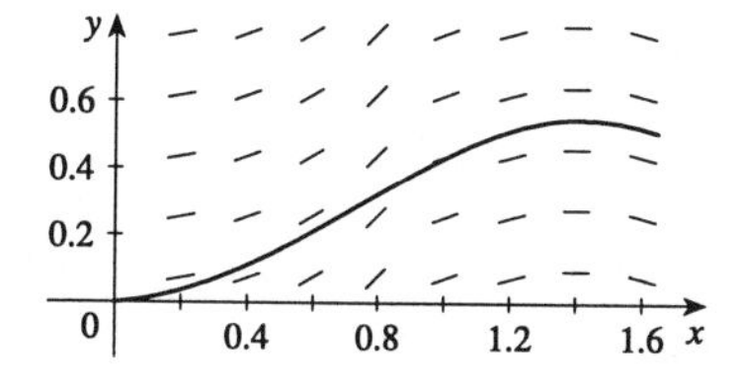

49. $v(t) = s'(t) = 3 - 2t \quad \Rightarrow \quad s(t) = 3t - t^2 + C \quad \Rightarrow \quad 4 = s(0) = C \quad \Rightarrow \quad s(t) = 3t - t^2 + 4$

51. $a(t) = v'(t) = 3t + 8 \quad \Rightarrow \quad v(t) = \frac{3}{2}t^2 + 8t + C \quad \Rightarrow \quad -2 = v(0) = C \quad \Rightarrow \quad v(t) = \frac{3}{2}t^2 + 8t - 2 \quad \Rightarrow$
$s(t) = \frac{1}{2}t^3 + 4t^2 - 2t + D \quad \Rightarrow \quad 1 = s(0) = D \quad \Rightarrow \quad s(t) = \frac{1}{2}t^3 + 4t^2 - 2t + 1$

53. $a(t) = v'(t) = t^2 - t \quad \Rightarrow \quad v(t) = \frac{1}{3}t^3 - \frac{1}{2}t^2 + C \quad \Rightarrow \quad s(t) = \frac{1}{12}t^4 - \frac{1}{6}t^3 + Ct + D \quad \Rightarrow \quad 0 = s(0) = D$
and $12 = s(6) = 108 - 36 + 6C + 0 \quad \Rightarrow \quad C = -10 \quad \Rightarrow \quad s(t) = \frac{1}{12}t^4 - \frac{1}{6}t^3 - 10t$

55. **(a)** $v'(t) = a(t) = -9.8 \quad \Rightarrow \quad v(t) = -9.8t + C$, but $C = v(0) = 0$, so $v(t) = -9.8t \quad \Rightarrow$
$s(t) = -4.9t^2 + D \quad \Rightarrow \quad D = s(0) = 450 \quad \Rightarrow \quad s(t) = 450 - 4.9t^2$

(b) It reaches the ground when $0 = s(t) = 450 - 4.9t^2 \quad \Rightarrow \quad t^2 = 450/4.9 \quad \Rightarrow \quad t = \sqrt{450/4.9} \approx 9.58$ s.

(c) $v = -9.8\sqrt{450/4.9} \approx -93.9$ m/s

57. **(a)** $v'(t) = -9.8 \quad \Rightarrow \quad v(t) = -9.8t + C \quad \Rightarrow \quad 5 = v(0) = C$, so $v(t) = 5 - 9.8t \quad \Rightarrow$
$s(t) = 5t - 4.9t^2 + D \quad \Rightarrow \quad D = s(0) = 450 \quad \Rightarrow \quad s(t) = 450 + 5t - 4.9t^2$

(b) It reaches the ground when $450 + 5t - 4.9t^2 = 0$. By the quadratic formula, the positive root of this equation is $t = \dfrac{5 + \sqrt{8845}}{9.8} \approx 10.1$ s

(c) $v = 5 - 9.8 \cdot \dfrac{5 + \sqrt{8845}}{9.8} \approx -94.0$ m/s

59. By Exercise 58, $s(t) = -4.9t^2 + v_0t + s_0$ and $v(t) = s'(t) = -9.8t + v_0$. So $[v(t)]^2 = (9.8)^2t^2 - 19.6v_0t + v_0^2$ and $v_0^2 - 19.6[s(t) - s_0] = v_0^2 - 19.6[-4.9t^2 + v_0t] = v_0^2 + (9.8)^2t^2 - 19.6v_0t = [v(t)]^2$

61. Marginal cost $= 1.92 - 0.002x = C'(x) \quad \Rightarrow \quad C(x) = 1.92x - 0.001x^2 + K$. But
$C(1) = 1.92 - 0.001 + K = 562 \quad \Rightarrow \quad K = 560.081$. Therefore $C(x) = 1.92x - 0.001x^2 + 560.081 \quad \Rightarrow$
$C(100) = 1.92(100) - 0.001(100)^2 + 560.081 = 742.081$, so the cost of producing 100 items is \$742.08.

63. Taking the upward direction to be positive we have that for $0 \le t \le 10$ (using the subscript 1 to refer to $0 \le t \le 10$), $a_1(t) = -9 + 0.9t = v_1'(t) \quad \Rightarrow \quad v_1(t) = -9t + 0.45t^2 + v_0$, but $v_1(0) = v_0 = -10 \quad \Rightarrow$
$v_1(t) = -9t + 0.45t^2 - 10 = s_1'(t) \quad \Rightarrow \quad s_1(t) = -\frac{9}{2}t^2 + 0.15t^3 - 10t + s_0$. But $s_1(0) = 500 = s_0 \quad \Rightarrow$
$s_1(t) = -\frac{9}{2}t^2 + 0.15t^3 - 10t + 500$. Now for $t > 10$, $a(t) = 0 = v'(t) \quad \Rightarrow$
$v(t) = \text{constant} = v_1(10) = -9(10) + 0.45(10)^2 - 10 = -55 \quad \Rightarrow \quad v(t) = -55 = s'(t) \quad \Rightarrow$
$s(t) = -55t + s_{10}$. But $s(10) = s_1(10) \quad \Rightarrow \quad -55(10) + s_{10} = 100 \quad \Rightarrow \quad s_{10} = 650 \quad \Rightarrow$
$s(t) = -55t + 650$. When the raindrop hits the ground we have that $s(t) = 0 \quad \Rightarrow \quad -55t + 650 = 0 \quad \Rightarrow$
$t = \dfrac{650}{55} = \dfrac{130}{11} \approx 11.8$ s.

65. $a(t) = a$ and the initial velocity is 30 mi/h $= 30 \cdot \frac{5280}{3600} = 44$ ft/s and final velocity
50 mi/h $= 50 \cdot \frac{5280}{3600} = \frac{220}{3}$ ft/s. So $v(t) = at + 44 \quad \Rightarrow \quad \frac{220}{3} = v(5) = 5a + 44 \quad \Rightarrow \quad a = \frac{88}{15} \approx 5.87$ ft/s^2.

67. The height at time t is $s(t) = -16t^2 + h$, where $h = s(0)$ is the height of the cliff. $v(t) = -32t = -120$ when $t = 3.75$, so $0 = s(3.75) = -16(3.75)^2 + h \quad \Rightarrow \quad h = 16(3.75)^2 = 225$ ft.

REVIEW EXERCISES FOR CHAPTER 3

1. False. For example, take $f(x) = x^3$, then $f'(x) = 3x^2$ and $f'(0) = 3(0)^2 = 0$, but $f(0) = 0$ is not a maximum or minimum; $(0, 0)$ is an inflection point.

3. False. For example, $f(x) = x$ is continuous on $(0, 1)$ but attains neither a maximum nor a minimum value on $(0, 1)$.

5. True, by the Test for Monotonic Functions.

7. False. $f(x) = g(x) + C$ by Corollary 3.2.7. For example, $f(x) = x + 2$, $g(x) = x + 1$ $\Rightarrow$ $f'(x) = g'(x) = 1$, but $f(x) \neq g(x)$.

9. True. The graph of one such function is sketched. [An example is $f(x) = e^{-x}$.]

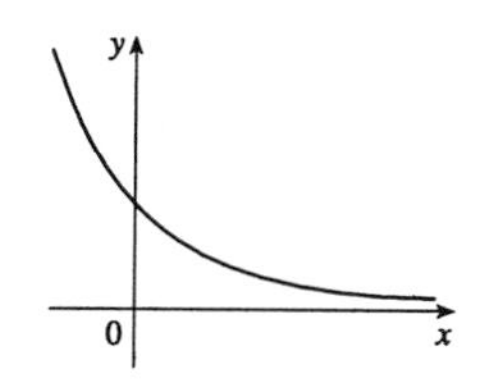

11. True. Let $x_1 < x_2$ where $x_1, x_2 \in I$. Then $f(x_1) < f(x_2)$ and $g(x_1) < g(x_2)$ (since f and g are increasing on I), so $(f + g)(x_1) = f(x_1) + g(x_1) < f(x_2) + g(x_2) = (f + g)(x_2)$.

13. False. Take $f(x) = x$ and $g(x) = x - 1$. Then both f and g are increasing on $[0, 1]$. But $f(x)g(x) = x(x - 1)$ is not increasing on $[0, 1]$.

15. True. Let $x_1, x_2 \in I$ and $x_1 < x_2$, then $f(x_1) < f(x_2)$ (f is increasing) $\Rightarrow$ $\dfrac{1}{f(x_1)} > \dfrac{1}{f(x_2)}$ (f is positive) $\Rightarrow$ $g(x_1) > g(x_2)$ $\Rightarrow$ $g(x) = \dfrac{1}{f(x)}$ is decreasing on I.

17. $f(x) = x^3 - 12x + 5$, $-5 \leq x \leq 3$. $f'(x) = 3x^2 - 12 = 0$ $\Rightarrow$ $x^2 = 4$ $\Rightarrow$ $x = \pm 2$. $f''(x) = 6x$ $\Rightarrow$ $f''(-2) = -12 < 0$, so $f(-2) = 21$ is a local maximum, and $f''(2) = 12 > 0$, so $f(2) = -11$ is a local minimum. Also $f(-5) = -60$ and $f(3) = -4$, so $f(-2) = 21$ is the absolute maximum and $f(-5) = -60$ is the absolute minimum.

19. $f(x) = \dfrac{x - 2}{x + 2}$, $0 \leq x \leq 4$. $f'(x) = \dfrac{(x + 2) - (x - 2)}{(x + 2)^2} = \dfrac{4}{(x + 2)^2} > 0$ $\Rightarrow$ f is increasing on $[0, 4]$, so f has no local extremum and $f(0) = -1$ is the absolute minimum and $f(4) = \frac{1}{3}$ is the absolute maximum.

21. $f(x) = x - \sqrt{2}\sin x$, $0 \leq x \leq \pi$. $f'(x) = 1 - \sqrt{2}\cos x = 0$ $\Rightarrow$ $\cos x = \frac{1}{\sqrt{2}}$ $\Rightarrow$ $x = \frac{\pi}{4}$. $f''\left(\frac{\pi}{4}\right) = \sqrt{2}\sin\frac{\pi}{4} = 1 > 0$, so $f\left(\frac{\pi}{4}\right) = \frac{\pi}{4} - 1$ is a local minimum. Also $f(0) = 0$ and $f(\pi) = \pi$, so the absolute minimum is $f\left(\frac{\pi}{4}\right) = \frac{\pi}{4} - 1$, the absolute maximum is $f(\pi) = \pi$.

23. $\displaystyle\lim_{x\to\infty} \frac{1 + 2x - x^2}{1 - x + 2x^2} = \lim_{x\to\infty} \frac{(1/x^2) + (2/x) - 1}{(1/x^2) - (1/x) + 2} = \frac{0 + 0 - 1}{0 - 0 + 2} = -\frac{1}{2}$

25. $\displaystyle\lim_{x\to\infty}\frac{\sqrt{x^2-9}}{2x-6}=\lim_{x\to\infty}\frac{\sqrt{1-9/x^2}}{2-6/x}=\frac{\sqrt{1-0}}{2-0}=\frac{1}{2}$

27. $\lim_{x\to\infty}\left(\sqrt[3]{x}-\frac{1}{3}x\right)=\lim_{x\to\infty}\sqrt[3]{x}\left(1-\frac{1}{3}x^{2/3}\right)=-\infty$, since $\sqrt[3]{x}\to\infty$ and $1-\frac{1}{3}x^{2/3}\to-\infty$.

29. $y=f(x)=1+x+x^3$ **A.** $D=\mathbb{R}$ **B.** y-intercept $=1$ **C.** No symmetry **D.** $\lim_{x\to\infty}(1+x+x^3)=\infty$, $\lim_{x\to-\infty}(1+x+x^3)=-\infty$, no asymptotes **E.** $f'(x)=1+3x^2 \Rightarrow f'(x)>0$, so f is increasing on $\mathbb{R}$ **F.** No local extremum **G.** $f''(x)=6x \Rightarrow f''(x)>0$ if $x>0$ and $f''(x)<0$ if $x<0$, so f is CU on $(0,\infty)$ and CD on $(-\infty,0)$. IP $(0,1)$

H.

31. $y=f(x)=\dfrac{1}{x(x-3)^2}$ **A.** $D=\{x\mid x\neq 0,3\}=(-\infty,0)\cup(0,3)\cup(3,\infty)$

B. No intercepts. **C.** No symmetry. **D.** $\displaystyle\lim_{x\to\pm\infty}\frac{1}{x(x-3)^2}=0$, so $y=0$ is a HA. $\displaystyle\lim_{x\to0^+}\frac{1}{x(x-3)^2}=\infty$, $\displaystyle\lim_{x\to0^-}\frac{1}{x(x-3)^2}=-\infty$, $\displaystyle\lim_{x\to3}\frac{1}{x(x-3)^2}=\infty$, so $x=0$ and $x=3$ are VA.

E. $f'(x)=-\dfrac{(x-3)^2+2x(x-3)}{x^2(x-3)^4}=\dfrac{3(1-x)}{x^2(x-3)^3} \Rightarrow f'(x)>0 \Leftrightarrow 1<x<3$, so f is increasing on $[1,3)$ and decreasing on $(-\infty,0)$, $(0,1]$, and $(3,\infty)$. **F.** $f(1)=\frac{1}{4}$ is a local minimum.

G. $f''(x)=\dfrac{6(2x^2-4x+3)}{x^3(x-3)^4}$. Note that $2x^2-4x+3>0$ for all x since it has negative discriminant. So $f''(x)>0$ $\Leftrightarrow$ $x>0$ $\Rightarrow$ f is CU on $(0,3)$ and $(3,\infty)$ and CD on $(-\infty,0)$. No IP

H.

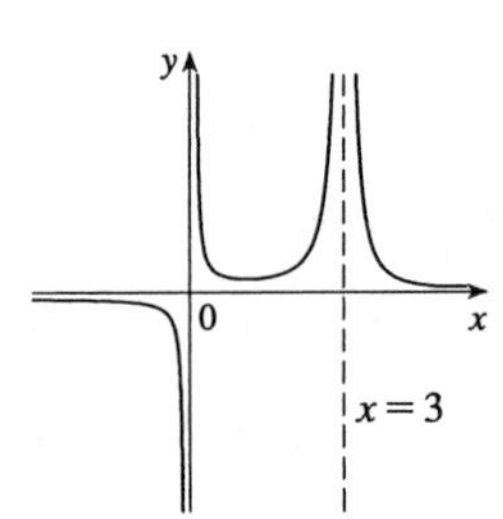

33. $y=f(x)=x\sqrt{5-x}$ **A.** $D=\{x\mid x\leq 5\}=(-\infty,5]$ **B.** x-intercepts $0,5$; y-intercept $=f(0)=0$

C. No symmetry **D.** $\lim_{x\to-\infty}x\sqrt{5-x}=-\infty$, no asymptotes

E. $f'(x)=\sqrt{5-x}-\dfrac{x}{2\sqrt{5-x}}=\dfrac{10-3x}{2\sqrt{5-x}}>0 \Leftrightarrow x<\frac{10}{3}$. So f is increasing on $\left(-\infty,\frac{10}{3}\right]$ and decreasing on $\left[\frac{10}{3},5\right]$. **F.** $f\left(\frac{10}{3}\right)=\frac{10\sqrt{5}}{3\sqrt{3}}$ is a local and absolute maximum.

G. $f''(x)=\dfrac{-6\sqrt{5-x}-(10-3x)\left(-1/\sqrt{5-x}\right)}{4(5-x)}$

$=\dfrac{3x-20}{4(5-x)^{3/2}}<0$ for all x in D,

so f is CD on $(-\infty,5)$.

H.

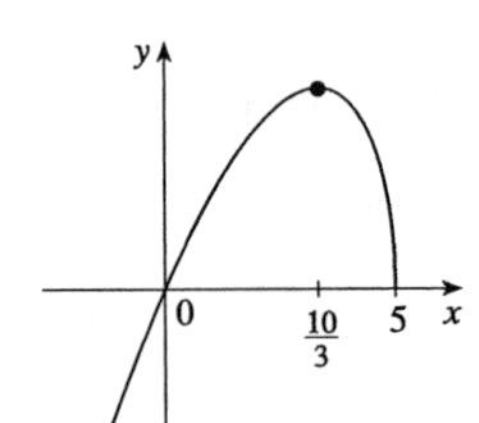

35. $y = f(x) = \dfrac{x^2}{x+8} = x - 8 + \dfrac{64}{x+8}$ **A.** $D = \{x \mid x \neq -8\}$ **B.** Intercepts are 0 **C.** No symmetry

D. $\displaystyle\lim_{x\to\infty} \frac{x^2}{x+8} = \infty$, but $f(x) - (x-8) = \dfrac{64}{x+8} \to 0$ as $x \to \infty$, so $y = x - 8$ is a slant asymptote.

$\displaystyle\lim_{x\to-8^+} \frac{x^2}{x+8} = \infty$ and $\displaystyle\lim_{x\to-8^-} \frac{x^2}{x+8} = -\infty$, so $x = -8$ is a VA. **E.** $f'(x) = 1 - \dfrac{64}{(x+8)^2} = \dfrac{x(x+16)}{(x+8)^2} > 0$

$\Leftrightarrow$ $x > 0$ or $x < -16$, so f is increasing on $(-\infty, -16]$ and $[0, \infty)$ and decreasing on $[-16, -8)$ and $(-8, 0]$. **F.** $f(-16) = -32$ is a local maximum, $f(0) = 0$ is a local minimum.

G. $f''(x) = 128/(x+8)^3 > 0 \quad \Leftrightarrow \quad x > -8$, so f is CU on $(-8, \infty)$ and CD on $(-\infty, -8)$. No IP

H.

$x = -8$ y 0 x $(-16, -32)$ $y = x - 8$

37. $y = f(x) = \sqrt{x} - \sqrt[3]{x}$ **A.** $D = [0, \infty)$ **B.** y-intercept 0, x-intercepts 0, 1 **C.** No symmetry

D. $\displaystyle\lim_{x\to\infty} \left(x^{1/2} - x^{1/3}\right) = \lim_{x\to\infty} \left[x^{1/3}\left(x^{1/6} - 1\right)\right] = \infty$, no asymptotes

E. $f'(x) = \frac{1}{2}x^{-1/2} - \frac{1}{3}x^{-2/3} = \dfrac{3x^{1/6} - 2}{6x^{2/3}} > 0 \quad \Leftrightarrow \quad 3x^{1/6} > 2 \quad \Leftrightarrow \quad x > \left(\frac{2}{3}\right)^6$, so f is increasing on $\left[\left(\frac{2}{3}\right)^6, \infty\right)$ and decreasing on $\left[0, \left(\frac{2}{3}\right)^6\right]$. **F.** $f\left(\left(\frac{2}{3}\right)^6\right) = -\frac{4}{27}$ is a local minimum.

G. $f''(x) = -\frac{1}{4}x^{-3/2} + \frac{2}{9}x^{-5/3}$

$= \dfrac{8 - 9x^{1/6}}{36x^{5/3}} > 0 \quad \Leftrightarrow \quad x^{1/6} < \frac{8}{9} \quad \Leftrightarrow$

$x < \left(\frac{8}{9}\right)^6$, so f is CU on $\left(0, \left(\frac{8}{9}\right)^6\right)$ and CD on $\left(\left(\frac{8}{9}\right)^6, \infty\right)$. IP $\left(\frac{8}{9}, -\frac{64}{729}\right)$

H.

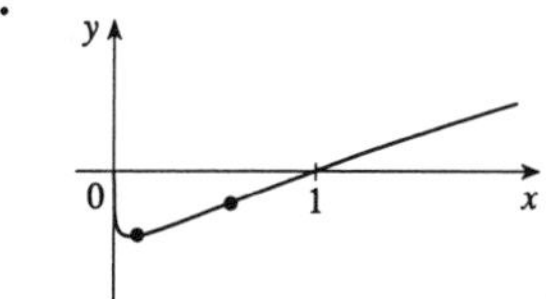

39. $f(x) = \dfrac{x^2 - 1}{x^3} \quad \Rightarrow \quad f'(x) = \dfrac{x^3(2x) - (x^2 - 1)3x^2}{x^6} = \dfrac{3 - x^2}{x^4} \quad \Rightarrow$

$f''(x) = \dfrac{x^4(-2x) - (3 - x^2)4x^3}{x^8} = \dfrac{2x^2 - 12}{x^5}$

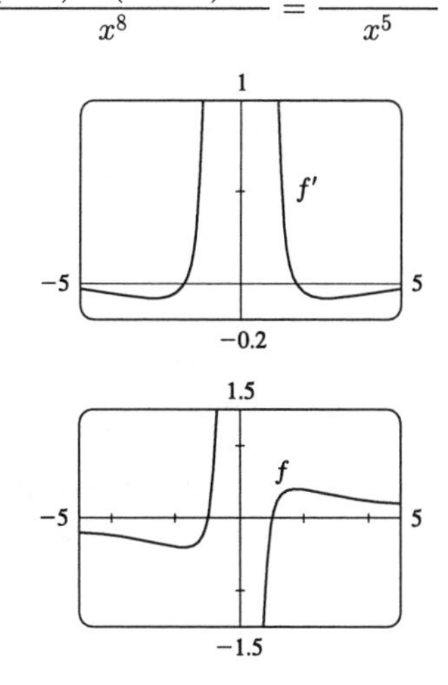

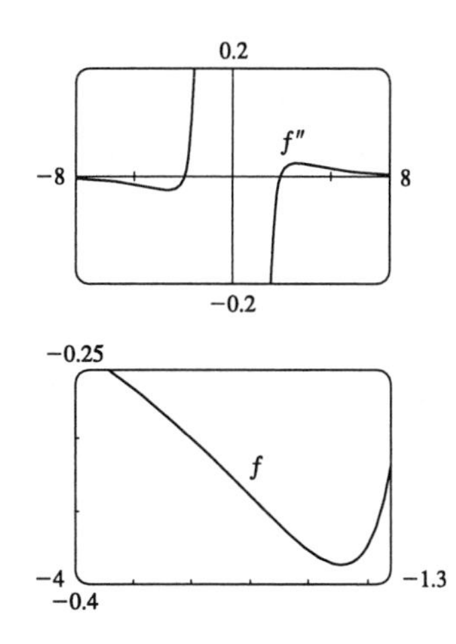

From the graphs of f' and f'', it appears that f is increasing on $[-1.73, 0)$ and $(0, 1.73]$ and decreasing on $(-\infty, -1.73]$ and $[1.73, \infty)$; f has a local maximum of about $f(1.73) = 0.38$ and a local minimum of about $f(-1.7) = -0.38$; f is CU on $(-2.45, 0)$ and $(2.45, \infty)$, and CD on $(-\infty, -2.45)$ and $(0, 2.45)$; and f has inflection points at about $(-2.45, -0.34)$ and $(2.45, 0.34)$. Now $f'(x) = \dfrac{3 - x^2}{x^4}$ is positive for $0 < x^2 < 3$, that is, f is increasing on $\left[-\sqrt{3}, 0\right)$ and $\left(0, \sqrt{3}\right]$; and $f'(x)$ is negative (and so f is decreasing) on $\left(-\infty, -\sqrt{3}\right]$ and $\left[\sqrt{3}, \infty\right)$. $f'(x) = 0$ when $x = \pm\sqrt{3}$. f' goes from positive to negative at $x = \sqrt{3}$, so f has a local maximum of $f\left(\sqrt{3}\right) = \dfrac{\left(\sqrt{3}\right)^2 - 1}{\left(\sqrt{3}\right)^3} = \dfrac{2\sqrt{3}}{9}$; and since f is odd, we know that maxima on the interval $[0, \infty)$ correspond to minima on $(-\infty, 0]$, so f has a local minimum of $f\left(-\sqrt{3}\right) = -\dfrac{2\sqrt{3}}{9}$. Also, $f''(x) = \dfrac{2x^2 - 12}{x^5}$ is positive (so f is CU) on $\left(-\sqrt{6}, 0\right)$ and $\left(\sqrt{6}, \infty\right)$, and negative (so f is CD) on $\left(-\infty, -\sqrt{6}\right)$ and $\left(0, \sqrt{6}\right)$. There are IP at $\left(\sqrt{6}, \frac{5\sqrt{6}}{36}\right)$ and $\left(-\sqrt{6}, -\frac{5\sqrt{6}}{36}\right)$.

41. $f(x) = 3x^6 - 5x^5 + x^4 - 5x^3 - 2x^2 + 3$, $f'(x) = 18x^5 - 25x^4 + 4x^3 - 15x^2 - 4x$, $f''(x) = 90x^4 - 100x^3 + 12x^2 - 30x - 4$

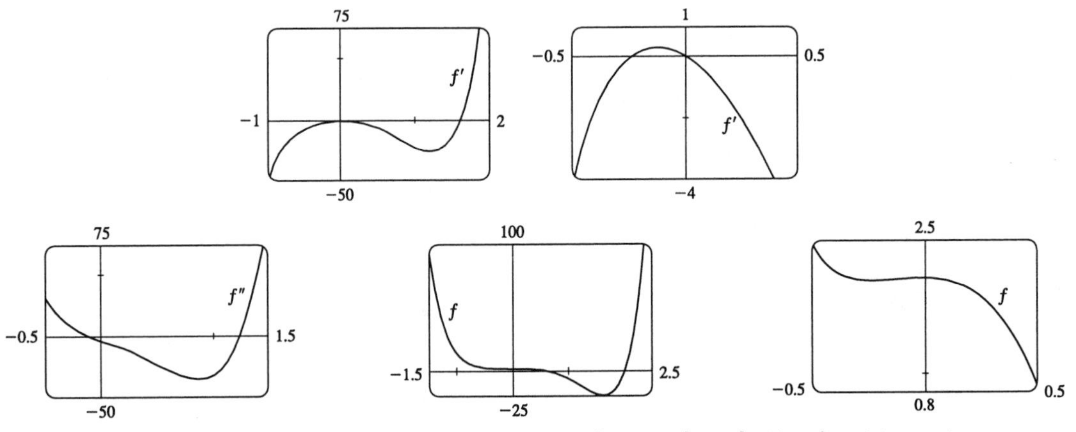

From the graphs of f' and f'', it appears that f is increasing on $[-0.23, 0]$ and $[1.62, \infty)$ and decreasing on $(-\infty, -0.23]$ and $[0, 1.62]$; f has a local maximum of about $f(0) = 2$ and local minima of about $f(-0.23) = 1.96$ and $f(1.62) = -19.2$; f is CU on $(-\infty, -0.12)$ and $(1.24, \infty)$ and CD on $(-0.12, 1.24)$; and f has inflection points at about $(-0.12, 1.98)$ and $(1.2, -12.1)$.

43. $f(x) = x^{101} + x^{51} + x - 1 = 0$. Since f is continuous and $f(0) = -1$ and $f(1) = 2$, the equation has at least one root in $(0, 1)$, by the Intermediate Value Theorem. Suppose the equation has two roots, a and b, with $a < b$. Then $f(a) = 0 = f(b)$, so by Rolle's Theorem, $f'(x) = 0$ has a root in (a, b). But this is impossible since $f'(x) = 101x^{100} + 51x^{50} + 1 \geq 1$ for all x.

45. Since f is continuous on $[32, 33]$ and differentiable on $(32, 33)$, then by the Mean Value Theorem there exists a number c in $(32, 33)$ such that $f'(c) = \frac{1}{5}c^{-4/5} = \dfrac{\sqrt[5]{33} - \sqrt[5]{32}}{33 - 32} = \sqrt[5]{33} - 2$, but $\frac{1}{5}c^{-4/5} > 0 \Rightarrow \sqrt[5]{33} - 2 > 0 \Rightarrow \sqrt[5]{33} > 2$. Also f' is decreasing, so that $f'(c) < f'(32) = \frac{1}{5}(32)^{-4/5} = 0.0125 \Rightarrow 0.0125 > f'(c) = \sqrt[5]{33} - 2 \Rightarrow \sqrt[5]{33} < 2.0125$. Therefore $2 < \sqrt[5]{33} < 2.0125$.

47.

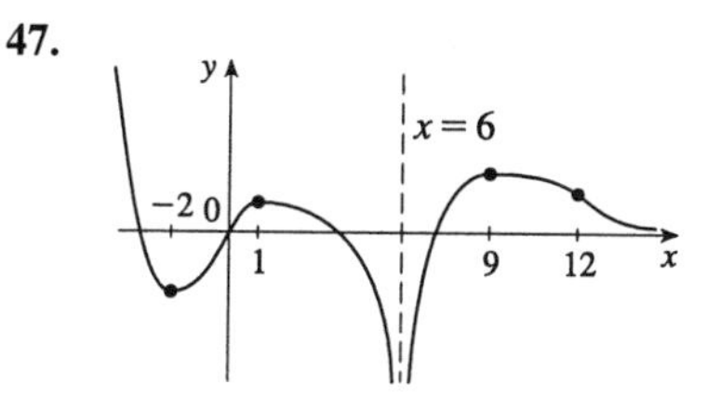

49. For $(1, 6)$ to be on the curve, we have that $6 = y(1) = 1^3 + a(1)^2 + b(1) + 1 = a + b + 2 \Rightarrow b = 4 - a$. Now $y' = 3x^2 + 2ax + b$ and $y'' = 6x + 2a$. Also, for $(1, 6)$ to be an inflection point it must be true that $y''(1) = 6(1) + 2a = 0 \Rightarrow a = -3 \Rightarrow b = 4 - (-3) = 7$.

51. **(a)** $g(x) = f(x^2) \Rightarrow g'(x) = 2xf'(x^2)$ by the Chain Rule. Since $f'(x) > 0$ for all $x \neq 0$, we must have $f'(x^2) > 0$ for $x \neq 0$, so $g'(x) = 0 \Leftrightarrow x = 0$. Now $g'(x)$ changes sign (from negative to positive) at $x = 0$, since one of its factors, namely $f'(x^2)$, is positive for all x, and its other factor, namely $2x$, changes from negative to positive at this point, so by the First Derivative Test, f has a local and absolute minimum at $x = 0$.

(b) $g'(x) = 2xf'(x^2) \Rightarrow g''(x) = 2[xf''(x^2)(2x) + f'(x^2)] = 4x^2f''(x^2) + 2f'(x^2)$ by the Product Rule and the Chain Rule. But $x^2 > 0$ for all $x \neq 0$ and thus $f''(x^2) > 0$ (since f is CU for $x > 0$) and $f'(x^2) > 0$ for all $x \neq 0$, so since all of its factors are positive, $g''(x) > 0$ for $x \neq 0$. Whether $g''(0)$ is positive or 0 doesn't matter (since the sign of g'' does not change there); g is concave up on $\mathbb{R}$.

53. If $B = 0$, the line is vertical and the distance from $x = -\dfrac{C}{A}$ to (x_1, y_1) is $\left|x_1 + \dfrac{C}{A}\right| = \dfrac{|Ax_1 + By_1 + C|}{\sqrt{A^2 + B^2}}$, so assume $B \neq 0$. The square of the distance from (x_1, y_1) to the line is $f(x) = (x - x_1)^2 + (y - y_1)^2$ where $Ax + By + C = 0$, so we minimize $f(x) = (x - x_1)^2 + \left(-\dfrac{A}{B}x - \dfrac{C}{B} - y_1\right)^2 \Rightarrow$

$f'(x) = 2(x - x_1) + 2\left(-\dfrac{A}{B}x - \dfrac{C}{B} - y_1\right)\left(-\dfrac{A}{B}\right)$. $f'(x) = 0 \Rightarrow x = \dfrac{B^2x_1 - ABy_1 - AC}{A^2 + B^2}$ and this gives a minimum since $f''(x) = 2\left(1 + \dfrac{A^2}{B^2}\right) > 0$. Substituting this value of x and simplifying gives

$f(x) = \dfrac{(Ax_1 + By_1 + C)^2}{A^2 + B^2}$, so the minimum distance is $\dfrac{|Ax_1 + By_1 + C|}{\sqrt{A^2 + B^2}}$.

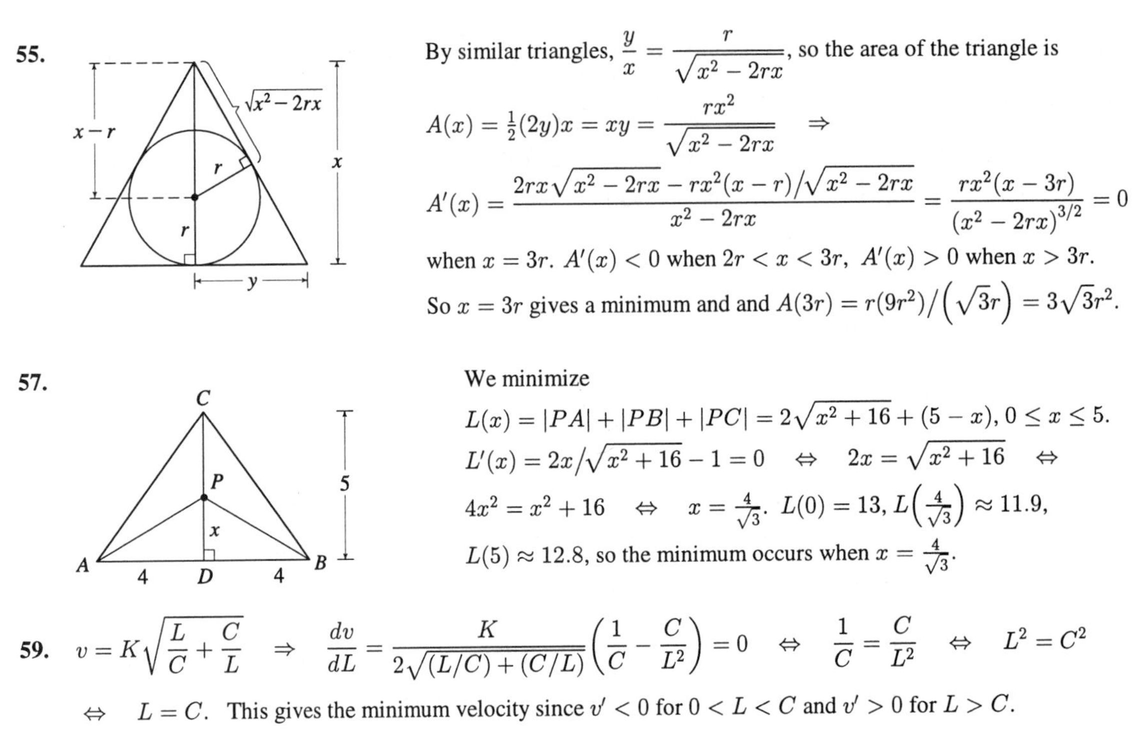

55. By similar triangles, $\dfrac{y}{x} = \dfrac{r}{\sqrt{x^2 - 2rx}}$, so the area of the triangle is

$A(x) = \frac{1}{2}(2y)x = xy = \dfrac{rx^2}{\sqrt{x^2 - 2rx}} \quad \Rightarrow$

$A'(x) = \dfrac{2rx\sqrt{x^2 - 2rx} - rx^2(x - r)/\sqrt{x^2 - 2rx}}{x^2 - 2rx} = \dfrac{rx^2(x - 3r)}{(x^2 - 2rx)^{3/2}} = 0$

when $x = 3r$. $A'(x) < 0$ when $2r < x < 3r$, $A'(x) > 0$ when $x > 3r$.

So $x = 3r$ gives a minimum and and $A(3r) = r(9r^2)/\left(\sqrt{3}r\right) = 3\sqrt{3}r^2$.

57. We minimize

$L(x) = |PA| + |PB| + |PC| = 2\sqrt{x^2 + 16} + (5 - x), 0 \le x \le 5.$

$L'(x) = 2x/\sqrt{x^2 + 16} - 1 = 0 \quad \Leftrightarrow \quad 2x = \sqrt{x^2 + 16} \quad \Leftrightarrow$

$4x^2 = x^2 + 16 \quad \Leftrightarrow \quad x = \frac{4}{\sqrt{3}}$. $L(0) = 13$, $L\left(\frac{4}{\sqrt{3}}\right) \approx 11.9$,

$L(5) \approx 12.8$, so the minimum occurs when $x = \frac{4}{\sqrt{3}}$.

59. $v = K\sqrt{\dfrac{L}{C} + \dfrac{C}{L}} \quad \Rightarrow \quad \dfrac{dv}{dL} = \dfrac{K}{2\sqrt{(L/C) + (C/L)}}\left(\dfrac{1}{C} - \dfrac{C}{L^2}\right) = 0 \quad \Leftrightarrow \quad \dfrac{1}{C} = \dfrac{C}{L^2} \quad \Leftrightarrow \quad L^2 = C^2$

$\Leftrightarrow \quad L = C$. This gives the minimum velocity since $v' < 0$ for $0 < L < C$ and $v' > 0$ for $L > C$.

61. Let x = selling price of ticket. Then $12 - x$ is the amount the ticket price has been lowered, so the number of tickets sold is $11{,}000 + 1000(12 - x) = 23{,}000 - 1000x$. The revenue is $R(x) = x(23{,}000 - 1000x) = 23{,}000x - 1000x^2$, so $R'(x) = 23{,}000 - 2000x = 0$ when $x = 11.5$. Since $R''(x) = -2000 < 0$, the maximum revenue occurs when the ticket prices are \$11.50.

63. $f'(x) = x - \sqrt[4]{x} = x - x^{1/4} \quad \Rightarrow \quad f(x) = \frac{1}{2}x^2 - \frac{4}{5}x^{5/4} + C$

65. $f'(x) = (1 + x)/\sqrt{x} = x^{-1/2} + x^{1/2} \quad \Rightarrow \quad f(x) = 2x^{1/2} + \frac{2}{3}x^{3/2} + C \quad \Rightarrow \quad 0 = f(1) = 2 + \frac{2}{3} + C \quad \Rightarrow$
$C = -\frac{8}{3} \quad \Rightarrow \quad f(x) = 2x^{1/2} + \frac{2}{3}x^{3/2} - \frac{8}{3}$

67. $f''(x) = x^3 + x \quad \Rightarrow \quad f'(x) = \frac{1}{4}x^4 + \frac{1}{2}x^2 + C \quad \Rightarrow \quad 1 = f'(0) = C \quad \Rightarrow \quad f'(x) = \frac{1}{4}x^4 + \frac{1}{2}x^2 + 1 \quad \Rightarrow$
$f(x) = \frac{1}{20}x^5 + \frac{1}{6}x^3 + x + D \quad \Rightarrow \quad -1 = f(0) = D \quad \Rightarrow \quad f(x) = \frac{1}{20}x^5 + \frac{1}{6}x^3 + x - 1$

69.

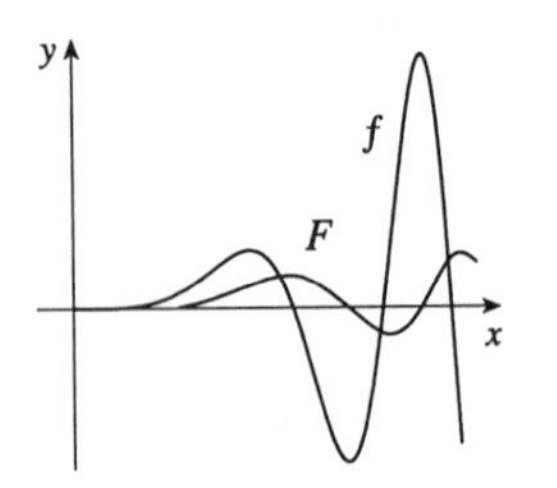

71. Choosing the positive direction to be upward, we have $a(t) = -9.8 \quad \Rightarrow \quad v(t) = -9.8t + v_0$, but $v(0) = 0 = v_0 \quad \Rightarrow \quad v(t) = -9.8t = s'(t) \quad \Rightarrow \quad s(t) = -4.9t^2 + s_0$, but $s(0) = s_0 = 500 \quad \Rightarrow$ $s(t) = -4.9t^2 + 500$. When $s = 0$, $-4.9t^2 + 500 = 0 \quad \Rightarrow \quad t = \sqrt{\frac{500}{4.9}} \quad \Rightarrow$ $v = -9.8\sqrt{\frac{500}{4.9}} \approx -98.995$ m/s. Therefore the canister will not burst.

73. **(a)** The cross-sectional area is

$A = 2x \cdot 2y = 4xy = 4x\sqrt{100 - x^2}$, $0 \le x \le 10$, so

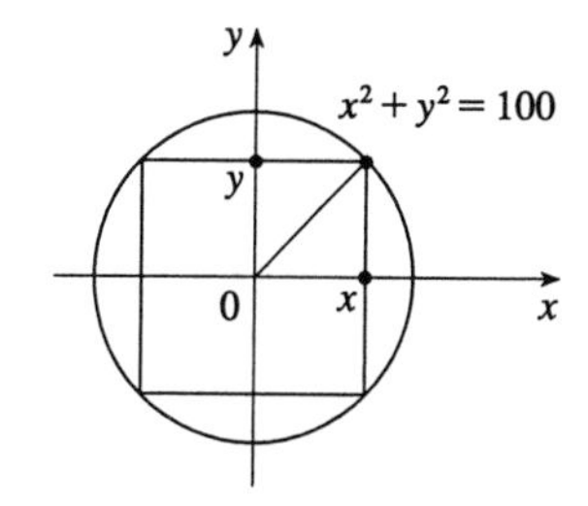

$$\frac{dA}{dx} = 4x\left(\tfrac{1}{2}\right)\left(100 - x^2\right)^{-1/2}(-2x) + \left(100 - x^2\right)^{1/2} \cdot 4$$

$$= \frac{-4x^2}{\left(100 - x^2\right)^{1/2}} + 4\left(100 - x^2\right)^{1/2} = 0 \text{ when}$$

$-4x^2 + 4(100 - x^2) = 0 \quad \Rightarrow \quad -8x^2 + 400 = 0$

$\Rightarrow \quad x^2 = 50 \quad \Rightarrow \quad x = \sqrt{50} \quad \Rightarrow \quad y = \sqrt{50}$. And $A(0) = A(10) = 0$. Therefore, the rectangle of maximum area is a square.

(b) $y = \sqrt{100 - x^2}$. The cross-sectional area of each rectangular plank is

$A = 2x\left(y - \sqrt{50}\right) = 2x\left[\sqrt{100 - x^2} - \sqrt{50}\right]$, $0 \le x \le \sqrt{50}$, so

$$\frac{dA}{dx} = 2\left[\sqrt{100 - x^2} - \sqrt{50}\right] + 2x\left(\tfrac{1}{2}\right)\left(100 - x^2\right)^{-1/2}(-2x)$$

$$= 2\left(100 - x^2\right)^{1/2} - 2\sqrt{50} - \frac{2x^2}{\left(100 - x^2\right)^{1/2}}.$$

Set $\frac{dA}{dx} = 0$: $\left(100 - x^2\right) - \sqrt{50}\left(100 - x^2\right)^{1/2} - x^2 = 0 \quad \Rightarrow \quad 100 - 2x^2 = \sqrt{50}\left(100 - x^2\right)^{1/2}$

$\Rightarrow \quad 10{,}000 - 400x^2 + 4x^4 = 50(100 - x^2) \quad \Rightarrow \quad 2500 - 175x^2 + 2x^4 = 0 \quad \Rightarrow$

$x^2 = \dfrac{175 \pm \sqrt{10{,}625}}{4} \approx 69.5$ or $17.98 \quad \Rightarrow \quad x \approx 8.3$ or 4.2. But $8.3 > \sqrt{50}$, so $x \approx 4.2 \quad \Rightarrow$

$y \approx \sqrt{100 - (4.2)^2} \approx 2.0$. Each plank should have dimensions about 8.4 inches by 2 inches.

(c) The strength is $S = k\,(2x)(2y)^2 = 8kxy^2 = 8kx(100 - x^2)$, $0 \le x \le 10$.

$\frac{dS}{dx} = 800k - 24kx^2 = 0$ when $24kx^2 = 800k \quad \Rightarrow \quad x^2 = \frac{100}{3} \quad \Rightarrow \quad x = \frac{10}{\sqrt{3}} \quad \Rightarrow$

$y = \sqrt{\frac{200}{3}} = \frac{10\sqrt{2}}{\sqrt{3}}$ and $S(0) = S(10) = 0$, so the maximum occurs when $x = \frac{10}{\sqrt{3}}$. The dimensions should be $\frac{20}{\sqrt{3}}$ inches by $\frac{20\sqrt{2}}{\sqrt{3}}$ inches.

APPLICATIONS PLUS (page 255)

1. **(a)** $I = \dfrac{k\cos\theta}{d^2} = \dfrac{k(h/d)}{d^2} = k\dfrac{h}{d^2} = k\dfrac{h}{\left(\sqrt{1600+h^2}\right)^3} = k\dfrac{h}{(1600+h^2)^{3/2}} \quad\Rightarrow$

$$\frac{dI}{dh} = k\frac{(1600+h^2)^{3/2} - kh\frac{3}{2}(1600+h^2)^{1/2}\cdot 2h}{(1600+h^2)^3} = \frac{k(1600+h^2)^{1/2}(1600+h^2-3h^2)}{(1600+h^2)^3}$$

$= \dfrac{k(1600-2h^2)}{(1600+h^2)^{5/2}}$. Set $\dfrac{dI}{dh} = 0$: $1600 - 2h^2 = 0 \quad\Rightarrow\quad h^2 = 800 \quad\Rightarrow\quad h = \sqrt{800} = 20\sqrt{2}$.

By the First Derivative Test, I has a relative maximum at $h = 20\sqrt{2} \approx 28$ ft.

(b)

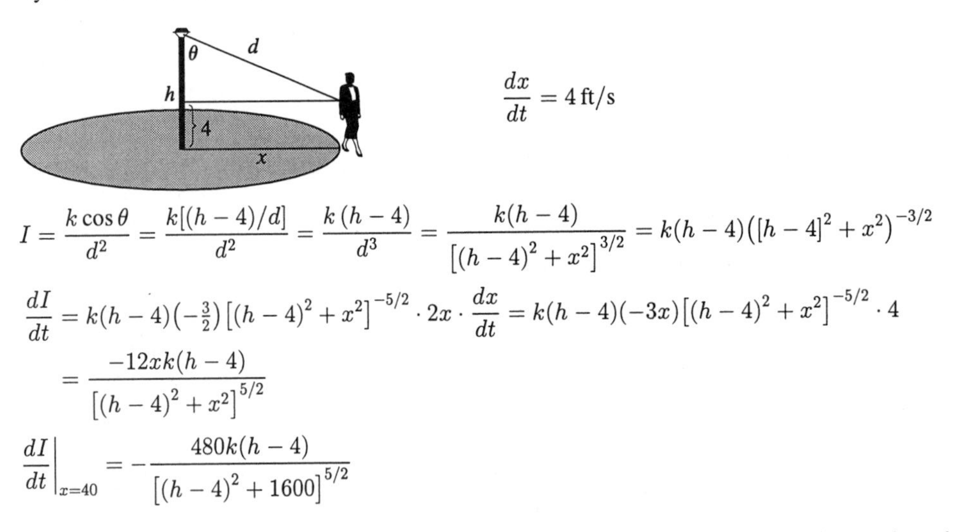

$\dfrac{dx}{dt} = 4$ ft/s

$$I = \frac{k\cos\theta}{d^2} = \frac{k[(h-4)/d]}{d^2} = \frac{k\,(h-4)}{d^3} = \frac{k(h-4)}{\left[(h-4)^2+x^2\right]^{3/2}} = k(h-4)\left([h-4]^2+x^2\right)^{-3/2}$$

$$\frac{dI}{dt} = k(h-4)\left(-\tfrac{3}{2}\right)\left[(h-4)^2+x^2\right]^{-5/2}\cdot 2x\cdot\frac{dx}{dt} = k(h-4)(-3x)\left[(h-4)^2+x^2\right]^{-5/2}\cdot 4$$

$$= \frac{-12xk(h-4)}{\left[(h-4)^2+x^2\right]^{5/2}}$$

$$\left.\frac{dI}{dt}\right|_{x=40} = -\frac{480k(h-4)}{\left[(h-4)^2+1600\right]^{5/2}}$$

3. We can assume without loss of generality that $\theta = 0$ at time $t = 0$, so that $\theta = 12\pi t$ rad. (The angular velocity of the wheel is 360 rpm $= 360\cdot 2\pi$ rad/60 s $= 12\pi$ rad/s.) Then the position of A as a function of time is $A = (40\cos\theta, 40\sin\theta) = (40\cos 12\pi t, 40\sin 12\pi t)$, so $\sin\alpha = \dfrac{40\sin\theta}{120} = \dfrac{\sin\theta}{3} = \frac{1}{3}\sin 12\pi t$.

(a) Differentiating the expression for $\sin\alpha$, we get $\cos\alpha\cdot\dfrac{d\alpha}{dt} = \frac{1}{3}\cdot 12\pi\cdot\cos 12\pi t = 4\pi\cos\theta$. When $\theta = \frac{\pi}{3}$, we have $\sin\alpha = \frac{1}{3}\sin\theta = \frac{\sqrt{3}}{6}$, so $\cos\alpha = \sqrt{1-\left(\frac{\sqrt{3}}{6}\right)^2} = \sqrt{\frac{11}{12}}$ and

$$\frac{d\alpha}{dt} = \frac{4\pi\cos\frac{\pi}{3}}{\cos\alpha} = \frac{2\pi}{\sqrt{11/12}} = \frac{4\pi\sqrt{3}}{\sqrt{11}}\text{ rad/s}.$$

(b) By the Law of Cosines, $|AP|^2 = |OA|^2 + |OP|^2 - 2|OA||OP|\cos\theta \quad\Rightarrow$

$120^2 = 40^2 + |OP|^2 - 2\cdot 40|OP|\cos\theta \quad\Rightarrow\quad |OP|^2 - (80\cos\theta)|OP| - 12{,}800 = 0 \quad\Rightarrow$

$|OP| = \frac{1}{2}\left(80\cos\theta \pm \sqrt{6400\cos^2\theta + 51{,}200}\right) = 40\cos\theta \pm 40\sqrt{\cos^2\theta + 8}$

$= 40\left(\cos\theta + \sqrt{8+\cos^2\theta}\right)$ cm (since $|OP| > 0$).

As a check, note that $|OP| = 160$ cm when $\theta = 0$ and $|OP| = 80\sqrt{2}$ cm when $\theta = \frac{\pi}{2}$.

(c) By part (b), the x-coordinate of P is given by $x = 40\left(\cos\theta + \sqrt{8+\cos^2\theta}\right)$, so

$$\frac{dx}{dt} = \frac{dx}{d\theta}\frac{d\theta}{dt} = 40\left(-\sin\theta - \frac{2\cos\theta\sin\theta}{2\sqrt{8+\cos^2\theta}}\right)\cdot 12\pi = -480\pi\sin\theta\left(1+\frac{\sin\theta\cos\theta}{\sqrt{8+\cos^2\theta}}\right)\text{cm/s.}$$ In particular, $dx/dt = 0$ cm/s when $\theta = 0$ and $dx/dt = -480\pi$ cm/s when $\theta = \frac{\pi}{2}$.

5. **(a)** $T = 2\pi\sqrt{\dfrac{L}{g}} = \dfrac{2\pi}{\sqrt{g}}L^{1/2} \quad\Rightarrow\quad dT = \dfrac{2\pi}{\sqrt{g}}\dfrac{1}{2}L^{-1/2}dL \quad\Rightarrow\quad \dfrac{dT}{2\pi/\sqrt{g}} = \dfrac{dL}{2\sqrt{L}} \quad\Rightarrow$

$\dfrac{dT}{(2\pi/\sqrt{g})\sqrt{L}} = \dfrac{dL}{2L} \quad\Rightarrow\quad \dfrac{dT}{T} = \dfrac{dL}{2L}$

(b) $dL = \dfrac{2L}{T}dT$. Set $dT = -15$ s, $T = 3600$ s. Then $dL = \dfrac{2L}{3600}\cdot(-15) = -\frac{30}{3600}L = -\frac{1}{120}L$. Thus, shorten the pendulum by $\frac{1}{120}L$.

(c) $T = \dfrac{2\pi\sqrt{L}}{\sqrt{g}} = 2\pi\sqrt{L}g^{-1/2} \quad\Rightarrow\quad dT = 2\pi\sqrt{L}\left(-\frac{1}{2}\right)t^{-3/2}dg$. Therefore, $dg = -\dfrac{g\sqrt{g}\,dT}{\pi\sqrt{L}}$.

7. **(a)** $T_1 = \dfrac{D}{c_1}$, $T_2 = \dfrac{2|PR|}{c_1} + \dfrac{|RS|}{c_2} = \dfrac{2h\sec\theta}{c_1} + \dfrac{D - 2h\tan\theta}{c_2}$, $T_3 = \dfrac{2\sqrt{h^2 + D^2/4}}{c_1} = \dfrac{\sqrt{4h^2+D^2}}{c_1}$.

(b) $\dfrac{dT_2}{d\theta} = \dfrac{2h}{c_1}\cdot\sec\theta\tan\theta - \dfrac{2h}{c_2}\sec^2\theta = 0$ when $2h\sec\theta\left[\dfrac{1}{c_1}\tan\theta - \dfrac{1}{c_2}\sec\theta\right] = 0 \quad\Rightarrow$

$\dfrac{1}{c_1}\dfrac{\sin\theta}{\cos\theta} - \dfrac{1}{c_2}\dfrac{1}{\cos\theta} = 0 \quad\Rightarrow\quad \sin\theta = \dfrac{c_1}{c_2}$. The First Derivative Test shows that this gives a minimum.

(c) Using part (a), we have $\dfrac{1}{4} = \dfrac{1}{c_1}$. Therefore, $c_1 = 4$. Also $\dfrac{3}{4\sqrt{5}} = \dfrac{\sqrt{4h^2+1}}{4} \quad\Rightarrow\quad \sqrt{4h^2+1} = \dfrac{3}{\sqrt{5}}$

$\Rightarrow\quad 4h^2 + 1 = \frac{9}{5} \quad\Rightarrow\quad h^2 = \frac{1}{5} \quad\Rightarrow\quad h = \frac{1}{\sqrt{5}}$. From (b), $\sin\theta = \dfrac{c_1}{c_2} = \dfrac{4}{c_2} \quad\Rightarrow$

$\sec\theta = \dfrac{c_2}{\sqrt{c_2^2 - 16}}$ and $\tan\theta = \dfrac{4}{\sqrt{c_2^2-16}}$. Thus $\dfrac{1}{3} = \dfrac{\dfrac{2}{\sqrt{5}}\cdot\dfrac{c_2}{\sqrt{c_2^2-16}}}{4} + \dfrac{1 - \dfrac{2}{\sqrt{5}}\cdot\dfrac{4}{\sqrt{c_2^2-16}}}{c_2} \quad\Rightarrow$

$4c_2 = \dfrac{6c_2^2}{\sqrt{5}\sqrt{c_2^2-16}} + 12\left(1 - \dfrac{8}{\sqrt{5}\sqrt{c_2^2-16}}\right) \quad\Rightarrow\quad 4c_2 - 12 = \dfrac{6(c_2^2-16)}{\sqrt{5}\sqrt{c_2^2-16}} = \frac{6}{\sqrt{5}}\sqrt{c_2^2-16}$

$\Rightarrow\quad 2c_2 - 6 = \frac{3}{\sqrt{5}}\sqrt{c_2^2-16} \quad\Rightarrow\quad 20c_2^2 - 120c_2 + 180 = 9c_2^2 - 144 \quad\Rightarrow$

$11c_2^2 - 120c_2 + 324 = (c_2 - 6)(11c_2 - 54) = 0 \Rightarrow c_2 = 6$ or $\frac{54}{11}$. But the root $\frac{54}{11}$ is inadmissible because if $\tan\theta = \dfrac{4}{\sqrt{\left(\frac{54}{11}\right)^2 - 16}} \approx 1.4$, then $\theta > 45°$, which is impossible (from the diagram). So $c_2 = 6$.

9. **(a)** Condition (i) will hold if and only if all of the following four conditions hold: (α) $P(0) = 0$; (β) $P'(0) = 0$ (for a smooth landing); (γ) $P'(\ell) = 0$ (since the plane is cruising horizontally when it begins its descent); and (δ) $P(\ell) = h$.

First of all, condition α implies that $P(0) = d = 0$, so $P(x) = ax^3 + bx^2 + cx \Rightarrow$ $P'(x) = 3ax^2 + 2bx + c$. But $P'(0) = c = 0$ by condition β. So $P'(\ell) = 3a\ell^2 + 2b\ell = \ell(3a\ell + 2b)$. Now by condition γ, $3a\ell + 2b = 0 \Rightarrow a = -\frac{2b}{3\ell}$. Therefore $P(x) = -\frac{2b}{3\ell}x^3 + bx^2$. Setting $P(\ell) = h$ for condition δ, we get $P(\ell) = -\frac{2b}{3\ell}\ell^3 + b\ell^2 = h \Rightarrow b = \frac{3h}{\ell^2} \Rightarrow a = -\frac{2h}{\ell^3}$. So $P(x) = -\frac{2h}{\ell^3}x^3 + \frac{3h}{\ell^2}x^2$.

(b) By condition (ii), $\frac{dx}{dt} = -v$ for all t, so $x(t) = \ell - vt$. Condition (iii) states that $\left|\frac{d^2y}{dt^2}\right| \le k$. By the Chain Rule, we have $\frac{dy}{dt} = -\frac{2h}{\ell^3}(3x^2)\frac{dx}{dt} + \frac{3h}{\ell^2}(2x)\frac{dx}{dt} = -\frac{6hx^2v}{\ell^3} + \frac{6hxv}{\ell^2}$ (for $x \le \ell$) $\Rightarrow$ $\frac{d^2y}{dt^2} = -\frac{6hv}{\ell^3}(2x)\frac{dx}{dt} + \frac{6hv}{\ell^2}\frac{dx}{dt} = -\frac{12hv^2}{\ell^3}x + \frac{6hv^2}{\ell^2}$. In particular, when $t = 0$, $x = \ell$ and so $\left.\frac{d^2y}{dt^2}\right|_{t=0} = -\frac{12hv^2}{\ell^3}\ell + \frac{6hv^2}{\ell^2} = -\frac{6hv^2}{\ell^2}$. Thus $\left|\frac{d^2y}{dt^2}\right|_{t=0} = \frac{6hv^2}{\ell^2} \le k$. (This condition also follows from taking $x = 0$.)

(c) We substitute $k = 860\text{ mi/h}^2$, $h = 35{,}000\text{ ft} \times \frac{1\text{ mi}}{5280\text{ ft}}$, and $v = 300\text{ mi/h}$ into the result of part (b):

$$\frac{6\left(35{,}000 \cdot \frac{1}{5280}\right)(300)^2}{\ell^2} \le 860 \quad \Leftrightarrow \quad \ell \ge 300\sqrt{6 \cdot \frac{35{,}000}{5280 \cdot 860}} \approx 64.5\text{ miles.}$$

CHAPTER FOUR

EXERCISES 4.1

1. $\sum_{i=1}^{5}\sqrt{i}=\sqrt{1}+\sqrt{2}+\sqrt{3}+\sqrt{4}+\sqrt{5}$

3. $\sum_{i=4}^{6}3^i=3^4+3^5+3^6$

5. $\sum_{k=0}^{4}\frac{2k-1}{2k+1}=-1+\frac{1}{3}+\frac{3}{5}+\frac{5}{7}+\frac{7}{9}$

7. $\sum_{i=1}^{n}i^{10}=1^{10}+2^{10}+3^{10}+\cdots+n^{10}$

9. $\sum_{j=0}^{n-1}(-1)^j=1-1+1-1+\cdots+(-1)^{n-1}$

11. $1+2+3+4+\cdots+10=\sum_{i=1}^{10}i$

13. $\frac{1}{2}+\frac{2}{3}+\frac{3}{4}+\frac{4}{5}+\cdots+\frac{19}{20}=\sum_{i=1}^{19}\frac{i}{i+1}$

15. $2+4+6+8+\cdots+2n=\sum_{i=1}^{n}2i$

17. $1+2+4+8+16+32=\sum_{i=0}^{5}2^i$

19. $x+x^2+x^3+\cdots+x^n=\sum_{i=1}^{n}x^i$

21. $\sum_{i=4}^{8}(3i-2)=10+13+16+19+22=80$

23. $\sum_{j=1}^{6}3^{j+1}=3^2+3^3+3^4+3^5+3^6+3^7=9+27+81+243+729+2187=3276$

(For a more general method, see Exercise 47.)

25. $\sum_{n=1}^{20}(-1)^n=-1+1-1+1-1+1-1+1-1+1-1+1-1+1-1+1-1+1-1+1=0$

27. $\sum_{i=0}^{4}(2^i+i^2)=(1+0)+(2+1)+(4+4)+(8+9)+(16+16)=61$

29. $\sum_{i=1}^{n}2i=2\sum_{i=1}^{n}i=n(n+1)$

31.
$$\sum_{i=1}^{n}(i^2+3i+4)=\sum_{i=1}^{n}i^2+3\sum_{i=1}^{n}i+\sum_{i=1}^{n}4=\frac{n(n+1)(2n+1)}{6}+\frac{3n(n+1)}{2}+4n$$
$$=\tfrac{1}{6}[(2n^3+3n^2+n)+(9n^2+9n)+24n]=\tfrac{1}{6}(2n^3+12n^2+34n)=\tfrac{1}{3}n(n^2+6n+17)$$

33.
$$\sum_{i=1}^{n}(i+1)(i+2)=\sum_{i=1}^{n}(i^2+3i+2)=\sum_{i=1}^{n}i^2+3\sum_{i=1}^{n}i+\sum_{i=1}^{n}2$$
$$=\frac{n(n+1)(2n+1)}{6}+\frac{3n(n+1)}{2}+2n=\frac{n(n+1)}{6}[(2n+1)+9]+2n$$
$$=\frac{n(n+1)}{3}(n+5)+2n=\frac{n}{3}[(n+1)(n+5)+6]=\frac{n}{3}(n^2+6n+11)$$

35. $\sum_{i=1}^{n}(i^3 - i - 2) = \sum_{i=1}^{n} i^3 - \sum_{i=1}^{n} i - \sum_{i=1}^{n} 2 = \left[\frac{n(n+1)}{2}\right]^2 - \frac{n(n+1)}{2} - 2n$
$= \frac{1}{4}n(n+1)[n(n+1) - 2] - 2n = \frac{1}{4}n(n+1)(n+2)(n-1) - 2n$
$= \frac{1}{4}n[(n+1)(n-1)(n+2) - 8] = \frac{1}{4}n[(n^2-1)(n+2) - 8] = \frac{1}{4}n(n^3 + 2n^2 - n - 10)$

37. By Theorem 2(a) and Example 3, $\sum_{i=1}^{n} c = c\sum_{i=1}^{n} 1 = cn$.

39. $\sum_{i=1}^{n}\left[(i+1)^4 - i^4\right] = (2^4 - 1^4) + (3^4 - 2^4) + (4^4 - 3^4) + \cdots + \left[(n+1)^4 - n^4\right]$
$= (n+1)^4 - 1^4 = n^4 + 4n^3 + 6n^2 + 4n$

On the other hand, $\sum_{i=1}^{n}\left[(i+1)^4 - i^4\right] = \sum_{i=1}^{n}(4i^3 + 6i^2 + 4i + 1) = 4\sum_{i=1}^{n} i^3 + 6\sum_{i=1}^{n} i^2 + 4\sum_{i=1}^{n} i + \sum_{i=1}^{n} 1$
$= 4S + n(n+1)(2n+1) + 2n(n+1) + n$ (where $S = \sum_{i=1}^{n} i^3$)
$= 4S + 2n^3 + 3n^2 + n + 2n^2 + 2n + n = 4S + 2n^3 + 5n^2 + 4n$

Thus $n^4 + 4n^3 + 6n^2 + 4n = 4S + 2n^3 + 5n^2 + 4n$, from which it follows that
$4S = n^4 + 2n^3 + n^2 = n^2(n^2 + 2n + 1) = n^2(n+1)^2$ and $S = \left[\frac{n(n+1)}{2}\right]^2$.

41. **(a)** $\sum_{i=1}^{n}\left(i^4 - (i-1)^4\right) = (1^4 - 0^4) + (2^4 - 1^4) + (3^4 - 2^4) + \cdots + \left[n^4 - (n-1)^4\right] = n^4 - 0 = n^4$

(b) $\sum_{i=1}^{100}(5^i - 5^{i-1}) = (5^1 - 5^0) + (5^2 - 5^1) + (5^3 - 5^2) + \cdots + (5^{100} - 5^{99}) = 5^{100} - 5^0 = 5^{100} - 1$

(c) $\sum_{i=3}^{99}\left(\frac{1}{i} - \frac{1}{i+1}\right) = \left(\frac{1}{3} - \frac{1}{4}\right) + \left(\frac{1}{4} - \frac{1}{5}\right) + \left(\frac{1}{5} - \frac{1}{6}\right) + \cdots + \left(\frac{1}{99} - \frac{1}{100}\right) = \frac{1}{3} - \frac{1}{100} = \frac{97}{300}$

(d) $\sum_{i=1}^{n}(a_i - a_{i-1}) = (a_1 - a_0) + (a_2 - a_1) + (a_3 - a_2) + \cdots + (a_n - a_{n-1}) = a_n - a_0$

43. $\lim_{n\to\infty}\sum_{i=1}^{n}\frac{1}{n}\left(\frac{i}{n}\right)^2 = \lim_{n\to\infty}\frac{1}{n^3}\sum_{i=1}^{n} i^2 = \lim_{n\to\infty}\frac{1}{n^3}\frac{n(n+1)(2n+1)}{6} = \lim_{n\to\infty}\frac{1}{6}\left(1 + \frac{1}{n}\right)\left(2 + \frac{1}{n}\right) = \frac{1}{6}(1)(2) = \frac{1}{3}$

45. $\lim_{n\to\infty}\sum_{i=1}^{n}\frac{2}{n}\left[\left(\frac{2i}{n}\right)^3 + 5\left(\frac{2i}{n}\right)\right] = \lim_{n\to\infty}\sum_{i=1}^{n}\left[\frac{16}{n^4}i^3 + \frac{20}{n^2}i\right] = \lim_{n\to\infty}\left[\frac{16}{n^4}\sum_{i=1}^{n} i^3 + \frac{20}{n^2}\sum_{i=1}^{n} i\right]$
$= \lim_{n\to\infty}\left[\frac{16}{n^4}\frac{n^2(n+1)^2}{4} + \frac{20}{n^2}\frac{n(n+1)}{2}\right] = \lim_{n\to\infty}\left[\frac{4(n+1)^2}{n^2} + \frac{10n(n+1)}{n^2}\right]$
$= \lim_{n\to\infty}\left[4\left(1 + \frac{1}{n}\right)^2 + 10\left(1 + \frac{1}{n}\right)\right] = 4 \cdot 1 + 10 \cdot 1 = 14$

47. Let $S = \sum_{i=1}^{n} ar^{i-1} = a + ar + ar^2 + \cdots + ar^{n-1}$. Then $rS = ar + ar^2 + \cdots + ar^{n-1} + ar^n$. Subtracting the first equation from the second, we find $(r-1)S = ar^n - a = a(r^n - 1)$, so $S = \frac{a(r^n - 1)}{r - 1}$.

49. $\displaystyle\sum_{i=1}^{n}(2i+2^i) = 2\sum_{i=1}^{n} i + \sum_{i=1}^{n} 2\cdot 2^{i-1} = 2\,\frac{n(n+1)}{2} + \frac{2(2^n-1)}{2-1} = 2^{n+1}+n^2+n-2.$

For the first sum we have used Theorem 3(c), and for the second, Exercise 47 with $a=r=2$.

51. By Theorem 3(c) we have that $\displaystyle\sum_{i=1}^{n} i = \frac{n(n+1)}{2} = 78 \;\Leftrightarrow\; n(n+1)=156 \;\Leftrightarrow\; n^2+n-156=0 \;\Leftrightarrow$

$(n+13)(n-12)=0 \;\Leftrightarrow\; n=12$ or -13. But $n=-13$ produces a negative answer for the sum, so $n=12$.

53. From Formula 18c in Appendix D, $\sin x \sin y = \frac{1}{2}[\cos(x-y)-\cos(x+y)]$, so

$2\sin u \sin v = \cos(u-v)-\cos(u+v)$ $\quad(\bigstar)$. Taking $u=\frac{1}{2}x$ and $v=ix$, we get

$2\sin\left(\frac{1}{2}x\right)\sin ix = \cos\left(\left(\frac{1}{2}-i\right)x\right)-\cos\left(\left(\frac{1}{2}+i\right)x\right) = \cos\left(\left(i-\frac{1}{2}\right)x\right)-\cos\left(\left(i+\frac{1}{2}\right)x\right)$. Thus

$$\begin{aligned}
2\sin\left(\tfrac{1}{2}x\right)\sum_{i=1}^{n}\sin ix &= \sum_{i=1}^{n} 2\sin\left(\tfrac{1}{2}x\right)\sin ix = \sum_{i=1}^{n}\left[\cos\left(\left(i-\tfrac{1}{2}\right)x\right)-\cos\left(\left(i+\tfrac{1}{2}\right)x\right)\right] \\
&= -\sum_{i=1}^{n}\left[\cos\left(\left(i+\tfrac{1}{2}\right)x\right)-\cos\left(\left(i-\tfrac{1}{2}\right)x\right)\right] = -\left[\cos\left(\left(n+\tfrac{1}{2}\right)x\right)-\cos\left(\tfrac{1}{2}x\right)\right] \quad \text{(telescoping sum)} \\
&= \cos\left(\tfrac{1}{2}(n+1)x-\tfrac{1}{2}nx\right)-\cos\left(\tfrac{1}{2}(n+1)x+\tfrac{1}{2}nx\right) \\
&= 2\sin\left(\tfrac{1}{2}(n+1)x\right)\sin\left(\tfrac{1}{2}nx\right) \quad [\text{by } (\bigstar) \text{ with } u=\tfrac{1}{2}(n+1)x \text{ and } v=\tfrac{1}{2}nx]
\end{aligned}$$

If x is not an integer multiple of 2π, then $\sin\left(\frac{1}{2}x\right)\neq 0$, so we can divide by $2\sin\left(\frac{1}{2}x\right)$ and get

$$\sum_{i=1}^{n}\sin ix = \frac{\sin\left(\frac{1}{2}nx\right)\sin\left(\frac{1}{2}(n+1)x\right)}{\sin\left(\frac{1}{2}x\right)}.$$

EXERCISES 4.2

1. **(a)** $\|P\| = \max\{1,1,1,1\} = 1$

(b) $\displaystyle\sum_{i=1}^{n} f(x_i^*)\Delta x_i = \sum_{i=1}^{4} f(i-1)\cdot 1 = 16+15+12+7 = 50$

(c)

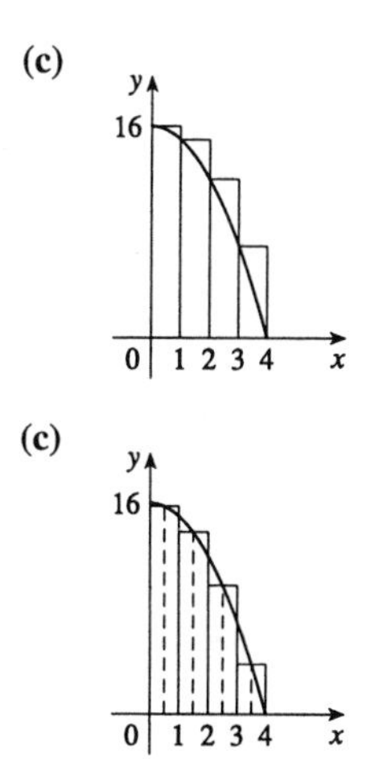

3. **(a)** $\|P\| = \max\{1,1,1,1\} = 1$

(b) $\displaystyle\sum_{i=1}^{n} f(x_i^*)\Delta x_i = \sum_{i=1}^{4} f\left(i-\tfrac{1}{2}\right)\cdot 1 = 15.75+13.75+9.75+3.75 = 43$

(c)

5. (a) $\|P\| = \max\{0.5, 0.5, 0.5, 0.5, 0.5, 0.5\} = 0.5$

(b) $$\sum_{i=1}^{6} f(x_i^*)\Delta x_i = [f(-0.5) + f(0) + f(0.5) + f(1) + f(1.5) + f(2)](0.5)$$
$$= \tfrac{1}{2}[1.875 + 2 + 2.125 + 3 + 5.375 + 10]$$
$$= \tfrac{1}{2}(24.375)$$
$$= 12.1875$$

(c)

7. (a) $\|P\| = \max\{\frac{\pi}{4}, \frac{\pi}{4}, \frac{\pi}{4}, \frac{\pi}{4}\} = \frac{\pi}{4}$

(b) $$\sum_{i=1}^{n} f(x_i^*)\Delta x_i = \sum_{i=1}^{4} f(x_i^*)\tfrac{\pi}{4} = \tfrac{\pi}{4}\left[f\left(\tfrac{\pi}{6}\right) + f\left(\tfrac{\pi}{3}\right) + f\left(\tfrac{2\pi}{3}\right) + f\left(\tfrac{5\pi}{6}\right)\right]$$
$$= \tfrac{\pi}{4}\left[1 + \sqrt{3} + \sqrt{3} + 1\right]$$
$$= \tfrac{\pi}{2}\left(1 + \sqrt{3}\right)$$

(c)

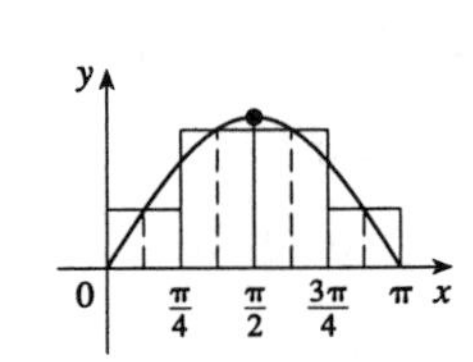

9. (a) $y = x^2 - 2x + 2 = (x-1)^2 + 1$. By counting squares, we estimate that the area under the curve is between 5 and 7 — perhaps near 6.

(b) $f(x) = y = x^2 - 2x + 2$ on $[0, 3]$ with partition points $x_i = 0 + \frac{3i}{n} = \frac{3i}{n}$, $\Delta x_i = \frac{3}{n}$ and $x_i^* = x_i$, so

$$R_n = \sum_{i=1}^{n} f(x_i^*)\Delta x_i = \frac{3}{n}\sum_{i=1}^{n} f\left(\frac{3i}{n}\right) = \frac{3}{n}\sum_{i=1}^{n}\left[\left(\frac{3i}{n}\right)^2 - 2\left(\frac{3i}{n}\right) + 2\right] = \frac{27}{n^3}\sum_{i=1}^{n} i^2 - \frac{18}{n^2}\sum_{i=1}^{n} i + 6$$
$$= \frac{27}{6}\left[\frac{n(n+1)(2n+1)}{n^3}\right] - \frac{18}{2}\left[\frac{n(n+1)}{n^2}\right] + 6 = \frac{9}{2}\left(2 + \frac{3}{n} + \frac{1}{n^2}\right) - 9\left(1 + \frac{1}{n}\right) + 6$$
$$= 6 + \frac{9}{2n} + \frac{9}{2n^2}.$$

(c) $R_6 = 6 + \frac{9}{2\cdot 6} + \frac{9}{2\cdot 36} = \frac{55}{8} = 6.875$, $R_{12} = 6 + \frac{9}{2\cdot 12} + \frac{9}{2\cdot 144} = \frac{205}{32} = 6.40625$,

$R_{24} = 6 + \frac{9}{2\cdot 24} + \frac{9}{2\cdot 576} = \frac{793}{128} = 6.1953125$

(d) Since $\|P\| \to 0$ as $n \to \infty$, the area is $A = \lim_{n\to\infty} R_n = \lim_{n\to\infty}\left(6 + \frac{9}{2n} + \frac{9}{2n^2}\right) = 6$.

11. $f(x) = x^2 + 1$ on $[0, 2]$ with partition points $x_i = 2i/n$ $(i = 0, 1, 2, \ldots, n)$, so $\Delta x_1 = \Delta x_2 = \cdots = \Delta x_n = 2/n$. $\|P\| = \max\{\Delta x_i\} = \frac{2}{n}$, so $\|P\| \to 0$ is equivalent to $n \to \infty$. Taking x_i^* to be the midpoint of $[x_{i-1}, x_i] = \left[2\frac{i-1}{n}, 2\frac{i}{n}\right]$, we get $x_i^* = \frac{2i-1}{n}$. Thus

$$A = \lim_{\|P\|\to 0} \sum_{i=1}^{n} f(x_i^*)\Delta x_i = \lim_{n\to\infty} \sum_{i=1}^{n} \left[\left(\frac{2i-1}{n}\right)^2 + 1 \right] \frac{2}{n} = \lim_{n\to\infty} \sum_{i=1}^{n} \left[\frac{8i^2}{n^3} - \frac{8i}{n^3} + \frac{2}{n^3} + \frac{2}{n} \right]$$

$$= \lim_{n\to\infty} \left[\frac{8}{n^3} \sum_{i=1}^{n} i^2 - \frac{8}{n^3} \sum_{i=1}^{n} i + \left(\frac{2}{n^3} + \frac{2}{n} \right) \sum_{i=1}^{n} 1 \right]$$

$$= \lim_{n\to\infty} \left[\frac{8}{n^3} \frac{n(n+1)(2n+1)}{6} - \frac{8}{n^3} \frac{n(n+1)}{2} + \left(\frac{2}{n^3} + \frac{2}{n} \right) n \right]$$

$$= \lim_{n\to\infty} \left[\frac{4}{3} \cdot 1 \left(1 + \frac{1}{n}\right) \left(2 + \frac{1}{n}\right) - \frac{4}{n} \cdot 1 \left(1 + \frac{1}{n}\right) + \frac{2}{n^2} + 2 \right]$$

$$= \left(\tfrac{4}{3} \cdot 1 \cdot 1 \cdot 2\right) - (0 \cdot 1 \cdot 1) + 0 + 2 = \tfrac{8}{3} + 2 = \tfrac{14}{3}.$$

13. $f(x) = 2x + 1$ on $[0, 5]$. $x_i^* = x_i = \dfrac{5i}{n}$ for $i = 1, \ldots, n$ and $\Delta x_i = \dfrac{5}{n}$.

$$A = \lim_{n\to\infty} \sum_{i=1}^{n} \left[2\left(\frac{5i}{n}\right) + 1 \right] \frac{5}{n} = \lim_{n\to\infty} \left[\frac{50}{n^2} \sum_{i=1}^{n} i + \frac{5}{n} \sum_{i=1}^{n} 1 \right]$$

$$= \lim_{n\to\infty} \left[\frac{50}{n^2} \frac{n(n+1)}{2} + \frac{5}{n} n \right] = \lim_{n\to\infty} \left[25 \cdot 1 \left(1 + \frac{1}{n}\right) + 5 \right]$$

$$= 25 + 5 = 30$$

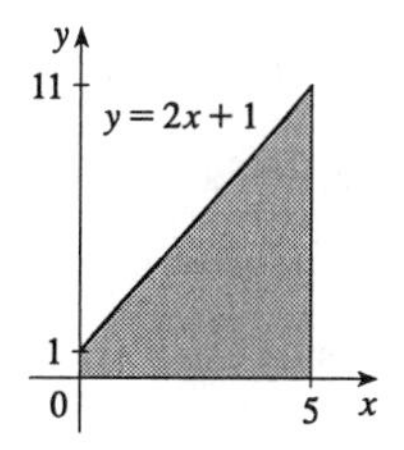

15. $f(x) = 2x^2 - 4x + 5$ on $[-3, 2]$. $x_i^* = x_i = -3 + \dfrac{5i}{n}$ $\quad (i = 1, 2, \ldots, n)$.

$$A = \lim_{n\to\infty} \sum_{i=1}^{n} \left[2\left(-3 + \frac{5i}{n}\right)^2 - 4\left(-3 + \frac{5i}{n}\right) + 5 \right] \frac{5}{n}$$

$$= \lim_{n\to\infty} \sum_{i=1}^{n} \left[\frac{50i^2}{n^2} - \frac{80i}{n} + 35 \right] \frac{5}{n}$$

$$= \lim_{n\to\infty} \left[\frac{250}{n^3} \sum_{i=1}^{n} i^2 - \frac{400}{n^2} \sum_{i=1}^{n} i + \frac{175}{n} \sum_{i=1}^{n} 1 \right]$$

$$= \lim_{n\to\infty} \left[\frac{250}{n^3} \frac{n(n+1)(2n+1)}{6} - \frac{400}{n^2} \frac{n(n+1)}{2} + \frac{175}{n} n \right]$$

$$= \lim_{n\to\infty} \left[\frac{125}{3} \cdot 1 \left(1 + \frac{1}{n}\right) \left(2 + \frac{1}{n}\right) - 200 \cdot 1 \left(1 + \frac{1}{n}\right) + 175 \right]$$

$$= \left(\tfrac{125}{3} \cdot 1 \cdot 1 \cdot 2\right) - (200 \cdot 1 \cdot 1) + 175 = \tfrac{175}{3}$$

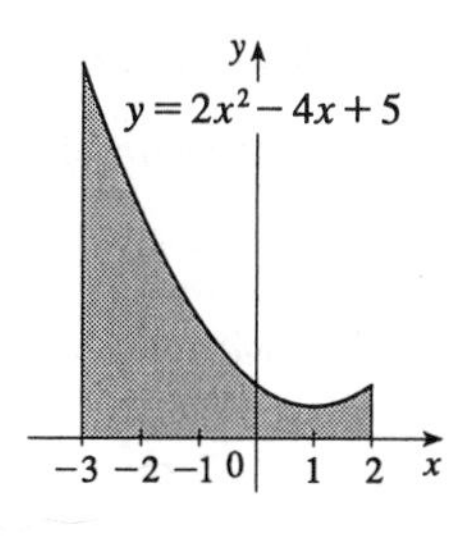

17. $f(x) = x^3 + 2x$ on $[0, 2]$. $x_i^* = x_i = \dfrac{2i}{n}$ for $i = 1, 2, \ldots, n$.

$$A = \lim_{n\to\infty} \sum_{i=1}^{n} \left[\left(\frac{2i}{n}\right)^3 + \frac{4i}{n} \right] \frac{2}{n} = \lim_{n\to\infty} \left[\frac{16}{n^4} \sum_{i=1}^{n} i^3 + \frac{8}{n^2} \sum_{i=1}^{n} i \right]$$

$$= \lim_{n\to\infty} \left[\frac{16}{n^4} \frac{n^2(n+1)^2}{4} + \frac{8}{n^2} \frac{n(n+1)}{2} \right]$$

$$= \lim_{n\to\infty} \left[4\left(1 + \frac{1}{n}\right)^2 + 4\left(1 + \frac{1}{n}\right) \right] = 4 \cdot 1 + 4 \cdot 1 = 8$$

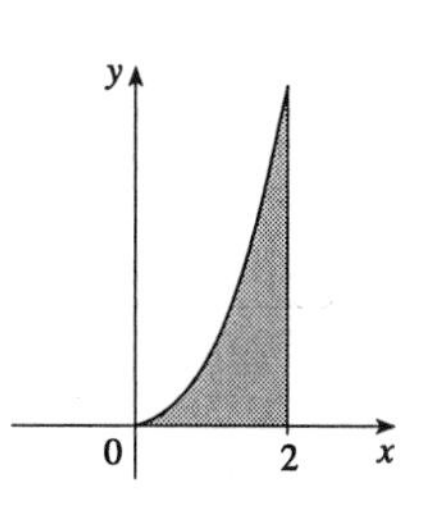

19. Here is one possible algorithm (ordered sequence of operations) for calculating the sums:

1 Let SUM $= 0$, let X-MIN $= 0$, let X-MAX $= \pi$, let STEP-SIZE $= \frac{\pi}{10}$ (or $\frac{\pi}{30}$ or $\frac{\pi}{50}$, depending on which sum we are calculating), and let RIGHT-ENDPOINT $=$ X-MIN $+$ STEP-SIZE.

2 Repeat steps 2a, 2b in sequence until RIGHT-ENDPOINT $>$ X-MAX.

2a Add sin(RIGHT-ENDPOINT) to SUM.

2b Add STEP-SIZE to RIGHT-ENDPOINT.

At the end of this procedure, the variable SUM is equal to the answer we are looking for. We find that

$$R_{10} = \frac{\pi}{10}\sum_{i=1}^{10} \sin\left(\frac{i\pi}{10}\right) \approx 1.9835,\ R_{30} = \frac{\pi}{30}\sum_{i=1}^{30} \sin\left(\frac{i\pi}{30}\right) \approx 1.9982, \text{ and } R_{50} = \frac{\pi}{50}\sum_{i=1}^{50} \sin\left(\frac{i\pi}{50}\right) \approx 1.9993.$$

It appears that the exact area is 2.

21. In Maple, we have to perform a number of steps before getting a numerical answer. After loading the `student` package [command: `with(student);`] we use the command `leftsum(x^(1/2),x=1..4,10` *[or 30, or 50]*`);` which gives us the expression in summation notation. To get a numerical approximation to the sum, we use `evalf(");`.

Mathematica does not have a special command for these sums, so we must type them in manually. For example, the first left sum is given by `(3/10)*Sum[Sqrt[1+3(i-1)/10],{i,1,10}]`, and we use the `N` command on the resulting output to get a numerical approximation.

In Derive, we use the `LEFT_RIEMANN` command to get the left sums, but must define the right sums ourselves.

(a) The left sums are $L_{10} = \frac{3}{10}\sum_{i=1}^{10}\sqrt{1+\frac{3(i-1)}{10}} \approx 4.5148$, $L_{30} = \frac{3}{30}\sum_{i=1}^{30}\sqrt{1+\frac{3(i-1)}{30}} \approx 4.6165$, $L_{50} = \frac{3}{50}\sum_{i=1}^{50}\sqrt{1+\frac{3(i-1)}{50}} \approx 4.6366$. The right sums are $R_{10} = \frac{3}{10}\sum_{i=1}^{10}\sqrt{1+\frac{3i}{10}} \approx 4.8148$, $R_{30} = \frac{3}{30}\sum_{i=1}^{30}\sqrt{1+\frac{3i}{30}} \approx 4.7165$, $R_{50} = \frac{3}{50}\sum_{i=1}^{50}\sqrt{1+\frac{3i}{50}} \approx 4.6966$.

(b) In Maple, we use the `leftbox` and `rightbox` commands (with the same arguments as `leftsum` and `rightsum` above) to generate the graphs.

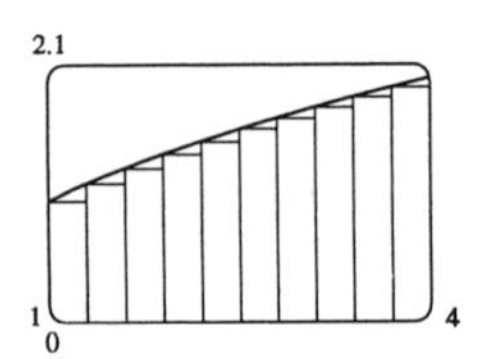

left endpoints, $n = 10$

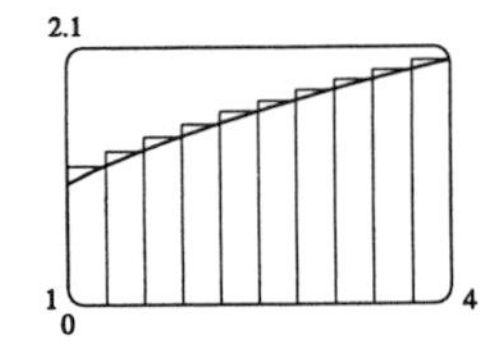

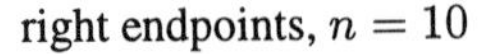
right endpoints, $n = 10$

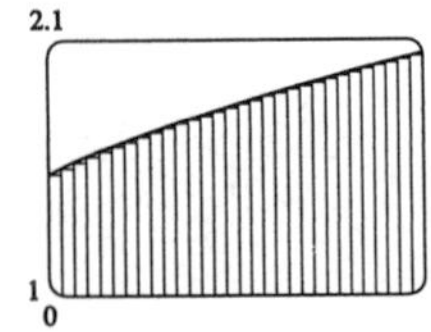

left endpoints, $n = 30$

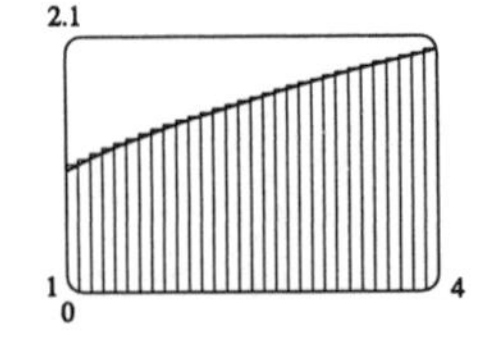

right endpoints, $n = 30$

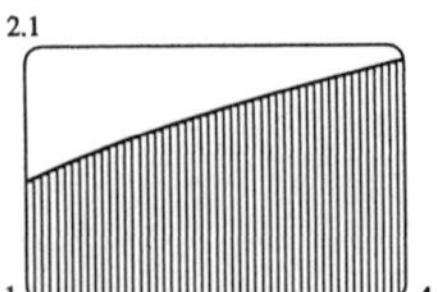

left endpoints, $n = 50$

right endpoints, $n = 50$

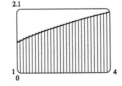

(c) We know that since $\sqrt{x}$ is an increasing function on $[1,4]$, all of the left sums are smaller than the actual area, and all of the right sums are larger than the actual area (see Example 3). Since the left sum with $n=50$ is about $4.637 > 4.6$ and the right sum with $n=50$ is about $4.697 < 4.7$, we conclude that $4.6 < L_{50} <$ actual area $< R_{50} < 4.7$, so the actual area is between 4.6 and 4.7.

23. $\lim\limits_{n\to\infty}\sum\limits_{i=1}^{n}\frac{\pi}{4n}\tan\frac{\pi i}{4n}$ can be interpreted as the area of the region lying under the graph of $y=\tan x$ on the interval $\left[0,\frac{\pi}{4}\right]$, since for $y=\tan x$ on $\left[0,\frac{\pi}{4}\right]$ with partition points $x_i=\left(\frac{\pi}{4}\right)\frac{i}{n}$, $\Delta x=\frac{\pi}{4n}$ and $x_i^*=x_i$, the expression for the area is $A=\lim\limits_{n\to\infty}\sum\limits_{i=1}^{n}f(x_i^*)\Delta x=\lim\limits_{n\to\infty}\sum\limits_{i=1}^{n}\tan\left(\frac{\pi i}{4n}\right)\frac{\pi}{4n}$. Note that this answer is not unique, since the expression for the area is the same for the function $y=\tan(k\pi+x)$ on the interval $\left[k\pi,k\pi+\frac{\pi}{4}\right]$ where k is any integer.

25. $f(x)=\sin x$. $\Delta x_i=\frac{\pi}{n}$ and $x_i^*=x_i=\frac{i\pi}{n}$. So

$$A=\lim_{n\to\infty}\sum_{i=1}^{n}\left[\sin\frac{i\pi}{n}\left(\frac{\pi}{n}\right)\right]=\lim_{n\to\infty}\frac{\pi}{n}\sum_{i=1}^{n}\sin\frac{i\pi}{n}=\lim_{n\to\infty}\frac{\pi}{n}\frac{\sin\left(\frac{n}{2}\cdot\frac{\pi}{n}\right)\sin\left(\frac{n+1}{2}\cdot\frac{\pi}{n}\right)}{\sin\left(\frac{1}{2}\frac{\pi}{n}\right)}\quad\text{(see Exercise 4.1.53)}$$

$$=\lim_{n\to\infty}\frac{\pi}{n}\frac{\sin\left(\frac{\pi}{2}+\frac{\pi}{2n}\right)}{\sin\left(\frac{\pi}{2n}\right)}=\lim_{n\to\infty}2\frac{\frac{\pi}{2n}}{\sin\left(\frac{\pi}{2n}\right)}\cdot\lim_{n\to\infty}\cos\left(\frac{\pi}{2n}\right)=2\cdot1\cdot1=2.$$

Here we have used the identity $\sin\left(\frac{\pi}{2}+x\right)=\cos x$.

EXERCISES 4.3

1. $f(x)=7-2x$

(a) $\|P\|=\max\{0.6,0.6,0.8,1.2,0.8\}=1.2$

(b) $\sum\limits_{i=1}^{5}f(x_i^*)\Delta x_i=f(1.3)(0.6)+f(1.9)(0.6)+f(2.6)(0.8)+f(3.6)(1.2)+f(4.6)(0.8)$

$=(4.4)(0.6)+(3.2)(0.6)+(1.8)(0.8)+(-0.2)(1.2)+(-2.2)(0.8)=4.$

3. $f(x)=2-x^2$

(a) $\|P\|=\max\{0.6,0.4,1,0.8,0.6,0.6\}=1$

(b) $\sum\limits_{i=1}^{6}f(x_i^*)\Delta x_i=f(-1.4)(0.6)+f(-1)(0.4)+f(0)(1)+f(0.8)(0.8)+f(1.4)(0.6)+f(2)(0.6)$

$=(0.04)(0.6)+(1)(0.4)+(2)(1)+(1.36)(0.8)+(0.04)(0.6)+(-2)(0.6)=2.336$

5. $f(x)=x^3$

(a) $\|P\|=\max\{0.5,0.5,0.5,0.5\}=0.5$

(b) $\sum\limits_{i=1}^{n}f(x_i^*)\Delta x_i=\frac{1}{2}\sum\limits_{i=1}^{4}f(x_i^*)=\frac{1}{2}\left[(-1)^3+(-0.4)^3+(0.2)^3+1^3\right]=-0.028$

7. **(a)** Using the right endpoints, we calculate

$$\int_0^8 f(x)\,dx \approx \sum_{i=1}^{4} f(x_i)\Delta x_i = 2[f(2)+f(4)+f(6)+f(8)] = 2(1+2-2+1) = 4$$

(b) Using the left endpoints, we calculate

$$\int_0^8 f(x)\,dx \approx \sum_{i=1}^{4} f(x_{i-1})\Delta x_i = 2[f(0)+f(2)+f(4)+f(6)] = 2(2+1+2-2) = 6$$

(c) Using the midpoint of each interval, we calculate

$$\int_0^8 f(x)\,dx \approx \sum_{i=1}^{4} f\left(\frac{x_i+x_{i-1}}{2}\right)\Delta x_i = 2[f(1)+f(3)+f(5)+f(7)] = 2(3+2+1-1) = 10$$

9. The width of the intervals is $\Delta x = (5-0)/5 = 1$ so the partition points are $0, 1, 2, 3, 4, 5$ and the midpoints are $0.5, 1.5, 2.5, 3.5, 4.5$. The Midpoint Rule gives
$\int_0^5 x^3\,dx \approx \sum_{i=1}^{5} f(\overline{x}_i)\Delta x = (0.5)^3 + (1.5)^3 + (2.5)^3 + (3.5)^3 + (4.5)^3 = 153.125.$

11. $\Delta x = (2-1)/10 = 0.1$ so the partition points are $1.0, 1.1, \ldots, 2.0$ and the midpoints are $1.05, 1.15, \ldots, 1.95$.

$$\int_1^2 \sqrt{1+x^2}\,dx \approx \sum_{i=1}^{10} f(\overline{x}_i)\Delta x = 0.1\left[\sqrt{1+(1.05)^2}+\sqrt{1+(1.15)^2}+\cdots+\sqrt{1+(1.95)^2}\right] \approx 1.8100$$

13. In Maple, we use the command `with(student);` to load the `sum` and `box` commands, then `m:=middlesum(sqrt(1+x^2),x=1..2,10);` which gives us the sum in summation notation, then `M:=evalf(m);` which gives $M_{10} \approx 1.81001414$, confirming the result of Exercise 11. The command `middlebox(sqrt(1+x^2),x=1..2,10);` generates the graph. Repeating for $n = 20$ and $n = 30$ gives $M_{20} \approx 1.81007263$ and $M_{30} \approx 1.81008347$.

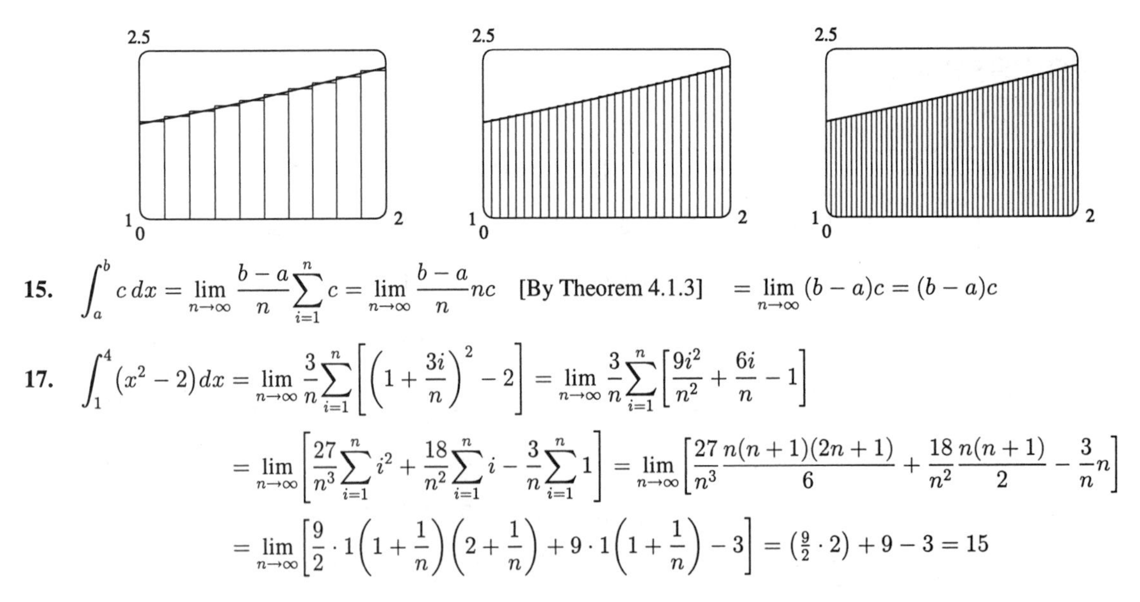

15. $$\int_a^b c\,dx = \lim_{n\to\infty} \frac{b-a}{n}\sum_{i=1}^{n} c = \lim_{n\to\infty} \frac{b-a}{n} nc \quad \text{[By Theorem 4.1.3]} \quad = \lim_{n\to\infty} (b-a)c = (b-a)c$$

17. $$\int_1^4 (x^2-2)\,dx = \lim_{n\to\infty} \frac{3}{n}\sum_{i=1}^{n}\left[\left(1+\frac{3i}{n}\right)^2 - 2\right] = \lim_{n\to\infty} \frac{3}{n}\sum_{i=1}^{n}\left[\frac{9i^2}{n^2}+\frac{6i}{n}-1\right]$$

$$= \lim_{n\to\infty}\left[\frac{27}{n^3}\sum_{i=1}^{n} i^2 + \frac{18}{n^2}\sum_{i=1}^{n} i - \frac{3}{n}\sum_{i=1}^{n} 1\right] = \lim_{n\to\infty}\left[\frac{27}{n^3}\frac{n(n+1)(2n+1)}{6} + \frac{18}{n^2}\frac{n(n+1)}{2} - \frac{3}{n}n\right]$$

$$= \lim_{n\to\infty}\left[\frac{9}{2}\cdot 1\left(1+\frac{1}{n}\right)\left(2+\frac{1}{n}\right) + 9\cdot 1\left(1+\frac{1}{n}\right) - 3\right] = \left(\tfrac{9}{2}\cdot 2\right) + 9 - 3 = 15$$

19. $$\int_0^b (x^3+4x)\,dx = \lim_{n\to\infty}\frac{b}{n}\sum_{i=1}^{n}\left[\left(\frac{bi}{n}\right)^3+4\left(\frac{bi}{n}\right)\right] = \lim_{n\to\infty}\left[\frac{b^4}{n^4}\sum_{i=1}^{n} i^3 + 4\frac{b^2}{n^2}\sum_{i=1}^{n} i\right]$$
$$= \lim_{n\to\infty}\left[\frac{b^4}{n^4}\frac{n^2(n+1)^2}{4}+\frac{4b^2}{n^2}\frac{n(n+1)}{2}\right] = \lim_{n\to\infty}\left[\frac{b^4}{4}\cdot 1^2\left(1+\frac{1}{n}\right)^2+2b^2\cdot 1\left(1+\frac{1}{n}\right)\right]$$
$$= \frac{b^4}{4}+2b^2$$

21. $$\int_a^b x\,dx = \lim_{n\to\infty}\frac{b-a}{n}\sum_{i=1}^{n}\left[a+\frac{b-a}{n}i\right] = \lim_{n\to\infty}\left[\frac{a(b-a)}{n}\sum_{i=1}^{n}1+\frac{(b-a)^2}{n^2}\sum_{i=1}^{n}i\right]$$
$$= \lim_{n\to\infty}\left[\frac{a(b-a)}{n}n+\frac{(b-a)^2}{n^2}\cdot\frac{n(n+1)}{2}\right] = a(b-a)+\lim_{n\to\infty}\frac{(b-a)^2}{2}\left(1+\frac{1}{n}\right)$$
$$= a(b-a)+\tfrac{1}{2}(b-a)^2 = (b-a)\left(a+\tfrac{1}{2}b-\tfrac{1}{2}a\right) = (b-a)\tfrac{1}{2}(b+a) = \tfrac{1}{2}(b^2-a^2)$$

23. $\int_1^3(1+2x)dx$ can be interpreted as the area under the graph of $f(x)=1+2x$ between $x=1$ and $x=3$. This is equal to the area of the rectangle plus the area of the triangle (see diagram) so $\int_1^3(1+2x)dx = A = 2\cdot 3+\frac{1}{2}\cdot 2\cdot 4 = 10$.
Or: Use the formula for the area of a trapezoid:
$a=\frac{1}{2}(2)(3+7)=10$.

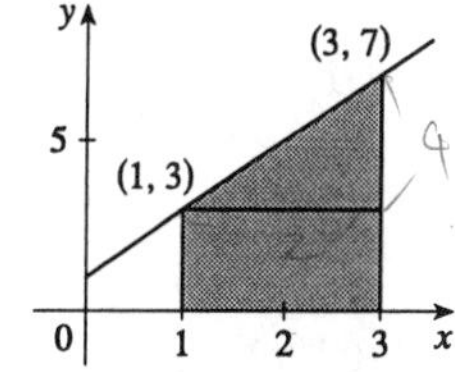

25. $\int_{-3}^{0}\left(1+\sqrt{9-x^2}\right)dx$ can be interpreted as the area under the graph of $f(x)=1+\sqrt{9-x^2}$ between $x=-3$ and $x=0$. This is equal to one-quarter the area of the circle with radius 3, plus the area of the rectangle (see diagram), so $\int_{-3}^{0}\left(1+\sqrt{9-x^2}\right)dx = \frac{1}{4}\pi 3^2+1\cdot 3 = 3+\frac{9}{4}\pi$.

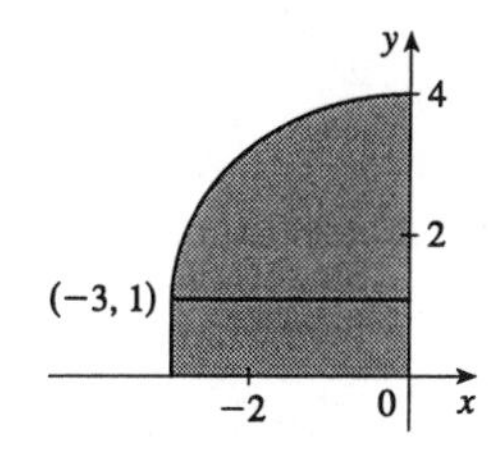

27. $\int_{-2}^{2}(1-|x|)dx$ can be interpreted as the area of the central triangle minus the areas of the outside ones (see diagram), so
$\int_{-2}^{2}(1-|x|)dx = \frac{1}{2}\cdot 2\cdot 1-2\cdot\frac{1}{2}\cdot 1\cdot 1 = 0$.

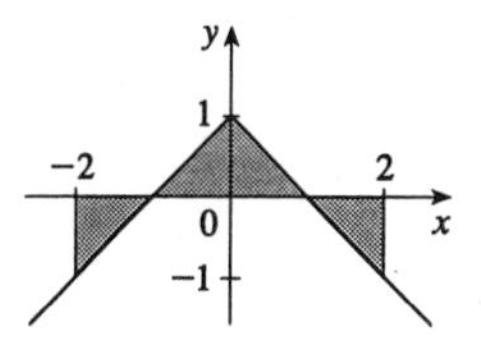

29. $\lim_{\|P\|\to 0}\sum_{i=1}^{n}\left[2(x_i^*)^2-5x_i^*\right]\Delta x_i = \int_0^1(2x^2-5x)\,dx$ by Definition 2.

31. $\lim_{\|P\|\to 0}\sum_{i=1}^{n}\cos x_i\,\Delta x_i = \int_0^{\pi}\cos x\,dx$

33. $\lim_{n\to\infty}\sum_{i=1}^{n}\frac{i^4}{n^5} = \lim_{n\to\infty}\frac{1}{n}\sum_{i=1}^{n}\left(\frac{i}{n}\right)^4 = \int_0^1 x^4\,dx$

35. $\lim_{n\to\infty}\sum_{i=1}^{n}\left[3\left(1+\frac{2i}{n}\right)^5-6\right]\frac{2}{n} = \int_1^3(3x^5-6)\,dx$

Note: To get started, notice that $\frac{2}{n}=\Delta x=\frac{b-a}{n}$ and $1+\frac{2i}{n}=a+\frac{b-a}{n}i$.

37. By the definition in Note 6, $\int_9^4\sqrt{t}\,dt = -\int_4^9\sqrt{t}\,dt = -\frac{38}{3}$.

39. $\int_{-4}^{-1}\sqrt{3}\,dx = \sqrt{3}(-1+4) = 3\sqrt{3}$

41. $\int_1^4 (2x^2 - 3x + 1)dx = 2\int_1^4 x^2\,dx - 3\int_1^4 x\,dx + \int_1^4 1\,dx$
$= 2 \cdot \frac{1}{3}(4^3 - 1^3) - 3 \cdot \frac{1}{2}(4^2 - 1^2) + 1(4 - 1) = \frac{45}{2} = 22.5$

43. $\int_{-1}^1 f(x)dx = \int_{-1}^0 f(x)dx + \int_0^1 f(x)dx = \int_{-1}^0 (-2x)dx + \int_0^1 3x^2\,dx$
$= -2\int_{-1}^0 x\,dx + 3\int_0^1 x^2\,dx = -2 \cdot \frac{1}{2}\left[0^2 - (-1)^2\right] + 3 \cdot \frac{1}{3}\left[1^3 - 0^3\right] = 2$

45. $\int_1^3 f(x)dx + \int_3^6 f(x)dx + \int_6^{12} f(x)dx = \int_1^6 f(x)dx + \int_6^{12} f(x)dx = \int_1^{12} f(x)dx$

47. $\int_2^{10} f(x)dx - \int_2^7 f(x)dx = \int_2^7 f(x)dx + \int_7^{10} f(x)dx - \int_2^7 f(x)dx = \int_7^{10} f(x)dx$

49. $0 \le \sin x < 1$ on $\left[0, \frac{\pi}{4}\right]$, so $\sin^3 x \le \sin^2 x$ on $\left[0, \frac{\pi}{4}\right]$. Hence $\int_0^{\pi/4} \sin^3 x\,dx \le \int_0^{\pi/4} \sin^2 x\,dx$ (Property 7).

51. $x \ge 4 \ge 8 - x$ on $[4, 6]$, so $\dfrac{1}{x} \le \dfrac{1}{8 - x}$ on $[4, 6]$, and $\displaystyle\int_4^6 \frac{1}{x}\,dx \le \int_4^6 \frac{1}{8 - x}\,dx.$

53. If $-1 \le x \le 1$, then $0 \le x^2 \le 1$ and $1 \le 1 + x^2 \le 2$, so $1 \le \sqrt{1 + x^2} \le \sqrt{2}$ and
$1[1 - (-1)] \le \int_{-1}^1 \sqrt{1 + x^2}\,dx \le \sqrt{2}[1 - (-1)]$ [Property 8]; that is, $2 \le \int_{-1}^1 \sqrt{1 + x^2}\,dx \le 2\sqrt{2}$.

55. If $1 \le x \le 2$, then $\dfrac{1}{2} \le \dfrac{1}{x} \le 1$, so $\frac{1}{2}(2 - 1) \le \displaystyle\int_1^2 \frac{1}{x}\,dx \le 1(2 - 1)$ or $\dfrac{1}{2} \le \displaystyle\int_1^2 \frac{1}{x}\,dx \le 1.$

57. If $f(x) = x^2 + 2x$, $-3 \le x \le 0$, then $f'(x) = 2x + 2 = 0$ when $x = -1$, and $f(-1) = -1$. At the endpoints, $f(-3) = 3$, $f(0) = 0$. Thus the absolute minimum is $m = -1$ and the absolute maximum is $M = 3$. Thus $-1[0 - (-3)] \le \int_{-3}^0 (x^2 + 2x)dx \le 3[0 - (-3)]$ or $-3 \le \int_{-3}^0 (x^2 + 2x)dx \le 9$.

59. For $-1 \le x \le 1$, $0 \le x^4 \le 1$ and $1 \le \sqrt{1 + x^4} \le \sqrt{2}$, so $1[1 - (-1)] \le \int_{-1}^1 \sqrt{1 + x^4}\,dx \le \sqrt{2}[1 - (-1)]$ or $2 \le \int_{-1}^1 \sqrt{1 + x^4}\,dx \le 2\sqrt{2}$.

61. $\sqrt{x^4 + 1} \ge \sqrt{x^4} = x^2$, so $\int_1^3 \sqrt{x^4 + 1}\,dx \ge \int_1^3 x^2\,dx = \frac{1}{3}(3^3 - 1^3) = \frac{26}{3}$.

63. $0 \le \sin x \le 1$ for $0 \le x \le \frac{\pi}{2}$, so $x \sin x \le x \quad \Rightarrow \quad \int_0^{\pi/2} x \sin x\,dx \le \int_0^{\pi/2} x\,dx = \frac{1}{2}\left[\left(\frac{\pi}{2}\right)^2 - 0^2\right] = \frac{\pi^2}{8}$.

65. Using a regular partition and right endpoints as in the proof of Property 2, we calculate

$$\int_a^b cf(x)dx = \lim_{n\to\infty} \sum_{i=1}^n cf(x_i)\Delta x_i = \lim_{n\to\infty} c\sum_{i=1}^n f(x_i)\Delta x_i = c\lim_{n\to\infty} \sum_{i=1}^n f(x_i)\Delta x_i = c\int_a^b f(x)dx.$$

67. By Property 7, the inequalities $-|f(x)| \le f(x) \le |f(x)|$ imply that $\int_a^b (-|f(x)|)dx \le \int_a^b f(x)dx \le \int_a^b |f(x)|dx$. By Property 3, the left-hand integral equals $-\int_a^b |f(x)|dx$. Thus $-M \le \int_a^b f(x)dx \le M$, where $M = \int_a^b |f(x)|dx$. [Notice that $M \ge 0$ by Property 6.] It follows that $\left|\int_a^b f(x)dx\right| \le M = \int_a^b |f(x)|dx$.

69. **(a)** $f(x) = x^2 \sin x$ is continuous on $[0, 2]$ and hence integrable by Theorem 4.

(b) $f(x) = \sec x$ is unbounded on $[0, 2]$, so f is not integrable (see the remarks following Theorem 4.)

(c) $f(x)$ is piecewise continuous on $[0, 2]$ with a single jump discontinuity at $x = 1$, so f is integrable.

(d) $f(x)$ has an infinite discontinuity at $x = 1$, so f is not integrable on $[0, 2]$.

71. f is bounded since $|f(x)| \le 1$ for all x in $[a, b]$. To see that f is not integrable on $[a, b]$, notice that $\sum_{i=1}^{n} f(x_i^*)\Delta x_i = 0$ if $x_1^*, \ldots, x_n^*$ are all chosen to be rational numbers, but $\sum_{i=1}^{n} f(x_i^*)\Delta x_i = \sum_{i=1}^{n} \Delta x_i = b - a$ if $x_1^*, \ldots, x_n^*$ are all chosen to be irrational numbers. This is true no matter how small $\|P\|$ is, since every interval $[x_{i-1}, x_i]$ with $x_{i-1} < x_i$ contains both rational and irrational numbers. $\sum_{i=1}^{n} f(x_i^*)\Delta x_i$ cannot approach both 0 and $b - a$ as $\|P\| \to 0$, so it has no limit as $\|P\| \to 0$.

73. Choose $x_i = 1 + \dfrac{i}{n}$ and $x_i^* = \sqrt{x_{i-1}\, x_i} = \sqrt{\left(1 + \dfrac{i-1}{n}\right)\left(1 + \dfrac{i}{n}\right)}$. Then

$$\int_1^2 x^{-2}\,dx = \lim_{n\to\infty} \frac{1}{n}\sum_{i=1}^{n} \frac{1}{\left(1 + \frac{i-1}{n}\right)\left(1 + \frac{i}{n}\right)} = \lim_{n\to\infty} n\sum_{i=1}^{n} \frac{1}{(n+i-1)(n+i)}$$

$$= \lim_{n\to\infty} n\sum_{i=1}^{n} \left[\frac{1}{n+i-1} - \frac{1}{n+1}\right] \quad \text{(by the hint)}$$

$$= \lim_{n\to\infty} n\left[\sum_{i=0}^{n-1} \frac{1}{n+i} - \sum_{i=1}^{n} \frac{1}{n+i}\right] = \lim_{n\to\infty} n\left[\frac{1}{n} - \frac{1}{2n}\right] = \lim_{n\to\infty}\left[1 - \frac{1}{2}\right] = \frac{1}{2}.$$

EXERCISES 4.4

1. (a) $g(0) = \int_0^0 f(t)dt = 0$, $g(1) = \int_0^1 f(t)dt = 1 \cdot 2 = 2$,

$g(2) = \int_0^2 f(t)dt = \int_0^1 f(t)dt + \int_1^2 f(t)dt = g(1) + \int_1^2 f(t)dt = 2 + 1 \cdot 2 + \frac{1}{2} \cdot 1 \cdot 2 = 5$,

$g(3) = \int_0^3 f(t)dt = g(2) + \int_2^3 f(t)dt = 5 + \frac{1}{2} \cdot 1 \cdot 4 = 7$,

$g(6) = g(3) + \int_3^6 f(t)dt = 7 + \left[-\left(\frac{1}{2} \cdot 2 \cdot 2 + 1 \cdot 2\right)\right] = 7 - 4 = 3$

(b) g is increasing on $[0, 3]$ because as x increases from 0 to 3, we keep adding more area.

(c) g has a maximum value when we start subtracting area, that is, at $x = 3$.

(d)

3.

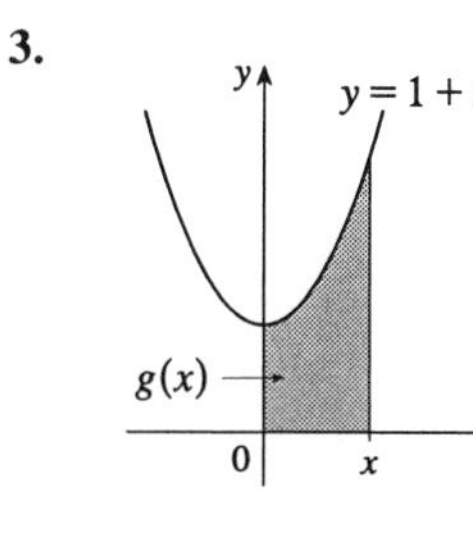

(a) By Part 1 of the Fundamental Theorem,

$g(x) = \int_0^x (1 + t^2)dt \quad \Rightarrow \quad g'(x) = f(x) = 1 + x^2$.

(b) By Part 2 of the Fundamental Theorem,

$$g(x) = \int_0^x (1 + t^2)dt = \left[t + \tfrac{1}{3}t^3\right]_0^x = \left(x + \tfrac{1}{3}x^3\right) - \left(0 + \tfrac{1}{3}0^3\right) = x + \tfrac{1}{3}x^3$$

$\Rightarrow \quad g'(x) = 1 + x^2$.

5. $g(x) = \int_1^x (t^2 - 1)^{20}dt \quad \Rightarrow \quad g'(x) = (x^2 - 1)^{20}$

7. $g(u) = \displaystyle\int_\pi^u \frac{1}{1 + t^4}\,dt \quad \Rightarrow \quad g'(u) = \frac{1}{1 + u^4}$

9. $F(x) = \int_x^2 \cos(t^2)dt = -\int_2^x \cos(t^2)dt \quad \Rightarrow \quad F'(x) = -\cos(x^2)$

11. Let $u = \dfrac{1}{x}$. Then $\dfrac{du}{dx} = -\dfrac{1}{x^2}$, so $\dfrac{d}{dx}\displaystyle\int_2^{1/x} \sin^4 t\,dt = \dfrac{d}{du}\int_2^u \sin^4 t\,dt \cdot \dfrac{du}{dx} = \sin^4 u\,\dfrac{du}{dx} = \dfrac{-\sin^4(1/x)}{x^2}$.

13. Let $u = \tan x$. Then $\dfrac{du}{dx} = \sec^2 x$, so

$$\frac{d}{dx}\int_{\tan x}^{17} \sin(t^4)\,dt = -\frac{d}{dx}\int_{17}^{\tan x} \sin(t^4)\,dt = -\frac{d}{du}\int_{17}^{u} \sin(t^4)\,dt \cdot \frac{du}{dx} = -\sin(u^4)\frac{du}{dx} = -\sin(\tan^4 x)\sec^2 x.$$

15. Let $t = 5x + 1$. Then $\dfrac{dt}{dx} = 5$, so

$$\frac{d}{dx}\int_0^{5x+1} \frac{1}{u^2-5}\,du = \frac{d}{dt}\int_0^t \frac{1}{u^2-5}\,du \cdot \frac{dt}{dx} = \frac{1}{t^2-5}\frac{dt}{dx} = \frac{5}{25x^2+10x-4}.$$

17. $\int_{-2}^4 (3x-5)dx = \left(3 \cdot \frac{1}{2}x^2 - 5x\right)\Big|_{-2}^4 = (3 \cdot 8 - 5 \cdot 4) - [3 \cdot 2 - (-10)] = -12$

19. $\int_0^1 (1 - 2x - 3x^2)dx = \left[x - 2 \cdot \frac{1}{2}x^2 - 3 \cdot \frac{1}{3}x^3\right]_0^1 = \left[x - x^2 - x^3\right]_0^1 = (1 - 1 - 1) - 0 = -1$

21. $\int_{-3}^0 (5y^4 - 6y^2 + 14)\,dy = \left[5\left(\frac{1}{5}y^5\right) - 6\left(\frac{1}{3}y^3\right) + 14y\right]_{-3}^0 = \left[y^5 - 2y^3 + 14y\right]_{-3}^0 = 0 - (-243 + 54 - 42) = 231$

23. $$\int_0^4 \sqrt{x}\,dx = \int_0^4 x^{1/2}\,dx = \left[\frac{x^{3/2}}{3/2}\right]_0^4 = \left[\tfrac{2}{3}x^{3/2}\right]_0^4 = \tfrac{2}{3}(4)^{3/2} - 0 = \tfrac{16}{3}$$

25. $$\int_1^3 \left[\frac{1}{t^2} - \frac{1}{t^4}\right]dt = \int_1^3 (t^{-2} - t^{-4})dt = \left[\frac{t^{-1}}{-1} - \frac{t^{-3}}{-3}\right]_1^3 = \left[\frac{1}{3t^3} - \frac{1}{t}\right]_1^3 = \left(\frac{1}{81} - \frac{1}{3}\right) - \left(\frac{1}{3} - 1\right) = \frac{28}{81}$$

27. $$\int_1^2 \frac{x^2+1}{\sqrt{x}}\,dx = \int_1^2 (x^{3/2} + x^{-1/2})\,dx = \left[\frac{x^{5/2}}{5/2} + \frac{x^{1/2}}{1/2}\right]_1^2 = \left[\tfrac{2}{5}x^{5/2} + 2x^{1/2}\right]_1^2$$
$$= \left(\tfrac{2}{5}4\sqrt{2} + 2\sqrt{2}\right) - \left(\tfrac{2}{5} + 2\right) = \frac{18\sqrt{2} - 12}{5} = \tfrac{6}{5}\left(3\sqrt{2} - 2\right)$$

29. $$\int_0^1 u(\sqrt{u} + \sqrt[3]{u})\,du = \int_0^1 (u^{3/2} + u^{4/3})\,du = \left[\frac{u^{5/2}}{5/2} + \frac{u^{7/3}}{7/3}\right]_0^1 = \left[\tfrac{2}{5}u^{5/2} + \tfrac{3}{7}u^{7/3}\right]_0^1 = \tfrac{2}{5} + \tfrac{3}{7} = \tfrac{29}{35}$$

31. $$\int_{-2}^3 |x^2 - 1|\,dx = \int_{-2}^{-1} (x^2 - 1)\,dx + \int_{-1}^1 (1 - x^2)\,dx + \int_1^3 (x^2 - 1)\,dx$$
$$= \left[\frac{x^3}{3} - x\right]_{-2}^{-1} + \left[x - \frac{x^3}{3}\right]_{-1}^1 + \left[\frac{x^3}{3} - x\right]_1^3$$
$$= \left(-\frac{1}{3} + 1\right) - \left(-\frac{8}{3} + 2\right) + \left(1 - \frac{1}{3}\right) - \left(-1 + \frac{1}{3}\right) + (9 - 3) - \left(\frac{1}{3} - 1\right) = \frac{28}{3}$$

33. $\int_3^3 \sqrt{x^5 + 2}\,dx = 0$ by the definition in Note 6 in Section 4.3.

35. $\displaystyle\int_{-4}^2 \frac{2}{x^6}\,dx$ does not exist since $f(x) = \dfrac{2}{x^6}$ has an infinite discontinuity at 0.

37. $$\int_1^4 \left(\sqrt{t} - \frac{2}{\sqrt{t}}\right)dt = \int_1^4 (t^{1/2} - 2t^{-1/2})\,dt = \left[\frac{t^{3/2}}{3/2} - 2\frac{t^{1/2}}{1/2}\right]_1^4 = \left[\tfrac{2}{3}t^{3/2} - 4t^{1/2}\right]_1^4$$
$$= \left[\tfrac{2}{3} \cdot 8 - 4 \cdot 2\right] - \left[\tfrac{2}{3} - 4\right] = \tfrac{2}{3}$$

39. $\displaystyle\int_{-1}^{0}(x+1)^3\,dx=\int_{-1}^{0}(x^3+3x^2+3x+1)\,dx=\left[\frac{x^4}{4}+3\frac{x^3}{3}+3\frac{x^2}{2}+x\right]_{-1}^{0}=0-\left[\frac{1}{4}-1+\frac{3}{2}-1\right]$
$\displaystyle=2-\frac{7}{4}=\frac{1}{4}$

41. $\displaystyle\int_{\pi/4}^{\pi/3}\sin t\,dt=[-\cos t]_{\pi/4}^{\pi/3}=-\cos\tfrac{\pi}{3}+\cos\tfrac{\pi}{4}=-\tfrac{1}{2}+\tfrac{1}{\sqrt{2}}=\tfrac{\sqrt{2}-1}{2}$

43. $\int_{\pi/2}^{\pi}\sec x\tan x\,dx$ does not exist since $\sec x\tan x$ has an infinite discontinuity at $\frac{\pi}{2}$.

45. $\int_{\pi/6}^{\pi/3}\csc^2\theta\,d\theta=[-\cot\theta]_{\pi/6}^{\pi/3}=-\cot\frac{\pi}{3}+\cot\frac{\pi}{6}=-\frac{1}{3}\sqrt{3}+\sqrt{3}=\frac{2}{3}\sqrt{3}$

47. $\displaystyle\int_0^1\left[\sqrt[4]{x^5}+\sqrt[5]{x^4}\right]dx=\int_0^1\left(x^{5/4}+x^{4/5}\right)dx=\left[\frac{x^{9/4}}{9/4}+\frac{x^{9/5}}{9/5}\right]_0^1=\left[\tfrac{4}{9}x^{9/4}+\tfrac{5}{9}x^{9/5}\right]_0^1=\tfrac{4}{9}+\tfrac{5}{9}-0=1$

49. $\int_{-1}^{2}(x-2|x|)dx=\int_{-1}^{0}3x\,dx+\int_0^2(-x)dx=3\left[\frac{1}{2}x^2\right]_{-1}^{0}-\left[\frac{1}{2}x^2\right]_0^2=\left(3\cdot 0-3\cdot\frac{1}{2}\right)-(2-0)=-\frac{7}{2}=-3.5$

51. $\int_0^2 f(x)dx=\int_0^1 x^4dx+\int_1^2 x^5dx=\frac{1}{5}x^5\big|_0^1+\frac{1}{6}x^6\big|_1^2=\left(\frac{1}{5}-0\right)+\left(\frac{64}{6}-\frac{1}{6}\right)=10.7$

53. From the graph, it appears that the area is about 60. The actual area is
$\int_0^{27}x^{1/3}dx=\left[\frac{3}{4}x^{4/3}\right]_0^{27}=\frac{3}{4}\cdot 81-0$
$=\frac{243}{4}=60.75$. This is $\frac{3}{4}$ of the area of the viewing rectangle.

55. It appears that the area under the graph is about $\frac{2}{3}$ of the area of the viewing rectangle, or about $\frac{2}{3}\pi\approx 2.1$. The actual area is
$\int_0^{\pi}\sin x\,dx=[-\cos x]_0^{\pi}=-\cos\pi+\cos 0$
$=-(-1)+1=2.$

57. By zooming in on the graph of $y=x+x^2-x^4$, we see that the graph has x-intercepts at $x=0$ and at $x\approx 1.32$. So the area of the region below the curve and above the y-axis is about
$\int_0^{1.32}\left(x+x^2-x^4\right)dx=\left[\frac{1}{2}x^2+\frac{1}{3}x^3-\frac{1}{5}x^5\right]_0^{1.32}$
$=\left[\frac{1}{2}(1.32)^2+\frac{1}{3}(1.32)^3-\frac{1}{5}(1.32)^5\right]-0\approx 0.84$

59. $\displaystyle\frac{d}{dx}\left[\frac{x}{a^2\sqrt{a^2-x^2}}+C\right]=\frac{1}{a^2}\frac{\sqrt{a^2-x^2}-x(-x)/\sqrt{a^2-x^2}}{a^2-x^2}=\frac{1}{a^2}\frac{(a^2-x^2)+x^2}{(a^2-x^2)^{3/2}}=\frac{1}{\sqrt{(a^2-x^2)^3}}$

61. $\displaystyle\frac{d}{dx}\left(\frac{x}{2}-\frac{\sin 2x}{4}+C\right)=\tfrac{1}{2}-\tfrac{1}{4}(\cos 2x)(2)+0=\tfrac{1}{2}-\tfrac{1}{2}\cos 2x=\tfrac{1}{2}-\tfrac{1}{2}(1-2\sin^2 x)=\sin^2 x$

63. $\int x\sqrt{x}\,dx=\int x^{3/2}\,dx=\frac{2}{5}x^{5/2}+C$

65. $\displaystyle\int\left(2-\sqrt{x}\right)^2dx=\int\left(4-4\sqrt{x}+x\right)dx=4x-4\frac{x^{3/2}}{3/2}+\frac{x^2}{2}+C=4x-\tfrac{8}{3}x^{3/2}+\tfrac{1}{2}x^2+C$

67. $\int(2x+\sec x\tan x)dx=x^2+\sec x+C$

69. **(a)** displacement $= \int_0^3 (3t-5)dt = \left[\frac{3}{2}t^2 - 5t\right]_0^3 = \frac{27}{2} - 15 = -\frac{3}{2}$ m

(b) distance traveled $= \int_0^3 |3t-5|dt = \int_0^{5/3}(5-3t)dt + \int_{5/3}^3 (3t-5)dt = \left[5t - \frac{3}{2}t^2\right]_0^{5/3} + \left[\frac{3}{2}t^2 - 5t\right]_{5/3}^3$
$= \frac{25}{3} - \frac{3}{2}\cdot\frac{25}{9} + \frac{27}{2} - 15 - \left(\frac{3}{2}\cdot\frac{25}{9} - \frac{25}{3}\right) = \frac{41}{6}$ m

71. **(a)** $v'(t) = a(t) = t+4 \quad\Rightarrow\quad v(t) = \frac{1}{2}t^2 + 4t + C \quad\Rightarrow\quad 5 = v(0) = C \quad\Rightarrow\quad v(t) = \frac{1}{2}t^2 + 4t + 5$ m/s

Or: $v(t) - v(0) = \int_0^t a(u)du = \int_0^t (u+4)du = \frac{1}{2}u^2 + 4u\Big|_0^t = \frac{1}{2}t^2 + 4t \;\Rightarrow\; v(t) = \frac{1}{2}t^2 + 4t + 5$ m/s.

(b) distance traveled $= \int_0^{10}|v(t)|\,dt = \int_0^{10}\left|\frac{1}{2}t^2 + 4t + 5\right|dt = \int_0^{10}\left(\frac{1}{2}t^2 + 4t + 5\right)dt$
$= \left[\frac{1}{6}t^3 + 2t^2 + 5t\right]_0^{10} = \frac{500}{3} + 200 + 50 = 416\frac{2}{3}$ m

73. Since $m'(x) = \rho(x)$, $m = \int_0^4 \rho(x)dx = \int_0^4 (9 + 2\sqrt{x})dx = \left[9x + \frac{4}{3}x^{3/2}\right]_0^4 = 36 + \frac{32}{3} - 0 = \frac{140}{3} = 46\frac{2}{3}$ kg.

75. Let s be the position of the car. We know from Equation 12 that $s(100) - s(0) = \int_0^{100} v(t)dt$. We use the Midpoint Rule for $0 \le t \le 100$ with $n = 5$. Note that the length of each of the five time intervals is 20 seconds $= \frac{1}{180}$ hour. So the distance traveled is

$\int_0^{100} v(t)dt \approx \frac{1}{180}[v(10) + v(30) + v(50) + v(70) + v(90)] = \frac{1}{180}(38 + 58 + 51 + 53 + 47) = \frac{247}{180} \approx 1.4$ miles

77. $$g(x) = \int_{2x}^{3x}\frac{u-1}{u+1}du = \int_{2x}^{0}\frac{u-1}{u+1}du + \int_0^{3x}\frac{u-1}{u+1}du = -\int_0^{2x}\frac{u-1}{u+1}du + \int_0^{3x}\frac{u-1}{u+1}du \quad\Rightarrow$$
$$g'(x) = -\frac{2x-1}{2x+1}\cdot\frac{d}{dx}(2x) + \frac{3x-1}{3x+1}\cdot\frac{d}{dx}(3x) = -2\cdot\frac{2x-1}{2x+1} + 3\cdot\frac{3x-1}{3x+1}$$

79. $$y = \int_{\sqrt{x}}^{x^3}\sqrt{t}\sin t\,dt = \int_{\sqrt{x}}^{1}\sqrt{t}\sin t\,dt + \int_1^{x^3}\sqrt{t}\sin t\,dt = -\int_1^{\sqrt{x}}\sqrt{t}\sin t\,dt + \int_1^{x^3}\sqrt{t}\sin t\,dt \quad\Rightarrow$$
$$y' = -\sqrt[4]{x}\left(\sin\sqrt{x}\right)\cdot\frac{d}{dx}\left(\sqrt{x}\right) + x^{3/2}\sin\left(x^3\right)\cdot\frac{d}{dx}\left(x^3\right) = -\frac{\sqrt[4]{x}\sin\sqrt{x}}{2\sqrt{x}} + x^{3/2}\sin\left(x^3\right)\left(3x^2\right)$$
$$= 3x^{7/2}\sin\left(x^3\right) - \left(\sin\sqrt{x}\right)/\left(2\sqrt[4]{x}\right)$$

81. $$F(x) = \int_1^x f(t)dt \quad\Rightarrow\quad F'(x) = f(x) = \int_1^{x^2}\frac{\sqrt{1+u^4}}{u}du \quad\Rightarrow$$
$$F''(x) = f'(x) = \frac{\sqrt{1+(x^2)^4}}{x^2}\cdot\frac{d}{dx}\left(x^2\right) = \frac{2\sqrt{1+x^8}}{x}.\quad \text{So } F''(2) = \sqrt{1+2^8} = \sqrt{257}.$$

83. **(a)** The Fresnel Function $S(x) = \int_0^x \sin\left(\frac{\pi}{2}t^2\right)dt$ has local maximum values where $0 = S'(x) = \sin\left(\frac{\pi}{2}x^2\right)$ and S' changes from positive to negative. For $x > 0$, this happens when $\frac{\pi}{2}x^2 = (2n-1)\pi \quad\Leftrightarrow\quad x = \sqrt{2(2n-1)}$, n any positive integer. For $x < 0$, S' changes from positive to negative where $x = -2\sqrt{n}$, since if $x < 0$, then as x increases, x^2 decreases. S' does not change sign at $x = 0$.

(b) S is concave upward on those intervals where $S''(x) \ge 0$. Differentiating our expression for $S'(x)$, we get $S''(x) = \cos\left(\frac{\pi}{2}x^2\right)\left(2\frac{\pi}{2}x\right) = \pi x\cos\left(\frac{\pi}{2}x^2\right)$. For $x > 0$, $S''(x) > 0$ where $\cos\left(\frac{\pi}{2}x^2\right) > 0 \;\Leftrightarrow\; 0 < \frac{\pi}{2}x^2 < \frac{\pi}{2}$ or $\left(2n - \frac{1}{2}\right)\pi < \frac{\pi}{2}x^2 < \left(2n + \frac{1}{2}\right)\pi$, n any integer $\quad\Leftrightarrow\quad 0 < x < 1$ or $\sqrt{4n-1} < x < \sqrt{4n+1}$, n any positive integer. For $x < 0$, as x increases, x^2 decreases, so the intervals of upward concavity for $x < 0$ are $\left(-\sqrt{4n-1}, -\sqrt{4n-3}\right)$, n any positive integer. To summarize: S is concave upward on the intervals $(0,1)$, $\left(-\sqrt{3}, -1\right)$, $\left(\sqrt{3}, \sqrt{5}\right)$, $\left(-\sqrt{7}, -\sqrt{5}\right)$, $\left(\sqrt{7}, 3\right)$, ….

(c) 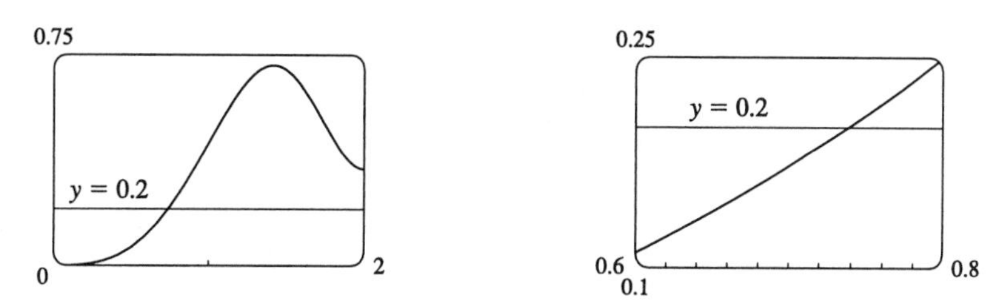

In Maple, we use `plot({int(sin(Pi*t^2/2),t=0..x),0.2},x=0..2);`. Note that Maple recognizes the Fresnel function, calling it `FresnelS(x)`. In Mathematica, we use `Plot[{Integrate[Sin[Pi*t^2/2],{t,0,x}],0.2},{x,0,2}]`. From the graphs, we see that $\int_0^x \sin\left(\frac{\pi}{2}t^2\right)dt = 0.2$ at $x \approx 0.74$.

85. (a) By the Fundamental Theorem of Calculus, $g'(x) = f(x)$. So $g'(x) = 0$ at $x = 1, 3, 5, 7$, and 9. g has local maxima at $x = 1$ and at $x = 5$ (since $f = g'$ changes from positive to negative there) and local minima at $x = 3$ and at $x = 7$. There is no local extremum at $x = 9$, since f is not defined for $x > 9$.

(b) We can see from the graph that $\left|\int_0^1 f\,dt\right| < \left|\int_1^3 f\,dt\right| < \cdots < \left|\int_7^9 f\,dt\right|$. So

$$g(9) = \int_0^9 f\,dt = \left|\int_0^1 f\,dt\right| - \left|\int_1^3 f\,dt\right| + \cdots - \left|\int_5^7 f\,dt\right| + \left|\int_7^9 f\,dt\right|$$

$$> g(5) = \int_0^5 f\,dt = \left|\int_0^1 f\,dt\right| - \cdots + \left|\int_3^5 f\,dt\right|,$$

which in turn is larger than $g(1)$. So the absolute maximum of $g(x)$ occurs at $x = 9$.

(c) g is concave downward on those intervals where $g'' < 0$. But $g'(x) = f(x)$, so $g''(x) = f'(x)$, which is negative on $\left(\frac{1}{2}, 2\right)$, $(4, 6)$ and $(8, 9)$. So g is concave down on these intervals.

(d) 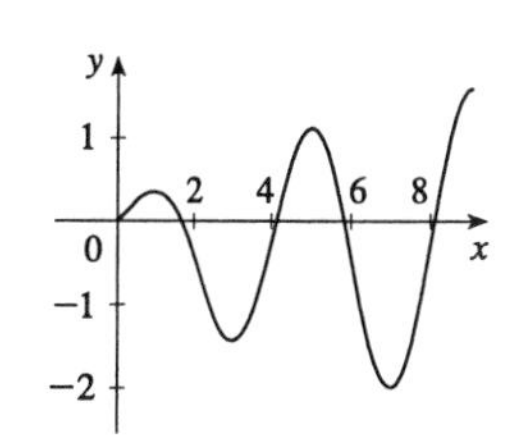

87. $\displaystyle\lim_{n\to\infty}\sum_{i=1}^{n}\frac{i^3}{n^4} = \lim_{n\to\infty}\frac{1-0}{n}\sum_{i=1}^{n}\left(\frac{i}{n}\right)^3 = \int_0^1 x^3dx = \left.\frac{x^4}{4}\right|_0^1 = \frac{1}{4}$

89. Suppose $h < 0$. Since f is continuous on $[x+h, x]$, the Extreme Value Theorem says that there are numbers u and v in $[x+h, x]$ such that $f(u) = m$ and $f(v) = M$, where m and M are the absolute minimum and maximum values of f on $[x+h, x]$. By Property 8 of integrals, $m(-h) \le \int_{x+h}^{x} f(t)dt \le M(-h)$; that is, $f(u)(-h) \le -\int_x^{x+h} f(t)dt \le f(v)(-h)$. Since $-h > 0$, we can divide this inequality by $-h$: $f(u) \le \dfrac{1}{h}\displaystyle\int_x^{x+h} f(t)dt \le f(v)$. By Equation 3, $\dfrac{g(x+h)-g(x)}{h} = \dfrac{1}{h}\displaystyle\int_x^{x+h} f(t)dt$ for $h \ne 0$, and hence $f(u) \le \dfrac{g(x+h)-g(x)}{h} \le f(v)$ which is Equation 4 in the case where $h < 0$.

91. **(a)** Let $f(x) = \sqrt{x} \Rightarrow f'(x) = 1/(2\sqrt{x}) > 0$ for $x > 0 \Rightarrow f$ is increasing on $[0, \infty)$. If $x \geq 0$, then $x^3 \geq 0$, so $1 + x^3 \geq 1$ and since f is increasing, this means that $f(1 + x^3) \geq f(1) \Rightarrow \sqrt{1 + x^3} \geq 1$ for $x \geq 0$. Next let $g(t) = t^2 - t \Rightarrow g'(t) = 2t - 1 \Rightarrow g'(t) > 0$ when $t \geq 1$. Thus g is increasing on $[1, \infty)$. And since $g(1) = 0$, $g(t) \geq 0$ when $t \geq 1$. Now let $t = \sqrt{1 + x^3}$, where $x \geq 0$. $\sqrt{1 + x^3} \geq 1$ (from above) $\Rightarrow t \geq 1 \Rightarrow g(t) \geq 0 \Rightarrow (1 + x^3) - \sqrt{1 + x^3} \geq 0$ for $x \geq 0$. Therefore $1 \leq \sqrt{1 + x^3} \leq 1 + x^3$ for $x \geq 0$.

(b) From part (a) and Property 7: $\int_0^1 1\,dx \leq \int_0^1 \sqrt{1 + x^3}\,dx \leq \int_0^1 (1 + x^3)dx \Leftrightarrow x\big|_0^1 \leq \int_0^1 \sqrt{1 + x^3}\,dx \leq \left[x + \frac{1}{4}x^4\right]_0^1 \Leftrightarrow 1 \leq \int_0^1 \sqrt{1 + x^3}\,dx \leq 1 + \frac{1}{4} = 1.25$.

93. If $w'(t)$ is the rate of change of weight in pounds per year, then $w(t)$ represents the weight in pounds of the child at age t. We know from the Fundamental Theorem of Calculus that $\int_5^{10} w'(t)dt = w(10) - w(5)$, so the integral represents the weight gained by the child between the ages of 5 and 10.

95. We differentiate both sides, using the Fundamental Theorem of Calculus, to get $\dfrac{f(x)}{x^2} = 2\dfrac{1}{2\sqrt{x}} \Leftrightarrow f(x) = x^{3/2}$. To find a, we substitute $x = a$ in the original equation to obtain $6 + \displaystyle\int_a^a \frac{f(t)}{t^2}\,dt = 2\sqrt{a} \Rightarrow 6 + 0 = 2\sqrt{a} \Rightarrow a = 9$.

97. $\int_4^8 (1/x)dx = [\ln x]_4^8 = \ln 8 - \ln 4 = \ln \frac{8}{4} = \ln 2$

99. $\displaystyle\int_8^9 2^t\,dt = \left[\frac{1}{\ln 2}2^t\right]_8^9 = \frac{1}{\ln 2}(2^9 - 2^8) = \frac{2^8}{\ln 2}$

101. $\displaystyle\int_1^{\sqrt{3}} \frac{6}{1 + x^2}\,dx = 6\left[\tan^{-1} x\right]_1^{\sqrt{3}} = 6\tan^{-1}\sqrt{3} - 6\tan^{-1} 1 = 6 \cdot \tfrac{\pi}{3} - 6 \cdot \tfrac{\pi}{4} = \tfrac{\pi}{2}$

103.
$$\int_1^e \frac{x^2 + x + 1}{x}\,dx = \int_1^e \left[x + 1 + \frac{1}{x}\right]dx = \left[\tfrac{1}{2}x^2 + x + \ln x\right]_1^e$$
$$= \left[\tfrac{1}{2}e^2 + e + \ln e\right] - \left[\tfrac{1}{2} + 1 + \ln 1\right] = \tfrac{1}{2}e^2 + e - \tfrac{1}{2}$$

105. $\displaystyle\int \left[x^2 + 1 + \frac{1}{x^2 + 1}\right]dx = \tfrac{1}{3}x^3 + x + \tan^{-1} x + C$

EXERCISES 4.5

1. Let $u = x^2 - 1$. Then $du = 2x\,dx$, so $\int x(x^2-1)^{99}\,dx = \int u^{99}\left(\frac{1}{2}\,du\right) = \frac{1}{2}\frac{u^{100}}{100} + C = \frac{1}{200}(x^2-1)^{100} + C$.

3. Let $u = 4x$. Then $du = 4\,dx$, so $\int \sin 4x\,dx = \int \sin u\left(\frac{1}{4}\,du\right) = \frac{1}{4}(-\cos u) + C = -\frac{1}{4}\cos 4x + C$

5. Let $u = x^2 + 6x$. Then $du = 2(x+3)dx$, so

$$\int \frac{x+3}{(x^2+6x)^2}\,dx = \frac{1}{2}\int \frac{du}{u^2} = \frac{1}{2}\int u^{-2}\,du = -\tfrac{1}{2}u^{-1} + C = -\frac{1}{2(x^2+6x)} + C.$$

7. Let $u = x^2 + x + 1$. Then $du = (2x+1)dx$, so

$\int (2x+1)(x^2+x+1)^3\,dx = \int u^3\,du = \frac{1}{4}u^4 + C = \frac{1}{4}(x^2+x+1)^4 + C$.

9. Let $u = x - 1$. Then $du = dx$, so $\int \sqrt{x-1}\,dx = \int u^{1/2}\,du = \frac{2}{3}u^{3/2} + C = \frac{2}{3}(x-1)^{3/2} + C$.

11. Let $u = 2 + x^4$. Then $du = 4x^3\,dx$, so $\int x^3\sqrt{2+x^4}\,dx = \int u^{1/2}\left(\frac{1}{4}\,du\right) = \frac{1}{4}\frac{u^{3/2}}{3/2} + C = \frac{1}{6}(2+x^4)^{3/2} + C$.

13. Let $u = t + 1$. Then $du = dt$, so $\int \frac{2}{(t+1)^6}\,dt = 2\int u^{-6}\,du = -\frac{2}{5}u^{-5} + C = -\frac{2}{5(t+1)^5} + C$.

15. Let $u = 1 - 2y$. Then $du = -2\,dy$, so

$$\int (1-2y)^{1.3}dy = \int u^{1.3}\left(-\tfrac{1}{2}\,du\right) = -\frac{1}{2}\left(\frac{u^{2.3}}{2.3}\right) + C = -\frac{(1-2y)^{2.3}}{4.6} + C.$$

17. Let $u = 2\theta$. Then $du = 2\,d\theta$, so $\int \cos 2\theta\,d\theta = \int \cos u\left(\frac{1}{2}\,du\right) = \frac{1}{2}\sin u + C = \frac{1}{2}\sin 2\theta + C$.

19. Let $u = x + 2$. Then $du = dx$, so $\int \frac{x}{\sqrt[4]{x+2}}\,dx = \int \frac{u-2}{\sqrt[4]{u}}\,du = \int \left(u^{3/4} - 2u^{-1/4}\right)du$

$= \frac{4}{7}u^{7/4} - 2\cdot\frac{4}{3}u^{3/4} + C = \frac{4}{7}(x+2)^{7/4} - \frac{8}{3}(x+2)^{3/4} + C$.

21. Let $u = t^2$. Then $du = 2t\,dt$, so $\int t\sin(t^2)dt = \int \sin u\left(\frac{1}{2}\,du\right) = -\frac{1}{2}\cos u + C = -\frac{1}{2}\cos(t^2) + C$.

23. Let $u = 1 - x^2$. Then $x^2 = 1 - u$ and $2x\,dx = -du$, so

$$\int x^3(1-x^2)^{3/2}\,dx = \int (1-x^2)^{3/2}x^2\cdot x\,dx = \int u^{3/2}(1-u)\left(-\tfrac{1}{2}\right)du = \tfrac{1}{2}\int \left(u^{5/2} - u^{3/2}\right)du$$

$$= \tfrac{1}{2}\left[\tfrac{2}{7}u^{7/2} - \tfrac{2}{5}u^{5/2}\right] + C = \tfrac{1}{7}(1-x^2)^{7/2} - \tfrac{1}{5}(1-x^2)^{5/2} + C.$$

25. Let $u = 1 + \sec x$. Then $du = \sec x\tan x\,dx$, so

$\int \sec x\tan x\sqrt{1+\sec x}\,dx = \int u^{1/2}\,du = \frac{2}{3}u^{3/2} + C = \frac{2}{3}(1+\sec x)^{3/2} + C$.

27. Let $u = \cos x$. Then $du = -\sin x\,dx$, so $\int \cos^4 x\sin x\,dx = \int u^4(-du) = -\frac{1}{5}u^5 + C = -\frac{1}{5}\cos^5 x + C$.

29. Let $u = 2x + 3$. Then $du = 2\,dx$, so

$\int \sin(2x+3)dx = \int \sin u\left(\frac{1}{2}\,du\right) = -\frac{1}{2}\cos u + C = -\frac{1}{2}\cos(2x+3) + C$.

31. Let $u = 3x$. Then $du = 3\,dx$, so

$\int(\sin 3\alpha - \sin 3x)dx = \int(\sin 3\alpha - \sin u)\frac{1}{3}\,du = \frac{1}{3}[(\sin 3\alpha)\,u + \cos u] + C = (\sin 3\alpha)\,x + \frac{1}{3}\cos 3x + C$.

33. Let $u = b + cx^{a+1}$. Then $du = (a+1)cx^a\,dx$, so

$$\int x^a\sqrt{b + cx^{a+1}}\,dx = \int u^{1/2}\frac{1}{(a+1)c}\,du = \frac{1}{(a+1)c}\left(\tfrac{2}{3}u^{3/2}\right) + C = \frac{2}{3c(a+1)}\left(b + cx^{a+1}\right)^{3/2} + C.$$

35. $f(x) = \dfrac{3x-1}{(3x^2-2x+1)^4}$. Let $u = 3x^2 - 2x + 1$.

Then $du = (6x-2)dx = 2(3x-1)dx$, so

$$\int \frac{3x-1}{(3x^2-2x+1)^4}\,dx = \int \frac{1}{u^4}\left(\tfrac{1}{2}\,du\right) = \tfrac{1}{2}\int u^{-4}du$$

$$= -\tfrac{1}{6}u^{-3} + C = -\frac{1}{6(3x^2-2x+1)^3} + C.$$

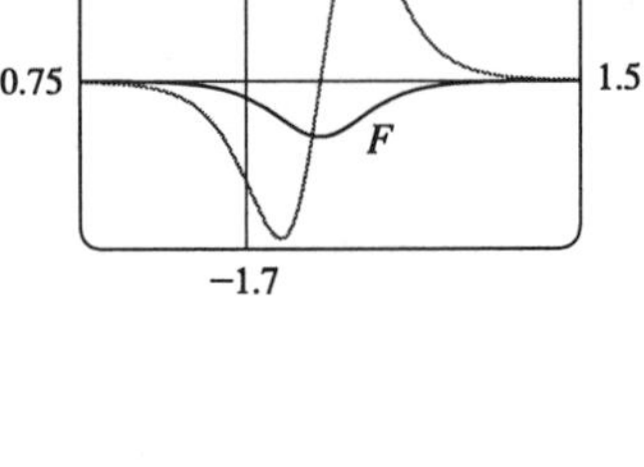

Notice that at $x = \frac{1}{3}$, the integrand goes from negative to positive, and the graph of the integral has a horizontal tangent (a local minimum).

37. $f(x) = \sin^3 x\cos x$. Let $u = \sin x$. Then $du = \cos x\,dx$, so

$\int \sin^3 x\cos x\,dx = \int u^3\,du = \frac{1}{4}u^4 + C = \frac{1}{4}\sin^4 x + C$.

Note that at $x = \frac{\pi}{2}$, the graph of the integrand crosses the x-axis from above, and the integral has a local maximum. Also, both f and F are periodic with period π, so at $x = 0$ and at $x = \pi$, the graph of the integrand crosses the x-axis from below, and the integral has local minima.

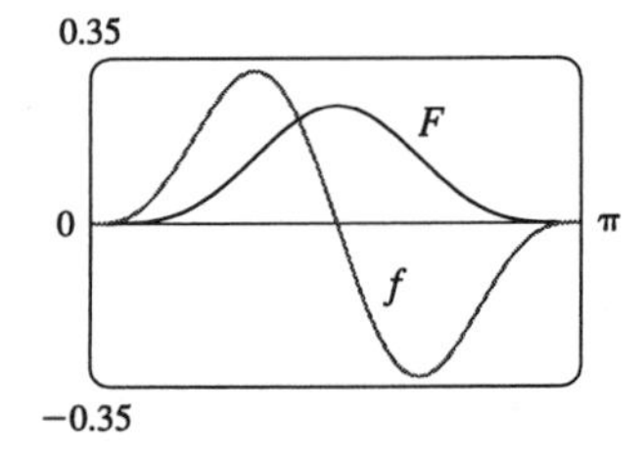

39. Let $u = 2x - 1$. Then $du = 2\,dx$, so $\int_0^1 (2x-1)^{100}dx = \int_{-1}^1 u^{100}\left(\frac{1}{2}\,du\right) = \int_0^1 u^{100}\,du$ [since the integrand is an even function] $= \left[\frac{1}{101}u^{101}\right]_0^1 = \frac{1}{101}$.

41. Let $u = x^4 + x$. Then $du = (4x^3+1)dx$, so $\displaystyle\int_0^1 \left(x^4+x\right)^5\left(4x^3+1\right)dx = \int_0^2 u^5 du = \left[\frac{u^6}{6}\right]_0^2 = \frac{2^6}{6} = \frac{32}{3}$.

43. Let $u = x - 1$. Then $du = dx$, so

$\int_1^2 x\sqrt{x-1}\,dx = \int_0^1 (u+1)\sqrt{u}\,du = \int_0^1 \left(u^{3/2} + u^{1/2}\right)du = \left[\frac{2}{5}u^{5/2} + \frac{2}{3}u^{3/2}\right]_0^1 = \dfrac{2}{5} + \dfrac{2}{3} = \dfrac{16}{15}$.

45. Let $u = \pi t$. Then $du = \pi\,dt$, so $\int_0^1 \cos\pi t\,dt = \int_0^\pi \cos u\left(\frac{1}{\pi}\,du\right) = \frac{1}{\pi}\sin u\big|_0^\pi = \frac{1}{\pi}(0-0) = 0$.

47. Let $u = 1 + \dfrac{1}{x}$. Then $du = -\dfrac{dx}{x^2}$, so

$\displaystyle\int_1^4 \frac{1}{x^2}\sqrt{1+\frac{1}{x}}\,dx = \int_2^{5/4} u^{1/2}(-du) = \int_{5/4}^2 u^{1/2}\,du = \left[\tfrac{2}{3}u^{3/2}\right]_{5/4}^2 = \tfrac{2}{3}\left[2\sqrt{2} - \tfrac{5\sqrt{5}}{8}\right] = \tfrac{4\sqrt{2}}{3} - \tfrac{5\sqrt{5}}{12}$.

49. Let $u = \cos\theta$. Then $du = -\sin\theta\,d\theta$, so

$$\int_0^{\pi/3} \frac{\sin\theta}{\cos^2\theta}\,d\theta = \int_1^{1/2} \frac{-du}{u^2} = \int_{1/2}^1 u^{-2}\,du = \left[-\frac{1}{u}\right]_{1/2}^1 = -1 + 2 = 1.$$

51. Let $u = 1 + 2x$. Then $du = 2\,dx$, so $\displaystyle\int_0^{13} \frac{dx}{\sqrt[3]{(1+2x)^2}} = \int_1^{27} u^{-2/3}\left(\tfrac{1}{2}\,du\right) = \tfrac{1}{2}\cdot 3u^{1/3}\Big|_1^{27} = \tfrac{3}{2}(3-1) = 3$.

53. $\displaystyle\int_0^4 \frac{dx}{(x-2)^3}$ does not exist since $\displaystyle\frac{1}{(x-2)^3}$ has an infinite discontinuity at $x=2$.

55. Let $u = x^2 + a^2$. Then $du = 2x\,dx$, so
$\int_0^a x\sqrt{x^2+a^2}\,dx = \int_{a^2}^{2a^2} u^{1/2}\left(\frac{1}{2}\,du\right) = \frac{1}{2}\left[\frac{2}{3}\,u^{3/2}\right]_{a^2}^{2a^2} = \left[\frac{1}{3}u^{3/2}\right]_{a^2}^{2a^2} = \frac{1}{3}\left(2\sqrt{2}-1\right)a^3.$

57. From the graph, it appears that the area under the curve is about $1 +$ a little more than $\frac{1}{2}\cdot 1\cdot 0.7$, or about 1.4. The exact area is given by $A = \int_0^1 \sqrt{2x+1}\,dx$. Let $u = 2x+1$, so $du = 2\,dx$, the limits change to $2\cdot 0+1=1$ and $2\cdot 1+1=3$, and
$A = \int_1^3 \sqrt{u}\left(\frac{1}{2}\,du\right) = \frac{1}{3}u^{3/2}\Big|_1^3 = \frac{1}{3}\left(3\sqrt{3}-1\right)$
$= \sqrt{3} - \frac{1}{3} \approx 1.399.$

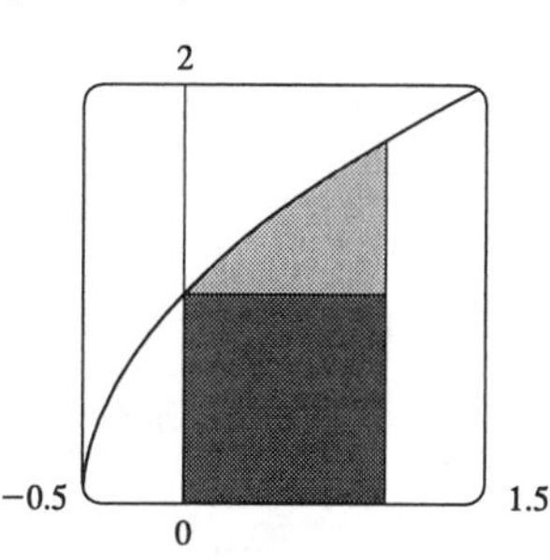

59. We split the integral: $\int_{-2}^2 (x+3)\sqrt{4-x^2}\,dx = \int_{-2}^2 x\sqrt{4-x^2}\,dx + \int_{-2}^2 3\sqrt{4-x^2}\,dx$. The first integral is 0 by Theorem 6, since $f(x) = x\sqrt{4-x^2}$ is an odd function and we are integrating from $x=-2$ to $x=2$. The second integral we interpret as three times the area of the half-circle with radius 2, so the original integral is equal to $0 + 3\cdot\frac{1}{2}(\pi\cdot 2^2) = 6\pi$.

61. The volume of inhaled air in the lungs at time t is
$V(t) = \int_0^t f(u)du = \int_0^t \frac{1}{2}\sin\left(\frac{2}{5}\pi u\right)du = \int_0^{2\pi t/5} \frac{1}{2}\sin v\left(\frac{5}{2\pi}\,dv\right)$ [We substitute $v = \frac{2\pi}{5}u \quad\Rightarrow\quad dv = \frac{2\pi}{5}\,du$]
$= \frac{5}{4\pi}(-\cos v)\big|_0^{2\pi t/5} = \frac{5}{4\pi}\left[-\cos\left(\frac{2}{5}\pi t\right)+1\right] = \frac{5}{4\pi}\left[1-\cos\left(\frac{2}{5}\pi t\right)\right]$ liters.

63. We make the substitution $u = 2x$ in $\int_0^2 f(2x)dx$. So $du = 2dx$ and the limits become $u = 2\cdot 0 = 2$ and $u = 2\cdot 2 = 4$. Hence $\int_0^4 f(u)\left(\frac{1}{2}\,du\right) = \frac{1}{2}\int_0^4 f(u)\,du = \frac{1}{2}(10) = 5$.

65. Let $u = -x$. Then $du = -dx$. When $x = a$, $u = -a$; when $x = b$, $u = -b$. So
$\int_a^b f(-x)dx = \int_{-a}^{-b} f(u)(-du) = \int_{-b}^{-a} f(u)du = \int_{-b}^{-a} f(x)dx.$
From the diagram, we see that the equality follows from the fact that we are reflecting the graph of f, and the limits of integration, about the y-axis.

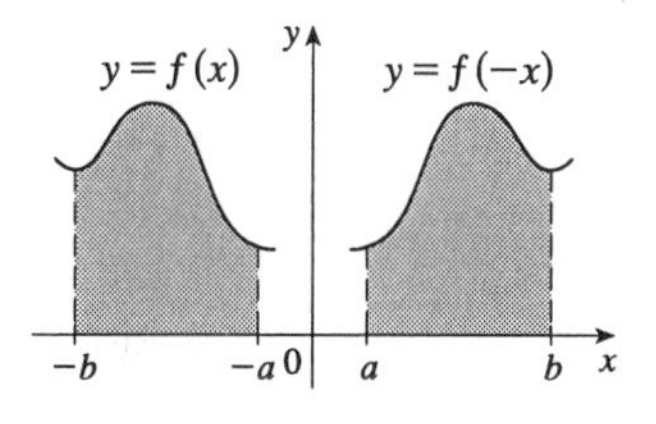

67. Let $u = 1-x$. Then $du = -dx$. When $x = 1$, $u = 0$ and when $x = 0$, $u = 1$. So
$\int_0^1 x^a(1-x)^b\,dx = -\int_1^0 (1-u)^a u^b\,du = \int_0^1 u^b(1-u)^a\,du = \int_0^1 x^b(1-x)^a\,dx.$

69. Let $u = 2x-1$. Then $du = 2\,dx$, so $\displaystyle\int\frac{dx}{2x-1} = \int\frac{\frac{1}{2}du}{u} = \frac{1}{2}\ln|u| + C = \frac{1}{2}\ln|2x-1| + C.$

71. Let $u = \ln x$. Then $du = \dfrac{dx}{x}$, so $\displaystyle\int\frac{(\ln x)^2}{x}\,dx = \int u^2du = \frac{1}{3}u^3 + C = \frac{1}{3}(\ln x)^3 + C.$

73. Let $u = 1+e^x$. Then $du = e^x\,dx$, so $\int e^x(1+e^x)^{10}\,dx = \int u^{10}\,du = \frac{1}{11}u^{11} + C = \frac{1}{11}(1+e^x)^{11} + C.$

75. Let $u = \ln x$. Then $du = \dfrac{dx}{x}$, so $\displaystyle\int\frac{dx}{x\ln x} = \int\frac{du}{u} = \ln|u| + C = \ln|\ln x| + C.$

77. $\displaystyle\int \frac{e^x+1}{e^x}\,dx = \int (1+e^{-x})dx = x - e^{-x} + C$ [Substitute $u = -x$.]

79. Let $u = x^2 + 2x$. Then $du = 2(x+1)dx$, so $\displaystyle\int \frac{x+1}{x^2+2x}\,dx = \int \frac{\frac{1}{2}\,du}{u} = \tfrac{1}{2}\ln|u| + C = \tfrac{1}{2}\ln|x^2+2x| + C.$

81. $\displaystyle\int \frac{1+x}{1+x^2}\,dx = \int \frac{1}{1+x^2}\,dx + \int \frac{x}{1+x^2}\,dx = \tan^{-1}x + \frac{1}{2}\int \frac{2x\,dx}{1+x^2} = \tan^{-1}x + \tfrac{1}{2}\ln(1+x^2) + C$

(At the last step, we evaluate $\int du/u$ where $u = 1 + x^2$.)

83. Let $u = 2x + 3$. Then $du = 2\,dx$, so

$$\int_0^3 \frac{dx}{2x+3} = \int_3^9 \frac{\frac{1}{2}\,du}{u} = \tfrac{1}{2}\ln u\big|_3^9 = \tfrac{1}{2}(\ln 9 - \ln 3) = \tfrac{1}{2}(\ln 3^2 - \ln 3) = \tfrac{1}{2}(2\ln 3 - \ln 3) = \tfrac{1}{2}\ln 3 \quad (\text{or } \ln\sqrt{3}).$$

85. Let $u = \ln x$. Then $du = \dfrac{dx}{x}$, so $\displaystyle\int_e^{e^4} \frac{dx}{x\sqrt{\ln x}} = \int_1^4 u^{-1/2}\,du = 2u^{1/2}\big|_1^4 = 2\cdot 2 - 2\cdot 1 = 2.$

87. $\dfrac{x\sin x}{1+\cos^2 x} = x \cdot \dfrac{\sin x}{2-\sin^2 x} = xf(\sin x)$, where $f(t) = \dfrac{t}{2-t^2}$. By Exercise 68,

$$\int_0^\pi \frac{x\sin x}{1+\cos^2 x}\,dx = \int_0^\pi xf(\sin x)dx = \frac{\pi}{2}\int_0^\pi f(\sin x)dx = \frac{\pi}{2}\int_0^\pi \frac{\sin x}{1+\cos^2 x}\,dx.$$

Let $u = \cos x$. Then $du = -\sin x\,dx$. When $x = \pi$, $u = -1$ and when $x = 0$, $u = 1$. So

$$\frac{\pi}{2}\int_0^\pi \frac{\sin x}{1+\cos^2 x}\,dx = -\frac{\pi}{2}\int_1^{-1} \frac{du}{1+u^2} = \frac{\pi}{2}\int_{-1}^1 \frac{du}{1+u^2} = \frac{\pi}{2}\left[\tan^{-1}u\right]_{-1}^1$$
$$= \tfrac{\pi}{2}\left[\tan^{-1}1 - \tan^{-1}(-1)\right] = \tfrac{\pi}{2}\left[\tfrac{\pi}{4} - \left(-\tfrac{\pi}{4}\right)\right] = \tfrac{\pi^2}{4}.$$

REVIEW EXERCISES FOR CHAPTER 4

1. True by Theorem 4.1.2 (b).

3. True by repeated application of Theorem 2.2.3 (b).

5. True by Property 2 of Integrals.

7. False. For example, let $f(x) = x^2$. Then $\int_0^1 \sqrt{x^2}\,dx = \int_0^1 x\,dx = \frac{1}{2}$, but $\sqrt{\int_0^1 x^2\,dx} = \sqrt{\frac{1}{3}} = \frac{1}{\sqrt{3}}$.

9. True by Property 7 of Integrals.

11. True. The integrand is an odd function that is continuous on $[-1, 1]$, so the result follows from Equation 4.5.6 (b).

13. False. The function $f(x) = 1/x^4$ is not bounded on the interval $[-2, 1]$. It has an infinite discontinuity at $x = 0$, so it is not integrable on the interval. (If the integral were to exist, a positive value would be expected by Property 6 of Integrals.)

15. False. For example, the function $y = |x|$ is continuous on $\mathbb{R}$, but has no derivative at $x = 0$.

17. First note that either a or b must be the graph of $\int_0^x f(t)dt$, since $\int_0^0 f(t)dt = 0$, and $c(0) \neq 0$. Now notice that $b > 0$ when c is increasing, and that $c > 0$ when a is increasing. It follows that c is the graph of $f(x)$, b is the graph of $f'(x)$, and a is the graph of $\int_0^x f(t)dt$.

19. $\sum_{i=1}^{n} f(x_i^*)\,\Delta x_i = \sum_{i=1}^{4} f\left(\frac{i-1}{2}\right)\cdot\frac{1}{2} = \frac{1}{2}\left[f(0) + f\left(\frac{1}{2}\right) + f(1) + f\left(\frac{3}{2}\right)\right]$.

$f(x) = 2 + (x-2)^2$, so $f(0) = 6$, $f\left(\frac{1}{2}\right) = 4.25$, $f(1) = 3$, and $f\left(\frac{3}{2}\right) = 2.25$. Thus $\sum_{i=1}^{n} f(x_i^*)\,\Delta x_i = \frac{1}{2}(15.5) = 7.75$.

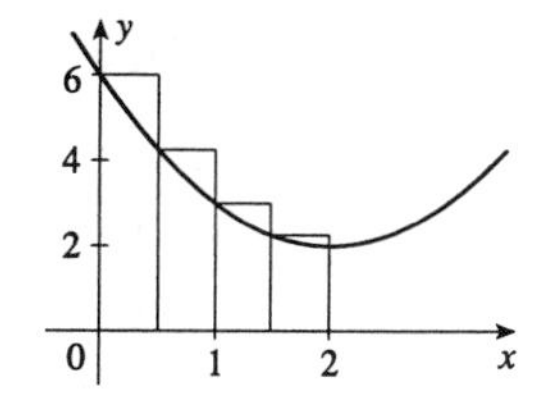

21. By Theorem 4.3.5, $\int_2^4 (3-4x)dx = \lim_{n\to\infty} \frac{2}{n}\sum_{i=1}^{n}\left[3 - 4\left(2 + \frac{2i}{n}\right)\right] = \lim_{n\to\infty} \frac{2}{n}\sum_{i=1}^{n}\left[-5 - \frac{8}{n}i\right]$

$= \lim_{n\to\infty} \frac{2}{n}\left[-5n - \frac{8}{n}\frac{n(n+1)}{2}\right] = \lim_{n\to\infty}\left[-10 - 8\cdot 1\left(1 + \frac{1}{n}\right)\right] = -10 - 8 = -18.$

23. $\int_0^5 (x^3 - 2x^2)dx = \frac{1}{4}x^4 - \frac{2}{3}x^3\Big|_0^5 = \frac{625}{4} - \frac{250}{3} = \frac{875}{12}$ (We did this integral in Exercise 22.)

25. $\int_0^1 (1 - x^9)dx = \left[x - \frac{1}{10}x^{10}\right]_0^1 = 1 - \frac{1}{10} = \frac{9}{10}$

27. $\int_1^8 \sqrt[3]{x}(x-1)dx = \int_1^8 (x^{4/3} - x^{1/3})dx = \frac{3}{7}x^{7/3} - \frac{3}{4}x^{4/3}\Big|_1^8 = \left(\frac{3}{7}\cdot 128 - \frac{3}{4}\cdot 16\right) - \left(\frac{3}{7} - \frac{3}{4}\right) = \frac{1209}{28}$

29. Let $u = 1 + 2x^3$. Then $du = 6x^2\,dx$, so $\int_0^2 x^2(1+2x^3)^3\,dx = \int_1^{17} u^3\left(\frac{1}{6}\,du\right) = \frac{1}{24}u^4\Big|_1^{17} = \frac{1}{24}(17^4 - 1) = 3480.$

31. Let $u = 2x + 3$. Then $du = 2\,dx$, so $\int_3^{11} \frac{dx}{\sqrt{2x+3}} = \int_9^{25} u^{-1/2}\left(\frac{1}{2}\,du\right) = u^{1/2}\Big|_9^{25} = 5 - 3 = 2.$

33. $\int_{-2}^{-1} \frac{dx}{(2x+3)^4}$ does not exist since the integrand has an infinite discontinuity at $x = -\frac{3}{2}$.

35. Let $u = 2 + x^5$. Then $du = 5x^4\,dx$, so

$$\int \frac{x^4\,dx}{(2+x^5)^6} = \int u^{-6}\left(\tfrac{1}{5}\,du\right) = \frac{1}{5}\left(\frac{u^{-5}}{-5}\right) + C = -\frac{1}{25u^5} + C = -\frac{1}{25(2+x^5)^5} + C.$$

37. Let $u = \pi x$. Then $du = \pi\,dx$, so $\int \sin \pi x\,dx = \int \frac{\sin u\,du}{\pi} = \frac{-\cos u}{\pi} + C = -\frac{\cos \pi x}{\pi} + C.$

39. Let $u = \frac{1}{t}$. Then $du = -\frac{1}{t^2}\,dt$, so $\int \frac{\cos(1/t)}{t^2}\,dt = \int \cos u\,(-du) = -\sin u + C = -\sin\left(\frac{1}{t}\right) + C.$

41. $\int_0^{2\pi} |\sin x|dx = \int_0^{\pi} \sin x\,dx - \int_{\pi}^{2\pi} \sin x\,dx = 2\int_0^{\pi} \sin x\,dx = -2\cos x\Big|_0^{\pi} = -2[(-1) - 1] = 4$

43. $f(x) = \frac{\cos x}{\sqrt{1+\sin x}}$. Let $u = 1 + \sin x$. Then $du = \cos x\,dx$, so

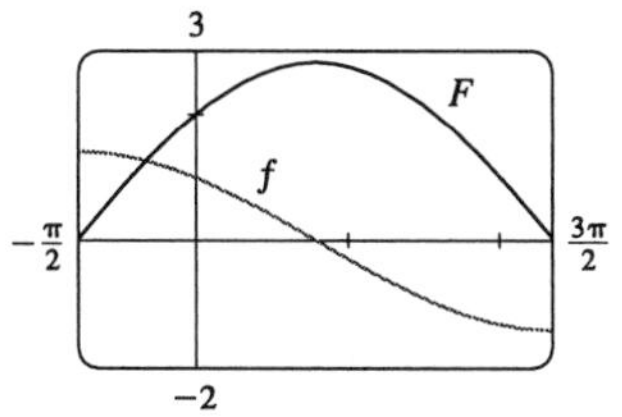

$$\int \frac{\cos x\,dx}{\sqrt{1+\sin x}} = \int u^{-1/2}\,du = 2u^{1/2} + C = 2\sqrt{1+\sin x} + C.$$

45.

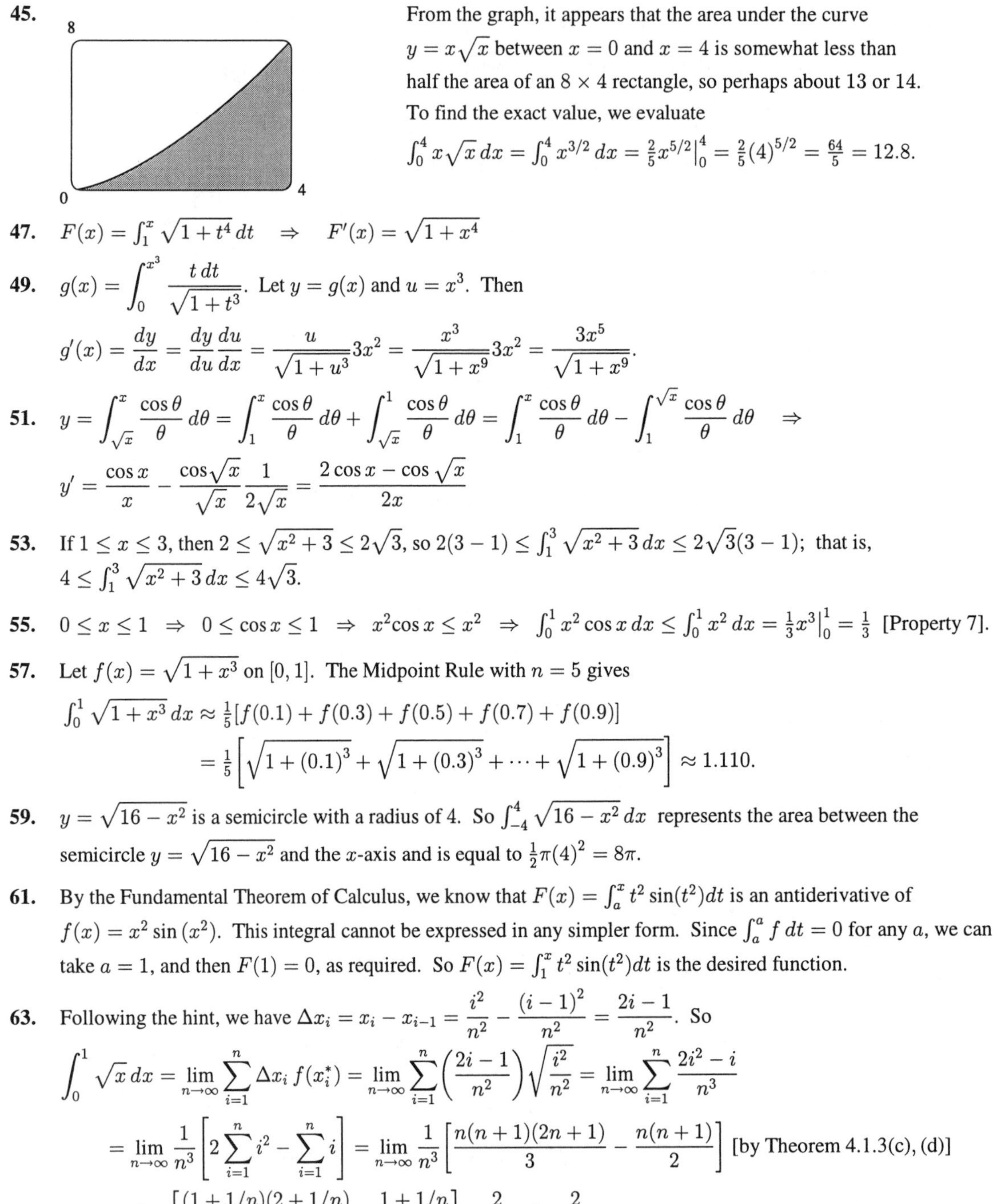

From the graph, it appears that the area under the curve $y = x\sqrt{x}$ between $x = 0$ and $x = 4$ is somewhat less than half the area of an 8×4 rectangle, so perhaps about 13 or 14. To find the exact value, we evaluate

$\int_0^4 x\sqrt{x}\,dx = \int_0^4 x^{3/2}\,dx = \frac{2}{5}x^{5/2}\Big|_0^4 = \frac{2}{5}(4)^{5/2} = \frac{64}{5} = 12.8.$

47. $F(x) = \int_1^x \sqrt{1+t^4}\,dt \quad \Rightarrow \quad F'(x) = \sqrt{1+x^4}$

49. $g(x) = \displaystyle\int_0^{x^3} \frac{t\,dt}{\sqrt{1+t^3}}$. Let $y = g(x)$ and $u = x^3$. Then

$$g'(x) = \frac{dy}{dx} = \frac{dy}{du}\frac{du}{dx} = \frac{u}{\sqrt{1+u^3}}3x^2 = \frac{x^3}{\sqrt{1+x^9}}3x^2 = \frac{3x^5}{\sqrt{1+x^9}}.$$

51. $y = \displaystyle\int_{\sqrt{x}}^x \frac{\cos\theta}{\theta}\,d\theta = \int_1^x \frac{\cos\theta}{\theta}\,d\theta + \int_{\sqrt{x}}^1 \frac{\cos\theta}{\theta}\,d\theta = \int_1^x \frac{\cos\theta}{\theta}\,d\theta - \int_1^{\sqrt{x}} \frac{\cos\theta}{\theta}\,d\theta \quad \Rightarrow$

$$y' = \frac{\cos x}{x} - \frac{\cos\sqrt{x}}{\sqrt{x}}\frac{1}{2\sqrt{x}} = \frac{2\cos x - \cos\sqrt{x}}{2x}$$

53. If $1 \le x \le 3$, then $2 \le \sqrt{x^2+3} \le 2\sqrt{3}$, so $2(3-1) \le \int_1^3 \sqrt{x^2+3}\,dx \le 2\sqrt{3}(3-1)$; that is, $4 \le \int_1^3 \sqrt{x^2+3}\,dx \le 4\sqrt{3}$.

55. $0 \le x \le 1 \;\Rightarrow\; 0 \le \cos x \le 1 \;\Rightarrow\; x^2\cos x \le x^2 \;\Rightarrow\; \int_0^1 x^2\cos x\,dx \le \int_0^1 x^2\,dx = \frac{1}{3}x^3\Big|_0^1 = \frac{1}{3}$ [Property 7].

57. Let $f(x) = \sqrt{1+x^3}$ on $[0,1]$. The Midpoint Rule with $n = 5$ gives

$$\begin{aligned}\int_0^1 \sqrt{1+x^3}\,dx &\approx \tfrac{1}{5}[f(0.1) + f(0.3) + f(0.5) + f(0.7) + f(0.9)] \\ &= \tfrac{1}{5}\left[\sqrt{1+(0.1)^3} + \sqrt{1+(0.3)^3} + \cdots + \sqrt{1+(0.9)^3}\right] \approx 1.110.\end{aligned}$$

59. $y = \sqrt{16-x^2}$ is a semicircle with a radius of 4. So $\int_{-4}^4 \sqrt{16-x^2}\,dx$ represents the area between the semicircle $y = \sqrt{16-x^2}$ and the x-axis and is equal to $\frac{1}{2}\pi(4)^2 = 8\pi$.

61. By the Fundamental Theorem of Calculus, we know that $F(x) = \int_a^x t^2\sin(t^2)dt$ is an antiderivative of $f(x) = x^2\sin(x^2)$. This integral cannot be expressed in any simpler form. Since $\int_a^a f\,dt = 0$ for any a, we can take $a = 1$, and then $F(1) = 0$, as required. So $F(x) = \int_1^x t^2\sin(t^2)dt$ is the desired function.

63. Following the hint, we have $\Delta x_i = x_i - x_{i-1} = \dfrac{i^2}{n^2} - \dfrac{(i-1)^2}{n^2} = \dfrac{2i-1}{n^2}$. So

$$\begin{aligned}\int_0^1 \sqrt{x}\,dx &= \lim_{n\to\infty}\sum_{i=1}^n \Delta x_i\,f(x_i^*) = \lim_{n\to\infty}\sum_{i=1}^n\left(\frac{2i-1}{n^2}\right)\sqrt{\frac{i^2}{n^2}} = \lim_{n\to\infty}\sum_{i=1}^n\frac{2i^2-i}{n^3} \\ &= \lim_{n\to\infty}\frac{1}{n^3}\left[2\sum_{i=1}^n i^2 - \sum_{i=1}^n i\right] = \lim_{n\to\infty}\frac{1}{n^3}\left[\frac{n(n+1)(2n+1)}{3} - \frac{n(n+1)}{2}\right] \text{ [by Theorem 4.1.3(c), (d)]} \\ &= \lim_{n\to\infty}\left[\frac{(1+1/n)(2+1/n)}{3} - \frac{1+1/n}{2n}\right] = \frac{2}{3} - 0 = \frac{2}{3}.\end{aligned}$$

65. Let $u = f(x)$ so $du = f'(x)dx$. So $2\int_a^b f(x)f'(x)dx = 2\int_{f(a)}^{f(b)} u\,du = u^2\Big|_{f(a)}^{f(b)} = [f(b)]^2 - [f(a)]^2$.

67. Let $u = 1-x$, then $du = -dx$, so $\int_0^1 f(1-x)dx = \int_1^0 f(u)(-du) = \int_0^1 f(u)du = \int_0^1 f(x)dx$.

PROBLEMS PLUS (page 304)

1. Differentiating both sides of the equation $x \sin \pi x = \int_0^{x^2} f(t)dt$ (using Part 1 of the Fundamental Theorem of Calculus for the right side) gives $\sin \pi x + \pi x \cos \pi x = 2x f(x^2)$. Putting $x = 2$, we obtain $\sin 2\pi + 2\pi \cos 2\pi = 4f(4)$, so $f(4) = \frac{1}{4}(0 + 2\pi \cdot 1) = \frac{\pi}{2}$.

3. For $1 \le x \le 2$, we have $x^4 \le 2^4 = 16$, so $1 + x^4 \le 17$ and $\dfrac{1}{1+x^4} \ge \dfrac{1}{17}$. Thus $\displaystyle\int_1^2 \frac{1}{1+x^4}\,dx \ge \int_1^2 \frac{1}{17}\,dx = \frac{1}{17}$. Also $1 + x^4 > x^4$ for $1 \le x \le 2$, so $\dfrac{1}{1+x^4} < \dfrac{1}{x^4}$ and $\displaystyle\int_1^2 \frac{1}{1+x^4}\,dx < \int_1^2 x^{-4}\,dx = \left[\frac{x^{-3}}{-3}\right]_1^2 = -\frac{1}{24} + \frac{1}{3} = \frac{7}{24}$. Thus we have the estimate $\displaystyle\frac{1}{17} \le \int_1^2 \frac{1}{1+x^4}\,dx \le \frac{7}{24}$.

5. Differentiating $x^2 + xy + y^2 = 12$ implicitly with respect to x gives $2x + y + x\dfrac{dy}{dx} + 2y\dfrac{dy}{dx} = 0$, so $\dfrac{dy}{dx} = -\dfrac{2x+y}{x+2y}$. At a highest or lowest point, $\dfrac{dy}{dx} = 0 \quad \Leftrightarrow \quad y = -2x$. Substituting this into the original equation gives $x^2 + x(-2x) + (-2x)^2 = 12$, so $3x^2 = 12$ and $x = \pm 2$. If $x = 2$, then $y = -2x = -4$, and if $x = -2$ then $y = 4$. Thus the highest and lowest points are $(-2, 4)$ and $(2, -4)$.

7. Such a function cannot exist. $f'(x) > 3$ for all x means that f is differentiable (and hence continuous) for all x. So by Part 2 of the Fundamental Theorem, $\int_1^4 f'(x)dx = f(4) - f(1) = 7 - (-1) = 8$. However, if $f'(x) > 3$ for all x, then $\int_1^4 f'(x)dx \ge 3 \cdot (4 - 1) = 9$ by Property 8 of integrals.

Alternate Solution: By the Mean Value Theorem there exists a number $c \in (1, 4)$ such that $f'(c) = \dfrac{f(4) - f(1)}{4 - 1} = \dfrac{7 - (-1)}{3} = \dfrac{8}{3} \quad \Rightarrow \quad 8 = 3f'(c)$. But $f'(x) > 3 \quad \Rightarrow \quad 3f'(c) > 9$, so such a function cannot exist.

9. $f(x) = \dfrac{1}{1+|x|} + \dfrac{1}{1+|x-2|}$

$$= \begin{cases} \dfrac{1}{1-x} + \dfrac{1}{1-(x-2)} & \text{if } x < 0 \\ \dfrac{1}{1+x} + \dfrac{1}{1-(x-2)} & \text{if } 0 \le x < 2 \\ \dfrac{1}{1+x} + \dfrac{1}{1+(x-2)} & \text{if } x \ge 2 \end{cases} \quad \Rightarrow \quad f'(x) = \begin{cases} \dfrac{1}{(1-x)^2} + \dfrac{1}{(3-x)^2} & \text{if } x < 0 \\ \dfrac{-1}{(1+x)^2} + \dfrac{1}{(3-x)^2} & \text{if } 0 < x < 2 \\ \dfrac{-1}{(1+x)^2} - \dfrac{1}{(x-1)^2} & \text{if } x > 2 \end{cases}$$

Clearly $f'(x) > 0$ for $x < 0$ and $f'(x) < 0$ for $x > 2$. For $0 < x < 2$, we have $f'(x) = \dfrac{1}{(3-x)^2} - \dfrac{1}{(x+1)^2} = \dfrac{(x^2+2x+1) - (x^2-6x+9)}{(3-x)^2(x+1)^2} = \dfrac{8(x-1)}{(3-x)^2(x+1)^2}$, so $f'(x) < 0$ for $x < 1$, $f'(1) = 0$ and $f'(x) > 0$ for $x > 1$. We have shown that $f'(x) > 0$ for $x < 0$; $f'(x) < 0$ for $0 < x < 1$; $f'(x) > 0$ for $1 < x < 2$; and $f'(x) < 0$ for $x > 2$. Therefore by the First Derivative Test, the local maxima of f are at $x = 0$ and $x = 2$, where f takes the value $\frac{4}{3}$. Therefore $\frac{4}{3}$ is the absolute maximum value of f.

11. We must find a value x_0 such that the normal lines to the parabola $y = x^2$ at $x = \pm x_0$ intersect at a point one unit from the points $(\pm x_0, x_0^2)$. The normals to $y = x^2$ at $x = \pm x_0$ have slopes $-\dfrac{1}{\pm 2x_0}$ and pass through $(\pm x_0, x_0^2)$ respectively, so the normals have the equations $y - x_0^2 = -\dfrac{1}{2x_0}(x - x_0)$ and $y - x_0^2 = \dfrac{1}{2x_0}(x + x_0)$. The common y-intercept is $x_0^2 + \frac{1}{2}$. We want to find the value of x_0 for which the distance from $\left(0, x_0^2 + \frac{1}{2}\right)$ to (x_0, x_0^2) equals 1. The square of the distance is $(x_0 - 0)^2 + \left[x_0^2 - \left(x_0^2 + \frac{1}{2}\right)\right]^2 = x_0^2 + \frac{1}{4} = 1 \quad \Leftrightarrow \quad x_0 = \pm\frac{\sqrt{3}}{2}$. For these values of x_0, the y-intercept is $x_0^2 + \frac{1}{2} = \frac{5}{4}$, so the center of the circle is at $\left(0, \frac{5}{4}\right)$.

Alternate Solution: Let the center of the circle be $(0, a)$. Then the equation of the circle is $x^2 + (y - a)^2 = 1$. Solving with the equation of the parabola, $y = x^2$, we get $x^2 + \left(x^2 - a\right)^2 = 1 \quad \Leftrightarrow \quad x^2 + x^4 - 2ax^2 + a^2 = 1$ $\Leftrightarrow \quad x^4 + (1 - 2a)x^2 + a^2 - 1 = 0$. The parabola and the circle will be tangent to each other when this quadratic equation in x^2 has equal roots, that is, when the discriminant is 0. Thus $(1 - 2a)^2 - 4(a^2 - a) = 0$ $\Leftrightarrow \quad 1 - 4a + 4a^2 - 4a^2 + 4 = 0 \quad \Leftrightarrow \quad 4a = 5$, so $a = \frac{5}{4}$. The center of the circle is $\left(0, \frac{5}{4}\right)$.

13. $f(x) = \displaystyle\int_0^{g(x)} \frac{1}{\sqrt{1 + t^3}}\,dt$, where $g(x) = \displaystyle\int_0^{\cos x} \left[1 + \sin\left(t^2\right)\right] dt$. Using Part 1 of the Fundamental Theorem of Calculus and the Chain Rule (twice) we have that

$f'(x) = \dfrac{1}{\sqrt{1 + [g(x)]^3}}\,g'(x) = \dfrac{1}{\sqrt{1 + [g(x)]^3}}\left[1 + \sin\left(\cos^2 x\right)\right](-\sin x)$. Now $g\left(\frac{\pi}{2}\right) = \int_0^0 [1 + \sin(t^2)]dt = 0$, so $f'\left(\frac{\pi}{2}\right) = \frac{1}{\sqrt{1+0}}(1 + \sin 0)(-1) = 1 \cdot 1 \cdot (-1) = -1$.

15. $f(x) = [\![x]\!] + \sqrt{x - [\![x]\!]}$. On each interval of the form $[n, n + 1)$, where n is an integer, we have $f(x) = n + \sqrt{x - n}$. It is easy to see that this function is continuous and increasing on $[n, n + 1)$. Also, the left-hand limit

$$\lim_{x \to (n+1)^-} f(x) = \lim_{x \to (n+1)^-} \left[[\![x]\!] + \sqrt{x - [\![x]\!]}\right] = \lim_{x \to (n+1)^-} [\![x]\!] + \sqrt{\lim_{x \to (n+1)^-} x - \lim_{x \to (n+1)^-} [\![x]\!]}$$
$$= n + \sqrt{n + 1 - n} = n + 1 = f(n + 1) = \lim_{x \to (n+1)^+} f(x),$$

so f is continuous and increasing everywhere.

17. $f(x) = 2 + x - x^2 = (-x + 2)(x + 1)$. So $f(x) = 0 \quad \Leftrightarrow \quad x = 2$ or $x = -1$, and $f(x) \geq 0$ for $x \in [-1, 2]$ and $f(x) < 0$ everywhere else. The integral $\int_a^b (2 + x - x^2)dx$ has a maximum on the interval where the integrand is positive, which is $[-1, 2]$. So $a = -1$, $b = 2$. (Any larger interval gives a smaller integral since $f(x) < 0$ outside $[-1, 2]$. Any smaller interval also gives a smaller integral since $f(x) \geq 0$ in $[-1, 2]$.)

19. $A = (x_1, x_1^2)$ and $B = (x_2, x_2^2)$, where x_1 and x_2 are the solutions of the quadratic equation $x^2 = mx + b$. Let $P = (x, x^2)$ and set $A_1 = (x_1, 0)$, $B_1 = (x_2, 0)$, and $P_1 = (x, 0)$. Let $f(x)$ denote the area of triangle PAB. Then $f(x)$ can be expressed in terms of the areas of three trapezoids as follows:

$$f(x) = \text{area}(A_1ABB_1) - \text{area}(A_1APP_1) - \text{area}(B_1BPP_1)$$
$$= \tfrac{1}{2}(x_2 - x_1)(x_1^2 + x_2^2) - \tfrac{1}{2}(x - x_1)(x_1^2 + x^2) - \tfrac{1}{2}(x_2 - x)(x^2 + x_2^2).$$

After expansion, cancelling of terms, and factoring, we find that $f(x) = \frac{1}{2}(x_2 - x_1)(x - x_1)(x_2 - x)$.

Note: Another way to get an expression for $f(x)$ is to use the formula for an area of a triangle in terms of the coordinates of the vertices: $f(x)=\frac{1}{2}[(x_2x_1^2-x_1x_2^2)+(x_1x^2-xx_1^2)+(xx_2^2-x_2x^2)]$. From our expression for $f(x)$, it follows that $f'(x)=\frac{1}{2}(x_2-x_1)(x_1+x_2-2x)$ and $f''(x)=-(x_2-x_1)<0$. Thus the area $f(x)$ is maximized when $x=\frac{1}{2}(x_1+x_2)$, and $f\left(\frac{1}{2}(x_1+x_2)\right)=\frac{1}{2}(x_2-x_1)\frac{1}{2}(x_2-x_1)\frac{1}{2}(x_2-x_1)=\frac{1}{8}(x_2-x_1)^3$. In terms of m and b, $x_1=\frac{1}{2}\left(m-\sqrt{m^2+4b}\right)$ and $x_2=\frac{1}{2}\left(m+\sqrt{m^2+4b}\right)$, so the maximal area is $\frac{1}{8}(m^2+4b)^{3/2}$ and it is attained at the point $P\left(\frac{1}{2}m,\frac{1}{4}m^2\right)$.

21. Since $[\![x]\!]\le x<[\![x]\!]+1$, we have $1\le\dfrac{x}{[\![x]\!]}\le 1+\dfrac{1}{[\![x]\!]}$ for $x>0$. As $x\to\infty$, $[\![x]\!]\to\infty$, so $\dfrac{1}{[\![x]\!]}\to 0$ and $1+\dfrac{1}{[\![x]\!]}\to 1$. Thus $\lim\limits_{x\to\infty}\dfrac{x}{[\![x]\!]}=1$ by the Squeeze Theorem.

23. Differentiating the equation $\int_0^x f(t)dt=[f(x)]^2$ using Part 1 of the Fundamental Theorem gives $f(x)=2f(x)f'(x)\quad\Rightarrow\quad f(x)[2f'(x)-1]=0$, so $f(x)=0$ or $f'(x)=\frac{1}{2}$. $f'(x)=\frac{1}{2}\quad\Rightarrow$ $f(x)=\frac{1}{2}x+C$. To find C we substitute into the original equation to get $\int_0^x\left(\frac{1}{2}t+C\right)dt=\left(\frac{1}{2}x+C\right)^2\quad\Leftrightarrow$ $\frac{1}{4}x^2+Cx=\frac{1}{4}x^2+Cx+C^2$. It follows that $C=0$ so $f(x)=\frac{1}{2}x$. Therefore $f(x)=0$ or $f(x)=\frac{1}{2}x$.

25. We find the tangent line to $f_n(x)=n\cos nx$ at the point $\left(\dfrac{\pi}{2n},0\right)$:

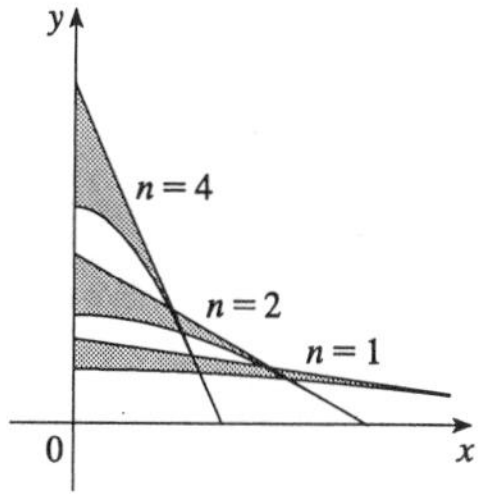

$\dfrac{df_n}{dx}=n(-\sin nx)n=-n^2\sin nx\quad\Rightarrow$

$f_n'\left(\dfrac{\pi}{2n}\right)=-n^2\sin\left(n\dfrac{\pi}{2n}\right)=-n^2(1)=-n^2$. So the equation of the tangent line to f_n is $y=t_n(x)=-n^2\left(x-\dfrac{\pi}{2n}\right)=\frac{\pi}{2}n-n^2x$.

Note that the tangent line lies above the curve on $\left(0,\dfrac{\pi}{2n}\right)$.

The area bounded by the graphs of $t_n(x)$ and $f_n(x)$ is

$$A_n=\int_0^{\pi/(2n)}[t_n(x)-f_n(x)]dx=\int_0^{\pi/(2n)}\left(\tfrac{\pi}{2}n-n^2x-n\cos nx\right)dx$$

$$=\left[\tfrac{\pi}{2}nx-\tfrac{1}{2}n^2x^2-\sin nx\right]_0^{\pi/(2n)}=\tfrac{\pi}{2}n\left(\frac{\pi}{2n}\right)-\tfrac{1}{2}n^2\left(\frac{\pi}{2n}\right)^2-\sin\left[n\left(\frac{\pi}{2n}\right)\right]$$

$$=\tfrac{1}{8}\pi^2-1\text{, a constant.}$$

27. $f(x)=(a^2+a-6)\cos 2x+(a-2)x+\cos 1\quad\Rightarrow\quad f'(x)=-(a^2+a-6)\sin 2x(2)+(a-2)$. The derivative exists for all x, so the only possible critical points will occur where $f'(x)=0\quad\Leftrightarrow$ $2(a-2)(a+3)\sin 2x=a-2\quad\Leftrightarrow\quad$ either $a=2$ or $2(a+3)\sin 2x=1$, with the latter implying that $\sin 2x=\dfrac{1}{2(a+3)}$. Since the range of $\sin 2x$ is $[-1,1]$, this equation has no solution whenever either $\dfrac{1}{2(a+3)}>1$ or $\dfrac{1}{2(a+3)}>1$. Solving these inequalities, we get $-\frac{7}{2}<a<-\frac{5}{2}$.

29. **(a)**

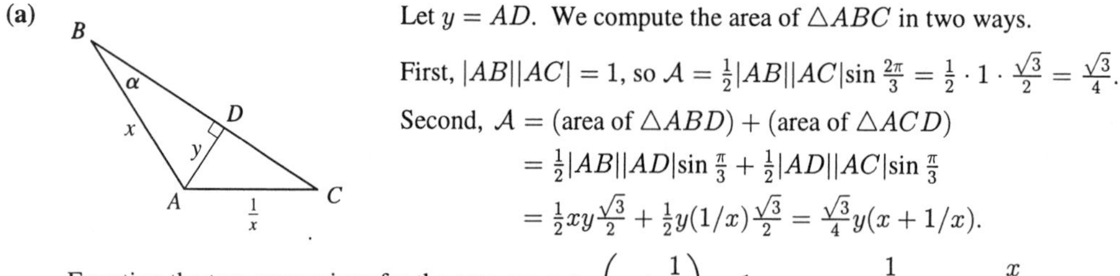

Let $y = AD$. We compute the area of $\triangle ABC$ in two ways.

First, $|AB||AC| = 1$, so $\mathcal{A} = \frac{1}{2}|AB||AC|\sin\frac{2\pi}{3} = \frac{1}{2}\cdot 1\cdot\frac{\sqrt{3}}{2} = \frac{\sqrt{3}}{4}$.

Second, $\mathcal{A} = (\text{area of } \triangle ABD) + (\text{area of } \triangle ACD)$

$$= \tfrac{1}{2}|AB||AD|\sin\tfrac{\pi}{3} + \tfrac{1}{2}|AD||AC|\sin\tfrac{\pi}{3}$$

$$= \tfrac{1}{2}xy\tfrac{\sqrt{3}}{2} + \tfrac{1}{2}y(1/x)\tfrac{\sqrt{3}}{2} = \tfrac{\sqrt{3}}{4}y(x+1/x).$$

Equating the two expressions for the area, we get $y\left(x + \dfrac{1}{x}\right) = 1$, or $y = \dfrac{1}{x + 1/x} = \dfrac{x}{x^2+1}$, $x > 0$.

Another Method: By using the Law of Sines on the triangles ABD and ABC, we can get $\dfrac{x}{y} = \frac{\sqrt{3}}{2}\cot\alpha + \frac{1}{2}$ and $\frac{\sqrt{3}}{2}\cot\alpha = x^2 + \frac{1}{2}$. Eliminating $\cot\alpha$ gives $\dfrac{x}{y} = \left(x^2 + \frac{1}{2}\right) + \frac{1}{2} \quad\Rightarrow$ $y = \dfrac{x}{x^2+1}$, $x > 0$.

(b) We differentiate our expression for y with respect to x to find the maximum:

$\dfrac{dy}{dx} = \dfrac{(x^2+1) - x(2x)}{(x^2+1)^2} = \dfrac{1-x^2}{(x^2+1)^2} = 0$ when $x = 1$. This indicates a maximum by the First Derivative Test, since $y'(x) > 0$ for $0 < x < 1$ and $y'(x) < 0$ for $x > 1$, so the maximum value of y is $y(1) = \frac{1}{2}$.

31.
$$\lim_{n\to\infty}\left(\frac{1}{\sqrt{n}\sqrt{n+1}} + \frac{1}{\sqrt{n}\sqrt{n+2}} + \cdots + \frac{1}{\sqrt{n}\sqrt{n+n}}\right) = \lim_{n\to\infty}\frac{1}{n}\left(\sqrt{\frac{n}{n+1}} + \sqrt{\frac{n}{n+2}} + \cdots + \sqrt{\frac{n}{n+n}}\right)$$

$$= \lim_{n\to\infty}\frac{1}{n}\left(\frac{1}{\sqrt{1+1/n}} + \frac{1}{\sqrt{1+2/n}} + \cdots + \frac{1}{\sqrt{1+1}}\right) = \lim_{n\to\infty}\frac{1}{n}\sum_{i=1}^{n} f\left(\frac{i}{n}\right) \quad \left(\text{where } f(x) = \frac{1}{\sqrt{1+x}}\right)$$

$$= \int_0^1 \frac{1}{\sqrt{1+x}}\,dx = \left[2\sqrt{1+x}\right]_0^1 = 2\left(\sqrt{2} - 1\right)$$

33. Let the roots have common difference d. If we let m be the average of the roots, then we can simplify the calculations by writing the roots as $m - 3a$, $m - a$, $m + a$, $m + 3a$, where $d = 2a$. The polynomial is then of the form $P(x) = c(x - m + 3a)(x - m + a)(x - m - a)(x - m - 3a)$. To simplify further, we make the change of variable $u = x - m$. Then $P(x) = c(u + 3a)(u + a)(u - a)(u - 3a) = c(u^2 - a^2)(u^2 - 9a^2)$.

So $\dfrac{dP}{dx} = \dfrac{dP}{du}\dfrac{du}{dx} = \dfrac{dP}{du} = c\left[2u\left(u^2 - 9a^2\right) + \left(u^2 - a^2\right)(2u)\right] = 2cu\left(2u^2 - 10a^2\right) = 4cu\left(u^2 - 5a^2\right) = 0$ when $u = 0, \pm\sqrt{5}a$. So the roots of P' are $m - \sqrt{5}a$, m, and $m + \sqrt{5}a$, which form an arithmetic sequence.

CHAPTER FIVE

EXERCISES 5.1

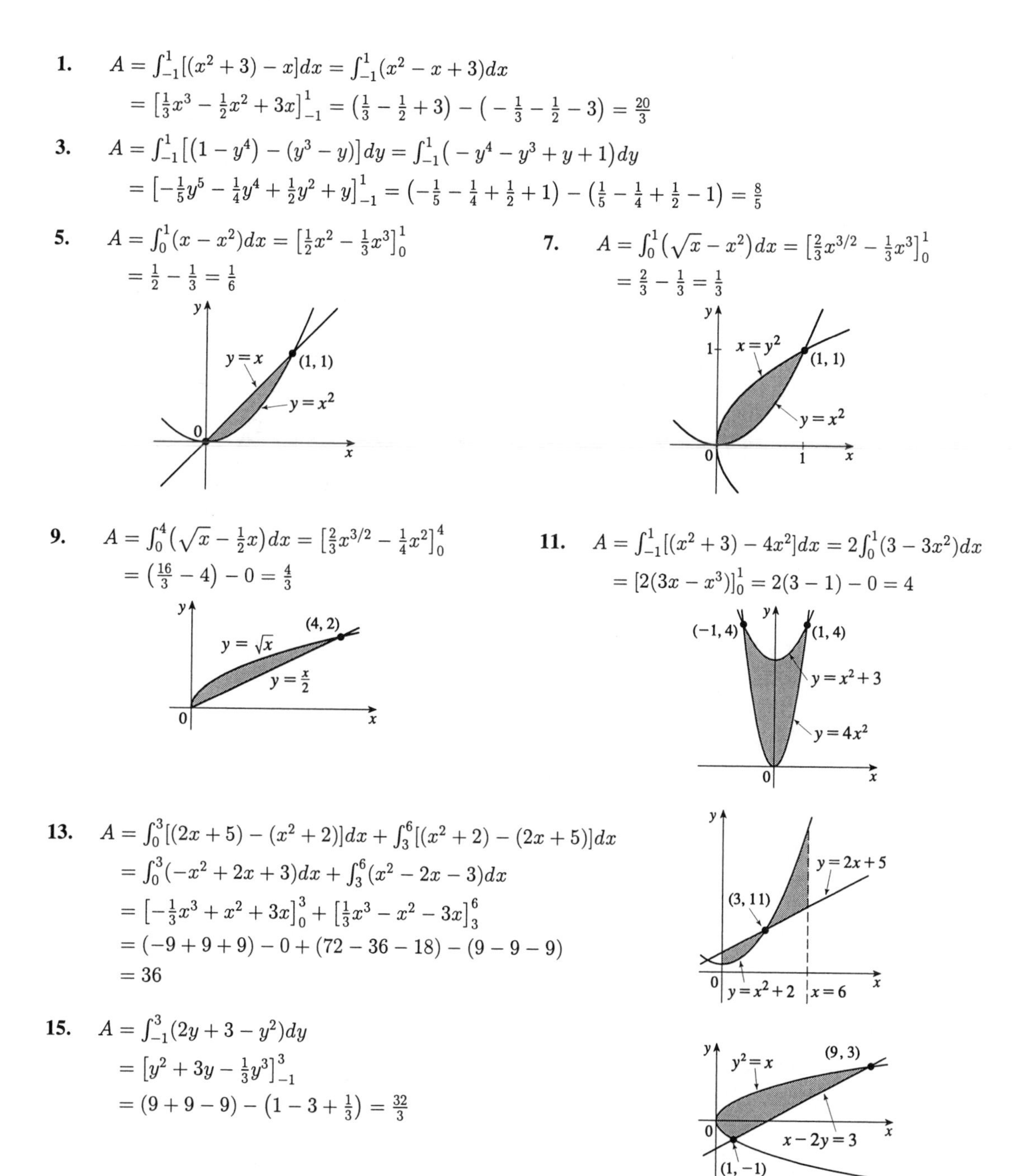

1. $A = \int_{-1}^{1}[(x^2+3) - x]dx = \int_{-1}^{1}(x^2 - x + 3)dx$
$= \left[\frac{1}{3}x^3 - \frac{1}{2}x^2 + 3x\right]_{-1}^{1} = \left(\frac{1}{3} - \frac{1}{2} + 3\right) - \left(-\frac{1}{3} - \frac{1}{2} - 3\right) = \frac{20}{3}$

3. $A = \int_{-1}^{1}\left[(1-y^4) - (y^3 - y)\right]dy = \int_{-1}^{1}\left(-y^4 - y^3 + y + 1\right)dy$
$= \left[-\frac{1}{5}y^5 - \frac{1}{4}y^4 + \frac{1}{2}y^2 + y\right]_{-1}^{1} = \left(-\frac{1}{5} - \frac{1}{4} + \frac{1}{2} + 1\right) - \left(\frac{1}{5} - \frac{1}{4} + \frac{1}{2} - 1\right) = \frac{8}{5}$

5. $A = \int_0^1 (x - x^2)dx = \left[\frac{1}{2}x^2 - \frac{1}{3}x^3\right]_0^1$
$= \frac{1}{2} - \frac{1}{3} = \frac{1}{6}$

7. $A = \int_0^1 \left(\sqrt{x} - x^2\right)dx = \left[\frac{2}{3}x^{3/2} - \frac{1}{3}x^3\right]_0^1$
$= \frac{2}{3} - \frac{1}{3} = \frac{1}{3}$

9. $A = \int_0^4 \left(\sqrt{x} - \frac{1}{2}x\right)dx = \left[\frac{2}{3}x^{3/2} - \frac{1}{4}x^2\right]_0^4$
$= \left(\frac{16}{3} - 4\right) - 0 = \frac{4}{3}$

11. $A = \int_{-1}^{1}[(x^2+3) - 4x^2]dx = 2\int_0^1 (3 - 3x^2)dx$
$= [2(3x - x^3)]_0^1 = 2(3-1) - 0 = 4$

13. $A = \int_0^3 [(2x+5) - (x^2+2)]dx + \int_3^6 [(x^2+2) - (2x+5)]dx$
$= \int_0^3 (-x^2 + 2x + 3)dx + \int_3^6 (x^2 - 2x - 3)dx$
$= \left[-\frac{1}{3}x^3 + x^2 + 3x\right]_0^3 + \left[\frac{1}{3}x^3 - x^2 - 3x\right]_3^6$
$= (-9 + 9 + 9) - 0 + (72 - 36 - 18) - (9 - 9 - 9)$
$= 36$

15. $A = \int_{-1}^{3}(2y + 3 - y^2)dy$
$= \left[y^2 + 3y - \frac{1}{3}y^3\right]_{-1}^{3}$
$= (9 + 9 - 9) - \left(1 - 3 + \frac{1}{3}\right) = \frac{32}{3}$

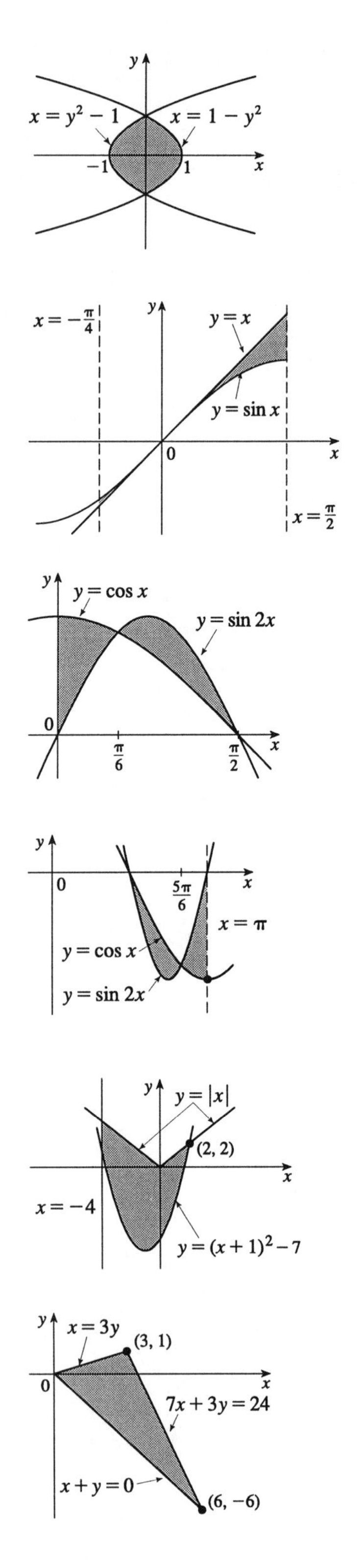

17. $A = \int_{-1}^{1}[(1-y^2) - (y^2-1)]dy$
$= \int_{-1}^{1} 2(1-y^2)dy = 4\int_0^1 (1-y^2)dy$
$= 4\left[y - \frac{1}{3}y^3\right]_0^1$
$= 4\left(1 - \frac{1}{3}\right)$
$= \frac{8}{3}$

19. $A = \int_{-\pi/4}^{0}(\sin x - x)dx + \int_0^{\pi/2}(x - \sin x)dx$
$= \left[-\cos x - \frac{1}{2}x^2\right]_{-\pi/4}^{0} + \left[\frac{1}{2}x^2 + \cos x\right]_0^{\pi/2}$
$= 1 - \left(-\frac{1}{\sqrt{2}} - \frac{\pi^2}{32}\right) + \frac{\pi^2}{8} - 1$
$= \frac{5}{32}\pi^2 + \frac{1}{\sqrt{2}} - 2$

21. Notice that $\cos x = \sin 2x = 2\sin x \cos x \quad \Leftrightarrow$
$2\sin x = 1$ or $\cos x = 0 \quad \Leftrightarrow \quad x = \frac{\pi}{6}$ or $\frac{\pi}{2}$.
$A = \int_0^{\pi/6}(\cos x - \sin 2x)dx + \int_{\pi/6}^{\pi/2}(\sin 2x - \cos x)dx$
$= \left[\sin x + \frac{1}{2}\cos 2x\right]_0^{\pi/6} + \left[-\frac{1}{2}\cos 2x - \sin x\right]_{\pi/6}^{\pi/2}$
$= \frac{1}{2} + \frac{1}{2}\cdot\frac{1}{2} - \left(0 + \frac{1}{2}\cdot 1\right)$
$+ \left(\frac{1}{2} - 1\right) - \left(-\frac{1}{2}\cdot\frac{1}{2} - \frac{1}{2}\right) = \frac{1}{2}$

23. $\cos x = \sin 2x = 2\sin x\cos x \quad \Leftrightarrow \quad \cos x = 0$ or
$\sin x = \frac{1}{2} \quad \Leftrightarrow \quad x = \frac{\pi}{2}$ or $\frac{5\pi}{6}$.
$A = \int_{\pi/2}^{5\pi/6}(\cos x - \sin 2x)dx + \int_{5\pi/6}^{\pi}(\sin 2x - \cos x)dx$
$= \left[\sin x + \frac{1}{2}\cos 2x\right]_{\pi/2}^{5\pi/6} - \left[\sin x + \frac{1}{2}\cos 2x\right]_{5\pi/6}^{\pi}$
$= \left(\frac{1}{2} + \frac{1}{2}\cdot\frac{1}{2}\right) - \left(1 - \frac{1}{2}\right) - \left(0 + \frac{1}{2}\right) + \left(\frac{1}{2} + \frac{1}{2}\cdot\frac{1}{2}\right) = \frac{1}{2}$

25. $A = \int_{-4}^{0}\left(-x - \left[(x+1)^2 - 7\right]\right)dx + \int_0^2\left(x - \left[(x+1)^2 - 7\right]\right)dx$
$= \int_{-4}^{0}(-x^2 - 3x + 6)dx + \int_0^2(-x^2 - x + 6)dx$
$= \left[-\frac{1}{3}x^3 - \frac{3}{2}x^2 + 6x\right]_{-4}^{0} + \left[-\frac{1}{3}x^3 - \frac{1}{2}x^2 + 6x\right]_0^2$
$= 0 - \left(\frac{64}{3} - 24 - 24\right) + \left(-\frac{8}{3} - 2 + 12\right) - 0 = 34$

27. $A = \int_0^3\left[\frac{1}{3}x - (-x)\right]dx + \int_3^6\left[\left(8 - \frac{7}{3}x\right) - (-x)\right]dx$
$= \int_0^3 \frac{4}{3}x\,dx + \int_3^6\left(-\frac{4}{3}x + 8\right)dx$
$= \left[\frac{2}{3}x^2\right]_0^3 + \left[-\frac{2}{3}x^2 + 8x\right]_3^6$
$= (6 - 0) + (24 - 18) = 12$

29. **(a)** $A = \displaystyle\int_{-4}^{-1} \left[(2x+4) + \sqrt{-4x}\right] dx + \int_{-1}^{0} 2\sqrt{-4x}\, dx$

$= \left[x^2 + 4x\right]_{-4}^{-1} + 2\int_{-4}^{-1} \sqrt{-x}\, dx + 4\int_{-1}^{0} \sqrt{-x}\, dx$

$= (-3 - 0) + 2\int_{1}^{4} \sqrt{u}\, du + 4\int_{0}^{1} \sqrt{u}\, du \quad (u = -x)$

$= -3 + \left[\frac{4}{3}u^{3/2}\right]_1^4 + \left[\frac{8}{3}u^{3/2}\right]_0^1 = -3 + \frac{28}{3} + \frac{8}{3} = 9$

(b) $A = \int_{-4}^{2} \left[-\frac{1}{4}y^2 - \left(\frac{1}{2}y - 2\right)\right] dy = \left[-\frac{1}{12}y^3 - \frac{1}{4}y^2 + 2y\right]_{-4}^{2}$

$= \left(-\frac{2}{3} - 1 + 4\right) - \left(\frac{16}{3} - 4 - 8\right) = 9$

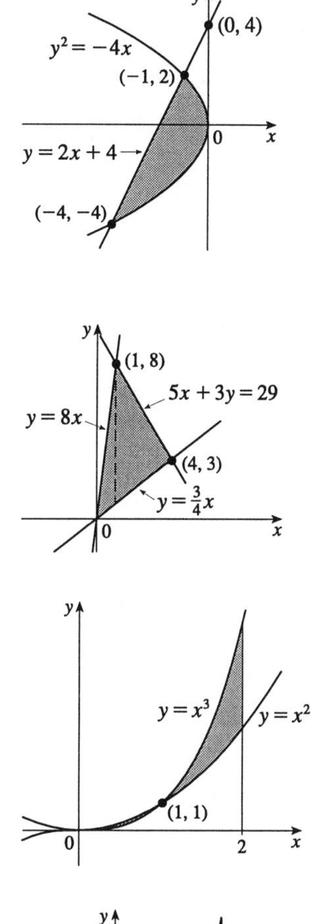

31. $A = \int_0^1 \left(8x - \frac{3}{4}x\right) dx + \int_1^4 \left[\left(-\frac{5}{3}x + \frac{29}{3}\right) - \frac{3}{4}x\right] dx$

$= \frac{29}{4}\int_0^1 x\, dx + \int_1^4 \left(-\frac{29}{12}x + \frac{29}{3}\right) dx$

$= \frac{29}{4}\left[\frac{1}{2}x^2\right]_0^1 - \frac{29}{12}\left[\frac{1}{2}x^2 - 4x\right]_1^4$

$= \frac{29}{8} - \frac{29}{12}\left(-8 - \frac{1}{2} + 4\right) = 14.5$

33. $\int_0^2 |x^2 - x^3|\, dx = \int_0^1 (x^2 - x^3)\, dx + \int_1^2 (x^3 - x^2)\, dx$

$= \left[\frac{1}{3}x^3 - \frac{1}{4}x^4\right]_0^1 + \left[\frac{1}{4}x^4 - \frac{1}{3}x^3\right]_1^2$

$= \frac{1}{3} - \frac{1}{4} + \left(4 - \frac{8}{3}\right) - \left(\frac{1}{4} - \frac{1}{3}\right) = 1.5$

35. $\int_{-1}^{2} x^3\, dx = \left[\frac{1}{4}x^4\right]_{-1}^{2} = 4 - \frac{1}{4} = \frac{15}{4} = 3.75$

37. Let $f(x) = \sqrt{1 + x^3} - (1 - x)$, $\Delta x = \dfrac{2 - 0}{4} = \dfrac{1}{2}$.

$A = \displaystyle\int_0^2 \left[\sqrt{1 + x^3} - (1 - x)\right] dx \approx \frac{1}{2}\left[f\left(\tfrac{1}{4}\right) + f\left(\tfrac{3}{4}\right) + f\left(\tfrac{5}{4}\right) + f\left(\tfrac{7}{4}\right)\right]$

$= \frac{1}{2}\left[\left(\frac{\sqrt{65}}{8} - \frac{3}{4}\right) + \left(\frac{\sqrt{91}}{8} - \frac{1}{4}\right) + \left(\frac{3\sqrt{21}}{8} + \frac{1}{4}\right) + \left(\frac{\sqrt{407}}{8} + \frac{3}{4}\right)\right]$

$= \frac{1}{16}\left(\sqrt{65} + \sqrt{91} + 3\sqrt{21} + \sqrt{407}\right) \approx 3.22$

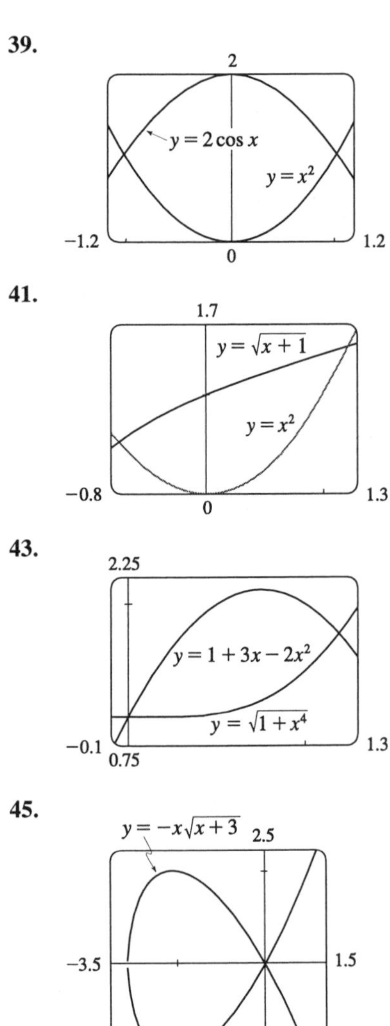

39. From zooming in on the graph or using a cursor, we see that the curves intersect at $x \approx \pm 1.02$, with $2\cos x > x^2$ on $[-1.02, 1.02]$. So the area between them is

$$A \approx \int_{-1.02}^{1.02} (2\cos x - x^2)\,dx = 2\int_0^{1.02} (2\cos x - x^2)\,dx$$
$$= 2\left[2\sin x - \tfrac{1}{3}x^3\right]_0^{1.02} \approx 2.70.$$

41. From the graph, we see that the curves intersect at $x \approx -0.72$ and at $x \approx 1.22$, with $\sqrt{x+1} > x^2$ on $[-0.72, 1.22]$. So the area between the curves is

$$A \approx \int_{-0.72}^{1.22} \left(\sqrt{x+1} - x^2\right) dx = \left[\tfrac{2}{3}(x+1)^{3/2} - \tfrac{1}{3}x^3\right]_{-0.72}^{1.22}$$
$$\approx 1.38.$$

43. From the graph, we see that the curves intersect at $x = 0$ and at $x \approx 1.19$, with $1 + 3x - 2x^2 > \sqrt{1 + x^4}$ on $[0, 1.19]$. So, using the Midpoint Rule with $f(x) = 1 + 3x - 2x^2 - \sqrt{1 + x^4}$ on $[0, 1.19]$ with $n = 4$, we calculate the approximate area between the curves:

$$A \approx \int_0^{1.19} \left(1 + 3x - 2x^2 - \sqrt{1 + x^4}\right) dx$$
$$\approx \tfrac{1.19}{4}\left[f\left(\tfrac{1.19}{8}\right) + f\left(\tfrac{3 \cdot 1.19}{8}\right) + f\left(\tfrac{5 \cdot 1.19}{8}\right) + f\left(\tfrac{7 \cdot 1.19}{8}\right)\right] \approx 0.83.$$

45. To graph this function, we must first express it as a combination of explicit functions of y; namely, $y = \pm x\sqrt{x+3}$. We can see from the graph that the loop extends from $x = -3$ to $x = 0$, and that by symmetry, the area we seek is just twice the area under the top half of the curve on this interval, the equation of the top half being $y = -x\sqrt{x+3}$. So the area is $A = 2\int_{-3}^{0} \left(-x\sqrt{x+3}\right) dx$. We substitute $u = x + 3$, so $du = dx$ and the limits change to 0 and 3, and we get

$$A = -2\int_0^3 \left[(u - 3)\sqrt{u}\right] du = -2\int_0^3 \left(u^{3/2} - 3u^{1/2}\right) du$$
$$= -2\left[\tfrac{2}{5}u^{5/2} - 2u^{3/2}\right]_0^3 = -2\left[\tfrac{2}{5}\left(3^2\sqrt{3}\right) - 2\left(3\sqrt{3}\right)\right] = \tfrac{24}{5}\sqrt{3}.$$

47. We first assume that $c > 0$, since c can be replaced by $-c$ in both equations without changing the graphs, and if $c = 0$ the curves do not enclose a region. We see from the graph that the enclosed area lies between $x = -c$ and $x = c$, and by symmetry, it is equal to twice the area under the top half of the graph (whose equation is $y = c^2 - x^2$). The enclosed area is

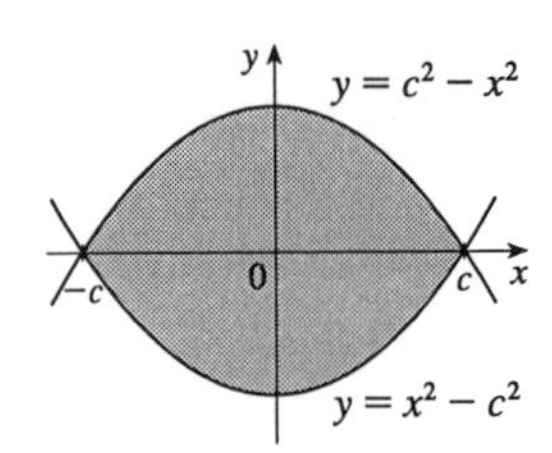

$$2\int_{-c}^{c} (c^2 - x^2)\,dx = 2\left[c^2x + \tfrac{1}{3}x^3\right]_{-c}^{c} = 2\left(\left[c^3 + \tfrac{1}{3}c^3\right] - \left[-c^3 + \tfrac{1}{3}(-c)^3\right]\right)$$
$$= \tfrac{8}{3}c^3, \text{ which is equal to 576 when } c = \sqrt[3]{216} = 6.$$

Note that $c = -6$ is another solution, since the graphs are the same.

49. By the symmetry of the problem, we consider only the first quadrant where $y = x^2 \quad \Rightarrow \quad x = \sqrt{y}$. We are looking for a number b such that $\int_0^4 x\,dy = 2\int_0^b x\,dy \quad \Rightarrow \quad \int_0^4 \sqrt{y}\,dy = 2\int_0^b \sqrt{y}\,dy \quad \Rightarrow$
$\frac{2}{3}\left[y^{3/2}\right]_0^4 = \frac{4}{3}\left[y^{3/2}\right]_0^b \quad \Rightarrow \quad \frac{2}{3}(8-0) = \frac{4}{3}\left(b^{3/2}-0\right) \quad \Rightarrow \quad b^{3/2} = 4 \quad \Rightarrow \quad b = 4^{2/3}$.

51. We know that the area under curve A between $t = 0$ and $t = x$ is $\int_0^x v_A(t)\,dt = s_A(x)$, where $v_A(t)$ is the velocity of car A and s_A is its displacement. Similarly, the area under curve B between $t = 0$ and $t = x$ is $\int_0^x v_B(t)\,dt = s_B(x)$.

(a) After one minute, the area under curve A is greater than the area under curve B. So A is ahead after one minute.

(b) After two minutes, car B is traveling faster than car A and has gained some ground, but the area under curve A from $t = 0$ to $t = 2$ is still greater than the corresponding area for curve B, so car A is still ahead.

(c) From the graph, it appears that the area between curves A and B for $0 \le t \le 1$ (when car A is going faster), which corresponds to the distance by which car A is ahead, seems to be about 3 squares. Therefore the cars will be side by side at the time x where the area between the curves for $1 \le t \le x$ (when car B is going faster) is the same as the area for $0 \le t \le 1$. From the graph, it appears that this time is $x \approx 2.2$. So the cars are side by side when $t \approx 2.2$ minutes.

53. $A = \int_1^2 \left(\frac{1}{x} - \frac{1}{x^2}\right)dx = \left[\ln x + \frac{1}{x}\right]_1^2$
$= \left(\ln 2 + \frac{1}{2}\right) - (\ln 1 + 1) = \ln 2 - \frac{1}{2}$

55. $A = 2\int_0^1 \left(\frac{2}{x^2+1} - x^2\right)dx = \left[4\tan^{-1}x - \frac{2}{3}x^3\right]_0^1$
$= 4 \cdot \frac{\pi}{4} - \frac{2}{3} = \pi - \frac{2}{3}$

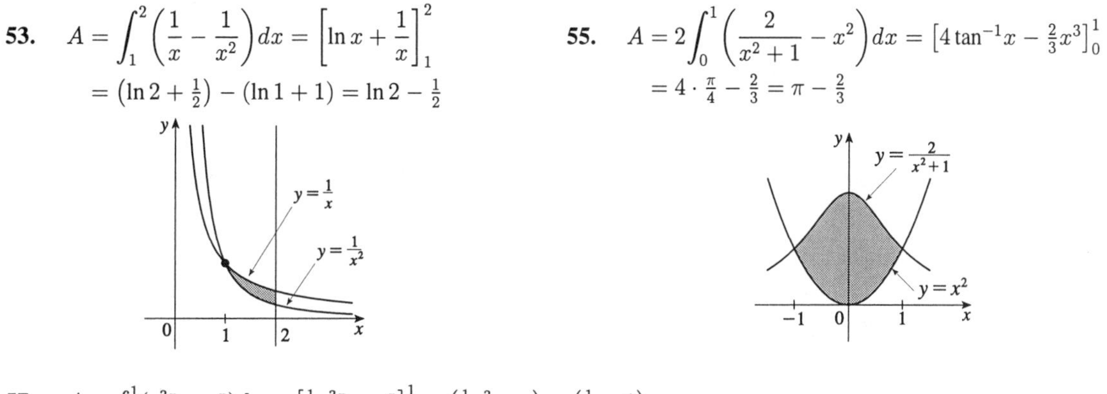

57. $A = \int_0^1 (e^{3x} - e^x)dx = \left[\frac{1}{3}e^{3x} - e^x\right]_0^1 = \left(\frac{1}{3}e^3 - e\right) - \left(\frac{1}{3} - 1\right)$
$= \frac{1}{3}e^3 - e + \frac{2}{3}$

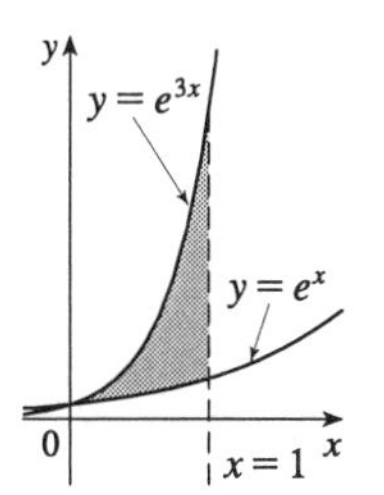

EXERCISES 5.2

1. $V = \int_0^1 \pi (x^2)^2\, dx = \pi \int_0^1 x^4\, dx = \pi\left[\frac{1}{5}x^5\right]_0^1 = \frac{\pi}{5}$

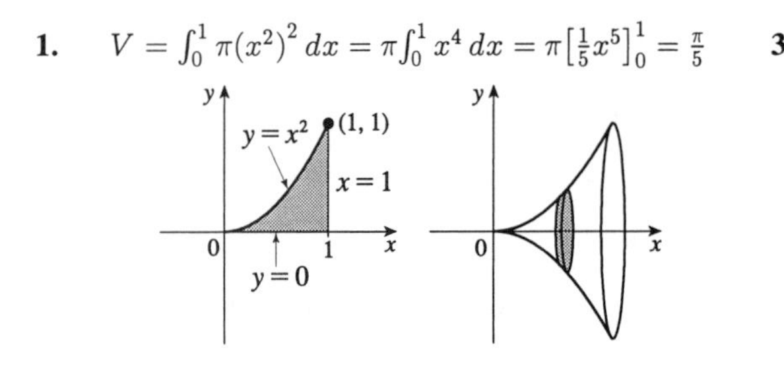

3. $V = \int_0^1 \pi(-x+1)^2\, dx = \pi\int_0^1 (x^2 - 2x + 1)dx$
$= \pi\left[\frac{1}{3}x^3 - x^2 + x\right]_0^1 = \pi\left(\frac{1}{3} - 1 + 1\right) = \frac{\pi}{3}$

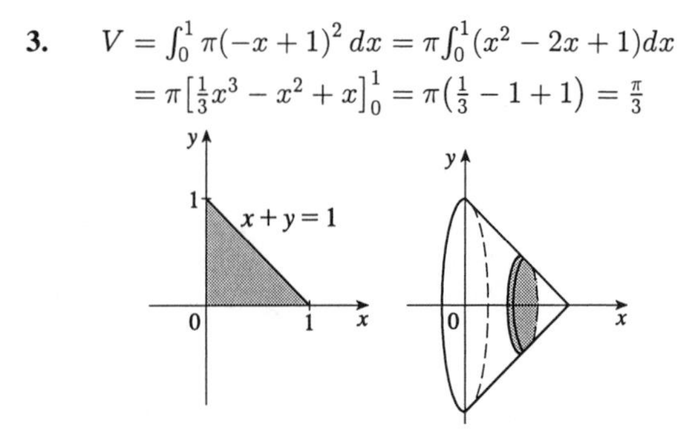

5. $V = \int_0^4 \pi\left(\sqrt{y}\right)^2 dy = \pi\int_0^4 y\, dy = \pi\left[\frac{1}{2}y^2\right]_0^4 = 8\pi$

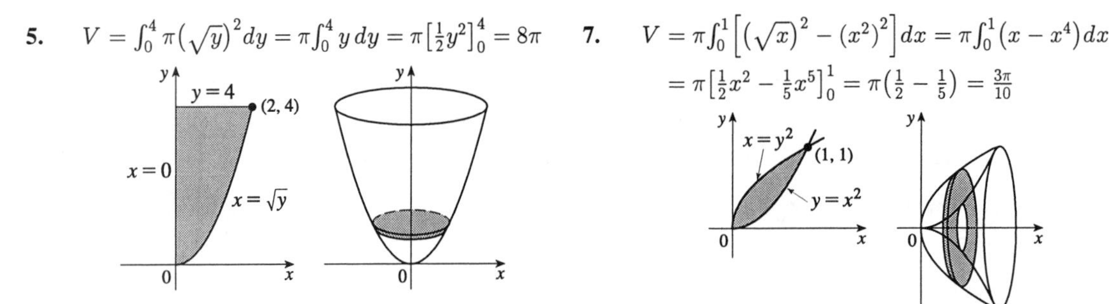

7. $V = \pi\int_0^1 \left[\left(\sqrt{x}\right)^2 - \left(x^2\right)^2\right]dx = \pi\int_0^1 (x - x^4)dx$
$= \pi\left[\frac{1}{2}x^2 - \frac{1}{5}x^5\right]_0^1 = \pi\left(\frac{1}{2} - \frac{1}{5}\right) = \frac{3\pi}{10}$

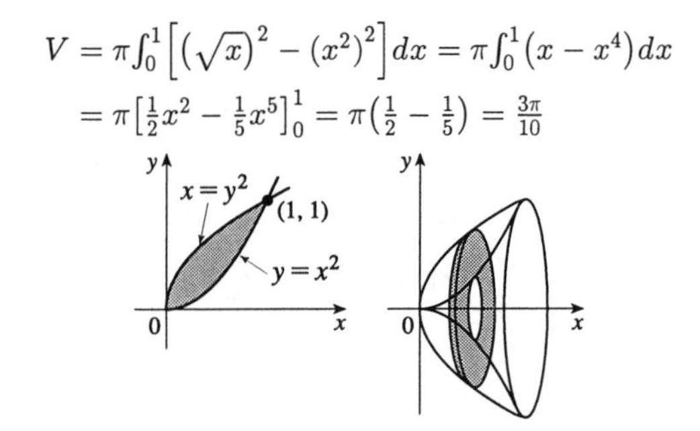

9. $V = \pi\int_0^2 \left[(2y)^2 - \left(y^2\right)^2\right]dy = \pi\int_0^2 (4y^2 - y^4)dy$
$= \pi\left[\frac{4}{3}y^3 - \frac{1}{5}y^5\right]_0^2 = \pi\left(\frac{32}{3} - \frac{32}{5}\right) = \frac{64\pi}{15}$

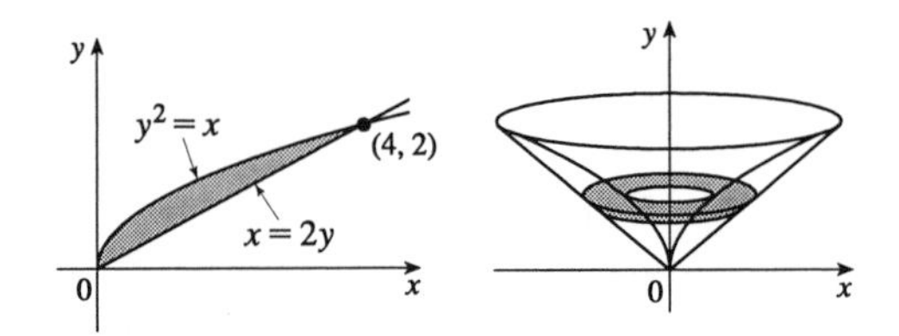

11. $V = \pi\int_{-1}^1 \left[\left(2 - x^4\right)^2 - 1^2\right]dx$
$= 2\pi\int_0^1 (3 - 4x^4 + x^8)dx$
$= 2\pi\left[3x - \frac{4}{5}x^5 + \frac{1}{9}x^9\right]_0^1$
$= 2\pi\left(3 - \frac{4}{5} + \frac{1}{9}\right) = \frac{208}{45}\pi$

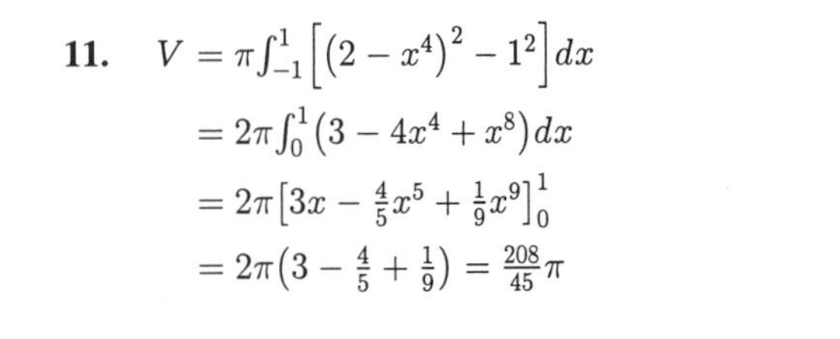

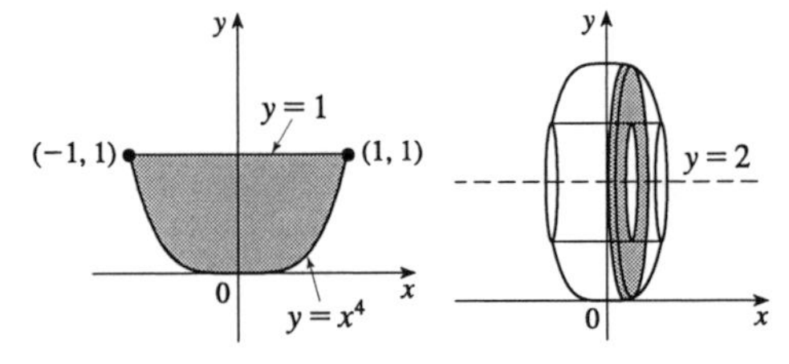

13. $V = \pi\int_0^8 \left(\frac{1}{4}x\right)^2 dx = \frac{\pi}{16}\left[\frac{1}{3}x^3\right]_0^8 = \frac{32}{3}\pi$

15. $V = \pi\int_0^2 (8-4y)^2 dy = \pi\left[64y - 32y^2 + \frac{16}{3}y^3\right]_0^2 = \pi\left(128 - 128 + \frac{128}{3}\right) = \frac{128}{3}\pi$

17. $V = \pi\int_0^8 \left[\left(\sqrt[3]{x}\right)^2 - \left(\frac{1}{4}x\right)^2\right] = \pi\int_0^8 \left(x^{2/3} - \frac{1}{16}x^2\right)dx = \pi\left[\frac{3}{5}x^{5/3} - \frac{1}{48}x^3\right]_0^8 = \pi\left(\frac{96}{5} - \frac{32}{3}\right) = \frac{128}{15}\pi$

19. $V = \pi\int_0^8 \left[\left(2 - \frac{1}{4}x\right)^2 - \left(2 - \sqrt[3]{x}\right)^2\right]dx = \pi\int_0^8 \left(-x + \frac{1}{16}x^2 + 4x^{1/3} - x^{2/3}\right)dx$
$= \pi\left[-\frac{1}{2}x^2 + \frac{1}{48}x^3 + 3x^{4/3} - \frac{3}{5}x^{5/3}\right]_0^8 = \pi\left(-32 + \frac{32}{3} + 48 - \frac{96}{5}\right) = \frac{112}{15}\pi$

21. $V = \pi\int_0^8 (2^2 - x^{2/3})\,dx = \pi\left[4x - \frac{3}{5}x^{5/3}\right]_0^8 = \pi\left(32 - \frac{96}{5}\right) = \frac{64}{5}\pi$

23. $V = \pi\int_0^8 \left(2 - \sqrt[3]{x}\right)^2 dx = \pi\int_0^8 \left(4 - 4x^{1/3} + x^{2/3}\right) dx$
$= \pi\left[4x - 3x^{4/3} + \frac{3}{5}x^{5/3}\right]_0^8 = \pi\left(32 - 48 + \frac{96}{5}\right) = \frac{16}{5}\pi$

25. $V = \pi\int_0^2 (x^2 - 1)^2 dx = \pi\int_0^2 \left(x^4 - 2x^2 + 1\right) dx = \pi\left[\frac{1}{5}x^5 - \frac{2}{3}x^3 + x\right]_0^2 = \pi\left(\frac{32}{5} - \frac{16}{3} + 2\right) = \frac{46}{15}\pi$

27. $V = \pi\int_{-1}^1 (\sec^2 x - 1^2)dx = \pi[\tan x - x]_{-1}^1 = \pi[(\tan 1 - 1) - (-\tan 1 + 1)] = 2\pi(\tan 1 - 1)$

29. $V = \pi\int_{-3}^{-2} (-x-2)^2 dx + \pi\int_{-2}^0 (x+2)^2 dx = \pi\int_{-3}^0 (x+2)^2 dx = \left[\frac{\pi}{3}(x+2)^3\right]_{-3}^0 = \frac{\pi}{3}[8 - (-1)] = 3\pi$

31. $V = \pi\int_0^{\pi/4} [1^2 - \tan^2 x]dx$

33. $x - 1 = (x-4)^2 + 1 \quad \Leftrightarrow \quad x^2 - 9x + 18 = 0 \quad \Leftrightarrow \quad x = 3$ or 6, so
$V = \pi\int_3^6 \left[\left[6 - (x-4)^2\right]^2 - (8-x)^2\right] dx = \pi\int_3^6 \left(x^4 - 16x^3 + 83x^2 - 144x + 36\right) dx.$

35. $V = \pi\int_0^{\pi/2} \left[(1 + \cos x)^2 - 1^2\right] dx = \pi\int_0^{\pi/2} (2\cos x + \cos^2 x)dx$

37. We see from the graph in Exercise 5.1.41 that the x-coordinates of the points of intersection are $x \approx -0.72$ and $x \approx 1.22$, with $\sqrt{x+1} > x^2$ on $[-0.72, 1.22]$, so the volume of revolution is about

$$\pi\int_{-0.72}^{1.22}\left[\left(\sqrt{x+1}\right)^2 - \left(x^2\right)^2\right]dx = \pi\int_{-0.72}^{1.22}\left(x + 1 - x^4\right)dx = \pi\left[\tfrac{1}{2}x^2 + x - \tfrac{1}{5}x^5\right]_{-0.72}^{1.22} \approx 5.80.$$

39. $V = \pi\int_0^1 3^2\,dx + \pi\int_1^4 1^2\,dx + \pi\int_4^5 3^2\,dx$
$= 9\pi + 3\pi + 9\pi = 21\pi$

41. The solid is obtained by rotating the region under the curve $y = \tan x$, from $x = 0$ to $x = \frac{\pi}{4}$, about the x-axis.

43. The solid is obtained by rotating the region between the curves $x = y$ and $x = \sqrt{y}$ about the y-axis.

45. The solid is obtained by rotating the region between the curves $y = 5 - 2x^2$ and $y = 5 - 2x$ about the x-axis.
Or: The solid is obtained by rotating the region bounded by the curves $y = 2x$ and $y = 2x^2$ about the line $y = 5$.

47.
$$V = \pi\int_0^h \left(-\frac{r}{h}y + r\right)^2 dy$$
$$= \pi\int_0^h \left[\frac{r^2}{h^2}y^2 - \frac{2r^2}{h}y + r^2\right] dy$$
$$= \pi\left[\frac{r^2}{3h^2}y^3 - \frac{r^2}{h}y^2 + r^2 y\right]_0^h = \tfrac{1}{3}\pi r^2 h$$

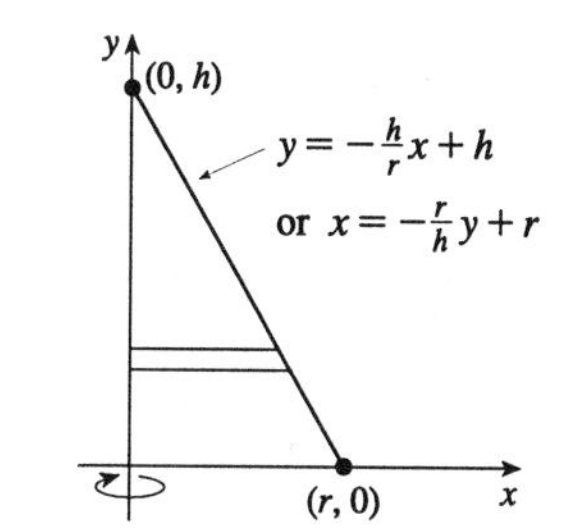

49. $V = \pi \int_{r-h}^{r} (r^2 - y^2)\,dy = \pi \left[r^2 y - \frac{y^3}{3} \right]_{r-h}^{r}$

$= \pi \left[\left(r^3 - \frac{r^3}{3} \right) - \left(r^2(r-h) - \frac{(r-h)^3}{3} \right) \right]$

$= \pi h^2 \left(r - \frac{h}{3} \right)$

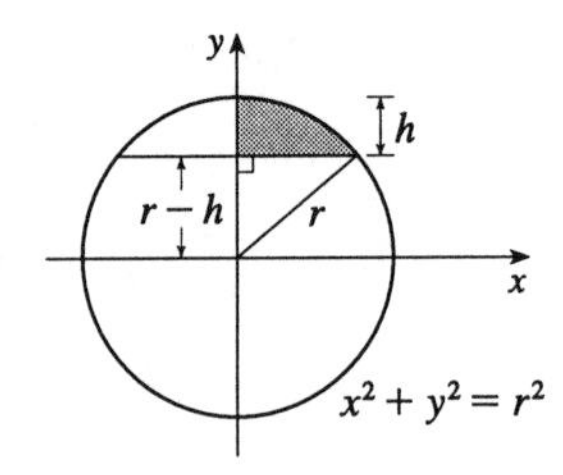

51. For a cross-section at height y, we see from similar triangles that $\frac{\alpha/2}{b/2} = \frac{h-y}{h}$, so $\alpha = b\left(1 - \frac{y}{h}\right)$. Similarly, $\beta = 2b\left(1 - \frac{y}{h}\right)$. So

$V = \int_0^h A(y)\,dy = \int_0^h 2b^2 \left(1 - \frac{y}{h}\right)^2 dy = 2b^2 \int_0^h \left(1 - \frac{2y}{h} + \frac{y^2}{h^2}\right) dy$

$= 2b^2 \left[y - \frac{y^2}{h} + \frac{y^3}{3h^2} \right]_0^h = 2b^2 \left[h - h + \tfrac{1}{3}h \right] = \tfrac{2}{3} b^2 h$

$\left(= \tfrac{1}{3} Bh\right.$, where B is the area of the base, as with any pyramid.$\left.\right)$

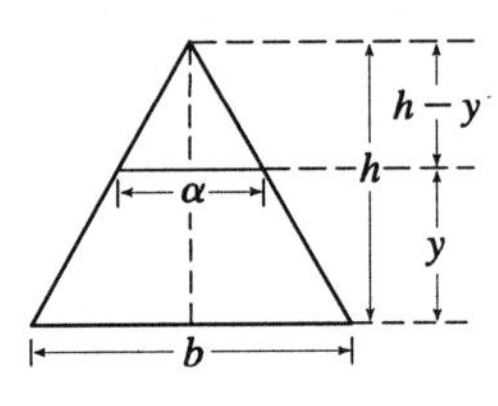

53. A cross-section at height z is a triangle similar to the base, so its area is

$A(z) = \frac{1}{2} \cdot 3\left(\frac{5-z}{5}\right) \cdot 4\left(\frac{5-z}{5}\right) = 6\left(1 - \frac{z}{5}\right)^2$, so

$V = \int_0^5 A(z)\,dz = 6 \int_0^5 \left(1 - \frac{z}{5}\right)^2 dz$

$= 6\left[(-5)\tfrac{1}{3}\left(1 - \tfrac{1}{5}z\right)^3 \right]_0^5 = -10(-1) = 10\,\text{cm}^3.$

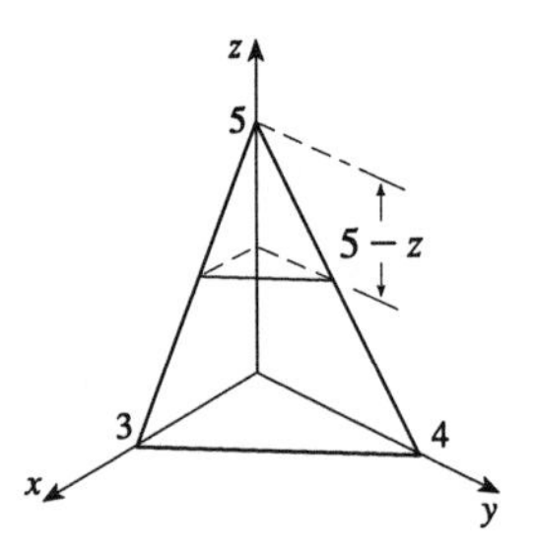

55. $V = \int_{-2}^{2} A(x)\,dx = 2\int_0^2 A(x)\,dx$

$= 2\int_0^2 \tfrac{1}{2}\left(\sqrt{2}y\right)^2 dx = 2\int_0^2 y^2\,dx$

$= \tfrac{1}{2} \int_0^2 (36 - 9x^2)\,dx = \tfrac{9}{2} \int_0^2 (4 - x^2)\,dx$

$= \frac{9}{2}\left[4x - \frac{x^3}{3} \right]_0^2 = \tfrac{9}{2}\left(8 - \tfrac{8}{3}\right) = 24$

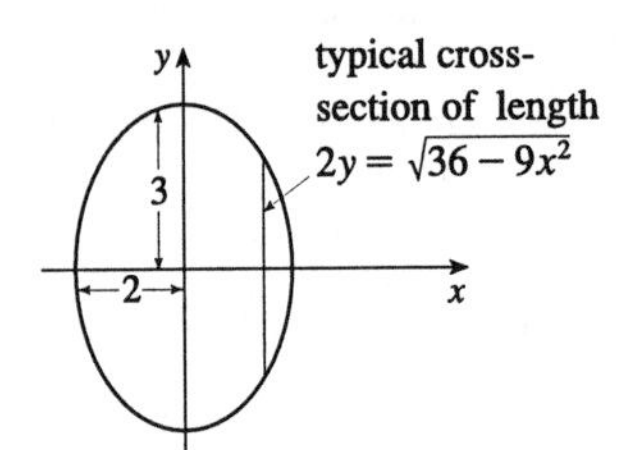

57. The cross section of the base corresponding to the coordinate y has length $2x = 2\sqrt{y}$, so

$V = \int_0^1 A(y)\,dy = \int_0^1 (2x)^2\,dy = \int_0^1 4x^2\,dy$

$= \int_0^1 4y\,dy = \left[2y^2\right]_0^1 = 2.$

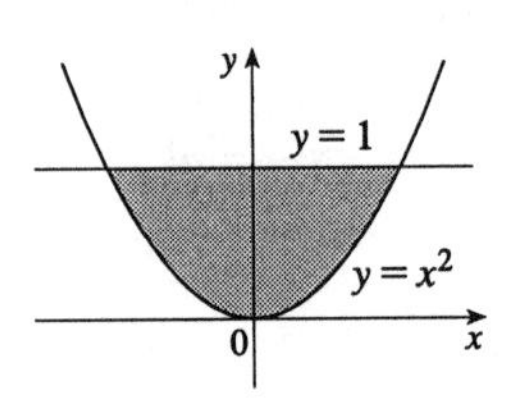

59. Assume that the base of each isosceles triangle lies in the base of S. Then its area is $A(x) = \frac{1}{2}bh = \frac{1}{2}\left(1 - \frac{1}{2}x\right)\left(1 - \frac{1}{2}x\right) = \frac{1}{2}\left(1 - \frac{1}{2}x\right)^2$, and the volume is $V = \int_0^2 A(x)\,dx = \int_0^2 \frac{1}{2}y^2\,dx = \frac{1}{2}\int_0^2 \left(1 - \frac{1}{2}x\right)^2 dx = \frac{1}{2}\left[\frac{2}{3}\left(\frac{1}{2}x - 1\right)^3\right]_0^2 = \frac{1}{3}$.

61. **(a)** The torus is obtained by rotating the circle $(x - R)^2 + y^2 = r^2$ about the y-axis. Solving for y, we see that the right half of the circle is given by $x = R + \sqrt{r^2 - y^2} = f(y)$ and the left half by $x = R - \sqrt{r^2 - y^2} = g(y)$.

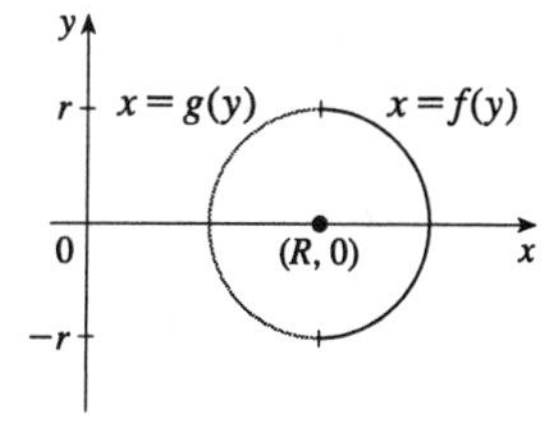

So
$$\begin{aligned} V &= \pi \int_{-r}^{r} \left([f(y)]^2 - [g(y)]^2\right) dy \\ &= 2\pi \int_0^r 4R\sqrt{r^2 - y^2}\,dy \\ &= 8\pi R \int_0^r \sqrt{r^2 - y^2}\,dy. \end{aligned}$$

(b) Observe that the integral represents a quarter of the area of a circle with radius r, so $8\pi R \int_0^r \sqrt{r^2 - y^2}\,dy = 8\pi R \frac{1}{4}(\pi r^2) = 2\pi^2 r^2 R$.

63. The volume is obtained by rotating the area common to two circles of radius r, as shown. The volume of the right half is

$$\begin{aligned} V_{\text{right}} &= \pi \int_0^{r/2} \left[r^2 - \left(\tfrac{1}{2}r + x\right)^2\right] dx \\ &= \pi \left[r^2 x - \tfrac{1}{3}\left(\tfrac{1}{2}r + x\right)^3\right]_0^{r/2} \\ &= \pi\left[\left(\tfrac{1}{2}r^3 - \tfrac{1}{3}r^3\right) - \left(0 - \tfrac{1}{24}r^3\right)\right] \\ &= \tfrac{5}{24}\pi r^3. \end{aligned}$$

So by symmetry, the total volume is twice this, or $\frac{5}{12}\pi r^3$.

Alternate Solution: We observe that the volume is the twice the volume of a cap of a sphere, so we can use the formula from Exercise 49 with $h = \frac{1}{2}r$: $V = 2\pi\left(\frac{1}{2}r\right)^2\left(r - \dfrac{r/2}{3}\right) = \frac{5}{12}\pi r^3$.

65. The cross-sections perpendicular to the y-axis in Figure 17 are rectangles. The rectangle corresponding to the coordinate y has a base of length $2\sqrt{16 - y^2}$ in the xy-plane and a height of $\frac{1}{\sqrt{3}}y$, since $\angle BAC = 30°$ and $|BC| = \frac{1}{\sqrt{3}}|AB|$. Thus $A(y) = \frac{2}{\sqrt{3}}y\sqrt{16 - y^2}$ and

$$\begin{aligned} V &= \int_0^4 A(y)\,dy = \int_0^4 A(y)\,dy = \tfrac{2}{\sqrt{3}}\int_0^4 \sqrt{16 - y^2}\,y\,dy \\ &= \tfrac{2}{\sqrt{3}}\int_{16}^0 u^{1/2}\left(-\tfrac{1}{2}\,du\right) \quad [\text{Put } u = 16 - y^2, \text{ so } du = -2y\,dy] \\ &= \tfrac{1}{\sqrt{3}}\int_0^{16} u^{1/2}\,du = \tfrac{1}{\sqrt{3}}\tfrac{2}{3}\left[u^{3/2}\right]_0^{16} = \tfrac{2}{3\sqrt{3}}(64) = \tfrac{128}{3\sqrt{3}}. \end{aligned}$$

67. Take the x-axis to be the axis of the cylindrical hole of radius r. A quarter of the cross-section through y, perpendicular to the y-axis, is the rectangle shown. Using Pythagoras twice, we see that the dimensions of this rectangle are $x = \sqrt{R^2 - y^2}$ and $z = \sqrt{r^2 - y^2}$, so $\frac{1}{4}A(y) = xz = \sqrt{r^2 - y^2}\sqrt{R^2 - y^2}$, and $V = \int_{-r}^{r} A(y)\,dy = \int_{-r}^{r} 4\sqrt{r^2-y^2}\sqrt{R^2-y^2}\,dy = 8\int_0^r \sqrt{r^2-y^2}\sqrt{R^2-y^2}\,dy.$

69. **(a)** Volume$(S_1) = \int_0^h A(z)dz =$ Volume(S_2) since the cross-sectional area $A(z)$ at height z is the same for both solids.

(b) By Cavalieri's Principle, the volume of the cylinder in the figure is the same as that of a right circular cylinder with radius r and height h, that is, $\pi r^2 h$.

71. **(a)** The radius of the barrel is the same at each end by symmetry, since the function $y = R - cx^2$ is even. Since the barrel is obtained by rotating the function y about the x-axis, this radius is equal to the value of y at $x = \frac{1}{2}h$, which is $R - c\left(\frac{1}{2}h\right)^2 = R - d = r$.

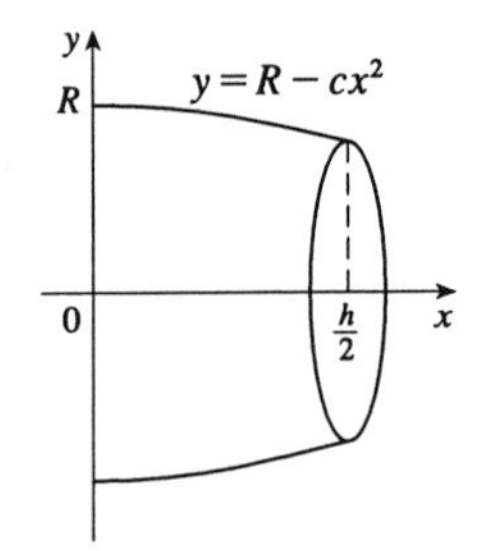

(b) The barrel is symmetric about the y-axis, so its volume is twice the volume of that part of the barrel for $x > 0$. Also, the barrel is a volume of rotation, so we use Formula 2:

$$V = 2\int_0^{h/2} \pi(y^2)dx = 2\pi\int_0^{h/2}\left(R - cx^2\right)^2 dx = 2\pi\left[R^2x - \tfrac{2}{3}Rcx^3 + \tfrac{1}{5}c^2x^5\right]_0^{h/2}$$
$$= 2\pi\left(\tfrac{1}{2}R^2h - \tfrac{1}{12}Rch^3 + \tfrac{1}{160}c^2h^5\right).$$

Trying to make this look more like the expression we want, we rewrite it as $V = \frac{1}{3}\pi h\left[2R^2 + \left(R^2 - \frac{1}{2}Rch^2 + \frac{3}{80}c^2h^4\right)\right]$. But $R^2 - \frac{1}{2}Rch^2 + \frac{3}{80}c^2h^4 = \left(R - \frac{1}{4}ch^2\right)^2 - \frac{1}{40}c^2h^4 = (R - d)^2 - \frac{2}{5}\left(\frac{1}{4}ch^2\right)^2 = r^2 - \frac{2}{5}d^2$. Substituting this back into V, we see that $V = \frac{1}{3}\pi h\left(2R^2 + r^2 - \frac{2}{5}d^2\right)$, as required.

73. We are given that the rate of change of the volume of water is $dV/dt = -kA(x)$, where k is some positive constant and $A(x)$ is the area of the surface when the water has depth x. Now we are concerned with the rate of change of the depth of the water with respect to time, that is, $\dfrac{dx}{dt}$. But by the Chain Rule, $\dfrac{dV}{dt} = \dfrac{dV}{dx}\dfrac{dx}{dt}$, so the first equation can be written $\dfrac{dV}{dx}\dfrac{dx}{dt} = -kA(x)$ (★). Also, we know that the total volume of water up to a depth x is $V(x) = \int_0^x A(s)\,ds$, where $A(s)$ is the area of a cross-section of the water at a depth s. Differentiating this equation with respect to x, we get $dV/dx = A(x)$. Substituting this into equation ★, we get $A(x)(dx/dt) = -kA(x) \quad\Rightarrow\quad dx/dt = -k$, a constant.

EXERCISES 5.3

1. $V = \int_1^2 2\pi x \cdot x^2\,dx = 2\pi \int_1^2 x^3\,dx = 2\pi\left[\frac{1}{4}x^4\right]_1^2 = 2\pi\left(\frac{15}{4}\right) = \frac{15}{2}\pi$

3. $V = \int_0^4 2\pi x\sqrt{4+x^2}\,dx = \pi\int_0^4 \sqrt{x^2+4}\,2x\,dx = \left[\pi\,\frac{2}{3}(x^2+4)^{3/2}\right]_0^4$
$= \left(\frac{2}{3}\pi\right)\left(20\sqrt{20}-8\right) = \frac{16}{3}\pi\left(5\sqrt{5}-1\right)$

5. $V = \int_0^2 2\pi x(4-x^2)dx = 2\pi\int_0^2(4x-x^3)dx = 2\pi\left[2x^2-\frac{1}{4}x^4\right]_0^2 = 2\pi(8-4) = 8\pi$

Note: If we integrated from -2 to 2, we would be generating the volume twice.

7. $V = \int_0^1 2\pi x(x^2-x^3)dx = 2\pi\int_0^1(x^3-x^4)dx = 2\pi\left[\frac{1}{4}x^4-\frac{1}{5}x^5\right]_0^1 = 2\pi\left(\frac{1}{4}-\frac{1}{5}\right) = \frac{1}{10}\pi$

9. $V = \int_0^{16} 2\pi y\sqrt[4]{y}\,dy = 2\pi\int_0^{16} y^{5/4}\,dy = 2\pi\left[\frac{4}{9}y^{9/4}\right]_0^{16} = \frac{8}{9}\pi(512-0) = \frac{4096}{9}\pi$

11. $V = \int_0^9 2\pi y\cdot 2\sqrt{y}\,dy = 4\pi\int_0^9 y^{3/2}\,dy = 4\pi\left[\frac{2}{5}y^{5/2}\right]_0^9 = \frac{8}{5}\pi(243-0) = \frac{1944}{5}\pi$

13. $V = \int_0^1 2\pi y[(2-y)-y^2]dy = 2\pi\left[y^2-\frac{1}{3}y^3-\frac{1}{4}y^4\right]_0^1 = 2\pi\left(1-\frac{1}{3}-\frac{1}{4}\right) = \frac{5}{6}\pi$

15. $V = \int_1^4 2\pi x\sqrt{x}\,dx = 2\pi\int_1^4 x^{3/2}\,dx$
$= 2\pi\left[\frac{2}{5}x^{5/2}\right]_1^4 = \frac{4}{5}\pi(32-1) = \frac{124}{5}\pi$

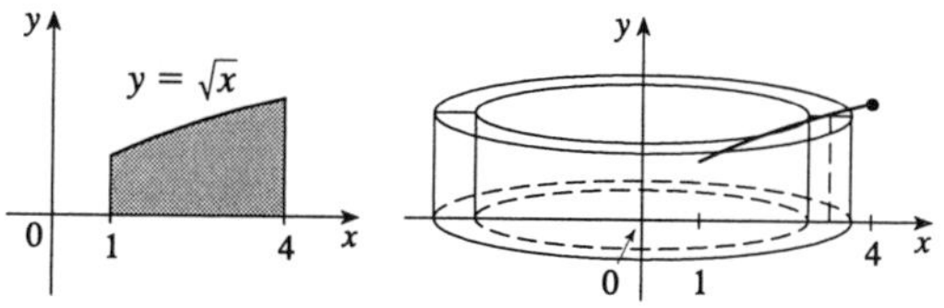

17. $V = \int_1^2 2\pi(x-1)\,x^2\,dx = 2\pi\left[\frac{1}{4}x^4-\frac{1}{3}x^3\right]_1^2$
$= 2\pi\left[\left(4-\frac{8}{3}\right)-\left(\frac{1}{4}-\frac{1}{3}\right)\right] = \frac{17}{6}\pi$

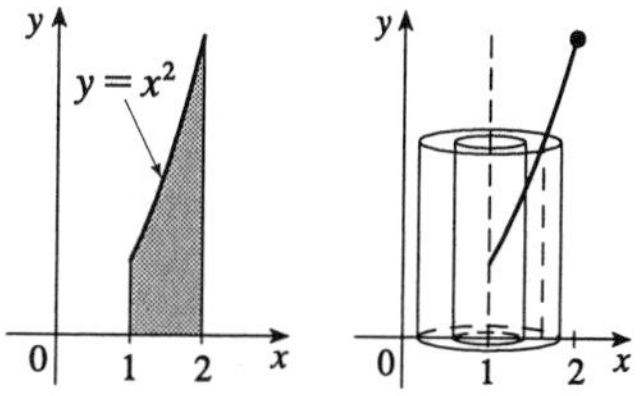

19. $V = \int_0^2 2\pi(3-y)(5-x)dy$
$= \int_0^2 2\pi(3-y)(5-y^2-1)dy$
$= \int_0^2 2\pi(12-4y-3y^2+y^3)\,dy$
$= 2\pi\left[12y-2y^2-y^3+\frac{1}{4}y^4\right]_0^2$
$= 2\pi(24-8-8+4) = 24\pi$

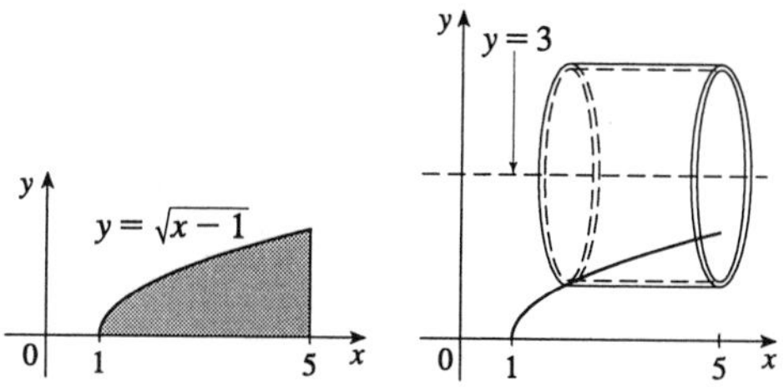

21. $V = \int_{2\pi}^{3\pi} 2\pi x\sin x\,dx$

23. $V = \int_0^{\pi/4} 2\pi y\cos y\;dy$

25. $V = \int_0^1 2\pi(x+1)\left(\sin\frac{\pi}{2}x - x^4\right)dx$

27. The solid is obtained by rotating the region bounded by the curve $y=\cos x$ and the line $y=0$, from $x=0$ to $x=\frac{\pi}{2}$, about the y-axis.

29. The solid is obtained by rotating the region in the first quadrant bounded by the curves $y = x^2$ and $y = x^6$ about the y-axis.

31.

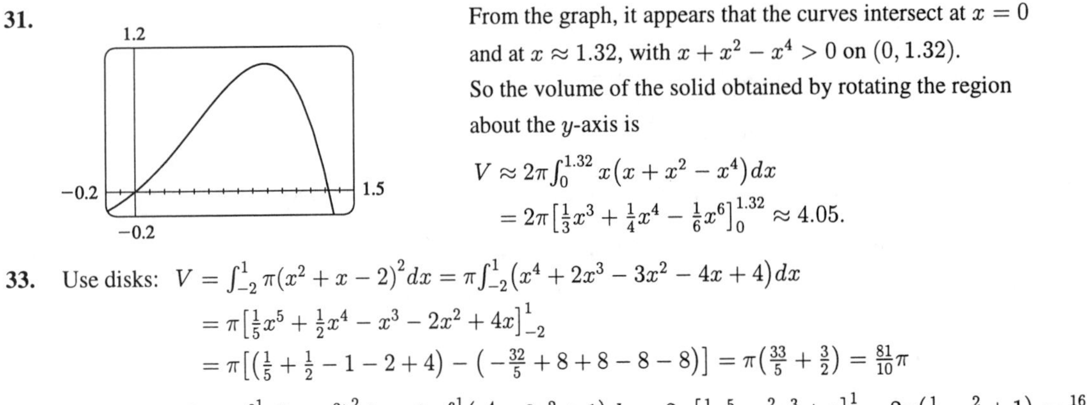

From the graph, it appears that the curves intersect at $x = 0$ and at $x \approx 1.32$, with $x + x^2 - x^4 > 0$ on $(0, 1.32)$. So the volume of the solid obtained by rotating the region about the y-axis is

$$V \approx 2\pi \int_0^{1.32} x\left(x + x^2 - x^4\right) dx = 2\pi\left[\tfrac{1}{3}x^3 + \tfrac{1}{4}x^4 - \tfrac{1}{6}x^6\right]_0^{1.32} \approx 4.05.$$

33. Use disks: $V = \int_{-2}^{1} \pi\left(x^2 + x - 2\right)^2 dx = \pi \int_{-2}^{1}\left(x^4 + 2x^3 - 3x^2 - 4x + 4\right) dx$

$$= \pi\left[\tfrac{1}{5}x^5 + \tfrac{1}{2}x^4 - x^3 - 2x^2 + 4x\right]_{-2}^{1}$$

$$= \pi\left[\left(\tfrac{1}{5} + \tfrac{1}{2} - 1 - 2 + 4\right) - \left(-\tfrac{32}{5} + 8 + 8 - 8 - 8\right)\right] = \pi\left(\tfrac{33}{5} + \tfrac{3}{2}\right) = \tfrac{81}{10}\pi$$

35. Use disks: $V = \pi \int_{-1}^{1}\left(1 - y^2\right)^2 dy = 2\pi \int_0^1 \left(y^4 - 2y^2 + 1\right) dy = 2\pi\left[\tfrac{1}{5}y^5 - \tfrac{2}{3}y^3 + y\right]_0^1 = 2\pi\left(\tfrac{1}{5} - \tfrac{2}{3} + 1\right) = \tfrac{16}{15}\pi$

37. Use disks: $V = \pi \displaystyle\int_0^2 \left[\sqrt{1 - (y-1)^2}\right]^2 dy = \pi \int_0^2 \left(2y - y^2\right) dy = \pi\left[y^2 - \tfrac{1}{3}y^3\right]_0^2 = \pi\left(4 - \tfrac{8}{3}\right) = \tfrac{4}{3}\pi$

39. $V = 2\int_0^r 2\pi x\sqrt{r^2 - x^2}\, dx = -2\pi \int_0^r \left(r^2 - x^2\right)^{1/2}(-2x)dx = \left[-2\pi \cdot \tfrac{2}{3}\left(r^2 - x^2\right)^{3/2}\right]_0^r = -\tfrac{4}{3}\pi\left(0 - r^3\right) = \tfrac{4}{3}\pi r^3$

41. $V = 2\pi \displaystyle\int_0^r x\left(-\frac{h}{r}x + h\right) dx = 2\pi h \int_0^r \left(-\frac{x^2}{r} + x\right) dx = 2\pi h\left[-\frac{x^3}{3r} + \frac{x^2}{2}\right]_0^r = 2\pi h \frac{r^2}{6} = \frac{\pi r^2 h}{3}$

43. If $a < b \leq 0$, then a typical cylindrical shell has radius $-x$ and height $f(x)$, so $V = \int_a^b 2\pi(-x)f(x)dx = -\int_a^b 2\pi x f(x)dx$.

45. $\Delta x = \dfrac{\pi/4 - 0}{4} = \dfrac{\pi}{16}$.

$V = \int_0^{\pi/4} 2\pi x \tan x\, dx \approx 2\pi \cdot \frac{\pi}{16}\left(\frac{\pi}{32}\tan\frac{\pi}{32} + \frac{3\pi}{32}\tan\frac{3\pi}{32} + \frac{5\pi}{32}\tan\frac{5\pi}{32} + \frac{7\pi}{32}\tan\frac{7\pi}{32}\right) \approx 1.142$

EXERCISES 5.4

1. By Equation 2, $W = Fd = (900)(8) = 7200$ J.

3. By Equation 4, $W = \int_a^b f(x)\, dx = \int_0^{10}\left(5x^2 + 1\right)dx = \left[\frac{5}{3}x^3 + x\right]_0^{10} = \frac{5000}{3} + 10 = \frac{5030}{3}$ ft-lb.

5. $10 = f(x) = kx = \frac{1}{3}k$ (4 inches $= \frac{1}{3}$ foot), so $k = 30$ (The units for k are pounds per foot.)

Now 6 inches $= \frac{1}{2}$ foot, so $W = \int_0^{1/2} 30x\, dx = \left[15x^2\right]_0^{1/2} = \frac{15}{4}$ ft-lb.

7. If $\int_0^{0.12} kx\, dx = 2$ J, then $2 = \left[\frac{1}{2}kx^2\right]_0^{0.12} = \frac{1}{2}k(0.0144) = 0.0072k$ and $k = \frac{2}{0.0072} = \frac{2500}{9} \approx 277.78$. Thus the work needed to stretch the spring from 35 cm to 40 cm is

$\int_{0.05}^{0.10} \frac{2500}{9}x\, dx = \left[\frac{1250}{9}x^2\right]_{1/20}^{1/10} = \frac{1250}{9}\left(\frac{1}{100} - \frac{1}{400}\right) = \frac{25}{24} \approx 1.04$ J.

9. $f(x) = kx$, so $30 = \frac{2500}{9}x$ and $x = \frac{270}{2500}$ m $= 10.8$ cm.

11. First notice that the exact height of the building does not matter. The portion of the rope from x ft to $(x + \Delta x)$ft below the top of the building weighs $\frac{1}{2}\Delta x$ lb and must be lifted x ft (approximately), so its contribution to the total work is $\frac{1}{2}x\Delta x$ ft-lb. The total work is $W = \int_0^{50} \frac{1}{2}x\,dx = \left[\frac{1}{4}x^2\right]_0^{50} = \frac{2500}{4} = 625$ ft-lb.

13. The work needed to lift the cable is $\int_0^{500} 2x\,dx = x^2\big|_0^{500} = 250{,}000$ ft-lb. The work needed to lift the coal is 800 lb $\cdot$ 500 ft $= 400{,}000$ ft-lb. Thus the total work required is $250{,}000 + 400{,}000 = 650{,}000$ ft-lb.

15. A "slice" of water Δx m thick and lying at a depth of x m (where $0 \le x \le \frac{1}{2}$) has volume $2\Delta x\text{ m}^3$, a mass of $2000\Delta x$ kg, weighs about $(9.8)(2000\Delta x) = 19{,}600\,\Delta x$ N, and thus requires about $19{,}600x\Delta x$ J of work for its removal. So $W \approx \int_0^{1/2} 19{,}600x\,dx = 9800x^2\big|_0^{1/2} = 2450$ J.

17. A "slice" of water Δx m thick and lying x ft above the bottom has volume $8x\Delta x\text{ m}^3$ and weighs about $(9.8 \times 10^3)(8x\Delta x)$ N. It must be lifted $(5 - x)$ m by the pump, so the work needed is about $(9.8 \times 10^3)(5 - x)(8x\Delta x)$ J. The total work required is

$$W \approx \int_0^3 (9.8 \times 10^3)(5 - x)8x\,dx = (9.8 \times 10^3)\int_0^3 (40x - 8x^2)dx$$
$$= (9.8 \times 10^3)\left[20x^2 - \tfrac{8}{3}x^3\right]_0^3 = (9.8 \times 10^3)(180 - 72) = (9.8 \times 10^3)(108) = 1058.4 \times 10^3$$
$$\approx 1.06 \times 10^6 \text{ J.}$$

19. Measure depth x downward from the flat top of the tank, so that $0 \le x \le 2$ ft. Then

$$\Delta W = (62.5)\left(2\sqrt{4 - x^2}\right)(8\Delta x)(x + 1) \text{ ft-lb, so}$$
$$W \approx (62.5)(16)\int_0^2 (x + 1)\sqrt{4 - x^2}\,dx = 1000\left(\int_0^2 x\sqrt{4 - x^2}\,dx + \int_0^2 \sqrt{4 - x^2}\,dx\right)$$
$$= 1000\left[\int_0^4 u^{1/2}\left(\tfrac{1}{2}\,du\right) + \tfrac{1}{4}\pi(2^2)\right] \qquad (\text{Put } u = 4 - x^2, \text{ so } du = -2x\,dx)$$
$$= 1000\left(\tfrac{1}{2} \cdot \tfrac{2}{3}u^{3/2}\big|_0^4 + \pi\right) = 1000\left(\tfrac{8}{3} + \pi\right) \approx 5.8 \times 10^3 \text{ ft-lb.}$$

Note: The second integral represents the area of a quarter-circle of radius 2.

21. If only 4.7×10^5 J of work is done, then only the water above a certain level (call it h) will be pumped out. So we use the same formula as in Exercise 17, except that the work is fixed, and we are trying to find the lower limit of integration: $4.7 \times 10^5 \approx \int_h^3 (9.8 \times 10^3)(5 - x)8x\,dx = (9.8 \times 10^3)\left[20x^2 - \frac{8}{3}x^3\right]_h^3 \Leftrightarrow$
$\frac{4.7}{9.8} \times 10^2 \approx 48 = \left(20 \cdot 3^2 - \frac{8}{3} \cdot 3^3\right) - \left(20h^2 - \frac{8}{3}h^3\right) \Leftrightarrow 2h^3 - 15h^2 + 45 = 0.$

To find the solution of this equation, we plot $2h^3 - 15h^2 + 45$ between $h = 0$ and $h = 3$. We see that the equation is satisfied for $h \approx 2.0$. So the depth of water remaining in the tank is about 2.0 m.

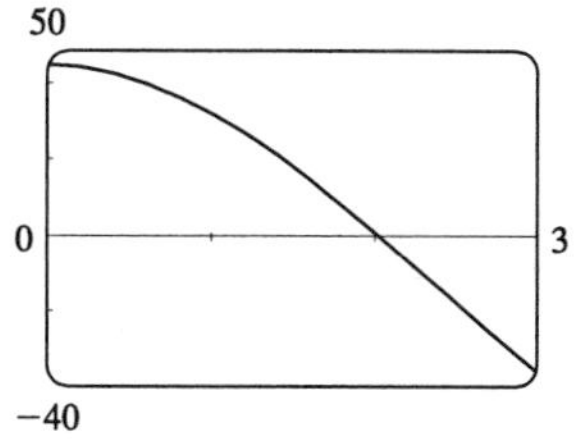

23. $V = \pi r^2 x$, so V is a function of x and P can also be regarded as a function of x. If $V_1 = \pi r^2 x_1$ and $V_2 = \pi r^2 x_2$, then $W = \int_{x_1}^{x_2} F(x)\,dx = \int_{x_1}^{x_2} \pi r^2 P(V(x))dx = \int_{x_1}^{x_2} P(V(x))dV(x)$ [Put $V(x) = \pi r^2 x$, so $dV(x) = \pi r^2\,dx$] $= \int_{V_1}^{V_2} P(V)dV$ by the Substitution Rule.

25. $W = \displaystyle\int_a^b F(r)dr = \int_a^b G\frac{m_1 m_2}{r^2}\,dr = Gm_1m_2\left[\frac{-1}{r}\right]_a^b = Gm_1m_2\left(\frac{1}{a} - \frac{1}{b}\right)$

EXERCISES 5.5

1. $f_{\text{ave}} = \frac{1}{3-0}\int_0^3 (x^2 - 2x)dx = \frac{1}{3}\left[\frac{1}{3}x^3 - x^2\right]_0^3 = \frac{1}{3}(9 - 9) = 0$

3. $f_{\text{ave}} = \frac{1}{1-(-1)}\int_{-1}^1 x^4\,dx = \frac{1}{2}\cdot 2\int_0^1 x^4\,dx = \left[\frac{1}{5}x^5\right]_0^1 = \frac{1}{5}$

5. $f_{\text{ave}} = \dfrac{1}{\frac{\pi}{4} - \left(-\frac{\pi}{2}\right)}\displaystyle\int_{-\pi/2}^{\pi/4} \sin^2 x \cos x\,dx = \frac{4}{3\pi}\int_{-\pi/2}^{\pi/4} \sin^2 x \cos x\,dx$

$= \frac{4}{3\pi}\int_{-1}^{1/\sqrt{2}} u^2\,du$ [Put $u = \sin x$, so $du = \cos x\,dx$] $= \frac{4}{3\pi}\left[\frac{1}{3}u^3\right]_{-1}^{1/\sqrt{2}}$

$= \frac{4}{9\pi}\left(\frac{1}{2\sqrt{2}} + 1\right) = \frac{4}{9\pi}\left(\frac{\sqrt{2}}{4} + 1\right) = \frac{\sqrt{2}+4}{9\pi}$

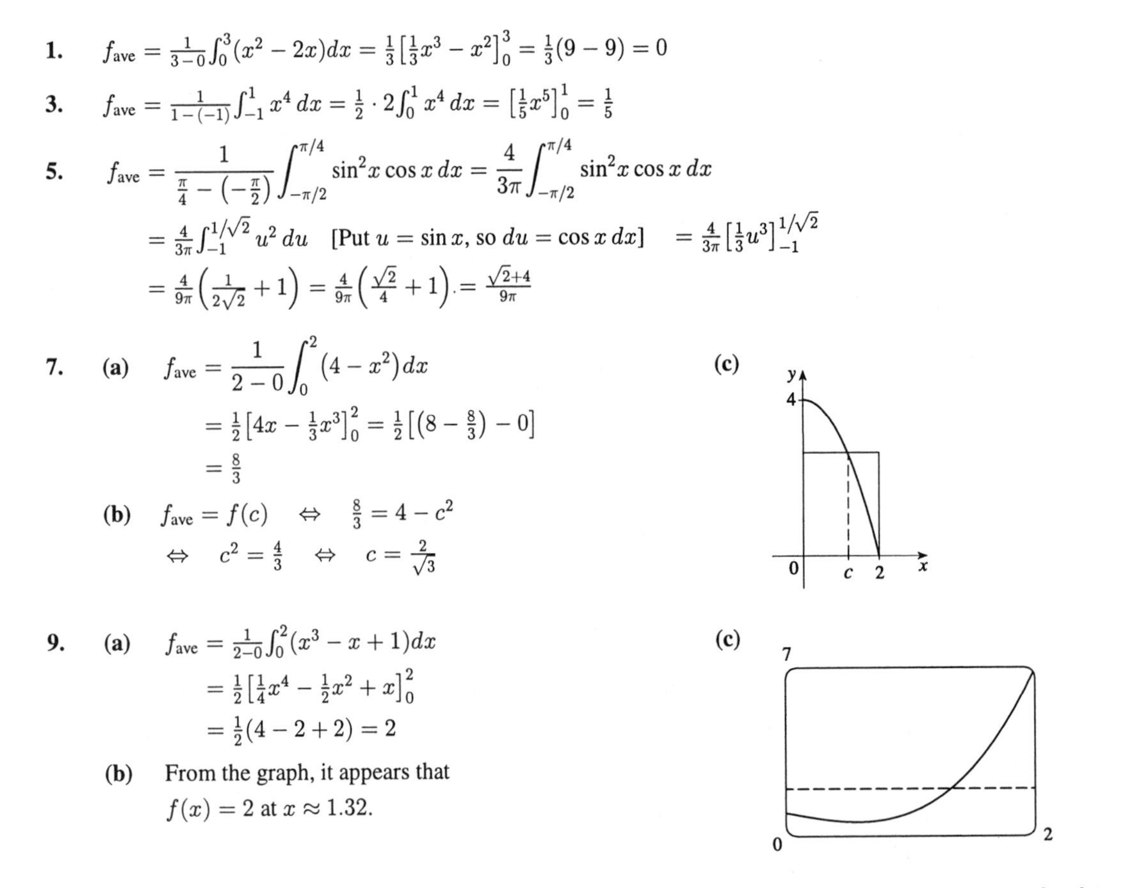

7. (a) $f_{\text{ave}} = \dfrac{1}{2-0}\displaystyle\int_0^2 (4 - x^2)dx$

$= \frac{1}{2}\left[4x - \frac{1}{3}x^3\right]_0^2 = \frac{1}{2}\left[\left(8 - \frac{8}{3}\right) - 0\right]$

$= \frac{8}{3}$

(b) $f_{\text{ave}} = f(c) \;\Leftrightarrow\; \frac{8}{3} = 4 - c^2$

$\Leftrightarrow\; c^2 = \frac{4}{3} \;\Leftrightarrow\; c = \frac{2}{\sqrt{3}}$

(c)

9. (a) $f_{\text{ave}} = \frac{1}{2-0}\int_0^2 (x^3 - x + 1)dx$

$= \frac{1}{2}\left[\frac{1}{4}x^4 - \frac{1}{2}x^2 + x\right]_0^2$

$= \frac{1}{2}(4 - 2 + 2) = 2$

(b) From the graph, it appears that $f(x) = 2$ at $x \approx 1.32$.

(c)

11. Since f is continuous on $[1, 3]$, by the Mean Value Theorem for Integrals there exists a number c in $[1, 3]$ such that $\int_1^3 f(x)\,dx = 8 = f(c)(3 - 1) = 2f(c)$; that is, there is a number c such that $f(c) = \frac{8}{2} = 4$.

13. $T_{\text{ave}} = \frac{1}{12}\int_0^{12}\left[50 + 14\sin\frac{1}{12}\pi t\right]dt = \frac{1}{12}\left[50t - 14\cdot\frac{12}{\pi}\cos\frac{1}{12}\pi t\right]_0^{12}$

$= \frac{1}{12}\left[50\cdot 12 + 14\cdot\frac{12}{\pi} + 14\cdot\frac{12}{\pi}\right] = \left(50 + \frac{28}{\pi}\right)^\circ\text{F} \approx 59^\circ\text{F}$

15. $\rho_{\text{ave}} = \dfrac{1}{8}\displaystyle\int_0^8 \frac{12}{\sqrt{x+1}}\,dx = \frac{3}{2}\int_0^8 (x+1)^{-1/2}\,dx = \left[3\sqrt{x+1}\right]_0^8 = 9 - 3 = 6\,\text{kg/m}$

17. $V_{\text{ave}} = \frac{1}{5}\int_0^5 V(t)dt = \frac{1}{5}\int_0^5 \frac{5}{4\pi}\left[1 - \cos\left(\frac{2}{5}\pi t\right)\right]dt = \frac{1}{4\pi}\int_0^5\left[1 - \cos\left(\frac{2}{5}\pi t\right)\right]dt$

$= \frac{1}{4\pi}\left[t - \frac{5}{2\pi}\sin\left(\frac{2}{5}\pi t\right)\right]_0^5 = \frac{1}{4\pi}[(5-0)-0] = \frac{5}{4\pi} \approx 0.4\,\text{L}$

19. Let $F(x) = \int_a^x f(t)\,dt$ for x in $[a,b]$. Then F is continuous on $[a,b]$ and differentiable on (a,b), so by the Mean Value Theorem there is a number c in (a,b) such that $F(b) - F(a) = F'(c)(b-a)$. But $F'(x) = f(x)$ by the Fundamental Theorem of Calculus. Therefore $\int_a^b f(t)dt - 0 = f(c)(b-a)$.

REVIEW EXERCISES FOR CHAPTER 5

1. **(a)** $A = \int_a^b [f(x) - g(x)]dx$ **(b)** $A = \int_c^d [u(y) - v(y)]dy$

(c) Here we use disks: $V = \pi\int_a^b \left([f(x)]^2 - [g(x)]^2\right)dx$

(d) Here we use cylindrical shells: $V = 2\pi\int_a^b x[f(x) - g(x)]dx$

(e) Use shells: $V = 2\pi\int_c^d y[u(y) - v(y)]dy$

(f) Use disks: $V = \pi\int_c^d \left([u(y)]^2 - [v(y)]^2\right)dy$

3. $A = \int_0^6 [(12x - 2x^2) - (x^2 - 6x)]dx = \int_0^6 (18x - 3x^2)dx = [9x^2 - x^3]_0^6 = 9\cdot 36 - 216 = 108$

5. By symmetry, $A = 2\int_0^1 (x^{1/3} - x^3)\,dx = 2\left[\frac{3}{4}x^{4/3} - \frac{1}{4}x^4\right]_0^1 = 2\left(\frac{3}{4} - \frac{1}{4}\right) = 1.$

7. $A = \int_0^\pi |\sin x - (-\cos x)|dx = \int_0^{3\pi/4}(\sin x + \cos x)dx - \int_{3\pi/4}^\pi(\sin x + \cos x)dx$

$= [\sin x - \cos x]_0^{3\pi/4} - [-\cos x + \sin x]_{3\pi/4}^\pi$

$= \left(\frac{1}{\sqrt{2}} + \frac{1}{\sqrt{2}}\right) - (0-1) - (1+0) + \left(\frac{1}{\sqrt{2}} + \frac{1}{\sqrt{2}}\right) = \sqrt{2} + 1 - 1 + \sqrt{2} = 2\sqrt{2}$

9. $V = \displaystyle\int_1^3 \pi\left(\sqrt{x-1}\right)^2 dx = \pi\int_1^3 (x-1)dx = \pi\left[\tfrac{1}{2}x^2 - x\right]_1^3 = \pi\left[\left(\tfrac{9}{2} - 3\right) - \left(\tfrac{1}{2} - 1\right)\right] = 2\pi$

11. $V = \int_1^3 2\pi y(-y^2 + 4y - 3)dy = 2\pi\int_1^3(-y^3 + 4y^2 - 3y)dy = 2\pi\left[-\frac{1}{4}y^4 + \frac{4}{3}y^3 - \frac{3}{2}y^2\right]_1^3$

$= 2\pi\left[\left(-\frac{81}{4} + 36 - \frac{27}{2}\right) - \left(-\frac{1}{4} + \frac{4}{3} - \frac{3}{2}\right)\right] = \frac{16\pi}{3}$

13. $V = \int_a^{a+h} 2\pi x\cdot 2\sqrt{x^2 - a^2}\,dx = 2\pi\int_0^{2ah+h^2} u^{1/2}\,du$ (Put $u = x^2 - a^2$, so $du = 2x\,dx$)

$= 2\pi\left[\frac{2}{3}u^{3/2}\right]_0^{2ah+h^2} = \frac{4}{3}\pi(2ah + h^2)^{3/2}$

15. $V = \displaystyle\int_0^1 \pi\left[\left(1 - x^3\right)^2 - \left(1 - x^2\right)^2\right]dx$

17. **(a)** $V = \int_0^1 \pi(x^2 - x^4)\,dx = \pi\left[\frac{1}{3}x^3 - \frac{1}{5}x^5\right]_0^1 = \pi\left[\frac{1}{3} - \frac{1}{5}\right] = \frac{2\pi}{15}$

Or: $V = \int_0^1 2\pi y(\sqrt{y} - y)\,dy = 2\pi\left[\frac{2}{5}y^{5/2} - \frac{1}{3}y^3\right]_0^1 = \frac{2\pi}{15}$

(b) $V = \int_0^1 \pi\left[(\sqrt{y})^2 - y^2\right]dy = \pi\left[\frac{1}{2}y^2 - \frac{1}{3}y^3\right]_0^1 = \pi\left[\frac{1}{2} - \frac{1}{3}\right] = \frac{\pi}{6}$

Or: $V = \int_0^1 2\pi x(x - x^2)dx = 2\pi\left[\frac{1}{3}x^3 - \frac{1}{4}x^4\right]_0^1 = \frac{\pi}{6}$

(c) $V = \int_0^1 \pi\left[(2 - x^2)^2 - (2 - x)^2\right]dx = \int_0^1 \pi(x^4 - 5x^2 + 4x)\,dx$

$= \pi\left[\frac{1}{5}x^5 - \frac{5}{3}x^3 + 2x^2\right]_0^1 = \pi\left[\frac{1}{5} - \frac{5}{3} + 2\right] = \frac{8\pi}{15}$

Or: $V = \int_0^1 2\pi(2 - y)(\sqrt{y} - y)\,dy = 2\pi\int_0^1 (y^2 - y^{3/2} - 2y + 2y^{1/2})\,dy$

$= 2\pi\left[\frac{1}{3}y^3 - \frac{2}{5}y^{5/2} - y^2 + \frac{4}{3}y^{3/2}\right]_0^1 = \frac{8\pi}{15}$

19. **(a)** Using the Midpoint Rule on $[0, 1]$ with $f(x) = \tan(x^2)$ and $n = 4$, we estimate

$A = \int_0^1 \tan(x^2)dx \approx \frac{1}{4}\left[\tan\left(\left(\frac{1}{8}\right)^2\right) + \tan\left(\left(\frac{3}{8}\right)^2\right) + \tan\left(\left(\frac{5}{8}\right)^2\right) + \tan\left(\left(\frac{7}{8}\right)^2\right)\right] \approx 0.38.$

(b) Using the Midpoint Rule on $[0, 1]$ with $f(x) = \pi\tan^2(x^2)$ (for disks) and $n = 4$, we estimate

$V = \int_0^1 f(x)dx \approx \frac{1}{4}\pi\left[\tan^2\left(\left(\frac{1}{8}\right)^2\right) + \tan^2\left(\left(\frac{3}{8}\right)^2\right) + \tan^2\left(\left(\frac{5}{8}\right)^2\right) + \tan^2\left(\left(\frac{7}{8}\right)^2\right)\right] \approx 0.87.$

21. The solid is obtained by rotating the region under the curve $y = \sin x$, above $y = 0$, from $x = 0$ to $x = \pi$, about the x-axis.

23. The solid is obtained by rotating the region in the first quadrant bounded by the curve $x = 4 - y^2$ and the coordinate axes about the x-axis.

25. Take the base to be the disk $x^2 + y^2 \le 9$. Then $V = \int_{-3}^{3} A(x)dx$, where $A(x_0)$ is the area of the isosceles right triangle whose hypotenuse lies along the line $x = x_0$ in the xy-plane. $A(x) = \frac{1}{4}\left(2\sqrt{9 - x^2}\right)^2 = 9 - x^2$, so $V = 2\int_0^3 A(x)dx = 2\int_0^3 (9 - x^2)dx = 2\left[9x - \frac{1}{3}x^3\right]_0^3 = 2(27 - 9) = 36.$

27. Equilateral triangles with sides measuring $\frac{1}{4}x$ meters have height $\frac{1}{4}x\sin 60° = \frac{\sqrt{3}}{8}x$. Therefore, $A(x) = \frac{1}{2}\cdot\frac{1}{4}x\cdot\frac{\sqrt{3}}{8}x = \frac{\sqrt{3}}{64}x^2$. $V = \int_0^{20} A(x)dx = \frac{\sqrt{3}}{64}\int_0^{20} x^2\,dx = \frac{\sqrt{3}}{64}\left[\frac{1}{3}x^3\right]_0^{20} = \frac{1000\sqrt{3}}{24} = \frac{125\sqrt{3}}{3}\text{ m}^3$

29. $30\text{ N} = f(x) = kx = k(0.03\text{ m})$, so $k = 30/0.03 = 1000\text{ N/m}$.

$W = \int_0^{0.08} kx\,dx = 1000\int_0^{0.08} x\,dx = 500[x^2]_0^{0.08} = 500(0.08)^2 = 3.2\text{ J}.$

31. **(a)** $W = \int_0^4 \pi\left(2\sqrt{y}\right)^2 62.5(4-y)dy = 250\pi \int_h^4 y(4-y)dy = 250\pi\left[2y^2 - \frac{1}{3}y^3\right]_0^4$
$= 250\pi\left(32 - \frac{64}{3}\right) = \frac{8000\pi}{3}$ ft-lb

(b) In part (a) we knew the final water level (0) but not the amount of work done. Here we use the same equation, except with the work fixed, and the lower limit of integration (that is, the final water level — call it h) unknown:

$W = 4000 = \int_h^4 \pi\left(2\sqrt{y}\right)^2 62.5(4-y)dy = 250\pi \int_h^4 y(4-y)dy = 250\pi\left[2y^2 - \frac{1}{3}y^3\right]_h^4$
$= 250\pi\left[\left(2\cdot 16 - \frac{1}{3}\cdot 64\right) - \left(2h^2 - \frac{1}{3}h^3\right)\right]$

$\Leftrightarrow \quad h^3 - 6h^2 + 32 - \frac{48}{\pi} = 0.$

We plot the graph of the function $f(h) = h^3 - 6h^2 + 32 - \frac{48}{\pi}$ on the interval $[0, 4]$ to see where it is 0.

From the graph, it appears that $f(h) = 0$ for $h \approx 2.1$. So the depth of water remaining is about 2.1 ft.

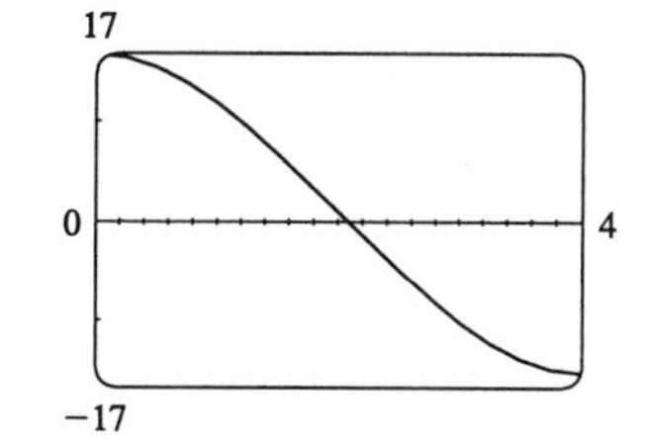

33. $\lim_{h\to 0} f_{\text{ave}} = \lim_{h\to 0} \frac{1}{h}\int_x^{x+h} f(t)dt = \lim_{h\to 0} \frac{F(x+h) - F(x)}{h}$, where $F(x) = \int_a^x f(t)dt$. But we recognize this limit as being $F'(x)$ by the definition of the derivative. Therefore $\lim_{h\to 0} f_{\text{ave}} = F'(x) = f(x)$ by Part 1 of the Fundamental Theorem of Calculus.

APPLICATIONS PLUS (page 340)

1. **(a)** By Formula 5.2.3, $V = \int_0^h \pi[f(y)]^2 dy$.

(b) $\dfrac{dV}{dt} = \dfrac{dV}{dh} \cdot \dfrac{dh}{dt} = \pi[f(h)]^2 \dfrac{dh}{dt}$

(c) $kA\sqrt{h} = \pi[f(h)]^2 \dfrac{dh}{dt}$. Set $\dfrac{dh}{dt} = C$: $\pi[f(h)]^2 C = kA\sqrt{h} \Rightarrow [f(h)]^2 = \dfrac{kA}{\pi C}\sqrt{h} \Rightarrow$

$f(h) = \sqrt{\dfrac{kA}{\pi C}} h^{1/4}$, that is, $f(y) = \sqrt{\dfrac{kA}{\pi C}} y^{1/4}$. The advantage of having $\dfrac{dh}{dt} = C$ is that the markings on the container will be equally spaced.

3. **(a)** First note that $90 \text{ mi/h} = 90 \times \frac{5280}{3600} \text{ ft/s} = 132 \text{ ft/s}$. Then $a(t) = 4 \text{ ft/s}^2 \Rightarrow v(t) = 4t = 132$ when $t = \frac{132}{4} = 33$ s. It takes 33 s to reach 132 ft/s. Therefore, taking $s(0) = 0$, we have $s(t) = 2t^2$, $0 \le t \le 33$. So $s(33) = 2178$ ft.

For $33 \le t \le 933$ we have $v(t) = 132$ ft/s $\Rightarrow s(t) = 132(t - 33) + C$ and $s(33) = 2178 \Rightarrow$ $C = 2178$, so $s(t) = 132(t - 33) + 2178$, $33 \le t \le 933$.

Therefore $s(933) = 132(933) + 2178 = 120{,}978 \text{ ft} = 22.9125 \text{ mi}$.

(b) As in part (a), the train accelerates for 33 s and travels 2178 ft while doing so. Similarly, it decelerates for 33 s and travels 2178 ft at the end of its trip. During the remaining $900 - 66 = 834$ s it travels at 132 ft/s, so the distance traveled is $132 \cdot 834 = 110{,}088$ ft. Thus the total distance is

$2178 + 110{,}088 + 2178 = 114{,}444 \text{ ft} = 21.675 \text{ mi}$.

5. **(a)** Let $F(t) = \int_0^t f(s)ds$. Then, by Part 1 of the Fundamental Theorem of Calculus, $F'(t) = f(t) =$ rate of depreciation, so $F(t)$ represents the loss in value over the interval $[0, t]$.

(b) $C(t) = [A + F(t)]/t$ represents the average expenditure over the interval $[0, t]$. The company wants to minimize average expenditure.

(c) $C(t) = \dfrac{1}{t}\left(A + \displaystyle\int_0^t f(s)ds\right)$. Using Part 1 of the Fundamental Theorem of Calculus, we have

$C'(t) = -\dfrac{1}{t^2}\left(A + \displaystyle\int_0^t f(s)ds\right) + \dfrac{1}{t}f(t) = 0$ when $tf(t) = A + \displaystyle\int_0^t f(s)ds \Rightarrow$

$f(t) = \dfrac{1}{t}\left(A + \displaystyle\int_0^t f(s)ds\right) = C(t)$.

7. **(a)** $P = \dfrac{\text{area under } y = L\sin\theta}{\text{area of rectangle}} = \dfrac{\int_0^\pi L\sin\theta\,d\theta}{\pi L} = \dfrac{-L\cos\theta|_0^\pi}{\pi L} = \dfrac{-(-1)+1}{\pi} = \dfrac{2}{\pi}$

(b) $P = \dfrac{\text{area under } y = \frac{1}{2}L\sin\theta}{\text{area of rectangle}} = \dfrac{\int_0^\pi \frac{1}{2}L\sin\theta\,d\theta}{\pi L} = \dfrac{\int_0^\pi \sin\theta\,d\theta}{2\pi} = \dfrac{-\cos\theta|_0^\pi}{2\pi} = \dfrac{2}{2\pi} = \dfrac{1}{\pi}$

(c) $P = \dfrac{\text{area under } y = \frac{1}{5}L\sin\theta}{\text{area of rectangle}} = \dfrac{\int_0^\pi \frac{1}{5}L\sin\theta\,d\theta}{\pi L} = \dfrac{\int_0^\pi \sin\theta\,d\theta}{5\pi} = \dfrac{2}{5\pi}$

9. **(a)** $F = ma = m\dfrac{dv}{dt}$, so by the Substitution Rule we have

$$\int_{t_0}^{t_1} F(t)dt = \int_{t_0}^{t_1} m\left(\frac{dv}{dt}\right)dt = m\int_{v_0}^{v_1} dv = mv\big|_{v_0}^{v_1} = mv_1 - mv_0 = p(t_1) - p(t_0).$$

(b) **(i)** We have $v_1 = 110\text{ mi/h} = \frac{110(5280)}{3600}\text{ ft/s} = 161.\overline{3}\text{ ft/s}$, $v_0 = -90\text{ mi/h} = -132\text{ ft/s}$, and $m = \frac{5/16}{32} = \frac{5}{512}$. So the change in momentum is

$p(t_1) - p(t_0) = mv_1 - mv_0 = \frac{5}{512}\left[161.\overline{3} - (-132)\right] = \frac{1466.\overline{6}}{512} \approx 2.86\text{ slug-ft/s}.$

(ii) From part (a) and part (b)(i) we have $\int_0^{0.01} F(t)dt = p(0.01) - p(0) \approx 2.86$, so the average force over the interval $[0, 0.01] = \frac{1}{0.01}\int_0^{0.01} F(t)dt \approx \frac{1}{0.01}(2.86) = 286\text{ lb}$.

11. **(a)** The segment is obtained by rotating the part of the circle $x^2 + y^2 = r^2$ given by $r - h \le y \le r$ about the y-axis. (Take the y-axis pointing downward.) So

$$\begin{aligned} V &= \textstyle\int_{r-h}^{r} \pi x^2\,dy = \int_{r-h}^{r} \pi(r^2 - y^2)dy = \pi\left[r^2 y - \frac{1}{3}y^3\right]_{r-h}^{r} = \pi\left[\left(r^3 - \frac{1}{3}r^3\right) - \left(r^2[r-h] - \frac{1}{3}[r-h]^3\right)\right] \\ &= \pi\left[\tfrac{2}{3}r^3 - \left(r^3 - r^2 h - \tfrac{1}{3}r^3 + r^2 h - rh^2 + \tfrac{1}{3}h^3\right)\right] = \pi\left[\tfrac{2}{3}r^3 - \left(\tfrac{2}{3}r^3 - rh^2 + h^3\right)\right] = \pi\left(rh^2 - \tfrac{1}{3}h^3\right) \\ &= \pi h^2\left(r - \tfrac{h}{3}\right) = \tfrac{1}{3}\pi h^2(3r - h) \end{aligned}$$

(b) The smaller segment has height $h = 1 - x$ and so by part (a) its volume is $V = \frac{1}{3}\pi(1-x)^2[3 - (1-x)] = \frac{1}{3}\pi(x-1)^2(x+2)$. This volume must be $\frac{1}{3}$ of the total volume of the sphere, which is $\frac{4}{3}\pi(1)^3$. So $\frac{1}{3}\pi(x-1)^2(x+2) = \frac{1}{3}\left(\frac{4}{3}\pi\right) \Rightarrow (x^2 - 2x + 1)(x+2) = \frac{4}{3} \Rightarrow x^3 - 3x + 2 = \frac{4}{3} \Rightarrow 3x^3 - 9x + 2 = 0$. Using Newton's method with $f(x) = 3x^3 - 9x + 2$, $f'(x) = 9x^2 - 9$, we get $x_{n+1} = x_n - \dfrac{3x_n^3 - 9x_n + 2}{9x_n^2 - 9}$. Taking $x_1 = 0$, we get $x_2 \approx 0.2222$, $x_3 \approx 0.2261$, and $x_4 \approx 0.2261$, so, correct to four decimal places, $x \approx 0.2261$.

(c) With $r = 0.5$ and $s = 0.75$, the given equation becomes $x^3 - 3(0.5)x^2 + 4(0.5)^3(0.75) = 0 \Rightarrow x^3 - \frac{3}{2}x^2 + 4\left(\frac{1}{8}\right)\frac{3}{4} = 0 \Rightarrow 8x^3 - 12x^2 + 3 = 0$. We use Newton's method with $f(x) = 8x^3 - 12x^2 + 3$, $f'(x) = 24x^2 - 24x$, so $x_{n+1} = x_n - \dfrac{8x_n^3 - 12x_n^2 + 3}{24x_n^2 - 24x_n}$. Take $x_1 = 0.5$. Then $x_2 \approx 0.6667$, $x_3 \approx 0.6736$, and $x_4 \approx 0.6736$. So to four decimals the depth is 0.6736 m.

(d) **(i)** From part (a), the volume of water in the bowl is

$V = \frac{1}{3}\pi h^2(3r - h) = \frac{1}{3}\pi h^2(15 - h) = 5\pi h^2 - \frac{1}{3}\pi h^3$. We are given that $\dfrac{dV}{dt} = 0.2\text{ m}^3/\text{s}$ and we want to find $\dfrac{dh}{dt}$ when $h = 3$. Now $\dfrac{dV}{dt} = 10\pi h\dfrac{dh}{dt} - \pi h^2\dfrac{dh}{dt}$, so $\dfrac{dh}{dt} = \dfrac{0.2}{\pi(10h - h^2)}$. When $h = 3$, we have $\dfrac{dh}{dt} = \dfrac{0.2}{\pi(10 \cdot 3 - 3^2)} = \dfrac{1}{105\pi} \approx 0.003\text{ in/s}$.

(ii) From part (a), the volume of water required to fill the bowl is

$V = \frac{1}{2} \cdot \frac{4}{3}\pi(5)^3 - \frac{1}{3}\pi(4)^2(15 - 4) = \frac{2}{3} \cdot 125\pi - \frac{16}{3} \cdot 11\pi = \frac{74}{3}\pi$. To find the time required to fill the bowl we divide this volume by the rate: Time $= \dfrac{74\pi/3}{0.2} = \dfrac{370\pi}{3} \approx 387\text{ s} \approx 6.5\text{ min}$.

CHAPTER SIX

EXERCISES 6.1

1. The diagram shows that there is a horizontal line which intersects the graph more than once, so the function is not one-to-one.

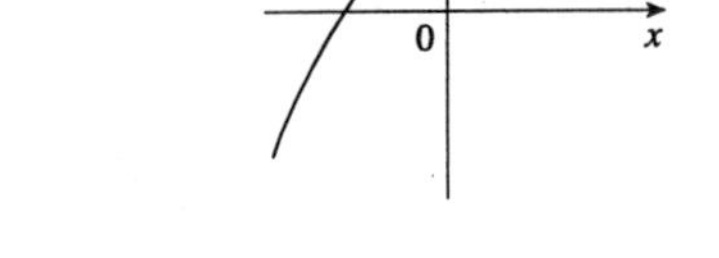

3. The function is one-to-one because no horizontal line intersects the graph more than once.

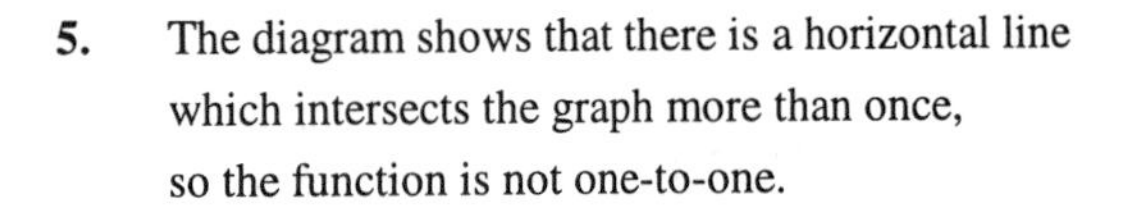

5. The diagram shows that there is a horizontal line which intersects the graph more than once, so the function is not one-to-one.

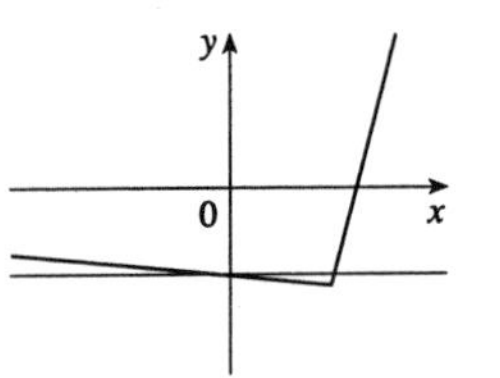

7. $x_1 \neq x_2 \quad \Rightarrow \quad 7x_1 \neq 7x_2 \quad \Rightarrow \quad 7x_1 - 3 \neq 7x_2 - 3 \quad \Rightarrow \quad f(x_1) \neq f(x_2)$, so f is 1-1.

9. $x_1 \neq x_2 \quad \Rightarrow \quad \sqrt{x_1} \neq \sqrt{x_2} \quad \Rightarrow \quad g(x_1) \neq g(x_2)$, so g is 1-1.

11. $h(x) = x^4 + 5 \quad \Rightarrow \quad h(1) = 6 = h(-1)$, so h is not 1-1.

13. $x_1 \neq x_2 \quad \Rightarrow \quad 4x_1 \neq 4x_2 \quad \Rightarrow \quad 4x_1 + 7 \neq 4x_2 + 7 \quad \Rightarrow \quad f(x_1) \neq f(x_2)$, so f is 1-1. $y = 4x + 7 \quad \Rightarrow$ $4x = y - 7 \quad \Rightarrow \quad x = (y-7)/4$. Interchange x and y: $y = (x-7)/4$. So $f^{-1}(x) = (x-7)/4$.

15. $f(x) = \dfrac{1+3x}{5-2x}$. If $f(x_1) = f(x_2)$, then $\dfrac{1+3x_1}{5-2x_1} = \dfrac{1+3x_2}{5-2x_2} \quad \Rightarrow$
$5 + 15x_1 - 2x_2 - 6x_1x_2 = 5 - 2x_1 + 15x_2 - 6x_1x_2 \quad \Rightarrow \quad 17x_1 = 17x_2 \quad \Rightarrow \quad x_1 = x_2$, so f is one-to-one.
$y = \dfrac{1+3x}{5-2x} \quad \Rightarrow \quad 5y - 2xy = 1 + 3x \quad \Rightarrow \quad x(3+2y) = 5y - 1 \quad \Rightarrow \quad x = \dfrac{5y-1}{2y+3}$.
Interchange x and y: $y = \dfrac{5x-1}{2x+3}$. So $f^{-1}(x) = \dfrac{5x-1}{2x+3}$.

17. $x_1 \neq x_2 \quad \Rightarrow \quad 5x_1 \neq 5x_2 \quad \Rightarrow \quad 2 + 5x_1 \neq 2 + 5x_2 \quad \Rightarrow \quad \sqrt{2+5x_1} \neq \sqrt{2+5x_2} \quad \Rightarrow$
$f(x_1) \neq f(x_2)$, so f is 1-1. $y = \sqrt{2+5x} \quad \Rightarrow \quad y^2 = 2 + 5x$ and $y \geq 0 \quad \Rightarrow \quad 5x = y^2 - 2 \quad \Rightarrow$
$x = \dfrac{y^2-2}{5}$, $y \geq 0$. Interchange x and y: $y = \dfrac{x^2-2}{5}$, $x \geq 0$. So $f^{-1}(x) = \dfrac{x^2-2}{5}$, $x \geq 0$.

19. **(a)** $x_1 \neq x_2 \Rightarrow 2x_1 \neq 2x_2 \Rightarrow 2x_1 + 1 \neq 2x_2 + 1 \Rightarrow f(x_1) \neq f(x_2)$, so f is 1-1.

(b) $f(1) = 3 \quad \Rightarrow \quad g(3) = 1$. Also $f'(x) = 2$, so $g'(3) = 1/f'(3) = \frac{1}{2}$.

(c) $y = 2x + 1 \quad \Rightarrow \quad x = \frac{1}{2}(y - 1)$.
Interchanging x and y gives
$y = \frac{1}{2}(x - 1)$, so $f^{-1}(x) = \frac{1}{2}(x - 1)$.
Domain(g) = range$(f) = \mathbb{R}$.
Range(g) = domain$(f) = \mathbb{R}$.

(d) $g(x) = \frac{1}{2}(x - 1) \quad \Rightarrow \quad g'(x) = \frac{1}{2}$
$\Rightarrow \quad g'(3) = \frac{1}{2}$ as in (b).

(e)

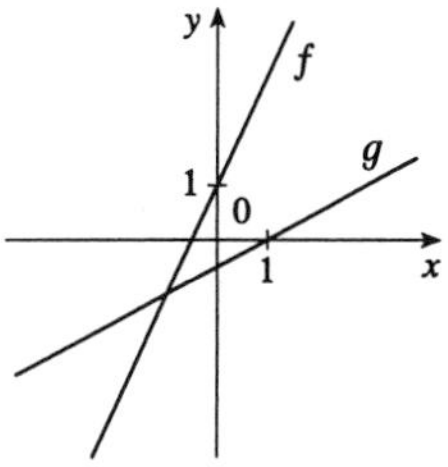

21. **(a)** $x_1 \neq x_2 \quad \Rightarrow \quad x_1^3 \neq x_2^3 \quad \Rightarrow \quad f(x_1) \neq f(x_2)$, so f is one-to-one.

(b) $f'(x) = 3x^2$ and $f(2) = 8 \quad \Rightarrow \quad g(8) = 2$, so $g'(8) = 1/f'(g(8)) = 1/f'(2) = \frac{1}{12}$.

(c) $y = x^3 \quad \Rightarrow \quad x = y^{1/3}$. Interchanging x and y gives $y = x^{1/3}$, so $f^{-1}(x) = x^{1/3}$.
Domain(g) = range$(f) = \mathbb{R}$.
Range(g) = domain$(f) = \mathbb{R}$.

(d) $g(x) = x^{1/3} \quad \Rightarrow$
$g'(x) = \frac{1}{3}x^{-2/3} \quad \Rightarrow$
$g'(8) = \frac{1}{3}\left(\frac{1}{4}\right) = \frac{1}{12}$ as in part (b).

(e)

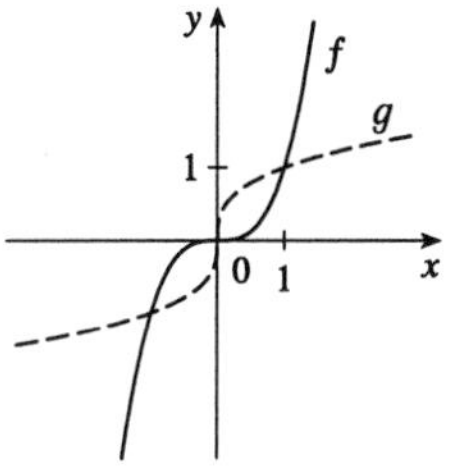

23. **(a)** Since $x \geq 0$, $x_1 \neq x_2 \Rightarrow x_1^2 \neq x_2^2 \Rightarrow 9 - x_1^2 \neq 9 - x_2^2 \Rightarrow f(x_1) \neq f(x_2)$, so f is 1-1.

(b) $f'(x) = -2x$ and $f(1) = 8 \Rightarrow g(8) = 1$, so $g'(8) = \dfrac{1}{f'(g(8))} = \dfrac{1}{f'(1)} = \dfrac{1}{(-2)} = -\dfrac{1}{2}$.

(c) $y = 9 - x^2 \quad \Rightarrow \quad x^2 = 9 - y \quad \Rightarrow$
$x = \sqrt{9 - x}$. Interchange x and y:
$y = \sqrt{9 - x}$, so $f^{-1}(x) = \sqrt{9 - x}$.
Domain(g) = range$(f) = [0, 9]$.
Range(g) = domain$(f) = [0, 3]$.

(d) $g'(x) = -1/\left(2\sqrt{9 - x}\right) \quad \Rightarrow$
$g'(8) = -\frac{1}{2}$ as in (b).

(e)

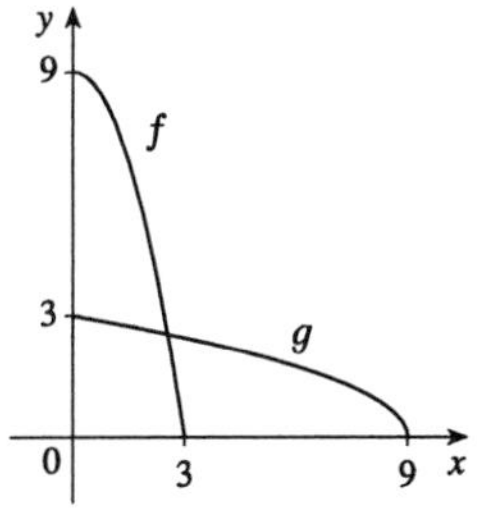

25. $f(0) = 1 \Rightarrow g(1) = 0$, and $f'(x) = 3x^2 + 1 \Rightarrow f'(0) = 1$. Therefore $g'(1) = \dfrac{1}{f'(g(1))} = \dfrac{1}{f'(0)} = \dfrac{1}{1} = 1$.

27. $f(0) = 3 \quad \Rightarrow \quad g(3) = 0$, and $f'(x) = 2x + \frac{\pi}{2}\sec^2(\pi x/2) \quad \Rightarrow \quad f'(0) = 1 \cdot \frac{\pi}{2} = \frac{\pi}{2}$. Thus
$$g'(3) = \frac{1}{f'(g(3))} = \frac{1}{f'(0)} = \frac{2}{\pi}.$$

29. $f(4) = 5 \quad \Rightarrow \quad g(5) = 4$. Therefore, $g'(5) = \dfrac{1}{f'(g(5))} = \dfrac{1}{f'(4)} = \dfrac{1}{2/3} = \dfrac{3}{2}$.

31. $y = 1 - \dfrac{2}{x^2} \quad \Rightarrow \quad 1 - y = \dfrac{2}{x^2} \quad \Rightarrow \quad x^2 = \dfrac{2}{1-y} \quad \Rightarrow \quad x = \sqrt{\dfrac{2}{1-y}}$, since $x > 0$. Interchange x and y:

$y = \sqrt{\dfrac{2}{1-x}}$. So $f^{-1}(x) = \sqrt{\dfrac{2}{1-x}}$.

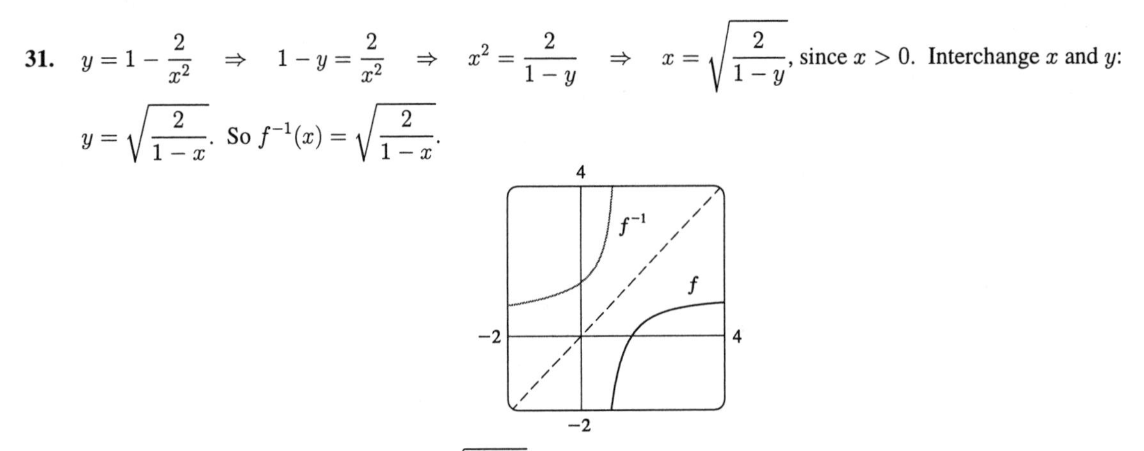

33. Since $f'(x) = \dfrac{2x}{2\sqrt{x^2+1}} - 1 = \dfrac{x - \sqrt{x^2+1}}{\sqrt{x^2+1}}$ is negative for all x, we know that f is a decreasing function on $\mathbb{R}$, and hence is 1-1. We could also use the Horizontal Line Test to show that f is 1-1. The parametric equations for the graph of f are $x = t$, $y = \sqrt{t^2+1} - t$; for the graph of f^{-1} they are $x = \sqrt{t^2+1} - t$, $y = t$.

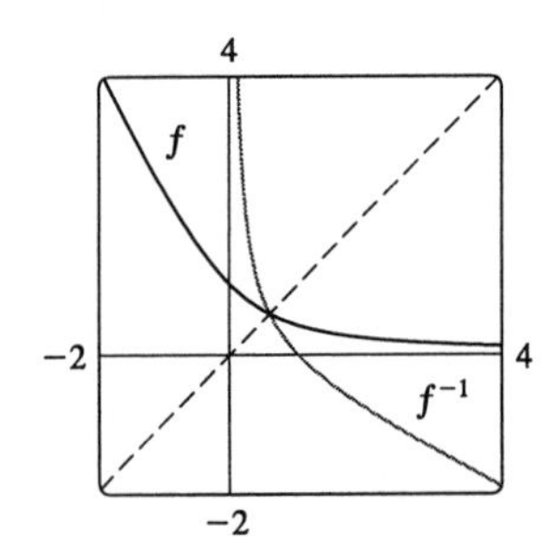

35. **(a)** $\sqrt[5]{x} - \sqrt[5]{y} = y \quad \Rightarrow \quad \sqrt[5]{x} = y + \sqrt[5]{y} \quad \Rightarrow \quad x = \left(y + \sqrt[5]{y}\right)^5$. Interchange x and y: $y = \left(x + \sqrt[5]{x}\right)^5$. So $f^{-1}(x) = \left(x + \sqrt[5]{x}\right)^5$.

(b) The parametric equations for the graphof f^{-1} are $x = t$, $y = \left(t + \sqrt[5]{t}\right)^5$. So the parametric equations for the graph of $f = \left(f^{-1}\right)^{-1}$ are $x = \left(t + \sqrt[5]{t}\right)^5$, $y = t$.

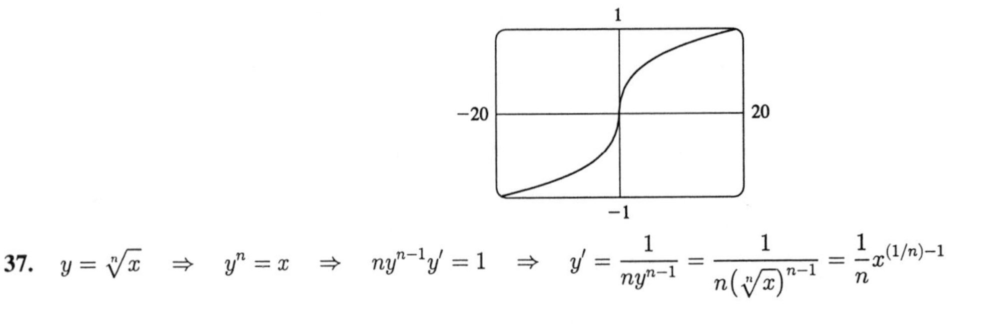

37. $y = \sqrt[n]{x} \quad \Rightarrow \quad y^n = x \quad \Rightarrow \quad ny^{n-1}y' = 1 \quad \Rightarrow \quad y' = \dfrac{1}{ny^{n-1}} = \dfrac{1}{n\left(\sqrt[n]{x}\right)^{n-1}} = \dfrac{1}{n}x^{(1/n)-1}$

39. Suppose that f is increasing. If $x_1 \neq x_2$, then either $x_1 < x_2$ or $x_2 < x_1$. If $x_1 < x_2$, then $f(x_1) < f(x_2)$. If $x_2 < x_1$, then $f(x_2) < f(x_1)$. In either case, $x_1 \neq x_2 \Rightarrow f(x_1) \neq f(x_2)$, so f is one-to-one.

EXERCISES 6.2

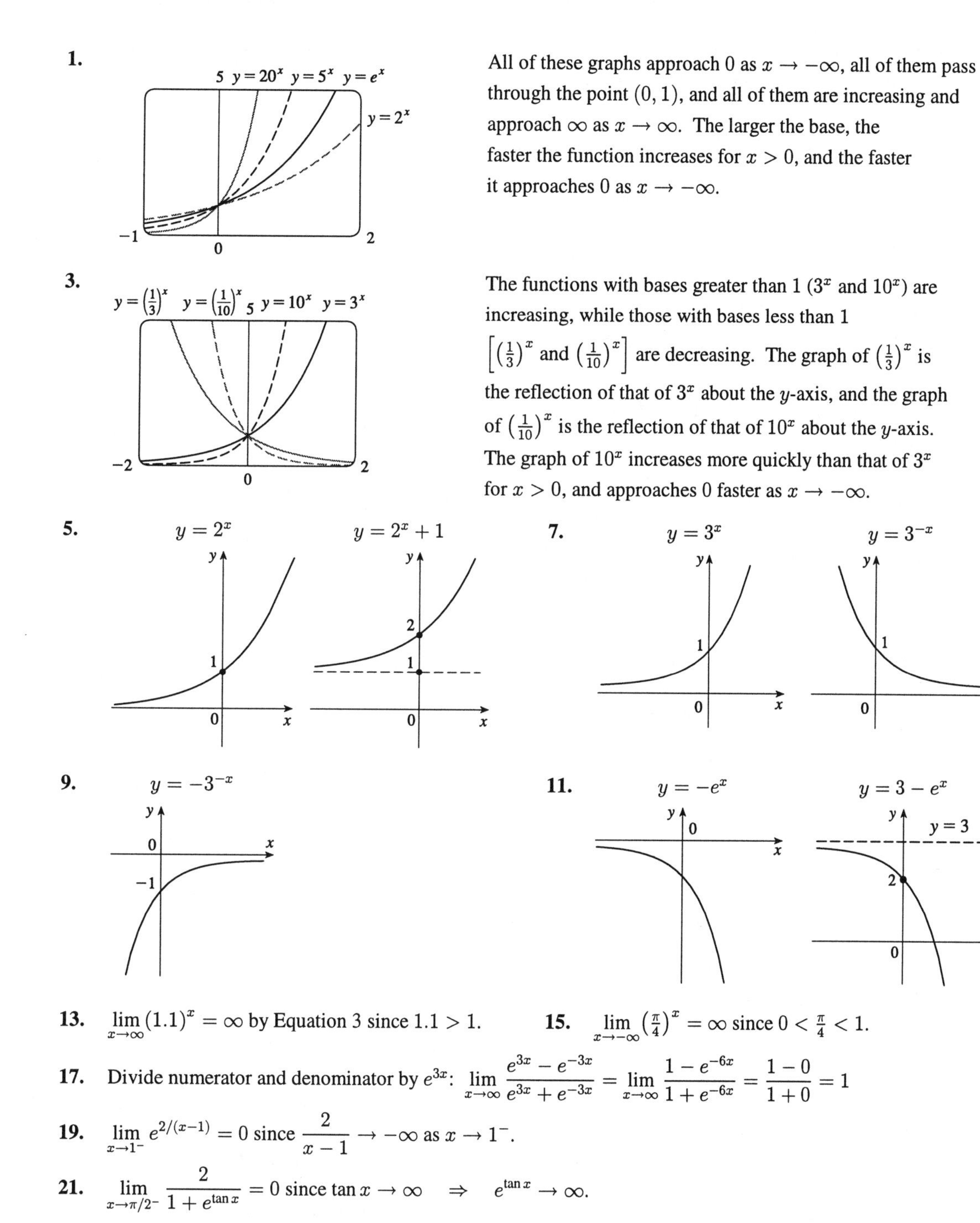

1. All of these graphs approach 0 as $x \to -\infty$, all of them pass through the point $(0, 1)$, and all of them are increasing and approach ∞ as $x \to \infty$. The larger the base, the faster the function increases for $x > 0$, and the faster it approaches 0 as $x \to -\infty$.

3. The functions with bases greater than 1 (3^x and 10^x) are increasing, while those with bases less than 1 $\left[\left(\frac{1}{3}\right)^x \text{ and } \left(\frac{1}{10}\right)^x\right]$ are decreasing. The graph of $\left(\frac{1}{3}\right)^x$ is the reflection of that of 3^x about the y-axis, and the graph of $\left(\frac{1}{10}\right)^x$ is the reflection of that of 10^x about the y-axis. The graph of 10^x increases more quickly than that of 3^x for $x > 0$, and approaches 0 faster as $x \to -\infty$.

5. $y = 2^x$ $y = 2^x + 1$

7. $y = 3^x$ $y = 3^{-x}$

9. $y = -3^{-x}$

11. $y = -e^x$ $y = 3 - e^x$

13. $\lim\limits_{x \to \infty} (1.1)^x = \infty$ by Equation 3 since $1.1 > 1$.

15. $\lim\limits_{x \to -\infty} \left(\frac{\pi}{4}\right)^x = \infty$ since $0 < \frac{\pi}{4} < 1$.

17. Divide numerator and denominator by e^{3x}: $\lim\limits_{x \to \infty} \dfrac{e^{3x} - e^{-3x}}{e^{3x} + e^{-3x}} = \lim\limits_{x \to \infty} \dfrac{1 - e^{-6x}}{1 + e^{-6x}} = \dfrac{1 - 0}{1 + 0} = 1$

19. $\lim\limits_{x \to 1^-} e^{2/(x-1)} = 0$ since $\dfrac{2}{x-1} \to -\infty$ as $x \to 1^-$.

21. $\lim\limits_{x \to \pi/2^-} \dfrac{2}{1 + e^{\tan x}} = 0$ since $\tan x \to \infty \quad \Rightarrow \quad e^{\tan x} \to \infty$.

23. $2 \text{ ft} = 24 \text{ in}$, $f(24) = 24^2 \text{ in} = 576 \text{ in} = 48 \text{ ft}$. $g(24) = 2^{24} \text{ in} = 2^{24}/(12 \cdot 5280) \text{mi} \approx 265 \text{ mi}$

25. **(a)** Let $f(h) = \dfrac{4^h - 1}{h}$. Then $f(0.1) \approx 1.487$, $f(0.01) \approx 1.396$, $f(0.001) \approx 1.387$, and $f(0.0001) \approx 1.386$.

These quantities represent the slopes of secant lines to the curve $y = 4^x$, through the points $(0,1)$ and $(h, 4^h)$ $\left(\text{since they are of the form } \dfrac{4^h - 4^0}{h - 0}\right)$.

(b) The value of the limit $\lim\limits_{h\to 0} \dfrac{4^h - 1}{h}$ is about 1.39, judging from the calculations in part (a).

(c) The limit in part (b) represents the slope of the tangent line to the curve $y = 4^x$ at $(0,1)$.

27. $f(x) = e^{\sqrt{x}} \quad\Rightarrow\quad f'(x) = e^{\sqrt{x}}\big/\left(2\sqrt{x}\right)$

29. $y = xe^{2x} \quad\Rightarrow\quad y' = e^{2x} + xe^{2x}(2) = e^{2x}(1 + 2x)$

31. $h(t) = \sqrt{1 - e^t} \quad\Rightarrow\quad h'(t) = -e^t\big/\left(2\sqrt{1 - e^t}\right)$

33. $y = e^{x\cos x} \quad\Rightarrow\quad y' = e^{x\cos x}(\cos x - x\sin x)$

35. $y = e^{-1/x} \quad\Rightarrow\quad y' = e^{-1/x}/x^2$

37. $y = \tan(e^{3x-2}) \quad\Rightarrow\quad y' = 3e^{3x-2}\sec^2(e^{3x-2})$

39. $y = \dfrac{e^{3x}}{1 + e^x} \quad\Rightarrow\quad y' = \dfrac{3e^{3x}(1 + e^x) - e^{3x}(e^x)}{(1 + e^x)^2} = \dfrac{3e^{3x} + 3e^{4x} - e^{4x}}{(1 + e^x)^2} = \dfrac{3e^{3x} + 2e^{4x}}{(1 + e^x)^2}$

41. $y = x^e \quad\Rightarrow\quad y' = ex^{e-1}$

43. $y = f(x) = e^{-x}\sin x \quad\Rightarrow\quad f'(x) = -e^{-x}\sin x + e^{-x}\cos x \quad\Rightarrow\quad f'(\pi) = e^{-\pi}(\cos\pi - \sin\pi) = -e^{-\pi}$, so the equation of the tangent at $(\pi, 0)$ is $y - 0 = -e^{-\pi}(x - \pi)$ or $x + e^{\pi}y = \pi$.

45. $\cos(x - y) = xe^x \quad\Rightarrow\quad -\sin(x - y)(1 - y') = e^x + xe^x \quad\Rightarrow\quad y' = 1 + \dfrac{e^x(1 + x)}{\sin(x - y)}$

47. $y = e^{2x} + e^{-3x} \quad\Rightarrow\quad y' = 2e^{2x} - 3e^{-3x} \quad\Rightarrow\quad y'' = 4e^{2x} + 9e^{-3x}$, so
$y'' + y' - 6y = (4e^{2x} + 9e^{-3x}) + (2e^{2x} - 3e^{-3x}) - 6(e^{2x} + e^{-3x}) = 0$.

49. $y = e^{rx} \quad\Rightarrow\quad y' = re^{rx} \quad\Rightarrow\quad y'' = r^2e^{rx}$, so $y'' + 5y' - 6y = r^2e^{rx} + 5re^{rx} - 6e^{rx}$
$= e^{rx}(r^2 + 5r - 6) = e^{rx}(r + 6)(r - 1) = 0 \quad\Rightarrow\quad (r + 6)(r - 1) = 0 \quad\Rightarrow\quad r = 1$ or -6.

51. $f(x) = e^{-2x} \quad\Rightarrow\quad f'(x) = -2e^{-2x} \quad\Rightarrow\quad f''(x) = (-2)^2e^{-2x} \quad\Rightarrow\quad f'''(x) = (-2)^3e^{-2x} \quad\Rightarrow\quad \cdots$
$\Rightarrow\quad f^{(8)}(x) = (-2)^8e^{-2x} = 256e^{-2x}$

53. **(a)** $f(x) = e^x + x$ is continuous on $\mathbb{R}$ and $f(-1) = e^{-1} - 1 < 0 < 1 = f(0)$, so by the Intermediate Value Theorem, $e^x + x = 0$ has a root in $(-1, 0)$.

(b) $f(x) = e^x + x \quad\Rightarrow\quad f'(x) = e^x + 1$, so $x_{n+1} = x_n - \dfrac{e^{x_n} + x_n}{e^{x_n} + 1}$. From Exercise 31 we know that there is a root between -1 and 0, so we take $x_1 = -0.5$. Then $x_2 \approx -0.566311$, $x_3 \approx -0.567143$, and $x_4 \approx -0.567143$, so the root is -0.567143 to six decimal places.

55. **(a)** $\lim\limits_{t\to\infty} p(t) = \lim\limits_{t\to\infty} \dfrac{1}{1+ae^{-kt}} = \dfrac{1}{1+a\cdot 0} = 1$, since $k>0 \quad\Rightarrow\quad -kt \to -\infty$.

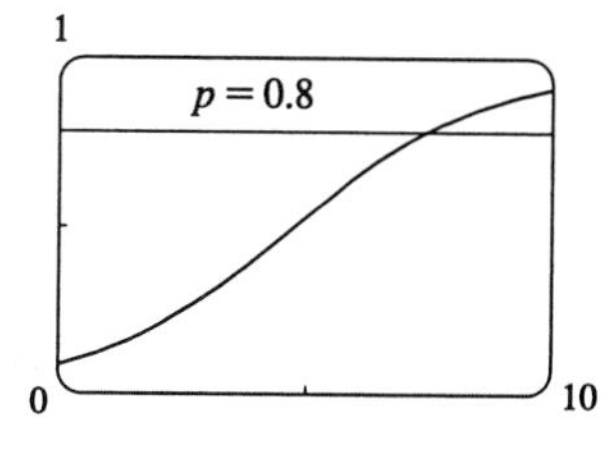

(b) $\dfrac{dp}{dt} = -\left(1+ae^{-kt}\right)^{-2}\left(-kae^{-kt}\right) = \dfrac{kae^{-kt}}{(1+ae^{-kt})^2}$

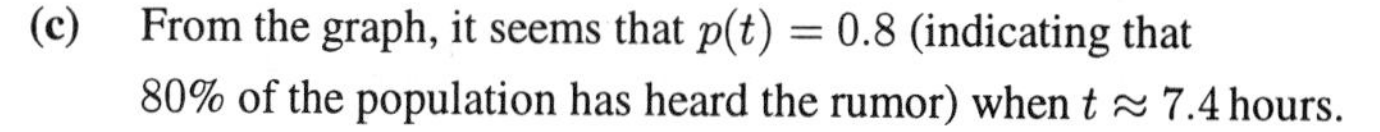

(c) From the graph, it seems that $p(t)=0.8$ (indicating that 80% of the population has heard the rumor) when $t\approx 7.4$ hours.

57. $x - e^x \quad\Rightarrow\quad f'(x) = 1 - e^x = 0 \quad\Leftrightarrow\quad e^x = 1 \quad\Leftrightarrow\quad x = 0$. Now $f'(x)>0$ for all $x<0$ and $f'(x)<0$ for all $x>0$, so the absolute maximum value is $f(0) = 0-1 = -1$.

59. **(a)** $f(x) = xe^x \quad\Rightarrow\quad f'(x) = e^x + xe^x = e^x(1+x) > 0 \quad\Leftrightarrow\quad 1+x>0 \quad\Leftrightarrow\quad x>-1$, so f is increasing on $[-1,\infty)$ and decreasing on $(-\infty,-1]$.

(b) $f''(x) = e^x(1+x) + e^x = e^x(2+x) > 0 \quad\Leftrightarrow\quad 2+x>0 \quad\Leftrightarrow\quad x>-2$, so f is CU on $(-2,\infty)$ and CD on $(-\infty,-2)$.

(c) f has an inflection point at $(-2,-2e^{-2})$.

61. $y = f(x) = e^{-1/(x+1)}$ **A.** $D = \{x \mid x\neq -1\} = (-\infty,-1)\cup(-1,\infty)$ **B.** No x-intercept; y-intercept $= f(0) = e^{-1}$ **C.** No symmetry **D.** $\lim\limits_{x\to\pm\infty} e^{-1/(x+1)} = 1$ since $-1/(x+1)\to 0$, so $y=1$ is a HA. $\lim\limits_{x\to -1^+} e^{-1/(x+1)} = 0$ since $-1/(x+1)\to-\infty$, $\lim\limits_{x\to -1^-} e^{-1/(x+1)} = \infty$ since $-1/(x+1)\to\infty$, so $x=-1$ is a VA. **E.** $f'(x) = e^{-1/(x+1)}/(x+1)^2 \quad\Rightarrow\quad f'(x)>0$ for all x except 1, so f is increasing on $(-\infty,-1)$ and $(-1,\infty)$. **F.** No extrema

G. $f''(x) = \dfrac{e^{-1/(x+1)}}{(x+1)^4} + \dfrac{e^{-1/(x+1)}(-2)}{(x+1)^3} = -\dfrac{e^{-1/(x+1)}(2x+1)}{(x+1)^4}$

$\Rightarrow\quad f''(x)>0 \quad\Leftrightarrow\quad 2x+1<0 \quad\Leftrightarrow\quad x<-\frac{1}{2}$, so f is CU on $(-\infty,-1)$ and $\left(-1,-\frac{1}{2}\right)$, and CD on $\left(-\frac{1}{2},\infty\right)$.

f has an IP at $\left(-\frac{1}{2}, e^{-2}\right)$.

H.

63. $y = 1/(1+e^{-x})$ **A.** $D=\mathbb{R}$ **B.** No x-intercepts; y-intercept $= f(0) = \frac{1}{2}$. **C.** No symmetry **D.** $\lim\limits_{x\to\infty} 1/(1+e^{-x}) = \frac{1}{1+0} = 1$ and $\lim\limits_{x\to-\infty} 1/(1+e^{-x}) = 0$ $\left(\text{since } \lim\limits_{x\to-\infty} e^{-x} = \infty\right)$, so f has horizontal asymptotes $y=0$ and $y=1$. **E.** $f'(x) = -(1+e^{-x})^{-2}(-e^{-x}) = e^{-x}/(1+e^{-x})^2$. This is positive for all x, so f is increasing on $\mathbb{R}$. **F.** No extrema

G. $f''(x) = \dfrac{(1+e^{-x})^2(-e^{-x}) - e^{-x}(2)(1+e^{-x})(-e^{-x})}{(1+e^{-x})^4}$

$= \dfrac{e^{-x}(e^{-x}-1)}{(1+e^{-x})^3}$. The second factor in the numerator is negative for $x>0$ and positive for $x<0$, and the other factors are always positive, so f is CU on $(-\infty,0)$ and CD on $(0,\infty)$. f has an inflection point at $\left(0,\frac{1}{2}\right)$.

H.

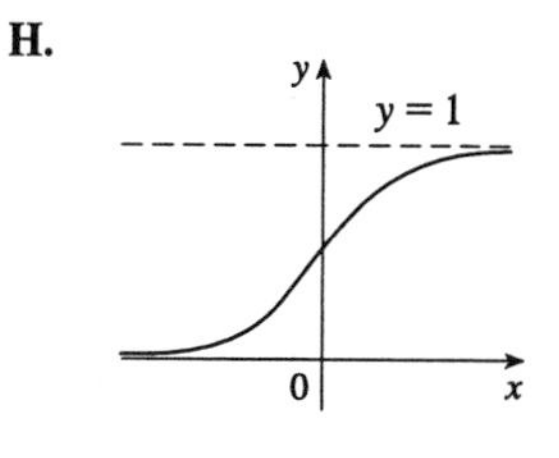

65.

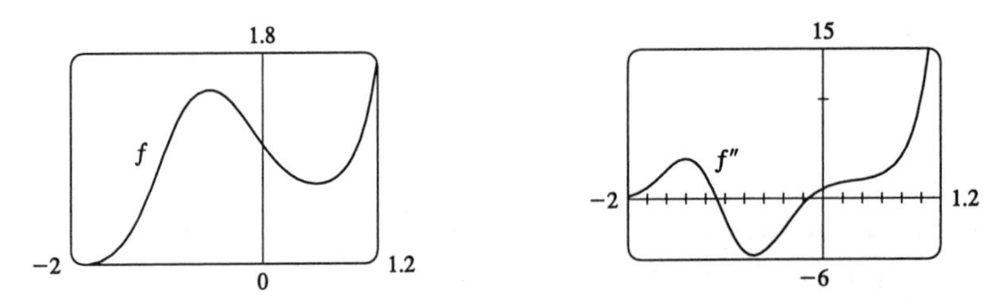

$f(x) = e^{x^3-x} \to 0$ as $x \to -\infty$, and $f(x) \to \infty$ as $x \to \infty$. From the graph, it appears that f has a local minimum of about $f(0.58) = 0.68$, and a local maximum of about $f(-0.58) = 1.47$. To find the exact values, we calculate $f'(x) = (3x^2 - 1)e^{x^3-x}$, which is 0 when $3x^2 - 1 = 0 \quad \Leftrightarrow \quad x = \pm\frac{1}{\sqrt{3}}$. The negative root corresponds to the local maximum $f\left(-\frac{1}{\sqrt{3}}\right) = e^{(-1/\sqrt{3})^3-(-1/\sqrt{3})} = e^{2\sqrt{3}/9}$, and the positive root corresponds to the local minimum $f\left(\frac{1}{\sqrt{3}}\right) = e^{(1/\sqrt{3})^3-(1/\sqrt{3})} = e^{-2\sqrt{3}/9}$. To estimate the inflection points, we calculate and graph $f''(x) = \dfrac{d}{dx}\left[(3x^2-1)e^{x^3-x}\right] = (3x^2-1)e^{x^3-x}(3x^2-1) + e^{x^3-x}(6x) = e^{x^3-x}(9x^4 - 6x^2 + 6x + 1)$. From the graph, it appears that $f''(x)$ changes sign (and thus f has inflection points) at $x \approx -0.15$ and $x \approx -1.09$. From the graph of f, we see that these x-values correspond to inflection points at about $(-0.15, 1.15)$ and $(-1.09, 0.82)$.

67. $u = -6x \quad \Rightarrow \quad du = -6\,dx$, so $\int e^{-6x}\,dx = -\frac{1}{6}\int e^u\,du = -\frac{1}{6}e^u + C = -\frac{1}{6}e^{-6x} + C$.

69. Let $u = 1 + e^x$. Then $du = e^x\,dx$, so $\int e^x(1+e^x)^{10}\,dx = \int u^{10}\,du = \frac{1}{11}u^{11} + C = \frac{1}{11}(1+e^x)^{11} + C$.

71. $\displaystyle\int \frac{e^x+1}{e^x}\,dx = \int (1 + e^{-x})dx = x - e^{-x} + C$

73. Let $u = -x^2$. Then $du = -2x\,dx$, so $\int xe^{-x^2}dx = -\frac{1}{2}\int e^u\,du = -\frac{1}{2}e^u + C = -\frac{1}{2}e^{-x^2} + C$.

75. Let $u = x^2 - 4x - 3$. Then $du = 2(x-2)dx$, so
$\int (x-2)e^{x^2-4x-3}\,dx = \frac{1}{2}\int e^u\,du = \frac{1}{2}e^u + C = \frac{1}{2}e^{x^2-4x-3} + C$.

77. Area $= \int_0^1 (e^{3x} - e^x)dx = \left[\frac{1}{3}e^{3x} - e^x\right]_0^1 = \left(\frac{1}{3}e^3 - e\right) - \left(\frac{1}{3} - 1\right) = \frac{1}{3}e^3 - e + \frac{2}{3} \approx 4.644$

79. $V = \int_0^1 \pi(e^x)^2\,dx = \int_0^1 \pi e^{2x}\,dx = \frac{1}{2}\left[\pi e^{2x}\right]_0^1 = \frac{\pi}{2}(e^2 - 1)$

81. We use Theorem 6.1.8. Note that $f(0) = 3 + 0 + e^0 = 4$, so $f^{-1}(4) = 0$. Also $f'(x) = 1 + e^x$. Therefore,
$(f^{-1})'(4) = \dfrac{1}{f'(f^{-1}(4))} = \dfrac{1}{f'(0)} = \dfrac{1}{1+e^0} = \dfrac{1}{2}$.

83. **(a)** Let $f(x) = e^x - 1 - x$. Now $f(0) = e^0 - 1 = 0$, and for $x \geq 0$, we have $f'(x) = e^x - 1 \geq 0$. Now, since $f(0) = 0$ and f is increasing on $[0, \infty)$, $f(x) \geq 0$ for $x \geq 0 \quad \Rightarrow \quad e^x - 1 - x \geq 0 \quad \Rightarrow$ $e^x \geq 1 + x$.

(b) For $0 \leq x \leq 1$, $x^2 \leq x$, so $e^{x^2} \leq e^x$ (since e^x is increasing.) Hence [from (a)] $1 + x^2 \leq e^{x^2} \leq e^x$. So $\frac{4}{3} = \int_0^1 (1+x^2)dx \leq \int_0^1 e^{x^2}dx \leq \int_0^1 e^x\,dx = e - 1 < e \quad \Rightarrow \quad \frac{4}{3} \leq \int_0^1 e^{x^2}dx \leq e$.

85. **(a)** By Exercise 83(a), the result holds for $n = 1$. Suppose that $e^x \geq 1 + x + \dfrac{x^2}{2!} + \cdots + \dfrac{x^k}{k!}$ for $x \geq 0$. Let $f(x) = e^x - 1 - x - \dfrac{x^2}{2!} - \cdots - \dfrac{x^k}{k!} - \dfrac{x^{k+1}}{(k+1)!}$. Then $f'(x) = e^x - 1 - x - \cdots - \dfrac{x^k}{k!} \geq 0$ by assumption. Hence $f(x)$ is increasing on $[0, \infty)$. So $0 \leq x$ implies that $0 = f(0) \leq f(x) = e^x - 1 - x - \cdots - \dfrac{x^k}{k!} - \dfrac{x^{k+1}}{(k+1)!}$, and hence $e^x \geq 1 + x + \cdots + \dfrac{x^k}{k!} + \dfrac{x^{k+1}}{(k+1)!}$ for $x \geq 0$. Therefore, for $x \geq 0$, $e^x \geq 1 + x + \dfrac{x^2}{2!} + \cdots + \dfrac{x^n}{n!}$ for every positive integer n, by mathematical induction.

(b) Taking $n = 4$ and $x = 1$ in (a), we have $e = e^1 \geq 1 + \dfrac{1}{2} + \dfrac{1}{6} + \dfrac{1}{24} = 2.708\overline{3} > 2.7$.

(c) $e^x \geq 1 + x + \cdots + \dfrac{x^k}{k!} + \dfrac{x^{k+1}}{(k+1)!} \quad \Rightarrow \quad \dfrac{e^x}{x^k} \geq \dfrac{1}{x^k} + \dfrac{1}{x^{k-1}} + \cdots + \dfrac{1}{k!} + \dfrac{x}{(k+1)!} \geq \dfrac{x}{(k+1)!}$. But $\lim\limits_{x\to\infty} \dfrac{x}{(k+1)!} = \infty$, so $\lim\limits_{x\to\infty} \dfrac{e^x}{x^k} = \infty$.

EXERCISES 6.3

1. $\log_2 64 = 6$ since $2^6 = 64$.

3. $\log_8 2 = \frac{1}{3}$ since $8^{1/3} = 2$.

5. $\log_3 \frac{1}{27} = -3$ since $3^{-3} = \frac{1}{27}$.

7. $\ln e^{\sqrt{2}} = \sqrt{2}$

9. $\log_{10} 1.25 + \log_{10} 80 = \log_{10}(1.25 \cdot 80) = \log_{10} 100 = 2$

11. $\log_8 6 - \log_8 3 + \log_8 4 = \log_8 \frac{6 \cdot 4}{3} = \log_8 8 = 1$

13. $2^{(\log_2 3 + \log_2 5)} = 2^{\log_2 15} = 15$

15. $\log_5 a + \log_5 b - \log_5 c = \log_5(ab/c)$

17. $2\ln 4 - \ln 2 = \ln 4^2 - \ln 2 = \ln 16 - \ln 2 = \ln \frac{16}{2} = \ln 8$

Or: $2\ln 4 - \ln 2 = 2\ln 2^2 - \ln 2 = 4\ln 2 - \ln 2 = 3\ln 2$

19. $\frac{1}{3}\ln x - 4\ln(2x+3) = \ln(x^{1/3}) - \ln(2x+3)^4 = \ln\left(x^{1/3}/(2x+3)^4\right)$

21. **(a)** $\log_2 5 = \dfrac{\ln 5}{\ln 2} \approx 2.321928$

(b) $\log_5 26.05 = \dfrac{\ln 26.05}{\ln 5} \approx 2.025563$

(c) $\log_3 e = \dfrac{1}{\ln 3} \approx 0.910239$

(d) $\log_{0.7} 14 = \dfrac{\ln 14}{\ln 0.7} \approx -7.399054$

23. 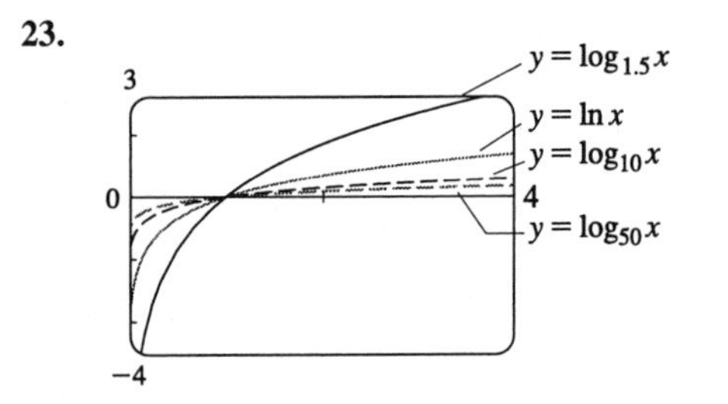

To graph these functions, we use $\log_{1.5} x = \dfrac{\ln x}{\ln 1.5}$ and $\log_{50} x = \dfrac{\ln x}{\ln 50}$. These graphs all approach $-\infty$ as $x \to 0^+$, and they all pass through the point $(1, 0)$. Also, they are all increasing, and all approach ∞ as $x \to \infty$. The functions with larger bases increase extremely slowly, and the ones with smaller bases do so somewhat more quickly. The functions with large bases approach the y-axis more closely as $x \to 0^+$.

25. $y = \log_{10} x$ $\qquad$ $y = \log_{10}(x+5)$

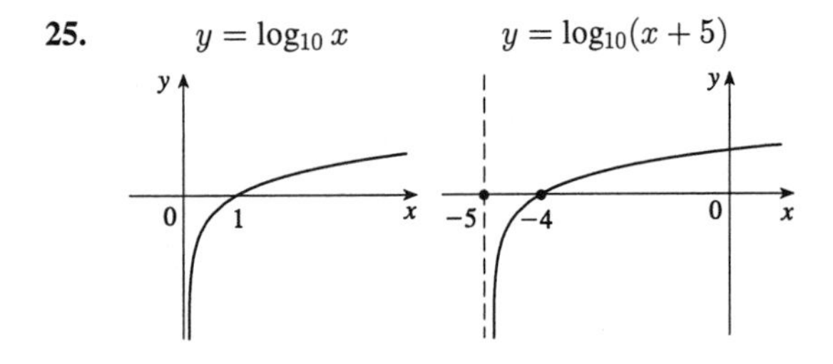

27. $y = \ln x$ $\qquad$ $y = -\ln x$

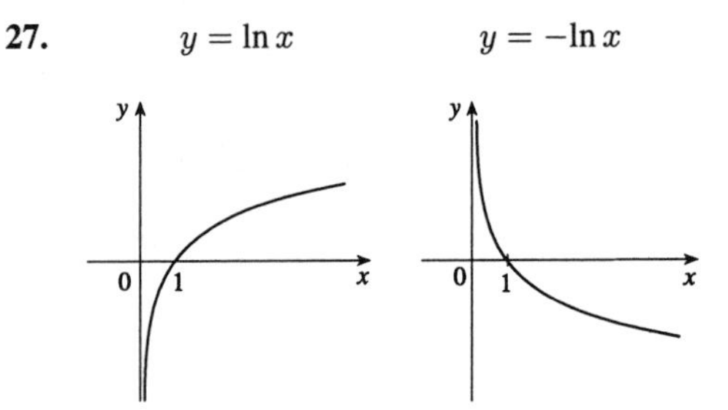

29. $y = \ln(-x)$ $\qquad$ $y = -\ln(-x)$

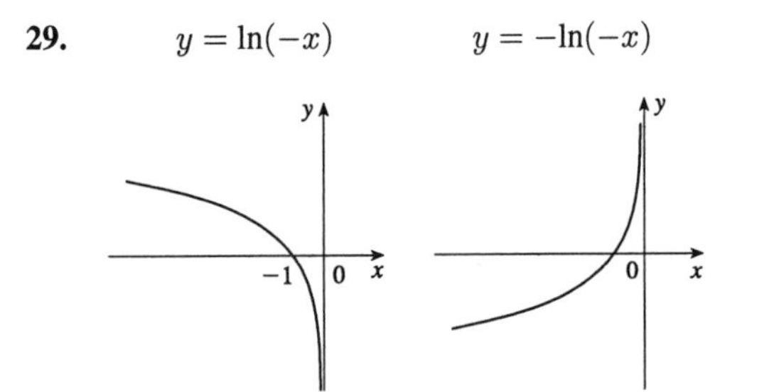

31. $y = \ln(x^2) = 2\ln|x|$

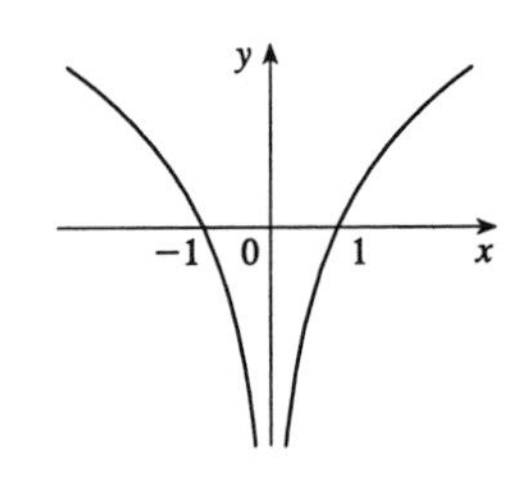

33. $y = \ln x$ $\qquad$ $y = \ln(x+3)$

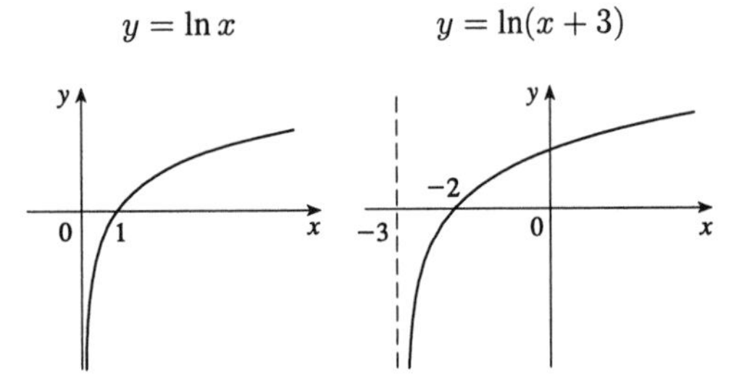

35. $\log_2 x = 3 \quad \Leftrightarrow \quad x = 2^3 = 8$

37. $e^x = 16 \quad \Leftrightarrow \quad \ln e^x = \ln 16 \quad \Leftrightarrow \quad x = \ln 16 \quad \Leftrightarrow \quad x = 4\ln 2$

39. $\ln(2x-1) = 3 \quad \Leftrightarrow \quad e^{\ln(2x-1)} = e^3 \quad \Leftrightarrow \quad 2x - 1 = e^3 \quad \Leftrightarrow \quad x = \frac{1}{2}(e^3 + 1)$

41. $3^{x+2} = m \quad \Leftrightarrow \quad \log_3 m = x + 2 \quad \Leftrightarrow \quad x = \log_3 m - 2$

43. $\ln x = \ln 5 + \ln 8 = \ln 40 \quad \Leftrightarrow \quad x = 40$

45. $5 = \ln(e^{2x-1}) = 2x - 1 \quad \Leftrightarrow \quad x = 3$

47. $\ln(\ln x) = 1 \quad \Leftrightarrow \quad e^{\ln(\ln x)} = e^1 \quad \Leftrightarrow \quad \ln x = e^1 = e \quad \Leftrightarrow \quad e^{\ln x} = e^e \quad \Leftrightarrow \quad x = e^e$

49. $2^{3^x} = 5 \quad \Leftrightarrow \quad 3^x = \log_2 5 \quad \Leftrightarrow \quad \log_3(\log_2 5) = x$. *Or:* $2^{3^x} = 5 \quad \Leftrightarrow \quad \ln 2^{3^x} = \ln 5 \quad \Leftrightarrow \quad 3^x \ln 2 = \ln 5$ $\quad \Leftrightarrow \quad 3^x = \dfrac{\ln 5}{\ln 2}$. Hence $\ln 3^x = x \ln 3 = \ln\left(\dfrac{\ln 5}{\ln 2}\right) \quad \Leftrightarrow \quad x = \dfrac{\ln(\ln 5/\ln 2)}{\ln 3}$.

51. $\ln(x+6) + \ln(x-3) = \ln 5 + \ln 2 \quad \Leftrightarrow \quad \ln[(x+6)(x-3)] = \ln 10 \quad \Leftrightarrow \quad (x+6)(x-3) = 10 \quad \Leftrightarrow$ $x^2 + 3x - 18 = 10 \quad \Leftrightarrow \quad x^2 + 3x - 28 = 0 \quad \Leftrightarrow \quad (x+7)(x-4) = 0 \quad \Leftrightarrow \quad x = -7$ or 4. However, $x = -7$ is not a solution since $\ln(-7+6)$ is not defined. So $x = 4$ is the only solution.

53. $e^{ax} = Ce^{bx} \quad \Leftrightarrow \ln e^{ax} = \ln(Ce^{bx}) \Leftrightarrow ax = \ln C + bx \Leftrightarrow (a-b)x = \ln C \Leftrightarrow x = \dfrac{\ln C}{a-b}$

55. $\ln(x-5) = 3 \quad \Rightarrow \quad x - 5 = e^3 \quad \Rightarrow \quad x = e^3 + 5 \approx 25.0855$

57. $e^{2-3x} = 20 \quad \Rightarrow \quad 2 - 3x = \ln 20 \quad \Rightarrow \quad x = \frac{1}{3}(2 - \ln 20) \approx -0.3319$

59. 3 ft = 36 in, so we need x such that $\log_2 x = 36 \quad \Leftrightarrow \quad x = 2^{36} = 68{,}719{,}476{,}736$. In miles, this is $68{,}719{,}476{,}736 \text{ in} \cdot \dfrac{1 \text{ ft}}{12 \text{ in}} \cdot \dfrac{1 \text{ mi}}{5280 \text{ ft}} \approx 1{,}084{,}587.7$ mi.

61. If I is the intensity of the 1989 San Francisco earthquake, then $\log_{10}(I/S) = 7.1 \quad \Rightarrow$ $\log_{10}(16I/S) = \log_{10}16 + \log_{10}(I/S) = \log_{10}16 + 7.1 \approx 8.3$.

63. $\lim\limits_{x \to 5^+} \ln(x-5) = -\infty$ since $x - 5 \to 0^+$ as $x \to 5^+$.

65. $\lim\limits_{x \to \infty} \log_2(x^2 - x) = \infty$ since $x^2 - x \to \infty$ as $x \to \infty$.

67. $\lim\limits_{x \to \pi/2^-} \log_{10}(\cos x) = -\infty$ since $\cos x \to 0^+$ as $x \to \frac{\pi}{2}^-$.

69. $\lim\limits_{x \to \infty} \ln(1 + e^{-x^2}) = \ln\left(1 + \lim\limits_{x \to \infty} e^{-x^2}\right) = \ln(1+0) = 0$

71. $f(x) = \log_{10}(1-x)$ Domain$(f) = \{x \mid 1 - x > 0\} = \{x \mid x < 1\} = (-\infty, 1)$. Range$(f) = \mathbb{R}$.

73. $F(t) = \sqrt{t}\ln(t^2 - 1)$ Domain$(F) = \{t \mid t \geq 0 \text{ and } t^2 - 1 > 0\} = \{t \mid t > 1\} = (1, \infty)$. Range$(F) = \mathbb{R}$.

75. $y = \ln(x+3) \quad \Rightarrow \quad e^y = e^{\ln(x+3)} = x + 3 \quad \Rightarrow \quad x = e^y - 3$.
Interchange x and y: the inverse function is $y = e^x - 3$.

77. $y = e^{\sqrt{x}} \quad \Rightarrow \quad \ln y = \ln e^{\sqrt{x}} = \sqrt{x} \quad \Rightarrow \quad x = (\ln y)^2$. Also note that $\sqrt{x} \geq 0 \quad \Rightarrow \quad y = e^{\sqrt{x}} \geq 1$.
Interchange x and y: the inverse function is $y = (\ln x)^2$, $x \geq 1$.

79. $y = \dfrac{10^x}{10^x + 1} \quad \Rightarrow \quad 10^x y + y = 10^x \quad \Rightarrow \quad 10^x(1 - y) = y \quad \Rightarrow \quad 10^x = \dfrac{y}{1-y} \quad \Rightarrow$ $x = \log_{10}\left(\dfrac{y}{1-y}\right)$. Interchange x and y: $y = \log_{10}\left(\dfrac{x}{1-x}\right)$ is the inverse function.

81. $y = e^x - 2e^{-x}$, so $y' = e^x + 2e^{-x}$, $y'' = e^x - 2e^{-x}$. $y'' > 0 \quad \Leftrightarrow \quad e^x - 2e^{-x} > 0 \quad \Leftrightarrow \quad e^x > 2e^{-x} \quad \Leftrightarrow$ $e^{2x} > 2 \quad \Leftrightarrow \quad 2x > \ln 2 \quad \Leftrightarrow \quad x > \frac{1}{2}\ln 2$. Therefore, y is concave upward on $\left(\frac{1}{2}\ln 2, \infty\right)$.

83. **(a)** We have to show that $-f(x) = f(-x)$.

$$-f(x) = -\ln\left(x + \sqrt{x^2+1}\right) = \ln\left[\left(x + \sqrt{x^2+1}\right)^{-1}\right] = \ln \frac{1}{x + \sqrt{x^2+1}}$$

$$= \ln\left(\frac{1}{x+\sqrt{x^2+1}} \cdot \frac{x - \sqrt{x^2+1}}{x - \sqrt{x^2+1}}\right) = \ln \frac{x - \sqrt{x^2+1}}{x^2 - x^2 - 1}$$

$= \ln\left(\sqrt{x^2+1} - x\right) = f(-x)$. Thus, f is an odd function.

(b) Let $y = \ln\left(x + \sqrt{x^2+1}\right)$, then $e^y = x + \sqrt{x^2+1} \quad \Leftrightarrow \quad (e^y - x)^2 = x^2 + 1 \quad \Leftrightarrow$

$e^{2y} - 2xe^y + x^2 = x^2 + 1 \quad \Leftrightarrow \quad 2xe^y = e^{2y} - 1 \quad \Leftrightarrow \quad x = \dfrac{e^{2y} - 1}{2e^y} = \frac{1}{2}(e^y - e^{-y})$. Thus, the inverse function is $f^{-1}(x) = \frac{1}{2}(e^x - e^{-x})$.

85. Let $x = \log_{10} 99$, $y = \log_9 82$. Then $10^x = 99 < 10^2 \quad \Rightarrow \quad x < 2$, and $9^y = 82 > 9^2 \quad \Rightarrow \quad y > 2$. Therefore $y = \log_9 82$ is larger.

87. **(a)** Let $\epsilon > 0$ be given. We need N such that $|a^x - 0| < \epsilon$ when $x < N$. But $a^x < \epsilon \quad \Leftrightarrow \quad x < \log_a \epsilon$. Let $N = \log_a \epsilon$. Then $x < N \quad \Rightarrow \quad x < \log_a \epsilon \quad \Rightarrow \quad |a^x - 0| = a^x < \epsilon$, so $\lim\limits_{x \to -\infty} a^x = 0$.

(b) Let $M > 0$ be given. We need N such that $a^x > M$ when $x > N$. But $a^x > M \quad \Leftrightarrow \quad x > \log_a M$. Let $N = \log_a M$. Then $x > N \quad \Rightarrow \quad x > \log_a M \quad \Rightarrow \quad a^x > M$, so $\lim\limits_{x \to \infty} a^x = \infty$.

89. $\ln(x^2 - 2x - 2) \le 0 \quad \Rightarrow \quad 0 < x^2 - 2x - 2 \le 1$. Now $x^2 - 2x - 2 \le 1$ gives $x^2 - 2x - 3 \le 0$ and hence $(x-3)(x+1) \le 0$. So $-1 \le x \le 3$. Now $0 < x^2 - 2x - 2 \quad \Rightarrow \quad x < 1 - \sqrt{3}$ or $x > 1 + \sqrt{3}$. Therefore $\ln(x^2 - 2x - 2) \le 0 \quad \Leftrightarrow \quad -1 \le x < 1 - \sqrt{3}$ or $1 + \sqrt{3} < x \le 3$.

EXERCISES 6.4

1. $f(x) = \ln(x+1) \quad \Rightarrow \quad f'(x) = 1/(x+1)$, $\text{Dom}(f) = \text{Dom}(f') = \{x \mid x + 1 > 0\}$
$= \{x \mid x > -1\} = (-1, \infty)$ [Note that, in general, $\text{Dom}(f') \subset \text{Dom}(f)$.]

3. $f(x) = x^2 \ln(1 - x^2) \quad \Rightarrow \quad f'(x) = 2x \ln(1 - x^2) + \dfrac{x^2(-2x)}{1 - x^2} = 2x \ln(1 - x^2) - \dfrac{2x^3}{1 - x^2}$,
$\text{Dom}(f) = \text{Dom}(f') = \{x \mid 1 - x^2 > 0\} = \{x \mid |x| < 1\} = (-1, 1)$

5. $f(x) = \log_3(x^2 - 4) \quad \Rightarrow \quad f'(x) = \dfrac{2x}{(x^2 - 4)\ln 3}$,
$\text{Dom}(f) = \text{Dom}(f') = \{x \mid x^2 - 4 > 0\} = \{x \mid |x| > 2\} = (-\infty, -2) \cup (2, \infty)$

7. $y = x \ln x \quad \Rightarrow \quad y' = \ln x + x(1/x) = \ln x + 1 \quad \Rightarrow \quad y'' = 1/x$

9. $y = \log_{10} x \quad \Rightarrow \quad y' = \dfrac{1}{x \ln 10} \quad \Rightarrow \quad y'' = -\dfrac{1}{x^2 \ln 10}$

11. $f(x) = \sqrt{x} \ln x \quad \Rightarrow \quad f'(x) = \dfrac{1}{2\sqrt{x}} \ln x + \sqrt{x}\left(\dfrac{1}{x}\right) = \dfrac{\ln x + 2}{2\sqrt{x}}$

13. $g(x) = \ln \dfrac{a-x}{a+x} = \ln(a-x) - \ln(a+x) \quad \Rightarrow \quad g'(x) = \dfrac{-1}{a-x} - \dfrac{1}{a+x} = \dfrac{-2a}{a^2-x^2}$

15. $F(x) = \ln\sqrt{x} = \frac{1}{2}\ln x \quad \Rightarrow \quad F'(x) = \dfrac{1}{2}\left(\dfrac{1}{x}\right) = \dfrac{1}{2x}$

17. $f(t) = \log_2(t^4 - t^2 + 1) \quad \Rightarrow \quad f'(t) = \dfrac{4t^3 - 2t}{(t^4 - t^2 + 1)\ln 2}$

19. $g(u) = \dfrac{1-\ln u}{1+\ln u} \quad \Rightarrow \quad g'(u) = \dfrac{(1+\ln u)(-1/u) - (1-\ln u)(1/u)}{(1+\ln u)^2} = -\dfrac{2}{u(1+\ln u)^2}$

21. $y = (\ln\sin x)^3 \quad \Rightarrow \quad y' = 3(\ln\sin x)^2 \dfrac{\cos x}{\sin x} = 3(\ln\sin x)^2 \cot x$

23. $y = \dfrac{\ln x}{1+x^2} \quad \Rightarrow \quad y' = \dfrac{(1+x^2)(1/x) - 2x\ln x}{(1+x^2)^2} = \dfrac{1+x^2-2x^2\ln x}{x(1+x^2)^2}$

25. $y = \ln|x^3 - x^2| \quad \Rightarrow \quad y' = \dfrac{1}{x^3-x^2}(3x^2-2x) = \dfrac{x(3x-2)}{x^2(x-1)} = \dfrac{3x-2}{x(x-1)}$

27. $F(x) = e^x \ln x \quad \Rightarrow \quad F'(x) = e^x\ln x + e^x\left(\dfrac{1}{x}\right) = e^x\left(\ln x + \dfrac{1}{x}\right)$

29. $f(t) = \pi^{-t} \quad \Rightarrow \quad f'(t) = \pi^{-t}(\ln\pi)(-1) = -\pi^{-t}\ln\pi$

31. $h(t) = t^3 - 3^t \quad \Rightarrow \quad h'(t) = 3t^2 - 3^t\ln 3$

33. $y = \ln[e^{-x}(1+x)] = \ln(e^{-x}) + \ln(1+x) = -x + \ln(1+x) \ \Rightarrow\ y' = -1 + \dfrac{1}{1+x} = -\dfrac{x}{1+x}$

35. $y = x^{\sin x} \ \Rightarrow\ \ln y = \sin x \ln x \ \Rightarrow\ \dfrac{y'}{y} = \cos x\ln x + \dfrac{\sin x}{x} \ \Rightarrow\ y' = x^{\sin x}\left[\cos x\ln x + \dfrac{\sin x}{x}\right]$

37. $y = x^{e^x} \quad \Rightarrow \quad \ln y = e^x\ln x \quad \Rightarrow \quad \dfrac{y'}{y} = e^x\ln x + \dfrac{e^x}{x} \quad \Rightarrow \quad y' = x^{e^x}e^x\left(\ln x + \dfrac{1}{x}\right)$

39. $y = (\ln x)^x \quad \Rightarrow \quad \ln y = x\ln\ln x \quad \Rightarrow \quad \dfrac{y'}{y} = \ln\ln x + x\cdot\dfrac{1}{\ln x}\cdot\dfrac{1}{x} \quad \Rightarrow \quad y' = (\ln x)^x\left(\ln\ln x + \dfrac{1}{\ln x}\right)$

41. $y = x^{1/\ln x} \quad \Rightarrow \quad \ln y = \left(\dfrac{1}{\ln x}\right)\ln x = 1 \quad \Rightarrow \quad y = e \quad \Rightarrow \quad y' = 0$

43. $y = \cos\left(x^{\sqrt{x}}\right) \quad \Rightarrow \quad y' = -\sin\left(x^{\sqrt{x}}\right)x^{\sqrt{x}}\left(\dfrac{\ln x + 2}{2\sqrt{x}}\right)$ by Example 16

45. $f(x) = \dfrac{x}{\ln x} \quad \Rightarrow \quad f'(x) = \dfrac{\ln x - x(1/x)}{(\ln x)^2} = \dfrac{\ln x - 1}{(\ln x)^2} \quad \Rightarrow \quad f'(e) = \dfrac{1-1}{1^2} = 0$

47. $f(x) = \sin x + \ln x \quad \Rightarrow$

$f'(x) = \cos x + \dfrac{1}{x}$

This is reasonable, because the graph shows that f increases when $f'(x)$ is positive.

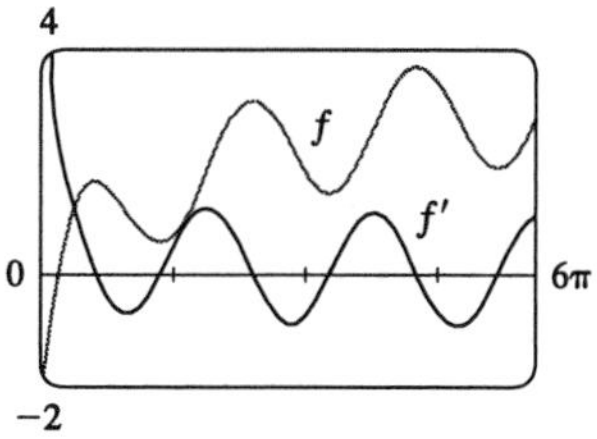

49. $y = f(x) = \ln\ln x \quad \Rightarrow \quad f'(x) = \dfrac{1}{\ln x}\left(\dfrac{1}{x}\right) \quad \Rightarrow \quad f'(e) = \dfrac{1}{e}$, so the equation of the tangent at $(e, 0)$ is

$y - 0 = \dfrac{1}{e}(x - e)$ or $x - ey = e$.

51. $y = \ln(x^2 + y^2) \;\Rightarrow\; y' = \dfrac{2x + 2yy'}{x^2 + y^2} \;\Rightarrow\; x^2y' + y^2y' = 2x + 2yy' \;\Rightarrow\; y' = \dfrac{2x}{x^2 + y^2 - 2y}$

53. $f(x) = \ln(x-1) \;\Rightarrow\; f'(x) = 1/(x-1) = (x-1)^{-1} \;\Rightarrow\; f''(x) = -(x-1)^{-2} \;\Rightarrow$
$f'''(x) = 2(x-1)^{-3} \;\Rightarrow\; f^{(4)}(x) = -2\cdot 3(x-1)^{-4} \;\Rightarrow\; \cdots \;\Rightarrow$
$f^{(n)}(x) = (-1)^{n-1}\cdot 2\cdot 3\cdot 4\cdot\cdots\cdot(n-1)(x-1)^{-n} = (-1)^{n-1}\dfrac{(n-1)!}{(x-1)^n}$

55.

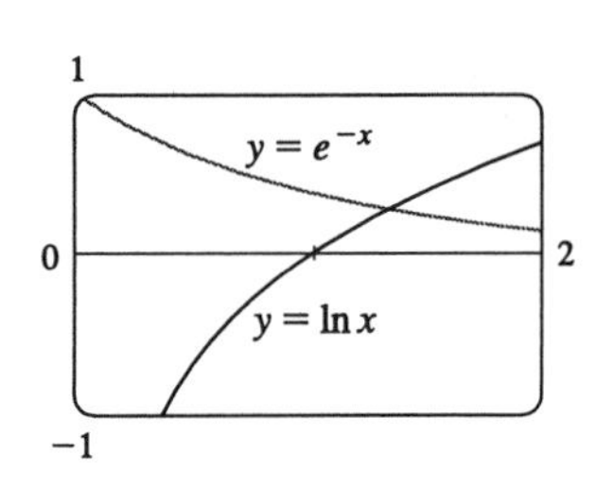

From the graph, it appears that the only root of the equation occurs at about $x = 1.3$. So we use Newton's Method with this as our initial approximation, and with $f(x) = \ln x - e^{-x} \;\Rightarrow\; f'(x) = 1/x + e^{-x}$. The formula is $x_{n+1} = x_n - f(x_n)/f'(x_n)$, and we calculate $x_1 = 1.3$, $x_2 \approx 1.309760$, $x_3 \approx x_4 \approx 1.309800$. So, correct to six decimal places, the root of the equation $\ln x = e^{-x}$ is $x = 1.309800$.

57. $f(x) = \dfrac{\ln x}{\sqrt{x}} \;\Rightarrow\; f'(x) = \dfrac{\sqrt{x}(1/x) - (\ln x)\left[1/(2\sqrt{x})\right]}{x} = \dfrac{2 - \ln x}{2x^{3/2}} \;\Rightarrow$
$f''(x) = \dfrac{2x^{3/2}(-1/x) - (2 - \ln x)(3x^{1/2})}{4x^3} = \dfrac{3\ln x - 8}{4x^{5/2}} > 0 \;\Leftrightarrow\; \ln x > \frac{8}{3} \;\Leftrightarrow\; x > e^{8/3}$, so f is CU on $(e^{8/3}, \infty)$ and CD on $(0, e^{8/3})$. The inflection point is $\left(e^{8/3}, \frac{8}{3}e^{-4/3}\right)$.

59. $y = f(x) = \ln(\cos x)$ **A.** $D = \{x \mid \cos x > 0\} = \left(-\frac{\pi}{2}, \frac{\pi}{2}\right) \cup \left(\frac{3\pi}{2}, \frac{5\pi}{2}\right) \cup \cdots$
$= \left\{x \mid 2n\pi - \frac{\pi}{2} < x < 2n\pi + \frac{\pi}{2}, n = 0, \pm1, \pm2, \ldots\right\}$ **B.** x-intercepts occur when $\ln(\cos x) = 0 \;\Leftrightarrow$ $\cos x = 1 \;\Leftrightarrow\; x = 2n\pi$, y-intercept $= f(0) = 0$. **C.** $f(-x) = f(x)$, so the curve is symmetric about the y-axis. $f(x + 2\pi) = f(x)$, f has period 2π, so in parts D-G we consider only $-\frac{\pi}{2} < x < \frac{\pi}{2}$.
D. $\lim\limits_{x\to\pi/2^-} \ln(\cos x) = -\infty$ and $\lim\limits_{x\to-\pi/2^+} \ln(\cos x) = -\infty$, so $x = \frac{\pi}{2}$ and $x = -\frac{\pi}{2}$ are VA. No HA.
E. $f'(x) = (1/\cos x)(-\sin x) = -\tan x > 0 \;\Leftrightarrow$ $-\frac{\pi}{2} < x < 0$, so f is increasing on $\left(-\frac{\pi}{2}, 0\right]$ and decreasing on $\left[0, \frac{\pi}{2}\right)$. **F.** $f(0) = 0$ is a local maximum.
G. $f''(x) = -\sec^2 x < 0 \;\Rightarrow\; f$ is CD on $\left(-\frac{\pi}{2}, \frac{\pi}{2}\right)$. No IP.

H.

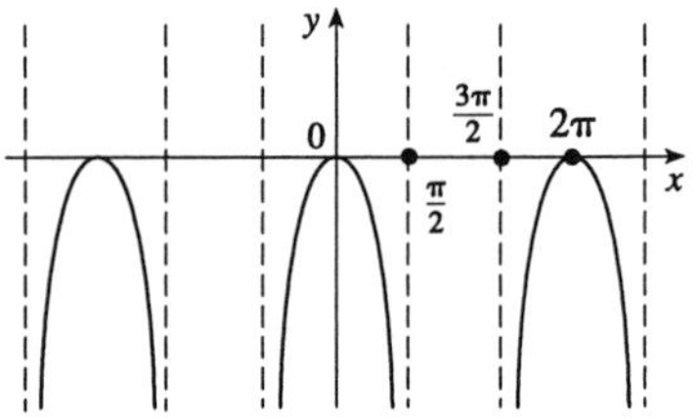

61. $y = f(x) = \ln(1 + x^2)$ **A.** $D = \mathbb{R}$ **B.** Both intercepts are 0. **C.** $f(-x) = f(x)$, so the curve is symmetric about the y-axis. **D.** $\lim\limits_{x\to\pm\infty} \ln(1 + x^2) = \infty$, no asymptotes. **E.** $f'(x) = \dfrac{2x}{1 + x^2} > 0 \;\Leftrightarrow\; x > 0$, so f is increasing on $[0, \infty)$ and decreasing on $(-\infty, 0]$.
F. $f(0) = 0$ is a local and absolute minimum.
G. $f''(x) = \dfrac{2(1 + x^2) - 2x(2x)}{(1 + x^2)^2} = \dfrac{2(1 - x^2)}{(1 + x^2)^2} > 0$
$\Leftrightarrow\; |x| < 1$, so f is CU on $(-1, 1)$, CD on $(-\infty, -1)$ and $(1, \infty)$. IP $(1, \ln 2)$ and $(-1, \ln 2)$.

H.

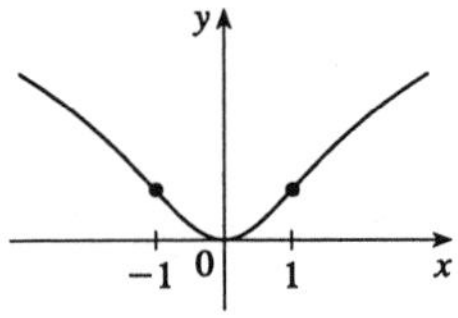

63. $y = f(x) = \ln(x^2 - x)$ **A.** $\{x \mid x^2 - x > 0\} = \{x \mid x < 0 \text{ or } x > 1\} = (-\infty, 0) \cup (1, \infty)$. **B.** x-intercepts occur when $x^2 - x = 1 \quad \Leftrightarrow \quad x^2 - x - 1 = 0 \quad \Leftrightarrow \quad x = \frac{1}{2}\left(1 \pm \sqrt{5}\right)$. No y-intercept **C.** No symmetry **D.** $\lim\limits_{x \to \infty} \ln(x^2 - x) = \infty$, no HA. $\lim\limits_{x \to 0^-} \ln(x^2 - x) = -\infty$, $\lim\limits_{x \to 1^+} \ln(x^2 - x) = -\infty$, so $x = 0$ and $x = 1$ are VA. **E.** $f'(x) = \dfrac{2x - 1}{x^2 - x} > 0$ when $x > 1$ and $f'(x) < 0$ when $x < 0$, so f is increasing on $(1, \infty)$ and decreasing on $(-\infty, 0)$. **F.** No extrema

G. $f''(x) = \dfrac{2(x^2 - x) - (2x - 1)^2}{(x^2 - x)^2} = \dfrac{-2x^2 + 2x - 1}{(x^2 - x)^2}$

$\Rightarrow \quad f''(x) < 0$ for all x since $-2x^2 + 2x - 1$ has a negative discriminant. So f is CD on $(-\infty, 0)$ and $(1, \infty)$. No IP.

H.

65.

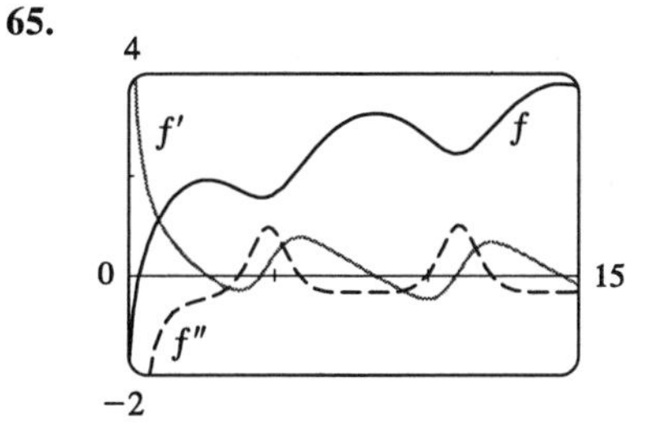

We use the CAS to calculate $f'(x) = \dfrac{2 + \sin x + x \cos x}{2x + x \sin x}$ and $f''(x) = \dfrac{2x^2 \sin x + 4 \sin x - \cos^2 x + x^2 + 5}{x^2(\cos^2 x - 4 \sin x - 5)}$. From the graphs, it seems that $f' > 0$ (and so f is increasing) on approximately the intervals $[0, 2.7]$, $[4.5, 8.2]$ and $[10.9, 14.3]$. It seems that f'' changes sign (indicating inflection points) at $x \approx 3.8$, 5.7, 10.0 and 12.0. Looking back at the graph of f, this implies that the inflection points have approximate coordinates $(3.8, 1.7)$, $(5.7, 2.1)$, $(10.0, 2.7)$, and $(12.0, 2.9)$.

67. $\displaystyle\int_4^8 \frac{1}{x}\, dx = \ln x\big|_4^8 = \ln 8 - \ln 4 = \ln \tfrac{8}{4} = \ln 2$

69. $\displaystyle\int_1^e \frac{x^2 + x + 1}{x}\, dx = \int_1^e \left(x + 1 + \frac{1}{x}\right) dx = \left[\tfrac{1}{2}x^2 + x + \ln x\right]_1^e = \left(\tfrac{1}{2}e^2 + e + 1\right) - \left(\tfrac{1}{2} + 1 + 0\right) = \tfrac{1}{2}e^2 + e - \tfrac{1}{2}$

71. Let $u = 2x - 1$. Then $du = 2\, dx$, so $\displaystyle\int \frac{dx}{2x - 1} = \frac{1}{2}\int \frac{du}{u} = \tfrac{1}{2}\ln|u| + C = \tfrac{1}{2}\ln|2x - 1| + C$.

73. Let $u = x^2 + 2x$. Then $du = 2(x + 1)dx$, so

$$\int \frac{x + 1}{x^2 + 2x}\, dx = \frac{1}{2}\int \frac{du}{u} = \tfrac{1}{2}\ln|u| + C = \tfrac{1}{2}\ln|x^2 + 2x| + C.$$

75. Let $u = \ln x$. Then $du = \dfrac{dx}{x} \quad \Rightarrow \quad \displaystyle\int \frac{(\ln x)^2}{x}\, dx = \int u^2\, du = \tfrac{1}{3}u^3 + C = \tfrac{1}{3}(\ln x)^3 + C$.

77. Let $u = 1 + \cos x$. Then $du = -\sin x\, dx$, so

$$\int \frac{\sin x}{1 + \cos x}\, dx = -\int \frac{du}{u} = -\ln|u| + C = -\ln(1 + \cos x) + C.$$

79. $\displaystyle\int_3^4 5^t\, dt = \left[\frac{5^t}{\ln 5}\right]_3^4 = \frac{5^4 - 5^3}{\ln 5} = \frac{500}{\ln 5}$

81. **(a)** $\dfrac{d}{dx}(\ln|\sin x| + C) = \dfrac{1}{\sin x}\cos x = \cot x$

(b) Let $u = \sin x$. Then $du = \cos x\, dx$, so $\displaystyle\int \cot x\, dx = \int \frac{\cos x}{\sin x}\, dx = \int \frac{du}{u} = \ln|u| + C = \ln|\sin x| + C$.

83. The cross-sectional area is $\pi\left(1/\sqrt{x+1}\right)^2 = \pi/(x+1)$. Therefore, the volume is
$$\int_0^1 \frac{\pi}{x+1}\,dx = \pi[\ln(x+1)]_0^1 = \pi\ln 2 - \ln 1 = \pi\ln 2$$

85. $y = (3x-7)^4(8x^2-1)^3 \quad\Rightarrow\quad \ln|y| = 4\ln|3x-7| + 3\ln|8x^2-1| \quad\Rightarrow\quad \dfrac{y'}{y} = \dfrac{12}{3x-7} + \dfrac{48x}{8x^2-1} \quad\Rightarrow$
$$y' = (3x-7)^4(8x^2-1)^3\left(\frac{12}{3x-7} + \frac{48x}{8x^2-1}\right)$$

87. $y = \dfrac{(x+1)^4(x-5)^3}{(x-3)^8} \quad\Rightarrow\quad \ln|y| = 4\ln|x+1| + 3\ln|x-5| - 8\ln|x-3| \quad\Rightarrow$
$$\frac{y'}{y} = \frac{4}{x+1} + \frac{3}{x-5} - \frac{8}{x-3} \quad\Rightarrow\quad y' = \frac{(x+1)^4(x-5)^3}{(x-3)^8}\left(\frac{4}{x+1} + \frac{3}{x-5} - \frac{8}{x-3}\right)$$

89. $y = \dfrac{e^x\sqrt{x^5+2}}{(x+1)^4(x^2+3)^2} \quad\Rightarrow\quad \ln y = x + \frac{1}{2}\ln(x^5+2) - 4\ln|x+1| - 2\ln(x^2+3) \quad\Rightarrow$
$$\frac{y'}{y} = 1 + \frac{5x^4}{2(x^5+2)} - \frac{4}{x+1} - \frac{4x}{x^2+3}. \text{ So } y' = \frac{e^x\sqrt{x^5+2}}{(x+1)^4(x^2+3)^2}\left[1 + \frac{5x^4}{2(x^5+2)} - \frac{4}{x+1} - \frac{4x}{x^2+3}\right].$$

91. The domain of $f(x) = 1/x$ is $(-\infty,0)\cup(0,\infty)$, so its general antiderivative is $F(x) = \begin{cases} \ln x + C_1 & \text{if } x > 0 \\ \ln|x| + C_2 & \text{if } x < 0 \end{cases}$

93. $f(x) = 2x + \ln x \quad\Rightarrow\quad f'(x) = 2 + 1/x$. If $g = f^{-1}$, then $f(1) = 2 \quad\Rightarrow\quad g(2) = 1$, so $g'(2) = 1/f'(g(2)) = 1/f'(1) = \frac{1}{3}$.

95. The curve and the line will determine a region when they intersect at two or more points. So we solve the equation $x/(x^2+1) = mx$ $\Rightarrow\quad x = 0$ or $mx^2 + m - 1 = 0 \quad\Rightarrow\quad x = 0$ or $x = \dfrac{\pm\sqrt{-4(m)(m-1)}}{2m} = \pm\sqrt{\dfrac{1}{m} - 1}$. Note that if $m = 1$, this has only the solution $x = 0$, and no region is determined. But if $1/m - 1 > 0 \quad\Leftrightarrow\quad 1/m > 1 \quad\Leftrightarrow\quad 0 < m < 1$, then there are two solutions. [Another way of seeing this is to observe that the slope of the tangent to $y = x/(x^2+1)$ at the origin is $y' = 1$ and therefore we must have $0 < m < 1$.] Note that we cannot just integrate between the positive and negative roots, since the curve and the line cross at the origin. Since mx and $x/(x^2+1)$ are both odd functions, the total area is twice the area between the curves on the interval $\left[0, \sqrt{1/m - 1}\right]$. So the total area enclosed is

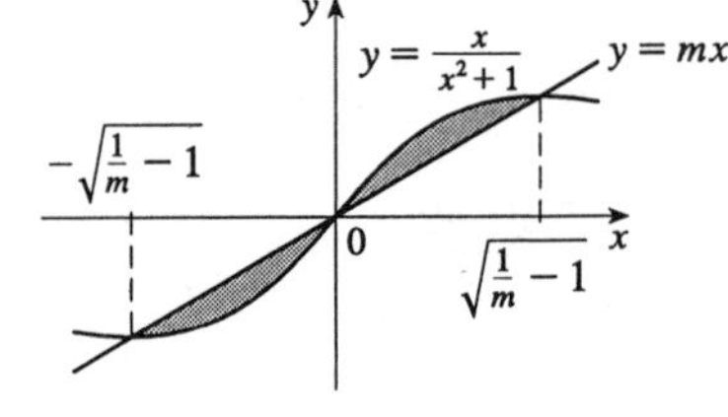

$$2\int_0^{\sqrt{1/m-1}}\left[\frac{x}{x^2+1} - mx\right]dx = 2\left[\tfrac{1}{2}\ln(x^2+1) - \tfrac{1}{2}mx^2\right]_0^{\sqrt{1/m-1}}$$
$$= \left[\ln\left(\frac{1}{m} - 1 + 1\right) - m\left(\frac{1}{m} - 1\right)\right] - (\ln 1 - 0) = \ln\left(\frac{1}{m}\right) + m - 1 = m - \ln m - 1.$$

97. If $f(x) = \ln(1+x)$, then $f'(x) = 1/(1+x)$, so $f'(0) = 1$. Thus
$$\lim_{x\to 0}\frac{\ln(1+x)}{x} = \lim_{x\to 0}\frac{f(x)}{x} = \lim_{x\to 0}\frac{f(x) - f(0)}{x - 0} = f'(0) = 1.$$

EXERCISES 6.2*

1. $\ln \dfrac{ab^2}{c} = \ln ab^2 - \ln c = \ln a + \ln b^2 - \ln c = \ln a + 2\ln b - \ln c$

3. $\ln \sqrt[3]{2xy} = \ln(2xy)^{1/3} = \frac{1}{3}\ln(2xy) = \frac{1}{3}(\ln 2 + \ln x + \ln y)$

5. $2\ln 4 - \ln 2 = \ln 4^2 - \ln 2 = \ln 16 - \ln 2 = \ln \frac{16}{2} = \ln 8$

Or: $2\ln 4 - \ln 2 = 2\ln 2^2 - \ln 2 = 4\ln 2 - \ln 2 = 3\ln 2$

7. $\frac{1}{3}\ln x - 4\ln(2x+3) = \ln(x^{1/3}) - \ln(2x+3)^4 = \ln\left[\sqrt[3]{x}/(2x+3)^4\right]$

9. $y = \ln x$ $\quad$ $y = -\ln x$ $\quad$ **11.** $y = \ln x$ $\quad$ $y = \ln(x+3)$

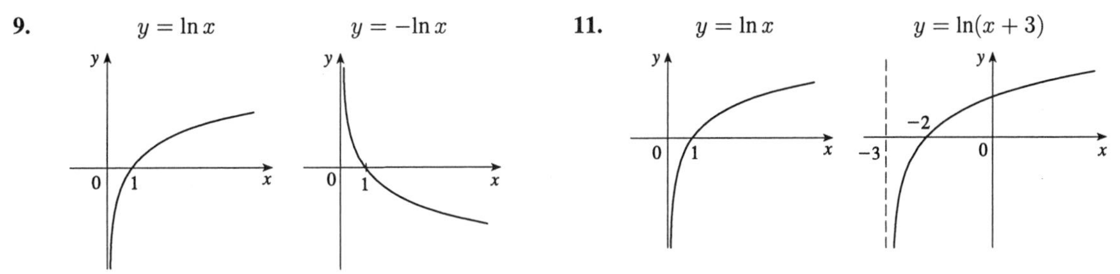

13. $f(x) = \ln(x+1) \quad\Rightarrow\quad f'(x) = 1/(x+1)$, $\text{Dom}(f) = \text{Dom}(f') = \{x \mid x+1 > 0\}$
$= \{x \mid x > -1\} = (-1, \infty)$. [Note that, in general, $\text{Dom}(f') \subset \text{Dom}(f)$.]

15. $f(x) = x^2\ln(1-x^2) \quad\Rightarrow\quad f'(x) = 2x\ln(1-x^2) + \dfrac{x^2(-2x)}{1-x^2} = 2x\ln(1-x^2) - \dfrac{2x^3}{1-x^2}$,
$\text{Dom}(f) = \text{Dom}(f') = \{x \mid 1-x^2 > 0\} = \{x \mid |x| < 1\} = (-1, 1)$

17. $y = x\ln x \quad\Rightarrow\quad y' = \ln x + x(1/x) = \ln x + 1 \quad\Rightarrow\quad y'' = 1/x$

19. $f(x) = \sqrt{x}\ln x \quad\Rightarrow\quad f'(x) = \dfrac{1}{2\sqrt{x}}\ln x + \sqrt{x}\left(\dfrac{1}{x}\right) = \dfrac{\ln x + 2}{2\sqrt{x}}$

21. $g(x) = \ln\dfrac{a-x}{a+x} = \ln(a-x) - \ln(a+x) \quad\Rightarrow\quad g'(x) = \dfrac{-1}{a-x} - \dfrac{1}{a+x} = \dfrac{-2a}{a^2-x^2}$

23. $F(x) = \ln\sqrt{x} = \frac{1}{2}\ln x \quad\Rightarrow\quad F'(x) = \dfrac{1}{2}\left(\dfrac{1}{x}\right) = \dfrac{1}{2x}$

25. $h(y) = \ln(y^3\sin y) = 3\ln y + \ln(\sin y) \quad\Rightarrow\quad h'(y) = \dfrac{3}{y} + \dfrac{1}{\sin y}(\cos y) = \dfrac{3}{y} + \cot y$

27. $g(u) = \dfrac{1-\ln u}{1+\ln u} \quad\Rightarrow\quad g'(u) = \dfrac{(1+\ln u)(-1/u) - (1-\ln u)(1/u)}{(1+\ln u)^2} = -\dfrac{2}{u(1+\ln u)^2}$

29. $y = (\ln\sin x)^3 \quad\Rightarrow\quad y' = 3(\ln\sin x)^2\dfrac{\cos x}{\sin x} = 3(\ln\sin x)^2\cot x$

31. $y = \dfrac{\ln x}{1+x^2} \quad\Rightarrow\quad y' = \dfrac{(1+x^2)(1/x) - 2x\ln x}{(1+x^2)^2} = \dfrac{1+x^2-2x^2\ln x}{x(1+x^2)^2}$

33. $y = \ln\left(\dfrac{x+1}{x-1}\right)^{3/5} = \frac{3}{5}[\ln(x+1) - \ln(x-1)] \quad\Rightarrow\quad y' = \dfrac{3}{5}\left(\dfrac{1}{x+1} - \dfrac{1}{x-1}\right) = \dfrac{-6}{5(x^2-1)}$

35. $y = \ln|x^3 - x^2| \quad \Rightarrow \quad y' = \dfrac{1}{x^3 - x^2}(3x^2 - 2x) = \dfrac{x(3x-2)}{x^2(x-1)} = \dfrac{3x-2}{x(x-1)}$

37. $f(x) = \dfrac{x}{\ln x} \quad \Rightarrow \quad f'(x) = \dfrac{\ln x - x(1/x)}{(\ln x)^2} = \dfrac{\ln x - 1}{(\ln x)^2} \quad \Rightarrow \quad f'(e) = \dfrac{1-1}{1^2} = 0$

39. $f(x) = \sin x + \ln x \quad \Rightarrow$
$f'(x) = \cos x + \dfrac{1}{x}$
This is reasonable, because the graph shows that f increases when $f'(x)$ is positive.

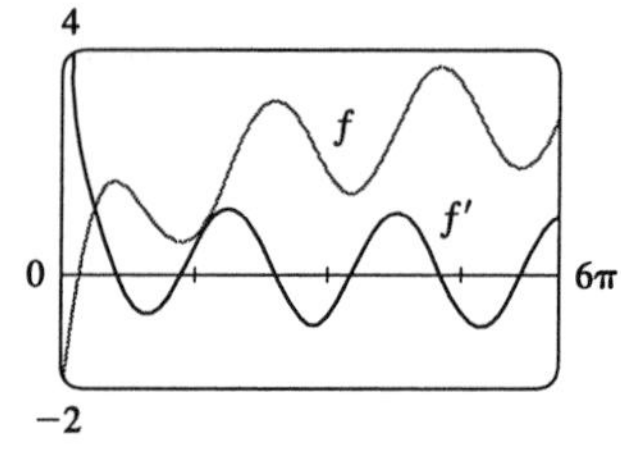

41. $y = f(x) = \ln \ln x \quad \Rightarrow \quad f'(x) = \dfrac{1}{\ln x}\left(\dfrac{1}{x}\right) \quad \Rightarrow \quad f'(e) = \dfrac{1}{e}$, so the equation of the tangent at $(e, 0)$ is $y - 0 = \dfrac{1}{e}(x - e)$ or $x - ey = e$.

43. $y = \ln(x^2 + y^2) \;\Rightarrow\; y' = \dfrac{2x + 2yy'}{x^2 + y^2} \;\Rightarrow\; x^2y' + y^2y' = 2x + 2yy' \;\Rightarrow\; y' = \dfrac{2x}{x^2 + y^2 - 2y}$

45. $f(x) = \ln(x-1) \quad \Rightarrow \quad f'(x) = 1/(x-1) = (x-1)^{-1} \quad \Rightarrow \quad f''(x) = -(x-1)^{-2} \quad \Rightarrow$
$f'''(x) = 2(x-1)^{-3} \quad \Rightarrow \quad f^{(4)}(x) = -2 \cdot 3(x-1)^{-4} \quad \Rightarrow \quad \cdots \quad \Rightarrow$
$f^{(n)}(x) = (-1)^{n-1} \cdot 2 \cdot 3 \cdot 4 \cdot \cdots \cdot (n-1)(x-1)^{-n} = (-1)^{n-1}\dfrac{(n-1)!}{(x-1)^n}$

47.

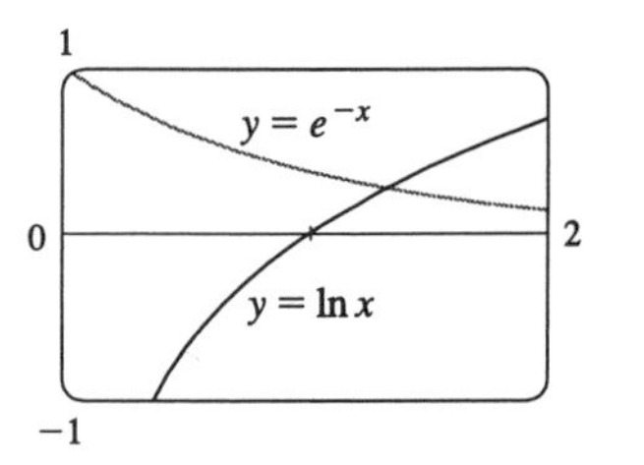

From the graph, it appears that the only root of the equation occurs at about $x = 1.3$. So we use Newton's Method with this as our initial approximation, and with $f(x) = \ln x - e^{-x} \quad \Rightarrow$ $f'(x) = 1/x + e^{-x}$. The formula is $x_{n+1} = x_n - f(x_n)/f'(x_n)$, and we calculate $x_1 = 1.3$, $x_2 \approx 1.309760$, $x_3 \approx x_4 \approx 1.309800$. So, correct to six decimal places, the root of the equation $\ln x = e^{-x}$ is $x = 1.309800$.

49. $y = f(x) = \ln(\cos x)$ **A.** $D = \{x \mid \cos x > 0\} = \left(-\frac{\pi}{2}, \frac{\pi}{2}\right) \cup \left(\frac{3\pi}{2}, \frac{5\pi}{2}\right) \cup \cdots$
$= \left\{x \mid 2n\pi - \frac{\pi}{2} < x < 2n\pi + \frac{\pi}{2}, n = 0, \pm1, \pm2, \ldots\right\}$ **B.** x-intercepts occur when $\ln(\cos x) = 0 \quad \Leftrightarrow$ $\cos x = 1 \quad \Leftrightarrow \quad x = 2n\pi$, y-intercept $= f(0) = 0$. **C.** $f(-x) = f(x)$, so the curve is symmetric about the y-axis. $f(x + 2\pi) = f(x)$, f has period 2π, so in parts D-G we consider only $-\frac{\pi}{2} < x < \frac{\pi}{2}$.

D. $\lim\limits_{x \to \pi/2^-} \ln(\cos x) = -\infty$ and $\lim\limits_{x \to -\pi/2^+} \ln(\cos x) = -\infty$, so $x = \frac{\pi}{2}$ and $x = -\frac{\pi}{2}$ are VA. No HA.
E. $f'(x) = (1/\cos x)(-\sin x) = -\tan x > 0 \quad \Leftrightarrow$ $-\frac{\pi}{2} < x < 0$, so f is increasing on $\left(-\frac{\pi}{2}, 0\right]$ and decreasing on $\left[0, \frac{\pi}{2}\right)$. **F.** $f(0) = 0$ is a local maximum.
G. $f''(x) = -\sec^2 x < 0 \quad \Rightarrow \quad f$ is CD on $\left(-\frac{\pi}{2}, \frac{\pi}{2}\right)$. No IP.

H.

y
0
$\frac{3\pi}{2}$
2π
$\frac{\pi}{2}$
x

51. $y = f(x) = \ln\left(x + \sqrt{1+x^2}\right)$ **A.** $x + \sqrt{1+x^2} > 0$ for all x since $1 + x^2 > x^2 \Rightarrow \sqrt{1+x^2} > |x|$, so $D = \mathbb{R}$. **B.** y-intercept $= f(0) = 0$, x–intercept occurs when $x + \sqrt{1+x^2} = 1 \Rightarrow \sqrt{1+x^2} = 1 - x \Rightarrow 1 + x^2 = 1 - 2x + x^2 \Rightarrow x = 0$. **C.** $\ln\left(-x + \sqrt{1+x^2}\right) = -\ln\left(x + \sqrt{1+x^2}\right)$ since $\left(\sqrt{1+x^2} - x\right)\left(\sqrt{1+x^2} + x\right) = 1$, so the curve is symmetric about the origin.

D. $\lim\limits_{x\to\infty} \ln\left(x + \sqrt{1+x^2}\right) = \infty$, $\lim\limits_{x\to-\infty} \ln\left(x + \sqrt{1+x^2}\right) = \lim\limits_{x\to-\infty} \ln \dfrac{1}{\sqrt{1+x^2} - x} = -\infty$, no HA.

E. $f'(x) = \dfrac{1}{x + \sqrt{1+x^2}}\left(1 + \dfrac{x}{\sqrt{1+x^2}}\right) = \dfrac{1}{\sqrt{1+x^2}} > 0$, so f is increasing on $(-\infty, \infty)$.

F. No extrema.

G. $f''(x) = -\dfrac{x}{(1+x^2)^{3/2}}$ $\Rightarrow f''(x) > 0 \Leftrightarrow x < 0$, so f is CU on $(-\infty, 0)$ and CD on $(0, \infty)$, and there is an IP at $(0, 0)$.

H.

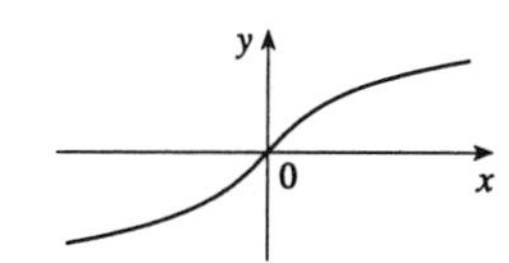

53. $y = f(x) = \ln(1 + x^2)$ **A.** $D = \mathbb{R}$ **B.** Both intercepts are 0. **C.** $f(-x) = f(x)$, so the curve is symmetric about the y-axis. **D.** $\lim\limits_{x\to\pm\infty} \ln(1 + x^2) = \infty$, no asymptotes. **E.** $f'(x) = \dfrac{2x}{1+x^2} > 0 \Leftrightarrow x > 0$, so f is increasing on $[0, \infty)$ and decreasing on $(-\infty, 0]$.

F. $f(0) = 0$ is a local and absolute minimum.

G. $f''(x) = \dfrac{2(1+x^2) - 2x(2x)}{(1+x^2)^2} = \dfrac{2(1-x^2)}{(1+x^2)^2} > 0$ $\Leftrightarrow |x| < 1$, so f is CU on $(-1, 1)$, CD on $(-\infty, -1)$ and $(1, \infty)$. IP $(1, \ln 2)$ and $(-1, \ln 2)$.

H.

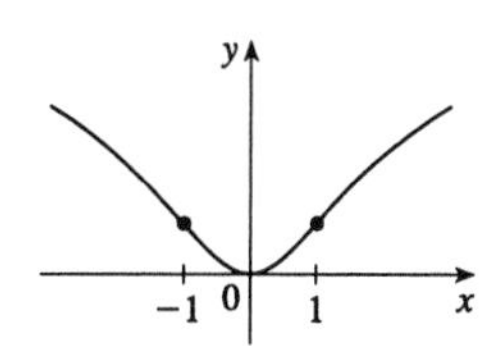

55.

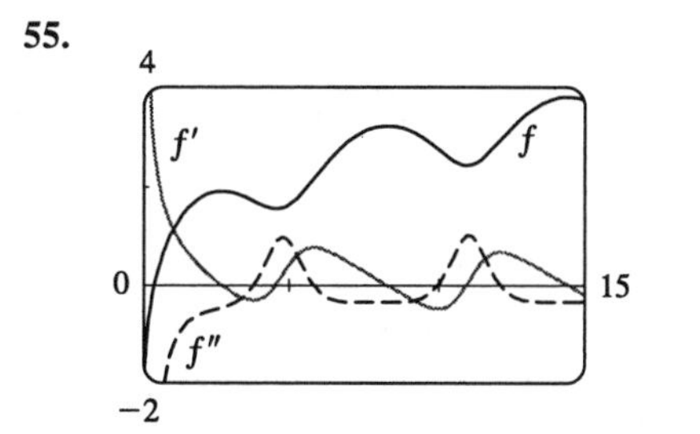

We use the CAS to calculate $f'(x) = \dfrac{2 + \sin x + x\cos x}{2x + x\sin x}$ and $f''(x) = \dfrac{2x^2\sin x + 4\sin x - \cos^2 x + x^2 + 5}{x^2(\cos^2 x - 4\sin x - 5)}$. From the graphs, it seems that $f' > 0$ (and so f is increasing) on approximately the intervals $[0, 2.7]$, $[4.5, 8.2]$ and $[10.9, 14.3]$. It seems that f'' changes sign (indicating inflection points) at $x \approx 3.8$, 5.7, 10.0 and 12.0. Looking back at the graph of f, this implies that the inflection points have approximate coordinates $(3.8, 1.7)$, $(5.7, 2.1)$, $(10.0, 2.7)$, and $(12.0, 2.9)$.

57. $\displaystyle\int_4^8 \frac{1}{x}\,dx = \ln x\Big|_4^8 = \ln 8 - \ln 4 = \ln\tfrac{8}{4} = \ln 2$

59. $\displaystyle\int_1^e \frac{x^2 + x + 1}{x}\,dx = \int_1^e \left(x + 1 + \frac{1}{x}\right)dx = \left[\tfrac{1}{2}x^2 + x + \ln x\right]_1^e = \left(\tfrac{1}{2}e^2 + e + 1\right) - \left(\tfrac{1}{2} + 1 + 0\right) = \tfrac{1}{2}e^2 + e - \tfrac{1}{2}$

61. Let $u = 2x - 1$. Then $du = 2\,dx$, so $\displaystyle\int \frac{dx}{2x - 1} = \frac{1}{2}\int \frac{du}{u} = \tfrac{1}{2}\ln|u| + C = \tfrac{1}{2}\ln|2x - 1| + C$.

63. Let $u = x^2 + 2x$. Then $du = 2(x+1)dx$, so
$$\int \frac{x+1}{x^2+2x}\,dx = \tfrac{1}{2}\int \frac{du}{u} = \tfrac{1}{2}\ln|u| + C = \tfrac{1}{2}\ln|x^2+2x| + C.$$

65. Let $u = \ln x$. Then $du = \dfrac{dx}{x}$ $\Rightarrow$ $\displaystyle\int \frac{(\ln x)^2}{x}\,dx = \int u^2\,du = \tfrac{1}{3}u^3 + C = \tfrac{1}{3}(\ln x)^3 + C.$

67. Let $u = 1 + \cos x$. Then $du = -\sin x\,dx$, so
$$\int \frac{\sin x}{1+\cos x}\,dx = -\int \frac{du}{u} = -\ln|u| + C = -\ln(1+\cos x) + C.$$

69. **(a)** $\dfrac{d}{dx}(\ln|\sin x| + C) = \dfrac{1}{\sin x}\cos x = \cot x$

(b) Let $u = \sin x$. Then $du = \cos x\,dx$, so $\displaystyle\int \cot x\,dx = \int \frac{\cos x}{\sin x}\,dx = \int \frac{du}{u} = \ln|u| + C = \ln|\sin x| + C.$

71. The cross-sectional area is $\pi\left(1/\sqrt{x+1}\right)^2 = \pi/(x+1)$. Therefore, the volume is
$$\int_0^1 \frac{\pi}{x+1}\,dx = \pi[\ln(x+1)]_0^1 = \pi \ln 2 - \ln 1 = \pi\ln 2$$

73. $y = (3x-7)^4(8x^2-1)^3$ $\Rightarrow$ $\ln|y| = 4\ln|3x-7| + 3\ln|8x^2-1|$ $\Rightarrow$ $\dfrac{y'}{y} = \dfrac{12}{3x-7} + \dfrac{48x}{8x^2-1}$ $\Rightarrow$
$$y' = (3x-7)^4(8x^2-1)^3\left(\frac{12}{3x-7} + \frac{48x}{8x^2-1}\right)$$

75. $y = \sqrt{\dfrac{x^2+1}{x+1}}$ $\Rightarrow$ $\ln y = \tfrac{1}{2}[\ln(x^2+1) - \ln(x+1)]$ $\Rightarrow$ $\dfrac{y'}{y} = \dfrac{1}{2}\left(\dfrac{2x}{x^2+1} - \dfrac{1}{x+1}\right)$ $\Rightarrow$
$$y' = \sqrt{\frac{x^2+1}{x+1}}\left[\frac{x}{x^2+1} - \frac{1}{2(x+1)}\right]$$

77. The domain of $f(x) = \dfrac{1}{x}$ is $(-\infty, 0) \cup (0, \infty)$, so its general antiderivative is $F(x) = \begin{cases} \ln x + C_1 & \text{if } x > 0 \\ \ln|x| + C_2 & \text{if } x < 0. \end{cases}$

79. $f(x) = 2x + \ln x$ $\Rightarrow$ $f'(x) = 2 + 1/x$. If $g = f^{-1}$, then $f(1) = 2$ $\Rightarrow$ $g(2) = 1$, so $g'(2) = 1/f'(g(2)) = 1/f'(1) = \tfrac{1}{3}$.

81. **(a)** We interpret $\ln 1.5$ as the area under the curve $y = 1/x$ from $x = 1$ to $x = 1.5$. The area of the rectangle $BCDE$ is $\tfrac{1}{2}\cdot\tfrac{2}{3} = \tfrac{1}{3}$. The area of the trapezoid $ABCD$ is $\tfrac{1}{2}\cdot\tfrac{1}{2}\left(1 + \tfrac{2}{3}\right) = \tfrac{5}{12}$. Thus, by comparing areas, we observe that $\tfrac{1}{3} < \ln 1.5 < \tfrac{5}{12}$.

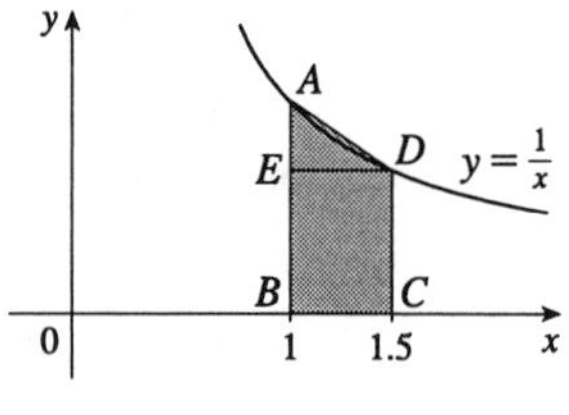

(b) With $f(t) = 1/t$, $n = 10$, and $\Delta x = 0.05$, we have
$$\ln 1.5 = \int_1^{1.5}(1/t)dt \approx (0.05)[f(1.025) + f(1.075) + \cdots + f(1.475)]$$
$$= (0.05)\left[\frac{1}{1.025} + \frac{1}{1.075} + \cdots + \frac{1}{1.475}\right] \approx 0.4054.$$

83.

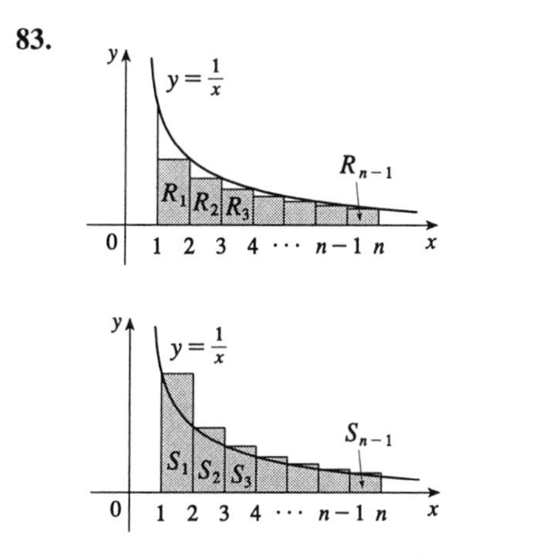

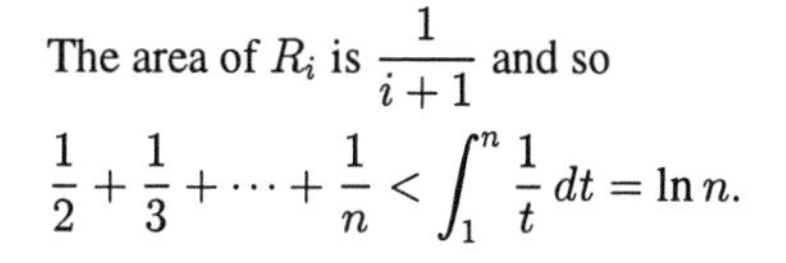

The area of R_i is $\dfrac{1}{i+1}$ and so

$$\frac{1}{2}+\frac{1}{3}+\cdots+\frac{1}{n}<\int_1^n \frac{1}{t}\,dt=\ln n.$$

The area of S_i is $\dfrac{1}{i}$ and so

$$1+\frac{1}{2}+\cdots+\frac{1}{n-1}>\int_1^n \frac{1}{t}\,dt=\ln n.$$

85. The curve and the line will determine a region when they intersect at two or more points.

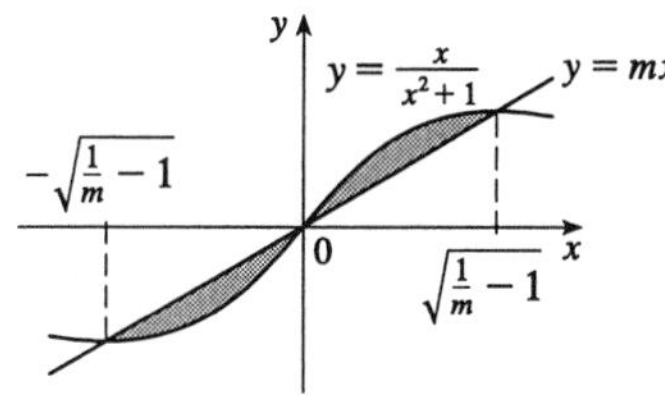

We solve the equation $\dfrac{x}{x^2+1}=mx \quad\Rightarrow\quad x=0$ or $mx^2+m-1=0 \quad\Rightarrow\quad x=0$ or $x=\dfrac{\pm\sqrt{-4(m)(m-1)}}{2m}=\pm\sqrt{\dfrac{1}{m}-1}$. Note that if $m=1$, this has only the solution $x=0$, and no region is determined. But if $\dfrac{1}{m}-1>0 \quad\Leftrightarrow\quad \dfrac{1}{m}>1 \quad\Leftrightarrow\quad 0<m<1$, then there are two solutions. [Another way of seeing this is to observe that the slope of the tangent to $y=\dfrac{x}{x^2+1}$ at the origin is $y'=1$ and therefore we must have $0<m<1$.] Note that we cannot just integrate between the positive and negative roots, since the curve and the line cross at the origin. Since mx and $\dfrac{x}{x^2+1}$ are both odd functions, the total area is twice the area between the curves on the interval $\left[0,\sqrt{\dfrac{1}{m}-1}\right]$. So the total area enclosed is

$$2\int_0^{\sqrt{1/m-1}}\left[\frac{x}{x^2+1}-mx\right]dx=2\left[\tfrac{1}{2}\ln(x^2+1)-\tfrac{1}{2}mx^2\right]_0^{\sqrt{1/m-1}}$$
$$=\left[\ln\left(\frac{1}{m}-1+1\right)-m\left(\frac{1}{m}-1\right)\right]-(\ln 1-0)=\ln\left(\frac{1}{m}\right)+m-1=m-\ln m-1.$$

87. If $f(x)=\ln(1+x)$, then $f'(x)=\dfrac{1}{1+x}$, so $f'(0)=1$. Thus

$$\lim_{x\to 0}\frac{\ln(1+x)}{x}=\lim_{x\to 0}\frac{f(x)}{x}=\lim_{x\to 0}\frac{f(x)-f(0)}{x-0}=f'(0)=1.$$

EXERCISES 6.3*

1. $\ln e^{\sqrt{2}} = \sqrt{2}$

3. $e^{3\ln 2} = \left(e^{\ln 2}\right)^3 = 2^3 = 8$

5. $\ln e^{\sin x} = \sin x$

7. $e^x = 16 \quad \Leftrightarrow \quad \ln e^x = \ln 16 \quad \Leftrightarrow \quad x = \ln 16 \quad \Leftrightarrow \quad x = 4\ln 2$

9. $\ln(2x-1) = 3 \quad \Leftrightarrow \quad e^{\ln(2x-1)} = e^3 \quad \Leftrightarrow \quad 2x - 1 = e^3 \quad \Leftrightarrow \quad x = \frac{1}{2}(e^3+1)$

11. $\ln(\ln x) = 1 \quad \Leftrightarrow \quad e^{\ln(\ln x)} = e^1 \quad \Leftrightarrow \quad \ln x = e^1 = e \quad \Leftrightarrow \quad e^{\ln x} = e^e \quad \Leftrightarrow \quad x = e^e$

13. $e^{ax} = Ce^{bx} \Leftrightarrow \ln e^{ax} = \ln\left(Ce^{bx}\right) \Leftrightarrow ax = \ln C + bx \Leftrightarrow (a-b)x = \ln C \Leftrightarrow x = \dfrac{\ln C}{a-b}$

15. $\ln(x-5) = 3 \quad \Rightarrow \quad x - 5 = e^3 \quad \Rightarrow \quad x = e^3 + 5 \approx 25.0855$

17. $y = e^x$ $\qquad$ $y = e^{-x}$ $\qquad$ **19.** $y = -e^x$ $\qquad$ $y = 3 - e^x$

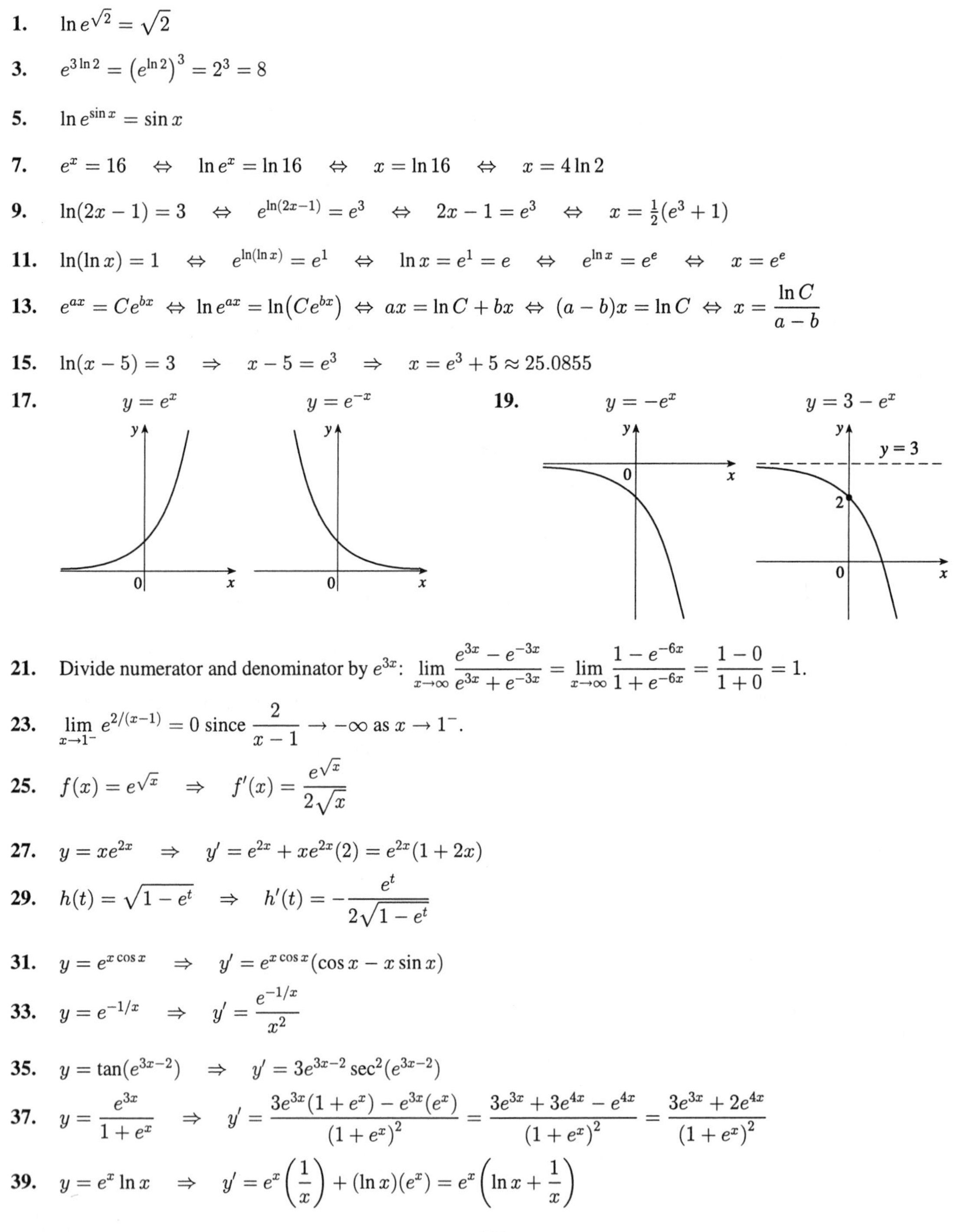

21. Divide numerator and denominator by e^{3x}: $\displaystyle\lim_{x\to\infty} \frac{e^{3x} - e^{-3x}}{e^{3x} + e^{-3x}} = \lim_{x\to\infty} \frac{1 - e^{-6x}}{1 + e^{-6x}} = \frac{1-0}{1+0} = 1.$

23. $\displaystyle\lim_{x\to 1^-} e^{2/(x-1)} = 0$ since $\dfrac{2}{x-1} \to -\infty$ as $x \to 1^-$.

25. $f(x) = e^{\sqrt{x}} \quad \Rightarrow \quad f'(x) = \dfrac{e^{\sqrt{x}}}{2\sqrt{x}}$

27. $y = xe^{2x} \quad \Rightarrow \quad y' = e^{2x} + xe^{2x}(2) = e^{2x}(1+2x)$

29. $h(t) = \sqrt{1-e^t} \quad \Rightarrow \quad h'(t) = -\dfrac{e^t}{2\sqrt{1-e^t}}$

31. $y = e^{x\cos x} \quad \Rightarrow \quad y' = e^{x\cos x}(\cos x - x\sin x)$

33. $y = e^{-1/x} \quad \Rightarrow \quad y' = \dfrac{e^{-1/x}}{x^2}$

35. $y = \tan(e^{3x-2}) \quad \Rightarrow \quad y' = 3e^{3x-2}\sec^2(e^{3x-2})$

37. $y = \dfrac{e^{3x}}{1+e^x} \quad \Rightarrow \quad y' = \dfrac{3e^{3x}(1+e^x) - e^{3x}(e^x)}{(1+e^x)^2} = \dfrac{3e^{3x} + 3e^{4x} - e^{4x}}{(1+e^x)^2} = \dfrac{3e^{3x} + 2e^{4x}}{(1+e^x)^2}$

39. $y = e^x \ln x \quad \Rightarrow \quad y' = e^x\left(\dfrac{1}{x}\right) + (\ln x)(e^x) = e^x\left(\ln x + \dfrac{1}{x}\right)$

41. $y = f(x) = e^{-x}\sin x \quad\Rightarrow\quad f'(x) = -e^{-x}\sin x + e^{-x}\cos x \quad\Rightarrow\quad f'(\pi) = e^{-\pi}(\cos\pi - \sin\pi) = -e^{-\pi}$, so the equation of the tangent at $(\pi, 0)$ is $y - 0 = -e^{-\pi}(x - \pi)$ or $x + e^{\pi}y = \pi$.

43. $\cos(x - y) = xe^x \quad\Rightarrow\quad -\sin(x-y)(1 - y') = e^x + xe^x \quad\Rightarrow\quad y' = 1 + \dfrac{e^x(1+x)}{\sin(x-y)}$

45. $y = e^{rx} \quad\Rightarrow\quad y' = re^{rx} \quad\Rightarrow\quad y'' = r^2e^{rx}$, so $y'' + 5y' - 6y = r^2e^{rx} + 5re^{rx} - 6e^{rx}$
$= e^{rx}(r^2 + 5r - 6) = e^{rx}(r+6)(r-1) = 0 \quad\Rightarrow\quad (r+6)(r-1) = 0 \quad\Rightarrow\quad r = 1$ or -6.

47. $f(x) = e^{-2x} \quad\Rightarrow\quad f'(x) = -2e^{-2x} \quad\Rightarrow\quad f''(x) = (-2)^2e^{-2x} \quad\Rightarrow\quad f'''(x) = (-2)^3e^{-2x} \quad\Rightarrow\quad \cdots$
$\Rightarrow\quad f^{(8)}(x) = (-2)^8e^{-2x} = 256e^{-2x}$

49. **(a)** $f(x) = e^x + x$ is continuous on $\mathbb{R}$ and $f(-1) = e^{-1} - 1 < 0 < 1 = f(0)$, so by the Intermediate Value Theorem, $e^x + x = 0$ has a root in $(-1, 0)$.

(b) $f(x) = e^x + x \quad\Rightarrow\quad f'(x) = e^x + 1$, so $x_{n+1} = x_n - \dfrac{e^{x_n} + x_n}{e^{x_n} + 1}$. From Exercise 31 we know that there is a root between -1 and 0, so we take $x_1 = -0.5$. Then $x_2 \approx -0.566311$, $x_3 \approx -0.567143$, and $x_4 \approx -0.567143$, so the root is -0.567143 to six decimal places.

51. **(a)** $\lim\limits_{t\to\infty} p(t) = \lim\limits_{t\to\infty} \dfrac{1}{1 + ae^{-kt}} = \dfrac{1}{1 + a\cdot 0} = 1$, since $k > 0 \quad\Rightarrow\quad -kt \to -\infty$.

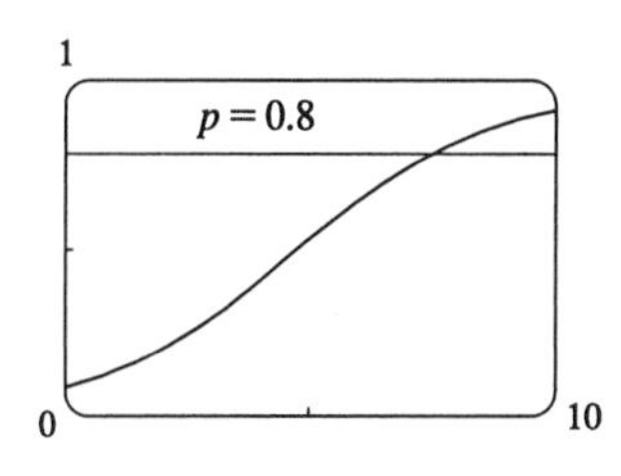

(b) $\dfrac{dp}{dt} = -(1 + ae^{-kt})^{-2}(-kae^{-kt}) = \dfrac{kae^{-kt}}{(1 + ae^{-kt})^2}$

(c) From the graph, it seems that $p(t) = 0.8$ (indicating that 80% of the population has heard the rumor) when $t \approx 7.4$ hours.

53. $x - e^x \quad\Rightarrow\quad f'(x) = 1 - e^x = 0 \quad\Leftrightarrow\quad e^x = 1 \quad\Leftrightarrow\quad x = 0$. Now $f'(x) > 0$ for all $x < 0$ and $f'(x) < 0$ for all $x > 0$, so the absolute maximum value is $f(0) = 0 - 1 = -1$.

55. $y = e^x - 2e^{-x}$, so $y' = e^x + 2e^{-x}$, $y'' = e^x - 2e^{-x}$. $y'' > 0 \quad\Leftrightarrow\quad e^x - 2e^{-x} > 0 \quad\Leftrightarrow\quad e^x > 2e^{-x} \quad\Leftrightarrow\quad e^{2x} > 2 \quad\Leftrightarrow\quad 2x > \ln 2 \quad\Leftrightarrow\quad x > \frac{1}{2}\ln 2$. Therefore, y is concave upward on $\left(\frac{1}{2}\ln 2, \infty\right)$.

57. $y = f(x) = e^{-1/(x+1)}$ **A.** $D = \{x \mid x \neq -1\} = (-\infty, -1) \cup (-1, \infty)$ **B.** No x-intercept; y-intercept $= f(0) = e^{-1}$ **C.** No symmetry **D.** $\lim\limits_{x\to\pm\infty} e^{-1/(x+1)} = 1$ since $-1/(x+1) \to 0$, so $y = 1$ is a HA. $\lim\limits_{x\to -1^+} e^{-1/(x+1)} = 0$ since $-1/(x+1) \to -\infty$, $\lim\limits_{x\to -1^-} e^{-1/(x+1)} = \infty$ since $-1/(x+1) \to \infty$, so $x = -1$ is a VA. **E.** $f'(x) = e^{-1/(x+1)}/(x+1)^2 \quad\Rightarrow\quad f'(x) > 0$ for all x except 1, so f is increasing on $(-\infty, -1)$ and $(-1, \infty)$. **F.** No extrema

G. $f''(x) = \dfrac{e^{-1/(x+1)}}{(x+1)^4} + \dfrac{e^{-1/(x+1)}(-2)}{(x+1)^3} = -\dfrac{e^{-1/(x+1)}(2x+1)}{(x+1)^4} \quad\Rightarrow$

$f''(x) > 0 \quad\Leftrightarrow\quad 2x + 1 < 0 \quad\Leftrightarrow\quad x < -\frac{1}{2}$, so f is CU on $(-\infty, -1)$ and $\left(-1, -\frac{1}{2}\right)$, and CD on $\left(-\frac{1}{2}, \infty\right)$. IP at $\left(-\frac{1}{2}, e^{-2}\right)$.

H.

59. $y = 1/(1+e^{-x})$ **A.** $D = \mathbb{R}$ **B.** No x-intercepts; y-intercept $= f(0) = \frac{1}{2}$. **C.** No symmetry **D.** $\lim\limits_{x\to\infty} 1/(1+e^{-x}) = \frac{1}{1+0} = 1$ and $\lim\limits_{x\to-\infty} 1/(1+e^{-x}) = 0$ $\left(\text{since } \lim\limits_{x\to-\infty} e^{-x} = \infty\right)$, so f has horizontal asymptotes $y = 0$ and $y = 1$. **E.** $f'(x) = -(1+e^{-x})^{-2}(-e^{-x}) = e^{-x}/(1+e^{-x})^2$. This is positive for all x, so f is increasing on $\mathbb{R}$. **F.** No extrema

G. $f''(x) = \dfrac{(1+e^{-x})^2(-e^{-x}) - e^{-x}(2)(1+e^{-x})(-e^{-x})}{(1+e^{-x})^4}$

$= \dfrac{e^{-x}(e^{-x}-1)}{(1+e^{-x})^3}$. The second factor in the numerator is negative for $x > 0$ and positive for $x < 0$, and the other factors are always positive, so f is CU on $(-\infty, 0)$ and CD on $(0, \infty)$. f has an inflection point at $\left(0, \frac{1}{2}\right)$.

H.

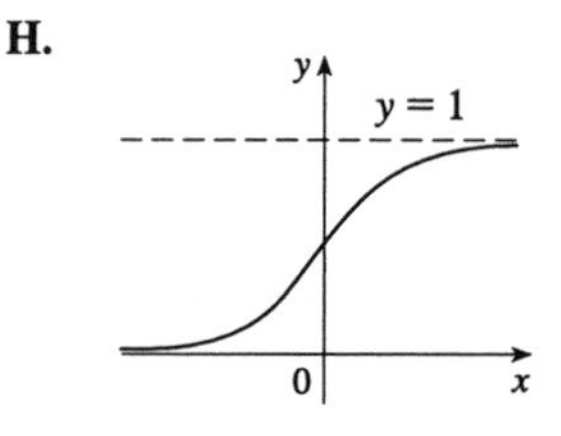

61.

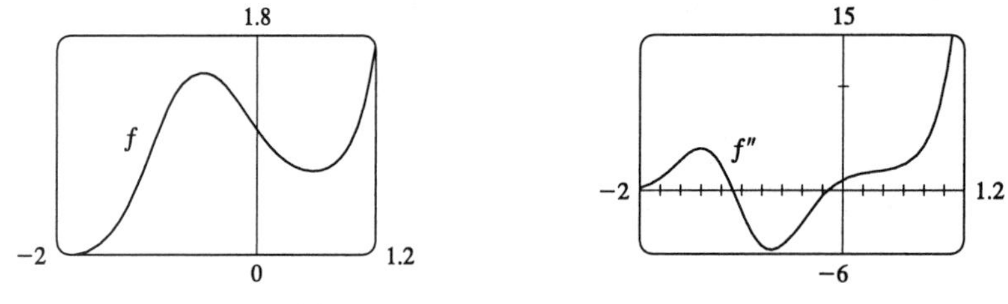

$f(x) = e^{x^3-x} \to 0$ as $x \to -\infty$, and $f(x) \to \infty$ as $x \to \infty$. From the graph, it appears that f has a local minimum of about $f(0.58) = 0.68$, and a local maximum of about $f(-0.58) = 1.47$. To find the exact values, we calculate $f'(x) = (3x^2-1)e^{x^3-x}$, which is 0 when $3x^2 - 1 = 0 \quad \Leftrightarrow \quad x = \pm\frac{1}{\sqrt{3}}$. The negative root corresponds to the local maximum $f\left(-\frac{1}{\sqrt{3}}\right) = e^{(-1/\sqrt{3})^3-(-1/\sqrt{3})} = e^{2\sqrt{3}/9}$, and the positive root corresponds to the local minimum $f\left(\frac{1}{\sqrt{3}}\right) = e^{(1/\sqrt{3})^3-(1/\sqrt{3})} = e^{-2\sqrt{3}/9}$. To estimate the inflection points, we calculate and graph $f''(x) = \dfrac{d}{dx}\left[(3x^2-1)e^{x^3-x}\right] = (3x^2-1)e^{x^3-x}(3x^2-1) + e^{x^3-x}(6x) = e^{x^3-x}(9x^4 - 6x^2 + 6x + 1)$. From the graph, it appears that $f''(x)$ changes sign (and thus f has inflection points) at $x \approx -0.15$ and $x \approx -1.09$. From the graph of f, we see that these x-values correspond to inflection points at about $(-0.15, 1.15)$ and $(-1.09, 0.82)$.

63. $u = -6x \quad \Rightarrow \quad du = -6\,dx$, so $\int e^{-6x}\,dx = -\frac{1}{6}\int e^u\,du = -\frac{1}{6}e^u + C = -\frac{1}{6}e^{-6x} + C$.

65. Let $u = 1 + e^x$. Then $du = e^x\,dx$, so $\int e^x(1+e^x)^{10}\,dx = \int u^{10}\,du = \frac{1}{11}u^{11} + C = \frac{1}{11}(1+e^x)^{11} + C$.

67. $\displaystyle\int \frac{e^x+1}{e^x}\,dx = \int (1+e^{-x})dx = x - e^{-x} + C$

69. Let $u = -x^2$. Then $du = -2x\,dx$, so $\int xe^{-x^2}dx = -\frac{1}{2}\int e^u\,du = -\frac{1}{2}e^u + C = -\frac{1}{2}e^{-x^2} + C$.

71. Let $u = x^2 - 4x - 3$. Then $du = 2(x-2)dx$, so $\int (x-2)e^{x^2-4x-3}\,dx = \frac{1}{2}\int e^u\,du = \frac{1}{2}e^u + C = \frac{1}{2}e^{x^2-4x-3} + C$.

73. Area $= \int_0^1 (e^{3x} - e^x)dx = \left[\frac{1}{3}e^{3x} - e^x\right]_0^1 = \left(\frac{1}{3}e^3 - e\right) - \left(\frac{1}{3} - 1\right) = \frac{1}{3}e^3 - e + \frac{2}{3} \approx 4.644$

75. $V = \int_0^1 \pi(e^x)^2\,dx = \int_0^1 \pi e^{2x}\,dx = \frac{1}{2}[\pi e^{2x}]_0^1 = \frac{\pi}{2}(e^2 - 1)$

77. $y = e^{\sqrt{x}} \quad \Rightarrow \quad \ln y = \ln e^{\sqrt{x}} = \sqrt{x}$

$\Rightarrow \quad x = (\ln y)^2$. Interchange x and y: the inverse function is $y = (\ln x)^2$. The domain of the inverse function is the range of the original function $y = e^{\sqrt{x}}$, that is, $[1, \infty)$.

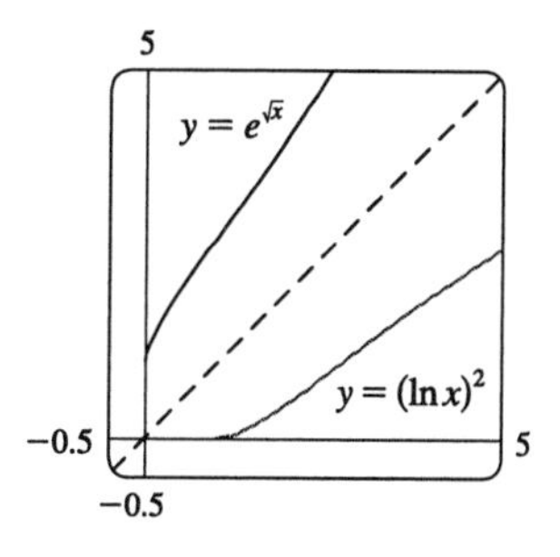

79. We use Theorem 6.1.8. Note that $f(0) = 3 + 0 + e^0 = 4$, so $f^{-1}(4) = 0$. Also $f'(x) = 1 + e^x$. Therefore, $(f^{-1})'(4) = \dfrac{1}{f'(f^{-1}(4))} = \dfrac{1}{f'(0)} = \dfrac{1}{1 + e^0} = \dfrac{1}{2}$.

81. Using the second law of logarithms and Equation 5, we have $\ln\left(\dfrac{e^x}{e^y}\right) = \ln e^x - \ln e^y = x - y = \ln(e^{x-y})$. Since ln is a one-to-one function, it follows that $\dfrac{e^x}{e^y} = e^{x-y}$.

83. **(a)** Let $f(x) = e^x - 1 - x$. Now $f(0) = e^0 - 1 = 0$, and for $x \ge 0$, we have $f'(x) = e^x - 1 \ge 0$. Now, since $f(0) = 0$ and f is increasing on $[0, \infty)$, $f(x) \ge 0$ for $x \ge 0 \;\Rightarrow\; e^x - 1 - x \ge 0 \;\Rightarrow\; e^x \ge 1 + x$.

(b) For $0 \le x \le 1$, $x^2 \le x$, so $e^{x^2} \le e^x$ (since e^x is increasing). Hence [from (a)] $1 + x^2 \le e^{x^2} \le e^x$. So $\frac{4}{3} = \int_0^1 (1 + x^2)dx \le \int_0^1 e^{x^2}dx \le \int_0^1 e^x\,dx = e - 1 < e \quad \Rightarrow \quad \frac{4}{3} \le \int_0^1 e^{x^2}dx \le e$.

85. **(a)** By Exercise 83(a), the result holds for $n = 1$. Suppose that $e^x \ge 1 + x + \dfrac{x^2}{2!} + \cdots + \dfrac{x^k}{k!}$ for $x \ge 0$. Let $f(x) = e^x - 1 - x - \dfrac{x^2}{2!} - \cdots - \dfrac{x^k}{k!} - \dfrac{x^{k+1}}{(k+1)!}$. Then $f'(x) = e^x - 1 - x - \cdots - \dfrac{x^k}{k!} \ge 0$ by assumption. Hence $f(x)$ is increasing on $[0, \infty)$. So $0 \le x$ implies that $0 = f(0) \le f(x) = e^x - 1 - x - \cdots - \dfrac{x^k}{k!} - \dfrac{x^{k+1}}{(k+1)!}$, and hence $e^x \ge 1 + x + \cdots + \dfrac{x^k}{k!} + \dfrac{x^{k+1}}{(k+1)!}$ for $x \ge 0$. Therefore, for $x \ge 0$, $e^x \ge 1 + x + \dfrac{x^2}{2!} + \cdots + \dfrac{x^n}{n!}$ for every positive integer n, by mathematical induction.

(b) Taking $n = 4$ and $x = 1$ in (a), we have $e = e^1 \ge 1 + \dfrac{1}{2} + \dfrac{1}{6} + \dfrac{1}{24} = 2.708\overline{3} > 2.7$.

(c) $e^x \ge 1 + x + \cdots + \dfrac{x^k}{k!} + \dfrac{x^{k+1}}{(k+1)!} \quad \Rightarrow \quad \dfrac{e^x}{x^k} \ge \dfrac{1}{x^k} + \dfrac{1}{x^{k-1}} + \cdots + \dfrac{1}{k!} + \dfrac{x}{(k+1)!} \ge \dfrac{x}{(k+1)!}$. But $\lim\limits_{x\to\infty} \dfrac{x}{(k+1)!} = \infty$, so $\lim\limits_{x\to\infty} \dfrac{e^x}{x^k} = \infty$.

EXERCISES 6.4*

1. $10^{\pi} = e^{\pi \ln 10}$

3. $2^{\cos x} = e^{(\cos x)\ln 2}$

5. $\log_2 64 = 6$ since $2^6 = 64$.

7. $2^{(\log_2 3 + \log_2 5)} = 2^{\log_2 15} = 15$

9.

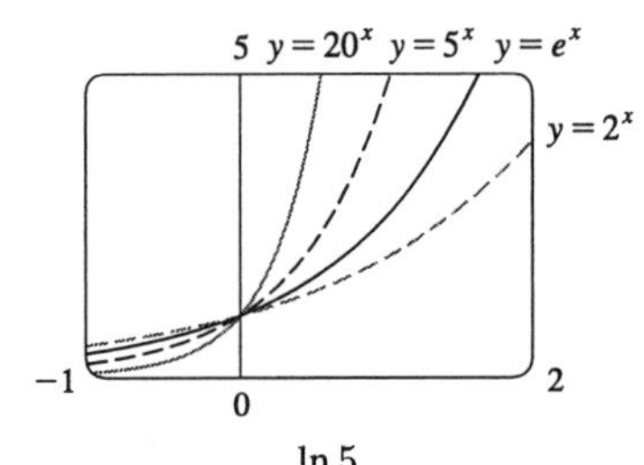

All of these graphs approach 0 as $x \to -\infty$, all of them pass through the point $(0, 1)$, and all of them are increasing and approach ∞ as $x \to \infty$. The larger the base, the faster the function increases for $x > 0$, and the faster it approaches 0 as $x \to -\infty$.

11.

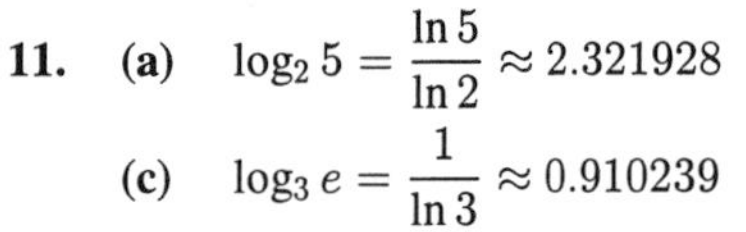

(a) $\log_2 5 = \dfrac{\ln 5}{\ln 2} \approx 2.321928$

(b) $\log_5 26.05 = \dfrac{\ln 26.05}{\ln 5} \approx 2.025563$

(c) $\log_3 e = \dfrac{1}{\ln 3} \approx 0.910239$

(d) $\log_{0.7} 14 = \dfrac{\ln 14}{\ln 0.7} \approx -7.399054$

13.

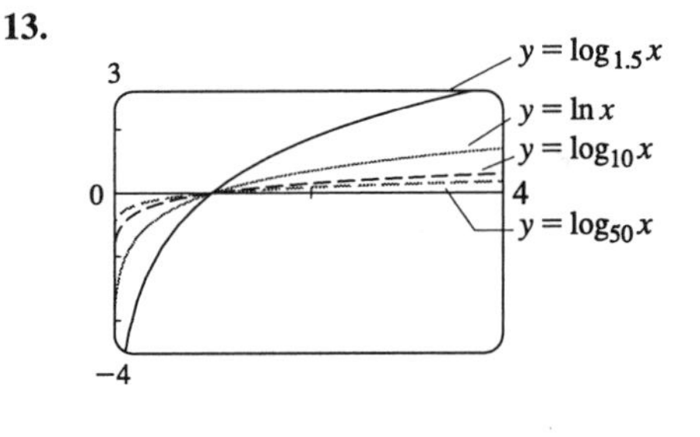

To graph these functions, we use $\log_{1.5} x = \dfrac{\ln x}{\ln 1.5}$ and $\log_{50} x = \dfrac{\ln x}{\ln 50}$. These graphs all approach $-\infty$ as $x \to 0^+$, and they all pass through the point $(1, 0)$. Also, they are all increasing, and all approach ∞ as $x \to \infty$. The functions with larger bases increase extremely slowly, and the ones with smaller bases do so somewhat more quickly. The functions with large bases approach the y-axis more closely as $x \to 0^+$.

15. **(a)** 2 ft = 24 in, so $f(24) = 24^2$ in $= 576$ in $= 48$ ft, and $g(24) = 2^{24}$ in $= 2^{24}/(12 \cdot 5280)$mi ≈ 265 mi.

(b) 3 ft = 36 in, so we need x such that $\log_2 x = 36 \quad \Leftrightarrow \quad x = 2^{36} = 68{,}719{,}476{,}736$. In miles, this is

$$68{,}719{,}476{,}736 \text{ in} \cdot \frac{1 \text{ mi}}{5280 \text{ ft}} \cdot \frac{1 \text{ ft}}{12 \text{ in}} \approx 1{,}084{,}587.7 \text{ mi}.$$

17. $\displaystyle\lim_{x \to \pi/2^-} \log_{10}(\cos x) = -\infty$ since $\cos x \to 0^+$ as $x \to \frac{\pi}{2}^-$.

19. $h(t) = t^3 - 3^t \quad \Rightarrow \quad h'(t) = 3t^2 - 3^t \ln 3$

21. $f(t) = \pi^{-t} \quad \Rightarrow \quad f'(t) = \pi^{-t}(\ln \pi)(-1) = -\pi^{-t} \ln \pi$

23. $f(x) = \log_3(x^2 - 4) \quad \Rightarrow \quad f'(x) = \dfrac{2x}{(\ln 3)(x^2 - 4)}$

25. $f(t) = \log_2(t^4 - t^2 + 1) \quad \Rightarrow \quad f'(t) = \dfrac{4t^3 - 2t}{(\ln 2)(t^4 - t^2 + 1)}$

27. $y = 2^{3^x} \quad \Rightarrow \quad y' = 2^{3^x}(\ln 2)3^x \ln 3 = (\ln 2)(\ln 3)3^x 2^{3^x}$

29. $y = x^{\sin x} \Rightarrow \ln y = \sin x \ln x \Rightarrow \dfrac{y'}{y} = \cos x \ln x + \dfrac{\sin x}{x} \Rightarrow y' = x^{\sin x}\left[\cos x \ln x + \dfrac{\sin x}{x}\right]$

31. $y = x^{e^x} \Rightarrow \ln y = e^x \ln x \Rightarrow \dfrac{y'}{y} = e^x \ln x + \dfrac{e^x}{x} \Rightarrow y' = x^{e^x} e^x\left(\ln x + \dfrac{1}{x}\right)$

33. $y = (\ln x)^x \Rightarrow \ln y = x \ln \ln x \Rightarrow \dfrac{y'}{y} = \ln \ln x + x \cdot \dfrac{1}{\ln x} \cdot \dfrac{1}{x} \Rightarrow y' = (\ln x)^x\left(\ln \ln x + \dfrac{1}{\ln x}\right)$

35. $y = x^{1/\ln x} \Rightarrow \ln y = \left(\dfrac{1}{\ln x}\right)\ln x = 1 \Rightarrow y = e \Rightarrow y' = 0$

37. $y = \cos\left(x^{\sqrt{x}}\right) \Rightarrow y' = -\sin\left(x^{\sqrt{x}}\right)x^{\sqrt{x}}\left(\dfrac{\ln x + 2}{2\sqrt{x}}\right)$ by Example ~~16~~ 4

39. $y = 10^x \Rightarrow y' = 10^x \ln 10$, so at $(1, 10)$, the slope of the tangent line is $10^1 \ln 10 = 10 \ln 10$, and its equation is $y - 10 = 10 \ln 10(x - 1)$, or $y = (10 \ln 10)x + 10(1 - \ln 10)$.

41. $\displaystyle\int_3^4 5^t\,dt = \left[\frac{5^t}{\ln 5}\right]_3^4 = \frac{5^4 - 5^3}{\ln 5} = \frac{500}{\ln 5}$

43. $\displaystyle\int \frac{\log_{10} x}{x}\,dx = \int \frac{(\ln x)/(\ln 10)}{x}\,dx = \frac{1}{\ln 10}\int \frac{\ln x}{x}\,dx$. Now put $u = \ln x$, so $du = \dfrac{1}{x}\,dx$, and the expression becomes $\displaystyle\frac{1}{\ln 10}\int u\,du = \frac{1}{\ln 10}\left(\tfrac{1}{2}u^2 + C_1\right) = \frac{1}{2\ln 10}(\ln x)^2 + C$.

Or: The substitution $u = \log_{10} x$ gives $du = \dfrac{dx}{x \ln 10}$ and we get $\displaystyle\int \frac{\log_{10} x}{x}\,dx = \tfrac{1}{2}\ln 10(\log_{10} x)^2 + C$.

45.
$$\begin{aligned}
A &= \int_{-1}^{0}(2^x - 5^x)dx + \int_0^1 (5^x - 2^x)dx \\
&= \left[\frac{2^x}{\ln 2} - \frac{5^x}{\ln 5}\right]_{-1}^{0} + \left[\frac{5^x}{\ln 5} - \frac{2^x}{\ln 2}\right]_0^1 \\
&= \left(\frac{1}{\ln 2} - \frac{1}{\ln 5}\right) - \left(\frac{1/2}{\ln 2} - \frac{1/5}{\ln 5}\right) + \left(\frac{5}{\ln 5} - \frac{2}{\ln 2}\right) - \left(\frac{1}{\ln 5} - \frac{1}{\ln 2}\right) \\
&= \frac{16}{5\ln 5} - \frac{1}{2\ln 2}
\end{aligned}$$

47. We see that the graphs of $y = 2^x$ and $y = 1 + 3^{-x}$ intersect at $x \approx 0.6$. We let $f(x) = 2^x - 1 - 3^{-x}$ and calculate $f'(x) = 2^x \ln 2 + 3^{-x}\ln 3$, and using the formula $x_{n+1} = x_n - f(x_n)/f'(x_n)$ (Newton's Method), we get $x_1 = 0.6$, $x_2 \approx x_3 \approx 0.600967$. So, correct to six decimal places, the root occurs at $x = 0.600967$.

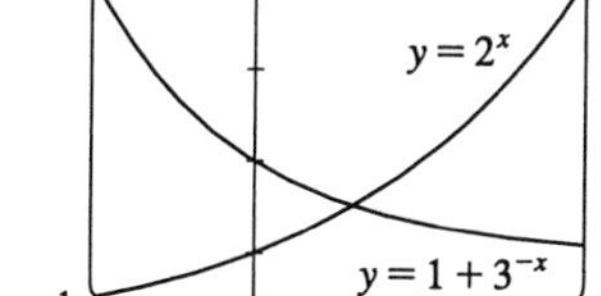

49. $y = \dfrac{10^x}{10^x + 1} \Leftrightarrow (10^x + 1)y = 10^x \Leftrightarrow y = 10^x(1 - y) \Leftrightarrow 10^x = \dfrac{y}{1 - y} \Leftrightarrow$

$\log_{10} 10^x = \log_{10}\left(\dfrac{y}{1 - y}\right) \Leftrightarrow x = \log_{10} y - \log_{10}(1 - y)$. Interchange x and y:

$y = \log_{10} x - \log_{10}(1 - x)$ is the inverse function.

51. If I is the intensity of the 1989 San Francisco earthquake, then $\log_{10}(I/S) = 7.1 \quad \Rightarrow$
$\log_{10}(16I/S) = \log_{10}16 + \log_{10}(I/S) = \log_{10}16 + 7.1 \approx 8.3$.

53. We find I with the loudness formula from Exercise 52, substituting $I_0 = 10^{-12}$ and $L = 50$: $50 = 10\log_{10}\dfrac{I}{10^{-12}}$
$\Leftrightarrow \quad 5 = \log_{10}\dfrac{I}{10^{-12}} \quad \Leftrightarrow \quad 10^5 = \dfrac{I}{10^{-12}} \quad \Leftrightarrow \quad I = 10^{-7}\ \text{watt}/\text{m}^2$. Now we differentiate L with respect to I: $L = 10\log_{10}\dfrac{I}{I_0} \quad \Rightarrow \quad \dfrac{dL}{dI} = 10\dfrac{1}{(I/I_0)\ln 10}\left(\dfrac{1}{I_0}\right) = \dfrac{10}{\ln 10}\left(\dfrac{1}{I}\right)$. Substituting $I = 10^{-7}$, we get
$L'(50) = \dfrac{10}{\ln 10}\left(\dfrac{1}{10^{-7}}\right) = \dfrac{10^8}{\ln 10} \approx 4.34 \times 10^7\ \dfrac{\text{dB}}{\text{watt}/\text{m}^2}$.

55. Using Definition 1 and the second law of exponents for e^x, we have
$a^{x-y} = e^{(x-y)\ln a} = e^{x\ln a - y\ln a} = e^{x\ln a}/e^{y\ln a} = a^x/a^y$.

57. Let $\log_a x = r$ and $\log_a y = s$. Then $a^r = x$ and $a^s = y$.

(a) $xy = a^r a^s = a^{r+s} \quad \Rightarrow \quad \log_a(xy) = r + s = \log_a x + \log_a y$

(b) $x/y = a^r/a^s = a^{r-s} \quad \Rightarrow \quad \log_a(x/y) = r - s = \log_a x - \log_a y$

(c) $x^y = (a^r)^y = a^{ry} \quad \Rightarrow \quad \log_a(x^y) = ry = y\log_a x$

EXERCISES 6.5

1. **(a)** By Theorem 2, $y(t) = y(0)e^{kt} = 100e^{kt} \quad \Rightarrow \quad y(\tfrac{1}{3}) = 100e^{k/3} = 200 \quad \Rightarrow$
$k/3 = \ln(200/100) = \ln 2 \quad \Rightarrow \quad k = 3\ln 2$. So $y(t) = 100e^{(3\ln 2)t} = 100 \cdot 2^{3t}$.

(b) $y(10) = 100 \cdot 2^{30} \approx 1.07 \times 10^{11}$ cells

(c) $y(t) = 100 \cdot 2^{3t} = 10{,}000 \quad \Rightarrow \quad 2^{3t} = 100 \quad \Rightarrow \quad 3t\ln 2 = \ln 100 \quad \Rightarrow \quad t = (\ln 100)/(3\ln 2) \approx 2.2$ h

3. **(a)** $y(t) = y(0)e^{kt} = 500\,e^{kt} \quad \Rightarrow \quad y(3) = 500e^{3k} = 8000 \quad \Rightarrow \quad e^{3k} = 16 \quad \Rightarrow \quad 3k = \ln 16 \quad \Rightarrow$
$y(t) = 500e^{(\ln 16)t/3} = 500 \cdot 16^{t/3}$

(b) $y(4) = 500 \cdot 16^{4/3} \approx 20{,}159$

(c) $y(t) = 500 \cdot 16^{t/3} = 30{,}000 \quad \Rightarrow \quad 16^{t/3} = 60 \quad \Rightarrow \quad \tfrac{1}{3}t\ln 16 = \ln 60 \quad \Rightarrow$
$t = 3(\ln 60)/(\ln 16) \approx 4.4$ h

5. **(a)** Let the population (in millions) in the year t be $P(t)$. Since the initial time is the year 1750, we substitute $t - 1750$ for t in Theorem 2, so the exponential model gives $P(t) = P(1750)e^{k(t-1750)}$. Then
$P(1800) = 906 = 728e^{k(1800-1750)} \quad \Rightarrow \quad \ln\frac{906}{728} = k(50) \quad \Rightarrow \quad k = \frac{1}{50}\ln\frac{906}{728} \approx 0.0043748$. So with this model, we estimate $P(1900) \approx P(1750)e^{k(1900-1750)} \approx 728e^{150(0.0043748)} \approx 1403$ million, and
$P(1950) \approx 728e^{200(0.0043748)} \approx 1746$ million. Both of these estimates are much too low.

(b) In this case, the exponential model gives $P(t) = P(1850)e^{k(t-1850)} \Rightarrow$
$P(1900) = 1608 = 1171e^{k(1900-1850)} \Rightarrow \ln\frac{1608}{1171} = k(50) \Rightarrow k = \frac{1}{50}\ln\frac{1608}{1171} \approx 0.006343$. So with this model, we estimate $P(1950) \approx 1171e^{100(0.006343)} \approx 2208$ million. This is still too low, but closer than the estimate of $P(1950)$ in part (a).

(c) The exponential model gives $P(t) = P(1900)e^{k(t-1900)} \Rightarrow P(1950) = 2517 = 1608e^{k(1950-1900)}$
$\Rightarrow \ln\frac{2517}{1608} = k(50) \Rightarrow k = \frac{1}{50}\ln\frac{2517}{1608} \approx 0.008962$. With this model, we estimate $P(1992) \approx 1608e^{0.008962(1992-1900)} \approx 3667$ million. This is much too low.

The discrepancy is explained by the fact that the world birth rate (average yearly number of births per person) is about the same as always, whereas the mortality rate (especially the infant mortality rate) is much lower, owing mostly to advances in medical science and to the wars in the first part of the 20th century. The exponential model assumes, among other things, that the birth and mortality rates will remain constant.

7. **(a)** If $y = [N_2O_5]$ then $\frac{dy}{dt} = -0.0005y \Rightarrow y(t) = y(0)e^{-0.0005t} = Ce^{-0.0005t}$.

(b) $y(t) = Ce^{-0.0005t} = 0.9C \Rightarrow e^{-0.0005t} = 0.9 \Rightarrow -0.0005t = \ln 0.9 \Rightarrow$
$t = -2000\ln 0.9 \approx 211$ s

9. **(a)** If $y(t)$ is the mass remaining after t days, then $y(t) = y(0)e^{kt} = 50e^{kt} \Rightarrow$
$y(0.00014) = 50e^{0.00014k} = 25 \Rightarrow e^{0.00014k} = \frac{1}{2} \Rightarrow k = -(\ln 2)/0.00014 \Rightarrow y(t) = 50$
$e^{-(\ln 2)t/0.00014} = 50 \cdot 2^{-t/0.00014}$

(b) $y(0.01) = 50 \cdot 2^{-0.01/0.00014} \approx 1.57 \times 10^{-20}$ mg

(c) $50e^{-(\ln 2)t/0.00014} = 40 \Rightarrow -(\ln 2)t/0.00014 = \ln 0.8 \Rightarrow t = -0.00014\frac{\ln 0.8}{\ln 2} \approx 4.5 \times 10^{-5}$ s

11. Let $y(t)$ be the level of radioactivity. Thus, $y(t) = y(0)e^{-kt}$ and k is determined by using the half-life:
$\frac{1}{2} = e^{-5730k} \Rightarrow k = -\frac{\ln\frac{1}{2}}{5730} = \frac{\ln 2}{5730}$. If 0.74 of the ^{14}C remains, then we know that $0.74 = e^{-t(\ln 2)/5730} \Rightarrow$
$\ln 0.74 = -\frac{t\ln 2}{5730} \Rightarrow t = -\frac{5730(\ln 0.74)}{\ln 2} \approx 2489 \approx 2500$ years.

13. Let $y(t) =$ temperature after t minutes. Then $\frac{dy}{dt} = -\frac{1}{10}[y(t) - 21]$. If $u(t) = y(t) - 21$, then $\frac{du}{dt} = -\frac{u}{10}$
$\Rightarrow u(t) = u(0)\,e^{-t/10} = 12\,e^{-t/10} \Rightarrow y(t) = 21 + u(t) = 21 + 12\,e^{-t/10}$.

15. **(a)** Let $y(t) =$ temperature after t minutes. Newton's Law of Cooling implies that $\frac{dy}{dt} = k(y - 75)$. Let
$u(t) = y(t) - 75$. Then $\frac{du}{dt} = ku$, so $u(t) = u(0)e^{kt} = 110e^{kt} \Rightarrow y(t) = 75 + 110e^{kt} \Rightarrow$
$y(30) = 75 + 110e^{30k} = 150 \Rightarrow e^{30k} = \frac{75}{110} = \frac{15}{22} \Rightarrow k = \frac{1}{30}\ln\frac{15}{22}$, so $y(t) = 75 + 110e^{\frac{1}{30}t\ln(\frac{15}{22})}$
and $y(45) = 75 + 110e^{\frac{45}{30}\ln(\frac{15}{22})} \approx 137$ °F.

(b) $y(t) = 75 + 110e^{\frac{1}{30}t\ln(\frac{15}{22})} = 100 \Rightarrow e^{\frac{1}{30}t\ln(\frac{15}{22})} = \frac{25}{110} \Rightarrow \frac{1}{30}t\ln\frac{15}{22} = \ln\frac{25}{110} \Rightarrow t = \frac{30\ln\frac{25}{110}}{\ln\frac{15}{22}} \approx 116$ min

17. **(a)** Let $P(h)$ be the pressure at altitude h. Then $dP/dh = kP \quad\Rightarrow\quad P(h) = P(0)e^{kh} = 101.3e^{kh} \quad\Rightarrow$ $P(1000) = 101.3e^{1000k} = 87.14 \quad\Rightarrow\quad 1000k = \ln\left(\frac{87.14}{101.3}\right) \quad\Rightarrow\quad P(h) = 101.3\, e^{\frac{1}{1000}h\ln\left(\frac{87.14}{101.3}\right)}$, so $P(3000) = 101.3e^{3\ln\left(\frac{87.14}{101.3}\right)} \approx 64.5\,\text{kPa}$.

(b) $P(6187) = 101.3\, e^{\frac{6187}{1000}\ln\left(\frac{87.14}{101.3}\right)} \approx 39.9\,\text{kPa}$

19. With the notation of Example 4, $A_0 = 3000$, $i = 0.05$, and $t = 5$.

(a) $n = 1$: $\quad A = 3000(1.05)^5 = \$3828.84$

(b) $n = 2$: $\quad A = 3000\left(1 + \frac{0.05}{2}\right)^{10} = \$3840.25$

(c) $n = 12$: $\quad A = 3000\left(1 + \frac{0.05}{12}\right)^{60} = \$3850.08$

(d) $n = 52$: $\quad A = 3000\left(1 + \frac{0.05}{52}\right)^{5\cdot 52} = \$3851.61$

(e) $n = 365$: $\quad A = 3000\left(1 + \frac{0.05}{365}\right)^{5\cdot 365} = \$3852.01$

(f) continuously: $\quad A = 3000e^{(0.05)5} = \$3852.08$

21. **(a)** If $y(t)$ is the amount of salt at time t, then $y(0) = 1500(0.3) = 450\,\text{kg}$. The rate of change of y is $\frac{dy}{dt} = -\left(\frac{y(t)}{1500}\,\frac{\text{kg}}{\text{L}}\right)\left(20\,\frac{\text{L}}{\text{min}}\right) = -\frac{1}{75}y(t)\,\frac{\text{kg}}{\text{min}}$, so $y(t) = y(0)e^{-t/75} = 450e^{-t/75} \quad\Rightarrow$ $y(30) = 450e^{-0.4} \approx 301.6\,\text{kg}$.

(b) When the concentration is $0.2\,\text{kg/L}$, the amount of salt is $1500(0.2) = 300\,\text{kg}$. So $y(t) = 450e^{-t/75} = 300$ $\Rightarrow\quad e^{-t/75} = \frac{2}{3} \quad\Rightarrow\quad -t/75 = \ln\frac{2}{3} \quad\Rightarrow\quad t = -75\ln\frac{2}{3} \approx 30.41\,\text{min}$.

EXERCISES 6.6

1. $\cos^{-1}(-1) = \pi$ since $\cos\pi = -1$.

3. $\tan^{-1}\sqrt{3} = \frac{\pi}{3}$ since $\tan\frac{\pi}{3} = \sqrt{3}$.

5. $\csc^{-1}\sqrt{2} = \frac{\pi}{4}$ since $\csc\frac{\pi}{4} = \sqrt{2}$.

7. $\cot^{-1}\left(-\sqrt{3}\right) = \frac{5\pi}{6}$ since $\cot\frac{5\pi}{6} = -\sqrt{3}$.

9. $\sin(\sin^{-1}0.7) = 0.7$

11. $\tan^{-1}\left(\tan\frac{4\pi}{3}\right) = \tan^{-1}\sqrt{3} = \frac{\pi}{3}$

13. Let $\theta = \cos^{-1}\frac{4}{5}$, so $\cos\theta = \frac{4}{5}$. Then $\sin\left(\cos^{-1}\frac{4}{5}\right) = \sin\theta = \sqrt{1 - \left(\frac{4}{5}\right)^2} = \sqrt{\frac{9}{25}} = \frac{3}{5}$.

15. $\arcsin\left(\sin\frac{5\pi}{4}\right) = \arcsin\left(-\frac{1}{\sqrt{2}}\right) = -\frac{\pi}{4}$

17. Let $\theta = \sin^{-1}\frac{5}{13}$. Then $\sin\theta = \frac{5}{13}$, so $\cos\left(2\sin^{-1}\frac{5}{13}\right) = \cos 2\theta = 1 - 2\sin^2\theta = 1 - 2\left(\frac{5}{13}\right)^2 = \frac{119}{169}$.

19. Let $y = \sin^{-1}x$. Then $-\frac{\pi}{2} \le y \le \frac{\pi}{2} \quad\Rightarrow\quad \cos y \ge 0$, so $\cos(\sin^{-1}x) = \cos y = \sqrt{1 - \sin^2 y} = \sqrt{1 - x^2}$

21. Let $y = \tan^{-1}x$. Then $\tan y = x$, so from the triangle we see that $\sin(\tan^{-1}x) = \sin y = \dfrac{x}{\sqrt{1+x^2}}$.

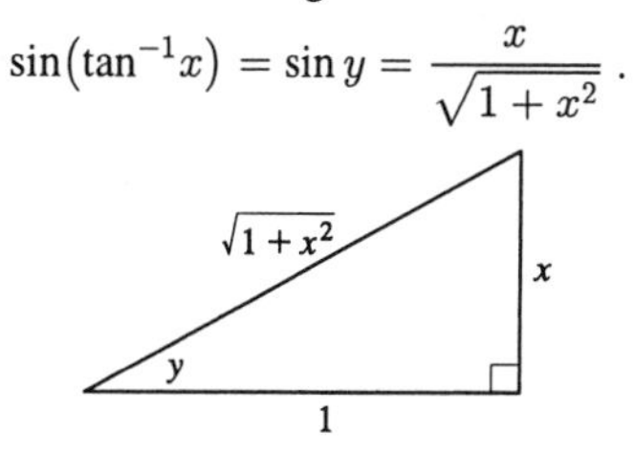

23. The graph of $\sin^{-1}x$ is the reflection of the graph of $\sin x$ about the line $y = x$.

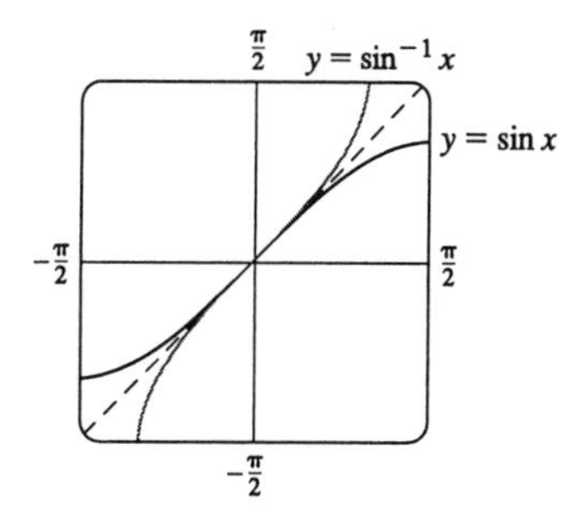

25. Let $y = \cos^{-1}x$. Then $\cos y = x$ and $0 \le y \le \pi \quad\Rightarrow\quad -\sin y\,\dfrac{dy}{dx} = 1 \quad\Rightarrow$

$$\frac{dy}{dx} = -\frac{1}{\sin y} = -\frac{1}{\sqrt{1-\cos^2 y}} = -\frac{1}{\sqrt{1-x^2}}$$ (Note that $\sin y \ge 0$ for $0 \le y \le \pi$.)

27. Let $y = \cot^{-1}x$. Then $\cot y = x \quad\Rightarrow\quad -\csc^2 y\,\dfrac{dy}{dx} = 1 \quad\Rightarrow\quad \dfrac{dy}{dx} = -\dfrac{1}{\csc^2 y} = -\dfrac{1}{1+\cot^2 y} = -\dfrac{1}{1+x^2}$.

29. Let $y = \csc^{-1}x$. Then $\csc y = x \quad\Rightarrow\quad -\csc y \cot y\,\dfrac{dy}{dx} = 1 \quad\Rightarrow$

$$\frac{dy}{dx} = -\frac{1}{\csc y \cot y} = -\frac{1}{\csc y\sqrt{\csc^2 y - 1}} = -\frac{1}{x\sqrt{x^2-1}}.$$ Note that $\cot y \ge 0$ on the domain of $\csc^{-1}x$.

31. $g(x) = \tan^{-1}(x^3) \quad\Rightarrow\quad g'(x) = \dfrac{1}{1+(x^3)^2}(3x^2) = \dfrac{3x^2}{1+x^6}$

33. $y = \sin^{-1}(x^2) \quad\Rightarrow\quad y' = \dfrac{1}{\sqrt{1-(x^2)^2}}(2x) = \dfrac{2x}{\sqrt{1-x^4}}$

35. $H(x) = (1+x^2)\arctan x \quad\Rightarrow\quad H'(x) = (2x)\arctan x + (1+x^2)\dfrac{1}{1+x^2} = 1 + 2x\arctan x$

37. $g(t) = \sin^{-1}\left(\dfrac{4}{t}\right) \quad\Rightarrow\quad g'(t) = \dfrac{1}{\sqrt{1-(4/t)^2}}\left(-\dfrac{4}{t^2}\right) = -\dfrac{4}{\sqrt{t^4-16t^2}}$

39. $G(t) = \cos^{-1}\sqrt{2t-1} \quad\Rightarrow\quad G'(t) = -\dfrac{1}{\sqrt{1-(2t-1)}}\dfrac{2}{2\sqrt{2t-1}} = -\dfrac{1}{\sqrt{2(-2t^2+3t-1)}}$

41. $y = \sec^{-1}\sqrt{1+x^2} \quad\Rightarrow\quad y' = \left[\dfrac{1}{\sqrt{1+x^2}\sqrt{(1+x^2)-1}}\right]\left[\dfrac{2x}{2\sqrt{1+x^2}}\right] = \dfrac{x}{(1+x^2)\sqrt{x^2}} = \dfrac{x}{(1+x^2)|x|}$

43. $y = \tan^{-1}(\sin x) \quad\Rightarrow\quad y' = \dfrac{\cos x}{1+\sin^2 x}$

45. $y = (\tan^{-1}x)^{-1} \quad\Rightarrow\quad y' = -(\tan^{-1}x)^{-2}\left(\dfrac{1}{1+x^2}\right) = -\dfrac{1}{(1+x^2)(\tan^{-1}x)^2}$

47. $y = x^2\cot^{-1}(3x) \quad\Rightarrow\quad y' = 2x\cot^{-1}(3x) + x^2\left[-\dfrac{1}{1+(3x)^2}\right](3) = 2x\cot^{-1}(3x) - \dfrac{3x^2}{1+9x^2}$

49. $y = \arccos\left(\dfrac{b + a\cos x}{a + b\cos x}\right) \quad \Rightarrow$

$$y' = -\frac{1}{\sqrt{1 - \left(\dfrac{b + a\cos x}{a + b\cos x}\right)^2}} \frac{(a + b\cos x)(-a\sin x) - (b + a\cos x)(-b\sin x)}{(a + b\cos x)^2}$$

$$= \frac{1}{\sqrt{a^2 + b^2\cos^2 x - b^2 - a^2\cos^2 x}} \frac{(a^2 - b^2)\sin x}{|a + b\cos x|}$$

$$= \frac{1}{\sqrt{a^2 - b^2}\sqrt{1 - \cos^2 x}} \frac{(a^2 - b^2)\sin x}{|a + b\cos x|}$$

$$= \frac{\sqrt{a^2 - b^2}}{|a + b\cos x|} \frac{\sin x}{|\sin x|}$$

But $0 \le x \le \pi$, so $|\sin x| = \sin x$. Also $a > b > 0 \quad \Rightarrow \quad b\cos x \ge -b > -a$, so $a + b\cos x > 0$.

Thus $y' = \dfrac{\sqrt{a^2 - b^2}}{a + b\cos x}$.

51. $g(x) = \sin^{-1}(3x + 1) \quad \Rightarrow \quad g'(x) = \dfrac{3}{\sqrt{1 - (3x + 1)^2}} = \dfrac{3}{\sqrt{-9x^2 - 6x}}$,

$\text{Dom}(g) = \{x \mid -1 \le 3x + 1 \le 1\} = \left\{x \mid -\frac{2}{3} \le x \le 0\right\} = \left[-\frac{2}{3}, 0\right]$,

$\text{Dom}(g') = \{x \mid -1 < 3x + 1 < 1\} = \left(-\frac{2}{3}, 0\right)$

53. $S(x) = \sin^{-1}(\tan^{-1}x) \quad \Rightarrow \quad S'(x) = \left[\sqrt{1 - (\tan^{-1}x)^2}(1 + x^2)\right]^{-1}$,

$\text{Dom}(S) = \{x \mid -1 \le \tan^{-1}x \le 1\} = \{x \mid \tan(-1) \le x \le \tan 1\} = [-\tan 1, \tan 1]$,

$\text{Dom}(S') = \{x \mid -1 < \tan^{-1}x < 1\} = (-\tan 1, \tan 1)$

55. $U(t) = 2^{\arctan t} \quad \Rightarrow \quad U'(t) = 2^{\arctan t} \cdot \dfrac{\ln 2}{1 + t^2}$, $\text{Dom}(U) = \text{Dom}(U') = \mathbb{R}$

57. $g(x) = x\sin^{-1}\left(\dfrac{x}{4}\right) + \sqrt{16 - x^2} \quad \Rightarrow \quad g'(x) = \sin^{-1}\left(\dfrac{x}{4}\right) + \dfrac{x}{4\sqrt{1 - (x/4)^2}} - \dfrac{x}{\sqrt{16 - x^2}} = \sin^{-1}\left(\dfrac{x}{4}\right)$

$\Rightarrow \quad g'(2) = \sin^{-1}\frac{1}{2} = \frac{\pi}{6}$

59. $f(x) = e^x - x^2\arctan x \quad \Rightarrow$

$$f'(x) = e^x - \left[x^2\left(\frac{1}{1 + x^2}\right) + 2x\arctan x\right]$$

$$= e^x - \frac{x^2}{1 + x^2} - 2x\arctan x$$

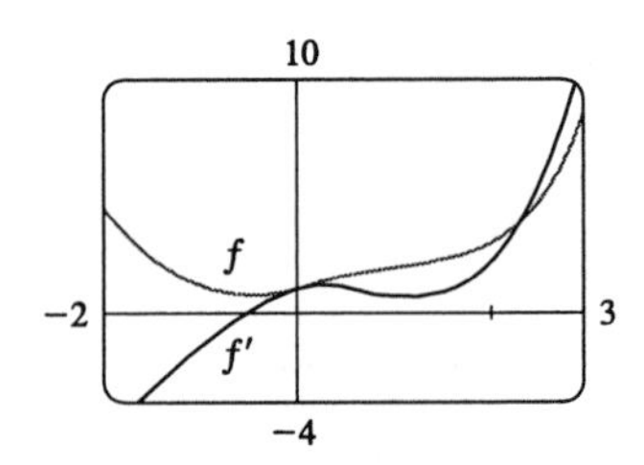

This is reasonable because the graphs show that f is increasing when $f'(x)$ is positive.

61. $\lim\limits_{x \to -1^+} \sin^{-1}x = \sin^{-1}(-1) = -\frac{\pi}{2}$

63. $\lim\limits_{x \to \infty} \tan^{-1}(x^2) = \frac{\pi}{2}$ since $x^2 \to \infty$ as $x \to \infty$.

65.

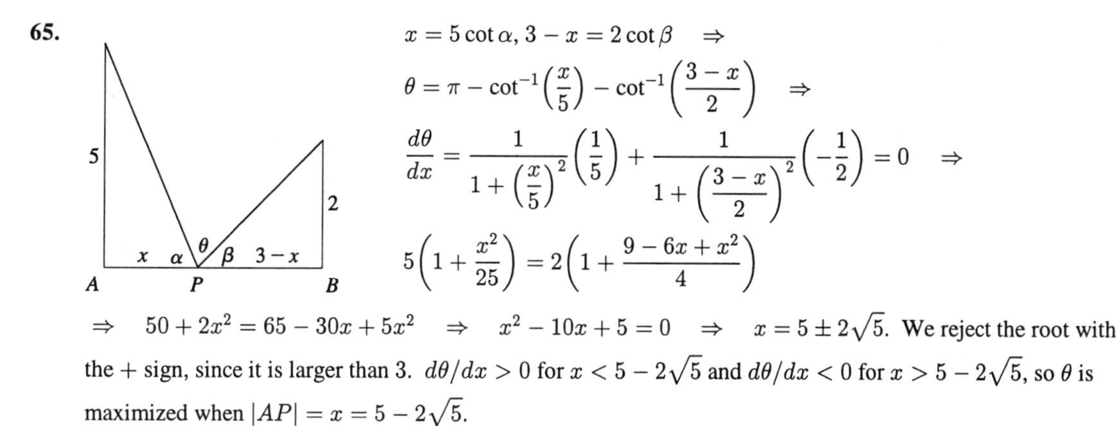

$x = 5\cot\alpha,\ 3 - x = 2\cot\beta \quad \Rightarrow$

$\theta = \pi - \cot^{-1}\left(\frac{x}{5}\right) - \cot^{-1}\left(\frac{3-x}{2}\right) \quad \Rightarrow$

$\frac{d\theta}{dx} = \frac{1}{1+\left(\frac{x}{5}\right)^2}\left(\frac{1}{5}\right) + \frac{1}{1+\left(\frac{3-x}{2}\right)^2}\left(-\frac{1}{2}\right) = 0 \quad \Rightarrow$

$5\left(1 + \frac{x^2}{25}\right) = 2\left(1 + \frac{9 - 6x + x^2}{4}\right)$

$\Rightarrow \quad 50 + 2x^2 = 65 - 30x + 5x^2 \quad \Rightarrow \quad x^2 - 10x + 5 = 0 \quad \Rightarrow \quad x = 5 \pm 2\sqrt{5}$. We reject the root with the $+$ sign, since it is larger than 3. $d\theta/dx > 0$ for $x < 5 - 2\sqrt{5}$ and $d\theta/dx < 0$ for $x > 5 - 2\sqrt{5}$, so θ is maximized when $|AP| = x = 5 - 2\sqrt{5}$.

67.

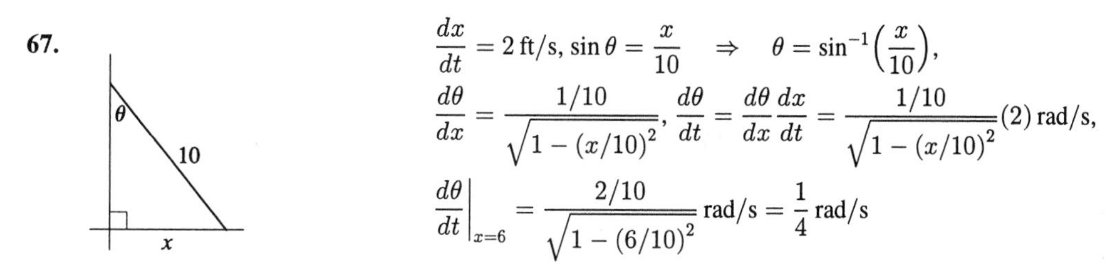

$\frac{dx}{dt} = 2\text{ ft/s},\ \sin\theta = \frac{x}{10} \quad \Rightarrow \quad \theta = \sin^{-1}\left(\frac{x}{10}\right),$

$\frac{d\theta}{dx} = \frac{1/10}{\sqrt{1-(x/10)^2}},\ \frac{d\theta}{dt} = \frac{d\theta}{dx}\frac{dx}{dt} = \frac{1/10}{\sqrt{1-(x/10)^2}}(2)\text{ rad/s},$

$\left.\frac{d\theta}{dt}\right|_{x=6} = \frac{2/10}{\sqrt{1-(6/10)^2}}\text{ rad/s} = \frac{1}{4}\text{ rad/s}$

69. By reflecting the graph of $y = \sec x$ (see Figure 11) about the line $y = x$, we get the graph of $y = \sec^{-1}x$.

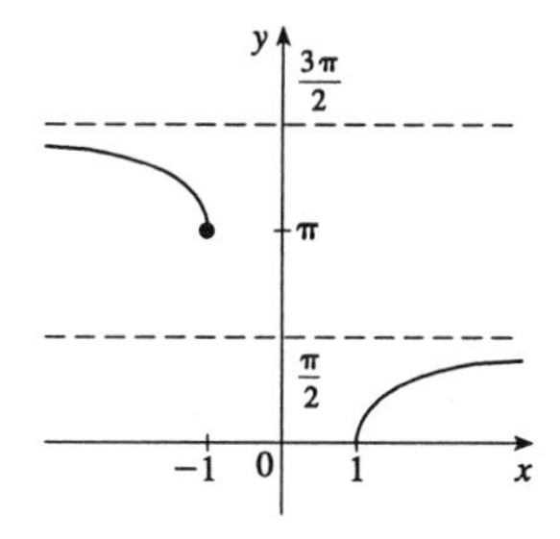

71. $y = f(x) = \sin^{-1}[x/(x+1)]$ **A.** $D = \{x \mid -1 \le x/(x+1) \le 1\}$. For $x > -1$ we have $-x - 1 \le x \le x + 1 \quad \Leftrightarrow \quad 2x \ge -1 \quad \Leftrightarrow \quad x \ge -\frac{1}{2}$, so $D = \left[-\frac{1}{2}, \infty\right)$. **B.** Intercepts are 0 **C.** No symmetry **D.** $\lim\limits_{x\to\infty} \sin^{-1}\left(\frac{x}{x+1}\right) = \lim\limits_{x\to\infty} \sin^{-1}\left(\frac{1}{1+1/x}\right) = \sin^{-1}1 = \frac{\pi}{2}$, so $y = \frac{\pi}{2}$ is a HA.

E. $f'(x) = \frac{1}{\sqrt{1-[x/(x+1)]^2}}\frac{(x+1)-x}{(x+1)^2} = \frac{1}{(x+1)\sqrt{2x+1}} > 0$, so f is increasing on $\left[-\frac{1}{2}, \infty\right)$. **F.** No local maximum or minimum, $f\left(-\frac{1}{2}\right) = \sin^{-1}(-1) = -\frac{\pi}{2}$ is an absolute minimum **G.** $f''(x) = \frac{\sqrt{2x+1} + (x+1)/\sqrt{2x+1}}{(x+1)^2(2x+1)}$

$= -\frac{3x+2}{(x+1)^2(2x+1)^{3/2}} < 0$ on D, so f is CD on $\left(-\frac{1}{2}, \infty\right)$.

H.

73. $y = f(x) = x - \tan^{-1}x$ **A.** $D = \mathbb{R}$ **B.** Intercepts are 0 **C.** $f(-x) = -f(x)$, so the curve is symmetric about the origin. **D.** $\lim\limits_{x\to\infty}(x - \tan^{-1}x) = \infty$ and $\lim\limits_{x\to-\infty}(x - \tan^{-1}x) = -\infty$, no HA.

But $f(x) - \left(x - \frac{\pi}{2}\right) = -\tan^{-1}x + \frac{\pi}{2} \to 0$ as $x \to \infty$, and $f(x) - \left(x + \frac{\pi}{2}\right) = -\tan^{-1}x - \frac{\pi}{2} \to 0$ as $x \to -\infty$, so $y = x \pm \frac{\pi}{2}$ are slant asymptotes. **E.** $f'(x) = 1 - \dfrac{1}{x^2+1} = \dfrac{x^2}{x^2+1} > 0$, so f is increasing on $\mathbb{R}$.

F. No extrema **G.** $f''(x) = \dfrac{(1+x^2)(2x) - x^2(2x)}{(1+x^2)^2} = \dfrac{2x}{(1+x^2)^2} > 0$

$\Leftrightarrow$ $x > 0$, so f is CU on $(0, \infty)$, CD on $(-\infty, 0)$. IP $(0, 0)$

H.

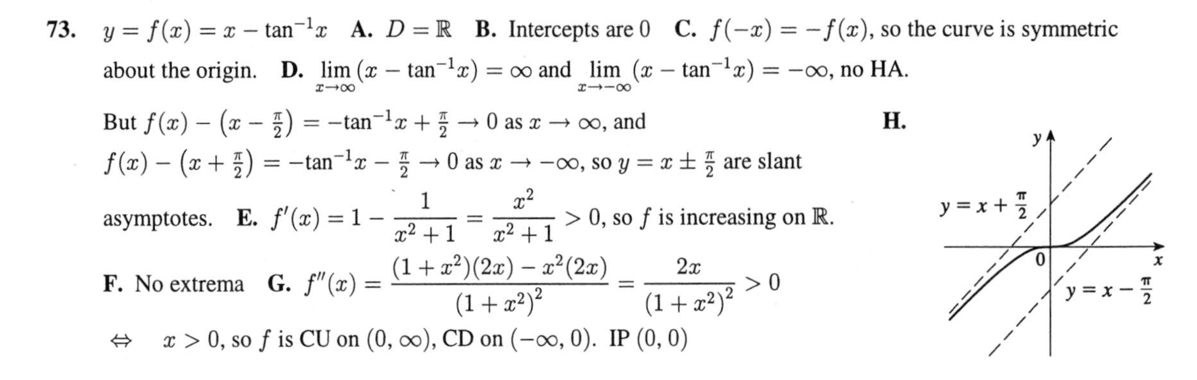

75. $f(x) = \arctan[\cos(3\arcsin x)]$. We use a CAS to compute f' and f'', and to graph f, f', and f'':

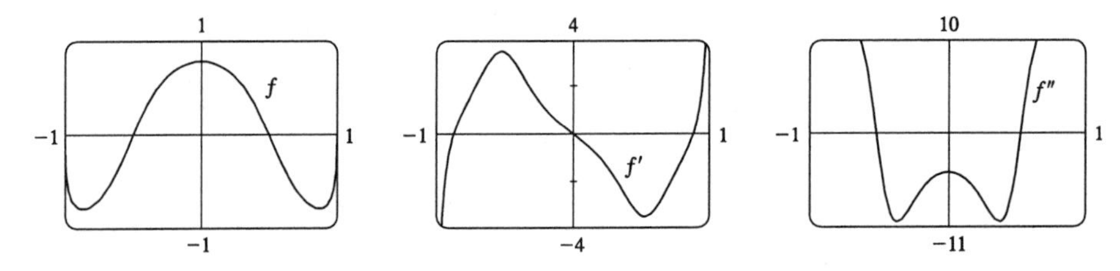

From the graph of f', it appears that the only maximum occurs at $x = 0$ and there are minima at $x = \pm 0.87$. From the graph of f'', it appears that there are inflection points at $x = \pm 0.52$.

77. $f(x) = 2x + 5/\sqrt{1-x^2} \quad \Rightarrow \quad F(x) = x^2 + 5\sin^{-1}x + C$

79. $\displaystyle\int_1^{\sqrt{3}} \frac{6}{1+x^2}\,dx = \left[6\tan^{-1}x\right]_1^{\sqrt{3}} = 6\left(\tan^{-1}\sqrt{3} - \tan^{-1}1\right) = 6\left(\tfrac{\pi}{3} - \tfrac{\pi}{4}\right) = \tfrac{\pi}{2}$

81. Let $u = x^3$. Then $du = 3x^2\,dx$, so $\displaystyle\int \frac{x^2}{\sqrt{1-x^6}}\,dx = \frac{1}{3}\int \frac{1}{\sqrt{1-u^2}}\,du = \tfrac{1}{3}\sin^{-1}u + C = \tfrac{1}{3}\sin^{-1}(x^3) + C.$

83. $\displaystyle\int \frac{x+9}{x^2+9}\,dx = \int \frac{x}{x^2+9}\,dx + 9\int \frac{1}{x^2+9}\,dx = \tfrac{1}{2}\ln(x^2+9) + 3\tan^{-1}\frac{x}{3} + C$

(Let $u = x^2 + 9$ in the first integral; use Equation 14 in the second.)

85. Let $u = 3x$. Then $du = 3\,dx$, so $\displaystyle\int \frac{dx}{1+9x^2} = \frac{1}{3}\int \frac{du}{1+u^2} = \tfrac{1}{3}\tan^{-1}u + C = \tfrac{1}{3}\tan^{-1}(3x) + C.$

87. Let $u = e^x$. Then $du = e^x\,dx$, so $\displaystyle\int \frac{e^x\,dx}{e^{2x}+1} = \int \frac{du}{u^2+1} = \tan^{-1}u + C = \tan^{-1}(e^x) + C.$

89. Let $u = \sin^{-1}x$. Then $du = \dfrac{1}{\sqrt{1-x^2}}\,dx$, so $\displaystyle\int_0^{1/2} \frac{\sin^{-1}x}{\sqrt{1-x^2}}\,dx = \int_0^{\pi/6} u\,du = \frac{u^2}{2}\bigg|_0^{\pi/6} = \frac{1}{2}\left(\frac{\pi}{6}\right)^2 = \frac{\pi^2}{72}.$

91. Let $u = x/a$. Then $du = dx/a$, so

$$\int \frac{dx}{\sqrt{a^2-x^2}} = \int \frac{dx}{a\sqrt{1-(x/a)^2}} = \int \frac{du}{\sqrt{1-u^2}} = \sin^{-1}u + C = \sin^{-1}\frac{x}{a} + C.$$

93.

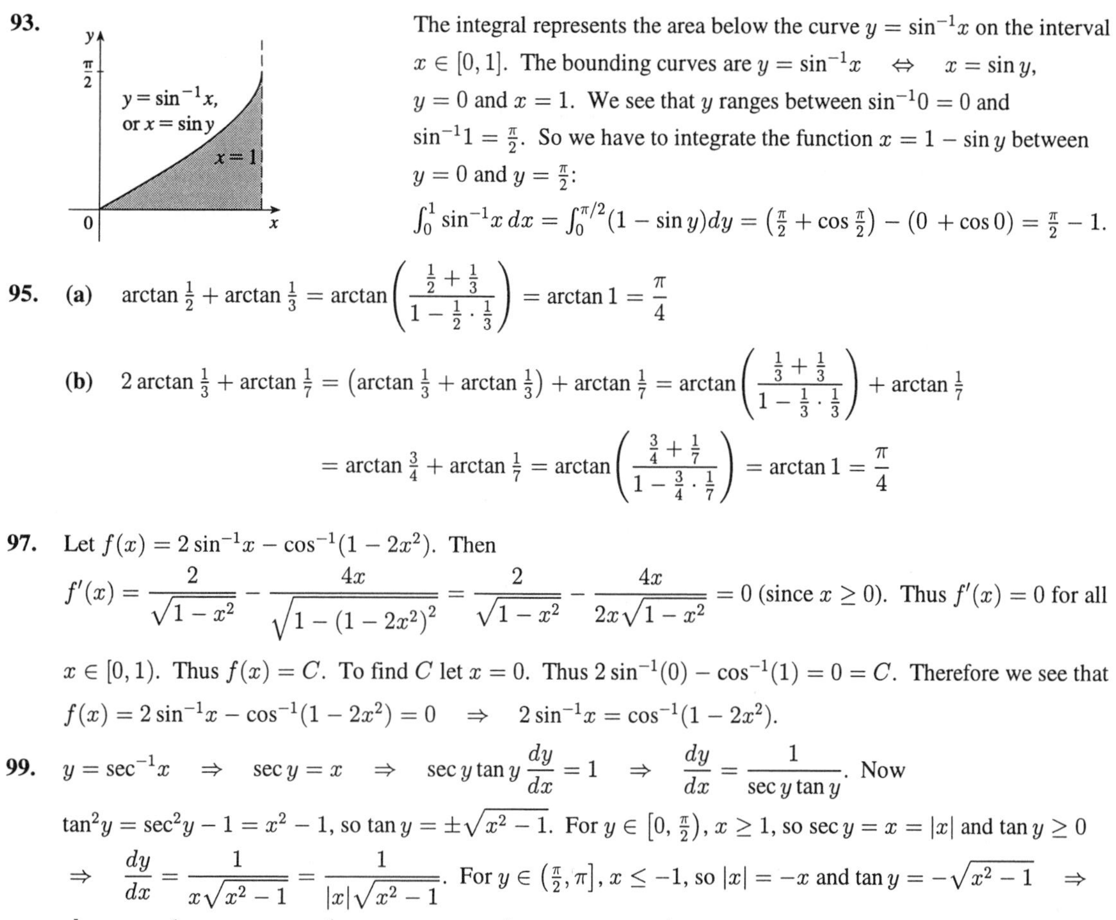

The integral represents the area below the curve $y = \sin^{-1}x$ on the interval $x \in [0, 1]$. The bounding curves are $y = \sin^{-1}x \quad \Leftrightarrow \quad x = \sin y$, $y = 0$ and $x = 1$. We see that y ranges between $\sin^{-1}0 = 0$ and $\sin^{-1}1 = \frac{\pi}{2}$. So we have to integrate the function $x = 1 - \sin y$ between $y = 0$ and $y = \frac{\pi}{2}$:

$\int_0^1 \sin^{-1}x\,dx = \int_0^{\pi/2}(1 - \sin y)dy = \left(\frac{\pi}{2} + \cos\frac{\pi}{2}\right) - (0 + \cos 0) = \frac{\pi}{2} - 1.$

95. **(a)** $\arctan\frac{1}{2} + \arctan\frac{1}{3} = \arctan\left(\dfrac{\frac{1}{2} + \frac{1}{3}}{1 - \frac{1}{2}\cdot\frac{1}{3}}\right) = \arctan 1 = \dfrac{\pi}{4}$

(b) $2\arctan\frac{1}{3} + \arctan\frac{1}{7} = \left(\arctan\frac{1}{3} + \arctan\frac{1}{3}\right) + \arctan\frac{1}{7} = \arctan\left(\dfrac{\frac{1}{3} + \frac{1}{3}}{1 - \frac{1}{3}\cdot\frac{1}{3}}\right) + \arctan\frac{1}{7}$

$$= \arctan\tfrac{3}{4} + \arctan\tfrac{1}{7} = \arctan\left(\dfrac{\frac{3}{4} + \frac{1}{7}}{1 - \frac{3}{4}\cdot\frac{1}{7}}\right) = \arctan 1 = \dfrac{\pi}{4}$$

97. Let $f(x) = 2\sin^{-1}x - \cos^{-1}(1 - 2x^2)$. Then

$f'(x) = \dfrac{2}{\sqrt{1 - x^2}} - \dfrac{4x}{\sqrt{1 - (1 - 2x^2)^2}} = \dfrac{2}{\sqrt{1 - x^2}} - \dfrac{4x}{2x\sqrt{1 - x^2}} = 0$ (since $x \geq 0$). Thus $f'(x) = 0$ for all $x \in [0, 1)$. Thus $f(x) = C$. To find C let $x = 0$. Thus $2\sin^{-1}(0) - \cos^{-1}(1) = 0 = C$. Therefore we see that $f(x) = 2\sin^{-1}x - \cos^{-1}(1 - 2x^2) = 0 \quad \Rightarrow \quad 2\sin^{-1}x = \cos^{-1}(1 - 2x^2)$.

99. $y = \sec^{-1}x \quad \Rightarrow \quad \sec y = x \quad \Rightarrow \quad \sec y \tan y\,\dfrac{dy}{dx} = 1 \quad \Rightarrow \quad \dfrac{dy}{dx} = \dfrac{1}{\sec y \tan y}$. Now

$\tan^2 y = \sec^2 y - 1 = x^2 - 1$, so $\tan y = \pm\sqrt{x^2 - 1}$. For $y \in \left[0, \frac{\pi}{2}\right)$, $x \geq 1$, so $\sec y = x = |x|$ and $\tan y \geq 0$

$\Rightarrow \quad \dfrac{dy}{dx} = \dfrac{1}{x\sqrt{x^2 - 1}} = \dfrac{1}{|x|\sqrt{x^2 - 1}}$. For $y \in \left(\frac{\pi}{2}, \pi\right]$, $x \leq -1$, so $|x| = -x$ and $\tan y = -\sqrt{x^2 - 1} \quad \Rightarrow$

$\dfrac{dy}{dx} = \dfrac{1}{\sec y \tan y} = \dfrac{1}{x\left(-\sqrt{x^2 - 1}\right)} = \dfrac{1}{(-x)\sqrt{x^2 - 1}} = \dfrac{1}{|x|\sqrt{x^2 - 1}}$.

EXERCISES 6.7

1. **(a)** $\sinh 0 = \frac{1}{2}(e^0 - e^0) = 0$ **(b)** $\cosh 0 = \frac{1}{2}(e^0 + e^0) = \frac{1}{2}(1 + 1) = 1$

3. **(a)** $\sinh(\ln 2) = \dfrac{e^{\ln 2} - e^{-\ln 2}}{2} = \dfrac{2 - \frac{1}{2}}{2} = \dfrac{3}{4}$ **(b)** $\sinh 2 = \frac{1}{2}(e^2 - e^{-2}) \approx 3.62686$

5. **(a)** $\operatorname{sech} 0 = \dfrac{1}{\cosh 0} = \dfrac{1}{1} = 1$ **(b)** $\cosh^{-1}1 = 0$ because $\cosh 0 = 1$.

7. $\sinh(-x) = \frac{1}{2}\left[e^{-x} - e^{-(-x)}\right] = \frac{1}{2}(e^{-x} - e^{x}) = -\frac{1}{2}(e^{x} - e^{-x}) = -\sinh x$

9. $\cosh x + \sinh x = \frac{1}{2}(e^x + e^{-x}) + \frac{1}{2}(e^x - e^{-x}) = \frac{1}{2}(2e^x) = e^x$

11. $\sinh x\cosh y+\cosh x\sinh y=\left[\frac{1}{2}(e^x-e^{-x})\right]\left[\frac{1}{2}(e^y+e^{-y})\right]+\left[\frac{1}{2}(e^x+e^{-x})\right]\left[\frac{1}{2}(e^y-e^{-y})\right]$

$=\frac{1}{4}[(e^{x+y}+e^{x-y}-e^{-x+y}-e^{-x-y})+(e^{x+y}-e^{x-y}+e^{-x+y}-e^{-x-y})]$

$=\frac{1}{4}(2e^{x+y}-2e^{-x-y})=\frac{1}{2}\left[e^{x+y}-e^{-(x+y)}\right]=\sinh(x+y)$

13. Divide both sides of the identity $\cosh^2x-\sinh^2x=1$ by $\sinh^2x$:

$$\frac{\cosh^2x}{\sinh^2x}-1=\frac{1}{\sinh^2x}\quad\Leftrightarrow\quad\coth^2x-1=\operatorname{csch}^2x.$$

15. By Exercise 11, $\sinh 2x=\sinh(x+x)=\sinh x\cosh x+\cosh x\sinh x=2\sinh x\cosh x$.

17. $\tanh(\ln x)=\dfrac{\sinh(\ln x)}{\cosh(\ln x)}=\dfrac{(e^{\ln x}-e^{-\ln x})/2}{(e^{\ln x}+e^{-\ln x})/2}=\dfrac{x-1/x}{x+1/x}=\dfrac{x^2-1}{x^2+1}$

19. By Exercise 9, $(\cosh x+\sinh x)^n=(e^x)^n=e^{nx}=\cosh nx+\sinh nx$.

21. $\tanh x=\frac{4}{5}>0$, so $x>0$. $\coth x=1/\tanh x=\frac{5}{4}$, $\operatorname{sech}^2x=1-\tanh^2x=1-\left(\frac{4}{5}\right)^2=\frac{9}{25}\quad\Rightarrow\quad\operatorname{sech}x=\frac{3}{5}$ (since $\operatorname{sech}x>0$), $\cosh x=1/\operatorname{sech}x=\frac{5}{3}$, $\sinh x=\tanh x\cosh x=\frac{4}{5}\cdot\frac{5}{3}=\frac{4}{3}$, and $\operatorname{csch}x=1/\sinh x=\frac{3}{4}$.

23. **(a)** $\displaystyle\lim_{x\to\infty}\tanh x=\lim_{x\to\infty}\frac{e^x-e^{-x}}{e^x+e^{-x}}=\lim_{x\to\infty}\frac{1-e^{-2x}}{1+e^{-2x}}=\frac{1-0}{1+0}=1$

(b) $\displaystyle\lim_{x\to-\infty}\tanh x=\lim_{x\to-\infty}\frac{e^x-e^{-x}}{e^x+e^{-x}}=\lim_{x\to-\infty}\frac{e^{2x}-1}{e^{2x}+1}=\frac{0-1}{0+1}=-1$

(c) $\displaystyle\lim_{x\to\infty}\sinh x=\lim_{x\to\infty}\frac{e^x-e^{-x}}{2}=\infty$

(d) $\displaystyle\lim_{x\to-\infty}\sinh x=\lim_{x\to-\infty}\frac{e^x-e^{-x}}{2}=-\infty$

(e) $\displaystyle\lim_{x\to\infty}\operatorname{sech}x=\lim_{x\to\infty}\frac{2}{e^x+e^{-x}}=0$

(f) $\displaystyle\lim_{x\to\infty}\coth x=\lim_{x\to\infty}\frac{e^x+e^{-x}}{e^x-e^{-x}}=\lim_{x\to\infty}\frac{1+e^{-2x}}{1-e^{-2x}}=\frac{1+0}{1-0}=1$ [*Or:* Use part (a)]

(g) $\displaystyle\lim_{x\to0^+}\coth x=\lim_{x\to0^+}\frac{\cosh x}{\sinh x}=\infty$, since $\sinh x\to0$ and $\coth x>0$.

(h) $\displaystyle\lim_{x\to0^-}\coth x=\lim_{x\to0^-}\frac{\cosh x}{\sinh x}=-\infty$, since $\sinh x\to0$ and $\coth x<0$.

(i) $\displaystyle\lim_{x\to-\infty}\operatorname{csch}x=\lim_{x\to-\infty}\frac{2}{e^x-e^{-x}}=0$

25. Let $y=\sinh^{-1}x$. Then $\sinh y=x$ and, by Example 1(a), $\cosh y=\sqrt{1+\sinh^2y}=\sqrt{1+x^2}$. So by Exercise 9, $e^y=\sinh y+\cosh y=x+\sqrt{1+x^2}\quad\Rightarrow\quad y=\ln\left(x+\sqrt{1+x^2}\right)$.

27. **(a)** Let $y=\tanh^{-1}x$. Then $x=\tanh y=\dfrac{e^y-e^{-y}}{e^y+e^{-y}}=\dfrac{e^{2y}-1}{e^{2y}+1}\quad\Rightarrow\quad xe^{2y}+x=e^{2y}-1\quad\Rightarrow$

$$e^{2y}=\frac{1+x}{1-x}\quad\Rightarrow\quad 2y=\ln\left(\frac{1+x}{1-x}\right)\quad\Rightarrow\quad y=\frac{1}{2}\ln\left(\frac{1+x}{1-x}\right).$$

(b) Let $y=\tanh^{-1}x$. Then $x=\tanh y$, so from Exercise 18 we have $e^{2y}=\dfrac{1+\tanh y}{1-\tanh y}=\dfrac{1+x}{1-x}\quad\Rightarrow$

$$2y=\ln\left(\frac{1+x}{1-x}\right)\quad\Rightarrow\quad y=\frac{1}{2}\ln\left(\frac{1+x}{1-x}\right).$$

29. **(a)** Let $y = \cosh^{-1}x$. Then $\cosh y = x$ and $y \geq 0 \;\Rightarrow\; \sinh y \,\dfrac{dy}{dx} = 1 \;\Rightarrow$

$$\frac{dy}{dx} = \frac{1}{\sinh y} = \frac{1}{\sqrt{\cosh^2 y - 1}} = \frac{1}{\sqrt{x^2 - 1}}$$ (since $\sinh y \geq 0$ for $y \geq 0$). *Or:* Use Formula 4.

(b) Let $y = \tanh^{-1}x$. Then $\tanh y = x \;\Rightarrow\; \operatorname{sech}^2 y \,\dfrac{dy}{dx} = 1 \;\Rightarrow\; \dfrac{dy}{dx} = \dfrac{1}{\operatorname{sech}^2 y} = \dfrac{1}{1 - \tanh^2 y} = \dfrac{1}{1 - x^2}$.

Or: Use Formula 5.

(c) Let $y = \operatorname{csch}^{-1}x$. Then $\operatorname{csch} y = x \;\Rightarrow\; -\operatorname{csch} y \coth y \,\dfrac{dy}{dx} = 1 \;\Rightarrow\; \dfrac{dy}{dx} = -\dfrac{1}{\operatorname{csch} y \coth y}$.

By Exercise 13, $\coth y = \pm\sqrt{\operatorname{csch}^2 y + 1} = \pm\sqrt{x^2 + 1}$. If $x > 0$, then $\coth y > 0$, so $\coth y = \sqrt{x^2 + 1}$. If $x < 0$, then $\coth y < 0$, so $\coth y = -\sqrt{x^2 + 1}$. In either case we have

$$\frac{dy}{dx} = -\frac{1}{\operatorname{csch} y \coth y} = -\frac{1}{|x|\sqrt{x^2 + 1}}.$$

(d) Let $y = \operatorname{sech}^{-1}x$. Then $\operatorname{sech} y = x \;\Rightarrow\; -\operatorname{sech} y \tanh y \,\dfrac{dy}{dx} = 1 \;\Rightarrow$

$$\frac{dy}{dx} = -\frac{1}{\operatorname{sech} y \tanh y} = -\frac{1}{\operatorname{sech} y \sqrt{1 - \operatorname{sech}^2 y}} = -\frac{1}{x\sqrt{1 - x^2}}.$$ (Note that $y > 0$ and so $\tanh y > 0$.)

(e) Let $y = \coth^{-1}x$. Then $\coth y = x \;\Rightarrow\; -\operatorname{csch}^2 y \,\dfrac{dy}{dx} = 1 \;\Rightarrow\; \dfrac{dy}{dx} = -\dfrac{1}{\operatorname{csch}^2 y} = \dfrac{1}{1 - \coth^2 y} = \dfrac{1}{1 - x^2}$

by Exercise 13.

31. $f(x) = \tanh 3x \;\Rightarrow\; f'(x) = 3\operatorname{sech}^2 3x$

33. $h(x) = \cosh(x^4) \;\Rightarrow\; h'(x) = \sinh(x^4)4x^3 = 4x^3\sinh(x^4)$

35. $G(x) = x^2 \operatorname{sech} x \;\Rightarrow\; G'(x) = 2x\operatorname{sech} x - x^2 \operatorname{sech} x \tanh x$

37. $H(t) = \tanh(e^t) \;\Rightarrow\; H'(t) = \operatorname{sech}^2(e^t)[e^t] = e^t \operatorname{sech}^2(e^t)$

39. $y = x^{\cosh x} \;\Rightarrow\; \ln y = \cosh x \ln x \;\Rightarrow\; \dfrac{y'}{y} = \sinh x \ln x + \dfrac{\cosh x}{x} \;\Rightarrow\; y' = x^{\cosh x}\left(\sinh x \ln x + \dfrac{\cosh x}{x}\right)$

41. $y = \cosh^{-1}(x^2) \;\Rightarrow\; y' = \left(1\Big/\sqrt{(x^2)^2 - 1}\right)(2x) = 2x/\sqrt{x^4 - 1}$

43. $y = x\ln(\operatorname{sech} 4x) \;\Rightarrow\; y' = \ln(\operatorname{sech} 4x) + x\,\dfrac{-\operatorname{sech} 4x \tanh 4x}{\operatorname{sech} 4x}(4) = \ln(\operatorname{sech} 4x) - 4x\tanh 4x$

45. $y = x\sinh^{-1}(x/3) - \sqrt{9 + x^2} \;\Rightarrow$

$$y' = \sinh^{-1}\left(\frac{x}{3}\right) + x\frac{1/3}{\sqrt{1 + (x/3)^2}} - \frac{2x}{2\sqrt{9 + x^2}} = \sinh^{-1}\left(\frac{x}{3}\right) + \frac{x}{\sqrt{9 + x^2}} - \frac{x}{\sqrt{9 + x^2}} = \sinh^{-1}\left(\frac{x}{3}\right)$$

47. $y = \coth^{-1}\sqrt{x^2 + 1} \;\Rightarrow\; y' = \dfrac{1}{1 - (x^2 + 1)}\,\dfrac{2x}{2\sqrt{x^2 + 1}} = -\dfrac{1}{x\sqrt{x^2 + 1}}$

49. The tangent to $y = \cosh x$ has slope 1 when $y' = \sinh x = 1 \;\Rightarrow\; x = \sinh^{-1} 1 = \ln\left(1 + \sqrt{2}\right)$, by Equation 3. Since $\sinh x = 1$ and $y = \cosh x = \sqrt{1 + \sinh^2 x}$, we have $\cosh x = \sqrt{2}$. The point is $\left(\ln\left(1 + \sqrt{2}\right), \sqrt{2}\right)$.

51. $u = 2x \;\Rightarrow\; du = 2\,dx$, so $\int \sinh 2x\,dx = \frac{1}{2}\int \sinh u\,du = \frac{1}{2}\cosh u + C = \frac{1}{2}\cosh 2x + C$.

53. Let $u = \sinh x$, so $du = \cosh x\,dx$, and $\displaystyle\int \coth x\,dx = \int \frac{\cosh x}{\sinh x}\,dx = \int \frac{du}{u} = \ln|u| + C = \ln|\sinh x| + C$.

55. Let $u = x/2$. Then $du = \frac{1}{2}\,dx \;\Rightarrow$

$$\int \frac{1}{\sqrt{4 + x^2}}\,dx = \frac{1}{2}\int \frac{1}{\sqrt{1 + (x/2)^2}}\,dx = \int \frac{1}{\sqrt{1 + u^2}}\,du = \sinh^{-1} u + C = \sinh^{-1}\left(\frac{x}{2}\right) + C.$$

57. $\displaystyle\int_0^{1/2} \frac{1}{1 - x^2}\,dx = \left[\tanh^{-1} x\right]_0^{1/2} = \tanh^{-1}\tfrac{1}{2} = \tfrac{1}{2}\ln\left(\frac{1 + 1/2}{1 - 1/2}\right)$ (from Equation 5) $= \frac{1}{2}\ln 3$.

59. **(a)**

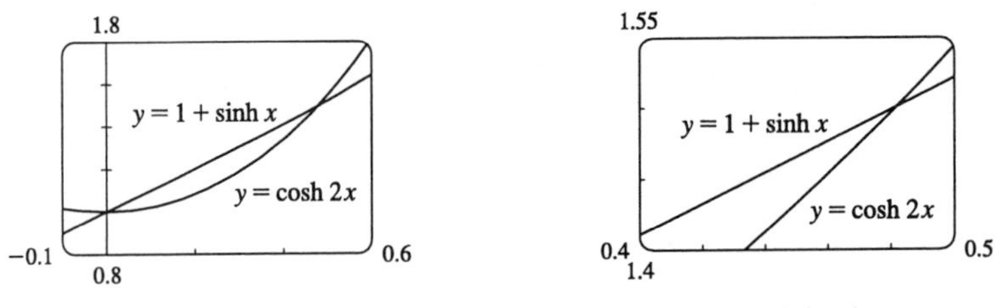

From the graphs, we estimate that the two curves $y = \cosh 2x$ and $y = 1 + \sinh x$ intersect at $x = 0$ and at $x \approx 0.481$.

(b) We have found the two roots of the equation $\cosh 2x = 1 + \sinh x$ to be $x = 0$ and $x \approx 0.481$. Note from the first graph that $1 + \sinh x > \cosh 2x$ on the interval $(0, 0.481)$, so the area between the two curves is

$A \approx \int_0^{0.481}(1 + \sinh x - \cosh 2x)dx = \left[x + \cosh x - \frac{1}{2}\sinh 2x\right]_0^{0.481}$

$= [0.481 + \cosh 0.481 - \frac{1}{2}\sinh(2 \cdot 0.481)] - [0 + \cosh 0 - \frac{1}{2}\sinh(2 \cdot 0)] \approx 0.0402.$

61. **(a)** $y = A\sinh mx + B\cosh mx \;\Rightarrow\; y' = mA\cosh mx + mB\sinh mx \;\Rightarrow$

$y'' = m^2 A\sinh mx + m^2 B\cosh mx = m^2 y$

(b) From part (a), a solution of $y'' = 9y$ is $y(x) = A\sinh 3x + B\cosh 3x$. So

$-4 = y(0) = A\sinh 0 + B\cosh 0 = B$, so $B = -4$. Now $y'(x) = 3A\cosh 3x - 12\sinh 3x \;\Rightarrow$

$6 = y'(0) = 3A \;\Rightarrow\; A = 2$, so $y = 2\sinh 3x - 4\cosh 3x$.

63. $\cosh x = \cosh[\ln(\sec\theta + \tan\theta)] = \frac{1}{2}\left[e^{\ln(\sec\theta + \tan\theta)} + e^{-\ln(\sec\theta + \tan\theta)}\right]$

$$= \frac{1}{2}\left[\sec\theta + \tan\theta + \frac{1}{\sec\theta + \tan\theta}\right] = \frac{1}{2}\left[\sec\theta + \tan\theta + \frac{\sec\theta - \tan\theta}{(\sec\theta + \tan\theta)(\sec\theta - \tan\theta)}\right]$$

$$= \frac{1}{2}\left[\sec\theta + \tan\theta + \frac{\sec\theta - \tan\theta}{\sec^2\theta - \tan^2\theta}\right] = \tfrac{1}{2}(\sec\theta + \tan\theta + \sec\theta - \tan\theta) = \sec\theta.$$

EXERCISES 6.8

NOTE: The use of l'Hospital's Rule is indicated by an H above the equal sign: $\overset{\text{H}}{=}$

1. $\displaystyle \lim_{x\to 2}\frac{x-2}{x^2-4}=\lim_{x\to 2}\frac{x-2}{(x-2)(x+2)}=\lim_{x\to 2}\frac{1}{x+2}=\frac{1}{4}$

3. $\displaystyle \lim_{x\to -1}\frac{x^6-1}{x^4-1}\overset{\text{H}}{=}\lim_{x\to -1}\frac{6x^5}{4x^3}=\frac{-6}{-4}=\frac{3}{2}$

5. $\displaystyle \lim_{x\to 0}\frac{e^x-1}{\sin x}\overset{\text{H}}{=}\lim_{x\to 0}\frac{e^x}{\cos x}=\frac{1}{1}=1$

7. $\displaystyle \lim_{x\to 0}\frac{\sin x}{x^3}\overset{\text{H}}{=}\lim_{x\to 0}\frac{\cos x}{3x^2}=\infty$

9. $\displaystyle \lim_{x\to 0}\frac{\tan x}{x+\sin x}\overset{\text{H}}{=}\lim_{x\to 0}\frac{\sec^2 x}{1+\cos x}=\frac{1}{1+1}=\frac{1}{2}$

11. $\displaystyle \lim_{x\to\infty}\frac{\ln x}{x}\overset{\text{H}}{=}\lim_{x\to\infty}\frac{1/x}{1}=0$

13. $\displaystyle \lim_{x\to\infty}\frac{e^x}{x^3}\overset{\text{H}}{=}\lim_{x\to\infty}\frac{e^x}{3x^2}\overset{\text{H}}{=}\lim_{x\to\infty}\frac{e^x}{6x}\overset{\text{H}}{=}\lim_{x\to\infty}\frac{e^x}{6}=\infty$

15. $\displaystyle \lim_{x\to a}\frac{x^{1/3}-a^{1/3}}{x-a}\overset{\text{H}}{=}\lim_{x\to a}\frac{(1/3)x^{-2/3}}{1}=\frac{1}{3a^{2/3}}$

17. $\displaystyle \lim_{x\to 0}\frac{e^x-1-x}{x^2}\overset{\text{H}}{=}\lim_{x\to 0}\frac{e^x-1}{2x}\overset{\text{H}}{=}\lim_{x\to 0}\frac{e^x}{2}=\frac{1}{2}$

19. $\displaystyle \lim_{x\to 0}\frac{\sin x}{e^x}=\frac{0}{1}=0$

21. $\displaystyle \lim_{x\to 0}\frac{1-\cos x}{x^2}\overset{\text{H}}{=}\lim_{x\to 0}\frac{\sin x}{2x}\overset{\text{H}}{=}\lim_{x\to 0}\frac{\cos x}{2}=\frac{1}{2}$

23. $\displaystyle \lim_{x\to 2^-}\frac{\ln x}{\sqrt{2-x}}=\infty$ since $\sqrt{2-x}\to 0$ but $\ln x\to \ln 2$

25. $\displaystyle \lim_{x\to\infty}\frac{\ln\ln x}{\sqrt{x}}\overset{\text{H}}{=}\lim_{x\to\infty}\frac{1/(x\ln x)}{1/(2\sqrt{x})}=\lim_{x\to\infty}\frac{2}{\sqrt{x}\ln x}=0$

27. $\displaystyle \lim_{x\to 0}\frac{\tan^{-1}(2x)}{3x}\overset{\text{H}}{=}\lim_{x\to 0}\frac{2/(1+4x^2)}{3}=\frac{2}{3}$

29. $\displaystyle \lim_{x\to 0}\frac{\tan\alpha x}{x}\overset{\text{H}}{=}\lim_{x\to 0}\frac{\alpha\sec^2\alpha x}{1}=\alpha$

31. $\displaystyle \lim_{x\to 0}\frac{\tan 2x}{\tanh 3x}\overset{\text{H}}{=}\lim_{x\to 0}\frac{2\sec^2 2x}{3\,\text{sech}^2 3x}=\frac{2}{3}$

33. $\displaystyle \lim_{x\to 0}\frac{x+\sin 3x}{x-\sin 3x}\overset{\text{H}}{=}\lim_{x\to 0}\frac{1+3\cos 3x}{1-3\cos 3x}=\frac{1+3}{1-3}=-2$

35. $\displaystyle \lim_{x\to 0}\frac{e^{4x}-1}{\cos x}=\frac{0}{1}=0$

37. $\displaystyle \lim_{x\to 0}\frac{\tan x-\sin x}{x^3}\overset{\text{H}}{=}\lim_{x\to 0}\frac{\sec^2 x-\cos x}{3x^2}\overset{\text{H}}{=}\lim_{x\to 0}\frac{2\sec^2 x\tan x+\sin x}{6x}$

$\displaystyle \overset{\text{H}}{=}\lim_{x\to 0}\frac{4\sec^2 x\tan^2 x+2\sec^4 x+\cos x}{6}=\frac{0+2+1}{6}=\frac{1}{2}$

39. $\displaystyle \lim_{x\to 0^+}\sqrt{x}\ln x=\lim_{x\to 0^+}\frac{\ln x}{x^{-1/2}}\overset{\text{H}}{=}\lim_{x\to 0^+}\frac{1/x}{-\frac{1}{2}x^{-3/2}}=\lim_{x\to 0^+}\left(-2\sqrt{x}\right)=0$

41. $\displaystyle \lim_{x\to\infty}e^{-x}\ln x=\lim_{x\to\infty}\frac{\ln x}{e^x}\overset{\text{H}}{=}\lim_{x\to\infty}\frac{1/x}{e^x}=\lim_{x\to\infty}\frac{1}{xe^x}=0$

43. $\displaystyle \lim_{x\to\infty}x^3e^{-x^2}=\lim_{x\to\infty}\frac{x^3}{e^{x^2}}\overset{\text{H}}{=}\lim_{x\to\infty}\frac{3x^2}{2xe^{x^2}}=\lim_{x\to\infty}\frac{3x}{2e^{x^2}}\overset{\text{H}}{=}\lim_{x\to\infty}\frac{3}{4xe^{x^2}}=0$

45. $\displaystyle \lim_{x\to\pi}(x-\pi)\cot x=\lim_{x\to\pi}\frac{x-\pi}{\tan x}\overset{\text{H}}{=}\lim_{x\to\pi}\frac{1}{\sec^2 x}=\frac{1}{(-1)^2}=1$

47. $\lim_{x\to 0}\left(\frac{1}{x^4}-\frac{1}{x^2}\right)=\lim_{x\to 0}\frac{1-x^2}{x^4}=\infty$

49. $\lim_{x\to 0}\left(\frac{1}{x}-\csc x\right)=\lim_{x\to 0}\left(\frac{1}{x}-\frac{1}{\sin x}\right)=\lim_{x\to 0}\frac{\sin x-x}{x\sin x}$

$$\overset{\text{H}}{=}\lim_{x\to 0}\frac{\cos x-1}{\sin x+x\cos x}\overset{\text{H}}{=}\lim_{x\to 0}\frac{-\sin x}{2\cos x-x\sin x}=\frac{0}{2}=0$$

51. $\lim_{x\to\infty}\left(x-\sqrt{x^2-1}\right)=\lim_{x\to\infty}\left(x-\sqrt{x^2-1}\right)\frac{x+\sqrt{x^2-1}}{x+\sqrt{x^2-1}}=\lim_{x\to\infty}\frac{x^2-(x^2-1)}{x+\sqrt{x^2-1}}=\lim_{x\to\infty}\frac{1}{x+\sqrt{x^2-1}}=0$

53. $\lim_{x\to\infty}\left[\frac{x^3}{x^2-1}-\frac{x^3}{x^2+1}\right]=\lim_{x\to\infty}\frac{x^3(x^2+1)-x^3(x^2-1)}{(x^2-1)(x^2+1)}=\lim_{x\to\infty}\frac{2x^3}{x^4-1}=\lim_{x\to\infty}\frac{2/x}{1-1/x^4}=0$

55. $y=x^{\sin x}\quad\Rightarrow\quad \ln y=\sin x\ln x$, so $\lim_{x\to 0^+}\ln y=\lim_{x\to 0^+}\sin x\ln x=\lim_{x\to 0^+}\frac{\ln x}{\csc x}\overset{\text{H}}{=}\lim_{x\to 0^+}\frac{1/x}{-\csc x\cot x}$

$=-\left[\lim_{x\to 0^+}\frac{\sin x}{x}\right]\left[\lim_{x\to 0^+}\tan x\right]=-1\cdot 0=0\quad\Rightarrow\quad \lim_{x\to 0^+}x^{\sin x}=\lim_{x\to 0^+}e^{\ln y}=e^0=1.$

57. $y=(1-2x)^{1/x}\quad\Rightarrow\quad \ln y=\frac{1}{x}\ln(1-2x)\quad\Rightarrow\quad \lim_{x\to 0}\ln y=\lim_{x\to 0}\frac{\ln(1-2x)}{x}\overset{\text{H}}{=}\lim_{x\to 0}\frac{-2/(1-2x)}{1}=-2$

$\Rightarrow\quad \lim_{x\to 0}(1-2x)^{1/x}=\lim_{x\to 0}e^{\ln y}=e^{-2}$

59. $y=\left(1+\frac{3}{x}+\frac{5}{x^2}\right)^x\quad\Rightarrow\quad \ln y=x\ln\left(1+\frac{3}{x}+\frac{5}{x^2}\right)\quad\Rightarrow$

$$\lim_{x\to\infty}\ln y=\lim_{x\to\infty}\frac{\ln\left(1+\frac{3}{x}+\frac{5}{x^2}\right)}{1/x}\overset{\text{H}}{=}\lim_{x\to\infty}\frac{\left(-\frac{3}{x^2}-\frac{10}{x^3}\right)\Big/\left(1+\frac{3}{x}+\frac{5}{x^2}\right)}{-1/x^2}=\lim_{x\to\infty}\frac{3+10/x}{1+3/x+5/x^2}=3,\text{ so}$$

$\lim_{x\to\infty}\left(1+\frac{3}{x}+\frac{5}{x^2}\right)^x=\lim_{x\to\infty}e^{\ln y}=e^3.$

61. $y=x^{1/x}\quad\Rightarrow\quad \ln y=(1/x)\ln x\quad\Rightarrow\quad \lim_{x\to\infty}\ln y=\lim_{x\to\infty}\frac{\ln x}{x}\overset{\text{H}}{=}\lim_{x\to\infty}\frac{1/x}{1}=0\quad\Rightarrow$

$\lim_{x\to\infty}x^{1/x}=\lim_{x\to\infty}e^{\ln y}=e^0=1$

63. $y=(\cot x)^{\sin x}\quad\Rightarrow\quad \ln y=\sin x\ln(\cot x)\quad\Rightarrow$

$\lim_{x\to 0^+}\ln y=\lim_{x\to 0^+}\frac{\ln(\cot x)}{\csc x}\overset{\text{H}}{=}\lim_{x\to 0^+}\frac{(-\csc^2 x)/\cot x}{-\csc x\cot x}=\lim_{x\to 0^+}\frac{\csc x}{\cot^2 x}$

$=\lim_{x\to 0^+}\frac{\sin x}{\cos^2 x}=0$, so $\lim_{x\to 0^+}(\cot x)^{\sin x}=\lim_{x\to 0^+}e^{\ln y}=e^0=1.$

65. $y=\left(\frac{x}{x+1}\right)^x\quad\Rightarrow\quad \ln y=x\ln\left(\frac{x}{x+1}\right)\quad\Rightarrow$

$\lim_{x\to\infty}\ln y=\lim_{x\to\infty}x\ln\left(\frac{x}{x+1}\right)=\lim_{x\to\infty}\frac{\ln x-\ln(x+1)}{1/x}\overset{\text{H}}{=}\lim_{x\to\infty}\frac{1/x-1/(x+1)}{-1/x^2}$

$=\lim_{x\to\infty}\left(-x+\frac{x^2}{x+1}\right)=\lim_{x\to\infty}\frac{-x}{x+1}=-1$, so $\lim_{x\to\infty}\left(\frac{x}{x+1}\right)^x=\lim_{x\to\infty}e^{\ln y}=e^{-1}$

Or: $\lim_{x\to\infty}\left(\frac{x}{x+1}\right)^x=\lim_{x\to\infty}\left[\left(\frac{x+1}{x}\right)^{-1}\right]^x=\left[\lim_{x\to\infty}\left(1+\frac{1}{x}\right)^x\right]^{-1}=e^{-1}$

67. Let $y = (-\ln x)^x$. Then $\ln y = x\ln(-\ln x) \;\Rightarrow\; \lim\limits_{x\to 0^+}\ln y = \lim\limits_{x\to 0^+} x\ln(-\ln x) = \lim\limits_{x\to 0^+}\dfrac{\ln(-\ln x)}{1/x}$

$\overset{\text{H}}{=} \lim\limits_{x\to 0^+}\dfrac{(1/-\ln x)(-1/x)}{-1/x^2} = \lim\limits_{x\to 0^+}\dfrac{-x}{\ln x} = 0 \;\Rightarrow\; \lim\limits_{x\to 0^+}(-\ln x)^x = e^0 = 1.$

69.

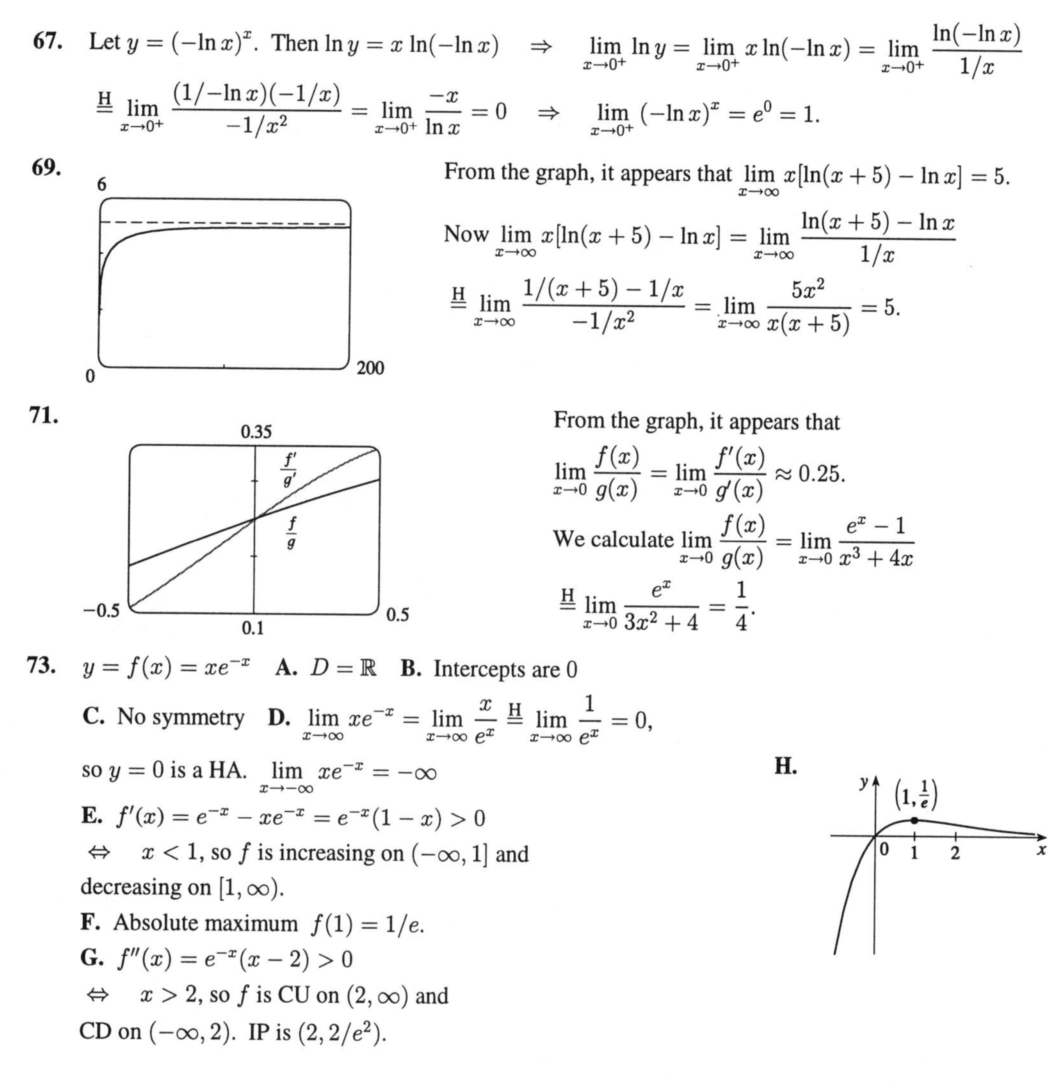

From the graph, it appears that $\lim\limits_{x\to\infty} x[\ln(x+5) - \ln x] = 5.$

Now $\lim\limits_{x\to\infty} x[\ln(x+5) - \ln x] = \lim\limits_{x\to\infty}\dfrac{\ln(x+5) - \ln x}{1/x}$

$\overset{\text{H}}{=} \lim\limits_{x\to\infty}\dfrac{1/(x+5) - 1/x}{-1/x^2} = \lim\limits_{x\to\infty}\dfrac{5x^2}{x(x+5)} = 5.$

71. From the graph, it appears that

$\lim\limits_{x\to 0}\dfrac{f(x)}{g(x)} = \lim\limits_{x\to 0}\dfrac{f'(x)}{g'(x)} \approx 0.25.$

We calculate $\lim\limits_{x\to 0}\dfrac{f(x)}{g(x)} = \lim\limits_{x\to 0}\dfrac{e^x - 1}{x^3 + 4x}$

$\overset{\text{H}}{=} \lim\limits_{x\to 0}\dfrac{e^x}{3x^2 + 4} = \dfrac{1}{4}.$

73. $y = f(x) = xe^{-x}$ **A.** $D = \mathbb{R}$ **B.** Intercepts are 0

C. No symmetry **D.** $\lim\limits_{x\to\infty} xe^{-x} = \lim\limits_{x\to\infty}\dfrac{x}{e^x} \overset{\text{H}}{=} \lim\limits_{x\to\infty}\dfrac{1}{e^x} = 0,$

so $y = 0$ is a HA. $\lim\limits_{x\to-\infty} xe^{-x} = -\infty$

E. $f'(x) = e^{-x} - xe^{-x} = e^{-x}(1 - x) > 0$

$\Leftrightarrow\; x < 1$, so f is increasing on $(-\infty, 1]$ and decreasing on $[1, \infty)$.

F. Absolute maximum $f(1) = 1/e$.

G. $f''(x) = e^{-x}(x - 2) > 0$

$\Leftrightarrow\; x > 2$, so f is CU on $(2, \infty)$ and CD on $(-\infty, 2)$. IP is $(2, 2/e^2)$.

H.

75. $y = f(x) = x\ln x$ **A.** $D = (0, \infty)$ **B.** x-intercept when $\ln x = 0 \;\Leftrightarrow\; x = 1$, no y-intercept

C. No symmetry **D.** $\lim\limits_{x\to\infty} x\ln x = \infty$, $\lim\limits_{x\to 0^+} x\ln x$

$= \lim\limits_{x\to 0^+}\dfrac{\ln x}{1/x} \overset{\text{H}}{=} \lim\limits_{x\to 0^+}\dfrac{1/x}{-1/x^2} = \lim\limits_{x\to 0^+}(-x) = 0$, no asymptotes. **E.** $f'(x) = \ln x + 1 = 0$ when $\ln x = -1$

$\Leftrightarrow\; x = e^{-1}$. $f'(x) > 0 \;\Leftrightarrow\; \ln x > -1 \;\Leftrightarrow\; x > e^{-1}$, so f is increasing on $[1/e, \infty)$ and decreasing on $(0, 1/e]$. **F.** $f(1/e) = -1/e$ is an absolute and local minimum. **G.** $f''(x) = 1/x > 0$, so f is CU on $(0, \infty)$. No IP

H.

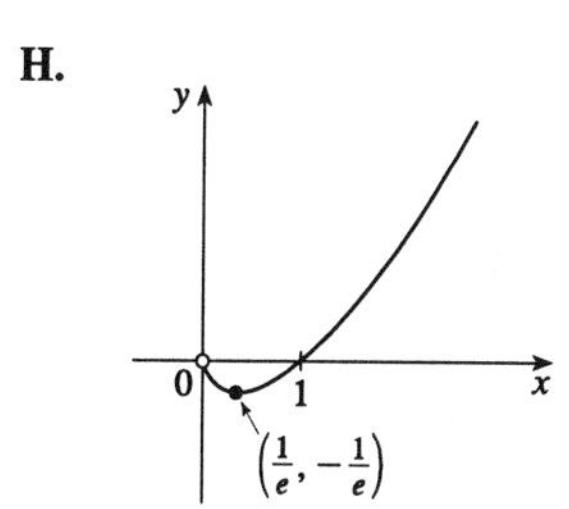

77. $y = f(x) = x^2 \ln x$ **A.** $D = (0, \infty)$ **B.** x-intercept when $\ln x = 0 \Leftrightarrow x = 1$, no y-intercept **C.** No symmetry **D.** $\lim_{x\to\infty} x^2 \ln x = \infty$, $\lim_{x\to 0^+} x^2 \ln x = \lim_{x\to 0^+} \dfrac{\ln x}{1/x^2} \overset{\text{H}}{=} \lim_{x\to 0^+} \dfrac{1/x}{-2/x^3} = \lim_{x\to 0^+}\left(-\dfrac{x^2}{2}\right) = 0$, no asymptote

E. $f'(x) = 2x\ln x + x = x(2\ln x + 1) > 0 \Leftrightarrow \ln x > -\frac{1}{2} \Leftrightarrow$ $x > e^{-1/2}$, so f is increasing on $\left[1/\sqrt{e}, \infty\right)$, decreasing on $\left(0, 1/\sqrt{e}\,\right]$.

F. $f\left(1/\sqrt{e}\right) = -1/(2e)$ is an absolute minimum.

G. $f''(x) = 2\ln x + 3 > 0 \Leftrightarrow \ln x > -\frac{3}{2} \Leftrightarrow x > e^{-3/2}$, so f is CU on $\left(e^{-3/2}, \infty\right)$ and CD on $\left(0, e^{-3/2}\right)$. IP is $\left(e^{-3/2}, -3/(2e^3)\right)$

H.

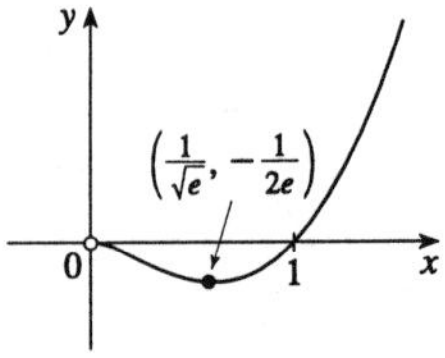

79. $y = f(x) = xe^{-x^2}$ **A.** $D = \mathbb{R}$ **B.** Intercepts are 0 **C.** $f(-x) = -f(x)$, so the curve is symmetric about the origin. **D.** $\lim_{x\to\pm\infty} xe^{-x^2} = \lim_{x\to\pm\infty} \dfrac{x}{e^{x^2}} \overset{\text{H}}{=} \lim_{x\to\pm\infty} \dfrac{1}{2xe^{x^2}} = 0$, so $y = 0$ is a HA.

E. $f'(x) = e^{-x^2} - 2x^2 e^{-x^2} = e^{-x^2}(1 - 2x^2) > 0 \Leftrightarrow x^2 < \frac{1}{2} \Leftrightarrow |x| < \frac{1}{\sqrt{2}}$, so f is increasing on $\left[-\frac{1}{\sqrt{2}}, \frac{1}{\sqrt{2}}\right]$ and decreasing on $\left(-\infty, -\frac{1}{\sqrt{2}}\right]$ and $\left[\frac{1}{\sqrt{2}}, \infty\right)$. **F.** $f\left(\frac{1}{\sqrt{2}}\right) = 1/\sqrt{2e}$ is a local maximum, $f\left(-\frac{1}{\sqrt{2}}\right) = -1/\sqrt{2e}$ is a local minimum.

G.
$$f''(x) = -2xe^{-x^2}(1 - 2x^2) - 4xe^{-x^2} = 2xe^{-x^2}(2x^2 - 3) > 0 \Leftrightarrow$$
$x > \sqrt{\frac{3}{2}}$ or $-\sqrt{\frac{3}{2}} < x < 0$, so f is CU on $\left(\sqrt{\frac{3}{2}}, \infty\right)$ and $\left(-\sqrt{\frac{3}{2}}, 0\right)$ and CD on $\left(-\infty, -\sqrt{\frac{3}{2}}\right)$ and $\left(0, \sqrt{\frac{3}{2}}\right)$.
IP are $(0, 0)$ and $\left(\pm\sqrt{\frac{3}{2}}, \pm\sqrt{\frac{3}{2}}e^{-3/2}\right)$.

H.

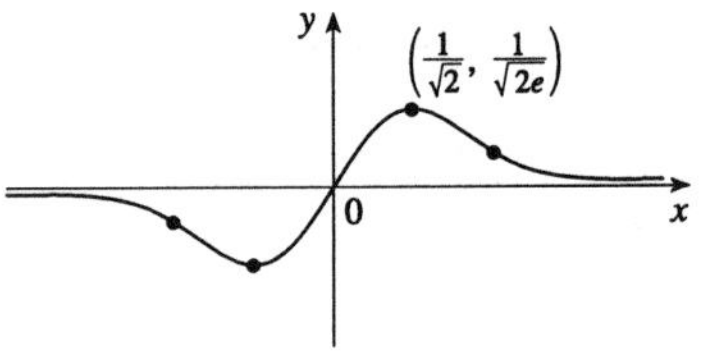

81. $y = f(x) = xe^{1/x}$ **A.** $D = \{x \mid x \neq 0\}$ **B.** No intercepts **C.** No symmetry **D.** $\lim_{x\to\infty} xe^{1/x} = \infty$, $\lim_{x\to-\infty} xe^{1/x} = -\infty$, no HA. $\lim_{x\to 0^+} xe^{1/x} = \lim_{x\to 0^+} \dfrac{e^{1/x}}{1/x} \overset{\text{H}}{=} \lim_{x\to 0^+} \dfrac{e^{1/x}(-1/x^2)}{-1/x^2} = \lim_{x\to 0^+} e^{1/x} = \infty$, so $x = 0$ is a VA.
Also $\lim_{x\to 0^-} xe^{1/x} = 0$ since $1/x \to -\infty \Rightarrow e^{1/x} \to 0$.

E. $f'(x) = e^{1/x} + xe^{1/x}(-1/x^2) = e^{1/x}(1 - 1/x) > 0 \Leftrightarrow 1/x < 1$ $\Leftrightarrow x < 0$ or $x > 1$, so f is increasing on $(-\infty, 0)$ and $[1, \infty)$, decreasing on $(0, 1]$. **F.** $f(1) = e$ is a local minimum.

G. $f''(x) = e^{1/x}(-1/x^2)(1 - 1/x) + e^{1/x}(1/x^2) = e^{1/x}/x^3 > 0$ $\Leftrightarrow x > 0$, so f is CU on $(0, \infty)$ and CD on $(-\infty, 0)$. No IP

H.

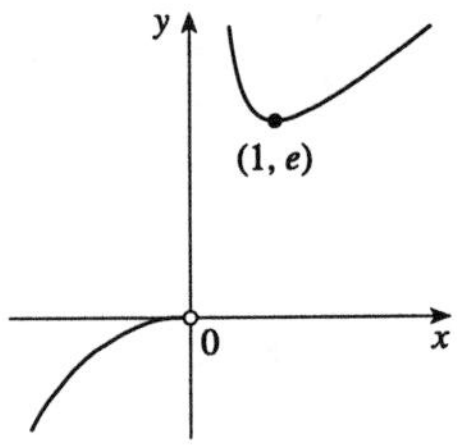

83. $y = f(x) = x - \ln(1 + x)$ **A.** $D = \{x \mid x > -1\} = (-1, \infty)$ **B.** Intercepts are 0 **C.** No symmetry
D. $\lim_{x\to -1^+} [x - \ln(1 + x)] = \infty$, so $x = -1$ is a VA. $\lim_{x\to\infty} [x - \ln(1 + x)] = \lim_{x\to\infty} x\left(1 - \dfrac{\ln(1 + x)}{x}\right) = \infty$, since $\lim_{x\to\infty} \dfrac{\ln(1 + x)}{x} \overset{\text{H}}{=} \lim_{x\to\infty} \dfrac{1/(1 + x)}{1} = 0$.

E. $f'(x) = 1 - 1/(1 + x) = x/(1 + x) > 0 \Leftrightarrow x > 0$ since $x + 1 > 0$. So f is increasing on $[0, \infty)$ and decreasing on $(-1, 0]$.

F. $f(0) = 0$ is an absolute minimum. **G.** $f''(x) = 1/(1 + x)^2 > 0$, so f is CU on $(-1, \infty)$.

H.

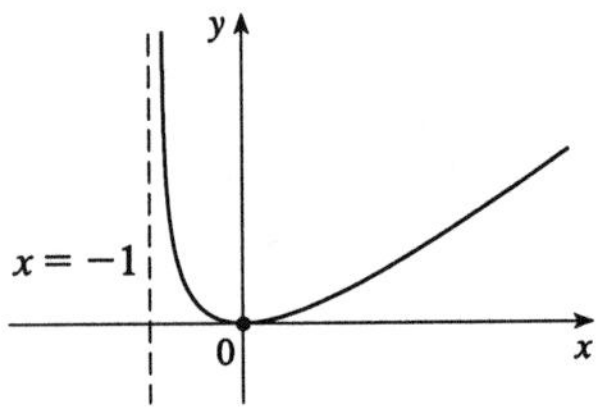

85. **(a)**

1.6

0 3

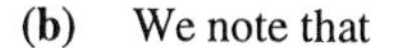

(b) We note that

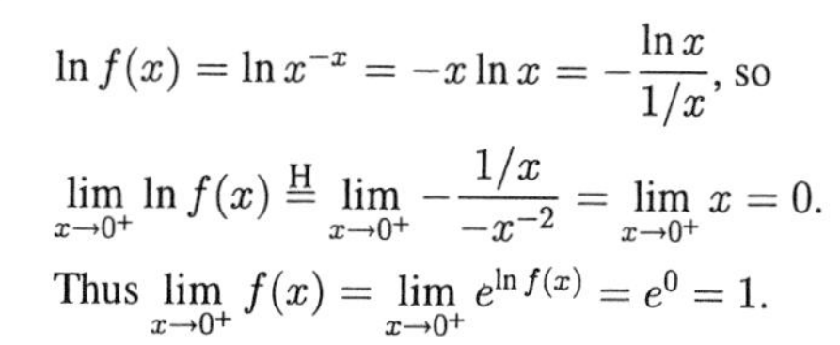

$$\ln f(x) = \ln x^{-x} = -x \ln x = -\frac{\ln x}{1/x}, \text{ so}$$

$$\lim_{x\to 0^+} \ln f(x) \overset{\text{H}}{=} \lim_{x\to 0^+} -\frac{1/x}{-x^{-2}} = \lim_{x\to 0^+} x = 0.$$

Thus $\lim_{x\to 0^+} f(x) = \lim_{x\to 0^+} e^{\ln f(x)} = e^0 = 1.$

(c) From the graph, it appears that there is a local and absolute maximum of about $f(0.37) \approx 1.44$. To find the exact value, we differentiate: $f(x) = x^{-x} = e^{-x\ln x} \quad \Rightarrow$

$$f'(x) = e^{-x\ln x}\left[-x\left(\frac{1}{x}\right) + \ln x(-1)\right] = -x^{-x}(1 + \ln x).$$ This is 0 only when $1 + \ln x = 0 \quad \Leftrightarrow$

$x = e^{-1}$. Also $f'(x)$ changes from positive to negative at e^{-1}.

So the maximum value is $f(1/e) = (1/e)^{-1/e} = e^{1/e}$.

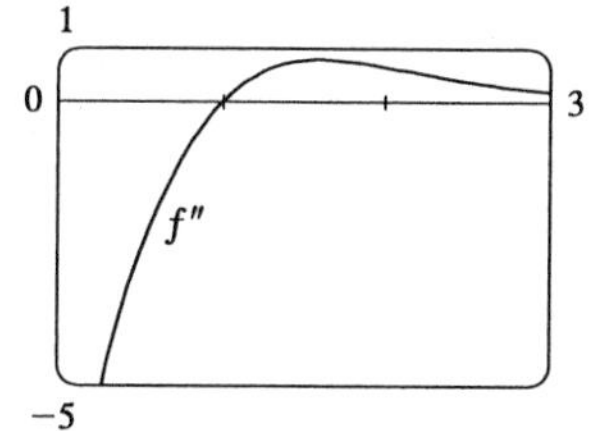

(d) We differentiate again to get

$$f''(x) = -x^{-x}(1/x) + (1 + \ln x)^2(x^{-x}) = x^{-x}\left[(1 + \ln x)^2 - 1/x\right].$$

From the graph of $f''(x)$, it seems that $f''(x)$ changes from negative to positive at $x = 1$, so we estimate that f has an IP at $x = 1$.

87. **(a)**

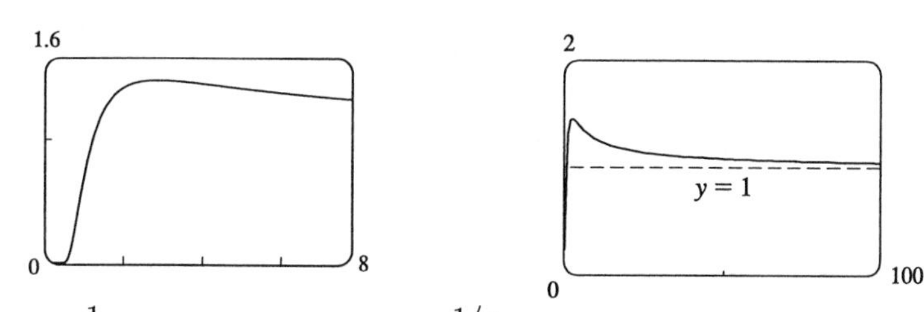

(b) $\ln f(x) = \ln x^{1/x} = \frac{1}{x}\ln x$, so $\lim_{x\to\infty} \ln f(x) \overset{\text{H}}{=} \lim_{x\to\infty} \frac{1/x}{1} = 0$. Therefore $\lim_{x\to\infty} f(x) = \lim_{x\to\infty} e^{\ln f(x)} = 1$.

Also $\lim_{x\to 0^+} \ln f(x) = \lim_{x\to 0^+}\left(\frac{1}{x}\ln x\right) = -\infty$. So $\lim_{x\to 0^+} f(x) = \lim_{x\to 0^+} e^{\ln f(x)} = 0$.

(c) From the graph, it appears that f has a local maximum at about $f(2.7) \approx 1.44$. To find the exact value, we differentiate: $f(x) = x^{1/x} = e^{(1/x)\ln x} \quad \Rightarrow$

$$f'(x) = e^{(1/x)\ln x}\left[\frac{1}{x}\left(\frac{1}{x}\right) + \ln x(-x^{-2})\right] = (1 - \ln x)x^{1/x}x^{-2}.$$ This is 0 only when $1 - \ln x = 0 \quad \Rightarrow$

$x = e$, and $f'(x)$ changes from positive to negative there. So the local maximum value is $f(e) = e^{1/e}$.

(d) We differentiate again: $f''(x) = (1 - \ln x)x^{1/x}(-2x^{-3}) + (1 - \ln x)^2 x^{1/x}x^{-4} + (-1/x)x^{1/x}x^{-2}$. From the graphs it appears that $f''(x)$ changes sign at $x \approx 0.58$ and at $x \approx 4.4$, so f has inflection points there.

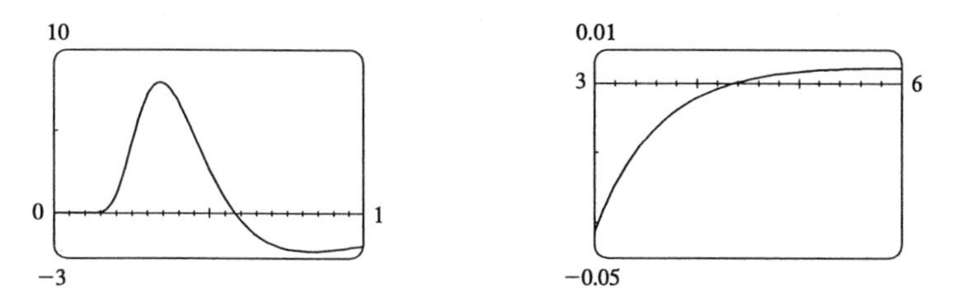

89. If $c < 0$, then $\lim\limits_{x\to-\infty} f(x) = \lim\limits_{x\to-\infty} \dfrac{x}{e^{cx}} \overset{\text{H}}{=} \lim\limits_{x\to-\infty} \dfrac{1}{ce^{cx}} = 0$, and $\lim\limits_{x\to\infty} f(x) = \infty$.

If $c > 0$, then $\lim\limits_{x\to-\infty} f(x) = -\infty$, and $\lim\limits_{x\to\infty} f(x) \overset{\text{H}}{=} \lim\limits_{x\to\infty} \dfrac{1}{ce^{cx}} = 0$.

If $c = 0$, then $f(x) = x$, so $\lim\limits_{x\to\pm\infty} f(x) = \pm\infty$ respectively.

So we see that $c = 0$ is a transitional value. We now exclude the case $c = 0$, since we know how the function behaves in that case.

To find the maxima and minima of f, we differentiate: $f(x) = xe^{-cx} \quad \Rightarrow$ $f'(x) = x(-ce^{-cx}) + e^{-cx} = (1 - cx)e^{-cx}$. This is 0 when $1 - cx = 0 \quad \Leftrightarrow \quad x = 1/c$. If $c < 0$ then this represents a minimum of $f(1/c) = 1/(ce)$, since $f'(x)$ changes from negative to positive at $x = 1/c$; and if $c > 0$, it represents a maximum. As $|c|$ increases, the extremum gets closer to the origin.

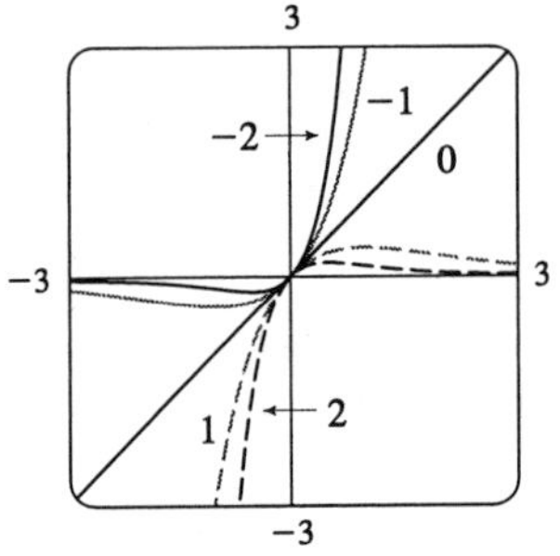

To find the inflection points, we differentiate again:

$f'(x) = e^{-cx}(1 - cx) \quad \Rightarrow$

$f''(x) = e^{-cx}(-c) + (1 - cx)(-ce^{-cx}) = (cx - 2)ce^{-cx}$.

This changes sign when $cx - 2 = 0 \quad \Leftrightarrow \quad x = 2/c$.

So as $|c|$ increases, the points of inflection get closer to the origin.

91. Both numerator and denominator approach 0 as $x \to 0$, so we use l'Hospital's Rule (and the Fundamental Theorem of Calculus, Part 1):

$$\lim_{x\to0} \frac{S(x)}{x^3} = \lim_{x\to0} \frac{\int_0^x \sin(\pi t^2/2)\,dt}{x^3} \overset{\text{H}}{=} \lim_{x\to0} \frac{\sin(\pi x^2/2)}{3x^2} \overset{\text{H}}{=} \lim_{x\to0} \frac{\pi x \cos(\pi x^2/2)}{6x} = \frac{\pi}{6} \cdot \cos 0 = \frac{\pi}{6}$$

93. Since $\lim\limits_{h\to0} [f(x+h) - f(x-h)] = f(x) - f(x) = 0$ (f is differentiable and hence continuous) and $\lim\limits_{h\to0} 2h = 0$, we use l'Hospital's Rule:

$$\lim_{h\to0} \frac{f(x+h) - f(x-h)}{2h} \overset{\text{H}}{=} \lim_{h\to0} \frac{f'(x+h) - f'(x-h)(-1)}{2} = \frac{f'(x) + f'(x)}{2} = \frac{2f'(x)}{2} = f'(x)$$

95. $\displaystyle \lim_{x\to\infty} \frac{e^x}{x^n} \overset{\text{H}}{=} \lim_{x\to\infty} \frac{e^x}{n\,x^{n-1}} \overset{\text{H}}{=} \lim_{x\to\infty} \frac{e^x}{n(n-1)x^{n-2}} \overset{\text{H}}{=} \cdots \overset{\text{H}}{=} \lim_{x\to\infty} \frac{e^x}{n!} = \infty$

97. $\displaystyle \lim_{x\to0^+} x^\alpha \ln x = \lim_{x\to0^+} \frac{\ln x}{x^{-\alpha}} \overset{\text{H}}{=} \lim_{x\to0^+} \frac{1/x}{-\alpha x^{-\alpha-1}} = \lim_{x\to0^+} \frac{x^\alpha}{-\alpha} = 0$ since $\alpha > 0$.

99. Let the radius of the circle be r. We see that $A(\theta)$ is just the area of the whole figure (a sector of the circle with radius 1), minus the area of $\triangle OPR$. But the area of the sector of the circle is $\frac{1}{2}r^2\theta$ (see endpapers), and the area of the triangle is $\frac{1}{2}r|PQ| = \frac{1}{2}r(r\sin\theta) = \frac{1}{2}r^2\sin\theta$. So we have $A(\theta) = \frac{1}{2}r^2\theta - \frac{1}{2}r^2\sin\theta = \frac{1}{2}r^2(\theta - \sin\theta)$. Now by elementary trigonometry, $B(\theta) = \frac{1}{2}|QR||PQ| = \frac{1}{2}r(1 - \cos\theta)(r\sin\theta)$. So the limit we want is

$$\lim_{\theta\to0^+} \frac{A(\theta)}{B(\theta)} = \lim_{\theta\to0^+} \frac{\frac{1}{2}r^2(\theta - \sin\theta)}{\frac{1}{2}r^2(1 - \cos\theta)\sin\theta} \overset{\text{H}}{=} \lim_{\theta\to0^+} \frac{1 - \cos\theta}{(1 - \cos\theta)\cos\theta + \sin\theta(\sin\theta)}$$

$$= \lim_{\theta\to0^+} \frac{1 - \cos\theta}{\cos\theta - \cos^2\theta + \sin^2\theta} \overset{\text{H}}{=} \lim_{\theta\to0^+} \frac{\sin\theta}{-\sin\theta + 4\sin\theta(\cos\theta)} = \frac{1}{-1 + 4\cos 0} = \frac{1}{3}.$$

101. **(a)** We show that $\lim\limits_{x\to 0}\dfrac{f(x)}{x^n}=0$ for every integer $n\geq 0$. Let $y=\dfrac{1}{x^2}$. Then

$$\lim_{x\to 0}\frac{f(x)}{x^{2n}}=\lim_{x\to 0}\frac{e^{-1/x^2}}{(x^2)^n}=\lim_{y\to\infty}\frac{y^n}{e^y}\overset{\text{H}}{=}\lim_{y\to\infty}\frac{ny^{n-1}}{e^y}\overset{\text{H}}{=}\cdots\overset{\text{H}}{=}\lim_{y\to\infty}\frac{n!}{e^y}=0\quad\Rightarrow$$

$$\lim_{x\to 0}\frac{f(x)}{x^n}=\lim_{x\to 0}x^n\frac{f(x)}{x^{2n}}=\lim_{x\to 0}x^n\lim_{x\to 0}\frac{f(x)}{x^{2n}}=0.\text{ Thus }f'(0)=\lim_{x\to 0}\frac{f(x)-f(0)}{x-0}=\lim_{x\to 0}\frac{f(x)}{x}=0.$$

(b) Using the Chain Rule and the Quotient Rule we see that $f^{(n)}(x)$ exists for $x\neq 0$. In fact, we prove by induction that for each $n\geq 0$, there is a polynomial p_n and a non-negative integer k_n with $f^{(n)}(x)=p_n(x)f(x)/x^{k_n}$ for $x\neq 0$. This is true for $n=0$; suppose it is true for the nth derivative. Then

$$\begin{aligned}f^{(n+1)}(x)&=\left[x^{k_n}\left[p_n'(x)f(x)+p_n(x)f'(x)\right]-k_nx^{k_n-1}p_n(x)f(x)\right]x^{-2k_n}\\&=\left[x^{k_n}p_n'(x)+p_n(x)\left(2/x^3\right)-k_nx^{k_n-1}p_n(x)\right]f(x)x^{-2k_n}\\&=\left[x^{k_n+3}p_n'(x)+2\,p_n(x)-k_nx^{k_n+2}p_n(x)\right]f(x)x^{-(2k_n+3)},\end{aligned}$$

which has the desired form.

Now we show by induction that $f^{(n)}(0)=0$ for all n. By (a), $f'(0)=0$. Suppose that $f^{(n)}(0)=0$. Then

$$\begin{aligned}f^{(n+1)}(0)&=\lim_{x\to 0}\frac{f^{(n)}(x)-f^{(n)}(0)}{x-0}=\lim_{x\to 0}\frac{f^{(n)}(x)}{x}=\lim_{x\to 0}\frac{p_n(x)f(x)/x^{k_n}}{x}\\&=\lim_{x\to 0}\frac{p_n(x)f(x)}{x^{k_n+1}}=\lim_{x\to 0}p_n(x)\lim_{x\to 0}\frac{f(x)}{x^{k_n+1}}=p_n(0)\cdot 0=0.\end{aligned}$$

REVIEW EXERCISES FOR CHAPTER 6

1. False. For example, $\cos\frac{\pi}{2}=\cos\left(-\frac{\pi}{2}\right)=0$, so $\cos x$ is not 1-1.

3. True, since $\ln x$ is an increasing function on $(0,\infty)$.

5. True, since $e^x\neq 0$ for all x.

7. False. For example, $(\ln e)^6=1^6=1$, but $6\ln e=6$. In fact $\ln(x^6)=6\ln x$.

9. False. $\ln 10$ is a constant, so its derivative is 0.

11. False. The "-1" is not an exponent; it is an indication of an inverse function. See Equation 6.6.4.

13. True. See Figure 2 in Section 6.7.

15. True. $\int_2^{16}(1/x)dx=\ln x\big|_2^{16}=\ln 16-\ln 2=\ln\frac{16}{2}=\ln 8=\ln 2^3=3\ln 2$.

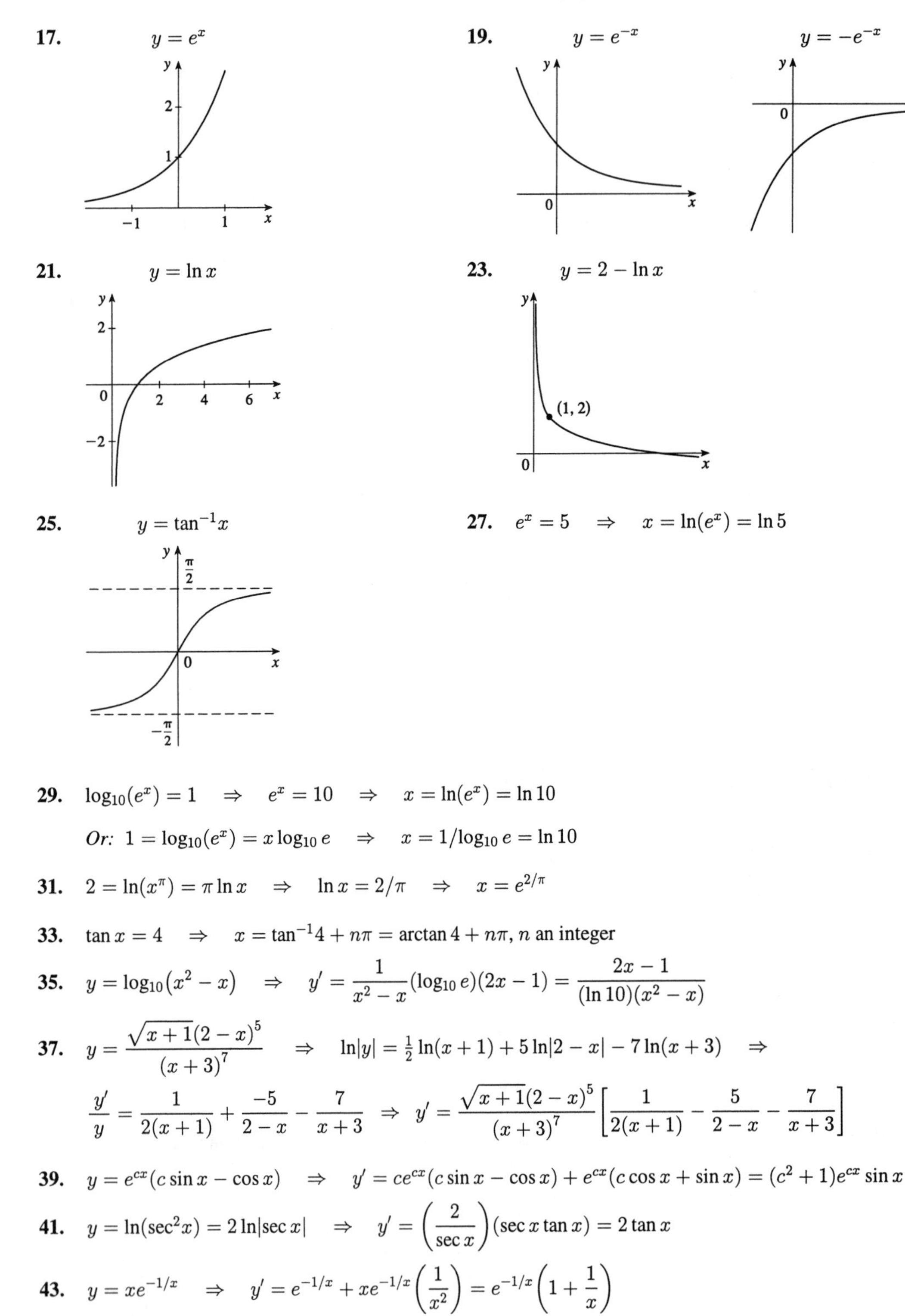

17. $y = e^x$

19. $y = e^{-x}$ $\quad$ $y = -e^{-x}$

21. $y = \ln x$

23. $y = 2 - \ln x$

25. $y = \tan^{-1}x$

27. $e^x = 5 \quad \Rightarrow \quad x = \ln(e^x) = \ln 5$

29. $\log_{10}(e^x) = 1 \quad \Rightarrow \quad e^x = 10 \quad \Rightarrow \quad x = \ln(e^x) = \ln 10$

Or: $1 = \log_{10}(e^x) = x \log_{10} e \quad \Rightarrow \quad x = 1/\log_{10} e = \ln 10$

31. $2 = \ln(x^\pi) = \pi \ln x \quad \Rightarrow \quad \ln x = 2/\pi \quad \Rightarrow \quad x = e^{2/\pi}$

33. $\tan x = 4 \quad \Rightarrow \quad x = \tan^{-1}4 + n\pi = \arctan 4 + n\pi$, n an integer

35. $y = \log_{10}(x^2 - x) \quad \Rightarrow \quad y' = \dfrac{1}{x^2 - x}(\log_{10} e)(2x - 1) = \dfrac{2x - 1}{(\ln 10)(x^2 - x)}$

37. $y = \dfrac{\sqrt{x+1}\,(2-x)^5}{(x+3)^7} \quad \Rightarrow \quad \ln|y| = \frac{1}{2}\ln(x+1) + 5\ln|2-x| - 7\ln(x+3) \quad \Rightarrow$

$$\frac{y'}{y} = \frac{1}{2(x+1)} + \frac{-5}{2-x} - \frac{7}{x+3} \quad \Rightarrow \quad y' = \frac{\sqrt{x+1}\,(2-x)^5}{(x+3)^7}\left[\frac{1}{2(x+1)} - \frac{5}{2-x} - \frac{7}{x+3}\right]$$

39. $y = e^{cx}(c\sin x - \cos x) \quad \Rightarrow \quad y' = ce^{cx}(c\sin x - \cos x) + e^{cx}(c\cos x + \sin x) = (c^2+1)e^{cx}\sin x$

41. $y = \ln(\sec^2 x) = 2\ln|\sec x| \quad \Rightarrow \quad y' = \left(\dfrac{2}{\sec x}\right)(\sec x \tan x) = 2\tan x$

43. $y = xe^{-1/x} \quad \Rightarrow \quad y' = e^{-1/x} + xe^{-1/x}\left(\dfrac{1}{x^2}\right) = e^{-1/x}\left(1 + \dfrac{1}{x}\right)$

45. $y = (\cos^{-1}x)^{\sin^{-1}x} \Rightarrow \ln y = \sin^{-1}x \ln(\cos^{-1}x) \Rightarrow$

$$\frac{y'}{y} = \frac{1}{\sqrt{1-x^2}}\ln(\cos^{-1}x) + (\sin^{-1}x)\left(\frac{1}{\cos^{-1}x}\right)\left(-\frac{1}{\sqrt{1-x^2}}\right) \Rightarrow$$

$$y' = (\cos^{-1}x)^{\sin^{-1}x - 1}\left[\frac{\cos^{-1}x \ln(\cos^{-1}x) - \sin^{-1}x}{\sqrt{1-x^2}}\right]$$

47. $y = e^{e^x} \Rightarrow y' = e^{e^x}e^x = e^{x+e^x}$

49. $y = \ln\dfrac{1}{x} + \dfrac{1}{\ln x} = -\ln x + (\ln x)^{-1} \Rightarrow y' = -\dfrac{1}{x} - \dfrac{1}{x(\ln x)^2}$

51. $y = 7^{\sqrt{2x}} \Rightarrow y' = 7^{\sqrt{2x}}(\ln 7)\left[1/(2\sqrt{2x})\right](2) = 7^{\sqrt{2x}}(\ln 7)/\sqrt{2x}$

53. $y = \ln(\cosh 3x) \Rightarrow y' = (1/\cosh 3x)(\sinh 3x)(3) = 3\tanh 3x$

55. $y = \cosh^{-1}(\sinh x) \Rightarrow y' = (\cosh x)/\sqrt{\sinh^2 x - 1}$

57. $y = \ln \sin x - \frac{1}{2}\sin^2 x \Rightarrow y' = \dfrac{\cos x}{\sin x} - \sin x \cos x = \cot x - \sin x \cos x$

59. $y = \sin^{-1}\left(\dfrac{x-1}{x+1}\right) \Rightarrow$

$$y' = \frac{1}{\sqrt{1 - [(x-1)/(x+1)]^2}}\,\frac{(x+1)-(x-1)}{(x+1)^2} = \frac{1}{\sqrt{(x+1)^2 - (x-1)^2}}\left(\frac{2}{x+1}\right)$$

$$= \frac{2}{\sqrt{4x}(x+1)} = \frac{1}{\sqrt{x}(x+1)}.$$ [Note that the domain of y is $x \geq 0$.]

61. $y = \frac{1}{4}\left[\ln(x^2+x+1) - \ln(x^2-x+1)\right] + \dfrac{1}{2\sqrt{3}}\left[\tan^{-1}\left(\dfrac{2x+1}{\sqrt{3}}\right) + \tan^{-1}\left(\dfrac{2x-1}{\sqrt{3}}\right)\right] \Rightarrow$

$$y' = \frac{1}{4}\left[\frac{2x+1}{x^2+x+1} - \frac{2x-1}{x^2-x+1}\right] + \frac{1}{2\sqrt{3}}\left[\frac{2/\sqrt{3}}{1+\left[(2x+1)/\sqrt{3}\right]^2} + \frac{2/\sqrt{3}}{1+\left[(2x-1)/\sqrt{3}\right]^2}\right]$$

$$= \frac{1}{4}\left[\frac{2x+1}{x^2+x+1} - \frac{2x-1}{x^2-x+1}\right] + \frac{1}{4(x^2+x+1)} + \frac{1}{4(x^2-x+1)}$$

$$= \frac{1}{2}\left[\frac{x+1}{x^2+x+1} - \frac{x-1}{x^2-x+1}\right] = \frac{1}{x^4+x^2+1}$$

63. $f(x) = 2^x \Rightarrow f'(x) = 2^x \ln 2 \Rightarrow f''(x) = 2^x(\ln 2)^2 \Rightarrow \cdots \Rightarrow f^{(n)}(x) = 2^x(\ln 2)^n$

65. We first show it is true for $n = 1$: $f'(x) = e^x + xe^x = (x+1)e^x$. We now assume it is true for $n = k$: $f^{(k)}(x) = (x+k)e^x$. With this assumption, we must show it is true for $n = k+1$:

$$f^{(k+1)}(x) = \frac{d}{dx}\left[f^{(k)}(x)\right] = \frac{d}{dx}[(x+k)e^x] = e^x + (x+k)e^x = [x+(k+1)]e^x.$$

Therefore $f^{(n)}(x) = (x+n)e^x$ by mathematical induction.

67. $y = f(x) = \ln(e^x + e^{2x}) \Rightarrow f'(x) = \dfrac{e^x + 2e^{2x}}{e^x + e^{2x}} \Rightarrow f'(0) = \frac{3}{2}$, so the tangent line at $(0, \ln 2)$ is $y - \ln 2 = \frac{3}{2}x$ or $3x - 2y + \ln 4 = 0$.

69.

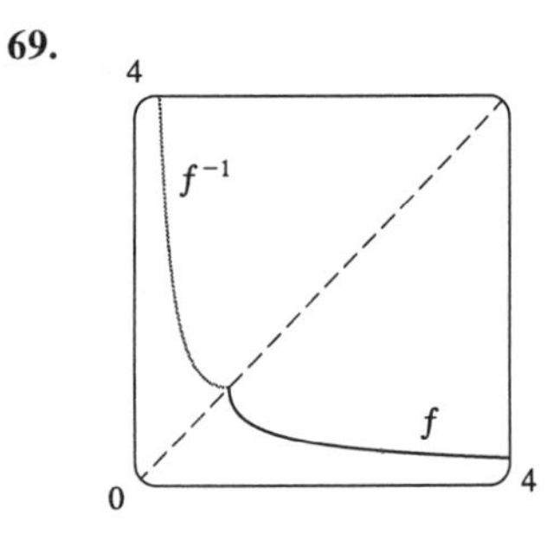

$f(x) = \sqrt{x} - \sqrt{x-1}$ has domain $[1, \infty)$. To see that f is 1-1, we can either graph the function and use the Horizontal Line Test, or we can calculate $f'(x) = \dfrac{1}{2\sqrt{x}} - \dfrac{1}{2\sqrt{x-1}} = \dfrac{\sqrt{x-1} - \sqrt{x}}{2\sqrt{x(x-1)}} < 0$, so f is decreasing and hence 1-1. The parametric equations of the graph of f are $x = t$, $y = \sqrt{t} - \sqrt{t-1}$, and so the parametric equations of the graph of f^{-1} are $x = \sqrt{t} - \sqrt{t-1}$, $y = t$.

71. $y = [\ln(x+4)]^2 \quad \Rightarrow \quad y' = 2\ln(x+4)/(x+4) = 0 \quad \Leftrightarrow \quad \ln(x+4) = 0 \quad \Leftrightarrow \quad x + 4 = 1 \quad \Leftrightarrow$ $x = -3$, so the tangent is horizontal at $(-3, 0)$.

73. The slope of the tangent at the point (a, e^a) is $\left[\dfrac{d}{dx}e^x\right]_{x=a} = e^a$. The equation of the tangent line is thus $y - e^a = e^a(x - a)$. We substitute $x = 0$, $y = 0$ into this equation, since we want the line to pass through the origin: $0 - e^a = e^a(0 - a) \quad \Leftrightarrow \quad -e^a = e^a(-a) \quad \Leftrightarrow \quad a = 1$. So the equation of the tangent is $y - e = e(x - 1)$, or $y = ex$.

75. $\lim\limits_{x \to -\infty} 10^{-x} = \infty$ since $-x \to \infty$ as $x \to -\infty$.

77. $\lim\limits_{x \to 0^+} \ln(\tan x) = -\infty$ since $\tan x \to 0^+$ as $x \to 0^+$.

79. $\lim\limits_{x \to -4^+} e^{1/(x+4)} = \infty$ since $\dfrac{1}{x+4} \to \infty$ as $x \to -4^+$.

81. $$\lim_{x \to \infty} \frac{e^x}{e^{2x} + e^{-x}} = \lim_{x \to \infty} \frac{e^{-x}}{1 + e^{-3x}} = \frac{0}{1+0} = 0$$

83. $$\lim_{x \to 1} \cos^{-1}\left(\frac{x}{x+1}\right) = \cos^{-1}\tfrac{1}{2} = \tfrac{\pi}{3}$$

85. $$\lim_{x \to \pi} \frac{\sin x}{x^2 - \pi^2} \overset{\text{H}}{=} \lim_{x \to \pi} \frac{\cos x}{2x} = -\frac{1}{2\pi}$$

87. $$\lim_{x \to \infty} \frac{\ln(\ln x)}{\ln x} \overset{\text{H}}{=} \lim_{x \to \infty} \frac{1/(x \ln x)}{1/x} = \lim_{x \to \infty} \frac{1}{\ln x} = 0$$

89. $$\lim_{x \to 0} \frac{\ln(1-x) + x + \frac{1}{2}x^2}{x^3} \overset{\text{H}}{=} \lim_{x \to 0} \frac{-\dfrac{1}{1-x} + 1 + x}{3x^2} \overset{\text{H}}{=} \lim_{x \to 0} \frac{-\dfrac{1}{(1-x)^2} + 1}{6x} \overset{\text{H}}{=} \lim_{x \to 0} \frac{-\dfrac{2}{(1-x)^3}}{6} = -\frac{2}{6} = -\frac{1}{3}$$

91. $$\lim_{x \to 0^+} \sin x (\ln x)^2 = \lim_{x \to 0^+} \frac{(\ln x)^2}{\csc x} \overset{\text{H}}{=} \lim_{x \to 0^+} \frac{2\ln x / x}{-\csc x \cot x} = -2 \lim_{x \to 0} \frac{\sin x}{x} \lim_{x \to 0} \frac{\ln x}{\cot x} = -2 \lim_{x \to 0} \frac{\ln x}{\cot x}$$
$$\overset{\text{H}}{=} -2 \lim_{x \to 0} \frac{1/x}{-\csc^2 x} = 2 \lim_{x \to 0} \frac{\sin^2 x}{x} = 2 \lim_{x \to 0} \frac{\sin x}{x} \lim_{x \to 0} \sin x$$
$$= 2 \cdot 1 \cdot 0 = 0$$

93. $$\lim_{x \to 1} (\ln x)^{\sin x} = (\ln 1)^{\sin 1} = 0^{\sin 1} = 0$$

95. $$\lim_{x \to 0^+} \frac{x^{1/3} - 1}{x^{1/4} - 1} = \frac{0 - 1}{0 - 1} = 1$$

97. $y = f(x) = \tan^{-1}(1/x)$ **A.** $D = \{x \mid x \neq 0\}$ **B.** No intercepts **C.** $f(-x) = -f(x)$, so the curve is symmetric about the origin. **D.** $\lim\limits_{x\to\pm\infty} \tan^{-1}(1/x) = \tan^{-1}0 = 0$, so $y = 0$ is a HA. $\lim\limits_{x\to 0^+} \tan^{-1}(1/x) = \frac{\pi}{2}$ and $\lim\limits_{x\to 0^-} \tan^{-1}(1/x) = -\frac{\pi}{2}$ since $\dfrac{1}{x} \to \pm\infty$ as $x \to 0^\pm$.

E. $f'(x) = \dfrac{1}{1 + (1/x)^2}\left(-1/x^2\right) = \dfrac{-1}{x^2+1} \quad\Rightarrow\quad f'(x) < 0$, so f is decreasing on $(-\infty, 0)$ and $(0, \infty)$. **F.** No extrema

G. $f''(x) = \dfrac{2x}{(x^2+1)^2} > 0 \quad\Leftrightarrow\quad x > 0$, so f is CU on $(0, \infty)$ and CD on $(-\infty, 0)$.

H.

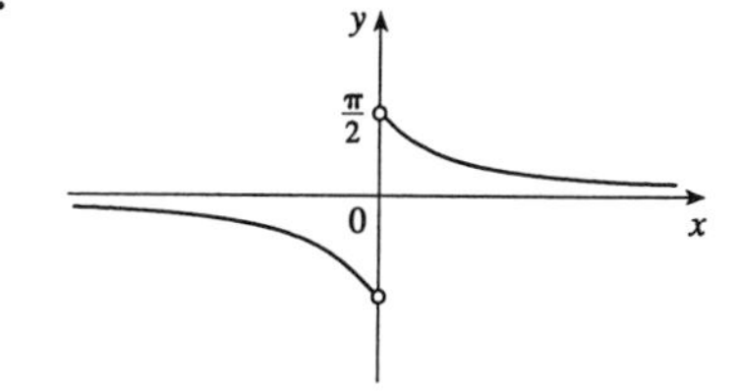

99. $y = f(x) = 2^{1/(x-1)}$ **A.** $D = \{x \mid x \neq 1\}$ **B.** No x-intercepts; y-intercept $= f(0) = 1/2$. **C.** No symmetry **D.** $\lim\limits_{x\to\pm\infty} 2^{1/(x-1)} = 2^0 = 1$, so $y = 1$ is a HA. $\lim\limits_{x\to 1^+} 2^{1/(x-1)} = \infty$, so $x = 1$ is a VA. Also $\lim\limits_{x\to 1^-} 2^{1/(x-1)} = 0$.

E. $f'(x) = 2^{1/(x-1)}(-\ln 2)/(x-1)^2 < 0$, so f is decreasing on $(-\infty, 1)$ and $(1, \infty)$. **F.** No extrema

G. $y'' = \dfrac{2^{1/(x-1)}(\ln 2)(2x - 2 + \ln 2)}{(x-1)^4} > 0$

$\Leftrightarrow \quad 2x - 2 + \ln 2 > 0 \quad\Leftrightarrow\quad x > 1 - \frac{1}{2}\ln 2$,

so f is CU on $\left(1 - \ln\sqrt{2}, 1\right)$ and $(1, \infty)$ and CD on $\left(-\infty, 1 - \ln\sqrt{2}\right)$. IP at $x = 1 - \ln\sqrt{2}$.

H.

101. $y = f(x) = e^x + e^{-3x}$ **A.** $D = \mathbb{R}$ **B.** No x-intercepts; y-intercept $= f(0) = 2$ **C.** No symmetry

D. $\lim\limits_{x\to\pm\infty} (e^x + e^{-3x}) = \infty$, no asymptote

E. $f'(x) = e^x - 3e^{-3x} = e^{-3x}\left(e^{4x} - 3\right) > 0 \quad\Leftrightarrow$ $e^{4x} > 3 \quad\Leftrightarrow\quad 4x > \ln 3 \quad\Leftrightarrow\quad x > \frac{1}{4}\ln 3$, so f is increasing on $\left[\frac{1}{4}\ln 3, \infty\right)$ and decreasing on $\left(-\infty, \frac{1}{4}\ln 3\right]$.

F. Absolute minimum $f\left(\frac{1}{4}\ln 3\right) = 3^{1/4} + 3^{-3/4}$.

G. $f''(x) = e^x + 9e^{-3x} > 0$, so f is CU on $(-\infty, \infty)$.

H.

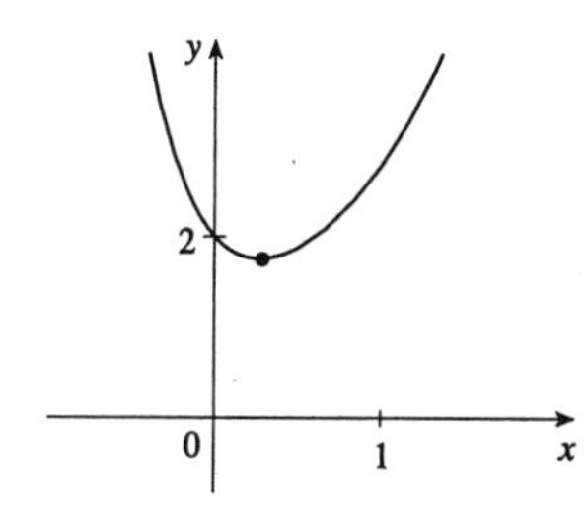

103.

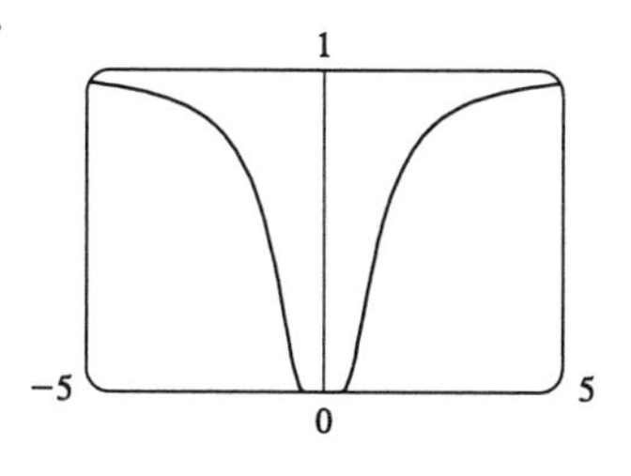

From the graph, we estimate the points of inflection to be about $(\pm 0.8, 0.2)$.

$f(x) = e^{-1/x^2} \quad\Rightarrow\quad f'(x) = 2x^{-3}e^{-1/x^2} \quad\Rightarrow$

$f''(x) = 2\left[x^{-3}\left(2x^{-3}\right)e^{-1/x^2} + e^{-1/x^2}\left(-3x^{-4}\right)\right]$

$= 2x^{-6}e^{-1/x^2}\left(2 - 3x^2\right)$. This is 0 when $2 - 3x^2 = 0$

$\Leftrightarrow \quad x = \pm\sqrt{\frac{2}{3}}$, so the inflection points are $\left(\pm\sqrt{\frac{2}{3}}, e^{-3/2}\right)$.

105. **(a)** $y(t) = y(0)e^{kt} = 1000e^{kt} \Rightarrow y(2) = 1000e^{2k} = 9000 \Rightarrow e^{2k} = 9 \Rightarrow 2k = \ln 9 \Rightarrow$ $k = \frac{1}{2}\ln 9 = \ln 3 \Rightarrow y(t) = 1000e^{(\ln 3)t} = 1000 \cdot 3^t$

(b) $y(3) = 1000 \cdot 3^3 = 27{,}000$

(c) $1000 \cdot 3^t = 2000 \Rightarrow 3^t = 2 \Rightarrow t\ln 3 = \ln 2 \Rightarrow t = (\ln 2)/(\ln 3) \approx 0.63\text{ h}$

107. Using the formula in Example 6.5.4, $A(t) = A_0\left(1 + \dfrac{i}{n}\right)^{nt}$, where $A_0 = 10{,}000.00$ and $i = 0.06$ then for:

(a) $n = 1$: $A(4) = 10{,}000(1 + 0.06)^{1 \cdot 4} = \$12{,}624.77$

(b) $n = 2$: $A(4) = 10{,}000\left(1 + \frac{0.06}{2}\right)^{2 \cdot 4} = \$12{,}667.70$

(c) $n = 4$: $A(4) = 10{,}000\left(1 + \frac{0.06}{4}\right)^{4 \cdot 4} = \$12{,}689.86$

(d) $n = 12$: $A(4) = 10{,}000\left(1 + \frac{0.06}{12}\right)^{12 \cdot 4} = \$12{,}704.89$

(e) $n = 365$: $A(4) = 10{,}000\left(1 + \frac{0.06}{365}\right)^{365 \cdot 4} = \$12{,}712.24$

(f) Using the formula for continuous interest, $A(t) = A_0e^{it}$, we have $A(4) = 10{,}000 \cdot e^{0.06 \cdot 4} = \$12{,}712.49$.

109. **(a)** $C'(t) = -kC(t) \Rightarrow C(t) = C(0)e^{-kt}$ by Theorem 9.5.4. But $C(0) = C_0$. Thus $C(t) = C_0e^{-kt}$.

(b) $C(30) = \frac{1}{2}C_0$ since the concentration is reduced by half. Thus, $\frac{1}{2}C_0 = C_0e^{-30k} \Rightarrow \ln\frac{1}{2} = -30k \Rightarrow$ $k = -\frac{1}{30}\ln\frac{1}{2} = \frac{1}{30}\ln 2$. Since 10% of the original concentration remains if 90% is eliminated, we want the value of t such that $C(t) = \frac{1}{10}C_0$. Therefore, $\frac{1}{10}C_0 = C_0e^{-t(\ln 2)/30} \Rightarrow t = -\dfrac{30}{\ln 2}\ln 0.1 \approx 100\text{ h}$.

111. $s(t) = Ae^{-ct}\cos(\omega t + \delta) \Rightarrow$

$$v(t) = s'(t) = -cAe^{-ct}\cos(\omega t + \delta) + Ae^{-ct}[-\omega\sin(\omega t + \delta)] = -Ae^{-ct}[c\cos(\omega t + \delta) + \omega\sin(\omega t + \delta)] \Rightarrow$$

$$a(t) = v'(t) = cAe^{-ct}[c\cos(\omega t + \delta) + \omega\sin(\omega t + \delta)] = -Ae^{-ct}\left[-\omega c\sin(\omega t + \delta) + \omega^2\cos(\omega t + \delta)\right]$$
$$= Ae^{-ct}\left[(c^2 - \omega^2)\cos(\omega t + \delta) + 2c\omega\sin(\omega t + \delta)\right]$$

113. $f(x) = e^{g(x)} \Rightarrow f'(x) = e^{g(x)}g'(x)$

115. $f(x) = \ln|g(x)| \Rightarrow f'(x) = g'(x)/g(x)$

117. $f(x) = \ln g(e^x) \Rightarrow f'(x) = \dfrac{1}{g(e^x)}\,g'(e^x)e^x$

119. $\displaystyle\int_0^{2\sqrt{3}} \frac{1}{x^2 + 4}\,dx = \left[\tfrac{1}{2}\tan^{-1}(x/2)\right]_0^{2\sqrt{3}} = \tfrac{1}{2}\left(\tan^{-1}\sqrt{3} - \tan^{-1}0\right) = \tfrac{1}{2} \cdot \tfrac{\pi}{3} = \tfrac{\pi}{6}$

121. $$\int_2^4 \frac{1 + x - x^2}{x^2}\,dx = \int_2^4 \left(x^{-2} + \frac{1}{x} - 1\right)dx = \left[-\frac{1}{x} + \ln x - x\right]_2^4$$
$$= \left(-\tfrac{1}{4} + \ln 4 - 4\right) - \left(-\tfrac{1}{2} + \ln 2 - 2\right) = \ln 2 - \tfrac{7}{4}$$

123. Let $u = e^x + 1$. Then $du = e^x\,dx$, so $\displaystyle\int \frac{e^x}{e^x + 1}\,dx = \int \frac{du}{u} = \ln|u| + C = \ln(e^x + 1) + C$.

125. Let $u = \sqrt{x}$. Then $du = \dfrac{dx}{2\sqrt{x}} \Rightarrow \displaystyle\int \frac{e^{\sqrt{x}}}{\sqrt{x}}\,dx = 2\int e^u\,du = 2e^u + C = 2e^{\sqrt{x}} + C$.

127. Let $u = \ln(\cos x)$. Then $du = \dfrac{-\sin x}{\cos x}\,dx = -\tan x\,dx \Rightarrow$
$\int \tan x\ln(\cos x)dx = -\int u\,du = -\frac{1}{2}u^2 + C = -\frac{1}{2}[\ln(\cos x)]^2 + C$.

129. Let $u = 1 + x^4$. Then $du = 4x^3\,dx \quad \Rightarrow \quad \displaystyle\int \frac{x^3}{1+x^4}\,dx = \frac{1}{4}\int \frac{1}{u}\,du = \tfrac{1}{4}\ln|u| + C = \tfrac{1}{4}\ln(1+x^4) + C.$

131. Let $u = 1 + \sec\theta$, so $du = \sec\theta\tan\theta\,d\theta \Rightarrow \displaystyle\int \frac{\sec\theta\tan\theta}{1+\sec\theta}\,d\theta = \int \frac{1}{u}\,du = \ln|u| + C = \ln|1+\sec\theta| + C.$

133. $u = 3t \quad \Rightarrow \quad \int \cosh 3t\,dt = \frac{1}{3}\int \cosh u\,du = \frac{1}{3}\sinh u + C = \frac{1}{3}\sinh 3t + C$

135. $\cos x \le 1 \quad \Rightarrow \quad e^x\cos x \le e^x \quad \Rightarrow \quad \int_0^1 e^x\cos x\,dx \le \int_0^1 e^x\,dx = [e^x]_0^1 = e - 1$

137. $f'(x) = \dfrac{d}{dx}\displaystyle\int_1^{\sqrt{x}} \frac{e^s}{s}\,ds = \frac{e^{\sqrt{x}}}{\sqrt{x}}\frac{d}{dx}\sqrt{x} = \frac{e^{\sqrt{x}}}{\sqrt{x}}\frac{1}{2\sqrt{x}} = \frac{e^{\sqrt{x}}}{2x}$

139. $f_{\text{ave}} = \frac{1}{4-1}\int_1^4 (1/x)dx = \left[\frac{1}{3}\ln x\right]_1^4 = \frac{1}{3}\ln 4$

141. $V = \displaystyle\int_0^1 \frac{2\pi x}{1+x^4}\,dx$ by cylindrical shells. Let $u = x^2 \quad \Rightarrow \quad du = 2x\,dx$. Then

$$V = \int_0^1 \frac{\pi}{1+u^2}\,du = \pi\left[\tan^{-1}u\right]_0^1 = \pi\left(\tan^{-1}1 - \tan^{-1}0\right) = \pi\left(\frac{\pi}{4}\right) = \frac{\pi^2}{4}.$$

143. $f(x) = \ln x + \tan^{-1}x \quad \Rightarrow \quad f(1) = \ln 1 + \tan^{-1}1 = \frac{\pi}{4} \quad \Rightarrow \quad g\left(\frac{\pi}{4}\right) = 1.$

$f'(x) = \dfrac{1}{x} + \dfrac{1}{1+x^2}$, so $g'\left(\frac{\pi}{4}\right) = \dfrac{1}{f'(1)} = \dfrac{1}{3/2} = \dfrac{2}{3}.$

145.

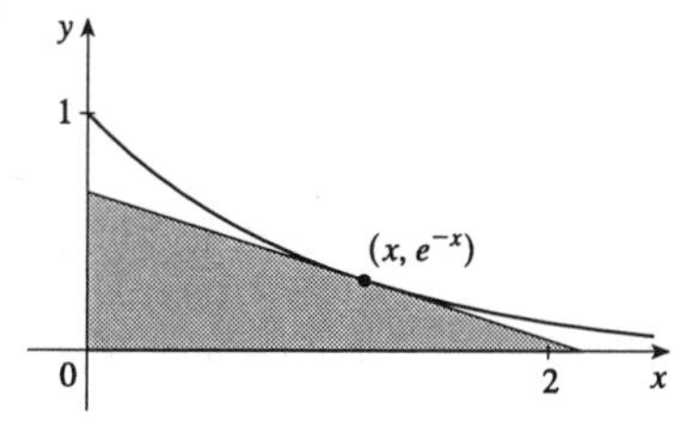

We find the equation of a tangent to the curve $y = e^{-x}$, so that we can find the x- and y-intercepts of this tangent, and then we can find the area of the triangle. The slope of the tangent at the point (a, e^{-a}) is given by $\left.\dfrac{d}{dx}e^{-x}\right|_{x=a} = -e^{-a}$, and so the equation of the tangent is $y - e^{-a} = -e^{-a}(x-a) \quad \Leftrightarrow$ $y = e^{-a}(a - x + 1)$. The y-intercept of this line is $y = e^{-a}(a - 0 + 1) = e^{-a}(a+1)$. To find the x-intercept we set $y = 0 \quad \Rightarrow \quad e^{-a}(a - x + 1) = 0 \quad \Rightarrow$ $x = a + 1$. So the area of the triangle is $A(a) = \frac{1}{2}[e^{-a}(a+1)](a+1) = \frac{1}{2}e^{-a}(a+1)^2$. We differentiate this with respect to a: $A'(a) = \frac{1}{2}\left[e^{-a}(2)(a+1) + (a+1)^2e^{-a}(-1)\right] = \frac{1}{2}e^{-a}(1-a^2)$. This is 0 at $a = \pm 1$, and the root $a = 1$ gives a maximum, by the First Derivative Test. So the maximum area of the triangle is $A(1) = \frac{1}{2}e^{-1}(1+1)^2 = 2e^{-1} = 2/e$.

147. $\displaystyle\lim_{x\to -1} F(x) = \lim_{x\to -1}\frac{b^{x+1} - a^{x+1}}{x+1} \overset{\text{H}}{=} \lim_{x\to -1}\frac{b^{x+1}\ln b - a^{x+1}\ln a}{1} = \ln b - \ln a = F(-1)$, so F is continuous at -1.

149. Differentiating both sides of the given equation, using the Fundamental Theorem for each side, gives $f(x) = e^{2x} + 2xe^{2x} + e^{-x}f(x)$. So $f(x)(1 - e^{-x}) = e^{2x} + 2xe^{2x}$. Hence $f(x) = \dfrac{e^{2x}(1+2x)}{1 - e^{-x}}$.

151. Let $y = \tan^{-1}x$. Then $\tan y = x$, so from the triangle we see that $\sin\left(\tan^{-1}x\right) = \sin y = \dfrac{x}{\sqrt{1+x^2}}$. Using this fact we have that $\sin\left(\tan^{-1}(\sinh x)\right) = \dfrac{\sinh x}{\sqrt{1+\sinh^2 x}} = \dfrac{\sinh x}{\cosh x} = \tanh x.$

Hence $\sin^{-1}(\tanh x) = \sin^{-1}(\sin(\tan^{-1}(\sinh x))) = \tan^{-1}(\sinh x)$.

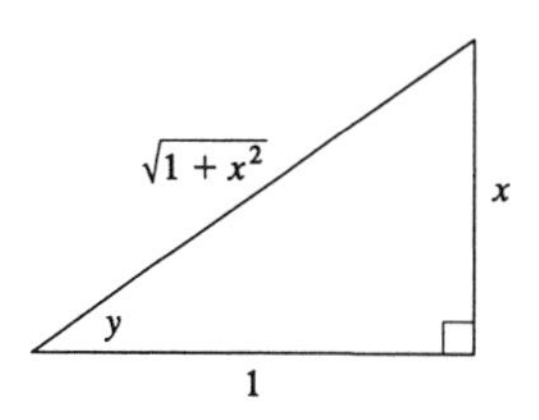

PROBLEMS PLUS (page 433)

1. Let $y = f(x) = e^{-x^2}$. The area of the rectangle under the curve from $-x$ to x is $A(x) = 2xe^{-x^2}$ where $x \geq 0$. We maximize $A(x)$: $A'(x) = 2e^{-x^2} - 4x^2e^{-x^2} = 2e^{-x^2}(1 - 2x^2) = 0 \Rightarrow x = \frac{1}{\sqrt{2}}$. This gives a maximum since $A'(x) > 0$ for $0 \leq x < \frac{1}{\sqrt{2}}$ and $A'(x) < 0$ for $x > \frac{1}{\sqrt{2}}$. We next determine the points of inflection of $f(x)$. Now $f'(x) = -2xe^{-x^2} = -A(x)$. So $f''(x) = -A'(x)$. So $f''(x) < 0$ for $-\frac{1}{\sqrt{2}} < x < \frac{1}{\sqrt{2}}$ and $f''(x) > 0$ for $x < -\frac{1}{\sqrt{2}}$ and $x > \frac{1}{\sqrt{2}}$. So $f(x)$ changes concavity at $x = \pm\frac{1}{\sqrt{2}}$. So the two vertices of the rectangle of largest area are at the inflection points.

3. We use proof by contradiction. Suppose that $\log_2 5$ is a rational number. Then $\log_2 5 = m/n$ where m and n are positive integers $\Rightarrow 2^{m/n} = 5 \Rightarrow 2^m = 5^n$. But this is impossible since 2^m is even and 5^n is odd. So $\log_2 5$ is irrational.

5. First notice that if we can prove the simpler inequality $\dfrac{x^2+1}{x} \geq 2$ for $x > 0$, then the desired inequality follows because $\dfrac{(x^2+1)(y^2+1)(z^2+1)}{xyz} = \left(\dfrac{x^2+1}{x}\right)\left(\dfrac{y^2+1}{y}\right)\left(\dfrac{z^2+1}{z}\right) \geq 2 \cdot 2 \cdot 2 = 8$. So we let $f(x) = \dfrac{x^2+1}{x} = x + \dfrac{1}{x}$, $x > 0$. Then $f'(x) = 1 - \dfrac{1}{x^2} = 0$ if $x = 1$, and $f'(x) < 0$ for $0 < x < 1$, $f'(x) > 0$ for $x > 1$. Thus the absolute minimum value of $f(x)$ for $x > 0$ is $f(1) = 2$. Therefore $\dfrac{x^2+1}{x} \geq 2$ for all positive x. (Or, without calculus, $(x^2+1)/x \geq 2 \Leftrightarrow x^2 + 1 \geq 2x \Leftrightarrow x^2 - 2x + 1 \geq 0 \Leftrightarrow (x-1)^2 \geq 0$, which is true.)

7. Consider the statement that $\left(d^n/dx^n\right)(e^{ax}\sin bx) = r^n e^{ax}\sin(bx + n\theta)$. For $n = 1$,

$$(d/dx)(e^{ax}\sin bx) = ae^{ax}\sin bx + be^{ax}\cos bx, \text{ and}$$
$$re^{ax}\sin(bx+\theta) = re^{ax}[\sin bx\cos\theta + \cos bx\sin\theta] = re^{ax}((a/r)\sin bx + (b/r)\cos bx)$$
$$= ae^{ax}\sin bx + be^{ax}\cos bx, \text{ since } \tan\theta = b/a \Rightarrow \sin\theta = b/r \text{ and } \cos\theta = a/r.$$

So the statement is true for $n = 1$. Assume it is true for $n = k$. Then

$$\frac{d^{k+1}}{dx^{k+1}}(e^{ax}\sin bx) = \frac{d}{dx}\left[r^k e^{ax}\sin(bx+k\theta)\right] = r^k ae^{ax}\sin(bx+k\theta) + r^k e^{ax} b\cos(bx+k\theta)$$
$$= r^k e^{ax}[a\sin(bx+k\theta) + b\cos(bx+k\theta)]. \text{ But}$$
$$\sin[bx+(k+1)\theta] = \sin[(bx+k\theta)+\theta] = \sin(bx+k\theta)\cos\theta + \sin\theta\cos(bx+k\theta)$$
$$= (a/r)\sin(bx+k\theta) + (b/r)\cos(bx+k\theta).$$

Hence $a\sin(bx+k\theta) + b\cos(bx+k\theta) = r\sin[bx+(k+1)\theta]$.

So $\left(d^{k+1}/dx^{k+1}\right)(e^{ax}\sin bx) = r^k e^{ax}[a\sin(bx+k\theta) + b\sin(bx+k\theta)] = r^k e^{ax}[r\sin(bx+(k+1)\theta)]$
$$= r^{k+1}e^{ax}[\sin(bx+(k+1)\theta)].$$

Therefore the statement is true for all n by mathematical induction.

9. The volume generated from $x = 0$ to $x = b$ is $\int_0^b \pi[f(x)]^2\,dx$. Hence we are given that $b^2 = \int_0^b \pi[f(x)]^2\,dx$ for all $b > 0$. Differentiating both sides of this equation using the Fundamental Theorem of Calculus gives $2b = \pi[f(b)]^2 \quad\Rightarrow\quad f(b) = \sqrt{2b/\pi}$, since f is positive. Therefore $f(x) = \sqrt{2x/\pi}$.

11. Let the line through A and B have equation $y = mx + b$. Now $x^2 = mx + b$ gives $x^2 - mx - b = 0$ and hence $x = \dfrac{m \pm \sqrt{m^2 + 4b}}{2}$. So A has x-coordinate $x_1 = \dfrac{1}{2}\left(m - \sqrt{m^2 + 4b}\right)$ and B has x-coordinate $x_2 = \frac{1}{2}\left(m + \sqrt{m^2 + 4b}\right)$. So the parabolic segment has area

$$\int_{x_1}^{x_2}[(mx + b) - x^2]dx = \left[\tfrac{1}{2}mx^2 + bx - \tfrac{1}{3}x^3\right]_{x_1}^{x_2} = \tfrac{1}{2}m(x_2^2 - x_1^2) + b(x_2 - x_1) - \tfrac{1}{3}(x_2^3 - x_1^3)$$

$$= (x_2 - x_1)\left[\tfrac{1}{2}m(x_2 + x_1) + b - \tfrac{1}{3}(x_2^2 + x_1x_2 + x_1^2)\right]$$

$$= \sqrt{m^2 + 4b}\left(\tfrac{1}{2}m^2 + b - \tfrac{1}{3}(mx_2 + b - b + mx_1 + b)\right) = \sqrt{m^2 + 4b}\left[\tfrac{1}{2}m^2 + b - \tfrac{1}{3}(m^2 + b)\right]$$

$$= \tfrac{1}{6}(m^2 + 4b)^{3/2}$$

Now since the line through C has slope m, we see that if C has x-coordinate c, then $2c = m$ $\left[\text{since } (x^2)' = 2x\right]$ and hence $c = m/2$. So C has coordinates $\left(\frac{1}{2}m, \frac{1}{4}m^2\right)$. The line through AC has slope $\dfrac{\frac{1}{4}m^2 - x_1^2}{\frac{1}{2}m - x_1} = \dfrac{m}{2} + x_1$ and equation $y - x_1^2 = \left(\frac{1}{2}m + x_1\right)(x - x_1)$ or $y = \left(\frac{1}{2}m + x_1\right)x - \frac{1}{2}mx_1$. Similarly, the equation of the line through BC is $y = \left(\frac{1}{2}m + x_2\right)x - \frac{1}{2}mx_2$. So the area of the triangular region is

$$\int_{x_1}^{m/2}\left[(mx + b) - \left[\left(\tfrac{1}{2}m + x_1\right)x - \tfrac{1}{2}mx_1\right]\right]dx + \int_{m/2}^{x_2}\left[(mx + b) - \left[\left(\tfrac{1}{2}m + x_2\right)x - \tfrac{1}{2}mx_2\right]\right]dx$$

$$= \int_{x_1}^{m/2}\left[\left(\tfrac{1}{2}m - x_1\right)x + \left(b + \tfrac{1}{2}mx_1\right)\right]dx + \int_{m/2}^{x_2}\left[\left(\tfrac{1}{2}m - x_2\right)x + \left(b + \tfrac{1}{2}mx_2\right)\right]dx$$

$$= \left[\tfrac{1}{2}\left(\tfrac{1}{2}m - x_1\right)x^2 + \left(b + \tfrac{1}{2}mx_1\right)x\right]_{x_1}^{m/2} + \left[\tfrac{1}{2}\left(\tfrac{1}{2}m - x_2\right)x^2 + \left(b + \tfrac{1}{2}mx_2\right)x\right]_{m/2}^{x_2}$$

$$= \left[\tfrac{1}{16}m^3 - \tfrac{1}{8}m^2x_1 + \tfrac{1}{2}bm + \tfrac{1}{4}m^2x_1\right] - \left[\tfrac{1}{4}mx_1^2 - \tfrac{1}{2}x_1^3 + bx_1 + \tfrac{1}{2}mx_1^2\right]$$
$$+ \left[\tfrac{1}{4}mx_2^2 - \tfrac{1}{2}x_2^3 + bx_2 + \tfrac{1}{2}mx_2^2\right] - \left[\tfrac{1}{16}m^3 - \tfrac{1}{8}m^2x_2 + \tfrac{1}{2}bm + \tfrac{1}{4}m^2x_2\right]$$

$$= \left(b - \tfrac{1}{8}m^2\right)(x_2 - x_1) + \tfrac{3}{4}m(x_2^2 - x_1^2) - \tfrac{1}{2}(x_2^3 - x_1^3)$$

$$= (x_2 - x_1)\left[\left(b - \tfrac{1}{8}m^2\right) + \tfrac{3}{4}m(x_2 + x_1) - \tfrac{1}{2}(x_2^2 + x_2x_1 + x_1^2)\right]$$

$$= \sqrt{m^2 + 4b}\left[\left(b - \tfrac{1}{8}m^2\right) + \tfrac{3}{4}m^2 - \tfrac{1}{2}(m^2 + b)\right] = \tfrac{1}{8}\sqrt{m^2 + 4b}(m^2 + 4b) = \tfrac{1}{8}(m^2 + 4b)^{3/2}.$$

The result follows since $\frac{4}{3}\left[\frac{1}{8}(m^2 + 4b)^{3/2}\right] = \frac{1}{6}(m^2 + 4b)^{3/2}$, the area of the parabolic segment.

Alternate Solution: Let $A = (a, a^2)$, $B = (b, b^2)$. Then $m_{AB} = (b^2 - a^2)/(b - a) = a + b$, so the equation of AB is $y - a^2 = (a + b)(x - a)$, or $y = (a + b)x - ab$, and the area of the parabolic segment is $\int_a^b[(a + b)x - ab - x^2]dx = \left[(a + b)\frac{1}{2}x^2 - abx - \frac{1}{3}x^3\right]_a^b = \frac{1}{2}(a + b)(b^2 - a^2) - ab(b - a) - \frac{1}{3}(b^3 - a^3)$ $= \frac{1}{6}(b - a)^3$. At C, $y' = 2x = b + a$, so the x-coordinate of C is $\frac{1}{2}(a + b)$. If we calculate the area of triangle ABC as in Problems Plus #19 after Chapter 4 (by subtracting areas of trapezoids) we find that the area is $\frac{1}{2}(b - a)\left[\frac{1}{2}(a + b) - a\right]\left[b - \frac{1}{2}(a + b)\right] = \frac{1}{2}(b - a)\frac{1}{2}(b - a)\frac{1}{2}(b - a) = \frac{1}{8}(b - a)^3$. This is $\frac{3}{4}$ of the area of the parabolic segment calculated above.

13. By l'Hospital's Rule and the Fundamental Theorem, using the notation $\exp(y) = e^y$,

$$\lim_{x\to 0}\frac{\int_0^x (1-\tan 2t)^{1/t}\,dt}{x} \overset{\text{H}}{=} \lim_{x\to 0}\frac{(1-\tan 2x)^{1/x}}{1} = \exp\left[\lim_{x\to 0}\frac{\ln(1-\tan 2x)}{x}\right]$$

$$\overset{\text{H}}{=} \exp\left(\lim_{x\to 0}\frac{-2\sec^2 2x}{1-\tan 2x}\right) = \exp\left(\frac{-2\cdot 1^2}{1-0}\right)$$

$$= e^{-2}.$$

15. We first show that $\dfrac{x}{1+x^2} < \tan^{-1}x$ for $x > 0$. Let $f(x) = \tan^{-1}x - \dfrac{x}{1+x^2}$. Then

$$f'(x) = \frac{1}{1+x^2} - \frac{1(1+x^2) - x(2x)}{(1+x^2)^2} = \frac{(1+x^2)-(1-x^2)}{(1+x^2)^2} = \frac{2x^2}{(1+x^2)^2} > 0 \text{ for } x > 0.$$ So $f(x)$ is increasing on $[0,\infty)$. Hence $0 < x \;\Rightarrow\; 0 = f(0) < f(x) = \tan^{-1}x - \dfrac{x}{1+x^2}$. So $\dfrac{x}{1+x^2} < \tan^{-1}x$ for $0 < x$. We next show that $\tan^{-1}x < x$ for $x > 0$. Let $h(x) = x - \tan^{-1}x$. Then

$$h'(x) = 1 - \frac{1}{1+x^2} = \frac{x^2}{1+x^2} > 0.$$ Hence $h(x)$ is increasing on $[0,\infty)$. So for $0 < x$, $0 = h(0) < h(x) = x - \tan^{-1}x$. Hence $\tan^{-1}x < x$ for $x > 0$.

17. By the Fundamental Theorem of Calculus, $f'(x) = \sqrt{1+x^3} > 0$ for $x > -1$. So f is increasing on $[-1,\infty)$ and hence is one-to-one. Note that $f(1) = 0$, so $f^{-1}(1) = 0 \;\Rightarrow\; (f^{-1})'(0) = 1/f'(1) = \frac{1}{\sqrt{2}}$.

19. Let $L = \lim_{x\to\infty}\left(\dfrac{x+a}{x-a}\right)^x$, so $\ln L = \lim_{x\to\infty}\ln\left(\dfrac{x+a}{x-a}\right)^x = \lim_{x\to\infty} x\ln\left(\dfrac{x+a}{x-a}\right) = \lim_{x\to\infty}\dfrac{\ln(x+a)-\ln(x-a)}{1/x}$

$$\overset{\text{H}}{=} \lim_{x\to\infty}\frac{\dfrac{1}{x+a} - \dfrac{1}{x-a}}{-1/x^2} = -\lim_{x\to\infty}\frac{(x-a)x^2 - (x+a)x^2}{(x+a)(x-a)} = -\lim_{x\to\infty}\frac{-2ax^2}{x^2-a^2} = \lim_{x\to\infty}\frac{2a}{1-a^2/x^2} = 2a.$$ Hence $\ln L = 2a$, so $L = e^{2a}$. Hence $L = e^1 \;\Rightarrow\; 2a = 1 \;\Rightarrow\; a = \frac{1}{2}$.

21. Note that $\dfrac{d}{dx}\left(\displaystyle\int_0^x\left[\int_0^u f(t)dt\right]du\right) = \displaystyle\int_0^x f(t)dt$ by Part 1 of the Fundamental Theorem of Calculus, while

$$\frac{d}{dx}\left[\int_0^x f(u)(x-u)du\right] = \frac{d}{dx}\left[x\int_0^x f(u)du\right] - \frac{d}{dx}\left[\int_0^x f(u)u\,du\right] = \int_0^x f(u)du + xf(x) - f(x)x$$

$$= \int_0^x f(u)du.$$

Hence $\int_0^x f(u)(x-u)du = \int_0^x\left[\int_0^u f(t)dt\right]du + C$. Setting $x = 0$ gives $C = 0$.

23. $\lim_{x\to\infty} x^c e^{-2x}\displaystyle\int_0^x e^{2t}\sqrt{t^2+1}\,dt = \lim_{x\to\infty}\dfrac{\int_0^x e^{2t}\sqrt{t^2+1}\,dt}{x^{-c}e^{2x}}$. By l'Hospital's Rule, this is equal to

$$\lim_{x\to\infty}\frac{e^{2x}\sqrt{x^2+1}}{-cx^{-c-1}e^{2x}+2x^{-c}e^{2x}} = \lim_{x\to\infty}\frac{\sqrt{x^2+1}}{2x^{-c}-cx^{-c-1}} = \lim_{x\to\infty}\frac{x^{c+1}\sqrt{x^2+1}}{2x-c}.$$ This limit is finite only for $c \le -1$, and is zero for $c < -1$. If $c = -1$, the limit is $\lim_{x\to\infty}\dfrac{\sqrt{x^2+1}}{2x+1} = \dfrac{1}{2}$.

25. Both sides of the inequality are positive, so $\cosh(\sinh x) < \sinh(\cosh x) \iff \cosh^2(\sinh x) < \sinh^2(\cosh x)$

$\iff \sinh^2(\sinh x) + 1 < \sinh^2(\cosh x) \iff 1 < [\sinh(\cosh x) - \sinh(\sinh x)][\sinh(\cosh x) + \sinh(\sinh x)]$

$$\iff \quad 1 < \left[\sinh\left(\frac{e^x + e^{-x}}{2}\right) - \sinh\left(\frac{e^x - e^{-x}}{2}\right)\right]\left[\sinh\left(\frac{e^x + e^{-x}}{2}\right) + \sinh\left(\frac{e^x - e^{-x}}{2}\right)\right]$$

$$\iff \quad 1 < \left[2\cosh\left(\frac{e^x}{2}\right)\sinh\left(\frac{e^{-x}}{2}\right)\right]\left[2\sinh\left(\frac{e^x}{2}\right)\cosh\left(\frac{e^{-x}}{2}\right)\right] \quad \text{(use the addition formulas and cancel)}$$

$$\iff \quad 1 < \left[2\sinh\left(\frac{e^x}{2}\right)\cosh\left(\frac{e^x}{2}\right)\right]\left[2\sinh\left(\frac{e^{-x}}{2}\right)\cosh\left(\frac{e^{-x}}{2}\right)\right]$$

$\iff \quad 1 < \sinh e^x \sinh e^{-x}$, by the half-angle formula. Now both e^x and e^{-x} are positive, and $\sinh y > y$ for $y > 0$, since $\sinh 0 = 0$ and $(\sinh y - y)' = \cosh y - 1 > 0$ for $x > 0$, so $1 = e^x e^{-x} < \sinh e^x \sinh e^{-x}$. So, following this chain of reasoning backward, we arrive at the desired result.

Another Method: Using Formula 6.7.3, we have

$\sinh^{-1}(\cosh(\sinh x)) = \ln\left(\cosh(\sinh x) + \sqrt{1 + \cosh^2(\sinh x)}\right) = \ln(\cosh(\sinh x) + \sinh(\cosh x))$

$= \ln\left(e^{\sinh x}\right) = \sinh x$. But $\sinh x < \cosh x$, so $\sinh^{-1}(\cosh(\sinh x)) < \cosh x$. Since sinh is an increasing function, we can apply it to both sides of the inequality and get $\cosh(\sinh x) < \sinh(\cosh x)$.

27. We must find expressions for the areas A and B, and then set them equal and see what this says about the curve C. If $P = (a, 2a^2)$, then area A is just $\int_0^a (2x^2 - x^2)dx = \int_0^a x^2\,dx = \frac{1}{3}a^3$. To find area B, we use y as the variable of integration. So we find the equation of the middle curve as a function of y: $y = 2x^2 \iff$ $x = \sqrt{y/2}$, since we are concerned with the first quadrant only. We can express area B as $\int_0^{2a^2}\left[\sqrt{y/2} - C(y)\right]dy = \left[\frac{4}{3}(y/2)^{3/2}\right]_0^{2a^2} - \int_0^{2a^2} C(y)dy = \frac{4}{3}a^3 - \int_0^{2a^2} C(y)dy$, where $C(y)$ is the function with graph C. Setting $A = B$, we get $\frac{1}{3}a^3 = \frac{4}{3}a^3 - \int_0^{2a^2} C(y)dy \iff \int_0^{2a^2} C(y)dy = a^3$. Now we differentiate this equation with respect to a using the Chain Rule and the Fundamental Theorem: $C(2a^2)(4a) = 3a^2 \Rightarrow$ $C(y) = \frac{3}{4}\sqrt{y/2}$, where $y = 2a^2$. Now we can solve explicitly for y: $x = \frac{3}{4}\sqrt{y/2} \Rightarrow x^2 = \frac{9}{16}(y/2) \Rightarrow$ $y = \frac{32}{9}x^2$.

29. Suppose that the curve $y = a^x$ intersects the line $y = x$. Then $a^{x_0} = x_0$ for some $x_0 > 0$, and hence $a = x_0^{1/x_0}$. We find the maximum value of $g(x) = x^{1/x}$, > 0, because if a is larger than the maximum value of this function, then the curve $y = a^x$ does not intersect the line $y = x$.

$g'(x) = e^{(1/x)\ln x}\left(-\frac{1}{x^2}\ln x + \frac{1}{x}\cdot\frac{1}{x}\right) = x^{1/x}\left(\frac{1}{x^2}\right)(1 - \ln x)$. This is 0 only where $x = e$, and for $0 < x < e$, $f'(x) > 0$, while for $x > e$, $f'(x) < 0$, so g has an absolute maximum of $g(e) = e^{1/e}$. So if $y = a^x$ intersects $y = x$, we must have $0 < a \le e^{1/e}$. Conversely, suppose that $0 < a \le e^{1/e}$. Then $a^e \le e$, so the graph of $y = a^x$ lies below or touches the graph of $y = x$ at $x = e$. Also $a^0 = 1 > 0$, so the graph of $y = a^x$ lies above that of $y = x$ at $x = 0$. Therefore, by the Intermediate Value Theorem, the graphs of $y = a^x$ and $y = x$ must intersect somewhere between $x = 0$ and $x = e$.

31. Note that $f(0)=0$, so for $x \neq 0$, $\left|\dfrac{f(x)-f(0)}{x-0}\right| = \left|\dfrac{f(x)}{x}\right| = \dfrac{|f(x)|}{|x|} \leq \dfrac{|\sin x|}{|x|} = \dfrac{\sin x}{x}$. Therefore

$|f'(0)| = \left|\lim\limits_{x\to 0}\dfrac{f(x)-f(0)}{x-0}\right| = \lim\limits_{x\to 0}\left|\dfrac{f(x)-f(0)}{x-0}\right| \leq \lim\limits_{x\to 0}\dfrac{\sin x}{x} = 1$. But

$f'(x) = a_1\cos x + 2a_2\cos 2x + \cdots + na_n\cos nx$, so $|f'(0)| = |a_1 + 2a_2 + \cdots + na_n| \leq 1$.

Another Solution: We are given that $\left|\sum\limits_{k=1}^{n} a_k \sin kx\right| \leq |\sin x|$. So for x close to 0, and $x \neq 0$, we have

$\left|\sum\limits_{k=1}^{n} a_k \dfrac{\sin kx}{\sin x}\right| \leq 1 \;\Rightarrow\; \lim\limits_{x\to 0}\left|\sum\limits_{k=1}^{n} a_k \dfrac{\sin kx}{\sin x}\right| \leq 1 \;\Rightarrow\; \left|\sum\limits_{k=1}^{n} a_k \lim\limits_{x\to 0}\dfrac{\sin kx}{\sin x}\right| \leq 1$. But by l'Hospital's Rule,

$\lim\limits_{x\to 0}\dfrac{\sin kx}{\sin x} = \lim\limits_{x\to 0}\dfrac{k\cos kx}{\cos x} = k$, so $\left|\sum\limits_{k=1}^{n} ka_k\right| \leq 1$.

33. The volume is $\int_0^{\sqrt{2}} \pi r^2\, ds$, where s is measured along the line $y=x$ from the origin to P. From the figure we have $r^2 + s^2 = d^2 = x^2 + x^4$, and from the distance formula we have

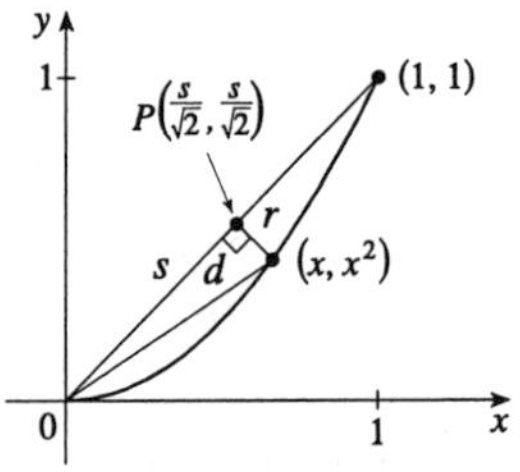

$$r^2 = \left(x - \tfrac{1}{\sqrt{2}}s\right)^2 + \left(x^2 - \tfrac{1}{\sqrt{2}}s\right)^2$$
$$= x^2 + x^4 + s^2 - \sqrt{2}s\left(x + x^2\right)$$

$\Rightarrow \quad x^2 + x^4 - s^2 = x^2 + x^4 + s^2 - \sqrt{2}s(x+x^2) \quad\Rightarrow\quad 2s^2 = \sqrt{2}s(x+x^2) \quad\Rightarrow\quad s = (x+x^2)/\sqrt{2}$

$\Rightarrow \quad r^2 = x^2 + x^4 - s^2 = x^2 + x^4 - \frac{1}{2}\left(x^2 + 2x^3 + x^4\right) = \frac{1}{2}x^4 - x^3 + \frac{1}{2}x^2$. Also, $ds = \frac{1}{\sqrt{2}}(1+2x)dx$, so

$$V = \int_0^{\sqrt{2}} \pi r^2\, ds = \pi\int_0^1 \left(\tfrac{1}{2}x^4 - x^3 + \tfrac{1}{2}x^2\right)\tfrac{1}{\sqrt{2}}(1+2x)dx$$
$$= \tfrac{\pi}{2\sqrt{2}}\int_0^1 \left(2x^5 - 3x^4 + x^2\right)dx = \tfrac{\pi}{2\sqrt{2}}\left(\tfrac{1}{3} - \tfrac{3}{5} + \tfrac{1}{3}\right) = \tfrac{\pi}{30\sqrt{2}}.$$

For a more general method, see Problems Plus 22 after Chapter 8.

CHAPTER SEVEN

EXERCISES 7.1

1. Let $u = x$, $dv = e^{2x}\,dx \quad \Rightarrow \quad du = dx$, $v = \frac{1}{2}e^{2x}$. Then by Equation 2,
$\int xe^{2x}\,dx = \frac{1}{2}xe^{2x} - \int \frac{1}{2}e^{2x}\,dx = \frac{1}{2}xe^{2x} - \frac{1}{4}e^{2x} + C.$

3. Let $u = x$, $dv = \sin 4x\,dx \quad \Rightarrow \quad du = dx$, $v = -\frac{1}{4}\cos 4x$. Then
$\int x\sin 4x\,dx = -\frac{1}{4}x\cos 4x - \int\left(-\frac{1}{4}\cos 4x\right)dx = -\frac{1}{4}x\cos 4x + \frac{1}{16}\sin 4x + C.$

5. Let $u = x^2$, $dv = \cos 3x\,dx \quad \Rightarrow \quad du = 2x\,dx$, $v = \frac{1}{3}\sin 3x$. Then
$I = \int x^2\cos 3x\,dx = \frac{1}{3}x^2\sin 3x - \frac{2}{3}\int x\sin 3x\,dx$ by Equation 2. Next let $U = x$, $dV = \sin 3x\,dx \quad \Rightarrow$
$dU = dx$, $V = -\frac{1}{3}\cos 3x$ to get $\int x\sin 3x\,dx = -\frac{1}{3}x\cos 3x + \frac{1}{3}\int\cos 3x\,dx = -\frac{1}{3}x\cos 3x + \frac{1}{9}\sin 3x + C_1$.
Substituting for $\int x\sin 3x\,dx$, we get $I = \frac{1}{3}x^2\sin 3x - \frac{2}{3}\left(-\frac{1}{3}x\cos 3x + \frac{1}{9}\sin 3x + C_1\right)$
$= \frac{1}{3}x^2\sin 3x + \frac{2}{9}x\cos 3x - \frac{2}{27}\sin 3x + C$, where $C = -\frac{2}{3}C_1$.

7. Let $u = (\ln x)^2$, $dv = dx \quad \Rightarrow \quad du = 2\ln x \cdot \frac{1}{x}\,dx$, $v = x$. Then $I = \int(\ln x)^2dx = x(\ln x)^2 - 2\int\ln x\,dx$.
Taking $U = \ln x$, $dV = dx \quad \Rightarrow \quad dU = 1/x\,dx$, $V = x$, we find that
$\int\ln x\,dx = x\ln x - \displaystyle\int x\cdot\frac{1}{x}\,dx = x\ln x - x + C_1$. Thus $I = x(\ln x)^2 - 2x\ln x + 2x + C$, where $C = -2C_1$.

9. $I = \int\theta\sin\theta\cos\theta\,d\theta = \frac{1}{4}\int 2\theta\sin 2\theta\,d\theta = \frac{1}{8}\int t\sin t\,dt$ (Put $t = 2\theta \quad \Rightarrow \quad dt = d\theta/2$.)
Let $u = t$, $dv = \sin t\,dt \quad \Rightarrow \quad du = dt$, $v = -\cos t$. Then
$I = \frac{1}{8}(-t\cos t + \int\cos t\,dt) = \frac{1}{8}(-t\cos t + \sin t) + C = \frac{1}{8}(\sin 2\theta - 2\theta\cos 2\theta) + C.$

11. Let $u = \ln t$, $dv = t^2\,dt \quad \Rightarrow \quad du = dt/t$, $v = \frac{1}{3}t^3$. Then
$\int t^2\ln t\,dt = \frac{1}{3}t^3\ln t - \int\frac{1}{3}t^3(1/t)\,dt = \frac{1}{3}t^3\ln t - \frac{1}{9}t^3 + C = \frac{1}{9}t^3(3\ln t - 1) + C.$

13. First let $u = \sin 3\theta$, $dv = e^{2\theta}d\theta \quad \Rightarrow \quad du = 3\cos 3\theta\,d\theta$, $v = \frac{1}{2}e^{2\theta}$. Then
$I = \int e^{2\theta}\sin 3\theta\,d\theta = \frac{1}{2}e^{2\theta}\sin 3\theta - \frac{3}{2}\int e^{2\theta}\cos 3\theta\,d\theta$. Next let $U = \cos 3\theta$, $dU = -3\sin 3\theta\,d\theta$, $dV = e^{2\theta}d\theta$,
$v = \frac{1}{2}e^{2\theta}$ to get $\int e^{2\theta}\cos 3\theta\,d\theta = \frac{1}{2}e^{2\theta}\cos 3\theta + \frac{3}{2}\int e^{2\theta}\sin 3\theta\,d\theta$. Substituting in the previous formula gives
$I = \frac{1}{2}e^{2\theta}\sin 3\theta - \frac{3}{4}e^{2\theta}\cos 3\theta - \frac{9}{4}\int e^{2\theta}\sin 3\theta\,d\theta$ or $\frac{13}{4}\int e^{2\theta}\sin 3\theta\,d\theta = \frac{1}{2}e^{2\theta}\sin 3\theta - \frac{3}{4}e^{2\theta}\cos 3\theta + C_1$.
Hence $\int e^{2\theta}\sin 3\theta\,d\theta = \frac{1}{13}e^{2\theta}(2\sin 3\theta - 3\cos 3\theta) + C$, where $C = \frac{4}{13}C_1$.

15. Let $u = y$, $dv = \sinh y\,dy \quad \Rightarrow \quad du = dy$, $v = \cosh y$. Then
$\int y\sinh y\,dy = y\cosh y - \int\cosh y\,dy = y\cosh y - \sinh y + C.$

17. Let $u = t$, $dv = e^{-t}\,dt \quad \Rightarrow \quad du = dt$, $v = -e^{-t}$. Then Formula 6 says $\int_0^1 te^{-t}\,dt = \left[-te^{-t}\right]_0^1 + \int_0^1$
$e^{-t}\,dt = -1/e + \left[-e^{-t}\right]_0^1 = -1/e - 1/e + 1 = 1 - 2/e.$

19. Let $u = x$, $dv = \cos 2x\,dx \quad \Rightarrow \quad du = dx$, $v = \frac{1}{2}\sin 2x\,dx$. Then
$\int_0^{\pi/2} x\cos 2x\,dx = \left[\frac{1}{2}x\sin 2x\right]_0^{\pi/2} - \frac{1}{2}\int_0^{\pi/2}\sin 2x\,dx = 0 + \left[\frac{1}{4}\cos 2x\right]_0^{\pi/2} = \frac{1}{4}(-1 - 1) = -\frac{1}{2}.$

21. Let $u = \cos^{-1}x$, $dv = dx \quad \Rightarrow \quad du = -\dfrac{dx}{\sqrt{1-x^2}}$, $v = x$. Then

$I = \displaystyle\int_0^{1/2} \cos^{-1}x\,dx = \left[x\cos^{-1}x\right]_0^{1/2} + \int_0^{1/2} \frac{x\,dx}{\sqrt{1-x^2}} = \tfrac{1}{2}\cdot\tfrac{\pi}{3} + \int_1^{3/4} t^{-1/2}\left[-\tfrac{1}{2}\,dt\right]$, where $t = 1 - x^2 \quad \Rightarrow$

$dt = -2x\,dx$. Thus $I = \frac{\pi}{6} + \frac{1}{2}\int_{3/4}^{1} t^{-1/2}\,dt = \left[\sqrt{t}\right]_{3/4}^{1} = \frac{\pi}{6} + 1 - \frac{\sqrt{3}}{2} = \frac{1}{6}\left(\pi + 6 - 3\sqrt{3}\right)$.

23. Let $u = \ln(\sin x)$, $dv = \cos x\,dx \quad \Rightarrow \quad du = \dfrac{\cos x}{\sin x}\,dx$, $v = \sin x$. Then

$I = \int \cos x \ln(\sin x)dx = \sin x \ln(\sin x) - \int \cos x\,dx = \sin x \ln(\sin x) - \sin x + C$.

Another Method: Substitute $t = \sin x$, so $dt = \cos x\,dx$. Then $I = \int \ln t\,dt = t\ln t - t + C$ (see Example 2) and so $I = \sin x(\ln \sin x - 1) + C$.

25. Let $u = 2x + 3$, $dv = e^x\,dx \quad \Rightarrow \quad du = 2\,dx$, $v = e^x$. Then

$\int (2x+3)e^x\,dx = (2x+3)e^x - \int e^x \cdot 2\,dx = (2x+3)e^x - 2e^x + C = (2x+1)e^x + C$.

27. Let $w = \ln x \quad \Rightarrow \quad dw = dx/x$. Then $x = e^w$ and $dx = e^w\,dw$, so

$\int \cos(\ln x)dx = \int e^w \cos w\,dw = \frac{1}{2}e^w(\sin w + \cos w) + C$ (by the method of Example 4)

$= \frac{1}{2}x[\sin(\ln x) + \cos(\ln x)] + C$.

29. $I = \int_1^4 \ln\sqrt{x}\,dx = \frac{1}{2}\int_1^4 \ln x\,dx = \frac{1}{2}[x\ln x - x]_1^4$ as in Example 2. So

$I = \frac{1}{2}[(4\ln 4 - 4) - (0 - 1)] = 4\ln 2 - \frac{3}{2}$.

31. Let $w = \sqrt{x}$, so that $x = w^2$ and $dx = 2w\,dw$. Then use $u = 2w$, $dv = \sin w\,dw$. Thus

$\int \sin\sqrt{x}\,dx = \int 2w\sin w\,dw = -2w\cos w + \int 2\cos w\,dw = -2w\cos w + 2\sin w + C$

$= -2\sqrt{x}\cos\sqrt{x} + 2\sin\sqrt{x} + C$.

33. $\int x^5 e^{x^2}\,dx = \int (x^2)^2 e^{x^2} x\,dx = \int t^2 e^t \,\frac{1}{2}\,dt$ (where $t = x^2 \Rightarrow \frac{1}{2}\,dt = x\,dx$)

$= \frac{1}{2}(t^2 - 2t + 2)e^t + C$ (by Example 3) $= \frac{1}{2}\left(x^4 - 2x^2 + 2\right)e^{x^2} + C$.

35. Let $u = x$, $dv = \cos \pi x\,dx \quad \Rightarrow \quad du = dx$,

$v = \displaystyle\int \cos\pi x\,dx = \frac{\sin \pi x}{\pi}$. Thus

$\displaystyle\int x\cos\pi x\,dx = x\cdot\frac{\sin\pi x}{\pi} - \int \frac{\sin\pi x}{\pi}\,dx = \frac{x\sin\pi x}{\pi} + \frac{\cos\pi x}{\pi^2} + C.$

We see from the graph that this is reasonable, since the antiderivative has extrema where the original function is 0.

1.2
f
−2
2
F
−1.2

37. **(a)** Take $n = 2$ in Example 6 to get $\int \sin^2 x\,dx = -\frac{1}{2}\cos x\sin x + \frac{1}{2}\int 1\,dx = \dfrac{x}{2} - \dfrac{\sin 2x}{4} + C$.

(b) $\int \sin^4 x\,dx = -\frac{1}{4}\cos x\sin^3 x + \frac{3}{4}\int \sin^2 x\,dx = -\frac{1}{4}\cos x\sin^3 x + \frac{3}{8}x - \frac{3}{16}\sin 2x + C$.

39. **(a)** $\int_0^{\pi/2} \sin^n x\,dx = \left[-\dfrac{\cos x\sin^{n-1}x}{n}\right]_0^{\pi/2} + \dfrac{n-1}{n}\displaystyle\int_0^{\pi/2} \sin^{n-2}x\,dx = \frac{n-1}{n}\int_0^{\pi/2} \sin^{n-2}x\,dx$

(b) $\int_0^{\pi/2} \sin^3 x\,dx = \frac{2}{3}\int_0^{\pi/2} \sin x\,dx = \left[-\frac{2}{3}\cos x\right]_0^{\pi/2} = \frac{2}{3}$; $\int_0^{\pi/2} \sin^5 x\,dx = \frac{4}{5}\int_0^{\pi/2} \sin^3 x\,dx = \frac{4}{5}\cdot\frac{2}{3} = \frac{8}{15}$

(c) The formula holds for $n = 1$ (that is, $2n + 1 = 3$) by (b). Assume it holds for some $k \geq 1$. Then

$$\int_0^{\pi/2} \sin^{2k+1} x\,dx = \frac{2 \cdot 4 \cdot 6 \cdot \cdots \cdot (2k)}{3 \cdot 5 \cdot 7 \cdot \cdots \cdot (2k+1)}.$$

By Example 6, $\displaystyle\int_0^{\pi/2} \sin^{2k+3} x\,dx = \frac{2k+2}{2k+3}\int_0^{\pi/2} \sin^{2k+1} x\,dx = \frac{2 \cdot 4 \cdot 6 \cdot \cdots \cdot [2(k+1)]}{2 \cdot 4 \cdot 6 \cdot \cdots \cdot [2(k+1)+1]}$ as desired.

By induction, the formula holds for all $n \geq 1$.

41. Let $u = (\ln x)^n$, $dv = dx \quad \Rightarrow \quad du = n(\ln x)^{n-1}(dx/x)$, $v = x$. Then
$\int (\ln x)^n\,dx = x(\ln x)^n - n\int (\ln x)^{n-1}\,dx$, by Equation 2.

43. Let $u = (x^2 + a^2)^n$, $dv = dx \quad \Rightarrow \quad du = n(x^2 + a^2)^{n-1} 2x\,dx$, $v = x$. Then

$$\begin{aligned}\int (x^2 + a^2)^n\,dx &= x(x^2 + a^2)^n - 2n\int x^2(x^2 + a^2)^{n-1}\,dx \\ &= x(x^2 + a^2)^n - 2n\left[\int (x^2 + a^2)^n\,dx - a^2\int (x^2 + a^2)^{n-1}\,dx\right] \quad [\text{since } x^2 = (x^2 + a^2) - a^2]\end{aligned}$$

$\Rightarrow \quad (2n+1)\int (x^2 + a^2)^n\,dx = x(x^2 + a^2)^n + 2na^2\int (x^2 + a^2)^{n-1}\,dx$, and

$$\int (x^2 + a^2)^n\,dx = \frac{x(x^2 + a^2)^n}{2n+1} + \frac{2na^2}{2n+1}\int (x^2 + a^2)^{n-1}\,dx \quad (\text{provided } 2n + 1 \neq 0).$$

45. Take $n = 3$ in Exercise 41 to get
$\int (\ln x)^3\,dx = x(\ln x)^3 - 3\int (\ln x)^2\,dx = x(\ln x)^3 - 3x(\ln x)^2 + 6x\ln x - 6x + C$ (by Exercise 7).

47. Let $u = \sin^{-1}x$, $dv = dx \quad \Rightarrow \quad du = \dfrac{dx}{\sqrt{1 - x^2}}$, $v = x$. Then

$$\text{area} = \int_0^{1/2} \sin^{-1}x\,dx = \left[x\sin^{-1}x\right]_0^{1/2} - \int_0^{1/2} \frac{x}{\sqrt{1 - x^2}}\,dx = \tfrac{1}{2}\left(\tfrac{\pi}{6}\right) + \left[\sqrt{1 - x^2}\right]_0^{1/2}$$
$$= \tfrac{\pi}{12} + \tfrac{\sqrt{3}}{2} - 1 = \tfrac{1}{12}\left(\pi + 6\sqrt{3} - 12\right).$$

49.

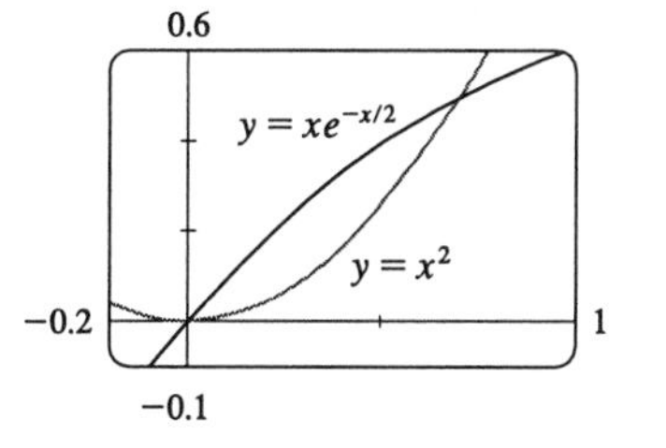

From the graph, we see that the curves intersect at approximately $x = 0$ and $x = 0.70$, with $xe^{-x/2} > x^2$ on $(0, 0.70)$. So the area bounded by the curves is approximately $A = \int_0^{0.70} (xe^{-x/2} - x^2)\,dx$. We separate this into two integrals, and evaluate the first one by parts with $u = x$, $dv = e^{-x/2}\,dx \quad \Rightarrow \quad du = dx$, $v = -2e^{-x/2}$:

$A = \left[-2xe^{-x/2}\right]_0^{0.70} - \int_0^{0.70} (-2e^{-x/2})\,dx - \left[\tfrac{1}{3}x^3\right]_0^{0.70} = [-2(0.70)e^{-0.35} - 0] - \left[4e^{-x/2}\right]_0^{0.70} - \tfrac{1}{3}[0.70^3 - 0]$
≈ 0.080.

51. Volume $= \int_{2\pi}^{3\pi} 2\pi x\sin x\,dx$. Let $u = x$, $dv = \sin x\,dx \quad \Rightarrow \quad du = dx$, $v = -\cos x \quad \Rightarrow$
$V = 2\pi[-x\cos x + \sin x]_{2\pi}^{3\pi} = 2\pi[(3\pi + 0) - (-2\pi + 0)] = 2\pi(5\pi) = 10\pi^2$.

53. Volume $= \int_{-1}^{0} 2\pi(1 - x)e^{-x}\,dx$. Let $u = 1 - x$, $dv = e^{-x}\,dx \quad \Rightarrow \quad du = -\,dx$, $v = -e^{-x} \quad \Rightarrow$
$V = 2\pi[xe^{-x}]_{-1}^{0} = 2\pi(0 + e) = 2\pi e$

55. Since $v(t) > 0$ for all t, the desired distance $s(t) = \int_0^t v(w)dw = \int_0^t w^2 e^{-w}\,dw$. Let $u = w^2$, $dv = e^{-w}\,dw$ $\Rightarrow$ $du = 2w\,dw$, $v = -e^{-w}$. Then $s(t) = \left[-w^2 e^{-w}\right]_0^t + 2\int_0^t we^{-w}\,dw$.

Now let $U = w$, $dV = e^{-w}\,dw$ $\Rightarrow$ $dU = dw$, $V = -e^{-w}$. Then
$s(t) = -t^2 e^{-t} + 2\left(\left[-we^{-w}\right]_0^t + \int_0^t e^{-w}\,dw\right) = -t^2 e^{-t} - 2te^{-t} - 2e^{-t} + 2 = 2 - e^{-t}(t^2 + 2t + 2)$ meters.

57. Take $g(x) = x$ in Equation 1.

59. By Exercise 58, $\int_1^e \ln x\,dx = e\ln e - 1\ln 1 - \int_{\ln 1}^{\ln e} e^y\,dy = e - \int_0^1 e^y\,dy = e - \left[e^y\right]_0^1 = e - (e-1) = 1$.

61. Using the formula for volumes of rotation (Equation 5.2.3) and the figure, we see that
Volume $= \int_0^d \pi b^2\,dy - \int_0^c \pi a^2\,dy - \int_c^d \pi[g(y)]^2\,dy = \pi b^2 d - \pi a^2 c - \int_c^d \pi[g(y)]^2\,dy$.
Let $y = f(x)$, which gives $dy = f'(x)dx$ and $g(y) = x$, so that $V = \pi b^2 d - \pi a^2 c - \pi\int_a^b x^2 f'(x)dx$. Now integrate by parts with $u = x^2$, and $dv = f'(x)dx$ $\Rightarrow$ $du = 2x\,dx$, $v = f(x)$, and
$\int_a^b x^2\,f'(x)dx = \left[x^2\,f(x)\right]_a^b - \int_a^b 2x\,f(x)dx = b^2\,f(b) - a^2\,f(a) - \int_a^b 2x\,f(x)dx$, but $f(a) = c$ and $f(b) = d$
$\Rightarrow$ $V = \pi b^2 d - \pi a^2 c - \pi\left[b^2 d - a^2 c - \int_a^b 2x f(x)dx\right] = \int_a^b 2\pi x f(x)dx$.

EXERCISES 7.2

1. $\int_0^{\pi/2} \sin^2 3x\,dx = \int_0^{\pi/2} \frac{1}{2}(1 - \cos 6x)dx = \left[\frac{1}{2}x - \frac{1}{12}\sin 6x\right]_0^{\pi/2} = \frac{\pi}{4}$

3. $\int \cos^4 x\,dx = \int\left[\frac{1}{2}(1 + \cos 2x)\right]^2 dx = \frac{1}{4}\int(1 + 2\cos 2x + \cos^2 2x)dx$
$= \frac{1}{4}x + \frac{1}{4}\sin 2x + \frac{1}{4}\int\frac{1}{2}(1 + \cos 4x)dx = \frac{1}{4}\left[x + \sin 2x + \frac{1}{2}x + \frac{1}{8}\sin 4x\right] + C$
$= \frac{3}{8}x + \frac{1}{4}\sin 2x + \frac{1}{32}\sin 4x + C$

5. Let $u = \cos x$ $\Rightarrow$ $du = -\sin x\,dx$. Then $\int \sin^3 x\cos^4 x\,dx = \int \cos^4 x(1 - \cos^2 x)\sin x\,dx$
$= \int u^4(1 - u^2)(-du) = \int(u^6 - u^4)du = \frac{1}{7}u^7 - \frac{1}{5}u^5 + C = \frac{1}{7}\cos^7 x - \frac{1}{5}\cos^5 x + C$.

7. $\int_0^{\pi/4} \sin^4 x\cos^2 x\,dx = \int_0^{\pi/4} \sin^2 x(\sin x\cos x)^2\,dx = \int_0^{\pi/4} \frac{1}{2}(1 - \cos 2x)\left(\frac{1}{2}\sin 2x\right)^2 dx$
$= \frac{1}{8}\int_0^{\pi/4}(1 - \cos 2x)\sin^2 2x\,dx = \frac{1}{8}\int_0^{\pi/4}\sin^2 2x\,dx - \frac{1}{8}\int_0^{\pi/4}\sin^2 2x\cos 2x\,dx$
$= \frac{1}{16}\int_0^{\pi/4}(1 - \cos 4x)dx - \frac{1}{16}\left[\frac{1}{3}\sin^3 2x\right]_0^{\pi/4} = \frac{1}{16}\left[x - \frac{1}{4}\sin 4x - \frac{1}{3}\sin^3 2x\right]_0^{\pi/4}$
$= \frac{1}{16}\left(\frac{\pi}{4} - 0 - \frac{1}{3}\right) = \frac{1}{192}(3\pi - 4)$

9. $\int(1 - \sin 2x)^2\,dx = \int(1 - 2\sin 2x + \sin^2 2x)dx = \int\left[1 - 2\sin 2x + \frac{1}{2}(1 - \cos 4x)\right]dx$
$= \int\left[\frac{3}{2} - 2\sin 2x - \frac{1}{2}\cos 4x\right]dx = \frac{3}{2}x + \cos 2x - \frac{1}{8}\sin 4x + C$

11. Let $u = \sin x$ $\Rightarrow$ $du = \cos x\,dx$. Then
$\int \cos^5 x\sin^5 x\,dx = \int u^5(1 - u^2)^2\,du = \int u^5(1 - 2u^2 + u^4)du = \int(u^5 - 2u^7 + u^9)du$
$= \frac{1}{10}u^{10} - \frac{1}{4}u^8 + \frac{1}{6}u^6 + C = \frac{1}{10}\sin^{10} x - \frac{1}{4}\sin^8 x + \frac{1}{6}\sin^6 x + C$.

Or: Let $v = \cos x$, $dv = -\sin x\,dx$. Then
$\int \cos^5 x\sin^5 x\,dx = \int v^5(1 - v^2)^2(-dv) = \int(-v^5 + 2v^7 - v^9)dv$
$= -\frac{1}{10}v^{10} + \frac{1}{4}v^8 - \frac{1}{6}v^6 + C = -\frac{1}{10}\cos^{10} x + \frac{1}{4}\cos^8 x - \frac{1}{6}\cos^6 x + C$.

13. Let $u = \cos x$, $du = -\sin x\,dx$. Then $\int \sin^3 x\,\sqrt{\cos x}\,dx = \int (1 - \cos^2 x)\sqrt{\cos x}\,\sin x\,dx$
$= \int (1 - u^2)u^{1/2}(-du) = \int (u^{5/2} - u^{1/2})du = \frac{2}{7}u^{7/2} - \frac{2}{3}u^{3/2} + C$
$= \frac{2}{7}(\cos x)^{7/2} - \frac{2}{3}(\cos x)^{3/2} + C = \left[\frac{2}{7}\cos^3 x - \frac{2}{3}\cos x\right]\sqrt{\cos x} + C.$

15. Let $u = \cos x \quad \Rightarrow \quad du = -\sin x\,dx$. Then $\displaystyle\int \cos^2 x \tan^3 x\,dx = \int \frac{\sin^3 x}{\cos x}\,dx$

$$= \int \frac{(1 - u^2)(-du)}{u} = \int \left[\frac{-1}{u} + u\right]du = -\ln|u| + \tfrac{1}{2}u^2 + C = \tfrac{1}{2}\cos^2 x - \ln|\cos x| + C.$$

17. $\displaystyle\int \frac{1 - \sin x}{\cos x}\,dx = \int (\sec x - \tan x)dx = \ln|\sec x + \tan x| - \ln|\sec x| + C$ (by Example 8)

$$= \ln|(\sec x + \tan x)\cos x| + C = \ln|1 + \sin x| + C = \ln(1 + \sin x) + C,$$

since $1 + \sin x \geq 0$.

Or: $\displaystyle\int \frac{1 - \sin x}{\cos x}\,dx = \int \frac{1 - \sin x}{\cos x}\cdot\frac{1 + \sin x}{1 + \sin x}\,dx = \int \frac{(1 - \sin^2 x)dx}{\cos x(1 + \sin x)} = \int \frac{\cos x\,dx}{1 + \sin x} = \int \frac{dw}{w}$

(where $w = 1 + \sin x$, $dw = \cos x\,dx$) $= \ln|w| + C = \ln|1 + \sin x| + C = \ln(1 + \sin x) + C.$

19. $\int \tan^2 x\,dx = \int (\sec^2 x - 1)dx = \tan x - x + C.$

21. $\int \sec^4 x\,dx = \int (\tan^2 x + 1)\sec^2 x\,dx = \int \tan^2 x \sec^2 x\,dx + \int \sec^2 x\,dx = \frac{1}{3}\tan^3 x + \tan x + C$

23. Let $u = \tan x \quad \Rightarrow \quad du = \sec^2 x\,dx$. Then $\int_0^{\pi/4} \tan^4 x \sec^2 x\,dx = \int_0^1 u^4\,du = \left[\frac{1}{5}u^5\right]_0^1 = \frac{1}{5}$.

25. Let $u = \sec x \quad \Rightarrow \quad du = \sec x \tan x\,dx$. Then
$\int \tan x \sec^3 x\,dx = \int \sec^2 x \sec x \tan x\,dx = \int u^2\,du = \frac{1}{3}u^3 + C = \frac{1}{3}\sec^3 x + C.$

27. $\int \tan^5 x\,dx = \int (\sec^2 x - 1)^2 \tan x\,dx = \int \sec^4 x \tan x\,dx - 2\int \sec^2 x \tan x\,dx + \int \tan x\,dx$
$= \int \sec^3 x \sec x \tan x\,dx - 2\int \tan x \sec^2 x\,dx + \int \tan x\,dx$
$= \frac{1}{4}\sec^4 x - \tan^2 x + \ln|\sec x| + C.$ *Or:* $\frac{1}{4}\sec^4 x - \sec^2 x + \ln|\sec x| + C.$

29. Let $u = \sec x \quad \Rightarrow \quad du = \sec x \tan x\,dx$. Then

$\int_0^{\pi/3} \tan^5 x \sec x\,dx = \int_0^{\pi/3} (\sec^2 x - 1)^2 \sec x \tan x\,dx = \int_1^2 (u^2 - 1)^2\,du$
$= \int_1^2 (u^4 - 2u^2 + 1)du = \left[\frac{1}{5}u^5 - \frac{2}{3}u^3 + u\right]_1^2 = \left[\frac{32}{5} - \frac{16}{3} + 2\right] - \left[\frac{1}{5} - \frac{2}{3} + 1\right] = \frac{38}{15}.$

31. Let $u = \tan x \quad \Rightarrow \quad du = \sec^2 x\,dx$. Then
$\displaystyle\int \frac{\sec^2 x}{\cot x}\,dx = \int \tan x \sec^2 x\,dx = \int u\,du = \tfrac{1}{2}u^2 + C = \tfrac{1}{2}\tan^2 x + C.$

33. $\int_{\pi/6}^{\pi/2} \cot^2 x\,dx = \int_{\pi/6}^{\pi/2} (\csc^2 x - 1)dx = [-\cot x - x]_{\pi/6}^{\pi/2} = \left(0 - \frac{\pi}{2}\right) - \left(-\sqrt{3} - \frac{\pi}{6}\right) = \sqrt{3} - \frac{\pi}{3}$

35. Let $u = \cot x \quad \Rightarrow \quad du = -\csc^2 x\,dx$. Then
$\int \cot^4 x \csc^4 x\,dx = \int u^4(u^2 + 1)(-du) = -\int (u^6 + u^4)du = -\frac{1}{7}u^7 - \frac{1}{5}u^5 + C = -\frac{1}{7}\cot^7 x - \frac{1}{5}\cot^5 x + C.$

37. $\displaystyle I = \int \csc x\,dx = \int \frac{\csc x(\csc x - \cot x)}{\csc x - \cot x}\,dx = \int \frac{-\csc x \cot x + \csc^2 x}{\csc x - \cot x}\,dx$. Let $u = \csc x - \cot x \quad \Rightarrow$
$du = (-\csc x \cot x + \csc^2 x)dx$. Then $I = \int du/u = \ln|u| = \ln|\csc x - \cot x| + C.$

39. $\int \sin 5x \sin 2x\,dx = \int \frac{1}{2}[\cos(5x-2x) - \cos(5x+2x)]dx = \frac{1}{2}\int(\cos 3x - \cos 7x)dx$
$= \frac{1}{6}\sin 3x - \frac{1}{14}\sin 7x + C$

41. $\int \cos 3x \cos 4x\,dx = \int \frac{1}{2}[\cos(3x-4x) + \cos(3x+4x)]dx = \frac{1}{2}\int(\cos x + \cos 7x)dx = \frac{1}{2}\sin x + \frac{1}{14}\sin 7x + C$

43. $\displaystyle\int \frac{1-\tan^2 x}{\sec^2 x}\,dx = \int\left(\cos^2 x - \sin^2 x\right)dx = \int \cos 2x\,dx = \tfrac{1}{2}\sin 2x + C$

45. Let $u = \cos x \quad\Rightarrow\quad du = -\sin x\,dx$. Then

$\int \sin^5 x\,dx = \int(1-\cos^2 x)^2 \sin x\,dx = \int(1-u^2)^2(-du)$
$= \int(-1+2u^2-u^4)du = -\frac{1}{5}u^5 + \frac{2}{3}u^3 - u + C$
$= -\frac{1}{5}\cos^5 x + \frac{2}{3}\cos^3 x - \cos x + C.$

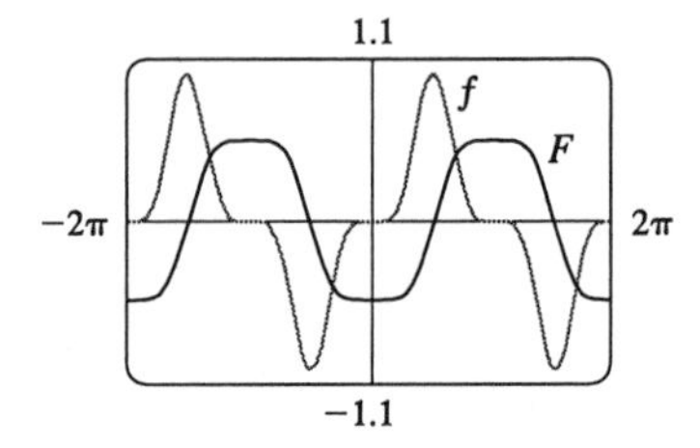

Notice that F is increasing when $f(x) > 0$, so the graphs serve as a check on our work.

47. $f_{\text{ave}} = \frac{1}{2\pi}\int_{-\pi}^{\pi}\sin^2 x\cos^3 x\,dx = \frac{1}{2\pi}\int_{-\pi}^{\pi}\sin^2 x(1-\sin^2 x)\cos x\,dx = \frac{1}{2\pi}\int_0^0 u^2(1-u^2)du$ (where $u = \sin x$) $= 0$

49. For $0 < x < \frac{\pi}{2}$, we have $0 < \sin x < 1$, so $\sin^3 x < \sin x$. Hence the area is
$\int_0^{\pi/2}(\sin x - \sin^3 x)dx = \int_0^{\pi/2}\sin x(1-\sin^2 x)dx = \int_0^{\pi/2}\cos^2 x\sin x\,dx$. Now let $u = \cos x \quad\Rightarrow$
$du = -\sin x\,dx$. Then area $= \int_1^0 u^2(-du) = \int_0^1 u^2\,du = \left[\frac{1}{3}u^3\right]_0^1 = \frac{1}{3}$.

51.

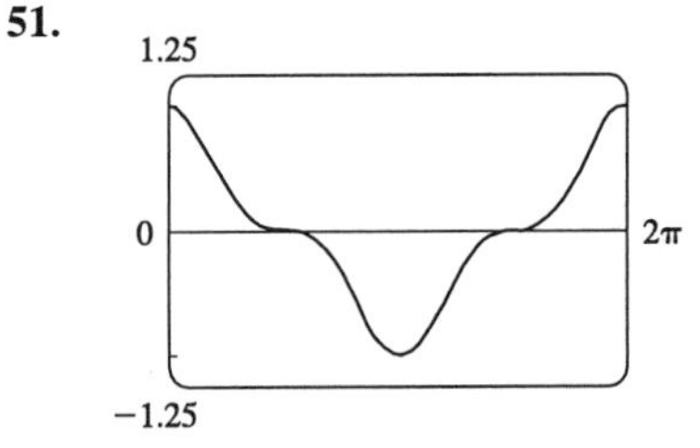

It seems from the graph that $\int_0^{2\pi}\cos^3 x\,dx = 0$, since the area below the x-axis and above the graph looks about equal to the area above the axis and below the graph. By Example 1, the integral is $\left[\sin x - \frac{1}{3}\sin^3 x\right]_0^{2\pi} = 0$. Note that due to symmetry, the integral of any odd power of $\sin x$ or $\cos x$ between limits which differ by $2n\pi$ (n any integer) is 0.

53. $V = \int_{\pi/2}^{\pi}\pi\sin^2 x\,dx = \pi\int_{\pi/2}^{\pi}\frac{1}{2}(1-\cos 2x)dx = \pi\left[\frac{1}{2}x - \frac{1}{4}\sin 2x\right]_{\pi/2}^{\pi} = \pi\left(\frac{\pi}{2} - 0 - \frac{\pi}{4} + 0\right) = \frac{\pi^2}{4}$

55. Volume $= \pi\int_0^{\pi/2}\left[(1+\cos x)^2 - 1^2\right]dx = \pi\int_0^{\pi/2}(2\cos x + \cos^2 x)dx$
$= \pi\left[2\sin x + \frac{1}{2}x + \frac{1}{4}\sin 2x\right]_0^{\pi/2} = \pi\left(2 + \frac{\pi}{4}\right) = 2\pi + \frac{\pi^2}{4}$

57. $s = f(t) = \int_0^t \sin\omega u\cos^2\omega u\,du$. Let $y = \cos\omega u \quad\Rightarrow\quad dy = -\omega\sin\omega u\,du$. Then
$$s = -\frac{1}{\omega}\int_1^{\cos\omega t} y^2\,dy = -\frac{1}{\omega}\left[\tfrac{1}{3}y^3\right]_1^{\cos\omega t} = \frac{1}{3\omega}\left(1-\cos^3\omega t\right).$$

59. Just note that the integrand is odd $[f(-x) = -f(x)]$.

Or: If $m \neq n$, calculate $\int_{-\pi}^{\pi}\sin mx\cos nx\,dx = \int_{-\pi}^{\pi}\frac{1}{2}[\sin(m-n)x + \sin(m+n)x]dx$
$= \dfrac{1}{2}\left[-\dfrac{\cos(m-n)x}{m-n} - \dfrac{\cos(m+n)x}{m+n}\right]_{-\pi}^{\pi} = 0$. If $m = n$, then the first term in each set of brackets is zero.

61. $\int_{-\pi}^{\pi}\cos mx\cos nx\,dx = \int_{-\pi}^{\pi}\frac{1}{2}[\cos(m-n)x + \cos(m+n)x]dx$. If $m \neq n$, this is equal to
$\dfrac{1}{2}\left[\dfrac{\sin(m-n)x}{m-n} + \dfrac{\sin(m+n)x}{m+n}\right]_{-\pi}^{\pi} = 0$. If $m = n$, we get
$\int_{-\pi}^{\pi}\frac{1}{2}[1+\cos(m+n)x]dx = \left[\frac{1}{2}x\right]_{-\pi}^{\pi} + \left[\dfrac{\sin(m+n)x}{2(m+n)}\right]_{-\pi}^{\pi} = \pi + 0 = \pi.$

EXERCISES 7.3

1. Let $x = \sin\theta$, where $-\frac{\pi}{2} \le \theta \le \frac{\pi}{2}$. Then $dx = \cos\theta\,d\theta$ and $\sqrt{1-x^2} = |\cos\theta| = \cos\theta$ (since $\cos\theta > 0$ for θ in $\left[-\frac{\pi}{2}, \frac{\pi}{2}\right]$). Thus

$$\int_{1/2}^{\sqrt{3}/2} \frac{dx}{x^2\sqrt{1-x^2}} = \int_{\pi/6}^{\pi/3} \frac{\cos\theta\,d\theta}{\sin^2\theta\cos\theta} = \int_{\pi/6}^{\pi/3} \csc^2\theta\,d\theta = \left[-\cot\theta\right]_{\pi/6}^{\pi/3}$$
$$= -\frac{1}{\sqrt{3}} - \left(-\sqrt{3}\right) = \frac{3}{\sqrt{3}} - \frac{1}{\sqrt{3}} = \frac{2}{\sqrt{3}}.$$

3. Let $u = 1 - x^2$. Then $du = -2x\,dx$, so $\displaystyle\int \frac{x}{\sqrt{1-x^2}}\,dx = -\frac{1}{2}\int \frac{du}{\sqrt{u}} = -\sqrt{u} + C = -\sqrt{1-x^2} + C.$

5. Let $2x = \sin\theta$, where $-\frac{\pi}{2} \le \theta \le \frac{\pi}{2}$. Then $x = \frac{1}{2}\sin\theta$, $dx = \frac{1}{2}\cos\theta\,d\theta$, and $\sqrt{1-4x^2} = \sqrt{1-(2x)^2} = \cos\theta$.

$$\int\sqrt{1-4x^2}\,dx = \int\cos\theta\left(\tfrac{1}{2}\cos\theta\right)d\theta = \tfrac{1}{4}\int(1+\cos 2\theta)d\theta$$
$$= \tfrac{1}{4}\left(\theta + \tfrac{1}{2}\sin 2\theta\right) + C = \tfrac{1}{4}(\theta + \sin\theta\cos\theta) + C$$
$$= \tfrac{1}{4}\left[\sin^{-1}(2x) + 2x\sqrt{1-4x^2}\right] + C$$

7. Let $x = 3\tan\theta$, where $-\frac{\pi}{2} < \theta < \frac{\pi}{2}$. Then $dx = 3\sec^2\theta\,d\theta$ and $\sqrt{9+x^2} = 3\sec\theta$.

$$\int_0^3 \frac{dx}{\sqrt{9+x^2}} = \int_0^{\pi/4} \frac{3\sec^2\theta\,d\theta}{3\sec\theta} = \int_0^{\pi/4}\sec\theta\,d\theta = \left[\ln|\sec\theta + \tan\theta|\right]_0^{\pi/4} = \ln\left(\sqrt{2}+1\right) - \ln 1 = \ln\left(\sqrt{2}+1\right)$$

9. Let $x = 4\sec\theta$, where $0 \le \theta < \frac{\pi}{2}$ or $\pi \le \theta < \frac{3\pi}{2}$. Then $dx = 4\sec\theta\tan\theta\,d\theta$ and $\sqrt{x^2-16} = 4|\tan\theta| = 4\tan\theta$. Thus

$$\int \frac{dx}{x^3\sqrt{x^2-16}} = \int \frac{4\sec\theta\tan\theta\,d\theta}{64\sec^3\theta\cdot 4\tan\theta} = \tfrac{1}{64}\int\cos^2\theta\,d\theta = \tfrac{1}{128}\int(1+\cos 2\theta)d\theta$$
$$= \tfrac{1}{128}\left(\theta + \tfrac{1}{2}\sin 2\theta\right) + C = \tfrac{1}{128}(\theta + \sin\theta\cos\theta) + C = \frac{1}{128}\left(\sec^{-1}\frac{x}{4} + \frac{4\sqrt{x^2-16}}{x^2}\right) + C$$

by the diagrams for $0 \le \theta < \frac{\pi}{2}$ and $\pi \le \theta < \frac{3\pi}{2}$, where the labels of the legs in the second diagram indicate the x-and y-coordinates of P rather than the lengths of those sides. Henceforth we omit the second diagram from our solutions.

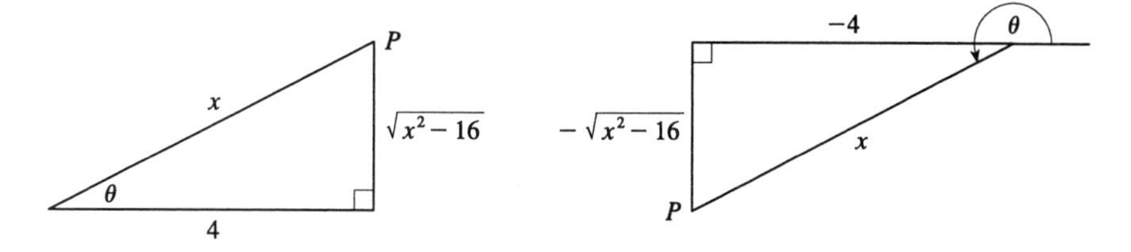

11. $9x^2 - 4 = (3x)^2 - 4$, so let $3x = 2\sec\theta$, where $0 \le \theta < \frac{\pi}{2}$ or $\pi \le \theta < \frac{3\pi}{2}$. Then $dx = \frac{2}{3}\sec\theta\tan\theta\,d\theta$ and $\sqrt{9x^2-4} = 2\tan\theta$.

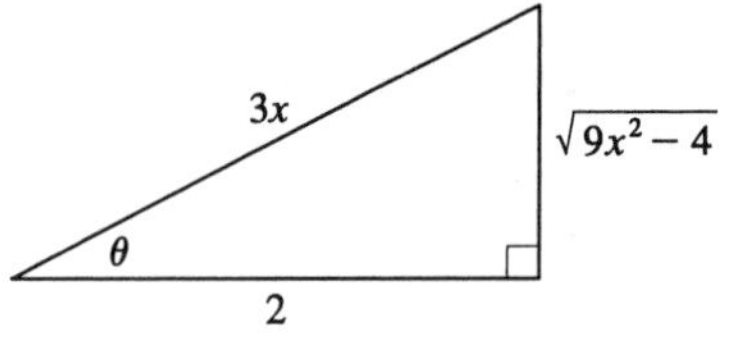

$$\int \frac{\sqrt{9x^2-4}}{x}\,dx = \int \frac{2\tan\theta}{\frac{2}{3}\sec\theta}\cdot\frac{2}{3}\sec\theta\tan\theta\,d\theta$$
$$= 2\int \tan^2\theta\,d\theta = 2\int(\sec^2\theta - 1)d\theta = 2(\tan\theta - \theta) + C$$
$$= \sqrt{9x^2-4} - 2\sec^{-1}\left(\frac{3x}{2}\right) + C$$

13. Let $x = a\sin\theta$, where $-\frac{\pi}{2} \le \theta \le \frac{\pi}{2}$. Then $dx = a\cos\theta\,d\theta$ and

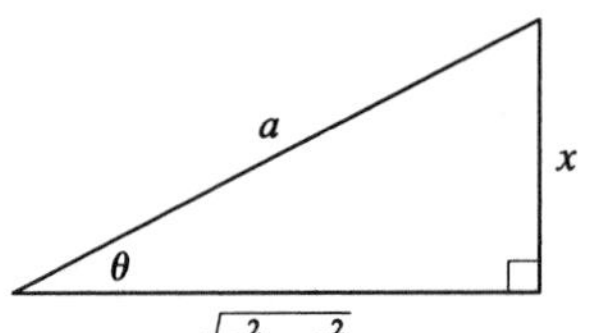

$$\int \frac{x^2\,dx}{(a^2-x^2)^{3/2}} = \int \frac{a^2\sin^2\theta\, a\cos\theta\,d\theta}{a^3\cos^3\theta} = \int \tan^2\theta\,d\theta$$
$$= \int(\sec^2\theta - 1)d\theta = \tan\theta - \theta + C$$
$$= \frac{x}{\sqrt{a^2-x^2}} - \sin^{-1}\frac{x}{a} + C.$$

15. Let $x = \sqrt{3}\tan\theta$, where $-\frac{\pi}{2} < \theta < \frac{\pi}{2}$. Then

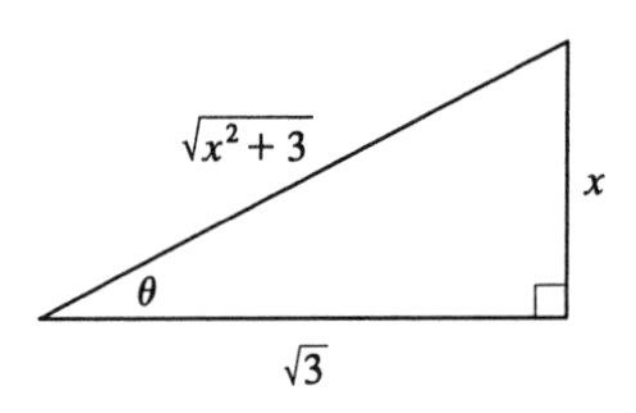

$$\int \frac{dx}{x\sqrt{x^2+3}} = \int \frac{\sqrt{3}\sec^2\theta\,d\theta}{\sqrt{3}\tan\theta\,\sqrt{3}\sec\theta} = \frac{1}{\sqrt{3}}\int\csc\theta\,d\theta$$
$$= \frac{1}{\sqrt{3}}\ln|\csc\theta - \cot\theta| + C = \frac{1}{\sqrt{3}}\ln\left|\frac{\sqrt{x^2+3}-\sqrt{3}}{x}\right| + C.$$

17. Let $u = 4 - 9x^2 \quad\Rightarrow\quad du = -18x\,dx$. Then $x^2 = \frac{1}{9}(4-u)$ and

$$\int_0^{2/3} x^3\sqrt{4-9x^2}\,dx = \int_4^0 \tfrac{1}{9}(4-u)u^{1/2}\left(-\tfrac{1}{18}\right)du = \tfrac{1}{162}\int_0^4\left(4u^{1/2} - u^{3/2}\right)du$$
$$= \tfrac{1}{162}\left[\tfrac{8}{3}u^{3/2} - \tfrac{2}{5}u^{5/2}\right]_0^4 = \tfrac{1}{162}\left[\tfrac{64}{3} - \tfrac{64}{5}\right] = \tfrac{64}{1215}.$$

Or: Let $3x = 2\sin\theta$, where $-\frac{\pi}{2} \le \theta \le \frac{\pi}{2}$.

19. Let $u = 1 + x^2$, $du = 2x\,dx$. Then $\int 5x\sqrt{1+x^2}\,dx = \frac{5}{2}\int u^{1/2}\,du = \frac{5}{3}u^{3/2} + C = \frac{5}{3}(1+x^2)^{3/2} + C$.

21. $2x - x^2 = -(x^2 - 2x + 1) + 1 = 1 - (x-1)^2$. Let $u = x - 1$. Then $du = dx$ and

$$\int\sqrt{2x-x^2}\,dx = \int\sqrt{1-u^2}\,du = \int\cos^2\theta\,d\theta \quad \text{(where } u = \sin\theta,\ -\tfrac{\pi}{2} \le \theta \le \tfrac{\pi}{2}\text{)}$$
$$= \tfrac{1}{2}\int(1+\cos 2\theta)d\theta = \tfrac{1}{2}\left(\theta + \tfrac{1}{2}\sin 2\theta\right) + C = \tfrac{1}{2}\left(\sin^{-1}u + u\sqrt{1-u^2}\right) + C$$
$$= \tfrac{1}{2}\left[\sin^{-1}(x-1) + (x-1)\sqrt{2x-x^2}\right] + C.$$

23. $9x^2 + 6x - 8 = (3x+1)^2 - 9$, so let $u = 3x + 1$, $du = 3\,dx$. Then $\displaystyle\int \frac{dx}{\sqrt{9x^2+6x-8}} = \int \frac{\frac{1}{3}\,du}{\sqrt{u^2-9}}$. Now let $u = 3\sec\theta$, where $0 \le \theta < \frac{\pi}{2}$ or $\pi \le \theta < \frac{3\pi}{2}$. Then $du = 3\sec\theta\tan\theta\,d\theta$ and $\sqrt{u^2-9} = 3\tan\theta$, so

$$\int \frac{\frac{1}{3}\,du}{\sqrt{u^2-9}} = \int \frac{\sec\theta\tan\theta\,d\theta}{3\tan\theta} = \tfrac{1}{3}\int\sec\theta\,d\theta = \tfrac{1}{3}\ln|\sec\theta + \tan\theta| + C_1$$
$$= \frac{1}{3}\ln\left|\frac{u+\sqrt{u^2-9}}{3}\right| + C_1 = \tfrac{1}{3}\ln\left|u + \sqrt{u^2-9}\right| + C = \tfrac{1}{3}\ln\left|3x + 1 + \sqrt{9x^2+6x-8}\right| + C.$$

25. $x^2+2x+2=(x+1)^2+1$. Let $u=x+1$, $du=dx$. Then

$$\int\frac{dx}{(x^2+2x+2)^2}=\int\frac{du}{(u^2+1)^2}=\int\frac{\sec^2\theta\,d\theta}{\sec^4\theta}\quad\begin{pmatrix}\text{where } u=\tan\theta,\ du=\sec^2\theta\,d\theta,\\ \text{and } u^2+1=\sec^2\theta\end{pmatrix}$$

$$=\int\cos^2\theta\,d\theta=\tfrac{1}{2}(\theta+\sin\theta\cos\theta)+C\quad\text{(as in Exercise 21)}$$

$$=\frac{1}{2}\left[\tan^{-1}u+\frac{u}{1+u^2}\right]+C=\frac{1}{2}\left[\tan^{-1}(x+1)+\frac{x+1}{x^2+2x+2}\right]+C.$$

27. Let $u=e^t\ \Rightarrow\ du=e^t\,dt$. Then $\int e^t\sqrt{9-e^{2t}}\,dt=\int\sqrt{9-u^2}\,du=\int(3\cos\theta)3\cos\theta\,d\theta$ (where $u=3\sin\theta$, $-\frac{\pi}{2}\le\theta\le\frac{\pi}{2}$) $=9\int\cos^2\theta\,d\theta=\frac{9}{2}(\theta+\sin\theta\cos\theta)+C$ (as in Exercise 21)

$$=\frac{9}{2}\left[\sin^{-1}\left(\frac{u}{3}\right)+\frac{u}{3}\cdot\frac{\sqrt{9-u^2}}{3}\right]+C=\tfrac{9}{2}\sin^{-1}\left(\tfrac{1}{3}e^t\right)+\tfrac{1}{2}e^t\sqrt{9-e^{2t}}+C.$$

29. **(a)** Let $x=a\tan\theta$, where $-\frac{\pi}{2}<\theta<\frac{\pi}{2}$. Then $\sqrt{x^2+a^2}=a\sec\theta$ and

$$\int\frac{dx}{\sqrt{x^2+a^2}}=\int\frac{a\sec^2\theta\,d\theta}{a\sec\theta}=\int\sec\theta\,d\theta=\ln|\sec\theta+\tan\theta|+C_1=\ln\left|\frac{\sqrt{x^2+a^2}}{a}+\frac{x}{a}\right|+C_1$$

$$=\ln\left(x+\sqrt{x^2+a^2}\right)+C,\text{ where } C=C_1-\ln|a|$$

(b) Let $x=a\sinh t$, so that $dx=a\cosh t\,dt$ and $\sqrt{x^2+a^2}=a\cosh t$. Then

$$\int\frac{dx}{\sqrt{x^2+a^2}}=\int\frac{a\cosh t\,dt}{a\cosh t}=t+C=\sinh^{-1}(x/a)+C.$$

31. Area of $\triangle POQ=\frac{1}{2}(r\cos\theta)(r\sin\theta)=\frac{1}{2}r^2\sin\theta\cos\theta$. Area of region $PQR=\int_{r\cos\theta}^{r}\sqrt{r^2-x^2}\,dx$.

Let $x=r\cos u\ \Rightarrow\ dx=-r\sin u\,du$ for $\theta\le u\le\frac{\pi}{2}$. Then we obtain

$$\int\sqrt{r^2-x^2}\,dx=\int r\sin u(-r\sin u)du=-r^2\int\sin^2u\,du$$

$$=-\tfrac{1}{2}r^2(u-\sin u\cos u)+C=-\tfrac{1}{2}r^2\cos^{-1}(x/r)+\tfrac{1}{2}x\sqrt{r^2-x^2}+C.\ \text{So}$$

area of region $PQR=\frac{1}{2}\left[-r^2\cos^{-1}(x/r)+x\sqrt{r^2-x^2}\right]_{r\cos\theta}^{r}=\frac{1}{2}[0-(-r^2\theta+r\cos\theta\, r\sin\theta)]$

$=\frac{1}{2}r^2\theta-\frac{1}{2}r^2\sin\theta\cos\theta$, so (area of sector POR) $=$ (area of $\triangle POQ$) $+$ (area of region PQR) $=\frac{1}{2}r^2\theta$.

33.

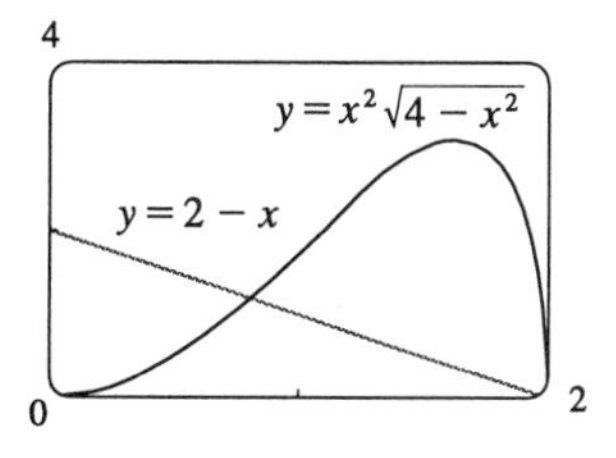

From the graph, it appears that the curve $y=x^2\sqrt{4-x^2}$ and the line $y=2-x$ intersect at about $x=0.81$ and $x=2$, with $x^2\sqrt{4-x^2}>2-x$ on $(0.81,2)$. So the area bounded by the curve and the line is

$$A\approx\int_{0.81}^{2}\left[x^2\sqrt{4-x^2}-(2-x)\right]dx=\int_{0.81}^{2}x^2\sqrt{4-x^2}\,dx-\left[2x-\tfrac{1}{2}x^2\right]_{0.81}^{2}.$$

To evaluate the integral, we put $x=2\sin\theta$, where $-\frac{\pi}{2}\le\theta\le\frac{\pi}{2}$. Then $dx=2\cos\theta\,d\theta$, $x=2\ \Rightarrow\ \theta=\sin^{-1}1=\frac{\pi}{2}$, and $x=0.81\ \Rightarrow\ \theta=\sin^{-1}0.405\approx0.417$. So

$$\int_{0.81}^{2}x^2\sqrt{4-x^2}\,dx\approx\int_{0.417}^{\pi/2}4\sin^2\theta(2\cos\theta)(2\cos\theta\,d\theta)=4\int_{0.417}^{\pi/2}\sin^2 2\theta\,d\theta$$

$$=4\int_{0.417}^{\pi/2}\tfrac{1}{2}(1-\cos4\theta)d\theta=2\left[\theta-\tfrac{1}{4}\sin4\theta\right]_{0.417}^{\pi/2}=2\left(\left[\tfrac{\pi}{2}-0\right]-\left[0.417-\tfrac{1}{4}(0.995)\right]\right)\approx2.81.\ \text{So}$$

$$A\approx2.81-\left[\left(2\cdot2-\tfrac{1}{2}\cdot2^2\right)-\left(2\cdot0.81-\tfrac{1}{2}\cdot0.81^2\right)\right]\approx2.10.$$

35. Let the equation of the large circle be $x^2 + y^2 = R^2$. Then the equation of the small circle is $x^2 + (y - b)^2 = r^2$, where $b = \sqrt{R^2 - r^2}$ is the distance between the centers of the circles. The desired area is

$$A = \int_{-r}^{r} \left[\left(b + \sqrt{r^2 - x^2}\right) - \sqrt{R^2 - x^2}\right] dx = 2\int_0^r \left(b + \sqrt{r^2 - x^2} - \sqrt{R^2 - x^2}\right) dx$$
$$= 2\int_0^r b\,dx + 2\int_0^r \sqrt{r^2 - x^2}\,dx - 2\int_0^r \sqrt{R^2 - x^2}\,dx.$$

The first integral is just $2br = 2r\sqrt{R^2 - r^2}$. To evaluate the other two integrals, note that

$$\int \sqrt{a^2 - x^2}\,dx = \int a^2 \cos^2\theta\,d\theta \quad (x = a\sin\theta,\ dx = a\cos\theta\,d\theta)$$
$$= \tfrac{1}{2}\left(\tfrac{1}{2}a^2\right)\int(1 + \cos 2\theta) = \tfrac{1}{2}a^2\left(\theta + \tfrac{1}{2}\sin 2\theta\right) + C = \tfrac{1}{2}a^2(\theta + \sin\theta\cos\theta) + C$$
$$= \frac{a^2}{2}\arcsin\left(\frac{x}{a}\right) + \frac{a^2}{2}\left(\frac{x}{a}\right)\frac{\sqrt{a^2 - x^2}}{a} + C = \frac{a^2}{2}\arcsin\left(\frac{x}{a}\right) + \frac{x}{2}\sqrt{a^2 - x^2} + C,$$ so the desired area is

$$A = 2r\sqrt{R^2 - r^2} + \left[r^2 \arcsin(x/r) + x\sqrt{r^2 - x^2}\right]_0^r - \left[R^2 \arcsin(x/R) + x\sqrt{R^2 - x^2}\right]_0^r$$
$$= 2r\sqrt{R^2 - r^2} + r^2\left(\tfrac{\pi}{2}\right) - \left[R^2 \arcsin(r/R) + r\sqrt{R^2 - r^2}\right] = r\sqrt{R^2 - r^2} + \frac{\pi}{2}r^2 - R^2\arcsin(r/R).$$

37. We use cylindrical shells and assume that $R > r$. $x^2 = r^2 - (y - R)^2 \quad\Rightarrow\quad x = \pm\sqrt{r^2 - (y - R)^2}$, so $g(y) = 2\sqrt{r^2 - (y - R)^2}$ in Formula 5.3.3, and

$$V = \int_{R-r}^{R+r} 2\pi y \cdot 2\sqrt{r^2 - (y - R)^2}\,dy = \int_{-r}^{r} 4\pi(u + R)\sqrt{r^2 - u^2}\,du \quad (\text{where } u = y - R)$$
$$= 4\pi\int_{-r}^{r} u\sqrt{r^2 - u^2}\,du + 4\pi R\int_{-r}^{r} \sqrt{r^2 - u^2}\,du \quad \left(\begin{array}{c}\text{Let } u = r\sin\theta,\ du = r\cos\theta\,d\theta\\ \text{in the second integral}\end{array}\right)$$
$$= 4\pi\left[-\tfrac{1}{3}(r^2 - u^2)^{3/2}\right]_{-r}^{r} + 4\pi R\int_{-\pi/2}^{\pi/2} r^2\cos^2\theta\,d\theta = -\tfrac{4\pi}{3}(0 - 0) + 4\pi R r^2\int_{-\pi/2}^{\pi/2}\cos^2 d\theta$$
$$= 2\pi R r^2\int_{-\pi/2}^{\pi/2}(1 + \cos 2\theta)d\theta = 2\pi R r^2\left[\theta + \tfrac{1}{2}\sin 2\theta\right]_{-\pi/2}^{\pi/2} = 2\pi^2 R r^2.$$

Another Method: Use washers instead of shells, so $V = 8\pi R\int_0^r \sqrt{r^2 - y^2}\,dy$ as in Exercise 5.2.61(a), but evaluate the integral using $y = r\sin\theta$.

EXERCISES 7.4

1. $$\frac{1}{(x-1)(x+2)} = \frac{A}{x-1} + \frac{B}{x+2}$$

3. $$\frac{x^2 + 3x - 4}{(2x-1)^2(2x+3)} = \frac{A}{2x-1} + \frac{B}{(2x-1)^2} + \frac{C}{2x+3}$$

5. $$\frac{1}{x^4 - x^3} = \frac{1}{x^3(x-1)} = \frac{A}{x} + \frac{B}{x^2} + \frac{C}{x^3} + \frac{D}{x-1}$$

7. $$\frac{x^2+1}{x^2-1} = 1 + \frac{2}{(x-1)(x+1)} = 1 + \frac{A}{x-1} + \frac{B}{x+1}$$

9. $$\frac{x^2-2}{x(x^2+2)} = \frac{A}{x} + \frac{Bx+C}{x^2+2}$$

11. $$\frac{x^4+x^2+1}{(x^2+1)(x^2+4)^2} = \frac{Ax+B}{x^2+1} + \frac{Cx+D}{x^2+4} + \frac{Ex+F}{(x^2+4)^2}$$

13. $\dfrac{x^4}{(x^2+9)^3} = \dfrac{Ax+B}{x^2+9} + \dfrac{Cx+D}{(x^2+9)^2} + \dfrac{Ex+F}{(x^2+9)^3}$

15. $\dfrac{x^3+x^2+1}{x^4+x^3+2x^2} = \dfrac{x^3+x^2+1}{x^2(x^2+x+2)} = \dfrac{A}{x} + \dfrac{B}{x^2} + \dfrac{Cx+D}{x^2+x+2}$

17. $\displaystyle\int \frac{x^2}{x+1}\,dx = \int \left(x - 1 + \frac{1}{x+1}\right) dx = \tfrac{1}{2}x^2 - x + \ln|x+1| + C$

19. $\dfrac{4x-1}{(x-1)(x+2)} = \dfrac{A}{x-1} + \dfrac{B}{x+2} \quad\Rightarrow\quad 4x - 1 = A(x+2) + B(x-1)$ Take $x = 1$ to get $3 = 3A$, then $x = -2$ to get $-9 = -3B \quad\Rightarrow\quad A = 1, B = 3$. Now

$$\int_2^4 \frac{4x-1}{(x-1)(x+2)}\,dx = \int_2^4 \left[\frac{1}{x-1} + \frac{3}{x+2}\right] dx = [\ln(x-1) + 3\ln(x+2)]_2^4$$
$$= \ln 3 + 3\ln 6 - \ln 1 - 3\ln 4 = 4\ln 3 - 3\ln 2 = \ln \tfrac{81}{8}.$$

21. $\displaystyle\int \frac{6x-5}{2x+3}\,dx = \int \left[3 - \frac{14}{2x+3}\right] dx = 3x - 7\ln|2x+3| + C$

23. $\dfrac{x^2+1}{x^2-x} = 1 + \dfrac{x+1}{x(x-1)} = 1 - \dfrac{1}{x} + \dfrac{2}{x-1}$, so

$$\int \frac{x^2+1}{x^2-x}\,dx = x - \ln|x| + 2\ln|x-1| + C = x + \ln\frac{(x-1)^2}{|x|} + C.$$

25. $\dfrac{2x+3}{(x+1)^2} = \dfrac{A}{x+1} + \dfrac{B}{(x+1)^2} \quad\Rightarrow\quad 2x + 3 = A(x+1) + B$. Take $x = -1$ to get $B = 1$, and equate coefficients of x to get $A = 2$. Now

$$\int_0^1 \frac{2x+3}{(x+1)^2}\,dx = \int_0^1 \left[\frac{2}{x+1} + \frac{1}{(x+1)^2}\right] dx = \left[2\ln(x+1) - \frac{1}{x+1}\right]_0^1$$
$$= 2\ln 2 - \tfrac{1}{2} - (2\ln 1 - 1) = 2\ln 2 + \tfrac{1}{2}.$$

27. $\dfrac{6x^2+5x-3}{x^3+2x^2-3x} = \dfrac{A}{x} + \dfrac{B}{x+3} + \dfrac{C}{x-1} \quad\Rightarrow$

$6x^2 + 5x - 3 = A(x+3)(x-1) + B(x)(x-1) + C(x)(x+3)$.

Set $x = 0$ to get $A = 1$, then take $x = -3$ to get $B = 3$, then set $x = 1$ to get $C = 2$:

$$\int_2^3 \frac{6x^2+5x-3}{x^3+2x^2-3x}\,dx = \int_2^3 \left[\frac{1}{x} + \frac{3}{x+3} + \frac{2}{x-1}\right] dx = [\ln x + 3\ln(x+3) + 2\ln(x-1)]_2^3$$
$$= (\ln 3 + 3\ln 6 + 2\ln 2) - (\ln 2 + 3\ln 5) = 4\ln 6 - 3\ln 5.$$

29. $\dfrac{1}{(x-1)^2(x+4)} = \dfrac{A}{x-1} + \dfrac{B}{(x-1)^2} + \dfrac{C}{x+4} \quad\Rightarrow\quad 1 = A(x-1)(x+4) + B(x+4) + C(x-1)^2$. Set $x = 1$ to get $B = \frac{1}{5}$ and take $x = -4$ to get $C = \frac{1}{25}$. Now equating the coefficients of x^2, we get $0 = Ax^2 + Cx^2$ or $A = -C = -\frac{1}{25} \quad\Rightarrow$

$$\int \frac{dx}{(x-1)^2(x+4)} = \int \left[\frac{-1/25}{x-1} + \frac{1/5}{(x-1)^2} + \frac{1/25}{x+4}\right] dx = -\frac{1}{25}\ln|x-1| - \frac{1}{5}\cdot\frac{1}{x-1} + \frac{1}{25}\ln|x+4| + C$$
$$= \frac{1}{25}\left[\ln\left|\frac{x+4}{x-1}\right| - \frac{5}{x-1}\right] + C.$$

31. $\dfrac{5x^2+3x-2}{x^3+2x^2} = \dfrac{5x^2+3x-2}{x^2(x+2)} = \dfrac{A}{x} + \dfrac{B}{x^2} + \dfrac{C}{x+2}$. Multiply by $x^2(x+2)$ to get

$5x^2+3x-2 = Ax(x+2) + B(x+2) + Cx^2$. Set $x=-2$ to get $C=3$, and take $x=0$ to get $B=-1$. Equating the coefficients of x^2 gives $5x^2 = Ax^2 + Cx^2$ or $A=2$. So

$$\int \frac{5x^2+3x-2}{x^3+2x^2}\,dx = \int \left[\frac{2}{x} - \frac{1}{x^2} + \frac{3}{x+2}\right] dx = 2\ln|x| + \frac{1}{x} + 3\ln|x+2| + C.$$

33. Let $u = x^3+3x^2+4$. Then $du = 3(x^2+2x)dx \Rightarrow \displaystyle\int \frac{x^2+2x}{x^3+3x^2+4}\,dx = \frac{1}{3}\int \frac{du}{u} = \tfrac{1}{3}\ln|x^3+3x^2+4| + C.$

35. $\dfrac{x^2}{(x+1)^3} = \dfrac{A}{x+1} + \dfrac{B}{(x+1)^2} + \dfrac{C}{(x+1)^3}$. Multiply by $(x+1)^3$ to get $x^2 = A(x+1)^2 + B(x+1) + C$.

Setting $x=-1$ gives $C=1$. Equating the coefficients of x^2 gives $A=1$, and setting $x=0$ gives $B=-2$.

Now $\displaystyle\int \frac{x^2\,dx}{(x+1)^3} = \int \left[\frac{1}{x+1} - \frac{2}{(x+1)^2} + \frac{1}{(x+1)^3}\right] dx = \ln|x+1| + \frac{2}{x+1} - \frac{1}{2(x+1)^2} + C.$

37. $\dfrac{1}{x^4-x^2} = \dfrac{1}{x^2(x-1)(x+1)} = \dfrac{A}{x} + \dfrac{B}{x^2} + \dfrac{C}{x-1} + \dfrac{D}{x+1}$. Multiply by $x^2(x-1)(x+1)$ to get

$1 = Ax(x-1)(x+1) + B(x-1)(x+1) + Cx^2(x+1) + Dx^2(x-1)$. Setting $x=1$ gives $C=\frac{1}{2}$, taking $x=-1$ gives $D=-\frac{1}{2}$. Equating the coefficients of x^3 gives $0 = A+C+D = A$. Finally, setting $x=0$ yields $B=-1$. Now $\displaystyle\int \frac{dx}{x^4-x^2} = \int \left[\frac{-1}{x^2} + \frac{1/2}{x-1} - \frac{1/2}{x+1}\right] dx = \frac{1}{x} + \frac{1}{2}\ln\left|\frac{x-1}{x+1}\right| + C.$

39. $\dfrac{x^3}{x^2+1} = \dfrac{(x^3+x)-x}{x^2+1} = x - \dfrac{x}{x^2+1}$, so $\displaystyle\int_0^1 \frac{x^3}{x^2+1}\,dx = \int_0^1 x\,dx - \int_0^1 \frac{x\,dx}{x^2+1}$

$= \left[\frac{1}{2}x^2\right]_0^1 - \dfrac{1}{2}\displaystyle\int_1^2 \frac{1}{u}\,du$ (where $u = x^2+1$, $du = 2x\,dx$) $= \frac{1}{2} - \left[\frac{1}{2}\ln u\right]_1^2 = \frac{1}{2} - \frac{1}{2}\ln 2 = \frac{1}{2}(1-\ln 2)$

41. Complete the square: $x^2+x+1 = \left(x+\frac{1}{2}\right)^2 + \frac{3}{4}$ and let $u = x+\frac{1}{2}$. Then

$$\int_0^1 \frac{x}{x^2+x+1}\,dx = \int_{1/2}^{3/2} \frac{u-1/2}{u^2+3/4}\,du = \int_{1/2}^{3/2} \frac{u}{u^2+3/4}\,du - \frac{1}{2}\int_{1/2}^{3/2} \frac{1}{u^2+3/4}\,du$$

$$= \tfrac{1}{2}\ln\left(u^2+\tfrac{3}{4}\right) - \tfrac{1}{2}\tfrac{1}{\sqrt{3/2}}\left[\tan^{-1}\left(\tfrac{2}{\sqrt{3}}u\right)\right]_{1/2}^{3/2} = \tfrac{1}{2}\ln 3 - \tfrac{1}{\sqrt{3}}\left(\tfrac{\pi}{3} - \tfrac{\pi}{6}\right) = \ln\sqrt{3} - \tfrac{\pi}{6\sqrt{3}}.$$

43. $\dfrac{3x^2-4x+5}{(x-1)(x^2+1)} = \dfrac{A}{x-1} + \dfrac{Bx+C}{x^2+1} \quad\Rightarrow\quad 3x^2-4x+5 = A(x^2+1) + (Bx+C)(x-1)$. Take $x=1$ to get $4=2A$ or $A=2$. Now $(Bx+C)(x-1) = 3x^2-4x+5-2(x^2+1) = x^2-4x+3$. Equating coefficients of x^2 and then comparing the constant terms, we get $B=1$ and $C=-3$. Hence

$$\int \frac{3x^2-4x+5}{(x-1)(x^2+1)}\,dx = \int \left[\frac{2}{x-1} + \frac{x-3}{x^2+1}\right] dx = 2\ln|x-1| + \int \frac{x\,dx}{x^2+1} - 3\int \frac{dx}{x^2+1}$$

$$= 2\ln|x-1| + \tfrac{1}{2}\ln(x^2+1) - 3\tan^{-1}x + C = \ln(x-1)^2 + \ln\sqrt{x^2+1} - 3\tan^{-1}x + C.$$

45. $\dfrac{1}{x^3-1} = \dfrac{1}{(x-1)(x^2+x+1)} = \dfrac{A}{x-1} + \dfrac{Bx+C}{x^2+x+1} \quad \Rightarrow \quad 1 = A(x^2+x+1) + (Bx+C)(x-1)$. Take $x=1$ to get $A=\frac{1}{3}$. Equate coefficients of x^2 and 1 to get $0 = \frac{1}{3}+B$, $1 = \frac{1}{3} - C$, so $B = -\frac{1}{3}$, $C = -\frac{2}{3} \quad \Rightarrow$

$$\int \frac{dx}{x^3-1} = \int \frac{1/3}{x-1}\,dx + \int \frac{(-1/3)x - 2/3}{x^2+x+1}\,dx = \tfrac{1}{3}\ln|x-1| - \frac{1}{3}\int \frac{x+2}{x^2+x+1}\,dx$$

$$= \tfrac{1}{3}\ln|x-1| - \frac{1}{3}\int \frac{x+1/2}{x^2+x+1}\,dx - \tfrac{1}{3}\int \frac{(3/2)dx}{(x+1/2)^2+3/4}$$

$$= \tfrac{1}{3}\ln|x-1| - \tfrac{1}{6}\ln(x^2+x+1) - \tfrac{1}{2}\left(\tfrac{2}{\sqrt{3}}\right)\tan^{-1}\left[\left(x+\tfrac{1}{2}\right)\Big/\left(\tfrac{\sqrt{3}}{2}\right)\right] + K$$

$$= \tfrac{1}{3}\ln|x-1| - \tfrac{1}{6}\ln(x^2+x+1) - \tfrac{1}{\sqrt{3}}\tan^{-1}\left[\tfrac{1}{\sqrt{3}}(2x+1)\right] + K.$$

47. $\dfrac{x^2-2x-1}{(x-1)^2(x^2+1)} = \dfrac{A}{x-1} + \dfrac{B}{(x-1)^2} + \dfrac{Cx+D}{x^2+1} \quad \Rightarrow$

$x^2-2x-1 = A(x-1)(x^2+1) + B(x^2+1) + (Cx+D)(x-1)^2$. Setting $x=1$ gives $B=-1$. Equating the coefficients of x^3 gives $A=-C$. Equating the constant terms gives $-1 = -A - 1 + D$, so $D = A$, and setting $x=2$ gives $-1 = 5A - 5 - 2A + A$ or $A = 1$. We have

$$\int \frac{x^2-2x-1}{(x-1)^2(x^2+1)}\,dx = \int \left[\frac{1}{x-1} - \frac{1}{(x-1)^2} - \frac{x-1}{x^2+1}\right]dx$$

$$= \ln|x-1| + \frac{1}{x-1} - \tfrac{1}{2}\ln(x^2+1) + \tan^{-1}x + C.$$

49. $\dfrac{3x^3-x^2+6x-4}{(x^2+1)(x^2+2)} = \dfrac{Ax+B}{x^2+1} + \dfrac{Cx+D}{x^2+2} \quad \Rightarrow$

$3x^3-x^2+6x-4 = (Ax+B)(x^2+2) + (Cx+D)(x^2+1)$. Equating the coefficients gives $A+C=3$, $B+D=-1$, $2A+C=6$, and $2B+D=-4 \quad \Rightarrow \quad A=3$, $C=0$, $B=-3$, and $D=2$. Now

$$\int \frac{3x^3-x^2+6x-4}{(x^2+1)(x^2+2)}\,dx = 3\int \frac{x-1}{x^2+1}\,dx + 2\int \frac{dx}{x^2+2} = \tfrac{3}{2}\ln(x^2+1) - 3\tan^{-1}x + \sqrt{2}\tan^{-1}\left(\frac{x}{\sqrt{2}}\right) + C.$$

51. $$\int \frac{x-3}{(x^2+2x+4)^2}\,dx = \int \frac{x-3}{\left[(x+1)^2+3\right]^2}\,dx = \int \frac{u-4}{(u^2+3)^2}\,du \quad (\text{with } u = x+1)$$

$$= \int \frac{u\,du}{(u^2+3)^2} - 4\int \frac{du}{(u^2+3)^2} = \frac{1}{2}\int \frac{dv}{v^2} - 4\int \frac{\sqrt{3}\sec^2\theta\,d\theta}{9\sec^4\theta} \quad \left[\begin{array}{c} v = u^2+3 \text{ in the first integral;} \\ u = \sqrt{3}\tan\theta \text{ in the second} \end{array}\right]$$

$$= \frac{-1}{(2v)} - \frac{4\sqrt{3}}{9}\int \cos^2\theta\,d\theta = \frac{-1}{2(u^2+3)} - \tfrac{2\sqrt{3}}{9}(\theta + \sin\theta\cos\theta) + C$$

$$= \frac{-1}{2(x^2+2x+4)} - \frac{2\sqrt{3}}{9}\left[\tan^{-1}\left(\frac{x+1}{\sqrt{3}}\right) + \frac{\sqrt{3}(x+1)}{x^2+2x+4}\right] + C$$

$$= \frac{-1}{2(x^2+2x+4)} - \frac{2\sqrt{3}}{9}\tan^{-1}\left(\frac{x+1}{\sqrt{3}}\right) - \frac{2(x+1)}{3(x^2+2x+4)} + C$$

53. Let $u = \sin^2 x - 3\sin x + 2$. Then $du = (2\sin x\cos x - 3\cos x)dx$, so

$$\int \frac{(2\sin x - 3)\cos x}{\sin^2 x - 3\sin x + 2}\,dx = \int \frac{du}{u} = \ln|u| + C = \ln\left|\sin^2 x - 3\sin x + 2\right| + C.$$

55. From the graph, we see that the integral will be negative, and we guess that the area is about the same as that of a rectangle with width 2 and height 0.3, so we estimate the integral to be $-(2 \cdot 0.3) = -0.6$.

Now $\dfrac{1}{x^2-2x-3} = \dfrac{1}{(x-3)(x+1)} = \dfrac{A}{x-3} + \dfrac{B}{x+1} \quad \Leftrightarrow$

$1 = (A+B)x + A - 3B$, so $A = -B$ and $A - 3B = 1 \quad \Leftrightarrow$

$A = \frac{1}{4}$ and $B = -\frac{1}{4}$, so the integral becomes

$$\int_0^2 \frac{dx}{x^2-2x-3} = \frac{1}{4}\int_0^2 \frac{dx}{x-3} - \frac{1}{4}\int_0^2 \frac{dx}{x+1} = \tfrac{1}{4}[\ln|x-3| - \ln|x+1|]_0^2 = \frac{1}{4}\left[\ln\left|\frac{x-3}{x+1}\right|\right]_0^2$$

$$= \tfrac{1}{4}\left(\ln\tfrac{1}{3} - \ln 3\right) = -\tfrac{1}{2}\ln 3 \approx -0.55.$$

57. If $|x| < a$, then $\displaystyle\int \frac{dx}{a^2-x^2} = \int \frac{a\,\text{sech}^2 u\,du}{a^2\,\text{sech}^2 u}$ (put $x = a\tanh u$) $= \dfrac{u}{a} + C = \dfrac{1}{a}\tanh^{-1}\left(\dfrac{x}{a}\right) + C$.

If $|x| > a$, then $\displaystyle\int \frac{dx}{a^2-x^2} = \int \frac{-a\,\text{csch}^2 u\,du}{-a^2\,\text{csch}^2 u}$ (put $x = a\coth u$) $= \dfrac{u}{a} + C = \dfrac{1}{a}\coth^{-1}\left(\dfrac{x}{a}\right) + C$.

59. $\displaystyle\int \frac{dx}{x^2-2x} = \int \frac{dx}{(x-1)^2-1} = \int \frac{du}{u^2-1}$ (put $u = x-1$)

$= \dfrac{1}{2}\ln\left|\dfrac{u-1}{u+1}\right| + C$ (by Equation 6) $= \dfrac{1}{2}\ln\left|\dfrac{x-2}{x}\right| + C$

61. $\displaystyle\int \frac{x\,dx}{x^2+x-1} = \frac{1}{2}\int \frac{(2x+1)dx}{x^2+x-1} - \frac{1}{2}\int \frac{dx}{\left(x+\frac{1}{2}\right)^2 - \frac{5}{4}} = \tfrac{1}{2}\ln|x^2+x-1| - \frac{1}{2}\int \frac{du}{u^2 - \left(\frac{\sqrt{5}}{2}\right)^2}$

$\left(\text{where } u = x + \tfrac{1}{2}\right) \quad = \tfrac{1}{2}\ln|x^2+x-1| - \dfrac{1}{2\sqrt{5}}\ln\left|\dfrac{u-\sqrt{5}/2}{u+\sqrt{5}/2}\right| + C$

$= \tfrac{1}{2}\ln|x^2+x-1| - \dfrac{1}{2\sqrt{5}}\ln\left|\dfrac{2x+1-\sqrt{5}}{2x+1+\sqrt{5}}\right| + C$

63. $\dfrac{x+1}{x-1} = 1 + \dfrac{2}{x-1} > 0$ for $2 \le x \le 3$, so

area $= \displaystyle\int_2^3 \left[1 + \frac{2}{x-1}\right]dx = [x + 2\ln|x-1|]_2^3 = (3 + 2\ln 2) - (2 + 2\ln 1) = 1 + 2\ln 2.$

65. In this case, we use cylindrical shells, so the volume is $V = 2\pi\displaystyle\int_0^1 \frac{x\,dx}{x^2+3x+2} = 2\pi\int_0^1 \frac{x}{(x+1)(x+2)}$. We use partial fractions to simplify the integrand: $\dfrac{x}{(x+1)(x+2)} = \dfrac{A}{x+1} + \dfrac{B}{x+2} \quad \Rightarrow$

$x = (A+B)x + 2A + B$. So $A + B = 1$ and $2A + B = 0 \quad \Rightarrow \quad A = -1$ and $B = 2$. So the volume is

$$2\pi\int_0^1 \left[\frac{-1}{x+1} + \frac{2}{x+2}\right]dx = 2\pi[-\ln|x+1| + 2\ln|x+2|]_0^1$$

$$= 2\pi(-\ln 2 + 2\ln 3 + \ln 1 - 2\ln 2) = 2\pi(2\ln 3 - 3\ln 2) = 2\pi\ln\tfrac{9}{8}.$$

67. **(a)** In Maple, we define $f(x)$, and then use `convert(f,parfrac,x);` to obtain

$$f(x)=\frac{24{,}110/4879}{5x+2}-\frac{668/323}{2x+1}-\frac{9438/80{,}155}{3x-7}+\frac{(22{,}098x+48{,}935)/260{,}015}{x^2+x+5}.$$ In Mathematica, we use the command `Apart`, and in Derive, we use `Expand`.

(b)
$$\begin{aligned}\int f(x)dx &= \tfrac{24{,}110}{4879}\cdot\tfrac{1}{5}\ln|5x+2|-\tfrac{668}{323}\cdot 2\ln|2x+1|-\tfrac{9438}{80{,}155}\cdot\tfrac{1}{3}\ln|3x-7|\\ &\quad+\frac{1}{260{,}015}\int\frac{22{,}098\left(x+\frac{1}{2}\right)+37{,}886}{\left(x+\frac{1}{2}\right)^2+\frac{19}{4}}\,dx+C\\ &= \tfrac{24{,}110}{4879}\cdot\tfrac{1}{5}\ln|5x+2|-\tfrac{668}{323}\cdot\tfrac{1}{2}\ln|2x+1|-\tfrac{9438}{80{,}155}\cdot\tfrac{1}{3}\ln|3x-7|\\ &\quad+\tfrac{1}{260{,}015}\left[22{,}098\cdot\tfrac{1}{2}\ln|x^2+x+5|+37{,}886\cdot\sqrt{\tfrac{4}{19}}\tan^{-1}\left(\tfrac{1}{\sqrt{19/4}}\left(x+\tfrac{1}{2}\right)\right)\right]+C\\ &= \tfrac{4822}{4879}\ln|5x+2|-\tfrac{334}{323}\ln|2x+1|-\tfrac{3146}{80{,}155}\ln|3x-7|+\tfrac{11{,}049}{260{,}015}\ln|x^2+x+5|\\ &\quad+\tfrac{75{,}772}{260{,}015\sqrt{19}}\tan^{-1}\left[\tfrac{1}{\sqrt{19}}(2x+1)\right]+C.\end{aligned}$$

If we tell Maple to `integrate(f,x);` we get

$$\begin{aligned}&\frac{4822\ln(5x+2)}{4879}-\frac{334\ln(2x+1)}{323}-\frac{3146\ln(3x-7)}{80{,}155}\\ &\quad+\frac{11{,}049\ln(x^2+x+5)}{260{,}015}+\frac{3988\sqrt{19}}{260{,}115}\tan^{-1}\left[\tfrac{\sqrt{19}}{19}(2x+1)\right].\end{aligned}$$

The main difference in Maple's answer is that the absolute value signs and the constant of integration have been omitted. Also, the fractions have been reduced and the denominators rationalized.

69. There are only finitely many values of x where $Q(x)=0$ (assuming that Q is not the zero polynomial). At all other values of x, $F(x)/Q(x)=G(x)/Q(x)$, so $F(x)=G(x)$. In other words, the values of F and G agree at all except perhaps finitely many values of x. By continuity of F and G, the polynomials F and G must agree at those values of x too.

More explicitly: if a is a value of x such that $Q(a)=0$, then $Q(x)\neq 0$ for all x sufficiently close to a. Thus
$$F(a)=\lim_{x\to a}F(x)\begin{bmatrix}\text{by continuity}\\ \text{of } F\end{bmatrix}=\lim_{x\to a}G(x)\begin{bmatrix}\text{since } F(x)=G(x)\\ \text{whenever } Q(x)\neq 0\end{bmatrix}=G(a)\begin{bmatrix}\text{by continuity}\\ \text{of } G\end{bmatrix}.$$

EXERCISES 7.5

1. Let $u = \sqrt{x}$. Then $x = u^2$, $dx = 2u\,du$ $\Rightarrow$

$$\int_0^1 \frac{dx}{1+\sqrt{x}} = \int_0^1 \frac{2u\,du}{1+u} = 2\int_0^1 \left[1 - \frac{1}{1+u}\right] du = 2[u - \ln(1+u)]_0^1 = 2(1 - \ln 2).$$

3. Let $u = \sqrt{x}$. Then $x = u^2$, $dx = 2u\,du$ $\Rightarrow$

$$\int \frac{\sqrt{x}\,dx}{x+1} = \int \frac{u \cdot 2u\,du}{u^2+1} = 2\int \left[1 - \frac{1}{u^2+1}\right] du = 2\left(u - \tan^{-1} u\right) + C = 2\left(\sqrt{x} - \tan^{-1}\sqrt{x}\right) + C.$$

5. Let $u = \sqrt[3]{x}$. Then $x = u^3$, $dx = 3u^2\,du$ $\Rightarrow$

$$\int \frac{dx}{x - \sqrt[3]{x}} = \int \frac{3u^2\,du}{u^3 - u} = 3\int \frac{u\,du}{u^2-1} = \tfrac{3}{2}\ln|u^2 - 1| + C = \tfrac{3}{2}\ln\left|x^{2/3} - 1\right| + C.$$

7. Let $u = \sqrt{x-1}$. Then $x = u^2 + 1$, $dx = 2u\,du$ $\Rightarrow$ $\displaystyle\int_5^{10} \frac{x^2\,dx}{\sqrt{x-1}} = \int_2^3 \frac{\left(u^2+1\right)^2 2u\,du}{u}$

$$= 2\textstyle\int_2^3 \left(u^4 + 2u^2 + 1\right) du = 2\left[\frac{1}{5}u^5 + \frac{2}{3}u^3 + u\right]_2^3 = 2\left(\frac{243}{5} + 18 + 3\right) - 2\left(\frac{32}{5} + \frac{16}{3} + 2\right) = \frac{1676}{15}.$$

9. Let $u = \sqrt{x}$. Then $x = u^2$, $dx = 2u\,du$ $\Rightarrow$

$$\int \frac{dx}{\sqrt{1+\sqrt{x}}} = \int \frac{2u\,du}{\sqrt{1+u}} = 2\int \frac{(v^2-1)2v\,dv}{v} \quad \left(\text{put } v = \sqrt{1+u},\ u = v^2 - 1,\ du = 2v\,dv\right)$$

$$= 4\textstyle\int (v^2 - 1)dv = \frac{4}{3}v^3 - 4v + C = \frac{4}{3}\left(1 + \sqrt{x}\right)^{3/2} - 4\sqrt{1 + \sqrt{x}} + C.$$

11. Let $u = \sqrt{x}$. Then $x = u^2$, $dx = 2u\,du$ $\Rightarrow$

$$\int \frac{\sqrt{x}+1}{\sqrt{x}-1}\,dx = \int \frac{u+1}{u-1}\,2u\,du = 2\int \frac{u^2+u}{u-1}\,du = 2\int \left[u + 2 + \frac{2}{u-1}\right] du$$

$$= u^2 + 4u + 4\ln|u - 1| + C = x + 4\sqrt{x} + 4\ln\left|\sqrt{x} - 1\right| + C.$$

13. Let $u = \sqrt[3]{x^2+1}$. Then $x^2 = u^3 - 1$, $2x\,dx = 3u^2\,du$ $\Rightarrow$

$$\int \frac{x^3\,dx}{\sqrt[3]{x^2+1}} = \int \frac{(u^3 - 1)\frac{3}{2}u^2\,du}{u} = \tfrac{3}{2}\textstyle\int \left(u^4 - u\right) du$$

$$= \tfrac{3}{10}u^5 - \tfrac{3}{4}u^2 + C = \tfrac{3}{10}\left(x^2+1\right)^{5/3} - \tfrac{3}{4}\left(x^2+1\right)^{2/3} + C.$$

15. Let $u = \sqrt[4]{x}$. Then $x = u^4$, $dx = 4u^3\,du$ $\Rightarrow$

$$\int \frac{dx}{\sqrt{x} + \sqrt[4]{x}} = \int \frac{4u^3\,du}{u^2+u} = 4\int \frac{u^2\,du}{u+1} = 4\int \left[u - 1 + \frac{1}{u+1}\right] du$$

$$= 2u^2 - 4u + 4\ln|u+1| + C = 2\sqrt{x} - 4\sqrt[4]{x} + 4\ln\left(\sqrt[4]{x} + 1\right) + C.$$

17. Let $u = \sqrt{x}$. Then $x = u^2$, $dx = 2u\,du$ $\Rightarrow$

$$\int \sqrt{\frac{1-x}{x}}\,dx = \int \frac{\sqrt{1-u^2}}{u}\,2u\,du = 2\int \sqrt{1-u^2}\,du = 2\int \cos^2\theta\,d\theta \quad (\text{put } u = \sin\theta)$$

$$= \theta + \sin\theta\cos\theta + C = \sin^{-1}\sqrt{x} + \sqrt{x(1-x)} + C.$$

Or: Let $u = \sqrt{\dfrac{1-x}{x}}$. This gives $I = \sqrt{x(1-x)} - \tan^{-1}\sqrt{\dfrac{1-x}{x}} + C$.

19. Let $u = e^x$. Then $x = \ln u$, $dx = \dfrac{du}{u} \quad \Rightarrow$

$$\int \frac{e^{2x}\,dx}{e^{2x}+3e^x+2} = \int \frac{u^2(du/u)}{u^2+3u+2} = \int \frac{u\,du}{(u+1)(u+2)} = \int \left[\frac{-1}{u+1} + \frac{2}{u+2}\right] du$$
$$= 2\ln|u+2| - \ln|u+1| + C = \ln\left[(e^x+2)^2/(e^x+1)\right] + C.$$

21. Let $u = e^x$. Then $x = \ln u$, $dx = \dfrac{du}{u} \quad \Rightarrow$

$$\int \sqrt{1-e^x}\,dx = \int \sqrt{1-u}\left(\frac{du}{u}\right) = \int \frac{\sqrt{1-u}\,du}{u} = \int \frac{v(-2v)dv}{1-v^2} \qquad \left[\begin{array}{c}\text{put } v = \sqrt{1-u},\ u = 1-v^2, \\ du = -2v\,dv)\end{array}\right]$$
$$= 2\int \left[1 + \frac{1}{v^2-1}\right] dv = 2\left[v + \frac{1}{2}\ln\left|\frac{v-1}{v+1}\right|\right] + C = 2\sqrt{1-e^x} + \ln\left(\frac{1-\sqrt{1-e^x}}{1+\sqrt{1-e^x}}\right) + C.$$

23. Let $t = \tan\left(\dfrac{x}{2}\right)$. Then, by Equation 1,

$$\int_0^{\pi/2} \frac{dx}{\sin x + \cos x} = \int_0^1 \frac{2\,dt}{2t+1-t^2} = -2\int_0^1 \frac{dt}{t^2-2t-1} = -2\int_0^1 \frac{dt}{(t-1)^2-2}$$
$$= -\frac{1}{\sqrt{2}}\ln\left|\frac{t-1-\sqrt{2}}{t-1+\sqrt{2}}\right|\Bigg|_0^1 = -\frac{1}{\sqrt{2}}\left[\ln 1 - \ln\frac{\sqrt{2}+1}{\sqrt{2}-1}\right] = \frac{1}{\sqrt{2}}\ln\left(\sqrt{2}+1\right)^2$$
$$= \sqrt{2}\ln\left(\sqrt{2}+1\right) \text{ or } -\sqrt{2}\ln\left(\sqrt{2}-1\right) \text{ since } \sqrt{2}+1 = \tfrac{1}{\sqrt{2}-1}$$
$$\text{or } \tfrac{1}{\sqrt{2}}\ln\left(3+2\sqrt{2}\right) \text{ since } \left(\sqrt{2}+1\right)^2 = 3+2\sqrt{2}.$$

25. Let $t = \tan(x/2)$. Then, by Equation 1,

$$\int \frac{dx}{3\sin x + 4\cos x} = \int \frac{2\,dt}{6t+4(1-t^2)} = \int \frac{-\,dt}{2t^2-3t-2} = -\int \left[\frac{-2/5}{2t+1} + \frac{1/5}{t-2}\right] dt$$
$$= \tfrac{1}{5}\ln\left|\frac{2t+1}{t-2}\right| + C = \tfrac{1}{5}\ln\left|\frac{2\tan(x/2)+1}{\tan(x/2)-2}\right| + C.$$

27. Let $t = \tan\left(\dfrac{x}{2}\right)$. Then $\displaystyle\int \frac{dx}{2\sin x + \sin 2x} = \frac{1}{2}\int \frac{dx}{\sin x + \sin x\cos x}$

$$= \frac{1}{2}\int \frac{2\,dt/(1+t^2)}{2t/(1+t^2) + 2t(1-t^2)/(1+t^2)^2} = \frac{1}{2}\int \frac{(1+t^2)dt}{t(1+t^2)+t(1-t^2)}$$
$$= \frac{1}{4}\int \frac{(1+t^2)dt}{t} = \frac{1}{4}\int \left[\frac{1}{t} + t\right] dt = \tfrac{1}{4}\ln|t| + \tfrac{1}{8}t^2 + C = \frac{1}{4}\ln\left|\tan\frac{x}{2}\right| + \frac{1}{8}\tan^2\frac{x}{2} + C.$$

29. Let $t = \tan(x/2)$. Then

$$\int \frac{dx}{a\sin x + b\cos x} = \int \frac{2\,dt}{a(2t)+b(1-t^2)} = -\frac{2}{b}\int \frac{dt}{t^2-2(a/b)t-1}$$
$$= -\frac{2}{b}\int \frac{dt}{(t-a/b)^2 - \left(1+a^2/b^2\right)} = -\frac{1}{b}\,\frac{b}{\sqrt{a^2+b^2}}\ln\left|\frac{t-a/b-\sqrt{a^2+b^2}/b}{t-a/b+\sqrt{a^2+b^2}/b}\right| + C$$
$$= \frac{1}{\sqrt{a^2+b^2}}\ln\left|\frac{b\tan(x/2)-a+\sqrt{a^2+b^2}}{b\tan(x/2)-a-\sqrt{a^2+b^2}}\right| + C.$$

31. **(a)** Let $t = \tan\left(\frac{x}{2}\right)$. Then $\int \sec x\,dx = \int \frac{dx}{\cos x} = \int \frac{2\,dt}{1-t^2} = \int \left[\frac{1}{1-t} + \frac{1}{1+t}\right] dt$

$= \ln|1+t| - \ln|1-t| + C = \ln\left|\frac{1+t}{1-t}\right| + C = \ln\left|\frac{1+\tan(x/2)}{1-\tan(x/2)}\right| + C.$

(b) $\tan\left(\frac{\pi}{4} + \frac{x}{2}\right) = \frac{\tan(\pi/4) + \tan(x/2)}{1 - \tan(\pi/4)\tan(x/2)} = \frac{1+\tan(x/2)}{1-\tan(x/2)}$. Substituting in the formula from part (a), we get $\int \sec x\,dx = \ln\left|\tan\left(\frac{1}{4}\pi + \frac{1}{2}x\right)\right| + C.$

33. According to Equation 7.2.1, $\int \sec x\,dx = \ln|\sec x + \tan x| + C$. Now

$$\frac{1+\tan(x/2)}{1-\tan(x/2)} = \frac{1+\sin(x/2)/\cos(x/2)}{1-\sin(x/2)/\cos(x/2)} = \frac{\cos(x/2)+\sin(x/2)}{\cos(x/2)-\sin(x/2)}$$

$$= \frac{[\cos(x/2)+\sin(x/2)]^2}{[\cos(x/2)-\sin(x/2)][\cos(x/2)+\sin(x/2)]} = \frac{1+2\cos(x/2)\sin(x/2)}{\cos^2(x/2)-\sin^2(x/2)}$$

$$= \frac{1+\sin x}{\cos x} \quad \text{(using identities from the endpapers)} \quad = \sec x + \tan x,$$

so $\ln\left|\frac{1+\tan(x/2)}{1-\tan(x/2)}\right| = \ln|\sec x + \tan x|$, and the formula in Exercise 31(a) agrees with (7.2.1).

EXERCISES 7.6

1. $\int \frac{2x+5}{x-3}\,dx = \int \frac{(2x-6)+11}{x-3}\,dx = \int \left[2 + \frac{11}{x-3}\right] dx = 2x + 11\ln|x-3| + C$

3. $\int \sin^2 x \cos^3 x\,dx = \int \sin^2 x(1-\sin^2 x)\cos x\,dx = \int u^2(1-u^2)\,du$ (put $u = \sin x$)

$= \int (u^2 - u^4)\,du = \frac{1}{3}u^3 - \frac{1}{5}u^5 + C = \frac{1}{3}\sin^3 x - \frac{1}{5}\sin^5 x + C$

5. Let $u = 1 - x^2$. Then $du = -2x\,dx \Rightarrow$

$$\int_0^{1/2} \frac{x\,dx}{\sqrt{1-x^2}} = -\int_1^{3/4} \frac{du}{2\sqrt{u}} = \int_{3/4}^1 \frac{du}{2\sqrt{u}} = \left[\sqrt{u}\right]_{3/4}^1 = 1 - \frac{\sqrt{3}}{2}.$$

7. Let $u = \sqrt{x-2}$. Then $x = u^2 + 2$, $dx = 2u\,du \Rightarrow \int \frac{\sqrt{x-2}}{x+2}\,dx = \int \frac{u \cdot 2u\,du}{u^2+4}$

$= 2\int \left(1 - \frac{4}{u^2+4}\right) du = 2u - \frac{8}{2}\tan^{-1}\left(\frac{1}{2}u\right) + C = 2\sqrt{x-2} - 4\tan^{-1}\left(\frac{1}{2}\sqrt{x-2}\right) + C.$

9. Use integration by parts: $u = \ln(1+x^2)$, $dv = dx \Rightarrow du = \frac{2x}{1+x^2}\,dx$, $v = x$, so

$\int \ln(1+x^2)\,dx = x\ln(1+x^2) - \int x \cdot \frac{2x\,dx}{1+x^2} = x\ln(1+x^2) - 2\int \left[1 - \frac{1}{1+x^2}\right] dx$

$= x\ln(1+x^2) - 2x + 2\tan^{-1}x + C.$

11. Let $u = 1 + \sqrt{x}$. Then $x = (u-1)^2$, $dx = 2(u-1)du \Rightarrow$

$\int_0^1 \left(1+\sqrt{x}\right)^8 dx = \int_1^2 u^8 \cdot 2(u-1)du = 2\int_1^2 (u^9 - u^8)du = \left[\frac{1}{5}u^{10} - 2 \cdot \frac{1}{9}u^9\right]_1^2 = \frac{1024}{5} - \frac{1024}{9} - \frac{1}{5} + \frac{2}{9} = \frac{4097}{45}.$

13. $\displaystyle\int \frac{x\,dx}{x^2-2x+2} = \frac{1}{2}\int \frac{(2x-2)dx}{x^2-2x+2} + \int \frac{dx}{(x-1)^2+1} = \tfrac{1}{2}\ln(x^2-2x+2) + \tan^{-1}(x-1) + C$

15. Let $u = \sqrt{9-x^2}$. Then $u^2 = 9 - x^2$, $u\,du = -x\,dx \quad \Rightarrow$

$$\int \frac{\sqrt{9-x^2}}{x}\,dx = \int \frac{\sqrt{9-x^2}}{x^2}\,x\,dx = \int \frac{u}{9-u^2}(-u)du = \int \left[1 - \frac{9}{9-u^2}\right]du$$

$$= u + 9\int \frac{du}{u^2-9} = u + \frac{9}{2\cdot 3}\ln\left|\frac{u-3}{u+3}\right| + C = \sqrt{9-x^2} + \frac{3}{2}\ln\left|\frac{\sqrt{9-x^2}-3}{\sqrt{9-x^2}+3}\right| + C$$

$$= \sqrt{9-x^2} + \frac{3}{2}\ln\frac{\left(\sqrt{9-x^2}-3\right)^2}{x^2} + C = \sqrt{9-x^2} + 3\ln\left|\frac{3-\sqrt{9-x^2}}{x}\right| + C.$$

Or: Put $x = 3\sin\theta$.

17. Integrate by parts: $u = x^2$, $dv = \cosh x\,dx \quad \Rightarrow \quad du = 2x\,dx$, $v = \sinh x$, so

$I = \int x^2\cosh x\,dx = x^2\sinh x - \int 2x\sinh x\,dx$.

Now let $U = x$, $dV = \sinh x\,dx \quad \Rightarrow \quad dU = dx$, $V = \cosh x$. So

$I = x^2\sinh x - 2(x\cosh x - \int\cosh x\,dx)$

$\quad = x^2\sinh x - 2[x\cosh x - \sinh x] = (x^2+2)\sinh x - 2x\cosh x + C$.

19. Let $u = \sin x$. Then $\displaystyle\int \frac{\cos x\,dx}{1+\sin^2 x} = \int \frac{du}{1+u^2} = \tan^{-1}u + C = \tan^{-1}(\sin x) + C$

21. $\int_0^1 \cos\pi x\tan\pi x\,dx = \int_0^1 \sin\pi x\,dx = -\frac{1}{\pi}\int_0^1(-\pi\sin\pi x)dx = -\frac{1}{\pi}[\cos\pi x]_0^1 = -\frac{1}{\pi}(-1-1) = \frac{2}{\pi}$.

23. Integrate by parts twice, first with $u = e^{3x}$, $dv = \cos 5x\,dx$:

$\int e^{3x}\cos 5x\,dx = \frac{1}{5}e^{3x}\sin 5x - \int\frac{3}{5}e^{3x}\sin 5x\,dx = \frac{1}{5}e^{3x}\sin 5x + \frac{3}{25}e^{3x}\cos 5x - \frac{9}{25}\int e^{3x}\cos 5x\,dx$, so

$\frac{34}{25}\int e^{3x}\cos 5x\,dx = \frac{1}{25}e^{3x}(5\sin 5x + 3\cos 5x) + C_1$ and $\int e^{3x}\cos 5x\,dx = \frac{1}{34}e^{3x}(5\sin 5x + 3\cos 5x) + C$.

25. $\displaystyle\int \frac{dx}{x^3+x^2+x+1} = \int \frac{dx}{(x+1)(x^2+1)} = \int\left[\frac{1/2}{x+1} - \frac{x/2-1/2}{x^2+1}\right]dx$

$$= \frac{1}{2}\int\left(\frac{1}{x+1} - \frac{x}{x^2+1} + \frac{1}{x^2+1}\right)dx = \tfrac{1}{2}\ln|x+1| - \tfrac{1}{4}\ln(x^2+1) + \tfrac{1}{2}\tan^{-1}x + C$$

27. Let $t = x^3$. Then $dt = 3x^2\,dx \quad \Rightarrow \quad I = \int x^5 e^{-x^3}\,dx = \frac{1}{3}\int te^{-t}\,dt$. Now integrate by parts with $u = t$, $dv = e^{-t}\,dt$: $I = -\frac{1}{3}te^{-t} + \frac{1}{3}\int e^{-t}\,dt = -\frac{1}{3}te^{-t} - \frac{1}{3}e^{-t} + C = -\frac{1}{3}e^{-x^3}(x^3+1) + C$.

29. Let $u = 3x + 2$. Then

$$\int \frac{dx}{\sqrt{9x^2+12x-5}} = \int \frac{dx}{\sqrt{(3x+2)^2-9}} = \frac{1}{3}\int \frac{du}{\sqrt{u^2-9}} = \tfrac{1}{3}\cosh^{-1}(u/3) + C_1$$

$= \frac{1}{3}\cosh^{-1}\left[\frac{1}{3}(3x+2)\right] + C_1 = \frac{1}{3}\ln\left|3x + 2 + \sqrt{9x^2+12x-5}\right| + C$ (by Equation 6.7.4).

Or: Substitute $u = 3\sec\theta$.

31. $\int x^{1/3}(1-x^{1/2})dx = \int(x^{1/3} - x^{5/6})dx = \frac{3}{4}x^{4/3} - \frac{6}{11}x^{11/6} + C$

33. Let $u = x^2 + 1$. Then $du = 2x\,dx$, so

$$\int \frac{x}{x^4+2x^2+10}\,dx = \int \frac{x}{(x^2+1)^2+9} = \frac{1}{2}\int \frac{du}{u^2+9} = \frac{1}{2}\cdot\frac{1}{3}\tan^{-1}\frac{u}{3} + C = \frac{1}{6}\tan^{-1}\frac{x^2+1}{3} + C.$$

35. $\int \sin^2 x \cos^4 x\,dx = \int (\sin x \cos x)^2 \cos^2 x\,dx = \int \frac{1}{4}\sin^2 2x\,\frac{1}{2}(1+\cos 2x)dx$
$= \frac{1}{8}\int \sin^2 2x\,dx + \frac{1}{8}\int \sin^2 2x \cos 2x\,dx = \frac{1}{16}\int (1-\cos 4x)dx + \frac{1}{16}\int \sin^2 2x(2\cos 2x)dx$
$= \frac{1}{16}x - \frac{1}{64}\sin 4x + \frac{1}{48}\sin^3 2x + C.$

Or: Write $\int \sin^2 x \cos^4 x\,dx = \frac{1}{8}\int (1-\cos 2x)(1+\cos 2x)^2\,dx.$

37. Let $u = 1 - x^2$. Then $du = -2x\,dx \quad \Rightarrow$

$$\int \frac{x\,dx}{1-x^2+\sqrt{1-x^2}} = -\frac{1}{2}\int \frac{du}{u+\sqrt{u}} = -\int \frac{v\,dv}{v^2+v} \quad \text{(put } v=\sqrt{u},\, u=v^2,\, du=2v\,dv\text{)}$$
$$= -\int \frac{dv}{v+1} = -\ln|v+1| + C = -\ln\left(\sqrt{1-x^2}+1\right) + C.$$

39. Let $u = e^x$. Then $x = \ln u$, $dx = du/u \quad \Rightarrow$

$$\int \frac{e^x\,dx}{e^{2x}-1} = \int \frac{u(du/u)}{u^2-1} = \int \frac{du}{u^2-1} = \frac{1}{2}\ln\left|\frac{u-1}{u+1}\right| + C = \frac{1}{2}\ln\left|\frac{e^x-1}{e^x+1}\right| + C.$$

41. $\int_{-1}^{1} x^5 \cosh x\,dx = 0$ by Theorem 4.5.6, since $x^5\cosh x$ is odd.

43. $\int_{-3}^{3}|x^3+x^2-2x|dx = \int_{-3}^{3}|(x+2)x(x-1)|dx$
$= -\int_{-3}^{-2}(x^3+x^2-2x)dx + \int_{-2}^{0}(x^3+x^2-2x)dx - \int_{0}^{1}(x^3+x^2-2x)dx + \int_{1}^{3}(x^3+x^2-2x)dx.$

Let $f(x) = \frac{1}{4}x^4 + \frac{1}{3}x^3 - x^2$. Then $f'(x) = x^3+x^2-2x$, so

$\int_{-3}^{3}|x^3+x^2-2x|dx = -f(-2)+f(-3)+f(0)-f(-2)-f(1)+f(0)+f(3)-f(1)$
$= f(-3) - 2f(-2) + 2f(0) - 2f(1) + f(3) = \frac{9}{4} - 2\left(-\frac{8}{3}\right) + 2\cdot 0 - 2\left(-\frac{5}{12}\right) + \frac{81}{4} = \frac{86}{3}.$

45. Let $u = \ln(\sin x)$. Then $du = \cot x\,dx \quad \Rightarrow \quad \int \cot x \ln(\sin x)dx = \int u\,du = \frac{1}{2}u^2 + C = \frac{1}{2}[\ln(\sin x)]^2 + C.$

47. $\dfrac{x}{(x^2+1)(x^2+4)} = \dfrac{Ax+B}{x^2+1} + \dfrac{Cx+D}{x^2+4} \quad \Rightarrow \quad x = (Ax+B)(x^2+4) + (Cx+D)(x^2+1) \quad \Rightarrow$

$0 = A+C$, $0 = B+D$, $1 = 4A+C$, and $0 = 4B+D \quad \Rightarrow \quad A = -C = \frac{1}{3}$, $B = D = 0$.

$$\int \frac{x\,dx}{(x^2+1)(x^2+4)} = \frac{1}{3}\int\left[\frac{x}{x^2+1} - \frac{x}{x^2+4}\right]dx = \frac{\ln(x^2+1)}{6} - \frac{\ln(x^2+4)}{6} + C = \frac{1}{6}\ln\frac{x^2+1}{x^2+4} + C.$$

49. Let $u = \sqrt[3]{x+c}$. Then $x = u^3 - c \quad \Rightarrow \quad \int x\sqrt[3]{x+c}\,dx = \int (u^3-c)u\cdot 3u^2\,du$
$= 3\int (u^6 - cu^3)du = \frac{3}{7}u^7 - \frac{3}{4}cu^4 + C = \frac{3}{7}(x+c)^{7/3} - \frac{3}{4}c(x+c)^{4/3} + C.$

51. Let $u = \sqrt{x+1}$. Then $x = u^2 - 1 \quad \Rightarrow \quad \displaystyle\int \frac{dx}{x+4+4\sqrt{x+1}} = \int \frac{2u\,du}{u^2+3+4u} = \int\left[\frac{-1}{u+1} + \frac{3}{u+3}\right]du$
$= 3\ln|u+3| - \ln|u+1| + C = 3\ln\left(\sqrt{x+1}+3\right) - \ln\left(\sqrt{x+1}+1\right) + C.$

53. Use parts twice. First let $u = x^2+4x-3$, $dv = \sin 2x\,dx \quad \Rightarrow \quad du = (2x+4)dx$, $v = -\frac{1}{2}\cos 2x$. Then
$I = \int (x^2+4x-3)\sin 2x\,dx = (x^2+4x-3)\left(-\frac{1}{2}\cos 2x\right) + \int (2x+4)\left(\frac{1}{2}\cos 2x\right)dx.$

Now let $U = 2x+4$, $dV = \frac{1}{2}\cos 2x\,dx \quad \Rightarrow \quad dU = 2\,dx$, $V = \frac{1}{4}\sin 2x$. Then
$I = (x^2+4x-3)\left(-\frac{1}{2}\cos 2x\right) + (2x+4)\left(\frac{1}{4}\sin 2x\right) - \frac{1}{2}\int \sin 2x\,dx$
$= -\frac{1}{2}(x^2+4x-3)\cos 2x + \frac{1}{2}(x+2)\sin 2x + \frac{1}{4}\cos 2x + C$
$= \frac{1}{2}(x+2)\sin 2x - \frac{1}{4}(2x^2+8x-7)\cos 2x + C.$

55. Let $u = x^2$. Then $du = 2x\,dx \Rightarrow$

$$\int \frac{x\,dx}{\sqrt{16 - x^4}} = \frac{1}{2}\int \frac{du}{\sqrt{16 - u^2}} = \tfrac{1}{2}\sin^{-1}\left(\tfrac{1}{4}u\right) + C = \tfrac{1}{2}\sin^{-1}\left(\tfrac{1}{4}x^2\right) + C.$$

57. Let $u = \csc 2x$. Then $du = -2\cot 2x \csc 2x\,dx \Rightarrow$

$$\int \cot^3 2x \csc^3 2x\,dx = \int \csc^2 2x(\csc^2 2x - 1)\cot 2x \csc 2x\,dx = \int u^2(u^2 - 1)\left(-\tfrac{1}{2}\,du\right)$$
$$= -\tfrac{1}{2}\int (u^4 - u^2)\,du = -\tfrac{1}{2}\left[\tfrac{1}{5}u^5 - \tfrac{1}{3}u^3\right] + C = \tfrac{1}{6}\csc^3 2x - \tfrac{1}{10}\csc^5 2x + C.$$

59. Let $u = \arctan x$. Then $du = \dfrac{dx}{1 + x^2} \Rightarrow \displaystyle\int \frac{e^{\arctan x}}{1 + x^2}\,dx = \int e^u\,du = e^u + C = e^{\arctan x} + C.$

61. Integrate by parts three times, first with $u = t^3$, $dv = e^{-2t}\,dt$:

$$\int t^3 e^{-2t}\,dt = -\tfrac{1}{2}t^3 e^{-2t} + \tfrac{1}{2}\int 3t^2 e^{-2t}\,dt = -\tfrac{1}{2}t^3 e^{-2t} - \tfrac{3}{4}t^2 e^{-2t} + \tfrac{1}{2}\int 3te^{-2t}\,dt$$
$$= -e^{-2t}\left[\tfrac{1}{2}t^3 + \tfrac{3}{4}t^2\right] - \tfrac{3}{4}te^{-2t} + \tfrac{3}{4}\int e^{-2t}\,dt = -e^{-2t}\left[\tfrac{1}{2}t^3 + \tfrac{3}{4}t^2 + \tfrac{3}{4}t + \tfrac{3}{8}\right] + C$$
$$= -\tfrac{1}{8}e^{-2t}(4t^3 + 6t^2 + 6t + 3) + C.$$

63. $\int \sin x \sin 2x \sin 3x\,dx = \int \sin x \cdot \tfrac{1}{2}[\cos(2x - 3x) - \cos(2x + 3x)]dx$

$$= \tfrac{1}{2}\int(\sin x \cos x - \sin x \cos 5x)dx = \tfrac{1}{4}\int \sin 2x\,dx - \tfrac{1}{2}\int \tfrac{1}{2}[\sin(x + 5x) + \sin(x - 5x)]dx$$
$$= -\tfrac{1}{8}\cos 2x - \tfrac{1}{4}\int(\sin 6x - \sin 4x)dx = -\tfrac{1}{8}\cos 2x + \tfrac{1}{24}\cos 6x - \tfrac{1}{16}\cos 4x + C$$

65. As in Example 5, $\displaystyle\int \sqrt{\frac{1 + x}{1 - x}}\,dx = \int \frac{1 + x}{\sqrt{1 - x^2}}\,dx = \int \frac{dx}{\sqrt{1 - x^2}} + \int \frac{x\,dx}{\sqrt{1 - x^2}} = \sin^{-1}x - \sqrt{1 - x^2} + C.$

Another Method: Substitute $u = \sqrt{(1 + x)/(1 - x)}$.

67. $$\int \frac{x + a}{x^2 + a^2}\,dx = \frac{1}{2}\int \frac{2x\,dx}{x^2 + a^2} + a\int \frac{dx}{x^2 + a^2} = \tfrac{1}{2}\ln(x^2 + a^2) + a \cdot \frac{1}{a}\tan^{-1}\left(\frac{x}{a}\right) + C$$
$$= \ln\sqrt{x^2 + a^2} + \tan^{-1}(x/a) + C.$$

69. Let $u = x^5$. Then $du = 5x^4\,dx \Rightarrow \displaystyle\int \frac{x^4\,dx}{x^{10} + 16} = \int \frac{\frac{1}{5}\,du}{u^2 + 16} = \tfrac{1}{5} \cdot \tfrac{1}{4}\tan^{-1}\left(\tfrac{1}{4}u\right) + C = \tfrac{1}{20}\tan^{-1}\left(\tfrac{1}{4}x^5\right) + C.$

71. Integrate by parts with $u = x$, $dv = \sec x \tan x\,dx \Rightarrow du = dx$, $v = \sec x$:

$\int x \sec x \tan x\,dx = x\sec x - \int \sec x\,dx = x\sec x - \ln|\sec x + \tan x| + C.$

73. $$\int \frac{dx}{\sqrt{x + 1} + \sqrt{x}} = \int\left(\sqrt{x + 1} - \sqrt{x}\right)dx = \frac{2}{3}\left[(x + 1)^{3/2} - x^{3/2}\right] + C.$$

75. Let $u = \sqrt{x}$. Then $du = dx/(2\sqrt{x}) \Rightarrow$

$$\int \frac{\arctan\sqrt{x}}{\sqrt{x}}\,dx = \int \tan^{-1}u\,2\,du = 2u\tan^{-1}u - \int \frac{2u\,du}{1 + u^2} \quad \text{(by parts)}$$
$$= 2u\tan^{-1}u - \ln(1 + u^2) + C = 2\sqrt{x}\tan^{-1}\sqrt{x} - \ln(1 + x) + C.$$

77. Let $u = e^x$. Then $x = \ln u$, $dx = du/u \Rightarrow$

$$\int \frac{dx}{e^{3x} - e^x} = \int \frac{du/u}{u^3 - u} = \int \frac{du}{(u - 1)u^2(u + 1)} = \int\left[\frac{1/2}{u - 1} - \frac{1}{u^2} - \frac{1/2}{u + 1}\right]du = \frac{1}{u} + \frac{1}{2}\ln\left|\frac{u - 1}{u + 1}\right| + C$$
$$= e^{-x} + \tfrac{1}{2}\ln|(e^x - 1)/(e^x + 1)| + C.$$

79. Let $u = \sqrt{2x - 25} \Rightarrow u^2 = 2x - 25 \Rightarrow 2u\,du = 2\,dx \Rightarrow$

$$\int \frac{dx}{x\sqrt{2x - 25}} = \int \frac{u\,du}{\frac{1}{2}(u^2 + 25) \cdot u} = 2\int \frac{du}{u^2 + 25} = \tfrac{2}{5}\tan^{-1}\left(\tfrac{1}{5}u\right) + C = \tfrac{2}{5}\tan^{-1}\left(\tfrac{1}{5}\sqrt{2x - 25}\right) + C.$$

EXERCISES 7.7

1. By Formula 99,

$$\int e^{-3x}\cos 4x\,dx = \frac{e^{-3x}}{(-3)^2+4^2}(-3\cos 4x + 4\sin 4x) + C = \frac{e^{-3x}}{25}(-3\cos 4x + 4\sin 4x) + C.$$

3. Let $u = 3x$. Then $du = 3\,dx$, so $\displaystyle\int \frac{\sqrt{9x^2-1}}{x^2}\,dx = \int \frac{\sqrt{u^2-1}}{u^2/9}\,\frac{du}{3} = 3\int \frac{\sqrt{u^2-1}}{u^2}\,du$

$$= -\frac{3\sqrt{u^2-1}}{u} + 3\ln\left|u + \sqrt{u^2-1}\right| + C \quad \text{(by Formula 42)} = -\frac{\sqrt{9x^2-1}}{x} + 3\ln\left|3x + \sqrt{9x^2-1}\right| + C.$$

5. $\int x^2 e^{3x}\,dx = \frac{1}{3}x^2e^{3x} - \frac{2}{3}\int xe^{3x}\,dx$ (Formula 97) $= \frac{1}{3}x^2e^{3x} - \frac{2}{3}\left[\frac{1}{9}(3x-1)e^{3x}\right] + C$ (Formula 96) $= \frac{1}{27}(9x^2 - 6x + 2)e^{3x} + C$

7. Let $u = x^2$. Then $du = 2x\,dx$, so $\int x\sin^{-1}(x^2)dx = \frac{1}{2}\int \sin^{-1}u\,du = \frac{1}{2}\left(u\sin^{-1}u + \sqrt{1-u^2}\right) + C$ (Formula 87) $= \frac{1}{2}\left(x^2\sin^{-1}(x^2) + \sqrt{1-x^4}\right) + C.$

9. Let $u = e^x$. Then $du = e^x\,dx$, so $\int e^x\,\text{sech}(e^x)dx = \int \text{sech}\,u\,du = \tan^{-1}|\sinh u| + C$ (Formula 107) $= \tan^{-1}[\sinh(e^x)] + C.$

11. Let $u = x + 2$. Then $\displaystyle\int \sqrt{5 - 4x - x^2}\,dx = \int \sqrt{9 - (x+2)^2}\,dx = \int \sqrt{9-u^2}\,du$

$$= \frac{u}{2}\sqrt{9-u^2} + \frac{9}{2}\sin^{-1}\frac{u}{3} + C \text{ (Formula 30)} = \frac{x+2}{2}\sqrt{5-4x-x^2} + \frac{9}{2}\sin^{-1}\frac{x+2}{3} + C.$$

13. $\int \sec^5 x\,dx = \frac{1}{4}\tan x\sec^3 x + \frac{3}{4}\int \sec^3 x\,dx$ (Formula 77)
$= \frac{1}{4}\tan x\sec^3 x + \frac{3}{4}\left(\frac{1}{2}\tan x\sec x + \frac{1}{2}\int \sec x\,dx\right)$ (Formula 77 again)
$= \frac{1}{4}\tan x\sec^3 x + \frac{3}{8}\tan x\sec x + \frac{3}{8}\ln|\sec x + \tan x| + C$ (Formula 14)

15. Let $u = \sin x$. Then $du = \cos x\,dx$, so $\int \sin^2 x\cos x\ln(\sin x)dx = \int u^2\ln u\,du$
$= \frac{1}{9}u^3(3\ln u - 1) + C$ (Formula 101) $= \frac{1}{9}\sin^3 x[3\ln(\sin x) - 1] + C.$

17. $\displaystyle\int \sqrt{2+3\cos x}\tan x\,dx = -\int \frac{\sqrt{2+3\cos x}}{\cos x}(-\sin x\,dx) = -\int \frac{\sqrt{2+3u}}{u}\,du$ (where $u = \cos x$)

$$= -2\sqrt{2+3u} - 2\int \frac{du}{u\sqrt{2+3u}} \quad \text{(Formula 58)} \quad = -2\sqrt{2+3u} - 2\cdot\frac{1}{\sqrt{2}}\ln\left|\frac{\sqrt{2+3u}-\sqrt{2}}{\sqrt{2+3u}+\sqrt{2}}\right| + C$$

$$\text{(Formula 57)} \quad = -2\sqrt{2+3\cos x} - \sqrt{2}\ln\left|\frac{\sqrt{2+3\cos x}-\sqrt{2}}{\sqrt{2+3\cos x}+\sqrt{2}}\right| + C$$

19. $\int_0^{\pi/2}\cos^5 x\,dx = \frac{1}{5}\left[\cos^4 x\sin x\right]_0^{\pi/2} + \frac{4}{5}\int_0^{\pi/2}\cos^3 x\,dx$ (Formula 74)
$= 0 + \frac{4}{5}\left[\frac{1}{3}(2 + \cos^2 x)\sin x\right]_0^{\pi/2}$ (Formula 68) $= \frac{4}{15}(2 - 0) = \frac{8}{15}$

21. Let $u = x^5$, $du = 5x^4\,dx$.

$$\int \frac{x^4\,dx}{\sqrt{x^{10}-2}} = \frac{1}{5}\int \frac{du}{\sqrt{u^2-2}} = \frac{1}{5}\ln\left|u + \sqrt{u^2-2}\right| + C \text{ (Formula 43)} = \frac{1}{5}\ln\left|x^5 + \sqrt{x^{10}-2}\right| + C.$$

23. Let $u = 1 + e^x$, so $du = e^x\,dx$. Then $\int e^x \ln(1 + e^x)dx = \int \ln u\,du = u \ln u - u + C$ (Formula 100)
$= (1 + e^x)\ln(1 + e^x) - e^x - 1 + C = (1 + e^x)\ln(1 + e^x) - e^x + C_1$.

25. Let $u = e^x \Rightarrow \ln u = x \Rightarrow dx = \dfrac{du}{u}$. Then $\displaystyle\int \sqrt{e^{2x} - 1}\,dx = \int \frac{\sqrt{u^2 - 1}}{u}\,du$
$= \sqrt{u^2 - 1} - \cos^{-1}(1/u) + C$ (Formula 41) $= \sqrt{e^{2x} - 1} - \cos^{-1}(e^{-x}) + C$.
Or: Let $u = \sqrt{e^{2x} - 1}$.

27. $\text{Volume} = \displaystyle\int_0^1 \frac{2\pi x}{(1 + 5x)^2}\,dx = 2\pi\left[\frac{1}{25(1 + 5x)} + \tfrac{1}{25}\ln|1 + 5x|\right]_0^1$ (Formula 51)
$= \frac{2\pi}{25}\left(\frac{1}{6} + \ln 6 - 1 - \ln 1\right) = \frac{2\pi}{25}\left(\ln 6 - \frac{5}{6}\right)$

29. **(a)** $\displaystyle\frac{d}{du}\left[\frac{1}{b^3}\left(a + bu - \frac{a^2}{a + bu} - 2a\ln|a + bu|\right) + C\right] = \frac{1}{b^3}\left[b + \frac{ba^2}{(a + bu)^2} - \frac{2ab}{(a + bu)}\right]$

$$= \frac{1}{b^3}\left[\frac{b(a + bu)^2 + ba^2 - (a + bu)2ab}{(a + bu)^2}\right] = \frac{1}{b^3}\left[\frac{b^3u^2}{(a + bu)^2}\right] = \frac{u^2}{(a + bu)^2}$$

(b) Let $t = a + bu \Rightarrow dt = b\,du$.

$$\int \frac{u^2\,du}{(a + bu)^2} = \frac{1}{b^3}\int \frac{(t - a)^2}{t^2}\,dt = \frac{1}{b^3}\int\left(1 - \frac{2a}{t} + \frac{a^2}{t^2}\right)dt = \frac{1}{b^3}\left(t - 2a\ln|t| - \frac{a^2}{t}\right) + C$$
$$= \frac{1}{b^3}\left(a + bu - \frac{a^2}{a + bu} - 2a\ln|a + bu|\right) + C$$

31. Maple, Mathematica and Derive all give $\int x^2\sqrt{5 - x^2}\,dx = -\frac{1}{4}x(5 - x^2)^{3/2} + \frac{5}{8}x\sqrt{5 - x^2} + \frac{25}{8}\sin^{-1}\left(\frac{1}{\sqrt{5}}x\right)$.
Using Formula 31, we get $\int x^2\sqrt{5 - x^2}\,dx = \frac{1}{8}x(2x^2 - 5)\sqrt{5 - x^2} + \frac{1}{8}(5^2)\sin^{-1}\left(\frac{1}{\sqrt{5}}x\right) + C$. But
$-\frac{1}{4}x(5 - x^2)^{3/2} + \frac{5}{8}x\sqrt{5 - x^2} = \frac{1}{8}x\sqrt{5 - x^2}[5 - 2(5 - x^2)] = \frac{1}{8}x(2x^2 - 5)\sqrt{5 - x^2}$, and the $\sin^{-1}$ terms are the same in each expression, so the answers are equivalent.

33. Maple and Derive both give $\int \sin^3 x\cos^2 x\,dx = -\frac{1}{5}\sin^2 x\cos^3 x - \frac{2}{15}\cos^3 x$ (although Derive factors the expression), and Mathematica gives $\int \sin^3 x\cos^2 x\,dx = -\frac{1}{8}\cos x - \frac{1}{48}\cos 3x + \frac{1}{80}\cos 5x$. We can use a CAS to show that both of these expressions are equal to $-\frac{1}{3}\cos^3 x + \frac{1}{5}\cos^5 x$. Using Formula 86, we write
$\int \sin^3 x\cos^2 x\,dx = -\frac{1}{5}\sin^2 x\cos^3 x + \frac{2}{5}\int \sin x\cos^2 x\,dx = -\frac{1}{5}\sin^2 x\cos^3 x + \frac{2}{5}\left(-\frac{1}{3}\cos^3 x\right) + C$
$= -\frac{1}{5}\sin^2 x\cos^3 x - \frac{2}{15}\cos^3 x + C$.

35. Maple gives $\int x\sqrt{1 + 2x}\,dx = \frac{1}{10}(1 + 2x)^{5/2} - \frac{1}{6}(1 + 2x)^{3/2}$, Mathematica gives $\sqrt{1 + 2x}\left(\frac{2}{5}x^2 + \frac{1}{15}x - \frac{1}{15}\right)$, and Derive gives $\frac{1}{15}(1 + 2x)^{3/2}(3x - 1)$. The first two expressions can be simplified to Derive's result. If we use Formula 54, we get $\int x\sqrt{1 + 2x}\,dx = \frac{2}{15(2)^2}(3 \cdot 2x - 2 \cdot 1)(1 + 2x)^{3/2} + C$
$= \frac{1}{30}(6x - 2)(1 + 2x)^{3/2} + C = \frac{1}{15}(3x - 1)(1 + 2x)^{3/2}$.

37. Maple gives $\int \tan^3 x\,dx = \frac{1}{2}\tan^2 x - \frac{1}{2}\ln(1 + \tan^2 x)$, while Mathematica and Derive both give $\ln\cos x + \frac{1}{2}\tan^2 x$. These expressions are equivalent, since $-\frac{1}{2}\ln(1 + \tan^2 x) = \ln\left[(\sec^2 x)^{-1/2}\right] = \ln\cos x$.
Using Formula 69, we get $\int \tan^3 x\,dx = \frac{1}{2}\tan^2 x + \ln|\cos x| + C$.

39. Maple gives the antiderivative

$$F(x) = \int \frac{x^2 - 1}{x^4 + x^2 + 1}\,dx = -\tfrac{1}{2}\ln(x^2 + x + 1) + \tfrac{1}{2}\ln(x^2 - x + 1).$$

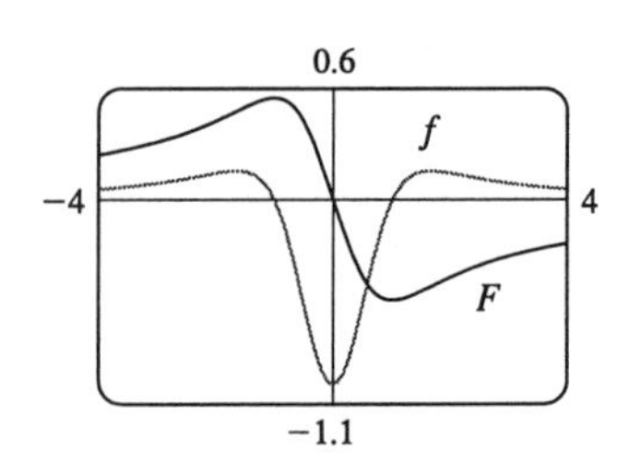

We can see that at 0, this antiderivative is 0. From the graphs, it appears that F has a maximum at $x = -1$ and a minimum at $x = 1$ [since $F'(x) = f(x)$ changes sign at these x-values], and that F has inflection points at $x \approx -1.7$, $x = 0$ and $x \approx 1.7$ [since $f(x)$ has extrema at these x-values].

41. Since f is everywhere positive, we know that its antiderivative F is increasing. The antiderivative given by Maple is

$$\int \sin^4 x \cos^6 x\,dx = -\tfrac{1}{10}\sin^3 x \cos^7 x - \tfrac{3}{80}\sin x \cos^7 x + \tfrac{1}{160}\cos^5 x \sin x + \tfrac{1}{128}\cos^3 x \sin x + \tfrac{3}{256}\cos x \sin x + \tfrac{3}{256}x,$$

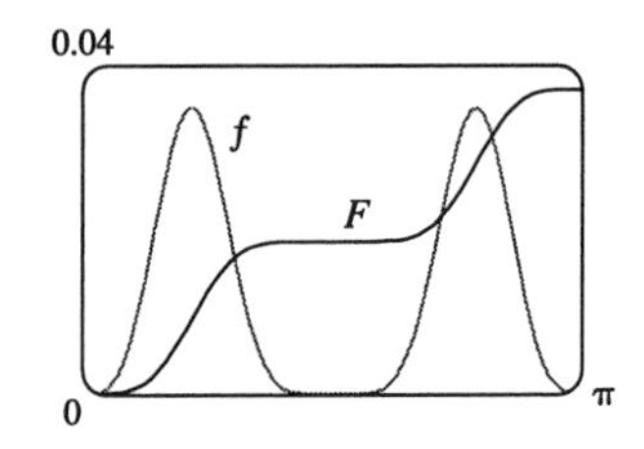

and this is 0 at $x = 0$.

EXERCISES 7.8

1. **(a)** $L_2 = \sum_{i=1}^{2} f(x_{i-1})\Delta x = 2\,f(0) + 2\,f(2) = 2(0.5) + 2(2.5) = 6$

$R_2 = \sum_{i=1}^{2} f(x_i)\Delta x = 2\,f(2) + 2\,f(4) = 2(2.5) + 2(3.5) = 12$

$M_2 = \sum_{i=1}^{2} f(\overline{x}_i)\Delta x = 2\,f(1) + 2\,f(3) \approx 2(1.7) + 2(3.2) = 9.8$

(b) L_2 is an underestimate, since the area under the small rectangles is less than the area under the curve, and R_2 is an overestimate, since the area under the large rectangles is greater than the area under the curve. It appears that M_2 is an overestimate, though it is fairly close to I. See the solution to Exercise 37 for a proof of the fact that if f is concave down on $[a, b]$, then the Midpoint Rule is an overestimate of $\int_a^b f(x)dx$.

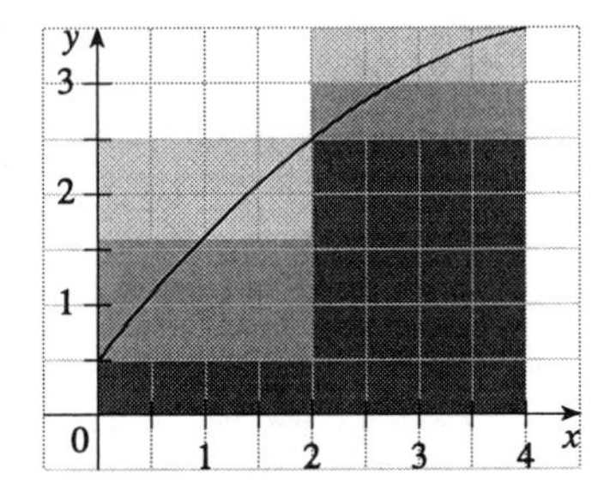

(c) $T_2 = \left(\tfrac{1}{2}\Delta x\right)[f(x_0) + 2f(x_1) + f(x_2)] = \tfrac{2}{2}[f(0) + 2\,f(2) + f(4)] = 0.5 + 2(2.5) + 3.5 = 9.$

This approximation is an underestimate, since the graph is concave down. See the solution to Exercise 37 for a general proof of this conclusion.

(d) For any n, we will have $L_n < T_n < I < M_n < R_n$.

3. $f(x) = \sqrt{1 + x^3}$, $\Delta x = \dfrac{1 - (-1)}{8} = \dfrac{1}{4}$

(a) $T_8 = \frac{0.25}{2}\left[f(-1) + 2f\left(-\frac{3}{4}\right) + 2f\left(-\frac{1}{2}\right) + \cdots + 2f\left(\frac{1}{2}\right) + 2f\left(\frac{3}{4}\right) + f(1)\right] \approx 1.913972$

(b) $S_8 = \frac{0.25}{3}\left[f(-1) + 4f\left(-\frac{3}{4}\right) + 2f\left(-\frac{1}{2}\right) + 4f\left(-\frac{1}{4}\right) + 2f(0) + 4f\left(\frac{1}{4}\right) + 2f\left(\frac{1}{2}\right) + 4f\left(\frac{3}{4}\right) + f(1)\right]$
≈ 1.934766

5. $f(x) = \dfrac{\sin x}{x}$, $\Delta x = \dfrac{\pi - \pi/2}{6} = \dfrac{\pi}{12}$

(a) $T_6 = \frac{\pi}{24}\left[f\left(\frac{\pi}{2}\right) + 2f\left(\frac{7\pi}{12}\right) + 2f\left(\frac{2\pi}{3}\right) + 2f\left(\frac{3\pi}{4}\right) + 2f\left(\frac{5\pi}{6}\right) + 2f\left(\frac{11\pi}{12}\right) + f(\pi)\right] \approx 0.481672$

(b) $S_6 = \frac{\pi}{36}\left[f\left(\frac{\pi}{2}\right) + 4f\left(\frac{7\pi}{12}\right) + 2f\left(\frac{2\pi}{3}\right) + 4f\left(\frac{3\pi}{4}\right) + 2f\left(\frac{5\pi}{6}\right) + 4f\left(\frac{11\pi}{12}\right) + f(\pi)\right] \approx 0.481172$

7. $f(x) = e^{-x^2}$, $\Delta x = \dfrac{1-0}{10} = 0.1$

(a) $T_{10} = \frac{0.1}{2}[f(0) + 2f(0.1) + 2f(0.2) + \cdots + 2f(0.8) + 2f(0.9) + f(1)] \approx 0.746211$

(b) $M_{10} = 0.1[f(0.05) + f(0.15) + f(0.25) + \cdots + f(0.75) + f(0.85) + f(0.95)] \approx 0.747131$

(c) $S_{10} = \frac{0.1}{3}[f(0) + 4f(0.1) + 2f(0.2) + 4f(0.3) + 2f(0.4) + 4f(0.5)$
$+ 2f(0.6) + 4f(0.7) + 2f(0.8) + 4f(0.9) + f(1)] \approx 0.746825$

9. $f(x) = \cos(e^x)$, $\Delta x = \dfrac{1/2 - 0}{8} = \dfrac{1}{16}$

(a) $T_8 = \frac{1}{32}\left[f(0) + 2f\left(\frac{1}{16}\right) + 2f\left(\frac{1}{8}\right) + \cdots + 2f\left(\frac{7}{16}\right) + f\left(\frac{1}{2}\right)\right] \approx 0.132465$

(b) $M_8 = \frac{1}{16}\left[f\left(\frac{1}{32}\right) + f\left(\frac{3}{32}\right) + f\left(\frac{5}{32}\right) + \cdots + f\left(\frac{15}{32}\right)\right] \approx 0.132857$

(c) $S_8 = \frac{1}{48}\left[f(0) + 4f\left(\frac{1}{16}\right) + 2f\left(\frac{1}{8}\right) + 4f\left(\frac{3}{16}\right) + 2f\left(\frac{1}{4}\right) + 4f\left(\frac{5}{16}\right) + 2f\left(\frac{3}{8}\right) + 4f\left(\frac{7}{16}\right) + f\left(\frac{1}{2}\right)\right]$
≈ 0.132727

11. $f(x) = x^5 e^x$, $\Delta x = \dfrac{1-0}{10} = \dfrac{1}{10}$

(a) $T_{10} = \frac{0.1}{2}[f(0) + 2f(0.1) + 2f(0.2) + \cdots + 2f(0.9) + f(1)] \approx 0.409140$

(b) $M_{10} = 0.1[f(0.05) + f(0.15) + f(0.25) + \cdots + f(0.95)] \approx 0.388849$

(c) $S_{10} = \frac{0.1}{3}[f(0) + 4f(0.1) + 2f(0.2) + 4f(0.3) + 2f(0.4) + 4f(0.5)$
$+ 2f(0.6) + 4f(0.7) + 2f(0.8) + 4f(0.9) + f(1)] \approx 0.395802$

13. $f(x) = e^{1/x}$, $\Delta x = \dfrac{2-1}{4} = \dfrac{1}{4}$

(a) $T_4 = \frac{1}{4\cdot 2}[f(1) + 2f(1.25) + 2f(1.5) + 2f(1.75) + f(2)] \approx 2.031893$

(b) $M_4 = \frac{1}{4}[f(1.125) + f(1.375) + f(1.625) + f(1.875)] \approx 2.014207$

(c) $S_4 = \frac{1}{4\cdot 3}[f(1) + 4f(1.25) + 2f(1.5) + 4f(1.75) + f(2)] \approx 2.020651$

15. $f(x) = \dfrac{1}{1+x^4}$, $\Delta x = \dfrac{3-0}{6} = \dfrac{1}{2}$

(a) $T_6 = \frac{1}{2\cdot 2}[f(0) + 2f(0.5) + 2f(1) + 2f(1.5) + 2f(2) + 2f(2.5) + f(3)] \approx 1.098004$

(b) $M_6 = \frac{1}{2}[f(0.25) + f(0.75) + f(1.25) + f(1.75) + f(2.25) + f(2.75)] \approx 1.098709$

(c) $S_6 = \frac{1}{2\cdot 3}[f(0) + 4f(0.5) + 2f(1) + 4f(1.5) + 2f(2) + 4f(2.5) + f(3)] = 1.109031$

17. $f(x) = e^{-x^2}$, $\Delta x = \dfrac{2-0}{10} = \dfrac{1}{5}$

(a) $T_{10} = \frac{1}{5\cdot 2}[f(0) + 2(f(0.2) + f(0.4) + \cdots + f(1.8)) + f(2)] \approx 0.881839$
$M_{10} = \frac{1}{5}[f(0.1) + f(0.3) + f(0.5) + \cdots + f(1.7) + f(1.9)] \approx 0.882202$

(b) $f(x) = e^{-x^2}$, $f'(x) = -2xe^{-x^2}$, $f''(x) = (4x^2 - 2)e^{-x^2}$, $f'''(x) = 4x(3 - 2x^2)e^{-x^2}$.
$f'''(x) = 0 \quad \Leftrightarrow \quad x = 0$ or $x = \pm\sqrt{3/2}$. So to find the maximum value of $|f''(x)|$ on $[0, 2]$, we need only consider its values at $x = 0$, $x = 2$, and $x = \sqrt{3/2}$. $|f''(0)| = 2$, $|f''(2)| \approx 0.2564$ and $\left|f''\left(\sqrt{3/2}\right)\right| = 4e^{-3/2} \approx 0.8925$. Thus, taking $K = 2$, $a = 0$, $b = 2$, and $n = 10$ in Theorem 5, we get $|E_T| \le 2 \cdot 2^3/\left[12(10)^2\right] = \frac{1}{75} = 0.01\overline{3}$, and $|E_M| \le 2 \cdot 2^3/\left[24(10)^2\right] = 0.00\overline{6}$.

19. **(a)** $T_{10} = \frac{1}{10 \cdot 2}[f(0) + 2(f(0.1) + f(0.2) + \cdots + f(0.9)) + f(1)] \approx 1.719713$

$S_{10} = \frac{1}{10 \cdot 3}[f(0) + 4f(0.1) + 2f(0.2) + 4f(0.3) + \cdots + 4f(0.9) + f(1)] \approx 1.7182828$

Since $\int_0^1 e^x\,dx = [e^x]_0^1 = e - 1 \approx 1.71828183$, $E_T \approx -0.00143166$ and $E_S \approx -0.00000095$.

(b) $f(x) = e^x \quad \Rightarrow \quad f''(x) = e^x \leq e$ for $0 \leq x \leq 1$. Taking $K = e$, $a = 0$, $b = 1$, and $n = 10$ in Theorem 5, we get $|E_T| \leq \dfrac{e(1)^3}{12(10)^2} \approx 0.002265 > 0.00143166$ [actual $|E_T|$ from (a)]. $f^{(4)}(x) = e^x < e$ for $0 \leq x \leq 1$. Using Theorem 7, we have $|E_S| \leq e(1)^5/\left[180(10)^4\right] \approx 0.0000015 > 0.00000095$ [actual $|E_S|$ from (a)]. We see that the actual errors are about two-thirds the size of the error estimates.

21. Take $K = 2$ (as in Exercise 17) in Theorem 5. $|E_T| \leq \dfrac{K(b-a)^3}{12n^2} \leq 10^{-5} \quad \Leftrightarrow \quad \dfrac{1}{6n^2} \leq 10^{-5} \quad \Leftrightarrow$

$6n^2 \geq 10^5 \quad \Leftrightarrow \quad n \geq 129.099\ldots \quad \Leftrightarrow \quad n \geq 130$. Take $n = 130$ in the trapezoidal method. For E_M, again take $K = 2$ in Theorem 5 to get $|E_M| \leq 2(1)^3/(24n^2) \leq 10^{-5} \quad \Leftrightarrow \quad n^2 \geq 2/\left[24(10^{-5})\right] \quad \Leftrightarrow \quad n \geq 91.3$

$\Rightarrow \quad n \geq 92$. Take $n = 92$ for M_n.

23. **(a)** Using the CAS, we differentiate $f(x) = e^{\cos x}$ twice, and find that $f''(x) = e^{\cos x}(\sin^2 x - \cos x)$. From the graph, we see that $|f''(x)| < 2.8$ on $[0, 2\pi]$. Other possible upper bounds for $|f''(x)|$ are $K = 3$ or $K = e$ (the actual maximum value.)

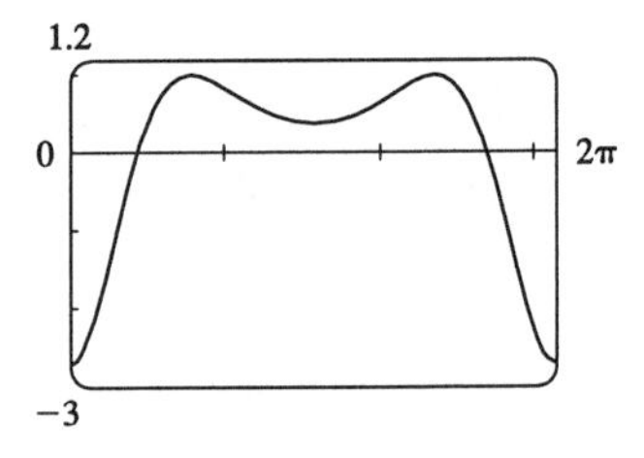

(b) A CAS gives $M_{10} \approx 7.954926518$.
(In Maple, use `student[middlesum]`.)

(c) Using Theorem 5 for the Midpoint Rule, with $K = 2.8$, we get $|E_M| \leq \dfrac{2.8(2\pi - 0)^3}{24 \cdot 10^2} \approx 0.287$.

(d) A CAS gives $I \approx 7.954926521$.

(e) The actual error is only about 3×10^{-9}, much less than the estimate in part (c).

(f) We use the CAS to differentiate twice more:
$f^{(4)}(x) = e^{\cos x}\left(\sin^4 x - 6\sin^2 x \cos x + 3 - 7\sin^2 x + \cos x\right)$
From the graph, it appears that $\left|f^{(4)}(x)\right| < 10.9$ on $[0, 2\pi]$. Another possible upper bound for $\left|f^{(4)}(x)\right|$ is $4e$ (the actual maximum value.)

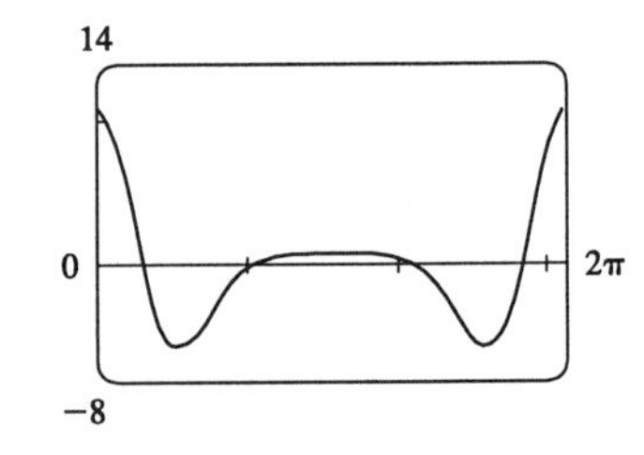

(g) A CAS gives $S_{10} \approx 7.953789422$.
(In Maple, use `student[simpson]`.)

(h) Using Theorem 7 with $K = 10.9$, we get $|E_S| \leq \dfrac{10.9(2\pi - 0)^5}{180 \cdot 10^4} \approx 0.0593$.

(i) The actual error is about $7.954926521 - 7.953789427 \approx 0.00114$. This is quite a bit smaller than the estimate in part (h), though the difference is not nearly as great as it was in the case of the Midpoint Rule.

(j) To ensure that $|E_S| \leq 0.0001$, we use Theorem 7: $|E_S| \leq \dfrac{10.9(2\pi)^5}{180 \cdot n^4} \leq 0.0001 \quad \Leftrightarrow \quad \dfrac{10.9(2\pi)^5}{180 \cdot 0.0001} \leq n^4$

$\Leftrightarrow \quad n^4 \geq 5{,}929{,}981 \quad \Leftrightarrow \quad n \geq 49.4$. So we must take $n \geq 50$ to ensure that $|I - S_n| \leq 0.0001$.

25. $\int_0^1 x^3\,dx = \left[\frac{1}{4}x^4\right]_0^1 = 0.25$. $f(x) = x^3$.

$n = 4$: $L_4 = \frac{1}{4}\left[0^3 + \left(\frac{1}{4}\right)^3 + \left(\frac{1}{2}\right)^3 + \left(\frac{3}{4}\right)^3\right] = 0.140625$

$R_4 = \frac{1}{4}\left[\left(\frac{1}{4}\right)^3 + \left(\frac{1}{2}\right)^3 + \left(\frac{3}{4}\right)^3 + 1^3\right] = 0.390625$,

$T_4 = \frac{1}{4\cdot 2}\left[0^3 + 2\left(\frac{1}{4}\right)^3 + 2\left(\frac{1}{2}\right)^3 + 2\left(\frac{3}{4}\right)^3 + 1^3\right] = \frac{17}{64} = 0.265625$,

$M_4 = \frac{1}{4}\left[\left(\frac{1}{8}\right)^3 + \left(\frac{3}{8}\right)^3 + \left(\frac{5}{8}\right)^3 + \left(\frac{7}{8}\right)^3\right] = 0.2421875$

$E_L = \int_0^1 x^3\,dx - L_4 = \frac{1}{4} - 0.140625 = 0.109375$, $E_R = \frac{1}{4} - 0.390625 = -0.140625$,

$E_T = \frac{1}{4} - 0.265625 = -0.015625$, $E_M = \frac{1}{4} - 0.2421875 = 0.0078125$

$n = 8$: $L_8 = \frac{1}{8}\left[f(0) + f\left(\frac{1}{8}\right) + f\left(\frac{2}{8}\right) + \cdots + f\left(\frac{7}{8}\right)\right] \approx 0.191406$

$R_8 = \frac{1}{8}\left[f\left(\frac{1}{8}\right) + f\left(\frac{2}{8}\right) + \cdots + f\left(\frac{7}{8}\right) + f(1)\right] \approx 0.316406$

$T_8 = \frac{1}{8\cdot 2}\left[f(0) + 2\left(f\left(\frac{1}{8}\right) + f\left(\frac{2}{8}\right) + \cdots + f\left(\frac{7}{8}\right)\right) + f(1)\right] \approx 0.253906$

$M_8 = \frac{1}{8}\left[f\left(\frac{1}{16}\right) + f\left(\frac{3}{16}\right) + \cdots + f\left(\frac{13}{16}\right) + f\left(\frac{15}{16}\right)\right] = 0.248047$

$E_L \approx \frac{1}{4} - 0.191406 \approx 0.058594$, $E_R \approx \frac{1}{4} - 0.316406 \approx -0.066406$,

$E_T \approx \frac{1}{4} - 0.253906 \approx -0.003906$, $E_M \approx \frac{1}{4} - 0.248047 \approx 0.001953$.

$n = 16$: $L_{16} = \frac{1}{16}\left[f(0) + f\left(\frac{1}{16}\right) + f\left(\frac{2}{16}\right) + \cdots + f\left(\frac{15}{16}\right)\right] \approx 0.219727$

$R_{16} = \frac{1}{16}\left[f\left(\frac{1}{16}\right) + f\left(\frac{2}{16}\right) + \cdots + f\left(\frac{15}{16}\right) + f(1)\right] \approx 0.282227$

$T_{16} = \frac{1}{16\cdot 2}\left(f(0) + 2\left[f\left(\frac{1}{16}\right) + f\left(\frac{2}{16}\right) + \cdots + f\left(\frac{15}{16}\right)\right] + f(1)\right) \approx 0.250977$

$M_{16} = \frac{1}{16}\left[f\left(\frac{1}{32}\right) + f\left(\frac{3}{32}\right) + \cdots + f\left(\frac{31}{32}\right)\right] \approx 0.249512$

$E_L \approx \frac{1}{4} - 0.219727 \approx 0.030273$, $E_R \approx \frac{1}{4} - 0.282227 \approx -0.032227$,

$E_T \approx \frac{1}{4} - 0.250977 \approx -0.000977$, $E_M \approx \frac{1}{4} - 0.249512 \approx 0.000488$.

n	L_n	R_n	T_n	M_n
4	0.140625	0.390625	0.265625	0.242188
8	0.191406	0.316406	0.253906	0.248047
16	0.219727	0.282227	0.250977	0.249512

n	E_L	E_R	E_T	E_M
4	0.109375	−0.140625	−0.015625	0.007813
8	0.058594	−0.066406	−0.003906	0.001953
16	0.030273	−0.032227	−0.000977	0.000488

Observations:

1. E_L and E_R are always opposite in sign, as are E_T and E_M.
2. As n is doubled, E_L and E_R are decreased by about a factor of 2, and E_T and E_M are decreased by a factor of about 4.
3. The Midpoint approximation is about twice as accurate as the Trapezoidal approximation.
4. All the approximations become more accurate as the value of n increases.
5. The Midpoint and Trapezoidal approximations are much more accurate than the endpoint approximations.

27. $\int_1^4 \sqrt{x}\,dx = \left[\frac{2}{3}x^{3/2}\right]_1^4 = \frac{2}{3}(8-1) = \frac{14}{3} \approx 4.666667$

$n = 6$: $\Delta x = (4-1)/6 = \frac{1}{2}$

$$T_6 = \frac{1}{2\cdot 2}\left[\sqrt{1} + 2\sqrt{1.5} + 2\sqrt{2} + 2\sqrt{2.5} + 2\sqrt{3} + 2\sqrt{3.5} + \sqrt{4}\right] \approx 4.661488$$

$$M_6 = \frac{1}{2}\left[\sqrt{1.25} + \sqrt{1.75} + \sqrt{2.25} + \sqrt{2.75} + \sqrt{3.25} + \sqrt{3.75}\right] \approx 4.669245$$

$$S_6 = \frac{1}{2\cdot 3}\left[\sqrt{1} + 4\sqrt{1.5} + 2\sqrt{2} + 4\sqrt{2.5} + 2\sqrt{3} + 4\sqrt{3.5} + \sqrt{4}\right] \approx 4.666563$$

$E_T \approx \frac{14}{3} - 4.661488 \approx 0.005178$, $E_M \approx \frac{14}{3} - 4.669245 \approx -0.002578$,

$E_S \approx \frac{14}{3} - 4.666563 \approx 0.000104$.

$n = 12$: $\Delta x = (4-1)/12 = \frac{1}{4}$

$$T_{12} = \frac{1}{4\cdot 2}(f(1) + 2[f(1.25) + f(1.5) + \cdots + f(3.5) + f(3.75)] + f(4)) \approx 4.665367$$

$$M_{12} = \frac{1}{4}[f(1.125) + f(1.375) + f(1.625) + \cdots + f(3.875)] \approx 4.667316$$

$$S_{12} = \frac{1}{4\cdot 3}[f(1) + 4\,f(1.25) + 2\,f(1.5) + 4\,f(1.75) + \cdots + 4\,f(3.75) + f(4)] \approx 4.666659$$

$E_T \approx \frac{14}{3} - 4.665367 \approx 0.001300$, $E_M \approx \frac{14}{3} - 4.667316 \approx -0.000649$,

$E_S \approx \frac{14}{3} - 4.666659 \approx 0.000007$.

Note: These errors were computed more precisely and then rounded to six places. That is, they were not computed by comparing the rounded values of T_n, M_n, and S_n with the rounded value of the actual definite integral.

n	T_n	M_n	S_n
6	4.661488	4.669245	4.666563
12	4.665367	4.667316	4.666659

n	E_T	E_M	E_S
6	0.005178	−0.002578	0.000104
12	0.001300	−0.000649	0.000007

Observations:

1. E_T and E_M are opposite in sign and decrease by a factor of about 4 as n is doubled.
2. The Simpson's approximation is much more accurate than the Midpoint and Trapezoidal approximations, and seems to decrease by a factor of about 16 as n is doubled.

29. $$\int_1^{3.2} y\,dx \approx \frac{0.2}{2}[4.9 + 2(5.4) + 2(5.8) + 2(6.2) + 2(6.7) + 2(7.0) + 2(7.3) + 2(7.5) + 2(8.0) + 2(8.2) + 2(8.3) + 8.3] = 15.4$$

31. $\Delta t = 1\text{ min} = \frac{1}{60}\text{ h}$, so distance $= \int_0^{1/6} v(t)dt \approx \frac{1/60}{3}[40 + 4(42) + 2(45) + 4(49) + 2(52) + 4(54) + 2(56) + 4(57) + 2(57) + 4(55) + 56] \approx 8.6\text{ mi}$.

33. $\Delta x = (4-0)/4 = 1$

(a) $T_4 = \frac{1}{2}[f(0) + 2f(1) + 2f(2) + 2f(3) + f(4)] \approx \frac{1}{2}[0 + 2(3) + 2(5) + 2(3) + 1] = 11.5$

(b) $M_4 = 1\cdot[f(.5) + f(1.5) + f(2.5) + f(3.5)] \approx 1 + 4.5 + 4.5 + 2 = 12$

(c) $S_4 = \frac{1}{3}[f(0) + 4f(1) + 2f(2) + 4f(3) + f(4)] \approx \frac{1}{3}[0 + 4(3) + 2(5) + 4(3) + 1] = 11.\overline{6}$

35. Volume $= \pi \int_0^2 \left(\sqrt[3]{1+x^3}\right)^2 dx = \pi \int_0^2 \left(1+x^3\right)^{2/3} dx$. $V \approx \pi \cdot S_{10}$ where $f(x) = \left(1+x^3\right)^{2/3}$ and $\Delta x = (2-0)/10 = \frac{1}{5}$. Therefore

$$V \approx \pi \cdot S_{10} = \pi \tfrac{1}{5 \cdot 3}\left[f(0) + 4f(0.2) + 2f(0.4) + 4f(0.6) + 2f(0.8) + 4f(1) + 2f(1.2) + 4f(1.4) + 2f(1.6) + 4f(1.8) + f(2)\right] \approx 12.325078.$$

37. Since the Trapezoidal and Midpoint approximations on the interval $[a, b]$ are the sums of the Trapezoidal and Midpoint approximations on the subintervals $[x_{i-1}, x_i]$, $i = 1, 2, \ldots, n$, we can focus our attention on one such interval. The condition $f''(x) < 0$ for $a \le x \le b$ means that the graph of f is concave down as in Figure 5. In that figure, T_n is the area of the trapezoid $AQRD$, $\int_a^b f(x)dx$ is the area of the region $AQPRD$, and M_n is the area of the trapezoid $ABCD$, so $T_n < \int_a^b f(x)dx < M_n$. In general, the condition $f'' < 0$ implies that the graph of f on $[a, b]$ lies above the chord joining the points $(a, f(a))$ and $(b, f(b))$. Thus $\int_a^b f(x)dx > T_n$. Since M_n is the area under a tangent to the graph, and since $f'' < 0$ implies that the tangent lies above the graph, we also have $M_n > \int_a^b f(x)dx$. Thus $T_n < \int_a^b f(x)dx < M_n$.

39. $T_n = \frac{1}{2}\Delta x[f(x_0) + 2f(x_1) + \cdots + 2f(x_{n-1}) + f(x_n)]$ and
$M_n = \Delta x[f(\overline{x}_1) + f(\overline{x}_2) + \cdots + f(\overline{x}_{n-1}) + f(\overline{x}_n)]$, where $\overline{x}_i = \frac{1}{2}(x_{i-1} + x_i)$. Now

$$T_{2n} = \tfrac{1}{2}\left(\tfrac{1}{2}\Delta x\right)[f(x_0) + 2f(\overline{x}_1) + 2f(x_1) + 2f(\overline{x}_2) + 2f(x_2) + \cdots + 2f(\overline{x}_{n-1}) + 2f(x_{n-1}) + 2f(\overline{x}_n) + f(x_n)],$$ so

$$\tfrac{1}{2}(T_n + M_n) = \tfrac{1}{4}\Delta x[f(x_0) + 2f(x_1) + \cdots + 2f(x_{n-1}) + f(x_n)] + \tfrac{1}{4}\Delta x[2f(\overline{x}_1) + 2f(\overline{x}_2) + \cdots + 2f(\overline{x}_{n-1}) + 2f(\overline{x}_n)] = T_{2n}.$$

EXERCISES 7.9

1. The area under the graph of $y = 1/x^3 = x^{-3}$ between $x = 1$ and $x = t$ is
$A(t) = \int_1^t x^{-3}\,dx = \left[-\frac{1}{2}x^{-2}\right]_1^t = \frac{1}{2} - 1/(2t^2)$. So the area for $0 \le x \le 10$ is $A(10) = 0.5 - 0.005 = 0.495$, the area for $0 \le x \le 100$ is $A(100) = 0.5 - 0.00005 = 0.49995$, and the area for $0 \le x \le 1000$ is $A(1000) = 0.5 - 0.0000005 = 0.4999995$. The total area under the curve for $x \ge 1$ is
$\int_1^\infty x^{-3}\,dx = \lim\limits_{t\to\infty}\left[-\frac{1}{2}t^{-2} - \left(-\frac{1}{2}\right)\right] = 0 - \left(-\frac{1}{2}\right) = \frac{1}{2}$.

3. $$\int_2^\infty \frac{dx}{\sqrt{x+3}} = \lim_{t\to\infty}\int_2^t \frac{dx}{\sqrt{x+3}} = \lim_{t\to\infty}\left[2\sqrt{x+3}\right]_2^t = \lim_{t\to\infty}\left(2\sqrt{t+3} - 2\sqrt{5}\right) = \infty.$$ Divergent

5. $$\int_{-\infty}^1 \frac{dx}{(2x-3)^2} = \lim_{t\to-\infty}\frac{1}{2}\int_t^1 \frac{2\,dx}{(2x-3)^2} = \lim_{t\to-\infty}\frac{1}{2}\left[-\frac{1}{2x-3}\right]_t^1 = \lim_{t\to-\infty}\left[\frac{1}{2} + \frac{1}{2(2t-3)}\right] = \frac{1}{2}$$

7. $\int_{-\infty}^\infty x\,dx = \int_{-\infty}^0 x\,dx + \int_0^\infty x\,dx$. $\int_{-\infty}^0 x\,dx = \lim\limits_{t\to-\infty}\left[\frac{1}{2}x^2\right]_t^0 = \lim\limits_{t\to-\infty}\left(-\frac{1}{2}t^2\right) = -\infty$. Divergent

9. $\int_0^\infty e^{-x}\,dx = \lim\limits_{t\to\infty}\int_0^t e^{-x}\,dx = \lim\limits_{t\to\infty}\left[-e^{-x}\right]_0^t = \lim\limits_{t\to\infty}\left(-e^{-t} + 1\right) = 1$

11. $\int_{-\infty}^{\infty} xe^{-x^2}\,dx = \int_{-\infty}^{0} xe^{-x^2}\,dx + \int_{0}^{\infty} xe^{-x^2}\,dx$, $\int_{-\infty}^{0} xe^{-x^2}\,dx = \lim_{t\to-\infty} -\frac{1}{2}\left[e^{-x^2}\right]_0^t = \lim_{t\to-\infty} -\frac{1}{2}\left(1 - e^{-t^2}\right) = -\frac{1}{2}$, and $\int_0^\infty xe^{-x^2}\,dx = \lim_{t\to\infty} -\frac{1}{2}\left[e^{-x^2}\right]_t^0 = \lim_{t\to\infty} -\frac{1}{2}\left(e^{-t^2} - 1\right) = \frac{1}{2}$. Therefore $\int_{-\infty}^{\infty} xe^{-x^2}\,dx = -\frac{1}{2} + \frac{1}{2} = 0$.

13. $$\int_0^\infty \frac{dx}{(x+2)(x+3)} = \lim_{t\to\infty}\int_0^t \left[\frac{1}{x+2} - \frac{1}{x+3}\right]dx = \lim_{t\to\infty}\left[\ln\left(\frac{x+2}{x+3}\right)\right]_0^t = \lim_{t\to\infty}\left[\ln\left(\frac{t+2}{t+3}\right) - \ln\frac{2}{3}\right]$$
$$= \ln 1 - \ln\tfrac{2}{3} = -\ln\tfrac{2}{3}$$

15. $\int_0^\infty \cos x\,dx = \lim_{t\to\infty}\left[\sin x\right]_0^t = \lim_{t\to\infty}\sin t$, which does not exist. Divergent

17. $$\int_0^\infty \frac{5\,dx}{2x+3} = \frac{5}{2}\lim_{t\to\infty}\int_0^t \frac{2\,dx}{2x+3} = \tfrac{5}{2}\lim_{t\to\infty}\left[\ln(2x+3)\right]_0^t = \tfrac{5}{2}\lim_{t\to\infty}\left[\ln(2t+3) - \ln 3\right] = \infty.$$ Divergent

19. $$\int_{-\infty}^1 xe^{2x}\,dx = \lim_{t\to-\infty}\int_t^1 xe^{2x}\,dx = \lim_{t\to-\infty}\left[\tfrac{1}{2}xe^{2x} - \tfrac{1}{4}e^{2x}\right]_t^1 \quad \text{(by parts)}$$
$$= \lim_{t\to-\infty}\left[\tfrac{1}{2}e^2 - \tfrac{1}{4}e^2 - \tfrac{1}{2}te^{2t} + \tfrac{1}{4}e^{2t}\right] = \tfrac{1}{4}e^2 - 0 + 0 = \tfrac{1}{4}e^2,$$
since $\lim_{t\to-\infty} te^{2t} = \lim_{t\to-\infty}\frac{t}{e^{-2t}} \overset{\text{H}}{=} \lim_{t\to-\infty}\frac{1}{-2e^{-2t}} = \lim_{t\to-\infty} -\tfrac{1}{2}e^{2t} = 0$.

21. $$\int_1^\infty \frac{\ln x}{x}\,dx = \lim_{t\to\infty}\left[\frac{(\ln x)^2}{2}\right]_1^t = \lim_{t\to\infty}\frac{(\ln t)^2}{2} = \infty.$$ Divergent

23. $$\int_{-\infty}^\infty \frac{x\,dx}{1+x^2} = \int_{-\infty}^0 \frac{x\,dx}{1+x^2} + \int_0^\infty \frac{x\,dx}{1+x^2} \text{ and}$$
$$\int_{-\infty}^0 \frac{x\,dx}{1+x^2} = \lim_{t\to-\infty}\left[\tfrac{1}{2}\ln(1+x^2)\right]_t^0 = \lim_{t\to-\infty}\left[0 - \tfrac{1}{2}\ln(1+t^2)\right] = -\infty.$$ Divergent

25. Integrate by parts with $u = \ln x$, $dv = dx/x^2 \quad\Rightarrow\quad du = dx/x$, $v = -1/x$.
$$\int_1^\infty \frac{\ln x}{x^2}\,dx = \lim_{t\to\infty}\int_1^t \frac{\ln x}{x^2}\,dx = \lim_{t\to\infty}\left[-\frac{\ln x}{x} - \frac{1}{x}\right]_1^t = \lim_{t\to\infty}\left[-\frac{\ln t}{t} - \frac{1}{t} + 0 + 1\right]$$
$$= -0 - 0 + 0 + 1 = 1, \text{ since } \lim_{t\to\infty}\frac{\ln t}{t} \overset{\text{H}}{=} \lim_{t\to\infty}\frac{1/t}{1} = 0.$$

27. $$\int_0^3 \frac{dx}{\sqrt{x}} = \lim_{t\to0^+}\int_t^3 \frac{dx}{\sqrt{x}} = \lim_{t\to0^+}\left[2\sqrt{x}\right]_t^3 = \lim_{t\to0^+}\left(2\sqrt{3} - 2\sqrt{t}\right) = 2\sqrt{3}$$

29. $$\int_{-1}^0 \frac{dx}{x^2} = \lim_{t\to0^-}\int_{-1}^t \frac{dx}{x^2} = \lim_{t\to0^-}\left[\frac{-1}{x}\right]_{-1}^t = \lim_{t\to0^-}\left[-\frac{1}{t} + \frac{1}{-1}\right] = \infty.$$ Divergent

31. $$\int_{-2}^3 \frac{dx}{x^4} = \int_{-2}^0 \frac{dx}{x^4} + \int_0^3 \frac{dx}{x^4} \text{ and } \int_{-2}^0 \frac{dx}{x^4} = \lim_{t\to0^-}\left[-\tfrac{1}{3}x^{-3}\right]_{-2}^t = \lim_{t\to0^-}\left[-\frac{1}{3t^3} - \frac{1}{24}\right] = \infty.$$ Divergent

33. $$\int_4^5 \frac{dx}{(5-x)^{2/5}} = \lim_{t\to5^-}\left[-\tfrac{5}{3}(5-x)^{3/5}\right]_4^t = \lim_{t\to5^-}\left[-\tfrac{5}{3}(5-t)^{3/5} + \tfrac{5}{3}\right] = 0 + \tfrac{5}{3} = \tfrac{5}{3}$$

35. $\int_{\pi/4}^{\pi/2}\tan^2 x\,dx = \lim_{t\to\pi/2^-}\int_{\pi/4}^t(\sec^2 x - 1)dx = \lim_{t\to\pi/2^-}\left[\tan x - x\right]_{\pi/4}^t$
$$= \tfrac{\pi}{4} - 1 + \lim_{t\to\pi/2^-}(\tan t - t) = \infty.$$ Divergent

37. $\int_0^\pi \sec x\,dx = \int_0^{\pi/2}\sec x\,dx + \int_{\pi/2}^\pi \sec x\,dx$. $\int_0^{\pi/2}\sec x\,dx = \lim_{t\to\pi/2^-}\int_0^t \sec x\,dx$
$$= \lim_{t\to\pi/2^-}\left[\ln|\sec x + \tan x|\right]_0^t = \lim_{t\to\pi/2^-}\ln|\sec t + \tan t| = \infty.$$ Divergent

39. $\displaystyle\int_{-2}^{2}\frac{dx}{x^2-1}=\int_{-2}^{-1}\frac{dx}{x^2-1}+\int_{-1}^{0}\frac{dx}{x^2-1}+\int_{0}^{1}\frac{dx}{x^2-1}+\int_{1}^{2}\frac{dx}{x^2-1}$, and

$\displaystyle\int\frac{dx}{x^2-1}=\int\frac{dx}{(x-1)(x+1)}=\frac{1}{2}\ln\left|\frac{x-1}{x+1}\right|+C$, so

$\displaystyle\int_{0}^{1}\frac{dx}{x^2-1}=\lim_{t\to1^-}\left[\frac{1}{2}\ln\left|\frac{x-1}{x+1}\right|\right]_0^t=\lim_{t\to1^-}\frac{1}{2}\ln\left|\frac{t-1}{t+1}\right|=-\infty$. Divergent

41. Integrate by parts with $u=\ln x$, $dv=x\,dx$:

$\int_0^1 x\ln x\,dx=\lim\limits_{t\to0^+}\int_t^1 x\ln x\,dx=\lim\limits_{t\to0^+}\left[\frac{1}{2}x^2\ln x-\frac{1}{4}x^2\right]_t^1=-\frac{1}{4}-\lim\limits_{t\to0^+}\frac{1}{2}t^2\ln t$

$\displaystyle=-\frac{1}{4}-\frac{1}{2}\lim_{t\to0^+}\frac{\ln t}{1/t^2}\overset{\text{H}}{=}-\frac{1}{4}-\frac{1}{2}\lim_{t\to0^+}\frac{1/t}{-2/t^3}=-\frac{1}{4}+\frac{1}{4}\lim_{t\to0^+}t^2=-\frac{1}{4}$

43.

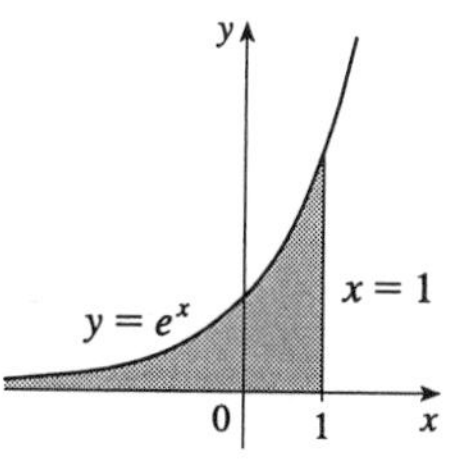

Area $=\int_{-\infty}^{1}e^x\,dx=\lim\limits_{t\to-\infty}\left[e^x\right]_t^1=e-\lim\limits_{t\to-\infty}e^t=e$

45.

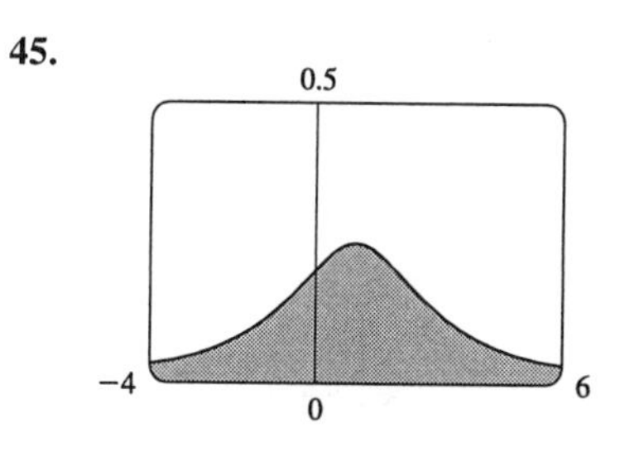

$\displaystyle\text{Area}=\int_{-\infty}^{\infty}\frac{dx}{x^2-2x+5}=\int_{-\infty}^{0}\frac{dx}{(x-1)^2+4}+\int_{0}^{\infty}\frac{dx}{(x-1)^2+4}$

$\displaystyle=\lim_{t\to-\infty}\left[\frac{1}{2}\tan^{-1}\left(\frac{x-1}{2}\right)\right]_t^0+\lim_{t\to\infty}\left[\frac{1}{2}\tan^{-1}\left(\frac{x-1}{2}\right)\right]_0^t$

$=\frac{1}{2}\tan^{-1}\left(-\frac{1}{2}\right)-\frac{1}{2}\left(-\frac{\pi}{2}\right)+\frac{1}{2}\left(\frac{\pi}{2}\right)-\frac{1}{2}\tan^{-1}\left(-\frac{1}{2}\right)=\frac{\pi}{2}$.

47.

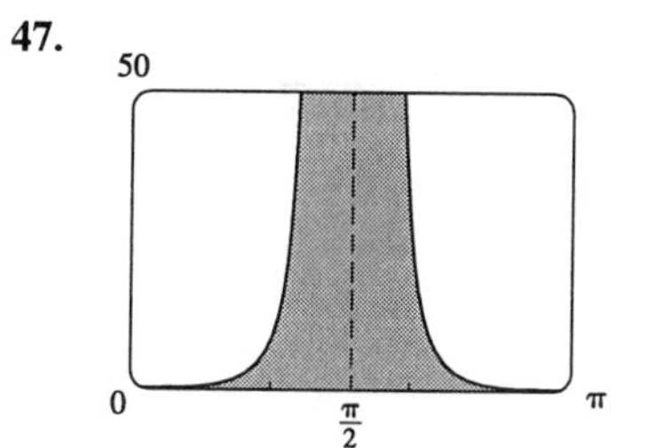

Area $=\int_0^\pi\tan^2x\sec^2x\,dx$

$=\int_0^{\pi/2}\tan^2x\sec^2x\,dx+\int_{\pi/2}^{\pi}\tan^2x\sec^2x\,dx$.

But $\int_0^{\pi/2}\tan^2x\sec^2x\,dx=\lim\limits_{t\to\pi/2^-}\left[\frac{1}{3}\tan^3x\right]_0^t=\infty$,

so the area is infinite.

49. $\displaystyle\frac{\sin^2x}{x^2}\le\frac{1}{x^2}$ on $[1,\infty)$. $\displaystyle\int_1^\infty\frac{dx}{x^2}$ is convergent by Example 4, so $\displaystyle\int_1^\infty\frac{\sin^2x}{x^2}\,dx$ is convergent by the Comparison Theorem.

51. For $x\ge1$, $x+e^{2x}>e^{2x}>0\ \Rightarrow\ \displaystyle\frac{1}{x+e^{2x}}\le\frac{1}{e^{2x}}=e^{-2x}$ on $[1,\infty)$.

$\int_1^\infty e^{-2x}\,dx=\lim\limits_{t\to\infty}\left[-\frac{1}{2}e^{-2x}\right]_1^t=\lim\limits_{t\to\infty}\left[-\frac{1}{2}e^{-2t}+\frac{1}{2}e^{-2}\right]=\frac{1}{2}e^{-2}$. Therefore $\int_1^\infty e^{-2x}\,dx$ is convergent, and by the Comparison Theorem, $\displaystyle\int_1^\infty\frac{dx}{x+e^{2x}}$ is also convergent.

53. $\dfrac{1}{x\sin x} \geq \dfrac{1}{x}$ on $\left(0, \frac{\pi}{2}\right]$ since $0 \leq \sin x \leq 1$. $\displaystyle\int_0^{\pi/2} \frac{dx}{x} = \lim_{t\to 0^+} \int_t^{\pi/2} \frac{dx}{x} = \lim_{t\to 0^+} [\ln x]_t^{\pi/2}$. But $\ln t \to -\infty$ as $t \to 0^+$, so $\displaystyle\int_0^{\pi/2} \frac{dx}{x}$ is divergent, and by the Comparison Theorem, $\displaystyle\int_0^{\pi/2} \frac{dx}{x\sin x}$ is also divergent.

55.
$$\int_0^\infty \frac{dx}{\sqrt{x}(1+x)} = \int_0^1 \frac{dx}{\sqrt{x}(1+x)} + \int_1^\infty \frac{dx}{\sqrt{x}(1+x)} = \lim_{t\to 0^+} \int_t^1 \frac{dx}{\sqrt{x}(1+x)} + \lim_{t\to\infty} \int_1^t \frac{dx}{\sqrt{x}(1+x)}$$
$$= \lim_{t\to 0^+} \int_{\sqrt{t}}^1 \frac{2\,du}{1+u^2} + \lim_{t\to\infty} \int_1^{\sqrt{t}} \frac{2\,du}{1+u^2} \quad [\text{put } u = \sqrt{x},\, x = u^2] = \lim_{t\to 0^+} \left[2\tan^{-1} u\right]_{\sqrt{t}}^1 + \lim_{t\to\infty} \left[2\tan^{-1} u\right]_1^{\sqrt{t}}$$
$$= \lim_{t\to 0^+} \left[2\left(\tfrac{\pi}{4}\right) - 2\tan^{-1}\sqrt{t}\right] + \lim_{t\to\infty} \left[2\tan^{-1}\sqrt{t} - 2\left(\tfrac{\pi}{4}\right)\right] = \tfrac{\pi}{2} - 0 + 2\left(\tfrac{\pi}{2}\right) - \tfrac{\pi}{2} = \pi$$

57. If $p = 1$, then $\displaystyle\int_0^1 \frac{dx}{x^p} = \lim_{t\to 0^+} [\ln x]_t^1 = \infty$. Divergent. If $p \neq 1$, then $\displaystyle\int_0^1 \frac{dx}{x^p} = \lim_{t\to 0^+} \int_t^1 \frac{dx}{x^p}$ (Note that the integral is not improper if $p < 0$) $\displaystyle = \lim_{t\to 0^+} \left[\frac{x^{-p+1}}{-p+1}\right]_t^1 = \lim_{t\to 0^+} \frac{1}{1-p}\left[1 - \frac{1}{t^{p-1}}\right]$. If $p > 1$, then $p - 1 > 0$, so $\dfrac{1}{t^{p-1}} \to \infty$ as $t \to 0^+$, and the integral diverges.

Finally, if $p < 1$, then $\displaystyle\int_0^1 \frac{dx}{x^p} = \frac{1}{1-p}\left[\lim_{t\to 0^+} \left(1 - t^{1-p}\right)\right] = \frac{1}{1-p}$.

Thus the integral converges if and only if $p < 1$, and in that case its value is $\dfrac{1}{1-p}$.

59. First suppose $p = -1$. Then

$\displaystyle\int_0^1 x^p \ln x\,dx = \int_0^1 \frac{\ln x}{x}\,dx = \lim_{t\to 0^+} \int_t^1 \frac{\ln x}{x}\,dx = \lim_{t\to 0^+} \left[\tfrac{1}{2}(\ln x)^2\right]_t^1 = -\tfrac{1}{2} \lim_{t\to 0^+} (\ln t)^2 = -\infty$, so the integral diverges. Now suppose $p \neq -1$. Then integration by parts gives

$\displaystyle\int x^p \ln x\,dx = \frac{x^{p+1}}{p+1}\ln x - \int \frac{x^p}{p+1}\,dx = \frac{x^{p+1}}{p+1}\ln x - \frac{x^{p+1}}{(p+1)^2} + C$. If $p < -1$, then $p + 1 < 0$, so

$\displaystyle\int_0^1 x^p \ln x\,dx = \lim_{t\to 0^+} \left[\frac{x^{p+1}}{p+1}\ln x - \frac{x^{p+1}}{(p+1)^2}\right]_t^1 = \frac{-1}{(p+1)^2} - \left(\frac{1}{p+1}\right) \lim_{t\to 0^+} \left[t^{p+1}\left(\ln t - \frac{1}{p+1}\right)\right] = \infty$. If $p > -1$, then $p + 1 > 0$ and

$$\int_0^1 x^p \ln x\,dx = \frac{-1}{(p+1)^2} - \left(\frac{1}{p+1}\right) \lim_{t\to 0^+} \frac{\ln t - 1/(p+1)}{t^{-(p+1)}}$$
$$\overset{\text{H}}{=} \frac{-1}{(p+1)^2} - \left(\frac{1}{p+1}\right) \lim_{t\to 0^+} \frac{1/t}{-(p+1)t^{-(p+2)}} = \frac{-1}{(p+1)^2} + \frac{1}{(p+1)^2} \lim_{t\to 0^+} t^{p+1} = \frac{-1}{(p+1)^2}.$$

Thus the integral converges to $-\dfrac{1}{(p+1)^2}$ if $p > -1$ and diverges otherwise.

61. **(a)** $\int_{-\infty}^\infty x\,dx = \int_{-\infty}^0 x\,dx + \int_0^\infty x\,dx$, and $\int_0^\infty x\,dx = \lim_{t\to\infty} \int_0^t x\,dx = \lim_{t\to\infty} \left[\tfrac{1}{2}t^2 - \tfrac{1}{2}(0^2)\right] = \infty$, so the integral is divergent.

(b) $\int_{-t}^t x\,dx = \left[\tfrac{1}{2}x^2\right]_{-t}^t = \tfrac{1}{2}t^2 - \tfrac{1}{2}t^2 = 0$, so $\lim_{t\to\infty} \int_{-t}^t x\,dx = 0$. Therefore $\int_{-\infty}^\infty x\,dx \neq \lim_{t\to\infty} \int_{-t}^t x\,dx$.

63. Volume $\displaystyle = \int_1^\infty \pi\left(\frac{1}{x}\right)^2 dx = \pi \lim_{t\to\infty} \int_1^t \frac{dx}{x^2} = \pi \lim_{t\to\infty} \left[-\frac{1}{x}\right]_1^t = \pi \lim_{t\to\infty} \left(1 - \frac{1}{t}\right) = \pi < \infty.$

65. Work $= \int_R^\infty F\,dr = \lim_{t\to\infty}\int_R^t \frac{GmM}{r^2}\,dr = \lim_{t\to\infty} GmM\left(\frac{1}{R}-\frac{1}{t}\right) = \frac{GmM}{R}$. The initial kinetic energy provides the work, so $\frac{1}{2}mv_0^2 = \frac{GmM}{R} \quad\Rightarrow\quad v_0 = \sqrt{\frac{2GM}{R}}$.

67. **(a)** $F(s) = \int_0^\infty f(t)e^{-st}\,dt = \int_0^\infty e^{-st}\,dt = \lim_{n\to\infty}\left[-\frac{e^{-st}}{s}\right]_0^n = \lim_{n\to\infty}\left(\frac{e^{-sn}}{-s}+\frac{1}{s}\right)$. This converges to $\frac{1}{s}$ only if $s>0$. Therefore $F(s)=\frac{1}{s}$ with domain $\{s \mid s>0\}$.

(b) $F(s) = \int_0^\infty f(t)e^{-st}\,dt = \int_0^\infty e^t e^{-st}\,dt = \lim_{n\to\infty}\int_0^n e^{t(1-s)}\,dt$

$$= \lim_{n\to\infty}\left[\frac{1}{1-s}e^{t(1-s)}\right]_0^n = \lim_{n\to\infty}\left(\frac{e^{(1-s)n}}{1-s}-\frac{1}{1-s}\right).$$

This converges only if $1-s<0 \quad\Rightarrow\quad s>1$, in which case $F(s)=\frac{1}{s-1}$ with domain $\{s \mid s>1\}$.

(c) $F(s) = \int_0^\infty f(t)e^{-st}\,dt = \lim_{n\to\infty}\int_0^n te^{-st}\,dt$. Use integration by parts: let $u=t$, $dv=e^{-st}\,dt \quad\Rightarrow$ $du=dt$, $v=-\frac{e^{-st}}{s}$. Then $F(s)=\lim_{n\to\infty}\left[-\frac{t}{s}e^{-st}-\frac{1}{s^2}e^{-st}\right]_0^n = \lim_{n\to\infty}\left(\frac{-n}{se^{sn}}-\frac{1}{s^2e^{sn}}+0+\frac{1}{s^2}\right)=\frac{1}{s^2}$ only if $s>0$. Therefore $F(s)=\frac{1}{s^2}$ and the domain of F is $\{s \mid s>0\}$.

69. $G(s)=\int_0^\infty f'(t)e^{-st}\,dt$. Integrate by parts with $u=e^{-st}$, $dv=f'(t)dt \quad\Rightarrow\quad du=-se^{-st}$, $v=f(t)$:
$G(s)=\lim_{n\to\infty}\left[f(t)e^{-st}\right]_0^n + s\int_0^\infty f(t)e^{-st}\,dt = \lim_{n\to\infty} f(n)e^{-sn}-f(0)+sF(s)$.
But $0\le f(t)\le Me^{at} \quad\Rightarrow\quad 0\le f(t)e^{-st}\le Me^{at}e^{-st}$ and $\lim_{t\to\infty} Me^{t(a-s)}=0$ for $s>a$. So by the Squeeze Theorem, $\lim_{t\to\infty} f(t)e^{-st}=0$ for $s>a \quad\Rightarrow\quad G(s)=0-f(0)+sF(s)=sF(s)-f(0)$ for $s>a$.

71. Use integration by parts: let $u=x$, $dv=xe^{-x^2}\,dx \quad\Rightarrow\quad du=dx$, $v=-\frac{1}{2}e^{-x^2}$. So
$\int_0^\infty x^2e^{-x^2}\,dx = \lim_{t\to\infty}\left[-\frac{x}{2}e^{-x^2}\right]_0^t + \frac{1}{2}\int_0^\infty e^{-x^2}\,dx = \lim_{t\to\infty} -\frac{t}{2e^{t^2}}+\frac{1}{2}\int_0^\infty e^{-x^2}\,dx = \frac{1}{2}\int_0^\infty e^{-x^2}\,dx$.
(The limit is 0 by l'Hospital's Rule.)

73. For the first part of the integral, let $x=2\tan\theta \quad\Rightarrow\quad dx=2\sec^2\theta\,d\theta$.
$\int\frac{1}{\sqrt{x^2+4}}\,dx=\int\sec\theta=\ln|\sec\theta+\tan\theta|$. But $\tan\theta=\frac{1}{2}x$, and
$\sec\theta=\sqrt{1+\tan^2\theta}=\sqrt{1+\frac{1}{4}x^2}=\frac{1}{2}\sqrt{x^2+4}$. So

$$\int_0^\infty\left(\frac{1}{\sqrt{x^2+4}}-\frac{C}{x+2}\right)dx=\lim_{t\to\infty}\left[\ln\left|\frac{\sqrt{x^2+4}}{2}+\frac{x}{2}\right|-C\ln|x+2|\right]_0^t$$

$$=\lim_{t\to\infty}\ln\left(\frac{\sqrt{t^2+4}+t}{2(t+2)^C}\right)-(\ln 1-C\ln 2)=\ln\left(\lim_{t\to\infty}\frac{t+\sqrt{t^2+4}}{(t+2)^C}\right)+\ln 2^{C-1}.$$

By l'Hospital's Rule, $\lim_{t\to\infty}\frac{t+\sqrt{t^2+4}}{(t+2)^C}=\lim_{t\to\infty}\frac{1+t/\sqrt{t^2+4}}{C(t+2)^{C-1}}=\frac{2}{C\lim_{t\to\infty}(t+2)^{C-1}}$.

If $C<1$, we get ∞ and the interval diverges. If $C=1$, we get 2, so the original integral converges to $\ln 2+\ln 2^0=\ln 2$. If $C>1$, we get 0, so the original integral diverges to $-\infty$.

75. We integrate by parts with $u = \dfrac{1}{\ln(1+x+t)}$, $dv = \sin t\,dt$, so $du = \dfrac{-1}{(1+x+t)[\ln(1+x+t)]^2}$ and $v = -\cos t$. The integral becomes

$$I = \int_0^\infty \frac{\sin t\,dt}{\ln(1+x+t)} = \lim_{b\to\infty}\left(\frac{-\cos t}{\ln(1+x+t)}\bigg|_0^b - \int_0^b \frac{\cos t\,dt}{(1+x+t)[\ln(1+x+t)]^2}\right)$$
$$= \lim_{b\to\infty}\frac{-\cos b}{\ln(1+x+b)} + \frac{1}{\ln(1+x)} + \int_0^\infty \frac{-\cos t\,dt}{(1+x+t)[\ln(1+x+t)]^2}$$
$$= \frac{1}{\ln(1+x)} + J, \text{ where } J = \int_0^\infty \frac{-\cos t\,dt}{(1+x+t)[\ln(1+x+t)]^2}.$$

Now $-1 \le -\cos t \le 1$ for all t; in fact, the inequality is strict except at isolated points. So

$$-\int_0^\infty \frac{dt}{(1+x+t)[\ln(1+x+t)]^2} < J < \int_0^\infty \frac{dt}{(1+x+t)[\ln(1+x+t)]^2} \quad\Leftrightarrow$$
$$-\frac{1}{\ln(1+x)} < J < \frac{1}{\ln(1+x)} \quad\Leftrightarrow\quad 0 < I < \frac{2}{\ln(1+x)}.$$

REVIEW EXERCISES FOR CHAPTER 7

1. False. Since the numerator has a higher degree than the denominator,

$$\frac{x(x^2+4)}{x^2-4} = x + \frac{8x}{x^2-4} = x + \frac{A}{x+2} + \frac{B}{x-2}.$$

3. False. It can be put in the form $\dfrac{A}{x} + \dfrac{B}{x^2} + \dfrac{C}{x-4}$.

5. False. This is an improper integral, since the denominator vanishes at $x = 1$.

$$\int_0^4 \frac{x}{x^2-1}\,dx = \int_0^1 \frac{x}{x^2-1}\,dx + \int_1^4 \frac{x}{x^2-1}\,dx \text{ and } \int_0^1 \frac{x}{x^2-1} = \lim_{t\to1^-}\int_0^t \frac{x}{x^2-1}\,dx$$
$$= \lim_{t\to1^-}\left[\tfrac{1}{2}\ln|x^2-1|\right]_0^t = \lim_{t\to1^-}\tfrac{1}{2}\ln|t^2-1| = \infty. \text{ So the integral diverges.}$$

7. False. See Exercise 61 in Section 7.9.

9. $\displaystyle\int \frac{x-1}{x+1}\,dx = \int\left[1 - \frac{2}{x+1}\right]dx = x - 2\ln|x+1| + C$

11. Let $u = \arctan x$. Then $du = dx/(1+x^2)$, so $\displaystyle\int \frac{(\arctan x)^5}{1+x^2}\,dx = \int u^5\,du = \tfrac{1}{6}u^6 + C = \tfrac{1}{6}(\arctan x)^6 + C.$

13. Let $u = \sin x$. Then $\displaystyle\int \frac{\cos x\,dx}{e^{\sin x}} = \int e^{-u}\,du = -e^{-u} + C = -\frac{1}{e^{\sin x}} + C.$

15. Use integration by parts with $u = \ln x$, $dv = x^4\,dx \;\Rightarrow\; du = dx/x$, $v = x^5/5$:

$\int x^4 \ln x\,dx = \frac{1}{5}x^5\ln x - \frac{1}{5}\int x^4\,dx = \frac{1}{5}x^5\ln x - \frac{1}{25}x^5 + C = \frac{1}{25}x^5(5\ln x - 1) + C.$

17. Let $u = x^2$. Then $du = 2x\,dx$, so $\int x\sin(x^2)dx = \frac{1}{2}\int \sin u\,du = -\frac{1}{2}\cos u + C = -\frac{1}{2}\cos(x^2) + C$.

19. $\displaystyle\int \frac{dx}{2x^2 - 5x + 2} = \int \left[\frac{-2/3}{2x-1} + \frac{1/3}{x-2}\right]dx = -\tfrac{1}{3}\ln|2x-1| + \tfrac{1}{3}\ln|x-2| + C = \tfrac{1}{3}\ln\left|\frac{x-2}{2x-1}\right| + C.$

21. Let $u = \sec x$. Then $du = \sec x\tan x\,dx$, so $\int \tan^7 x\sec^3 x\,dx = \int \tan^6 x\sec^2 x\sec x\tan x\,dx$
$= \int (u^2-1)^3 u^2\,du = \int (u^8 - 3u^6 + 3u^4 - u^2)du = \frac{1}{9}u^9 - \frac{3}{7}u^7 + \frac{3}{5}u^5 - \frac{1}{3}u^3 + C$
$= \frac{1}{9}\sec^9 x - \frac{3}{7}\sec^7 x + \frac{3}{5}\sec^5 x - \frac{1}{3}\sec^3 x + C.$

23. Let $u = \sqrt{1+2x}$. Then $x = \frac{1}{2}(u^2-1)$, $dx = u\,du$, so $\displaystyle\int \frac{dx}{\sqrt{1+2x}+3} = \int \frac{u\,du}{u+3}$
$\displaystyle= \int \left[1 - \frac{3}{u+3}\right]du = u - 3\ln|u+3| + C = \sqrt{1+2x} - 3\ln\left(\sqrt{1+2x}+3\right) + C.$

25. $\displaystyle u = \sqrt{x} \quad\Rightarrow\quad du = \frac{dx}{2\sqrt{x}} \quad\Rightarrow\quad \int \frac{e^{\sqrt{x}}\,dx}{\sqrt{x}} = 2\int e^u\,du = 2e^u + C = 2e^{\sqrt{x}} + C$

27. Let $x = \sec\theta$. Then

$$\int \frac{dx}{(x^2-1)^{3/2}} = \int \frac{\sec\theta\tan\theta}{\tan^3\theta}\,d\theta = \int \frac{\sec\theta}{\tan^2\theta}\,d\theta = \int \frac{\cos\theta\,d\theta}{\sin^2\theta} = -\frac{1}{\sin\theta} + C = -\frac{x}{\sqrt{x^2-1}} + C.$$

29. $\displaystyle\int \frac{dx}{x^3+x} = \int \left(\frac{1}{x} - \frac{x}{x^2+1}\right)dx = \ln|x| - \tfrac{1}{2}\ln(x^2+1) + C$

31. $\int \cot^2 x\,dx = \int (\csc^2 x - 1)dx = -\cot x - x + C$

33. $\displaystyle\int \frac{2x^2+3x+11}{x^3+x^2+3x-5}\,dx = \int \left(\frac{2}{x-1} - \frac{1}{x^2+2x+5}\right)dx = 2\ln|x-1| - \int \frac{dx}{(x+1)^2+4}$

$$= 2\ln|x-1| - \frac{1}{2}\tan^{-1}\left(\frac{x+1}{2}\right) + C$$

35. Let $u = \cot 4x$. Then $du = -4\csc^2 4x\,dx \quad\Rightarrow\quad \displaystyle\int \csc^4 4x\,dx = \int (\cot^2 4x + 1)\csc^2 4x\,dx$
$= \int (u^2+1)\left(-\frac{1}{4}\,du\right) = -\frac{1}{4}\left(\frac{1}{3}u^3 + u\right) + C = -\frac{1}{12}(\cot^3 4x + 3\cot 4x) + C.$

37. Let $u = \ln x$. Then $\displaystyle\int \frac{\ln(\ln x)}{x}\,dx = \int \ln u\,du$. Now use parts with $w = \ln u$, $dv = du \quad\Rightarrow\quad dw = du/u$,
$v = u \quad\Rightarrow\quad \int \ln u\,du = u\ln u - u + C = (\ln x)[\ln(\ln x) - 1] + C.$

39. Let $u = 2x+1$. Then $du = 2\,dx \quad\Rightarrow$

$$\int \frac{dx}{\sqrt{4x^2+4x+5}} = \int \frac{(1/2)du}{\sqrt{u^2+4}} = \frac{1}{2}\int \frac{2\sec^2\theta\,d\theta}{2\sec\theta} \quad (\text{put } u = 2\tan\theta,\ du = 2\sec^2\theta\,d\theta)$$

$$= \frac{1}{2}\int \sec\theta\,d\theta = \tfrac{1}{2}\ln|\sec\theta + \tan\theta| + C_1 = \frac{1}{2}\ln\left|\frac{\sqrt{u^2+4}}{2} + \frac{u}{2}\right| + C_1$$

$$= \tfrac{1}{2}\ln\left(u + \sqrt{u^2+4}\right) + C = \tfrac{1}{2}\ln\left(2x+1+\sqrt{4x^2+4x+5}\right) + C.$$

41. $\int (\cos x + \sin x)^2\cos 2x\,dx = \int (\cos^2 x + 2\sin x\cos x + \sin^2 x)\cos 2x\,dx$
$= \int (1 + \sin 2x)\cos 2x\,dx = \int \cos 2x\,dx + \frac{1}{2}\int \sin 4x\,dx = \frac{1}{2}\sin 2x - \frac{1}{8}\cos 4x + C.$

Or: $\int (\cos x + \sin x)^2\cos 2x\,dx = \int (\cos x + \sin x)^2(\cos^2 x - \sin^2 x)dx$
$= \int (\cos x + \sin x)^3(\cos x - \sin x)dx = \frac{1}{4}(\cos x + \sin x)^4 + C_2.$

43. $\int_0^{\pi/2} \cos^3 x \sin 2x\, dx = \int_0^{\pi/2} 2\cos^4 x \sin x\, dx = \left[-\frac{2}{5}\cos^5 x\right]_0^{\pi/2} = \frac{2}{5}$

45. $\displaystyle\int_0^3 \frac{dx}{x^2-x-2} = \int_0^3 \frac{dx}{(x+1)(x-2)} = \int_0^2 \frac{dx}{(x+1)(x-2)} + \int_2^3 \frac{dx}{(x+1)(x-2)}$, and

$$\int_2^3 \frac{dx}{x^2-x-2} = \lim_{t\to 2^+}\int_t^3\left[\frac{-1/3}{x+1}+\frac{1/3}{x-2}\right]dx = \lim_{t\to 2^+}\left[\frac{1}{3}\ln\left|\frac{x-2}{x+1}\right|\right]_t^3 = \lim_{t\to 2^+}\left[\frac{1}{3}\ln\frac{1}{4} - \frac{1}{3}\ln\left|\frac{t-2}{t+1}\right|\right] = \infty.$$

Divergent

47. $\displaystyle\int_0^1 \frac{t^2-1}{t^2+1}\,dt = \int_0^1\left[1-\frac{2}{t^2+1}\right]dt = \left[t-2\tan^{-1}t\right]_0^1 = \left(1-2\cdot\frac{\pi}{4}\right)-0 = 1-\frac{\pi}{2}$

49. $\displaystyle\int_0^\infty \frac{dx}{(x+2)^4} = \lim_{t\to\infty}\left[\frac{-1}{3(x+2)^3}\right]_0^t = \lim_{t\to\infty}\left[\frac{1}{3\cdot 2^3}-\frac{1}{3(t+2)^3}\right] = \frac{1}{24}$

51. Let $u = \ln x$. Then

$$\int_1^e \frac{dx}{x\sqrt{\ln x}} = \lim_{t\to 1^+}\int_t^e \frac{dx}{x\sqrt{\ln x}} = \lim_{t\to 1^+}\int_{\ln t}^1 \frac{du}{\sqrt{u}} = \lim_{t\to 1^+}\left[2\sqrt{u}\right]_{\ln t}^1 = \lim_{t\to 1^+}\left(2-2\sqrt{\ln t}\right) = 2.$$

53. Let $u = \sqrt{x}+2$. Then $x = (u-2)^2$, $dx = 2(u-2)du$, so

$$\int_1^4 \frac{\sqrt{x}\,dx}{\sqrt{x}+2} = \int_3^4 \frac{2(u-2)^2\,du}{u} = \int_3^4\left[2u-8+\frac{8}{u}\right]du = \left[u^2-8u+8\ln u\right]_3^4$$

$$= (16-32+8\ln 4)-(9-24+8\ln 3) = -1+8\ln 4-8\ln 3 = 8\ln\tfrac{4}{3}-1.$$

55. Let $u = 2x+1$. Then $\displaystyle\int_{-\infty}^\infty \frac{dx}{4x^2+4x+5} = \int_{-\infty}^\infty \frac{\frac{1}{2}du}{u^2+4} = \frac{1}{2}\int_{-\infty}^0 \frac{du}{u^2+4} + \frac{1}{2}\int_0^\infty \frac{du}{u^2+4}$

$= \frac{1}{2}\lim\limits_{t\to-\infty}\left[\frac{1}{2}\tan^{-1}\left(\frac{1}{2}u\right)\right]_t^0 + \frac{1}{2}\lim\limits_{t\to\infty}\left[\frac{1}{2}\tan^{-1}\left(\frac{1}{2}u\right)\right]_0^t = \frac{1}{4}\left[0-\left(-\frac{\pi}{2}\right)\right]+\frac{1}{4}\left[\frac{\pi}{2}-0\right] = \frac{\pi}{4}.$

57. Let $x = \sec\theta$. Then $\displaystyle\int_1^2 \frac{\sqrt{x^2-1}}{x}\,dx = \int_0^{\pi/3}\frac{\tan\theta}{\sec\theta}\sec\theta\tan\theta\,d\theta = \int_0^{\pi/3}\tan^2\theta\,d\theta$

$= \int_0^{\pi/3}(\sec^2\theta-1)d\theta = \left[\tan\theta-\theta\right]_0^{\pi/3} = \sqrt{3}-\frac{\pi}{3}.$

59. $\int_0^\infty e^{ax}\cos bx\,dx = \lim\limits_{t\to\infty}\int_0^t e^{ax}\cos bx\,dx$. Integrate by parts twice:

$\displaystyle\int e^{ax}\cos bx\,dx = \frac{1}{b}e^{ax}\sin bx - \frac{a}{b}\int e^{ax}\sin bx\,dx = \frac{1}{b}e^{ax}\sin bx + \frac{a}{b^2}e^{ax}\cos bx - \frac{a^2}{b^2}\int e^{ax}\cos bx\,dx$, so

$\displaystyle\left(1+\frac{a^2}{b^2}\right)\int e^{ax}\cos bx\,dx = \frac{1}{b}e^{ax}\sin bx + \frac{a}{b^2}e^{ax}\cos bx + C_1$. Thus

$\displaystyle\int e^{ax}\cos bx\,dx = \frac{e^{ax}}{a^2+b^2}(b\sin bx + a\cos bx) + C$. Now

$$\int_0^\infty e^{ax}\cos bx\,dx = \lim_{t\to\infty}\left[\frac{e^{ax}}{a^2+b^2}(a\cos bx + b\sin bx)\right]_0^t = \lim_{t\to\infty}\frac{e^{at}}{a^2+b^2}(a\cos bt + b\sin bt) - \frac{a}{a^2+b^2}.$$

If $a \geq 0$, the limit does not exist and the integral is divergent. If $a < 0$, the limit is 0 (since $|e^{at}\cos bt| \leq e^{at}$ and $|e^{at}\sin bt| \leq e^{at}$), so the integral converges to $-a/(a^2+b^2)$.

61. We first make the substitution $t = x + 1$, so $\ln(x^2 + 2x + 2) = \ln\left[(x+1)^2 + 1\right] = \ln(t^2 + 1)$. Then we use parts with $u = \ln(t^2 + 1)$, $dv = dt$:

$$\begin{aligned}\int \ln(t^2+1)\,dt &= t\ln(t^2+1) - \int \frac{t(2t)dt}{t^2+1} = t\ln(t^2+1) - 2\int \frac{t^2\,dt}{t^2+1} \\ &= t\ln(t^2+1) - 2\int\left(1 - \frac{1}{t^2+1}\right)dt = t\ln(t^2+1) - 2t + 2\arctan t + C \\ &= (x+1)\ln(x^2+2x+2) - 2x + 2\arctan(x+1) + K.\end{aligned}$$

[Alternately, we could have integrated by parts immediately with $u = \ln(x^2 + 2x + 2)$]
Notice from the graph that $f = 0$ where F has a horizontal tangent. Also, F is always increasing, and $f \geq 0$.

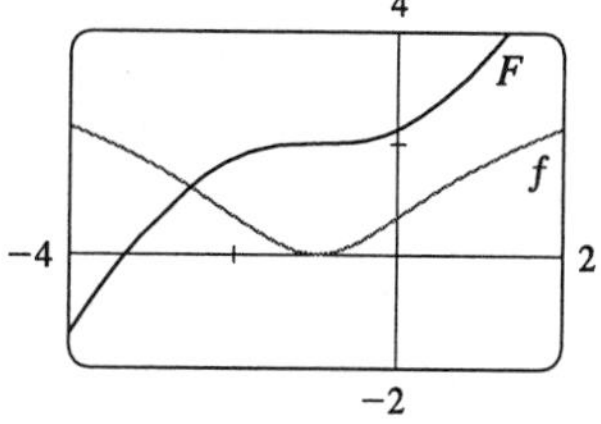

63. Let $u = e^x$. Then $du = e^x\,dx$, so $\int e^x\sqrt{1 - e^{2x}}\,dx = \int \sqrt{1-u^2}\,du = \frac{1}{2}u\sqrt{1-u^2} + \frac{1}{2}\sin^{-1}u + C$

(Formula 30) $= \frac{1}{2}\left[e^x\sqrt{1-e^{2x}} + \sin^{-1}(e^x)\right] + C.$

65. Let $u = x + \frac{1}{2}$. Then $du = dx$, so $\displaystyle\int \sqrt{x^2+x+1}\,dx = \int \sqrt{\left(x+\tfrac{1}{2}\right)^2 + \tfrac{3}{4}}\,dx$

$$= \int \sqrt{u^2 + \left(\tfrac{\sqrt{3}}{2}\right)^2}\,du = \tfrac{1}{2}u\sqrt{u^2+\tfrac{3}{4}} + \tfrac{3}{8}\ln\left|u + \sqrt{u^2+\tfrac{3}{4}}\right| + C \quad \text{(Formula 21)}$$

$$= \frac{2x+1}{4}\sqrt{x^2+x+1} + \tfrac{3}{8}\ln\left|x + \tfrac{1}{2} + \sqrt{x^2+x+1}\right| + C.$$

67. **(a)** $$\frac{d}{du}\left[-\frac{1}{u}\sqrt{a^2-u^2} - \sin^{-1}\left(\frac{u}{a}\right) + C\right] = \frac{1}{u^2}\sqrt{a^2-u^2} + \frac{1}{\sqrt{a^2-u^2}} - \frac{1}{\sqrt{1-u^2/a^2}}\cdot\frac{1}{a}$$

$$= \left(a^2-u^2\right)^{-1/2}\left[\frac{1}{u^2}\left(a^2-u^2\right) + 1 - 1\right] = \frac{\sqrt{a^2-u^2}}{u^2}.$$

(b) Let $u = a\sin\theta \;\Rightarrow\; du = a\cos\theta\,d\theta,\ a^2 - u^2 = a^2(1 - \sin^2\theta) = a^2\cos^2\theta.$

$$\begin{aligned}\int \frac{\sqrt{a^2-u^2}}{u^2}\,du &= \int \frac{a^2\cos^2\theta}{a^2\sin^2\theta}\,d\theta = \int \frac{1-\sin^2\theta}{\sin^2\theta}\,d\theta \\ &= \int\left(\csc^2\theta - 1\right)d\theta = -\cot\theta - \theta + C = -\frac{\sqrt{a^2-u^2}}{u} - \sin^{-1}\left(\frac{u}{a}\right) + C.\end{aligned}$$

69. $f(x) = \sqrt{1+x^4}$, $\Delta x = \dfrac{b-a}{n} = \dfrac{1-0}{10} = \dfrac{1}{10}$

(a) $T_{10} = \frac{0.1}{2}[f(0) + 2f(0.1) + 2f(0.2) + \cdots + 2f(0.8) + 2f(0.9) + f(1)] \approx 1.090608$

(b) $M_{10} = 0.1\left[f\left(\frac{1}{20}\right) + f\left(\frac{3}{20}\right) + f\left(\frac{5}{20}\right) + \cdots + f\left(\frac{19}{20}\right)\right] \approx 1.088840$

(c) $$\begin{aligned}S_{10} = \tfrac{0.1}{3}\big[&f(0) + 4f(0.1) + 2f(0.2) + 4f(0.3) + 2f(0.4) + 4f(0.5) \\ &+ 2f(0.6) + 4f(0.7) + 2f(0.8) + 4f(0.9) + f(1)\big] \approx 1.089429\end{aligned}$$

71. $f(x) = (1+x^4)^{1/2}$, $f'(x) = \frac{1}{2}(1+x^4)^{-1/2}(4x^3) = 2x^3(1+x^4)^{-1/2}$, $f''(x) = (2x^6+6x^2)(1+x^4)^{-3/2}$. Thus $|f''(x)| \le 8 \cdot 1^{-3/2} = 8$ on $[0,1]$. By taking $K = 8$, we find that the error in Exercise 69(a) is bounded by $K\dfrac{(b-a)^3}{12n^2} = \dfrac{8}{1200} = \dfrac{1}{150} < 0.0067$, and in (b) by $K\dfrac{(b-a)^3}{24n^2} = \dfrac{1}{300} = 0.00\overline{3}$.

73. **(a)** $f(x) = \sin(\sin x)$. A CAS gives

$$f^{(4)}(x) = \sin(\sin x)\cos^4 x + 6\cos(\sin x)\cos^2 x \sin x + 3\sin(\sin x) + \sin(\sin x)\cos^2 x + \cos(\sin x)\sin x.$$

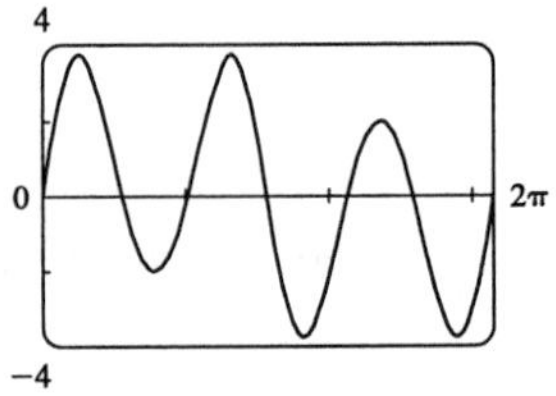

From the graph, we see that $f^{(4)}(x) < 3.8$ for $x \in [0, \pi]$.

(b) We use Simpson's Rule with $f(x) = \sin(\sin x)$ and $\Delta x = \frac{\pi}{10}$:

$\int_0^\pi f(x)dx \approx \frac{\pi}{10 \cdot 3}\left[f(0) + 4f\left(\frac{\pi}{10}\right) + 2f\left(\frac{2\pi}{10}\right) + \cdots + 2f\left(\frac{8\pi}{10}\right) + 4f\left(\frac{9\pi}{10}\right) + f(\pi)\right] \approx 1.7867$. From part (a), we know that $f^{(4)}(x) < 3.8$ on $[0,\pi]$, so we use Theorem 7.8.7 with $K = 3.8$, and estimate the error as $|E_S| \le 3.8(\pi - 0)^5/\left[180(10)^4\right] \approx 0.000646$.

(c) If we want the error to be less than 0.00001, we must have $|E_S| \le 3.8\pi^5/(180n^4) \le 0.00001$, so $n^4 \ge 3.8\pi^5/[180(0.00001)] \approx 646{,}041.5 \quad \Rightarrow \quad n \ge 28.35$. Since n must be even for Simpson's Rule, we must have $n \ge 30$ to ensure the desired accuracy.

75. $\dfrac{x^3}{x^5+2} \le \dfrac{x^3}{x^5} = \dfrac{1}{x^2}$ for x in $[1,\infty)$. $\displaystyle\int_1^\infty \frac{1}{x^2}\,dx$ is convergent by (7.9.2) with $p = 2 > 1$. Thus $\displaystyle\int_1^\infty \frac{x^3}{x^5+2}\,dx$ is convergent by the Comparison Theorem.

77. For x in $\left[0, \frac{\pi}{2}\right]$, $0 \le \cos^2 x \le \cos x$. For x in $\left[\frac{\pi}{2}, \pi\right]$, $\cos x \le 0 \le \cos^2 x$. Thus

area $= \int_0^{\pi/2}(\cos x - \cos^2 x)dx + \int_{\pi/2}^{\pi}(\cos^2 x - \cos x)dx$

$= \left[\sin x - \frac{1}{2}x - \frac{1}{4}\sin 2x\right]_0^{\pi/2} + \left[\frac{1}{2}x + \frac{1}{4}\sin 2x - \sin x\right]_{\pi/2}^{\pi} = \left[\left(1 - \frac{\pi}{4}\right) - 0\right] + \left[\frac{\pi}{2} - \left(\frac{\pi}{4} - 1\right)\right] = 2.$

79. Using the formula for disks, the volume is

$V = \int_0^{\pi/2} \pi[f(x)]^2\,dx = \pi\int_0^{\pi/2}(\cos^2 x)^2\,dx = \pi\int_0^{\pi/2}\left[\frac{1}{2}(1+\cos 2x)\right]^2 dx = \frac{\pi}{4}\int_0^{\pi/2}(1 + \cos^2 2x + 2\cos 2x)dx$

$= \frac{\pi}{4}\int_0^{\pi/2}\left[1 + \frac{1}{2}(1+\cos 4x) + 2\cos 2x\right]dx = \frac{\pi}{4}\left[\frac{3}{2}x + \frac{1}{2}\left(\frac{1}{4}\sin 4x\right) + 2\left(\frac{1}{2}\sin 2x\right)\right]_0^{\pi/2}$

$= \frac{\pi}{4}\left[\left(\frac{3\pi}{4} + \frac{1}{8}\cdot 0 + 0\right) - 0\right] = \frac{3\pi^2}{16}.$

81. For $n \ge 0$, $\int_0^\infty x^n\,dx = \lim\limits_{t\to\infty}\left[x^{n+1}/(n+1)\right]_0^t = \infty$. For $n < 0$, $\int_0^\infty x^n\,dx = \int_0^1 x^n\,dx + \int_1^\infty x^n\,dx$. Both integrals are improper. By (7.9.2), the second integral diverges if $-1 \le n < 0$. By Exercise 7.9.57, the first integral diverges if $n \le -1$. Thus $\int_0^\infty x^n\,dx$ is divergent for all values of n.

83. By the Fundamental Theorem of Calculus,

$\int_0^\infty f'(x)dx = \lim\limits_{t\to\infty}\int_0^t f'(x)dx = \lim\limits_{t\to\infty}\left[f(t) - f(0)\right] = \lim\limits_{t\to\infty} f(t) - f(0) = 0 - f(0) = -f(0).$

85. Let $u = 1/x \quad \Rightarrow \quad x = 1/u \quad \Rightarrow \quad dx = -(1/u^2)du$.

$$\int_0^\infty \frac{\ln x}{1+x^2}\,dx = \int_\infty^0 \frac{\ln(1/u)}{1+1/u^2}\left(-\frac{du}{u^2}\right) = \int_\infty^0 \frac{-\ln u}{u^2+1}(-du) = \int_\infty^0 \frac{\ln u}{1+u^2}\,du$$

$$= -\int_0^\infty \frac{\ln u}{1+u^2}\,du. \text{ Therefore } \int_0^\infty \frac{\ln x}{1+x^2}\,dx = -\int_0^\infty \frac{\ln x}{1+x^2}\,dx = 0.$$

APPLICATIONS PLUS (page 499)

1. **(a)** Coefficient of inequality $= \dfrac{\text{area between Lorenz curve and straight line}}{\text{area under straight line}}$

$$= \frac{\int_0^1 [x - L(x)]dx}{\int_0^1 x\,dx} = \frac{\int_0^1 [x - L(x)]dx}{[x^2/2]_0^1} = \frac{\int_0^1 [x - L(x)]dx}{1/2} = 2\int_0^1 [x - L(x)]dx$$

(b) $L(x) = \frac{5}{12}x^2 + \frac{7}{12}x \quad \Rightarrow \quad L\left(\frac{1}{2}\right) = \frac{5}{48} + \frac{7}{24} = \frac{19}{48} = 0.3958\overline{3}$, so the bottom 50% of the households receive about 40% of the income.

Coefficient of inequality $= 2\int_0^1 \left[x - \frac{5}{12}x^2 - \frac{7}{12}x\right]dx = 2\int_0^1 \frac{5}{12}(x - x^2)dx = \frac{5}{6}\left(\frac{1}{2}x^2 - \frac{1}{3}x^3\right)\Big|_0^1 = \frac{5}{36}$

(c) Coefficient of inequality $= 2\displaystyle\int_0^1 [x - L(x)]dx = 2\int_0^1 \left(x - \frac{5x^3}{4 + x^2}\right)dx$

$$= 2\int_0^1 \left[x - \left(5x - \frac{20x}{x^2 + 4}\right)\right]dx = 2\int_0^1 \left(-4x + \frac{20x}{x^2 + 4}\right)dx$$

$$= 2\left[-2x^2 + 10\ln(x^2 + 4)\right]_0^1 = 2(-2 + 10\ln 5 - 10\ln 4)$$

$$= -4 + 20\ln\tfrac{5}{4} \approx 0.46$$

3. **(a)** The tangent to the curve $y = f(x)$ at $x = x_0$ has the equation $y - f(x_0) = f'(x_0)(x - x_0)$.

The y-intercept of this tangent line is $f(x_0) - f'(x_0)x_0$. Thus L is the distance from the point $(0, f(x_0) - f'(x_0)x_0)$ to the point $(x_0, f'(x_0))$. That is, $L^2 = x_0^2 + [f'(x_0)]^2 x_0^2$, so $[f'(x_0)]^2 = \dfrac{L^2 - x_0^2}{x_0^2}$ and $f'(x_0) = -\dfrac{\sqrt{L^2 - x_0^2}}{x_0}$ for each $0 < x_0 < L$.

(b) $\dfrac{dy}{dx} = -\dfrac{\sqrt{L^2 - x^2}}{x} \quad \Rightarrow$

$$y = \int -\frac{\sqrt{L^2 - x^2}}{x}\,dx = \int \frac{-L\cos\theta\, L\cos\theta\, d\theta}{L\sin\theta} \quad \text{(where } x = L\sin\theta\text{)}$$

$$= L\int \frac{\sin^2\theta - 1}{\sin\theta}\,d\theta = L\int(\sin\theta - \csc\theta)d\theta = -L\cos\theta + L\ln|\csc\theta + \cot\theta| + C$$

$$= -\sqrt{L^2 - x^2} + L\ln\left(\frac{L}{x} + \frac{\sqrt{L^2 - x^2}}{x}\right) + C$$

When $x = L$, $0 = y = -0 + L\ln(1 + 0) + C$, so $C = 0$. Therefore

$$y = -\sqrt{L^2 - x^2} + L\ln\left(\frac{L + \sqrt{L^2 - x^2}}{x}\right).$$

5. **(a)** $\dfrac{dC}{dt} = r - kC = k\left(\dfrac{r}{k} - C\right)$. Let $u(t) = \dfrac{r}{k} - C(t)$. Then $\dfrac{du}{dt} = -\dfrac{dC}{dt}$. Therefore $\dfrac{du}{dt} = -ku$, and by Theorem 6.5.2, the solution to this equation is $u(t) = u(0)e^{-kt} = \left[\dfrac{r}{k} - C(0)\right]e^{-kt}$. Therefore

$$\frac{r}{k} - C(t) = \left(\frac{r}{k} - C_0\right)e^{-kt} \quad \Rightarrow \quad C(t) = \frac{r}{k} - \left(\frac{r}{k} - C_0\right)e^{-kt} = C_0e^{-kt} + \frac{r}{k}\left(1 - e^{-kt}\right).$$

(b) If $C_0 < r/k$, then the first formula for $C(t)$ shows that $C(t)$ increases monotonically and $\lim_{t\to\infty} C(t) = r/k$. The second expression for $C(t)$ shows how the role of C_0 steadily diminishes as that of r/k increases.

7. **(a)** Here we have a differential equation of the form $\dfrac{dv}{dt} = kv$, so by Theorem 6.5.2, the solution is $v(t) = v(0)e^{kt}$. In this case $k = -\frac{1}{10}$ and $v(0) = 100$ ft/s, so $v(t) = 100e^{-t/10}$. We are interested in the time that the ball takes to travel 280 ft, so we find the distance function

$s(t) = \int_0^t v(x)dx = \int_0^t 100e^{-x/10}\,dx = 100\left[-10e^{-x/10}\right]_0^t = -1000\left(e^{-t/10} - 1\right) = 1000\left(1 - e^{-t/10}\right)$.

Now we set $s(t) = 280$ and solve for t: $280 = 1000\left(1 - e^{-t/10}\right) \quad \Rightarrow \quad 1 - e^{-t/10} = \frac{7}{25} \quad \Rightarrow$

$-\frac{1}{10}t = \ln\left(1 - \frac{7}{25}\right) \approx -0.3285$, so $t \approx 3.285$ seconds.

(b) Let x be the distance of the shortstop from home plate. We calculate the time for the ball to reach home plate as a function of x, then differentiate with respect to x to find the value of x which corresponds to the minimum time.

The total time that it takes the ball to reach home is the sum of the times of the two throws, plus the relay time $\left(\frac{1}{2}\text{ s}\right)$. The distance from the fielder to the shortstop is $280 - x$, so to find the time t_1 taken by the first throw, we solve the equation $s_1(t_1) = 280 - x \;\Leftrightarrow\; 1 - e^{-t_1/10} = \dfrac{280 - x}{1000} \;\Leftrightarrow\; t_1 = -10\ln\dfrac{720 + x}{1000}$.

We find the time t_2 taken by the second throw if the shortstop throws with velocity w, since we see that this velocity varies in the rest of the problem. We use $v = we^{-t/10}$ and isolate t_2 in the equation

$s(t_2) = 10w\left(1 - e^{-t_2/10}\right) = x \quad \Leftrightarrow \quad e^{-t_2/10} = 1 - \dfrac{x}{10w} \quad \Leftrightarrow \quad t_2 = -10\ln\dfrac{10w - x}{10w}$, so the total time is $t_w(x) = \dfrac{1}{2} - 10\left[\ln\dfrac{720 + x}{1000} + \ln\dfrac{10w - x}{10w}\right]$. To find the minimum, we differentiate:

$\dfrac{dt_w}{dx} = -10\left[\dfrac{1}{720 + x} - \dfrac{1}{10w - x}\right]$, which changes from negative to positive when $720 + x = 10w - x$

$\Leftrightarrow \quad x = 5w - 360$. So by the First Derivative Test, t_w has a minimum at this distance from the shortstop to home plate. So if the shortstop throws at $w = 105$ ft/s, the minimum time is

$t_{105}(165) = -10\ln\frac{720 + 165}{1000} + \frac{1}{2} - 10\ln\frac{1050 - 165}{1050} \approx 3.431$ seconds. This is longer than the time taken in part (a), so in this case the manager should encourage a direct throw.

If $w = 115$ ft/s, the minimum time is $t_{115}(215) = -10\ln\frac{720 + 215}{1000} + \frac{1}{2} - 10\ln\frac{1150 - 215}{1150} \approx 3.242$ seconds. This is less than the time taken in part (a), so in this case, the manager should encourage a relayed throw.

(c) In general, the minimum time is

$$t_w(5w-360)=\frac{1}{2}-10\left[\ln\frac{360+5w}{1000}+\ln\frac{360+5w}{10w}\right]=\frac{1}{2}-10\ln\frac{(w+72)^2}{400w}.$$

We want to find out when this is about 3.285 seconds, the same time as the direct throw.

From the graph, we estimate that this is the case for $w\approx 112.8$ ft/s. So if the shortstop can throw the ball with this velocity, then a relayed throw takes the same time as a direct throw.

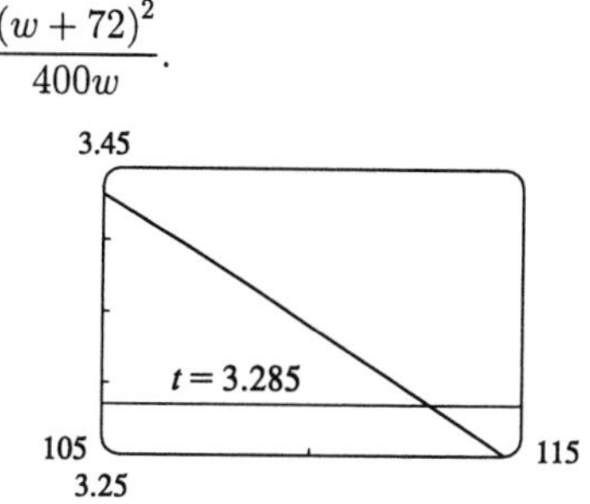

9. **(a)** $|VP|=9+x\cos\alpha$, $|PT|=35-(4+x\sin\alpha)=31-x\sin\alpha$, and $|PB|=(4+x\sin\alpha)-10=x\sin\alpha-6$.

So using the Pythagorean Theorem, we have

$$|VT|=\sqrt{|VP|^2+|PT|^2}=\sqrt{(9+x\cos\alpha)^2+(31-x\sin\alpha)^2}=a,\text{ and}$$

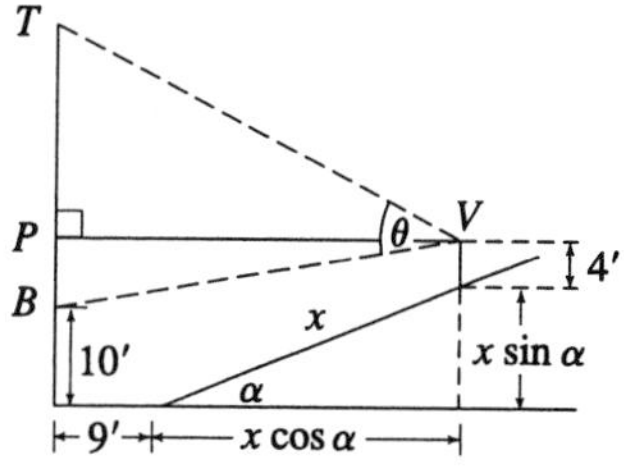

$|VB|=\sqrt{|VP|^2+|PB|^2}=\sqrt{(9+x\cos\alpha)^2+(x\sin\alpha-6)^2}=b$. Using the Law of Cosines on $\triangle VBT$, we get $25^2=a^2+b^2-2ab\cos\theta \quad\Leftrightarrow$

$$\cos\theta=\frac{a^2+b^2-625}{2ab}\quad\Leftrightarrow\quad\theta=\arccos\left(\frac{a^2+b^2-625}{2ab}\right),\text{ as required.}$$

(b) From the graph, it appears that the value of x which maximizes θ is $x\approx 8.25$ ft. The row closest to this value of x is the fourth row, at $x=9$ ft, and from the graph, the viewing angle in this row seems to be about 0.85 radians, or about 49°.

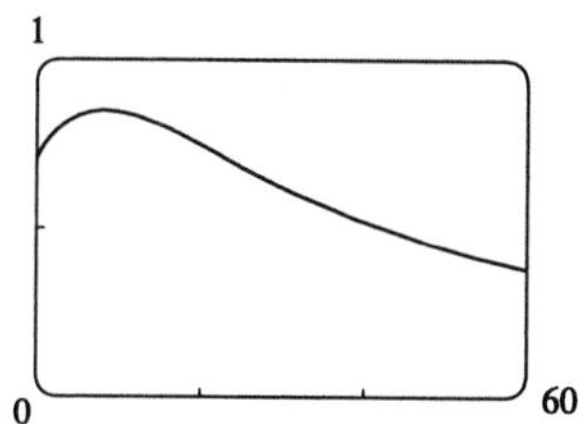

(c) With a CAS, we type in the definition of θ (calling it T), substitute in the proper values of a and b in terms of x and $\alpha=20°=\frac{\pi}{9}$ radians, and then use the differentiation command (`diff` in Maple) to find the derivative. We use a numerical root finder (`fsolve` in Maple) and find that the root of the equation $d\theta/dx=0$ is $x\approx 8.25306209$, as approximated above.

(d) From the graph in part (b), it seems that the average value of the function on the interval $[0,60]$ is about 0.6. We can use a CAS to approximate $\frac{1}{60}\int_0^{60}\theta(x)dx\approx 0.625\approx 36°$. (The calculation is faster if we reduce the number of digits of accuracy required.) The minimum value is $\theta(60)\approx 0.38$ and, from part (b), the maximum value is about 0.85.

CHAPTER EIGHT

EXERCISES 8.1

1. $\dfrac{dy}{dx} = y^2 \Rightarrow \dfrac{dy}{y^2} = dx \ (y \neq 0) \Rightarrow \displaystyle\int \frac{dy}{y^2} = \int dx \Rightarrow -\frac{1}{y} = x + C \Rightarrow -y = \frac{1}{x+C} \Rightarrow y = \frac{-1}{x+C}$, and $y = 0$ is also a solution.

3. $yy' = x \Rightarrow \int y\,dy = \int x\,dx \Rightarrow \dfrac{y^2}{2} = \dfrac{x^2}{2} + C_1 \Rightarrow y^2 = x^2 + 2C_1 \Rightarrow x^2 - y^2 = C$ (where $C = -2C_1$). This represents a family of hyperbolas.

5. $x^2y' + y = 0 \Rightarrow \dfrac{dy}{dx} = -\dfrac{y}{x^2} \Rightarrow \displaystyle\int \frac{dy}{y} = \int \frac{-dx}{x^2} \ (y \neq 0) \Rightarrow \ln|y| = \frac{1}{x} + K \Rightarrow$ $|y| = e^K e^{1/x} \Rightarrow y = Ce^{1/x}$, where now we allow C to be any constant.

7. $\dfrac{du}{dt} = e^{u+2t} = e^u e^{2t} \Rightarrow \int e^{-u}\,du = \int e^{2t}\,dt \Rightarrow -e^{-u} = \frac{1}{2}e^{2t} + C_1 \Rightarrow e^{-u} = -\frac{1}{2}e^{2t} + C$ (where $C = -C_1$ and the right-hand side is positive, since $e^{-u} > 0$) $\Rightarrow -u = \ln\left(C - \frac{1}{2}e^{2t}\right) \Rightarrow u = -\ln\left(C - \frac{1}{2}e^{2t}\right)$

9. $e^y y' = \dfrac{3x^2}{1+y}$, $y(2) = 0$. $\displaystyle\int e^y(1+y)dy = \int 3x^2\,dx \Rightarrow ye^y = x^3 + C$. $y(2) = 0$, so $0 = 2^3 + C$ and $C = -8$. Thus $ye^y = x^3 - 8$.

11. $xe^{-t}\dfrac{dx}{dt} = t$, $x(0) = 1$. $\int x\,dx = \int te^t dt \Rightarrow \frac{1}{2}x^2 = (t-1)e^t + C$. $x(0) = 1$, so $\frac{1}{2} = (0-1)e^0 + C$ and $C = \frac{3}{2}$. Thus $x^2 = 2(t-1)e^t + 3 \Rightarrow x = \sqrt{2(t-1)e^t + 3}$.

13. $\dfrac{du}{dt} = \dfrac{2t+1}{2(u-1)}$, $u(0) = -1$. $\int 2(u-1)du = \int (2t+1)dt \Rightarrow u^2 - 2u = t^2 + t + C$. $u(0) = -1$ so $(-1)^2 - 2(-1) = 0^2 + 0 + C$ and $C = 3$. Thus $u^2 - 2u = t^2 + t + 3$; the quadratic formula gives $u = 1 - \sqrt{t^2 + t + 4}$.

15. Let $y = f(x)$. Then $\dfrac{dy}{dx} = x^3 y$ and $y(0) = 1$. $\dfrac{dy}{y} = x^3\,dx$ (if $y \neq 0$), so $\displaystyle\int \frac{dy}{y} = \int x^3\,dx$ and $\ln|y| = \frac{1}{4}x^4 + C$; $y(0) = 1 \Rightarrow C = 0$, so $\ln|y| = \frac{1}{4}x^4$, $|y| = e^{x^4/4}$ and $y = f(x) = e^{x^4/4}$ [since $y(0) = 1$].

17. $\dfrac{dy}{dx} = 4x^3 y$, $y(0) = 7$. $\dfrac{dy}{y} = 4x^3\,dx$ (if $y \neq 0$) $\Rightarrow \displaystyle\int \frac{dy}{y} = \int 4x^3\,dx \Rightarrow \ln|y| = x^4 + C \Rightarrow$ $y = Ae^{x^4}$; $y(0) = 7 \Rightarrow A = 7 \Rightarrow y = 7e^{x^4}$.

19. $y' = e^{x-y}$, $y(0) = 1$.

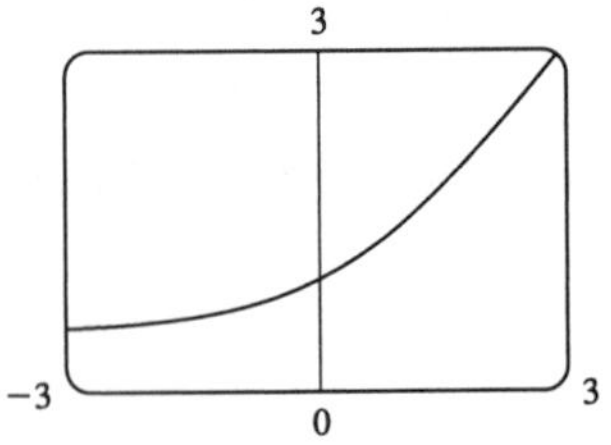

So $\dfrac{dy}{dx} = e^x e^{-y} \Leftrightarrow \displaystyle\int e^y dy = \int e^x dx \Leftrightarrow$ $e^y = e^x + C$. From the initial condition, we must have $e^1 = e^0 + C \Rightarrow C = e - 1$. So the solution is $e^y = e^x + e - 1 \Rightarrow y = \ln(e^x + e - 1)$.

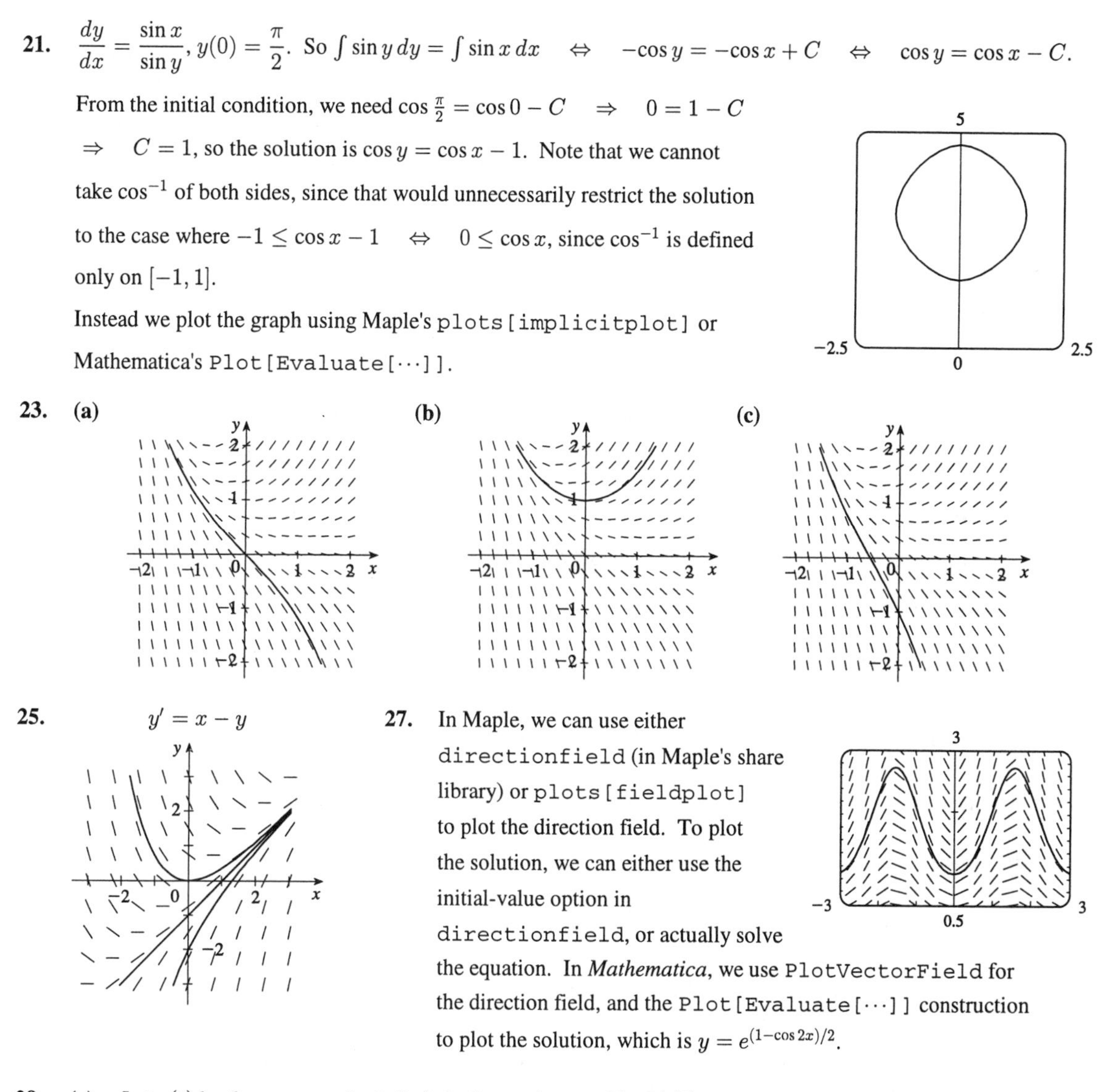

21. $\dfrac{dy}{dx} = \dfrac{\sin x}{\sin y}$, $y(0) = \dfrac{\pi}{2}$. So $\int \sin y\,dy = \int \sin x\,dx \quad \Leftrightarrow \quad -\cos y = -\cos x + C \quad \Leftrightarrow \quad \cos y = \cos x - C$.

From the initial condition, we need $\cos\frac{\pi}{2} = \cos 0 - C \quad \Rightarrow \quad 0 = 1 - C$ $\quad \Rightarrow \quad C = 1$, so the solution is $\cos y = \cos x - 1$. Note that we cannot take $\cos^{-1}$ of both sides, since that would unnecessarily restrict the solution to the case where $-1 \le \cos x - 1 \quad \Leftrightarrow \quad 0 \le \cos x$, since $\cos^{-1}$ is defined only on $[-1, 1]$.

Instead we plot the graph using Maple's `plots[implicitplot]` or Mathematica's `Plot[Evaluate[···]]`.

23. (a) (b) (c)

25. $y' = x - y$

27. In Maple, we can use either `directionfield` (in Maple's share library) or `plots[fieldplot]` to plot the direction field. To plot the solution, we can either use the initial-value option in `directionfield`, or actually solve the equation. In *Mathematica*, we use `PlotVectorField` for the direction field, and the `Plot[Evaluate[···]]` construction to plot the solution, which is $y = e^{(1-\cos 2x)/2}$.

29. **(a)** Let $y(t)$ be the amount of salt (in kg) after t minutes. Then $y(0) = 15$. The amount of liquid in the tank is 1000 L at all times, so the concentration at time t (in minutes) is $y(t)/1000$ kg/L and

$\dfrac{dy}{dt} = -\left[\dfrac{y(t)}{1000}\,\dfrac{\text{kg}}{\text{L}}\right]\left(10\,\dfrac{\text{L}}{\text{min}}\right) = -\dfrac{y(t)}{100}\,\dfrac{\text{kg}}{\text{min}}$. $\displaystyle\int \frac{dy}{y} = -\frac{1}{100}\int dt \quad \Rightarrow \quad \ln y = -\frac{t}{100} + C$, and

$y(0) = 15 \quad \Rightarrow \quad \ln 15 = C$, so $\ln y = \ln 15 - \dfrac{t}{100}$. It follows that $\ln\left(\dfrac{y}{15}\right) = -\dfrac{t}{100}$ and $\dfrac{y}{15} = e^{-t/100}$,

so $y = 15e^{-t/100}$ kg.

(b) After 20 min., $y = 15e^{-20/100} = 15e^{-0.2} \approx 12.3$ kg.

31. $\frac{dx}{dt} = k(a-x)(b-x)$, $a \neq b$. $\int \frac{dx}{(a-x)(b-x)} = \int k\,dt \quad \Rightarrow \quad \frac{1}{b-a}\int\left(\frac{1}{a-x} - \frac{1}{b-x}\right)dx = \int k\,dt$

$\Rightarrow \quad \frac{1}{b-a}(-\ln|a-x| + \ln|b-x|) = kt + C \quad \Rightarrow \quad \ln\left|\frac{b-x}{a-x}\right| = (b-a)(kt+C)$. Here the concentrations $[A] = a - x$ and $[B] = b - x$ cannot be negative, so $\frac{b-x}{a-x} \geq 0$ and $\left|\frac{b-x}{a-x}\right| = \frac{b-x}{a-x}$. We now have $\ln\left(\frac{b-x}{a-x}\right) = (b-a)(kt+C)$. Since $x(0) = 0$, $\ln\left(\frac{b}{a}\right) = (b-a)C$. Hence $\ln\left(\frac{b-x}{a-x}\right) = (b-a)kt + \ln\left(\frac{b}{a}\right)$, $\frac{b-x}{a-x} = \frac{b}{a}e^{(b-a)kt}$, and $x = \frac{b\left[e^{(b-a)kt} - 1\right]}{be^{(b-a)kt}/a - 1} = \frac{ab\left[e^{(b-a)kt} - 1\right]}{be^{(b-a)kt} - a}$ moles/L.

33. **(a)** Let $P(t)$ be the world population in the year t. Then $dP/dt = 0.02P$, so $\int (1/P)dP = \int 0.02\,dt$ and $\ln P = 0.02t + C \quad \Rightarrow \quad P(t) = Ae^{0.02t}$. $P(1986) = 5 \times 10^9 \quad \Rightarrow \quad P(t) = 5 \times 10^9 e^{0.02(t-1986)}$.

(b) **(i)** The predicted population in 2000 is $P(2000) = 5e^{0.28} \times 10^9 \approx 6.6$ billion.

(ii) The predicted population in 2100 is $P(2100) = 5e^{2.28} \times 10^9 \approx 49$ billion.

(iii) The predicted population in 2500 is $P(2500) = 5e^{10.28} \times 10^9 \approx 146$ trillion.

(c) According to this model, in 2000 the area per person will be $\frac{1.8 \times 10^{15}}{6.6 \times 10^9} \approx 270{,}000\text{ ft}^2$. In 2100 it will be $\frac{1.8 \times 10^{15}}{49 \times 10^9} \approx 37{,}000\text{ ft}^2$, and in 2500 it will be $\frac{1.8 \times 10^{15}}{146 \times 10^{12}} \approx 12\text{ ft}^2$. (!)

35. **(a)** Our assumption is that $\frac{dy}{dt} = ky(1-y)$, where y is the fraction of the population that has heard the rumor.

(b) Take $M = 1$ in (11) to get $y = \frac{y_0}{y_0 + (1-y_0)e^{-kt}}$.

(c) Let t be the number of hours since 8 A.M. Then $y_0 = y(0) = \frac{80}{1000} = 0.08$ and $y(4) = \frac{1}{2}$, so $\frac{1}{2} = y(4) = \frac{0.08}{0.08 + 0.92e^{-4k}}$. Thus $0.08 + 0.92e^{-4k} = 0.16$, $e^{-4k} = \frac{0.08}{0.92} = \frac{2}{23}$, and $e^{-k} = \left(\frac{2}{23}\right)^{1/4}$, so $y = \frac{0.08}{0.08 + 0.92(2/23)^{t/4}} = \frac{2}{2 + 23(2/23)^{t/4}}$ and $\left(\frac{2}{23}\right)^{t/4} = \frac{2}{23} \cdot \frac{1-y}{y}$ or $\left(\frac{2}{23}\right)^{t/4-1} = \frac{1-y}{y}$. It follows that $\frac{t}{4} - 1 = \frac{\ln[(1-y)/y]}{\ln(2/23)}$, so $t = 4\left[1 + \frac{\ln[(1-y)/y]}{\ln(2/23)}\right]$. When $y = 0.9$, $\frac{1-y}{y} = \frac{1}{9}$, so $t = 4\left[1 - \frac{\ln 9}{\ln 23}\right] \approx 7.6$ h or 7 h 36 min. Thus 90% of the population will have heard the rumor by 3:36 P.M..

37. y increases most rapidly when y' is maximal, that is, when $y'' = 0$. But $y' = ky(M-y) \quad \Rightarrow$ $y'' = ky'(M-y) + ky(-y') = ky'(M-2y) = k^2y(M-y)(M-2y)$. Since $0 < y < M$, we see that $y'' = 0$ $\Leftrightarrow \quad y = M/2$.

39. At $t = 0$, the exponential model $y = e^{0.1t}$ has derivative $y' = 0.1e^0 = 0.1$. From the original differential equation, the logistic model has derivative $y' = ky(M - y)$. At $t = 0$, this is equal to $ky_0(M - y_0) = 9k$. So the two derivatives are equal at $t = 0$ if $9k = 0.1 \quad \Leftrightarrow \quad k = 0.1/9 = \frac{1}{90}$. We graph both models, and see that for small values of t they agree closely, but for large values of t, the exponential model increases rapidly, while the logistic model levels off and approaches the line $y = 10$.

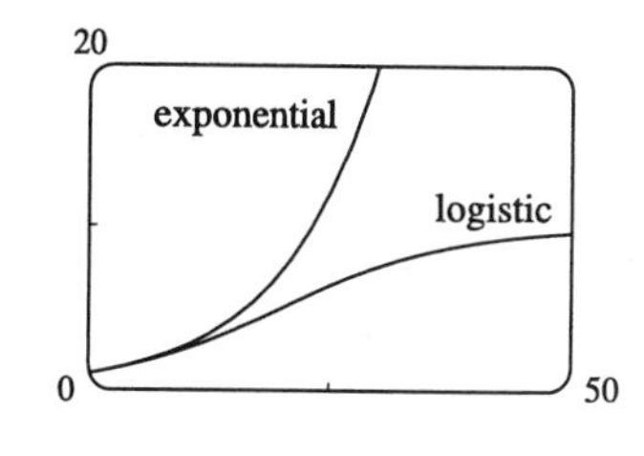

41. **(a)** The rate of growth of the area is jointly proportional to $\sqrt{A(t)}$ and $M - A(t)$; that is, the rate is proportional to the product of those two quantities. So for some constant k, $dA/dt = k\sqrt{A}(M - A)$. We are interested in the maximum of the function dA/dt, so we differentiate, using the Chain Rule and then substituting for dA/dt from the differential equation:

$$\frac{d}{dt}\left(\frac{dA}{dt}\right) = k\left[\tfrac{1}{2}A^{-1/2}(M - A)\frac{dA}{dt} + \sqrt{A}(-1)\frac{dA}{dt}\right] = \tfrac{1}{2}kA^{-1/2}\frac{dA}{dt}[(M - A) - 2A]$$

$= \frac{1}{2}k^2(M - A)(M - 3A)$. This is 0 when $M - A = 0$ [this situation never actually occurs, since the graph of $A(t)$ is asymptotic to the line $y = M$, as in the logistic model] and when $M - 3A = 0$ $\Leftrightarrow \quad A(t) = M/3$. This represents a maximum by the First Derivative Test, since $\dfrac{d}{dt}\left(\dfrac{dA}{dt}\right)$ goes from positive to negative when $A(t) = M/3$.

(b) To solve the differential equation, we separate variables in our expression for dA/dt and integrate: $\displaystyle\int \frac{dA}{\sqrt{A}(M - A)} = \int k\,dt = kt + C_1$. To evaluate the LHS, we make the substitution $x = \sqrt{A}$, so $A = x^2$ and $dA = 2x\,dx$. The LHS becomes

$$2\int \frac{dx}{M - x^2} = \int \left(\frac{1/\sqrt{M}}{\sqrt{M} + x} + \frac{1/\sqrt{M}}{\sqrt{M} - x}\right)dx \quad \text{(difference of squares; partial fractions)}$$

$$= \frac{1}{\sqrt{M}}\left(\ln\left|\sqrt{M} + x\right| - \ln\left|\sqrt{M} - x\right|\right) = \frac{1}{\sqrt{M}}\ln\frac{\sqrt{M} + \sqrt{A}}{\sqrt{M} - \sqrt{A}}.$$

So, multiplying by $\sqrt{M}$ and exponentiating both sides of the equation, we get $\dfrac{\sqrt{M} + \sqrt{A}}{\sqrt{M} - \sqrt{A}} = Ce^{\sqrt{M}kt}$, where $C = e^{\sqrt{M}C_1}$. Solving for A: $\sqrt{M} + \sqrt{A} = Ce^{\sqrt{M}kt}\left(\sqrt{M} - \sqrt{A}\right) \quad \Leftrightarrow$

$$\sqrt{A}\left(Ce^{\sqrt{M}kt} + 1\right) = \sqrt{M}\left(Ce^{\sqrt{M}kt} - 1\right) \quad \Leftrightarrow \quad A = M\left(\frac{Ce^{\sqrt{M}kt} - 1}{Ce^{\sqrt{M}kt} + 1}\right)^2.$$

To get C in terms of the initial area A_0 and the maximum area M, we substitute $t = 0$ and $A = A_0$:

$$A_0 = M\left(\frac{C - 1}{C + 1}\right)^2 \quad \Leftrightarrow \quad (C + 1)\sqrt{A_0} = (C - 1)\sqrt{M} \quad \Leftrightarrow \quad C = \frac{\sqrt{M} + \sqrt{A_0}}{\sqrt{M} - \sqrt{A_0}}.$$

(Notice that if $A_0 = 0$, then $C = 1$.)

43. $RI + LI'(t) = V$, $I(0) = 0$. $LI' = V - RI \;\Rightarrow\; L\dfrac{dI}{dt} = V - RI \;\Rightarrow\; \displaystyle\int \frac{L\,dI}{V - RI} = \int dt \;\Rightarrow$

$-(L/R)\ln|V - RI| = t + C \quad \Rightarrow \quad V - RI = Ae^{-Rt/L} \quad \Rightarrow \quad I = V/R - (A/R)e^{-Rt/L}$.

$I(0) = 0 \quad \Rightarrow \quad 0 = V/R - (A/R) \cdot e^0 \quad \Rightarrow \quad A = V$. So $I = (V/R)\left(1 - e^{-Rt/L}\right)$.

EXERCISES 8.2

1. $L = \int_{-1}^{3} \sqrt{1 + \left(\frac{dy}{dx}\right)^2}\,dx = \int_{-1}^{3} \sqrt{1 + 2^2}\,dx = \sqrt{5}[3 - (-1)] = 4\sqrt{5}.$

The arc length can be calculated using the distance formula, since the curve is a straight line, so
$L = [\text{distance from } (-1,-1) \text{ to } (3,7)] = \sqrt{[3-(-1)]^2 + [7-(-1)]^2} = \sqrt{80} = 4\sqrt{5}.$

3. $y^2 = (x-1)^3,\ y = (x-1)^{3/2} \ \Rightarrow\ \frac{dy}{dx} = \frac{3}{2}(x-1)^{1/2} \ \Rightarrow\ 1 + \left(\frac{dy}{dx}\right)^2 = 1 + \frac{9}{4}(x-1).$

So $L = \int_1^2 \sqrt{1 + \frac{9}{4}(x-1)}\,dx = \int_1^2 \sqrt{\frac{9}{4}x - \frac{5}{4}}\,dx = \left[\frac{4}{9}\cdot\frac{2}{3}\left(\frac{9}{4}x - \frac{5}{4}\right)^{3/2}\right]_1^2 = \frac{13\sqrt{13}-8}{27}.$

5. $y = \frac{1}{3}(x^2+2)^{3/2} \ \Rightarrow\ dy/dx = \frac{1}{2}(x^2+2)^{1/2}(2x) = x\sqrt{x^2+2} \ \Rightarrow$
$1 + (dy/dx)^2 = 1 + x^2(x^2+2) = (x^2+1)^2$. So $L = \int_0^1 (x^2+1)dx = \left[\frac{1}{3}x^3 + x\right]_0^1 = \frac{4}{3}.$

7. $y = \frac{x^4}{4} + \frac{1}{8x^2} \ \Rightarrow\ \frac{dy}{dx} = x^3 - \frac{1}{4x^3} \ \Rightarrow\ 1 + \left(\frac{dy}{dx}\right)^2 = 1 + x^6 - \frac{1}{2} + \frac{1}{16x^6} = x^6 + \frac{1}{2} + \frac{1}{16x^6}.$

So $L = \int_1^3 \left(x^3 + \frac{1}{4}x^{-3}\right)dx = \left[\frac{1}{4}x^4 - \frac{1}{8}x^{-2}\right]_1^3 = \left(\frac{81}{4} - \frac{1}{72}\right) - \left(\frac{1}{4} - \frac{1}{8}\right) = \frac{181}{9}.$

9. $y = \ln(\cos x) \ \Rightarrow\ y' = \frac{1}{\cos x}(-\sin x) = -\tan x \ \Rightarrow\ 1 + (y')^2 = 1 + \tan^2 x = \sec^2 x.$

So $L = \int_0^{\pi/4} \sec x\,dx = \ln(\sec x + \tan x)\big|_0^{\pi/4} = \ln\left(\sqrt{2}+1\right).$

11. $y = \ln(1-x^2) \ \Rightarrow\ \frac{dy}{dx} = \frac{-2x}{1-x^2} \ \Rightarrow\ 1 + \left(\frac{dy}{dx}\right)^2 = 1 + \frac{4x^2}{(1-x^2)^2} = \frac{(1+x^2)^2}{(1-x^2)^2}.$ So

$$L = \int_0^{1/2} \frac{1+x^2}{1-x^2}\,dx = \int_0^{1/2}\left[-1 + \frac{2}{(1-x)(1+x)}\right]dx = \int_0^{1/2}\left[-1 + \frac{1}{1+x} + \frac{1}{1-x}\right]dx$$
$$= \left[-x + \ln(1+x) - \ln(1-x)\right]_0^{1/2} = -\tfrac{1}{2} + \ln\tfrac{3}{2} - \ln\tfrac{1}{2} - 0 = \ln 3 - \tfrac{1}{2}.$$

13. $y = e^x \ \Rightarrow\ y' = e^x \ \Rightarrow\ 1 + (y')^2 = 1 + e^{2x}$. So

$$L = \int_0^1 \sqrt{1+e^{2x}}\,dx = \int_1^e \sqrt{1+u^2}\,\frac{du}{u} \quad [\text{where } u = e^x, \text{ so } x = \ln u,\ dx = du/u]$$
$$= \int_1^e \frac{\sqrt{1+u^2}}{u^2}\,u\,du = \int_{\sqrt{2}}^{\sqrt{1+e^2}} \frac{v}{v^2-1}\,v\,dv \quad \left[\text{where } v = \sqrt{1+u^2}, \text{ so } v^2 = 1+u^2,\ v\,dv = u\,du\right]$$
$$= \int_{\sqrt{2}}^{\sqrt{1+e^2}} \left(1 + \frac{1/2}{v-1} - \frac{1/2}{v+1}\right)dv = \left[v + \tfrac{1}{2}\ln\frac{v-1}{v+1}\right]_{\sqrt{2}}^{\sqrt{1+e^2}}$$
$$= \sqrt{1+e^2} - \sqrt{2} + \tfrac{1}{2}\ln\frac{\sqrt{1+e^2}-1}{\sqrt{1+e^2}+1} - \tfrac{1}{2}\ln\frac{\sqrt{2}-1}{\sqrt{2}+1}$$
$$= \sqrt{1+e^2} - \sqrt{2} + \ln\left(\sqrt{1+e^2}-1\right) - 1 - \ln\left(\sqrt{2}-1\right)$$

Or: Use Formula 23 for $\int\left(\sqrt{1+u^2}/u\right)du$, or substitute $u = \tan\theta$.

15. $y = \cosh x \ \Rightarrow\ y' = \sinh x \ \Rightarrow\ 1 + (y')^2 = 1 + \sinh^2 x = \cosh^2 x.$
So $L = \int_0^1 \cosh x\,dx = [\sinh x]_0^1 = \sinh 1 = \frac{1}{2}(e - 1/e).$

17. $y = x^3 \quad \Rightarrow \quad y' = 3x^2 \quad \Rightarrow \quad 1 + (y')^2 = 1 + 9x^4$. So $L = \int_0^1 \sqrt{1 + 9x^4}\, dx$.

19. $y = e^x \cos x \quad \Rightarrow \quad y' = e^x(\cos x - \sin x) \quad \Rightarrow$
$1 + (y')^2 = 1 + e^{2x}(\cos^2 x - 2\cos x \sin x + \sin^2 x) = 1 + e^{2x}(1 - \sin 2x)$. So
$L = \int_0^{\pi/2} \sqrt{1 + e^{2x}(1 - \sin 2x)}\, dx$.

21. $y = x^3 \quad \Rightarrow \quad 1 + (y')^2 = 1 + (3x^2)^2 = 1 + 9x^4$. So $L = \int_0^1 \sqrt{1 + 9x^4}\, dx$. Let $f(x) = \sqrt{1 + 9x^4}$.
Then by Simpson's Rule with $n = 10$,

$$L \approx \tfrac{1/10}{3}\big[f(0) + 4f(0.1) + 2f(0.2) + 4f(0.3) + 2f(0.4) + 4f(0.5) + 2f(0.6) + 4f(0.7) + 2f(0.8) + 4f(0.9) + f(1)\big] \approx 1.548.$$

23. $y = \sin x$, $1 + (dy/dx)^2 = 1 + \cos^2 x$, $L = \int_0^\pi \sqrt{1 + \cos^2 x}\, dx$. Let $g(x) = \sqrt{1 + \cos^2 x}$. Then

$$L \approx \tfrac{\pi/10}{3}\big[g(0) + 4g(\tfrac{\pi}{10}) + 2g(\tfrac{\pi}{5}) + 4g(\tfrac{3\pi}{10}) + 2g(\tfrac{2\pi}{5}) + 4g(\tfrac{\pi}{2}) + 2g(\tfrac{3\pi}{5}) + 4g(\tfrac{7\pi}{10}) + 2g(\tfrac{4\pi}{5}) + 4g(\tfrac{9\pi}{10}) + g(\pi)\big] \approx 3.820.$$

25. **(a)** **(b)**

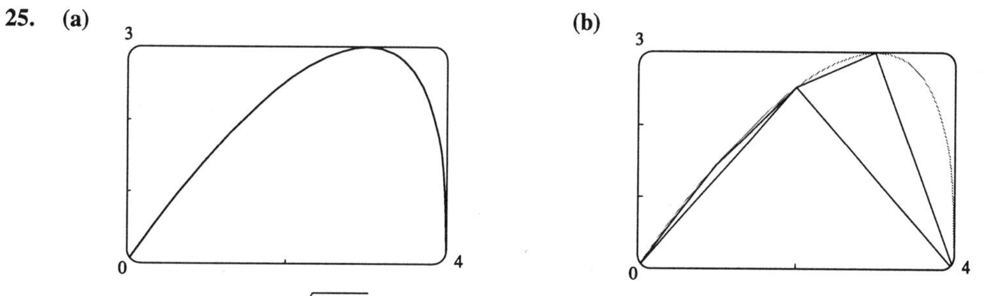

Let $f(x) = y = x\sqrt[3]{4 - x}$. The polygon with one side is just the line segment joining the points $(0, f(0)) = (0, 0)$ and $(4, f(4)) = (4, 0)$, and its length is 4. The polygon with two sides joins the points $(0, 0)$, $(2, f(2)) = \left(2, 2\sqrt[3]{2}\right)$ and $(4, 0)$. Its length is

$\sqrt{(2-0)^2 + (2\sqrt[3]{2} - 0)^2} + \sqrt{(4-2)^2 + (0 - 2\sqrt[3]{2})^2} = 2\sqrt{4 + 2^{8/3}} \approx 6.43$. Similarly, the inscribed polygon with four sides joins the points $(0, 0)$, $\left(1, \sqrt[3]{3}\right)$, $\left(2, 2\sqrt[3]{2}\right)$, $(3, 3)$, and $(4, 0)$, so its length is

$\sqrt{1 + (\sqrt[3]{3})^2} + \sqrt{1 + (2\sqrt[3]{2} - \sqrt[3]{3})^2} + \sqrt{1 + (3 - 2\sqrt[3]{2})^2} + \sqrt{1 + 9} \approx 7.50$.

(c) Using the arc length formula with $\dfrac{dy}{dx} = x\left[\frac{1}{3}(4 - x)^{-2/3}(-1)\right] + \sqrt[3]{4 - x} = \dfrac{12 - 4x}{3(4 - x)^{2/3}}$, the length of the curve is $L = \displaystyle\int_0^4 \sqrt{1 + \left(\frac{dy}{dx}\right)^2}\, dx = \int_0^4 \sqrt{1 + \left[\frac{12 - 4x}{3(4 - x)^{2/3}}\right]^2}\, dx$.

(d) According to a CAS, the length of the curve is $L \approx 7.7988$. The actual value is larger than any of the approximations in part (b). This is always true, since any approximating straight line between two points on the curve is shorter than the length of the curve between the two points.

27. $y = 2x^{3/2} \quad \Rightarrow \quad y' = 3x^{1/2} \quad \Rightarrow \quad 1 + (y')^2 = 1 + 9x$. The arc length function with starting point $P_0(1, 2)$ is
$s(x) = \int_1^x \sqrt{1 + 9t}\, dt = \left[\frac{2}{27}(1 + 9t)^{3/2}\right]_1^x = \frac{2}{27}\left[(1 + 9x)^{3/2} - 10\sqrt{10}\right]$.

29. $y^{2/3} = 1 - x^{2/3} \quad \Rightarrow \quad y = \left(1 - x^{2/3}\right)^{3/2} \quad \Rightarrow$

$\dfrac{dy}{dx} = \frac{3}{2}\left(1 - x^{2/3}\right)^{1/2}\left(-\frac{2}{3}x^{-1/3}\right) = -x^{-1/3}\left(1 - x^{2/3}\right)^{1/2} \quad \Rightarrow$

$\left(\dfrac{dy}{dx}\right)^2 = x^{-2/3}\left(1 - x^{2/3}\right) = x^{-2/3} - 1$. Thus

$L = 4\int_0^1 \sqrt{1 + (x^{-2/3} - 1)}\, dx = 4\int_0^1 x^{-1/3}\, dx = 4 \lim_{t \to 0^+} \left[\frac{3}{2}x^{2/3}\right]_t^1 = 6.$

31. $y = 4500 - \dfrac{x^2}{8000} \quad \Rightarrow \quad \dfrac{dy}{dx} = -\dfrac{x}{4000} \quad \Rightarrow \quad \left(\dfrac{dy}{dx}\right)^2 = \dfrac{x^2}{16{,}000{,}000}$. When $y = 4500$ m, $x = 0$ m. When $y = 0$ m, $x = 6000$ m. Therefore

$L = \int_0^{6000} \sqrt{1 + (x/4000)^2}\, dx = \int_0^{3/2} \sqrt{1 + u^2}\, 4000\, du \quad$ (put $u = \frac{1}{4000}x$)

$= 4000\left[\frac{1}{2}u\sqrt{1 + u^2} + \frac{1}{2}\ln\left(u + \sqrt{1 + u^2}\right)\right]_0^{3/2} \quad$ (Formula 21 or $u = \tan\theta$)

$= 4000\left[\frac{3}{4}\sqrt{\frac{13}{4}} + \frac{1}{2}\ln\left(\frac{3}{2} + \sqrt{\frac{13}{4}}\right)\right] = 1500\sqrt{13} + 2000\ln\frac{3 + \sqrt{13}}{2} \approx 7798$ m.

33. The sine wave has amplitude 1 and period 14, since it goes through two periods in a distance of 28 in., so its equation is $y = \sin\left(\frac{2\pi}{14}x\right) = \sin\left(\frac{\pi}{7}x\right)$. The width w of the flat metal sheet needed to make the panel is the arc length of the sine curve from $x = 0$ to $x = 28$. We set up the integral to evaluate w using the arc length formula with $\dfrac{dy}{dx} = \frac{\pi}{7}\cos\left(\frac{\pi}{7}x\right)$: $L = \displaystyle\int_0^{28} \sqrt{1 + \left[\tfrac{\pi}{7}\cos\left(\tfrac{\pi}{7}x\right)\right]^2}\, dx = 2\int_0^{14} \sqrt{1 + \left[\tfrac{\pi}{7}\cos\left(\tfrac{\pi}{7}x\right)\right]^2}\, dx$. This integral would be very difficult to evaluate exactly, so we use a CAS, and find that $L \approx 29.36$ inches.

35. $y = \displaystyle\int_1^x \sqrt{t^3 - 1}\, dt \quad \Rightarrow \quad \dfrac{dy}{dx} = \sqrt{x^3 - 1}$ (by the Fundamental Theorem, Part 1) $\Rightarrow \quad 1 + \left(\dfrac{dy}{dx}\right)^2 = x^3$

$\Rightarrow \quad L = \int_1^4 x^{3/2}\, dx = \frac{2}{5}\left[x^{5/2}\right]_1^4 = \frac{2}{5}(32 - 1) = \frac{62}{5} = 12.4.$

EXERCISES 8.3

1. $y = \sqrt{x} \quad \Rightarrow \quad 1 + \left(\dfrac{dy}{dx}\right)^2 = 1 + \left(\dfrac{1}{2\sqrt{x}}\right)^2 = 1 + \dfrac{1}{4x}$. So

$S = \displaystyle\int_4^9 2\pi y\sqrt{1 + \left(\frac{dy}{dx}\right)^2}\, dx = \int_4^9 2\pi\sqrt{x}\sqrt{1 + \frac{1}{4x}}\, dx = 2\pi\int_4^9 \sqrt{x + \tfrac{1}{4}}\, dx$

$= 2\pi\left[\frac{2}{3}\left(x + \frac{1}{4}\right)^{3/2}\right]_4^9 = \frac{4\pi}{3}\left[\frac{1}{8}(4x + 1)^{3/2}\right]_4^9 = \frac{\pi}{6}\left(37\sqrt{37} - 17\sqrt{17}\right).$

3. $y = x^3 \quad \Rightarrow \quad y' = 3x^2$. So $S = \displaystyle\int_0^2 2\pi y\sqrt{1 + (y')^2}\, dx = 2\pi\int_0^2 x^3\sqrt{1 + 9x^4}\, dx$ (Let $u = 1 + 9x^4$, so $du = 36x^3\, dx$) $= \frac{2\pi}{36}\int_1^{145} \sqrt{u}\, du = \frac{\pi}{18}\left[\frac{2}{3}u^{3/2}\right]_1^{145} = \frac{\pi}{27}\left(145\sqrt{145} - 1\right)$.

5. $y = \sin x \quad \Rightarrow \quad 1 + \left(\dfrac{dy}{dx}\right)^2 = 1 + \cos^2 x$. So

$$S = 2\pi \int_0^\pi \sin x \sqrt{1 + \cos^2 x}\, dx = 2\pi \int_{-1}^1 \sqrt{1 + u^2}\, du \quad (\text{put } u = -\cos x \Rightarrow du = \sin x\, dx)$$
$$= 4\pi \int_0^1 \sqrt{1 + u^2}\, du = 4\pi \int_0^{\pi/4} \sec^3\theta\, d\theta \quad (\text{put } u = \tan\theta \quad \Rightarrow \quad du = \sec^2\theta\, d\theta)$$
$$= 2\pi[\sec\theta\tan\theta + \ln|\sec\theta + \tan\theta|]_0^{\pi/4} = 2\pi\left[\sqrt{2} + \ln\left(\sqrt{2} + 1\right)\right].$$

7. $y = \cosh x \quad \Rightarrow \quad 1 + \left(\dfrac{dy}{dx}\right)^2 = 1 + \sinh^2 x = \cosh^2 x$. So $S = 2\pi \int_0^1 \cosh x \cosh x\, dx$

$$= 2\pi \int_0^1 \tfrac{1}{2}(1 + \cosh 2x)dx = \pi\left[x + \tfrac{1}{2}\sinh 2x\right]_0^1 = \pi\left(1 + \tfrac{1}{2}\sinh 2\right) \text{ or } \pi\left[1 + \tfrac{1}{4}(e^2 - e^{-2})\right].$$

9. $x = \frac{1}{3}(y^2 + 2)^{3/2} \quad \Rightarrow \quad dx/dy = \frac{1}{2}(y^2 + 2)^{1/2}(2y) = y\sqrt{y^2 + 2} \quad \Rightarrow$

$1 + (dx/dy)^2 = 1 + y^2(y^2 + 2) = (y^2 + 1)^2$. So

$$S = 2\pi \int_1^2 y(y^2 + 1)dy = 2\pi\left[\tfrac{1}{4}y^4 + \tfrac{1}{2}y^2\right]_1^2 = 2\pi\left(4 + 2 - \tfrac{1}{4} - \tfrac{1}{2}\right) = \tfrac{21\pi}{2}.$$

11. $y = \sqrt[3]{x} \quad \Rightarrow \quad x = y^3 \quad \Rightarrow \quad 1 + (dx/dy)^2 = 1 + 9y^4$. So

$$S = 2\pi \int_1^2 x\sqrt{1 + (dx/dy)^2}\, dy = 2\pi \int_1^2 y^3\sqrt{1 + 9y^4}\, dy = \frac{2\pi}{36}\int_1^2 \sqrt{1 + 9y^4}\, 36y^3\, dy$$
$$= \tfrac{\pi}{18}\left[\tfrac{2}{3}\left(1 + 9y^4\right)^{3/2}\right]_1^2 = \tfrac{\pi}{27}\left(145\sqrt{145} - 10\sqrt{10}\right).$$

13. $x = e^{2y} \quad \Rightarrow \quad 1 + (dx/dy)^2 = 1 + 4e^{4y}$. So

$$S = 2\pi \int_0^{1/2} e^{2y}\sqrt{1 + (2e^{2y})^2}\, dy = 2\pi \int_2^{2e} \sqrt{1 + u^2}\, \tfrac{1}{4}\, du \quad [\text{put } u = 2e^{2y},\ du = 4e^{2y}\, dy]$$
$$= \tfrac{\pi}{2}\int_2^{2e} \sqrt{1 + u^2}\, du = \tfrac{\pi}{2}\left[\tfrac{1}{2}u\sqrt{1 + u^2} + \tfrac{1}{2}\ln\left|u + \sqrt{1 + u^2}\right|\right]_2^{2e} \quad [\text{put } u = \tan\theta \text{ or use Formula 21}]$$
$$= \tfrac{\pi}{2}\left[e\sqrt{1 + 4e^2} + \tfrac{1}{2}\ln\left(2e + \sqrt{1 + 4e^2}\right) - \sqrt{5} - \tfrac{1}{2}\ln\left(2 + \sqrt{5}\right)\right]$$
$$= \frac{\pi}{4}\left[2e\sqrt{1 + 4e^2} - 2\sqrt{5} + \ln\left(\frac{2e + \sqrt{1 + 4e^2}}{2 + \sqrt{5}}\right)\right].$$

15. $x = \frac{1}{2\sqrt{2}}(y^2 - \ln y) \quad \Rightarrow \quad \dfrac{dx}{dy} = \dfrac{1}{2\sqrt{2}}\left(2y - \dfrac{1}{y}\right) \quad \Rightarrow \quad 1 + \left(\dfrac{dx}{dy}\right)^2 = 1 + \dfrac{1}{8}\left(2y - \dfrac{1}{y}\right)^2$

$$= 1 + \frac{1}{8}\left(4y^2 - 4 + \frac{1}{y^2}\right) = \frac{1}{8}\left(4y^2 + 4 + \frac{1}{y^2}\right) = \left[\frac{1}{2\sqrt{2}}\left(2y + \frac{1}{y}\right)\right]^2. \text{ So}$$
$$S = 2\pi \int_1^2 \frac{1}{2\sqrt{2}}(y^2 - \ln y)\frac{1}{2\sqrt{2}}\left(2y + \frac{1}{y}\right)dy = \frac{\pi}{4}\int_1^2 \left(2y^3 + y - 2y\ln y - \frac{\ln y}{y}\right)dy$$
$$= \tfrac{\pi}{4}\left[\tfrac{1}{2}y^4 + \tfrac{1}{2}y^2 - y^2\ln y + \tfrac{1}{2}y^2 - \tfrac{1}{2}(\ln y)^2\right]_1^2 = \tfrac{\pi}{8}\left[y^4 + 2y^2 - 2y^2\ln y - (\ln y)^2\right]_1^2$$
$$= \tfrac{\pi}{8}\left[16 + 8 - 8\ln 2 - (\ln 2)^2 - 1 - 2\right] = \tfrac{\pi}{8}\left[21 - 8\ln 2 - (\ln 2)^2\right].$$

17. $S = 2\pi \int_0^1 x^4\sqrt{1 + (4x^3)^2}\, dx = 2\pi \int_0^1 x^4\sqrt{16x^6 + 1}\, dx$

$$\approx 2\pi\tfrac{1/10}{3}\left[f(0) + 4f(0.1) + 2f(0.2) + 4f(0.3) + 2f(0.4) + 4f(0.5) + 2f(0.6)\right.$$
$$\left. + 4f(0.7) + 2f(0.8) + 4f(0.9) + f(1)\right] \approx 3.44. \text{ Here } f(x) = x^4\sqrt{16x^6 + 1}.$$

19. The curve $8y^2 = x^2(1-x^2)$ actually consists of two loops in the region described by the inequalities $|x| \le 1$, $|y| \le \frac{\sqrt{2}}{8}$. $\left(\text{The maximum value of } |y| \text{ is attained when } |x| = \frac{1}{\sqrt{2}}.\right)$ If we consider the loop in the region $x \ge 0$, the surface area S it generates when rotated about the x-axis is calculated as follows: $16y\dfrac{dy}{dx} = 2x - 4x^3$, so

$$\left(\frac{dy}{dx}\right)^2 = \left(\frac{x-2x^3}{8y}\right)^2 = \frac{x^2(1-2x^2)^2}{64y^2} = \frac{x^2(1-2x^2)^2}{8x^2(1-x^2)} = \frac{(1-2x^2)^2}{8(1-x^2)}$$ for $x \ne 0, \pm 1$. The formula also holds for $x = 0$ by continuity. $1 + \left(\dfrac{dy}{dx}\right)^2 = 1 + \dfrac{(1-2x^2)^2}{8(1-x^2)} = \dfrac{9-12x^2+4x^4}{8(1-x^2)} = \dfrac{(3-2x^2)^2}{8(1-x^2)}$. So

$$S = 2\pi\int_0^1 \frac{\sqrt{x^2(1-x^2)}}{2\sqrt{2}} \cdot \frac{3-2x^2}{2\sqrt{2}\sqrt{1-x^2}}\,dx = \tfrac{\pi}{4}\int_0^1 x(3-2x^2)dx = \tfrac{\pi}{4}\left[\tfrac{3}{2}x^2 - \tfrac{1}{2}x^4\right]_0^1 = \tfrac{\pi}{4}\left(\tfrac{3}{2}-\tfrac{1}{2}\right) = \tfrac{\pi}{4}.$$

21. $$S = 2\pi\int_1^\infty y\sqrt{1+\left(\frac{dy}{dx}\right)^2}\,dx = 2\pi\int_1^\infty \frac{1}{x}\sqrt{1+\frac{1}{x^4}}\,dx = 2\pi\int_1^\infty \frac{\sqrt{x^4+1}}{x^3}\,dx$$
$$> 2\pi\int_1^\infty \frac{x^2}{x^3}\,dx = 2\pi\int_1^\infty \frac{dx}{x} = 2\pi\lim_{t\to\infty}\left[\ln x\right]_1^t = 2\pi\lim_{t\to\infty}\ln t = \infty.$$

23. $$\frac{x^2}{a^2} + \frac{y^2}{b^2} = 1 \quad\Rightarrow\quad \frac{y(dy/dx)}{b^2} = -\frac{x}{a^2} \quad\Rightarrow\quad \frac{dy}{dx} = -\frac{b^2x}{a^2y} \quad\Rightarrow$$
$$1 + \left(\frac{dy}{dx}\right)^2 = 1 + \frac{b^4x^2}{a^4y^2} = \frac{b^4x^2 + a^4y^2}{a^4y^2} = \frac{b^4x^2 + a^4b^2(1-x^2/a^2)}{a^4b^2(1-x^2/a^2)}$$
$$= \frac{a^4b^2 + b^4x^2 - a^2b^2x^2}{a^4b^2 - a^2b^2x^2} = \frac{a^4 + b^2x^2 - a^2x^2}{a^4 - a^2x^2} = \frac{a^4 - (a^2-b^2)x^2}{a^2(a^2-x^2)}.$$

The ellipsoid's surface area is twice the area generated by rotating the first quadrant portion of the ellipse about the x-axis. Thus

$$S = 2\int_0^a 2\pi y\sqrt{1+\left(\frac{dy}{dx}\right)^2}\,dx = 4\pi\int_0^a \frac{b}{a}\sqrt{a^2-x^2}\,\frac{\sqrt{a^4-(a^2-b^2)x^2}}{a\sqrt{a^2-x^2}}\,dx = \frac{4\pi b}{a^2}\int_0^a \sqrt{a^4-(a^2-b^2)x^2}\,dx$$
$$= \frac{4\pi b}{a^2}\int_0^{a\sqrt{a^2-b^2}} \sqrt{a^4-u^2}\,\frac{du}{\sqrt{a^2-b^2}} \qquad \left(\text{put } u = \sqrt{a^2-b^2}\,x\right)$$
$$= \frac{4\pi b}{a^2\sqrt{a^2-b^2}}\left[\frac{u}{2}\sqrt{a^4-u^2} + \frac{a^4}{2}\sin^{-1}\frac{u}{a^2}\right]_0^{a\sqrt{a^2-b^2}} \qquad \text{(Formula 30)}$$
$$= \frac{4\pi b}{a^2\sqrt{a^2-b^2}}\left[\frac{a\sqrt{a^2-b^2}}{2}\sqrt{a^4-a^2(a^2-b^2)} + \frac{a^4}{2}\sin^{-1}\frac{\sqrt{a^2-b^2}}{a}\right] = 2\pi\left[b^2 + \frac{a^2b\sin^{-1}\dfrac{\sqrt{a^2-b^2}}{a}}{\sqrt{a^2-b^2}}\right]$$

25. In the derivation of (4), we computed a typical contribution to the surface area to be $2\pi\dfrac{y_{i-1}+y_i}{2}|P_{i-1}P_i|$, the area of a frustum of a cone. When $f(x)$ is not necessarily positive, the approximations $y_i = f(x_i) \approx f(x_i^*)$ and $y_{i-1} = f(x_{i-1}) \approx f(x_i^*)$ must be replaced by $y_i = |f(x_i)| \approx |f(x_i^*)|$ and $y_{i-1} = |f(x_{i-1})| \approx |f(x_i^*)|$. Thus $2\pi\dfrac{y_{i-1}+y_i}{2}|P_{i-1}P_i| \approx 2\pi|f(x_i^*)|\sqrt{1+[f'(x_i^*)]^2}\,\Delta x_i$. Continuing with the rest of the derivation as before, we obtain $S = \displaystyle\int_a^b 2\pi|f(x)|\sqrt{1+[f'(x)]^2}\,dx$.

27. For the upper semicircle, $f(x) = \sqrt{r^2 - x^2}$, $f'(x) = -x/\sqrt{r^2 - x^2}$. The surface area generated is

$$S_1 = \int_{-r}^{r} 2\pi\left(r - \sqrt{r^2 - x^2}\right)\sqrt{1 + \frac{x^2}{r^2 - x^2}}\, dx = 4\pi\int_0^r \left(r - \sqrt{r^2 - x^2}\right)\frac{r}{\sqrt{r^2 - x^2}}dx$$

$$= 4\pi\int_0^r \left(\frac{r^2}{\sqrt{r^2 - x^2}} - r\right)dx.$$

For the lower semicircle, $f(x) = -\sqrt{r^2 - x^2}$ and $f'(x) = \dfrac{x}{\sqrt{r^2 - x^2}}$, so $S_2 = 4\pi\displaystyle\int_0^r \left(\frac{r^2}{\sqrt{r^2 - x^2}} + r\right)dx.$

Thus the total area is $S = S_1 + S_2 = 8\pi\displaystyle\int_0^r \left(\frac{r^2}{\sqrt{r^2 - x^2}}\right)dx = 8\pi\left[r^2 \sin^{-1}\left(\frac{x}{r}\right)\right]_0^r = 8\pi r^2\left(\frac{\pi}{2}\right) = 4\pi^2 r^2.$

EXERCISES 8.4

1. $m_1 = 4$, $m_2 = 8$; $P_1(-1, 2)$, $P_2(2, 4)$. $m = m_1 + m_2 = 12$. $M_x = 4 \cdot 2 + 8 \cdot 4 = 40$;
$M_y = 4 \cdot (-1) + 8 \cdot 2 = 12$; $\overline{x} = M_y/m = 1$ and $\overline{y} = M_x/m = \frac{10}{3}$, so the center of mass is $(\overline{x}, \overline{y}) = \left(1, \frac{10}{3}\right)$.

3. $m = m_1 + m_2 + m_3 = 4 + 2 + 5 = 11$. $M_x = 4 \cdot (-2) + 2 \cdot 4 + 5 \cdot (-3) = -15$;
$M_y = 4 \cdot (-1) + 2 \cdot (-2) + 5 \cdot 5 = 17$, $(\overline{x}, \overline{y}) = \left(\frac{17}{11}, -\frac{15}{11}\right)$.

5. $A = \int_0^2 x^2\, dx = \left[\frac{1}{3}x^3\right]_0^2 = \frac{8}{3}$, $\overline{x} = A^{-1}\int_0^2 x \cdot x^2\, dx = \frac{3}{8}\left[\frac{1}{4}x^4\right]_0^2 = \frac{3}{8} \cdot 4 = \frac{3}{2}$,
$\overline{y} = A^{-1}\int_0^2 \frac{1}{2}(x^2)^2\, dx = \frac{3}{8} \cdot \frac{1}{2}\left[\frac{1}{5}x^5\right]_0^2 = \frac{3}{16} \cdot \frac{32}{5} = \frac{6}{5}$. Centroid $(\overline{x}, \overline{y}) = \left(\frac{3}{2}, \frac{6}{5}\right) = (1.5, 1.2)$.

7. $A = \int_{-1}^2 (3x + 5)dx = \left[\frac{3}{2}x^2 + 5x\right]_{-1}^2 = (6 + 10) - \left(\frac{3}{2} - 5\right) = 16 + \frac{7}{2} = \frac{39}{2}$,
$\overline{x} = A^{-1}\int_{-1}^2 x(3x + 5)dx = \frac{2}{39}\int_{-1}^2 (3x^2 + 5x)dx = \frac{2}{39}\left[x^3 + \frac{5}{2}x^2\right]_{-1}^2$
$= \frac{2}{39}\left[(8 + 10) - \left(-1 + \frac{5}{2}\right)\right] = \frac{2}{39}\left(\frac{36-3}{2}\right) = \frac{11}{13}$,
$\overline{y} = A^{-1}\int_{-1}^2 \frac{1}{2}(3x + 5)^2\, dx = \frac{1}{39}\int_{-1}^2 (9x^2 + 30x + 25)dx = \frac{1}{39}[3x^3 + 15x^2 + 25x]_{-1}^2$
$= \frac{1}{39}[(24 + 60 + 50) - (-3 + 15 - 25)] = \frac{147}{39} = \frac{49}{13}$. $(\overline{x}, \overline{y}) = \left(\frac{11}{13}, \frac{49}{13}\right)$.

9. By symmetry, $\overline{x} = 0$ and $A = 2\int_0^{\pi/4} \cos 2x\, dx = \sin 2x|_0^{\pi/4} = 1$,
$\overline{y} = A^{-1}\int_{-\pi/4}^{\pi/4} \frac{1}{2}\cos^2 2x\, dx = \int_0^{\pi/4} \cos^2 2x\, dx = \frac{1}{2}\int_0^{\pi/4}(1 + \cos 4x)dx = \frac{1}{2}\left[x + \frac{1}{4}\sin 4x\right]_0^{\pi/4}$
$= \frac{1}{2}\left(\frac{\pi}{4} + \frac{1}{4} \cdot 0\right) = \frac{\pi}{8}$. $(\overline{x}, \overline{y}) = \left(0, \frac{\pi}{8}\right)$.

11. $A = \int_0^1 e^x\, dx = [e^x]_0^1 = e - 1$,

$$\overline{x} = \frac{1}{A}\int_0^1 xe^x\, dx = \frac{1}{e-1}[xe^x - e^x]_0^1 \quad \text{(integration by parts)} \quad = \frac{1}{e-1}[0 - (-1)] = \frac{1}{e-1},$$

$$\overline{y} = \frac{1}{A}\int_0^1 \frac{(e^x)^2}{2}\, dx = \frac{1}{e-1} \cdot \tfrac{1}{4}[e^{2x}]_0^1 = \frac{1}{4(e-1)}(e^2 - 1) = \frac{e+1}{4}. \quad (\overline{x}, \overline{y}) = \left(\frac{1}{e-1}, \frac{e+1}{4}\right).$$

13. $A = \int_0^1 \left(\sqrt{x} - x\right) dx = \left[\frac{2}{3}x^{3/2} - \frac{1}{2}x^2\right]_0^1 = \frac{2}{3} - \frac{1}{2} = \frac{1}{6}$,

$\overline{x} = A^{-1}\int_0^1 x\left(\sqrt{x} - x\right) dx = 6\int_0^1 \left(x^{3/2} - x^2\right) dx = 6\left[\frac{2}{5}x^{5/2} - \frac{1}{3}x^3\right]_0^1 = 6\left(\frac{2}{5} - \frac{1}{3}\right) = \frac{2}{5}$,

$\overline{y} = A^{-1}\int_0^1 \frac{1}{2}\left[\left(\sqrt{x}\right)^2 - x^2\right] dx = 3\int_0^1 (x - x^2) dx = 3\left[\frac{1}{2}x^2 - \frac{1}{3}x^3\right]_0^1 = 3\left(\frac{1}{2} - \frac{1}{3}\right) = \frac{1}{2}$.

$(\overline{x}, \overline{y}) = \left(\frac{2}{5}, \frac{1}{2}\right) = (0.4, 0.5)$.

15. $A = \int_0^{\pi/4} (\cos x - \sin x) dx = [\sin x + \cos x]_0^{\pi/4} = \sqrt{2} - 1$,

$\overline{x} = A^{-1}\int_0^{\pi/4} x(\cos x - \sin x) dx = A^{-1}[x(\sin x + \cos x) + \cos x - \sin x]_0^{\pi/4}$ [integration by parts]

$= A^{-1}\left(\frac{\pi}{4}\sqrt{2} - 1\right) = \dfrac{\frac{1}{4}\pi\sqrt{2} - 1}{\sqrt{2} - 1}$,

$$\overline{y} = A^{-1}\int_0^{\pi/4} \frac{1}{2}\left(\cos^2 x - \sin^2 x\right) dx = \frac{1}{2A}\int_0^{\pi/4} \cos 2x \, dx = \frac{1}{4A}[\sin 2x]_0^{\pi/4} = \frac{1}{4A} = \frac{1}{4(\sqrt{2} - 1)}.$$

$$(\overline{x}, \overline{y}) = \left(\frac{\pi\sqrt{2} - 4}{4(\sqrt{2} - 1)}, \frac{1}{4(\sqrt{2} - 1)}\right).$$

17. By symmetry, $M_y = 0$ and $\overline{x} = 0$. $A = \frac{1}{2}bh = \frac{1}{2} \cdot 2 \cdot 2 = 2$.

$M_x = 2\rho\int_0^1 \frac{1}{2}(2 - 2x)^2 \, dx = 4\int_0^1 (1 - x)^2 \, dx$

$= 4\left[-\frac{1}{3}(1 - x)^3\right]_0^1 = 4 \cdot \frac{1}{3} = \frac{4}{3}$.

$\overline{y} = \dfrac{1}{\rho A}M_x = \dfrac{2}{3}$. $(\overline{x}, \overline{y}) = \left(0, \frac{2}{3}\right)$.

19. By symmetry, $M_y = 0$ and $\overline{x} = 0$. $A =$ area of triangle $+$ area of square $= 1 + 4 = 5$, so $m = \rho A = 4 \cdot 5 = 20$.

$M_x = \rho \cdot 2\int_0^1 \frac{1}{2}\left[(1 - x)^2 - (-2)^2\right] dx = 4\int_0^1 (x^2 - 2x - 3) dx$

$= 4\left[\frac{1}{3}x^3 - x^2 - 3x\right]_0^1 = 4\left(\frac{1}{3} - 1 - 3\right) = 4\left(-\frac{11}{3}\right) = -\frac{44}{3}$. $\overline{y} = M_x/m = \frac{1}{20}\left(-\frac{44}{3}\right) = -\frac{11}{15}$. $(\overline{x}, \overline{y}) = \left(0, -\frac{11}{15}\right)$.

21. Choose x- and y-axes so that the base (one side of the triangle) lies along the x-axis with the other vertex along the positive y-axis as shown. From geometry, we know the medians intersect at a point $\frac{2}{3}$ of the way from each vertex (along the median) to the opposite side.

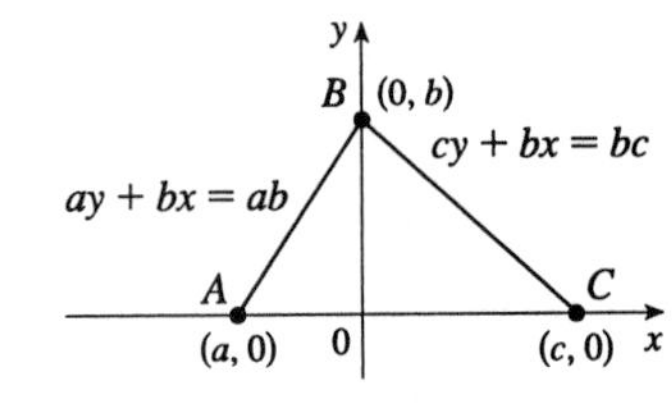

The median from B goes to the midpoint $\left(\frac{1}{2}(a + c), 0\right)$ of side AC, so the point of intersection of the medians is $\left(\frac{2}{3} \cdot \frac{1}{2}(a + c), \frac{1}{3}b\right) = \left(\frac{1}{3}(a + c), \frac{1}{3}b\right)$. This can also be verified by finding the equations of two medians, and solving them simultaneously to find their point of intersection. Now let us compute the location of the centroid of the triangle. The area is $A = \frac{1}{2}(c - a)b$.

$$\overline{x} = \frac{1}{A}\left[\int_a^0 x \cdot \frac{b}{a}(a - x) dx + \int_0^c x \cdot \frac{b}{c}(c - x) dx\right] = \frac{1}{A}\left[\frac{b}{a}\int_a^0 \left(ax - x^2\right) dx + \frac{b}{c}\int_0^c \left(cx - x^2\right) dx\right]$$

$$= \frac{b}{Aa}\left[\tfrac{1}{2}ax^2 - \tfrac{1}{3}x^3\right]_a^0 + \frac{b}{Ac}\left[\tfrac{1}{2}cx^2 - \tfrac{1}{3}x^3\right]_0^c = \frac{b}{Aa}\left[-\tfrac{1}{2}a^3 + \tfrac{1}{3}a^3\right] + \frac{b}{Ac}\left[\tfrac{1}{2}c^3 - \tfrac{1}{3}c^3\right]$$

$$= \frac{2}{a(c - a)} \cdot \frac{-a^3}{6} + \frac{2}{c(c - a)} \cdot \frac{c^3}{6} = \frac{1}{3(c - a)}\left(c^2 - a^2\right) = \frac{a + c}{3}, \text{ and}$$

$$\overline{y} = \frac{1}{A}\left[\int_a^0 \tfrac{1}{2}\left(\frac{b}{a}(a-x)\right)^2 dx + \int_0^c \tfrac{1}{2}\left(\frac{b}{c}(c-x)\right)^2 dx\right]$$

$$= \frac{1}{A}\left[\frac{b^2}{2a^2}\int_a^0 (a^2 - 2ax + x^2)\,dx + \frac{b^2}{2c^2}\int_0^c (c^2 - 2cx + x^2)\,dx\right]$$

$$= \frac{1}{A}\left[\frac{b^2}{2a^2}\left[a^2x - ax^2 + \tfrac{1}{3}x^3\right]_a^0 + \frac{b^2}{2c^2}\left[c^2x - cx^2 + \tfrac{1}{3}x^3\right]_0^c\right]$$

$$= \frac{1}{A}\left[\frac{b^2}{2a^2}\left(-a^3 + a^3 - \tfrac{1}{3}a^3\right) + \frac{b^2}{2c^2}\left(c^3 - c^3 + \tfrac{1}{3}c^3\right)\right] = \frac{1}{A}\left[\frac{b^2}{6}(-a+c)\right] = \frac{2}{(c-a)b}\cdot\frac{(c-a)b^2}{6} = \frac{b}{3}.$$

Thus $(\overline{x}, \overline{y}) = \left(\dfrac{a+c}{3}, \dfrac{b}{3}\right)$ as claimed.

Remarks: Actually the computation of $\overline{y}$ is all that is needed. By considering each side of the triangle in turn to be the base, we see that the centroid is $\frac{1}{3}$ of the way from each side to the opposite vertex and must therefore be the intersection of the medians.

The computation of $\overline{y}$ in this problem (and many others) can be simplified by using horizontal rather than vertical approximating rectangles. If the length of a thin rectangle at coordinate y is $\ell(y)$, then its area is $\ell(y)\Delta y$, its mass is $\rho\ell(y)\Delta y$, and its moment about the x-axis is $\Delta M_x = \rho y\ell(y)\Delta y$. Thus $M_x = \int \rho y\ell(y)dy$ and

$\overline{y} = \dfrac{\int \rho y\ell(y)dy}{\rho A} = \dfrac{1}{A}\displaystyle\int y\ell(y)dy$. In this problem,

$\ell(y) = \dfrac{c-a}{b}(b-y)$ by similar triangles, so

$$\overline{y} = \frac{1}{A}\int_0^b \frac{c-a}{b}y(b-y)dy = \frac{2}{b^2}\int_0^b (by - y^2)\,dy = \frac{2}{b^2}\left[\tfrac{1}{2}by^2 - \tfrac{1}{3}y^3\right]_0^b = \frac{2}{b^2}\cdot\frac{b^3}{6} = \frac{b}{3}.$$

Notice that only one integral is needed when this method is used.

Since the position of a centroid is independent of density when the density is constant, we will assume for convenience that $\rho = 1$ in Exercises 23 and 25.

23. Divide the lamina into two triangles and one rectangle with respective masses of 2, 2 and 4, so that the total mass is 8. Using the result of Exercise 21, the triangles have centroids $\left(-1, \frac{2}{3}\right)$ and $\left(1, \frac{2}{3}\right)$. The centroid of the rectangle (its center) is $\left(0, -\frac{1}{2}\right)$. So, using Formulas 5 and 7, we have

$\overline{y} = \dfrac{\sum m_i y_i}{m} = \frac{2}{8}\left(\frac{2}{3}\right) + \frac{2}{8}\left(\frac{2}{3}\right) + \frac{4}{8}\left(-\frac{1}{2}\right) = \frac{1}{12}$, and $\overline{x} = 0$, since the lamina is symmetric about the line $x = 0$.

Therefore $(\overline{x}, \overline{y}) = \left(0, \frac{1}{12}\right)$.

25. Suppose first that the large rectangle were complete, so that its mass would be $6 \cdot 3 = 18$. Its centroid would be $\left(1, \frac{3}{2}\right)$. The mass removed from this object to create the one being studied is 3. The centroid of the cut-out piece is $\left(\frac{3}{2}, \frac{3}{2}\right)$. Therefore, for the actual lamina, whose mass is 15, $\overline{x} = \frac{18}{15}(1) - \frac{3}{15}\left(\frac{3}{2}\right) = \frac{9}{10}$, and $\overline{y} = \frac{3}{2}$, since the lamina is symmetric about the line $y = \frac{3}{2}$. Therefore $(\overline{x}, \overline{y}) = \left(\frac{9}{10}, \frac{3}{2}\right)$.

27. A cone of height h and radius r can be generated by rotating a right triangle about one of its legs as shown. By Exercise 21, $\overline{x} = \frac{1}{3}r$, so by the Theorem of Pappus, the volume of the cone is $V = Ad = \frac{1}{2}rh \cdot 2\pi\left(\frac{1}{3}r\right) = \frac{1}{3}\pi r^2 h$.

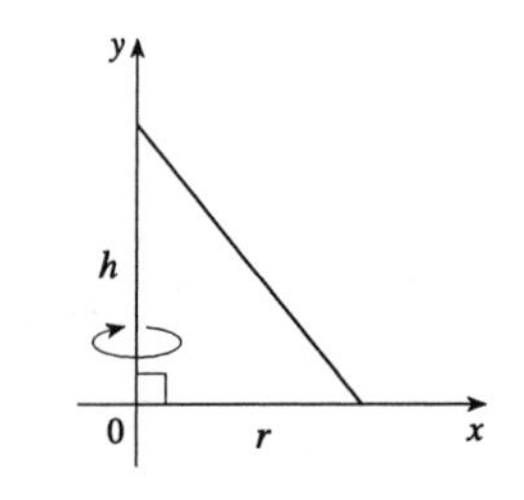

29. Suppose the region lies between two curves $y = f(x)$ and $y = g(x)$ where $f(x) \geq g(x)$, as illustrated in Figure 9. Take a partition P by points x_i with $a = x_0 < x_1 < \cdots < x_n = b$ and choose x_i^* to be the midpoint of the ith subinterval; that is, $x_i^* = \frac{1}{2}(x_{i-1} + x_i)$. Then the centroid of the ith approximating rectangle R_i is its center $C_i = \left(x_i^*, \frac{1}{2}[f(x_i^*) + g(x_i^*)]\right)$. Its area is $[f(x_i^*) - g(x_i^*)]\Delta x_i$, so its mass is $\rho[f(x_i^*) - g(x_i^*)]\Delta x_i$. Thus
$M_y(R_i) = \rho[f(x_i^*) - g(x_i^*)]\Delta x_i \cdot x_i^* = \rho x_i^*[f(x_i^*) - g(x_i^*)]\Delta x_i$ and
$M_x(R_i) = \rho[f(x_i^*) - g(x_i^*)]\Delta x_i \cdot \frac{1}{2}[f(x_i^*) + g(x_i^*)] = \rho \cdot \frac{1}{2}\left[f(x_i^*)^2 - g(x_i^*)^2\right]\Delta x_i$.

Summing over i and taking the limit as $\|P\| \to 0$, we get
$M_y = \lim_{\|P\| \to 0} \sum_i \rho x_i^*[f(x_i^*) - g(x_i^*)]\Delta x_i = \rho \int_a^b x[f(x) - g(x)]\,dx$ and
$M_x = \lim_{\|P\| \to 0} \sum_i \rho \cdot \frac{1}{2}\left[f(x_i^*)^2 - g(x_i^*)^2\right]\Delta x_i = \rho \int_a^b \frac{1}{2}\left[f(x)^2 - g(x)^2\right]dx$. Thus

$$\overline{x} = \frac{M_y}{m} = \frac{M_y}{\rho A} = \frac{1}{A}\int_a^b x[f(x) - g(x)]\,dx \text{ and } \overline{y} = \frac{M_x}{m} = \frac{M_x}{\rho A} = \frac{1}{A}\int_a^b \tfrac{1}{2}\left[f(x)^2 - g(x)^2\right]dx.$$

EXERCISES 8.5

1. **(a)** $P = \rho g d = (1000\,\text{kg/m}^3)(9.8\,\text{m/s}^2)(1\,\text{m}) = 9800\,\text{Pa} = 9.8\,\text{kPa}$

(b) $F = \rho g d A = PA = (9800\,\text{N/m}^2)(2\,\text{m}^2) = 1.96 \times 10^4\,\text{N}$

(c) $F = \int_0^1 \rho g x \cdot 1\,dx = 9800 \int_0^1 x\,dx = 4900x^2\big|_0^1 = 4.90 \times 10^3\,\text{N}$

3. $F = \int_0^{10} \rho g x \cdot 2\sqrt{100 - x^2}\,dx = 9.8 \times 10^3 \int_0^{10} \sqrt{100 - x^2}\,2x\,dx$
$= 9.8 \times 10^3 \int_{100}^0 u^{1/2}(-du)$ (put $u = 100 - x^2$)
$= 9.8 \times 10^3 \int_0^{100} u^{1/2}\,du = 9.8 \times 10^3 \left[\frac{2}{3}u^{3/2}\right]_0^{100}$
$= \frac{2}{3} \cdot 9.8 \times 10^6 \approx 6.5 \times 10^6\,\text{N}$

5. $F = \int_{-r}^{r} \rho g(x + r) \cdot 2\sqrt{r^2 - x^2}\,dx = \rho g \int_{-r}^{r} \sqrt{r^2 - x^2}\,2x\,dx + 2\rho g r \int_{-r}^{r} \sqrt{r^2 - x^2}\,dx$. The first integral is 0 because the integrand is an odd function. The second integral can be interpreted as the area of a semicircular disk with radius r, or we could make the trigonometric substitution $x = r\sin\theta$. Continuing:
$F = \rho g \cdot 0 + 2\rho g r \cdot \frac{1}{2}\pi r^2 = \rho g \pi r^3 = 1000 g \pi r^3\,\text{N}$ (SI units assumed).

7. $F = \int_0^6 \delta x \cdot \frac{2x}{3}\,dx = \left[\frac{2}{9}\delta x^3\right]_0^6$
$= 48\delta \approx 48 \times 62.5 = 3000\,\text{lb}$

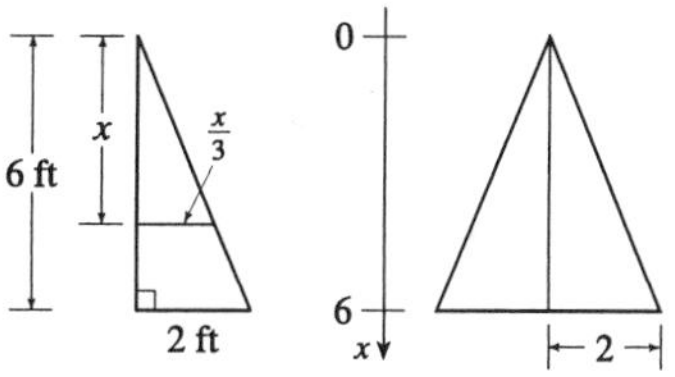

6 ft, x, $\frac{x}{3}$, 2 ft, 0, 6, x, 2

9. $F = \int_2^6 \delta(x-2)\frac{2}{3}x\,dx = \frac{2}{3}\delta\int_2^6 (x^2 - 2x)dx$
$= \frac{2}{3}\delta\left[\frac{1}{3}x^3 - x^2\right]_2^6 = \frac{2}{3}\delta\left[36 - \left(-\frac{4}{3}\right)\right]$
$= \frac{224}{9}\delta \approx 1.56 \times 10^3\,\text{lb}$

11. $F = \int_0^8 \delta x \cdot (12 + x)dx = \delta\int_0^8 (12x + x^2)dx$
$= \delta\left[6x^2 + \frac{x^3}{3}\right]_0^8 = \delta\left(384 + \frac{512}{3}\right)$
$= (62.5)\frac{1664}{3} \approx 3.47 \times 10^4\,\text{lb}$

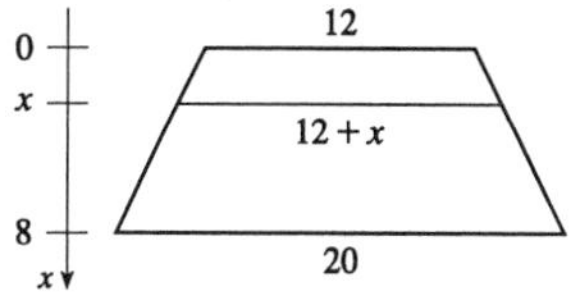

0, x, 8, x, 12, $12 + x$, 20

13. $F = \int_0^{4\sqrt{3}} \rho g\left(4\sqrt{3} - x\right)\frac{2x}{\sqrt{3}}\,dx = 8\rho g\int_0^{4\sqrt{3}} x\,dx - \frac{2\rho g}{\sqrt{3}}\int_0^{4\sqrt{3}} x^2\,dx$
$= 4\rho g\left[x^2\right]_0^{4\sqrt{3}} - \frac{2\rho g}{3\sqrt{3}}\left[x^3\right]_0^{4\sqrt{3}} = 192\rho g - \frac{2\rho g}{3\sqrt{3}}64 \cdot 3\sqrt{3}$
$= 192\rho g - 128\rho g = 64\rho g \approx 64(840)(9.8) \approx 5.27 \times 10^5\,\text{N}$

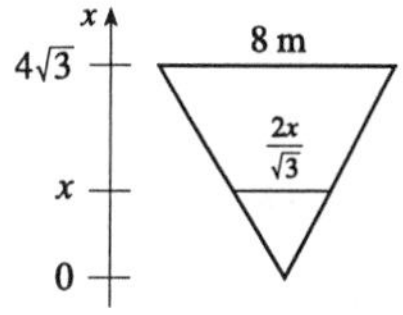

x, $4\sqrt{3}$, x, 0, 8 m, $\frac{2x}{\sqrt{3}}$

15. **(a)** $F = \rho g d A \approx (1000)(9.8)(0.8)(0.2)^2$
$\approx 314\,\text{N}$

(b) $F = \int_{0.8}^1 \rho g x(0.2)dx = 0.2\rho g\left[\frac{1}{2}x^2\right]_{0.8}^1$
$= (0.2\rho g)(0.18) = 0.036\rho g$
$\approx 353\,\text{N}$

17. $F = \int_0^2 \rho g x \cdot 3 \cdot \sqrt{2}\,dx = 3\sqrt{2}\rho g\int_0^2 x\,dx$
$= 3\sqrt{2}\rho g\left[\frac{1}{2}x^2\right]_0^2 = 6\sqrt{2}\rho g \approx 8.32 \times 10^4\,\text{N}$

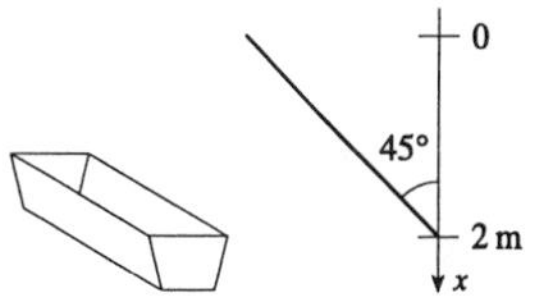

0, 45°, 2 m, x

19. Assume that the pool is filled with water.

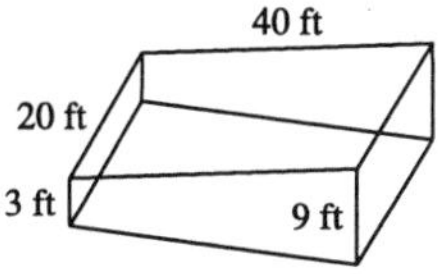

(a) $F = \int_0^3 \delta x\,20\,dx = 20\delta\left[\frac{1}{2}x^2\right]_0^3 = 20\delta \cdot \frac{9}{2} = 90\delta \approx 5625\,\text{lb} \approx 5.63 \times 10^3\,\text{lb}$

(b) $F = \int_0^9 \delta x 20\,dx = 20\delta\left[\frac{1}{2}x^2\right]_0^9 = 810\delta \approx 50625\,\text{lb} \approx 5.06 \times 10^4\,\text{lb}.$

40 ft, 20 ft, 3 ft, 9 ft

(c) $F = \int_0^3 \delta x\,40\,dx + \int_3^9 \delta x(40)\frac{9-x}{6}\,dx = 40\delta\left[\frac{1}{2}x^2\right]_0^3 + \frac{20}{3}\delta\int_3^9 (9x - x^2)dx$
$= 180\delta + \frac{20}{3}\delta\left[\frac{9}{2}x^2 - \frac{1}{3}x^3\right]_3^9 = 180\delta + \frac{20}{3}\delta\left[\left(\frac{729}{2} - 243\right) - \left(\frac{81}{2} - 9\right)\right] = 780\delta \approx 4.88 \times 10^4\,\text{lb}$

(d) $F = \int_3^9 \delta x\,20\frac{\sqrt{409}}{3}\,dx$
$= \frac{1}{3}\left(20\sqrt{409}\right)\delta\left[\frac{1}{2}x^2\right]_3^9$
$= \frac{1}{3} \cdot 10\sqrt{409}\delta(81 - 9)$
$\approx 3.03 \times 10^5\,\text{lb}$

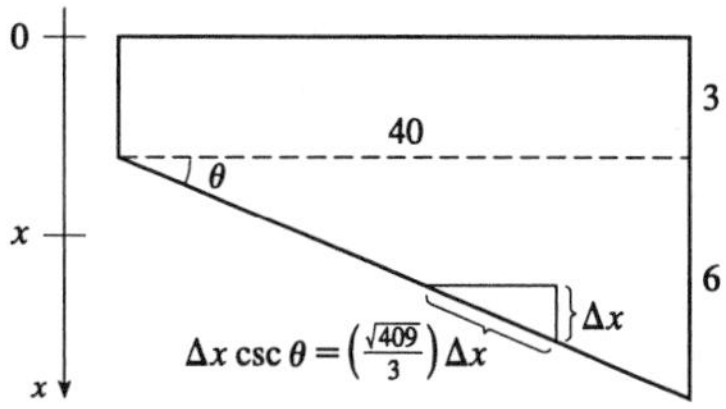

0, x, x, 40, 3, θ, 6, Δx, $\Delta x \csc\theta = \left(\frac{\sqrt{409}}{3}\right)\Delta x$

21. $\overline{x} = A^{-1}\int_a^b xw(x)dx$ (Equation 1) $\Rightarrow$ $A\overline{x} = \int_a^b xw(x)dx$ $\Rightarrow$ $(\rho g\overline{x})A = \int_a^b \rho g x w(x)dx = F$ by Exercise 20.

EXERCISES 8.6

1. $C(2000) = C(0) + \int_0^{2000} C'(x)dx = 1{,}500{,}000 + \int_0^{2000} (0.006x^2 - 1.5x + 8)dx$
$= 1{,}500{,}000 + \left[0.002x^3 - 0.75x^2 + 8x\right]_0^{2000} = \$14{,}516{,}000$

3. $C(5000) - C(3000) = \int_{3000}^{5000} (140 - 0.5x + 0.012x^2)dx = [140x - 0.25x^2 + 0.004x^3]_{3000}^{5000}$
$= 494{,}450{,}000 - 106{,}170{,}000 = \$388{,}280{,}000$

5. $p(x) = 20 = \dfrac{1000}{x+20} \quad \Rightarrow \quad x + 20 = 50 \quad \Rightarrow \quad x = 30.$

Consumer surplus $= \displaystyle\int_0^{30} [p(x) - 20]dx = \int_0^{30} \left(\frac{1000}{x+20} - 20\right)dx = [1000\ln(x+20) - 20x]_0^{30}$
$= 1000\ln\left(\frac{50}{20}\right) - 600 = 1000\ln\left(\frac{5}{2}\right) - 600 \approx \$316.29.$

7. $P = p(x) = 10 = 5 + \frac{1}{10}\sqrt{x} \quad \Rightarrow \quad 50 = \sqrt{x} \quad \Rightarrow \quad x = 2500.$

Producer surplus $= \int_0^{2500} [P - p(x)]dx = \int_0^{2500} \left(10 - 5 - \frac{1}{10}\sqrt{x}\right)dx = \left[5x - \frac{1}{15}x^{3/2}\right]_0^{2500} \approx \$4166.67$

9. The demand function is linear, with slope $\frac{-10}{100}$ and $p(1000) = 450$. So its equation is $p - 450 = -\frac{1}{10}(x - 1000)$ or $p = -\frac{1}{10}x + 550$. A selling price of \$400 $\Rightarrow \quad 400 = -\frac{1}{10}x + 550 \quad \Rightarrow \quad x = 1500.$

Consumer surplus $= \int_0^{1500} \left(550 - \frac{1}{10}x - 400\right)dx = \left[150x - \frac{1}{20}x^2\right]_0^{1500} = \$112{,}500$

11. Pretend that it is five years later. Then the fund will start in five years and continue for 15 years, so the present value (five years from now) is $\int_5^{20} 12{,}000e^{-0.06t}\,dt = -\frac{12{,}000}{0.06}\left[e^{-0.06t}\right]_5^{20} = \frac{12{,}000}{0.06}\left(e^{-0.3} - e^{-1.2}\right) \approx \$87{,}924.80.$

13. **(a)** $f(t) = A$, so present value $= \int_0^\infty Ae^{-rt}\,dt = \lim\limits_{x\to\infty} \int_0^x Ae^{-rt}\,dt = \lim\limits_{x\to\infty} -(A/r)\left[e^{-rt}\right]_0^x$
$= \lim\limits_{x\to\infty} -(A/r)[e^{-rx} - 1] = A/r \quad$ (since $r > 0$, $e^{-rx} \to 0$ as $x \to \infty$.)

(b) $r = 0.08$, $A = 5000$, so present value $= \frac{5000}{0.08} = \$62{,}500$ [by part (a)].

15. $f(8) - f(4) = \int_4^8 f'(t)dt = \int_4^8 \sqrt{t}\,dt = \frac{2}{3}t^{3/2}\Big|_4^8 = \frac{2}{3}\left(16\sqrt{2} - 8\right) = \frac{16(2\sqrt{2}-1)}{3} \approx \$9.75$ million

17. $F = \dfrac{\pi PR^4}{8\eta\ell} = \dfrac{\pi(4000)(0.008)^4}{8(0.027)(2)} \approx 1.19 \times 10^{-4}\ \text{cm}^3/\text{s}$

19. $\int_0^{12} c(t)dt = \int_0^{12} \frac{1}{4}t(12-t)dt = \left[\frac{3}{2}t^2 - \frac{1}{12}t^3\right]_0^{12} = \frac{144}{2} = 72\ \text{mg}\cdot\text{s/L}$. Therefore,
$F = A/72 = \frac{8}{72} = \frac{1}{9}\ \text{L/s} = \frac{60}{9}\ \text{L/min}.$

REVIEW EXERCISES FOR CHAPTER 8

1. $y^2(dy/dx) = x + \sin x \quad \Rightarrow \quad \int y^2\,dy = \int (x + \sin x)dx \quad \Rightarrow \quad \frac{1}{3}y^3 = \frac{1}{2}x^2 - \cos x + C \quad \Rightarrow$
$y^3 = \frac{3}{2}x^2 - 3\cos x + K$ (where $K = 3C$) $\quad \Rightarrow \quad y = \sqrt[3]{\frac{3}{2}x^2 - 3\cos x + K}$

3. $y' = \dfrac{1}{x^2y - 2x^2 + y - 2} \quad \Rightarrow \quad \dfrac{dy}{dx} = \dfrac{1}{(x^2+1)(y-2)} \quad \Rightarrow \quad \displaystyle\int (y-2)dy = \int \frac{dx}{x^2+1} \quad \Rightarrow$
$\frac{1}{2}y^2 - 2y = \tan^{-1}x + K \quad \Rightarrow \quad y = 2 \pm \sqrt{2\tan^{-1}x + C}$, where $C = 4 + 2K$.

5. $xyy' = \ln x$, $y(1) = 2$. $y\,dy = \dfrac{\ln x}{x}\,dx \quad \Rightarrow \quad \displaystyle\int y\,dy = \int \frac{\ln x}{x}\,dx$ (Make the substitution $u = \ln x$; then $du = dx/x$.) So $\int y\,dy = \int u\,du \;\Rightarrow\; \frac{1}{2}y^2 = \frac{1}{2}u^2 + C \;\Rightarrow\; \frac{1}{2}y^2 = \frac{1}{2}(\ln x)^2 + C$. $y(1) = 2 \quad \Rightarrow$
$\frac{1}{2}2^2 = \frac{1}{2}(\ln 1)^2 + C = C \quad \Leftrightarrow \quad C = 2$. Therefore, $\frac{1}{2}y^2 = \frac{1}{2}(\ln x)^2 + 2$, or $y = \sqrt{(\ln x)^2 + 4}$. The negative square root is inadmissible, since $y(1) > 0$.

7. $3x = 2(y-1)^{3/2}$, $2 \le y \le 5$. $x = \frac{2}{3}(y-1)^{3/2}$, so $dx/dy = (y-1)^{1/2}$ and the arc length formula gives
$L = \int_2^5 \sqrt{1 + (dx/dy)^2}\,dy = \int_2^5 \sqrt{1 + (y-1)}\,dy = \int_2^5 \sqrt{y}\,dy = \left[\frac{2}{3}y^{3/2}\right]_2^5 = \frac{2}{3}\left(5\sqrt{5} - 2\sqrt{2}\right)$.

9. **(a)** $y = \frac{1}{6}x^3 + 1/(2x)$, $1 \le x \le 2 \quad \Rightarrow \quad y' = \frac{1}{2}(x^2 - 1/x^2) \quad \Rightarrow \quad (y')^2 = \frac{1}{4}(x^4 - 2 + 1/x^4) \quad \Rightarrow$
$1 + (y')^2 = \frac{1}{4}(x^4 + 2 + 1/x^4) = \frac{1}{4}(x^2 + 1/x^2)^2 \quad \Rightarrow$
$$L = \int_1^2 \sqrt{1 + (y')^2}\,dy = \frac{1}{2}\int_1^2 \left(x^2 + \frac{1}{x^2}\right)dx = \frac{1}{2}\left[\frac{x^3}{3} - \frac{1}{x}\right]_1^2 = \frac{1}{2}\left(\frac{17}{6}\right) = \frac{17}{12}.$$

(b) $S = \int_1^2 2\pi y\sqrt{1 + (dy/dx)^2}\,dx = 2\pi\int_1^2 \left[\frac{1}{6}x^3 + 1/(2x)\right]\frac{1}{2}(x^2 + 1/x^2)\,dx$
$= \pi\int_1^2 \left(\frac{1}{6}x^5 + \frac{2}{3}x + \frac{1}{2}x^{-3}\right)dx = \pi\left[\frac{1}{36}x^6 + \frac{1}{3}x^2 - \frac{1}{4}x^{-2}\right]_1^2$
$= \pi\left[\left(\frac{64}{36} + \frac{4}{3} - \frac{1}{16}\right) - \left(\frac{1}{36} + \frac{1}{3} - \frac{1}{4}\right)\right] = \frac{47\pi}{16}$

11. $y = \dfrac{1}{x^2}$, $1 \le x \le 2$. $\dfrac{dy}{dx} = -\dfrac{2}{x^3}$, so $1 + \left(\dfrac{dy}{dx}\right)^2 = 1 + \dfrac{4}{x^6}$. $L = \displaystyle\int_1^2 \sqrt{1 + \frac{4}{x^6}}\,dx$. By Simpson's Rule with
$n = 10$, $L \approx \frac{1/10}{3}[f(1) + 4f(1.1) + 2f(1.2) + 4f(1.3) + 2f(1.4) + 4f(1.5)$
$+ 2f(1.6) + 4f(1.7) + 2f(1.8) + 4f(1.9) + f(2)] \approx 1.297$. Here $f(x) = \sqrt{1 + 4/x^6}$.

13. The loop lies between $x = 0$ and $x = 3a$ and is symmetric about the x-axis. We can assume without loss of generality that $a > 0$, since if $a = 0$, the graph is the parallel lines $x = 0$ and $x = 3a$, so there is no loop. The upper half of the loop is given by $y = \dfrac{1}{3\sqrt{a}}\sqrt{x}(3a - x) = \sqrt{a}x^{1/2} - \dfrac{x^{3/2}}{3\sqrt{a}}$, $0 \le x \le 3a$. The desired surface area is twice the area generated by the upper half of the loop, that is, $S = 2(2\pi)\int_0^{3a} x\sqrt{1 + (dy/dx)^2}\,dx$.
$\dfrac{dy}{dx} = \dfrac{\sqrt{a}}{2}x^{-1/2} - \dfrac{x^{1/2}}{2\sqrt{a}} \quad \Rightarrow \quad 1 + \left(\dfrac{dy}{dx}\right)^2 = \dfrac{a}{4x} + \dfrac{1}{2} + \dfrac{x}{4a}$. Therefore
$$S = 2(2\pi)\int_0^{3a} x\left(\frac{\sqrt{a}}{2}x^{-1/2} + \frac{x^{1/2}}{2\sqrt{a}}\right)dx = 2\pi\int_0^{3a}\left(\sqrt{a}x^{1/2} + \frac{x^{3/2}}{\sqrt{a}}\right)dx$$
$$= 2\pi\left[\frac{2\sqrt{a}}{3}x^{3/2} + \frac{2}{5\sqrt{a}}x^{5/2}\right]_0^{3a} = 2\pi\left[\frac{2\sqrt{a}}{3}3a\sqrt{3a} + \frac{2}{5\sqrt{a}}9a^2\sqrt{3a}\right] = \frac{56\sqrt{3}\pi a^2}{5}.$$

15. $A = \int_{-2}^{1}[(4-x^2)-(x+2)]dx = \int_{-2}^{1}(2-x-x^2)dx = \left[2x - \frac{1}{2}x^2 - \frac{1}{3}x^3\right]_{-2}^{1}$
$= \left(2 - \frac{1}{2} - \frac{1}{3}\right) - \left(-4 - 2 + \frac{8}{3}\right) = \frac{9}{2} \quad \Rightarrow$

$\overline{x} = A^{-1}\int_{-2}^{1} x(2-x-x^2)dx = \frac{2}{9}\int_{-2}^{1}(2x - x^2 - x^3)dx = \frac{2}{9}\left[x^2 - \frac{1}{3}x^3 - \frac{1}{4}x^4\right]_{-2}^{1}$
$= \frac{2}{9}\left[\left(1 - \frac{1}{3} - \frac{1}{4}\right) - \left(4 + \frac{8}{3} - 4\right)\right] = -\frac{1}{2}$ and

$\overline{y} = A^{-1}\int_{-2}^{1}\frac{1}{2}\left[\left(4-x^2\right)^2 - (x+2)^2\right]dx = \frac{1}{9}\int_{-2}^{1}\left(x^4 - 9x^2 - 4x + 12\right)dx$
$= \frac{1}{9}\left[\frac{1}{5}x^5 - 3x^3 - 2x^2 + 12x\right]_{-2}^{1} = \frac{1}{9}\left[\left(\frac{1}{5} - 3 - 2 + 12\right) - \left(-\frac{32}{5} + 24 - 8 - 24\right)\right] = \frac{12}{5}.$

So $(\overline{x}, \overline{y}) = \left(-\frac{1}{2}, \frac{12}{5}\right)$.

17. The equation of the line passing through $(0,0)$ and $(3,2)$ is $y = \frac{2}{3}x$. $A = \frac{1}{2}\cdot 3 \cdot 2 = 3$. Therefore,
$\overline{x} = \frac{1}{3}\int_0^3 x\left(\frac{2}{3}x\right)dx = \frac{2}{27}\left[x^3\right]_0^3 = 2$, and $\overline{y} = \frac{1}{3}\int_0^3 \frac{1}{2}\left(\frac{2}{3}x\right)^2 dx = \frac{2}{81}\left[x^3\right]_0^3 = \frac{2}{3}$. $(\overline{x}, \overline{y}) = \left(2, \frac{2}{3}\right)$.
Or: Use Exercise 8.4.21.

19. The centroid of this circle, $(1,0)$, travels a distance $2\pi(1)$ when the lamina is rotated about the y-axis. The area of the circle is $\pi(1)^2$. So by the Theorem of Pappus, $V = A2\pi\overline{x} = \pi(1)^2 2\pi(1) = 2\pi^2$.

21. As in Example 1 of Section 8.5, $F = \int_0^2 \rho g x(5-x)dx = \rho g\left[\frac{5}{2}x^2 - \frac{1}{3}x^3\right]_0^2 = \rho g\frac{22}{3} = \frac{22}{3}\delta \approx \frac{22}{3}\cdot 62.5 \approx 458$ lb.

23. $x = 100 \quad \Rightarrow \quad P = 2000 - 0.1(100) - 0.01(100)^2 = 1890$
Consumer surplus $= \int_0^{100}[p(x) - P]dx = \int_0^{100}(2000 - 0.1x - 0.01x^2 - 1890)dx$
$= \left[110x - 0.05x^2 - \frac{0.01}{3}x^3\right]_0^{100} = 11{,}000 - 500 - \frac{10{,}000}{3} \approx \$7166.67$

25. **(a)** $\dfrac{dL}{dt} \propto L_\infty - L \quad \Rightarrow \quad \dfrac{dL}{dt} = k(L_\infty - L) \quad \Rightarrow \quad \displaystyle\int \frac{dL}{L_\infty - L} = \int k\,dt \quad \Rightarrow \quad -\ln|L_\infty - L| = kt + C$
$\Rightarrow \quad L_\infty - L = Ae^{-kt} \quad \Rightarrow \quad L = L_\infty - Ae^{-kt}$. At $t = 0$, $L = L(0) = L_\infty - A \quad \Rightarrow$
$A = L_\infty - L(0) \quad \Rightarrow \quad L(t) = L_\infty - [L_\infty - L(0)]e^{-kt}$

(b) $L_\infty = 53$ cm, $L(0) = 10$ cm and $k = 0.2$. So $L(t) = 53 - (53-10)e^{-0.2t} = 53 - 43e^{-0.2t}$.

27. Let P be the population and I be the number of infected people. The rate of spread dI/dt is jointly proportional to I and to $P - I$, so for some constant k, $\dfrac{dI}{dt} = kI(P - I) \quad \Rightarrow \quad I = \dfrac{I_0 P}{I_0 + (P - I_0)e^{-kPt}}$ (from the discussion of logistic growth in Section 8.1).

Now, measuring t in days, we substitute $t = 7$, $P = 5000$, $I_0 = 160$ and $I(7) = 1200$ to find k:
$1200 = \dfrac{160\cdot 5000}{160 + (5000 - 160)e^{-5000\cdot 7\cdot k}} \quad \Leftrightarrow \quad k \approx 0.00006448$. So, putting $I = 5000 \times 80\% = 4000$, we solve for t: $4000 = \dfrac{160\cdot 5000}{160 + (5000-160)e^{-0.00006448\cdot 5000\cdot t}} \quad \Leftrightarrow \quad 160 + 4840e^{-0.3224t} = 200 \quad \Leftrightarrow$
$-0.3224t = \ln\dfrac{40}{4840} \quad \Leftrightarrow \quad t \approx 14.9$. So it takes about 15 days for 80% of the population to be infected.

29. **(a)** We are given that $V = \frac{1}{3}\pi r^2 h$, $dV/dt = 60{,}000\pi\ \text{ft}^3/\text{h}$, and $r = 1.5h = \frac{3}{2}h$. So $V = \frac{1}{3}\pi\left(\frac{3}{2}h\right)^2 \Rightarrow$ $\dfrac{dV}{dt} = \frac{3}{4}\pi \cdot 3h^2 \dfrac{dh}{dt} = \frac{9}{4}\pi h^2 \dfrac{dh}{dt}$. Therefore, $\dfrac{dh}{dt} = \dfrac{4(dV/dt)}{9\pi h^2} = \dfrac{240{,}000\pi}{9\pi h^2} = \dfrac{80{,}000}{3h^2}$ ($\bigstar$) $\Rightarrow$

$\int 3h^2\,dh = \int 80{,}000\,dt \Rightarrow h^3 = 80{,}000t + C$. When $t = 0$, $h = 60$. Therefore, $C = 60^3 = 216{,}000$, so $h^3 = 80{,}000t + 216{,}000$. Let $h = 100$. Then $100^3 = 1{,}000{,}000 = 80{,}000t + 216{,}000 \Rightarrow$ $80{,}000t = 784{,}000 \Rightarrow t = 9.8$, so the time required is 9.8 hours.

(b) The floor area of the silo is $F = \pi \cdot 200^2 = 40{,}000\pi\ \text{ft}^2$, and the area of the base of the pile is $A = \pi r^2 = \pi\left(\frac{3}{2}h\right)^2 = \frac{9\pi}{4}h^2 = 8100\pi\ \text{ft}^2$. So the area of the floor which is not covered is $F - A = 31{,}900\pi \approx 100{,}000\ \text{ft}^2$.

Now $A = \frac{9\pi}{4}h^2 \Rightarrow \dfrac{dA}{dt} = \dfrac{9\pi}{4} \cdot 2h\dfrac{dh}{dt}$, and from ($\bigstar$) in part (a) we know that when $h = 60$, $\dfrac{dh}{dt} = \dfrac{80{,}000}{3(60)^2} = \dfrac{200}{27}\dfrac{\text{ft}}{\text{h}}$. Therefore $\dfrac{dA}{dt} = \dfrac{9\pi}{4}(2)(60)\left(\dfrac{200}{27}\right) \approx 6283\,\dfrac{\text{ft}^2}{\text{h}}$.

(c) At $h = 90$ ft, $\dfrac{dV}{dt} = 60{,}000\pi - 20{,}000\pi = 40{,}000\pi\ \text{ft}^3/\text{h}$. From ($\bigstar$) in (a), $\dfrac{dh}{dt} = \dfrac{4(dV/dt)}{9\pi h^2} = \dfrac{4(40{,}000\pi)}{9\pi h^2} = \dfrac{160{,}000}{9h^2} \Rightarrow \int 9h^2\,dh = \int 160{,}000\,dt \Rightarrow$ $3h^3 = 160{,}000t + C$. When $t = 0$, $h = 90$; therefore, $C = 3 \cdot 729{,}000 = 2{,}187{,}000$. So $3h^3 = 160{,}000t + 2{,}187{,}000$. At the top, $h = 100$, so $3(100)^3 = 160{,}000t + 2{,}187{,}000 \Rightarrow$ $t = \dfrac{813{,}000}{160{,}000} \approx 5.1$. The pile reaches the top after about 5.1 h.

PROBLEMS PLUS (page 544)

1. By symmetry, the problem can be reduced to finding the line $x = c$ such that the shaded area is one-third of the area of the quarter-circle. The equation of the circle is $y = \sqrt{49 - x^2}$, so we require that $\int_0^c \sqrt{49 - x^2}\,dx = \frac{1}{3} \cdot \frac{1}{4}\pi(7)^2 \quad \Leftrightarrow$
$\left[\frac{1}{2}x\sqrt{49 - x^2} + \frac{49}{2}\sin^{-1}(x/7)\right]_0^c = \frac{49}{12}\pi$ (Formula 30) $\Leftrightarrow$
$\frac{1}{2}c\sqrt{49 - c^2} + \frac{49}{2}\sin^{-1}(c/7) = \frac{49}{12}\pi$.

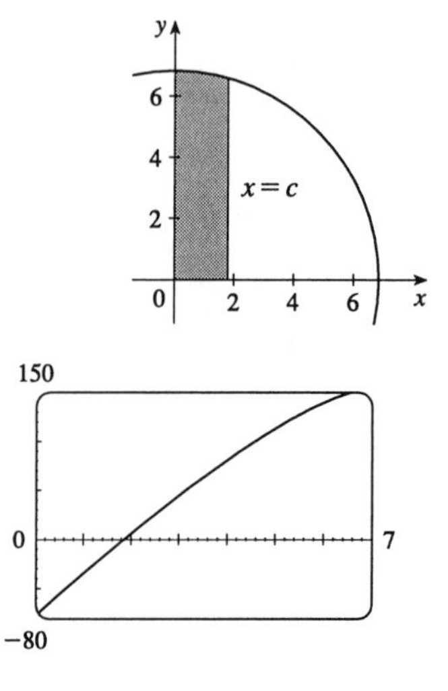

This equation would be difficult to solve exactly, so we plot the left-hand side as a function of c, and find that the equation holds for $c \approx 1.85$. So the cuts should be made at distances of about 1.85 inches from the center of the pizza.

3. $$f'(x) = \lim_{h\to 0} \frac{f(x+h) - f(x)}{h} = \lim_{h\to 0} \frac{f(x)[f(h) - 1]}{h} \quad [\text{since } f(x+h) = f(x)f(h)]$$
$$= f(x) \lim_{h\to 0} \frac{f(h) - 1}{h} = f(x) \lim_{h\to 0} \frac{f(h) - f(0)}{h - 0} = f(x)f'(0) = f(x)$$

Therefore, $f'(x) = f(x)$ for all x. In Leibniz notation, we can either solve the differential equation $\frac{dy}{dx} = y$ $\Rightarrow \quad \int \frac{dy}{y} = \int dx \quad \Rightarrow \quad \ln|y| = x + C \quad \Rightarrow \quad y = Ae^x$, or we can use Theorem 6.5.2 to get the same result. Now $f(0) = 1 \quad \Rightarrow \quad A = 1 \quad \Rightarrow \quad f(x) = e^x$.

5. First we show that $x(1 - x) \le \frac{1}{4}$ for all x. Let $f(x) = x(1 - x) = x - x^2$. Then $f'(x) = 1 - 2x$. This is 0 when $x = \frac{1}{2}$ and $f'(x) > 0$ for $x < \frac{1}{2}$, $f'(x) < 0$ for $x > \frac{1}{2}$, so the absolute maximum of f is $f\left(\frac{1}{2}\right) = \frac{1}{4}$. Thus $x(1 - x) \le \frac{1}{4}$ for all x.

Now suppose that the given assertion is false, that is, $a(1 - b) > \frac{1}{4}$ and $b(1 - a) > \frac{1}{4}$. Multiply these inequalities: $a(1 - b)b(1 - a) > \frac{1}{16} \quad \Rightarrow \quad [a(1 - a)][b(1 - b)] > \frac{1}{16}$. But we know that $a(1 - a) \le \frac{1}{4}$ and $b(1 - b) \le \frac{1}{4} \quad \Rightarrow \quad [a(1 - a)][b(1 - b)] \le \frac{1}{16}$. Thus we have a contradiction, so the given assertion is proved.

7. Let $F(x) = \int_a^x f(t)dt - \int_x^b f(t)dt$. Then $F(a) = -\int_a^b f(t)dt$ and $F(b) = \int_a^b f(t)dt$. Also F is continuous by the Fundamental Theorem. So by the Intermediate Value Theorem, there is a number c in $[a, b]$ such that $F(c) = 0$. For that number c, we have $\int_a^c f(t)dt = \int_c^b f(t)dt$.

9. The given integral represents the difference of the shaded areas, which appears to be 0. It can be calculated by integrating with respect to either x or y, so we find x in terms of y for each curve:

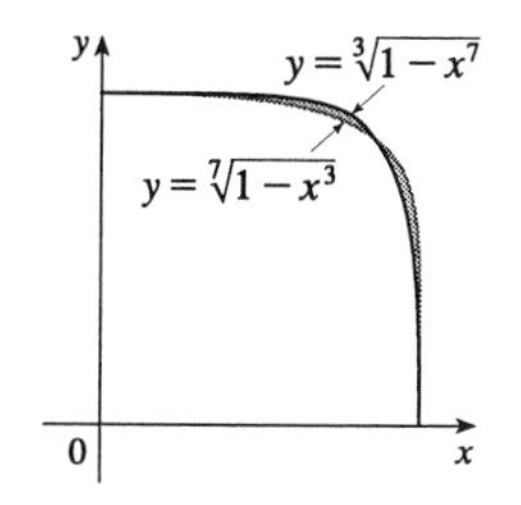

$y = \sqrt[3]{1-x^7} \Rightarrow x = \sqrt[7]{1-y^3}$ and $y = \sqrt[7]{1-x^3} \Rightarrow x = \sqrt[3]{1-y^7}$, so

$$\int_0^1 \left(\sqrt[3]{1-y^7} - \sqrt[7]{1-y^3}\right) dy = \int_0^1 \left(\sqrt[7]{1-x^3} - \sqrt[3]{1-x^7}\right) dx.$$

But this equation is of the form $z = -z$. So $\int_0^1 \left(\sqrt[3]{1-x^7} - \sqrt[7]{1-x^3}\right) dx = 0$.

11. First we find the domain explicitly. $5x^2 \geq x^4 + 4 \quad \Leftrightarrow \quad x^4 - 5x^2 + 4 \leq 0 \quad \Leftrightarrow \quad (x^2-4)(x^2-1) \leq 0$ $\Leftrightarrow \quad (x-2)(x-1)(x+1)(x+2) \leq 0 \quad \Leftrightarrow \quad x \in [-2,-1]$ or $[1,2]$. Therefore, the domain is $\{x \mid 5x^2 \geq x^4 + 4\} = [-2,-1] \cup [1,2]$. Now f is decreasing on its domain, since $f(x) = 3x - 2x^3 \quad \Rightarrow$ $f'(x) = 3 - 6x^2 \leq -3$. So f's maximum value is either $f(-2)$ or $f(1)$. $f(-2) = 10$ and $f(1) = 1$, so the maximum value of f is 10.

13. Recall that $\cos A \cos B = \frac{1}{2}[\cos(A+B) + \cos(A-B)]$. So

$$\begin{aligned} f(x) &= \textstyle\int_0^\pi \cos t \cos(x-t)\,dt = \frac{1}{2}\int_0^\pi [\cos(t+x-t) + \cos(t-x+t)]\,dt \\ &= \textstyle\frac{1}{2}\int_0^\pi [\cos x + \cos(2t-x)]\,dt = \frac{1}{2}\left[t\cos x + \frac{1}{2}\sin(2t-x)\right]_0^\pi \\ &= \textstyle\frac{\pi}{2}\cos x + \frac{1}{4}\sin(2\pi - x) - \frac{1}{4}\sin(-x) = \frac{\pi}{2}\cos x + \frac{1}{4}\sin(-x) - \frac{1}{4}\sin(-x) \\ &= \textstyle\frac{\pi}{2}\cos x. \end{aligned}$$

The minimum of $\cos x$ on this domain is -1, so the minimum value of $f(x)$ is $f(\pi) = -\frac{\pi}{2}$.

15. $0 < a < b$. $\displaystyle\int_0^1 [bx + a(1-x)]^t\,dx = \int_a^b \frac{u^t}{(b-a)}\,du$ [put $u = bx + a(1-x)$]

$$= \left[\frac{u^{t+1}}{(t+1)(b-a)}\right]_a^b = \frac{b^{t+1} - a^{t+1}}{(t+1)(b-a)}. \text{ Now let } y = \lim_{t\to 0}\left[\frac{b^{t+1} - a^{t+1}}{(t+1)(b-a)}\right]^{1/t}.$$

Then $\ln y = \displaystyle\lim_{t\to 0}\left[\frac{1}{t}\ln\frac{b^{t+1} - a^{t+1}}{(t+1)(b-a)}\right]$. This limit is of the form $\frac{0}{0}$, so we can apply l'Hospital's Rule to get

$$\ln y = \lim_{t\to 0}\left[\frac{b^{t+1}\ln b - a^{t+1}\ln a}{b^{t+1} - a^{t+1}} - \frac{1}{t+1}\right] = \frac{b\ln b - a\ln a}{b-a} - 1 = \frac{b\ln b}{b-a} - \frac{a\ln a}{b-a} - \ln e = \ln\frac{b^{b/(b-a)}}{ea^{a/(b-a)}}.$$

Therefore, $y = e^{-1}\left(\dfrac{b^b}{a^a}\right)^{1/(b-a)}$.

17. In accordance with the hint, we let $I_k = \int_0^1 (1-x^2)^k\,dx$, and we find an expression for I_{k+1} in terms of I_k. We integrate I_{k+1} by parts with $u = (1-x^2)^{k+1} \quad \Rightarrow \quad du = (k+1)(1-x^2)^k(-2x)$, $dv = dx \quad \Rightarrow \quad v = x$, and then split the remaining integral into identifiable quantities:

$$\begin{aligned} I_{k+1} &= x(1-x^2)^{k+1}\Big|_0^1 + 2(k+1)\textstyle\int_0^1 x^2(1-x^2)^k\,dx = (2k+2)\int_0^1 (1-x^2)^k[1-(1-x^2)]\,dx \\ &= (2k+2)(I_k - I_{k+1}). \text{ So } I_{k+1}[1 + (2k+2)] = (2k+2)I_k \end{aligned}$$

$$\Rightarrow \quad I_{k+1} = \frac{2k+2}{2k+3}I_k.$$

Now to complete the proof, we use induction: $I_0 = 1 = \dfrac{2^0(0!)^2}{1!}$, so the formula holds for $n = 0$. Now suppose it holds for $n = k$. Then

$$I_{k+1} = \frac{2k+2}{2k+3} I_k = \frac{2k+2}{2k+3}\left[\frac{2^{2k}(k!)^2}{(2k+1)!}\right] = \frac{2(k+1)2^{2k}(k!)^2}{(2k+3)(2k+1)!} = \frac{2(k+1)}{2k+2}\cdot\frac{2(k+1)2^{2k}(k!)^2}{(2k+3)(2k+1)!}$$

$$= \frac{[2(k+1)]^2 2^{2k}(k!)^2}{(2k+3)(2k+2)(2k+1)!} = \frac{2^{2(k+1)}[(k+1)!]^2}{[2(k+1)+1]!}.$$

So by induction, the formula holds for all integers $n \geq 0$.

19. We use the Fundamental Theorem of Calculus to differentiate the given equation:

$[f(x)]^2 = 100 + \int_0^x \left([f(t)]^2 + [f'(t)]^2\right) dt \quad\Rightarrow\quad 2f(x)f'(x) = [f(x)]^2 + [f'(x)]^2 \quad\Rightarrow$

$[f(x)]^2 + [f'(x)]^2 - 2f(x)f'(x) = [f(x) - f'(x)]^2 = 0 \quad\Leftrightarrow\quad f(x) = f'(x)$. We can solve this as a separable equation, or else use Theorem 6.5.2, which says that the only solutions are $f(x) = Ce^x$. Now $[f(0)]^2 = 100$, so $f(0) = C = \pm 10$, and hence $f(x) = \pm 10e^x$ are the only functions satisfying the given equation.

21. To find the height of the pyramid, we use similar triangles. The first figure shows a cross-section of the pyramid passing through the top and through two opposite corners of the square base. Now $|BD| = b$, since it is a radius of the sphere, which has diameter $2b$ since it is tangent to the opposite sides of the square base. Also, $|AD| = b$ since $\triangle ADB$ is isosceles. So the height is $|AB| = \sqrt{b^2 + b^2} = \sqrt{2}b$.

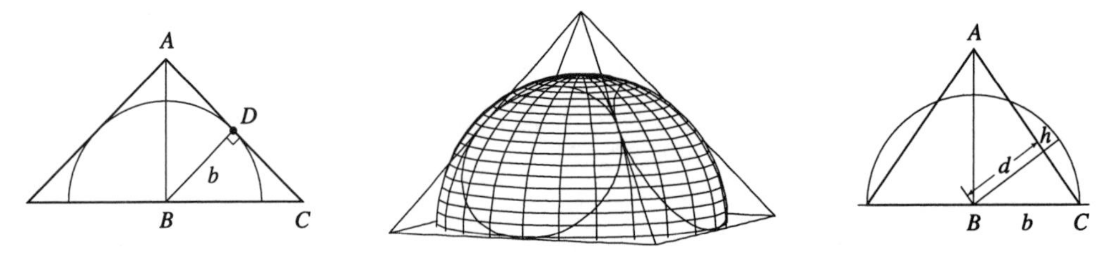

We observe that the shared volume is equal to half the volume of the sphere, minus the sum of the four equal volumes (caps of the sphere) cut off by the triangular faces of the pyramid. See Exercise 5.2.49 for a derivation of the formula for the volume of a cap of a sphere. To use the formula, we need to find the perpendicular distance h of each triangular face from the surface of the sphere. We first find the distance d from the center of the sphere to one of the triangular faces. The third figure shows a cross-section of the pyramid through the top and through the midpoints of opposite sides of the square base. From similar triangles we find that

$\dfrac{d}{b} = \dfrac{|AB|}{|AC|} = \dfrac{\sqrt{2}b}{\sqrt{b^2 + \left(\sqrt{2}b\right)^2}} \quad\Rightarrow\quad d = \dfrac{\sqrt{2}b^2}{\sqrt{3b^2}} = \frac{\sqrt{6}}{3}b$. So $h = b - \frac{\sqrt{6}}{3}b = \frac{3-\sqrt{6}}{3}b$. So, using the formula from Exercise 5.2.49 with $r = b$, we find that the volume of each of the caps is

$\pi\left(\frac{3-\sqrt{6}}{3}b\right)^2\left(b - \frac{3-\sqrt{6}}{3\cdot 3}b\right) = \frac{15-6\sqrt{6}}{9}\cdot\frac{6+\sqrt{6}}{9}\pi b^3 = \left(\frac{2}{3} - \frac{7}{27}\sqrt{6}\right)\pi b^3$. So, using our first observation, the shared volume is $V = \frac{1}{2}\left(\frac{4}{3}\pi b^3\right) - 4\left(\frac{2}{3} - \frac{7}{27}\sqrt{6}\right)\pi b^3 = \left(\frac{28}{27}\sqrt{6} - 2\right)\pi b^3$.

CHAPTER NINE

EXERCISES 9.1

1. **(a)**

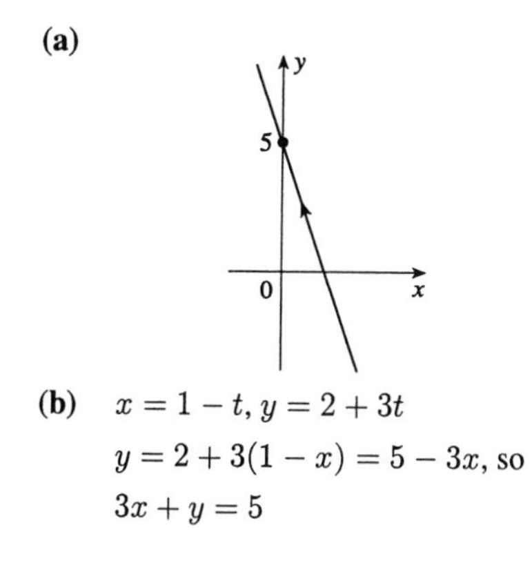

(b) $x = 1 - t,\ y = 2 + 3t$

$y = 2 + 3(1 - x) = 5 - 3x$, so

$3x + y = 5$

3. **(a)**

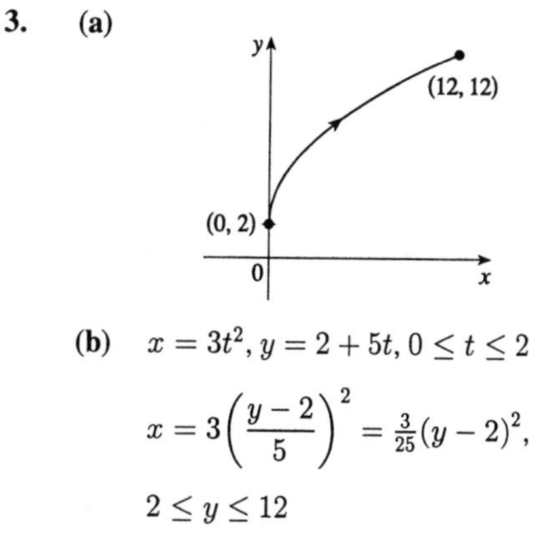

(b) $x = 3t^2,\ y = 2 + 5t,\ 0 \le t \le 2$

$$x = 3\left(\frac{y-2}{5}\right)^2 = \tfrac{3}{25}(y-2)^2,$$

$2 \le y \le 12$

5. **(a)**

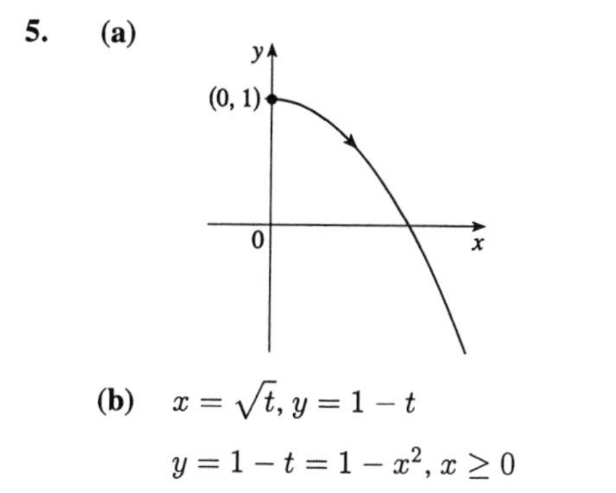

(b) $x = \sqrt{t},\ y = 1 - t$

$y = 1 - t = 1 - x^2,\ x \ge 0$

7. **(a)**

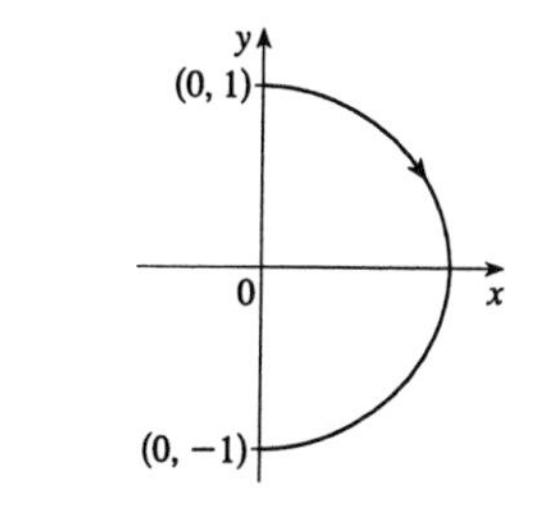

(b) $x = \sin\theta,\ y = \cos\theta,\ 0 \le \theta \le \pi$

$x^2 + y^2 = \sin^2\theta + \cos^2\theta = 1,\ 0 \le x \le 1$

9. **(a)**

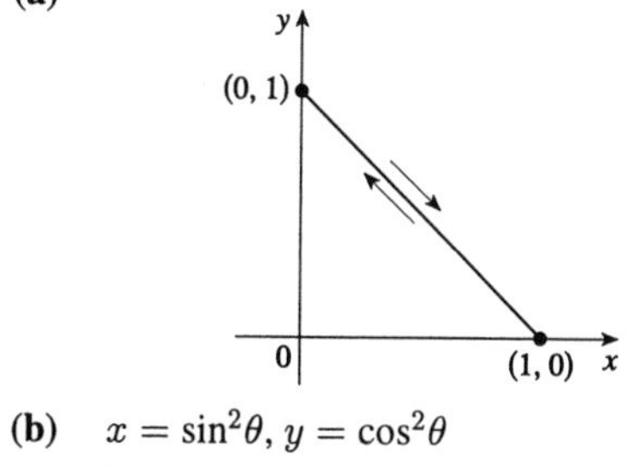

(b) $x = \sin^2\theta,\ y = \cos^2\theta$

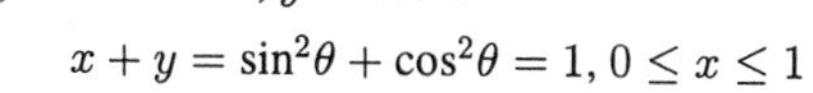

$x + y = \sin^2\theta + \cos^2\theta = 1,\ 0 \le x \le 1$

11. **(a)**

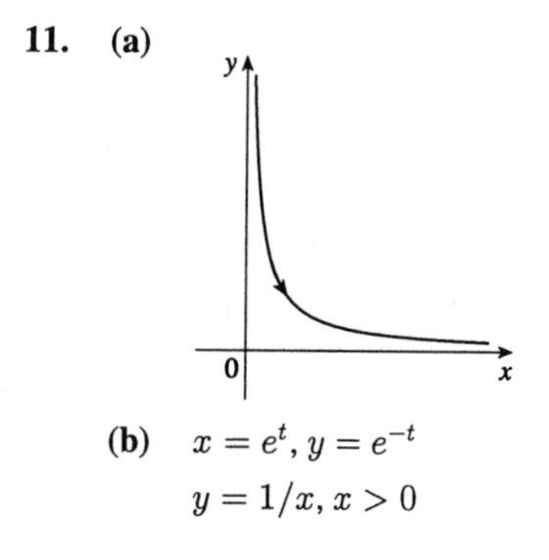

(b) $x = e^t,\ y = e^{-t}$

$y = 1/x,\ x > 0$

13. (a)

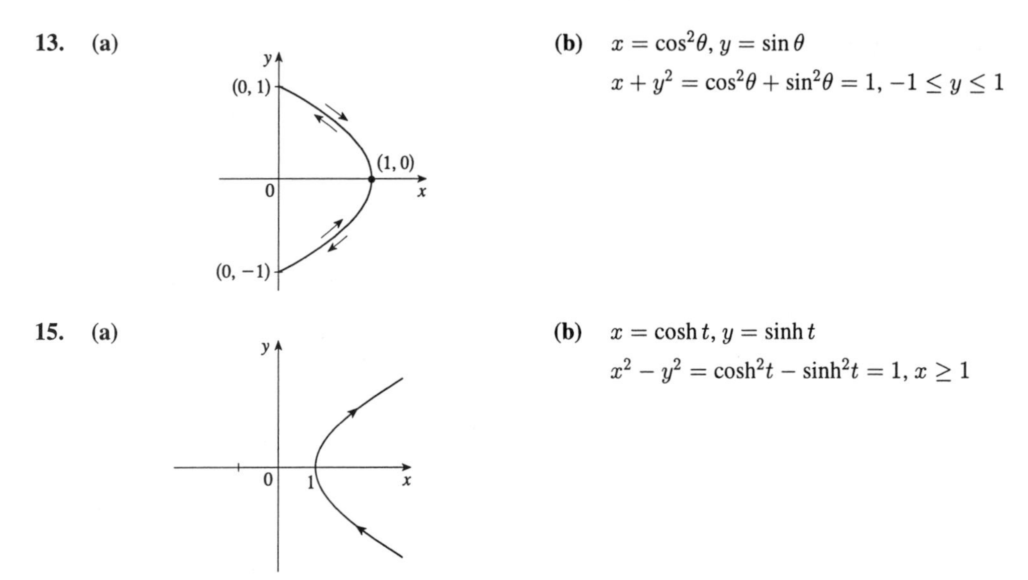

(b) $x = \cos^2\theta,\ y = \sin\theta$

$x + y^2 = \cos^2\theta + \sin^2\theta = 1,\ -1 \le y \le 1$

15. (a)

(b) $x = \cosh t,\ y = \sinh t$

$x^2 - y^2 = \cosh^2 t - \sinh^2 t = 1,\ x \ge 1$

17. $x^2 + y^2 = \cos^2\pi t + \sin^2\pi t = 1,\ 1 \le t \le 2$, so the particle moves counterclockwise along the circle $x^2 + y^2 = 1$ from $(-1, 0)$ to $(1, 0)$, along the lower half of the circle.

19. $x = 8t - 3,\ y = 2 - t,\ 0 \le t \le 1 \quad \Rightarrow \quad x = 8(2 - y) - 3 = 13 - 8y$, so the particle moves along the line $x + 8y = 13$ from $(-3, 2)$ to $(5, 1)$.

21. $\left(\frac{1}{2}x\right)^2 + \left(\frac{1}{3}y\right)^2 = \sin^2 t + \cos^2 t = 1$, so the particle moves once clockwise along the ellipse $\frac{1}{4}x^2 + \frac{1}{9}y^2 = 1$, starting and ending at $(0, 3)$.

23.

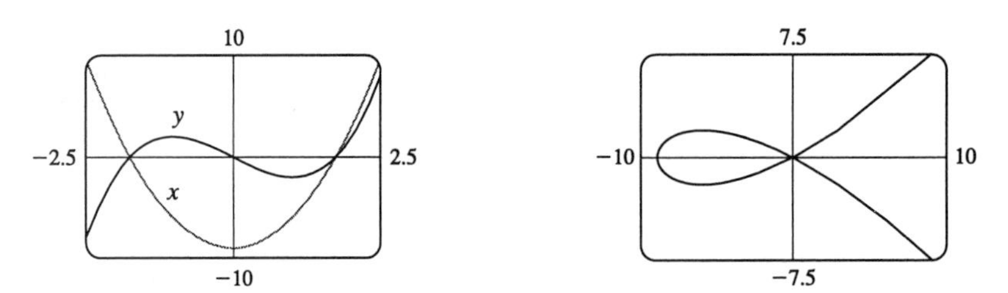

From the graphs, it seems that as $t \to -\infty$, $x \to \infty$ and $y \to -\infty$. So the point $(x(t), y(t))$ will move from far out in the fourth quadrant as t increases. At $t \approx -1.7$, both x and y are 0, so the graph passes through the origin. After that the graph passes through the second quadrant (x is negative, y is positive), then intersects the x-axis at $x \approx -9$ when $t = 0$. After this, the graph passes through the third quadrant, going through the origin again at $t \approx 1.7$, and then as $t \to \infty$, $x \to \infty$ and $y \to \infty$. Note that for every point $(x(t), y(t)) = (3(t^2 - 3), t^3 - 3t)$, we can substitute $-t$ to get the corresponding point $(x(-t), y(-t)) = \left(3\left[(-t)^2 - 3\right], (-t)^3 - 3(-t)\right)$ $= (x(t), -y(t))$, and so the graph is symmetric about the x-axis.

25.

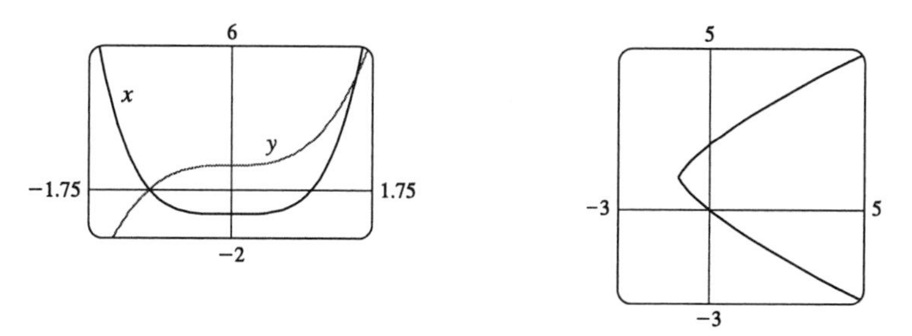

As $t \to -\infty$, $x \to \infty$ and $y \to -\infty$. The graph passes through the origin at $t = -1$, and then goes through the second quadrant (x negative, y positive), passing through the point $(-1, 1)$ at $t = 0$. As t increases, the graph passes through the point $(0, 2)$ at $t = 1$, and then as $t \to \infty$, both x and y approach ∞.

27. Clearly the curve passes through (x_1, y_1) when $t = 0$ and through (x_2, y_2) when $t = 1$. For $0 < t < 1$, x is strictly between x_1 and x_2 and y is strictly between y_1 and y_2. For every value of t, x and y satisfy the relation $y - y_1 = \dfrac{y_2 - y_1}{x_2 - x_1}(x - x_1)$, which is the equation of the straight line through (x_1, y_1) and (x_2, y_2).

Finally, any point (x, y) on that line satisfies $\dfrac{y - y_1}{y_2 - y_1} = \dfrac{x - x_1}{x_2 - x_1}$; if we call that common value t, then the given parametric equations yield the point (x, y); and any (x, y) on the line between (x_1, y_1) and (x_2, y_2) yields a value of t in $[0, 1]$. So the given parametric equations exactly specify the line segment from (x_1, y_1) to (x_2, y_2).

29. The case $\frac{\pi}{2} < \theta < \pi$ is illustrated. C has coordinates $(r\theta, r)$ as before, and Q has coordinates $(r\theta, r + r\cos(\pi - \theta)) = (r\theta, r(1 - \cos\theta))$, so P has coordinates $(r\theta - r\sin(\pi - \theta), r(1 - \cos\theta)) = (r(\theta - \sin\theta), r(1 - \cos\theta))$. Again we have the parametric equations $x = r(\theta - \sin\theta)$, $y = r(1 - \cos\theta)$.

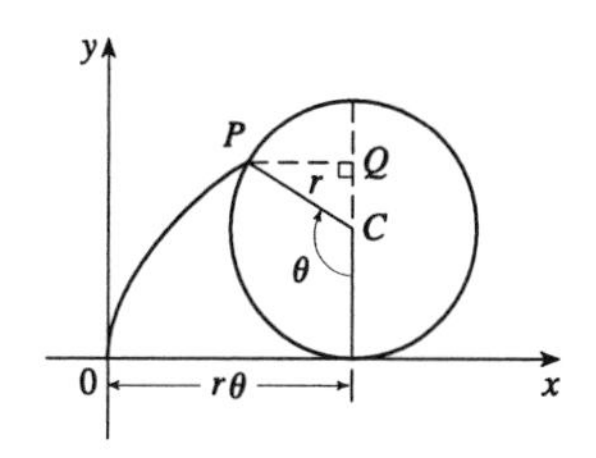

31.

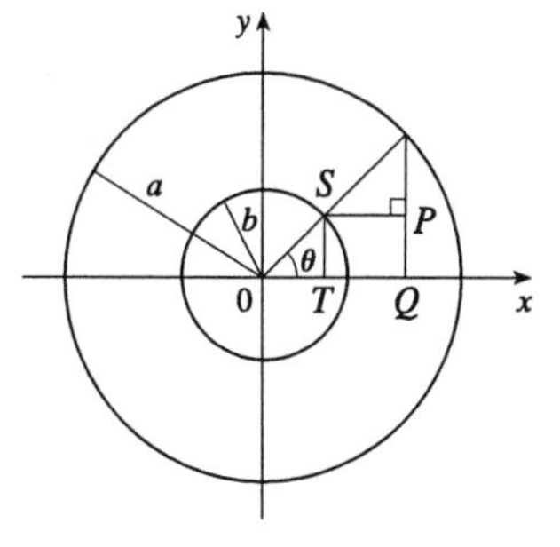

It is apparent that $x = |OQ|$ and $y = |QP| = |ST|$. From the diagram, $x = |OQ| = a\cos\theta$ and $y = |ST| = b\sin\theta$. Thus the parametric equations are $x = a\cos\theta$ and $y = b\sin\theta$. To eliminate θ we rearrange: $\sin\theta = y/b \;\Rightarrow\; \sin^2\theta = (y/b)^2$ and $\cos\theta = x/a \;\Rightarrow\; \cos^2\theta = (x/a)^2$. Adding the two equations: $\sin^2\theta + \cos^2\theta = 1 = x^2/a^2 + y^2/b^2$. Thus we have an ellipse.

33. **(a)** The center Q of the smaller circle has coordinates $((a-b)\cos\theta, (a-b)\sin\theta)$. Arc PS on circle C has length $a\theta$ since it is equal in length to arc AS (the smaller circle rolls without slipping against the larger). Thus $\angle PQS = \frac{a}{b}\theta$ and $\angle PQT = \frac{a}{b}\theta - \theta$, so P has coordinates

$$x = (a-b)\cos\theta + b\cos(\angle PQT) = (a-b)\cos\theta + b\cos\left(\frac{a-b}{b}\theta\right),$$

and $y = (a-b)\sin\theta - b\sin(\angle PQT) = (a-b)\sin\theta - b\sin\left(\frac{a-b}{b}\theta\right)$.

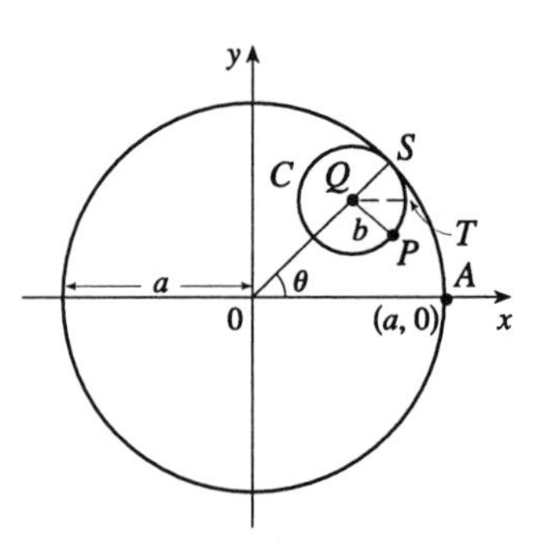

(b) If $b = \frac{a}{4}$, then $a - b = \frac{3a}{4}$ and $\frac{a-b}{b} = 3$, so

$$x = \frac{3a}{4}\cos\theta + \frac{a}{4}\cos 3\theta = \frac{3a}{4}\cos\theta + \frac{a}{4}\left(4\cos^3\theta - 3\cos\theta\right) = a\cos^3\theta \text{ and}$$

$$y = \frac{3a}{4}\sin\theta - \frac{a}{4}\sin 3\theta = \frac{3a}{4}\sin\theta - \frac{a}{4}\left(3\sin\theta - 4\sin^3\theta\right) = a\sin^3\theta.$$

The curve is symmetric about the origin.

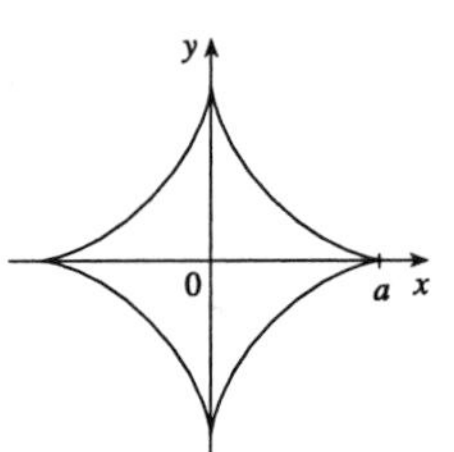

(c)

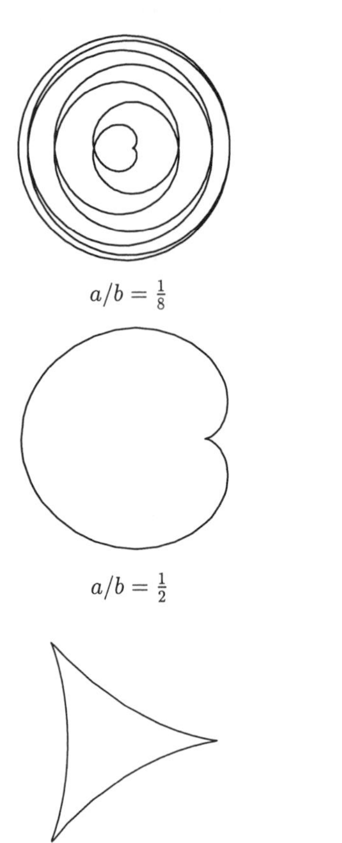

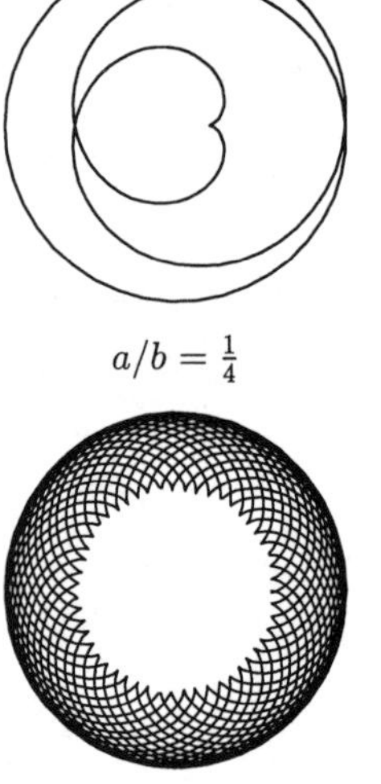

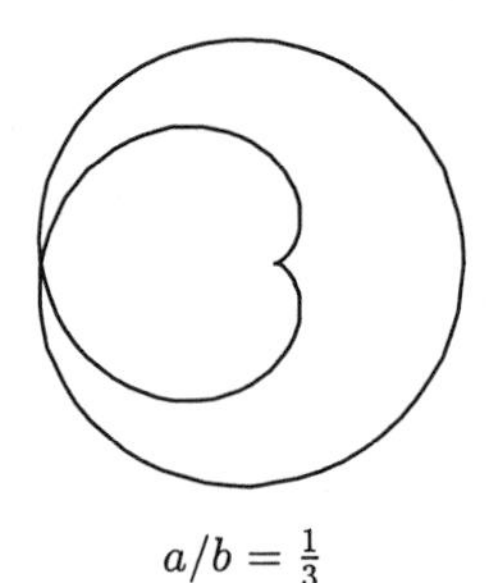

$a/b = \frac{1}{8}$ $a/b = \frac{1}{4}$ $a/b = \frac{1}{3}$

$a/b = \frac{1}{2}$ $a/b = e - 2, 0 \le t \le 446$

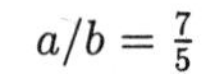

$a/b = \frac{7}{5}$

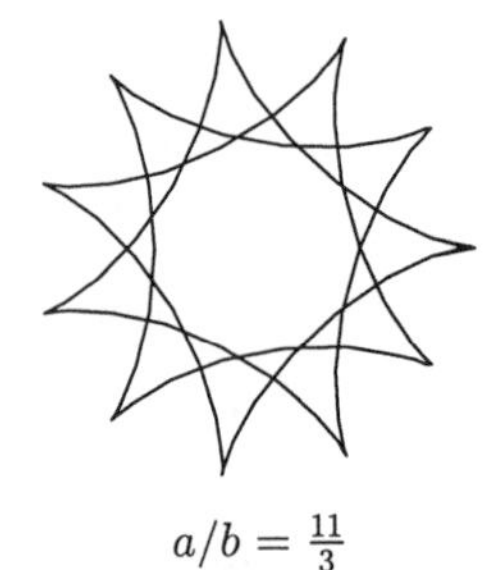

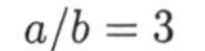

$a/b = 3$ $a/b = \frac{11}{3}$ $a/b = 23$

35. $C = (2a\cot\theta, 2a)$, so the x-coordinate of P is $x = 2a\cot\theta$. Let $B = (0, 2a)$. Then $\angle OAB$ is a right angle and $\angle OBA = \theta$, so $|OA| = 2a\sin\theta$ and $A = (2a\sin\theta\cos\theta, 2a\sin^2\theta)$.
Thus the y-coordinate of P is $y = 2a\sin^2\theta$.

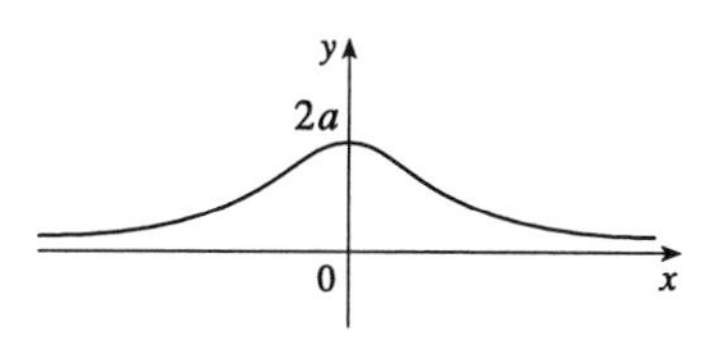

37. $x = t^2, y = t^3 - ct$. We use a graphing device to produce the graphs for various values of c.

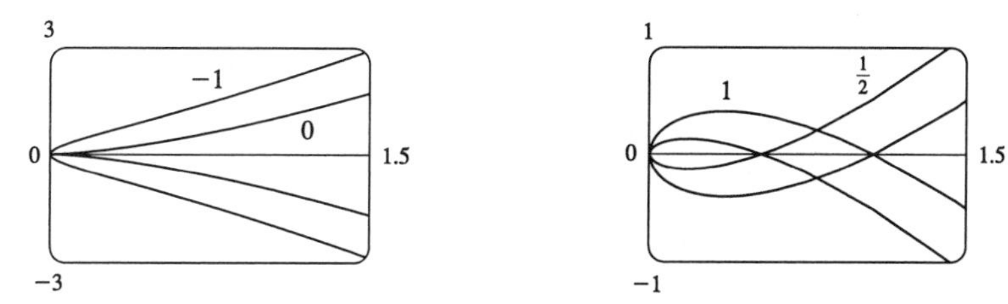

Note that all the members of the family are symmetric about the x-axis. For $c < 0$, the graph does not cross itself, but for $c = 0$ it has a cusp at $(0, 0)$ and for $c > 0$ the graph crosses itself at $x = c$, so the loop grows larger as c increases.

39. Note that all the Lissajous figures are symmetric about the x-axis. The parameters a and b simply stretch the graph in the x- and y-directions respectively. For $a = b = n = 1$ the graph is simply a circle with radius 1. For $n = 2$ the graph crosses itself at the origin and there are loops above and below the x-axis. In general, the figures have $n - 1$ points of intersection, all of which are on the y-axis, and a total of n closed loops.

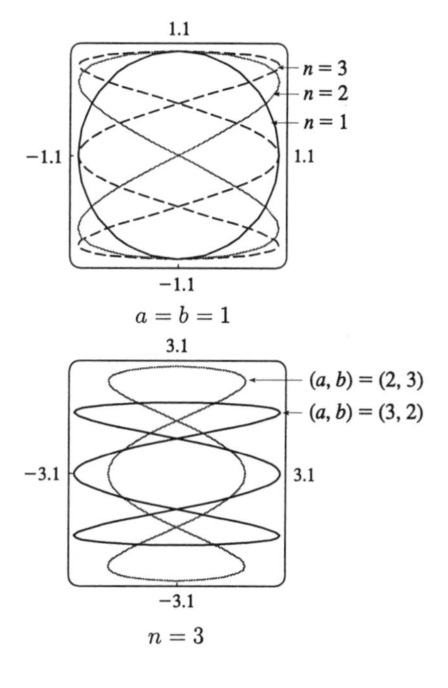

$a = b = 1$

$n = 3$

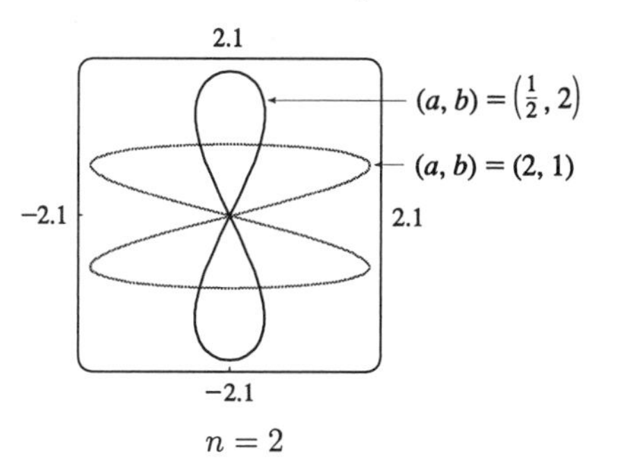

$n = 2$

EXERCISES 9.2

1. $x = t^2 + t, y = t^2 - t; t = 0.$ $\dfrac{dy}{dt} = 2t - 1, \dfrac{dx}{dt} = 2t + 1$, so $\dfrac{dy}{dx} = \dfrac{dy/dt}{dx/dt} = \dfrac{2t-1}{2t+1}$. When $t = 0, x = y = 0$ and $\dfrac{dy}{dx} = -1$. The tangent is $y - 0 = (-1)(x - 0)$ or $y = -x$.

3. $x = \ln t, y = te^t; t = 1.$ $\dfrac{dy}{dt} = (t+1)e^t, \dfrac{dx}{dt} = \dfrac{1}{t}, \dfrac{dy}{dx} = \dfrac{dy/dt}{dx/dt} = t(t+1)e^t$. When $t = 1, (x, y) = (0, e)$ and $\dfrac{dy}{dx} = 2e$, so an equation of the tangent is $y - e = 2e(x - 0)$ or $2ex - y + e = 0$.

5. **(a)** $x = 2t + 3, y = t^2 + 2t; (5, 3).$ $\dfrac{dy}{dt} = 2t + 2, \dfrac{dx}{dt} = 2$, and $\dfrac{dy}{dx} = \dfrac{dy/dt}{dx/dt} = t + 1$.

At $(5, 3)$, $t = 1$ and $\dfrac{dy}{dx} = 2$, so the tangent is $y - 3 = 2(x - 5)$ or $2x - y - 7 = 0$.

(b) $y = t^2 + 2t = \left(\dfrac{x-3}{2}\right)^2 + 2\left(\dfrac{x-3}{2}\right) = \dfrac{(x-3)^2}{4} + x - 3$, so $\dfrac{dy}{dx} = \dfrac{x-3}{2} + 1$.

When $x = 5$, $\dfrac{dy}{dx} = 2$, so an equation of the tangent is $2x - y - 7 = 0$, as before.

7. $x = 2\sin 2t, y = 2\sin t; \left(\sqrt{3}, 1\right)$.

$\dfrac{dy}{dx} = \dfrac{dy/dt}{dx/dt} = \dfrac{2\cos t}{2 \cdot 2\cos 2t} = \dfrac{\cos t}{2\cos 2t}$. The point $\left(\sqrt{3}, 1\right)$ corresponds to $t = \frac{\pi}{6}$, so the slope of the tangent at that point is

$\dfrac{\cos\frac{\pi}{6}}{2\cos\frac{\pi}{3}} = \dfrac{\sqrt{3}/2}{2 \cdot \frac{1}{2}} = \dfrac{\sqrt{3}}{2}$. An equation of the tangent is therefore

$(y - 1) = \frac{\sqrt{3}}{2}\left(x - \sqrt{3}\right)$ or $\sqrt{3}x - 2y - 1 = 0$.

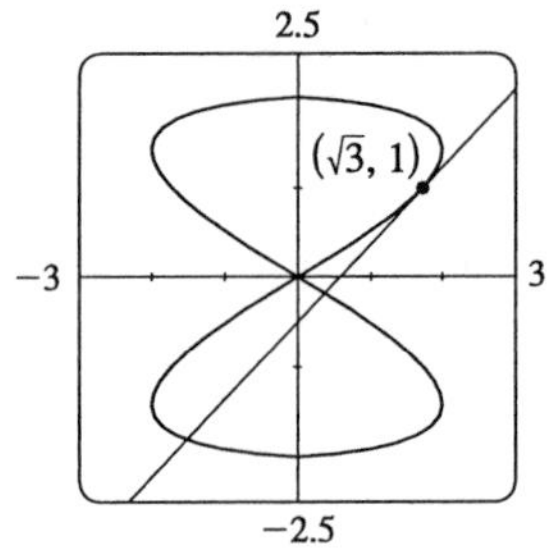

9. $x = t^2 + t, y = t^2 + 1.$ $\dfrac{dy}{dx} = \dfrac{dy/dt}{dx/dt} = \dfrac{2t}{2t+1} = 1 - \dfrac{1}{2t+1}; \dfrac{d}{dt}\left(\dfrac{dy}{dx}\right) = \dfrac{2}{(2t+1)^2}$;

$$\frac{d^2y}{dx^2} = \frac{d}{dx}\left(\frac{dy}{dx}\right) = \frac{d(dy/dx)/dt}{dx/dt} = \frac{2}{(2t+1)^3}$$

11. $x = \sin \pi t, y = \cos \pi t.$ $\dfrac{dy}{dx} = \dfrac{dy/dt}{dx/dt} = \dfrac{-\pi \sin \pi t}{\pi \cos \pi t} = -\tan \pi t$;

$$\frac{d^2y}{dx^2} = \frac{d}{dx}\left(\frac{dy}{dx}\right) = \frac{d(dy/dx)/dt}{dx/dt} = \frac{-\pi \sec^2 \pi t}{\pi \cos \pi t} = -\sec^3 \pi t$$

13. $x = e^{-t}, y = te^{2t}.$ $\dfrac{dy}{dx} = \dfrac{dy/dt}{dx/dt} = \dfrac{(2t+1)e^{2t}}{-e^{-t}} = -(2t+1)e^{3t}$;

$$\frac{d}{dt}\left(\frac{dy}{dx}\right) = -3(2t+1)e^{3t} - 2e^{3t} = -(6t+5)e^{3t};$$

$$\frac{d^2y}{dx^2} = \frac{d}{dx}\left(\frac{dy}{dx}\right) = \frac{d(dy/dx)/dt}{dx/dt} = \frac{-(6t+5)e^{3t}}{-e^{-t}} = (6t+5)e^{4t}$$

15. $x = t(t^2-3) = t^3 - 3t$, $y = 3(t^2-3)$. $\dfrac{dx}{dt} = 3t^2 - 3 = 3(t-1)(t+1)$; $\dfrac{dy}{dt} = 6t$. $\dfrac{dy}{dt} = 0 \Leftrightarrow t = 0 \Leftrightarrow (x,y) = (0,-9)$. $\dfrac{dx}{dt} = 0 \Leftrightarrow t = \pm 1 \Leftrightarrow (x,y) = (-2,-6)$ or $(2,-6)$. So there is a horizontal tangent at $(0,-9)$ and there are vertical tangents at $(-2,-6)$ and $(2,-6)$.

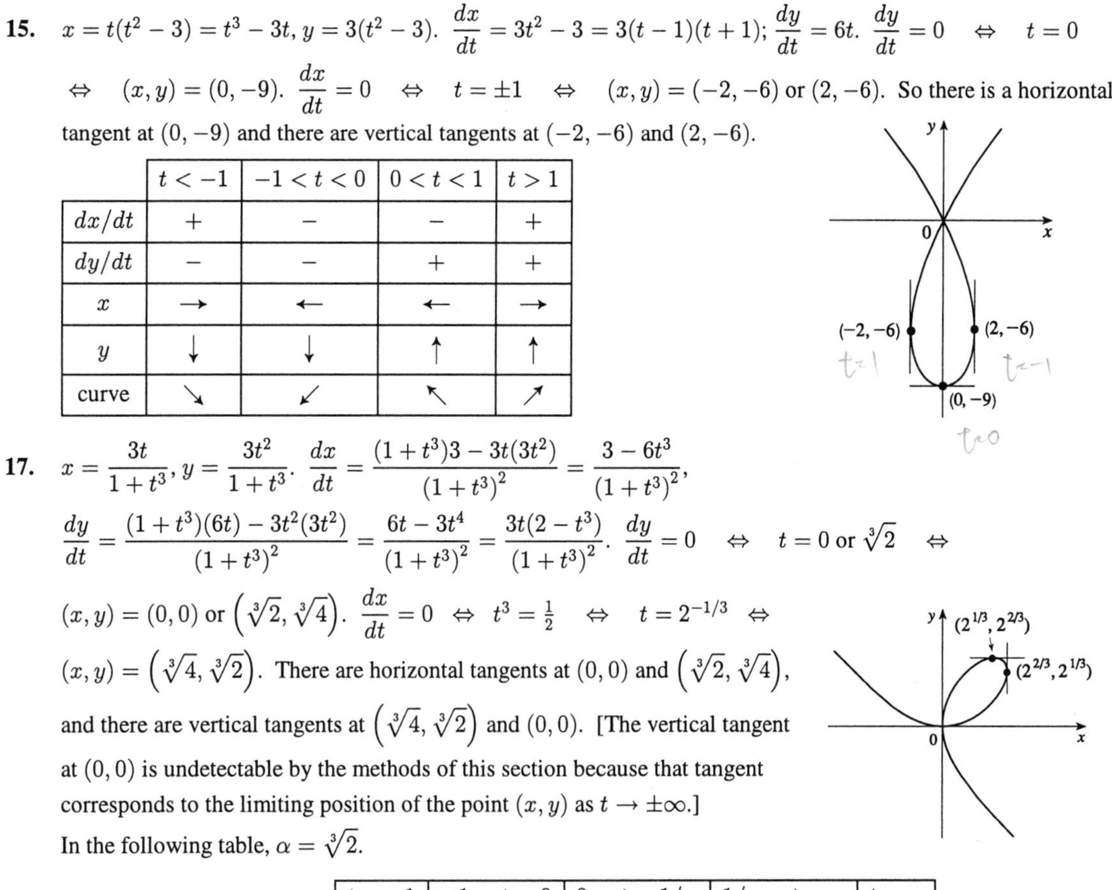

	$t < -1$	$-1 < t < 0$	$0 < t < 1$	$t > 1$
dx/dt	+	−	−	+
dy/dt	−	−	+	+
x	→	←	←	→
y	↓	↓	↑	↑
curve	↘	↙	↖	↗

17. $x = \dfrac{3t}{1+t^3}$, $y = \dfrac{3t^2}{1+t^3}$. $\dfrac{dx}{dt} = \dfrac{(1+t^3)3 - 3t(3t^2)}{(1+t^3)^2} = \dfrac{3-6t^3}{(1+t^3)^2}$,

$\dfrac{dy}{dt} = \dfrac{(1+t^3)(6t) - 3t^2(3t^2)}{(1+t^3)^2} = \dfrac{6t-3t^4}{(1+t^3)^2} = \dfrac{3t(2-t^3)}{(1+t^3)^2}$. $\dfrac{dy}{dt} = 0 \Leftrightarrow t = 0$ or $\sqrt[3]{2} \Leftrightarrow$

$(x,y) = (0,0)$ or $\left(\sqrt[3]{2}, \sqrt[3]{4}\right)$. $\dfrac{dx}{dt} = 0 \Leftrightarrow t^3 = \frac{1}{2} \Leftrightarrow t = 2^{-1/3} \Leftrightarrow$

$(x,y) = \left(\sqrt[3]{4}, \sqrt[3]{2}\right)$. There are horizontal tangents at $(0,0)$ and $\left(\sqrt[3]{2}, \sqrt[3]{4}\right)$, and there are vertical tangents at $\left(\sqrt[3]{4}, \sqrt[3]{2}\right)$ and $(0,0)$. [The vertical tangent at $(0,0)$ is undetectable by the methods of this section because that tangent corresponds to the limiting position of the point (x,y) as $t \to \pm\infty$.]

In the following table, $\alpha = \sqrt[3]{2}$.

	$t < -1$	$-1 < t < 0$	$0 < t < 1/\alpha$	$1/\alpha < t < \alpha$	$t > \alpha$
dx/dt	+	+	+	−	−
dy/dt	−	−	+	+	−
x	→	→	→	←	←
y	↓	↓	↑	↑	↓
curve	↘	↘	↗	↖	↙

19. From the graph, it appears that the leftmost point on the curve $x = t^4 - t^2$, $y = t + \ln t$ is about $(-0.25, 0.36)$. To find the exact coordinates, we find the value of t for which the graph has a vertical tangent, that is, $0 = dx/dt = 4t^3 - 2t$

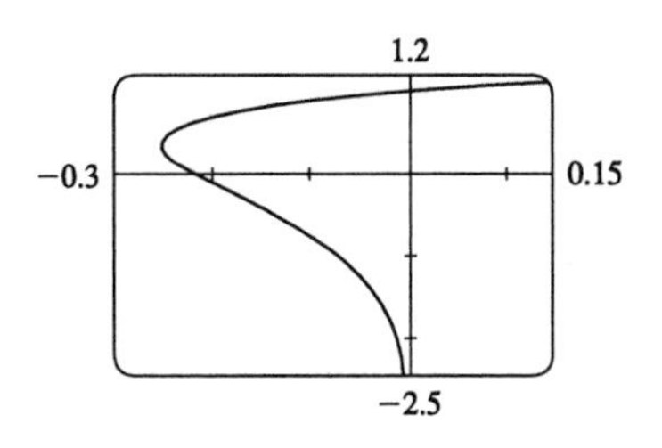

$\Leftrightarrow 2t(2t^2-1) = 0 \Leftrightarrow 2t\left(\sqrt{2}t+1\right)\left(\sqrt{2}t-1\right) = 0$

$\Leftrightarrow t = 0$ or $\pm\frac{1}{\sqrt{2}}$. The negative and 0 roots are inadmissible since $y(t)$ is only defined for $t > 0$, so the leftmost point must be

$$\left(x\left(\tfrac{1}{\sqrt{2}}\right), y\left(\tfrac{1}{\sqrt{2}}\right)\right) = \left(\left(\tfrac{1}{\sqrt{2}}\right)^4 - \left(\tfrac{1}{\sqrt{2}}\right)^2, \tfrac{1}{\sqrt{2}} + \ln\tfrac{1}{\sqrt{2}}\right) = \left(-\tfrac{1}{4}, \tfrac{1}{\sqrt{2}} - \tfrac{1}{2}\ln 2\right).$$

21. We graph the curve $x = t^4 - 2t^3 - 2t^2$, $y = t^3 - t$ in the viewing rectangle $[-2, 1.1]$ by $[-0.5, 0.5]$. This rectangle corresponds approximately to $t \in [-1, 0.8]$. We estimate that the curve has horizontal tangents at about $(-1, -0.4)$ and $(-0.17, 0.39)$ and vertical tangents at about $(0, 0)$ and $(-0.19, 0.37)$. We calculate $\frac{dy}{dx} = \frac{dy/dt}{dx/dt} = \frac{3t^2 - 1}{4t^3 - 6t^2 - 4t}$. The horizontal tangents occur when $dy/dt = 3t^2 - 1 = 0 \quad \Leftrightarrow \quad t = \pm\frac{1}{\sqrt{3}}$, so both horizontal tangents are shown in our graph. The vertical tangents occur when $dx/dt = 2t(2t^2 - 3t - 2) = 0 \quad \Leftrightarrow \quad 2t(2t+1)(t-2) = 0 \quad \Leftrightarrow \quad t = 0, -\frac{1}{2}$ or 2. It seems that we have missed one vertical tangent, and indeed if we plot the curve on the t-interval $[-1.2, 2.2]$ we see that there is another vertical tangent at $(-8, 6)$.

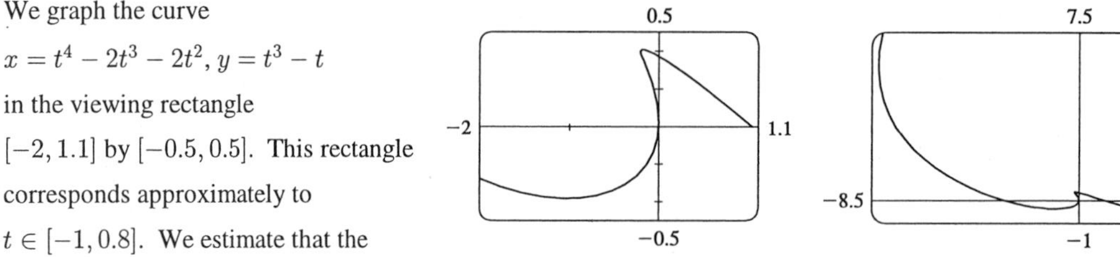

23. $x = \cos t$, $y = \sin t \cos t$. $\frac{dx}{dt} = -\sin t$, $\frac{dy}{dt} = -\sin^2 t + \cos^2 t = \cos 2t$. $(x, y) = (0, 0) \Leftrightarrow \quad \cos t = 0$ $\Leftrightarrow \quad t$ is an odd multiple of $\frac{\pi}{2}$.

When $t = \frac{\pi}{2}$, $\frac{dx}{dt} = -1$ and $\frac{dy}{dt} = -1$, so $\frac{dy}{dx} = 1$.

When $t = \frac{3\pi}{2}$, $\frac{dx}{dt} = 1$ and $\frac{dy}{dt} = -1$. So $\frac{dy}{dx} = -1$.

Thus $y = x$ and $y = -x$ are both tangent to the curve at $(0, 0)$.

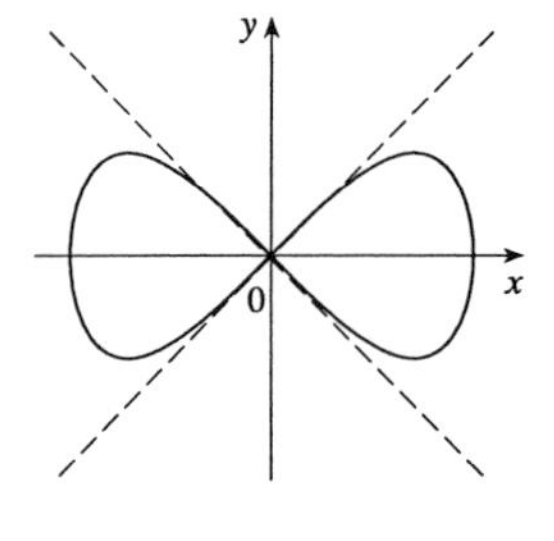

25. **(a)** $x = r\theta - d\sin\theta$, $y = r - d\cos\theta$; $\frac{dx}{d\theta} = r - d\cos\theta$, $\frac{dy}{d\theta} = d\sin\theta$. So $\frac{dy}{dx} = \frac{d\sin\theta}{r - d\cos\theta}$.

(b) If $0 < d < r$, then $|d\cos\theta| \le d < r$, so $r - d\cos\theta \ge r - d > 0$. This shows that $dx/d\theta$ never vanishes, so the trochoid can have no vertical tangents if $d < r$.

27. The line with parametric equations $x = -7t$, $y = 12t - 5$ is $y = 12\left(-\frac{1}{7}x\right) - 5$, which has slope $-\frac{12}{7}$. The curve $x = t^3 + 4t$, $y = 6t^2$ has slope $\frac{dy}{dx} = \frac{dy/dt}{dx/dt} = \frac{12t}{3t^2 + 4}$. This equals $-\frac{12}{7} \quad \Leftrightarrow \quad 3t^2 + 4 = -7t \quad \Leftrightarrow$ $(3t+4)(t+1) = 0 \quad \Leftrightarrow \quad t = -1$ or $t = -\frac{4}{3} \quad \Leftrightarrow \quad (x, y) = (-5, 6)$ or $\left(-\frac{208}{27}, \frac{32}{3}\right)$.

29. By symmetry of the ellipse about the x- and y-axes,

$$A = 4\int_0^a y\,dx = 4\int_{\pi/2}^0 b\sin\theta(-a\sin\theta)d\theta = 4ab\int_0^{\pi/2}\sin^2\theta\,d\theta = 4ab\int_0^{\pi/2}\tfrac{1}{2}(1 - \cos 2\theta)d\theta$$

$$= 2ab\left[\theta - \tfrac{1}{2}\sin 2\theta\right]_0^{\pi/2} = 2ab\left(\tfrac{\pi}{2}\right) = \pi ab.$$

31. $A = \int_0^1 (y-1)dx = \int_{\pi/2}^0 (e^t - 1)(-\sin t)dt = \int_0^{\pi/2}(e^t \sin t - \sin t)dt = \left[\frac{1}{2}e^t(\sin t - \cos t) + \cos t\right]_0^{\pi/2}$

$= \frac{1}{2}\left(e^{\pi/2} - 1\right)$ (Formula 98)

33. $A = \int_0^{2\pi r} y\,dx = \int_0^{2\pi}(r - d\cos\theta)(r - d\cos\theta)d\theta = \int_0^{2\pi}(r^2 - 2dr\cos\theta + d^2\cos^2\theta)d\theta$
$= \left[r^2\theta - 2dr\sin\theta + \frac{1}{2}d^2\left(\theta + \frac{1}{2}\sin 2\theta\right)\right]_0^{2\pi} = 2\pi r^2 + \pi d^2$

35. We plot the curve $x = t^3 - 12t$, $y = 3t^2 + 2t + 5$ in the parameter interval $t \in [-4, 3.5]$. In order to find the area of the loop, we need to estimate the two t-values corresponding to the point at which the curve crosses itself. By zooming in, we estimate the y-coordinate of the point of intersection to be 39.633, and so the two t-values at the point of intersection are approximately the two solutions of the equation $y = 3t^2 + 2t + 5 = 39.633$, which are $t = -\frac{1}{3} \pm \frac{\sqrt{419.596}}{6} \approx -3.7473$ or 3.0807. Now as in Example 4, we can evaluate the area of the loop simply by integrating $y\,dx$ between these two t-values, since this integral represents the area under the upper part of the loop for t between the first t-value and t_A, minus the area under the bottom part between t_A and t_B, plus the area under the top part between t_B and the final t-value. So since $dx = (3t^2 - 12)dt$, the area of the loop is $A \approx \int_{-3.7473}^{3.0807}(3t^2 + 2t + 5)(3t^2 - 12)dt \approx 741$.

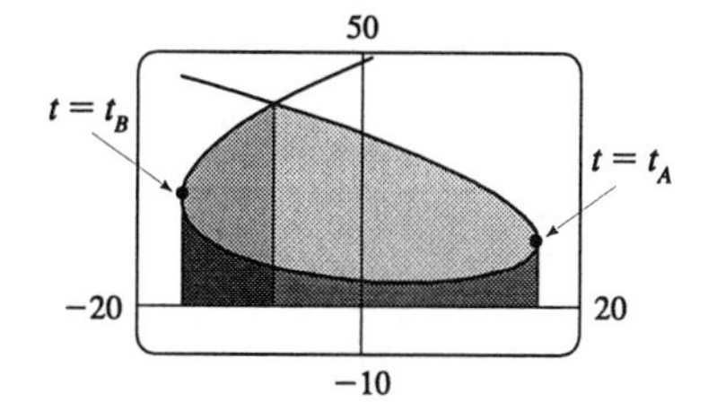

37. **(a)** The equations for line segment P_0P_1 are $x = x_0 + (x_1 - x_0)t$, $y = y_0 + (y_1 - y_0)t$, and those of the other line segments are similar, as shown in Exercise 9.1.27.

(b) It suffices to show that the slope of the tangent at P_0 is the same as that of line segment P_0P_1, namely $\dfrac{y_1 - y_0}{x_1 - x_0}$. We calculate the slope of the tangent to the Bézier curve:

$$\frac{dy/dt}{dx/dt} = \frac{-3y_0(1-t)^2 + 3y_1\left[-2t(1-t) + (1-t)^2\right] + 3y_2[-t^2 + (2t)(1-t)] + 3y_3t^2}{-3x_0(1-t)^2 + 3x_1\left[-2t(1-t) + (1-t)^2\right] + 3x_2[-t^2 + (2t)(1-t)] + 3x_3t^2}.$$

At point P_0, $t = 0$, so the slope of the tangent is $\dfrac{-3y_0 + 3y_1}{-3x_0 + 3x_1} = \dfrac{y_1 - y_0}{x_1 - x_0}$. So the tangent to the curve at P_0 passes through P_1. Similarly, the slope of the tangent at point P_3 (where $t = 1$) is $\dfrac{-3y_2 + 3y_3}{-3x_2 + 3x_3} = \dfrac{y_3 - y_2}{x_3 - x_2}$, which is also the slope of line P_2P_3.

(c) It appears that if P_1 were to the right of P_2, a loop would appear. We try setting $P_1 = (110, 30)$, and the resulting curve does indeed have a loop.

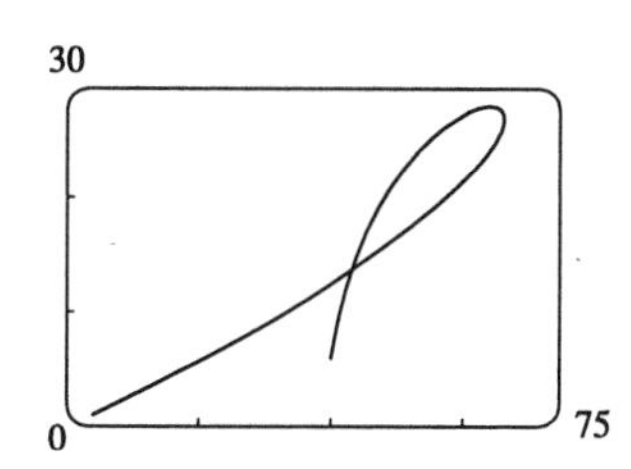

39. The coordinates of T are $(r\cos\theta, r\sin\theta)$. Since TP was unwound from arc TA, TP has length $r\theta$. Also $\angle PTQ = \angle PTR - \angle QTR = \frac{1}{2}\pi - \theta$, so P has coordinates $x = r\cos\theta + r\theta\cos\left(\frac{1}{2}\pi - \theta\right) = r(\cos\theta + \theta\sin\theta)$, $y = r\sin\theta - r\theta\sin\left(\frac{1}{2}\pi - \theta\right) = r(\sin\theta - \theta\cos\theta)$.

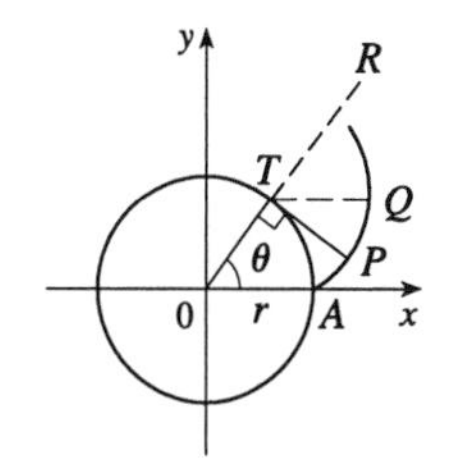

EXERCISES 9.3

1. $L = \int_0^1 \sqrt{(dx/dt)^2 + (dy/dt)^2}\,dt$ and $dx/dt = 3t^2$, $dy/dt = 4t^3 \quad \Rightarrow$

$L = \int_0^1 \sqrt{9t^4 + 16t^6}\,dt = \int_0^1 t^2\sqrt{9 + 16t^2}\,dt$

3. $\dfrac{dx}{dt} = \sin t + t\cos t$ and $\dfrac{dy}{dt} = \cos t - t\sin t \quad \Rightarrow$

$L = \int_0^{\pi/2} \sqrt{(\sin t + t\cos t)^2 + (\cos t - t\sin t)^2}\,dt = \int_0^{\pi/2} \sqrt{1 + t^2}\,dt$

5. $x = t^3$, $y = t^2$, $0 \le t \le 4$. $(dx/dt)^2 + (dy/dt)^2 = (3t^2)^2 + (2t)^2 = 9t^4 + 4t^2$

$L = \int_0^4 \sqrt{(dx/dt)^2 + (dy/dt)^2}\,dt = \int_0^4 \sqrt{9t^4 + 4t^2}\,dt = \int_0^4 t\sqrt{9t^2 + 4}\,dt = \frac{1}{18}\int_4^{148} \sqrt{u}\,du$

(where $u = 9t^2 + 4$). So $L = \frac{1}{18}\left(\frac{2}{3}\right)\left[u^{3/2}\right]_4^{148} = \frac{1}{27}\left(148^{3/2} - 4^{3/2}\right) = \frac{8}{27}\left(37^{3/2} - 1\right)$.

7. $x = 2 - 3\sin^2\theta$, $y = \cos 2\theta$, $0 \le \theta \le \frac{\pi}{2}$.

$(dx/d\theta)^2 + (dy/d\theta)^2 = (-6\sin\theta\cos\theta)^2 + (-2\sin 2\theta)^2 = (-3\sin 2\theta)^2 + (-2\sin 2\theta)^2 = 13\sin^2 2\theta \quad \Rightarrow$

$L = \int_0^{\pi/2} \sqrt{13}\sin 2\theta\,d\theta = \left[-\frac{\sqrt{13}}{2}\cos 2\theta\right]_0^{\pi/2} = -\frac{\sqrt{13}}{2}(-1 - 1) = \sqrt{13}$

9. $x = e^t\cos t$, $y = e^t\sin t$, $0 \le t \le \pi$.

$$\left(\frac{dx}{dt}\right)^2 + \left(\frac{dy}{dt}\right)^2 = \left[e^t(\cos t - \sin t)\right]^2 + \left[e^t(\sin t + \cos t)\right]^2$$
$$= e^{2t}\left(2\cos^2 t + 2\sin^2 t\right) = 2e^{2t} \quad \Rightarrow$$

$L = \int_0^{\pi} \sqrt{2}\,e^t\,dt = \sqrt{2}(e^{\pi} - 1)$

11. $x = \ln t$ and $y = e^{-t} \quad \Rightarrow \quad \dfrac{dx}{dt} = \dfrac{1}{t}$ and $\dfrac{dy}{dt} = -e^{-t} \quad \Rightarrow \quad L = \int_1^2 \sqrt{t^{-2} + e^{-2t}}\,dt$. Using Simpson's Rule with $n = 10$, $\Delta x = (2 - 1)/10 = 0.1$ and $f(t) = \sqrt{t^{-2} + e^{-2t}}$ we get

$L \approx \frac{0.1}{3}[f(1.0) + 4f(1.1) + 2f(1.2) + \cdots + 2f(1.8) + 4f(1.9) + f(2.0)] \approx 0.7314$.

13. $x = \sin^2\theta$, $y = \cos^2\theta$, $0 \le \theta \le 3\pi$.

$\left(\dfrac{dx}{d\theta}\right)^2 + \left(\dfrac{dy}{d\theta}\right)^2 = (2\sin\theta\cos\theta)^2 + (-2\cos\theta\sin\theta)^2 = 8\sin^2\theta\cos^2\theta = 2\sin^2 2\theta \quad \Rightarrow$

Distance $= \int_0^{3\pi} \sqrt{2}\,|\sin 2\theta|\,d\theta = 6\sqrt{2}\int_0^{\pi/2} \sin 2\theta\,d\theta$ (by symmetry)

$= \left[-3\sqrt{2}\cos 2\theta\right]_0^{\pi/2} = -3\sqrt{2}(-1 - 1) = 6\sqrt{2}$

The full curve is traversed as θ goes from 0 to $\frac{\pi}{2}$, because the curve is the segment of $x + y = 1$ that lies in the first quadrant (since $x, y \ge 0$), and this segment is completely traversed as θ goes from 0 to $\frac{\pi}{2}$. Thus $L = \int_0^{\pi/2} \sin 2\theta\,d\theta = \sqrt{2}$, as above.

15. $x = a\sin\theta,\ y = b\cos\theta,\ 0 \le \theta \le 2\pi$.

$$\begin{aligned}(dx/d\theta)^2 + (dy/d\theta)^2 &= (a\cos\theta)^2 + (-b\sin\theta)^2 = a^2\cos^2\theta + b^2\sin^2\theta \\ &= a^2(1-\sin^2\theta) + b^2\sin^2\theta = a^2 - (a^2-b^2)\sin^2\theta \\ &= a^2 - c^2\sin^2\theta = a^2\left(1 - \frac{c^2}{a^2}\sin^2\theta\right) = a^2(1-e^2\sin^2\theta)\end{aligned}$$

So $L = 4\int_0^{\pi/2} \sqrt{a^2(1-e^2\sin^2\theta)}\,d\theta$ (by symmetry) $= 4a\int_0^{\pi/2}\sqrt{1-e^2\sin^2\theta}\,d\theta$

17. **(a)** Notice that $0 \le t \le 2\pi$ does not give the complete curve because $x(0) \ne x(2\pi)$. In fact, we must take $t \in [0, 4\pi]$ in order to obtain the complete curve, since the first term in each of the parametric equations has period 2π and the second has period $\frac{4\pi}{11}$, and the least common integer multiple of these two numbers is 4π.

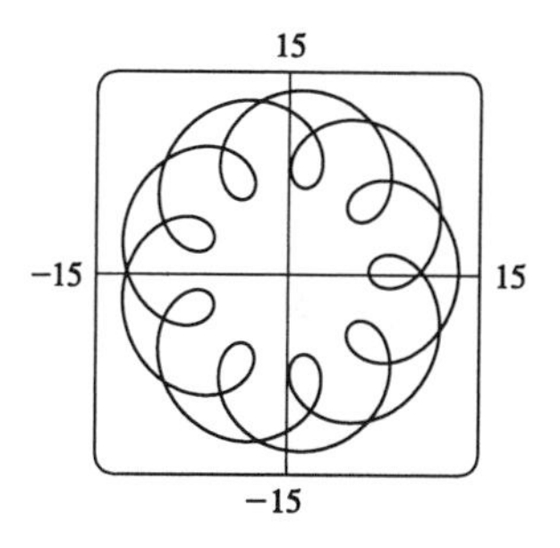

(b) We use the CAS to find the derivatives dx/dt and dy/dt, and then use Theorem 4 to find the arc length. Maple cannot do the integral exactly, so we use the command `evalf(Int(sqrt(diff(x,t)^2+diff(y,t)^2),t=0..4*Pi),4);` to estimate the length to be about 294. The 4 in the Maple command indicates that we want only four digits of accuracy; this speeds up the otherwise glacial calculation.

19. $x = t^3$ and $y = t^4 \quad\Rightarrow\quad dx/dt = 3t^2$ and $dy/dt = 4t^3$. So $S = \int_0^1 2\pi t^4\sqrt{9t^4+16t^6}\,dt = \int_0^1 2\pi t^6\sqrt{9+16t^2}\,dt$.

21. $x = t^3,\ y = t^2,\ 0 \le t \le 1$. $(dx/dt)^2 + (dy/dt)^2 = (3t^2)^2 + (2t)^2 = 9t^4 + 4t^2$.

$$\begin{aligned}S &= \int_0^1 2\pi y\sqrt{(dx/dt)^2 + (dy/dt)^2}\,dt = \int_0^1 2\pi t^2\sqrt{9t^4+4t^2}\,dt \\ &= 2\pi\int_4^{13}\frac{u-4}{9}\sqrt{u}\left(\tfrac{1}{18}du\right) \text{ (where } u = 9t^2+4) = \tfrac{\pi}{81}\left[\tfrac{2}{5}u^{5/2} - \tfrac{8}{3}u^{3/2}\right]_4^{13} = \tfrac{2\pi}{1215}\left(247\sqrt{13}+64\right)\end{aligned}$$

23. $x = a\cos^3\theta,\ y = a\sin^3\theta,\ 0 \le \theta \le \frac{\pi}{2}$

$(dx/d\theta)^2 + (dy/d\theta)^2 = \left(-3a\cos^2\theta\sin\theta\right)^2 + \left(3a\sin^2\theta\cos\theta\right)^2 = 9a^2\sin^2\theta\cos^2\theta$

$S = \int_0^{\pi/2} 2\pi a\sin^3\theta\, 3a\sin\theta\cos\theta\,d\theta = 6\pi a^2\int_0^{\pi/2}\sin^4\theta\cos\theta\,d\theta = \frac{6}{5}\pi a^2[\sin^5\theta]_0^{\pi/2} = \frac{6}{5}\pi a^2$

25. $(dx/dt)^2 + (dy/dt)^2 = (6t)^2 + (6t^2)^2 = 36t^2(1+t^2) \quad\Rightarrow$

$$\begin{aligned}S &= \int_0^5 2\pi x\sqrt{(dx/dt)^2 + (dy/dt)^2}\,dt = \int_0^5 2\pi(3t^2)6t\sqrt{1+t^2}\,dt = 18\pi\int_0^5 t^2\sqrt{1+t^2}\,2t\,dt \\ &= 18\pi\int_1^{26}(u-1)\sqrt{u}\,du \text{ (where } u = 1+t^2) = 18\pi\int_1^{26}(u^{3/2}-u^{1/2})\,du = 18\pi\left[\tfrac{2}{5}u^{5/2} - \tfrac{2}{3}u^{3/2}\right]_1^{26} \\ &= 18\pi\left[\left(\tfrac{2}{5}\cdot 676\sqrt{26} - \tfrac{2}{3}\cdot 26\sqrt{26}\right) - \left(\tfrac{2}{5} - \tfrac{2}{3}\right)\right] = \tfrac{24}{5}\pi\left(949\sqrt{26}+1\right)\end{aligned}$$

27. $x = a\cos\theta,\ y = b\sin\theta,\ 0 \le \theta \le 2\pi.$

$$(dx/d\theta)^2 + (dy/d\theta)^2 = (-a\sin\theta)^2 + (b\cos\theta)^2 = a^2\sin^2\theta + b^2\cos^2\theta = a^2(1-\cos^2\theta) + b^2\cos^2\theta$$

$$= a^2 - (a^2-b^2)\cos^2\theta = a^2 - c^2\cos^2\theta = a^2\left(1 - \frac{c^2}{a^2}\cos^2\theta\right) = a^2(1 - e^2\cos^2\theta)$$

(a) $S = \int_0^\pi 2\pi b\sin\theta\, a\sqrt{1-e^2\cos^2\theta}\,d\theta = 2\pi ab\int_{-e}^{e}\sqrt{1-u^2}(1/e)du$ [where $u = -e\cos\theta$, $du = e\sin\theta\,d\theta$]

$$= \frac{4\pi ab}{e}\int_0^e (1-u^2)^{1/2}\,du = \frac{4\pi ab}{e}\int_0^{\sin^{-1}e}\cos^2 v\,dv \text{ (where } u = \sin v) = \frac{2\pi ab}{e}\int_0^{\sin^{-1}e}(1+\cos 2v)dv$$

$$= \frac{2\pi ab}{e}\left[v + \tfrac{1}{2}\sin 2v\right]_0^{\sin^{-1}e} = \frac{2\pi ab}{e}\left[v + \sin v\cos v\right]_0^{\sin^{-1}e} = \frac{2\pi ab}{e}\left(\sin^{-1}e + e\sqrt{1-e^2}\right).$$

But $\sqrt{1-e^2} = \sqrt{1 - \frac{c^2}{a^2}} = \sqrt{\frac{a^2-c^2}{a^2}} = \sqrt{\frac{b^2}{a^2}} = \frac{b}{a}$, so $S = \frac{2\pi ab}{e}\sin^{-1}e + 2\pi b^2$.

(b) $S = \int_{-\pi/2}^{\pi/2} 2\pi a\cos\theta\, a\sqrt{1-e^2\cos^2\theta}\,d\theta = 4\pi a^2\int_0^{\pi/2}\cos\theta\sqrt{(1-e^2)+e^2\sin^2\theta}\,d\theta$

$$= \frac{4\pi a^2(1-e^2)}{e}\int_0^{\pi/2}\frac{e}{\sqrt{1-e^2}}\cos\theta\sqrt{1 + \left(\frac{e\sin\theta}{\sqrt{1-e^2}}\right)^2}\,d\theta$$

$$= \frac{4\pi a^2(1-e^2)}{e}\int_0^{e/\sqrt{1-e^2}}\sqrt{1+u^2}\,du \quad \left(\text{where } u = \frac{e\sin\theta}{\sqrt{1-e^2}}\right)$$

$$= \frac{4\pi a^2(1-e^2)}{e}\int_0^{\sin^{-1}e}\sec^3 v\,dv \quad (\text{where } u = \tan v,\ du = \sec^2 v\,dv)$$

$$= \frac{2\pi a^2(1-e^2)}{e}\left[\sec v\tan v + \ln|\sec v + \tan v|\right]_0^{\sin^{-1}e}$$

$$= \frac{2\pi a^2(1-e^2)}{e}\left[\frac{1}{\sqrt{1-e^2}}\frac{e}{\sqrt{1-e^2}} + \ln\left|\frac{1}{\sqrt{1-e^2}} + \frac{e}{\sqrt{1-e^2}}\right|\right]$$

$$= 2\pi a^2 + \frac{2\pi a^2(1-e^2)}{e}\ln\sqrt{\frac{1+e}{1-e}} = 2\pi a^2 + \frac{2\pi b^2}{e}\frac{1}{2}\ln\left(\frac{1+e}{1-e}\right) \quad \left(\text{since } 1-e^2 = \frac{b^2}{a^2}\right)$$

$$= 2\pi\left[a^2 + \frac{b^2}{2e}\ln\frac{1+e}{1-e}\right]$$

29. **(a)** $\phi = \tan^{-1}\left(\frac{dy}{dx}\right) \Rightarrow \frac{d\phi}{dt} = \frac{d}{dt}\tan^{-1}\left(\frac{dy}{dx}\right) = \frac{1}{1+(dy/dx)^2}\left[\frac{d}{dt}\left(\frac{dy}{dx}\right)\right]$. But $\frac{dy}{dx} = \frac{dy/dt}{dx/dt} = \frac{\dot{y}}{\dot{x}}$

$$\Rightarrow \frac{d}{dt}\left(\frac{dy}{dx}\right) = \frac{d}{dt}\left(\frac{\dot{y}}{\dot{x}}\right) = \frac{\ddot{y}\dot{x} - \ddot{x}\dot{y}}{\dot{x}^2} \Rightarrow \frac{d\phi}{dt} = \frac{1}{1+(\dot{y}/\dot{x})^2}\left[\frac{\ddot{y}\dot{x} - \ddot{x}\dot{y}}{\dot{x}^2}\right] = \frac{\dot{x}\ddot{y} - \ddot{x}\dot{y}}{\dot{x}^2 + \dot{y}^2}.$$

Using the Chain Rule, and the fact that $s = \int_0^t\sqrt{\left(\frac{dx}{dt}\right)^2 + \left(\frac{dy}{dt}\right)^2}\,dt \Rightarrow$

$\frac{ds}{dt} = \sqrt{\left(\frac{dx}{dt}\right)^2 + \left(\frac{dy}{dt}\right)^2} = (\dot{x}^2 + \dot{y}^2)^{1/2}$, we have that

$\frac{d\phi}{ds} = \frac{d\phi/dt}{ds/dt} = \left(\frac{\dot{x}\ddot{y} - \ddot{x}\dot{y}}{\dot{x}^2 + \dot{y}^2}\right)\frac{1}{(\dot{x}^2+\dot{y}^2)^{1/2}} = \frac{\dot{x}\ddot{y} - \ddot{x}\dot{y}}{(\dot{x}^2+\dot{y}^2)^{3/2}}$. So

$$\kappa = \left|\frac{d\phi}{ds}\right| = \left|\frac{\dot{x}\ddot{y} - \ddot{x}\dot{y}}{(\dot{x}^2+\dot{y}^2)^{3/2}}\right| = \frac{|\dot{x}\ddot{y} - \ddot{x}\dot{y}|}{(\dot{x}^2+\dot{y}^2)^{3/2}}.$$

(b) $x = x$ and $y = f(x) \quad \Rightarrow \quad \dot{x} = 1, \ddot{x} = 0$ and $\dot{y} = \dfrac{dy}{dx}, \ddot{y} = \dfrac{d^2y}{dx^2}$.

So $\kappa = \dfrac{|1 \cdot (d^2y/dx^2) - 0 \cdot (dy/dx)|}{\left[1 + (dy/dx)^2\right]^{3/2}} = \dfrac{|d^2y/dx^2|}{\left[1 + (dy/dx)^2\right]^{3/2}}$.

31. $x = \theta - \sin\theta \Rightarrow \dot{x} = 1 - \cos\theta \Rightarrow \ddot{x} = \sin\theta$, and $y = 1 - \cos\theta \Rightarrow \dot{y} = \sin\theta \Rightarrow \ddot{y} = \cos\theta$. Therefore $\kappa = \dfrac{|\cos\theta - \cos^2\theta - \sin^2\theta|}{\left[(1-\cos\theta)^2 + \sin^2\theta\right]^{3/2}} = \dfrac{|\cos\theta - (\cos^2\theta + \sin^2\theta)|}{(1 - 2\cos\theta + \cos^2\theta + \sin^2\theta)^{3/2}} = \dfrac{|\cos\theta - 1|}{(2 - 2\cos\theta)^{3/2}}$. The top of the arch is characterized by a horizontal tangent, and from Example 1 of Section 9.2 the tangent is horizontal when $\theta = (2n-1)\pi$, so take $n = 1$ and substitute $\theta = \pi$ into the expression for κ:

$$\kappa = \frac{|\cos\pi - 1|}{(2 - 2\cos\pi)^{3/2}} = \frac{|-1-1|}{[2 - 2(-1)]^{3/2}} = \frac{1}{4}.$$

EXERCISES 9.4

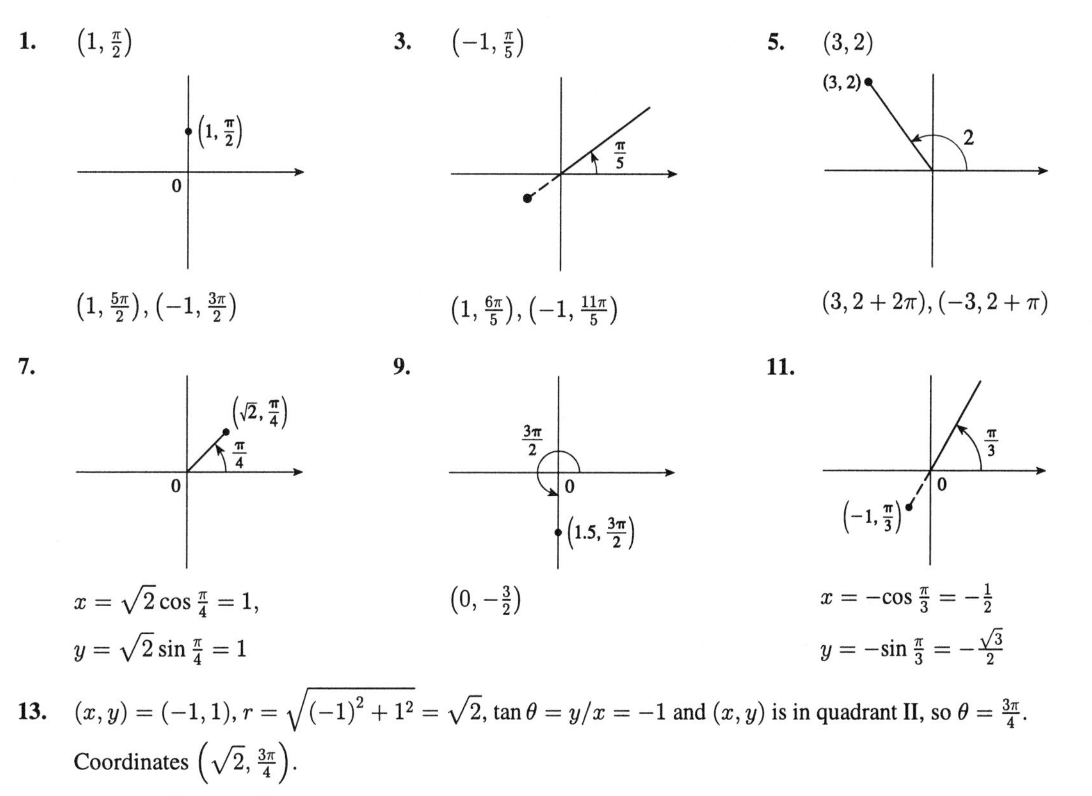

1. $(1, \frac{\pi}{2})$
$(1, \frac{\pi}{2})$
0
$(1, \frac{5\pi}{2}), (-1, \frac{3\pi}{2})$
3. $(-1, \frac{\pi}{5})$
$\frac{\pi}{5}$
$(1, \frac{6\pi}{5}), (-1, \frac{11\pi}{5})$
5. $(3, 2)$
$(3, 2)$
2
$(3, 2 + 2\pi), (-3, 2 + \pi)$
7.
$(\sqrt{2}, \frac{\pi}{4})$
$\frac{\pi}{4}$
0
$x = \sqrt{2}\cos\frac{\pi}{4} = 1$,
$y = \sqrt{2}\sin\frac{\pi}{4} = 1$
9.
$\frac{3\pi}{2}$
0
$(1.5, \frac{3\pi}{2})$
$(0, -\frac{3}{2})$
11.
$\frac{\pi}{3}$
0
$(-1, \frac{\pi}{3})$
$x = -\cos\frac{\pi}{3} = -\frac{1}{2}$
$y = -\sin\frac{\pi}{3} = -\frac{\sqrt{3}}{2}$

13. $(x, y) = (-1, 1)$, $r = \sqrt{(-1)^2 + 1^2} = \sqrt{2}$, $\tan\theta = y/x = -1$ and (x, y) is in quadrant II, so $\theta = \frac{3\pi}{4}$. Coordinates $\left(\sqrt{2}, \frac{3\pi}{4}\right)$.

15. $(x, y) = \left(2\sqrt{3}, -2\right)$. $r = \sqrt{12 + 4} = 4$, $\tan\theta = y/x = -\frac{1}{\sqrt{3}} \quad \Rightarrow \quad (x, y)$ is in quadrant IV, so $\theta = \frac{11\pi}{6}$. The polar coordinates are $\left(4, \frac{11\pi}{6}\right)$.

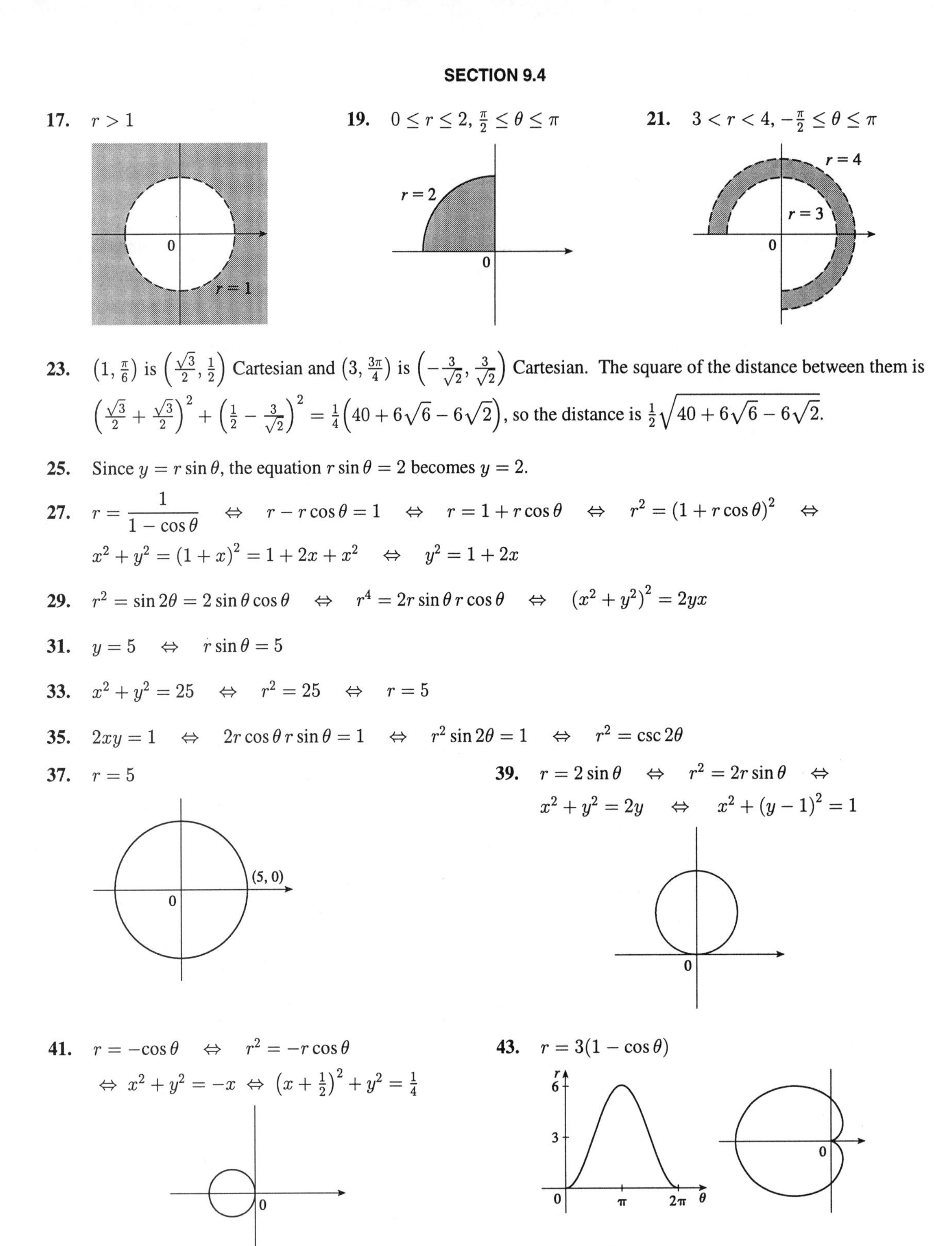

17. $r > 1$

19. $0 \le r \le 2,\ \frac{\pi}{2} \le \theta \le \pi$

21. $3 < r < 4,\ -\frac{\pi}{2} \le \theta \le \pi$

23. $\left(1, \frac{\pi}{6}\right)$ is $\left(\frac{\sqrt{3}}{2}, \frac{1}{2}\right)$ Cartesian and $\left(3, \frac{3\pi}{4}\right)$ is $\left(-\frac{3}{\sqrt{2}}, \frac{3}{\sqrt{2}}\right)$ Cartesian. The square of the distance between them is $\left(\frac{\sqrt{3}}{2} + \frac{\sqrt{3}}{2}\right)^2 + \left(\frac{1}{2} - \frac{3}{\sqrt{2}}\right)^2 = \frac{1}{4}\left(40 + 6\sqrt{6} - 6\sqrt{2}\right)$, so the distance is $\frac{1}{2}\sqrt{40 + 6\sqrt{6} - 6\sqrt{2}}$.

25. Since $y = r\sin\theta$, the equation $r\sin\theta = 2$ becomes $y = 2$.

27. $r = \dfrac{1}{1 - \cos\theta} \quad \Leftrightarrow \quad r - r\cos\theta = 1 \quad \Leftrightarrow \quad r = 1 + r\cos\theta \quad \Leftrightarrow \quad r^2 = (1 + r\cos\theta)^2 \quad \Leftrightarrow$

$x^2 + y^2 = (1 + x)^2 = 1 + 2x + x^2 \quad \Leftrightarrow \quad y^2 = 1 + 2x$

29. $r^2 = \sin 2\theta = 2\sin\theta\cos\theta \quad \Leftrightarrow \quad r^4 = 2r\sin\theta\, r\cos\theta \quad \Leftrightarrow \quad (x^2 + y^2)^2 = 2yx$

31. $y = 5 \quad \Leftrightarrow \quad r\sin\theta = 5$

33. $x^2 + y^2 = 25 \quad \Leftrightarrow \quad r^2 = 25 \quad \Leftrightarrow \quad r = 5$

35. $2xy = 1 \quad \Leftrightarrow \quad 2r\cos\theta\, r\sin\theta = 1 \quad \Leftrightarrow \quad r^2\sin 2\theta = 1 \quad \Leftrightarrow \quad r^2 = \csc 2\theta$

37. $r = 5$

39. $r = 2\sin\theta \quad \Leftrightarrow \quad r^2 = 2r\sin\theta \quad \Leftrightarrow$

$x^2 + y^2 = 2y \quad \Leftrightarrow \quad x^2 + (y - 1)^2 = 1$

41. $r = -\cos\theta \quad \Leftrightarrow \quad r^2 = -r\cos\theta$

$\Leftrightarrow\ x^2 + y^2 = -x\ \Leftrightarrow\ \left(x + \frac{1}{2}\right)^2 + y^2 = \frac{1}{4}$

43. $r = 3(1 - \cos\theta)$

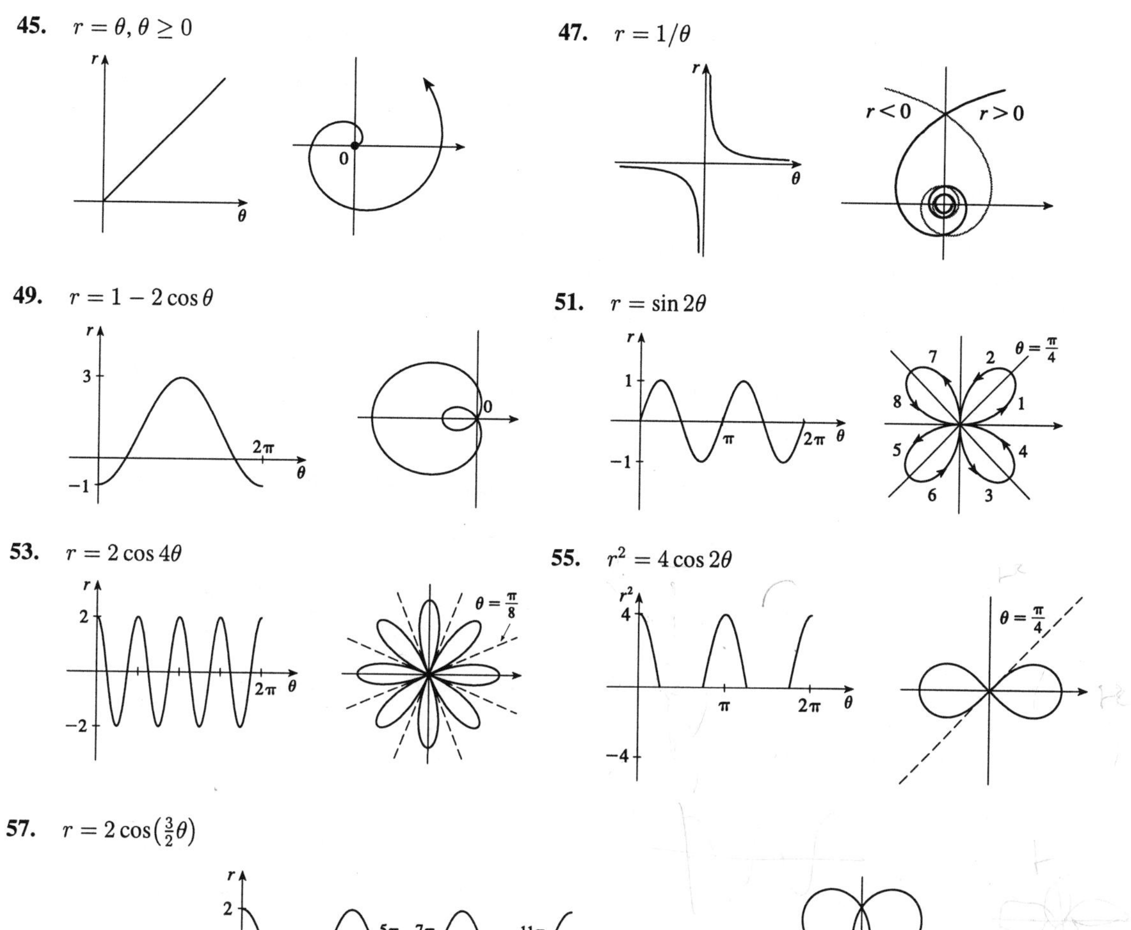

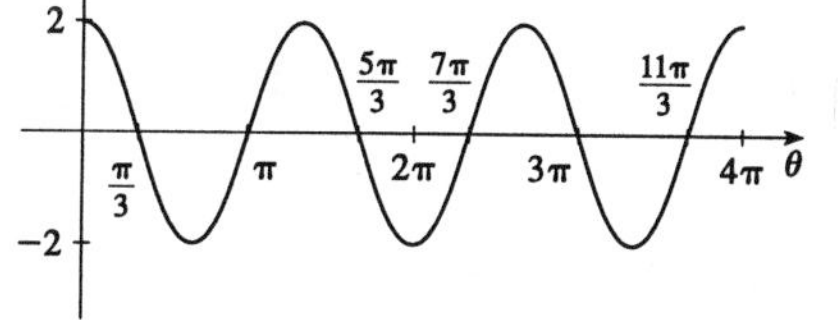

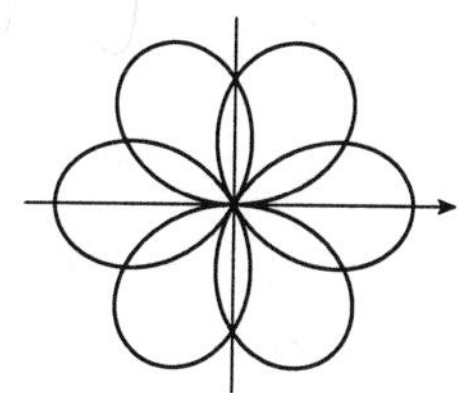

59. $x = r\cos\theta = 4\cos\theta + 2\sec\theta\cos\theta = 4\cos\theta + 2$. Now, $r \to \infty \Rightarrow (4 + 2\sec\theta) \to \infty \Rightarrow \theta \to \frac{\pi}{2}^-$ (since we need only consider $0 \leq \theta < 2\pi$), so $\lim\limits_{r\to\infty} x = \lim\limits_{\theta\to\pi/2^-} (4\cos\theta + 2) = 2$. Also, $r \to -\infty \Rightarrow (4 + 2\sec\theta) \to -\infty \Rightarrow \theta \to \frac{\pi}{2}^+$, so $\lim\limits_{r\to-\infty} x = \lim\limits_{\theta\to\pi/2^+} (4\cos\theta + 2) = 2$. Therefore $\lim\limits_{r\to\pm\infty} x = 2 \Rightarrow x = 2$ is a vertical asymptote.

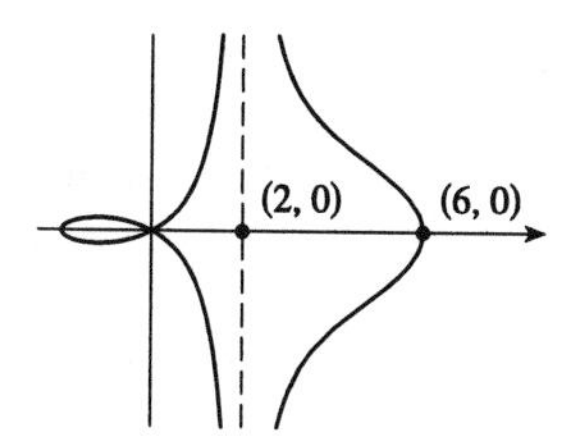

61. To show that $x = 1$ is an asymptote we must prove $\lim_{r\to\pm\infty} x = 1$. $x = r\cos\theta = \sin\theta\tan\theta\cos\theta = \sin^2\theta$. Now, $r \to \infty \;\Rightarrow\; \sin\theta\tan\theta \to \infty \;\Rightarrow\; \theta \to \frac{\pi}{2}^-$, so $\lim_{r\to\infty} x = \lim_{\theta\to\pi/2^-} \sin^2\theta = 1$. Also, $r \to -\infty \;\Rightarrow$ $\sin\theta\tan\theta \to -\infty \Rightarrow\; \theta \to \pi/2^+$, so $\lim_{r\to-\infty} x = \lim_{\theta\to\pi/2^+} \sin^2\theta = 1$.

Therefore $\lim_{r\to\pm\infty} x = 1 \;\;\Rightarrow\;\; x = 1$ is a vertical asymptote.

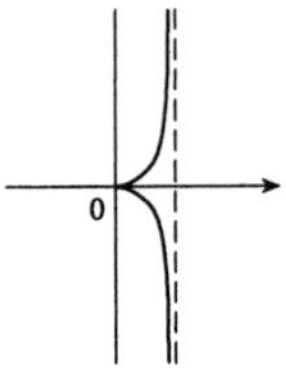

Also notice that $x = \sin^2\theta \geq 0$ for all θ, and $x = \sin^2\theta \leq 1$ for all θ.

And $x \neq 1$, since the curve is not defined at odd multiples of $\frac{\pi}{2}$.

Therefore the curve lies entirely within the vertical strip $0 \leq x < 1$.

0

63. $$\frac{dy}{dx} = \frac{dy/d\theta}{dx/d\theta} = \frac{(dr/d\theta)(\sin\theta) + r\cos\theta}{(dr/d\theta)(\cos\theta) - r\sin\theta} = \frac{-3\sin\theta\sin\theta + 3\cos\theta\cos\theta}{-3\sin\theta\cos\theta - 3\cos\theta\sin\theta} = \frac{3\cos 2\theta}{-3\sin 2\theta} = -\cot 2\theta$$
$$= \frac{1}{\sqrt{3}} \text{ when } \theta = \frac{\pi}{3}$$

Alternate Solution: $r = 3\cos\theta \;\;\Rightarrow\;\; x = r\cos\theta = 3\cos^2\theta,\ y = r\sin\theta = 3\sin\theta\cos\theta \;\;\Rightarrow$
$$\frac{dy}{dx} = \frac{dy/d\theta}{dx/d\theta} = \frac{-3\sin^2\theta + 3\cos^2\theta}{-6\cos\theta\sin\theta} = \frac{\cos 2\theta}{-\sin 2\theta} = -\cot 2\theta = \frac{1}{\sqrt{3}} \text{ when } \theta = \frac{\pi}{3}$$

65. $r = \theta \;\Rightarrow\; x = r\cos\theta = \theta\cos\theta,\ y = r\sin\theta = \theta\sin\theta \;\Rightarrow\; \dfrac{dy}{dx} = \dfrac{dy/d\theta}{dx/d\theta} = \dfrac{\sin\theta + \theta\cos\theta}{\cos\theta - \theta\sin\theta} = -\dfrac{2}{\pi}$ when $\theta = \dfrac{\pi}{2}$

67. $r = 1 + \cos\theta \;\;\Rightarrow\;\; x = r\cos\theta = \cos\theta + \cos^2\theta,\ y = r\sin\theta = \sin\theta + \sin\theta\cos\theta \;\;\Rightarrow$
$$\frac{dy}{dx} = \frac{dy/d\theta}{dx/d\theta} = \frac{\cos\theta + \cos^2\theta - \sin^2\theta}{-\sin\theta - 2\cos\theta\sin\theta} = \frac{\cos\theta + \cos 2\theta}{-\sin\theta - \sin 2\theta} = -1 \text{ when } \theta = \frac{\pi}{6}$$

69. $r = 3\cos\theta \;\;\Rightarrow\;\; x = r\cos\theta = 3\cos\theta\cos\theta,\ y = r\sin\theta = 3\cos\theta\sin\theta \;\;\Rightarrow$
$dy/d\theta = -3\sin^2\theta + 3\cos^2\theta = 3\cos 2\theta = 0 \;\;\Rightarrow\;\; 2\theta = \frac{\pi}{2}$ or $\frac{3\pi}{2} \;\;\Leftrightarrow\;\; \theta = \frac{\pi}{4}$ or $\frac{3\pi}{4}$. So the tangent is horizontal at $\left(\frac{3}{\sqrt{2}}, \frac{\pi}{4}\right)$ and $\left(-\frac{3}{\sqrt{2}}, \frac{3\pi}{4}\right)$ $\left[\text{same as } \left(\frac{3}{\sqrt{2}}, -\frac{\pi}{4}\right)\right]$. $dx/d\theta = -6\sin\theta\cos\theta = -3\sin 2\theta = 0 \;\;\Rightarrow$ $2\theta = 0$ or $\pi \;\;\Leftrightarrow\;\; \theta = 0$ or $\frac{\pi}{2}$. So the tangent is vertical at $(3, 0)$ and $\left(0, \frac{\pi}{2}\right)$.

71. $r = \cos 2\theta \;\;\Rightarrow\;\; x = r\cos\theta = \cos 2\theta\cos\theta,\ y = r\sin\theta = \cos 2\theta\sin\theta \;\;\Rightarrow$
$dy/d\theta = -2\sin 2\theta\sin\theta + \cos 2\theta\cos\theta = -4\sin^2\theta\cos\theta + \left(\cos^3\theta - \sin^2\theta\cos\theta\right)$
$= \cos\theta(\cos^2\theta - 5\sin^2\theta) = \cos\theta(1 - 6\sin^2\theta) = 0 \;\;\Rightarrow$
$\cos\theta = 0$ or $\sin\theta = \pm\frac{1}{\sqrt{6}} \;\;\Rightarrow\;\; \theta = \frac{\pi}{2}, \frac{3\pi}{2}, \alpha, \pi - \alpha, \pi + \alpha$, or $2\pi - \alpha$ $\left[\text{where } \alpha = \sin^{-1}\frac{1}{\sqrt{6}}\right]$.
So the tangent is horizontal at $\left(1, \frac{3\pi}{2}\right), \left(1, \frac{\pi}{2}\right), \left(\frac{2}{3}, \alpha\right), \left(\frac{2}{3}, \pi - \alpha\right), \left(\frac{2}{3}, \pi + \alpha\right)$, and $\left(\frac{2}{3}, 2\pi - \alpha\right)$.
$dx/d\theta = -2\sin 2\theta\cos\theta - \cos 2\theta\sin\theta = -4\sin\theta\cos^2\theta - (2\cos^2\theta - 1)\sin\theta = \sin\theta(1 - 6\cos^2\theta) = 0 \;\;\Rightarrow$
$\sin\theta = 0$ or $\cos\theta = \pm\frac{1}{\sqrt{6}} \;\;\Rightarrow\;\; \theta = 0, \pi, \frac{\pi}{2} - \alpha, \frac{\pi}{2} + \alpha, \frac{3\pi}{2} - \alpha$, or $\frac{3\pi}{2} + \alpha$. So the tangent is vertical at $(1, 0), (1, \pi), \left(\frac{2}{3}, \frac{3\pi}{2} - \alpha\right), \left(\frac{2}{3}, \frac{3\pi}{2} + \alpha\right), \left(\frac{2}{3}, \frac{\pi}{2} - \alpha\right)$, and $\left(\frac{2}{3}, \frac{\pi}{2} + \alpha\right)$.

73. $r = 1 + \cos\theta \;\;\Rightarrow\;\; x = r\cos\theta = \cos\theta(1 + \cos\theta),\ y = r\sin\theta = \sin\theta(1 + \cos\theta) \;\;\Rightarrow$
$dy/d\theta = (1 + \cos\theta)\cos\theta - \sin^2\theta = 2\cos^2\theta + \cos\theta - 1 = (2\cos\theta - 1)(\cos\theta + 1) = 0 \;\;\Rightarrow\;\; \cos\theta = \frac{1}{2}$ or $-1 \;\;\Rightarrow\;\; \theta = \frac{\pi}{3}, \pi$, or $\frac{5\pi}{3} \;\;\Rightarrow\;\;$ horizontal tangent at $\left(\frac{3}{2}, \frac{\pi}{3}\right), (0, \pi)$, and $\left(\frac{3}{2}, \frac{5\pi}{3}\right)$.

$dx/d\theta = -(1+\cos\theta)\sin\theta - \cos\theta\sin\theta = -\sin\theta(1+2\cos\theta) = 0 \quad\Rightarrow\quad \sin\theta = 0$ or $\cos\theta = -\frac{1}{2} \quad\Rightarrow$ $\theta = 0, \pi, \frac{2\pi}{3}$, or $\frac{4\pi}{3} \quad\Rightarrow\quad$ vertical tangent at $(2,0)$, $\left(\frac{1}{2},\frac{2\pi}{3}\right)$, and $\left(\frac{1}{2},\frac{4\pi}{3}\right)$. Note that the tangent is horizontal, not vertical when $\theta = \pi$, since $\displaystyle\lim_{\theta\to\pi}\frac{dy/d\theta}{dx/d\theta} = 0$.

75. $r = a\sin\theta + b\cos\theta \quad\Rightarrow\quad r^2 = ar\sin\theta + br\cos\theta \quad\Rightarrow\quad x^2+y^2 = ay+bx \quad\Rightarrow$ $\left(x-\frac{1}{2}b\right)^2 + \left(y-\frac{1}{2}a\right)^2 = \frac{1}{4}(a^2+b^2)$, and this is a circle with center $\left(\frac{1}{2}b, \frac{1}{2}a\right)$ and radius $\frac{1}{2}\sqrt{a^2+b^2}$.

77. **(a)** We see that the curve crosses itself at the origin, where $r = 0$ (in fact the inner loop corresponds to negative r-values,) so we solve the equation of the limaçon for $r = 0 \quad\Leftrightarrow\quad c\sin\theta = -1 \quad\Leftrightarrow$ $\sin\theta = -1/c$. Now if $|c| < 1$, then this equation has no solution and hence there is no inner loop. But if $c < -1$, then on the interval $(0, 2\pi)$ the equation has the two solutions $\theta = \sin^{-1}(-1/c)$ and $\theta = \pi - \sin^{-1}(-1/c)$, and if $c > 1$, the solutions are $\theta = \pi + \sin^{-1}(1/c)$ and $\theta = 2\pi - \sin^{-1}(1/c)$. In each case, $r < 0$ for θ between the two solutions, indicating a loop.

(b) For $0 < c < 1$, the dimple (if it exists) is characterized by the fact that y has a local maximum at $\theta = \frac{3\pi}{2}$. So we determine for what c-values $\dfrac{d^2y}{d\theta^2}$ is negative at $\theta = \frac{3\pi}{2}$, since by the Second Derivative Test this indicates a maximum: $y = r\sin\theta = \sin\theta + c\sin^2\theta \;\Rightarrow\; \dfrac{dy}{d\theta} = \cos\theta + 2c\sin\theta\cos\theta = \cos\theta + c\sin 2\theta$ $\Rightarrow\; \dfrac{d^2y}{d\theta^2} = -\sin\theta + 2c\cos 2\theta$. At $\theta = \frac{3\pi}{2}$, this is equal to $-(-1) + 2c(-1) = 1 - 2c$, which is negative only for $c > \frac{1}{2}$. A similar argument shows that for $-1 < c < 0$, y only has a local *minimum* at $\theta = \frac{\pi}{2}$ (indicating a dimple) for $c < -\frac{1}{2}$.

Note for Exercises 79-82: Maple is able to plot polar curves using the `polarplot` command, or using the `coords=polar` option in a regular `plot` command. In Mathematica, use `PolarPlot`. If your graphing device cannot plot polar equations, you must convert to parametric equations. For example, in Exercise 79, $x = r\cos\theta = [1+2\sin(\theta/2)]\cos\theta$, $y = r\sin\theta = [1+2\sin(\theta/2)]\sin\theta$.

79. $r = 1 + 2\sin(\theta/2)$

The correct parameter interval is $[0, 4\pi]$.

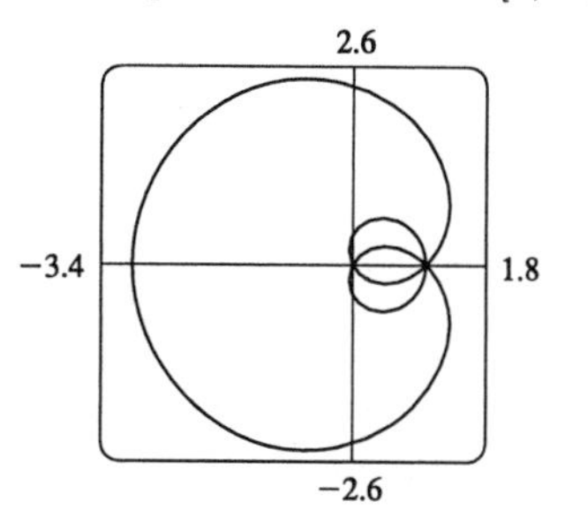

81. $r = \sin(9\theta/4)$

The correct parameter interval is $[0, 8\pi]$.

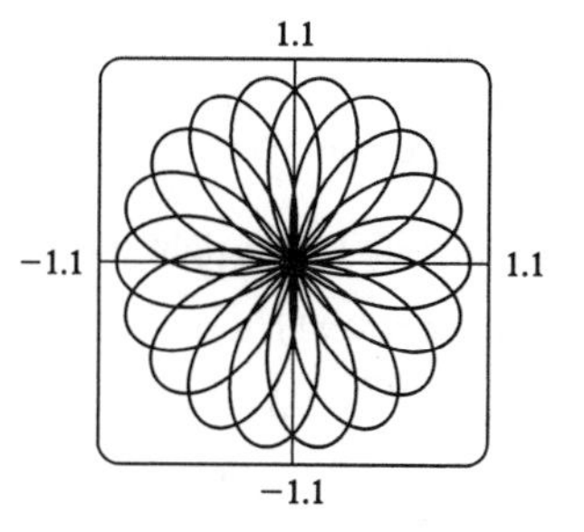

83.

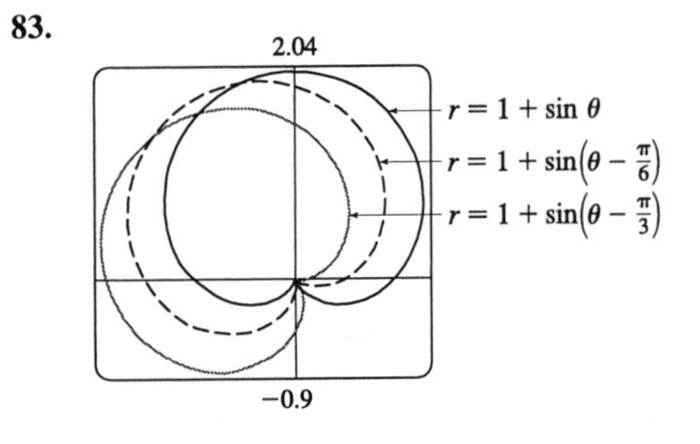

It appears that the graph of $r = 1 + \sin\left(\theta - \frac{\pi}{6}\right)$ is the same shape as the graph of $r = 1 + \sin\theta$, but rotated counterclockwise about the origin by $\frac{\pi}{6}$. Similarly, the graph of $r = 1 + \sin\left(\theta - \frac{\pi}{3}\right)$ is rotated by $\frac{\pi}{3}$. In general, the graph of $r = f(\theta - \alpha)$ is the same shape as that of $r = f(\theta)$, but rotated counterclockwise through α about the origin. That is, for any point (r_0, θ_0) on the curve $r = f(\theta)$, the point $(r_0, \theta_0 + \alpha)$ is on the curve the curve $r = f(\theta - \alpha)$, since $r_0 = f(\theta_0) = f((\theta_0 + \alpha) - \alpha)$.

85. **(a)** $r = \sin n\theta$. From the graphs, it seems that when n is even, the number of loops in the curve (called a rose) is $2n$, and when n is odd, the number of loops is simply n. This is because in the case of n odd, every point on the graph is traversed twice, due to the fact that

$$r(\theta + \pi) = \sin[n(\theta + \pi)] = \sin n\theta \cos n\pi + \cos n\theta \sin n\pi = \begin{cases} \sin n\theta & n \text{ even} \\ -\sin n\theta & n \text{ odd.} \end{cases}$$

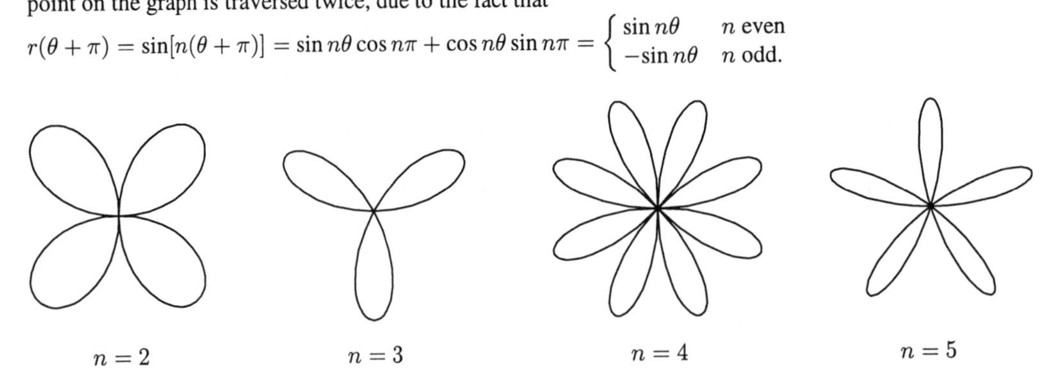

(b) The graph of $r = |\sin n\theta|$ has $2n$ loops whether n is odd or even, since $r(\theta + \pi) = r(\theta)$.

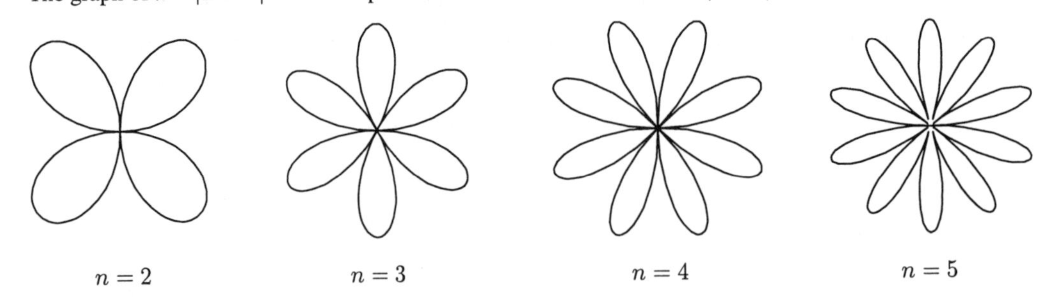

87. $r = \dfrac{1 - a\cos\theta}{1 + a\cos\theta}$. We start with $a = 0$, since in this case the curve is simply the circle $r = 1$. As a increases, the graph moves to the left, and its right side becomes flattened. As a increases through about 0.4, the right side seems to grow a dimple, which upon closer investigation (with narrower θ-ranges) seems to appear at $a \approx 0.42$ (the actual value is $\sqrt{2} - 1$.) As $a \to 1$, this dimple becomes more pronounced, and the curve begins to stretch out horizontally, until at $a = 1$ the denominator vanishes at $\theta = \pi$, and the dimple becomes an actual cusp.

For $a > 1$ we must choose our parameter interval carefully, since $r \to \infty$ as $1 + a\cos\theta \to 0 \quad \Leftrightarrow$ $\theta \to \pm\cos^{-1}(-1/a)$. As a increases from 1, the curve splits into two parts. The left part has a loop, which grows larger as a increases, and the right part grows broader vertically, and its left tip develops a dimple when $a \approx 2.42$ (actually, $\sqrt{2}+1$). As a increases, the dimple grows more and more pronounced.

If $a < 0$, we get the same graph as we do for the corresponding positive a-value, but with a rotation through π about the pole, as happened when c was replaced with $-c$ in Exercise 86.

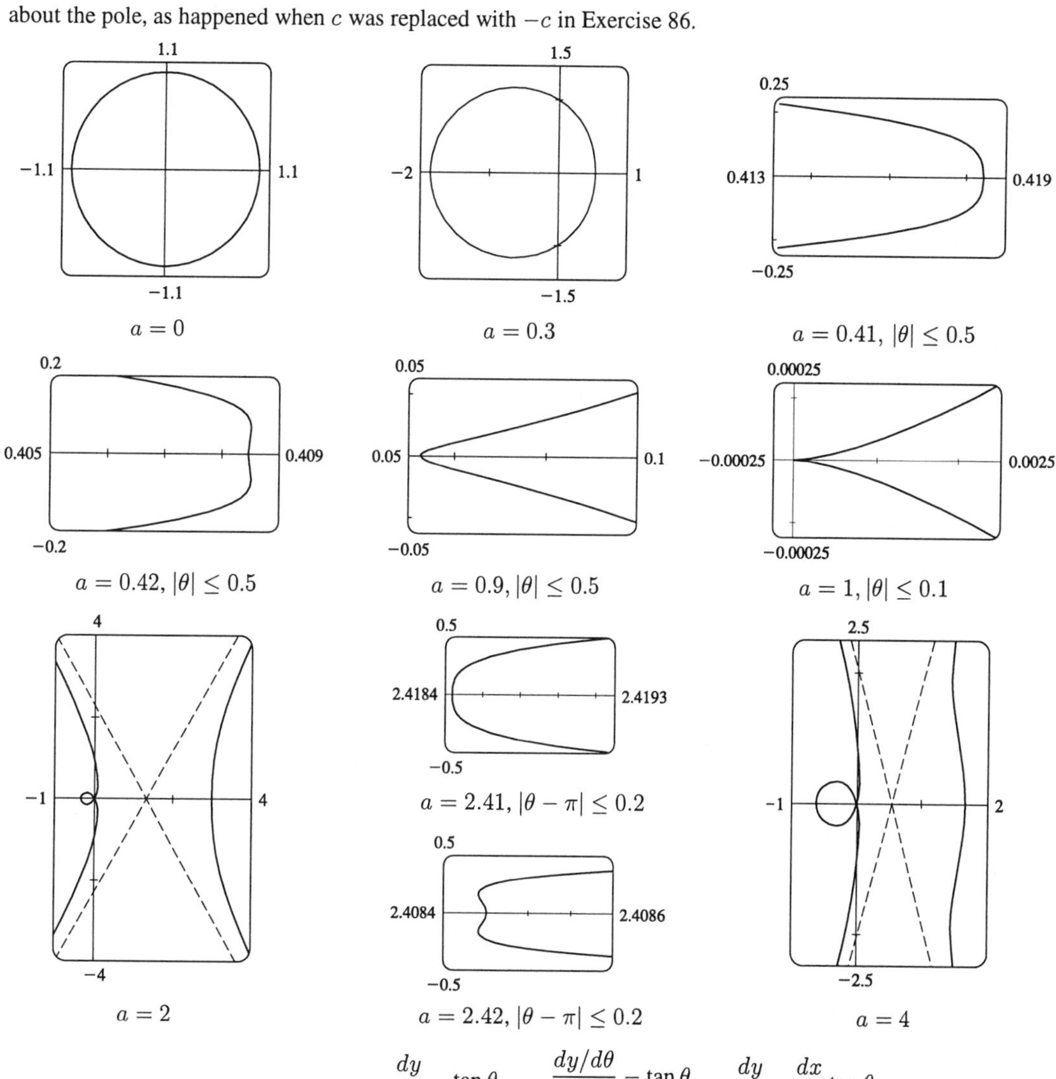

$a = 0$ $\quad$ $a = 0.3$ $\quad$ $a = 0.41, |\theta| \le 0.5$

$a = 0.42, |\theta| \le 0.5$ $\quad$ $a = 0.9, |\theta| \le 0.5$ $\quad$ $a = 1, |\theta| \le 0.1$

$a = 2.41, |\theta - \pi| \le 0.2$

$a = 2$ $\quad$ $a = 2.42, |\theta - \pi| \le 0.2$ $\quad$ $a = 4$

89. $\tan\psi = \tan(\phi - \theta) = \dfrac{\tan\phi - \tan\theta}{1 + \tan\phi\tan\theta} = \dfrac{\dfrac{dy}{dx} - \tan\theta}{1 + \dfrac{dy}{dx}\tan\theta} = \dfrac{\dfrac{dy/d\theta}{dx/d\theta} - \tan\theta}{1 + \dfrac{dy/d\theta}{dx/d\theta}\tan\theta} = \dfrac{\dfrac{dy}{d\theta} - \dfrac{dx}{d\theta}\tan\theta}{\dfrac{dx}{d\theta} + \dfrac{dy}{d\theta}\tan\theta}$

$$= \frac{\left(\dfrac{dr}{d\theta}\sin\theta + r\cos\theta\right) - \tan\theta\left(\dfrac{dr}{d\theta}\cos\theta - r\sin\theta\right)}{\left(\dfrac{dr}{d\theta}\cos\theta - r\sin\theta\right) + \tan\theta\left(\dfrac{dr}{d\theta}\sin\theta + r\cos\theta\right)} = \frac{r\cos^2\theta + r\sin^2\theta}{\dfrac{dr}{d\theta}\cos^2\theta + \dfrac{dr}{d\theta}\sin^2\theta} = \frac{r}{dr/d\theta}$$

EXERCISES 9.5

1. $A = \int_0^{\pi} \frac{1}{2} r^2\, d\theta = \int_0^{\pi} \frac{1}{2}\theta^2\, d\theta = \left[\frac{1}{6}\theta^3\right]_0^{\pi} = \frac{1}{6}\pi^3$

3. $A = \int_0^{\pi/6} \frac{1}{2}(2\cos\theta)^2\, d\theta = \int_0^{\pi/6}(1+\cos 2\theta)d\theta = \left[\theta + \frac{1}{2}\sin 2\theta\right]_0^{\pi/6} = \frac{\pi}{6} + \frac{\sqrt{3}}{4}$

5. $A = \int_0^{\pi/6} \frac{1}{2}\sin^2 2\theta\, d\theta = \frac{1}{4}\int_0^{\pi/6}(1-\cos 4\theta)d\theta = \left[\frac{1}{4}\theta - \frac{1}{16}\sin 4\theta\right]_0^{\pi/6} = \frac{4\pi - 3\sqrt{3}}{96}$

7. $A = \int_0^{\pi} \frac{1}{2}(5\sin\theta)^2\, d\theta$

$= \frac{25}{4}\int_0^{\pi}(1-\cos 2\theta)d\theta$

$= \frac{25}{4}\left[\theta - \frac{1}{2}\sin 2\theta\right]_0^{\pi}$

$= \frac{25}{4}\pi$

9. $A = 4\int_0^{\pi/4} \frac{1}{2} r^2\, d\theta$

$= 8\int_0^{\pi/4}\cos 2\theta\, d\theta$

$= [4\sin 2\theta]_0^{\pi/4} = 4$

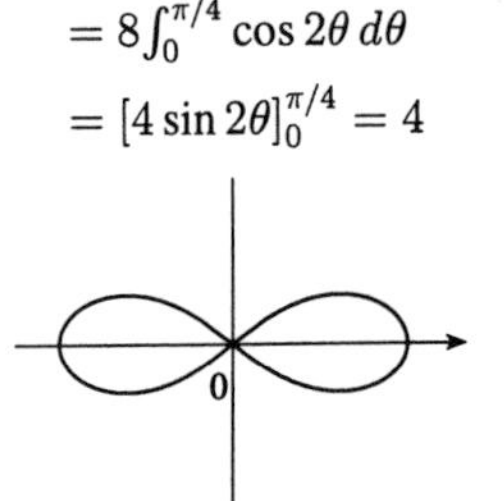

11. $A = 2\int_{-\pi/2}^{\pi/2} \frac{1}{2}(4-\sin\theta)^2\, d\theta$

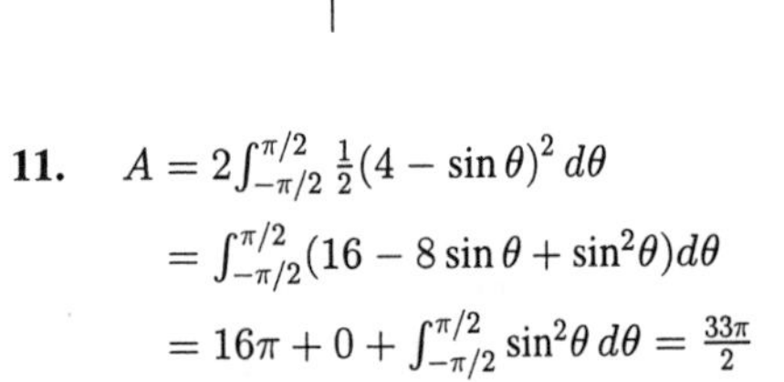

$= \int_{-\pi/2}^{\pi/2}(16 - 8\sin\theta + \sin^2\theta)d\theta$

$= 16\pi + 0 + \int_{-\pi/2}^{\pi/2}\sin^2\theta\, d\theta = \frac{33\pi}{2}$

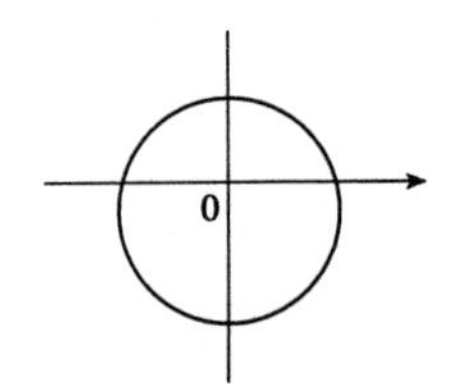

13. By symmetry, the total area is twice the area enclosed above the polar axis, so

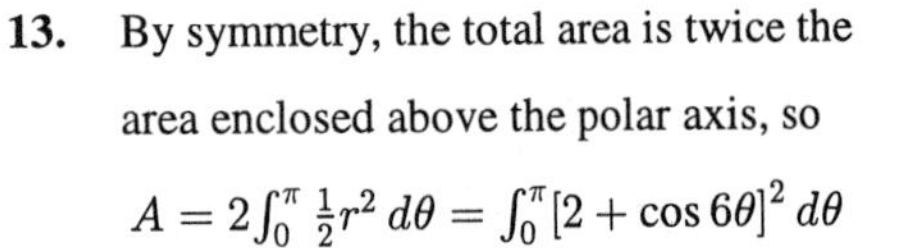

$A = 2\int_0^{\pi} \frac{1}{2} r^2\, d\theta = \int_0^{\pi}[2+\cos 6\theta]^2\, d\theta$

$= \left[4\theta + 4\left(\frac{1}{6}\sin 6\theta\right) + \left(\frac{1}{24}\sin 12\theta + \frac{1}{2}\theta\right)\right]_0^{\pi}$

$= 4\pi + \frac{\pi}{2} = \frac{9\pi}{2}.$

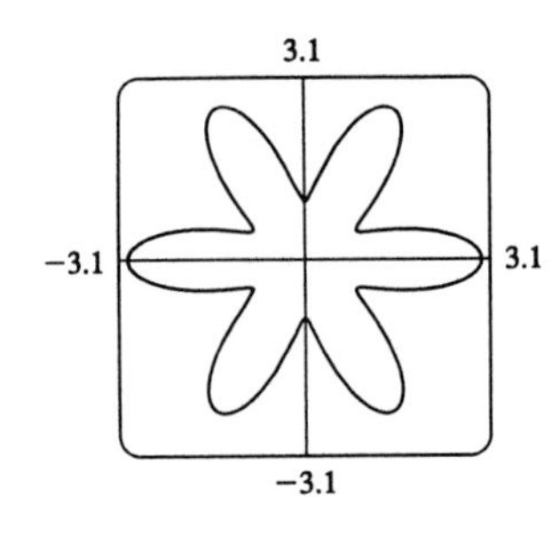

15. $A = 2\int_0^{\pi/6} \frac{1}{2}\cos^2 3\theta\, d\theta = \frac{1}{2}\int_0^{\pi/6}(1+\cos 6\theta)d\theta = \frac{1}{2}\left[\theta + \frac{1}{6}\sin 6\theta\right]_0^{\pi/6} = \frac{\pi}{12}$

17. $A = \int_0^{\pi/5} \frac{1}{2}\sin^2 5\theta\, d\theta = \frac{1}{4}\int_0^{\pi/5}(1-\cos 10\theta)d\theta = \frac{1}{4}\left[\theta - \frac{1}{10}\sin 10\theta\right]_0^{\pi/5} = \frac{\pi}{20}$

19. This is a limaçon, with inner loop traced out between $\theta = \frac{7\pi}{6}$ and $\frac{11\pi}{6}$.

$A = 2\int_{7\pi/6}^{3\pi/2} \frac{1}{2}(1+2\sin\theta)^2\, d\theta = \int_{7\pi/6}^{3\pi/2}(1 + 4\sin\theta + 4\sin^2\theta)d\theta$

$= [\theta - 4\cos\theta + 2\theta - \sin 2\theta]_{7\pi/6}^{3\pi/2} = \pi - \frac{3\sqrt{3}}{2}$

21. $1 - \cos\theta = \frac{3}{2} \quad\Leftrightarrow\quad \cos\theta = -\frac{1}{2} \quad\Rightarrow\quad \theta = \frac{2\pi}{3} \text{ or } \frac{4\pi}{3} \quad\Rightarrow$

$$A = \int_{2\pi/3}^{4\pi/3} \frac{1}{2}\left[(1-\cos\theta)^2 - \left(\tfrac{3}{2}\right)^2\right] d\theta$$
$$= \tfrac{1}{2}\int_{2\pi/3}^{4\pi/3}\left(-\tfrac{5}{4} - 2\cos\theta + \cos^2\theta\right) d\theta$$
$$= \tfrac{1}{2}\left[-\tfrac{5}{4}\theta - 2\sin\theta\right]_{2\pi/3}^{4\pi/3} + \tfrac{1}{2}\int_{2\pi/3}^{4\pi/3} \frac{1+\cos 2\theta}{2}\, d\theta$$
$$= -\tfrac{5}{12}\pi + \sqrt{3} + \tfrac{1}{4}\left[\theta + \tfrac{1}{2}\sin 2\theta\right]_{2\pi/3}^{4\pi/3} = \tfrac{9\sqrt{3}}{8} - \tfrac{1}{4}\pi$$

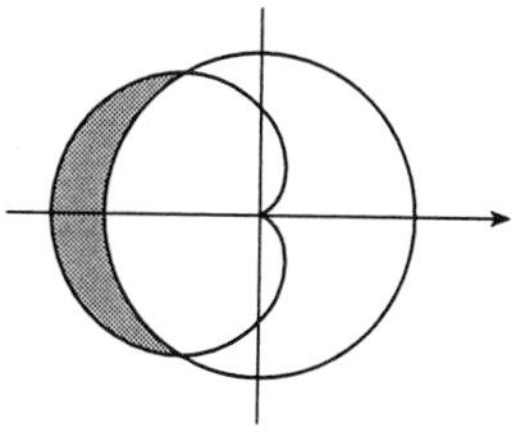

23. $4\sin\theta = 2 \quad\Leftrightarrow\quad \sin\theta = \frac{1}{2} \quad\Rightarrow\quad \theta = \frac{\pi}{6} \text{ or } \frac{5\pi}{6} \quad\Rightarrow$

$$A = 2\int_{\pi/6}^{\pi/2} \tfrac{1}{2}\left[(4\sin\theta)^2 - 2^2\right] d\theta$$
$$= \int_{\pi/6}^{\pi/2}(16\sin^2\theta - 4)\,d\theta = \int_{\pi/6}^{\pi/2}[8(1-\cos 2\theta) - 4]\,d\theta$$
$$= [4\theta - 4\sin 2\theta]_{\pi/6}^{\pi/2} = \tfrac{4}{3}\pi + 2\sqrt{3}$$

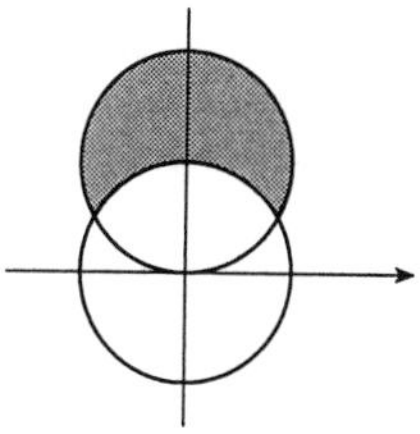

25. $3\cos\theta = 1 + \cos\theta \quad\Leftrightarrow\quad \cos\theta = \frac{1}{2} \quad\Rightarrow\quad \theta = \frac{\pi}{3} \text{ or } -\frac{\pi}{3}.$

$$A = 2\int_0^{\pi/3} \tfrac{1}{2}\left[(3\cos\theta)^2 - (1+\cos\theta)^2\right] d\theta$$
$$= \int_0^{\pi/3}(8\cos^2\theta - 2\cos\theta - 1)\,d\theta$$
$$= \int_0^{\pi/3}[4(1+\cos 2\theta) - 2\cos\theta - 1]\,d\theta$$
$$= [3\theta + 2\sin 2\theta - 2\sin\theta]_0^{\pi/3} = \pi + \sqrt{3} - \sqrt{3} = \pi$$

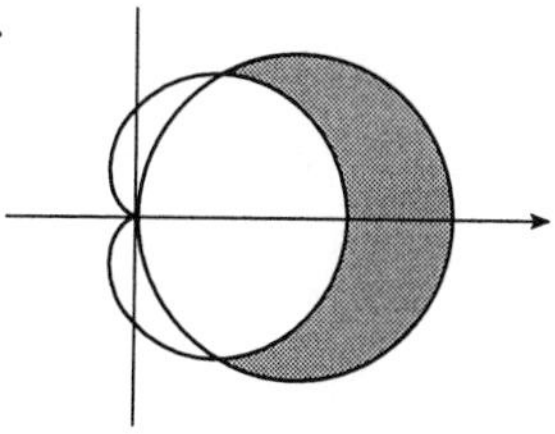

27. $$A = 2\int_0^{\pi/4} \tfrac{1}{2}\sin^2\theta\, d\theta = \int_0^{\pi/4} \frac{1-\cos 2\theta}{2}\, d\theta$$
$$= \left[\tfrac{1}{2}\theta - \tfrac{1}{4}\sin 2\theta\right]_0^{\pi/4} = \tfrac{1}{8}\pi - \tfrac{1}{4}$$

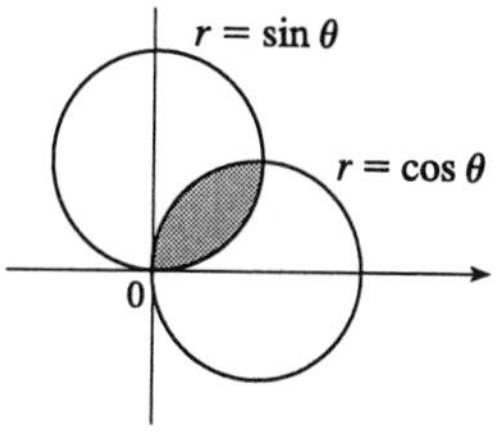

r = sin θ
r = cos θ
0

29. $\sin 2\theta = \cos 2\theta \quad\Rightarrow\quad \tan 2\theta = 1 \quad\Rightarrow\quad 2\theta = \frac{\pi}{4} \quad\Rightarrow\quad \theta = \frac{\pi}{8}$

$$\Rightarrow\quad A = 16\int_0^{\pi/8} \tfrac{1}{2}\sin^2 2\theta\, d\theta$$
$$= 4\int_0^{\pi/8}(1-\cos 4\theta)\,d\theta = 4\left[\theta - \tfrac{1}{4}\sin 4\theta\right]_0^{\pi/8}$$
$$= \tfrac{1}{2}\pi - 1$$

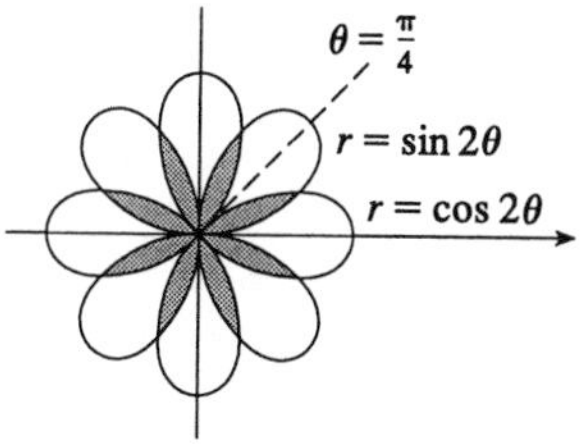

θ = π/4
r = sin 2θ
r = cos 2θ

31. $$A = 2\left[\int_{-\pi/2}^{-\pi/6} \tfrac{1}{2}(3+2\sin\theta)^2\, d\theta + \int_{-\pi/6}^{\pi/2} \tfrac{1}{2}2^2\, d\theta\right]$$
$$= \int_{-\pi/2}^{-\pi/6}(9 + 12\sin\theta + 4\sin^2\theta)\,d\theta + [4\theta]_{-\pi/6}^{\pi/2}$$
$$= [9\theta - 12\cos\theta + 2\theta - \sin 2\theta]_{-\pi/2}^{-\pi/6} + \tfrac{8\pi}{3}$$
$$= \tfrac{19\pi}{3} - \tfrac{11\sqrt{3}}{2}$$

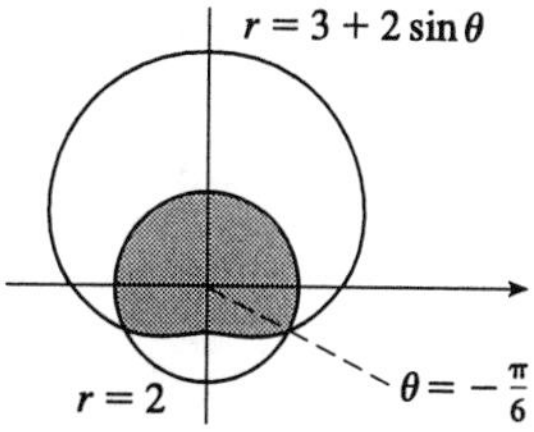

r = 3 + 2 sin θ
r = 2
θ = −π/6

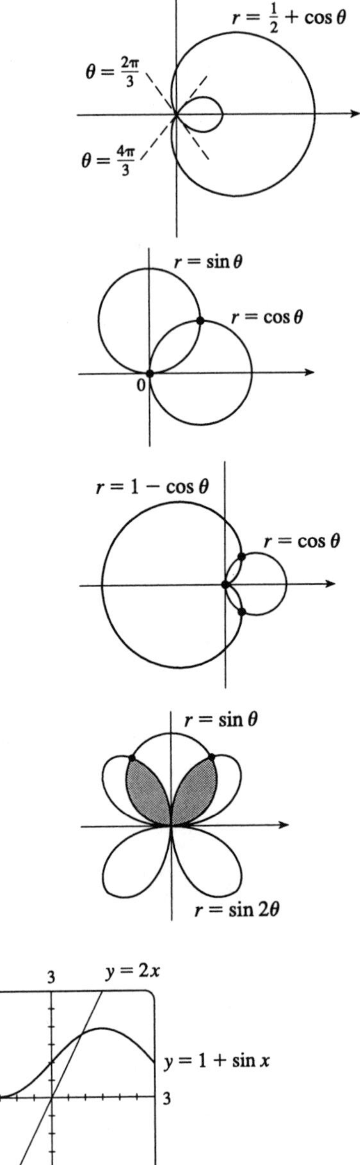

33. $A = 2\left[\int_0^{2\pi/3} \frac{1}{2}\left(\frac{1}{2} + \cos\theta\right)^2 d\theta - \int_{2\pi/3}^{\pi} \frac{1}{2}\left(\frac{1}{2} + \cos\theta\right)^2 d\theta\right]$

$$= \left[\frac{\theta}{4} + \sin\theta + \frac{\theta}{2} + \frac{\sin 2\theta}{4}\right]_0^{2\pi/3} - \left[\frac{\theta}{4} + \sin\theta + \frac{\theta}{2} + \frac{\sin 2\theta}{4}\right]_{2\pi/3}^{\pi}$$

$$= \frac{1}{4}\left(\pi + 3\sqrt{3}\right)$$

35. The two circles intersect at the pole since $(0, 0)$ satisfies the first equation and $\left(0, \frac{\pi}{2}\right)$ the second. The other intersection point $\left(\frac{1}{\sqrt{2}}, \frac{\pi}{4}\right)$ occurs where $\sin\theta = \cos\theta$.

37. The curves intersect at the pole since $\left(0, \frac{\pi}{2}\right)$ satisfies the first equation and $(0, 0)$ the second.
$\cos\theta = 1 - \cos\theta \quad \Rightarrow \quad \cos\theta = \frac{1}{2} \quad \Rightarrow$
$\theta = \frac{\pi}{3}$ or $\frac{5\pi}{3} \quad \Rightarrow \quad$ the other intersection points are $\left(\frac{1}{2}, \frac{\pi}{3}\right)$ and $\left(\frac{1}{2}, \frac{5\pi}{3}\right)$.

39. The pole is a point of intersection.
$\sin\theta = \sin 2\theta = 2\sin\theta\cos\theta \quad \Leftrightarrow$
$\sin\theta(1 - 2\cos\theta) = 0 \quad \Leftrightarrow \quad \sin\theta = 0$ or $\cos\theta = \frac{1}{2}$
$\Rightarrow \quad \theta = 0, \pi, \frac{\pi}{3}, -\frac{\pi}{3} \quad \Rightarrow \quad \left(\frac{\sqrt{3}}{2}, \frac{\pi}{3}\right)$ and $\left(\frac{\sqrt{3}}{2}, \frac{2\pi}{3}\right)$ are the other intersection points.

41.

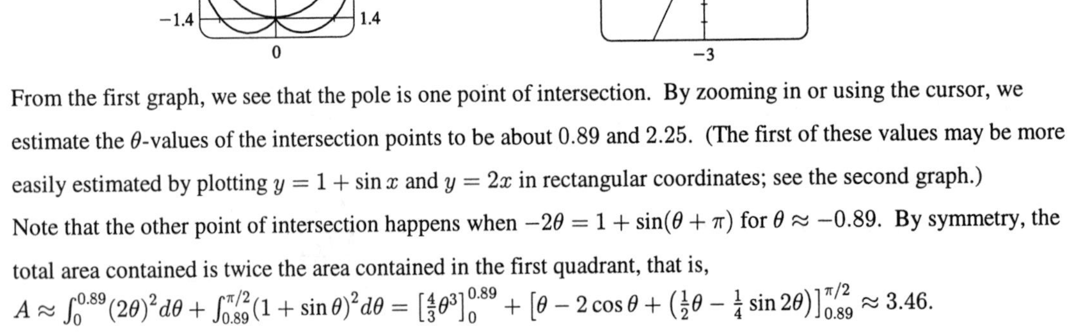

From the first graph, we see that the pole is one point of intersection. By zooming in or using the cursor, we estimate the θ-values of the intersection points to be about 0.89 and 2.25. (The first of these values may be more easily estimated by plotting $y = 1 + \sin x$ and $y = 2x$ in rectangular coordinates; see the second graph.) Note that the other point of intersection happens when $-2\theta = 1 + \sin(\theta + \pi)$ for $\theta \approx -0.89$. By symmetry, the total area contained is twice the area contained in the first quadrant, that is,
$A \approx \int_0^{0.89} (2\theta)^2 d\theta + \int_{0.89}^{\pi/2} (1 + \sin\theta)^2 d\theta = \left[\frac{4}{3}\theta^3\right]_0^{0.89} + \left[\theta - 2\cos\theta + \left(\frac{1}{2}\theta - \frac{1}{4}\sin 2\theta\right)\right]_{0.89}^{\pi/2} \approx 3.46.$

43. $L = \int_0^{3\pi/4} \sqrt{r^2 + (dr/d\theta)^2}\, d\theta = \int_0^{3\pi/4} \sqrt{(5\cos\theta)^2 + (-5\sin\theta)^2}\, d\theta$

$= 5\int_0^{3\pi/4} \sqrt{\cos^2\theta + \sin^2\theta}\, d\theta = 5\int_0^{3\pi/4} d\theta = \frac{15}{4}\pi$

45. $L = \int_0^{2\pi} \sqrt{(2^\theta)^2 + [(\ln 2)2^\theta]^2}\, d\theta = \int_0^{2\pi} 2^\theta \sqrt{1 + \ln^2 2}\, d\theta = \left[\sqrt{1+\ln^2 2}\left(\frac{2^\theta}{\ln 2}\right)\right]_0^{2\pi} = \frac{\sqrt{1+\ln^2 2}(2^{2\pi} - 1)}{\ln 2}$

47. $L = \int_0^{2\pi} \sqrt{(\theta^2)^2 + (2\theta)^2}\, d\theta = \int_0^{2\pi} \theta\sqrt{\theta^2 + 4}\, d\theta = \frac{1}{2} \cdot \frac{2}{3}\left[(\theta^2 + 4)^{3/2}\right]_0^{2\pi} = \frac{8}{3}\left[(\pi^2 + 1)^{3/2} - 1\right]$

49. $L = 2\int_0^{2\pi} \sqrt{\cos^8(\theta/4) + \cos^6(\theta/4)\sin^2(\theta/4)}\, d\theta$

$= 2\int_0^{2\pi} |\cos^3(\theta/4)| \sqrt{\cos^2(\theta/4) + \sin^2(\theta/4)}\, d\theta$

$= 2\int_0^{2\pi} |\cos^3(\theta/4)| d\theta = 8\int_0^{\pi/2} \cos^3 u\, du \quad (\text{where } u = \frac{1}{4}\theta)$

$= 8\left[\sin u - \frac{1}{3}\sin^3 u\right]_0^{\pi/2} = \frac{16}{3}$

Note that the curve is retraced after every interval of length 4π.

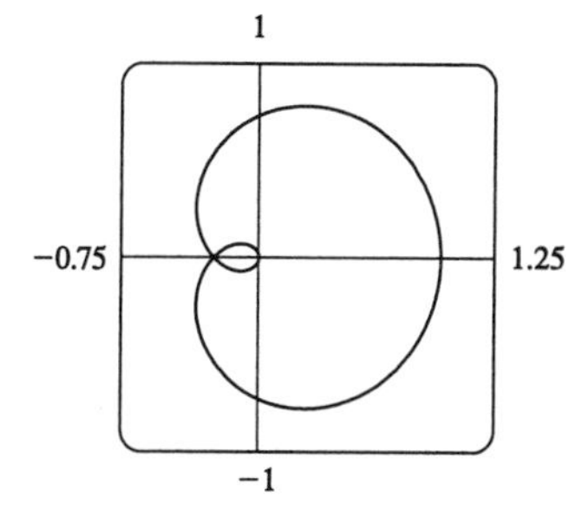

51. From Figure 4 it is apparent that one loop lies between $\theta = -\frac{\pi}{4}$ and $\theta = \frac{\pi}{4}$. Therefore $L = \int_{-\pi/4}^{\pi/4} \sqrt{\cos^2 2\theta + (-2\sin 2\theta)^2}\, d\theta = 2\int_0^{\pi/4} \sqrt{\cos^2 2\theta + 4\sin^2 2\theta}\, d\theta = 2\int_0^{\pi/4} \sqrt{1 + 3\sin^2 2\theta}\, d\theta$. Using Simpson's Rule with $n = 4$, $\Delta x = \frac{\pi/4}{4} = \frac{\pi}{16}$ and $f(\theta) = \sqrt{1 + 3\sin^2 2\theta}$, we get $L \approx 2 \cdot S_4 = 2\frac{\pi}{16 \cdot 3}\left[f(0) + 4f\left(\frac{\pi}{16}\right) + 2f\left(\frac{\pi}{8}\right) + 4f\left(\frac{3\pi}{16}\right) + f\left(\frac{\pi}{4}\right)\right] \approx 2.4228$. Therefore the length of one loop of the four-leaved rose is approximately 2.42.

53. **(a)** From (9.3.5), $S = \int_a^b 2\pi y \sqrt{(dx/d\theta)^2 + (dy/d\theta)^2}\, d\theta = \int_a^b 2\pi y \sqrt{r^2 + (dr/d\theta)^2}\, d\theta$

(see the derivation of Equation 5) $= \int_a^b 2\pi r \sin\theta \sqrt{r^2 + (dr/d\theta)^2}\, d\theta$.

(b) $r^2 = \cos 2\theta \quad \Rightarrow \quad 2r\frac{dr}{d\theta} = -2\sin 2\theta \quad \Rightarrow \quad \left(\frac{dr}{d\theta}\right)^2 = \frac{\sin^2 2\theta}{r^2} = \frac{\sin^2 2\theta}{\cos 2\theta}$

$S = 2\int_0^{\pi/4} 2\pi\sqrt{\cos 2\theta}\sin\theta \sqrt{\cos 2\theta + (\sin^2 2\theta)/\cos 2\theta}\, d\theta = 4\pi\int_0^{\pi/4} \sin\theta\, d\theta$

$= [-4\pi\cos\theta]_0^{\pi/4} = -4\pi\left(\frac{1}{\sqrt{2}} - 1\right) = 2\pi\left(2 - \sqrt{2}\right)$.

EXERCISES 9.6

1. $x^2 = -8y$. $4p = -8$, so $p = -2$. The vertex is $(0, 0)$, the focus is $(0, -2)$, and the directrix is $y = 2$.

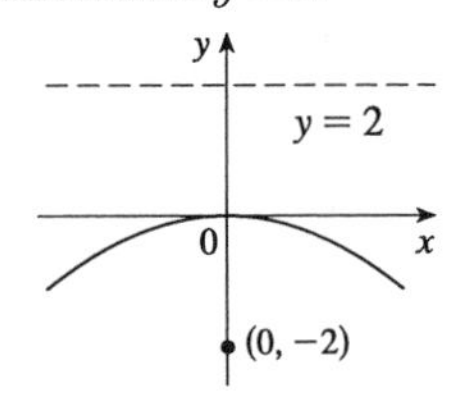

3. $y^2 = x$. $p = \frac{1}{4}$ and the vertex is $(0, 0)$, so the focus is $\left(\frac{1}{4}, 0\right)$, and the directrix is $x = -\frac{1}{4}$.

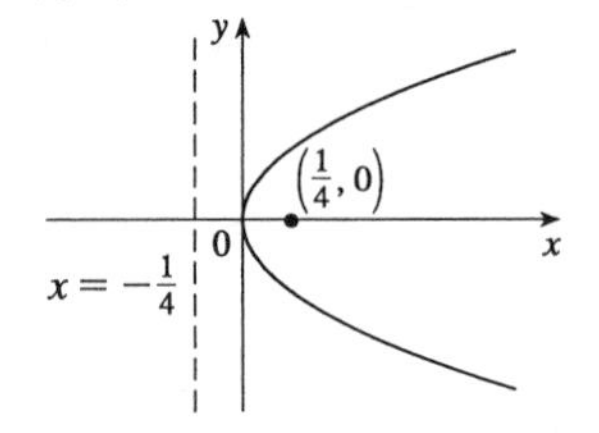

5. $x + 1 = 2(y - 3)^2 \quad\Rightarrow$ $(y - 3)^2 = \frac{1}{2}(x + 1) \quad\Rightarrow\quad p = \frac{1}{8} \quad\Rightarrow$ vertex $(-1, 3)$, focus $\left(-\frac{7}{8}, 3\right)$, directrix $x = -\frac{9}{8}$

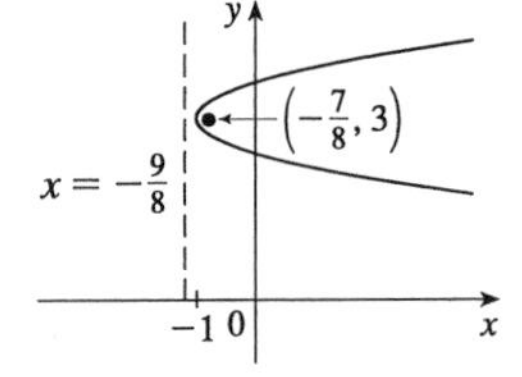

7. $2x + y^2 - 8y + 12 = 0 \;\Rightarrow\; (y - 4)^2 = -2(x - 2)$ $\Rightarrow\quad p = -\frac{1}{2} \quad\Rightarrow\quad$ vertex $(2, 4)$, focus $\left(\frac{3}{2}, 4\right)$, directrix $x = \frac{5}{2}$

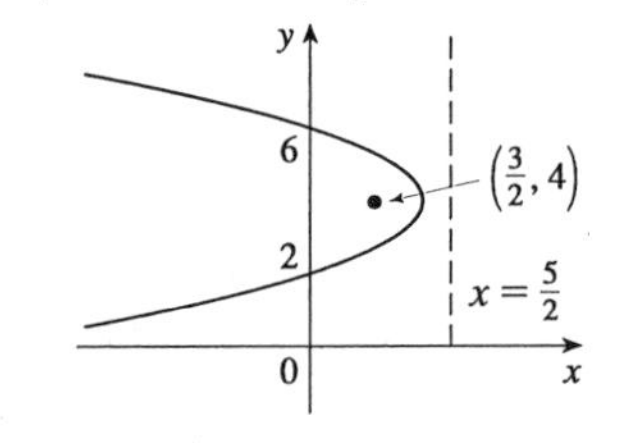

9. $x^2/16 + y^2/4 = 1 \quad\Rightarrow\quad a = 4, b = 2,$ $c = \sqrt{16 - 4} = 2\sqrt{3} \;\Rightarrow\;$ center $(0, 0)$, vertices $(\pm 4, 0)$, foci $\left(\pm 2\sqrt{3}, 0\right)$

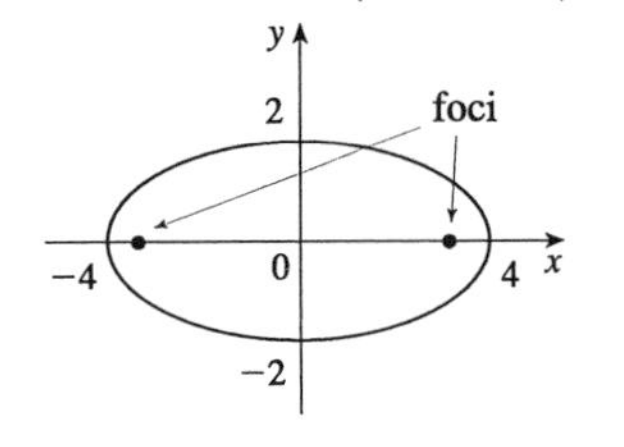

11. $25x^2 + 9y^2 = 225 \quad\Leftrightarrow\quad \frac{1}{9}x^2 + \frac{1}{25}y^2 = 1$ $\Rightarrow\; a = 5, b = 3, c = 4 \quad\Rightarrow$ center $(0, 0)$, vertices $(0, \pm 5)$, foci $(0, \pm 4)$

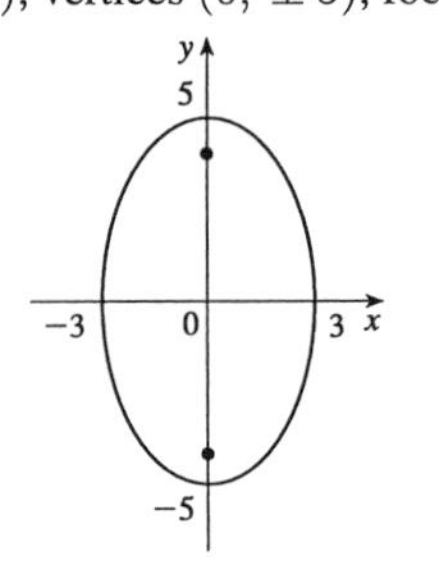

13. $x^2/144 - y^2/25 = 1 \quad\Rightarrow\quad a = 12, b = 5,$ $c = \sqrt{144 + 25} = 13 \Rightarrow$ center $(0, 0)$, vertices $(\pm 12, 0)$, foci $(\pm 13, 0)$, asymptotes $y = \pm \frac{5}{12}x$

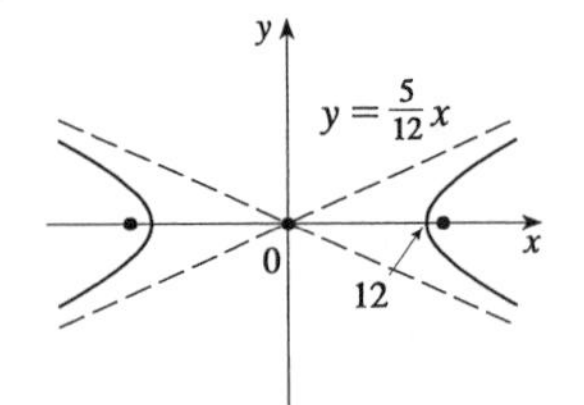

15. $9y^2 - x^2 = 9 \;\Rightarrow\; y^2 - \frac{1}{9}x^2 = 1 \;\Rightarrow\; a = 1,$ $b = 3, c = \sqrt{10} \;\Rightarrow\;$ center $(0, 0)$, vertices $(0, \pm 1)$, foci $\left(0, \pm\sqrt{10}\right)$, asymptotes $y = \pm \frac{1}{3}x$

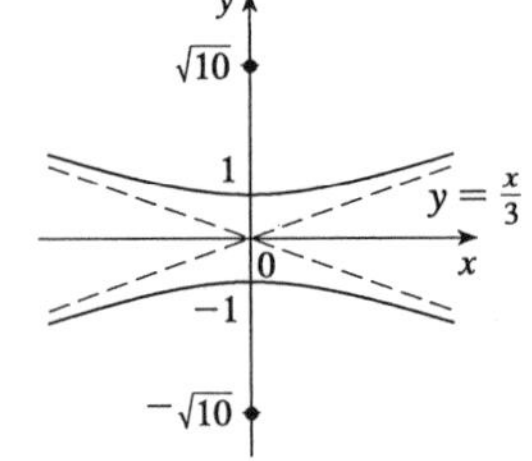

17. $9x^2 - 18x + 4y^2 = 27 \Leftrightarrow$
$\frac{(x-1)^2}{4} + \frac{y^2}{9} = 1 \Rightarrow a = 3, b = 2,$
$c = \sqrt{5} \Rightarrow$ center $(1, 0)$,
vertices $(1, \pm 3)$, foci $\left(1, \pm\sqrt{5}\right)$

19. $2y^2 - 4y - 3x^2 + 12x = -8 \Leftrightarrow$
$\frac{(x-2)^2}{6} - \frac{(y-1)^2}{9} = 1 \Rightarrow a = \sqrt{6},$
$b = 3, c = \sqrt{15} \Rightarrow$ center $(2, 1)$,
vertices $\left(2 \pm \sqrt{6}, 1\right)$, foci $\left(2 \pm \sqrt{15}, 1\right)$,
asymptotes $y - 1 = \pm\frac{3}{\sqrt{6}}(x - 2)$

21. Vertex at $(0, 0)$, $p = 3$, opens upward $\Rightarrow x^2 = 4py = 12y$

23. Vertex at $(2, 0)$, $p = 1$, opens to right $\Rightarrow y^2 = 4p(x-2) = 4(x-2)$

25. The parabola must have equation $y^2 = 4px$, so $(-4)^2 = 4p(1) \Rightarrow p = 4 \Rightarrow y^2 = 16x$.

27. Center $(0, 0)$, $c = 1$, $a = 2 \Rightarrow b = \sqrt{2^2 - 1^2} = \sqrt{3} \Rightarrow \frac{1}{4}x^2 + \frac{1}{3}y^2 = 1$

29. Center $(3, 0)$, $c = 1$, $a = 3 \Rightarrow b = \sqrt{8} = 2\sqrt{2} \Rightarrow \frac{1}{8}(x-3)^2 + \frac{1}{9}y^2 = 1$

31. Center $(2, 2)$, $c = 2$, $a = 3 \Rightarrow b = \sqrt{5} \Rightarrow \frac{1}{9}(x-2)^2 + \frac{1}{5}(y-2)^2 = 1$

33. Center $(0, 0)$, vertical axis, $c = 3$, $a = 1 \Rightarrow b = \sqrt{8} = 2\sqrt{2} \Rightarrow y^2 - \frac{1}{8}x^2 = 1$

35. Center $(4, 3)$, horizontal axis, $c = 3$, $a = 2 \Rightarrow b = \sqrt{5} \Rightarrow \frac{1}{4}(x-4)^2 - \frac{1}{5}(y-3)^2 = 1$

37. Center $(0, 0)$, horizontal axis, $a = 3$, $\frac{b}{a} = 2 \Rightarrow b = 6 \Rightarrow \frac{1}{9}x^2 - \frac{1}{36}y^2 = 1$

39. In Figure 8, we see that the point on the ellipse closest to a focus is the closer vertex (which is a distance $a - c$ from it) while the farthest point is the other vertex (at a distance of $a + c$). So for this lunar orbit, $(a - c) + (a + c) = 2a = (1728 + 110) + (1728 + 314)$, or $a = 1940$; and $(a + c) - (a - c) = 2c = 314 - 110$, or $c = 102$. Thus $b^2 = a^2 - c^2 = 3{,}753{,}196$, and the equation is
$\frac{x^2}{3{,}763{,}600} + \frac{y^2}{3{,}753{,}196} = 1.$

41. **(a)** Set up the coordinate system so that A is $(-200, 0)$ and B is $(200, 0)$.
$|PA| - |PB| = (1200)(980) = 1{,}176{,}000 \text{ ft} = \frac{2450}{11} \text{ mi} = 2a \Rightarrow a = \frac{1225}{11}$, and $c = 200$ so
$b^2 = c^2 - a^2 = \frac{3{,}339{,}375}{121} \Rightarrow \frac{121x^2}{1{,}500{,}625} - \frac{121y^2}{3{,}339{,}375} = 1.$

(b) Due north of $B \Rightarrow x = 200 \Rightarrow \frac{(121)(200)^2}{1{,}500{,}625} - \frac{121y^2}{3{,}339{,}375} = 1 \Rightarrow y = \frac{133{,}575}{539} \approx 248$ mi

43. The function whose graph is the upper branch of this hyperbola is concave upward. The function is

$y = f(x) = a\sqrt{1 + \dfrac{x^2}{b^2}} = \dfrac{a}{b}\sqrt{b^2 + x^2}$, so $y' = \dfrac{a}{b}x\left(b^2 + x^2\right)^{-1/2}$ and

$y'' = \dfrac{a}{b}\left[\left(b^2 + x^2\right)^{-1/2} - x^2\left(b^2 + x^2\right)^{-3/2}\right] = ab\left(b^2 + x^2\right)^{-3/2} > 0$ for all x, and so f is concave upward.

45. **(a)** ellipse **(b)** hyperbola **(c)** empty graph (no curve)

(d) In case (a), $a^2 = k$, $b^2 = k - 16$, and $c^2 = a^2 - b^2 = 16$, so the foci are at $(\pm 4, 0)$. In case (b), $k - 16 < 0$, so $a^2 = k$, $b^2 = 16 - k$, and $c^2 = a^2 + b^2 = 16$, and so again the foci are at $(\pm 4, 0)$.

47. Use the parametrization $x = 2\cos t$, $y = \sin t$, $0 \le t \le 2\pi$ to get

$L = 4\displaystyle\int_0^{\pi/2} \sqrt{(dx/dt)^2 + (dy/dt)^2}\,dt = 4\int_0^{\pi/2} \sqrt{4\sin^2 t + \cos^2 t}\,dt = 4\int_0^{\pi/2} \sqrt{3\sin^2 t + 1}\,dt$. Using Simpson's Rule with $n = 10$, $L \approx \frac{4}{3}\left(\frac{\pi}{20}\right)\left[f(0) + 4f\left(\frac{\pi}{20}\right) + 2f\left(\frac{\pi}{10}\right) + \cdots + 2f\left(\frac{2\pi}{5}\right) + 4f\left(\frac{9\pi}{20}\right) + f\left(\frac{\pi}{2}\right)\right]$, with $f(t) = \sqrt{3\sin^2 t + 1}$, so $L \approx 9.69$.

49. $\dfrac{x^2}{a^2} + \dfrac{y^2}{b^2} = 1 \;\Rightarrow\; \dfrac{2x}{a^2} + \dfrac{2yy'}{b^2} = 0 \;\Rightarrow\; y' = -\dfrac{b^2 x}{a^2 y}$ $(y \ne 0)$. Thus the slope of the tangent line at P is $-\dfrac{b^2 x_1}{a^2 y_1}$. The slope of F_1P is $\dfrac{y_1}{x_1 + c}$ and of F_2P is $\dfrac{y_1}{x_1 - c}$. By the formula from Problem 23 of Problems Plus after Chapter 2, we have

$$\tan\alpha = \frac{\dfrac{y_1}{x_1 + c} + \dfrac{b^2 x_1}{a^2 y_1}}{1 - \dfrac{b^2 x_1 y_1}{a^2 y_1 (x_1 + c)}} = \frac{a^2 y_1^2 + b^2 x_1 (x_1 + c)}{a^2 y_1 (x_1 + c) - b^2 x_1 y_1} = \frac{a^2 b^2 + b^2 c x_1}{c^2 x_1 y_1 + a^2 c y_1} \quad \left[\begin{array}{c}\text{using } b^2 x_1^2 + a^2 y_1^2 = a^2 b^2 \\ \text{and } a^2 - b^2 = c^2\end{array}\right]$$

$$= \frac{b^2 (c x_1 + a^2)}{c y_1 (c x_1 + a^2)} = \frac{b^2}{c y_1}, \text{ and}$$

$$\tan\beta = \frac{-\dfrac{y_1}{x_1 - c} - \dfrac{b^2 x_1}{a^2 y_1}}{1 - \dfrac{b^2 x_1 y_1}{a^2 y_1 (x_1 - c)}} = \frac{-a^2 y_1^2 - b^2 x_1 (x_1 - c)}{a^2 y_1 (x_1 - c) - b^2 x_1 y_1} = \frac{-a^2 b^2 + b^2 c x_1}{c^2 x_1 y_1 - a^2 c y_1} = \frac{b^2 (c x_1 - a^2)}{c y_1 (c x_1 - a^2)} = \frac{b^2}{c y_1}.$$ So $\alpha = \beta$.

EXERCISES 9.7

1. $r = \dfrac{ed}{1 + e\cos\theta} = \dfrac{\frac{2}{3} \cdot 3}{1 + \frac{2}{3}\cos\theta} = \dfrac{6}{3 + 2\cos\theta}$

3. $r = \dfrac{ed}{1 + e\sin\theta} = \dfrac{1 \cdot 2}{1 + \sin\theta} = \dfrac{2}{1 + \sin\theta}$

5. $r = 5\sec\theta \;\Leftrightarrow\; x = r\cos\theta = 5$, so $r = \dfrac{ed}{1 + e\cos\theta} = \dfrac{4 \cdot 5}{1 + 4\cos\theta} = \dfrac{20}{1 + 4\cos\theta}$

7. Focus $(0, 0)$, vertex $\left(5, \frac{\pi}{2}\right) \;\Rightarrow\;$ directrix $y = 10 \;\Rightarrow\; r = \dfrac{ed}{1 + e\sin\theta} = \dfrac{10}{1 + \sin\theta}$

e=1 r = d/(1±sinθ) exist at θ=π/2
D.N.E at θ=3π/2
→ d/(1+sinθ) θ=π/2 sinθ=1 r=d/2 θ=3π/2 sinθ=−1 r=D.N.E.

9. $e = 3 \;\Rightarrow\;$ hyperbola; $ed = 4 \;\Rightarrow\; d = \frac{4}{3} \;\Rightarrow$ directrix $x = \frac{4}{3}$; vertices $(1, 0)$ and $(-2, \pi) = (2, 0)$; center $\left(\frac{3}{2}, 0\right)$; asymptotes parallel to $\theta = \pm\cos^{-1}\left(-\frac{1}{3}\right)$

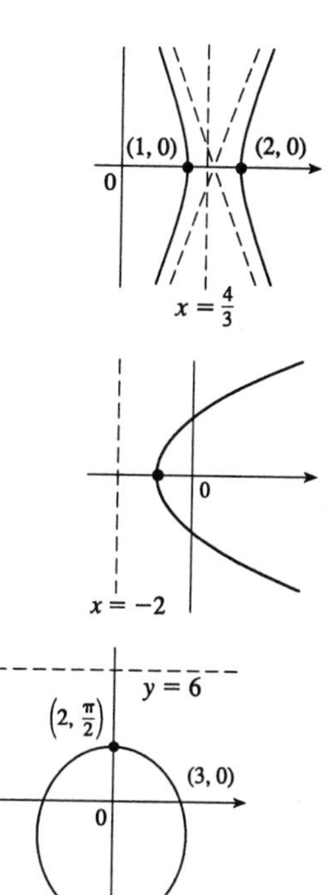

11. $e = 1 \;\Rightarrow\;$ parabola; $ed = 2 \;\Rightarrow\; d = 2$ $\Rightarrow$ directrix $x = -2$; vertex $(-1, 0) = (1, \pi)$

13. $r = \dfrac{3}{1 + \frac{1}{2}\sin\theta} \;\Rightarrow\; e = \frac{1}{2} \;\Rightarrow\;$ ellipse; $ed = 3 \;\Rightarrow\; d = 6 \;\Rightarrow\;$ directrix $y = 6$; vertices $\left(2, \frac{\pi}{2}\right)$ and $\left(6, \frac{3\pi}{2}\right)$; center $\left(2, \frac{3\pi}{2}\right)$

15. $r = \dfrac{7/2}{1 - \frac{5}{2}\sin\theta} \;\Rightarrow\; e = \frac{5}{2} \;\Rightarrow\;$ hyperbola; $ed = \frac{7}{2} \;\Rightarrow\; d = \frac{7}{5} \;\Rightarrow\;$ directrix $y = -\frac{7}{5}$; center $\left(\frac{5}{3}, \frac{3\pi}{2}\right)$; vertices $\left(-\frac{7}{3}, \frac{\pi}{2}\right) = \left(\frac{7}{3}, \frac{3\pi}{2}\right)$ and $\left(1, \frac{3\pi}{2}\right)$.

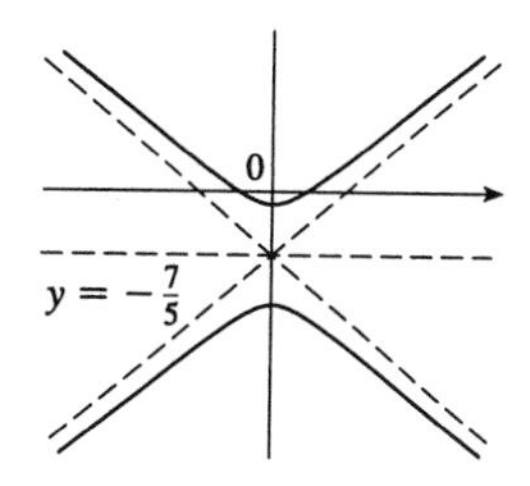

17. **(a)** The equation is $r = \dfrac{1}{4 - 3\cos\theta} = \dfrac{1/4}{1 - \frac{3}{4}\cos\theta}$, so $e = \frac{3}{4}$ and $ed = \frac{1}{4} \;\Rightarrow\; d = \frac{1}{3}$. The conic is an ellipse, and the equation of its directrix is $x = r\cos\theta = -\frac{1}{3} \;\Rightarrow\; r = -\dfrac{1}{3\cos\theta}$. We must be careful in our choice of parameter values in this equation ($-1 \le \theta \le 1$ works well.)

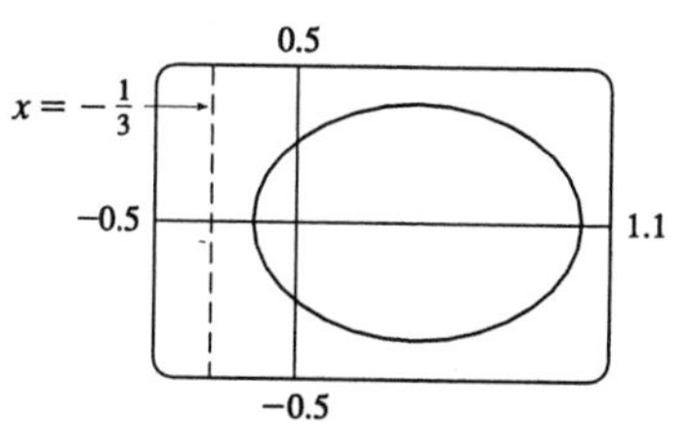

(b) The equation is obtained by replacing θ with $\theta - \frac{\pi}{3}$ in the equation of the original conic (see Example 4), so

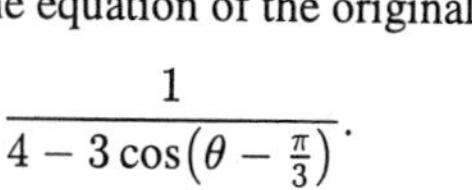

$$r = \frac{1}{4 - 3\cos\left(\theta - \frac{\pi}{3}\right)}.$$

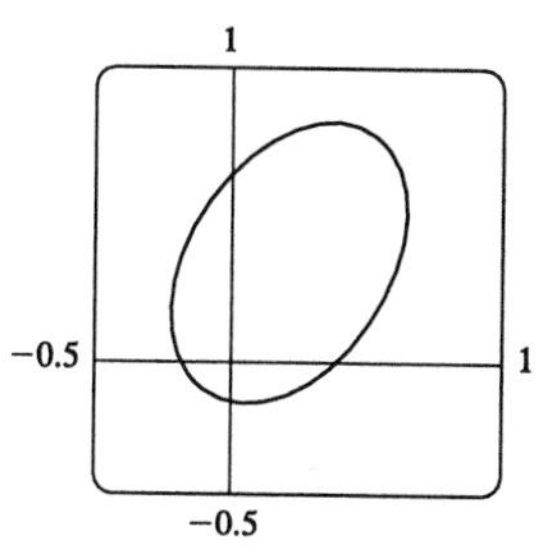

19. For $e < 1$ the curve is an ellipse. It is nearly circular when e is close to 0. As e increases, the graph is stretched out to the right, and grows larger (that is, its right-hand focus moves to the right while its left-hand focus remains at the origin.) At $e = 1$, the curve becomes a parabola with focus at the origin.

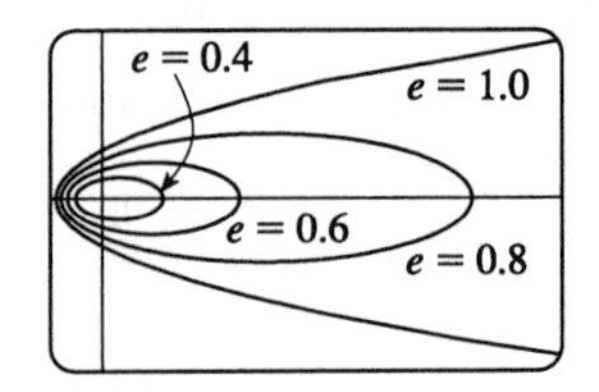

21. $|PF| = e|Pl| \Rightarrow$
$r = e[d - r\cos(\pi - \theta)] = e(d + r\cos\theta) \Rightarrow$
$r(1 - e\cos\theta) = ed \Rightarrow r = \dfrac{ed}{1 - e\cos\theta}$

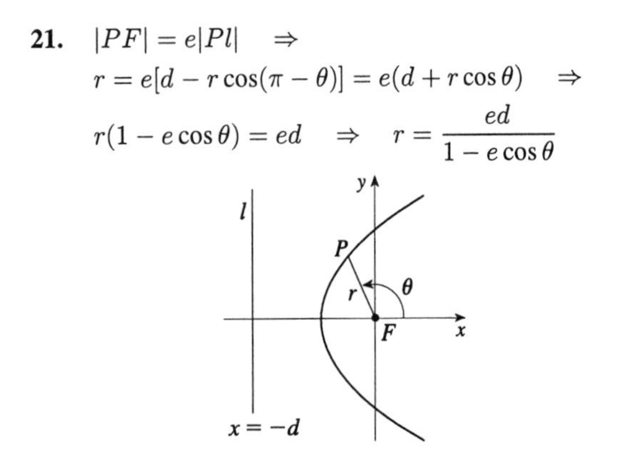

23. $|PF| = e|Pl| \Rightarrow$
$r = e[d - r\sin(\theta - \pi)] = e(d + r\sin\theta)$
$\Rightarrow r(1 - e\sin\theta) = ed \Rightarrow r = \dfrac{ed}{1 - e\sin\theta}$

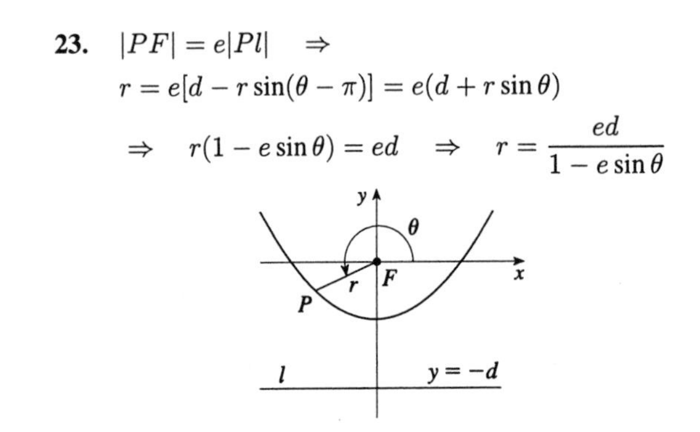

25. **(a)** If the directrix is $x = -d$, then $r = \dfrac{ed}{1 - e\cos\theta}$ [see Figure 8(b)], and, from (4), $a^2 = \dfrac{e^2d^2}{(1 - e^2)^2} \Rightarrow$
$ed = a(1 - e^2)$. Therefore, $r = \dfrac{a(1 - e^2)}{1 - e\cos\theta}$.

(b) $e = 0.017$ and the major axis $= 2a = 2.99 \times 10^8 \Rightarrow a = 1.495 \times 10^8$.
Therefore $r = \dfrac{1.495 \times 10^8\left[1 - (0.017)^2\right]}{1 - 0.017\cos\theta} \approx \dfrac{1.49 \times 10^8}{1 - 0.017\cos\theta}$.

27. The minimum distance is at perihelion where $4.6 \times 10^7 = r = a(1 - e) = a(1 - 0.206) = a(0.794) \Rightarrow$
$a = 4.6 \times 10^7/0.794$. So the maximum distance, which is at aphelion, is
$r = a(1 + e) = \left(4.6 \times 10^7/0.794\right) \times 10^7(1.206) \approx 7.0 \times 10^7$ km.

29. From Exercise 27, we have $e = 0.206$ and $a(1 - e) = 4.6 \times 10^7$ km. Thus $a = 4.6 \times 10^7/0.794$. From Exercise 25, we can write the equation of Mercury's orbit as $r = a\dfrac{1 - e^2}{1 - e\cos\theta}$. So since $\dfrac{dr}{d\theta} = \dfrac{-a(1 - e^2)e\sin\theta}{(1 - e\cos\theta)^2} \Rightarrow$
$r^2 + \left(\dfrac{dr}{d\theta}\right)^2 = \dfrac{a^2(1 - e^2)^2}{(1 - e\cos\theta)^2} + \dfrac{a^2(1 - e^2)^2 e^2\sin^2\theta}{(1 - e\cos\theta)^4} = \dfrac{a^2(1 - e^2)^2}{(1 - e\cos\theta)^4}\left(1 - 2e\cos\theta + e^2\right)$, the length of the orbit is $L = \displaystyle\int_0^{2\pi}\sqrt{r^2 + (dr/d\theta)^2}\,d\theta = a\left(1 - e^2\right)\int_0^{2\pi}\frac{\sqrt{1 + e^2 - 2e\cos\theta}}{(1 - e\cos\theta)^2}\,d\theta \approx 3.6 \times 10^8$ km.

This seems reasonable, since Mercury's orbit is nearly circular, and the circumference of a circle of radius a is $2\pi a \approx 3.6 \times 10^8$ km.

REVIEW EXERCISES FOR CHAPTER 9

1. $x = 1 - t^2, y = 1 - t \quad (-1 \le t \le 1)$

$x = 1 - (1 - y)^2 = 2y - y^2 \ (0 \le y \le 2)$

3. $x = 1 + \sin t, y = 2 + \cos t \quad \Rightarrow$

$(x - 1)^2 + (y - 2)^2 = \sin^2 t + \cos^2 t = 1$

5. $r = 1 + 3\cos\theta$

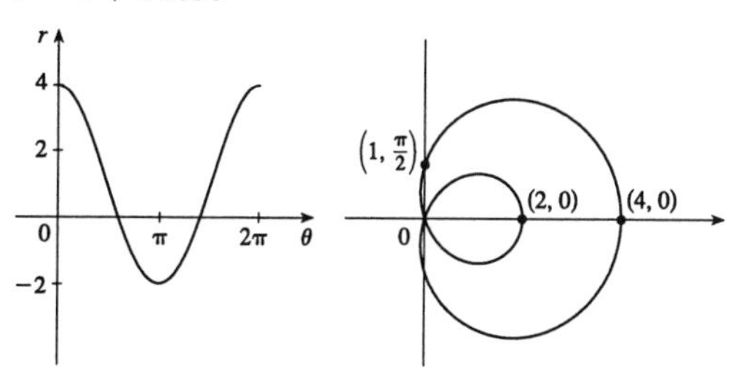

7. $r^2 = \sec 2\theta \quad \Rightarrow \quad r^2 \cos 2\theta = 1 \quad \Rightarrow$

$r^2(\cos^2\theta - \sin^2\theta) = 1 \quad \Rightarrow \quad x^2 - y^2 = 1,$

a hyperbola

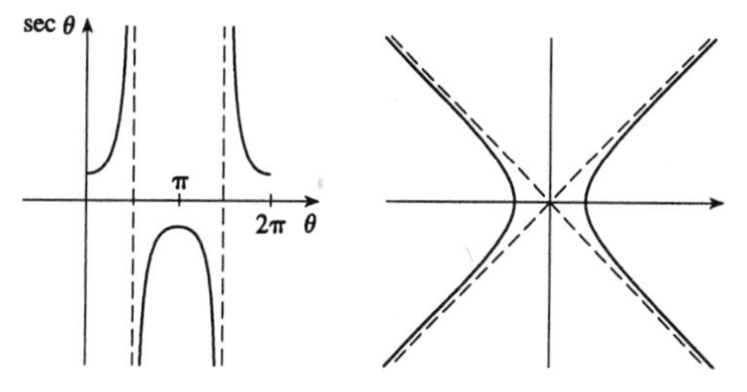

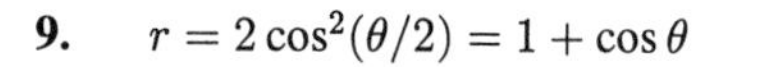

9. $r = 2\cos^2(\theta/2) = 1 + \cos\theta$

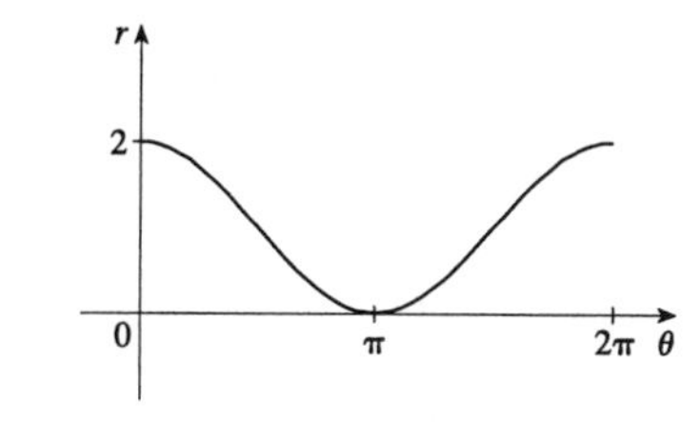

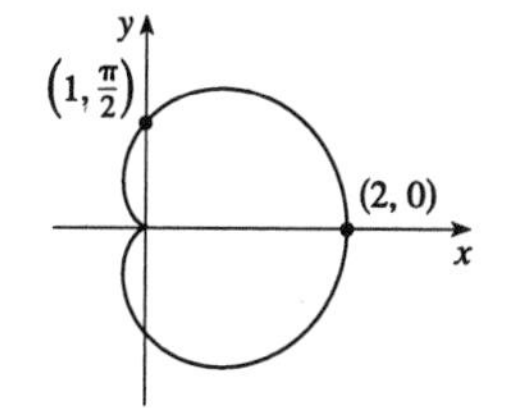

11. $r = \dfrac{1}{1 + \cos\theta} \quad \Rightarrow \quad e = 1 \quad \Rightarrow$ parabola; $d = 1 \quad \Rightarrow$ directrix $x = 1$ and vertex $\left(\frac{1}{2}, 0\right)$; y-intercepts are $\left(1, \frac{\pi}{2}\right)$ and $\left(1, \frac{3\pi}{2}\right)$.

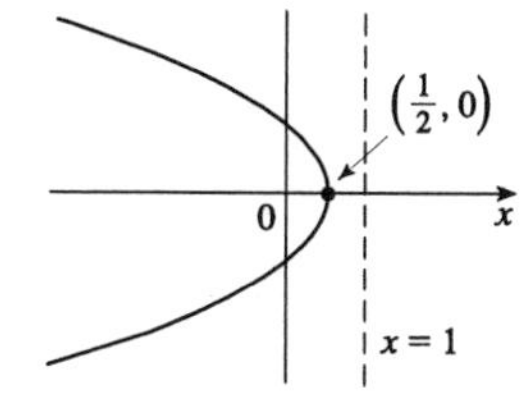

13. $x^2 + y^2 = 4x \quad \Leftrightarrow \quad r^2 = 4r\cos\theta \quad \Leftrightarrow \quad r = 4\cos\theta$

15. $r = (\sin\theta)/\theta$. As $\theta \to \pm\infty$, $r \to 0$.

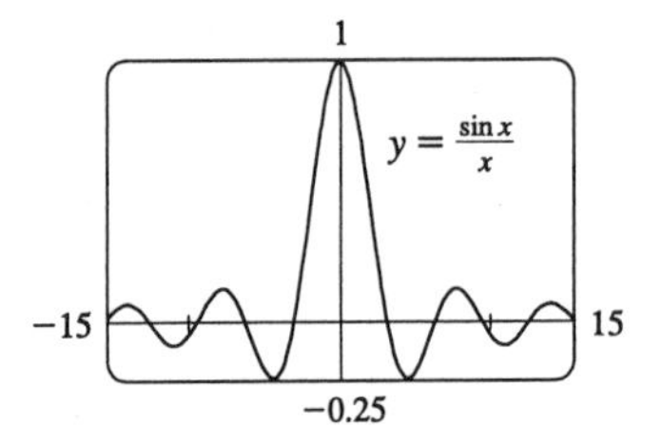

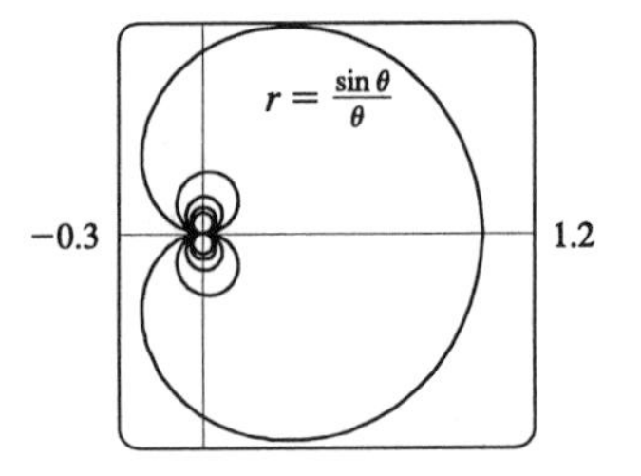

17. $x = t^2 + 2t$, $y = t^3 - t$. $\dfrac{dy}{dx} = \dfrac{dy/dt}{dx/dt} = \dfrac{3t^2 - 1}{2t + 2} = \dfrac{1}{2}$ when $t = 1$

19. $\dfrac{dy}{dx} = \dfrac{(dr/d\theta)\sin\theta + r\cos\theta}{(dr/d\theta)\cos\theta - r\sin\theta} = \dfrac{\sin\theta + \theta\cos\theta}{\cos\theta - \theta\sin\theta} = \dfrac{\frac{1}{\sqrt{2}} + \frac{\pi}{4}\cdot\frac{1}{\sqrt{2}}}{\frac{1}{\sqrt{2}} - \frac{\pi}{4}\cdot\frac{1}{\sqrt{2}}} = \dfrac{4+\pi}{4-\pi}$ when $\theta = \dfrac{\pi}{4}$

21. $\dfrac{dy}{dx} = \dfrac{dy/dt}{dx/dt} = \dfrac{t\cos t + \sin t}{-t\sin t + \cos t}$. $\dfrac{d^2y}{dx^2} = \dfrac{\frac{d}{dt}\left(\frac{dy}{dx}\right)}{dx/dt}$, where

$$\frac{d}{dt}\left(\frac{dy}{dx}\right) = \frac{(-t\sin t + \cos t)(-t\sin t + 2\cos t) - (t\cos t + \sin t)(-t\cos t - 2\sin t)}{(-t\sin t + \cos t)^2} = \frac{t^2 + 2}{(-t\sin t + \cos t)^2}$$

$$\Rightarrow \quad \frac{d^2y}{dx^2} = \frac{t^2 + 2}{(-t\sin t + \cos t)^3}$$

23. We graph the curve for $-2.2 \le t \le 1.2$. By zooming in or using a cursor, we find that the lowest point is about $(1.4, 0.75)$. To find the exact values, we find the t-value at which $dy/dt = 2t + 1 = 0 \quad \Leftrightarrow$ $t = -\frac{1}{2} \quad \Leftrightarrow \quad (x, y) = \left(\frac{11}{8}, \frac{3}{4}\right)$.

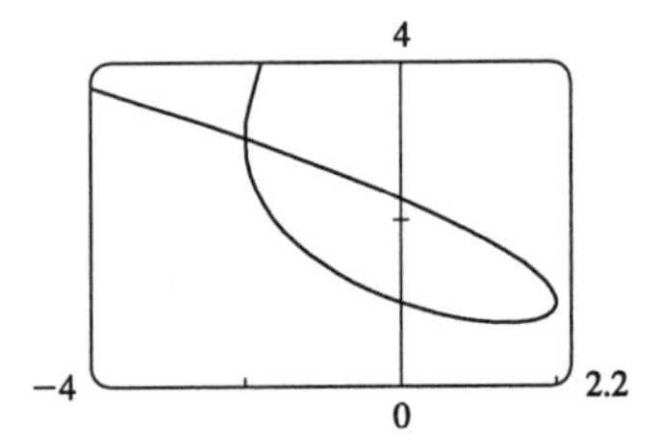

25. $dx/dt = -2a\sin t + 2a\sin 2t = 2a\sin t(2\cos t - 1) = 0 \quad \Leftrightarrow \quad \sin t = 0$ or $\cos t = \frac{1}{2} \quad \Rightarrow \quad t = 0, \frac{\pi}{3}, \pi$, or $\frac{5\pi}{3}$.

$dy/dt = 2a\cos t - 2a\cos 2t = 2a(1 + \cos t - 2\cos^2 t) = 2a(1 - \cos t)(1 + 2\cos t) = 0 \quad \Rightarrow \quad t = 0, \frac{2\pi}{3}$, or $\frac{4\pi}{3}$.

Thus the graph has vertical tangents where $t = \frac{\pi}{3}$, π and $\frac{5\pi}{3}$, and horizontal tangents where $t = \frac{2\pi}{3}$ and $\frac{4\pi}{3}$. To determine what the slope is where $t = 0$, we use l'Hospital's Rule to evaluate $\lim\limits_{t\to 0}\dfrac{dy/dt}{dx/dt} = 0$, so there is a horizontal tangent there.

t	x	y
0	a	0
$\frac{\pi}{3}$	$\frac{3}{2}a$	$\frac{\sqrt{3}}{2}a$
$\frac{2\pi}{3}$	$-\frac{1}{2}a$	$\frac{3\sqrt{3}}{2}a$
π	$-3a$	0
$\frac{4\pi}{3}$	$-\frac{1}{2}a$	$-\frac{3\sqrt{3}}{2}a$
$\frac{5\pi}{3}$	$\frac{3}{2}a$	$-\frac{\sqrt{3}}{2}a$

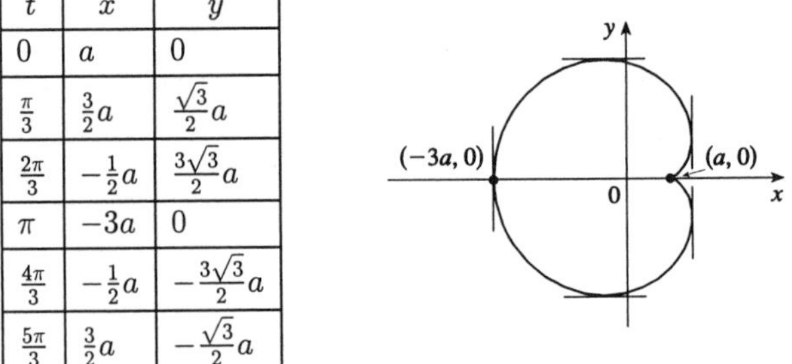

27. This curve has 10 "petals." For instance, for $-\frac{\pi}{10} \le \theta \le \frac{\pi}{10}$, there are two petals, one with $r > 0$ and one with $r < 0$. $A = 10\int_{-\pi/10}^{\pi/10} \frac{1}{2}r^2\,d\theta = 5\int_{-\pi/10}^{\pi/10} 9\cos 5\theta\,d\theta = 90\int_0^{\pi/10}\cos 5\theta\,d\theta = \left[18\sin 5\theta\right]_0^{\pi/10} = 18$

29. The curves intersect where $4\cos\theta = 2$; that is, at $\left(2, \frac{\pi}{3}\right)$ and $\left(2, -\frac{\pi}{3}\right)$.

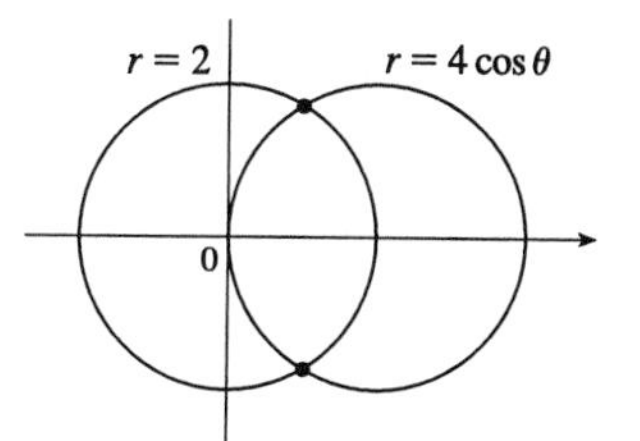

31. The curves intersect where $2\sin\theta = \sin\theta + \cos\theta \quad\Rightarrow\quad \sin\theta = \cos\theta \quad\Rightarrow$ $\theta = \frac{\pi}{4}$, and also at the origin (at which $\theta = \frac{3\pi}{4}$ on the second curve.)

$$A = \int_0^{\pi/4} \tfrac{1}{2}(2\sin\theta)^2\,d\theta + \int_{\pi/4}^{3\pi/4} \tfrac{1}{2}(\sin\theta + \cos\theta)^2\,d\theta$$

$$= \int_0^{\pi/4}(1 - \cos 2\theta)d\theta + \tfrac{1}{2}\int_{\pi/4}^{3\pi/4}(1 + \sin 2\theta)d\theta$$

$$= \left[\theta - \tfrac{1}{2}\sin 2\theta\right]_0^{\pi/4} + \left[\tfrac{1}{2}\theta - \tfrac{1}{4}\cos 2\theta\right]_{\pi/4}^{3\pi/4} = \tfrac{1}{2}(\pi - 1)$$

33. $x = 3t^2$, $y = 2t^3$. $L = \displaystyle\int_0^2 \sqrt{(dx/dt)^2 + (dy/dt)^2}\,dt = \int_0^2 \sqrt{(6t)^2 + (6t^2)^2}\,dt$

$$= 6\int_0^2 t\sqrt{1+t^2}\,dt = \left[2(1+t^2)^{3/2}\right]_0^2 = 2\left(5\sqrt{5} - 1\right)$$

35. $L = \displaystyle\int_\pi^{2\pi} \sqrt{r^2 + (dr/d\theta)^2}\,d\theta = \int_\pi^{2\pi} \sqrt{(1/\theta)^2 + (-1/\theta^2)^2}\,d\theta = \int_\pi^{2\pi} \frac{\sqrt{\theta^2+1}}{\theta^2}\,d\theta$

$$= \left[-\frac{\sqrt{\theta^2+1}}{\theta} + \ln\left|\theta + \sqrt{\theta^2+1}\right|\right]_\pi^{2\pi} \quad \text{(Formula 24)} \quad = \frac{\sqrt{\pi^2+1}}{\pi} - \frac{\sqrt{4\pi^2+1}}{2\pi} + \ln\left|\frac{2\pi + \sqrt{4\pi^2+1}}{\pi + \sqrt{\pi^2+1}}\right|$$

37. $S = \displaystyle\int_1^4 2\pi y\sqrt{(dx/dt)^2 + (dy/dt)^2}\,dt = \int_1^4 2\pi\left(\tfrac{1}{3}t^3 + \tfrac{1}{2}t^{-2}\right)\sqrt{\left(2/\sqrt{t}\right)^2 + (t^2 - t^{-3})^2}\,dt$

$$= 2\pi\int_1^4 \left(\tfrac{1}{3}t^3 + \tfrac{1}{2}t^{-2}\right)\sqrt{(t^2 + t^{-3})^2}\,dt = 2\pi\int_1^4 \left(\tfrac{1}{3}t^5 + \tfrac{5}{6} + \tfrac{1}{2}t^{-5}\right)dt = 2\pi\left[\tfrac{1}{18}t^6 + \tfrac{5}{6}t - \tfrac{1}{8}t^{-4}\right]_1^4 = \tfrac{471{,}295}{1024}\pi$$

39. For all c except -1, the curve is asymptotic to the line $x = 1$. For $c < -1$, the curve bulges to the right near $y = 0$. As c increases, the bulge becomes smaller, until at $c = -1$ the curve is the straight line $x = 1$. As c continues to increase, the curve bulges to the left, until at $c = 0$ there is a cusp at the origin. For $c > 0$, there is a loop to the left of the origin, whose size and roundness increase as c increases. Note that the x-intercept of the curve is always $-c$.

41. Ellipse, center $(0, 0)$, $a = 3$, $b = 2\sqrt{2}$, $c = 1 \quad\Rightarrow\quad$ foci $(\pm 1, 0)$, vertices $(\pm 3, 0)$

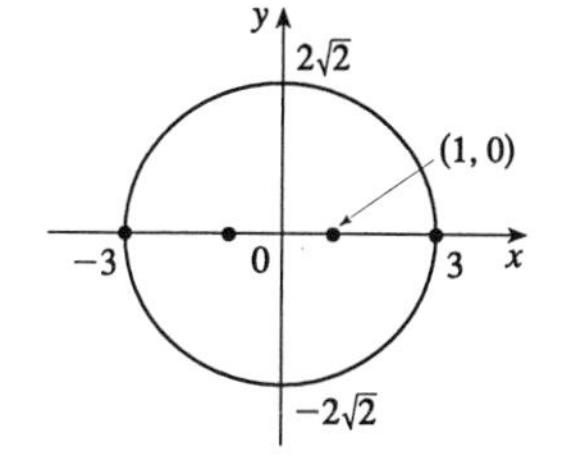

43. $6(y^2-6y+9)=-(x+1) \quad \Leftrightarrow$
$(y-3)^2=-\frac{1}{6}(x+1)$, a parabola
with vertex $(-1,3)$, opening to the left,
$p=-\frac{1}{24} \quad \Rightarrow \quad$ focus $\left(-\frac{25}{24},3\right)$ and
directrix $x=-\frac{23}{24}$.

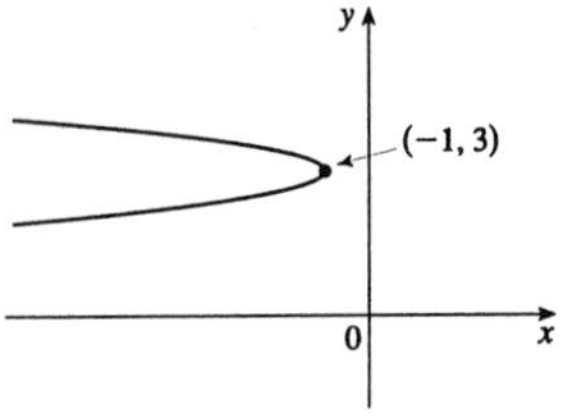

y $(-1,3)$ 0 x

45. The parabola opens upward with vertex $(0,4)$ and $p=2$, so its equation is $(x-0)^2=4\cdot 2(y-4) \quad \Leftrightarrow$
$x^2=8(y-4)$.

47. The hyperbola has center $(0,0)$ and foci on the x-axis. $c=3$ and $b/a=\frac{1}{2}$ (from the asymptotes) $\quad \Rightarrow$
$9=c^2=a^2+b^2=(2b)^2+b^2=5b^2 \quad \Rightarrow \quad b=\frac{3}{\sqrt{5}} \quad \Rightarrow \quad a=\frac{6}{\sqrt{5}} \quad \Rightarrow \quad$ the equation is $\dfrac{x^2}{36/5}-\dfrac{y^2}{9/5}=1$
$\Leftrightarrow \quad 5x^2-20y^2=36$.

49. $x^2=-y+100$ has its vertex at $(0,100)$, so one of the vertices of the ellipse is $(0,100)$. Another form of the equation of a parabola is $x^2=4p(y-100)$ so $4p(y-100)=-y+100 \quad \Rightarrow \quad 4py-4p(100)=100-y \quad \Rightarrow$
$4p=\dfrac{100-y}{y-100} \quad \Rightarrow \quad p=-\frac{1}{4}$. Therefore the shared focus is found at $\left(0,\frac{399}{4}\right)$ so $2c=\frac{399}{4}-0 \quad \Rightarrow \quad c=\frac{399}{8}$
and the center of the ellipse is $\left(0,\frac{399}{8}\right)$. So $a=100-\frac{399}{8}=\frac{401}{8}$ and $b^2=a^2-c^2=\frac{401^2-399^2}{8^2}=25$. So the equation of the ellipse is $\dfrac{x^2}{b^2}+\dfrac{\left(y-\frac{399}{8}\right)^2}{a^2}=1 \quad \Rightarrow \quad \dfrac{x^2}{25}+\dfrac{\left(y-\frac{399}{8}\right)^2}{\left(\frac{401}{8}\right)^2}=1$ or $\dfrac{x^2}{25}+\dfrac{(8y-399)^2}{160{,}801}=1$.

51. Directrix $x=4 \quad \Rightarrow \quad d=4$, so $e=\frac{1}{3} \quad \Rightarrow \quad r=\dfrac{ed}{1+e\cos\theta}=\dfrac{4}{3+\cos\theta}$.

53. In polar coordinates an equation for the circle is $r=2a\sin\theta$. Thus the coordinates of Q are
$x=r\cos\theta=2a\sin\theta\cos\theta$ and $y=r\sin\theta=2a\sin^2\theta$. The coordinates of R are $x=2a\cot\theta$ and $y=2a$.
Since P is the midpoint of QR, we use the midpoint formula to get $x=a(\sin\theta\cos\theta+\cot\theta)$ and
$y=a(1+\sin^2\theta)$.

APPLICATIONS PLUS (page 593)

1. **(a)**

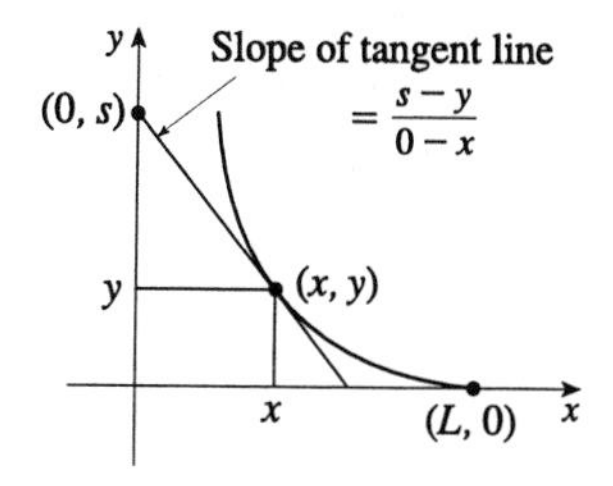

While running from $(L, 0)$ to (x, y), the dog travels a distance $s = \int_x^L \sqrt{1 + (dy/dx)^2}\, dx$. The dog and rabbit run at the same speed, so the rabbit's position when the dog has traveled a distance s is $(0, s)$. Since the dog runs straight for the rabbit, $\frac{dy}{dx} = \frac{s-y}{0-x}$ (see the figure). Thus $s = y - x\frac{dy}{dx}$ and $-\sqrt{1 + \left(\frac{dy}{dx}\right)^2} = \frac{ds}{dx} = \frac{dy}{dx} - x\frac{d^2y}{dx^2} - 1\frac{dy}{dx} = -x\frac{d^2y}{dx^2}$. Hence $x\frac{d^2y}{dx^2} = \sqrt{1 + \left(\frac{dy}{dx}\right)^2}$, as claimed.

(b) Letting $z = \frac{dy}{dx}$, we obtain the differential equation $x\frac{dz}{dx} = \sqrt{1+z^2}$, or $\frac{dz}{\sqrt{1+z^2}} = \frac{dx}{x}$. Integrating:

$\ln x = \int \frac{dz}{\sqrt{1+z^2}} \quad (z = \tan\theta,\, dz = \sec^2\theta\, d\theta) \quad = \int (\sec\theta) d\theta = \ln|\sec\theta + \tan\theta| + C$

$= \ln\left|\sqrt{1+z^2} + z\right| + C.$

When $x = L$, $z = \frac{dy}{dx} = 0$, so $\ln L = \ln 1 + C$. Therefore $C = \ln L$, so

$\ln x = \ln\left(\sqrt{1+z^2} + z\right) + \ln L = \ln\left[L\left(\sqrt{1+z^2} + z\right)\right]$ and $x = L\left(\sqrt{1+z^2} + z\right)$.

$\sqrt{1+z^2} = \frac{x}{L} - z \quad \Rightarrow \quad 1 + z^2 = \left(\frac{x}{L}\right)^2 - \frac{2xz}{L} + z^2 \quad \Rightarrow \quad \left(\frac{x}{L}\right)^2 - 2z\left(\frac{x}{L}\right) - 1 = 0 \quad \Rightarrow$

$z = \frac{(x/L)^2 - 1}{2(x/L)} = \frac{x^2 - L^2}{2Lx} = \frac{x}{2L} - \frac{L}{2}\frac{1}{x}$ (for $x > 0$).

Since $z = \frac{dy}{dx}$, $y = \frac{x^2}{4L} - \frac{L}{2}\ln x + C'$. But $y = 0$ when $x = L$. Therefore $0 = \frac{L}{4} - \frac{L}{2}\ln L + C'$ and $C' = \frac{L}{2}\ln L - \frac{L}{4}$. Therefore $y = \frac{x^2}{4L} - \frac{L}{2}\ln x + \frac{L}{2}\ln L - \frac{L}{4} = \frac{x^2 - L^2}{4L} - \frac{L}{2}\ln\left(\frac{x}{L}\right)$.

(c) As $x \to 0$, $y \to \infty$, so the dog never catches the rabbit.

3. **(a)** The two spherical zones, whose surface areas we will call S_1 and S_2, are generated by rotation about the y-axis of circular arcs, as indicated in the figure. The arcs are the upper and lower portions of the circle $x^2 + y^2 = r^2$ that are obtained when the circle is cut with the line $y = d$. The portion of the upper arc in the first quadrant is sufficient to generate the upper spherical zone. That portion of the arc can be described by the relation $x = \sqrt{r^2 - y^2}$ for $d \le y \le r$. Thus $\dfrac{dy}{dx} = \dfrac{-y}{\sqrt{r^2 - y^2}}$ and

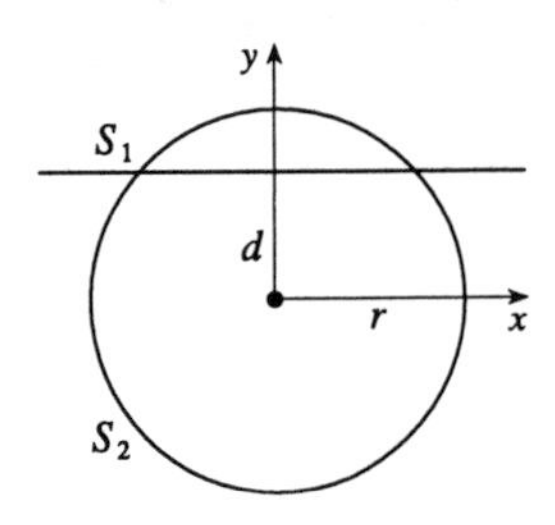

$$ds = \sqrt{1 + \left(\frac{dx}{dy}\right)^2}\,dy = \sqrt{1 + \frac{y^2}{r^2 - y^2}}\,dy = \sqrt{\frac{r^2}{r^2 - y^2}}\,dy = \frac{r\,dy}{\sqrt{r^2 - y^2}}.$$ From Formula 8.3.8 we have

$$S_1 = \int_d^r 2\pi x \sqrt{1 + \left(\frac{dx}{dy}\right)^2}\,dy = \int_d^r 2\pi\sqrt{r^2 - y^2}\,\frac{r\,dy}{\sqrt{r^2 - y^2}} = \int_d^r 2\pi r\,dy = 2\pi r(r - d).$$

Similarly, we can compute

$$S_2 = \int_{-r}^d 2\pi x \sqrt{1 + \left(\frac{dx}{dy}\right)^2}\,dy = \int_{-r}^d 2\pi r\,dy = 2\pi r(r + d).$$

Note that $S_1 + S_2 = 4\pi r^2$, the surface area of the entire sphere.

(b) $r = 3960$ mi and $d = r(\sin 75°) \approx 3825$ mi, so the surface area of the Arctic Ocean is about $2\pi r(r - d) \approx 2\pi(3960)(135) \approx 3.36 \times 10^6\ \text{mi}^2$.

(c) The area on the sphere lies between planes $y = y_1$ and $y = y_2$, where $y_2 - y_1 = h$. Thus we compute the surface area on the sphere to be $S = \displaystyle\int_{y_1}^{y_2} 2\pi x \sqrt{1 + \left(\frac{dx}{dy}\right)^2}\,dy = \int_{y_1}^{y_2} 2\pi r\,dy = 2\pi r(y_2 - y_1) = 2\pi rh.$

This equals the lateral area of a cylinder of radius r and height h, since such a cylinder is obtained by rotating the line $x = r$ about the y-axis, so the surface area of the cylinder between the planes $y = y_1$ and $y = y_2$ is

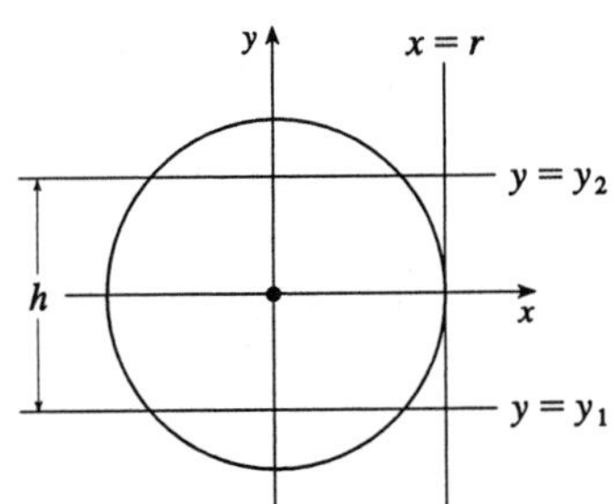

$$A = \int_{y_1}^{y_2} 2\pi x \sqrt{1 + \left(\frac{dx}{dy}\right)^2}\,dy = \int_{y_1}^{y_2} 2\pi r\,\sqrt{1 + 0^2}\,dy$$
$$= 2\pi r y\Big|_{y=y_1}^{y_2} = 2\pi r(y_2 - y_1) = 2\pi rh.$$

(d) $h = 2r\sin 23.45° \approx 3152$ mi, so the surface area of the Torrid Zone is $2\pi rh \approx 2\pi(3960)(3152) \approx 7.84 \times 10^7\ \text{mi}^2$.

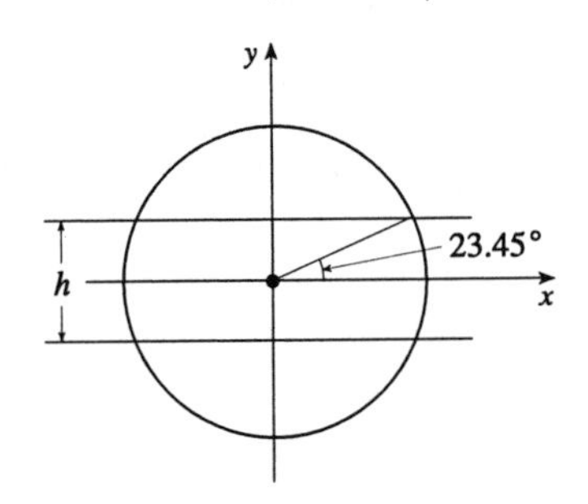

5. **(a)** $\dfrac{dy}{dt} = ky^{1+\varepsilon} \quad \Rightarrow \quad y^{-1-\varepsilon}\,dy = k\,dt \quad \Rightarrow \quad \dfrac{y^{-\varepsilon}}{-\varepsilon} = kt + C$. Since $y(0) = y_0$, we have $C = \dfrac{y_0^{-\varepsilon}}{-\varepsilon}$. Thus $\dfrac{y^{-\varepsilon}}{-\varepsilon} = kt + \dfrac{y_0^{-\varepsilon}}{-\varepsilon}$, or $y^{-\varepsilon} = y_0^{-\varepsilon} - \varepsilon kt$. So $y^{\varepsilon} = \dfrac{1}{y_0^{-\varepsilon} - \varepsilon kt} = \dfrac{y_0^{\varepsilon}}{1 - \varepsilon y_0^{\varepsilon}kt}$ and $y(t) = \dfrac{y_0}{\left(1 - \varepsilon y_0^{\varepsilon}kt\right)^{1/\varepsilon}}$.

(b) $y(t) \to \infty$ as $1 - \varepsilon y_0^{\varepsilon}kt \to 0$, that is, as $t \to \dfrac{1}{\varepsilon y_0^{\varepsilon}k}$. Define $T = \dfrac{1}{\varepsilon y_0^{\varepsilon}k}$. Then $\lim\limits_{t\to T^-} y(t) = \infty$.

(c) According to the data given, we have $\varepsilon = 0.01$, $y(0) = 2$, and $y(3) = 16$, where the time t is given in months. Thus $y_0 = 2$ and $16 = y(3) = \dfrac{y_0}{\left(1 - \varepsilon y_0^{\varepsilon}k \cdot 3\right)^{1/\varepsilon}}$. We could solve for k, but it is easier and more helpful to solve for $\varepsilon y_0^{\varepsilon}k$. $\left(k \text{ turns out to be } \dfrac{1 - 8^{-0.01}}{(0.03)(2^{0.01})} \approx 0.68125.\right)$

$16 = \dfrac{2}{\left(1 - 3\varepsilon y_0^{\varepsilon}k\right)^{100}}$, so $1 - 3\varepsilon y_0^{\varepsilon}k = \left(\frac{1}{8}\right)^{0.01} = 8^{-0.01}$ and $\varepsilon y_0^{\varepsilon}k = \frac{1}{3}\left(1 - 8^{-0.01}\right)$. Thus doomsday occurs when $t = T = \dfrac{1}{\varepsilon y_0^{\varepsilon}k} = \dfrac{3}{1 - 8^{-0.01}} \approx 145.77$ months or 12.15 years.

7. **(a)** If $\tan\theta = \sqrt{\dfrac{y}{C - y}}$, then $\tan^2\theta = \dfrac{y}{C - y}$, so $C\tan^2\theta - y\tan^2\theta = y$ and

$y = \dfrac{C\tan^2\theta}{1 + \tan^2\theta} = \dfrac{C\tan^2\theta}{\sec^2\theta} = C\tan^2\theta\cos^2\theta = C\sin^2\theta = \dfrac{C}{2}(1 - \cos 2\theta)$. Now

$dx = \sqrt{\dfrac{y}{C - y}}\,dy = \tan\theta \cdot \dfrac{C}{2} \cdot 2\sin 2\theta\,d\theta = C\tan\theta \cdot 2\sin\theta\cos\theta\,d\theta = 2C\sin^2\theta\,d\theta = C(1 - \cos 2\theta)d\theta$.

Thus $x = C\left(\theta - \frac{1}{2}\sin 2\theta\right) + K$ for some constant K. When $\theta = 0$, we have $y = 0$. We require that $x = 0$ when $\theta = 0$ so that the curve passes through the origin when $\theta = 0$. This yields $K = 0$. We now have $x = \frac{1}{2}C(2\theta - \sin 2\theta)$, $y = \frac{1}{2}C(1 - \cos 2\theta)$.

(b) Setting $\phi = 2\theta$ and $r = \frac{1}{2}C$, we get $x = r(\phi - \sin\phi)$, $y = r(1 - \cos\phi)$. Comparison with Equations 9.1.1 shows that the curve is a cycloid.

9. **(a)** Choose a vertical x-axis pointing downward with its origin at the surface. In order to calculate the pressure at depth z, consider a partition P of the interval $[0, z]$ by points x_i and choose a point $x_i^* \in [x_{i-1}, x_i]$ for each i. The thin layer of water lying between depth x_{i-1} and depth x_i has a density of approximately $\rho(x_i^*)$, so the weight of a piece of that layer with unit cross-sectional area would be $\rho(x_i^*)g\,\Delta x_i$, where $\Delta x_i = x_i - x_{i-1}$. The total weight of a column of water extending from the surface to depth z (with unit cross-sectional area) would be approximately $\sum_i \rho(x_i^*)g\,\Delta x_i$. The estimate becomes exact if we take the limit as $\|P\| \to 0$: weight (or force) per unit area at depth z is $W = \lim\limits_{\|P\|\to 0} \sum_i \rho(x_i^*)g\,\Delta x_i$. In other words, $P(z) = \int_0^z \rho(x)g\,dx$. More generally, if we make no assumptions about the location of the origin, then $P(z) = P_0 + \int_0^z \rho(x)g\,dx$, where P_0 is the pressure at $x = 0$. Differentiating, we get $dP/dz = \rho(z)g$.

(b) $F = \int_{-r}^{r} P(L+x) \cdot 2\sqrt{r^2 - x^2}\,dx$

$= \int_{-r}^{r} \left(P_0 + \int_0^{L+x} \rho_0 e^{z/H} g\,dz\right) \cdot 2\sqrt{r^2 - x^2}\,dx$

$= P_0 \int_{-r}^{r} 2\sqrt{r^2 - x^2}\,dx + \rho_0 g H \int_{-r}^{r} \left(e^{(L+x)/H} - 1\right) \cdot 2\sqrt{r^2 - x^2}\,dx$

$= (P_0 - \rho_0 g H) \int_{-r}^{r} 2\sqrt{r^2 - x^2}\,dx + \rho_0 g H \int_{-r}^{r} e^{(L+x)/H} \cdot 2\sqrt{r^2 - x^2}\,dx$

$= (P_0 - \rho_0 g H)(\pi r^2) + \rho_0 g H e^{L/H} \int_{-r}^{r} e^{x/H} \cdot 2\sqrt{r^2 - x^2}\,dx$

L

r

Notice that the result of Exercise 8.5.21 does not apply here since pressure does not increase linearly with depth the way it does for a fluid of constant density.

11. (a) $\dfrac{d^2y}{dx^2} = k\sqrt{1 + \left(\dfrac{dy}{dx}\right)^2}$. Setting $z = \dfrac{dy}{dx}$, we get $\dfrac{dz}{dx} = k\sqrt{1+z^2} \Rightarrow \dfrac{dz}{\sqrt{1+z^2}} = k\,dx$. Using Formula 25 gives $\ln\left(z + \sqrt{1+z^2}\right) = kx + c \Rightarrow z + \sqrt{1+z^2} = Ce^{kx}$ (where $C = e^c$) $\Rightarrow$ $\sqrt{1+z^2} = Ce^{kx} - z \Rightarrow 1 + z^2 = C^2 e^{2kx} - 2Ce^{kx}z + z^2 \Rightarrow 2Ce^{kx}z = C^2 e^{2kx} - 1 \Rightarrow$ $z = \dfrac{C}{2}e^{kx} - \dfrac{1}{2C}e^{-kx}$. Now $\dfrac{dy}{dx} = \dfrac{C}{2}e^{kx} - \dfrac{1}{2C}e^{-kx} \Rightarrow y = \dfrac{C}{2k}e^{kx} + \dfrac{1}{2Ck}e^{-kx} + C'$. From the diagram in the text, we see that $y(0) = a$ and $y(\pm b) = h$. $a = y(0) = \dfrac{C}{2k} + \dfrac{1}{2Ck} + C' \Rightarrow$ $C' = a - \dfrac{C}{2k} - \dfrac{1}{2Ck} \Rightarrow y = \dfrac{C}{2k}\left(e^{kx} - 1\right) + \dfrac{1}{2Ck}\left(e^{-kx} - 1\right) + a$. From $h = y(\pm b)$, we find $h = \dfrac{C}{2k}\left(e^{kb} - 1\right) + \dfrac{1}{2Ck}\left(e^{-kb} - 1\right) + a$ and $h = \dfrac{C}{2k}\left(e^{-kb} - 1\right) + \dfrac{1}{2Ck}\left(e^{kb} - 1\right) + a$. Subtracting the second equation from the first, we get $0 = \dfrac{C}{k}\dfrac{e^{kb} - e^{-kb}}{2} - \dfrac{1}{Ck}\dfrac{e^{kb} - e^{-kb}}{2} = \dfrac{1}{k}\left(C - \dfrac{1}{C}\right)\sinh(kb)$. Now $k > 0$ and $b > 0$, so $\sinh(kb) > 0$ and $C = \pm 1$. If $C = 1$, then

$y = \dfrac{1}{2k}\left(e^{kx} - 1\right) + \dfrac{1}{2k}\left(e^{-kx} - 1\right) + a = \dfrac{1}{k}\dfrac{e^{kx} + e^{-kx}}{2} - \dfrac{1}{k} + a = a + \dfrac{1}{k}(\cosh kx - 1)$. If $C = -1$, then $y = \dfrac{-1}{2k}\left(e^{kx} - 1\right) - \dfrac{1}{2k}\left(e^{-kx} - 1\right) + a = \dfrac{-1}{k}\dfrac{e^{kx} + e^{-kx}}{2} + \dfrac{1}{k} + a = a - \dfrac{1}{k}(\cosh kx - 1)$.

Since $k > 0$, $\cosh kx \geq 1$, and $y \geq a$, we conclude that $C = 1$ and $y = a + \dfrac{1}{k}(\cosh kx - 1)$, where $h = y(b) = a + \dfrac{1}{k}(\cosh kb - 1)$. Since $\cosh(kb) = \cosh(-kb)$, there is no further information to extract from the condition that $y(b) = y(-b)$. However, we could replace a with the expression $h - \dfrac{1}{k}(\cosh kb - 1)$, obtaining $y = h + \dfrac{1}{k}(\cosh kx - \cosh kb)$. It would be better still to keep a in the expression for y, and use the expression for h to solve for k in terms of a, b, and h. That would enable us to express y in terms of x and the given parameters a, b, and h. Sadly, it is not possible to solve for k in closed form. That would have to be done by numerical methods when specific parameter values are given.

(b) The length of the cable is $L = \int_{-b}^{b} \sqrt{1 + (dy/dx)^2}\,dx = \int_{-b}^{b} \sqrt{1 + \sinh^2 kx}\,dx$

$= \int_{-b}^{b} \cosh kx\,dx = \dfrac{1}{k}\sinh kx\Big|_{-b}^{b} = \dfrac{1}{k}[\sinh(kb) - \sinh(-kb)] = \dfrac{2}{k}\sinh(kb)$.

CHAPTER TEN

EXERCISES 10.1

1. $a_n = \dfrac{n}{2n+1}$, so the sequence is $\left\{\dfrac{1}{3}, \dfrac{2}{5}, \dfrac{3}{7}, \dfrac{4}{9}, \dfrac{5}{11}, \dots\right\}$.

3. $a_n = \dfrac{1 \cdot 3 \cdot 5 \cdot \cdots \cdot (2n-1)}{n!}$, so the sequence is $\left\{1, \dfrac{3}{2}, \dfrac{5}{2}, \dfrac{35}{8}, \dfrac{63}{8}, \dots\right\}$.

5. $a_n = \sin \dfrac{n\pi}{2}$, so the sequence is $\{1, 0, -1, 0, 1, \dots\}$.

7. $a_n = \dfrac{1}{2^n}$

9. $a_n = 3n - 2$

11. $a_n = (-1)^{n+1}\left(\dfrac{3}{2}\right)^n$

13. $\lim\limits_{n\to\infty} \dfrac{1}{4n^2} = \dfrac{1}{4} \lim\limits_{n\to\infty} \dfrac{1}{n^2} = \dfrac{1}{4} \cdot 0 = 0$. Convergent

15. $\lim\limits_{n\to\infty} \dfrac{n^2-1}{n^2+1} = \lim\limits_{n\to\infty} \dfrac{1-1/n^2}{1+1/n^2} = 1$. Convergent

17. $\{a_n\}$ diverges since $\dfrac{n^2}{n+1} = \dfrac{n}{1+1/n} \to \infty$ as $n \to \infty$.

19. $\lim\limits_{n\to\infty} |a_n| = \lim\limits_{n\to\infty} \dfrac{n^2}{1+n^3} = \lim\limits_{n\to\infty} \dfrac{1/n}{(1/n^3)+1} = 0$, so by Theorem 5, $\lim\limits_{n\to\infty} (-1)^n \left(\dfrac{n^2}{1+n^3}\right) = 0$.

21. $\{a_n\} = \{0, -1, 0, 1, 0, -1, 0, 1, \dots\}$. This sequence oscillates among 0, -1, and 1 and so diverges.

23. $a_n = \left(\frac{\pi}{3}\right)^n$ so $\{a_n\}$ diverges by Equation 7 with $r = \frac{\pi}{3} > 1$.

25. $0 < \dfrac{3+(-1)^n}{n^2} \le \dfrac{4}{n^2}$ and $\lim\limits_{n\to\infty} \dfrac{4}{n^2} = 0$, so $\left\{\dfrac{3+(-1)^n}{n^2}\right\}$ converges to 0 by the Squeeze Theorem.

27. $\lim\limits_{x\to\infty} \dfrac{\ln(x^2)}{x} = \lim\limits_{x\to\infty} \dfrac{2\ln x}{x} \overset{\text{H}}{=} \lim\limits_{x\to\infty} \dfrac{2/x}{1} = 0$, so by Theorem 2, $\left\{\dfrac{\ln(n^2)}{n}\right\}$ converges to 0.

29. $\sqrt{n+2} - \sqrt{n} = \left(\sqrt{n+2} - \sqrt{n}\right) \dfrac{\sqrt{n+2}+\sqrt{n}}{\sqrt{n+2}+\sqrt{n}} = \dfrac{2}{\sqrt{n+2}+\sqrt{n}} < \dfrac{2}{2\sqrt{n}} = \dfrac{1}{\sqrt{n}} \to 0$ as $n \to \infty$. So by the Squeeze Theorem $\left\{\sqrt{n+2} - \sqrt{n}\right\}$ converges to 0.

31. $\lim\limits_{x\to\infty} \dfrac{x}{2^x} \overset{\text{H}}{=} \lim\limits_{x\to\infty} \dfrac{1}{(\ln 2)2^x} = 0$, so by Theorem 2 $\{a_n\}$ converges to 0.

33. Let $y = x^{-1/x}$. Then $\ln y = -(\ln x)/x$ and $\lim\limits_{x\to\infty} (\ln y) \overset{\text{H}}{=} \lim\limits_{x\to\infty} -(1/x)/1 = 0$, so $\lim\limits_{x\to\infty} y = e^0 = 1$, and so $\{a_n\}$ converges to 1.

35. $0 \le \dfrac{\cos^2 n}{2^n} \le \dfrac{1}{2^n}$ $\left[\text{since } 0 \le \cos^2 n \le 1\right]$, so since $\lim\limits_{n\to\infty} \dfrac{1}{2^n} = 0$, $\{a_n\}$ converges to 0 by the Squeeze Theorem.

37. The series converges, since $a_n = \dfrac{1+2+3+\cdots+n}{n^2} = \dfrac{n(n+1)/2}{n^2}$ [Theorem 3]

$= \dfrac{n+1}{2n} = \dfrac{1+1/n}{2} \to \dfrac{1}{2}$ as $n \to \infty$.

39. $a_n = \frac{1}{2} \cdot \frac{2}{2} \cdot \frac{3}{2} \cdot \cdots \cdot \frac{(n-1)}{2} \cdot \frac{n}{2} \geq \frac{1}{2} \cdot \frac{n}{2} = \frac{n}{4} \to \infty$ as $n \to \infty$, so $\{a_n\}$ diverges.

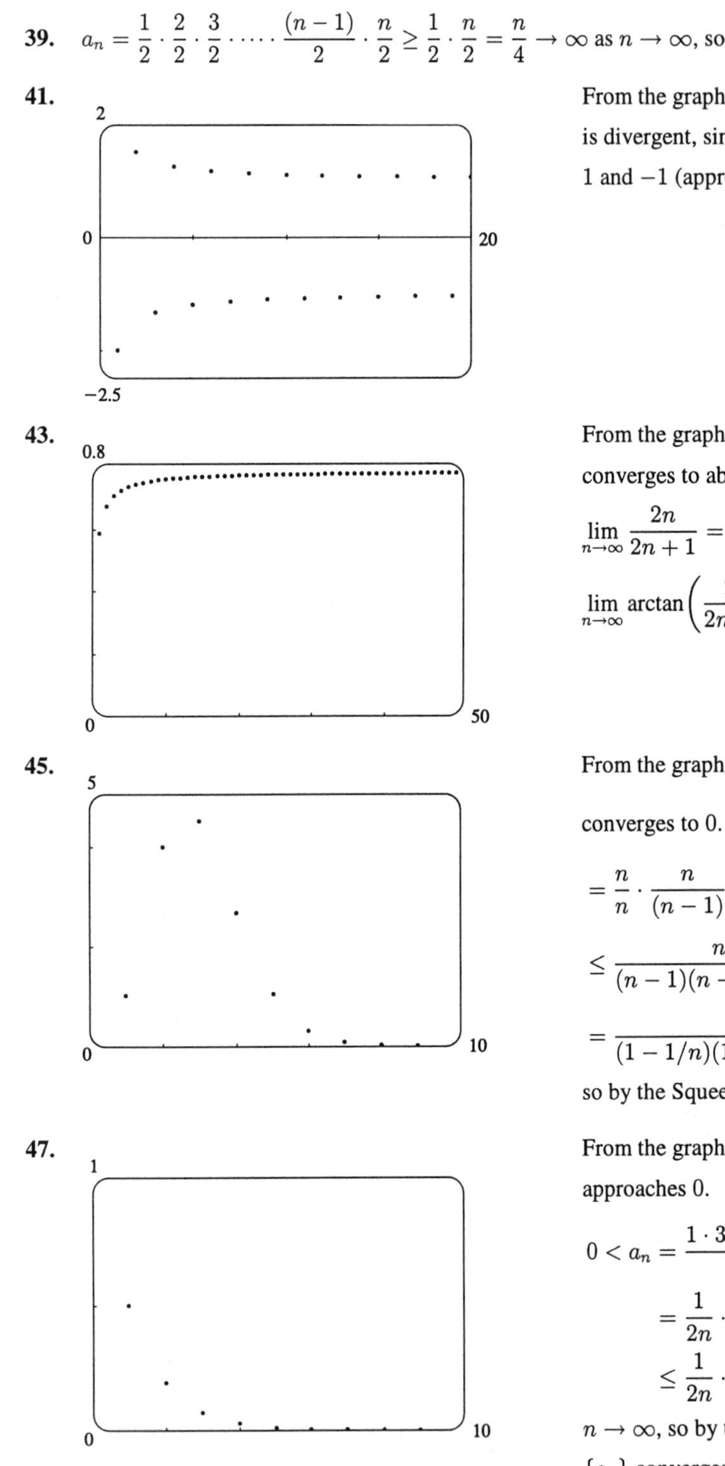

41. From the graph, we see that the sequence is divergent, since it oscillates between 1 and −1 (approximately).

43. From the graph, it appears that the sequence converges to about 0.78.

$\lim\limits_{n\to\infty} \frac{2n}{2n+1} = \lim\limits_{n\to\infty} \frac{2}{2+1/n} = 1$, so

$\lim\limits_{n\to\infty} \arctan\left(\frac{2n}{2n+1}\right) = \arctan 1 = \frac{\pi}{4}$.

45. From the graph, it appears that the sequence converges to 0. $0 < a_n = \frac{n^3}{n!}$

$$= \frac{n}{n} \cdot \frac{n}{(n-1)} \cdot \frac{n}{(n-2)} \cdot \frac{1}{(n-3)} \cdot \cdots \cdot \frac{1}{3} \cdot \frac{1}{2} \cdot \frac{1}{1}$$

$$\leq \frac{n^2}{(n-1)(n-2)(n-3)} \quad \text{(for } n \geq 4\text{)}$$

$$= \frac{1/n}{(1-1/n)(1-2/n)(1-3/n)} \to 0 \text{ as } n \to \infty,$$

so by the Squeeze Theorem, $\{a_n\}$ converges to 0.

47. From the graph, it appears that the sequence approaches 0.

$$0 < a_n = \frac{1 \cdot 3 \cdot 5 \cdot \cdots \cdot (2n-1)}{(2n)^n}$$

$$= \frac{1}{2n} \cdot \frac{3}{2n} \cdot \frac{5}{2n} \cdot \cdots \cdot \frac{2n-1}{2n}$$

$$\leq \frac{1}{2n} \cdot (1) \cdot (1) \cdot \cdots \cdot (1) = \frac{1}{2n} \to 0 \text{ as}$$

$n \to \infty$, so by the Squeeze Theorem $\{a_n\}$ converges to 0.

49. If $|r| \geq 1$, then $\{r^n\}$ diverges by (7), so $\{nr^n\}$ diverges also since $|nr^n| = n|r^n| \geq |r^n|$. If $|r| < 1$ then $\lim_{x\to\infty} xr^x = \lim_{x\to\infty} \frac{x}{r^{-x}} \overset{\text{H}}{=} \lim_{x\to\infty} \frac{1}{(-\ln r)r^{-x}} = \lim_{x\to\infty} \frac{r^x}{-\ln r} = 0$, so $\lim_{n\to\infty} nr^n = 0$, and hence $\{nr^n\}$ converges whenever $|r| < 1$.

51. $3(n+1)+5 > 3n+5$ so $\frac{1}{3(n+1)+5} < \frac{1}{3n+5} \quad \Leftrightarrow \quad a_{n+1} < a_n$ so $\{a_n\}$ is decreasing.

53. $\left\{\frac{n-2}{n+2}\right\}$ is increasing since $a_n < a_{n+1} \quad \Leftrightarrow \quad \frac{n-2}{n+2} < \frac{(n+1)-2}{(n+1)+2} \quad \Leftrightarrow \quad (n-2)(n+3) < (n+2)(n-1)$ $\Leftrightarrow \quad n^2+n-6 < n^2+n-2 \quad \Leftrightarrow \quad -6 < -2$, which is of course true.

55. $a_1 = 0 > a_2 = -1 < a_3 = 0$, so the sequence is not monotonic.

57. $\left\{\frac{n}{n^2+n-1}\right\}$ is decreasing since $a_{n+1} < a_n \quad \Leftrightarrow \quad \frac{n+1}{(n+1)^2+(n+1)-1} < \frac{n}{n^2+n-1} \quad \Leftrightarrow$ $(n+1)(n^2+n-1) < n(n^2+3n+1) \quad \Leftrightarrow \quad n^3+2n^2-1 < n^3+3n^2+n \quad \Leftrightarrow$ $0 < n^2+n+1 = \left(n+\frac{1}{2}\right)^2 + \frac{3}{4}$, which is obviously true.

59. $a_1 = 2^{1/2}, a_2 = 2^{3/4}, a_3 = 2^{7/8}, \ldots, a_n = 2^{(2^n-1)/2^n} = 2^{(1-1/2^n)}$. $\lim_{n\to\infty} a_n = \lim_{n\to\infty} 2^{(1-1/2^n)} = 2^1 = 2$.

Alternate Solution: Let $L = \lim_{n\to\infty} a_n$ (We could show the limit exists by showing that $\{a_n\}$ is bounded and increasing.) So L must satisfy $L = \sqrt{2 \cdot L} \Rightarrow L^2 = 2L \Rightarrow L(L-2) = 0$ ($L \neq 0$ since the sequence increases) so $L = 2$.

61. We show by induction that $\{a_n\}$ is increasing and bounded above by 3. Let $P(n)$ be the proposition that $a_{n+1} > a_n$ and $0 < a_n < 3$. Clearly $P(1)$ is true. Assume $P(n)$ is true. Then $a_{n+1} > a_n \quad \Rightarrow \quad \frac{1}{a_{n+1}} < \frac{1}{a_n}$ $\Rightarrow \quad -\frac{1}{a_{n+1}} > -\frac{1}{a_n} \quad \Rightarrow \quad a_{n+2} = 3 - \frac{1}{a_{n+1}} > 3 - \frac{1}{a_n} = a_{n+1} \quad \Leftrightarrow \quad P(n+1)$. This proves that $\{a_n\}$ is increasing and bounded above by 3, so $1 = a_1 < a_n < 3$, that is, $\{a_n\}$ is bounded, and hence convergent by Theorem 10.
If $L = \lim_{n\to\infty} a_n$, then $\lim_{n\to\infty} a_{n+1} = L$ also, so L must satisfy $L = 3 - \frac{1}{L}$, so $L^2 - 3L + 1 = 0$ and the quadratic formula gives $L = \frac{3\pm\sqrt{5}}{2}$. But $L > 1$, so $L = \frac{3+\sqrt{5}}{2}$.

63. **(a)** Let a_n be the number of rabbit pairs in the nth month. Clearly $a_1 = 1 = a_2$. In the nth month, each pair that is 2 or more months old (that is, a_{n-2} pairs) will have a pair of children to add to the a_{n-1} pairs already present. Thus $a_n = a_{n-1} + a_{n-2}$, so that $\{a_n\} = \{f_n\}$, the Fibonacci sequence.

(b) $a_{n-1} = \frac{f_n}{f_{n-1}} = \frac{f_{n-1}+f_{n-2}}{f_{n-1}} = 1 + \frac{f_{n-2}}{f_{n-1}} = 1 + \frac{1}{a_{n-2}}$. If $L = \lim_{n\to\infty} a_n$, then L must satisfy $L = 1 + \frac{1}{L}$ or $L^2 - L - 1 = 0$, so $L = \frac{1+\sqrt{5}}{2}$ (since L must be positive.)

65. **(a)**

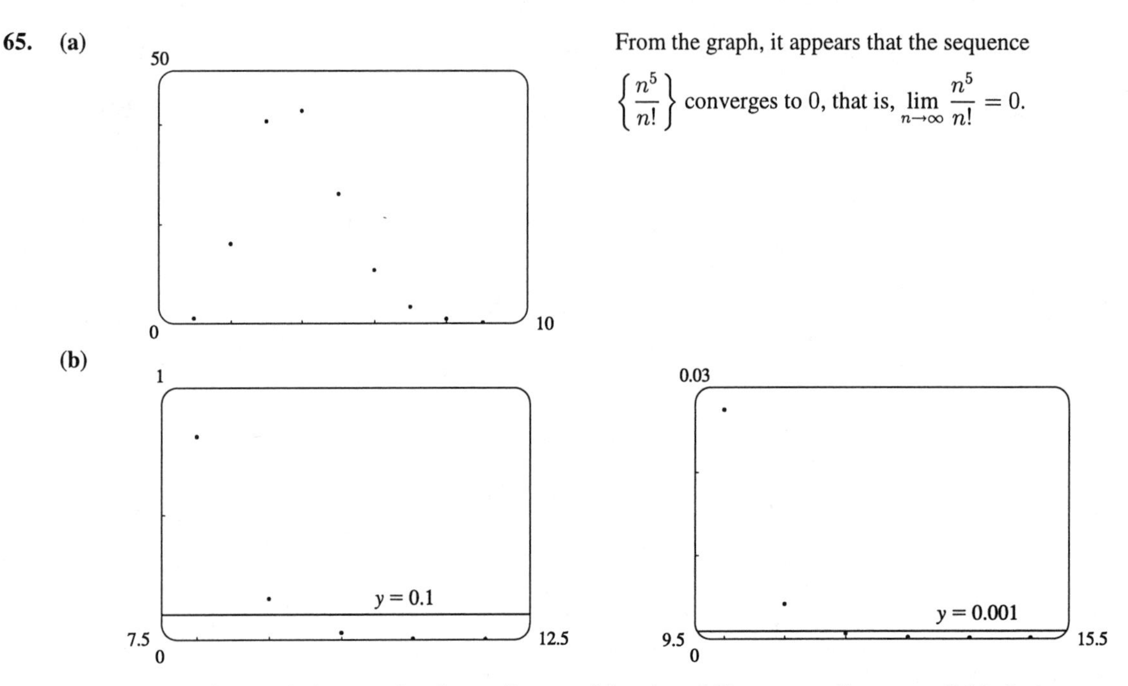

From the graph, it appears that the sequence $\left\{\frac{n^5}{n!}\right\}$ converges to 0, that is, $\lim_{n\to\infty} \frac{n^5}{n!} = 0$.

(b)

From the first graph, it seems that the smallest possible value of N corresponding to $\epsilon = 0.1$ is 9, since $n^5/n! < 0.1$ whenever $n \geq 10$, but $9^5/9! > 0.1$. From the second graph, it seems that for $\epsilon = 0.001$, the smallest possible value for N is 11.

67. If $\lim_{n\to\infty} |a_n| = 0$ then $\lim_{n\to\infty} -|a_n| = 0$, and since $-|a_n| \leq a_n \leq |a_n|$, we have that $\lim_{n\to\infty} a_n = 0$ by the Squeeze Theorem.

69. **(a)** First we show that $a > a_1 > b_1 > b$.

$$a_1 - b_1 = \frac{a+b}{2} - \sqrt{ab} = \tfrac{1}{2}\left(a - 2\sqrt{ab} + b\right) = \tfrac{1}{2}\left(\sqrt{a} - \sqrt{b}\right)^2 > 0 \quad (\text{since } a > b) \quad \Rightarrow \quad a_1 > b_1.$$

Also $a - a_1 = a - \frac{1}{2}(a+b) = \frac{1}{2}(a-b) > 0$ and $b - b_1 = b - \sqrt{ab} = \sqrt{b}\left(\sqrt{b} - \sqrt{a}\right) < 0$, so $a > a_1 > b_1 > b$. In the same way we can show that $a_1 > a_2 > b_2 > b_1$ and so the given assertion is true for $n = 1$. Suppose it is true for $n = k$, that is, $a_k > a_{k+1} > b_{k+1} > b_k$. Then

$$a_{k+2} - b_{k+2} = \tfrac{1}{2}(a_{k+1} + b_{k+1}) - \sqrt{a_{k+1}b_{k+1}} = \tfrac{1}{2}\left(a_{k+1} - 2\sqrt{a_{k+1}b_{k+1}} + b_{k+1}\right)$$

$= \frac{1}{2}\left(\sqrt{a_{k+1}} - \sqrt{b_{k+1}}\right)^2 > 0$ and $a_{k+1} - a_{k+2} = a_{k+1} - \frac{1}{2}(a_{k+1} + b_{k+1}) = \frac{1}{2}(a_{k+1} - b_{k+1}) > 0$,

$$b_{k+1} - b_{k+2} = b_{k+1} - \sqrt{a_{k+1}b_{k+1}} = \sqrt{b_{k+1}}\left(\sqrt{b_{k+1}} - \sqrt{a_{k+1}}\right) < 0 \quad \Rightarrow \quad a_{k+1} > a_{k+2} > b_{k+2} > b_{k+1},$$

so the assertion is true for $n = k+1$. Thus it is true for all n by mathematical induction.

(b) From part (a) we have $a > a_n > a_{n+1} > b_{n+1} > b_n > b$, which shows that both sequences are monotonic and bounded. So they are both convergent by Theorem 10.

(c) Let $\lim_{n\to\infty} a_n = \alpha$ and $\lim_{n\to\infty} b_n = \beta$. Then $\lim_{n\to\infty} a_{n+1} = \lim_{n\to\infty} \frac{a_n + b_n}{2} \quad \Rightarrow \quad \alpha = \frac{\alpha + \beta}{2} \quad \Rightarrow$

$2\alpha = \alpha + \beta \quad \Rightarrow \quad \alpha = \beta$.

71. **(a)** $2\cos\theta - 1 = \dfrac{1+2\cos 2\theta}{1+2\cos\theta} \quad\Leftrightarrow\quad (2\cos\theta+1)(2\cos\theta-1) = 1+2\cos 2\theta$ (provided $\cos\theta \neq -1$), and this is certainly true since the LHS $= 4\cos^2\theta - 1$ and the RHS $= 1+2(2\cos^2\theta-1) = 4\cos^2\theta - 1$.

(b) By part (a), we can write each a_k as $2\cos(\theta/2^k) - 1 = \dfrac{1+2\cos(\theta/2^{k-1})}{1+2\cos(\theta/2^k)}$, so we get

$$b_n = \frac{1+2\cos\theta}{1+2\cos(\theta/2)}\cdot\frac{1+2\cos(\theta/2)}{1+2\cos(\theta/4)}\cdot\cdots\cdot\frac{1+2\cos(\theta/2^{n-2})}{1+2\cos(\theta/2^{n-1})}\cdot\frac{1+2\cos(\theta/2^{n-1})}{1+2\cos(\theta/2^n)} = \frac{1+2\cos\theta}{1+2\cos(\theta/2^n)}$$

(telescoping product). So $\lim\limits_{n\to\infty} b_n = \lim\limits_{n\to\infty}\dfrac{1+2\cos\theta}{1+2\cos(\theta/2^n)} = \dfrac{1+2\cos\theta}{1+2\cos 0} = \frac{1}{3}(1+2\cos\theta)$.

EXERCISES 10.2

1.

n	s_n
1	3.33333
2	4.44444
3	4.81481
4	4.93827
5	4.97942
6	4.99314
7	4.99771
8	4.99924
9	4.99975
10	4.99992
11	4.99997
12	4.99999

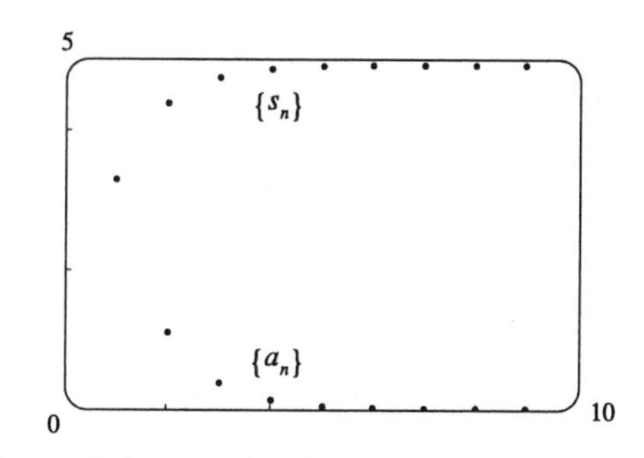

From the graph, it seems that the series converges. In fact, it is a geometric series with $a = \frac{10}{3}$ and $r = \frac{1}{3}$, so its sum is

$$\sum_{n=1}^{\infty}\frac{10}{3^n} = \frac{10/3}{1-1/3} = 5.$$

3.

n	s_n
1	0.50000
2	1.16667
3	1.91667
4	2.71667
5	3.55000
6	4.40714
7	5.28214
8	6.17103
9	7.07103
10	7.98012

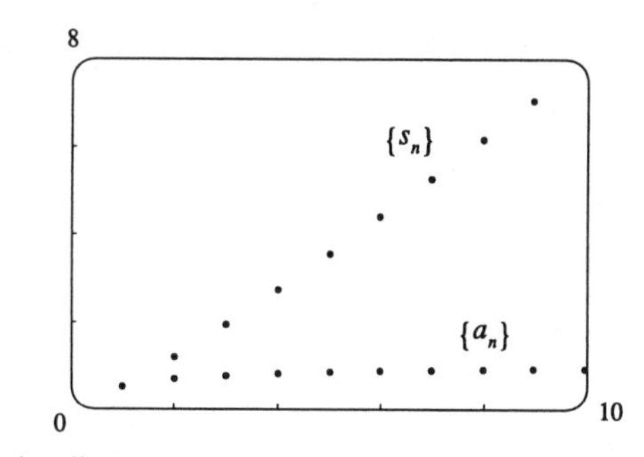

The series diverges, since its terms do not approach 0.

5.

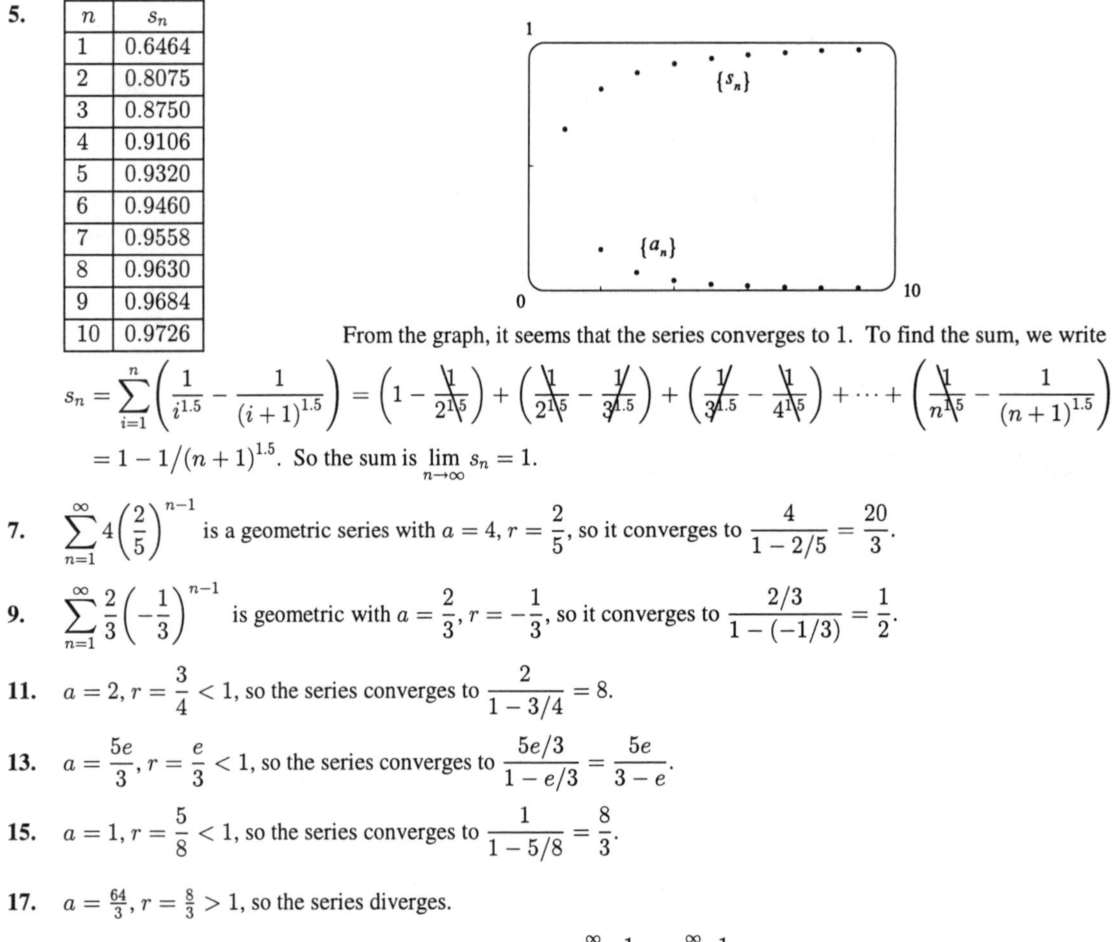

n	s_n
1	0.6464
2	0.8075
3	0.8750
4	0.9106
5	0.9320
6	0.9460
7	0.9558
8	0.9630
9	0.9684
10	0.9726

From the graph, it seems that the series converges to 1. To find the sum, we write

$$s_n = \sum_{i=1}^{n}\left(\frac{1}{i^{1.5}} - \frac{1}{(i+1)^{1.5}}\right) = \left(1 - \frac{1}{2^{1.5}}\right) + \left(\frac{1}{2^{1.5}} - \frac{1}{3^{1.5}}\right) + \left(\frac{1}{3^{1.5}} - \frac{1}{4^{1.5}}\right) + \cdots + \left(\frac{1}{n^{1.5}} - \frac{1}{(n+1)^{1.5}}\right)$$

$= 1 - 1/(n+1)^{1.5}$. So the sum is $\lim\limits_{n\to\infty} s_n = 1$.

7. $\displaystyle\sum_{n=1}^{\infty} 4\left(\frac{2}{5}\right)^{n-1}$ is a geometric series with $a = 4$, $r = \dfrac{2}{5}$, so it converges to $\dfrac{4}{1-2/5} = \dfrac{20}{3}$.

9. $\displaystyle\sum_{n=1}^{\infty} \frac{2}{3}\left(-\frac{1}{3}\right)^{n-1}$ is geometric with $a = \dfrac{2}{3}$, $r = -\dfrac{1}{3}$, so it converges to $\dfrac{2/3}{1-(-1/3)} = \dfrac{1}{2}$.

11. $a = 2$, $r = \dfrac{3}{4} < 1$, so the series converges to $\dfrac{2}{1-3/4} = 8$.

13. $a = \dfrac{5e}{3}$, $r = \dfrac{e}{3} < 1$, so the series converges to $\dfrac{5e/3}{1-e/3} = \dfrac{5e}{3-e}$.

15. $a = 1$, $r = \dfrac{5}{8} < 1$, so the series converges to $\dfrac{1}{1-5/8} = \dfrac{8}{3}$.

17. $a = \frac{64}{3}$, $r = \frac{8}{3} > 1$, so the series diverges.

19. This series diverges, since if it converged, so would $2 \cdot \displaystyle\sum_{n=1}^{\infty} \frac{1}{2n} = \sum_{n=1}^{\infty} \frac{1}{n}$ [by Theorem 8(a)], which we know diverges (Example 7).

21. Converges. $s_n = \displaystyle\sum_{i=1}^{n} \frac{1}{(3i-2)(3i+1)} = \sum_{i=1}^{n}\left[\frac{1/3}{3i-2} - \frac{1/3}{3i+1}\right]$ (partial fractions)

$$= \left[\frac{1}{3}\cdot 1 - \frac{1}{3}\cdot\frac{1}{4}\right] + \left[\frac{1}{3}\cdot\frac{1}{4} - \frac{1}{3}\cdot\frac{1}{7}\right] + \left[\frac{1}{3}\cdot\frac{1}{7} - \frac{1}{3}\cdot\frac{1}{10}\right] + \cdots$$

$$+ \left[\frac{1}{3}\cdot\frac{1}{3n-2} - \frac{1}{3}\cdot\frac{1}{3n+1}\right] = \frac{1}{3} - \frac{1}{3(3n+1)} \quad \text{(telescoping series)}$$

$$\Rightarrow \quad \lim_{n\to\infty} s_n = \frac{1}{3} \quad \Rightarrow \quad \sum_{n=1}^{\infty} \frac{1}{(3n-2)(3n+1)} = \frac{1}{3}$$

23. Converges by Theorem 8.

$$\sum_{n=1}^{\infty}[2(0.1)^n + (0.2)^n] = 2\sum_{n=1}^{\infty}(0.1)^n + \sum_{n=1}^{\infty}(0.2)^n = 2\left(\frac{0.1}{1-0.1}\right) + \frac{0.2}{1-0.2} = \frac{2}{9} + \frac{1}{4} = \frac{17}{36}$$

25. Diverges by the Test for Divergence. $\lim_{n\to\infty} \dfrac{n}{\sqrt{1+n^2}} = \lim_{n\to\infty} \dfrac{1}{\sqrt{1+1/n^2}} = 1 \neq 0.$

27. Converges. $s_n = \sum_{i=1}^{n} \dfrac{1}{i(i+2)} = \sum_{i=1}^{n} \left[\dfrac{1/2}{i} - \dfrac{1/2}{i+2}\right]$ (partial fractions)

$$= \left[\frac{1}{2} - \frac{1}{6}\right] + \left[\frac{1}{4} - \frac{1}{8}\right] + \left[\frac{1}{6} - \frac{1}{10}\right] + \cdots + \left[\frac{1}{2n-2} - \frac{1}{2n+2}\right] + \left[\frac{1}{2n} - \frac{1}{2n+4}\right]$$

$$= \frac{1}{2} + \frac{1}{4} - \frac{1}{2n+2} - \frac{1}{2n+4} \quad \text{(telescoping series).}$$

Thus $\sum_{n=1}^{\infty} \dfrac{1}{n(n+2)} = \lim_{n\to\infty} \left[\dfrac{1}{2} + \dfrac{1}{4} - \dfrac{1}{2n+2} - \dfrac{1}{2n+4}\right] = \dfrac{3}{4}.$

29. Converges. $\sum_{n=1}^{\infty} \dfrac{3^n + 2^n}{6^n} = \sum_{n=1}^{\infty} \left[\left(\dfrac{1}{2}\right)^n + \left(\dfrac{1}{3}\right)^n\right] = \dfrac{1/2}{1-1/2} + \dfrac{1/3}{1-1/3} = \dfrac{3}{2}$

31. Converges. $s_n = \left(\sin 1 - \sin \dfrac{1}{2}\right) + \left(\sin \dfrac{1}{2} - \sin \dfrac{1}{3}\right) + \cdots + \left[\sin \dfrac{1}{n} - \sin \dfrac{1}{n+1}\right] = \sin 1 - \sin \dfrac{1}{n+1}$, so

$\sum_{n=1}^{\infty} \left(\sin \dfrac{1}{n} - \sin \dfrac{1}{n+1}\right) = \lim_{n\to\infty} s_n = \sin 1 - \sin 0 = \sin 1$

33. Diverges since $\lim_{n\to\infty} \arctan n = \frac{\pi}{2} \neq 0.$

35. $s_n = (\ln 1 - \ln 2) + (\ln 2 - \ln 3) + (\ln 3 - \ln 4) + \cdots + [\ln n - \ln(n+1)] = \ln 1 - \ln(n+1) = -\ln(n+1)$ (telescoping series). Thus $\lim_{n\to\infty} s_n = -\infty$, so the series is divergent.

37. $0.\overline{5} = 0.5 + 0.05 + 0.005 + \cdots = \dfrac{0.5}{1-0.1} = \dfrac{5}{9}$

39. $0.\overline{307} = 0.307 + 0.000307 + 0.000000307 + \cdots = \dfrac{0.307}{1-0.001} = \dfrac{307}{999}$

41. $0.123\overline{456} = \dfrac{123}{1000} + \dfrac{0.000456}{1-0.001} = \dfrac{123}{1000} + \dfrac{456}{999{,}000} = \dfrac{123{,}333}{999{,}000} = \dfrac{41{,}111}{333{,}000}$

43. $\sum_{n=0}^{\infty} (x-3)^n$ is a geometric series with $r = x - 3$, so it converges whenever $|x-3| < 1 \Rightarrow$ $-1 < x - 3 < 1 \Leftrightarrow 2 < x < 4$. The sum is $\dfrac{1}{1-(x-3)} = \dfrac{1}{4-x}$.

45. $\sum_{n=2}^{\infty} \left(\dfrac{x}{5}\right)^n$ is a geometric series with $r = \dfrac{x}{5}$, so converges whenever $\left|\dfrac{x}{5}\right| < 1 \Leftrightarrow -5 < x < 5$. The sum is $\dfrac{(x/5)^2}{1-x/5} = \dfrac{x^2}{25-5x}$.

47. $\sum_{n=0}^{\infty} (2\sin x)^n$ is geometric so converges whenever $|2\sin x| < 1 \Leftrightarrow -\frac{1}{2} < \sin x < \frac{1}{2} \Leftrightarrow$ $n\pi - \frac{\pi}{6} < x < n\pi + \frac{\pi}{6}$, where the sum is $\dfrac{1}{1-2\sin x}$.

49. After defining f, We use `convert(f,parfrac);` in Maple or `Apart` in Mathematica to find that the general term is $\dfrac{1}{(4n+1)(4n-3)} = -\dfrac{1/4}{4n+1} + \dfrac{1/4}{4n-3}$. So the nth partial sum is

$$s_n = \sum_{k=1}^{n}\left(-\frac{1/4}{4k+1} + \frac{1/4}{4k-3}\right)$$

$$= \frac{1}{4}\left[\left(-\frac{1}{5}+1\right) + \left(-\frac{1}{9}+\frac{1}{5}\right) + \left(-\frac{1}{13}+\frac{1}{9}\right) + \cdots + \left(-\frac{1}{4n+1}+\frac{1}{4n-3}\right)\right] = \frac{1}{4}\left(1-\frac{1}{4n+1}\right).$$

The series converges to $\lim_{n\to\infty} s_n = \frac{1}{4}$. This can be confirmed by directly computing the sum using `sum(f,1..infinity);` (in Maple) or `Sum[f,{n,0,Infinity}]` (in Mathematica).

51. Plainly $a_1 = 0$ since $s_1 = 0$. For $n \neq 1$, $a_n = s_n - s_{n-1} = \dfrac{n-1}{n+1} - \dfrac{(n-1)-1}{(n-1)+1}$

$= \dfrac{(n-1)n-(n+1)(n-2)}{(n+1)n} = \dfrac{2}{n(n+1)}$. Also $\displaystyle\sum_{n=1}^{\infty} a_n = \lim_{n\to\infty} s_n = \lim_{n\to\infty}\frac{1-1/n}{1+1/n} = 1$.

53. **(a)** The first step in the chain occurs when the local government spends D dollars. The people who receive it spend a fraction c of those D dollars, that is, Dc dollars. Those who receive the Dc dollars spend a fraction c of it, that is, Dc^2 dollars. Continuing in this way, we see that the total spending after n transactions is $S_n = D + Dc + Dc^2 + \cdots + Dc^{n-1} = \dfrac{D(1-c^n)}{1-c}$ by (3).

(b) $\displaystyle\lim_{n\to\infty} S_n = \lim_{n\to\infty}\frac{D(1-c^n)}{1-c} = \frac{D}{1-c}\lim_{n\to\infty}(1-c^n) = \frac{D}{1-c} = \frac{D}{s} = kD$, since $0 < c < 1 \Rightarrow \displaystyle\lim_{n\to\infty} c^n = 0$. If $c = 0.8$, then $s = 1 - c = 0.2$ and the multiplier is $k = 1/s = 5$.

55. $\sum_{n=2}^{\infty}(1+c)^{-n}$ is a geometric series with $a = (1+c)^{-2}$ and $r = (1+c)^{-1}$, so the series converges when $\left|(1+c)^{-1}\right| < 1 \Rightarrow |1+c| > 1 \Rightarrow 1+c > 1$ or $1+c < -1 \Rightarrow c > 0$ or $c < -2$. We calculate the sum of the series and set it equal to 2: $\dfrac{(1+c)^{-2}}{1-(1+c)^{-1}} = 2 \Leftrightarrow \left(\dfrac{1}{1+c}\right)^2 = 2 - 2\left(\dfrac{1}{1+c}\right) \Leftrightarrow$ $1 - 2(1+c)^2 + 2(1+c) = 0 \Leftrightarrow 2c^2 + 2c - 1 = 0 \Leftrightarrow c = \frac{-2\pm\sqrt{12}}{4} = \frac{\pm\sqrt{3}-1}{2}$. However, the negative root is inadmissible because $-2 < \frac{-\sqrt{3}-1}{2} < 0$. So $c = \frac{\sqrt{3}-1}{2}$.

57. Let d_n be the diameter of C_n. We draw lines from the centers of the C_i to the center of D (or C), and using the Pythagorean Theorem, we can write

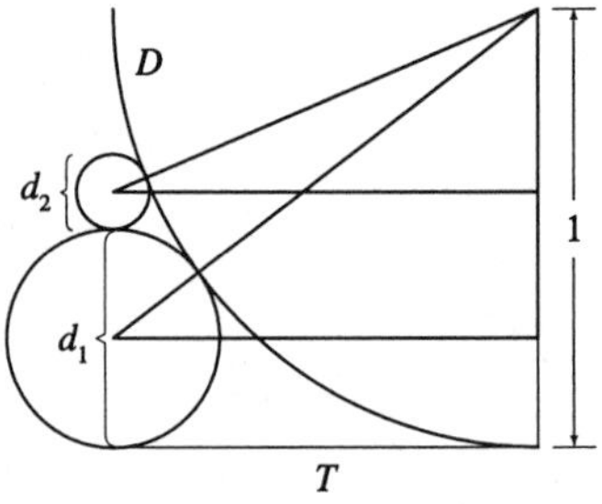

$1^2 + \left(1 - \frac{1}{2}d_1\right)^2 = \left(1 + \frac{1}{2}d_1\right)^2 \Leftrightarrow$

$1 = \left(1 + \frac{1}{2}d_1\right)^2 - \left(1 - \frac{1}{2}d_1\right)^2 = 2d_1$ (difference of squares)

$\Rightarrow d_1 = \frac{1}{2}$. Similarly,

$1 = \left(1 + \frac{1}{2}d_2\right)^2 - \left(1 - d_1 - \frac{1}{2}d_2\right)^2 = (2 - d_1)(d_1 + d_2)$

D
d_2
d_1
1
T

$\Leftrightarrow d_2 = \dfrac{(1-d_1)^2}{2-d_1}$, $1 = \left(1+\frac{1}{2}d_3\right)^2 - \left(1 - d_1 - d_2 - \frac{1}{2}d_3\right)^2 \Leftrightarrow d_3 = \dfrac{[1-(d_1+d_2)]^2}{2-(d_1+d_2)}$, and in general,

$$d_{n+1} = \frac{\left(1 - \sum_{i=1}^{n} d_i\right)^2}{2 - \sum_{i=1}^{n} d_i}.$$

If we actually calculate d_2 and d_3 from the formulas above, we find that they are $\frac{1}{6} = \frac{1}{2 \cdot 3}$ and $\frac{1}{12} = \frac{1}{3 \cdot 4}$ respectively, so we suspect that in general, $d_n = \frac{1}{n(n+1)}$. To prove this, we use induction: assume that for all $k \le n$, $d_k = \frac{1}{k(k+1)} = \frac{1}{k} - \frac{1}{k+1}$. Then $\sum_{i=1}^{n} d_i = 1 - \frac{1}{n+1} = \frac{n}{n+1}$ (telescoping sum). Substituting this into our formula for d_{n+1}, we get $d_{n+1} = \frac{\left[1 - \frac{n}{n+1}\right]^2}{2 - \left(\frac{n}{n+1}\right)} = \frac{\frac{1}{(n+1)^2}}{\frac{n+2}{n+1}} = \frac{1}{(n+1)(n+2)}$, and the induction is complete.

Now, we observe that the partial sums $\sum_{i=1}^{n} d_i$ of the diameters of the circles approach 1 as $n \to \infty$; that is, $\sum_{n=1}^{\infty} d_n = \sum_{n=1}^{\infty} \frac{1}{n(n+1)} = 1$, which is what we wanted to prove.

59. The series $1 - 1 + 1 - 1 + 1 - 1 + \cdots$ diverges (geometric series with $r = -1$) so we cannot say $0 = 1 - 1 + 1 - 1 + 1 - 1 + \cdots$.

61. $\sum_{n=1}^{\infty} ca_n = \lim_{n\to\infty} \sum_{i=1}^{n} ca_i = \lim_{n\to\infty} c \sum_{i=1}^{n} a_i = c \lim_{n\to\infty} \sum_{i=1}^{n} a_i = c \sum_{n=1}^{\infty} a_n$, which exists by hypothesis.

63. Suppose on the contrary that $\sum(a_n + b_n)$ converges. Then by Theorem 8(c), so would $\sum[(a_n + b_n) - a_n] = \sum b_n$, a contradiction.

65. The partial sums $\{s_n\}$ form an increasing sequence, since $s_n - s_{n-1} = a_n > 0$ for all n. Also, the sequence $\{s_n\}$ is bounded since $s_n \le 1000$ for all n. So by Theorem 10.1.10 , the sequence of partial sums converges, that is, the series $\sum a_n$ is convergent.

67. **(a)** At the first step, only the interval $\left(\frac{1}{3}, \frac{2}{3}\right)$ (length $\frac{1}{3}$) is removed. At the second step, we remove the intervals $\left(\frac{1}{9}, \frac{2}{9}\right)$ and $\left(\frac{7}{9}, \frac{8}{9}\right)$, which have a total length of $2 \cdot \left(\frac{1}{3}\right)^2$. At the third step, we remove 2^2 intervals, each of length $\left(\frac{1}{3}\right)^3$. In general, at the nth step we remove 2^{n-1} intervals, each of length $\left(\frac{1}{3}\right)^n$, for a length of $2^{n-1} \cdot \left(\frac{1}{3}\right)^n = \frac{1}{3}\left(\frac{2}{3}\right)^{n-1}$. Thus, the total length of all removed intervals is $\sum_{n=1}^{\infty} \frac{1}{3}\left(\frac{2}{3}\right)^{n-1} = \frac{1/3}{1 - 2/3} = 1$ (geometric series with $a = \frac{1}{3}$ and $r = \frac{2}{3}$).

Notice that at the nth step, the leftmost interval that is removed is $\left(\left(\frac{1}{3}\right)^n, \left(\frac{2}{3}\right)^n\right)$, so we never remove 0, so 0 is in the Cantor set. Also, the rightmost interval removed is $\left(1 - \left(\frac{2}{3}\right)^n, 1 - \left(\frac{1}{3}\right)^n\right)$, so 1 is never removed. Some other numbers in the Cantor set are $\frac{1}{3}$, $\frac{2}{3}$, $\frac{1}{9}$, $\frac{2}{9}$, $\frac{7}{9}$, and $\frac{8}{9}$.

(b) The area removed at the first step is $\frac{1}{9}$; at the second step, $2^3 \cdot \left(\frac{1}{9}\right)^2$; at the third step, $\left(2^3\right)^2 \cdot \left(\frac{1}{9}\right)^3$. In general, the area removed at the nth step is $\left(2^3\right)^{n-1}\left(\frac{1}{9}\right)^n = \frac{1}{9}\left(\frac{8}{9}\right)^{n-1}$, so the total area of all removed squares is $\sum_{n=1}^{\infty} \frac{1}{8}\left(\frac{8}{9}\right)^{n-1} = \frac{1/9}{1 - 8/9} = 1$.

69. **(a)** $s_1 = \frac{1}{1\cdot 2} = \frac{1}{2}$, $s_2 = \frac{1}{2} + \frac{1}{1\cdot 2\cdot 3} = \frac{5}{6}$, $s_3 = \frac{5}{6} + \frac{3}{1\cdot 2\cdot 3\cdot 4} = \frac{23}{24}$, $s_4 = \frac{23}{24} + \frac{4}{1\cdot 2\cdot 3\cdot 4\cdot 5} = \frac{119}{120}$.

The denominators are $(n+1)!$ so a guess would be $s_n = \frac{(n+1)!-1}{(n+1)!}$.

(b) For $n=1$, $s_1 = \frac{1}{2} = \frac{2!-1}{2!}$, so the formula holds for $n=1$. Assume $s_k = \frac{(k+1)!-1}{(k+1)!}$. Then

$$s_{k+1} = \frac{(k+1)!-1}{(k+1)!} + \frac{k+1}{(k+2)!} = \frac{(k+1)!-1}{(k+1)!} + \frac{k+1}{(k+1)!(k+2)} = \frac{(k+2)!-(k+2)+k+1}{(k+2)!}$$
$$= \frac{(k+2)!-1}{(k+2)!}$$

Thus the formula is true for $n=k+1$. So by induction, the guess is correct.

(c) $\lim_{n\to\infty} s_n = \lim_{n\to\infty} \frac{(n+1)!-1}{(n+1)!} = \lim_{n\to\infty}\left[1 - \frac{1}{(n+1)!}\right] = 1$ and so $\sum_{n=0}^{\infty} \frac{n}{(n+1)!} = 1$.

EXERCISES 10.3

1. $\sum_{n=1}^{\infty} \frac{2}{\sqrt[3]{n}} = 2\sum_{n=1}^{\infty} \frac{1}{n^{1/3}}$, which is a p-series, $p = \frac{1}{3} < 1$, so it diverges.

3. $\sum_{n=5}^{\infty} \frac{1}{n^{1.0001}}$ is a p-series, $p = 1.0001 > 1$, so it converges.

5. $\sum_{n=5}^{\infty} \frac{1}{(n-4)^2} = \sum_{n=1}^{\infty} \frac{1}{n^2}$ is a p-series, $p = 2 > 1$, so it converges.

7. Since $\frac{1}{\sqrt{x}+1}$ is continuous, positive, and decreasing on $[0,\infty)$ we can apply the Integral Test.

$$\int_1^{\infty} \frac{1}{\sqrt{x}+1}\,dx = \lim_{t\to\infty}\left[2\sqrt{x} - 2\ln(\sqrt{x}+1)\right]_1^t \quad \left[\begin{array}{c}\text{using the substitution } u = \sqrt{x}+1, \text{ so} \\ x = (u-1)^2 \text{ and } dx = 2(u-1)du\end{array}\right]$$
$$= \lim_{t\to\infty}\left(\left[2\sqrt{t} - 2\ln\left(\sqrt{t}+1\right)\right] - (2 - 2\ln 2)\right).$$

Now $2\sqrt{t} - 2\ln\left(\sqrt{t}+1\right) = 2\ln\left(\frac{e^{\sqrt{t}}}{\sqrt{t}+1}\right)$ and so $\lim_{t\to\infty}\left[2\sqrt{t} - 2\ln\left(\sqrt{t}+1\right)\right] = \infty$ (using l'Hospital's Rule) so both the integral and the original series diverge.

9. $f(x) = xe^{-x^2}$ is continuous and positive on $[1,\infty)$, and since $f'(x) = e^{-x^2}(1-2x^2) < 0$ for $x > 1$, f is decreasing as well. We can use the Integral Test. $\int_1^{\infty} xe^{-x^2}\,dx = \lim_{t\to\infty}\left[-\frac{1}{2}e^{-x^2}\right]_1^t = 0 - \left(-\frac{e^{-1}}{2}\right) = \frac{1}{2e}$ so the series converges.

11. $f(x) = \frac{x}{x^2+1}$ is continuous and positive on $[1,\infty)$, and since $f'(x) = \frac{1-x^2}{(x^2+1)^2} < 0$ for $x > 1$, f is also decreasing. Using the Integral Test, $\int_1^{\infty} \frac{x}{x^2+1}\,dx = \lim_{t\to\infty}\left[\frac{\ln(x^2+1)}{2}\right]_1^t = \infty$, so the series diverges.

13. $f(x) = \dfrac{1}{x \ln x}$ is continuous and positive on $[2, \infty)$, and also decreasing since $f'(x) = -\dfrac{1 + \ln x}{x^2 (\ln x)^2} < 0$ for $x > 2$, so we can use the Integral Test. $\displaystyle\int_2^\infty \frac{1}{x \ln x}\,dx = \lim_{t\to\infty} [\ln(\ln x)]_2^t = \lim_{t\to\infty} [\ln(\ln t) - \ln(\ln 2)] = \infty$, so the series diverges.

15. $f(x) = \dfrac{\arctan x}{1 + x^2}$ is continuous and positive on $[1, \infty)$. $f'(x) = \dfrac{1 - 2x \arctan x}{(1 + x^2)^2} < 0$ for $x > 1$, since $2x \arctan x \geq \frac{\pi}{2} > 1$ for $x \geq 1$. So f is decreasing and we can use the Integral Test.

$$\int_1^\infty \frac{\arctan x}{1 + x^2}\,dx = \lim_{t\to\infty} \left[\tfrac{1}{2} \arctan^2 x\right]_1^t = \frac{(\pi/2)^2}{2} - \frac{(\pi/4)^2}{2} = \frac{3\pi^2}{32},$$ so the series converges.

17. $f(x) = \dfrac{1}{x^2 + 2x + 2}$ is continuous and positive on $[1, \infty)$, and $f'(x) = -\dfrac{2x + 2}{(x^2 + 2x + 2)^2} < 0$ for $x \geq 1$, so f is decreasing and we can use the Integral Test. $\displaystyle\int_1^\infty \frac{1}{x^2 + 2x + 2}\,dx = \int_1^\infty \frac{1}{(x+1)^2 + 1}\,dx$

$= \displaystyle\lim_{t\to\infty} [\arctan(x + 1)]_1^t = \frac{\pi}{2} - \arctan 2$, so the series converges also.

19. We have already shown that when $p = 1$ the series diverges (in Exercise 13 above), so assume $p \neq 1$.

$f(x) = \dfrac{1}{x(\ln x)^p}$ is continuous and positive on $[2, \infty)$, and $f'(x) = -\dfrac{p + \ln x}{x^2 (\ln x)^{p+1}} < 0$ if $x > e^{-p}$, so that f is eventually decreasing and we can use the Integral Test.

$$\int_2^\infty \frac{1}{x(\ln x)^p}\,dx = \lim_{t\to\infty} \left[\frac{(\ln x)^{1-p}}{1 - p}\right]_2^t \quad (\text{for } p \neq 1) \quad = \lim_{t\to\infty} \left[\frac{(\ln t)^{1-p}}{1 - p}\right] - \frac{(\ln 2)^{1-p}}{1 - p}.$$

This limit exists whenever $1 - p < 0 \quad \Leftrightarrow \quad p > 1$, so the series converges for $p > 1$.

21. Clearly the series cannot converge if $p \geq -\frac{1}{2}$, because then $\lim\limits_{n\to\infty} n(1 + n^2)^p \neq 0$. Also, if $p = -1$ the series diverges (see Exercise 11 above.) So assume $p < -\frac{1}{2}$, $p \neq -1$. Then $f(x) = x(1 + x^2)^p$ is continuous, positive, and eventually decreasing on $[1, \infty)$, and we can use the Integral Test.

$$\int_1^\infty x(1 + x^2)^p\,dx = \lim_{t\to\infty} \left[\frac{1}{2} \cdot \frac{(1 + x^2)^{p+1}}{p + 1}\right]_1^t = \lim_{t\to\infty} \frac{1}{2} \cdot \frac{(1 + t^2)^{p+1}}{p + 1} - \frac{2^p}{p + 1}.$$ This limit exists and is finite $\Leftrightarrow \ p + 1 < 0 \ \Leftrightarrow \ p < -1$, so the series converges whenever $p < -1$.

23. Since this is a p-series with $p = x$, $\zeta(x)$ is defined when $x > 1$.

25. **(a)** $f(x) = \dfrac{1}{x^2}$ is positive and continuous and $f'(x) = \dfrac{-2}{x^3}$ is negative for $x > 1$, and so the Integral Test applies. $\displaystyle\sum_{n=1}^\infty \frac{1}{n^2} \approx s_{10} = \frac{1}{1^2} + \frac{1}{2^2} + \frac{1}{3^2} + \cdots + \frac{1}{10^2} \approx 1.54977.$

$$R_{10} \leq \int_{10}^\infty \frac{1}{x^2}\,dx = \lim_{t\to\infty} \left[\frac{-1}{x}\right]_{10}^t = \lim_{t\to\infty} \left(-\frac{1}{t} + \frac{1}{10}\right) = \frac{1}{10},$$ so the error is at most 0.1.

(b) $\displaystyle s_{10} + \int_{11}^\infty \frac{1}{x^2}\,dx \leq s \leq s_{10} + \int_{10}^\infty \frac{1}{x^2}\,dx \quad \Rightarrow \quad s_{10} + \frac{1}{11} \leq s \leq s_{10} + \frac{1}{10} \quad \Rightarrow$

$1.549768 + 0.090909 = 1.640677 \leq s \leq 1.549768 + 0.1 = 1.649768$, so we get $s \approx 1.64522$ with error ≤ 0.005.

(c) $R_n \leq \int_n^\infty (1/x^2)\,dx = 1/n$. So $R_n < 0.001$ for $n > 1000$.

27. $f(x) = x^{-3/2}$ is positive and continuous and $f'(x) = -\frac{3}{2}x^{-5/2}$ is negative for $x > 1$, so the Integral Test applies. Using (5), we need n such that

$$0.01 > \frac{1}{2}\left(\int_n^{\infty} x^{-3/2}\,dx - \int_{n+1}^{\infty} x^{-3/2}\,dx\right) = \frac{1}{2}\left(\lim_{t\to\infty}\left[\frac{-2}{\sqrt{x}}\right]_n^t - \lim_{t\to\infty}\left[\frac{-2}{\sqrt{x}}\right]_{n+1}^t\right) = \frac{1}{\sqrt{n}} - \frac{1}{\sqrt{n+1}} \quad \Leftrightarrow$$

$n > 13$. Then, again from (5),
$s \approx s_{14} + \frac{1}{2}\left(\int_{14}^{\infty} x^{-3/2}\,dx + \int_{15}^{\infty} x^{-3/2}\,dx\right) = 2.0872 + \frac{1}{\sqrt{14}} + \frac{1}{\sqrt{15}} \approx 2.6127$. Any larger value of n will also work. For instance, $s \approx s_{30} + \frac{1}{\sqrt{30}} + \frac{1}{\sqrt{31}} \approx 2.6124$.

29. **(a)** From (1), with $f(x) = \frac{1}{x}$, $\frac{1}{2} + \frac{1}{3} + \frac{1}{4} + \cdots + \frac{1}{n} \leq \int_1^n \frac{1}{x}\,dx = \ln n$, so

$$s_n = 1 + \frac{1}{2} + \frac{1}{3} + \frac{1}{4} + \cdots + \frac{1}{n} \leq 1 + \ln n.$$

(b) By part (a), $s_{10^6} \leq 1 + \ln 10^6 \approx 14.82 < 15$ and $s_{10^9} \leq 1 + \ln 10^9 \approx 21.72 < 22$.

31. $b^{\ln n} = e^{\ln b \ln n} = n^{\ln b} = \dfrac{1}{n^{-\ln b}}$. This is a p-series, which converges for all b such that $-\ln b > 1 \quad \Leftrightarrow$ $\ln b < -1$, so for $b < 1/e$.

EXERCISES 10.4

1. $\dfrac{1}{n^3 + n^2} < \dfrac{1}{n^3}$ since $n^3 + n^2 > n^3$ for all n, and since $\displaystyle\sum_{n=1}^{\infty} \frac{1}{n^3}$ is a convergent p-series ($p = 3 > 1$), $\displaystyle\sum_{n=1}^{\infty} \frac{1}{n^3 + n^2}$ converges also by the Comparison Test [part (a).]

3. $\dfrac{3}{n2^n} \leq \dfrac{3}{2^n}$. $\displaystyle\sum_{n=1}^{\infty} \frac{3}{2^n}$ is a geometric series with $|r| = \frac{1}{2} < 1$, and hence converges, so $\displaystyle\sum_{n=1}^{\infty} \frac{3}{n2^n}$ converges also, by the Comparison Test.

5. $\dfrac{1 + 5^n}{4^n} > \dfrac{5^n}{4^n} = \left(\dfrac{5}{4}\right)^n$. $\displaystyle\sum_{n=0}^{\infty} \left(\frac{5}{4}\right)^n$ is a divergent geometric series ($|r| = \frac{5}{4} > 1$) so $\displaystyle\sum_{n=0}^{\infty} \frac{1 + 5^n}{4^n}$ diverges by the Comparison Test.

7. $\dfrac{3}{n(n+3)} < \dfrac{3}{n^2}$. $\displaystyle\sum_{n=1}^{\infty} \frac{3}{n^2} = 3\sum_{n=1}^{\infty} \frac{1}{n^2}$ is a convergent p-series ($p = 2 > 1$) so $\displaystyle\sum_{n=1}^{\infty} \frac{3}{n(n+3)}$ converges by the Comparison Test.

9. $\dfrac{\sqrt{n}}{n-1} > \dfrac{\sqrt{n}}{n} = \dfrac{1}{n^{1/2}}$. $\displaystyle\sum_{n=2}^{\infty} \frac{1}{n^{1/2}}$ is a divergent p-series ($p = \frac{1}{2} < 1$) so $\displaystyle\sum_{n=2}^{\infty} \frac{\sqrt{n}}{n-1}$ diverges by the Comparison Test.

11. $n^3+1>n^3 \quad\Rightarrow\quad \dfrac{1}{n^3+1}<\dfrac{1}{n^3} \quad\Rightarrow\quad \dfrac{n}{n^3+1}<\dfrac{n}{n^3} \quad\Rightarrow\quad \dfrac{n-1}{n^3+1}<\dfrac{n}{n^3}=\dfrac{1}{n^2}$. Now $\sum_{n=1}^{\infty}\dfrac{1}{n^2}$ is a convergent p-series ($p=2>1$) so $\sum_{n=1}^{\infty}\dfrac{n-1}{n^3+1}$ converges by the Comparison Test.

13. $\dfrac{3+\cos n}{3^n}\le\dfrac{4}{3^n}$ since $\cos n\le 1$. $\sum_{n=1}^{\infty}\dfrac{4}{3^n}$ is a geometric series with $|r|=\frac{1}{3}<1$ so it converges, and so $\sum_{n=1}^{\infty}\dfrac{3+\cos n}{3^n}$ converges by the Comparison Test.

15. $\dfrac{n}{\sqrt{n^5+4}}<\dfrac{n}{\sqrt{n^5}}=\dfrac{1}{n^{3/2}}$. $\sum_{n=1}^{\infty}\dfrac{1}{n^{3/2}}$ is a convergent p-series ($p=\frac{3}{2}>1$) so $\sum_{n=1}^{\infty}\dfrac{n}{\sqrt{n^5+4}}$ converges by the Comparison Test.

17. $\dfrac{2^n}{1+3^n}<\dfrac{2^n}{3^n}=\left(\dfrac{2}{3}\right)^n$. $\sum_{n=1}^{\infty}\left(\dfrac{2}{3}\right)^n$ is a convergent geometric series ($|r|=\frac{2}{3}<1$), so $\sum_{n=1}^{\infty}\dfrac{2^n}{1+3^n}$ converges by the Comparison Test.

19. Let $a_n=\dfrac{1}{1+\sqrt{n}}$ and $b_n=\dfrac{1}{\sqrt{n}}$. Then $\lim\limits_{n\to\infty}\dfrac{a_n}{b_n}=\lim\limits_{n\to\infty}\dfrac{\sqrt{n}}{1+\sqrt{n}}=1>0$. Since $\sum_{n=1}^{\infty}\dfrac{1}{\sqrt{n}}$ is a divergent p-series ($p=\frac{1}{2}<1$), $\sum_{n=1}^{\infty}\dfrac{1}{1+\sqrt{n}}$ also diverges by the Limit Comparison Test.

21. Let $a_n=\dfrac{n^2+1}{n^4+1}$ and $b_n=\dfrac{1}{n^2}$. Then $\lim\limits_{n\to\infty}\dfrac{a_n}{b_n}=\lim\limits_{n\to\infty}\dfrac{n^4+n^2}{n^4+1}=1$. Since $\sum_{n=1}^{\infty}\dfrac{1}{n^2}$ is a convergent p-series ($p=2>1$), so is $\sum_{n=1}^{\infty}\dfrac{n^2+1}{n^4+1}$ by the Limit Comparison Test.

23. Let $a_n=\dfrac{n^2-n+2}{\sqrt[4]{n^{10}+n^5+3}}$ and $b_n=\dfrac{1}{\sqrt{n}}$. Then

$\lim\limits_{n\to\infty}\dfrac{a_n}{b_n}=\lim\limits_{n\to\infty}\dfrac{n^{5/2}-n^{3/2}+2n^{1/2}}{\sqrt[4]{n^{10}+n^5+3}}=\lim\limits_{n\to\infty}\dfrac{1-n^{-1}+2n^{-2}}{\sqrt[4]{1+n^{-5}+3n^{-10}}}=1$. Since $\sum_{n=1}^{\infty}\dfrac{1}{\sqrt{n}}$ is a divergent p-series ($p=\frac{1}{2}<1$), $\sum_{n=1}^{\infty}\dfrac{n^2-n+2}{\sqrt[4]{n^{10}+n^5+3}}$ diverges by the Limit Comparison Test.

25. Let $a_n=\dfrac{n+1}{n2^n}$ and $b_n=\dfrac{1}{2^n}$. Then $\lim\limits_{n\to\infty}\dfrac{a_n}{b_n}=\lim\limits_{n\to\infty}\dfrac{n+1}{n}=1$. Since $\sum_{n=1}^{\infty}\dfrac{1}{2^n}$ is a convergent geometric series ($|r|=\frac{1}{2}<1$), $\sum_{n=1}^{\infty}\dfrac{n+1}{n2^n}$ converges by the Limit Comparison Test.

27. Let $a_n=\dfrac{\ln n}{n^3}$ and $b_n=\dfrac{1}{n^2}$. Then $\lim\limits_{n\to\infty}\dfrac{a_n}{b_n}=\lim\limits_{n\to\infty}\dfrac{\ln n}{n}=\lim\limits_{n\to\infty}\dfrac{1/n}{1}=0$. So since $\sum_{n=1}^{\infty}\dfrac{1}{n^2}$ converges (p-series, $p=2>1$), so does $\sum_{n=1}^{\infty}\dfrac{\ln n}{n^3}$ by part (b) of the Limit Comparison Test.

29. Clearly $n! = n(n-1)(n-2)\cdots(3)(2) \geq 2 \cdot 2 \cdot 2 \cdot \cdots \cdot 2 \cdot 2 = 2^{n-1}$, so $\dfrac{1}{n!} \leq \dfrac{1}{2^{n-1}}$. $\displaystyle\sum_{n=1}^{\infty} \frac{1}{2^{n-1}}$ is a convergent geometric series ($|r| = \frac{1}{2} < 1$) so $\displaystyle\sum_{n=1}^{\infty} \frac{1}{n!}$ converges by the Comparison Test.

31. Let $a_n = \sin\left(\dfrac{1}{n}\right)$ and $b_n = \dfrac{1}{n}$. Then $\displaystyle\lim_{n\to\infty} \frac{a_n}{b_n} = \lim_{n\to\infty} \frac{\sin(1/n)}{1/n} = \lim_{\theta\to 0} \frac{\sin\theta}{\theta} = 1$, so since $\displaystyle\sum_{n=1}^{\infty} b_n$ is the harmonic series (which diverges), $\displaystyle\sum_{n=1}^{\infty} \sin\left(\frac{1}{n}\right)$ diverges as well by the Limit Comparison Test.

33. $\displaystyle\sum_{n=1}^{10} \frac{1}{n^4+n^2} = \frac{1}{2} + \frac{1}{20} + \frac{1}{90} + \cdots + \frac{1}{10{,}100} \approx 0.567975$. Now $\dfrac{1}{n^4+n^2} < \dfrac{1}{n^4}$, so using the reasoning and notation of Example 7, the error is $\displaystyle R_{10} \leq T_{10} = \sum_{n=11}^{\infty} \frac{1}{n^4} \leq \int_{10}^{\infty} \frac{dx}{x^4} = \lim_{t\to\infty}\left[-\frac{x^{-3}}{3}\right]_{10}^{t} = \frac{1}{3000} = 0.000\overline{3}$.

35. $\displaystyle\sum_{n=1}^{10} \frac{1}{1+2^n} = \frac{1}{3} + \frac{1}{5} + \frac{1}{9} + \cdots + \frac{1}{1025} \approx 0.76352$. Now $\dfrac{1}{1+2^n} < \dfrac{1}{2^n}$, so the error is

$\displaystyle R_{10} \leq T_{10} = \sum_{n=11}^{\infty} \frac{1}{2^n} = \frac{1/2^{11}}{1-1/2}$ (geometric series) ≈ 0.00098.

37. Since $\dfrac{d_n}{10^n} \leq \dfrac{9}{10^n}$ for each n, and since $\displaystyle\sum_{n=1}^{\infty} \frac{9}{10^n}$ is a convergent geometric series ($|r| = \frac{1}{10} < 1$),

$\displaystyle 0.d_1d_2d_3\ldots = \sum_{n=1}^{\infty} \frac{d_n}{10^n}$ will always converge by the Comparison Test.

39. Since $\sum a_n$ converges, $\displaystyle\lim_{n\to\infty} a_n = 0$, so there exists N such that $|a_n - 0| < 1$ for all $n > N$ $\Rightarrow$ $0 \leq a_n < 1$ for all $n > N$ $\Rightarrow$ $0 \leq a_n^2 \leq a_n$. Since $\sum a_n$ converges, so does $\sum a_n^2$ by the Comparison Test.

41. We wish to prove that if $\displaystyle\lim_{n\to\infty} \frac{a_n}{b_n} = \infty$ and $\sum b_n$ diverges, then so does $\sum a_n$. So suppose on the contrary that $\sum a_n$ converges. Since $\displaystyle\lim_{n\to\infty} \frac{a_n}{b_n} = \infty$, we have that $\displaystyle\lim_{n\to\infty} \frac{b_n}{a_n} = 0$, so by part (b) of the Limit Comparison Test (proved in Exercise 40), if $\sum a_n$ converges, so must $\sum b_n$. But this contradicts our hypothesis, so $\sum a_n$ must diverge.

43. $\displaystyle\lim_{n\to\infty} na_n = \lim_{n\to\infty} \frac{a_n}{1/n}$, so we apply the Limit Comparison Test with $b_n = \dfrac{1}{n}$. Since $\displaystyle\lim_{n\to\infty} na_n > 0$ we know that either both series converge or both series diverge, and we also know that $\displaystyle\sum_{n=0}^{\infty} \frac{1}{n}$ diverges (p-series with $p = 1$). Therefore $\sum a_n$ must be divergent.

45. Yes. Since $\sum a_n$ converges, its terms approach 0 as $n \to \infty$, so $\displaystyle\lim_{n\to\infty} \frac{\sin a_n}{a_n} = 1$ by Theorem 2.4.4. Thus $\sum \sin a_n$ converges by the Limit Comparison Test.

EXERCISES 10.5

1. $\sum_{n=1}^{\infty}(-1)^{n-1}\frac{3}{n+4}$. $b_n = \frac{3}{n+4} > 0$ and $b_{n+1} < b_n$ for all n; $\lim_{n\to\infty} b_n = 0$ so the series converges by the Alternating Series Test.

3. $\sum_{n=1}^{\infty}(-1)^{n}\frac{n}{n+1}$. $\lim_{n\to\infty}\frac{n}{n+1} = 1$ so $\lim_{n\to\infty}(-1)^n\frac{n}{n+1}$ does not exist and the series diverges by the Test for Divergence.

5. $\sum_{n=1}^{\infty}(-1)^{n-1}\frac{1}{n^2}$. $b_n = \frac{1}{n^2} > 0$ and $b_{n+1} < b_n$ for all n, and $\lim_{n\to\infty}\frac{1}{n^2} = 0$, so the series converges by the Alternating Series Test.

7. $\sum_{n=1}^{\infty}(-1)^{n+1}\frac{n}{5n+1}$. $\lim_{n\to\infty}\frac{n}{5n+1} = \frac{1}{5}$ so $\lim_{n\to\infty}(-1)^{n+1}\frac{n}{5n+1}$ does not exist and the series diverges by the Test for Divergence.

9. $\sum_{n=1}^{\infty}(-1)^{n}\frac{n}{n^2+1}$. $b_n = \frac{n}{n^2+1} > 0$ for all n. $b_{n+1} < b_n \quad\Leftrightarrow\quad \frac{n+1}{(n+1)^2+1} < \frac{n}{n^2+1} \quad\Leftrightarrow$
$(n+1)(n^2+1) < \left[(n+1)^2+1\right]n \quad\Leftrightarrow\quad n^3+n^2+n+1 < n^3+2n^2+2n \quad\Leftrightarrow\quad 0 < n^2+n-1$, which is true for all $n \geq 1$. Also $\lim_{n\to\infty}\frac{n}{n^2+1} = \lim_{n\to\infty}\frac{1/n}{1+1/n^2} = 0$. Therefore the series converges by the Alternating Series Test.

11. $\sum_{n=1}^{\infty}(-1)^{n-1}\frac{\sqrt{n}}{n+4}$. $b_n = \frac{\sqrt{n}}{n+4} > 0$ for all n. Let $f(x) = \frac{\sqrt{x}}{x+4}$. Then $f'(x) = \frac{4-x}{2\sqrt{x}(x+4)^2} < 0$ if $x > 4$, so $\{b_n\}$ is decreasing after $n = 4$. $\lim_{n\to\infty}\frac{\sqrt{n}}{n+4} = \lim_{n\to\infty}\frac{1}{\sqrt{n}+4/\sqrt{n}} = 0$. So the series converges by the Alternating Series Test.

13. $\sum_{n=2}^{\infty}(-1)^{n}\frac{n}{\ln n}$. $\lim_{n\to\infty}\frac{n}{\ln n} = \lim_{n\to\infty}\frac{1}{1/n} = \infty$ so the series diverges by the Test for Divergence.

15. $\sum_{n=1}^{\infty}\frac{\cos n\pi}{n^{3/4}} = \sum_{n=1}^{\infty}\frac{(-1)^n}{n^{3/4}}$. $b_n = \frac{1}{n^{3/4}}$ is decreasing and positive, and $\lim_{n\to\infty}\frac{1}{n^{3/4}} = 0$ so the series converges by the Alternating Series Test.

17. $\sum_{n=1}^{\infty}(-1)^{n}\sin\left(\frac{\pi}{n}\right)$. $b_n = \sin\left(\frac{\pi}{n}\right) > 0$ for $n \geq 2$ and $\sin\left(\frac{\pi}{n}\right) \geq \sin\left(\frac{\pi}{n+1}\right)$, and $\lim_{n\to\infty}\sin\left(\frac{\pi}{n}\right) = \sin 0 = 0$, so the series converges by the Alternating Series Test.

19. $\frac{n^n}{n!} = \frac{n\cdot n\cdot\cdots\cdot n}{1\cdot 2\cdot\cdots\cdot n} \geq n \quad\Rightarrow\quad \lim_{n\to\infty}\frac{n^n}{n!} = \infty \quad\Rightarrow\quad \lim_{n\to\infty}\frac{(-1)^n n^n}{n!}$ does not exist. So the series diverges by the Test for Divergence.

21. Let $\sum b_n$ be the series for which $b_n = 0$ if n is odd and $b_n = 1/n^2$ if n is even. Then $\sum b_n = \sum 1/(2n)^2$ clearly converges (by comparison with the p-series for $p = 2$). So suppose that $\sum(-1)^{n-1}b_n$ converges. Then by Theorem 10.2.8(b), so does $\sum\left[(-1)^{n-1}b_n + b_n\right] = 1 + \frac{1}{3} + \frac{1}{5} + \cdots = \sum \frac{1}{2n-1}$. But this diverges by comparison with the harmonic series, a contradiction. Therefore $\sum(-1)^{n-1}b_n$ must diverge. The Alternating Series Test does not apply since $\{b_n\}$ is not decreasing.

23. Clearly $b_n = \dfrac{1}{n+p}$ is decreasing and eventually positive and $\lim_{n\to\infty} b_n = 0$ for any p. So the series will converge (by the Alternating Series Test) for any p for which every b_n is defined, that is, $n + p \neq 0$ for $n \geq 1$, or p is not a negative integer.

25. If $b_n = \dfrac{1}{n^2}$, then $b_{11} = \dfrac{1}{121} < 0.01$, so by Theorem 1, $\displaystyle\sum_{n=1}^{\infty}\frac{1}{n^2} \approx \sum_{n=1}^{10}\frac{1}{n^2} \approx 0.82$.

27. $\displaystyle\sum_{n=0}^{\infty}(-1)^n\frac{2^n}{n!}$. Since $\dfrac{2}{n} < \dfrac{2}{3}$ for $n \geq 4$, $0 < \dfrac{2^n}{n!} < \dfrac{2}{1}\cdot\dfrac{2}{2}\cdot\dfrac{2}{3}\cdot\left(\dfrac{2}{3}\right)^{n-3} \to 0$ as $n \to \infty$, so by the Squeeze Theorem, $\lim_{n\to\infty}\dfrac{2^n}{n!} = 0$, and hence $\displaystyle\sum_{n=0}^{\infty}(-1)^n\frac{2^n}{n!}$ is a convergent alternating series. $\dfrac{2^8}{8!} = \dfrac{256}{40{,}320} < 0.01$, so $\sum_{n=0}^{\infty}(-1)^n\frac{2^n}{n!} \approx \sum_{n=0}^{7}(-1)^n\frac{2^n}{n!} \approx 0.13$.

29. $\displaystyle\sum_{n=1}^{\infty}\frac{(-1)^{n-1}}{(2n-1)!}$. $b_5 = \dfrac{1}{(2\cdot 5-1)!} = \dfrac{1}{362{,}880} < 0.00001$, so $\displaystyle\sum_{n=1}^{\infty}\frac{(-1)^{n-1}}{(2n-1)!} \approx \sum_{n=1}^{4}\frac{(-1)^{n-1}}{(2n-1)!} \approx 0.8415$.

31. $\displaystyle\sum_{n=0}^{\infty}\frac{(-1)^n}{2^n n!}$. $b_6 = \dfrac{1}{2^6 6!} = \dfrac{1}{46{,}080} < 0.000022$, so $\displaystyle\sum_{n=0}^{\infty}\frac{(-1)^n}{2^n n!} \approx \sum_{n=0}^{5}\frac{(-1)^n}{2^n n!} \approx 0.6065$.

33. $\displaystyle\sum_{n=1}^{\infty}\frac{(-1)^{n-1}}{n} = 1 - \frac{1}{2} + \frac{1}{3} - \frac{1}{4} + \cdots + \frac{1}{49} - \frac{1}{50} + \frac{1}{51} - \frac{1}{52} + \cdots$. The 50th partial sum of this series is an underestimate, since $\displaystyle\sum_{n=1}^{\infty}\frac{(-1)^{n-1}}{n} = s_{50} + \left(\frac{1}{51} - \frac{1}{52}\right) + \left(\frac{1}{53} - \frac{1}{54}\right) + \cdots$, and the terms in parentheses are all positive. The result can be seen geometrically in Figure 1.

35. **(a)** We will prove this by induction. Let $P(n)$ be the proposition that $s_{2n} = h_{2n} - h_n$. $P(1)$ is true by an easy calculation. So suppose that $P(n)$ is true. We will show that $P(n+1)$ must be true as a consequence.

$$h_{2n+2} - h_{n+1} = \left(h_{2n} + \frac{1}{2n+1} + \frac{1}{2n+2}\right) - \left(h_n + \frac{1}{n+1}\right)$$
$$= (h_{2n} - h_n) + \frac{1}{2n+1} - \frac{1}{2n+2} = s_{2n} + \frac{1}{2n+1} - \frac{1}{2n+2} = s_{2n+2},$$

which is $P(n+1)$, and proves that $s_{2n} = h_{2n} - h_n$ for all n.

(b) We know that $h_{2n} - \ln 2n \to \gamma$ and $h_n - \ln n \to \gamma$ as $n \to \infty$. So $s_{2n} = h_{2n} - h_n = (h_{2n} - \ln 2n) - (h_n - \ln n) + (\ln 2n - \ln n)$, and $\lim_{n\to\infty} s_{2n} = \gamma - \gamma + \lim_{n\to\infty}[\ln 2n - \ln n] = \lim_{n\to\infty}(\ln 2 + \ln n - \ln n) = \ln 2$.

EXERCISES 10.6

1. $\displaystyle\sum_{n=1}^{\infty}\frac{1}{n\sqrt{n}}=\sum_{n=1}^{\infty}\frac{1}{n^{3/2}}$ is a convergent p-series ($p=\frac{3}{2}>1$), so the given series is absolutely convergent.

3. $\displaystyle\lim_{n\to\infty}\left|\frac{a_{n+1}}{a_n}\right|=\lim_{n\to\infty}\left|\frac{(-3)^{n+1}/(n+1)^3}{(-3)^n/n^3}\right|=3\lim_{n\to\infty}\left(\frac{n}{n+1}\right)^3=3>1$, so the series diverges by the Ratio Test.

5. $\displaystyle\sum_{n=1}^{\infty}\frac{1}{2n+1}$ diverges (use the Integral Test or the Limit Comparison Test with $b_n=1/n$), but since

$\displaystyle\lim_{n\to\infty}\frac{1}{2n+1}=0$, $\displaystyle\sum_{n=1}^{\infty}\frac{(-1)^{n+1}}{2n+1}$ converges by the Alternating Series Test, and so is conditionally convergent.

7. $\displaystyle\lim_{n\to\infty}\left|\frac{a_{n+1}}{a_n}\right|=\lim_{n\to\infty}\frac{1/(2n+1)!}{1/(2n-1)!}=\lim_{n\to\infty}\frac{1}{(2n+1)2n}=0$, so by the Ratio Test the series is absolutely convergent.

9. $\displaystyle\sum_{n=1}^{\infty}\frac{n}{n^2+4}$ diverges (use the Limit Comparison Test with $b_n=1/n$). But since $\displaystyle 0\le\frac{n+1}{(n+1)^2+4}<\frac{n}{n^2+4}$

$\Leftrightarrow\quad n^3+n^2+4n+4<n^3+2n^2+5n\quad\Leftrightarrow\quad 0<n^2+n-4$ (which is true for $n\ge 2$), and since

$\displaystyle\lim_{n\to\infty}\frac{n}{n^2+4}=0$, $\displaystyle\sum_{n=1}^{\infty}(-1)^n\frac{n}{n^2+4}$ converges (conditionally) by the Alternating Series Test.

11. $\displaystyle\lim_{n\to\infty}\frac{2n}{3n-4}=\frac{2}{3}$, so $\displaystyle\sum_{n=1}^{\infty}(-1)^n\frac{2n}{3n-4}$ diverges by the Test for Divergence.

13. $\displaystyle\left|\frac{\sin 2n}{n^2}\right|\le\frac{1}{n^2}$ and $\displaystyle\sum_{n=1}^{\infty}\frac{1}{n^2}$ converges (p-series, $p=2>1$), so $\displaystyle\sum_{n=1}^{\infty}\frac{\sin 2n}{n^2}$ converges absolutely by the Comparison Test.

15. $\displaystyle\lim_{n\to\infty}\left|\frac{a_{n+1}}{a_n}\right|=\lim_{n\to\infty}\left|\frac{2^{n+1}/\left[(n+1)3^{n+2}\right]}{2^n/(n3^{n+1})}\right|=\frac{2}{3}\lim_{n\to\infty}\frac{n}{n+1}=\frac{2}{3}<1$ so the series converges absolutely by the Ratio Test.

17. $\displaystyle\lim_{n\to\infty}\left|\frac{a_{n+1}}{a_n}\right|=\lim_{n\to\infty}\frac{(n+2)5^{n+1}/\left[(n+1)3^{2(n+1)}\right]}{(n+1)5^n/(n3^{2n})}=\lim_{n\to\infty}\frac{5n(n+2)}{9(n+1)^2}=\frac{5}{9}<1$ so the series converges absolutely by the Ratio Test.

19. $\displaystyle\lim_{n\to\infty}\left|\frac{a_{n+1}}{a_n}\right|=\lim_{n\to\infty}\frac{(n+1)!/10^{n+1}}{n!/10^n}=\lim_{n\to\infty}\frac{n+1}{10}=\infty$, so the series diverges by the Ratio Test.

21. $\displaystyle\frac{|\cos(n\pi/3)|}{n!}\le\frac{1}{n!}$ and $\displaystyle\sum_{n=1}^{\infty}\frac{1}{n!}$ converges (Exercise 10.4.29), so the given series converges absolutely by the Comparison Test.

23. $\displaystyle\lim_{n\to\infty}\left|\frac{a_{n+1}}{a_n}\right|=\lim_{n\to\infty}\frac{(n+1)^{n+1}/5^{2n+5}}{n^n/5^{2n+3}}=\lim_{n\to\infty}\frac{1}{25}\left(\frac{n+1}{n}\right)^n(n+1)=\infty$, so the series diverges by the Ratio Test.

25. $\lim\limits_{n\to\infty} \sqrt[n]{|a_n|} = \lim\limits_{n\to\infty} \left|\dfrac{1-3n}{3+4n}\right| = \frac{3}{4} < 1$, so the series converges absolutely by the Root Test.

27. $\lim\limits_{n\to\infty} \left|\dfrac{a_{n+1}}{a_n}\right| = \lim\limits_{n\to\infty} \dfrac{(n+1)!/[1\cdot 3\cdot 5\cdot\cdots\cdot(2n+1)]}{n!/[1\cdot 3\cdot 5\cdot\cdots\cdot(2n-1)]} = \lim\limits_{n\to\infty} \dfrac{n+1}{2n+1} = \frac{1}{2} < 1$, so the series converges absolutely by the Ratio Test.

29. $\sum\limits_{n=1}^{\infty} \dfrac{2\cdot 4\cdot 6\cdot\cdots\cdot(2n)}{n!} = \sum\limits_{n=1}^{\infty} \dfrac{2^n n!}{n!} = \sum\limits_{n=1}^{\infty} 2^n$ which diverges by the Test for Divergence since $\lim\limits_{n\to\infty} 2^n = \infty$.

31. $\lim\limits_{n\to\infty} \left|\dfrac{a_{n+1}}{a_n}\right| = \lim\limits_{n\to\infty} \dfrac{(n+3)!/[(n+1)!\,10^{n+1}]}{(n+2)!/(n!\,10^n)} = \frac{1}{10}\lim\limits_{n\to\infty} \dfrac{n+3}{n+1} = \frac{1}{10} < 1$ so the series converges absolutely by the Ratio Test.

33. By the recursive definition, $\lim\limits_{n\to\infty} \left|\dfrac{a_{n+1}}{a_n}\right| = \lim\limits_{n\to\infty} \left|\dfrac{5n+1}{4n+3}\right| = \frac{5}{4} > 1$, so the series diverges by the Ratio Test.

35. **(a)** $\lim\limits_{n\to\infty} \left|\dfrac{1/(n+1)^3}{1/n^3}\right| = \lim\limits_{n\to\infty} \dfrac{n^3}{(n+1)^3} = \lim\limits_{n\to\infty} \dfrac{1}{(1+1/n)^3} = 1$. So inconclusive.

(b) $\lim\limits_{n\to\infty} \left|\dfrac{(n+1)}{2^{n+1}}\cdot\dfrac{2^n}{n}\right| = \lim\limits_{n\to\infty} \dfrac{n+1}{2n} = \lim\limits_{n\to\infty} \left(\dfrac{1}{2} + \dfrac{1}{2n}\right) = \dfrac{1}{2}$. So conclusive (convergent).

(c) $\lim\limits_{n\to\infty} \left|\dfrac{(-3)^n}{\sqrt{n+1}}\cdot\dfrac{\sqrt{n}}{(-3)^{n-1}}\right| = 3\lim\limits_{n\to\infty} \sqrt{\dfrac{n}{n+1}} = 3\lim\limits_{n\to\infty} \sqrt{\dfrac{1}{1+1/n}} = 3$. So conclusive (divergent).

(d) $\lim\limits_{n\to\infty} \left|\dfrac{\sqrt{n+1}}{1+(n+1)^2}\cdot\dfrac{1+n^2}{\sqrt{n}}\right| = \lim\limits_{n\to\infty} \left[\sqrt{1+\dfrac{1}{n}}\cdot\dfrac{1/n^2+1}{1/n^2+(1+1/n)^2}\right] = 1$. So inconclusive.

37. **(a)** $\lim\limits_{n\to\infty} \left|\dfrac{a_{n+1}}{a_n}\right| = \lim\limits_{n\to\infty} \dfrac{|x|^{n+1}/(n+1)!}{|x|^n/n!} = |x|\lim\limits_{n\to\infty} \dfrac{1}{n+1} = 0$, so by the Ratio Test the series converges for all x.

(b) Since the series of part (a) always converges, we must have $\lim\limits_{n\to\infty} \dfrac{x^n}{n!} = 0$ by Theorem 10.2.6.

39. **(a)** $s_5 = \sum\limits_{n=1}^{5} \dfrac{1}{n2^n} = \dfrac{1}{2} + \dfrac{1}{8} + \dfrac{1}{24} + \dfrac{1}{64} + \dfrac{1}{160} = \dfrac{661}{960} \approx 0.68854$. Now the ratios

$r_n = \dfrac{a_{n+1}}{a_n} = \dfrac{n2^n}{(n+1)2^{n+1}} = \dfrac{n}{2(n+1)}$ form an increasing sequence, since

$r_{n+1} - r_n = \dfrac{n+1}{2(n+2)} - \dfrac{n}{2(n+1)} = \dfrac{(n+1)^2 - n(n+2)}{2(n+1)(n+2)} = \dfrac{1}{2(n+1)(n+2)} > 0$. So by Exercise 38(b), the error is less than $\dfrac{a_6}{1 - \lim\limits_{n\to\infty} r_n} = \dfrac{1/(6\cdot 2^6)}{1-1/2} = \dfrac{1}{192} \approx 0.00521$.

(b) The error in using s_n as an approximation to the sum is $R_n = \dfrac{a_{n+1}}{1/2} = \dfrac{2}{(n+1)2^{n+1}}$. We want $R_n < 0.00005 \quad\Leftrightarrow\quad \dfrac{1}{(n+1)2^n} < 0.00005 \quad\Leftrightarrow\quad (n+1)2^n > 20{,}000$. To find such an n we can use trial and error or a graph. We calculate $(11+1)2^{11} = 24{,}576$, so $s_{11} = \sum\limits_{n=1}^{11} \dfrac{1}{n2^n} \approx 0.693109$ is within 0.00005 of the actual sum.

41. By the Triangle Inequality (see Exercise 4.1.44) we have $\left|\sum_{i=1}^{n} a_i\right| \le \sum_{i=1}^{n} |a_i| \quad\Rightarrow\quad -\sum_{i=1}^{n}|a_i| \le \sum_{i=1}^{n} a_i \le \sum_{i=1}^{n}|a_i|$

$\Rightarrow\quad -\lim_{n\to\infty}\sum_{i=1}^{n}|a_i| \le \lim_{n\to\infty}\sum_{i=1}^{n} a_i \le \lim_{n\to\infty}\sum_{i=1}^{n}|a_i| \;\Rightarrow\; -\sum_{n=1}^{\infty}|a_n| \le \sum_{n=1}^{\infty} a_n \le \sum_{n=1}^{\infty}|a_n| \;\Rightarrow\; \left|\sum_{n=1}^{\infty} a_n\right| \le \sum_{n=1}^{\infty}|a_n|.$

43. **(a)** Since $\sum a_n$ is absolutely convergent, and since $|a_n^+| \le |a_n|$ and $|a_n^-| \le |a_n|$ (because a_n^+ and a_n^- each equal either a_n or 0), we conclude by the Comparison Test that both $\sum a_n^+$ and $\sum a_n^-$ must be absolutely convergent. (Or use Theorem 10.2.8.)

(b) We will show by contradiction that both $\sum a_n^+$ and $\sum a_n^-$ must diverge. For suppose that $\sum a_n^+$ converged. Then so would $\sum\left(a_n^+ - \frac{1}{2}a_n\right)$ by Theorem 10.2.8. But

$\sum\left(a_n^+ - \frac{1}{2}a_n\right) = \sum\left[\frac{1}{2}(a_n + |a_n|) - \frac{1}{2}a_n\right] = \frac{1}{2}\sum|a_n|$, which diverges because $\sum a_n$ is only conditionally convergent. Hence $\sum a_n^+$ can't converge. Similarly, neither can $\sum a_n^-$.

EXERCISES 10.7

1. Use the Comparison Test, with $a_n = \dfrac{\sqrt{n}}{n^2+1}$ and $b_n = \dfrac{1}{n^{3/2}}$: $\dfrac{\sqrt{n}}{n^2+1} < \dfrac{\sqrt{n}}{n^2} = \dfrac{1}{n^{3/2}}$, and $\sum_{n=1}^{\infty}\dfrac{1}{n^{3/2}}$ is a convergent p-series ($p = \frac{3}{2} > 1$), so $\sum_{n=1}^{\infty} a_n = \sum_{n=1}^{\infty}\dfrac{\sqrt{n}}{n^2+1}$ converges as well.

3. $\sum_{n=1}^{\infty}\dfrac{4^n}{3^{2n-1}} = 3\sum_{n=1}^{\infty}\left(\dfrac{4}{9}\right)^n$ which is a convergent geometric series ($|r| = \frac{4}{9} < 1$.)

5. The series converges by the Alternating Series Test, since $a_n = \dfrac{1}{(\ln n)^2}$ is decreasing ($\ln x$ is an increasing function) and $\lim_{n\to\infty} a_n = 0$.

7. $\sum_{k=1}^{\infty}\dfrac{1}{k^{1.7}}$ is a convergent p-series ($p = 1.7 > 1$).

9. $\lim_{n\to\infty}\left|\dfrac{a_{n+1}}{a_n}\right| = \lim_{n\to\infty}\dfrac{(n+1)/e^{n+1}}{n/e^n} = \dfrac{1}{e}\lim_{n\to\infty}\dfrac{n+1}{n} = \dfrac{1}{e} < 1$, so the series converges by the Ratio Test.

11. Use the Limit Comparison Test with $a_n = \dfrac{n^3+1}{n^4-1}$ and $b_n = \dfrac{1}{n}$. $\lim_{n\to\infty}\dfrac{a_n}{b_n} = \lim_{n\to\infty}\dfrac{n^4+n}{n^4-1} = \lim_{n\to\infty}\dfrac{1+1/n^3}{1-1/n^4} = 1$, and since $\sum_{n=2}^{\infty} b_n$ diverges (harmonic series), so does $\sum_{n=2}^{\infty}\dfrac{n^3+1}{n^4-1}$.

13. Let $f(x) = \dfrac{2}{x(\ln x)^3}$. $f(x)$ is clearly positive and decreasing for $x \ge 2$, so we apply the Integral Test.

$\int_2^{\infty}\dfrac{2}{x(\ln x)^3}\,dx = \lim_{t\to\infty}\left[\dfrac{-1}{(\ln x)^2}\right]_2^t = 0 - \dfrac{-1}{(\ln 2)^2}$, which is finite, so $\sum_{n=2}^{\infty}\dfrac{2}{n(\ln n)^3}$ converges.

15. $\lim\limits_{n\to\infty}\left|\dfrac{a_{n+1}}{a_n}\right| = \lim\limits_{n\to\infty}\dfrac{3^{n+1}(n+1)^2/(n+1)!}{3^n n^2/n!} = 3\lim\limits_{n\to\infty}\dfrac{n+1}{n^2} = 0$, so the series converges by the Ratio Test.

17. $\dfrac{3^n}{5^n+n} \le \dfrac{3^n}{5^n} = \left(\dfrac{3}{5}\right)^n$. Since $\sum\limits_{n=1}^{\infty}\left(\dfrac{3}{5}\right)^n$ is a convergent geometric series ($|r| = \frac{3}{5} < 1$), $\sum\limits_{n=1}^{\infty}\dfrac{3^n}{5^n+n}$ converges by the Comparison Test.

19. $\lim\limits_{n\to\infty}\left|\dfrac{a_{n+1}}{a_n}\right| = \lim\limits_{n\to\infty}\dfrac{(n+1)!/[2\cdot5\cdot8\cdot\cdots\cdot(3n+5)]}{n!/[2\cdot5\cdot8\cdot\cdots\cdot(3n+2)]} = \lim\limits_{n\to\infty}\dfrac{n+1}{3n+5} = \dfrac{1}{3} < 1$, so the series converges by the Ratio Test.

21. Use the Limit Comparison Test with $a_i = \dfrac{1}{\sqrt{i(i+1)}}$ and $b_i = \dfrac{1}{i}$. $\lim\limits_{i\to\infty}\dfrac{a_i}{b_i} = \lim\limits_{i\to\infty}\dfrac{i}{\sqrt{i(i+1)}}$ $= \lim\limits_{i\to\infty}\dfrac{1}{\sqrt{1+1/i}} = 1$. Since $\sum\limits_{i=1}^{\infty} b_i$ diverges (harmonic series) so does $\sum\limits_{i=1}^{\infty}\dfrac{1}{\sqrt{i(i+1)}}$.

23. $\lim\limits_{n\to\infty} 2^{1/n} = 2^0 = 1$, so $\lim\limits_{n\to\infty}(-1)^n 2^{1/n}$ does not exist and the series diverges by the Test for Divergence.

25. Let $f(x) = \dfrac{\ln x}{\sqrt{x}}$. Then $f'(x) = \dfrac{2-\ln x}{2x^{3/2}} < 0$ when $\ln x > 2$ or $x > e^2$, so $\dfrac{\ln n}{\sqrt{n}}$ is decreasing for $n > e^2$. By l'Hospital's Rule, $\lim\limits_{n\to\infty}\dfrac{\ln n}{\sqrt{n}} = \lim\limits_{n\to\infty}\dfrac{1/n}{1/(2\sqrt{n})} = \lim\limits_{n\to\infty}\dfrac{2}{\sqrt{n}} = 0$, so the series converges by the Alternating Series Test.

27. The series diverges since it is a geometric series with $r = -\pi$ and $|r| = \pi > 1$. (Or use the Test for Divergence.)

29. $\sum\limits_{n=1}^{\infty}\dfrac{(-2)^{2n}}{n^n} = \sum\limits_{n=1}^{\infty}\left(\dfrac{4}{n}\right)^n$. $\lim\limits_{n\to\infty}\sqrt[n]{|a_n|} = \lim\limits_{n\to\infty}\dfrac{4}{n} = 0$, so the series converges by the Root Test.

31. $\displaystyle\int_2^{\infty}\frac{\ln x}{x^2}\,dx = \lim_{t\to\infty}\left[-\frac{\ln x}{x} - \frac{1}{x}\right]_1^t$ (using integration by parts) $= 1$ (by L'Hospital's Rule). So $\sum\limits_{n=1}^{\infty}\dfrac{\ln n}{n^2}$ converges by the Integral Test, and since $\dfrac{k\ln k}{(k+1)^3} < \dfrac{k\ln k}{k^3} = \dfrac{\ln k}{k^2}$, the given series converges by the Comparison Test.

33. $\lim\limits_{n\to\infty}\left|\dfrac{a_{n+1}}{a_n}\right| = \lim\limits_{n\to\infty}\dfrac{2^{n+1}/(2n+3)!}{2^n/(2n+1)!} = 2\lim\limits_{n\to\infty}\dfrac{1}{(2n+3)(2n+2)} = 0$, so the series converges by the Ratio Test.

35. $0 < \dfrac{\tan^{-1} n}{n^{3/2}} < \dfrac{\pi/2}{n^{3/2}}$. $\sum\limits_{n=1}^{\infty}\dfrac{\pi/2}{n^{3/2}} = \dfrac{\pi}{2}\sum\limits_{n=1}^{\infty}\dfrac{1}{n^{3/2}}$ which is a convergent p-series ($p = \frac{3}{2} > 1$), so $\sum\limits_{n=1}^{\infty}\dfrac{\tan^{-1} n}{n^{3/2}}$ converges by the Comparison Test.

37. $\lim\limits_{n\to\infty}\sqrt[n]{|a_n|} = \lim\limits_{n\to\infty}\left(\dfrac{n}{n+1}\right)^{n^2/n} = \lim\limits_{n\to\infty}\dfrac{1}{[(n+1)/n]^n} = \dfrac{1}{\lim\limits_{n\to\infty}(1+1/n)^n} = \dfrac{1}{e} < 1$ (see Equation 6.4.9 or 6.4*.8), so the series converges by the Root Test.

39. $\lim\limits_{n\to\infty}\sqrt[n]{|a_n|} = \lim\limits_{n\to\infty}(2^{1/n} - 1) = 1 - 1 = 0$, so the series converges by the Root Test.

EXERCISES 10.8

Note: "R" stands for "radius of convergence" and "I" stands for "interval of convergence" in this section.

1. **(a)** We are given that the power series $\sum_{n=0}^{\infty} c_n x^n$ is convergent for $x = 4$. So by Theorem 3 it must converge for at least $-4 < x \le 4$. In particular it converges when $x = -2$, that is, $\sum_{n=0}^{\infty} c_n(-2)^n$ is convergent.

(b) But it does not follow that $\sum_{n=0}^{\infty} c_n(-4)^n$ is necessarily convergent. [See the comments after Theorem 3. An example is $c_n = (-1)^n/(n4^n)$.]

3. If $a_n = \dfrac{x^n}{n+2}$, then $\lim_{n\to\infty}\left|\dfrac{a_{n+1}}{a_n}\right| = \lim_{n\to\infty}\left|\dfrac{x^{n+1}}{n+3}\cdot\dfrac{n+2}{x^n}\right| = |x|\lim_{n\to\infty}\dfrac{n+2}{n+3} = |x| < 1$ for convergence (by the Ratio Test). So $R = 1$. When $x = 1$, the series is $\sum_{n=0}^{\infty}\dfrac{1}{n+2}$ which diverges (Integral Test or Comparison Test), and when $x = -1$, it is $\sum_{n=0}^{\infty}\dfrac{(-1)^n}{n+2}$ which converges (Alternating Series Test), so $I = [-1, 1)$.

5. If $a_n = nx^n$, then $\lim_{n\to\infty}\left|\dfrac{a_{n+1}}{a_n}\right| = \lim_{n\to\infty}\left|\dfrac{(n+1)x^{n+1}}{nx^n}\right| = |x|\lim_{n\to\infty}\dfrac{n+1}{n} = |x| < 1$ for convergence (by the Ratio Test). So $R = 1$. When $x = 1$ or -1, $\lim_{n\to\infty} nx^n$ does not exist, so $\sum_{n=0}^{\infty} nx^n$ diverges for $x = \pm 1$. So $I = (-1, 1)$.

7. If $a_n = \dfrac{x^n}{n!}$, then $\lim_{n\to\infty}\left|\dfrac{a_{n+1}}{a_n}\right| = \lim_{n\to\infty}\left|\dfrac{x^{n+1}/(n+1)!}{x^n/n!}\right| = |x|\lim_{n\to\infty}\dfrac{1}{n+1} = 0 < 1$ for all x. So, by the Ratio Test, $R = \infty$, and $I = (-\infty, \infty)$.

9. If $a_n = \dfrac{(-1)^n x^n}{n2^n}$, then $\lim_{n\to\infty}\left|\dfrac{a_{n+1}}{a_n}\right| = \lim_{n\to\infty}\left|\dfrac{x^{n+1}/[(n+1)2^{n+1}]}{x^n/(n2^n)}\right| = \left|\dfrac{x}{2}\right|\lim_{n\to\infty}\dfrac{n}{n+1} = \left|\dfrac{x}{2}\right| < 1$ for convergence, so $|x| < 2$ and $R = 2$. When $x = 2$, $\sum_{n=1}^{\infty}\dfrac{(-1)^n x^n}{n2^n} = \sum_{n=1}^{\infty}\dfrac{(-1)^n}{n}$ which converges by the Alternating Series Test. When $x = -2$, $\sum_{n=1}^{\infty}\dfrac{(-1)^n x^n}{n2^n} = \sum_{n=1}^{\infty}\dfrac{1}{n}$ which diverges (harmonic series), so $I = (-2, 2]$.

11. If $a_n = \dfrac{3^n x^n}{(n+1)^2}$, then $\lim_{n\to\infty}\left|\dfrac{a_{n+1}}{a_n}\right| = \lim_{n\to\infty}\left|\dfrac{3^{n+1}x^{n+1}}{(n+2)^2}\cdot\dfrac{(n+1)^2}{3^n x^n}\right| = 3|x|\lim_{n\to\infty}\left(\dfrac{n+1}{n+2}\right)^2 = 3|x| < 1$ for convergence, so $|x| < \frac{1}{3}$ and $R = \frac{1}{3}$. When $x = \frac{1}{3}$, $\sum_{n=0}^{\infty}\dfrac{3^n x^n}{(n+1)^2} = \sum_{n=0}^{\infty}\dfrac{1}{(n+1)^2} = \sum_{n=1}^{\infty}\dfrac{1}{n^2}$ which is a convergent p-series $(p = 2 > 1)$. When $x = -\frac{1}{3}$, $\sum_{n=0}^{\infty}\dfrac{3^n x^n}{(n+1)^2} = \sum_{n=0}^{\infty}\dfrac{(-1)^n}{(n+1)^2}$ which converges by the Alternating Series Test, so $I = \left[-\frac{1}{3}, \frac{1}{3}\right]$.

13. If $a_n = \dfrac{x^n}{\ln n}$, then $\lim\limits_{n\to\infty}\left|\dfrac{a_{n+1}}{a_n}\right| = \lim\limits_{n\to\infty}\left|\dfrac{x^{n+1}}{\ln(n+1)}\cdot\dfrac{\ln n}{x^n}\right| = |x|\lim\limits_{n\to\infty}\dfrac{\ln n}{\ln(n+1)} = |x|$ (using l'Hospital's Rule), so $R = 1$. When $x = 1$, $\sum\limits_{n=2}^{\infty}\dfrac{x^n}{\ln n} = \sum\limits_{n=2}^{\infty}\dfrac{1}{\ln n}$ which diverges because $\dfrac{1}{\ln n} > \dfrac{1}{n}$ and $\sum\limits_{n=2}^{\infty}\dfrac{1}{n}$ is the divergent harmonic series. When $x = -1$, $\sum\limits_{n=2}^{\infty}\dfrac{x^n}{\ln n} = \sum\limits_{n=2}^{\infty}\dfrac{(-1)^n}{\ln n}$ which converges by the Alternating Series Test.

So $I = [-1, 1)$.

15. If $a_n = \dfrac{n}{4^n}(2x-1)^n$, then $\left|\dfrac{a_{n+1}}{a_n}\right| = \left|\dfrac{(n+1)(2x-1)^{n+1}}{4^{n+1}}\cdot\dfrac{4^n}{n(2x-1)^n}\right| = \left|\dfrac{2x-1}{4}\left(1+\dfrac{1}{n}\right)\right| \to \frac{1}{2}\left|x-\frac{1}{2}\right|$ as $n\to\infty$. For convergence, $\frac{1}{2}\left|x-\frac{1}{2}\right| < 1 \Rightarrow \left|x-\frac{1}{2}\right| < 2 \Rightarrow R = 2$ and $-2 < x - \frac{1}{2} < 2 \Rightarrow -\frac{3}{2} < x < \frac{5}{2}$. If $x = -\frac{3}{2}$, the series becomes $\sum\limits_{n=0}^{\infty}\dfrac{n}{4^n}(-4)^n = \sum\limits_{n=0}^{\infty}(-1)^n n$ which is divergent by the Test for Divergence.

If $x = \frac{5}{2}$, the series is $\sum\limits_{n=0}^{\infty}\dfrac{n}{4^n}4^n = \sum\limits_{n=0}^{\infty}n$, also divergent by the Test for Divergence. So $I = \left(-\frac{3}{2}, \frac{5}{2}\right)$.

17. If $a_n = \dfrac{(-1)^n(x-1)^n}{\sqrt{n}}$, then $\lim\limits_{n\to\infty}\left|\dfrac{a_{n+1}}{a_n}\right| = \lim\limits_{n\to\infty}\left|\dfrac{(x-1)^{n+1}}{\sqrt{n+1}}\cdot\dfrac{\sqrt{n}}{(x-1)^n}\right| = |x-1|\lim\limits_{n\to\infty}\sqrt{\dfrac{n}{n+1}} = |x-1| < 1$ for convergence, or $0 < x < 2$, and $R = 1$. When $x = 0$, $\sum\limits_{n=1}^{\infty}\dfrac{(-1)^n(x-1)^n}{\sqrt{n}} = \sum\limits_{n=1}^{\infty}\dfrac{1}{\sqrt{n}}$ which is a divergent p-series ($p = \frac{1}{2} < 1$). When $x = 2$, the series is $\sum\limits_{n=1}^{\infty}\dfrac{(-1)^n}{\sqrt{n}}$ which converges by the Alternating Series Test. So

$I = (0, 2]$.

19. If $a_n = \dfrac{(x-2)^n}{n^n}$, then $\lim\limits_{n\to\infty}\sqrt[n]{|a_n|} = \lim\limits_{n\to\infty}\dfrac{x-2}{n} = 0$, so the series converges for all x (by the Root Test).

$R = \infty$ and $I = (-\infty, \infty)$.

21. If $a_n = \dfrac{2^n(x-3)^n}{n+3}$, then $\lim\limits_{n\to\infty}\left|\dfrac{a_{n+1}}{a_n}\right| = \lim\limits_{n\to\infty}\left|\dfrac{2^{n+1}(x-3)^{n+1}}{n+4}\cdot\dfrac{n+3}{2^n(x-3)^n}\right| = 2|x-3|\lim\limits_{n\to\infty}\dfrac{n+3}{n+4}$

$= 2|x-3| < 1$ for convergence, or $|x-3| < \frac{1}{2} \;\Leftrightarrow\; \frac{5}{2} < x < \frac{7}{2}$, and $R = \frac{1}{2}$. When $x = \frac{5}{2}$,

$\sum\limits_{n=0}^{\infty}\dfrac{2^n(x-3)^n}{n+3} = \sum\limits_{n=0}^{\infty}\dfrac{(-1)^n}{n+3}$ which converges by the Alternating Series Test. When $x = \dfrac{7}{2}$,

$\sum\limits_{n=0}^{\infty}\dfrac{2^n(x-3)^n}{n+3} = \sum\limits_{n=0}^{\infty}\dfrac{1}{n+3} = \sum\limits_{n=3}^{\infty}\dfrac{1}{n}$, the harmonic series, which diverges. So $I = \left[\frac{5}{2}, \frac{7}{2}\right)$.

23. If $a_n = \left(\dfrac{n}{2}\right)^n(x+6)^n$, then $\lim\limits_{n\to\infty}\sqrt[n]{|a_n|} = \lim\limits_{n\to\infty}\dfrac{n(x+6)}{2} = \infty$ unless $x = -6$, in which case the limit is 0. So by the Root Test, the series converges only for $x = -6$. $R = 0$ and $I = \{-6\}$.

25. If $a_n = \dfrac{(2x-1)^n}{n^3}$, then $\lim\limits_{n\to\infty}\left|\dfrac{a_{n+1}}{a_n}\right| = |2x-1|\lim\limits_{n\to\infty}\left(\dfrac{n}{n+1}\right)^3 = |2x-1| < 1$ for convergence, so

$\left|x-\frac{1}{2}\right| < \frac{1}{2} \;\Leftrightarrow\; 0 < x < 1$, and $R = \frac{1}{2}$. The series $\sum\limits_{n=1}^{\infty}\dfrac{(2x-1)^n}{n^3}$ converges both for $x = 0$ and $x = 1$ (in the first case because of the Alternating Series Test and in the second case because we get a p-series with $p = 3 > 1$). So $I = [0, 1]$.

27. If $a_n = \dfrac{x^n}{(\ln n)^n}$ then $\lim\limits_{n\to\infty} \sqrt[n]{|a_n|} = \lim\limits_{n\to\infty} \dfrac{|x|}{\ln n} = 0 < 1$ for all x, so $R = \infty$ and $I = (-\infty, \infty)$ by the Root Test.

29. If $a_n = \dfrac{(n!)^k}{(kn)!}x^n$, then $\lim\limits_{n\to\infty}\left|\dfrac{a_{n+1}}{a_n}\right| = \lim\limits_{n\to\infty} \dfrac{[(n+1)!]^k (kn)!}{(n!)^k [k(n+1)]!}|x|$

$$= \lim_{n\to\infty} \frac{(n+1)^k}{(kn+k)(kn+k-1)\cdots(kn+2)(kn+1)}|x| = \lim_{n\to\infty}\left[\frac{(n+1)}{(kn+1)}\frac{(n+1)}{(kn+2)}\cdots\frac{(n+1)}{(kn+k)}\right]|x|$$

$$= \lim_{n\to\infty}\left[\frac{n+1}{kn+1}\right]\lim_{n\to\infty}\left[\frac{n+1}{kn+2}\right]\cdots\lim_{n\to\infty}\left[\frac{n+1}{kn+k}\right]|x| = \left(\frac{1}{k}\right)^k |x| < 1 \quad\Leftrightarrow\quad |x| < k^k \text{ for convergence, and}$$

the radius of convergence is $R = k^k$.

31. **(a)** If $a_n = \dfrac{(-1)^n x^{2n+1}}{n!(n+1)!\,2^{2n+1}}$, then $\lim\limits_{n\to\infty}\left|\dfrac{a_{n+1}}{a_n}\right| = \left(\dfrac{x}{2}\right)^2 \lim\limits_{n\to\infty} \dfrac{1}{(n+1)(n+2)} = 0$ for all x. So $J_1(x)$ converges for all x; the domain is $(-\infty, \infty)$.

(b), (c) The initial terms of $J_1(x)$ up to $n = 5$ are

$$a_0 = \frac{x}{2},\ a_1 = -\frac{x^3}{16},\ a_2 = \frac{x^5}{384},\ a_3 = -\frac{x^7}{18{,}432},\ a_4 = \frac{x^9}{1{,}474{,}560},\ a_5 = -\frac{x^{11}}{176{,}947{,}200}.$$

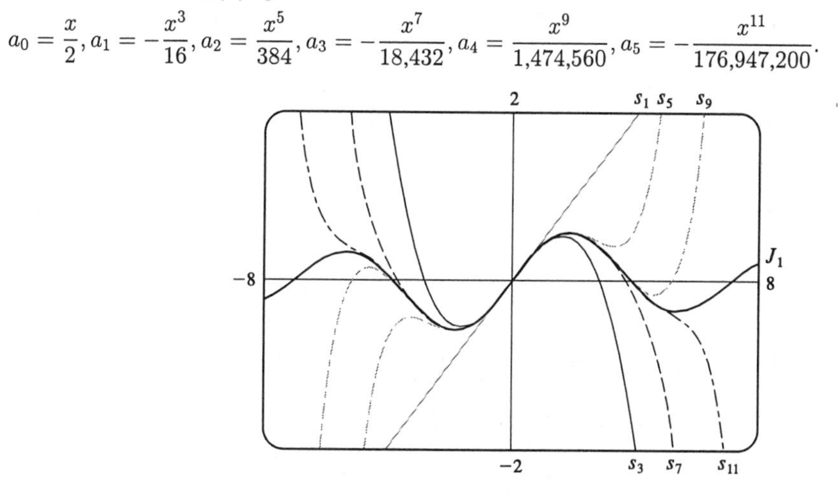

The partial sums seem to approximate $J_1(x)$ well near the origin, but as $|x|$ increases, we need to take a large number of terms to get a good approximation.

33. $s_{2n-1} = 1 + 2x + x^2 + 2x^3 + \cdots + x^{2n-2} + 2x^{2n-1} = (1+2x)(1 + x^2 + x^4 + \cdots + x^{2n-2})$

$$= (1+2x)\frac{1-x^{2n}}{1-x^2} \to \frac{1+2x}{1-x^2} \text{ as } n\to\infty, \text{ when } |x| < 1.$$

Also $s_{2n} = s_{2n-1} + x^{2n} \to \dfrac{1+2x}{1-x^2}$ since $x^{2n} \to 0$ for $|x| < 1$. Therefore $s_n \to \dfrac{1+2x}{1-x^2}$ by Exercise 10.1.70(a).

Thus the interval of convergence is $(-1, 1)$ and $f(x) = \dfrac{1+2x}{1-x^2}$.

35. We use the Root Test on the series $\sum c_n x^n$. $\lim\limits_{n\to\infty} \sqrt[n]{|c_n x^n|} = |x| \lim\limits_{n\to\infty} \sqrt[n]{|c_n|} = c|x| < 1$ for convergence, or $|x| < 1/c$, so $R = 1/c$.

37. $\sum(c_n + d_n)x^n = \sum c_n x^n + \sum d_n x^n$ on the interval $(-2, 2)$, since both series converge there. So the radius of convergence must be at least 2. Now since $\sum c_n x^n$ has $R = 2$, it must diverge either at $x = -2$ or at $x = 2$. So by Exercise 10.2.63, $\sum(c_n + d_n)x^n$ diverges either at $x = -2$ or at $x = 2$, and so its radius of convergence is 2.

EXERCISES 10.9

Note: "R" stands for "radius of convergence" and "I" stands for "interval of convergence" in this section.

1. $f(x) = \dfrac{1}{1+x} = \dfrac{1}{1-(-x)} = \displaystyle\sum_{n=0}^{\infty}(-1)^n x^n$ with $|-x| < 1 \Leftrightarrow |x| < 1$ so $R = 1$ and $I = (-1, 1)$.

3. $f(x) = \dfrac{1}{1+4x^2} = \displaystyle\sum_{n=0}^{\infty}(-1)^n (4x^2)^n$ $\left(\begin{array}{c}\text{substituting } 4x^2 \text{ for } x \text{ in the} \\ \text{series from Exercise 1}\end{array}\right)$ $= \displaystyle\sum_{n=0}^{\infty}(-1)^n 4^n x^{2n}$, with $|4x^2| < 1$ so $x^2 < \frac{1}{4} \Leftrightarrow |x| < \frac{1}{2}$, and so $R = \frac{1}{2}$ and $I = \left(-\frac{1}{2}, \frac{1}{2}\right)$.

5. $f(x) = \dfrac{1}{4+x^2} = \dfrac{1}{4}\left(\dfrac{1}{1+x^2/4}\right) = \dfrac{1}{4}\displaystyle\sum_{n=0}^{\infty}(-1)^n\left(\dfrac{x^2}{4}\right)^n$ (using Exercise 1)

$= \displaystyle\sum_{n=0}^{\infty}\dfrac{(-1)^n x^{2n}}{4^{n+1}}$, with $\left|\dfrac{x^2}{4}\right| < 1 \Leftrightarrow x^2 < 4 \Leftrightarrow |x| < 2$, so $R = 2$ and $I = (-2, 2)$.

7. $\dfrac{x}{x-3} = 1 + \dfrac{3}{x-3} = 1 - \dfrac{1}{1-x/3} = 1 - \displaystyle\sum_{n=0}^{\infty}\left(\dfrac{x}{3}\right)^n = -\displaystyle\sum_{n=1}^{\infty}\left(\dfrac{x}{3}\right)^n$. For convergence, $\dfrac{|x|}{3} < 1 \Leftrightarrow$ $|x| < 3$, so $R = 3$ and $I = (-3, 3)$.

Another Method: $\dfrac{x}{x-3} = -\dfrac{x}{3(1-x/3)} = -\dfrac{x}{3}\displaystyle\sum_{n=0}^{\infty}\left(\dfrac{x}{3}\right)^n = -\displaystyle\sum_{n=0}^{\infty}\dfrac{x^{n+1}}{3^{n+1}} = -\displaystyle\sum_{n=1}^{\infty}\dfrac{x^n}{3^n}$

9. $\dfrac{3x-2}{2x^2-3x+1} = \dfrac{3x-2}{(2x-1)(x-1)} = \dfrac{A}{2x-1} + \dfrac{B}{x-1} \Leftrightarrow A + 2B = 3$ and $-A - B = -2 \Leftrightarrow$ $A = B = 1$, so $f(x) = \dfrac{3x-2}{2x^2-3x+1} = \dfrac{1}{2x-1} + \dfrac{1}{x-1} = -\displaystyle\sum_{n=0}^{\infty}(2x)^n - \displaystyle\sum_{n=0}^{\infty}x^n = -\displaystyle\sum_{n=0}^{\infty}(2^n+1)x^n$, with $R = \frac{1}{2}$. At $x = \pm\frac{1}{2}$, the series diverges by the Test for Divergence, so $I = \left(-\frac{1}{2}, \frac{1}{2}\right)$.

11. $f(x) = \dfrac{1}{(1+x)^2} = -\dfrac{d}{dx}\left(\dfrac{1}{1+x}\right) = -\dfrac{d}{dx}\left(\displaystyle\sum_{n=0}^{\infty}(-1)^n x^n\right)$ (from Exercise 1)

$= \displaystyle\sum_{n=1}^{\infty}(-1)^{n+1} n x^{n-1} = \displaystyle\sum_{n=0}^{\infty}(-1)^n (n+1) x^n$ with $R = 1$.

13. $f(x) = \dfrac{1}{(1+x)^3} = -\dfrac{1}{2}\dfrac{d}{dx}\left[\dfrac{1}{(1+x)^2}\right] = -\dfrac{1}{2}\dfrac{d}{dx}\left(\displaystyle\sum_{n=0}^{\infty}(-1)^n (n+1) x^n\right)$ (from Exercise 11)

$= -\dfrac{1}{2}\displaystyle\sum_{n=1}^{\infty}(-1)^n (n+1) n x^{n-1} = \dfrac{1}{2}\displaystyle\sum_{n=0}^{\infty}(-1)^n (n+2)(n+1) x^n$ with $R = 1$.

15. $f(x) = \ln(5-x) = -\displaystyle\int\dfrac{dx}{5-x} = -\dfrac{1}{5}\displaystyle\int\dfrac{dx}{1-x/5} = -\dfrac{1}{5}\displaystyle\int\left[\sum_{n=0}^{\infty}\left(\dfrac{x}{5}\right)^n\right]dx$

$= C - \dfrac{1}{5}\displaystyle\sum_{n=0}^{\infty}\dfrac{x^{n+1}}{5^n(n+1)} = C - \displaystyle\sum_{n=1}^{\infty}\dfrac{x^n}{n5^n}$

Putting $x = 0$, we get $C = \ln 5$. The series converges for $|x/5| < 1 \Leftrightarrow |x| < 5$. So $R = 5$.

17. $f(x) = \ln(1+x) - \ln(1-x) = \displaystyle\int \frac{dx}{1+x} + \int \frac{dx}{1-x} = \int \left[\sum_{n=0}^{\infty} (-1)^n x^n + \sum_{n=0}^{\infty} x^n \right] dx$

$= \displaystyle\int \sum_{n=0}^{\infty} 2x^{2n}\, dx = \sum_{n=0}^{\infty} \frac{2x^{2n+1}}{2n+1} + C.$

But $f(0) = \ln 1 - \ln 1 = 0$, so $C = 0$ and we have $f(x) = \displaystyle\sum_{n=0}^{\infty} \frac{2x^{2n+1}}{2n+1}$ with $R = 1$.

19. $f(x) = \ln(3+x) = \displaystyle\int \frac{dx}{3+x} = \frac{1}{3}\int \frac{dx}{1+x/3} = \frac{1}{3}\int \sum_{n=0}^{\infty} (-1)^n \left(\frac{x}{3}\right)^n dx$ (from Exercise 1)

$= C + \displaystyle\frac{1}{3}\sum_{n=0}^{\infty} \frac{(-1/3)^n}{n+1} x^{n+1} = \ln 3 + \frac{1}{3}\sum_{n=1}^{\infty} \frac{(-1/3)^{n-1}}{n} x^n = \ln 3 + \sum_{n=1}^{\infty} \frac{(-1)^{n-1}}{n3^n}$ $[C = f(0) = \ln 3]$

with $R = 3$. The terms of the series are $a_0 = \ln 3$, $a_1 = \dfrac{x}{3}$, $a_2 = -\dfrac{x^2}{18}$, $a_3 = \dfrac{x^3}{81}$, $a_4 = -\dfrac{x^4}{324}$, $a_5 = \dfrac{x^5}{1215}$, ...

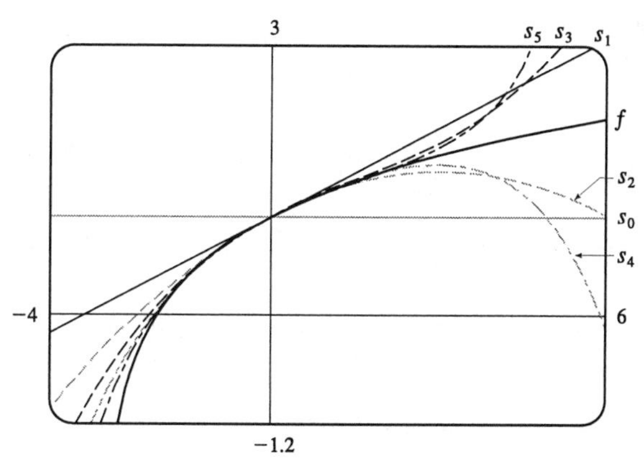

As n increases, $s_n(x)$ approximates f better on the interval of convergence, which is $(-3, 3)$.

21. $\displaystyle\int \frac{dx}{1+x^4} = \int \sum_{n=0}^{\infty} (-1)^n x^{4n}\, dx = C + \sum_{n=0}^{\infty} \frac{(-1)^n x^{4n+1}}{4n+1}$ with $R = 1$.

23. By Example 7, $\arctan x = \displaystyle\sum_{n=0}^{\infty} (-1)^n \frac{x^{2n+1}}{2n+1}$, so

$\displaystyle\int \frac{\arctan x}{x}\, dx = \int \sum_{n=0}^{\infty} (-1)^n \frac{x^{2n}}{2n+1}\, dx = C + \sum_{n=0}^{\infty} (-1)^n \frac{x^{2n+1}}{(2n+1)^2}$ with $R = 1$.

25. We use the representation $\displaystyle\int \frac{dx}{1+x^4} = C + \sum_{n=0}^{\infty} \frac{(-1)^n x^{4n+1}}{4n+1}$ from Exercise 21, with $C = 0$. So

$\displaystyle\int_0^{0.2} \frac{dx}{1+x^4} = \left[x - \frac{x^5}{5} + \frac{x^9}{9} - \frac{x^{13}}{13} + \cdots \right]_0^{0.2} = 0.2 - \frac{0.2^5}{5} + \frac{0.2^9}{9} - \frac{0.2^{13}}{13} + \cdots$. Since the series is alternating, the error in the nth-order approximation is less than the first neglected term by Theorem 10.5.1. If we use only the first two terms of the series, then the error is at most $0.2^9/9 \approx 5.7 \times 10^{-8}$. So, to six decimal places, $\displaystyle\int_0^{0.2} \frac{dx}{1+x^4} \approx 0.2 - \frac{0.2^5}{5} \approx 0.199936$.

27. We substitute x^4 for x in Example 7, and find that $\int x^2 \tan^{-1}(x^4)\,dx = \int x^2 \sum_{n=0}^{\infty} (-1)^n \frac{(x^4)^{2n+1}}{2n+1}\,dx$

$= \int \sum_{n=0}^{\infty} (-1)^n \frac{x^{8n+6}}{2n+1}\,dx = C + \sum_{n=0}^{\infty} (-1)^n \frac{x^{8n+7}}{(2n+1)(8n+7)}$. So

$\int_0^{1/3} x^2 \tan^{-1}(x^4)\,dx = \left[\frac{x^7}{7} - \frac{x^{15}}{45} + \cdots\right]_0^{1/3} = \frac{1}{7 \cdot 3^7} - \frac{1}{45 \cdot 3^{15}} + \cdots$. The series is alternating, so if we use only one term, the error is at most $1/(45 \cdot 3^{15}) \approx 1.5 \times 10^{-9}$. So $\int_0^{1/3} x^2 \tan^{-1}(x^4)\,dx \approx 1/(7 \cdot 3^7) \approx 0.000065$ to six decimal places.

29. Using the result of Example 6 with $x = -0.1$, we have

$\ln 1.1 = \ln[1 - (-0.1)] = 0.1 - \frac{0.01}{2} + \frac{0.001}{3} - \frac{0.0001}{4} + \frac{0.00001}{5} - \cdots$. If we use only the first four terms, the error is at most $\frac{0.00001}{5} = 0.000002$. So $\ln 1.1 \approx 0.1 - \frac{0.01}{2} + \frac{0.001}{3} - \frac{0.0001}{4} \approx 0.09531$.

31. **(a)** $J_0(x) = \sum_{n=0}^{\infty} \frac{(-1)^n x^{2n}}{2^{2n}(n!)^2}$, $J_0'(x) = \sum_{n=1}^{\infty} \frac{(-1)^n 2n x^{2n-1}}{2^{2n}(n!)^2}$, and $J_0''(x) = \sum_{n=1}^{\infty} \frac{(-1)^n 2n(2n-1) x^{2n-2}}{2^{2n}(n!)^2}$, so

$$x^2 J_0''(x) + x J_0'(x) + x^2 J_0(x) = \sum_{n=1}^{\infty} \frac{(-1)^n 2n(2n-1)x^{2n}}{2^{2n}(n!)^2} + \sum_{n=1}^{\infty} \frac{(-1)^n 2n x^{2n}}{2^{2n}(n!)^2} + \sum_{n=0}^{\infty} \frac{(-1)^n x^{2n+2}}{2^{2n}(n!)^2}$$

$$= \sum_{n=1}^{\infty} \frac{(-1)^n 2n(2n-1)x^{2n}}{2^{2n}(n!)^2} + \sum_{n=1}^{\infty} \frac{(-1)^n 2n x^{2n}}{2^{2n}(n!)^2} + \sum_{n=1}^{\infty} \frac{(-1)^{n-1} x^{2n}}{2^{2n-2}[(n-1)!]^2}$$

$$= \sum_{n=1}^{\infty} (-1)^n \left[\frac{2n(2n-1) + 2n - 2^2 n^2}{2^{2n}(n!)^2}\right] x^{2n} = \sum_{n=1}^{\infty} (-1)^n \left[\frac{4n^2 - 2n + 2n - 4n^2}{2^{2n}(n!)^2}\right] x^{2n} = 0.$$

(b) $$\int_0^1 J_0(x)\,dx = \int_0^1 \left[\sum_{n=0}^{\infty} \frac{(-1)^n x^{2n}}{2^{2n}(n!)^2}\right] dx = \int_0^1 dx - \int_0^1 \frac{x^2}{4}\,dx + \int_0^1 \frac{x^4}{64}\,dx - \int_0^1 \frac{x^6}{2304}\,dx + \cdots$$

$$= \left[x - \frac{x^3}{3 \cdot 4} + \frac{x^5}{5 \cdot 64} - \frac{x^7}{7 \cdot 2304} + \cdots\right]_0^1 = 1 - \frac{1}{12} + \frac{1}{320} - \frac{1}{16{,}128} + \cdots.$$

Since $\frac{1}{16{,}128} \approx 0.000062$, it follows from Theorem 10.5.1 that, correct to three decimal places, $\int_0^1 J_0(x)\,dx \approx 1 - \frac{1}{12} + \frac{1}{320} \approx 0.920$.

33. **(a)** We calculate $f'(x) = \sum_{n=1}^{\infty} \frac{n x^{n-1}}{n!} = \sum_{n=1}^{\infty} \frac{x^{n-1}}{(n-1)!} = \sum_{n=0}^{\infty} \frac{x^n}{n!} = f(x)$.

(b) By Theorem 6.5.2, the only solutions to the differential equation $\frac{df(x)}{dx} = f(x)$ are $f(x) = Ke^x$, but $f(0) = 1$, so $K = 1$ and $f(x) = e^x$.

Or: We could solve the equation $\frac{df(x)}{dx} = f(x)$ as a separable differential equation.

35. If $a_n = \dfrac{x^n}{n^2}$, then $\lim_{n\to\infty}\left|\dfrac{a_{n+1}}{a_n}\right| = |x|\lim_{n\to\infty}\left(\dfrac{n}{n+1}\right)^2 = |x| < 1$ for convergence, so $R = 1$. When $x = \pm 1$, $\sum_{n=1}^{\infty}\left|\dfrac{x^n}{n^2}\right| = \sum_{n=1}^{\infty}\dfrac{1}{n^2}$ which is a convergent p-series ($p = 2 > 1$), so the interval of convergence for f is $[-1,1]$. By Theorem 10.9.2, the radii of convergence of f' and f'' are both 1, so we need only check the endpoints. $f'(x) = \sum_{n=1}^{\infty}\dfrac{nx^{n-1}}{n^2} = \sum_{n=0}^{\infty}\dfrac{x^n}{n+1}$, and this series diverges for $x = 1$ (harmonic series) and converges for $x = -1$ (Alternating Series Test), so the interval of convergence is $[-1,1)$. $f''(x) = \sum_{n=1}^{\infty}\dfrac{nx^{n-1}}{n+1}$ diverges at both 1 and -1 (Test for Divergence) since $\lim_{n\to\infty}\dfrac{n}{n+1} = 1 \neq 0$, so its interval of convergence is $(-1,1)$.

EXERCISES 10.10

1.

n	$f^{(n)}(x)$	$f^{(n)}(0)$
0	$\cos x$	1
1	$-\sin x$	0
2	$-\cos x$	-1
3	$\sin x$	0
4	$\cos x$	1
$\cdots$	$\cdots$	$\cdots$

$$\cos x = f(0) + f'(0)x + \frac{f''(0)}{2!}x^2 + \frac{f^{(3)}(0)}{3!}x^3 + \frac{f^{(4)}(0)}{4!}x^4 + \cdots$$
$$= 1 - \frac{x^2}{2!} + \frac{x^4}{4!} - \cdots = \sum_{n=0}^{\infty}\frac{(-1)^n x^{2n}}{(2n)!}$$

If $a_n = \dfrac{(-1)^n x^{2n}}{(2n)!}$, then

$$\lim_{n\to\infty}\left|\frac{a_{n+1}}{a_n}\right| = x^2\lim_{n\to\infty}\frac{1}{(2n+2)(2n+1)} = 0 < 1 \text{ for all } x.$$

So $R = \infty$.

3.

n	$f^{(n)}(x)$	$f^{(n)}(0)$
0	$(1+x)^{-2}$	1
1	$-2(1+x)^{-3}$	-2
2	$2\cdot 3(1+x)^{-4}$	$2\cdot 3$
3	$-2\cdot 3\cdot 4(1+x)^{-5}$	$-2\cdot 3\cdot 4$
4	$2\cdot 3\cdot 4\cdot 5(1+x)^{-6}$	$2\cdot 3\cdot 4\cdot 5$
$\cdots$	$\cdots$	$\cdots$

So $f^{(n)}(0) = (-1)^n(n+1)!$ and

$$\frac{1}{(1+x)^2} = \sum_{n=0}^{\infty}\frac{(-1)^n(n+1)!}{n!}x^n = \sum_{n=0}^{\infty}(-1)^n(n+1)x^n.$$

If $a_n = (-1)^n(n+1)x^n$, then

$\lim_{n\to\infty}\left|\dfrac{a_{n+1}}{a_n}\right| = |x|$, so $R = 1$.

5.

n	$f^{(n)}(x)$	$f^{(n)}(0)$
0	$\sinh x$	0
1	$\cosh x$	1
2	$\sinh x$	0
3	$\cosh x$	1
4	$\sinh x$	0
$\cdots$	$\cdots$	$\cdots$

So $f^{(n)}(0) = \begin{cases} 0 & \text{if } n \text{ is even} \\ 1 & \text{if } n \text{ is odd} \end{cases}$

and $\sinh x = \sum_{n=0}^{\infty}\dfrac{x^{2n+1}}{(2n+1)!}$. If $a_n = \dfrac{x^{2n+1}}{(2n+1)!}$ then

$$\lim_{n\to\infty}\left|\frac{a_{n+1}}{a_n}\right| = x^2\lim_{n\to\infty}\frac{1}{(2n+3)(2n+2)} = 0 < 1$$

for all x, so $R = \infty$.

7.

n	$f^{(n)}(x)$	$f^{(n)}(\pi/4)$
0	$\sin x$	$\sqrt{2}/2$
1	$\cos x$	$\sqrt{2}/2$
2	$-\sin x$	$-\sqrt{2}/2$
3	$-\cos x$	$-\sqrt{2}/2$
4	$\sin x$	$\sqrt{2}/2$
$\cdots$	$\cdots$	$\cdots$

$$\begin{aligned}\sin x &= f\left(\tfrac{\pi}{4}\right) + f'\left(\tfrac{\pi}{4}\right)\left(x - \tfrac{\pi}{4}\right) + \frac{f''\left(\frac{\pi}{4}\right)}{2!}\left(x - \tfrac{\pi}{4}\right)^2 \\ &\quad + \frac{f^{(3)}\left(\frac{\pi}{4}\right)}{3!}\left(x - \tfrac{\pi}{4}\right)^3 + \frac{f^{(4)}\left(\frac{\pi}{4}\right)}{4!}\left(x - \tfrac{\pi}{4}\right)^4 + \cdots \\ &= \tfrac{\sqrt{2}}{2}\left[1 + \left(x - \tfrac{\pi}{4}\right) - \tfrac{1}{2!}\left(x - \tfrac{\pi}{4}\right)^2 - \tfrac{1}{3!}\left(x - \tfrac{\pi}{4}\right)^3 + \tfrac{1}{4!}\left(x - \tfrac{\pi}{4}\right)^4 + \cdots\right] \\ &= \frac{\sqrt{2}}{2}\sum_{n=0}^{\infty}\frac{(-1)^{n(n-1)/2}\left(x - \frac{\pi}{4}\right)^n}{n!}\end{aligned}$$

If $a_n = \dfrac{(-1)^{n(n-1)/2}\left(x - \frac{\pi}{4}\right)^n}{n!}$, then $\lim_{n\to\infty}\left|\dfrac{a_{n+1}}{a_n}\right| = \lim_{n\to\infty}\dfrac{\left|x - \frac{\pi}{4}\right|}{n+1} = 0 < 1$ for all x, so $R = \infty$.

9.

n	$f^{(n)}(x)$	$f^{(n)}(1)$
0	x^{-1}	1
1	$-x^{-2}$	-1
2	$2x^{-3}$	2
3	$-3 \cdot 2x^{-4}$	$-3 \cdot 2$
4	$4 \cdot 3 \cdot 2x^{-5}$	$4 \cdot 3 \cdot 2$
$\cdots$	$\cdots$	$\cdots$

So $f^{(n)}(1) = (-1)^n n!$, and

$$\frac{1}{x} = \sum_{n=0}^{\infty}\frac{(-1)^n n!}{n!}(x-1)^n = \sum_{n=0}^{\infty}(-1)^n(x-1)^n.$$

If $a_n = (-1)^n(x-1)^n$ then

$\lim_{n\to\infty}\left|\dfrac{a_{n+1}}{a_n}\right| = |x - 1| < 1$ for convergence,

so $0 < x < 2$ and $R = 1$.

11. Clearly $f^{(n)}(x) = e^x$, so $f^{(n)}(3) = e^3$ and $e^x = \sum_{n=0}^{\infty}\dfrac{e^3}{n!}(x-3)^n$. If $a_n = \dfrac{e^3}{n!}(x-3)^n$ then

$\lim_{n\to\infty}\left|\dfrac{a_{n+1}}{a_n}\right| = \lim_{n\to\infty}\dfrac{|x-3|}{n+1} = 0$ for all x, so $R = \infty$.

13. If $f(x) = \cos x$, then by Formula 9, $R_n(x) = \dfrac{f^{(n+1)}(z)}{(n+1)!}x^{n+1}$, where $0 < |z| < |x|$. But $f^{(n+1)}(z) = \pm\sin z$ or $\pm\cos z$. In each case, $\left|f^{(n+1)}(z)\right| \le 1$, so $|R_n(x)| \le \dfrac{1}{(n+1)!}x^{n+1} \to 0$ as $n \to \infty$ by Equation 11. So $\lim_{n\to\infty} R_n(x) = 0$ and, by Theorem 8, the series in Exercise 1 represents $\cos x$ for all x.

15. If $f(x) = \sinh x$, then $R_n(x) = \dfrac{f^{(n+1)}(z)}{(n+1)!}x^{n+1}$, where $0 < |z| < |x|$. But for all n,

$\left|f^{(n+1)}(z)\right| \le \cosh z \le \cosh x$ (since all derivatives are either sinh or cosh, $|\sinh z| < |\cosh z|$ for all z, and $|z| < |x| \Rightarrow \cosh z < \cosh x$), so $|R_n(z)| \le \dfrac{\cosh x}{(n+1)!}x^{n+1} \to 0$ as $n \to \infty$ (by Equation 11). So by Theorem 9, the series represents $\sinh x$ for all x.

17. $e^{3x} = \sum_{n=0}^{\infty}\dfrac{(3x)^n}{n!} = \sum_{n=0}^{\infty}\dfrac{3^n x^n}{n!}$, with $R = \infty$

19. $x^2\cos x = x^2\sum_{n=0}^{\infty}\dfrac{(-1)^n x^{2n}}{(2n)!} = \sum_{n=0}^{\infty}\dfrac{(-1)^n x^{2n+2}}{(2n)!}$, $R = \infty$

21. $x\sin\left(\dfrac{x}{2}\right) = x\sum_{n=0}^{\infty}\dfrac{(-1)^n(x/2)^{2n+1}}{(2n+1)!} = \sum_{n=0}^{\infty}\dfrac{(-1)^n x^{2n+2}}{(2n+1)!2^{2n+1}}$, with $R = \infty$.

23. $\sin^2 x = \frac{1}{2}[1 - \cos 2x] = \dfrac{1}{2}\left[1 - \sum_{n=0}^{\infty}\dfrac{(-1)^n(2x)^{2n}}{(2n)!}\right] = \dfrac{1}{2}\left[1 - 1 - \sum_{n=1}^{\infty}\dfrac{(-1)^n(2x)^{2n}}{(2n)!}\right] = \sum_{n=1}^{\infty}\dfrac{(-1)^{n+1}2^{2n-1}x^{2n}}{(2n)!}$,

with $R = \infty$.

25. $\dfrac{\sin x}{x} = \dfrac{1}{x}\sum_{n=0}^{\infty}\dfrac{(-1)^n x^{2n+1}}{(2n+1)!} = \sum_{n=0}^{\infty}\dfrac{(-1)^n x^{2n}}{(2n+1)!}$ and this series also gives the required value at $x = 0$, so $R = \infty$.

27.

n	$f^{(n)}(x)$	$f^{(n)}(0)$
0	$(1+x)^{1/2}$	1
1	$\frac{1}{2}(1+x)^{-1/2}$	$\frac{1}{2}$
2	$-\frac{1}{4}(1+x)^{-3/2}$	$-\frac{1}{4}$
3	$\frac{3}{8}(1+x)^{-5/2}$	$\frac{3}{8}$
4	$-\frac{15}{16}(1+x)^{-7/2}$	$-\frac{15}{16}$
...	...	...

So $f^{(n)}(0) = \dfrac{(-1)^{n-1}1\cdot 3\cdot 5\cdots(2n-3)}{2^n}$ for $n \geq 2$, and

$\sqrt{1+x} = 1 + \dfrac{x}{2} + \sum_{n=2}^{\infty}\dfrac{(-1)^{n-1}1\cdot 3\cdot 5\cdots(2n-3)}{2^n n!}x^n.$

If $a_n = \dfrac{(-1)^{n+1}1\cdot 3\cdot 5\cdot\cdots\cdot(2n-3)}{2^n n!}x^n$, then

$\lim_{n\to\infty}\left|\dfrac{a_{n+1}}{a_n}\right| = \dfrac{|x|}{2}\lim_{n\to\infty}\dfrac{2n-1}{n+1} = |x| < 1$ for convergence, so $R = 1$.

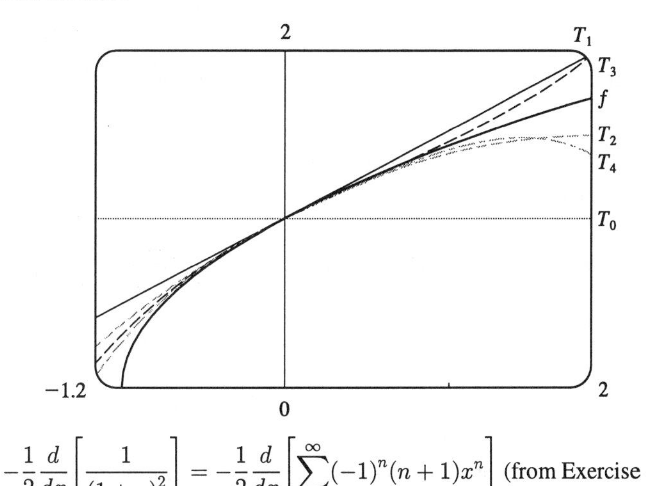

29. $f(x) = (1+x)^{-3} = -\dfrac{1}{2}\dfrac{d}{dx}\left[\dfrac{1}{(1+x)^2}\right] = -\dfrac{1}{2}\dfrac{d}{dx}\left[\sum_{n=0}^{\infty}(-1)^n(n+1)x^n\right]$ (from Exercise 5)

$$= -\frac{1}{2}\sum_{n=1}^{\infty}(-1)^n n(n+1)x^{n-1} = \sum_{n=0}^{\infty}\frac{(-1)^n(n+1)(n+2)x^n}{2},$$

with $R = 1$ since that is the R in Exercise 5.

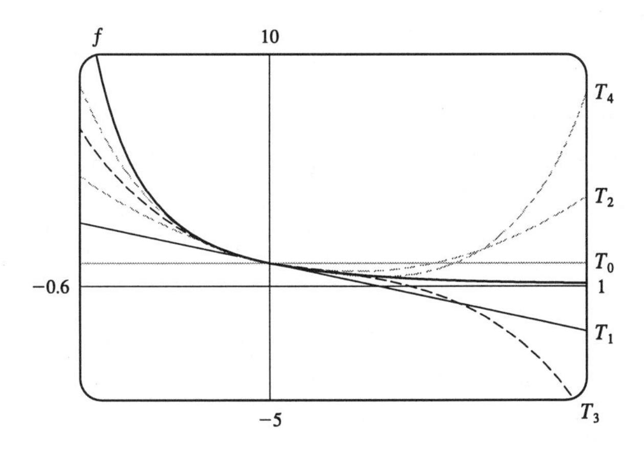

31. $\ln(1+x) = \int \frac{dx}{1+x} = \int \sum_{n=0}^{\infty} (-1)^n x^n \, dx = \sum_{n=1}^{\infty} \frac{(-1)^{n-1} x^n}{n}$ with $R = 1$, so $\ln(1.1) = \sum_{n=1}^{\infty} \frac{(-1)^{n-1} (0.1)^n}{n}$.

This is an alternating series with $b_5 = \frac{(0.1)^5}{5} = 0.000002$, so to five decimals,

$\ln(1.1) \approx \sum_{n=1}^{4} \frac{(-1)^{n-1} (0.1)^n}{n} \approx 0.09531.$

33. $\int \sin(x^2)\, dx = \int \sum_{n=0}^{\infty} (-1)^n \frac{(x^2)^{2n+1}}{(2n+1)!} \, dx = \int \sum_{n=0}^{\infty} \frac{(-1)^n x^{4n+2}}{(2n+1)!} \, dx = C + \sum_{n=0}^{\infty} \frac{(-1)^n x^{4n+3}}{(4n+3)(2n+1)!}$

35. Using the series we obtained in Exercise 27, we get

$\sqrt{x^3+1} = 1 + \frac{x^3}{2} + \sum_{n=2}^{\infty} \frac{(-1)^{n-1} 1 \cdot 3 \cdot 5 \cdot \cdots \cdot (2n-3)}{2^n \, n!} x^{3n}$, so

$$\int \sqrt{x^3+1}\, dx = \int \left(1 + \frac{x^3}{2} + \sum_{n=2}^{\infty} \frac{(-1)^{n-1} 1 \cdot 3 \cdot 5 \cdot \cdots \cdot (2n-3)}{2^n \, n!} x^{3n} \right) dx$$

$$= C + x + \frac{x^4}{8} + \sum_{n=2}^{\infty} \frac{(-1)^{n-1} 1 \cdot 3 \cdot 5 \cdot \cdots \cdot (2n-3)}{2^n n! \, (3n+1)} x^{3n+1}.$$

37. Using our series from Exercise 33, we get $\int_0^1 \sin(x^2)\, dx = \sum_{n=0}^{\infty} \left[\frac{(-1)^n x^{4n+3}}{(4n+3)(2n+1)!} \right]_0^1 = \sum_{n=0}^{\infty} \frac{(-1)^n}{(4n+3)(2n+1)!}$

and $|c_3| = \frac{1}{75{,}600} < 0.000014$, so by Theorem 10.5.1, we have

$\sum_{n=0}^{2} \frac{(-1)^n}{(4n+3)(2n+1)!} \approx \frac{1}{3} - \frac{1}{42} + \frac{1}{1320} \approx 0.310.$

39. We first find a series representation for $f(x) = (1+x)^{-1/2}$, and then substitute.

n	$f^{(n)}(x)$	$f^{(n)}(0)$
0	$(1+x)^{-1/2}$	1
1	$-\frac{1}{2}(1+x)^{-3/2}$	$-\frac{1}{2}$
2	$\frac{3}{4}(1+x)^{-5/2}$	$3/4$
3	$-\frac{15}{8}(1+x)^{-7/2}$	$-15/8$
$\cdots$	$\cdots$	$\cdots$

$$\frac{1}{\sqrt{1+x}} = 1 - \frac{x}{2} + \frac{3}{4}\left(\frac{x^2}{2!}\right) - \frac{15}{8}\left(\frac{x^3}{3!}\right) + \cdots \quad \Rightarrow$$

$$\frac{1}{\sqrt{1+x^3}} = 1 - \tfrac{1}{2}x^3 + \tfrac{3}{8}x^6 - \tfrac{5}{16}x^9 + \cdots \quad \Rightarrow$$

$$\int_0^{0.1} \frac{dx}{\sqrt{1+x^3}} = \left[x - \tfrac{1}{8}x^4 + \tfrac{3}{56}x^7 - \tfrac{1}{32}x^{10} + \cdots \right]_0^{0.1}$$

$\approx (0.1) - \frac{1}{8}(0.1)^4$ by Theorem 10.28

since $\frac{3}{56}(0.1)^7 \approx 0.0000000054 < 10^{-8}$.

Therefore $\int_0^{0.1} \frac{dx}{\sqrt{1+x^3}} \approx 0.09998750.$

41. As in Example 8(a), we have $e^{-x^2} = 1 - \frac{x^2}{1!} + \frac{x^4}{2!} + \frac{x^6}{3!} + \cdots$ and we know that $\cos x = 1 - \frac{x^2}{2!} + \frac{x^4}{4!} - \cdots$ from Equation 16. Therefore $e^{-x^2} \cos x = \left(1 - x^2 + \frac{1}{2}x^4 - \cdots\right)\left(1 - \frac{1}{2}x^2 + \frac{1}{24}x^4 - \cdots\right)$

$= 1 - \frac{1}{2}x^2 + \frac{1}{24}x^4 - x^2 + \frac{1}{2}x^4 + \frac{1}{2}x^4 + \cdots = 1 - \frac{3}{2}x^2 + \frac{25}{4}x^4 + \cdots$

43. From Example 6 in Section 10.9, we have

$\ln(1-x) = -x - \frac{1}{2}x^2 - \frac{1}{3}x^3 - \cdots,$

$|x| < 1$. Therefore

$$y = \frac{\ln(1-x)}{e^x} = \frac{-x - \frac{1}{2}x^2 - \frac{1}{3}x^3 - \cdots}{1 + x + \frac{1}{2}x^2 + \frac{1}{6}x^3 + \cdots}.$$

So by the long division at right,

$$\frac{\ln(1-x)}{e^x} = -x + \frac{x^2}{2} - \frac{x^3}{3} + \cdots, \ |x| < 1.$$

$$\begin{array}{r|l} & -x + \frac{1}{2}x^2 - \frac{1}{3}x^3 + \cdots \\ \hline 1 + x + \frac{1}{2}x^2 + \frac{1}{6}x^3 - \cdots & -x - \frac{1}{2}x^2 - \frac{1}{3}x^3 - \cdots \\ & -x - \quad x^2 - \frac{1}{2}x^3 - \cdots \\ \hline & \frac{1}{2}x^2 + \frac{1}{6}x^3 - \cdots \\ & \frac{1}{2}x^2 + \frac{1}{2}x^3 + \cdots \\ \hline & -\frac{1}{3}x^3 + \cdots \\ & -\frac{1}{3}x^3 + \cdots \\ \hline & \cdots \end{array}$$

45. $\displaystyle\sum_{n=0}^{\infty}(-1)^n \frac{x^{4n}}{n!} = \sum_{n=0}^{\infty} \frac{\left(-x^4\right)^n}{n!} = e^{-x^4}$ by (12).

47. $\displaystyle\sum_{n=0}^{\infty} \frac{(-1)^n \pi^{2n+1}}{4^{2n+1}(2n+1)!} = \sum_{n=0}^{\infty} \frac{(-1)^n (\pi/4)^{2n+1}}{(2n+1)!} = \sin\frac{\pi}{4} = \frac{1}{\sqrt{2}}$ by (15).

49. $\displaystyle\sum_{n=0}^{\infty} \frac{x^{n+1}}{(n+1)!} = \frac{x}{1!} + \frac{x^2}{2!} + \frac{x^3}{3!} + \cdots = \left(1 + \frac{x}{1!} + \frac{x^2}{2!} + \frac{x^3}{3!} + \cdots\right) - 1 = e^x - 1$ by (12).

51. By (12), $e^x = 1 + x + \dfrac{x^2}{2!} + \dfrac{x^3}{3!} + \dfrac{x^4}{4!} + \cdots$, but for $x > 0$, all of the terms after the first two on the RHS are positive, so $e^x > 1 + x$ for $x > 0$.

53.
$$\lim_{x\to 0} \frac{\sin x - x + \frac{1}{6}x^3}{x^5} = \lim_{x\to 0} \frac{\left(x - \frac{1}{6}x^3 + \frac{1}{5!}x^5 - \frac{1}{7!}x^7 + \cdots\right) - x + \frac{1}{6}x^3}{x^5}$$
$$= \lim_{x\to 0} \frac{\frac{1}{5!}x^5 - \frac{1}{7!}x^7 + \cdots}{x^5} = \lim_{x\to 0}\left(\frac{1}{5!} - \frac{x^2}{7!} + \frac{x^4}{9!} - \cdots\right) = \frac{1}{5!} = \frac{1}{120},$$

since power series are continuous functions.

55. We must show that f equals its Taylor series expansion on I; that is, we must show that $\lim\limits_{n\to\infty} |R_n(x)| = 0$. For $x \in I$, $|R_n(x)| = \left|\dfrac{f^{(n+1)}(z)}{(n+1)!}(x-a)^{n+1}\right| \le \dfrac{M \cdot R^{n+1}}{(n+1)!} \to 0$ as $n \to \infty$ by (11).

EXERCISES 10.11

1. $(1+x)^{1/2} = \sum_{n=0}^{\infty} \binom{1/2}{n} x^n = 1 + \left(\frac{1}{2}\right)x + \frac{\left(\frac{1}{2}\right)\left(-\frac{1}{2}\right)}{2!}x^2 + \frac{\left(\frac{1}{2}\right)\left(-\frac{1}{2}\right)\left(-\frac{3}{2}\right)}{3!}x^3 + \cdots$

$$= 1 + \frac{x}{2} - \frac{x^2}{2^2 \cdot 2!} + \frac{1 \cdot 3 \cdot x^3}{2^3 \cdot 3!} - \frac{1 \cdot 3 \cdot 5 \cdot x^4}{2^4 \cdot 4!} + \cdots$$

$$= 1 + \frac{x}{2} + \sum_{n=2}^{\infty} \frac{(-1)^{n-1} 1 \cdot 3 \cdot 5 \cdot \cdots \cdot (2n-3)x^n}{2^n \cdot n!}, \; R = 1$$

3. $[1+(2x)]^{-4} = 1 + (-4)(2x) + \frac{(-4)(-5)}{2!}(2x)^2 + \frac{(-4)(-5)(-6)}{3!}(2x)^3 + \cdots$

$$= 1 + \sum_{n=1}^{\infty} \frac{(-1)^n 2^n 4 \cdot 5 \cdot 6 \cdots (n+3)}{n!} x^n = \sum_{n=0}^{\infty} (-1)^n \frac{2^n (n+1)(n+2)(n+3)}{6} x^n,$$

and for convergence $|2x| < 1 \quad \Leftrightarrow \quad |x| < \frac{1}{2}$ so $R = \frac{1}{2}$.

5. $[1+(-x)]^{-1/2} = \sum_{n=0}^{\infty} \binom{-1/2}{n} (-x)^n = 1 + \left(-\frac{1}{2}\right)(-x) + \frac{\left(-\frac{1}{2}\right)\left(-\frac{3}{2}\right)}{2!}(-x)^2 + \cdots$

$$= 1 + \frac{x}{2} + \frac{1 \cdot 3}{2^2 2!}x^2 + \frac{1 \cdot 3 \cdot 5}{2^3 3!}x^3 + \frac{1 \cdot 3 \cdot 5 \cdot 7}{2^4 4!}x^4 + \cdots = 1 + \sum_{n=1}^{\infty} \frac{1 \cdot 3 \cdot 5 \cdot \cdots \cdot (2n-1)}{2^n \, n!} x^n,$$

so $\frac{x}{\sqrt{1-x}} = x + \sum_{n=1}^{\infty} \frac{1 \cdot 3 \cdot 5 \cdot \cdots \cdot (2n-1)}{2^n \, n!} x^{n+1}$ with $R = 1$.

7. $\left(1 - x^4\right)^{1/4} = 1 + \left(\frac{1}{4}\right)\left(-x^4\right) + \frac{\left(\frac{1}{4}\right)\left(-\frac{3}{4}\right)}{2!}\left(-x^4\right)^2 + \frac{\left(\frac{1}{4}\right)\left(-\frac{3}{4}\right)\left(-\frac{7}{4}\right)}{3!}\left(-x^4\right)^3 + \cdots$

$$= 1 - \frac{x^4}{4} - \sum_{n=2}^{\infty} \frac{3 \cdot 7 \cdot 11 \cdot \cdots \cdot (4n-5)}{4^n \cdot n!} x^{4n} \text{ with } R = 1.$$

9. $(1-x)^{-5} = 1 + (-5)(-x) + \frac{(-5)(-6)}{2!}(-x)^2 + \frac{(-5)(-6)(-7)}{3!}(-x)^3 + \cdots$

$$= 1 + \sum_{n=1}^{\infty} \frac{5 \cdot 6 \cdot 7 \cdots (n+4)}{n!} x^n = \sum_{n=0}^{\infty} \frac{(n+4)!}{4! \cdot n!} x^n \quad \Rightarrow$$

$$\frac{x^5}{(1-x)^5} = \sum_{n=0}^{\infty} \frac{(n+4)!}{4! \cdot n!} x^{n+5} \quad \left(\text{or } \sum_{n=0}^{\infty} \frac{(n+1)(n+2)(n+3)(n+4)}{24} x^{n+5}\right) \text{ with } R = 1.$$

11. $(8+x)^{-1/3} = \frac{1}{2}\left(1 + \frac{x}{8}\right)^{-1/3}$

$$= \frac{1}{2}\left[1 + \left(-\frac{1}{3}\right)\left(\frac{x}{8}\right) + \frac{\left(-\frac{1}{3}\right)\left(-\frac{4}{3}\right)}{2!}\left(\frac{x}{8}\right)^2 + \cdots\right]$$

$$= \frac{1}{2}\left[1 + \sum_{n=1}^{\infty} \frac{(-1)^n 1 \cdot 4 \cdot 7 \cdot \cdots \cdot (3n-2)}{3^n \cdot n! \, 8^n} x^n\right]$$

and $|x/8| < 1 \;\Leftrightarrow\; |x| < 8$, so $R = 8$. The first three Taylor polynomials are $T_1(x) = \frac{1}{2} - \frac{1}{48}x$, $T_2(x) = \frac{1}{2} - \frac{1}{48}x + \frac{1}{576}x^2$, and $T_3(x) = \frac{1}{2} - \frac{1}{48}x + \frac{1}{576}x^2 - \frac{4 \cdot 7}{27 \cdot 6 \cdot 512}x^3$
$= \frac{1}{2} - \frac{1}{48}x + \frac{1}{576}x^2 - \frac{7}{41{,}472}x^3$.

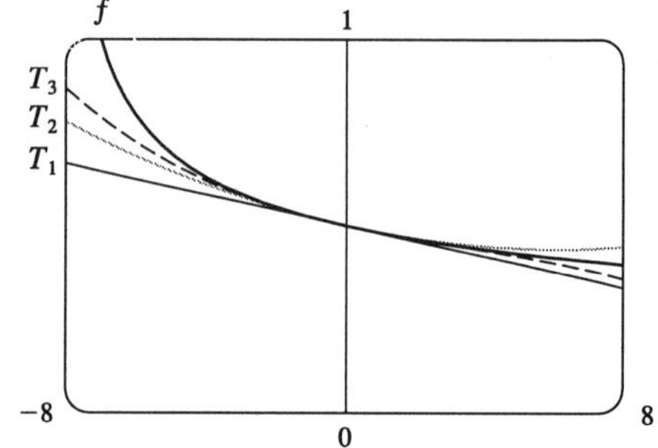

13. **(a)** $\left(1-x^2\right)^{-1/2}=1+\left(-\frac{1}{2}\right)\left(-x^2\right)+\frac{\left(-\frac{1}{2}\right)\left(-\frac{3}{2}\right)}{2!}\left(-x^2\right)^2+\frac{\left(-\frac{1}{2}\right)\left(-\frac{3}{2}\right)\left(-\frac{5}{2}\right)}{3!}\left(-x^2\right)^3+\cdots$

$$=1+\sum_{n=1}^{\infty}\frac{1\cdot 3\cdot 5\cdot\cdots\cdot(2n-1)}{2^n\cdot n!}x^{2n}$$

(b) $\sin^{-1}x=\int\frac{1}{\sqrt{1-x^2}}\,dx=C+x+\sum_{n=1}^{\infty}\frac{1\cdot 3\cdot 5\cdot\cdots\cdot(2n-1)}{(2n+1)2^n\cdot n!}x^{2n+1}$

$$=x+\sum_{n=1}^{\infty}\frac{1\cdot 3\cdot 5\cdots(2n-1)}{(2n+1)2^n\cdot n!}x^{2n+1}\text{ since }0=\sin^{-1}0=C.$$

15. **(a)** $(1+x)^{-1/2}=1+\left(-\frac{1}{2}\right)x+\frac{\left(-\frac{1}{2}\right)\left(-\frac{3}{2}\right)}{2!}x^2+\frac{\left(-\frac{1}{2}\right)\left(-\frac{3}{2}\right)\left(-\frac{5}{2}\right)}{3!}x^3+\cdots$

$$=1+\sum_{n=1}^{\infty}\frac{(-1)^n 1\cdot 3\cdot 5\cdot\cdots\cdot(2n-1)}{2^n\cdot n!}x^n$$

(b) Take $x=0.1$ in the above series. $\frac{1\cdot 3\cdot 5\cdot 7}{2^4\,4!}(0.1)^4<0.00003$, so

$\frac{1}{\sqrt{1.1}}\approx 1-\frac{0.1}{2}+\frac{1\cdot 3}{2^2\cdot 2!}(0.1)^2-\frac{1\cdot 3\cdot 5}{2^3\cdot 3!}(0.1)^3\approx 0.953.$

17. **(a)** $(1-x)^{-2}=1+(-2)(-x)+\frac{(-2)(-3)}{2!}(-x)^2+\cdots=\sum_{n=0}^{\infty}(n+1)x^n$, so

$$\frac{x}{(1-x)^2}=\sum_{n=0}^{\infty}(n+1)x^{n+1}=\sum_{n=1}^{\infty}nx^n.$$

(b) With $x=\frac{1}{2}$ in part (a), we have $\sum_{n=1}^{\infty}\frac{n}{2^n}=\frac{1/2}{(1-1/2)^2}=2.$

19. **(a)** $\left(1+x^2\right)^{1/2}=1+\left(\frac{1}{2}\right)x^2+\frac{\left(\frac{1}{2}\right)\left(-\frac{1}{2}\right)}{2!}\left(x^2\right)^2+\frac{\left(\frac{1}{2}\right)\left(-\frac{1}{2}\right)\left(-\frac{3}{2}\right)}{3!}\left(x^2\right)^3+\cdots$

$$=1+\frac{x^2}{2}+\sum_{n=2}^{\infty}\frac{(-1)^{n-1}1\cdot 3\cdot 5\cdot\cdots\cdot(2n-3)}{2^n\cdot n!}x^{2n}$$

(b) The coefficient of x^{10} in the above Maclaurin series is $\frac{f^{(10)}(0)}{10!}$, so $f^{(10)}(0)=10!\left(\frac{1\cdot 3\cdot 5\cdot 7}{2^5\cdot 5!}\right)=99{,}225.$

21. **(a)** $g'(x)=\sum_{n=1}^{\infty}\binom{k}{n}nx^{n-1}.$

$$(1+x)g'(x)=(1+x)\sum_{n=1}^{\infty}\binom{k}{n}nx^{n-1}=\sum_{n=1}^{\infty}\binom{k}{n}nx^{n-1}+\sum_{n=1}^{\infty}\binom{k}{n}nx^n$$

$$=\sum_{n=0}^{\infty}\binom{k}{n+1}(n+1)x^n+\sum_{n=0}^{\infty}\binom{k}{n}nx^n$$

$$=\sum_{n=0}^{\infty}(n+1)\frac{k(k-1)(k-2)\cdots(k-n)}{(n+1)!}x^n+\sum_{n=0}^{\infty}\left((n)\frac{k(k-1)(k-2)\cdots(k-n+1)}{n!}\right)x^n$$

$$=\sum_{n=0}^{\infty}\frac{(n+1)k(k-1)(k-2)\cdots(k-n+1)}{(n+1)!}[(k-n)+n]x^n$$

$$=\sum_{n=0}^{\infty}\frac{k^2(k-1)(k-2)\cdots(k-n+1)}{n!}x^n=k\sum_{n=0}^{\infty}\binom{k}{n}x^n=kg(x).\text{ So }g'(x)=\frac{kg(x)}{1+x}.$$

(b) $h'(x)=-k(1+x)^{-k-1}g(x)+(1+x)^{-k}g'(x)=-k(1+x)^{-k-1}g(x)+(1+x)^{-k}\frac{kg(x)}{1+x}$

$$=-k(1+x)^{-k-1}g(x)+k(1+x)^{-k-1}g(x)=0$$

(c) From part (b) we see that $h(x)$ must be constant for $x\in(-1,1)$, so $h(x)=h(0)=1$ for $x\in(-1,1)$.

Thus $h(x)=1=(1+x)^{-k}g(x)\quad\Leftrightarrow\quad g(x)=(1+x)^k$ for $x\in(-1,1)$.

EXERCISES 10.12

1.

n	$f^{(n)}(x)$	$f^{(n)}\left(\frac{\pi}{6}\right)$
0	$\sin x$	$\frac{1}{2}$
1	$\cos x$	$\frac{\sqrt{3}}{2}$
2	$-\sin x$	$-\frac{1}{2}$
3	$-\cos x$	$-\frac{\sqrt{3}}{2}$

$$T_3(x) = \sum_{n=0}^{3} \frac{f^{(n)}\left(\frac{\pi}{6}\right)}{n!}\left(x - \tfrac{\pi}{6}\right)^n$$
$$= \tfrac{1}{2} + \tfrac{\sqrt{3}}{2}\left(x - \tfrac{\pi}{6}\right) - \tfrac{1}{4}\left(x - \tfrac{\pi}{6}\right)^2 - \tfrac{\sqrt{3}}{12}\left(x - \tfrac{\pi}{6}\right)^3$$

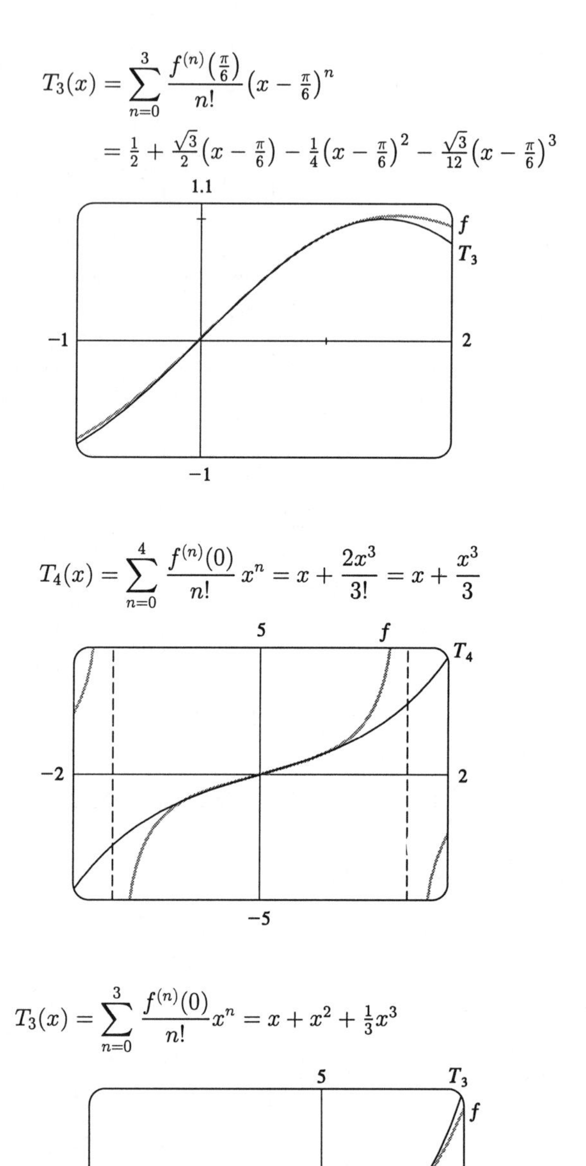

3.

n	$f^{(n)}(x)$	$f^{(n)}(0)$
0	$\tan x$	0
1	$\sec^2 x$	1
2	$2\sec^2 x \tan x$	0
3	$4\sec^2 x \tan^2 x + 2\sec^4 x$	2
4	$8\sec^2 x \tan^3 x + 16\sec^4 x \tan x$	0

$$T_4(x) = \sum_{n=0}^{4} \frac{f^{(n)}(0)}{n!} x^n = x + \frac{2x^3}{3!} = x + \frac{x^3}{3}$$

5.

n	$f^{(n)}(x)$	$f^{(n)}(0)$
0	$e^x \sin x$	0
1	$e^x(\sin x + \cos x)$	1
2	$2e^x \cos x$	2
3	$2e^x(\cos x - \sin x)$	2

$$T_3(x) = \sum_{n=0}^{3} \frac{f^{(n)}(0)}{n!} x^n = x + x^2 + \tfrac{1}{3}x^3$$

5

T3

f

−2.5

1.5

−1.5

7.

n	$f^{(n)}(x)$	$f^{(n)}(8)$
0	$x^{-1/3}$	$\frac{1}{2}$
1	$-\frac{1}{3}x^{-4/3}$	$-\frac{1}{48}$
2	$\frac{4}{9}x^{-7/3}$	$\frac{1}{288}$
3	$-\frac{28}{27}x^{-10/3}$	$-\frac{7}{6912}$

$$T_3(x) = \sum_{n=0}^{3} \frac{f^{(n)}(8)}{n!}(x-8)^n$$
$$= \tfrac{1}{2} - \tfrac{1}{48}(x-8) + \tfrac{1}{576}(x-8)^2 - \tfrac{7}{41{,}472}(x-8)^3$$

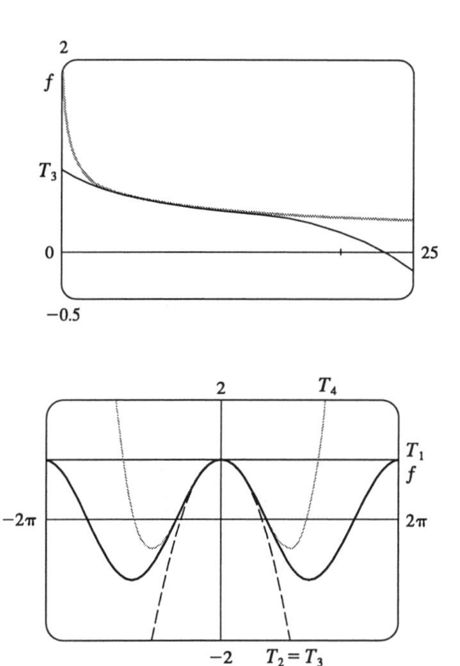

9.

n	$f^{(n)}(x)$	$f^{(n)}(0)$	$T_n(x)$
0	$\cos x$	1	1
1	$-\sin x$	0	1
2	$-\cos x$	-1	$1 - \frac{1}{2}x^2$
3	$\sin x$	0	$1 - \frac{1}{2}x^2$
4	$\cos x$	1	$1 - \frac{1}{2}x^2 + \frac{1}{24}x^4$

11. In Maple, we can find the Taylor polynomials by the following method: first define `f:=sec(x);` and then set `T2:=convert(taylor(f,x=0,3),polynom);`, `T4:=convert(taylor(f,x=0,5),polynom);`, etc. (The third argument in the `taylor` function is one more than the degree of the desired polynomial). We must `convert` to the type `polynom` because the output of the `taylor` function contains an error term which we do not want. In Mathematica, we use `Tn:=Normal[Series[f,{x,0,n}]]`, with `n=2`, 4, etc. Note that in Mathematica, the "degree" argument is the same as the degree of the desired polynomial. The eighth Taylor polynomial is $T_8(x) = 1 + \frac{1}{2}x^2 + \frac{5}{24}x^4 + \frac{61}{720}x^6 + \frac{277}{8064}x^8$.

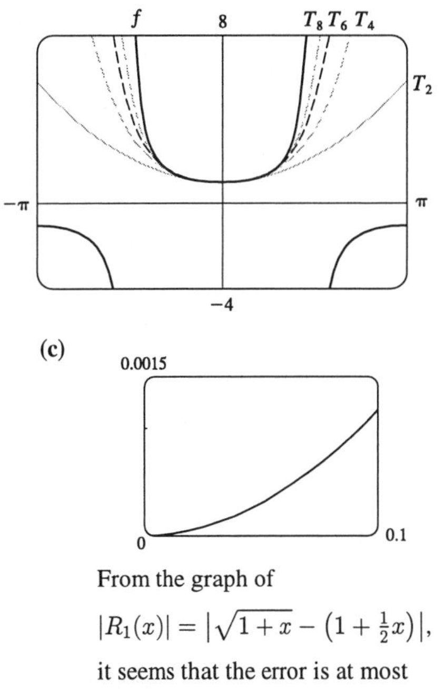

13. $f(x) = (1+x)^{1/2}$ $\quad f(0) = 1$

$f'(x) = \frac{1}{2}(1+x)^{-1/2}$ $\quad f'(0) = \frac{1}{2}$

$f''(x) = -\frac{1}{4}(1+x)^{-3/2}$

(a) $(1+x)^{1/2} \approx T_1(x) = 1 + \frac{1}{2}x$

(b) By Taylor's Formula, the remainder is

$$R_1(x) = \frac{f''(z)}{2!}x^2 = -\frac{1}{8(1+z)^{3/2}}x^2,$$ where z lies between 0 and x. Now $0 \le x \le 0.1 \;\Rightarrow\; 0 \le x^2 \le 0.01$ and $0 < z < 0.1 \;\Rightarrow\; 1 < 1 + z < 1.1$, so $|R_1(x)| < \frac{0.01}{8.1} = 0.00125$.

(c) From the graph of $|R_1(x)| = \left|\sqrt{1+x} - \left(1 + \frac{1}{2}x\right)\right|$, it seems that the error is at most 0.0013 on $(0, 0.1)$.

15. $f(x) = \sin x \qquad f\left(\frac{\pi}{4}\right) = \frac{\sqrt{2}}{2}$

$f'(x) = \cos x \qquad f'\left(\frac{\pi}{4}\right) = \frac{\sqrt{2}}{2}$

$f''(x) = -\sin x \qquad f''\left(\frac{\pi}{4}\right) = -\frac{\sqrt{2}}{2}$

$f'''(x) = -\cos x \qquad f'''\left(\frac{\pi}{4}\right) = -\frac{\sqrt{2}}{2}$

$f^{(4)}(x) = \sin x \qquad f^{(4)}\left(\frac{\pi}{4}\right) = \frac{\sqrt{2}}{2}$

$f^{(5)}(x) = \cos x \qquad f^{(5)}\left(\frac{\pi}{4}\right) = \frac{\sqrt{2}}{2}$

$f^{(6)}(x) = -\sin x$

(a) $\sin x \approx T_5(x) = \frac{\sqrt{2}}{2} + \frac{\sqrt{2}}{2}\left(x - \frac{\pi}{4}\right) - \frac{\sqrt{2}}{4}\left(x - \frac{\pi}{4}\right)^2 - \frac{\sqrt{2}}{12}\left(x - \frac{\pi}{4}\right)^3 + \frac{\sqrt{2}}{48}\left(x - \frac{\pi}{4}\right)^4 + \frac{\sqrt{2}}{240}\left(x - \frac{\pi}{4}\right)^5$

(b) The remainder is $R_5(x) = \frac{1}{6!}f^{(6)}(z)\left(x - \frac{\pi}{4}\right)^6 = \frac{1}{720}(-\sin z)\left(x - \frac{\pi}{4}\right)^6$, where z lies between $\frac{\pi}{4}$ and x. Since $0 \le x \le \frac{\pi}{2}$, $-\frac{\pi}{4} \le x - \frac{\pi}{4} \le \frac{\pi}{4} \quad \Rightarrow \quad 0 \le \left(x - \frac{\pi}{4}\right)^6 \le \left(\frac{\pi}{4}\right)^6$, and since $0 < z < \frac{\pi}{2}$, $0 < \sin z < 1$, so $|R_5(x)| < \frac{1}{720}\left(\frac{\pi}{4}\right)^6 \approx 0.00033$.

(c)

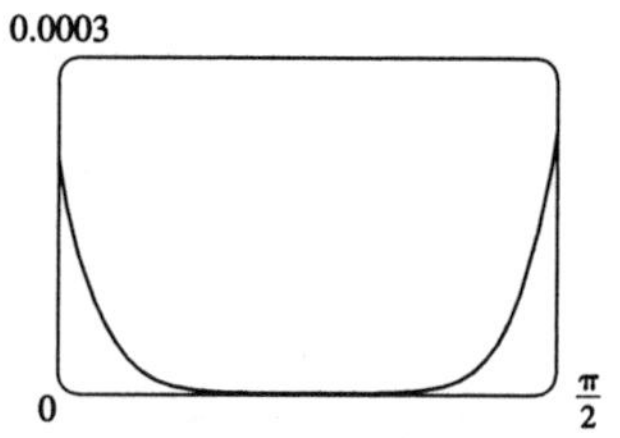

From the graph, it seems that the error is less than 0.00026 on $\left(0, \frac{\pi}{2}\right)$.

17. $f(x) = \tan x \qquad f(0) = 0$

$f'(x) = \sec^2 x \qquad f'(0) = 1$

$f''(x) = 2\sec^2 x \tan x \qquad f''(0) = 0$

$f'''(x) = 4\sec^2 x \tan^2 x + 2\sec^4 x \qquad f'''(0) = 2$

$f^{(4)}(x) = 8\sec^2 x \tan^3 x + 16\sec^4 x \tan x$

(a) $\tan x \approx T_3(x) = x + \frac{1}{3}x^3$

(b) $R_3(x) = \dfrac{f^{(4)}(z)}{4!}x^4 = \dfrac{8\sec^2 z \tan^3 z + 16\sec^4 z \tan z}{4!}x^4$

$= \dfrac{\sec^2 z \tan^3 z + 2\sec^4 z \tan z}{3}$ where z lies between 0 and x. Now $0 \le x^4 \le \left(\dfrac{\pi}{6}\right)^4$ and $0 < z < \dfrac{\pi}{6}$

$\Rightarrow \quad \sec^2 z < \frac{4}{3}$ and $\tan z < \frac{\sqrt{3}}{3}$, so $|R_3(x)| < \dfrac{\frac{4}{3}\cdot\frac{1}{3\sqrt{3}} + 2\cdot\frac{16}{9}\cdot\frac{1}{\sqrt{3}}}{3}\left(\dfrac{\pi}{6}\right)^4 = \dfrac{4\sqrt{3}}{9}\left(\dfrac{\pi}{6}\right)^4 < 0.06$.

(c)

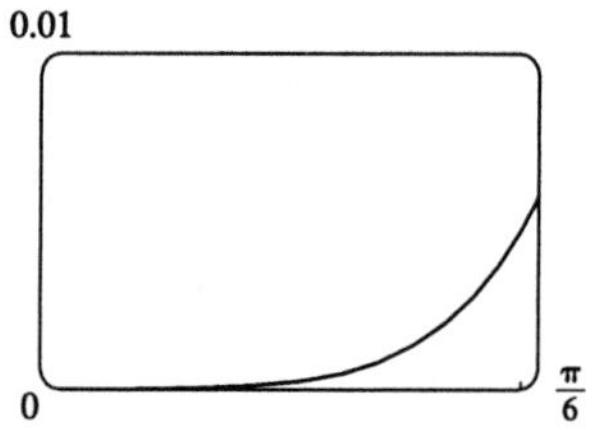

From the graph, it seems that the error is less than 0.006 on $(0, \pi)$

19. $f(x) = e^{x^2} \qquad f(0) = 1$

$f'(x) = e^{x^2}(2x) \qquad f'(0) = 0$

$f''(x) = e^{x^2}(2 + 4x^2) \qquad f''(0) = 2$

$f'''(x) = e^{x^2}(12x + 8x^3) \qquad f'''(0) = 0$

$f^{(4)}(x) = e^{x^2}(12 + 48x^2 + 16x^4)$

(a) $e^{x^2} \approx T_3(x) = 1 + x^2$

(b) $R_3(x) = \dfrac{f^{(4)}(z)}{4!}x^4 = \dfrac{e^{z^2}(3 + 12z^2 + 4z^4)}{6}x^4$, where

z lies between 0 and x. $0 \le x \le 0.1 \quad \Rightarrow \quad |R_3(x)| < \dfrac{e^{0.01}(3 + 0.12 + 0.0004)}{6}(0.0001) < 0.00006$.

(c)

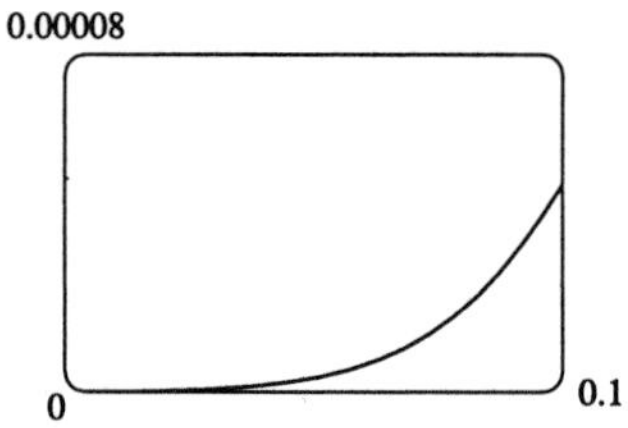

It appears that the error is less than 0.00005 on $(0, 0.1)$.

21. $f(x) = x^{3/4}$ $\qquad f(16) = 8$

$f'(x) = \frac{3}{4}x^{-1/4}$ $\qquad f'(16) = \frac{3}{8}$

$f''(x) = -\frac{3}{16}x^{-5/4}$ $\qquad f''(16) = -\frac{3}{512}$

$f'''(x) = \frac{15}{64}x^{-9/4}$ $\qquad f'''(16) = \frac{15}{32{,}768}$

$f^{(4)}(x) = -\frac{135}{256}x^{-13/4}$

(a) $x^{3/4} \approx T_3(x)$

$$= 8 + \tfrac{3}{8}(x-16) - \tfrac{3}{1024}(x-16)^2 + \tfrac{5}{65{,}536}(x-16)^3$$

(b) $R_3(x) = \dfrac{f^{(4)}(z)}{4!}(x-16)^4 = -\dfrac{135(x-16)^4}{256 \cdot 4!\, z^{13/4}}$, where z lies between 16 and x. $|x - 16| \le 1$ and $z > 15$ $\Rightarrow$ $|R_3(x)| < \dfrac{135}{256 \cdot 24 \cdot 15^{13/4}} < 0.0000034.$

(c)

5×10^{-6}

15   0   17

It appears that the error is less than 3×10^{-6} on $(15, 17)$.

23. From Exercise 1, $\sin x = \frac{1}{2} + \frac{\sqrt{3}}{2}\left(x - \frac{\pi}{6}\right) - \frac{1}{4}\left(x - \frac{\pi}{6}\right)^2 - \frac{\sqrt{3}}{12}\left(x - \frac{\pi}{6}\right)^3 + R_3(x)$, where $R_3(x) = \dfrac{\sin z}{4!}\left(x - \frac{\pi}{6}\right)^4$ and z lies between $\frac{\pi}{6}$ and x. Now $35^\circ = \left(\frac{\pi}{6} + \frac{\pi}{36}\right)$ radians, so the error is $\left|R_3\left(\frac{\pi}{36}\right)\right| < \frac{(\pi/36)^4}{4!} < 0.000003$. Therefore, to five decimal places, $\sin 35^\circ \approx \frac{1}{2} + \frac{\sqrt{3}}{2}\left(\frac{\pi}{36}\right) - \frac{1}{4}\left(\frac{\pi}{36}\right)^2 + \frac{\sqrt{3}}{12}\left(\frac{\pi}{36}\right)^3 \approx 0.57358$.

25. All derivatives of e^x are e^x, so the remainder term is $R_n(x) = \dfrac{e^z}{(n+1)!}x^{n+1}$, where $0 < z < 0.1$. So we want $R_n(0.1) \le \dfrac{e^{0.1}}{(n+1)!}(0.1)^{n+1} < 0.00001$, and we find that $n = 3$ satisfies this inequality. [In fact $R_3(0.1) < 0.0000046$.]

27. $\sin x = x - \frac{1}{3!}x^3 + \frac{1}{5!}x^5 - \cdots$. By the Alternating Series Estimation Theorem, the error in the approximation $\sin x = x - \frac{1}{3!}x^3$ is less than $\left|\frac{1}{5!}x^5\right| < 0.01 \quad \Leftrightarrow \quad |x^5| < 1.2 \quad \Leftrightarrow \quad |x| < (1.2)^{1/5} \approx 1.037$.

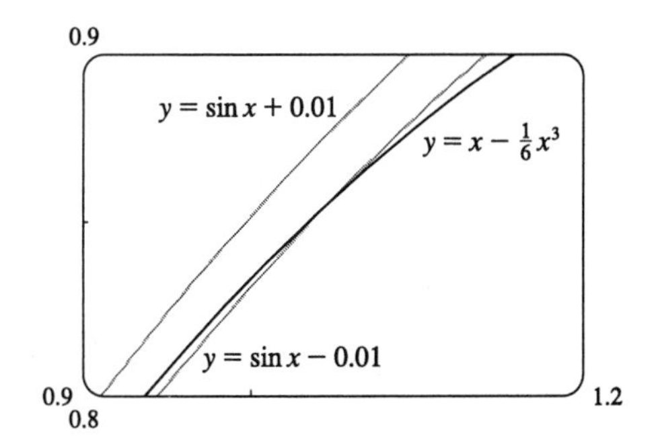

The graph confirms our estimate. Since both the sine function and the given approximation are odd functions, we only need to check the estimate for $x > 0$.

29. Let $s(t)$ be the position function of the car, and for convenience set $s(0) = 0$. The velocity of the car is $v(t) = s'(t)$ and the acceleration is $a(t) = s''(t)$, so the second degree Taylor polynomial is $T_2(t) = s(0) + v(0)t + \frac{a(0)}{2}t^2 = 20t + t^2$. We estimate the distance travelled during the next second to be $s(1) \approx T_2(1) = 20 + 1 = 21$ m. The function $T_2(t)$ would not be accurate over a full minute, since the car could not possibly maintain an acceleration of 2 m/s^2 for that long (if it did, its final speed would be $140\text{ m/s} \approx 315\text{ mi/h}$!)

(b) To convert to μm, we substitute $\lambda/10^6$ for λ in both laws. We can see that the two laws are very different for short wavelengths (Planck's Law gives a maximum at $\lambda \approx 0.5\ \mu\text{m}$; the Rayleigh-Jeans Law gives no extremum.) The two laws are similar for large λ.

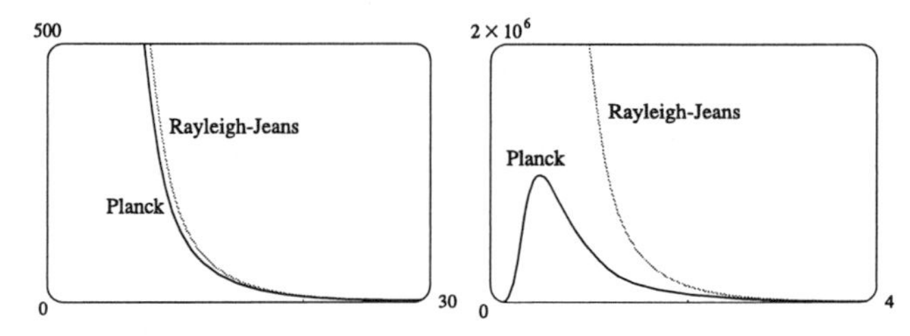

31. $E = \frac{q}{D^2} - \frac{q}{(D+d)^2} = \frac{q}{D^2} - \frac{q}{D^2(1+d/D)^2} = \frac{q}{D^2}\left[1 - \left(1 + \frac{d}{D}\right)^{-2}\right]$.

We use the Binomial Series to expand $(1 + d/D)^{-2}$:

$$E = \frac{q}{D^2}\left[1 - \left(1 - 2\left(\frac{d}{D}\right) + \frac{2\cdot 3}{2!}\left(\frac{d}{D}\right)^2 - \frac{2\cdot 3\cdot 4}{3!}\left(\frac{d}{D}\right)^3 + \cdots\right)\right]$$

$$= \frac{q}{D^2}\left[2\left(\frac{d}{D}\right) - 3\left(\frac{d}{D}\right)^2 + 4\left(\frac{d}{D}\right)^3 - \cdots\right] \approx 2qd \cdot \frac{1}{D^3} \text{ when } D \text{ is much larger than } d.$$

33. **(a)** If the water is deep, then $2\pi d/L$ is large, and we know that $\tanh x \to 1$ as $x \to \infty$. So we can approximate $\tanh(2\pi d/L) \approx 1$, and so $v^2 \approx gL/(2\pi) \quad\Leftrightarrow\quad v \approx \sqrt{gL/(2\pi)}$.

(b) From the calculations at right, the first term in the Maclaurin series of $\tanh x$ is x, so if the water is shallow, we can approximate $\tanh\frac{2\pi d}{L} \approx \frac{2\pi d}{L}$, and so $v^2 \approx \frac{gL}{2\pi}\cdot\frac{2\pi d}{L} \quad\Leftrightarrow\quad v \approx \sqrt{gd}$.

$f(x) = \tanh x \qquad f(0) = 0$

$f'(x) = \operatorname{sech}^2 x \qquad f'(0) = 1$

$f''(x) = -2\operatorname{sech}^2 x \tanh x \qquad f''(0) = 0$

$f'''(x) = 2\operatorname{sech}^2 x(3\tanh^2 x - 1) \qquad f'''(0) = -2$

(c) Since $\tanh x$ is an odd function, its Maclaurin series is alternating, so the error in the approximation $\tanh\frac{2\pi d}{L} \approx \frac{2\pi d}{L}$ is less than the first neglected term, which is $\frac{|f'''(0)|}{3!}\left(\frac{2\pi d}{L}\right)^3 = \frac{1}{3}\left(\frac{2\pi d}{L}\right)^3$. If $L > 10d$, then $\frac{1}{3}\left(\frac{2\pi d}{L}\right)^3 < \frac{1}{3}\left(2\pi\cdot\frac{1}{10}\right)^3 = \frac{\pi^3}{375}$, so the error in the approximation $v^2 = gd$ is less than $\frac{gL}{2\pi}\cdot\frac{\pi^3}{375} \approx 0.0132gL$.

35. Using Taylor's Formula with $n = 1$, $a = x_n$, $x = r$, we get $f(r) = f(x_n) + f'(x_n)(r - x_n) + R_1(x)$, where $R_1(x) = \frac{1}{2}f''(z)(r - x_n)^2$ and z lies between x_n and r. But r is a root, so $f(r) = 0$ and Taylor's Formula becomes $0 = f(x_n) + f'(x_n)(r - x_n) + \frac{1}{2}f''(z)(r - x_n)^2$. Taking the first two terms to the left side and dividing by $f'(x_n)$, we have $x_n - r - \dfrac{f(x_n)}{f'(x_n)} = \dfrac{1}{2}\dfrac{f''(z)}{f'(x_n)}|x_n - r|^2$. By the formula for Newton's Method, we have $|x_{n+1} - r| = \left|x_n - \dfrac{f(x_n)}{f'(x_n)} - r\right| = \dfrac{1}{2}\dfrac{|f''(z)|}{|f'(x_n)|}|x_n - r|^2 \le \dfrac{M}{2K}|x_n - r|^2$ since $|f''(z)| \le M$ and $|f'(x_n)| \ge K$.

REVIEW EXERCISES FOR CHAPTER 10

1. False. See the warning in Note 2 after Theorem 10.2.6.

3. False. For example, take $a_n = (-1)^n/(n6^n)$.

5. False, since $\lim\limits_{n\to\infty}\left|\dfrac{a_{n+1}}{a_n}\right| = \lim\limits_{n\to\infty}\left|\dfrac{n^3}{(n+1)^3}\right| = \lim\limits_{n\to\infty}\dfrac{1}{(1+1/n)^3} = 1$.

7. False. See the remarks after Example 3 in Section 10.4.

9. False. A power series has the form $a_0 + a_1x + a_2x^2 + a_3x^3 + \cdots$.

11. True. See Example 8 in Section 10.1.

13. True. By Theorem 10.10.5 the coefficient of x^3 is $\dfrac{f'''(0)}{3!} = \dfrac{1}{3} \quad\Rightarrow\quad f'''(0) = 2$.

Or: Use Theorem 10.9.2 to differentiate f three times.

15. False. For example, let $a_n = b_n = (-1)^n$. Then $\{a_n\}$ and $\{b_n\}$ are divergent, but $a_nb_n = 1$, so $\{a_nb_n\}$ is convergent.

17. True by Theorem 10.6.3. [$\sum(-1)^n a_n$ is absolutely convergent and hence convergent.]

19. $\lim\limits_{n\to\infty}\dfrac{n}{2n+5} = \lim\limits_{n\to\infty}\dfrac{1}{2+5/n} = \dfrac{1}{2}$ and the sequence is convergent.

21. $\{2n+5\}$ is divergent since $2n + 5 \to \infty$ as $n \to \infty$.

23. $\{\sin n\}$ is divergent since $\lim\limits_{n\to\infty}\sin n$ does not exist.

25. $\left\{\left(1+\dfrac{3}{n}\right)^{4n}\right\}$ is convergent. Let $y = \left(1+\dfrac{3}{x}\right)^{4x}$. Then

$$\lim_{x\to\infty}\ln y = \lim_{x\to\infty}4x\ln(1+3/x) = \lim_{x\to\infty}\frac{\ln(1+3/x)}{1/(4x)} \overset{\text{H}}{=} \lim_{x\to\infty}\frac{\dfrac{1}{1+3/x}\left(-\dfrac{3}{x^2}\right)}{-1/(4x^2)} = \lim_{x\to\infty}\frac{12}{1+3/x} = 12,\text{ so}$$

$\lim\limits_{x\to\infty}y = \lim\limits_{n\to\infty}(1+3/n)^{4n} = e^{12}$.

27. We use induction, hypothesizing that $a_{n-1} < a_n < 2$. Note first that $1 < a_2 = \frac{1}{3}(1+5) = \frac{5}{3} < 2$, so the hypothesis holds for $n = 2$. Now assume that $a_{k-1} < a_k < 2$. Then $a_k = \frac{1}{3}(a_{k-1}+4) < \frac{1}{3}(a_k+4) < \frac{1}{3}(2+4) = 2$. So $a_k < a_{k+1} < 2$, and the induction is complete. To find the limit of the sequence, we note that $L = \lim_{n\to\infty} a_n = \lim_{n\to\infty} a_{n+1} \quad\Rightarrow\quad L = \frac{1}{3}(L+4) \quad\Rightarrow\quad L = 2$.

29. Use the Limit Comparison Test with $a_n = \dfrac{n^2}{n^3+1}$ and $b_n = \dfrac{1}{n}$. $\lim_{n\to\infty} \dfrac{a_n}{b_n} = \lim_{n\to\infty} \dfrac{n^2/(n^3+1)}{1/n}$ $= \lim_{n\to\infty} \dfrac{1}{1+1/n^3} = 1$. Since $\sum_{n=1}^{\infty} \dfrac{1}{n}$ (the harmonic series) diverges, $\sum_{n=1}^{\infty} \dfrac{n^2}{n^3+1}$ diverges also.

31. An alternating series with $a_n = \dfrac{1}{n^{1/4}}$, $a_n > 0$ for all n, and $a_n > a_{n+1}$. $\lim_{n\to\infty} a_n = \lim_{n\to\infty} \dfrac{1}{n^{1/4}} = 0$, so the series converges by the Alternating Series Test.

33. $\lim_{n\to\infty} \sqrt[n]{|a_n|} = \lim_{n\to\infty} [n/(3n+1)] = \frac{1}{3} < 1$, so series converges by the Root Test.

35. $\dfrac{|\sin n|}{1+n^2} \le \dfrac{1}{1+n^2} < \dfrac{1}{n^2}$ and since $\sum_{n=1}^{\infty} \dfrac{1}{n^2}$ converges (p-series with $p = 2 > 1$), so does $\sum_{n=1}^{\infty} \dfrac{|\sin n|}{1+n^2}$ by the Comparison Test.

37. $\lim_{n\to\infty} \left|\dfrac{a_{n+1}}{a_n}\right| = \lim_{n\to\infty} \dfrac{1\cdot 3\cdot 5\cdot \cdots \cdot (2n-1)(2n+1)}{5^{n+1}(n+1)!} \cdot \dfrac{5^n\, n!}{1\cdot 3\cdot 5\cdot \cdots \cdot (2n-1)} = \lim_{n\to\infty} \dfrac{2n+1}{5(n+1)}$ $= \frac{2}{5} < 1$, so the series converges by the Ratio Test.

39. $\lim_{n\to\infty} \left|\dfrac{a_{n+1}}{a_n}\right| = \lim_{n\to\infty} \dfrac{4^{n+1}}{(n+1)3^{n+1}} \cdot \dfrac{n3^n}{4^n} = \frac{4}{3} \lim_{n\to\infty} \dfrac{n}{n+1} = \frac{4}{3} > 1$ so the series diverges by the Ratio Test.

41. Consider the series of absolute values: $\sum_{n=1}^{\infty} n^{-1/3}$ is a p-series with $p = \frac{1}{3} < 1$ and is therefore divergent. But if we apply the Alternating Series Test we see that $a_{n+1} < a_n$ and $\lim_{n\to\infty} n^{-1/3} = 0$. Therefore $\sum_{n=1}^{\infty} (-1)^{n-1} n^{-1/3}$ is conditionally convergent.

43. $\left|\dfrac{a_{n+1}}{a_n}\right| = \left|\dfrac{(-1)^{n+1}(n+2)3^{n+1}}{2^{2n+3}} \cdot \dfrac{2^{2n+1}}{(-1)^n(n+1)3^n}\right| = \dfrac{n+2}{n+1} \cdot \dfrac{3}{4} = \dfrac{1+(2/n)}{1+(1/n)} \cdot \dfrac{3}{4} \to \dfrac{3}{4} < 1$ as $n \to \infty$ so by the Ratio Test, $\sum_{n=1}^{\infty} \dfrac{(-1)^n(n+1)3^n}{2^{2n+1}}$ is absolutely convergent.

45. Convergent geometric series. $\sum_{n=1}^{\infty} \dfrac{2^{2n+1}}{5^n} = 2\sum_{n=1}^{\infty} \dfrac{4^n}{5^n} = 2\left(\dfrac{4/5}{1-4/5}\right) = 8$.

47.
$$\sum_{n=1}^{\infty} [\tan^{-1}(n+1) - \tan^{-1} n] = \lim_{n\to\infty} [(\tan^{-1} 2 - \tan^{-1} 1) + (\tan^{-1} 3 - \tan^{-1} 2) + \cdots + (\tan^{-1}(n+1) - \tan^{-1} n)]$$
$$= \lim_{n\to\infty} [\tan^{-1}(n+1) - \tan^{-1} 1] = \tfrac{\pi}{2} - \tfrac{\pi}{4} = \tfrac{\pi}{4}$$

49. $1.2 + 0.0\overline{345} = \frac{12}{10} + \frac{345/10{,}000}{1-1/1000} = \frac{12}{10} + \frac{345}{9990} = \frac{4111}{3330}$

51. $\sum_{n=1}^{\infty} \dfrac{(-1)^{n+1}}{n^5} = 1 - \dfrac{1}{32} + \dfrac{1}{243} - \dfrac{1}{1024} + \dfrac{1}{3125} - \dfrac{1}{7776} + \dfrac{1}{16{,}807} - \dfrac{1}{32{,}768} + \cdots$.

Since $\dfrac{1}{32{,}768} < 0.000031$, $\sum_{n=1}^{\infty} \dfrac{(-1)^{n+1}}{n^5} \approx \sum_{n=1}^{7} \dfrac{(-1)^{n+1}}{n^5} \approx 0.9721$.

53. $\sum_{n=1}^{\infty} \frac{1}{2+5^n} \approx \sum_{n=1}^{8} \frac{1}{2+5^n} \approx 0.18976224$. To estimate the error, note that $\frac{1}{2+5^n} < \frac{1}{5^n}$, so the remainder term is $R_8 = \sum_{n=9}^{\infty} \frac{1}{2+5^n} < \sum_{n=9}^{\infty} \frac{1}{5^n} = \frac{1/5^9}{1-1/5} \approx 6.4 \times 10^{-7}$ (geometric series with $a = 1/5^9$ and $r = \frac{1}{5}$).

55. Use the Limit Comparison Test. $\lim_{n\to\infty} \left| \frac{\left(\frac{n+1}{n}\right) a_n}{a_n} \right| = \lim_{n\to\infty} \frac{n+1}{n} = \lim_{n\to\infty} \left(1 + \frac{1}{n}\right) = 1 > 0$. Since $\sum |a_n|$ is convergent, so is $\sum \left| \left(\frac{n+1}{n}\right) a_n \right|$ by the Limit Comparison Test.

57. $\lim_{n\to\infty} \left| \frac{a_{n+1}}{a_n} \right| = \lim_{n\to\infty} \left| \frac{x^{n+1}}{3^{n+1}(n+1)^3} \cdot \frac{3^n n^3}{x^n} \right| = \frac{|x|}{3} \lim_{n\to\infty} \left(\frac{n}{n+1}\right)^3 = \frac{|x|}{3} < 1$ for convergence (Ratio Test) $\Rightarrow$ $|x| < 3$ and the radius of convergence is 3. When $x = \pm 3$, $\sum_{n=1}^{\infty} |a_n| = \sum_{n=1}^{\infty} \frac{1}{n^3}$ which is a convergent p-series ($p = 3 > 1$), so the interval of convergence is $[-3, 3]$.

59. $\lim_{n\to\infty} \left| \frac{a_{n+1}}{a_n} \right| = \lim_{n\to\infty} \left| \frac{2^{n+1}(x-3)^{n+1}}{\sqrt{n+4}} \cdot \frac{\sqrt{n+3}}{2^n (x-3)^n} \right| = 2|x-3| \lim_{n\to\infty} \sqrt{\frac{n+3}{n+4}} = 2|x-3| < 1 \quad \Leftrightarrow$ $|x-3| < \frac{1}{2}$ so the radius of convergence is $\frac{1}{2}$. For $x = \frac{7}{2}$ the series becomes $\sum_{n=0}^{\infty} \frac{1}{\sqrt{n+3}} = \sum_{n=3}^{\infty} \frac{1}{n^{1/2}}$ which diverges ($p = \frac{1}{2} < 1$), but for $x = \frac{5}{2}$ we get $\sum_{n=0}^{\infty} \frac{(-1)^n}{\sqrt{n+3}}$ which is a convergent alternating series, so the interval of convergence is $\left[\frac{5}{2}, \frac{7}{2}\right)$.

61.

$$\begin{aligned}
f(x) &= \sin x & f\left(\tfrac{\pi}{6}\right) &= \tfrac{1}{2} \\
f'(x) &= \cos x & f'\left(\tfrac{\pi}{6}\right) &= \tfrac{\sqrt{3}}{2} \\
f''(x) &= -\sin x & f''\left(\tfrac{\pi}{6}\right) &= -\tfrac{1}{2} \\
f'''(x) &= -\cos x & f'''\left(\tfrac{\pi}{6}\right) &= -\tfrac{\sqrt{3}}{2} \\
f^{(4)}(x) &= \sin x & f^{(4)}\left(\tfrac{\pi}{6}\right) &= \tfrac{1}{2} \\
&\cdots & &\cdots
\end{aligned}$$

$f^{(2n)}\left(\frac{\pi}{6}\right) = (-1)^n \cdot \frac{1}{2}$ and $f^{(2n+1)}\left(\frac{\pi}{6}\right) = (-1)^n \cdot \frac{\sqrt{3}}{2}$.

$$\sin x = \sum_{n=0}^{\infty} \frac{f^{(n)}\left(\frac{\pi}{6}\right)}{n!} \left(x - \tfrac{\pi}{6}\right)^n = \sum_{n=0}^{\infty} \frac{(-1)^n}{2(2n)!} \left(x - \tfrac{\pi}{6}\right)^{2n} + \sum_{n=0}^{\infty} \frac{(-1)^n \sqrt{3}}{2(2n+1)!} \left(x - \tfrac{\pi}{6}\right)^{2n+1}$$

63. $\frac{1}{1+x} = \frac{1}{1-(-x)} = \sum_{n=0}^{\infty} (-1)^n x^n$ for $|x| < 1 \quad \Rightarrow \quad \frac{x^2}{1+x} = \sum_{n=0}^{\infty} (-1)^n x^{n+2}$ with $R = 1$.

65. $\frac{1}{1-x} = \sum_{n=0}^{\infty} x^n$ for $|x| < 1 \quad \Rightarrow \quad \ln(1-x) = -\int \frac{dx}{1-x} = -\int \sum_{n=0}^{\infty} x^n \, dx = C - \sum_{n=0}^{\infty} \frac{x^{n+1}}{n+1}$.

$\ln(1-0) = C - 0 \quad \Rightarrow \quad C = 0 \quad \Rightarrow \quad \ln(1-x) = -\sum_{n=0}^{\infty} \frac{x^{n+1}}{n+1} = \sum_{n=1}^{\infty} \frac{-x^n}{n}$ with $R = 1$.

67. $\sin x = \sum_{n=0}^{\infty} \frac{(-1)^n x^{2n+1}}{(2n+1)!} \Rightarrow \sin(x^4) = \sum_{n=0}^{\infty} \frac{(-1)^n (x^4)^{2n+1}}{(2n+1)!} = \sum_{n=0}^{\infty} \frac{(-1)^n x^{8n+4}}{(2n+1)!}$ for all x, so the radius of convergence is ∞.

69. $(16-x)^{-1/4} = \frac{1}{2}\left(1-\frac{1}{16}x\right)^{-1/4} = \frac{1}{2}\left[1+\left(-\frac{1}{4}\right)\left(-\frac{x}{16}\right)+\frac{\left(-\frac{1}{4}\right)\left(-\frac{5}{4}\right)}{2!}\left(-\frac{x}{16}\right)^2+\cdots\right]$

$= \frac{1}{2}+\sum_{n=1}^{\infty}\frac{1\cdot 5\cdot 9\cdot\cdots\cdot(4n-3)}{2\cdot 4^n\cdot n!\cdot 16^n}x^n = \frac{1}{2}+\sum_{n=1}^{\infty}\frac{1\cdot 5\cdot 9\cdot\cdots\cdot(4n-3)}{2^{6n+1}\,n!}x^n$ for $\left|-\frac{x}{16}\right|<1 \quad\Rightarrow\quad R=16.$

71. $e^x = \sum_{n=0}^{\infty}\frac{x^n}{n!}$ so $\frac{e^x}{x} = \frac{1}{x}+\sum_{n=1}^{\infty}\frac{x^{n-1}}{n!}$ and $\int\frac{e^x}{x}\,dx = C+\ln|x|+\sum_{n=1}^{\infty}\frac{x^n}{n\cdot n!}$

73. (a)

$f(x)=x^{1/2}$ $\qquad f(1)=1$

$f'(x)=\frac{1}{2}x^{-1/2}$ $\qquad f'(1)=\frac{1}{2}$

$f''(x)=-\frac{1}{4}x^{-3/2}$ $\qquad f''(1)=-\frac{1}{4}$

$f'''(x)=\frac{3}{8}x^{-5/2}$ $\qquad f'''(1)=\frac{3}{8}$

$f^{(4)}(x)=-\frac{15}{16}x^{-7/2}$

$\sqrt{x}\approx T_3(x)=1+\frac{1}{2}(x-1)-\frac{1}{8}(x-1)^2+\frac{1}{16}(x-1)^3$

(b)

(c) By Taylor's Formula,

$$R_3(x)=\frac{f^{(4)}(z)}{4!}(x-1)^4=-\frac{5(x-1)^4}{128z^{7/2}},$$

with z between x and 1. If $0.9\le x\le 1.1$ then $0\le|x-1|\le 0.1$ and $z^{7/2}>(0.9)^{7/2}$ so

$$|R_3(x)|<\frac{5(0.1)^4}{128(0.9)^{7/2}}<0.000006.$$

(d)

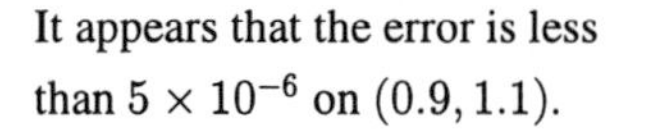

It appears that the error is less than 5×10^{-6} on $(0.9, 1.1)$.

75. $e^x=\sum_{n=0}^{\infty}\frac{x^n}{n!} \quad\Rightarrow\quad e^{-1/x^2}=\sum_{n=0}^{\infty}\frac{\left(-1/x^2\right)^n}{n!}=1-\frac{1}{x^2}+\frac{1}{2x^4}-\cdots \quad\Rightarrow$

$x^2\left(1-e^{-1/x^2}\right)=x^2\left(\frac{1}{x^2}-\frac{1}{2x^4}+\cdots\right)=1-\frac{1}{2x^2}+\cdots\to 1$ as $x\to\infty$.

77. $f(x)=\sum_{n=0}^{\infty}c_nx^n \quad\Rightarrow\quad f(-x)=\sum_{n=0}^{\infty}c_n(-x)^n=\sum_{n=0}^{\infty}(-1)^nc_nx^n$

(a) If f is an odd function, then $f(-x)=-f(x) \quad\Rightarrow\quad \sum_{n=0}^{\infty}(-1)^nc_nx^n=\sum_{n=0}^{\infty}-c_nx^n$. The coefficients of any power series are uniquely determined (by Theorem 10.10.5), so $(-1)^nc_n=-c_n$. If n is even, then $(-1)^n=1$, so $c_n=-c_n \quad\Rightarrow\quad 2c_n=0 \quad\Rightarrow\quad c_n=0$. Thus all even coefficients are 0.

(b) If f is even, then $f(-x)=f(x) \quad\Rightarrow\quad \sum_{n=0}^{\infty}(-1)^nc_nx^n=\sum_{n=0}^{\infty}c_nx^n \quad\Rightarrow\quad (-1)^nc_n=c_n$. If n is odd, then $(-1)^n=-1$, so $-c_n=c_n \quad\Rightarrow\quad 2c_n=0 \quad\Rightarrow\quad c_n=0$. Thus all odd coefficients are 0.

79. Let $f(x)=\sum_{m=0}^{\infty}c_mx^m$ and $g(x)=e^{f(x)}=\sum_{n=0}^{\infty}d_nx^n$. Then $g'(x)=\sum_{n=0}^{\infty}nd_nx^{n-1}$, so nd_n is the coefficient of x^{n-1}. Also

$g'(x)=e^{f(x)}f'(x)=\left(\sum_{n=0}^{\infty}d_nx^n\right)\left(\sum_{m=1}^{\infty}mc_mx^{m-1}\right)$

$=\left(d_0+d_1x+d_2x^2+\cdots+d_{n-1}x^{n-1}+\cdots\right)\left(c_1+2c_2x+3c_3x^2+\cdots+nc_nx^{n-1}+\cdots\right)$, so

the coefficient of x^{n-1} is $c_1d_{n-1}+2c_2d_{n-2}+3c_3d_{n-3}+\cdots+nc_nd_0=\sum_{i=1}^{n}ic_id_{n-i}$. Thus $nd_n=\sum_{i=1}^{n}ic_id_{n-i}$.

PROBLEMS PLUS (page 678)

1. It would be far too much work to compute 15 derivatives of f. The key idea is to remember that $f^{(n)}(0)$ occurs in the coefficient of x^n in the Maclaurin series of f. We start with the Maclaurin series for sin: $\sin x = x - \frac{x^3}{3!} + \frac{x^5}{5!} - \cdots$. Then $\sin(x^3) = x^3 - \frac{x^9}{3!} + \frac{x^{15}}{5!} - \cdots$ and so the coefficient of x^{15} is $\frac{f^{(15)}(0)}{15!} = \frac{1}{5!}$. Therefore, $f^{(15)}(0) = \frac{15!}{5!} = 6 \cdot 7 \cdot 8 \cdot 9 \cdot 10 \cdot 11 \cdot 12 \cdot 13 \cdot 14 \cdot 15 = 10{,}897{,}286{,}400$.

3. **(a)** From Formula 14a in Appendix D, with $x = y = \theta$, we get $\tan 2\theta = \frac{2\tan\theta}{1 - \tan^2\theta}$, so $\cot 2\theta = \frac{1 - \tan^2\theta}{2\tan\theta}$ $\Rightarrow$ $2\cot 2\theta = \frac{1 - \tan^2\theta}{\tan\theta} = \cot\theta - \tan\theta$. Replacing θ by $\frac{1}{2}x$, we get $2\cot x = \cot\frac{1}{2}x - \tan\frac{1}{2}x$, or $\tan\frac{1}{2}x = \cot\frac{1}{2}x - 2\cot x$.

(b) From part (a) we have $\tan\frac{x}{2^n} = \cot\frac{x}{2^n} - 2\cot\frac{x}{2^{n-1}}$, so the nth partial sum of the given series is

$$
\begin{aligned}
s_n &= \frac{\tan(x/2)}{2} + \frac{\tan(x/4)}{4} + \frac{\tan(x/8)}{8} + \cdots + \frac{\tan(x/2^n)}{2^n} \\
&= \left[\frac{\cot(x/2)}{2} - \cot x\right] + \left[\frac{\cot(x/4)}{4} - \frac{\cot(x/2)}{2}\right] + \left[\frac{\cot(x/8)}{8} - \frac{\cot(x/4)}{4}\right] \\
&\qquad + \cdots + \left[\frac{\cot(x/2^n)}{2^n} - \frac{\cot(x/2^{n-1})}{2^{n-1}}\right] \\
&= -\cot x + \frac{\cot(x/2^n)}{2^n} \quad \text{(telescoping sum).}
\end{aligned}
$$

Now $\frac{\cot(x/2^n)}{2^n} = \frac{\cos(x/2^n)}{2^n\sin(x/2^n)} = \frac{\cos(x/2^n)}{x} \cdot \frac{x/2^n}{\sin(x/2^n)} \to \frac{1}{x} \cdot 1 = \frac{1}{x}$ as $n \to \infty$ since $\frac{x}{2^n} \to 0$ for $x \neq 0$. Therefore, if $x \neq 0$ and $x \neq n\pi$, then $\sum_{n=1}^{\infty} \frac{1}{2^n}\tan\frac{x}{2^n} = \lim_{n\to\infty}\left(-\cot x + \frac{1}{2^n}\cot\frac{x}{2^n}\right) = -\cot x + \frac{1}{x}$.

If $x = 0$, then all terms in the series are 0, so the sum is 0.

5. **(a)** At each stage, each side is replaced by four shorter sides, each of length $\frac{1}{3}$ of the side length at the preceding stage. Writing s_0 and ℓ_0 for the number of sides and the length of the side of the initial triangle, we generate the table at right. In general, we have $s_n = 3 \cdot 4^n$ and $\ell_n = \left(\frac{1}{3}\right)^n$, so the length of the perimeter at the nth stage of construction is $p_n = s_n\ell_n = 3 \cdot 4^n \cdot \left(\frac{1}{3}\right)^n = 3 \cdot \left(\frac{4}{3}\right)^n$.

$s_0 = 3$	$\ell_0 = 1$
$s_1 = 3 \cdot 4$	$\ell_1 = \frac{1}{3}$
$s_2 = 3 \cdot 4^2$	$\ell_2 = \frac{1}{3^2}$
$s_3 = 3 \cdot 4^3$	$\ell_3 = \frac{1}{3^3}$
$\cdots$	$\cdots$

(b) $p_n = \frac{4^n}{3^{n-1}} = 4\left(\frac{4}{3}\right)^{n-1}$. Since $\frac{4}{3} > 1$, $p_n \to \infty$ as $n \to \infty$.

(c) The area of each of the small triangles added at a given stage is one-ninth of the area of the triangle added at the preceding stage. Let a be the area of the original triangle. Then the area a_n of each of the small triangles added at stage n is $a_n = a \cdot \frac{1}{9^n} = \frac{a}{9^n}$. Since a small triangle is added to each side at every stage, it follows that the total area A_n added to the figure at the nth stage is

$A_n = s_{n-1} \cdot a_n = 3 \cdot 4^{n-1} \cdot \frac{a}{9^n} = a \cdot \frac{4^{n-1}}{3^{2n-1}}$. Then the total area enclosed by the snowflake curve is

$A = a + A_1 + A_2 + A_3 + \cdots = a + a \cdot \frac{1}{3} + a \cdot \frac{4}{3^3} + a \cdot \frac{4^2}{3^5} + a \cdot \frac{4^3}{3^7} + \cdots$. After the first term, this is a geometric series with common ratio $\frac{4}{9}$, so $A = a + \frac{a/3}{1 - \frac{4}{9}} = a + \frac{a}{3} \cdot \frac{9}{5} = \frac{8a}{5}$. But the area of the original equilateral triangle with side 1 is $a = \frac{1}{2} \cdot 1 \cdot \sin\frac{\pi}{3} = \frac{\sqrt{3}}{4}$.

So the area enclosed by the snowflake curve is $\frac{8}{5} \cdot \frac{\sqrt{3}}{4} = \frac{2\sqrt{3}}{5}$.

7. $x^2 + y^2 \le 4y \quad \Leftrightarrow \quad x^2 + (y-2)^2 \le 4$, so S is part of a circle, as shown in the diagram. The area of S is

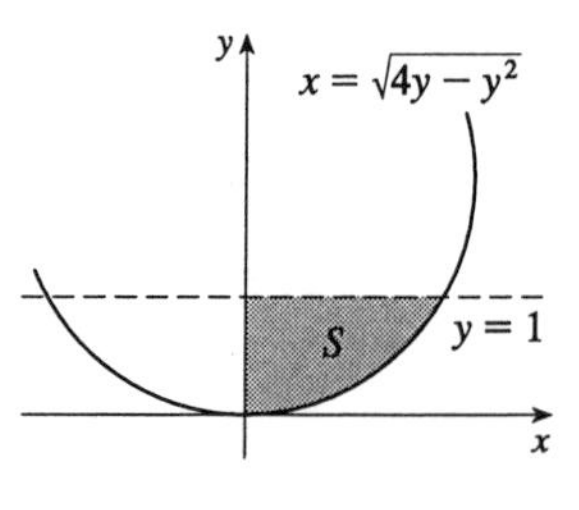

$\int_0^1 \sqrt{4y - y^2}\,dy = \int_{-2}^{-1} \sqrt{4 - v^2}\,dv \quad$ (put $v = y - 2$)

$= \left[\frac{1}{2}v\sqrt{4 - v^2} + \frac{1}{2}(4)\sin^{-1}\left(\frac{1}{2}v\right)\right]_{-2}^{-1} \quad$ (Formula 30) $\quad = \frac{2\pi}{3} - \frac{\sqrt{3}}{2}$

Another Method (without calculus): Note that $\theta = \angle ABC = \frac{\pi}{3}$, so the area is (area of sector AOC) − (area of $\triangle ABC$) $= \frac{1}{2}(2^2)\frac{\pi}{3} - \frac{1}{2}(1)\sqrt{3} = \frac{2\pi}{3} - \frac{\sqrt{3}}{2}$.

9. $a_{n+1} = \dfrac{a_n + b_n}{2}$, $b_{n+1} = \sqrt{b_n a_{n+1}}$. So $a_1 = \cos\theta$, $b_1 = 1 \Rightarrow a_2 = \dfrac{1 + \cos\theta}{2} = \cos^2\dfrac{\theta}{2}$,

$b_2 = \sqrt{b_1 a_2} = \sqrt{\cos^2\frac{\theta}{2}} = \cos\frac{\theta}{2}$ since $-\frac{\pi}{2} \le \theta \le \frac{\pi}{2}$. Then

$$a_3 = \frac{1}{2}\left(\cos\frac{\theta}{2} + \cos^2\frac{\theta}{2}\right) = \cos\frac{\theta}{2} \cdot \frac{1}{2}\left(1 + \cos\frac{\theta}{2}\right) = \cos\frac{\theta}{2}\cos^2\frac{\theta}{4} \quad \Rightarrow$$

$$b_3 = \sqrt{b_2 a_3} = \sqrt{\cos\frac{\theta}{2}\cos\frac{\theta}{2}\cos^2\frac{\theta}{4}} = \cos\frac{\theta}{2}\cos\frac{\theta}{4} \quad \Rightarrow$$

$$a_4 = \frac{1}{2}\left(\cos\frac{\theta}{2}\cos^2\frac{\theta}{4} + \cos\frac{\theta}{2}\cos\frac{\theta}{4}\right) = \cos\frac{\theta}{2}\cos\frac{\theta}{4} \cdot \frac{1}{2}\left(1 + \cos\frac{\theta}{4}\right) = \cos\frac{\theta}{2}\cos\frac{\theta}{4}\cos^2\frac{\theta}{8} \quad \Rightarrow$$

$b_4 = \sqrt{\cos\frac{\theta}{2}\cos\frac{\theta}{4}\cos\frac{\theta}{2}\cos\frac{\theta}{4}\cos^2\frac{\theta}{8}} = \cos\frac{\theta}{2}\cos\frac{\theta}{4}\cos\frac{\theta}{8}$. By now we see the pattern:

$b_n = \cos\frac{\theta}{2}\cos\frac{\theta}{2^2}\cos\frac{\theta}{2^3} \cdot \cdots \cdot \cos\frac{\theta}{2^{n-1}}$ and $a_n = b_n \cos\frac{\theta}{2^{n-1}}$. (This could be proved by mathematical induction.) By Exercise 8(a), $\sin\theta = 2^{n-1}\sin\frac{\theta}{2^{n-1}}\cos\frac{\theta}{2}\cos\frac{\theta}{4}\cdots\cos\frac{\theta}{2^{n-1}}$.

So $b_n = \cos\frac{\theta}{2}\cos\frac{\theta}{2^2}\cos\frac{\theta}{2^3}\cdots\cos\frac{\theta}{2^{n-1}} \to \frac{\sin\theta}{\theta}$ as $n \to \infty$ by Exercise 8(b), and

$a_n = b_n \cos\frac{\theta}{2^{n-1}} \to \frac{\sin\theta}{\theta} \cdot 1 = \frac{\sin\theta}{\theta}$ as $n \to \infty$. So $\lim\limits_{n\to\infty} a_n = \lim\limits_{n\to\infty} b_n = \frac{\sin\theta}{\theta}$.

11. We start with the geometric series $\sum_{n=0}^{\infty} x^n = \dfrac{1}{1-x}$, $|x| < 1$, and differentiate:

$$\sum_{n=1}^{\infty} nx^{n-1} = \frac{d}{dx}\left(\sum_{n=0}^{\infty} x^n\right) = \frac{d}{dx}\left(\frac{1}{1-x}\right) = \frac{1}{(1-x)^2} \text{ for } |x|<1 \quad \Rightarrow \quad \sum_{n=1}^{\infty} nx^n = x\sum_{n=1}^{\infty} nx^{n-1} = \frac{x}{(1-x)^2}$$

for $|x| < 1$. Differentiate again: $\displaystyle\sum_{n=1}^{\infty} n^2x^{n-1} = \frac{d}{dx}\frac{x}{(1-x)^2} = \frac{(1-x)^2 - x\cdot 2(1-x)(-1)}{(1-x)^4} = \frac{x+1}{(1-x)^3} \quad \Rightarrow$

$$\sum_{n=1}^{\infty} n^2x^n = \frac{x^2+x}{(1-x)^3} \quad \Rightarrow \quad \sum_{n=1}^{\infty} n^3x^{n-1} = \frac{d}{dx}\frac{x^2+x}{(1-x)^3}$$

$$= \frac{(1-x)^3(2x+1) - (x^2+x)3(1-x)^2(-1)}{(1-x)^6} = \frac{x^2+4x+1}{(1-x)^4} \quad \Rightarrow \quad \sum_{n=1}^{\infty} n^3x^n = \frac{x^3+4x^2+x}{(1-x)^4},\ |x|<1.$$

The radius of convergence is 1 because that is the radius of convergence for the geometric series we started with. If $x = \pm 1$, the series is $\sum n^3(\pm 1)^n$, which diverges by the Test For Divergence, so the interval of convergence is $(-1, 1)$.

13. **(a)** Let $a = \arctan x$ and $b = \arctan y$. Then, from the endpapers,

$$\tan(a-b) = \frac{\tan a - \tan b}{1 + \tan a \tan b} = \frac{\tan(\arctan x) - \tan(\arctan y)}{1 + \tan(\arctan x)\tan(\arctan y)} \quad \Rightarrow \quad \tan(a-b) = \frac{x-y}{1+xy} \quad \Rightarrow$$

$$\arctan x - \arctan y = a - b = \arctan\frac{x-y}{1+xy} \text{ since } -\frac{\pi}{2} < \arctan x - \arctan y < \frac{\pi}{2}.$$

(b) From part (a) we have $\arctan\frac{120}{119} - \arctan\frac{1}{239} = \arctan\dfrac{\frac{120}{119} - \frac{1}{239}}{1 + \frac{120}{119}\cdot\frac{1}{239}} = \arctan\dfrac{\frac{28{,}561}{28{,}441}}{\frac{28{,}561}{28{,}441}} = \arctan 1 = \frac{\pi}{4}$.

(c) Replacing y by $-y$ in the formula of part (a), we get $\arctan x + \arctan y = \arctan\dfrac{x+y}{1-xy}$. So

$$4\arctan\tfrac{1}{5} = 2\left(\arctan\tfrac{1}{5} + \arctan\tfrac{1}{5}\right) = 2\arctan\frac{\frac{1}{5}+\frac{1}{5}}{1 - \frac{1}{5}\cdot\frac{1}{5}} = 2\arctan\tfrac{5}{12}$$

$$= \arctan\tfrac{5}{12} + \arctan\tfrac{5}{12} = \arctan\frac{\frac{5}{12}+\frac{5}{12}}{1-\frac{5}{12}\cdot\frac{5}{12}} = \arctan\tfrac{120}{119}.$$

Thus, from part (b), we have $4\arctan\frac{1}{5} - \arctan\frac{1}{239} = \arctan\frac{120}{119} - \arctan\frac{1}{239} = \frac{\pi}{4}$.

(d) From Example 7 in Section 10.9 we have $\arctan x = x - \dfrac{x^3}{3} + \dfrac{x^5}{5} - \dfrac{x^7}{7} + \dfrac{x^9}{9} - \dfrac{x^{11}}{11} + \cdots$, so

$\arctan\dfrac{1}{5} = \dfrac{1}{5} - \dfrac{1}{3\cdot 5^3} + \dfrac{1}{5\cdot 5^5} - \dfrac{1}{7\cdot 5^7} + \dfrac{1}{9\cdot 5^9} - \dfrac{1}{11\cdot 5^{11}} + \cdots$. This is an alternating series and the size of the terms decreases to 0, so by Theorem 10.5.1, the sum lies between s_5 and s_6, that is, $0.197395560 < \arctan\frac{1}{5} < 0.197395562$.

(e) From the series in part (d) we get $\arctan\dfrac{1}{239} = \dfrac{1}{239} - \dfrac{1}{3\cdot 239^3} + \dfrac{1}{5\cdot 239^5} - \cdots$. The third term is less than 2.6×10^{-13}, so by Theorem 10.5.1 we have, to nine decimal places, $\arctan\frac{1}{239} \approx s_2 \approx 0.004184076$. Thus $0.004184075 < \arctan\frac{1}{239} < 0.004184077$.

(f) From part (c) we have $\pi = 16\arctan\frac{1}{5} - 4\arctan\frac{1}{239}$, so from parts (d) and (e) we have

$16(0.197395560) - 4(0.004184077) < \pi < 16(0.197395562) - 4(0.004184075) \quad \Rightarrow$

$3.141592652 < \pi < 3.141592692$. So, to 7 decimal places, $\pi \approx 3.1415927$.

15. $u = 1 + \frac{x^3}{3!} + \frac{x^6}{6!} + \frac{x^9}{9!} + \cdots$, $v = x + \frac{x^4}{4!} + \frac{x^7}{7!} + \frac{x^{10}}{10!} + \cdots$, $w = \frac{x^2}{2!} + \frac{x^5}{5!} + \frac{x^8}{8!} + \cdots$.

The key idea is to differentiate: $\frac{du}{dx} = \frac{3x^2}{3!} + \frac{6x^5}{6!} + \frac{9x^8}{9!} + \cdots = \frac{x^2}{2!} + \frac{x^5}{5!} + \frac{x^8}{8!} + \cdots = w$.

Similarly, $\frac{dv}{dx} = 1 + \frac{x^3}{3!} + \frac{x^6}{6!} + \frac{x^9}{9!} + \cdots = u$, and $\frac{dw}{dx} = x + \frac{x^4}{4!} + \frac{x^7}{7!} + \frac{x^{10}}{10!} + \cdots = v$.

So $u' = w$, $v' = u$, and $w' = v$. Now differentiate the left hand side of the desired equation:

$$\frac{d}{dx}(u^3 + v^3 + w^3 - 3uvw) = 3u^2u' + 3v^3v' + 3w^3w' - 3(u'vw + uv'w + uvw')$$

$$= 3u^2w + 3v^2u + 3w^2v - 3(vw^2 + u^2w + uv^2) = 0 \quad \Rightarrow \quad u^3 + v^3 + w^3 - 3uvw = C.$$

To find the value of the constant C, we put $x = 1$ in the equation and get $1^3 + 0 + 0 - 3(1 \cdot 0 \cdot 0) = C \quad \Rightarrow$ $C = 1$, so $u^3 + v^3 + w^3 - 3uvw = 1$.

17. $(a^n + b^n + c^n)^{1/n} = \left(c^n\left[\left(\frac{a}{c}\right)^n + \left(\frac{b}{c}\right)^n + 1\right]\right)^{1/n} = c\left[\left(\frac{a}{c}\right)^n + \left(\frac{b}{c}\right)^n + 1\right]^{1/n}$. Since $0 \le a \le c$, we have $0 \le a/c \le 1$, so $(a/c)^n \to 0$ or 1 as $n \to \infty$. Similarly, $(b/c)^n \to 0$ or 1 as $n \to \infty$. Thus $\left(\frac{a}{c}\right)^n + \left(\frac{b}{c}\right)^n + 1 \to d$, where $d = 1, 2$, or 3 and so $\left[\left(\frac{a}{c}\right)^n + \left(\frac{b}{c}\right)^n + 1\right]^{1/n} \to 1$. Therefore $\lim\limits_{n\to\infty}(a^n + b^n + c^n)^{1/n} = c$.

19. As in Section 10.9 we have to integrate the function x^x by integrating a series. Writing $x^x = \left(e^{\ln x}\right)^x = e^{x\ln x}$ and using the Maclaurin series for e^x, we have $x^x = e^{x\ln x} = \sum\limits_{n=0}^{\infty}\frac{(x\ln x)^n}{n!} = \sum\limits_{n=0}^{\infty}\frac{x^n(\ln x)^n}{n!}$. As with power series, we can integrate this series term-by-term: $\int_0^1 x^x\,dx = \sum\limits_{n=0}^{\infty}\int_0^1\frac{x^n(\ln x)^n}{n!}\,dx = \sum\limits_{n=0}^{\infty}\frac{1}{n!}\int_0^1 x^n(\ln x)^n\,dx$. We integrate by parts with $u = (\ln x)^n$, $dv = x^n\,dx$, so $du = \frac{n(\ln x)^{n-1}}{x}\,dx$ and $v = \frac{x^{n+1}}{n+1}$:

$$\int_0^1 x^n(\ln x)^n\,dx = \lim_{t\to 0^+}\int_t^1 x^n(\ln x)^n\,dx = \lim_{t\to 0^+}\left[\frac{x^{n+1}}{n+1}(\ln x)^n\right]_t^1 - \lim_{t\to 0^+}\int_t^1\frac{n}{n+1}x^n(\ln x)^{n-1}\,dx$$

$$= 0 - \frac{n}{n+1}\int_0^1 x^n(\ln x)^{n-1}\,dx$$ (where l'Hospital's Rule was used to help evaluate the first limit).

Further integration by parts gives $\int_0^1 x^n(\ln x)^k\,dx = -\frac{k}{n+1}\int_0^1 x^n(\ln x)^{k-1}\,dx$ and, combining these steps, we get $\int_0^1 x^n(\ln x)^n\,dx = \frac{(-1)^n\,n!}{(n+1)^n}\int_0^1 x^n\,dx = \frac{(-1)^n\,n!}{(n+1)^{n+1}} \quad \Rightarrow$

$$\int_0^1 x^x\,dx = \sum_{n=0}^{\infty}\frac{1}{n!}\int_0^1 x^n(\ln x)^n\,dx = \sum_{n=0}^{\infty}\frac{1}{n!}\frac{(-1)^n\,n!}{(n+1)^{n+1}} = \sum_{n=0}^{\infty}\frac{(-1)^n}{(n+1)^{n+1}} = \sum_{n=1}^{\infty}\frac{(-1)^{n-1}}{n^n}.$$

21. Call the series S. We group the terms according to the number of digits in their denominators:

$$S = \underbrace{\left(1 + \frac{1}{2} + \cdots + \frac{1}{8} + \frac{1}{9}\right)}_{g_1} + \underbrace{\left(\frac{1}{11} + \cdots + \frac{1}{99}\right)}_{g_2} + \underbrace{\left(\frac{1}{111} + \cdots + \frac{1}{999}\right)}_{g_3} + \cdots$$

Now in the group g_n, there are 9^n terms, since we have 9 choices for each of the n digits in the denominator. Furthermore, each term in g_n is less than $\frac{1}{10^{n-1}}$. So $g_n < 9^n \cdot \frac{1}{10^{n-1}} = 9\left(\frac{9}{10}\right)^{n-1}$.

Now $\sum_{n=1}^{\infty} 9\left(\frac{9}{10}\right)^{n-1}$ is a geometric series with $a = 9$ and $r = \frac{9}{10} < 1$. Therefore, by the Comparison Test,

$$S = \sum_{n=1}^{\infty} g_n < \sum_{n=1}^{\infty} 9\left(\frac{9}{10}\right)^{n-1} = \frac{9}{1-\frac{9}{10}} = 90.$$

23. If L is the length of a side of the equilateral triangle, then the area is $A = \frac{1}{2}L \cdot \frac{\sqrt{3}}{2}L = \frac{\sqrt{3}}{4}L^2$ and so $L^2 = \frac{4}{\sqrt{3}}A$. Let r be the radius of one of the circles when there are n rows of circles. The figure shows that

$$L = \sqrt{3}r + r + (n-2)(2r) + r + \sqrt{3}r$$
$$= r\left(2n - 2 + 2\sqrt{3}\right), \text{ so } r = \frac{L}{2\left(n + \sqrt{3} - 1\right)}.$$

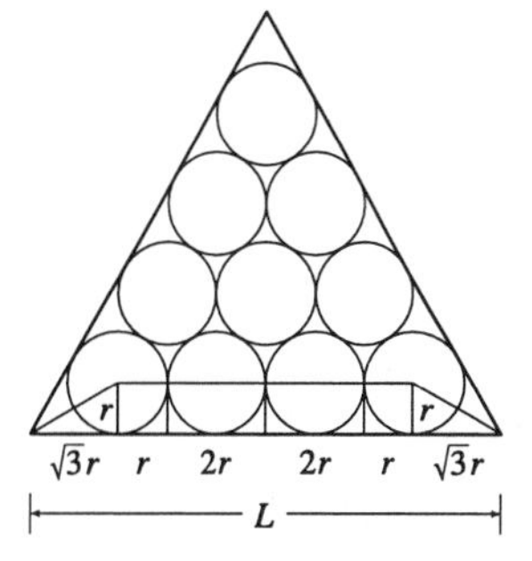

The number of circles is $1 + 2 + \cdots + n = \frac{n(n+1)}{2}$ and so the total area of the circles is

$$A_n = \frac{n(n+1)}{2}\pi r^2 = \frac{n(n+1)}{2}\pi\frac{L^2}{4\left(n+\sqrt{3}-1\right)^2} = \frac{n(n+1)}{2}\pi\frac{4A/\sqrt{3}}{4\left(n+\sqrt{3}-1\right)^2} = \frac{n(n+1)}{\left(n+\sqrt{3}-1\right)^2}\frac{\pi A}{2\sqrt{3}}$$

$$\Rightarrow \quad \frac{A_n}{A} = \frac{n(n+1)}{\left(n+\sqrt{3}-1\right)^2}\frac{\pi}{2\sqrt{3}} = \frac{1+1/n}{\left[1+\left(\sqrt{3}-1\right)/n\right]^2}\frac{\pi}{2\sqrt{3}} \to \frac{\pi}{2\sqrt{3}} \text{ as } n \to \infty.$$

25. **(a)** f is continuous when $x \neq 0$ since x, $\sin x$, and π/x are continuous when $x \neq 0$. Also $|\sin(\pi/x)| \leq 1$ $\Rightarrow$ $|x\sin(\pi/x)| \leq |x|$ and $\lim_{x\to 0}|x| = 0$, so by the Squeeze Theorem we have $\lim_{x\to 0}|x\sin(\pi/x)| = 0$, so $\lim_{x\to 0} f(x) = \lim_{x\to 0} x\sin(\pi/x) = 0 = f(0)$. Therefore f is continuous at 0, and so f is continuous on $(-1, 1)$.

(b) Note that $f(x) = 0$ when $x = 0$ and when $\pi/x = n\pi$ $\Rightarrow$ $x = 1/n$, n an integer. Since $-1 \leq \sin(\pi/x) \leq 1$, the graph of f lies between the lines $y = x$ and $y = -x$ and touches these lines when $\frac{\pi}{x} = \frac{\pi}{2} + n\pi$ $\Rightarrow$ $x = \frac{1}{n+1/2}$.

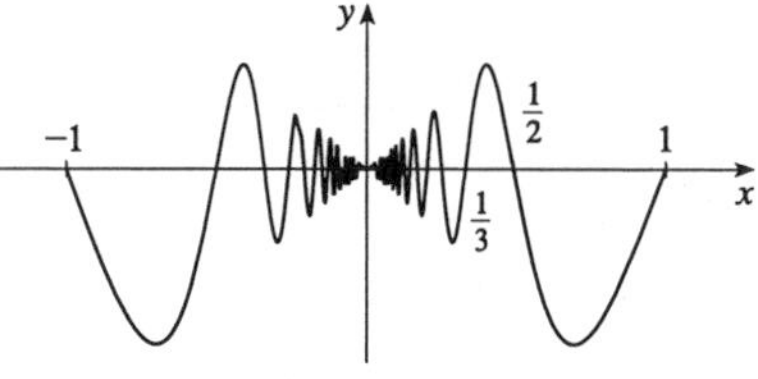

(c) The enlargement of the portion of the graph between $x = \frac{1}{n}$ and $x = \frac{1}{n-1}$ (the case where n is odd is illustrated) shows that the arc length from $x = \frac{1}{n}$ to $x = \frac{1}{n-1}$ is greater than $|PQ| = \frac{1}{n-1/2} = \frac{2}{2n-1}$.

Thus the total length of the graph is greater than $2\sum_{n=1}^{\infty}\frac{2}{2n-1}$.

This is a divergent series (by comparison with the harmonic series), so the graph has infinite length.

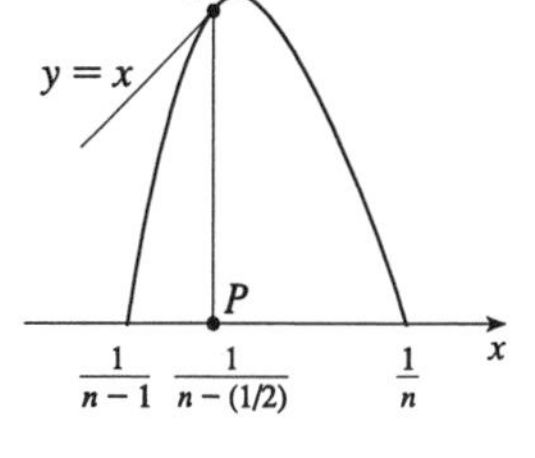

27. We use a method similar to that for Exercise 10.5.35.

Let s_n be the nth partial sum for the given series, and let h_n be the nth partial sum for the harmonic series.

From Exercise 10.3.32 we know that $h_n - \ln n \to \gamma$ as $n \to \infty$. So

$$\begin{aligned} s_{3n} &= 1 + \frac{1}{2} - \frac{2}{3} + \frac{1}{4} + \frac{1}{5} - \frac{2}{6} + \cdots + \frac{1}{3n-2} + \frac{1}{3n-1} - \frac{2}{3n} \\ &= \left(1 + \frac{1}{2} + \frac{1}{3} + \frac{1}{4} + \cdots + \frac{1}{3n}\right) - \left(\frac{3}{3} + \frac{3}{6} + \frac{3}{9} + \cdots + \frac{3}{3n}\right) = h_{3n} - h_n \\ &= \ln 3 + (h_{3n} - \ln 3n) - (h_n - \ln n) \to \ln 3 + \gamma - \gamma = \ln 3 \end{aligned}$$

Note: The method suggested in the first printing of the text doesn't quite work. We can differentiate to get

$$\begin{aligned} f'(x) &= 1 + x - 2x^2 + x^3 + x^4 - 2x^5 + \cdots \\ &= (1 + x - 2x^2) + x^3(1 + x - 2x^2) + x^6(1 + x - 2x^2) + \cdots \\ &= (1 + x - 2x^2)(1 + x^3 + x^6 + x^9 + \cdots) \\ &= \frac{1 + x - 2x^2}{1 - x^3} \quad (\text{if } |x| < 1) \\ &= \frac{(1-x)(1+2x)}{(1-x)(1+x+x^2)} \\ &= \frac{1+2x}{1+x+x^2}. \end{aligned}$$

Then $f(x) = \ln(1 + x + x^2) + C$. But $f(0) = 0$, so $C = 0$ and we have shown that

$x + \frac{x^2}{2} - \frac{2x^3}{3} + \frac{x^4}{4} + \frac{x^5}{5} - \frac{2x^6}{6} + \cdots = \ln(1 + x + x^2)$ for $-1 < x < 1$. As $x \to 1$, the limit of the right side is $\ln 3$.

However, it is not so easy to show that the limit of the left side is $1 + \frac{1}{2} - \frac{2}{3} + \frac{1}{4} + \frac{1}{5} - \frac{2}{6} + \cdots$.

29.

$$\begin{aligned} &\int_0^1 \left(\frac{-x-1}{100}\right)\left(\frac{1}{x+1} + \frac{1}{x+2} + \frac{1}{x+3} \cdots + \frac{1}{x+100}\right) dx \\ &= \int_0^1 \frac{(-x-1)(-x-2)(-x-3)\cdots(-x-100)}{100!}\left(\frac{1}{x+1} + \cdots + \frac{1}{x+100}\right) dx \\ &= \frac{1}{100!}\int_0^1 (x+1)(x+2)(x+3)\cdots(x+100)\left(\frac{1}{x+1} + \cdots + \frac{1}{x+100}\right) dx \\ &= \frac{1}{100!}(x+1)(x+2)(x+3)\cdots(x+100)\Big|_0^1 \quad (\text{the Product Rule in reverse}) \\ &= \frac{2 \cdot 3 \cdot \cdots \cdot 101}{100!} - \frac{100!}{100!} = 101 - 1 = 100. \end{aligned}$$

31. We can write the sum as

$$\sum_{n=1}^{\infty}\frac{1}{n}\left[\sum_{m=1}^{\infty}\frac{1}{m}\left(\frac{1}{m+(n+2)}\right)\right]=\sum_{n=1}^{\infty}\frac{1}{n}\left[\frac{1}{n+2}\sum_{m=1}^{\infty}\left(\frac{1}{m}-\frac{1}{m+(n+2)}\right)\right] \quad \text{(partial fractions)}$$

$$=\frac{1}{2}\sum_{n=1}^{\infty}\left[\left(\frac{1}{n}-\frac{1}{n+2}\right)\sum_{m=1}^{n+2}\left(\frac{1}{m}\right)\right] \quad \left(\begin{array}{c}\text{partial fractions in the outer sum; all terms}\\ \text{beyond the }(n+2)\text{th cancel in the inner sum}\end{array}\right)$$

$$=\frac{1}{2}\left[\sum_{n=1}^{\infty}\left(\frac{1}{n}\sum_{m=1}^{n+2}\frac{1}{m}\right)-\sum_{n=3}^{\infty}\left(\frac{1}{n}\sum_{m=1}^{n}\frac{1}{m}\right)\right] \quad \text{(change the index)}$$

$$=\frac{1}{2}\left[\sum_{n=1}^{2}\left(\frac{1}{n}\sum_{m=1}^{n+2}\frac{1}{m}\right)+\sum_{n=3}^{\infty}\left(\frac{1}{n}\sum_{m=1}^{n+2}\frac{1}{m}-\frac{1}{n}\sum_{m=1}^{n}\frac{1}{m}\right)\right]$$

$$=\frac{1}{2}\left[1\left(1+\frac{1}{2}+\frac{1}{3}\right)+\left(\frac{1}{2}\right)\left(1+\frac{1}{2}+\frac{1}{3}+\frac{1}{4}\right)+\sum_{n=3}^{\infty}\frac{1}{n}\left(\frac{1}{n+1}+\frac{1}{n+2}\right)\right]$$

$$=\frac{1}{2}\left[\frac{11}{6}+\frac{25}{24}+\sum_{n=3}^{\infty}\left(\frac{1}{n}-\frac{1}{n-1}\right)+\sum_{n=3}^{\infty}\left(\frac{1/2}{n}-\frac{1/2}{n-2}\right)\right] \quad \text{(partial fractions)}$$

$$=\frac{1}{2}\left[\frac{11}{6}+\frac{25}{24}+\frac{1}{3}+\frac{1}{2}\left(\frac{1}{3}+\frac{1}{4}\right)\right] \quad \text{(both series telescope)} \quad =\frac{7}{4}.$$

33. **(a)** We prove by induction that $1 < a_{n+1} < a_n$. For $n = 1$,

$$a_{1+1}=\frac{3\left(\frac{3}{2}\right)^2+4\left(\frac{3}{2}\right)-3}{4\left(\frac{3}{2}\right)^2}=\frac{3\left(\frac{9}{4}\right)+6-3}{9}=\frac{13}{12},$$ so $1 < a_2 < a_1$. Assume for k. Then

$$a_{k+1}=\frac{3a_k^2+4a_k-3}{4a_k^2}>1 \quad\Leftrightarrow\quad 3a_k^2+4a_k-3>4a_k^2 \quad\Leftrightarrow\quad 0>a_k^2-4a_k+3=(a_k-3)(a_k-1)$$

$\Leftrightarrow \quad 1 < a_k < 3$, which is true by the induction hypothesis. So we have the first inequality. For the second, $a_{k+1}=\dfrac{3a_k^2+4a_k-3}{4a_k^2}<a_k \quad\Leftrightarrow\quad 3a_k^2+4a_k-3<4a_k^3 \quad\Leftrightarrow$

$0<4a_k^3-3a_k^2-4a_k+3=(4a_k-3)\left(a_k^2-1\right)>0$ since $a_k>1$. So both results hold by induction.

(b) $\{a_n\}$ converges by part (a) and Theorem 10.1.10, so let $\lim\limits_{n\to\infty}a_n=L$. Then

$$L=\lim_{n\to\infty}\frac{3a_n^2+4a_n-3}{4a_n^2}=\frac{3\left(\lim\limits_{n\to\infty}a_n\right)^2+4\left(\lim\limits_{n\to\infty}a_n\right)-3}{4\left(\lim\limits_{n\to\infty}a_n\right)^2}=\frac{3L^2+4L-3}{4L^2} \quad\Rightarrow$$

$4L^3-3L^2-4L+3=0 \quad\Rightarrow\quad (4L-3)\left(L^2-1\right)=(4L-3)(L-1)(L+1)=0$. Since $a_n>1$ for all n, $L\geq 1$, but 1 is the only such root of this polynomial, so $L=1$.

(c) Observe that $a_{n+1}=\dfrac{3a_n^2+4a_n-3}{4a_n^2} \quad\Leftrightarrow\quad 4a_n^2\cdot a_{n+1}-3a_n^2=4a_n-3 \quad\Leftrightarrow$

$a_n^2(4a_{n+1}-3)=4a_n-3$. Substituting this for $n=2$ into the same for $n=1$ gives

$a_1^2a_2^2(4a_3-3)=(4a_1-3)$. If we carry on these substitutions we get $a_1^2a_2^2\cdots a_n^2(4a_{n+1}-3)=(4a_1-3)$

$\Leftrightarrow \quad a_1a_2\cdots a_n=\sqrt{\dfrac{4a_1-3}{4a_{n+1}-3}}$. So $\lim\limits_{n\to\infty}a_1a_2\cdots a_n=\lim\limits_{n\to\infty}\sqrt{\dfrac{4a_1-3}{4a_{n+1}-3}}=\sqrt{\dfrac{4\left(\frac{3}{2}\right)-3}{4(1)-3}}=\sqrt{3}.$

EXERCISES A

1. $|5 - 23| = |-18| = 18$

3. $|-\pi| = \pi$ because $\pi > 0$.

5. $\left|\sqrt{5} - 5\right| = -\left(\sqrt{5} - 5\right) = 5 - \sqrt{5}$ because $\sqrt{5} - 5 < 0$.

7. For $x < 2$, $x - 2 < 0$, so $|x - 2| = -(x - 2) = 2 - x$.

9. $|x + 1| = \begin{cases} x + 1 & \text{for } x + 1 \geq 0 \quad \Leftrightarrow \quad x \geq -1 \\ -(x + 1) & \text{for } x + 1 < 0 \quad \Leftrightarrow \quad x < -1 \end{cases}$

11. $|x^2 + 1| = x^2 + 1$ (since $x^2 + 1 \geq 0$ for all x).

13. $2x + 7 > 3 \quad \Leftrightarrow \quad 2x > -4 \quad \Leftrightarrow \quad x > -2$, so $x \in (-2, \infty)$.

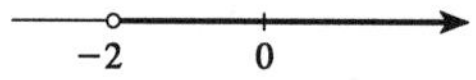

−2 0

15. $1 - x \leq 2 \quad \Leftrightarrow \quad -x \leq 1 \quad \Leftrightarrow \quad x \geq -1$, so $x \in [-1, \infty)$.

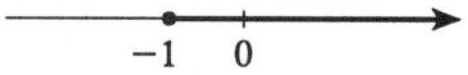

−1 0

17. $2x + 1 < 5x - 8 \quad \Leftrightarrow \quad 9 < 3x \quad \Leftrightarrow \quad 3 < x$, so $x \in (3, \infty)$.

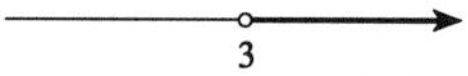

3

19. $-1 < 2x - 5 < 7 \quad \Leftrightarrow \quad 4 < 2x < 12 \quad \Leftrightarrow \quad 2 < x < 6$, so $x \in (2, 6)$.

2 6

21. $0 \leq 1 - x < 1 \quad \Leftrightarrow \quad -1 \leq -x < 0 \quad \Leftrightarrow \quad 1 \geq x > 0$, so $x \in (0, 1]$.

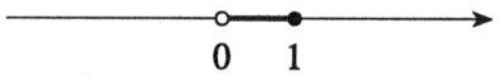

0 1

23. $4x < 2x + 1 \leq 3x + 2$. So $4x < 2x + 1 \quad \Leftrightarrow \quad 2x < 1 \quad \Leftrightarrow \quad x < \frac{1}{2}$, and $2x + 1 \leq 3x + 2 \quad \Leftrightarrow \quad -1 \leq x$. Thus $x \in \left[-1, \frac{1}{2}\right)$.

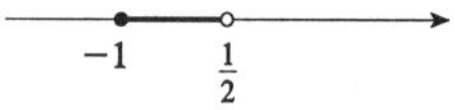

−1 $\frac{1}{2}$

25. $1 - x \geq 3 - 2x \geq x - 6$. So $1 - x \geq 3 - 2x \quad \Leftrightarrow \quad x \geq 2$, and $3 - 2x \geq x - 6 \quad \Leftrightarrow \quad 9 \geq 3x \quad \Leftrightarrow \quad 3 \geq x$. Thus $x \in [2, 3]$.

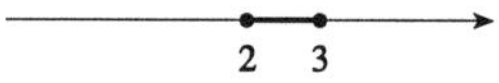

2 3

27. $(x - 1)(x - 2) > 0$. *Case 1:* $x - 1 > 0 \quad \Leftrightarrow \quad x > 1$, and $x - 2 > 0 \quad \Leftrightarrow \quad x > 2$, so $x \in [1, \infty)$. *Case 2:* $x - 1 < 0 \quad \Leftrightarrow \quad x < 1$, and $x - 2 < 0 \quad \Leftrightarrow \quad x < 2$, so $x \in (-\infty, 1)$. Thus the solution set is $(-\infty, 1) \cup (2, \infty)$.

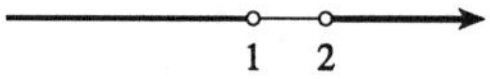

1 2

29. $2x^2 + x \leq 1 \quad \Leftrightarrow \quad 2x^2 + x - 1 \leq 0 \quad \Leftrightarrow \quad (2x - 1)(x + 1) \leq 0$. *Case 1:* $2x - 1 \geq 0 \quad \Leftrightarrow \quad x \geq \frac{1}{2}$, and $x + 1 \leq 0 \quad \Leftrightarrow \quad x \leq -1$, which is impossible. *Case 2:* $2x - 1 \leq 0 \quad \Leftrightarrow \quad x \leq \frac{1}{2}$, and $x + 1 \geq 0 \quad \Leftrightarrow \quad x \geq -1$, so $x \in \left[-1, \frac{1}{2}\right]$. Thus the solution set is $\left[-1, \frac{1}{2}\right]$.

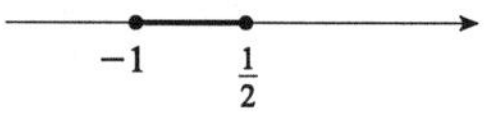

−1 $\frac{1}{2}$

31. $x^2 + x + 1 > 0 \quad \Leftrightarrow \quad x^2 + x + \frac{1}{4} + \frac{3}{4} > 0 \quad \Leftrightarrow \quad \left(x + \frac{1}{2}\right)^2 + \frac{3}{4} > 0$. But since $\left(x + \frac{1}{2}\right)^2 \geq 0$ for every real x, the original inequality will be true for all real x as well. Thus, the solution set is $(-\infty, \infty)$.

33. $x^2 < 3 \quad \Leftrightarrow \quad x^2 - 3 < 0 \quad \Leftrightarrow \quad \left(x - \sqrt{3}\right)\left(x + \sqrt{3}\right) < 0$. *Case 1:* $x > \sqrt{3}$ and $x < -\sqrt{3}$, which is impossible. *Case 2:* $x < \sqrt{3}$ and $x > -\sqrt{3}$. Thus the solution set is $\left(-\sqrt{3}, \sqrt{3}\right)$.

Another Method: $x^2 < 3 \quad \Leftrightarrow \quad |x| < \sqrt{3} \quad \Leftrightarrow \quad -\sqrt{3} < x < \sqrt{3}$.

35. $x^3 - x^2 \le 0 \quad \Leftrightarrow \quad x^2(x-1) \le 0$. Since $x^2 \ge 0$ for all x, the inequality is satisfied when $x - 1 \le 0 \quad \Leftrightarrow$ $x \le 1$. Thus the solution set is $(-\infty, 1]$.

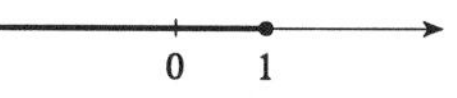

37. $x^3 > x \quad \Leftrightarrow \quad x^3 - x > 0 \quad \Leftrightarrow \quad x(x^2-1) > 0 \quad \Leftrightarrow \quad x(x-1)(x+1) > 0$. Constructing a table:

Interval	x	$x-1$	$x+1$	$x(x-1)(x+1)$
$x < -1$	$-$	$-$	$-$	$-$
$-1 < x < 0$	$-$	$-$	$+$	$+$
$0 < x < 1$	$+$	$-$	$+$	$-$
$x > 1$	$+$	$+$	$+$	$+$

Since $x^3 > x$ when the last column is positive, the solution set is $(-1, 0) \cup (1, \infty)$.

39. $1/x < 4$. This is clearly true for $x < 0$. So suppose $x > 0$. then $1/x < 4 \quad \Leftrightarrow \quad 1 < 4x \quad \Leftrightarrow \quad \frac{1}{4} < x$. Thus the solution set is $(-\infty, 0) \cup \left(\frac{1}{4}, \infty\right)$.

41. Multiply both sides by x. *Case 1:* If $x > 0$, then $4/x < x \quad \Leftrightarrow \quad 4 < x^2 \quad \Leftrightarrow \quad 2 < x$. *Case 2:* If $x < 0$, then $4/x < x \quad \Leftrightarrow \quad 4 > x^2 \quad \Leftrightarrow \quad -2 < x < 0$. Thus the solution set is $(-2, 0) \cup (2, \infty)$.

43. $\dfrac{2x+1}{x-5} < 3$. *Case 1:* If $x - 5 > 0$ (that is, $x > 5$), then $2x + 1 < 3(x-5) \quad \Leftrightarrow \quad 16 < x$, so $x \in (16, \infty)$.

Case 2: If $x - 5 < 0$ (that is, $x < 5$), then $2x + 1 > 3(x - 5) \quad \Leftrightarrow \quad 16 > x$, so in this case $x \in (-\infty, 5)$.

Combining the two cases, the solution set is $(-\infty, 5) \cup (16, \infty)$.

45. $\dfrac{x^2-1}{x^2+1} \ge 0$. Since $x^2 + 1 \ge 0$ for all real x, this inequality will hold whenever $x^2 - 1 \ge 0 \quad \Leftrightarrow$ $(x-1)(x+1) \ge 0$. *Case 1:* $x \ge 1$ and $x \ge -1$, so $x \in [1, \infty)$. *Case 2:* $x \le 1$ and $x \le -1$, so $x \in (-\infty, -1]$. Thus the solution set is $(-\infty, -1] \cup [1, \infty)$.

Another Method: $x^2 \ge 1 \quad \Leftrightarrow \quad |x| \ge 1 \quad \Leftrightarrow \quad x \ge 1$ or $x \le -1$.

47. $C = \frac{5}{9}(F - 32) \quad \Rightarrow \quad F = \frac{9}{5}C + 32$. So $50 \le F \le 95 \quad \Rightarrow \quad 50 \ge \frac{9}{5}C + 32 \le 95 \quad \Rightarrow \quad 18 \le \frac{9}{5}C \le 63$ $\Rightarrow \quad 10 \le C \le 35$. So the interval is $[10, 35]$.

49. **(a)** Let T represent the temperature in degrees Celsius and h the height in km. $T = 20$ when $h = 0$ and T decreases by $10°$ C for every km. Thus $T = 20 - 10h$ when $0 \le h \le 12$.

(b) From (a), $T = 20 - 10h \quad \Rightarrow \quad h = 2 - T/10$. So $0 \le h \le 5 \quad \Rightarrow \quad 0 \le 2 - T/10 \le 5 \quad \Rightarrow$ $-2 \le -T/10 \le 3 \quad \Rightarrow \quad -30 \le T \le 20$. Thus the range of temperatures to be expected is $[-30, 20]$.

51. $|2x| = 3 \quad \Leftrightarrow \quad$ either $2x = 3$ or $2x = -3 \quad \Leftrightarrow \quad x = \frac{3}{2}$ or $x = -\frac{3}{2}$.

53. $|x + 3| = |2x + 1| \quad \Leftrightarrow \quad$ either $x + 3 = 2x + 1$ or $x + 3 = -(2x + 1)$. In the first case, $x = 2$, and in the second case, $3x = -4 \quad \Leftrightarrow \quad x = -\frac{4}{3}$.

55. By (6), Property 5, $|x| < 3 \quad \Leftrightarrow \quad -3 < x < 3$, so $x \in (-3, 3)$.

57. $|x - 4| < 1 \quad \Leftrightarrow \quad -1 < x - 4 < 1 \quad \Leftrightarrow \quad 3 < x < 5$, so $x \in (3, 5)$.

59. $|x + 5| \ge 2 \quad \Leftrightarrow \quad x + 5 \ge 2$ or $x + 5 \le -2 \quad \Leftrightarrow \quad x \ge -3$ or $x \le -7$, so $x \in (-\infty, -7] \cup [-3, \infty)$.

61. $|2x - 3| \le 0.4 \quad \Leftrightarrow \quad -0.4 \le 2x - 3 \le 0.4 \quad \Leftrightarrow \quad 2.6 \le 2x \le 3.4 \quad \Leftrightarrow \quad 1.3 \le x \le 1.7$, so $x \in [1.3, 1.7]$.

63. $1 \le |x| \le 4$. So either $1 \le x \le 4$ or $1 \le -x \le 4 \quad \Leftrightarrow \quad -1 \ge x \ge -4$. Thus $x \in [-4, -1] \cup [1, 4]$.

65. $|x| > |x - 1|$. Since $|x|, |x - 1| \ge 0$, $|x| > |x - 1| \quad \Leftrightarrow \quad |x|^2 > |x - 1|^2 \quad \Leftrightarrow$
$x^2 > (x - 1)^2 = x^2 - 2x + 1 \quad \Leftrightarrow \quad 0 > -2x + 1 \quad \Leftrightarrow \quad x > \frac{1}{2}$, so $x \in \left(\frac{1}{2}, \infty\right)$.

67. $\left|\dfrac{x}{2 + x}\right| < 1 \quad \Leftrightarrow \quad \left(\dfrac{x}{2 + x}\right)^2 < 1 \quad \Leftrightarrow \quad x^2 < (2 + x)^2 \quad \Leftrightarrow \quad x^2 < 4 + 4x + x^2 \quad \Leftrightarrow \quad 0 < 4 + 4x$

$\Leftrightarrow \quad -1 < x$, so $x \in (-1, \infty)$.

69. $a(bx - c) \ge bc \quad \Leftrightarrow \quad bx - c \ge \dfrac{bc}{a} \quad \Leftrightarrow \quad bx \ge \dfrac{bc}{a} + c = \dfrac{bc + ac}{a} \quad \Leftrightarrow \quad x \ge \dfrac{bc + ac}{ab}$

71. $ax + b < c \quad \Leftrightarrow \quad ax < c - b \quad \Leftrightarrow \quad x > \dfrac{c - b}{a}$ (since $a < 0$)

73. $|(x + y) - 5| = |(x - 2) + (y - 3)| \le |x - 2| + |y - 3| < 0.01 + 0.04 = 0.05$

75. If $a < b$ then $a + a < a + b$ and $a + b < b + b$. So $2a < a + b < 2b$. Dividing by 2, $a < \frac{1}{2}(a + b) < b$.

77. $|ab| = \sqrt{(ab)^2} = \sqrt{a^2b^2} = \sqrt{a^2}\sqrt{b^2} = |a||b|$

79. If $0 < a < b$, then $a \cdot a < a \cdot b$ and $a \cdot b < b \cdot b$ [using (2), Rule 3]. So $a^2 < ab < b^2$ and hence $a^2 < b^2$.

81. Observe that the sum, difference and product of two integers is always an integer. Let the rational numbers be represented by $r = m/n$ and $s = p/q$ (where m, n, p and q are integers with $n \ne 0$, $q \ne 0$). Now $r + s = \dfrac{m}{n} + \dfrac{p}{q} = \dfrac{mq + pn}{nq}$, but $mq + pn$ and nq are both integers, so $\dfrac{mq + pn}{nq} = r + s$ is a rational number by definition. Similarly, $r - s = \dfrac{m}{n} - \dfrac{p}{q} = \dfrac{mq - pn}{nq}$ is a rational number. Finally, $r \cdot s = \dfrac{m}{n} \cdot \dfrac{p}{q} = \dfrac{mp}{nq}$ but mp and nq are both integers, so $\dfrac{mp}{nq} = r \cdot s$ is a rational number by definition.

EXERCISES B

1. From the Distance Formula (1) with $x_1 = 1$, $x_2 = 4$, $y_1 = 1$, $y_2 = 5$, we find the distance to be $\sqrt{(4-1)^2 + (5-1)^2} = \sqrt{3^2 + 4^2} = \sqrt{25} = 5$.

3. $\sqrt{(-1-6)^2 + [3-(-2)]^2} = \sqrt{(-7)^2 + 5^2} = \sqrt{74}$

5. $\sqrt{(4-2)^2 + (-7-5)^2} = \sqrt{2^2 + (-12)^2} = \sqrt{148} = 2\sqrt{37}$

7. From (2), the slope is $\dfrac{11-5}{4-1} = \dfrac{6}{3} = 2$.

9. $m = \dfrac{-6-3}{-1-(-3)} = -\dfrac{9}{2}$

11. Since $|AC| = \sqrt{(-4-0)^2 + (3-2)^2} = \sqrt{(-4)^2 + 1^2} = \sqrt{17}$ and $|BC| = \sqrt{[-4-(-3)]^2 + [3-(-1)]^2} = \sqrt{(-1)^2 + 4^2} = \sqrt{17}$, the triangle has two sides of equal length, and so is isosceles.

13. Label the points A, B, C, and D respectively. Then

$|AB| = \sqrt{[4-(-2)]^2 + (6-9)^2} = \sqrt{6^2 + (-3)^2} = 3\sqrt{5}$,

$|BC| = \sqrt{(1-4)^2 + (0-6)^2} = \sqrt{(-3)^2 + (-6)^2} = 3\sqrt{5}$,

$|CD| = \sqrt{(-5-1)^2 + (3-0)^2} = \sqrt{(-6)^2 + 3^2} = 3\sqrt{5}$, and

$|DA| = \sqrt{[-2-(-5)]^2 + (9-3)^2} = \sqrt{3^2 + 6^2} = 3\sqrt{5}$. So all sides are of equal length. Moreover, $m_{AB} = \dfrac{6-9}{4-(-2)} = -\dfrac{1}{2}$, $m_{BC} = \dfrac{0-6}{1-4} = 2$, $m_{CD} = \dfrac{3-0}{-5-1} = -\dfrac{1}{2}$, and $m_{DA} = \dfrac{9-3}{-2-(-5)} = 2$, so the sides are perpendicular. Thus, it is a square.

15. The slope of the line segment AB is $\dfrac{4-1}{7-1} = \dfrac{1}{2}$, the slope of CD is $\dfrac{7-10}{-1-5} = \dfrac{1}{2}$, the slope of BC is $\dfrac{10-4}{5-7} = -3$, and the slope of DA is $\dfrac{1-7}{1-(-1)} = -3$. So AB is parallel to CD and BC is parallel to DA. Hence $ABCD$ is a parallelogram.

17. $x = 3$

19. $xy = 0 \quad \Leftrightarrow \quad x = 0$ or $y = 0$

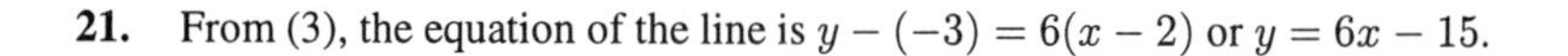

21. From (3), the equation of the line is $y - (-3) = 6(x-2)$ or $y = 6x - 15$.

23. $y - 7 = \frac{2}{3}(x - 1)$ or $2x - 3y + 19 = 0$

25. The slope is $m = \dfrac{6-1}{1-2} = -5$, so the equation of the line is $y - 1 = -5(x - 2)$ or $5x + y = 11$.

27. From (4), the equation is $y = 3x - 2$.

29. Since the line passes through $(1, 0)$ and $(0, -3)$, its slope is $m = \dfrac{-3-0}{0-1} = 3$, so its equation is $y = 3x - 3$.

31. Since $m = 0$, $y - 5 = 0(x - 4)$ or $y = 5$.

33. Putting the line $x + 2y = 6$ into its slope-intercept form $y = -\frac{1}{2}x + 3$, we see that this line has slope $-\frac{1}{2}$. So we want the line of slope $-\frac{1}{2}$ that passes through the point $(1, -6)$: $y - (-6) = -\frac{1}{2}(x - 1) \quad \Leftrightarrow \quad y = -\frac{1}{2}x - \frac{11}{2}$ or $x + 2y + 11 = 0$.

35. $2x + 5y + 8 = 0 \quad \Leftrightarrow \quad y = -\frac{2}{5}x - \frac{8}{5}$. Since this line has slope $-\frac{2}{5}$, a line perpendicular to it would have slope $\frac{5}{2}$, so the required line is $y - (-2) = \frac{5}{2}[x - (-1)] \quad \Leftrightarrow \quad y = \frac{5}{2}x + \frac{1}{2}$ or $5x - 2y + 1 = 0$.

37. $x + 3y = 0 \quad \Leftrightarrow \quad y = -\frac{1}{3}x$, so the slope is $-\frac{1}{3}$ and the y-intercept is 0.

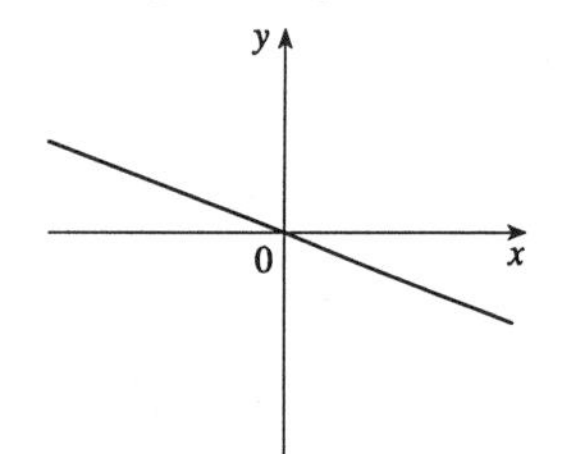

39. $y = -2$ is a horizontal line with slope 0 and y-intercept -2.

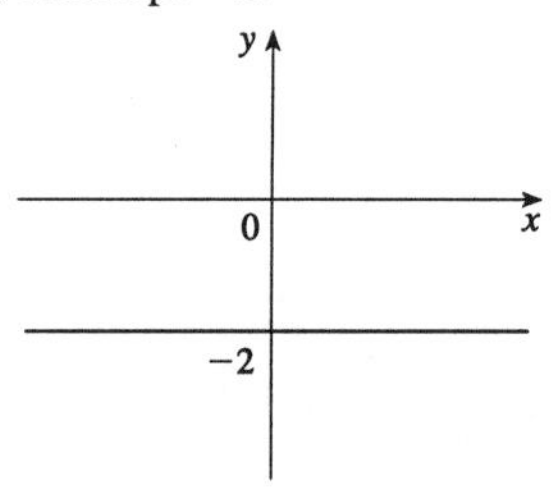

41. $3x - 4y = 12 \quad \Leftrightarrow \quad y = \frac{3}{4}x - 3$, so the slope is $\frac{3}{4}$ and the y-intercept is -3.

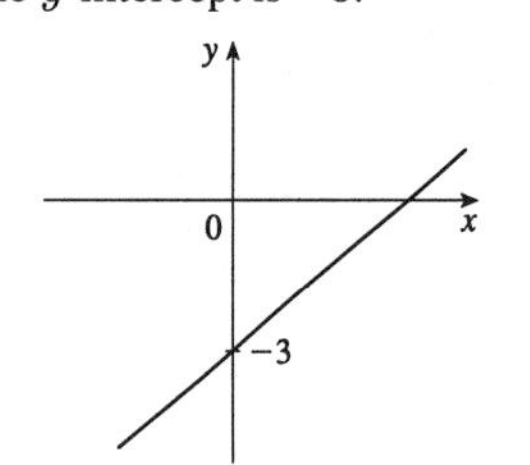

43. $\{(x, y) \mid x < 0\}$

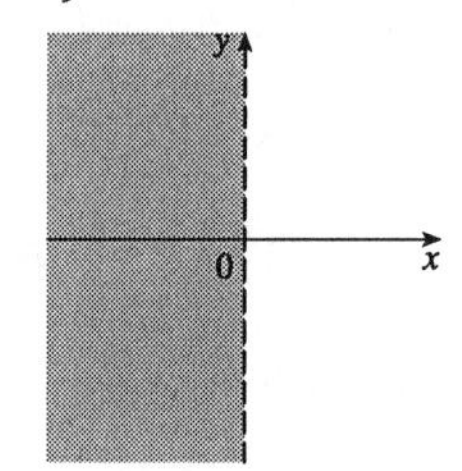

45. $\{(x, y) \mid xy < 0\} = \{(x, y) \mid x < 0 \text{ and } y > 0\} \cup \{(x, y) \mid x > 0 \text{ and } y < 0\}$

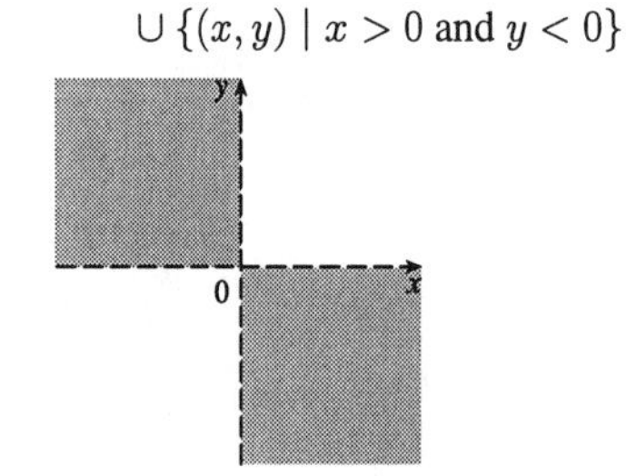

47. $\{(x, y) \mid |x| \leq 2\} = \{(x, y) \mid -2 \leq x \leq 2\}$

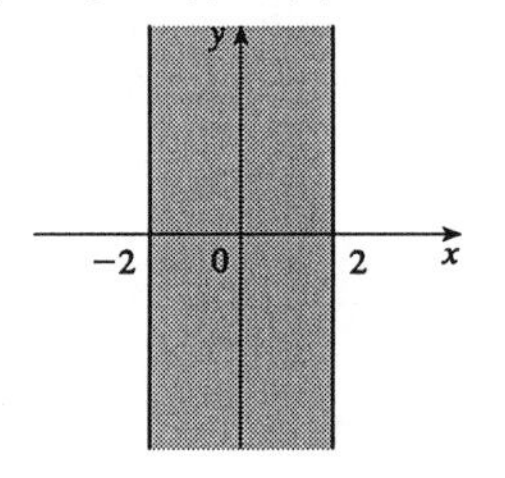

49. $\{(x,y) \mid 0 \le y \le 4,\, x \le 2\}$

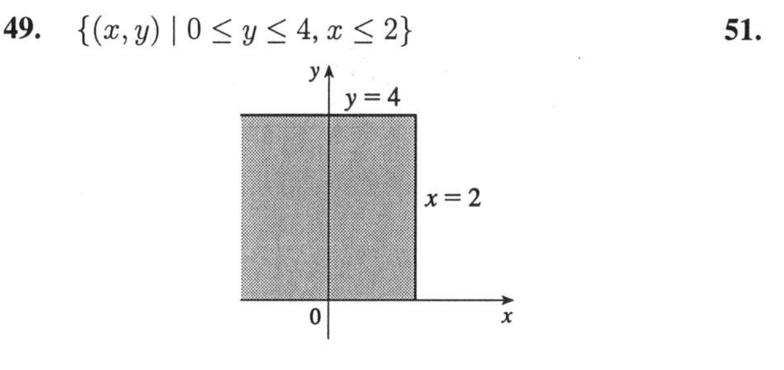

51. $\{(x,y) \mid 1 + x \le y \le 1 - 2x\}$

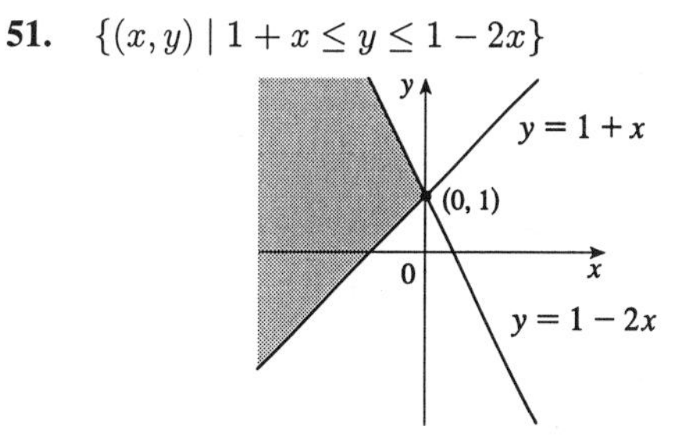

53. Let $P(0,y)$ be a point on the y-axis. The distance from P to $(5,-5)$ is $\sqrt{(5-0)^2 + (-5-y)^2} = \sqrt{5^2 + (y+5)^2}$. The distance from P to $(1,1)$ is $\sqrt{(1-0)^2 + (1-y)^2} = \sqrt{1^2 + (y-1)^2}$. We want these distances to be equal: $\sqrt{5^2 + (y+5)^2} = \sqrt{1^2 + (y-1)^2} \quad \Leftrightarrow \quad 5^2 + (y+5)^2 = 1^2 + (y-1)^2 \quad \Leftrightarrow$ $25 + (y^2 + 10y + 25) = 1 + (y^2 - 2y + 1) \;\Leftrightarrow\; 12y = -48 \;\Leftrightarrow\; y = -4$. So the desired point is $(0,-4)$.

55. Using the midpoint formula of Exercise 54, we get

(a) $\left(\dfrac{1+7}{2}, \dfrac{3+15}{2}\right) = (4,9)$ **(b)** $\left(\dfrac{-1+8}{2}, \dfrac{6-12}{2}\right) = \left(\dfrac{7}{2}, -3\right)$

57. $2x - y = 4 \quad \Leftrightarrow \quad y = 2x - 4 \quad \Rightarrow \quad m_1 = 2$ and $6x - 2y = 10 \quad \Leftrightarrow \quad 2y = 6x - 10 \quad \Leftrightarrow \quad y = 3x - 5$ $\Rightarrow \quad m_2 = 3$. Since $m_1 \ne m_2$, the two lines are not parallel [by 6(a)]. To find the point of intersection: $2x - 4 = 3x - 5 \quad \Leftrightarrow \quad x = 1 \quad \Rightarrow \quad y = -2$. Thus, the point of intersection is $(1,-2)$.

59. The slope of the segment AB is $\dfrac{-2-4}{7-1} = -1$, so its perpendicular bisector has slope 1. The midpoint of AB is $\left(\dfrac{1+7}{2}, \dfrac{4-2}{2}\right) = (4,1)$, so the equation of the perpendicular bisector is $y - 1 = 1(x-4)$ or $y = x - 3$.

61. **(a)** Since the x-intercept is a, the point $(a,0)$ is on the line, and similarly since the y-intercept is b, $(0,b)$ is on the line. Hence the slope of the line is $m = \dfrac{b-0}{0-a} = -\dfrac{b}{a}$. Substituting into $y = mx + b$ gives $y = -\dfrac{b}{a}x + b \quad \Leftrightarrow \quad y + \dfrac{b}{a}x = b \quad \Leftrightarrow \quad \dfrac{y}{b} + \dfrac{x}{a} = 1$.

(b) Letting $a = 6$ and $b = -8$ gives $\dfrac{y}{-8} + \dfrac{x}{6} = 1 \quad \Leftrightarrow \quad 6y - 8x = -48 \quad \Leftrightarrow \quad 8x - 6y - 48 = 0 \quad \Leftrightarrow$ $4x - 3y - 24 = 0$.

EXERCISES C

1. From (1), the equation is $(x-3)^2+(y+1)^2=25$.

3. The equation has the form $x^2+y^2=r^2$. Since $(4,7)$ lies on the circle, we have $4^2+7^2=r^2 \quad\Rightarrow\quad r^2=65$. So the required equation is $x^2+y^2=65$.

5. $x^2+y^2-4x+10y+13=0 \quad\Leftrightarrow\quad x^2-4x+y^2+10y=-13 \quad\Leftrightarrow$
$(x^2-4x+4)+(y^2+10y+25)=-13+4+25=16 \quad\Leftrightarrow\quad (x-2)^2+(y+5)^2=4^2$. Thus, we have a circle with center $(2,-5)$ and radius 4.

7. $x^2+y^2+x=0 \quad\Leftrightarrow\quad \left(x^2+x+\frac{1}{4}\right)+y^2=\frac{1}{4} \quad\Leftrightarrow\quad \left(x+\frac{1}{2}\right)^2+y^2=\left(\frac{1}{2}\right)^2$. Thus, we have a circle with center $\left(-\frac{1}{2},0\right)$ and radius $\frac{1}{2}$.

9. $2x^2+2y^2-x+y=1 \quad\Leftrightarrow\quad 2\left(x^2-\frac{1}{2}x+\frac{1}{16}\right)+2\left(y^2+\frac{1}{2}y+\frac{1}{16}\right)=1+\frac{1}{8}+\frac{1}{8} \quad\Leftrightarrow$
$2\left(x-\frac{1}{4}\right)^2+2\left(y+\frac{1}{4}\right)^2=\frac{5}{4} \quad\Leftrightarrow\quad \left(x-\frac{1}{4}\right)^2+\left(y+\frac{1}{4}\right)^2=\frac{5}{8}$. Thus, we have a circle with center $\left(\frac{1}{4},-\frac{1}{4}\right)$ and radius $\frac{\sqrt{5}}{2\sqrt{2}}=\frac{\sqrt{10}}{4}$.

11. $y=-x^2$. Parabola

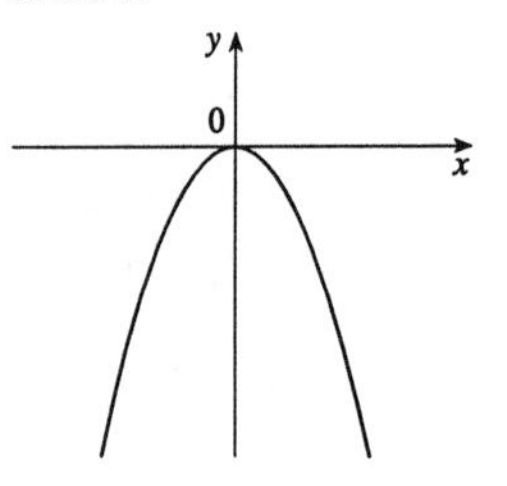

13. $x^2+4y^2=16 \quad\Leftrightarrow\quad \dfrac{x^2}{16}+\dfrac{y^2}{4}=1$. Ellipse

15. $16x^2-25y^2=400 \quad\Leftrightarrow\quad x^2/25-y^2/16=1$. Hyperbola

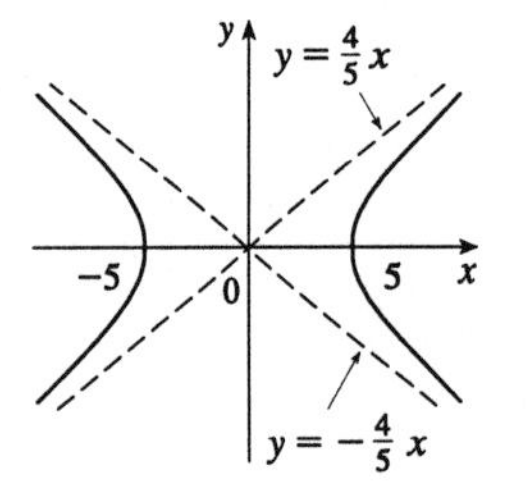

17. $4x^2+y^2=1 \quad\Leftrightarrow\quad \dfrac{x^2}{1/4}+y^2=1$. Ellipse

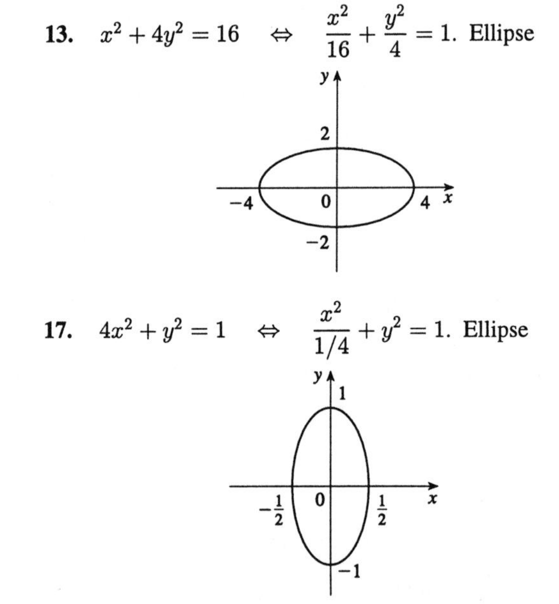

19. $x=y^2-1$. Parabola with vertex at $(-1,0)$

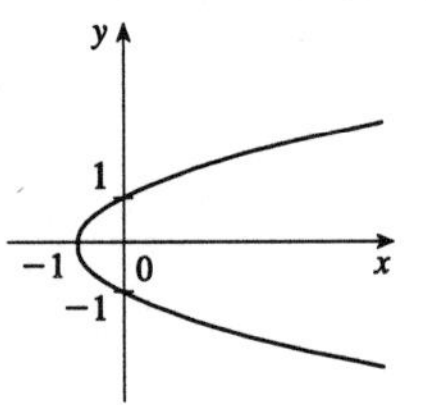

21. $9y^2-x^2=9 \quad\Leftrightarrow\quad y^2-x^2/9=1$. Hyperbola

23. $xy = 4$. Hyperbola

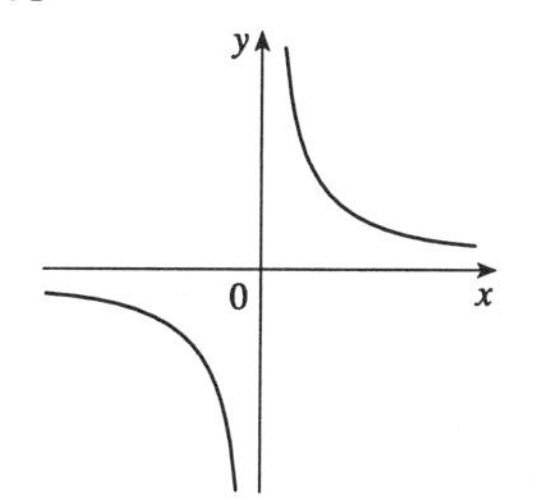

25. $9(x-1)^2 + 4(y-2)^2 = 36 \quad \Leftrightarrow$

$\dfrac{(x-1)^2}{4} + \dfrac{(y-2)^2}{9} = 1$. Ellipse centered at $(1,2)$

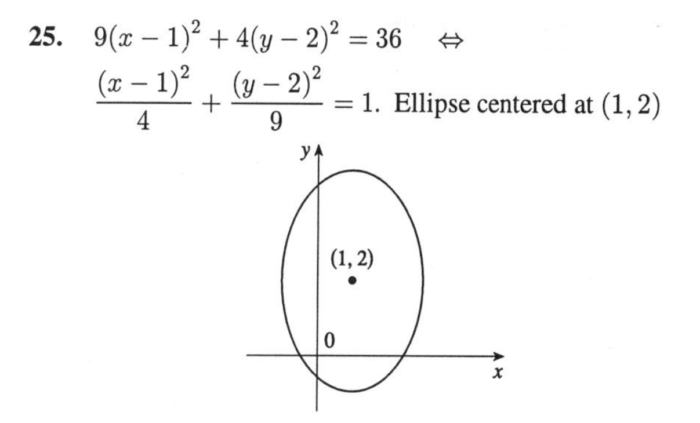

27. $y = x^2 - 6x + 13 = (x^2 - 6x + 9) + 4$

$= (x-3)^2 + 4$. Parabola with vertex at $(3,4)$

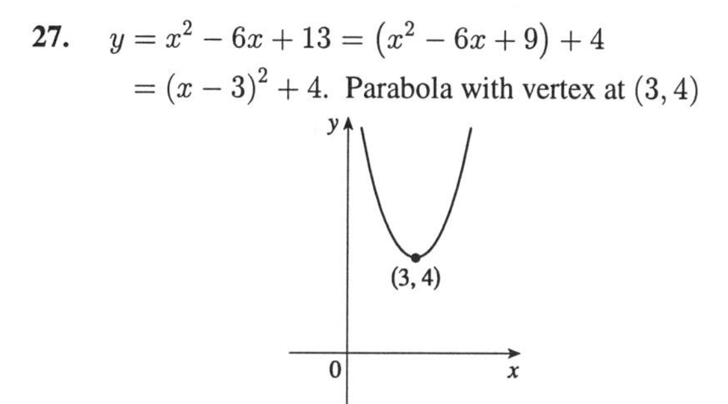

29. $x = -y^2 + 4$. Parabola with vertex at $(4,0)$

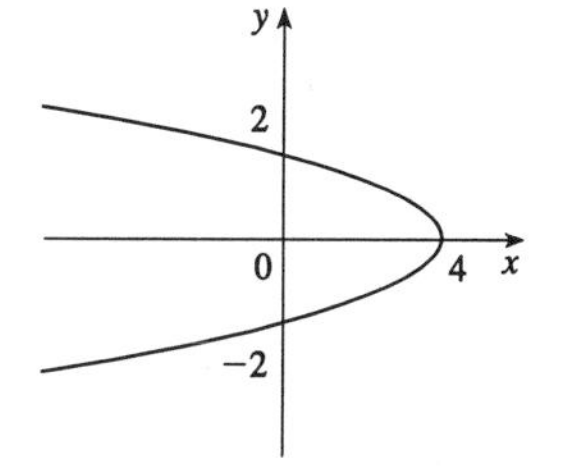

31. $x^2 + 4y^2 - 6x + 5 = 0 \quad \Leftrightarrow$

$(x^2 - 6x + 9) + 4y^2 = -5 + 9 = 4 \quad \Leftrightarrow$

$\dfrac{(x-3)^2}{4} + y^2 = 1$. Ellipse centered at $(3,0)$

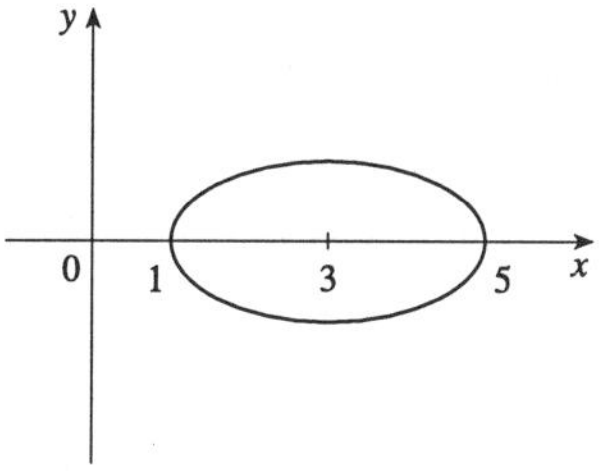

33. $y = 3x$ and $y = x^2$ intersect where $3x = x^2$

$\Leftrightarrow \quad 0 = x^2 - 3x = x(x-3)$,

that is, at $(0,0)$ and $(3,9)$.

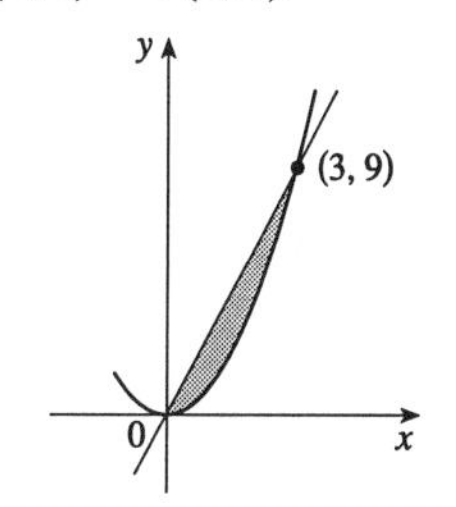

35. The parabola must have an equation of the form $y = a(x-1)^2 - 1$. Substituting $x = 3$ and $y = 3$ into the equation gives $3 = a(3-1)^2 - 1$, so $a = 1$, and the equation is $y = (x-1)^2 - 1 = x^2 - 2x$.

Note that using the other point $(-1,3)$ would have given the same value for a, and hence the same equation.

37. $\{(x,y) \mid x^2 + y^2 \le 1\}$

39. $\{(x,y) \mid y \ge x^2 - 1\}$

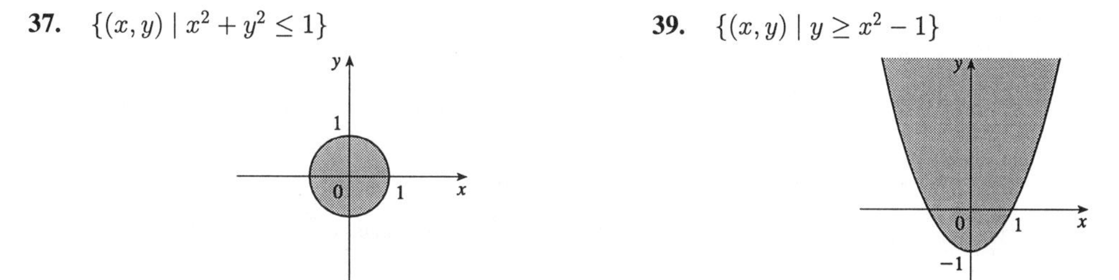

EXERCISES D

1. $210° = 210\left(\frac{\pi}{180}\right) = \frac{7\pi}{6}$ rad

3. $9° = 9\left(\frac{\pi}{180}\right) = \frac{\pi}{20}$ rad

5. $900° = 900\left(\frac{\pi}{180}\right) = 5\pi$ rad

7. 4π rad $= 4\pi\left(\frac{180}{\pi}\right) = 720°$

9. $\frac{5\pi}{12}$ rad $= \frac{5\pi}{12}\left(\frac{180}{\pi}\right) = 75°$

11. $-\frac{3\pi}{8}$ rad $= -\frac{3\pi}{8}\left(\frac{180}{\pi}\right) = -67.5°$

13. Using Formula 3, $a = r\theta = \frac{36\pi}{12} = 3\pi$ cm.

15. Using Formula 3, $\theta = \frac{1}{1.5} = \frac{2}{3}$ rad $= \frac{2}{3}\left(\frac{180}{\pi}\right) = \left(\frac{120}{\pi}\right)^{\circ}$.

17.

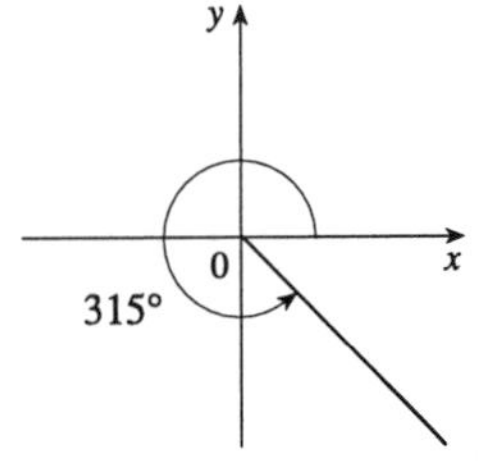

19. **21.**

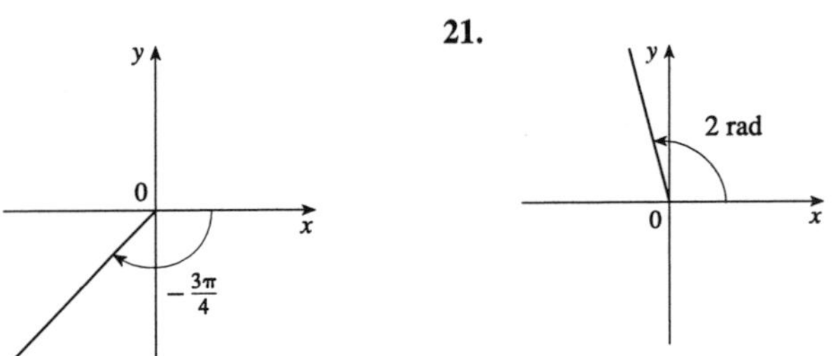

23.

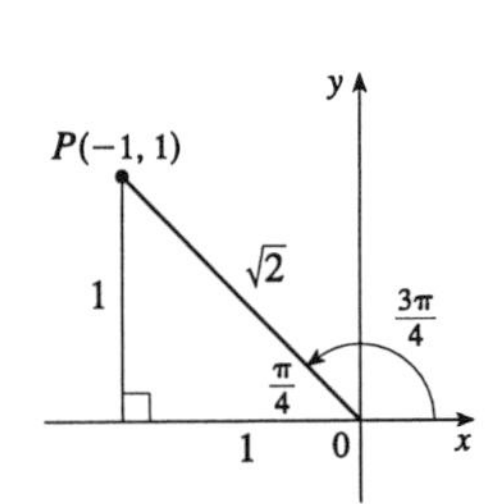

From the diagram we see that a point on the terminal line is $P(-1, 1)$. Therefore taking $x = -1$, $y = 1$, $r = \sqrt{2}$ in the definitions of the trigonometric ratios, we have

$\sin\frac{3\pi}{4} = \frac{1}{\sqrt{2}}$, $\cos\frac{3\pi}{4} = -\frac{1}{\sqrt{2}}$, $\tan\frac{3\pi}{4} = -1$,

$\csc\frac{3\pi}{4} = \sqrt{2}$, $\sec\frac{3\pi}{4} = -\sqrt{2}$, and $\cot\frac{3\pi}{4} = -1$.

25.

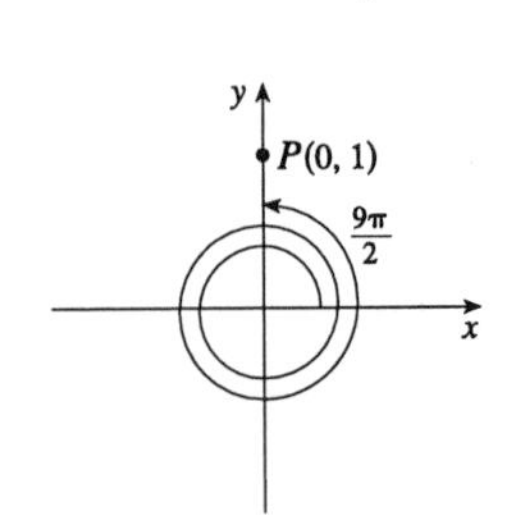

From the diagram we see that a point on the terminal line is $P(0, 1)$. Therefore taking $x = 0$, $y = 1$, $r = 1$ in the definitions of the trigonometric ratios, we have $\sin\frac{9\pi}{2} = 1$, $\cos\frac{9\pi}{2} = 0$, $\tan\frac{9\pi}{2} = y/x$ is undefined since $x = 0$, $\csc\frac{9\pi}{2} = 1$, $\sec\frac{9\pi}{2} = r/x$ is undefined since $x = 0$, and $\cot\frac{9\pi}{2} = 0$.

27.

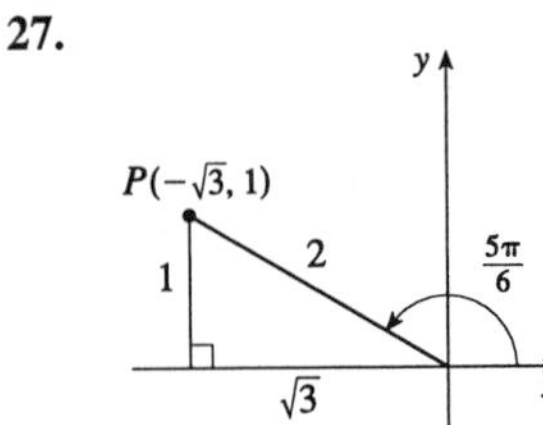

Using Figure 8 we see that a point on the terminal line is $P\left(-\sqrt{3}, 1\right)$. Therefore taking $x = -\sqrt{3}$, $y = 1$, $r = 2$ in the definitions of the trigonometric ratios, we have

$\sin\frac{5\pi}{6} = \frac{1}{2}$, $\cos\frac{5\pi}{6} = -\frac{\sqrt{3}}{2}$, $\tan\frac{5\pi}{6} = -\frac{1}{\sqrt{3}}$,

$\csc\frac{5\pi}{6} = 2$, $\sec\frac{5\pi}{6} = -\frac{2}{\sqrt{3}}$, and $\cot\frac{5\pi}{6} = -\sqrt{3}$.

29. $\sin\theta = y/r = \frac{3}{5} \quad\Rightarrow\quad y = 3$, $r = 5$, and $x = \sqrt{r^2 - y^2} = 4$ (since $0 < \theta < \frac{\pi}{2}$). Therefore taking $x = 4$, $y = 3$, $r = 5$ in the definitions of the trigonometric ratios, we have $\cos\theta = \frac{4}{5}$, $\tan\theta = \frac{3}{4}$, $\csc\theta = \frac{5}{3}$, $\sec\theta = \frac{5}{4}$, and $\cot\theta = \frac{4}{3}$.

31. $\frac{\pi}{2} < \phi < \pi \quad \Rightarrow \quad \phi$ is in the second quadrant, where x is negative and y is positive. Therefore $\sec\phi = r/x = -1.5 = -\frac{3}{2} \quad \Rightarrow \quad r = 3$, $x = -2$, and $y = \sqrt{r^2 - x^2} = \sqrt{5}$. Taking $x = -2$, $y = \sqrt{5}$, and $r = 3$ in the definitions of the trigonometric ratios, we have $\sin\phi = \frac{\sqrt{5}}{3}$, $\cos\phi = -\frac{2}{3}$, $\tan\phi = -\frac{\sqrt{5}}{2}$, $\csc\phi = \frac{3}{\sqrt{5}}$, and $\cot\theta = -\frac{2}{\sqrt{5}}$.

33. $\pi < \beta < 2\pi$ means that β is in the third or fourth quadrant where y is negative. Also since $\cot\beta = x/y = 3$ which is positive, x must also be negative. Therefore $\cot\beta = x/y = \frac{3}{1} \quad \Rightarrow \quad x = -3$, $y = -1$, and $r = \sqrt{x^2 + y^2} = \sqrt{10}$. Taking $x = -3$, $y = -1$ and $r = \sqrt{10}$ in the definitions of the trigonometric ratios, we have $\sin\beta = -\frac{1}{\sqrt{10}}$, $\cos\beta = -\frac{3}{\sqrt{10}}$, $\tan\beta = \frac{1}{3}$, $\csc\beta = -\sqrt{10}$, and $\sec\beta = -\frac{\sqrt{10}}{3}$.

35. $\sin 35^\circ = \dfrac{x}{10} \quad \Rightarrow \quad x = 10\sin 35^\circ \approx 5.73576$ cm

37. $\tan\frac{2\pi}{5} = \dfrac{x}{8} \quad \Rightarrow \quad x = 8\tan\frac{2\pi}{5} \approx 24.62147$ cm

39. **(a)** From the diagram we see that

$$\sin\theta = \frac{y}{r} = \frac{a}{c}, \text{ and } \sin(-\theta) = -\frac{a}{c} = -\sin\theta.$$

(b) Again from the diagram we see that

$$\text{that } \cos\theta = \frac{x}{r} = \frac{b}{c} = \cos(-\theta).$$

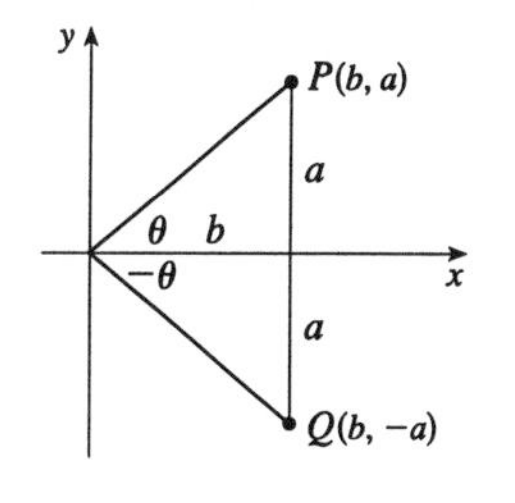

41. **(a)** Using (12a) and (13a), we have $\frac{1}{2}[\sin(x+y) + \sin(x-y)]$

$= \frac{1}{2}[\sin x\cos y + \cos x\sin y + \sin x\cos y - \cos x\sin y] = \frac{1}{2}(2\sin x\cos y) = \sin x\cos y$.

(b) This time, using (12b) and (13b), we have $\frac{1}{2}[\cos(x+y) + \cos(x-y)]$

$= \frac{1}{2}[\cos x\cos y - \sin x\sin y + \cos x\cos y + \sin x\sin y] = \frac{1}{2}(2\cos x\cos y) = \cos x\cos y$.

(c) Again using (12b) and (13b), we have $\frac{1}{2}[\cos(x-y) - \cos(x+y)]$

$= \frac{1}{2}[\cos x\cos y + \sin x\sin y - \cos x\cos y + \sin x\sin y] = \frac{1}{2}(2\sin x\sin y) = \sin x\sin y$.

43. Using (12a), $\sin\left(\frac{\pi}{2} + x\right) = \sin\frac{\pi}{2}\cos x + \cos\frac{\pi}{2}\sin x = 1\cdot\cos x + 0\cdot\sin x = \cos x$.

45. Using (6), $\sin\theta\cot\theta = \sin\theta\cdot\dfrac{\cos\theta}{\sin\theta} = \cos\theta$.

47. $\sec y - \cos y = \dfrac{1}{\cos y} - \cos y$ [by (6)] $= \dfrac{1-\cos^2 y}{\cos y} = \dfrac{\sin^2 y}{\cos y}$ [by (7)] $= \dfrac{\sin y}{\cos y}\sin y = \tan y\sin y$ [by (6)]

49. $\cot^2\theta + \sec^2\theta = \dfrac{\cos^2\theta}{\sin^2\theta} + \dfrac{1}{\cos^2\theta}$ [by (6)] $= \dfrac{\cos^2\theta\cos^2\theta + \sin^2\theta}{\sin^2\theta\cos^2\theta} = \dfrac{(1-\sin^2\theta)(1-\sin^2\theta) + \sin^2\theta}{\sin^2\theta\cos^2\theta}$ [by (7)]

$= \dfrac{1 - \sin^2\theta + \sin^4\theta}{\sin^2\theta\cos^2\theta} = \dfrac{\cos^2\theta + \sin^4\theta}{\sin^2\theta\cos^2\theta}$ [by (7)] $= \dfrac{1}{\sin^2\theta} + \dfrac{\sin^2\theta}{\cos^2\theta} = \csc^2\theta + \tan^2\theta$ [by (6)]

51. Using (14a), we have $\tan 2\theta = \tan(\theta+\theta) = \dfrac{\tan\theta + \tan\theta}{1 - \tan\theta\tan\theta} = \dfrac{2\tan\theta}{1-\tan^2\theta}$.

53. Using (15a) and (16a), $\sin x\sin 2x + \cos x\cos 2x = \sin x(2\sin x\cos x) + \cos x(2\cos^2 x - 1)$

$= 2\sin^2 x\cos x + 2\cos^3 x - \cos x = 2(1-\cos^2 x)\cos x + 2\cos^3 x - \cos x$ [by (7)]

$= 2\cos x - 2\cos^3 x + 2\cos^3 x - \cos x = \cos x$.

55. $\dfrac{\sin\phi}{1-\cos\phi} = \dfrac{\sin\phi}{1-\cos\phi}\cdot\dfrac{1+\cos\phi}{1+\cos\phi} = \dfrac{\sin\phi(1+\cos\phi)}{1-\cos^2\phi} = \dfrac{\sin\phi(1+\cos\phi)}{\sin^2\phi}$ [by (7)] $= \dfrac{1+\cos\phi}{\sin\phi}$

$= \dfrac{1}{\sin\phi} + \dfrac{\cos\phi}{\sin\phi} = \csc\phi + \cot\phi$ [by (6)]

57. Using (12a), $\sin 3\theta + \sin\theta = \sin(2\theta+\theta) + \sin\theta = \sin 2\theta\cos\theta + \cos 2\theta\sin\theta + \sin\theta$

$= \sin 2\theta\cos\theta + (2\cos^2\theta - 1)\sin\theta + \sin\theta$ [by (16a)] $= \sin 2\theta\cos\theta + 2\cos^2\theta\sin\theta - \sin\theta + \sin\theta$

$= \sin 2\theta\cos\theta + \sin 2\theta\cos\theta$ [by (15a)] $= 2\sin 2\theta\cos\theta$.

59. Since $\sin x = \frac{1}{3}$ we can label the opposite side as having length 1, the hypotenuse as having length 3, and use the Pythagorean Theorem to get that the adjacent side has length $\sqrt{8}$. Then, from the diagram, $\cos x = \frac{\sqrt{8}}{3}$. Similarly we have that $\sin y = \frac{3}{5}$.

Now use (12a): $\sin(x+y) = \sin x\cos y + \cos x\sin y = \frac{1}{3}\cdot\frac{4}{5} + \frac{\sqrt{8}}{3}\cdot\frac{3}{5} = \frac{4}{15} + \frac{3\sqrt{8}}{15} = \frac{4+6\sqrt{2}}{15}$.

61. Using (13b) and the values for $\cos x$ and $\sin y$ obtained in Exercise 59, we have $\cos(x-y) = \cos x\cos y + \sin x\sin y = \frac{\sqrt{8}}{3}\cdot\frac{4}{5} + \frac{1}{3}\cdot\frac{3}{5} = \frac{8\sqrt{2}+3}{15}$.

63. Using (15a) and the value for $\sin y$ obtained in Exercise 59, we have

$\sin 2y = 2\sin y\cos y = \dfrac{2\sin y}{\sec y} = 2\left(\frac{3}{5}\right)\left(\frac{4}{5}\right) = \frac{24}{25}$.

65. $2\cos x - 1 = 0 \quad\Leftrightarrow\quad \cos x = \frac{1}{2} \quad\Rightarrow\quad x = \frac{\pi}{3}, \frac{5\pi}{3}$

67. $2\sin^2 x = 1 \quad\Leftrightarrow\quad \sin^2 x = \frac{1}{2} \quad\Leftrightarrow\quad \sin x = \pm\frac{1}{\sqrt{2}} \quad\Rightarrow\quad x = \frac{\pi}{4}, \frac{3\pi}{4}, \frac{5\pi}{4}, \frac{7\pi}{4}$.

69. Using (15a), $\sin 2x = \cos x \;\Rightarrow\; 2\sin x\cos x - \cos x = 0 \;\Leftrightarrow\; \cos x(2\sin x - 1) = 0 \;\Leftrightarrow\; \cos x = 0$ or $2\sin x - 1 = 0 \;\Rightarrow\; x = \frac{\pi}{2}, \frac{3\pi}{2}$ or $\sin x = \frac{1}{2} \;\Rightarrow\; x = \frac{\pi}{6}$ or $\frac{5\pi}{6}$. Therefore the solutions are $x = \frac{\pi}{6}, \frac{\pi}{2}, \frac{5\pi}{6}, \frac{3\pi}{2}$.

71. $\sin x = \tan x \quad\Leftrightarrow\quad \sin x - \tan x = 0 \quad\Leftrightarrow\quad \sin x - \dfrac{\sin x}{\cos x} = 0 \quad\Leftrightarrow\quad \sin x\left(1 - \dfrac{1}{\cos x}\right) = 0 \quad\Leftrightarrow$

$\sin x = 0$ or $1 - \dfrac{1}{\cos x} = 0 \quad\Rightarrow\quad x = 0, \pi, 2\pi$ or $1 = \dfrac{1}{\cos x} \quad\Rightarrow\quad \cos x = 1 \quad\Rightarrow\quad x = 0, 2\pi$. Therefore the solutions are $x = 0, \pi, 2\pi$.

73. We know that $\sin x = \frac{1}{2}$ when $x = \frac{\pi}{6}$ or $\frac{5\pi}{6}$, and from Figure 13(a), we see that $\sin x \le \frac{1}{2} \quad\Rightarrow\quad 0 \le x \le \frac{\pi}{6}$ or $\frac{5\pi}{6} \le x \le 2\pi$.

75. $\tan x = -1$ when $x = \frac{3\pi}{4}, \frac{7\pi}{4}$, and $\tan x = 1$ when $x = \frac{\pi}{4}$ or $\frac{5\pi}{4}$. From Figure 14 we see that $-1 < \tan x < 1$ $\Rightarrow\quad 0 \le x < \frac{\pi}{4}, \frac{3\pi}{4} < x < \frac{5\pi}{4}$, and $\frac{7\pi}{4} < x \le 2\pi$.

77. $y = \cos\left(x - \frac{\pi}{3}\right)$. We start with the graph of $y = \cos x$ and shift it $\frac{\pi}{3}$ units to the right.

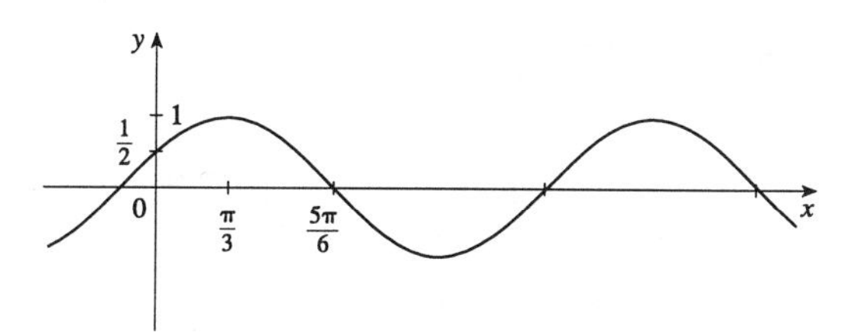

79. $y = \frac{1}{3}\tan\left(x - \frac{\pi}{2}\right)$. We start with the graph of $y = \tan x$, shift it $\frac{\pi}{2}$ units to the right and compress it to $\frac{1}{3}$ of its original vertical size.

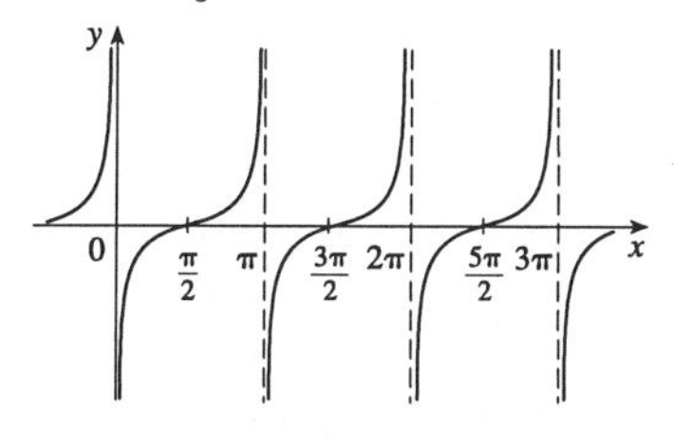

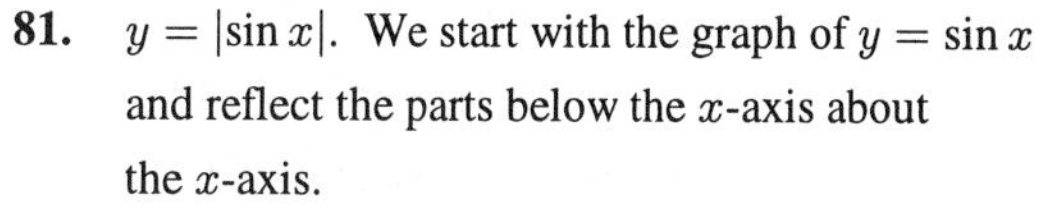

81. $y = |\sin x|$. We start with the graph of $y = \sin x$ and reflect the parts below the x-axis about the x-axis.

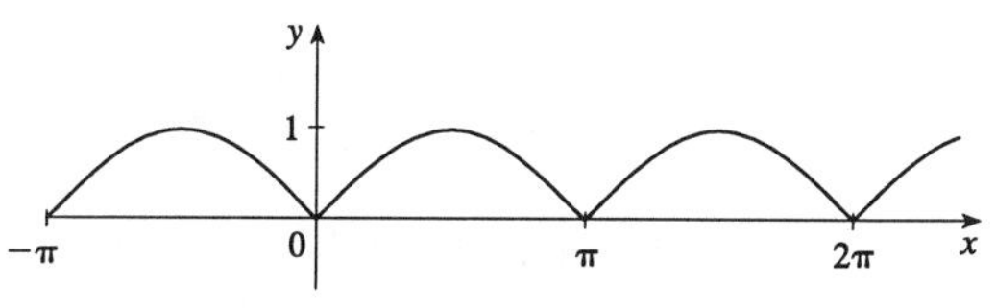

83. From the figure we see that $x = b\cos\theta$, $y = b\sin\theta$, and from the distance formula we have that the distance c from (x, y) to $(a, 0)$ is $c = \sqrt{(x-a)^2 + (y-0)^2} \quad \Rightarrow \quad c^2 = (b\cos\theta - a)^2 + (b\sin\theta)^2$

$= b^2\cos^2\theta - 2ab\cos\theta + a^2 + b^2\sin^2\theta = a^2 + b^2(\cos^2\theta + \sin^2\theta) - 2ab\cos\theta = a^2 + b^2 - 2ab\cos\theta$ [by (7)].

85. Using the Law of Cosines, we have $c^2 = 1^2 + 1^2 - 2(1)(1)\cos(\alpha - \beta) = 2[1 - \cos(\alpha - \beta)]$. Now, using the distance formula, $c^2 = |AB|^2 = (\cos\alpha - \cos\beta)^2 + (\sin\alpha - \sin\beta)^2$. Equating these two expressions for c^2, we get $2[1 - \cos(\alpha - \beta)] = \cos^2\alpha + \sin^2\alpha + \cos^2\beta + \sin^2\beta - 2\cos\alpha\cos\beta - 2\sin\alpha\sin\beta \quad \Rightarrow$

$1 - \cos(\alpha - \beta) = 1 - \cos\alpha\cos\beta - \sin\alpha\sin\beta \quad \Rightarrow \quad \cos(\alpha - \beta) = \cos\alpha\cos\beta + \sin\alpha\sin\beta$.

87. In Exercise 86 we used the subtraction formula for cosine to prove the addition formula for cosine. Using that formula with $x = \frac{\pi}{2} - \alpha$, $y = \beta$, we get $\cos\left[\left(\frac{\pi}{2} - \alpha\right) + \beta\right] = \cos\left(\frac{\pi}{2} - \alpha\right)\cos\beta - \sin\left(\frac{\pi}{2} - \alpha\right)\sin\beta \quad \Rightarrow$

$\cos\left[\frac{\pi}{2} - (\alpha - \beta)\right] = \cos\left(\frac{\pi}{2} - \alpha\right)\cos\beta - \sin\left(\frac{\pi}{2} - \alpha\right)\sin\beta$. Now we use the identities given in the problem to get $\sin(\alpha - \beta) = \sin\alpha\cos\beta - \cos\alpha\sin\beta$.

89. Using the formula derived in Exercise 88, the area of the triangle is $\frac{1}{2}(10)(3)\sin 107° \approx 14.34457$.

EXERCISES E

1. Let S_n be the statement that $2^n > n$.

1. S_1 is true because $2^1 = 2 > 1$.
2. Assume S_k is true, that is, $2^k > k$. Then $2^{k+1} = 2^k \cdot 2 > k \cdot 2 = 2k$ (since $2^k > k$). But $k > 1 \quad \Rightarrow$ $k + k > k + 1 \quad \Rightarrow \quad 2k > k + 1$, so that $2^{k+1} > 2k > k + 1$, which shows that S_{k+1} is true.
3. Therefore, by mathematical induction, $2^n > n$ for every positive integer n.

3. Let S_n be the statement that $(1 + x)^n \geq 1 + nx$.

1. S_1 is true because $(1 + x)^1 = 1 + (1)x$.
2. Assume S_k is true, that is, $(1 + x)^k \geq 1 + kx$. Then $(1 + x)^{k+1} = (1 + x)^k (1 + x) \geq (1 + kx)(1 + x)$ [since $(1 + x)^k \geq 1 + kx$] $\quad = 1 + x + kx + kx^2 \geq 1 + x + kx = 1 + (k + 1)x$, and so S_{k+1} is true.
3. Therefore, by mathematical induction, $(1 + x)^n \geq 1 + nx$ for every positive integer n.

5. Let S_n be the statement that $7^n - 1$ is divisible by 6.

1. S_1 is true because $7^1 - 1 = 6$ is divisible by 6.
2. Assume S_k is true, that is, $7^k - 1$ is divisible by 6; in other words $7^k - 1 = 6m$ for some positive integer m. Then $7^{k+1} - 1 = 7^k \cdot 7 - 1 = (6m + 1) \cdot 7 - 1 = 6(7m + 1)$, which is divisible by 6, so S_{k+1} is true.
3. Therefore, by mathematical induction, $7^n - 1$ is divisible by 6 for every positive integer n.

7. Let S_n be the statement that $1 + 3 + 5 + \cdots + (2n - 1) = n^2$.

1. S_1 is true because $[2(1) - 1] = 1 = 1^2$.
2. Assume S_k is true, that is $1 + 3 + 5 + \cdots + (2k - 1) = k^2$. Then $1 + 3 + 5 + \cdots + [2(k + 1) - 1] = 1 + 3 + 5 + \cdots + (2k - 1) + (2k + 1) = k^2 + (2k + 1) = (k + 1)^2$, which shows that S_{k+1} is true.
3. Therefore, by mathematical induction, $1 + 3 + 5 + \cdots + (2n - 1) = n^2$ for every positive integer n.

9. Let S_n be the statement that $\frac{1}{2} + \frac{1}{6} + \frac{1}{12} + \cdots + \frac{1}{n(n+1)} = \frac{n}{n+1}$.

1. S_1 is true because $\frac{1}{1(1+1)} = \frac{1}{1+1}$.
2. Assume S_k is true, that is, $\frac{1}{2} + \frac{1}{6} + \frac{1}{12} + \cdots + \frac{1}{k(k+1)} = \frac{k}{k+1}$. Then

$$\frac{1}{2} + \frac{1}{6} + \frac{1}{12} + \cdots + \frac{1}{(k+1)[(k+1)+1]} = \frac{1}{2} + \frac{1}{6} + \frac{1}{12} + \cdots + \frac{1}{(k+1)(k+2)}$$
$$= \frac{1}{2} + \frac{1}{6} + \frac{1}{12} + \cdots + \frac{1}{k(k+1)} + \frac{1}{(k+1)(k+2)} = \frac{k}{k+1} + \frac{1}{(k+1)(k+2)} = \frac{k(k+2)+1}{(k+1)(k+2)}$$
$$= \frac{k^2+2k+1}{(k+1)(k+2)} = \frac{(k+1)(k+1)}{(k+1)(k+2)} = \frac{(k+1)}{(k+1)+1}$$ which shows that S_{k+1} is true.

3. Therefore, by mathematical induction, $\frac{1}{2} + \frac{1}{6} + \frac{1}{12} + \cdots + \frac{1}{n(n+1)} = \frac{n}{n+1}$ for every positive integer n.

EXERCISES G

1. The computer results are from Maple, with `Digits:=16`. The last column shows the values of the sixth-degree Taylor polynomial for $f(x) = \csc^2 x - x^{-2}$ near $x = 0$. Note that the second arrangement of Taylor's polynomial is easier to use with a calculator:

$$T_6(x) = \tfrac{1}{3} + \tfrac{1}{15}x^2 + \tfrac{2}{189}x^4 + \tfrac{1}{675}x^6 = \left[\left(\tfrac{1}{675}x^2 + \tfrac{2}{189}\right)x^2 + \tfrac{1}{15}\right]x^2 + \tfrac{1}{3}$$

x	$f(x)_{\text{calculator}}$	$f(x)_{\text{computer}}$	$f(x)_{\text{Taylor}}$
0.1	0.33400107	0.3340010596845	0.33400106
0.01	0.333341	0.33334000010	0.33334000
0.001	0.3334	0.333333400	0.33333340
0.0001	0.34	0.3333334	0.33333333
0.00001	2.0	0.33334	0.33333333
0.000001	100 or 200	0.334	0.33333333
0.0000001	10,000 or 20,000	0.4	0.33333333
0.00000001	1,000,000	0	0.33333333

We see that the calculator results start to deteriorate seriously at $x = 0.0001$, and for smaller x, they are entirely meaningless. The different results "100 or 200" etc. depended on whether we calculated $\left[(\sin x)^2\right]^{-1}$ or $\left[(\sin x)^{-1}\right]^2$. With Maple, the result is off by more than 10% when $x = 0.0000001$ (compare with the calculator result!) A detailed analysis reveals that the values of the function are always greater than $\frac{1}{3}$, but the computer eventually gives results less than $\frac{1}{3}$.

The polynomial $T_6(x)$ was obtained by patient simplification of the expression for $f(x)$, starting with $\sin^2(x) = \frac{1}{2}(1 - \cos 2x)$, where $\cos 2x = 1 - \dfrac{(2x)^2}{2!} + \dfrac{(2x)^4}{4!} - \cdots - \dfrac{(2x)^{10}}{10!} + R_{12}(x)$.
Consequently, the exact value of the limit is $T_6(0) = \frac{1}{3}$. It can also be obtained by several applications of l'Hospital's Rule to the expression $f(x) = \dfrac{x^2 - \sin^2 x}{x^2 \sin^2 x}$ with intermediate simplifications.

3. From $f(x) = \dfrac{x^{25}}{(1.0001)^x}$ (we may assume $x > 0$; why?), we have $\ln f(x) = 25 \ln x - x \ln(1.0001)$ and $\dfrac{f'(x)}{f(x)} = \dfrac{25}{x} - \ln(1.0001)$. This derivative, as well as the derivative $f'(x)$ itself, is positive for

$0 < x < x_0 = \dfrac{25}{\ln(1.0001)} \approx 249{,}971.015$, and negative for $x > x_0$. Hence the maximum value of $f(x)$ is

$f(x_0) = \dfrac{x_0^{25}}{(1.0001)^{x_0}}$, a number too large to be calculated directly. Using decimal logarithms, $\log_{10} f(x_0) \approx 124.08987757$, so that $f(x_0) \approx 1.229922 \times 10^{124}$. The actual value of the limit is $\lim_{x\to\infty} f(x) = 0$; it would be wasteful and inelegant to use l'Hospital's Rule twenty-five times since we can transform $f(x)$ into

$f(x) = \left[\dfrac{x}{(1.0001)^{x/25}}\right]^{25}$, and the inside expression needs just one application of l'Hospital's Rule to give 0.

5. For $f(x) = \ln \ln x$ with $x \in [a, b]$, $a = 10^9$, and $b = 10^9 + 1$, we need $f'(x) = \dfrac{1}{x \ln x}$, $f''(x) = -\dfrac{\ln x + 1}{x^2 (\ln x)^2}$.

(a) $f'(b) < D < f'(a)$, where $f'(a) \approx 4.8254942434 \times 10^{-11}$, $f'(b) \approx 4.8254942383 \times 10^{-11}$.

(b) Let us estimate $f'(b) - f'(a) = (b - a) f''(c_1) = f''(c_1)$. Since f'' increases (its absolute value decreases), we have $|f'(b) - f'(a)| < |f''(a)| \approx 5.0583 \times 10^{-20}$.

7. **(a)** The 11-digit calculator value of $192 \sin \frac{\pi}{96}$ is 6.2820639018, while the value (on the same device) of p before rationalization is 6.282063885, which is 1.68×10^{-8} less than the trigonometric result.

(b) $p = \dfrac{96}{\sqrt{2+\sqrt{3}} \cdot \sqrt{2+\sqrt{2+\sqrt{3}}} \cdot \sqrt{2+\sqrt{2+\sqrt{2+\sqrt{3}}}} \cdot \sqrt{2+\sqrt{2+\sqrt{2+\sqrt{2+\sqrt{3}}}}}}$, but of course we can avoid repetitious calculations by storing intermediate results in a memory: $p_1 = \sqrt{2 + \sqrt{3}}$, $p_2 = \sqrt{2 + p_1}$, $p_3 = \sqrt{2 + p_2}$, $p_4 = \sqrt{2 + p_3}$, and so $p = \dfrac{96}{p_1 p_2 p_3 p_4}$. According to this formula, a calculator gives $p \approx 6.2820639016$, which is within 2×10^{-10} of the trigonometric result. With `Digits:=16;` Maple gives $p \approx 6.282063901781030$ before rationalization (off the trig result by about 1.1×10^{-14}) and $p \approx 6.282063901781018$ after rationalization (error of about 1.7×10^{-15}), a gain of about 1 digit of accuracy for rationalizing. If we set `Digits:=100;`, the difference between Maple's calculation of $192 \sin \frac{\pi}{96}$ and the radical is only about 4×10^{-99}.

9. **(a)** Let $A = \left[\frac{1}{2}\left(27q + \sqrt{729q^2 + 108p^3}\right)\right]^{1/3}$ and $B = \left[\frac{1}{2}\left(27q - \sqrt{729q^2 + 108p^3}\right)\right]^{1/3}$. Then $A^3 + B^3 = 27q$ and $AB = \frac{1}{4}[729q^2 - (729q^2 + 108p^3)]^{1/3} = -3p$. Substitute into the formula $A + B = \dfrac{A^3 + B^3}{A^2 - AB + B^2}$ where we replace B by $-\dfrac{3p}{A}$:

$$x = \tfrac{1}{3}(A + B) = \frac{27q/3}{\left[\frac{1}{2}\left(27q + \sqrt{729q^2 + 108p^3}\right)\right]^{2/3} + 3p + 9p^2 \left[\frac{1}{2}\left(27q + \sqrt{729q^2 + 108p^3}\right)\right]^{-2/3}}$$

which almost yields the given formula; since replacing q by $-q$ results in replacing x by $-x$, a simple discussion of the cases $q > 0$ and $q < 0$ allows us to replace q by $|q|$ in the denominator, so that it involves only positive numbers. The problems mentioned in the introduction to this exercise have disappeared.

(b) A direct attack works best here. To save space, let $\alpha = 2 + \sqrt{5}$, so we can rationalize, using $\alpha^{-1} = -2 + \sqrt{5}$ and $\alpha - \alpha^{-1} = 4$ (check it!):

$$u = \frac{4}{\alpha^{2/3} + 1 + \alpha^{-2/3}} \cdot \frac{\alpha^{1/3} - \alpha^{-1/3}}{\alpha^{1/3} - \alpha^{-1/3}} = \frac{4\left(\alpha^{1/3} - \alpha^{-1/3}\right)}{\alpha - \alpha^{-1}} = \alpha^{1/3} - \alpha^{-1/3}$$

and we cube the expression for u: $u^3 = \alpha - 3\alpha^{1/3} + 3\alpha^{-1/3} - \alpha^{-1} = 4 - 3u$, $u^3 + 3u - 4 = (u - 1)(u^2 + u + 4) = 0$, so that the only real root is $u = 1$. A check using the formula from part (a): $p = 3$, $q = -4$, so $729q^2 + 108p^3 = 14{,}580 = 54^2 \times 5$, and $x = \dfrac{36}{\left(54 + 27\sqrt{5}\right)^{2/3} + 9 + 81\left(54 + 27\sqrt{5}\right)^{-2/3}}$, which simplifies to the given form after reduction by 9.

11. Proof that $\lim_{n\to\infty} a_n = 0$: From $1 \le e^{1-x} \le e$ it follows that $x^n \le e^{1-x}x^n \le x^n e$, and integration gives $\frac{1}{n+1} = \int_0^1 x^n\,dx \le \int_0^1 e^{1-x}x^n\,dx \le \int_0^1 x^n e\,dx = \frac{e}{n+1}$, that is, $\frac{1}{n+1} \le a_n \le \frac{e}{n+1}$, and since $\lim_{n\to\infty}\frac{1}{n+1} = \lim_{n\to\infty}\frac{e}{n+1} = 0$, it follows from the Squeeze Theorem that $\lim_{n\to\infty} a_n = 0$. Of course, the expression $1/(n+1)$ on the left side could have been replaced by 0 and the proof would still be correct.

Calculations: Using the formula $a_n = \left[e - 1 - \left(\frac{1}{1!} + \frac{1}{2!} + \cdots + \frac{1}{n!}\right)\right] n!$ with an 11-digit pocket calculator:

n	a_n	n	a_n	n	a_n
0	1.7182818284	7	0.1404151360	14	−5.07636992
1	0.7182818284	8	0.1233210880	15	−77.1455488
2	0.4365636568	9	0.1098897920	16	−1235.3287808
3	0.3096909704	10	0.0988979200	17	−21001.589274
4	0.2387638816	11	0.0878771200	18	−378029.60693
5	0.1938194080	12	0.0545254400	19	−7182563.5317
6	0.1629164480	13	−0.2911692800	20	−143651271.63

It is clear that the values calculated from the direct reduction formula will diverge to $-\infty$. If we instead calculate a_n using the reduction formula in Maple (with `Digits:=16`), we get some odd results : $a_{20} = -1000$, $a_{28} = 10^{14}$, $a_{29} = 0$, and $a_{30} = 10^{17}$, for example. But for larger n, the results are at least small and positive (for example, $a_{1000} \approx 0.001$.) For $n > 32{,}175$, we get the delightful `object too large` error message. If, instead of using the reduction formula, we integrate directly with Maple, the results are much better.

13. We can start by expressing e^x and e^{-x} in terms of $E(x) = (e^x - 1)/x$ $(x \ne 0)$, where $E(0) = 1$ to make E continuous at 0 (by L'Hospital's Rule). Namely, $e^x = 1 + xE(x)$, $e^{-x} = 1 - xE(-x)$ and $\sinh x = \frac{1 + xE(x) - [1 - xE(-x)]}{2} = \frac{1}{2}x[E(x) + E(-x)]$, where the addition involves only positive numbers $E(x)$ and $E(-x)$, thus presenting no loss of accuracy due to subtraction.

Another form, which calls the function E only once: we write $\sinh x = \frac{(e^x)^2 - 1}{2e^x} = \frac{[1 + xE(x)]^2 - 1}{2[1 + xE(x)]} = \frac{x\left[1 + \frac{1}{2}|x|E(|x|)\right]E(|x|)}{1 + |x|E(|x|)}$, taking advantage of the fact that $\frac{\sinh x}{x}$ is an even function, so replacing x by $|x|$ does not change its value.

EXERCISES H

1. $(3+2i)+(7-3i)=(3+7)+(2-3)i=10-i$

3. $(3-i)(4+i)=12+3i-4i-(-1)=13-i$

5. $\overline{12+7i}=12-7i$

7. $\dfrac{2+3i}{1-5i}=\dfrac{2+3i}{1-5i}\cdot\dfrac{1+5i}{1+5i}=\dfrac{2+10i+3i+15(-1)}{1-25(-1)}=\dfrac{-13+13i}{26}=-\frac{1}{2}+\frac{1}{2}i$

9. $\dfrac{1}{1+i}=\dfrac{1}{1+i}\cdot\dfrac{1-i}{1-i}=\dfrac{1-i}{1-(-1)}=\dfrac{1-i}{2}=\frac{1}{2}-\frac{1}{2}i$

11. $i^3=i^2\cdot i=(-1)\,i=-i$

13. $\sqrt{-25}=\sqrt{25}\,i=5i$

15. $\overline{3+4i}=3-4i$, $|3+4i|=\sqrt{3^2+4^2}=\sqrt{25}=5$

17. $\overline{-4i}=\overline{0-4i}=0+4i=4i$, $|-4i|=\sqrt{0^2+(-4)^2}=4$

19. $4x^2+9=0 \quad\Leftrightarrow\quad 4x^2=-9 \quad\Leftrightarrow\quad x^2=-\frac{9}{4} \quad\Leftrightarrow\quad x=\pm\sqrt{-\frac{9}{4}}=\pm\sqrt{\frac{9}{4}}i=\pm\frac{3}{2}i.$

21. By the quadratic formula, $x^2-8x+17=0 \quad\Leftrightarrow\quad x=\dfrac{8\pm\sqrt{8^2-4(1)(17)}}{2(1)}=\dfrac{8\pm\sqrt{-4}}{2}=\dfrac{8\pm 2i}{2}=4\pm i.$

23. By the quadratic formula, $z^2+z+2=0 \quad\Leftrightarrow\quad z=\dfrac{-1\pm\sqrt{1-4(1)(2)}}{2(1)}=\dfrac{-1\pm\sqrt{-7}}{2}=-\frac{1}{2}\pm\frac{\sqrt{7}}{2}i.$

25. $r=\sqrt{(-3)^2+3^2}=3\sqrt{2}$, $\tan\theta=\frac{3}{-3}=-1 \quad\Rightarrow\quad \theta=\frac{3}{4}\pi$ (since the given number is in the second quadrant). Therefore $-3+3i=3\sqrt{2}\left(\cos\frac{3\pi}{4}+i\sin\frac{3\pi}{4}\right)$.

27. $r=\sqrt{3^2+4^2}=5$, $\tan\theta=\frac{4}{3} \quad\Rightarrow\quad \theta=\tan^{-1}\frac{4}{3}$ (since the given number is in the second quadrant). Therefore $3+4i=5\left[\cos\left(\tan^{-1}\frac{4}{3}\right)+i\sin\left(\tan^{-1}\frac{4}{3}\right)\right]$.

29. For $z=\sqrt{3}+i$, $r=\sqrt{\left(\sqrt{3}\right)^2+1^2}=2$, and $\tan\theta=\frac{1}{\sqrt{3}} \quad\Rightarrow\quad \theta=\frac{\pi}{6}$ so that $z=2\left(\cos\frac{\pi}{6}+i\sin\frac{\pi}{6}\right)$. For $w=1+\sqrt{3}i$, $r=2$, and $\tan\theta=\sqrt{3} \quad\Rightarrow\quad \theta=\frac{\pi}{3}$ so that $w=2\left(\cos\frac{\pi}{3}+i\sin\frac{\pi}{3}\right)$. Therefore
$zw=2\cdot2\left[\cos\left(\frac{\pi}{6}+\frac{\pi}{3}\right)+i\sin\left(\frac{\pi}{6}+\frac{\pi}{3}\right)\right]=4\left(\cos\frac{\pi}{2}+i\sin\frac{\pi}{2}\right)$,
$z/w=\frac{2}{2}\left[\cos\left(\frac{\pi}{6}-\frac{\pi}{3}\right)+i\sin\left(\frac{\pi}{6}-\frac{\pi}{3}\right)\right]=\cos\left(-\frac{\pi}{6}\right)+i\sin\left(-\frac{\pi}{6}\right)$, and $1=1+0i=\cos 0+i\sin 0 \quad\Rightarrow$
$1/z=\frac{1}{2}\left[\cos\left(0-\frac{\pi}{6}\right)+i\sin\left(0-\frac{\pi}{6}\right)\right]=\frac{1}{2}\left[\cos\left(-\frac{\pi}{6}\right)+i\sin\left(-\frac{\pi}{6}\right)\right]$.

31. For $z=2\sqrt{3}-2i$, $r=4$, $\tan\theta=\frac{-2}{2\sqrt{3}}=-\frac{1}{\sqrt{3}} \quad\Rightarrow\quad \theta=-\frac{\pi}{6} \quad\Rightarrow\quad z=4\left[\cos\left(-\frac{\pi}{6}\right)+i\sin\left(-\frac{\pi}{6}\right)\right]$. For $w=-1+i$, $r=\sqrt{2}$, $\tan\theta=\frac{1}{-1}=-1 \quad\Rightarrow\quad \theta=\frac{3\pi}{4} \quad\Rightarrow\quad z=\sqrt{2}\left(\cos\frac{3\pi}{4}+i\sin\frac{3\pi}{4}\right)$. Therefore
$zw=4\sqrt{2}\left[\cos\left(-\frac{\pi}{6}+\frac{3\pi}{4}\right)+i\sin\left(-\frac{\pi}{6}+\frac{3\pi}{4}\right)\right]=4\sqrt{2}\left(\cos\frac{7\pi}{12}+i\sin\frac{7\pi}{12}\right)$,
$z/w=\frac{4}{\sqrt{2}}\left[\cos\left(-\frac{\pi}{6}-\frac{3\pi}{4}\right)+i\sin\left(-\frac{\pi}{6}-\frac{3\pi}{4}\right)\right]=\frac{4}{\sqrt{2}}\left[\cos\left(-\frac{11\pi}{12}\right)+i\sin\left(-\frac{11\pi}{12}\right)\right]=2\sqrt{2}\left(\cos\frac{13\pi}{12}+i\sin\frac{13\pi}{12}\right)$,
and $1=1+0i=\cos 0+i\sin 0 \quad\Rightarrow\quad 1/z=\frac{1}{4}\left[\cos\left(0-\left(-\frac{\pi}{6}\right)\right)+i\sin\left(0-\left(-\frac{\pi}{6}\right)\right)\right]=\frac{1}{4}\left(\cos\frac{\pi}{6}+i\sin\frac{\pi}{6}\right)$.

33. For $z = 1 + i$, $r = \sqrt{2}$, $\tan\theta = \frac{1}{1} = 1 \quad\Rightarrow\quad \theta = \frac{\pi}{4} \quad\Rightarrow\quad 1 + i = \sqrt{2}\left(\cos\frac{\pi}{4} + i\sin\frac{\pi}{4}\right)$. So by De Moivre's Theorem, $(1+i)^{20} = \left[\sqrt{2}\left(\cos\frac{\pi}{4} + i\sin\frac{\pi}{4}\right)\right]^{20} = \left(2^{1/2}\right)^{20}\left(\cos\frac{20\pi}{4} + i\sin\frac{20\pi}{4}\right) = 2^{10}(\cos 5\pi + i\sin 5\pi)$
$= 2^{10}[-1 + i(0)] = -2^{10} = -1024$.

35. For $z = 2\sqrt{3} + 2i$, $r = 4$, $\tan\theta = \frac{2}{2\sqrt{3}} = \frac{1}{\sqrt{3}} \quad\Rightarrow\quad \theta = \frac{\pi}{6} \quad\Rightarrow\quad 2\sqrt{3} + 2i = 4\left(\cos\frac{\pi}{6} + i\sin\frac{\pi}{6}\right)$. So by De Moivre's Theorem,

$$\left(2\sqrt{3} + 2i\right)^5 = \left[4\left(\cos\frac{\pi}{6} + i\sin\frac{\pi}{6}\right)\right]^5 = 4^5\left(\cos\frac{5\pi}{6} + i\sin\frac{5\pi}{6}\right) = 4^5\left[-\frac{\sqrt{3}}{2} + i(0.5)\right] = -512\sqrt{3} + 512i.$$

37. $1 = 1 + 0i = \cos 0 + i\sin 0$. Using Equation 3 with $r = 1$, $n = 8$, and $\theta = 0$ we have

$$w_k = 1^{1/8}\left[\cos\left(\frac{0 + 2k\pi}{8}\right) + i\sin\left(\frac{0 + 2k\pi}{8}\right)\right] = \cos\frac{k\pi}{4} + i\sin\frac{k\pi}{4}, \text{ where } k = 0, 1, 2, \ldots, 7.$$

$w_0 = (\cos 0 + i\sin 0) = 1$ $\qquad w_4 = (\cos\pi + i\sin\pi) = -1$

$w_1 = \left(\cos\frac{\pi}{4} + i\sin\frac{\pi}{4}\right) = \frac{1}{\sqrt{2}} + \frac{1}{\sqrt{2}}i$ $\qquad w_5 = \left(\cos\frac{5\pi}{4} + i\sin\frac{5\pi}{4}\right) = -\frac{1}{\sqrt{2}} - \frac{1}{\sqrt{2}}i$

$w_2 = \left(\cos\frac{\pi}{2} + i\sin\frac{\pi}{2}\right) = i$ $\qquad w_6 = \left(\cos\frac{3\pi}{2} + i\sin\frac{3\pi}{2}\right) = -i$

$w_3 = \left(\cos\frac{3\pi}{4} + i\sin\frac{3\pi}{4}\right) = -\frac{1}{\sqrt{2}} + \frac{1}{\sqrt{2}}i$ $\qquad w_7 = \left(\cos\frac{7\pi}{4} + i\sin\frac{7\pi}{4}\right) = \frac{1}{\sqrt{2}} - \frac{1}{\sqrt{2}}i$

39. $0 = 0 + i = \cos\frac{\pi}{2} + i\sin\frac{\pi}{2}$. Using Equation 3 with $r = 1$, $n = 3$, and $\theta = \frac{\pi}{2}$, we have

$$w_k = 1^{1/3}\left[\cos\left(\frac{\pi/2 + 2k\pi}{3}\right) + i\sin\left(\frac{\pi/2 + 2k\pi}{3}\right)\right], \text{ where } k = 0, 1, 2.$$

$w_0 = \left(\cos\frac{\pi}{6} + i\sin\frac{\pi}{6}\right) = \frac{\sqrt{3}}{2} + \frac{1}{2}i$, $w_1 = \left(\cos\frac{5\pi}{6} + i\sin\frac{5\pi}{6}\right) = -\frac{\sqrt{3}}{2} + \frac{1}{2}i$

$w_2 = \left(\cos\frac{9\pi}{6} + i\sin\frac{9\pi}{6}\right) = -i$

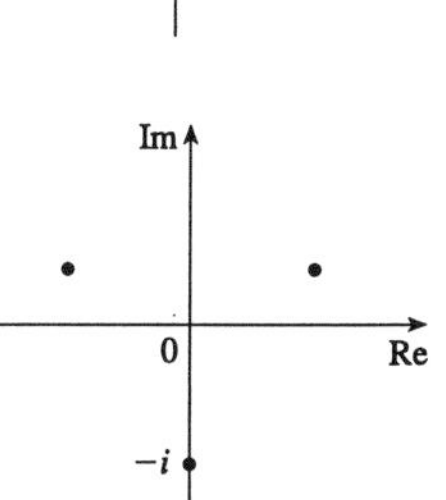

41. Using Euler's formula (6) with $y = \frac{\pi}{2}$, $e^{i\pi/2} = \cos\frac{\pi}{2} + i\sin\frac{\pi}{2} = i$.

43. Using Euler's formula with $y = \frac{3\pi}{4}$, $e^{i3\pi/4} = \cos\frac{3\pi}{4} + i\sin\frac{3\pi}{4} = -\frac{1}{\sqrt{2}} + \frac{1}{\sqrt{2}}i$.

45. Using Equation 7 with $x = 2$ and $y = \pi$, $e^{2+i\pi} = e^2 e^{i\pi} = e^2(\cos\pi + i\sin\pi) = e^2(-1 + 0) = -e^2$.

47. Take $r = 1$ and $n = 3$ in De Moivre's Theorem to get

$[1(\cos\theta + i\sin\theta)]^3 = 1^3(\cos 3\theta + i\sin 3\theta) \quad\Rightarrow\quad (\cos\theta + i\sin\theta)^3 = \cos 3\theta + i\sin 3\theta \quad\Rightarrow$

$\cos^3\theta + 3(\cos^2\theta)(i\sin\theta) + 3(\cos\theta)(i\sin\theta)^2 + (i\sin\theta)^3 = \cos 3\theta + i\sin 3\theta \quad\Rightarrow$

$(\cos^3\theta - 3\sin^2\theta\cos\theta) + (3\sin\theta\cos^2\theta - \sin^3\theta)i = \cos 3\theta + i\sin 3\theta$. Equating real and imaginary parts gives

$\cos 3\theta = \cos^3\theta - 3\sin^2\theta\cos\theta$ and $\sin 3\theta = 3\sin\theta\cos^2\theta - \sin^3\theta$.

49. $F(x) = e^{rx} = e^{(a+bi)x} = e^{ax+bxi} = e^{ax}(\cos bx + i\sin bx) = e^{ax}\cos bx + i(e^{ax}\sin bx) \quad\Rightarrow$

$F'(x) = (e^{ax}\cos bx)' + i(e^{ax}\sin bx)' = (ae^{ax}\cos bx - be^{ax}\sin bx) + i(ae^{ax}\sin bx + be^{ax}\cos bx)$

$= a[e^{ax}(\cos bx + i\sin bx)] + b[e^{ax}(-\sin bx + i\cos bx)] = ae^{rx} + b\left[e^{ax}\left(i^2\sin bx + i\cos bx\right)\right]$

$= ae^{rx} + bi[e^{ax}(\cos bx + i\sin bx)] = ae^{rx} + bie^{rx} = (a + bi)e^{rx} = re^{rx}$.

CUT ALONG DOTTED LINE

FOLD HERE

NO POSTAGE
NECESSARY
IF MAILED
IN THE
UNITED STATES

BUSINESS REPLY MAIL

FIRST CLASS PERMIT NO. 358 PACIFIC GROVE, CA

POSTAGE WILL BE PAID BY ADDRESSEE

ATTN: MARKETING

Brooks/Cole Publishing Company
511 Forest Lodge Road
Pacific Grove, California 93950-9968

FOLD HERE

CUT ALONG DOTTED LINE

FOLD HERE

NO POSTAGE
NECESSARY
IF MAILED
IN THE
UNITED STATES

BUSINESS REPLY MAIL

FIRST CLASS PERMIT NO. 358 PACIFIC GROVE, CA

POSTAGE WILL BE PAID BY ADDRESSEE

ATTN: MARKETING

Brooks/Cole Publishing Company
511 Forest Lodge Road
Pacific Grove, California 93950-9968

FOLD HERE

CUT ALONG DOTTED LINE

FOLD HERE

NO POSTAGE
NECESSARY
IF MAILED
IN THE
UNITED STATES

BUSINESS REPLY MAIL

FIRST CLASS PERMIT NO. 358 PACIFIC GROVE, CA

POSTAGE WILL BE PAID BY ADDRESSEE

ATTN: MARKETING

Brooks/Cole Publishing Company
511 Forest Lodge Road
Pacific Grove, California 93950-9968

FOLD HERE